Fundamentals of
Asymmetric Catalysis

Fundamentals of Asymmetric Catalysis

Patrick J. Walsh
Marisa C. Kozlowski
University of Pennsylvania
Department of Chemistry

University Science Books
Sausalito, California

University Science Books
www.uscibooks.com

Production Manager: Mark Ong
Manuscript Editor: John Murdzek
Design: Mark Ong
Compositor: Publishers' Design and Production Services, Inc.
Cover design: Genette Itoko McGrew
Printer & Binder: Maple Vail Book Manufacturing Group

This book is printed on acid-free paper.

Library of Congress Cataloging-in-Publication Data

Walsh, Patrick J., 1964–
    Fundamentals of asymmetric catalysis / Patrick J. Walsh, Marisa C. Kozlowski.
        p. cm.
    Includes bibliographical references and index.
    ISBN 978-1-891389-54-2 (alk. paper)
    1. Enantioselective catalysis. 2. Asymmetry synthesis. I. Kozlowski, Marisa C., 1967–   II. Title.
    QD505.W35 2008
    541'.395—dc22
                                                                        2008006608

Printed in the United States of America

10 9 8 7 6 5 4 3 2 1

Contents

Foreword

It is a pleasure to provide a foreword to this book titled the *Fundamentals of Asymmetric Catalysis*.

In reviewing this text, I am reminded of an earlier landmark monograph, *Asymmetric Organic Reactions*, written by Morrison and Moser. This text, published in 1971, appeared at a time when few chemists were engaged in the development of enantioselective processes. In fact, not a single non-enzymic enantioselective process was even discussed at that time. As a young academician, I read this treatise with great interest; however, I was slow in getting the message! In the intervening 37 years the field of asymmetric catalysis has emerged as one of the principal activities of reaction development! It is my projection that the Walsh & Kozlowski text will become the "Bible" for those who wish to master the concepts that are the underpinnings to the research advances in the area of asymmetric catalysis. The chapters are arranged by topic, not reactions, and provide a solid foundation of the concepts in asymmetric catalysis. Nonetheless, the reader is exposed to many reaction types. The authors have also done an excellent job at integrating metal catalyzed reactions with organocatalysis.

The first chapter deals with "Modes of Asymmetric Induction." The discussion integrates simple resolution, kinetic resolution, and the application of the Curtin–Hammett Principle in a series of case studies. Both metal- and organo-catalysts are nicely integrated into the discussion.

The second chapter deals with "Lewis Acid and Lewis Base Catalysis." This chapter covers both of these important topics in a balanced fashion. Again, all forms of catalysis are covered.

Chapter 3, "Beyond Lewis Acid and Lewis Base Catalyzed Activation," provides a discussion of chiral cations and anions along with a treatment of organo-catalysts that rely on H-bonding organizational principles. Other topics include group transfer catalysis, alkene metathesis, and alkene activation by π-coordination. This discussion represents an important catch-all of topics needed in the ensuing discussion.

In Chapter 4, "Asymmetric Induction in Enantioselective Catalysis," the fundamental issues of asymmetric induction are presented, again illustrated with current examples from the literature. A general discussion of the transmission of asymmetry, ligand architecture, and the importance of the trans influence are covered.

In Chapter 5, "Nonclassical Two-Point Catalyst Substrate Interactions," catalysts based on the pairing of a primary (Lewis acid) with a secondary weak interaction are discussed. This secondary interaction includes H-bonding, and π-stacking organizational interactions. In many of the cases in this chapter, the nature of the secondary catalyst-substrate is still under active debate. The reader is presented with the facts. Conclusions will follow in due course.

Chapter 6 addresses alternative approaches to asymmetric catalysis, including chiral poisoning, chiral activation, and chiral catalyst optimization by modification of achiral ligands.

In Chapters 7 through 9, a concise discussion of kinetic, parallel kinetic, and dynamic kinetic resolution is presented along with a host of case studies.

Chapter 10, "Desymmetrization Reactions," addresses the fundamental concepts associated with enantioselective desymmetrization. Topics from earlier chapters are integrated into this monologue.

In Chapter 11, the authors address the issue of non-linear effects. This important chapter ultimately contributes to an enhanced understanding of reaction mechanism as the investigation of non-linear effects is now a routine exercise that is incorporated into the presentation of a catalysis model.

Chapter 12, "Bifunctional, Dual, and Multifunctional Catalyst Systems," addresses the combined effects of many of the individual catalytic principles presented in earlier chapters. This integration of the individual concepts is one of the principal strengths of this treatise.

This monograph could easily have been terminated at this point as the fundamental principles of asymmetric catalysis have been thoroughly covered. Nonetheless, succeeding chapters including, "Asymmetric Catalysis with Enantiopure Substrates: Double Diastereoselection;" "Multistep Asymmetric Catalysis," "Supported Chiral Catalysts," and "Applications of Asymmetric Catalysis in Synthesis," are included along with critical discussions.

Imagine yourself as a professor who is faced with presenting a topical course in asymmetric catalysis. This task is currently not a trivial undertaking since there is no integrated discussion of the various topics one might wish to cover. This book provides the intellectual framework for a detailed discussion of the intricacies of catalytic processes and assists its readership in providing a strong bridge to the primary scientific literature. The discussion nicely tracks the intellectual perspective between that of a text and a monograph.

I predict that this book will be widely used as a text in advanced courses and as a primary reference text for the research scientist. The authors are to be congratulated for their accomplishment.

David A. Evans
Harvard University

Introduction to Fundamentals of Asymmetric Catalysis

The importance of stereochemistry[1] in chemical interactions is crucial in several fields of chemistry including pharmaceutical, biological, agricultural, fragrance, and materials chemistries. This is probably best appreciated in the context of drug-receptor interactions, because most biological targets are chiral entities. Hence, there is enormous demand to devise viable and practical methods for preparing chiral compounds as single enantiomers. It is this demand that drives the field of asymmetric synthesis.

An integral part of asymmetric synthesis is asymmetric catalysis, which is dedicated to the development and applications of enantioenriched catalysts to transform prochiral and racemic substances into valuable enantioenriched synthetic building blocks. Since the first reports appeared in the 1960s, a wide variety of chiral organometallic complexes and organocatalysts (catalysts without a reactive metal component) have been identified as asymmetric catalysts.[2-8] These catalysts not only effect useful reactions with high levels of enantioselectivity, but often do so with a broad range of substrates.[9] The Royal Swedish Academy of Sciences awarded the 2001 Nobel Prize in Chemistry to pioneering figures in the field of asymmetric catalysis, William S. Knowles, Ryoji Noyori, and K. Barry Sharpless, in recognition of these achievements: for "the development of catalytic asymmetric synthesis." In the following chapters, we outline the fundamental principles of asymmetric catalysis, a rapidly evolving field on the cutting edge of organic synthesis and organometallic chemistry.

Goals and Intended Audience

This text was written to introduce the fundamental concepts of asymmetric catalysis to chemists with some graduate level education or to advanced undergraduate students. It provides a treatment beyond the introductory level. This book is intended to be a useful source of information to professionals working in organic synthesis who have occasion to employ enantioselective catalysts and desire to increase their level of understanding and chances of a successful outcome.

Organization of this Book

This book is organized differently than most books on asymmetric catalysis. Rather than classifying by reaction type, the chapters are organized by concepts. Although the choice of reactions and catalysts to include was primarily based on their educational significance, examples have been selected so that most of the commonly encountered reactions, ligands, and catalysts in asymmetric catalysis are represented. With an emphasis on instructional value, the coverage of historical significance and the latest developments in catalyst optimization for a given reaction have been included in the references and can be found in more specialized and/or comprehensive collections and reviews. For comprehensive presentation by reaction type, the reader is directed to other excellent texts.[4,6,8]

Each chapter can be read separately and provides the underlying concepts along with selected examples. A detailed appendix defines the key terms, types of chirality, and nomenclature of asymmetric transformations (prochiral, etc.) and is intended as a reference source.

Some of the mechanisms discussed herein are mechanistic proposals and are presented as a framework for understanding the material and discussion. Further exploration of many of these mechanisms is expected and there will undoubtedly be changes as the understanding of the community evolves.

In this book, enantiomeric excess (ee) is used rather than enantiomeric ratios (er) as this in the current state found in most publications. Nonetheless, a compelling argument for the use of enantiomeric ratios can be made.[10] Before this can occur, a convention must be established as to how the enantiomeric ratios are reported (x:1 vs a fraction summing to 100%, i.e. 95:5). We support the latter since it is the most readily correlated to the difference in energy accounting for kinetic control of the product ratios in asymmetric reactions (see **Equation A.2** and **Equation 1.1**).

Challenges in Asymmetric Catalysis

Despite significant effort, the state of the art still falls short in many ways. The ideal catalytic asymmetric transformation would proceed in 100% yield, *and* would provide complete chemocontrol, regiocontrol, and stereocontrol (diastereoselectivity and enantioselectivity). It would be conducted with a minimum of solvent and additives, generate no wasteful byproducts, and employ an inexpensive recoverable catalyst at low loadings. For large scale processes, the turnover number (TON) and turnover frequency (TOF) are particularly important. The TON describes the catalyst loading as the number of reaction cycles each catalyst molecule completes. The TOF is the number of reaction cycles per unit time. The latter is particularly important given the high cost per hour of operating large scale chemical reactors. Few catalytic asymmetric transformations meet these rigorous criteria. As a result, many industrial processes

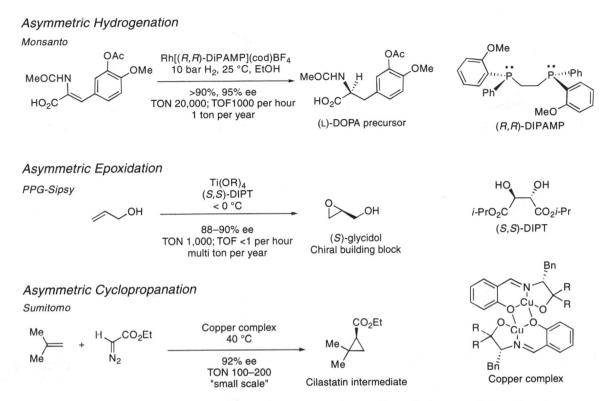

Figure. Examples of highly efficient catalytic asymmetric reactions that are performed on large scale.

for the synthesis of enantiomerically pure materials still rely on classical resolution of racemic mixtures or employ enzymatic resolution. The most progress toward the above goals has been made with catalytic asymmetric reductions and oxidations.[11-13] Illustrative examples include the asymmetric hydrogenation of a precursor en route to (L)-DOPA,[14,15] an anti-Parkinson's disease agent, and the asymmetric epoxidation of allyl alcohol to produce enantioenriched glycidol, an important chiral building block[16] (see **Figure**). While some catalytic asymmetric carbon-carbon bond forming transformations[13] approach the above criteria, as is the case with hydroformylation[17,18] or π–allylation,[19] the economics for utilizing these processes for large scale production are not yet favorable. An example of one of the few production scale asymmetric carbon-carbon bond forming reactions is the asymmetric cyclopropanation of 2-methylpropene to yield an intermediate for cilastatin,[20,21] an agent that is coadminstered with the antibiotic imipenem to prevent its degradation by renal dehydropeptidase. In this book, the fundamentals of asymmetric catalysis are examined to provide a basis for the further development of practical asymmetric catalysts that can meet the above criteria.

Topics Covered

Chapter 1 introduces the major types of asymmetric transformations (asymmetric catalysis with prochiral substrates, kinetic resolution and dynamic kinetic resolution of racemates) along with their corresponding energy diagrams. The relationship between energy and asymmetric induction is described. The effects of competing background processes are also discussed in terms of energy diagrams. Chapter 2 (Lewis acid and Lewis base) and Chapter 3 (Brønsted acid, Brønsted base, ion-pairing, group-transfer, cross-coupling, and π-activation) outline most of the chemical activation modes found in asymmetric catalysts. Many catalysts combine these activation features leading to bifunctional, dual, and even multifunctional catalysts (Chapter 12). The transfer of stereochemical information from the catalyst to the reacting substrates is discussed in Chapter 4 (classical interactions such as steric occlusion) and Chapter 5 (non-classical interactions such as cation-π). Methods for chiral catalyst optimization by means of additives (chiral activation, chiral deactivation, and modification with achiral ligands) are surveyed in Chapter 6. Chapters 7–9 describe the various forms of resolution: kinetic resolution, parallel kinetic resolution, dynamic kinetic resolution, and dynamic kinetic asymmetric transformations. As shown in Chapter 11, impure chiral catalysts can be highly advantageous (nonlinear effects) and are key to explaining the evolution of enantiopurity in nature (autoinduction and autocatalysis). Chapter 15 surveys the different types of supported chiral catalysts that combine the best features of homogeneous and heterogeneous catalysts.

Most practical asymmetric catalysts have focused on the generation of small chiral molecules containing one to two stereogenic units in highly enantioenriched form. Such compounds can be used as the starting point in the synthesis where all further stereogenic units are introduced diastereoselectively, thereby maximizing the contribution of the often valuable chiral component. The development of tandem or cascade processes (Chapter 14) that combine these events allows the rapid synthesis of highly complex structures. A different tactic is the use of desymmetrization (Chapter 10) to rapidly generate complex stereochemical arrays. The concept of using enantioenriched catalysts not only to create smaller subunits but also to assemble chiral subunits in transformations that generate additional stereogenic units has not received widespread acceptance. However, such double diastereoselective or triple diastereoselective processes hold much potential (Chapter 13). If inexpensive catalysts that exercise complete stereocontrol were available for every possible diastereomeric combination for a given process, then the assembly of complex stereochemical arrays would be highly simplified and could be practiced even by non-experts.

Alternatively, the convergent assembly of higher order structures with multiple stereogenic units can be undertaken from several different small chiral components (Chapter 16). By using enantiopure components, a convergent synthesis can provide a flexible approach to most if not all of the diastereomeric congeners. This approach has been the most successful when the convergent assembly transforms do not create further stereogenic units thereby removing the need for diastereoselectivity optimization. Another area that has received less attention is the use of chiral catalysis in transformations of complex or high value substrates. Most often chiral catalysts are employed to generate small chiral structures and can often do so with a high degree of substrate generality. The use of chiral catalysts with complex substrates containing multiple functional groups, however, is less explored (Chapter 16). Here, selective functionalization is crucial. Only chiral catalysts that are highly reliable can be employed in such situations where the substrates are valuable. Clearly, there are still many challenges facing the field of asymmetric catalyst development.

As a prerequisite to any useful discussion of asymmetric catalysis, a common means of describing various asymmetric groups and chirality types is required. In the Appendix, the definitions of chiral substances, their precursors, and their nomenclature are outlined. Different types of chiral compounds possessing a variety of stereogenic units, including central, axial, and planar chirality, are outlined. Achiral precursors that are prochiral, meaning that they can be transformed into enantiomerically pure chiral compounds, are delineated. The generation of pure enantiomers from racemic mixtures is also presented. Examples of the creation of each type of enantiomerically pure stereogenic unit are outlined. The Appendix concludes with strategies for the creation of multiple stereogenic units in a controlled manner to generate single enantiomers and diastereomers.

Asymmetric catalysis is a challenging but exciting area of endeavor. The ability to control the generation of nearly identical chemical compounds has captured the imagination of many outstanding scientists. In some ways, the field is beginning to mature (nearly 50 years old!). Even so, the stimulating discoveries within the last few years reveal the incredible potential in this field. Many challenges remain and, undoubtedly, many ingenious methods will be devised to address them. We hope that the community finds this text useful in the pursuit of this endeavor. We would like to thank the many individuals who have provided inspiration to us in the writing of this book. We are especially grateful to those who have read and commented on the text allowing it to reach its current form. In particular, Mukund Sibi (North Dakota State University) who read the entire text and Jennifer Love (University of British Columbia), Mathew Sigman (University of Utah), Jeffrey Johnson (University of North Caroline, Chapel Hill) who read large sections. Additional experts were consulted for individual chapters, including John Brown (Oxford University), David Glueck (Dartmouth College), Chris Vanderwal (University of California, Irvine), Huw Davies (SUNY Buffalo), Greg Cook (North Dakota State University) and Tomislav Rovis (Colorado State University). Their suggestions were valuable and insightful. We are grateful to Patrick Carroll (University of Pennsylvania) for assistance with the crystal structure graphics.

Patrick J. Walsh
Marisa C. Kozlowski
University of Pennsylvania

References

(1) Eliel, E. L.; Wilen, S. H.; Mander, L. N. *Stereochemistry of Organic Compounds*; Wiley-Interscience: New York, 1994.

(2) *Asymmetric Synthesis*; Aitken, R. A.; Kilényi, S. N., Eds.; Chapman & Hall: London, 1992.

(3) Koskinen, A. *Asymmetric Synthesis of Natural Products*; John Wiley & Sons: Chichester, 1993.

(4) *Catalytic Asymmetric Synthesis*; Ojima, I., Ed.; VCH: New York, 1993.

(5) Noyori, R. *Asymmetric Catalysis in Organic Synthesis*; Wiley-Interscience: New York, 1994.

(6) Nógrádi, M. *Stereoselective Synthesis*; VCH: Weinheim, 1995.

(7) Gawley, R. E.; Aube, J. *Principles of Asymmetric Synthesis*; Pergamon: Oxford, 1996; Vol. 14.

(8) *Comprehensive Asymmetric Catalysis*; Jacobsen, E. N.; Pfaltz, A.; Yamamoto, H., Eds.; Springer Verlag: New York, 1999; Vol. 1-3.

(9) Yoon, T. P.; Jacobsen, E. N. Privileged Chiral Catalysts. *Science* **2003**, *299*, 1691-1693.

(10) Gawley, R. E. Do the Terms "% ee" and "% de" Make Sense as Expressions of Stereoisomer Composition or Stereoselectivity? *J. Org. Chem.* **2006**, *71*, 2411–2416.

(11) Blaser, H. U.; Spindler, F.; Studer, M. Enantioselective Catalysis in Fine Chemicals Production. *Applied Catalysis A: General* **2001**, *221*, 119–143.

(12) *Asymmetric Catalysis on Industrial Scale: Challenges, Approaches and Solutions*; Blaser, H. U.; Schmidt, E., Eds.; Wiley: New York, 2004.

(13) Farina, V.; Reeves, J. T.; Senanayake, C. H.; Song, J. J. Asymmetric Synthesis of Active Pharmaceutical Ingredients. *Chem. Rev.* **2006**, *106*, 2734–2793.

(14) Vineyard, B. D.; Knowles, W. S.; Sabacky, M. J.; Bachman, G. L.; Weinkauff, D. J. Asymmetric Hydrogenation. Rhodium Chiral Bisphosphine Catalyst. *J. Am. Chem. Soc.* **1977**, *99*, 5946–5952.

(15) Knowles, W. S. Asymmetric Hydrogenation. *Acc. Chem. Res.* **1983**, *16*, 106–112.

(16) Shum, W.; Cannarsa, M. In *Chirality in Industry II: Developments in the Commercial Manufacture and Applications of Optically Active Compounds*; Collins, A. N., Sheldrake, G., Crosby, J., Eds.; Wiley: New York, 1997; pp 363.

(17) Chapuis, C.; Jacoby, D. Catalysis in the Preparation of Fragrances and Flavours. *Applied Catalysis A: General* **2001**, *221*, 93–117.

(18) Breit, B.; Seiche, W. Recent Advances on Chem-, Regio-, and Stereoselective Hydroformylation. *Synthesis* **2001**, 1–36.

(19) Trost, B. M.; Crawley, M. L. Asymmetric Transition-Metal-Catalyzed Allylic Alkylations: Applications in Total Synthesis. *Chem. Rev.* **2003**, *103*, 2921–2943.

(20) Aratani, T. Catalytic Asymmetric-Synthesis of Cyclopropane-Carboxylic Acids—an Application of Chiral Copper Carbenoid Reaction. *Pure Appl. Chem.* **1985**, *57*, 1839–1844.

(21) Aratani, T. In *Comprehensive Asymmetric Catalysis*; Jacobsen, E. N., Pfaltz, A., Yamamoto, H., Eds.; Springer Verlag: New York, 1999; Vol. 3; pp 1451–1460.

1

Modes of Asymmetric Induction

Enantioselective or asymmetric catalysis generally refers to asymmetric synthesis by means of substoichiometric chiral catalysts. However, there are actually several distinct modes of operation for asymmetric catalysts. Most commonly, asymmetric reactions commence with a prochiral substrate that may, in the presence of a chiral catalyst, be subject to two diastereomeric reaction pathways with, ideally, much different activation energies. The presence of competing achiral pathways and/or equilibrating intermediates can complicate the interpretation and optimization of such processes. Alternatively, chiral racemic substrates may be treated with chiral catalysts leading again to two diastereomeric reaction pathways. However, such kinetic resolution reactions have an upper yield of 50% resolved product (or 50% resolved starting material), unless some pathway allows equilibration of the starting material or a relevant intermediate (dynamic kinetic resolution, dynamic kinetic asymmetric transformation). Below, these different reaction paradigms are presented in terms of their characteristic features and reaction coordinate energy diagrams.

1.1 Simple Asymmetric Induction: No Catalyzed Background Reaction

The simplest form of asymmetric catalysis commences with an achiral substrate that is prochiral. Addition of a stoichiometric reagent then transforms the prochiral substrate (see Appendix A.2 for examples) into the two enantiomeric products. Ideally, this reagent causes no reaction in the absence of a catalyst (i.e., the activation energy for the uncatalyzed process is very high—see **Figure 1.1**). Addition of a substoichiometric chiral ligand then results in formation of a new catalyst species that causes substantial acceleration of the reaction. This effect is termed ligand accelerated catalysis.[1]

In the reaction illustrated in **Figure 1.1**, the prochiral substrate is the aldehyde and the stoichiometric reagent is diethylzinc.[2,3] Alone, these two species undergo little or no reaction at 0 °C to room temperature (see the plain line in the reaction coordinate energy diagram). A small (substoichiometric) amount of a chiral β-amino alcohol ligand reacts with the diethyl zinc to make a new chiral metal species, the β-amino zinc alkoxide, with evolution of ethane.[4,5,6] This new metal species does not transfer its ethyl group to the aldehyde, but does catalyze the addition of an ethyl group from another molecule of diethyl zinc to the aldehyde (see the transition structure in **Figure 1.1**). The chiral environment afforded by the chiral metal complex causes the approach of the activated diethylzinc to one prochiral face of the aldehyde to be more favorable than to the other face. As a result, the lower-energy pathway in the reaction coordinate energy diagram (the bold line in **Figure 1.1**) produces the (S) product, whereas the higher-energy pathway in the reaction coordinate energy diagram (the dashed line in **Figure 1.1**) produces the (R) product. In this particular example, all the accessible pathways to the product require the β-amino zinc alkoxide catalyst. Thus, the uncatalyzed reaction will not affect the product enantioselectivity; *only* the differences in energy between the catalyzed pathways are relevant to the asymmetric induction. The difference in energy between the two traversed pathways in **Figure 1.1** (i.e., the bold line and the dashed line) is 2.7 kcal/mol, which translates into a ~99:1 ratio for the two product enantiomers using $\Delta G = -RT\ln K$ (where K is the enantiomeric ratio). Since $\Delta G^{\ddagger}$ represents an activation energy, $\Delta\Delta G^{\ddagger}$ may not always correlate to enantiomeric ratios (see Section 1.5). This is especially true with computed transition states arising from different conformers of catalyst substrate adducts. As such, ΔG (i.e., $\Delta G_{TS1} - \Delta G_{TS2}$) is used here to represent the energy difference between the highest transition state from each pathway.

A more complete discussion of this reaction mechanism, which actually involves several competing diastereomeric pathways, is presented in the next section. In Sections 1.1–1.4, the energy diagrams depict only the starting materials, the key enantioselectivity-determining transition state, and the products. This simplified

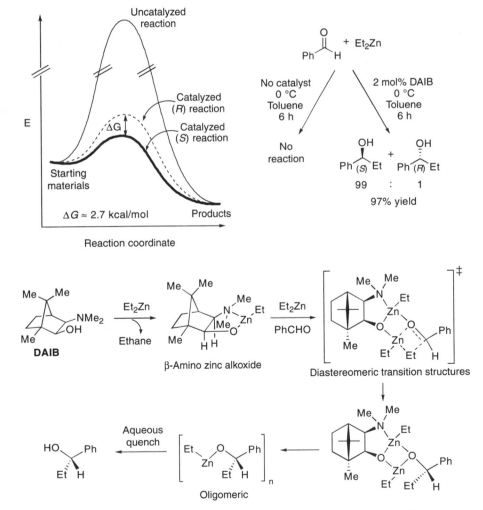

Figure 1.1. β-Amino-alcohol-catalyzed alkylation of aldehydes. An example of ligand accelerated catalysis with no background reaction.

rendering allows correlation of enantioselection directly to the illustrated ΔG and assumes that any intermediates (i.e., catalyst-coordinated substrate adducts) can equilibrate prior to the transition state (see Section 1.5).

The β-amino alcohol in this scheme is referred to as the ligand and is sometimes mistakenly termed the catalyst. While the ligand can be recovered from the reaction after hydrolysis of the zinc alkoxides present, the true catalyst, which remains unchanged after each turnover, is the β-amino zinc alkoxide. As an alternative to this ligand-accelerated catalysis, wherein a ligand alters one of the reagents and forms a catalytic species (or a more active catalytic species), catalysts may be employed that do not arise from modification of any of the reagents, but that function directly.

For example, the silyl ketene acetals in **Figure 1.2** undergo almost no reaction when treated with acetic anhydride (60 h, < 2% conversion).[7] Thus, there is essentially no background reaction and the potential for an energy profile similar to that in **Figure 1.1** arises. In fact, addition of a substoichiometric amount of a planar chiral 4-dimethylaminopyridine (DMAP) derivative gives rise to the much lower barrier pathway illustrated in **Figure 1.2** (0.3 h, 100% conversion). In the last step of the reaction, which is enantioselectivity-determining, the *N*-acyl–catalyst adduct can approach the top or bottom faces of the enolate, giving rise to the enantiomeric products (attack to the top face dominates). The difference in energy between these two diastereomeric pathways thus dictates the observed selectivity. Unlike the case above, the *unmodified* chiral DMAP derivative is the actual catalyst here. Like the case above, the absence of any meaningful background reaction simplifies the development of useful chiral catalysts.

The situations described above are considered ideal because the uncatalyzed background reaction essentially does not occur and thus does not need to be considered when assessing the level of enantioinduction. The most important factor effecting the

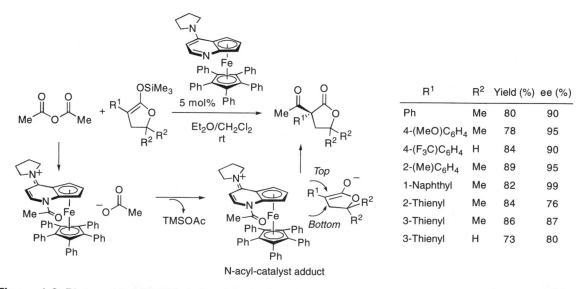

R¹	R²	Yield (%)	ee (%)
Ph	Me	80	90
4-(MeO)C₆H₄	Me	78	95
4-(F₃C)C₆H₄	H	84	90
2-(Me)C₆H₄	Me	89	95
1-Naphthyl	Me	82	99
2-Thienyl	Me	84	76
3-Thienyl	Me	86	87
3-Thienyl	H	73	80

N-acyl-catalyst adduct

Figure 1.2. Planar chiral DMAP derivatives in the enantioselective alkylation of substituted silyl ketene acetals to yield quaternary stereocenters. An example of catalytic asymmetric transformation with no background reaction.

enantiomeric excess is the difference in energy between the *accessible* pathways. No-tably, these energy differences are small (0.5–5.0 kcal/mol), especially when com-pared to the strengths of C–C bonds (80–90 kcal/mol) or metal ligand bonds (10–50 kcal/mol).

1.1.1 Simple Temperature Effects

If this energy difference between diastereomeric reaction pathways is relatively large (> 2.0 kcal/mol), then high enantiomeric excess can be observed even at room temper-ature. However, highly selective reactions are possible with even small energy differ-ences (1.0 kcal/mol) if low temperatures are employed since $\Delta G = RT\ln K$, where K is the enantiomeric ratio and ΔG is the difference in transition-state energies. **Table 1.1** illustrates the relationships between this ΔG, reaction temperature, enantiomeric ratio, and enantiomeric excess.

Importantly, care must be taken in extrapolating an enantiomeric excess value from one temperature to another according to **Table 1.1**; this table is based on a two-state model (only two diastereomeric pathways). If multiple diastereomeric pathways come into play, if different reaction pathways occur at different temperatures, or if the

				Reaction temperature (°C)				
ΔG (kcal/mol) between reaction pathways to two compounds		60	40	25	0	−20	−50	−78	−100
	0.1	54:46 er 8% ee	54:46 8	54:46 8	54:46 9	55:45 10	56:44 11	56:44 12	57:43 14
	0.2	57:43 15	58:42 16	58:42 17	59:41 18	60:40 20	61:39 22	63:37 25	64:36 28
	0.5	68:32 36	69:31 38	70:30 40	72:28 43	73:27 46	76:24 51	78:22 57	81:19 62
	1.0	82:18 64	83:17 67	84:16 69	86:14 73	88:12 76	91:9 81	93:7 86	**95:5 90**
	1.4	89:11 78	90:10 81	91:9 83	93:7 86	94:6 88	**96:4 92**	97:3 95	98:2 97
	1.8	94:6 88	**95:5 90**	95:5 91	97:3 93	97:3 95	98:2 97	99:1 98	99.5:0.5 99
	2.2	97:3 93	97:3 94	98:2 95	98:2 97	99:1 98	99.3:0.7 98.6	99.7:0.3 99.4	99.8:0.2 99.6
	2.6	98:2 96	99:1 97	99:1 98	99:1 98	99.4:0.6 98.8	99.7:0.3 99.4	99.9:0.1 99.8	99.9:0.1 99.8
	3.0	99:1 98	99:1 98	99.4:0.6 98.8	99.6:0.4 99.2	99.7:0.3 99.4	99.9:0.1 99.8	> 99.9:0.1 99.9	> 99.9:0.1 > 99.9
	3.4	99.4:0.6 98.8	99.6:0.4 99.2	99.7:0.3 99.4	99.8:0.2 99.6	99.9:0.1 99.8	>99.9:0.1 99.9	> 99.9:0.1 > 99.9	> 99.9:0.1 > 99.9

Table 1.1. Enantiomeric ratios and enantiomeric excesses (%) at different temperatures correspon-ding to several energy differences (ΔG) between two diastereomeric reaction pathways.

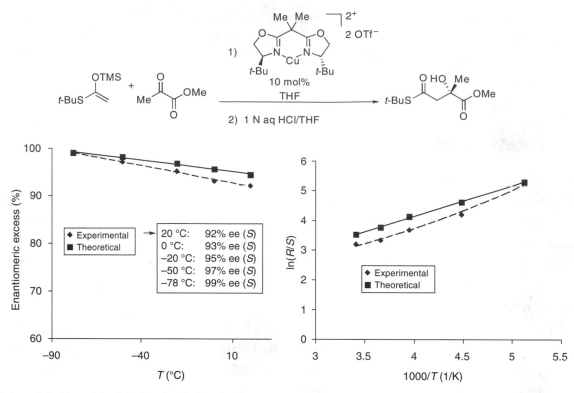

Figure 1.3. Near-ideal Arrhenius behavior for a two-state model in the chiral copper-bisoxazoline-catalyzed Mukaiyama aldol reaction with pyruvate.

rate-determining step shifts with temperature, then the observed enantiomeric excess values may depart from these trends (see Section 1.1.2).

An example of near-ideal behavior upon varying temperature in a reaction can be found in the additions of silyl ketene acetals to pyruvate esters as catalyzed by copper bisoxazoline complexes (**Figure 1.3**).[8] In a plot of temperature versus enantiomeric excess, a theoretical plot (the solid line), assuming a two-state scenario, is calculated from the intial −78 °C result (99% ee) according to **Table 1.1**. Even at room temperature, the agreement between experiment (92% ee, the dashed line) and the extrapolation (94%, the solid line) is very strong, indicating that a two-diastereomeric-pathway model can account for most of the behavior. Apparently, the competing thermal reaction and the reaction from free Cu(OTf)$_2$ (see Section 1.2) account for a very small percentage of the product obtained.

1.1.2 Temperature Effects in Multiple Pathway Reactions

While trends described above for temperature versus enantiomeric excess do not hold strictly for all reactions, higher enantiomeric excesses are often observed at lower temperatures even if multiple diastereomeric pathways intervene. If there is sufficient information about the composition of the diastereomeric pathways, it is even possible to determine the effect of temperature by generating a Boltzmann distribution based upon relative energy.

For example, in the addition of dialkyl zinc reagents to aldehydes with different catalysts (**Figure 1.4**), more than two diastereomeric reaction pathways (**Figure 1.5**) have been found to participate.[9,10] Thus, the reaction coordinate diagram is more complex and must take into account all the pathways sufficiently low in energy to contribute to the ultimate product distribution. A sample energy coordinate diagram for the reaction of Et$_2$Zn with butanal using the **DAIB**-derived catalyst is illustrated in **Figure 1.4**. Here, four of the eight lowest-energy diastereomeric pathways are illustrated.[11] Selectivity is dictated by a Boltzmann distribution over all the different pathways.

Such distributions can be calculated for any given temperature provided that the relative energies of the different pathways are known via experimental measurement or computation. For example, with **DAIB** and butanal, three of the eight lowest energy pathways contribute to the final product distribution and moderate selectivity (60% ee) is observed (**Figure 1.5**).[11] In the additions to benzaldehyde, higher selectivities are observed (82–99% ee) and only two reaction pathways are relevant. Notably, the two dominant pathways differ for each of the different catalysts (derived from **DAIB**, **A**, and **B**). Thus, even apparently similar highly selective reactions do not necessarily proceed via identical pathways. In fact, the basis for stereoselectivity is markedly different in these cases, which can further complicate the development of new chiral catalysts.

There is also no constraint that highly stereoselective processes proceed predominantly through two pathways, one leading to the (*S*) enantiomer and one to the (*R*) enantiomer. For example, it is entirely plausible that the major enantiomer is produced via several different low-energy pathways.

This situation is further complicated when different diastereomers, as well as enantiomers, are possible in a given reaction pathway. In an ideal asymmetric

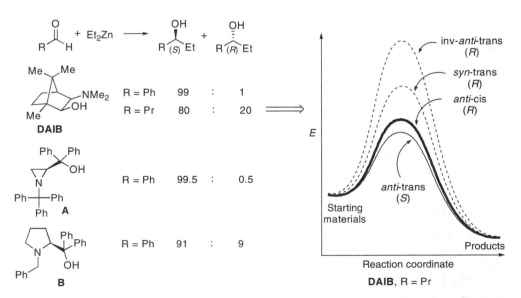

Figure 1.4. Energy diagram of the multiple diastereomeric pathways in the β-amino-alcohol-catalyzed alkylation of aldehydes.

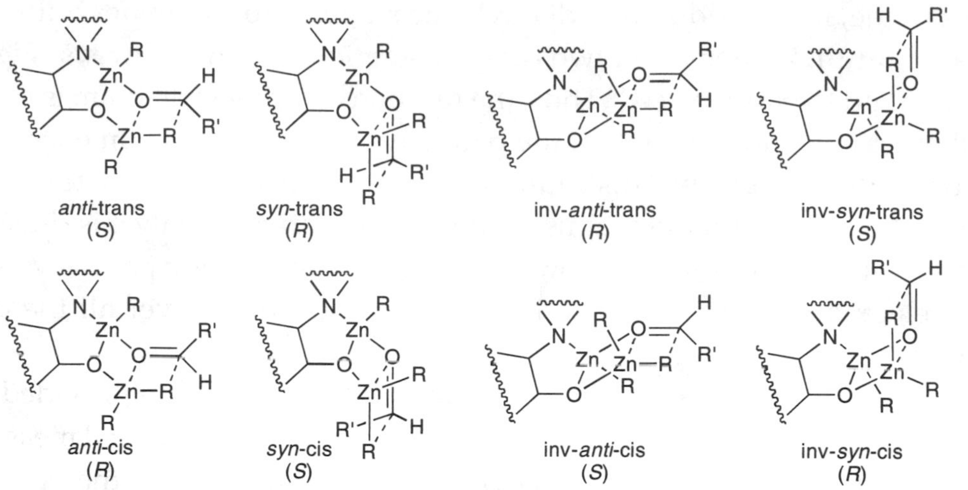

anti-trans (S)   syn-trans (R)   inv-anti-trans (R)   inv-syn-trans (S)

anti-cis (R)   syn-cis (S)   inv-anti-cis (S)   inv-syn-cis (R)

Catalyst, substrate	anti-trans (S)	inv-anti-trans (R)	syn-trans (R)	anti-cis (R)	
DAIB, R = Ph	0	> 7.0	3.0	4.5	kcal/mol
	99.7	0	0.3	0	% at 273 K
DAIB, R = Pr	0	> 7.0	4	1	
	85.3	0	0.1	14.6	
A, R = Ph	0	5.0	> 7.0	3.3	
	99.8	0	0	0.2	
B, R = Ph	0	0.5	> 7.0	4.3	
	70.7	29.3	0	0	

Figure 1.5. Multiple diastereomeric pathways in β-amino-alcohol-catalyzed alkylation of aldehydes from **Figure 1.4**. Structures, relative energies, and product distributions from Q2MM force field calculations are shown.[11]

catalytic process, only a single product is obtained. However, a process is considered useful if one compound predominates over the other possible products to a large extent (> 80–100% of the product material obtained). For example, in the (S)-proline-catalyzed intermolecular aldol reaction between cyclohexanone and isobutyralde-hyde, high diastereoselection (only the *anti* isomer is seen) and enantioselection (97% ee) are observed (**Figure 1.6**).[12] In contrast, the benzaldehyde version of the same reaction displays poor diastereoselection (50:50 *anti:syn*) and moderate enantioselection (*syn* 85% ee; *anti* 76% ee). In these reactions, at least four diastereomeric pathways are available, as illustrated in **Figure 1.6**. In fact, more pathways differing from the four canonical transition structures in the conformational disposition of the R group are likely. Thus, an efficient *and* selective catalyst must either stabilize one set of pathways or destabilize all other pathways to a large extent. Since different factors influence the relative stability of the major pathways, catalysts must be capable of multiple different stereodirecting roles in such situations. Thus, as the number of pathways increases, optimization of the reaction variables, such as temperature, can become more difficult, since the pathways are likely to be interdependent.

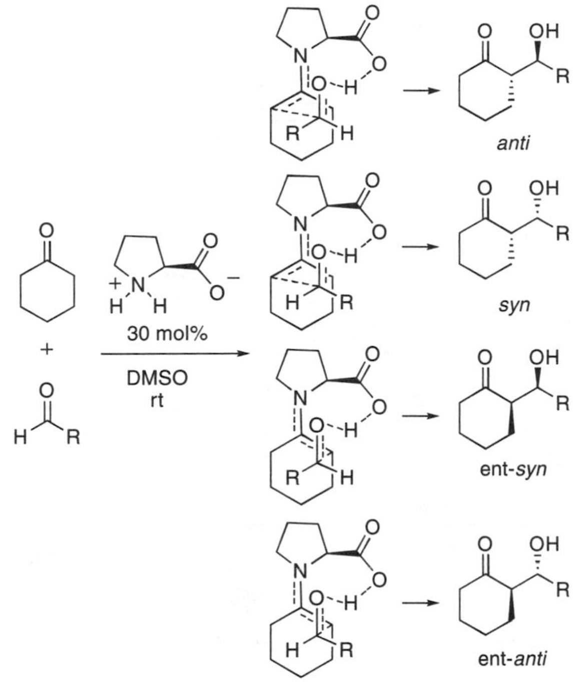

	R = Ph		R = i-Pr	
	Experimental ΔG_{exp} / %	Calculated ΔH_{298} / %	Experimental ΔG_{exp} / %	Calculated ΔH_{298} / %
anti	0.0 kcal/mol 45–47%	0.0 kcal/mol 50–80%	0.0 kcal/mol 97–100%	0.0 kcal/mol > 99%
syn	0.03 ± 0.05 43–45%	0.4 ± 0.4 20–50%	> 4.1 ± 0.03 < 1%	6.7 ± 0.4 < 1%
ent-syn	1.2 ± 0.05 5–7%	3.6 ± 0.4 < 1%	>4.1 ± 0.03 < 1%	7.8 ± 0.4 < 1%
ent-anti	1.4 ± 0.05 3–5%	2.3 ± 0.4 1–4%	2.5 ± 0.03 0–3%	4.6 ± 0.4 < 1%

Figure 1.6. (*S*)-Proline-catalyzed intermolecular aldol reaction. Multiple diastereomeric transition states give rise to diastereomeric and enantiomeric products. Relative heats and free energies of formation of the transition structures are given in kcal/mol. Experimental free energies are calculated from the experimentally measured distributions. Calculated enthalpies and distributions were obtained with B3LYP/6-31G*.

1.1.3 Nonideal Temperature Effects

A nonideal temperature versus enantioselectivity profile illustrating the hazards of multiple competing reaction pathways is presented in **Figure 1.7** (c.f. **Figure 1.3**). In the Michael addition reaction catalyzed by the illustrated bifunctional salen catalyst, good levels of enantioselection are observed at low temperature (–40 °C, 90% ee).[13] Upon raising the temperature, the enantioselection (dashed line in **Figure 1.7**) follows that projected from an extrapolation (solid line; see **Table 1.1**) of the –40 °C result up to –20 °C. In this regime, two diastereomeric pathways appear to account for the behavior. At 0 °C and 25 °C, the enantioselection deviates from the extrapolation, indicating that a two-state paradigm is no longer operative. Presumably racemic background reaction pathways (i.e., from residual Cs_2CO_3 or $CsHCO_3$) become accessible at these higher temperatures.

Further examples of unexpected temperature versus enantioselectivity trends can be found in the asymmetric enamide hydrogenation as catalyzed by bisphosphine–rhodium adducts (**Table 1.2** and **Figure 1.8**).

In some of the enamide hydrogenation reactions (entries 1–6), the enantioselection follows the general trend described above—namely, enantioselection falls as the

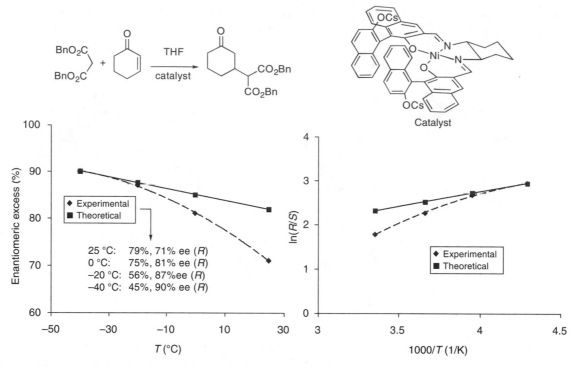

Figure 1.7. Non-Arrhenius behavior of a (salen)Ni–Cs naphthoxide catalyst in the Michael addition reaction.

Table 1.2. Catalysts and conditions utilized in the asymmetric enamide hydrogenations presented in **Figures 1.8**, **1.9**, and **1.10**.

Entry	R^1	R^2	R^3	Catalyst (mol%)	Catalyst	L*	Solvent	Product configuration	Reference
1	H	CH_2CO_2H	H	0.3	$Rh(L^*)_2BF_4$	(+)-Diop	n-BuOH/PhCH$_3$	(R)	14
2	H	NHAc	H	0.3	$Rh(L^*)_2BF_4$	(+)-Diop	n-BuOH/PhCH$_3$	(S)	14
3	H	NHAc	Ph	4.0	$Rh(cod)L^*ClO_4$	(−)-Diop	EtOH or n-BuOH	(R)	15
4	H	NHAc	Ph	4.0	$Rh(cod)L^*ClO_4$	(R,R)-Dioxop	EtOH or n-BuOH	(S)	15
5	Me	NHAc	Ph	4.0	$Rh(cod)L^*ClO_4$	(−)-Diop	EtOH or n-BuOH	(R)	15
6	Me	NHAc	Ph	4.0	$Rh(cod)L^*ClO_4$	(R,R)-Dioxop	EtOH or n-BuOH	(S)	15
7	H	NHAc	H	1.0	$Rh(nbd)L^*PF_6$	Diphosphinite	EtOH	(S)	16
8	H	NHAc	H	1.7	$Rh(cod)L^*Cl$	MABP	MeOH	(R)	17
9	H	NHAc	Ph	1.0	$Rh(nbd)L^*BF_4$	Chiraphos	EtOH	(R)	18
10	H	NHAc	Ph	1.0	$Rh(nbd)L^*BF_4$	BDPOP	EtOH	(R)	18
11	H	NHCOPh	Ph	0.8	$Rh(nbd)L^*PF_6$	Prophos	n-BuOH	(S)	19
12	H	NHCOPh	Ph	0.8	$Rh(nbd)L^*PF_6$	Cycphos	n-BuOH	(S)	19
13	H	NHAc	Ph	1.0	$Rh(nbd)L^*ClO_4$	cDDTHP	EtOH/PhH	(S)	20
14	H	NHAc	Ph	1.0	$Rh(nbd)L^*ClO_4$	tDDTHP	EtOH/PhH	(S)	20
15	H	NHCOPh	Ph	1.0	$Rh(cod)L^*ClO_4$	BPPM	EtOH	(R)	21
16	H	NHAc	Ph	0.05	$Rh(cod)L^*BF_4$	BBMPP	MeOH	(S)	22

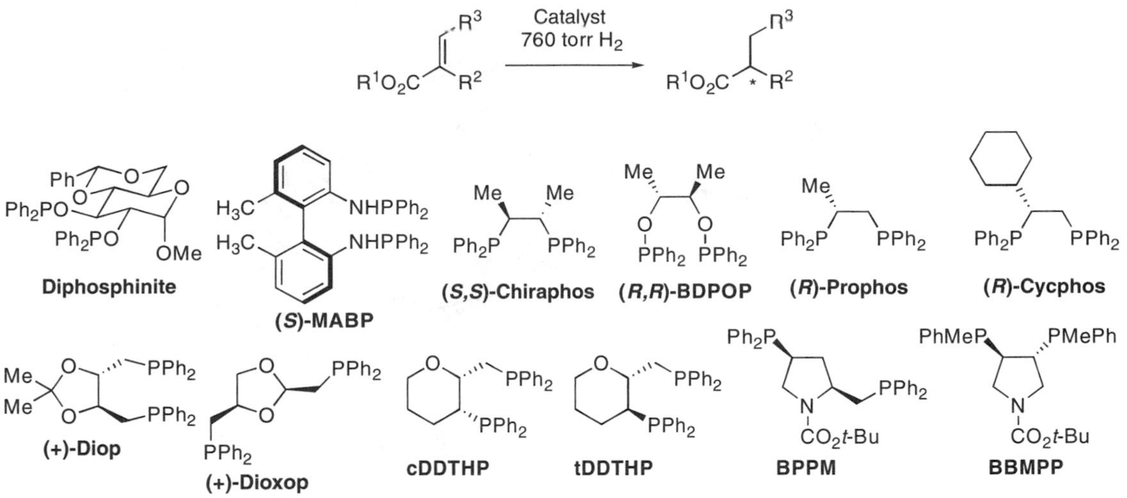

Figure 1.8. Bisphosphine ligands (L*) used in asymmetric rhodium-catalyzed enamide hydrogenations. See **Table 1.2** for reaction conditions.

reaction temperature is increased (**Figure 1.9a**). Furthermore, the behavior in many of these reactions can be accounted for using a two-state model as outlined above (see **Table 1.1**).

In other enamide hydrogenation reactions, the enantioselection paradoxically climbs as the temperature is raised (**Figure 1.9b**). In yet other reactions, the enantioselection displays a maximum at an intermediate temperature (**Figure 1.9c**). In these latter two cases, a shift in the mechanism most likely explains the results. This shift can occur gradually over a large temperature range, accounting for the data in **Figure 1.9b**, or can be sudden, accounting for the data in **Figure 1.9c**. Enantioselectivities should be measured at several temperatures when optimizing catalytic asymmetric processes. In addition, a simple temperature profile can provide insight into the basic mechanism, which is useful in further optimization. For example, an Arrhenius plot (**Equation 1.1**) can provide the activation energy difference between the two enantiomeric pathways, or an Eyring plot (**Equation 1.2**) can provide the corresponding $\Delta\Delta H^{\ddagger}$, $\Delta\Delta S^{\ddagger}$, and $\Delta\Delta G^{\ddagger}$. If the mechanism remains the same over a temperature range, the plot of the natural logarithm of the relative rates of formation of the two enantiomers (which is the same as the enantiomeric ratio; see **Equation 1.1**) versus $1/T$ should be linear.

$$\ln \frac{k_R}{k_S} = \frac{E_a(S) - E_a(R)}{RT} + \ln \frac{A_R}{A_S} \qquad \textit{Arrhenius} \qquad \textbf{Equation 1.1}$$

$$\frac{k_R}{k_S} = \frac{R}{S} = \text{er (enantiomeric ratio)}$$

$$\ln \frac{\kappa_S k_R}{\kappa_R k_S} = \frac{-\Delta\Delta H^{\ddagger}}{RT} + \frac{\Delta\Delta S^{\ddagger}}{R} \qquad \textit{Eyring} \qquad \textbf{Equation 1.2}$$

$$\kappa = \text{transmission coefficient} \approx 1$$

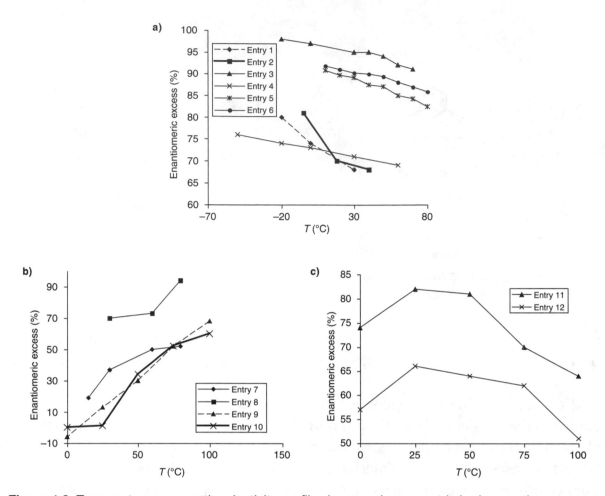

Figure 1.9. Temperature vs. enantioselectivity profiles in several asymmetric hydrogenation reactions with rhodium catalysts (see **Figure 1.8** and **Table 1.2**). a) Enantioselectivity *decreases* as temperature increases. b) Enantioselectivity *increases* as temperature increases. c) Nonlinear relationship between enantioselectivity and temperature.

The presence of a mechanism shift can be ascertained using this analysis as illustrated in **Figure 1.10** with the data from **Figure 1.9c**. When two roughly linear regions are observed, there is a difference in the activation enthalpy ($\Delta\Delta H^{\ddagger}$) and activation entropy ($\Delta\Delta S^{\ddagger}$) before and after the inflection point. This inflection point is termed the inversion temperature (T_{inv}).[23,24] Because temperature does not have an isolated effect, it can lead to a reweighting of the enthalpy/entropy dominance if more than one step contributes to the selectivity, as is the case here (see **Figure 1.24** and the accompanying text in Section 1.5 for further discussion).

An example of a mechanism shift with temperature can be found in the chiral platinum-catalyzed hydroformylation reaction shown in **Figure 1.11**. At 40 °C the (*S*)-branched enantiomer is predominant (60% ee), while at 100 °C the opposite—the (*R*) enantiomer-predominates slightly (10% ee).[25] This 70% ee difference is explained by a change in the enantioselectivity-determining step. At lower temperature, insertion of styrene into the Pt–H bond is largely irreversible and enantioselection is established in this early step. At higher temperatures, Pt–H insertion becomes reversible

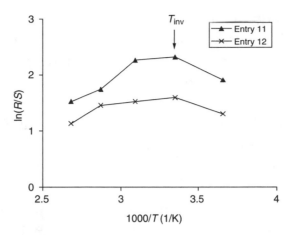

Figure 1.10. An Eyring plot of the temperature-dependent enantiomeric excess values from the systems in **Figure 1.9c** and **Table 1.2** measured at constant pressure.

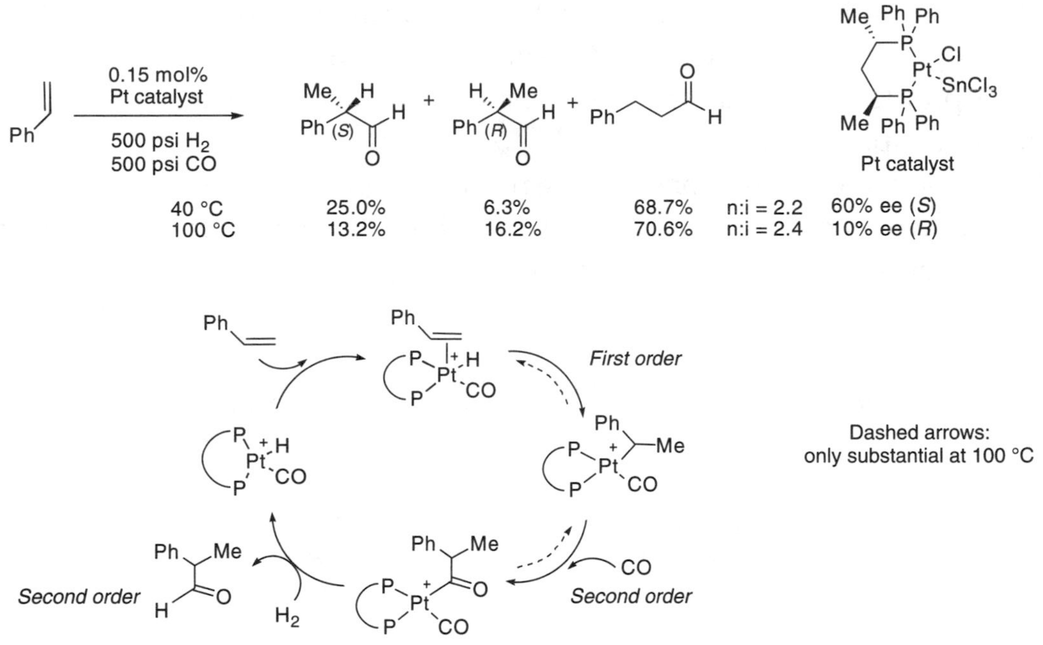

Figure 1.11. Changes in temperature resulting in reversal in the branched aldehyde enantioselection in asymmetric platinum-catalyzed hydroformylation.

(the dashed arrows in **Figure 1.11**). Under these circumstances, enantioselectivity is not completely determined until the final hydrogenolysis of the acyl platinum intermediate, which is slower for the (S) enantiomer.

In this case, a rationale for the temperature-dependent mechanism shift can be made based on the temperature dependence of a first-order reaction ($\Delta H^{\ddagger}$ = larger, $\Delta S^{\ddagger} \approx 0$) versus a competing second-order reaction ($\Delta H^{\ddagger}$ = smaller, $\Delta S^{\ddagger} << 0$). The first-order β-hydride elimination of the platinum alkyl back to styrene at 40 °C is slower than its second-order reaction with CO to give an acyl complex that is converted to an aldehyde. At 40 °C, selectivity is then largely set by the first-order formation of platinum alkyls. Heating will accelerate the first-order reaction as more molecules will have sufficient energy to surmount the reaction barrier. For the second-order process, the disfavorable entropy counters to this effect, causing a lesser degree of acceleration (or even a deceleration). This causes the reverse reactions (the dashed arrows in **Figure 1.11**) to become important at 100 °C. At this temperature, the first-order β-hydride elimination of the platinum alkyl back to styrene is accelerated more than the second-order acyl formation or the second-order hydrogenolysis of the acyl complex; the selectivity is then not determined until the second-order hydrogenolysis of the acyl intermediates.

The effect of temperature on the enantioselectivity of the asymmetric conjugate addition in **Figure 1.12** provides a cautionary example against assuming temperature-dependent trends, even with related catalyst systems.[26] While both catalysts are optimally selective at ambient temperature, *no enantioselectivity is seen at the commonly employed −78 °C*. In addition, there is a steady increase in enantioselectivity with temperature for ligand **A**. On the other hand, ligand **B** can provide either enantiomer in ~60% ee depending on the temperature employed. An examination of the [ln(*R/S*)] versus 1/*T* profile reveals that the mechanism of the reaction is considerably changed at different temperatures.

These observations have considerable practical significance in the development of enantioselective processes. Enantioselectivities should be measured at several temperatures to determine the necessary parameters for optimization.

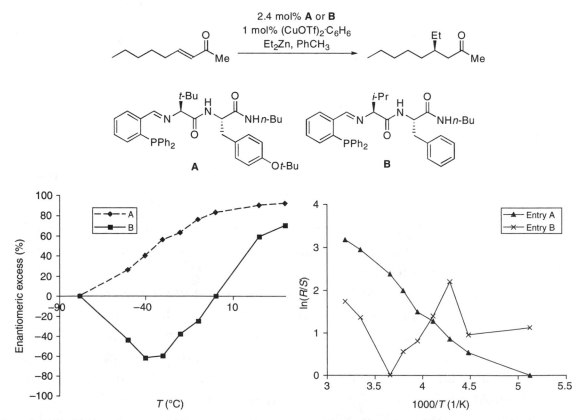

Figure 1.12. Changes in temperature resulting in unexpected enantioselectivity changes in an asymmetric conjugate addition.

1.1.4 Other Factors: Additives, Solvent, Concentration/Pressure

Other parameters, aside from temperature, can affect enantioselection, including additives, solvent, and concentration/pressure. For a discussion of additives and their effect on enantiomeric excess, see Chapter 6 and Section A.3 of Appendix A, **Figure A.72**.

The choice of solvent is often crucial for the optimal selectivity and conversion in any chemical reaction. Solvent effects are rather unpredictable since solvents can play several roles. Most often, their bulk properties (i.e., dielectric constant, polarity, etc.) are a primary consideration. However, solvents can also play discrete roles in asymmetric catalysis as metal ligands, hydrogen bond donors/acceptors, shuttles for reactive intermediates, and reactive elements (i.e., water in ketone/enamine chemistry).

In considering bulk properties, there are often no simple trends and screening of several solvents must be undertaken to empirically determine the optimum choice. Even so, there are several general rules of thumb. Typically, the most polar noncoordinating solvent accommodated by the reaction conditions is superior, allowing the greatest degree of solubility. For example, in the asymmetric addition of dialkyl zinc reagents to aldehydes (**Figure 1.1**), toluene is often superior to hexanes or pentane, although both will function if the components are soluble. In contrast, strongly coordinating solvents, such as tetrahydrofuran, partially or completely inhibit the catalyst and result in much lower enantioselection. Other asymmetric reactions, such as the asymmetric acylation of silyl ketene acetals, require more polar systems for solubility and can tolerate moderately coordinating solvents (**Figure 1.2**). Here, dichloromethane and diethyl ether are good candidates.

An example that shows dramatic solvent effects is illustrated in **Figure 1.13**. The rate and stereoselectivity of the Diels–Alder reaction are significantly influenced by the solvent polarity.[27] In particular, the rate of the reaction is faster in the polar CH_3NO_2 and the *endo*-selectivity is somewhat diminished from that seen in CH_2Cl_2. However, the level of asymmetric induction is dramatically affected; the cycloaddition process is essentially stereorandom in CH_2Cl_2, whereas in CH_3NO_2, much higher enantioselectivity is observed. Further experiments supported two different

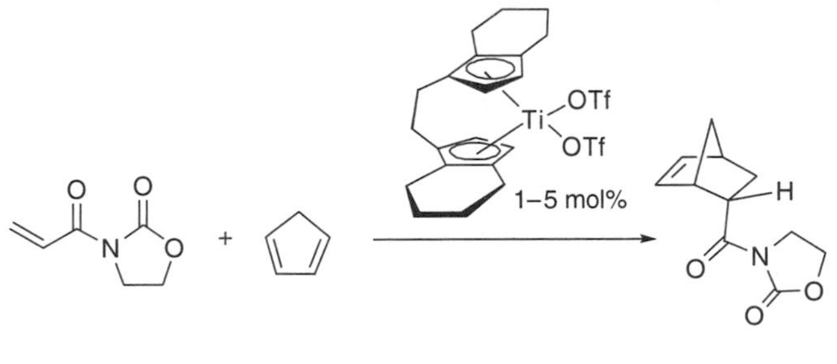

Entry	Solvent	T (°C)	Time (h)	Yield (%)	endo:exo	ee (%)
1	CH_2Cl_2	0	8	81	9:1	0
2	CH_3NO_2	0	0.5	94	7:1	88

Figure 1.13. Dramatic enantioselectivity changes in a Diels–Alder reaction with solvent.

coordination forms between the dienophile substrate and the catalyst. In CH_3NO_2, the complex leading to the more stereoselective pathway is present to a greater extent than in CH_2Cl_2.

For other reactions, such as the Mukaiyama aldol reaction of pyruvates with silyl ketene acetals (**Figure 1.3**), the use of polar coordinating solvents, such as tetrahydrofuran, is favorable, because the solvent serves to aid in the transfer of charged species (in this case, the trimethylsilyl cation) and accelerates the reaction. In cases involving the formation of enamines, imines, or related species via exchange, such as in the (S)-proline-catalyzed intermolecular aldol (**Figure 1.6**), highly polar solvent systems with trace amounts of water are to allow formation of the key charged intermediates.

Concentration is another reaction parameter that is important in catalytic enantioselective processes. In terms of reaction rate, the highest concentrations at which all components are soluble are best. These conditions also minimize waste. Usually the concentration of a reaction does not dramatically influence the enantioselection. However, enantioselection often erodes at lower concentrations when competing racemic background pathways contribute to a greater portion of the product. Thus, the greatest efficiency is achieved under solvent-free or highly concentrated conditions.[28] Under these conditions, catalyst saturation occurs and the fraction of the reaction that proceeds via the catalyzed pathway is maximal. Exceptions to this general trend can occur if there are competing unimolecular and bimolecular pathways. If the unimolecular pathway leads to the desired product, then the optimal outcome will be achieved at lower concentrations.

When a gas is employed in a reaction, the relationship between pressure (concentration) and enantioselection does not necessarily follow any simple trend. For example, all relationships between hydrogen pressure and enantioselection have been documented in the asymmetric hydrogenation of alkenes with rhodium catalysts (**Figures 1.8** and **1.14**). Upon increasing hydrogen pressure, entries 4, 6, and 15 from **Table 1.2** exhibit a decrease in the amount of major enantiomer. In the case of entry 15, the product is formed with overall lower enantioselection at higher pressures. For en-

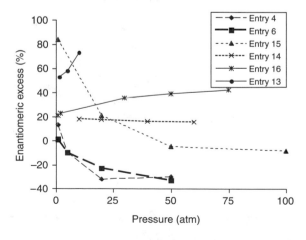

Figure 1.14. Impact of pressure on asymmetric hydrogenation reactions. See **Table 1.2** and **Figure 1.8** for conditions and catalysts.

tries 4 and 6, high pressure causes the opposite enantiomer to predominate. The enantioselection in some reactions is practically unaffected by pressure (c.f. entry 14). Entries 13 and 16 exhibit an increase in the amount of the major enantiomer with increasing hydrogen pressure. Changes in selectivity with pressure of a reagent arise due to a change in the rates of steps involving uptake or release of the reagent. In the rhodium-catalyzed hydrogenations described previously, increasing the hydrogen pressure increases the rate of the rate-limiting addition of hydrogen to the metal center (see **Figure 1.24**). However, early steps that are partially selectivity-determining, such as substrate coordination, remain unchanged. At higher hydrogen pressure, the oxidative addition of dihydrogen is now sufficiently fast that the substrate metal adducts no longer equilibrate, resulting in lower product enantioselectivity.

1.2 Simple Asymmetric Induction: Inherent Complicating Factors

While several complicating factors have been observed in even the simple asymmetric induction paradigms described so far, further complications occur when there are competing background processes arising from the thermal uncatalyzed reaction, ligand decelerated catalysis, product inhibition, and competing pathways.

For example, asymmetric catalysis in Grignard reagent and alkyl lithium carbonyl additions have generally failed. In contrast, the stoichiometric chiral versions can work well.[29,30] The problem is that achiral reagents also add quite effectively (i.e., background thermal reaction) and often even more quickly than the ligated adducts (i.e., ligand deccelerated catalysis), as illustrated in **Figure 1.15**.[30] Thus, even though

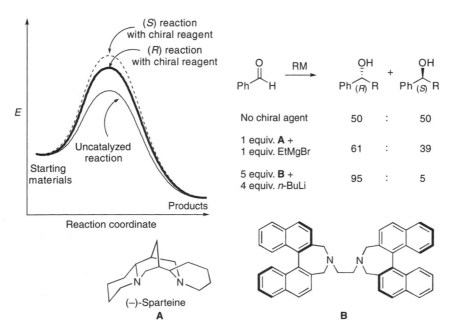

Figure 1.15. Enantioselective addition of Grignard reagents and alkyl lithium reagents to aldehydes using stoichiometric chiral additives.

the use of Grignard or organolithium reagents for this type of asymmetric catalytic transformation is highly desirable, due to their availability, development of such a process remains a challenging problem.

A problem encountered with many transition metal catalysts, especially Lewis acids, is ligand deccelerated catalysis. Such inhibition can result from coordination of solvent, product, or even counterions to the catalyst. For example, in the Diels–Alder reactions of acroleins with cyclopentadiene, a number of chiral cationic copper–pyridine–bisoxazoline adducts have been examined (**Figure 1.16**).[31,32] While the data from the Cu(OTf)$_2$ adduct indicate that the complex affords the architecture necessary for asymmetric induction, the reaction must be conducted at lower temperature (–20 °C) to circumvent the thermal cycloaddition reaction that competes at even 0 °C.[33] As a consequence, long reaction times (116 h) are necessary, which severely limit practical applications of this system.

The corresponding SbF$_6$ catalysts with less coordinating counterions presumably results in a more Lewis acidic catalyst. The net result is greater reactivity without sacrificing enantioselectivity. Under these circumstances, the presence of the highly reactive Cu(SbF$_6$)$_2$ must be avoided to prevent a racemic background reaction. Notably, this background reaction is different from the thermal background reaction. Here a background reaction can occur as the result of even minor impurities in the catalyst. For example, free Cu(OTf)$_2$ is a highly competent catalyst in this reaction, providing the product even more quickly at –20 °C (1 h) than the catalyst containing the pyri-

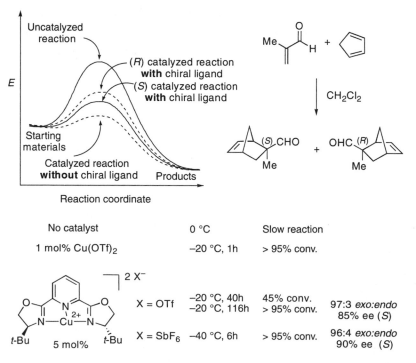

Figure 1.16. The copper–pyridine–bisoxazoline-catalyzed Diels–Alder reaction of methacrolein. An example with both an uncatalyzed background reaction (thermal) and competing catalysis from achiral metal species.

dine bisoxazoline Cu(OTf)$_2$ complex (116 h). Presumably the Cu(SbF$_6$)$_2$ is also a more reactive catalyst than its ligated adduct. In practice, this type of background reaction is suppressed by utilizing a small excess of the chiral ligand, which alone is an innocent bystander. Since complexation is highly favorable, this measure ensures that no uncomplexed Cu(OTf)$_2$ or Cu(SbF$_6$)$_2$ is present (**Equation 1.3**). In the development of a catalyst system, this situation is a primary consideration. Adjustment of the ligand:metal ratio must take into account the favorability of such an equilibrium, as well as the possibility that other contaminants or components may form active metal complexes (i.e., see the asymmetric dihydroxylation in **Figure 1.17**).

Equation 1.3

X = OTf or SbF$_6$

t-BuPybox + Cu(X)$_2$ ⇌ t-BuPyboxCu(X)$_2$

More reactive — *Less reactive*

↓ ↓

Racemic reaction — Asymmetric reaction

t-BuPybox

The Sharpless asymmetric dihydroxylation reaction[34] (see **Figure A.71** in Section A.3 of Appendix A) provides an example where the exact nature of the available metal species is crucial to the enantioselection observed (**Figure 1.17**). In the absence of an appropriate ligand L (an amine), the reaction is sluggish. If a chiral amine ligand L is present, then the reaction can proceed with catalytic amounts of OsO$_4$ via the primary catalytic cycle (ligand-accelerated catalysis). However, a secondary catalytic cycle can also intervene when certain oxidants, such as N-methylmorpholine N-oxide (NMO), are employed as the co-oxidant. The chiral ligand L is not involved in this

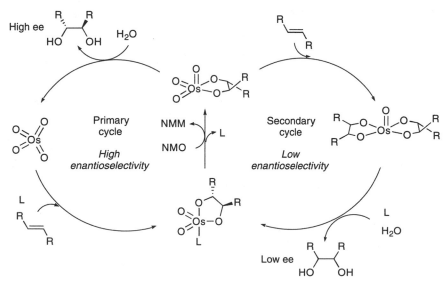

Figure 1.17. Catalytic cycle of the asymmetric dihydroxylation reaction (see **Figure A.71** in Section A.3 of Appendix A).

secondary catalytic cycle. Even though the oxidizing species is chiral (it contains the chiral diol produced in primary cycle) enantioinduction from this complex is low. In this situation, the enantiomeric excesses of the diol products obtained under the *catalytic* conditions (the secondary cycle participates) are lower than those produced by the *stoichiometric* reaction (only the primary cycle participates).

In order to generate a highly enantioselective dihydroxylation, the primary cycle must be fast at all stages of the reaction. Three discoveries permitted development of such a reaction that was catalytic in both the osmium species and the chiral ligand.[34] First, the participation of the secondary catalytic cycle can be virtually eliminated by performing the reaction under two-phase conditions with $K_3Fe(CN)_6$ as the stoichiometric reoxidant. Under these conditions, the only oxidizing species to remain in the organic layer is OsO_4, in contrast to the homogeneous NMO conditions. Second, hydrolysis of the osmium glycolate product can be accelerated considerably by $MeSO_2NH_2$, thereby preventing it from participating in the secondary cycle. Third, the discovery of ligands with two independent cinchona alkaloid units attached to a heterocyclic spacer led to a considerable increase in both the enantioselectivity and the scope of the reaction. Due to these improvements, it is now possible to obtain high enantioselectivities with a broad range of alkenes (**Figure A.71**).

1.3 Kinetic Resolution

Aside from asymmetric induction in the reactions of prochiral substrates to generate homochiral products, several other distinct protocols have been established to generate enantioenriched materials from racemic mixtures. In catalytic kinetic resolution, one enantiomer of a racemic substrate undergoes a more rapid reaction in the presence of a chiral catalyst. This topic is discussed in detail in Chapter 7. Conceptually, this process is outlined for a generic case in energy terms in **Figure 1.18**. **Figure 1.18a** compares directly the rates of the three relevant processes: 1) the uncatalyzed reaction, 2) the reaction of the (*R*) substrate with the chiral catalyst, and 3) the reaction of the (*S*) substrate with the chiral catalyst. Ideally, one of these last two processes will have an energy barrier much lower than the remaining two, such that only one reaction pathway is transited under the conditions employed. In such a situation, the reaction will proceed until one enantiomer of the substrate is completely consumed, resulting in a 50% yield of resolved substrate and a 50% yield of resolved product. In reality, the two catalyzed processes are often close in energy and the reaction needs to be limited by employing 0.5 equivalents of a reacting achiral reagent or by halting the reaction after 50% of the racemic substrate is consumed. Even so, a perfect resolution may not be possible if the two catalyzed pathways are too close in energy (< 2.8 kcal/mol apart for a room temperature process).

Figure 1.18b provides an energy coordinate diagram of just the two reactions involving the chiral catalyst and illustrates the biggest drawback to a catalytic kinetic

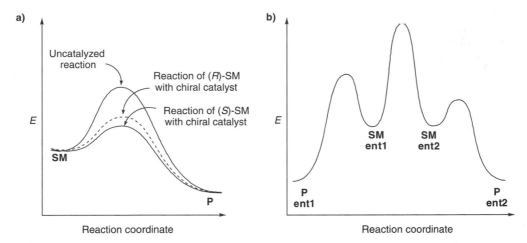

Figure 1.18. Kinetic resolution.

resolution. Since the two substrate enantiomers cannot interconvert, there is a maximum 50% yield for the resolved product (or 50% yield of resolved unreacted starting material).

An example of the utility of catalytic kinetic resolution can be found in the synthesis of (+)-cyclophellitol, an inhibitor of the HIV virus. A kinetic resolution of conduritol B via a catalytic asymmetric π-allylation is undertaken with a pivalate nucleophile to provide the resolved product (44% yield, 97% ee) along with the resolved conduritol B (50%, 88% ee) (**Figure 1.19**).[35] In this transformation, the enantiodifferentiation of the conduritol B enantiomers occurs during formation of the diastereomeric π-allyl intermediates. For the left enantiomer, this process occurs rapidly with the chiral catalyst leading to an *achiral* π-allyl (see **Figure 1.21**). A subsequent asymmetric reaction with the pivalate nucleophile controlled by the chiral palladium portion provides the resolved product. For the right enantiomer, reaction with the chiral calalyst to form the π-allyl intermediate is slow; unreacted conduritol B accumulates, leading to its resolution. This case is more complex than that diagrammed in **Figure 1.18**; kinetic resolutions are possible if any step (first, middle, or last) in the catalytic cycle displays rate differences between the two enantiomers, provided that all subsequent steps are faster (see Section 1.5 for a discussion of Curtin–Hammett effects).

1.4 Dynamic Kinetic Resolution

Catalytic dynamic kinetic resolution is similar to catalytic kinetic resolution in that one enantiomer of a racemic substrate undergoes a more rapid reaction due to the presence of a chiral catalyst. The difference is that 100% of the starting substrate can be converted to the desired product enantiomer. This topic is discussed in further detail in Chapter 9.

In order to achieve this 100% conversion to a single enantiomer, there must be a path that allows interconversion of the two enantiomers of the starting substrate under the reaction conditions. **Figure 1.20** illustrates this difference; in a kinetic

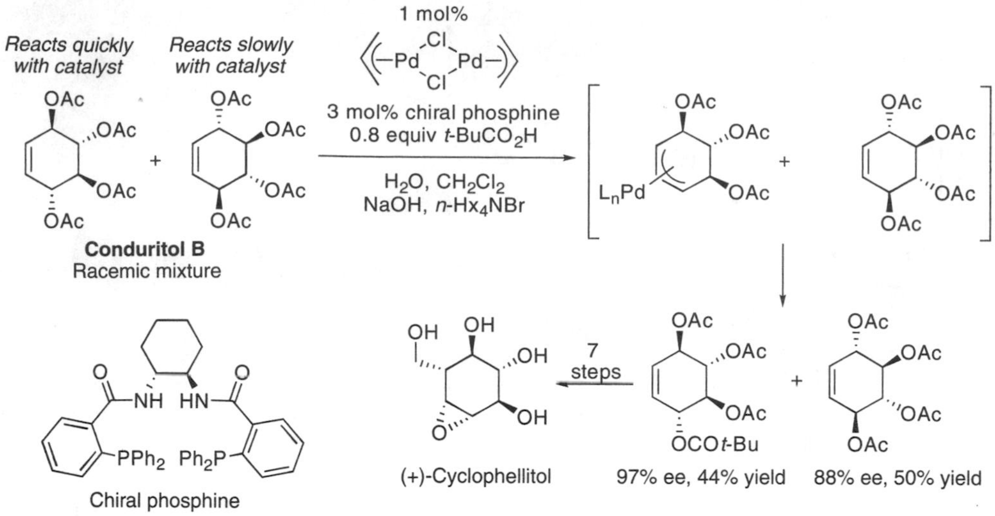

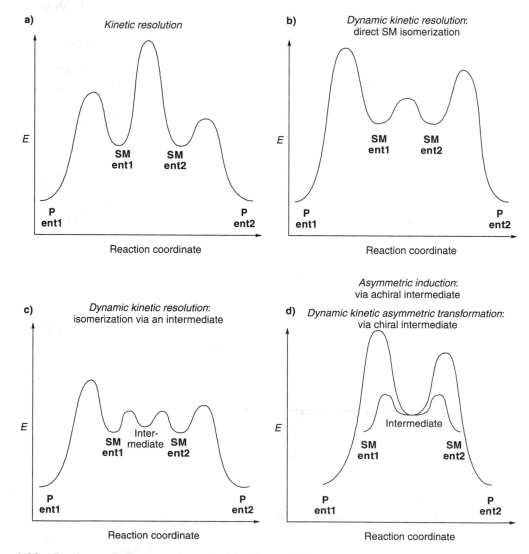

Figure 1.19. Kinetic resolution in asymmetric π-allylation.

Figure 1.20. Kinetic resolution vs. dynamic kinetic resolution.

resolution (**Figure 1.20a**), the barrier to intercoversion between the two substrate enantiomers is so high as to be inaccessible. In a dynamic kinetic resolution (**Figures 1.20b** and **1.20c**), the substrate enantiomers can interconvert. Since one of these enantiomers (**SM ent2**) is removed by a more rapid transformation to product (**P ent2**), Le Châtelier's principle dictates that the remaining enantiomer (**SM ent1**) will re-equilibrate, thereby replenishing **SM ent2**.

The mechanism for the interconversion can be direct (**Figure 1.20b**). For example, a thermal isomerization between two atropisomeric forms may occur (see **Figure A.69** in Section A.2.2.c and **Figure A.47** in Section A.2.1.b, both in Appendix A). Alternatively, enantiomer interconversion may require formation of an intermediate (**Figure 1.20c**). For example, racemization of a ketone with a stereocenter adjacent to the carbonyl proceeds via an enolic form and can occur on the time scale of a catalytic transformation, allowing a dynamic kinetic resolution (see **Figure A.68** in Section A.2.2.c of Appendix A). A separate added catalyst may even be employed, the sole purpose of which is to catalyze the substrate enantiomer interconversion (see Chapter 9, **Figures 14.35** and **14.36**).

Reactions of a racemic mixture that proceed via a single intermediate often take on the appearance of a dynamic kinetic resolution in that a racemic mixture is converted to a single enantiomer. From an energetic analysis, such reactions are distinct in that the relative reaction rates of two enantiomeric compounds with a chiral catalyst do not determine the enantioselection. Rather, the racemic material converges onto one intermediate. This intermediate is chiral by virtue of the incorporated asymmetric catalyst. In this dynamic kinetic asymmetric transformation (DyKAT, see Chapter 9), the relative energies of the diastereomeric reaction pathways arising from this single enantiomer intermediate give rise to the asymmetry-inducing event (**Figure 1.20d**, via the chiral intermediate).

An example can be found in the π-allylation protocol illustrated in **Figure 1.21**, which provides a key intermediate to (−)-cyclophellitol.[36] In contrast to the similar case in **Figure 1.19**, both enantiomers of the substrate undergo ionization with the

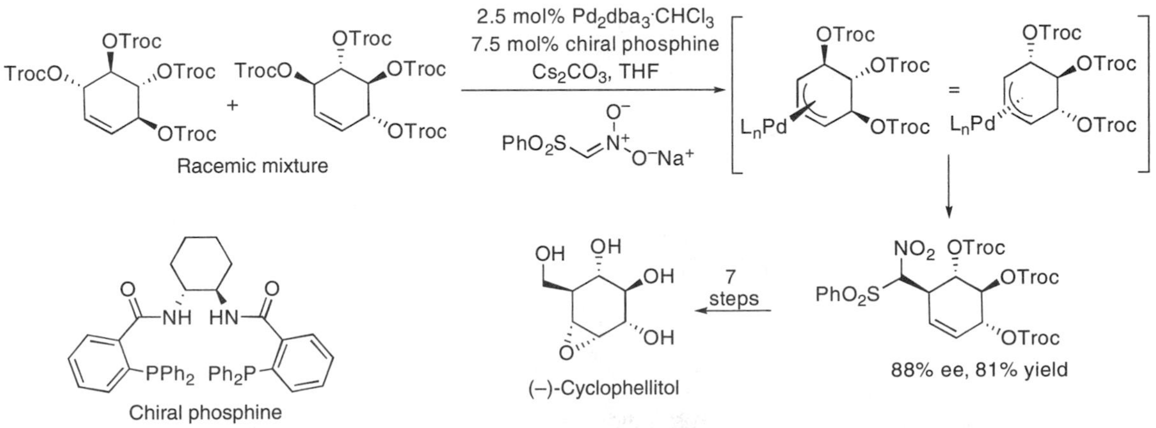

Figure 1.21. Dynamic kinetic asymmetric transformation (DyKAT) in asymmetric π-allylation.

chiral catalyst, leading to the *same* achiral allyl cation. However, due to the chiral palladium moiety, the coordinated complex is chiral. Differentiation of the termini of the allyl system caused by the chiral palladium gives rise to selective attack by the nucleophile (phenylsulfonylnitromethane anion), resulting in formation of the chiral product in 81% yield and 88% ee.

Alternatively, an apparent dynamic kinetic resolution can occur if both enantiomers of a racemic mixture converge onto an achiral intermediate. Then, the recognition of prochiral faces or groups in the achiral intermediate is the relevant asymmetry-inducing event and simple asymmetric induction pertains (**Figure 1.20d**, but with an achiral intermediate). An example of this type can be found in the chiral phase-transfer-catalyzed reaction of chiral ketones shown in **Figure A.48** (Section A.2.1.b in Appendix A). In this case, the substrate can readily racemize and it actually all converges upon an achiral enolate, which *then* forms a chiral enolate, with the chiral ammonium catalyst. The key stereochemistry step occurs by differentiation of the chiral faces of this intermediate.

A related case involving decarboxylation of a racemic mixture is illustrated in **Figure 1.22**.[37] Here, mechanistic studies provide evidence for the interchange of the chiral palladium allyl counterion from one deuterium-labeled enolate to another. As

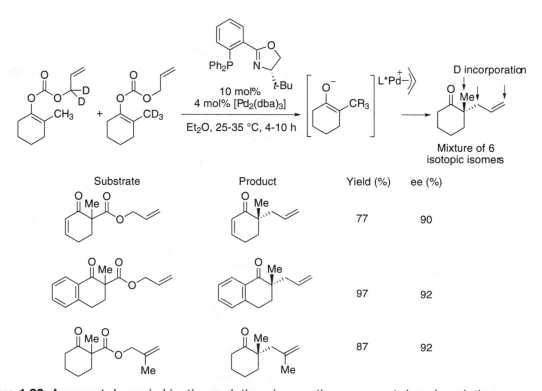

Figure 1.22. Apparent dynamic kinetic resolution via enantioconvergent decarboxylation.

such, there is strong support for formation of an achiral enolate. Subsequent differentiation of the enolate prochiral faces (simple enantioselection) by the chiral palladium catalyst gives rise to the enantioselection, resulting in an enantioconvergent process.

1.5 Curtin–Hammett: Equilibrating Intermediate States

In multistep reactions, a situation can occur that gives rise to an energy diagram similar to that encountered in dynamic kinetic resolution (**Figure 1.20** vs. **Figure 1.23**). Rather than equilibrating chiral starting materials, equilibrating diastereomeric intermediates (I_1 and I_2) that devolve from the same achiral starting material (**SM**) are present. In this Curtin–Hammett paradigm, I_1 and I_2 may rapidly exchange either by direct equilibration or by reverting through the starting material. For example, I_1 and I_2 may be two conformational forms of the substrate bound to the catalyst; interconversion by bond rotation could occur directly. Alternatively, I_1 and I_2 may be reaction intermediates that cannot directly interconvert, but that do undergo facile reversion to the starting material. The overall enantioselection in such processes is determined by the difference in the *relative* heights of the turnover-limiting barriers (ΔG). Thus, the activation energies for the conversion of intermediates I_1 and I_2 to products P_1 and P_2, as well as the relative energies of the intermediates, contribute to the enantioselection.

An illustration of this paradigm can be found in asymmetric catalytic hydrogenation, an extremely important process used in the manufacture of various pharmaceutical agents on an industrial scale, such as L-DOPA (Parkinson's disease) and naproxen (Alleve®, analgesic, anti-inflammatory). In an elegant example of mechanistic elucidation in transition metal catalysis, this process was studied and the

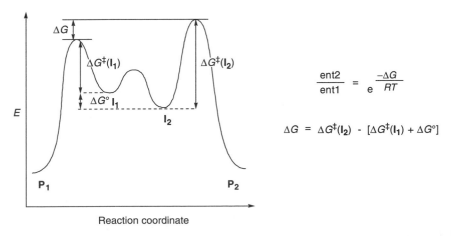

$$\frac{ent2}{ent1} = e^{\frac{-\Delta G}{RT}}$$

$$\Delta G = \Delta G^{\ddagger}(I_2) - [\Delta G^{\ddagger}(I_1) + \Delta G^{\circ}]$$

Figure 1.23. Energy diagram of an asymmetric reaction with equilibrating intermediates prior to the rate determining step. Both I_1 and I_2 form from the same **SM** (not shown).

following was demonstrated for the reaction illustrated in **Figure 1.24**:[38] 1) the product is obtained in ~95% ee, favoring P_1, 2) spectrosopic studies showed that the equilibrium ratio of $I_2{:}I_1$ is greater than 95:5, and 3) the conversion of I_1 to H_1 (and I_2 to H_2) is essentially irreversible and turnover-limiting.

In the preceding example, the selectivity of the transformation is ultimately dictated by a combination of two factors: the thermodynamic stabilities of the intermediates *and* the kinetic reactivity of the intermediates. From the energy coordinate diagram in **Figure 1.23**, the selectivity is related to the relative heights of the highest two barriers. In this case, the two factors counter one another, but this is not always the case (see below). This case cautions us that isolable intermediates (i.e., I_1 and I_2) and their ratios do not provide the whole story. Often, isolable intermediates are not catalytically relevant in that they are too stable. For a further discussion of this example, see **Figure 3.48** and the accompanying text.

This Curtin–Hammett analysis can also be applied to instances where the intermediate forms are not equilibrating. An example is seen in the highly effective systems for the kinetic resolution of benzylic alcohols by enantioselective oxidation with a palladium–sparteine catalyst system.[39,40,41] The proposed mechanism for this reaction involved four steps: alcohol binding to the palladium catalyst, alkoxide formation, β-hydride elimination, and palladium catalyst reoxidation (**Figure 1.25**).

Under conditions with saturating base (B:), mechanistic studies[42,43] reveal that the rate-determining step for the reaction shown in **Figure 1.25** is β-hydride elimination. However, further studies showed that the difference in the rates of formation of the two product enantiomers (intrinsic selectivity = intrinsic k_{rel} = 11) based on the β-hydride elimination step alone was two-fold lower than that observed experimentally with racemic substrate using saturating base (racemic k_{rel} = 25). The difference between these two values is a consequence of uneven partitioning of the small

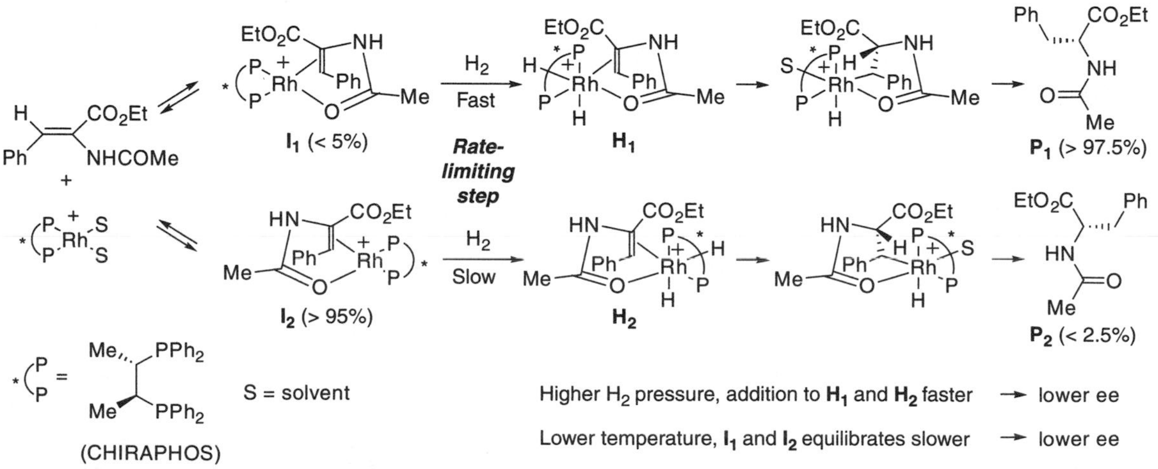

Figure 1.24. Curtin–Hammett effect in asymmetric hydrogenation.

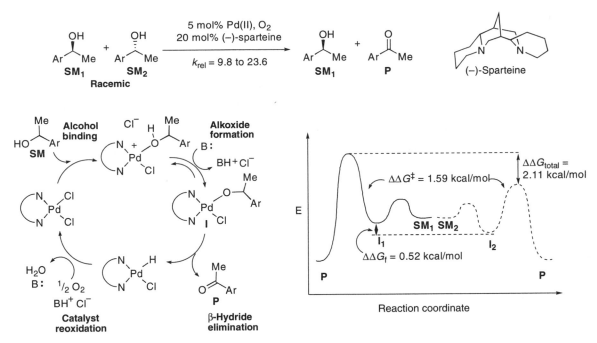

Figure 1.25. Asymmetric oxidative kinetic resolution of chiral alcohols using a chiral palladium catalyst.

amount of palladium catalyst (1 mol%) between excess amounts of each starting material enantiomer. In other words, there are unequal amounts of I_1 and I_2 after alcohol binding and alkoxide formation. If the amounts of these two compounds are taken into account, as represented by their energy difference (0.52 kcal/mol), then the picture shown in **Figure 1.25** evolves. Now, the selectivity (k_{rel}) for the racemic substrate is a consequence of the different energies of I_1 and I_2 ($\Delta\Delta G_f = 0.52$ kcal/mol) *and* of the different activation energies of I_1 and I_2 ($\Delta\Delta G^{\ddagger} = 1.59$ kcal/mol). In contrast to the hydrogenation case above, both of these factors act in concert here. As such, the energy difference between the two highest points on the energy diagram ($\Delta\Delta G_{total} = 2.11$ kcal/mol), which corresponds to the observed racemic k_{rel}, is the sum of these two values. This case also differs from the hydrogenation case described above in that the two intermediates, I_1 and I_2, are not strictly interconverting. However, the same type of analysis can be applied, at least during the initial stages of the reaction. The small amount of catalyst versus the excess of each SM_1 and SM_2 is proposed to allow thermodynamic control in the formation of I_1 and I_2. Understanding this mechanism has allowed rational modification of the reaction conditions to improve selectivity and reduce the amount of chiral sparteine required.[44,45]

In addition to these mechanistic studies, compelling evidence has been presented[46,47] that the C_1 symmetry of (–)-sparteine is crucial to the sterochemical course of this reaction.

Outlook

Most of the distinct methods for catalytic asymmetric synthesis (i.e., simple asymmetric induction, kinetic resolution, dynamic kinetic resolution, and dynamic kinetic asymmetric transformations) have yielded to development, providing highly valuable processes for converting achiral compounds or racemic compounds into enantiomerically pure materials. While each of these types of processes is characterized by specific concerns, there are many parameters that are important to any process (e.g., background reactions, multiple reaction pathways, temperature, additives, solvent, concentration, and multiple catalyst species). Despite the potential complexities introduced by these factors, it is possible to optimize many processes to useful levels. Understanding the mechanism increases our understanding of the various pitfalls and provides avenues for rationally improving catalytic asymmetric processes.

References

(1) Berrisford, D. J.; Bolm, C.; Sharpless, K. B. Ligand-Accelerated Catalysis. *Angew. Chem., Int. Ed. Engl.* **1995**, *34*, 1059–1070.

(2) Kitamura, M.; Suga, S.; Kawai, K.; Noyori, R. Catalytic Asymmetric Induction. Highly Enantioselective Addition of Dialkylzincs to Aldehydes. *J. Am. Chem. Soc.* **1986**, *108*, 6071–6072.

(3) Noyori, R.; Suga, S.; Kawai, K.; Okada, S.; Kitamura, M. Enantioselective Addition of Diorganozincs to Aldehydes Catalyzed by β-Amino Alcohols. *J. Organomet. Chem.* **1990**, *382*, 19–37.

(4) Kitamura, M.; Oka, H.; Noyori, R. Asymmetric Addition of Dialkylzincs to Benzaldehyde Derivatives Catalyzed by Chiral β-Amino Alcohols. Evidence for the Monomeric Alkylzinc Aminoalkoxide as Catalyst. *Tetrahedron* **1999**, *55*, 3605–3614.

(5) Noyori, R.; Suga, S.; Kawai, K.; Okada, S.; Kitamura, M. Enantioselective Alkylation of Carbonyl Compounds. From Stoichiometric to Catalytic Asymmetric Induction. *Pure Appl. Chem.* **1988**, *60*, 1597–1606.

(6) Noyori, R.; Kitamura, M. Enantioselective Addition of Organometallic Reagents to Carbonyl Compounds: Chirality Transfer, Multiplication, and Amplification. *Angew. Chem., Int. Ed. Engl.* **1991**, *30*, 49–69.

(7) Mermerian, A. H.; Fu, G. C. Catalytic Enantioselective Synthesis of Quarternary Stereocenters via Intermolecular C-Acylation of Silyl Ketene Acetals: Dual Activation of the Electrophile and Nucleophile. *J. Am. Chem. Soc.* **2003**, *125*, 4050–4051.

(8) Evans, D. A.; Burgey, C. S.; Kozlowski, M. C.; Tregay, S. W. C_2-Symmetric Copper(II) Complexes as Chiral Lewis Acids. Scope and Mechanism of the Catalytic Enantioselective Aldol Additions of Enolsilanes to Pyruvate Esters. *J. Am. Chem. Soc.* **1999**, *121*, 686–699.

(9) Yamakawa, M.; Noyori, R. An Ab Initio Molecular Orbital Study on the Amino Alcohol-Promoted Reaction of Dialkylzincs and Aldehydes. *J. Am. Chem. Soc.* **1995**, *117*, 6327–6335.

(10) Yamakawa, M.; Noyori, R. Asymmetric Addition of Dimethylzinc to Benzaldehyde Catalyzed by (2S)-3-exo-(Dimethylamino)isoborneol. A Theoretical Study on the Origin of Enantioselection. *Organometallics* **1999**, *18*, 128–133.

(11) Rasmussen, T.; Norrby, P.-O. Modeling the Stereoselectivity of the β-Amino Alcohol-Promoted Addition of Dialkylzinc to Aldehydes. *J. Am. Chem. Soc.* **2003**, *125*, 5130–5138.

(12) Bahmanyar, S.; Houk, K. N.; Martin, H. J.; List, B. Quantum Mechanical Predictions of the Stereoselectivities of Proline-Catalyzed Asymmetric Intermolecular Aldol Reactions. *J. Am. Chem. Soc.* **2003**, *125*, 2475–2479.

(13) Annamalai, V.; DiMauro, E. F.; Carroll, P. J.; Kozlowski, M. C. Catalysis of the Michael Addition Reaction by Late Transition Metal Complexes of BINOL-Derived Salens. *J. Org. Chem.* **2003**, *68*, 1973–1981.

(14) James, B. R.; Mahajan, D. Kinetic and Mechanistic Aspects of the Binding of Dihydrogen by Bis(ditertiaryphosphine)rhodium(I) Tetrafluoroborate Complexes, and Activity of the Dihydrides for Catalytic Asymmetric Hydrogenation of Prochiral Olefinic Acids. *J. Organomet. Chem.* **1985**, *279*, 31–48.

(15) Sinou, D. Hydrogenation Asymmetrique a L'Aide du Complexe DIOXOP-Rh(I) Voie Dihydro ou Voie Insaturee. *Tetrahedron Lett.* **1981**, *22*, 2987–2990.

(16) Cullen, W. R.; Sugi, Y. Asymmetric Hydrogenation Catalyzed by Diphosphinite Rhodium Complexes Derived from a Sugar. *Tetrahedron Lett.* **1978**, 1635–1636.

(17) Uehara, A.; Kubota, T.; Tsuchiya, R. New Atropisomeric Chiral Bisphosphine, (S)-6,6'-Dimethyl-2,2'-bis(diphenylphosphinamino)biphenyl, and Asymmetric Hydrogenation Using the Rh(I) Complex Thereof. *Chem. Lett.* **1983**, 441–444.

(18) Bakos, J.; Toth, I.; Heil, B.; Marko, L. A Facile Method for the Preparation of 2,4-Bis(diphenylphosphino)pentane (BDPP) Enantiomers and Their Application in Asymmetric Hydrogenation. *J. Organomet. Chem.* **1985**, *279*, 23–29.

(19) Riley, D. P.; Shumate, R. E. 1,2-Bis(diphenylphosphino)-1-cyclohexylethane. A New Chiral Phosphine Ligand for Catalytic Chiral Hydrogenations. *J. Org. Chem.* **1980**, *45*, 5187–5193.

(20) Sunjic, V.; Habus, I. Chiroptical Properties and Enantioselectivity in Hydrogenation with Rhodium(I) Complexes of Chiral Bis-diphenylphospines Derived from D-Glucose and D-Galactose. *J. Organomet. Chem.* **1989**, *370*, 295–304.

(21) Ojima, I.; Kogure, T.; Yoda, N. Asymmetric Hydrogenation of Prochiral Olefins Catalyzed by Rhodium Complexes with Chiral Pyrrolidinophosphines. Crucial Factors for the Effective Asymmetric Induction. *J. Org. Chem.* **1980**, *45*, 4728–4739.

(22) Nagel, U.; Rieger, B. Enantioselective Catalysis. 6. The Catalytic Hydrogenation of α-(Acetylamino)cinnamic Acid with Rhodium(I)-Bis(phospine) Complexes. On the Origin of Enantioselection. *Organometallics* **1989**, *8*, 1534–1538.

(23) Buschmann, H.; Scharf, H.-D.; Hoffmann, N.; Esser, P. The Isoinversion Principle—A General Model of Chemical Selectivity. *Angew. Chem., Int. Ed. Engl.* **1991**, *30*, 477–515.

(24) Heller, D.; Buschmann, H.; Scharf, H.-D. Nonlinear Temperature Behavior of Product Ratios in Selection Processes. *Angew. Chem., Int. Ed. Engl.* **1996**, *35*, 1852–1854.

(25) Casey, C. P.; Martins, S. C.; Fagan, M. A. Reversal of Enantioselectivity in the Hydroformylation of Styrene with [2S,4S-BDPP]Pt(SnCl₃)Cl at High Temperature Arises from a Change in the Enantioselective-Determining Step. *J. Am. Chem. Soc.* **2004**, *126*, 5585–5592.

(26) Hoveyda, A. H.; Hird, A. W.; Kacprzynski, M. A. Small Peptides as Ligands for Catalytic Asymmetric Alkylations of Olefins. Rational Design of Catalysts or of Searches That Lead to Them? *J. Chem. Soc., Chem. Commun.* **2004**, 1779–1785.

(27) Jaquith, J. B.; Guan, J.; Wang, S.; Collins, S. Asymmetric Induction in the Diels–Alder Reaction Catalyzed by Chiral Metallocene Triflate Complexes: Dramatic Effect of Solvent Polarity. *Organometallics* **1995**, *14*, 1079–1081.

(28) Walsh, P. J.; Li, H.; Anaya de Parrodi, C. A Green Approach to Asymmetric Catalysis: Solvent-Free and Highly Concentrated Reactions. *Chem. Rev.* **2007**, *107*, 2503–2545.

(29) Nozaki, H.; Aratani, T.; Toraya, T. Asymmetric Carbinol Synthesis by Means of (−)-Sparteine-Modified Organometallic Reagents. *Tetrahedron Lett.* **1968**, 4097–4098.

(30) Mazaleyrat, J. P.; Cram, D. J. Chiral Catalysis of Additions of Alkyllithiums to Aldehydes. *J. Am. Chem. Soc.* **1981**, *103*, 4585–4586.

(31) Evans, D. A.; Murry, J. A.; von Matt, P.; Norcross, R. D.; Miller, S. J. C_2-Symmetric Cationic Copper(II) Complexes as Chiral Lewis Acids: Counterion Effects in the Enantioselective Diels–Alder Reaction. *Angew. Chem., Int. Ed. Engl.* **1995**, *34*, 798–800.

(32) Evans, D. A.; Barnes, D. M.; Johnson, J. S.; Lectka, T.; von Matt, P.; Miller, S. J.; Murry, J. A.; Norcross, R. D.; Shaughnessy, E. A.; Campos, K. R. Bis(oxazoline) and Bis(oxazolinyl)pyridine Copper Complexes as Enantioselective Diels–Alder Catalysts: Reaction Scope and Synthetic Applications. *J. Am. Chem. Soc.* **1999**, *121*, 7582–7594.

(33) Mellor, J. M.; Webb, C. F. Stereochemistry of the Diels–Alder Reaction: Steric Effects of the Dienophile on Endo-Selectivity. *J. Chem. Soc., Perkin Trans. 2* **1974**, 17–22.

(34) Kolb, H. C.; VanNieuwenhze, M. S.; Sharpless, K. B. Catalytic Asymmetric Dihydroxylation. *Chem. Rev.* **1994**, *94*, 2483–2547.

(35) Trost, B. M.; Hembre, E. J. Pd Catalyzed Kinetic Resolution of Conduritol B. Asymmetric Synthesis of (+)-Cyclophellitol. *Tetrahedron Lett.* **1999**, *40*, 219–222.

(36) Trost, B. M.; Patterson, D. E.; Hembre, E. J. AAA in KAT/DYKAT Processes: First- and Second-Generation Asymmetric Syntheses of (+)- and (−)-Cyclophellitor. *Chem.—Eur. J.* **2001**, 3768–3775.

(37) Mohr, J. T.; Behenna, D. C.; Harned, A. M.; Stoltz, B. M. Deracemization of Quaternary Stereocenters by Pd-Catalyzed Enantioconvergent Decarboxylative Allylation of Racemic β-Ketoesters. *Angew. Chem., Int. Ed. Engl.* **2005**, *44*, 6924–6927.

(38) Halpern, J. Mechanism and Stereoselectivity of Asymmetric Hydrogenation. *Science* **1982**, *217*, 401–407.

(39) Jensen, D. R.; Pugsley, J. S.; Sigman, M. S. Palladium-Catalyzed Enantioselective Oxidations of Alcohols Using Molecular Oxygen. *J. Am. Chem. Soc.* **2001**, *123*, 7475–7476.

(40) Mandal, S. K.; Jensen, D. R.; Pugsley, J. S.; Sigman, M. S. Scope of Enantioselective Palladium(II)-Catalyzed Aerobic Alcohol Oxidations with (−)-Sparteine. *J. Org. Chem.* **2003**, *68*, 4600–4603.

(41) Bagdanoff, J. T.; Stoltz, B. M. Palladium-Catalyzed Oxidative Kinetic Resolution with Ambient Air as the Stoichiometric Oxidation Gas. *Angew. Chem., Int. Ed.* **2004**, *43*, 353–357.

(42) Mueller, J. A.; Jensen, D. R.; Sigman, M. S. Dual Role of (−)-Sparteine in the Palladium-Catalyzed Aerobic Oxidative Kinetic Resolution of Secondary Alcohols. *J. Am. Chem. Soc.* **2002**, *124*, 8202–8203.

(43) Mueller, J. A.; Sigman, M. S. Mechanistic Investigations of the Palladium-Catalyzed Aerobic Oxidative Kinetic Resolution of Secondary Alcohols Using (−)-Sparteine. *J. Am. Chem. Soc.* **2003**, *125*, 7005–7013.

(44) Jensen, D. R.; Sigman, M. S. Palladium Catalysts for Aerobic Oxidative Kinetic Resolution of Secondary Alcohols Based on Mechanistic Insight. *Org. Lett.* **2003**, *5*, 63–65.

(45) Mandal, S. K.; Sigman, M. S. Palladium-Catalyzed Aerobic Oxidative Kinetic Resolution of Alcohols with an Achiral Exogenous Base. *J. Org. Chem.* **2003**, *68*, 7535–7537.

(46) Trend, R. M.; Stoltz, B. M. An Experimentally Derived Model for Stereoselectivity in the Aerobic Oxidative Kinetic Resolution of Secondary Alcohols by (Sparteine)PdCl$_2$. *J. Am. Chem. Soc.* **2004**, *126*, 4482–4483.

(47) Nielsen, R. J.; Keith, J. M.; Stoltz, B. M.; Goddard, W. A., III. A Computational Model Relating Structure and Reactivity in Enantioselective Oxidations of Secondary Alcohols by (−)-Sparteine-Pd^{II} Complexes. *J. Am. Chem. Soc.* **2004**, *126*, 7967–7974.

Lewis Acid and Lewis Base Catalysis

There are several distinctly different types of catalytic modes to asymmetric induction. Commonly encountered asymmetric catalysts can be classified as Lewis acids, Lewis bases, Brønsted acids, Brønsted bases, ionic species, group transfer agents, cross-coupling agents, and π-activating agents. Each of these different catalyst types affects asymmetric activation through a unique mode. In this chapter, the substrate functional groups needed for a productive catalytic interaction and other prerequisites needed for Lewis acid and Lewis base catalysis are defined. The remaining types of asymmetric catalysts are discussed in Chapter 3.

2.1 Lewis Acid Catalysts

By far the most common means of activation of a substrate functional group for a catalytic asymmetric transformation involves the use of chiral Lewis acids. In these cases, a Lewis acid metal or metalloid species is typically merged with a chiral Lewis base ligand to generate a chiral Lewis acid (**Figure 2.1**). There is a continuum of reactivity displayed by these catalysts that combine Lewis acid and Lewis base groups (**Figure 2.1**). When addition of a chiral Lewis base ligand gives rise to a more reactive catalyst (right side of **Figure 2.1**) relative to the parent Lewis acid, then ligand-accelerated catalysis occurs.[1] This situation is optimal because an excess of the Lewis base is not needed to prevent background reaction from the Lewis acid without a ligand. Under these circumstances, the chiral Lewis acid adduct may even be formed by reversible coordination of a chiral Lewis base to an achiral Lewis acid in situ. Such cases are formally examples of Lewis base asymmetric catalysis. In this chapter, reactions with substoichiometric Lewis acid are covered here in Section 2.1, whereas those with stoichiomet-

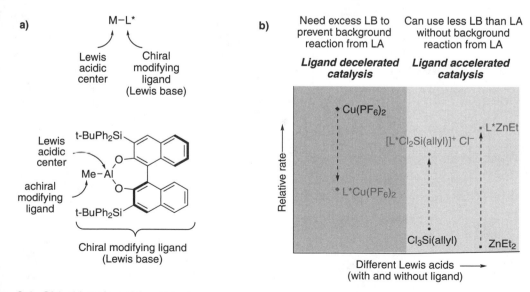

Figure 2.1. Chiral Lewis acid catalysts.

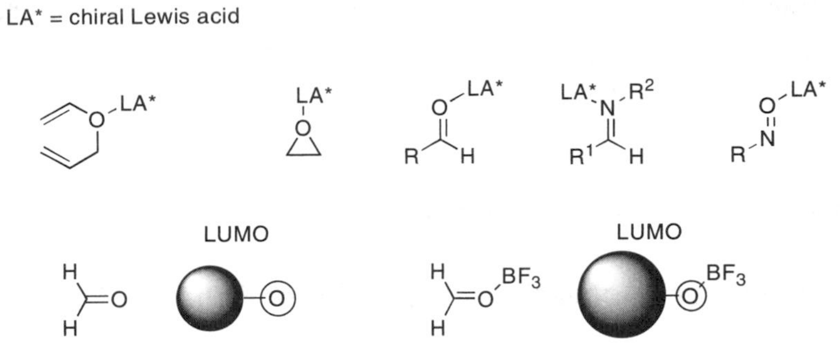

LA* = chiral Lewis acid

Figure 2.2. Activation by chiral Lewis acid catalysts.

ric achiral Lewis acid and substoichiometric chiral Lewis base are treated in Section 2.2.1.

In the cases where chiral Lewis base ligand gives rise to a slower catalyst (i.e., ligand-decelerated catalysis; see the left side of **Figure 2.1**), it is imperative that the amount of unliganded Lewis acid be limited to prevent incursion of a competing racemic pathway (see Section 1.2).

The presence of Lewis basic coordinating groups (ethers, epoxides, esters, ketones, aldehydes, imines, nitrones, nitro groups, nitrosyls, sulfonyls, phosphonyls, alcohols, sulfides, etc.) in many organic reaction substrates accounts for the widespread use of chiral Lewis acid catalysts (**Figure 2.2**). Lewis acids perturb the orbital levels and electron distribution of coordinated substrates. Usually, the resultant lowering of the lowest unoccupied molecular orbital (LUMO) energy level brings it closer in energy to the highest occupied molecular orbital (HOMO) of a nucleophilic partner. As a result, Lewis acids provide a high level of electrophilic activation. For example, BF_3-coordinated formaldehyde has a lower-energy LUMO compared to formaldehyde (**Figure 2.2**). As a result, the carbonyl carbon of the coordinated formaldehyde is more electrophilic.

Because of the large number of chiral Lewis acid catalysts that have been described in a diverse array of transformations, most of the variants in activation (direct activation, indirect activation, one-point substrate binding, two-point substrate binding, etc.) are well documented. Several examples are outlined in the rest of Section 2.1, highlighting the most common modes of Lewis acid activation.

2.1.1 Lewis Acid Direct Activation

Direct activation is used here to refer to Lewis acid coordination to at least one of the reacting centers of an electrophile (**Figure 2.3**). In contrast, indirect activation involves coordination of Lewis acid to a nonreacting atom. In direct or indirect activation, there may be one or more points of contact between the substrate and the Lewis acid center, as well as further contacts between the substrate and other portions of the catalyst (see Chapter 5).

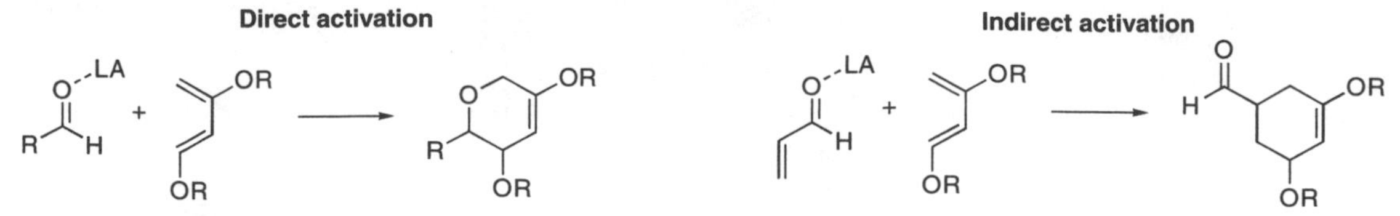

Figure 2.3. Direct vs. indirect Lewis acid activation.

2.1.1.a Lewis Acid Direct Activation via One-Point Binding

One-point binding entails formation of one coordination bond between the Lewis acid and electrophile in the activated complex. Cases in which the catalytic mode is unambiguously due to direct Lewis acid activation (vs. bifunctional activation) via one-point binding (vs. secondary catalyst–substrate interactions) are actually not as common as you might believe. An example in an intramolecular process can be found in the ene reaction in **Figure 2.4**.[2,3] Here, activation occurs via one-point coordination to the aldehyde, causing polarization of the C=O bond and lowering of the LUMO energy. As a consequence, the activation energy is lowered relative to the background reaction. The lowest energy pathways to the two enantiomers of the trans product are now diastereomeric (see Chapter 1), resulting in the observed selectivity.

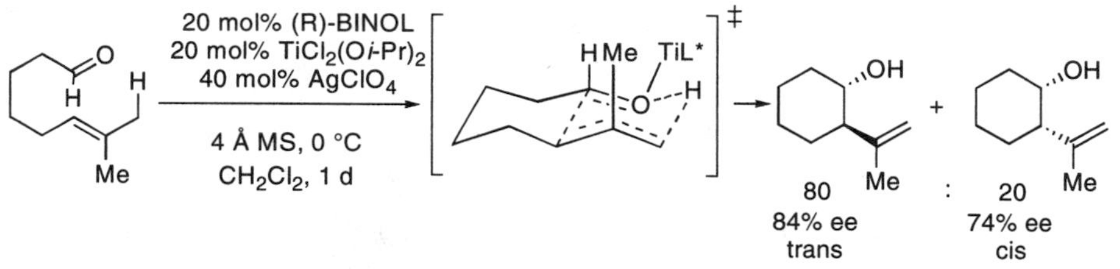

Figure 2.4. Catalytic Lewis acid catalyst in a unimolecular ene reaction.

Intermolecular cases are also known, as illustrated in the asymmetric titanium[3–5] catalyzed ene reaction in **Figure 2.5**.

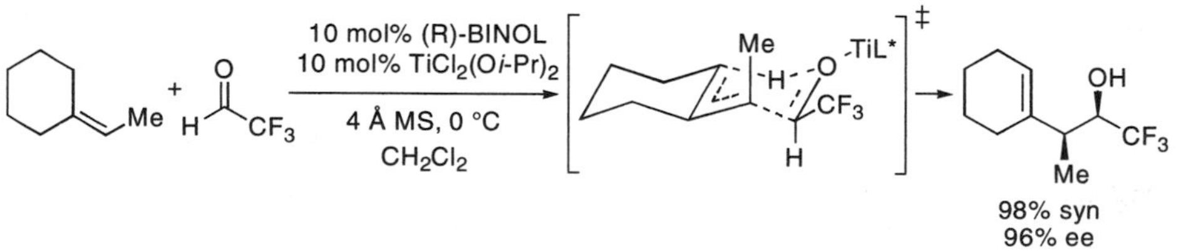

Figure 2.5. Catalytic Lewis acid catalysts in an intermolecular ene reaction.

A cycloaddition involving direct Lewis acid activation of an enal is illustrated in **Figure 2.6**.[6] Here, the chromium Schiff base adduct coordinates the aldehyde activating it as well as the conjugated alkene (compare to indirect activation below) for the hetero-Diels–Alder reaction. While a two-step mechanism involving conjugate addition of the enol ether to the α,β-unsaturated system cannot be excluded, the high diastereoselection observed (> 97.5:2.5) points to either a concerted, asynchronous transition state or highly ordered transition states in the two-step processes (i.e., both steps incorporating the chromium catalyst). From X-ray crystallography, the chromium catalyst exists as the illustrated dimer bridged via a water molecule. On the basis of solution molecular-weight and kinetic studies, it appears that this dimeric structure is maintained in the catalytic cycle. With each chromium bearing water at the terminal centers, dissociation of a water molecule is needed to open a coordination site for substrate complexation and may explain the crucial role of molecular sieves in this process. In contrast to the above examples, this case illustrates how the activated center (the carbonyl oxygen) can undergo a bond-forming as well as a bond-breaking process while being activated by a Lewis acid catalyst.

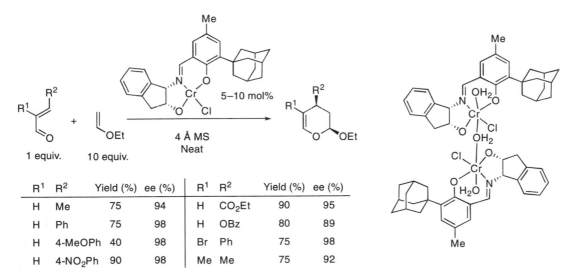

R¹	R²	Yield (%)	ee (%)	R¹	R²	Yield (%)	ee (%)
H	Me	75	94	H	CO₂Et	90	95
H	Ph	75	98	H	OBz	80	89
H	4-MeOPh	40	98	Br	Ph	75	98
H	4-NO₂Ph	90	98	Me	Me	75	92

Figure 2.6. Chiral Lewis acid catalyst in a hetero–Diels–Alder reactions.

Direct Lewis acid activation by one-point binding is possible with many other functional groups besides carbonyls (see **Figure 2.2**), making this activation mode highly prevalent in asymmetric catalysis.

2.1.1.b Lewis Acid Direct Activation via Two-Point Binding

Conceptually, two-point binding of a substrate to a catalyst is advantageous in that the resultant adduct will be conformationally restricted and well-defined relative to one-point binding adducts, where rotation about the Lewis acid coordination bond may be possible, leading to multiple reactive conformations. The restricted conformations with the chelating substrate greatly reduce the number of possible transition states. Furthermore, additional electrophilic activation of the substrate may be possible via the second point of connection between the Lewis acid and substrate. The chelation effect can also lead to a greater equilibrium constant between the unbound substrate and catalyst–substrate adduct, leading to a greater concentration of the activated adduct. In such instances, increased catalytic activity (e.g., greater turnover frequency, greater turnover number, or reaction at lower temperature) can be observed.

On the other hand, the prerequisite of a neighboring group to form the chelate limits the generality of the process.

For example, substrate activation via two-point binding was crucial in the development of the asymmetric Mukaiyama aldol reaction illustrated in **Figure 2.7**[7,8] The benzyloxy group of α-benzyloxyacetaldehyde was necessary for the high selectivity observed. This trend is in accord with binding of both the benzyloxy and aldehyde oxygens to the cationic copper catalyst. Support for this activation mode was found in the crystallographic structure of the cationic PhPyBox–copper species complexed to the aldehyde substrate (**Figure 2.7**). Importantly, the enantioselection anticipated from this isolated adduct correlated with the behavior of the catalyzed reaction process.

The Mukaiyama aldol process shown in **Figure 2.7** entails a singular Lewis acid catalyst, as supported by crossover experiments. With the copper catalyst there is complete scrambling of the silicon groups (**Figure 2.8**) from the silyl ketene acetal substrate, indicating that there is no intramolecular silyl transfer. This experiment rules out Lewis base coordination of the substrate silicon with the catalyst and subsequent intramolecular shuttling (**Figure 2.9**). Strikingly, the reaction proceeds with high enantioselectivity, even though an achiral pathway may arise from the silyl cation that transfers intermolecularly (via SbF_6^- or substrate). This situation stands in stark contrast to that observed with a $(BINOLate)TiCl_2$ catalyst in the same reaction.[9] In that case, one-point binding is inferred based upon similar high levels of selectivity with substrates that cannot chelate. Furthermore, the silyl group undergoes intramolecular transfer (**Figure 2.8**). A potential role of the catalyst to activate the silicon group via coordination to the BINOLate oxygens is illustrated in **Figure 2.9**.

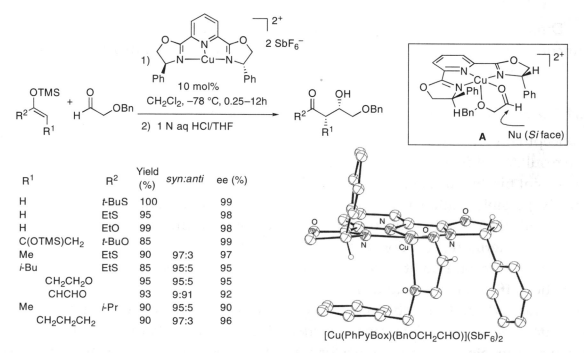

R^1	R^2	Yield (%)	syn:anti	ee (%)
H	t-BuS	100		99
H	EtS	95		98
H	EtO	99		98
C(OTMS)CH$_2$	t-BuO	85		99
Me	EtS	90	97:3	97
i-Bu	EtS	85	95:5	95
CH$_2$CH$_2$O		95	95:5	95
CHCHO		93	9:91	92
Me	i-Pr	90	95:5	90
CH$_2$CH$_2$CH$_2$		90	97:3	96

[Cu(PhPyBox)(BnOCH$_2$CHO)](SbF$_6$)$_2$

Figure 2.7. The asymmetric Mukaiyama aldol reaction of α-benzyloxyacetaldehyde with a PhPyBox–copper catalyst. The crystal structure of the substrate catalyst adduct is illustrated.

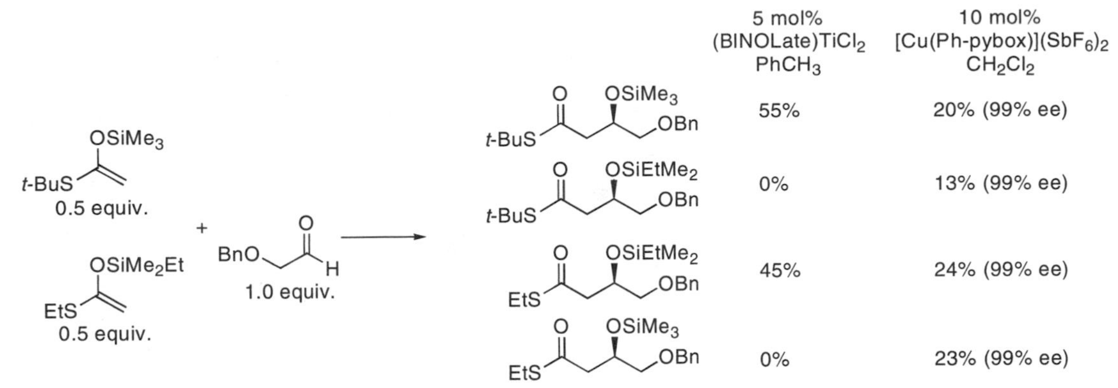

	5 mol% (BINOLate)TiCl$_2$ PhCH$_3$	10 mol% [Cu(Ph-pybox)](SbF$_6$)$_2$ CH$_2$Cl$_2$
t-BuS—C(=O)—CH$_2$—CH(OSiMe$_3$)—CH$_2$OBn	55%	20% (99% ee)
t-BuS—C(=O)—CH$_2$—CH(OSiEtMe$_2$)—CH$_2$OBn	0%	13% (99% ee)
EtS—C(=O)—CH$_2$—CH(OSiEtMe$_2$)—CH$_2$OBn	45%	24% (99% ee)
EtS—C(=O)—CH$_2$—CH(OSiMe$_3$)—CH$_2$OBn	0%	23% (99% ee)

Figure 2.8. Crossover experiments with doubly labeled silylketene acetals in the (BINOLate)TiCl$_2$ and copper PyBox catalyzed (from **Figure 2.7**) benzyloxyacetaldehyde Mukaiyama aldol reaction.

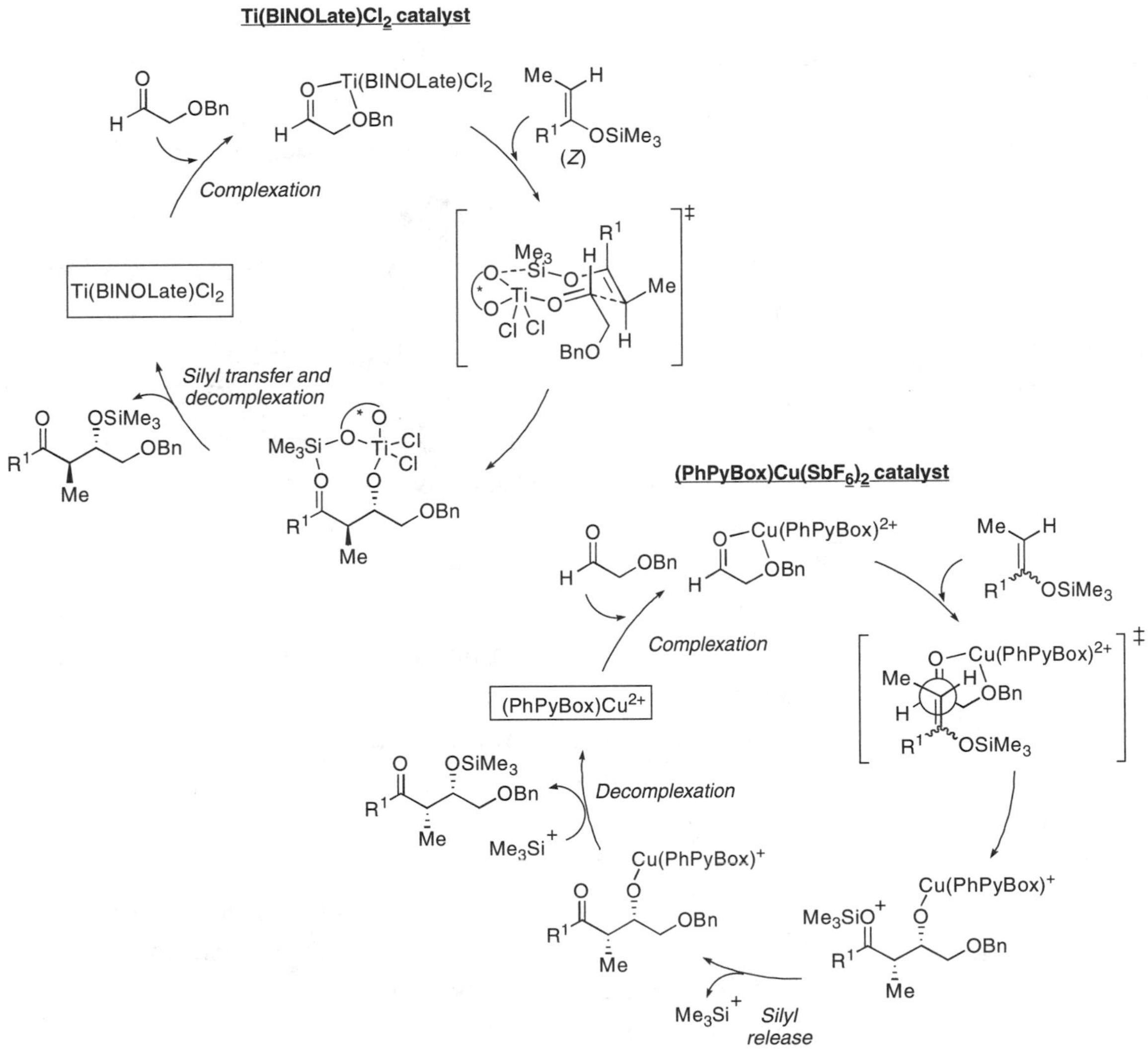

Figure 2.9. Mechanism of the (BINOLate)TiCl$_2$ and copper PyBox catalyzed (from **Figure 2.7**) benzyloxyacetaldehyde Mukaiyama aldol reaction.

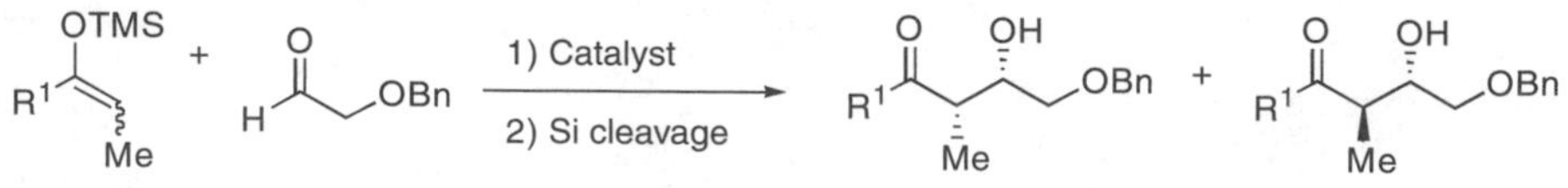

Figure 2.10. Different stereochemical outcomes from two-point and one-point substrate coordination to catalyst.

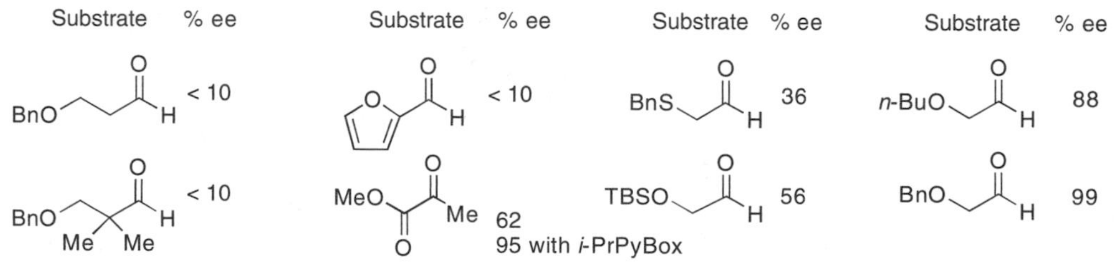

Catalyst	R^1	E:Z	Yield (%)	syn:anti	ee (%)	
10 mol% (PhPyBox)Cu(SbF$_6$)$_2$	EtS	5:95	90	97:3	97	*Stereoconvergent*
	EtS	99:1	48	86:14	85	
5 mol% (BINOLate)TiCl$_2$	t-BuS	7:93	72	8:92	90	*Stereodivergent*
	EtS	77:23	85	72:28	90	

Figure 2.11. Electrophile scope in the two-point binding copper PyBox catalysts (Figure 2.7: R^1 = H, R^2 = t-BuS).

On the other hand, one-point versus two-point substrate binding can allow access to different stereochemical arrays. For example, the chelation imposed in the two-point binding in the PyBox–copper-catalyzed reactions, combined with an open transition state, leads to the *syn* stereochemistry, regardless of the silyl ketene acetal geometry (**Figure 2.10**). In contrast, the one-point binding in the BINOLate–titanium-catalyzed reaction and the closed transition state allows formation of the *anti* isomer from the (Z)-silyl ketene acetal.

In practice, the two-point binding requirement places restrictions on the substrates that limit the scope of such transformations, as was seen with the aldehyde component in the PhPyBox–copper-catalyzed Mukaiyama aldol reactions (**Figure 2.11**).

A chelation approach has been applied to numerous other bidentate substrates. For example, the ketophosphonates illustrated in **Figure 2.12**[10,11] form five-membered chelates with chiral Lewis acids. From these adducts the activated carbonyl can participate in an asynchronous, concerted hetero-Diels–Alder to form a variety of dihydropyran derivatives.

Intramolecular reactions are also viable with this activation mode. For example, the allyl ethers formed from the enolic tautomers of α-ketoesters undergo two-point binding to chiral copper catalysts (**Figure 2.13**). These activated adducts subsequently undergo facile, enantioselective Claisen rearrangement.[12]

Direct Lewis acid activation by two-point binding is possible with many functional groups (see **Figure 2.2**), making this activation mode very powerful. However,

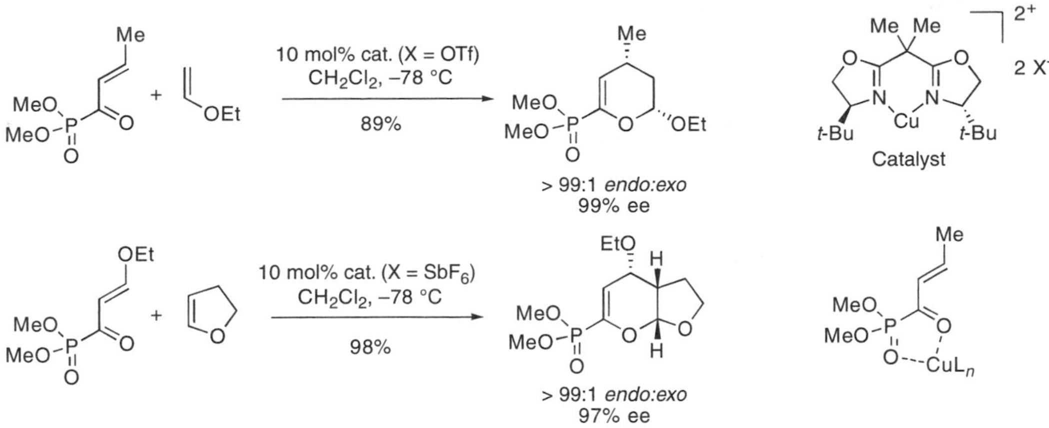

Figure 2.12. Hetero-Diels–Alder reactions utilizing two-point binding substrates.

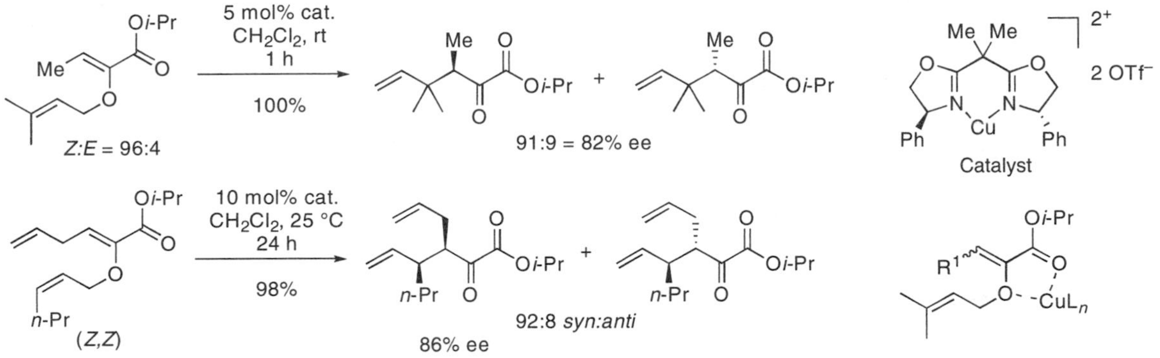

Figure 2.13. Claisen rearrangement with two-point binding.

it is important to note that selectivity is not necessarily higher with two-point binding; other factors in one-point binding scenarios (e.g., restricted rotation about the Lewis acid coordination bond) often allow high facial differentiation of a substrate and hence high stereoselection. Finally, both "points" in two-point binding do not have to occur via the same type of interaction (e.g., in the preceding cases with a Lewis acid). For a further discussion of nonclassical two-point catalyst–substrate interactions, see Chapter 5.

2.1.2 Lewis Acid Activation via a Conjugated System

In contrast to direct activation, a catalyst may also activate the reacting centers of a substrate indirectly. This type of activation is most commonly accomplished by way of a conjugated π system (see **Figure 2.3**). As was the case for direct activation, one or more points of contact, mediated by covalent, dative, or noncovalent (see Chapter 5) interactions, may exist between the catalyst and substrate.

2.1.2.a Lewis Acid Activation via a Conjugated System with One-Point Binding

In many instances, none of the atoms from the bonds formed or broken during the reaction are directly bound to the catalyst. Rather, indirect activation of the reactive sites is instigated via a π system. This phenomenon is most commonly observed in the Lewis acid activation of an α,β-unsaturated carbonyl, which lowers the LUMO energy of the electrophilic partner, thereby decreasing the HOMO–LUMO energy gap and facilitating reaction (**Figure 2.14**). Transformations that commonly utilize this activation mode include the Diels–Alder reaction, as well as numerous conjugate additions including Michael-type reactions.

The conformational mobility allowed in a one-point binding mode can create problems in indirect activation. These issues are diagrammed for an α,β-unsaturated carbonyl system in **Figure 2.15**. Not only is the position of the Lewis acid flexible, but the relationship of the alkene to the carbonyl, and hence to the Lewis acid, can vary.

For most α,β-unsaturated carbonyls, singular Lewis acid binding modes can be achieved, allowing for selective asymmetric catalysis (**Figure 2.16**). Greater control is possible under two-point binding conditions (e.g., the imide in **Figure 2.16**), as will be discussed in the next section. With ketones, control of Lewis acid binding is diffi-

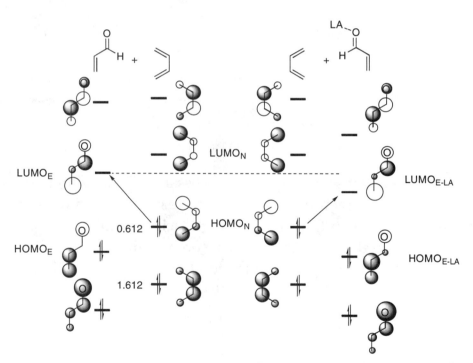

Figure 2.14. Hückel MO diagrams for a Diels–Alder reaction with and without Lewis acid activation of a conjugated carbonyl.

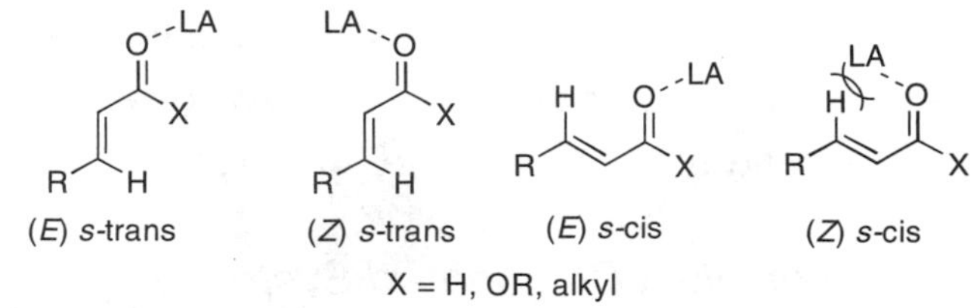

Figure 2.15. Conformers encountered in binding of a Lewis acid to an acylic α,β-unsaturated carbonyl.

cult, especially if the size of the groups flanking the carbonyl are similar. For an alternate approach that provides a solution in this system, see Section 2.2.2.c.

Examples of successful asymmetric catalysis via indirect activation with one-point binding have been realized for the first three examples from **Figure 2.16** (**Figure 2.17**). For instance, the Lewis acid aluminum catalysts derived from a vaulted biaryl was shown to be highly selective in the reaction of methacrolein with cyclopentadiene.[13]

Indirect activation of α,β-unsaturated esters has also been reported, as illustrated in **Figure 2.17**.[14] Here a boron Lewis acid coordinates the carbonyl and activates the dienophile. A secondary interaction between the naphthalene ring of the catalyst and

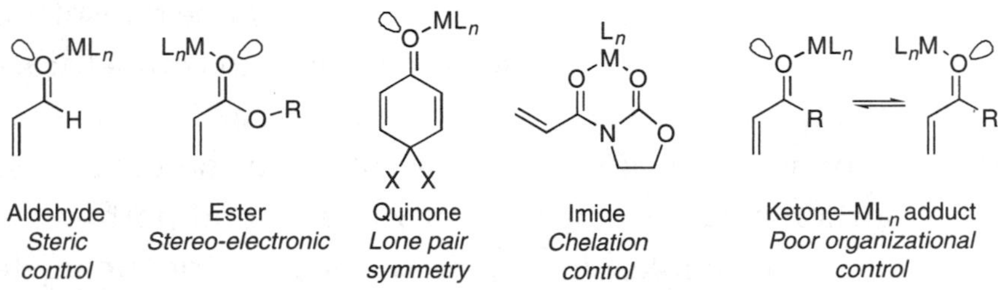

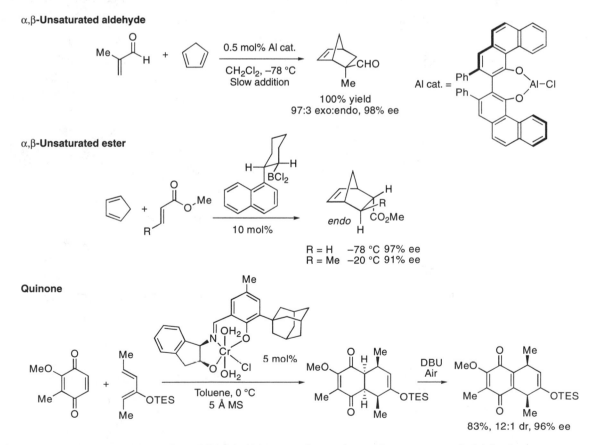

Figure 2.16. Successful and unsuccessful Lewis acid catalysis with various α,β-unsaturated carbonyl compounds.

Figure 2.17. Lewis acid catalyzed Diels–Alder reactions of an α,β-unsaturated aldehyde, an α,β-unsaturated ester, and a quinone.

the carbonyl of the substrate is also proposed to stabilize the reactive pathway, as well as provide stereochemical control. This type of secondary interaction is discussed further in Chapter 5 (see **Equation 5.1** and **Figure 5.3**) and is distinct from the two-point binding described in Section 2.1.2.b, where a Lewis acid acts upon two Lewis basic portions of a substrate.

Indirect activation of quinones via one-point binding can be found in the Diels–Alder reaction outlined in **Figure 2.17**.[15] In contrast to the related Diels–Alder reaction in **Figure 2.6**, here the Lewis acid takes the form of the monomer indicated. The dimeric form was significantly slower than the monomer, which was prepared selectively under acidic conditions. The product formed via Lewis acid activation and [4+2] addition is unstable and was converted directly to the quinone for isolation. The reaction selectivity is high, even with substituted dienes and unsymmetrical quinones.

These examples may give the impression that indirect activation is only viable for the Diels–Alder reaction. In fact, any reaction involving a conjugated system, where activation of one portion enhances the reactivity of a second portion, can be subject to indirect activation. For example, highly efficient asymmetric conjugate addition of silyl ketene thioacetals to α,β-unsaturated ketones can be catalyzed with a chiral boron Lewis acid that coordinates the ketone moiety (**Figure 2.18**).[16,17]

While the preceding examples have focused on α,β-unsaturated carbonyl derivatives, the same principles apply to many similar substrates of the type X=Y–Z, including those listed in **Figure 2.19**. In practice, much less attention has been given to these related compounds.

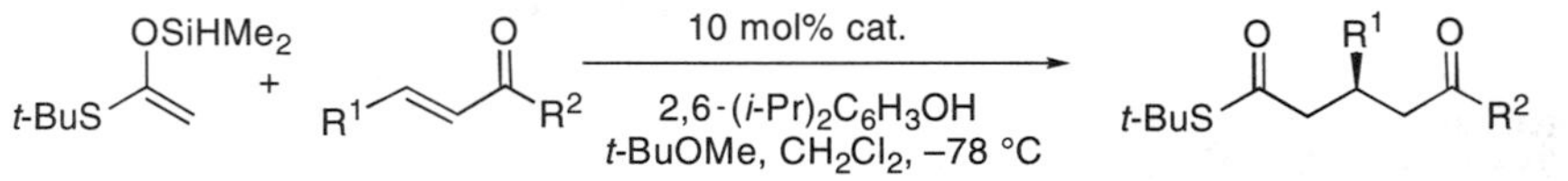

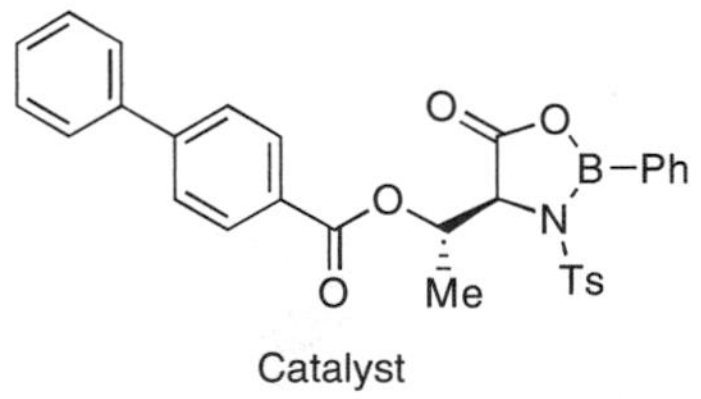

R^1	R^2	Yield (%)	ee (%)
Ph	Me	83	95
4-MeOPh	Me	54	92
4-CF$_3$Ph	Me	81	92
Me	Me	75	98
BnOCH$_2$	Me	60	90
Ph	i-Pr	64	88

Figure 2.18. Lewis acid catalyzed asymmetric conjugate addition of silyl ketene thioacetals to α,β-unsaturated ketones.

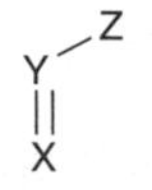

X = C, NR, etc.
Y = C, N, etc.
Z = COH, COR, CO$_2$R, CONR$_2$, CN, NO$_2$, POR$_2$, SOR, SO$_2$R, etc.

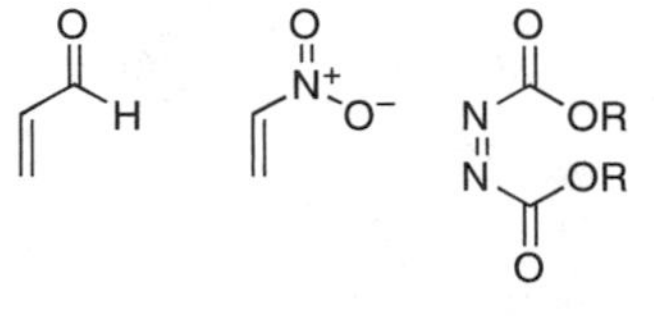

Figure 2.19. Substrates amenable to indirect Lewis acid activation.

2.1.2.b Lewis Acid Activation via a Conjugated System with Two-Point Binding

All of the advantages and disadvantages of two-point binding that apply to direct activation (Section 2.1.1.b) pertain to indirect activation as well. In addition, the greater number of conformational isomers in conjugated systems (see **Figure 2.15**) make the catalyst conformational restriction available with two-point binding particularly attractive (see **Figure 2.16**). A seminal example of two-point binding in asymmetric catalysis can be found in the oxazolidinone Diels–Alder reaction (**Figure 2.20**).[18] The α,β-unsaturated oxazolidinone confers two advantages. First, upon bidentate coordination to a Lewis acid, the orientation of the amide carbonyl is constrained, which enforces an s-cis olefin conformation as the s-trans conformation encounters $A^{1,3}$ strain with the five-membered oxazolidinone ring (see the model in **Figure 4.33**). Second, the bidentate coordination to the Lewis acid affords a greater degree of activation than monodentate coordination to the corresponding α,β-unsaturated amide. Evidence for the bidentate coordination is based upon the x-ray crystallographic structure of a closely related adduct[19] (see **Figures 4.31–4.33** for more detail).

The titanium catalyst/α,β-unsaturated oxazolidinone combination described above has been found to be useful in further reactions, including [2+2][20] and [3+2][21] cycloadditions. However, numerous titanium–ligand–substrate adduct geometries are possible besides that shown (see Section 4.3.2), providing multiple potential pathways to the minor enantiomers and eroding selectivity. As a consequence, high enantioselectivies are confined to a small number of substrates. In addition, the titanium catalysts are sensitive to air and water.

The discovery of bisoxazoline catalysts in conjunction with α,β-unsaturated oxazolidinones provided a solution to these problems. Numerous metal–bisoxazoline catalysts have been found selective in this same Diels–Alder reaction.[22–24] The copper catalysts outlined in **Figure 2.21** have been explored in detail.[25–28] Upon bidentate coordination, the rigid bisoxazoline creates a well-defined asymmetric environment, where the *tert*-butyl groups block one face of the reacting olefin (**Figure 2.21**). The net result is a highly efficient system for the Diels–Alder reaction of a much wider variety of dienes with several α,β-unsaturated oxazolidinone derivatives (**Figure 2.21**).

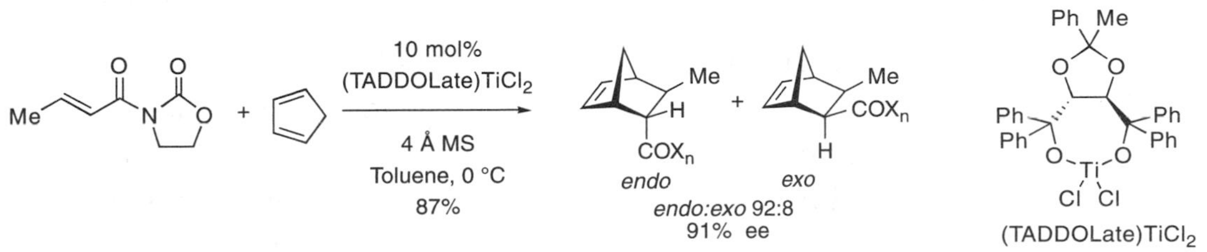

Figure 2.20. α,β-Unsaturated oxazolidinone Diels–Alder reaction with a TADDOLate–titanium catalyst.

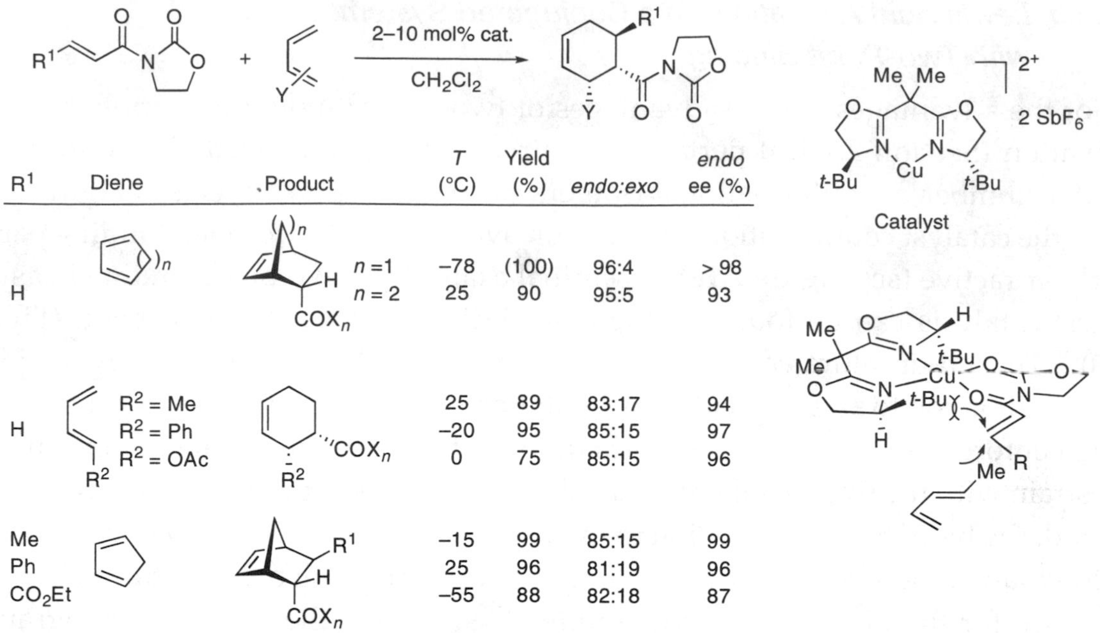

Figure 2.21. α,β-Unsaturated oxazolidinone Diels–Alder reaction with a bisoxazoline catalyst.

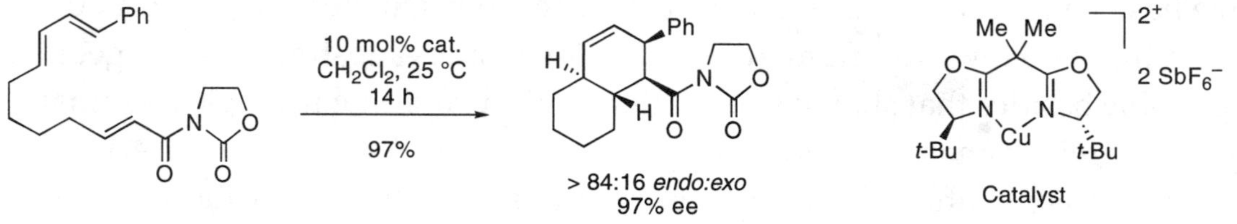

R¹	Diene	Product		T (°C)	Yield (%)	endo:exo	endo ee (%)
H			n = 1	−78	(100)	96:4	> 98
			n = 2	25	90	95:5	93
H	R² = Me			25	89	83:17	94
	R² = Ph			−20	95	85:15	97
	R² = OAc			0	75	85:15	96
Me				−15	99	85:15	99
Ph				25	96	81:19	96
CO₂Et				−55	88	82:18	87

Figure 2.22. Intramolecular oxazolidinone Diels–Alder reaction.

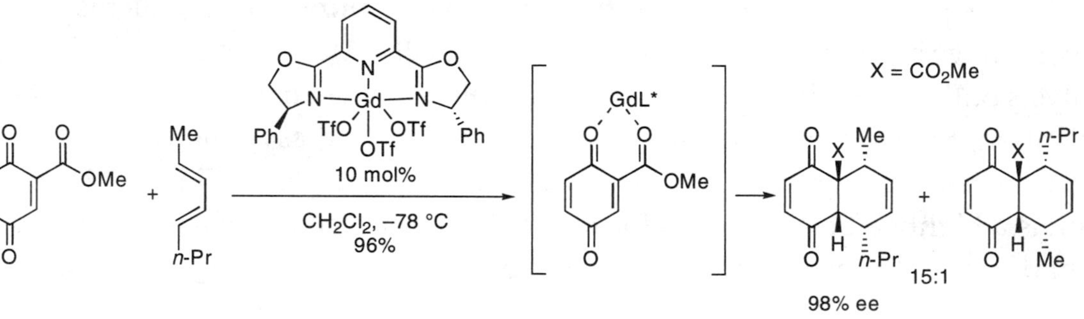

Figure 2.23. Diels–Alder reaction with two-point binding quinone substrates.

Unimolecular variants based upon the above principles are also possible. For example, the intramolecular Diels–Alder reactions with the oxazolidinone-derived substrates[28] are highly successful (**Figure 2.22**).

In general, the bisoxazoline- and pyridine–bisoxazoline-derived catalysts are very useful with chelating substrates.[22–24,29] They have been applied to numerous bidentate substrates, with those forming six-membered[30] (**Figure 2.23**) and five-membered[31] (**Figure 2.24**) chelates being the most successful.

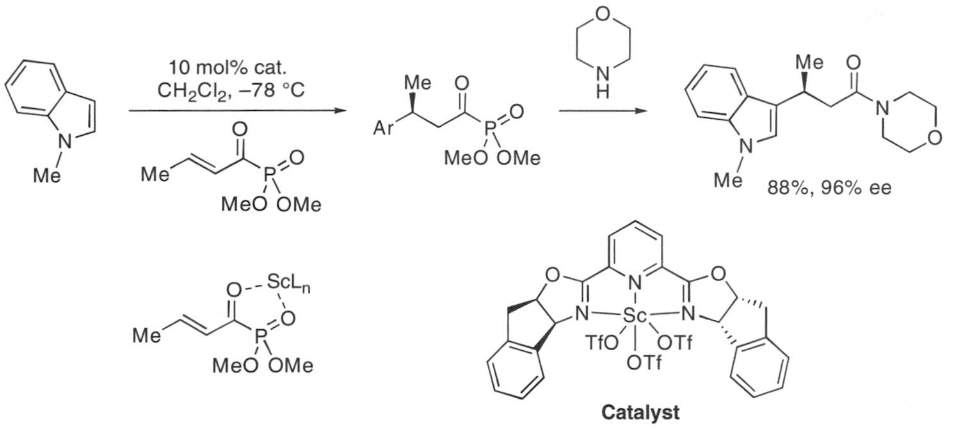

Figure 2.24. Friedel-Crafts reaction with two-point binding α-ketophosphonate substrates.

The case in **Figure 2.24** illustrates that this type of activation mode is not restricted to Diels–Alder reactions.[31] Good results have been obtained in the conjugate additions of a variety of systems, including silyl ketene acetals to α,β-unsaturated oxazolidinones[32] and diazo compounds.[33]

2.2 Lewis Base Catalysis

Due to the host of organic reactions involving Lewis basic functional groups (ethers, epoxides, esters, ketones, aldehydes, imines, nitrones, nitro groups, nitrosyls, sulfonyls, phosphonyls, alcohols, sulfides, etc.), the use of chiral Lewis acids has become widespread in asymmetric catalysis. The well-understood activating function of Lewis acids (see Section 2.1) allows the design and rapid screening of chiral catalytic versions. Even so, there are downsides to chiral Lewis acid catalysis. High catalyst loadings (or cost of the catalyst, including the chiral ligand, per turnover) are employed compared to the most efficient catalyst systems available (i.e., transition-metal catalysts for hydrogenation). Furthermore, reactions not amenable to Lewis acid activation can not be employed.

While initially less common than Lewis acid catalysis, Lewis base catalysis has become quite prominent, permitting access to a whole new range of chemistries. Lewis base catalysts can function via two different modes involving covalent (**Figure 2.25a** and **2.25b**) or noncovalent (dative) substrate modification (**Figure 2.25c** and **2.25d**). The first activation mode (**Figure 2.25a** and **2.25b**) is far more prevalent, where the chiral Lewis base (LB) undergoes a reaction with one of the substrates (S1,S2), forming a covalent bond to yield a reactive chiral adduct (S1′–LB). If the S1′ portion is achiral, then a chirality-transfer process (see Section A.1.1.g in Appendix A) between S1′–LB and another substrate, S2, can ensue (**Figure 2.25a**). In this type of reaction, the LB is displaced while transmitting asymmetry to the new stereocenter(s)

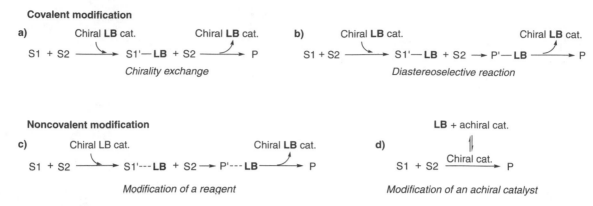

Figure 2.25. Activation modes for Lewis base catalysts.

formed upon combination of S1′ and S2. The result is an enantioenriched product, P. More commonly, a diastereoselective reaction (**Figure 2.25b**) occurs between either S1 and LB or S1′–LB and S2; the LB portion controls the newly established stereocenter(s) that ultimately emerge in the product (P). In both of these cases, turnover of the chiral Lewis base by facile cleavage of a covalent bond is a key feature of a useful catalyst.

An alternate, and less common, mode of Lewis base catalysis involves a substoichiometric amount of a chiral Lewis base, forming a noncovalent adduct during a reaction (**Figure 2.25c** and **2.25d**). When a stoichiometric reagent undergoes Lewis base activation (**Figure 2.25c**), it is crucial that the S1′–LB adduct be much more reactive than the reagent (S1) alone. In other words, ligand-accelerated catalysis (i.e., the far right of the plot in **Figure 2.1**) is needed to prevent a racemic background reaction. The LB must also be able to undergo facile exchange from P′–LB to S1′–LB in order for turnover to occur.

When the Lewis base activates a substoichiometric reagent (i.e., an achiral catalyst), the scenario reverts to a chiral Lewis acid catalyst (**Figure 2.25d**) which is covered in Section 2.1.

2.2.1 Lewis Base Catalysis via Noncovalent Substrate Modification

Because typical organic compounds possess few functional groups that can productively interact with Lewis bases in a *noncovalent* manner, chiral Lewis base catalysts have not enjoyed the same widespread utility as chiral Lewis acid catalysts. However, some organic systems, such as organosilanes, are known to interact with Lewis bases, producing species in which the ordinary number of bonds (four for neutral silicon) is exceeded. These hypercoordinate species often exhibit enhanced reactivity.[34] If turnover of a substoichiometric chiral Lewis base can occur, then a reaction that is catalytic in the Lewis base can be developed.

The most important consideration in this approach is the greater reactivity of the chiral hypercoordinate silicon species with respect to its corresponding achiral tetracoordinate form. If there is not a large differential, then the background reaction from

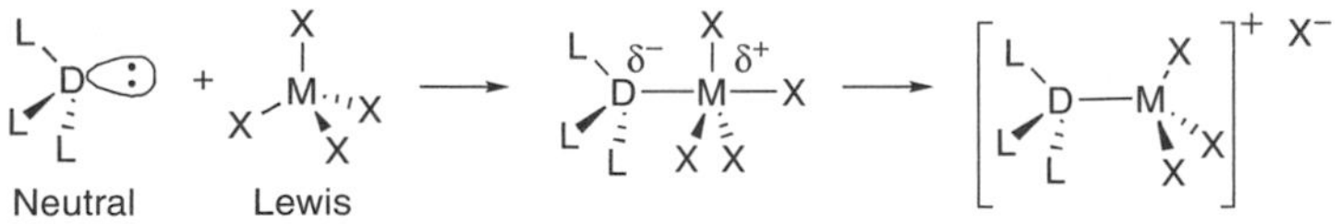

Figure 2.26. Polarization in a Lewis acid–Lewis base adduct.

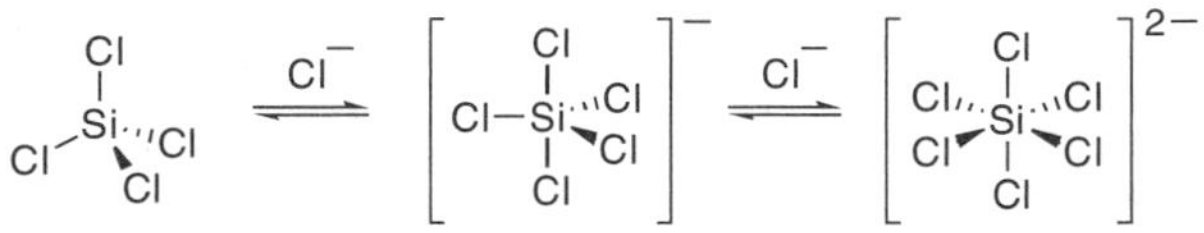

Charges: Si +0.178   Si +0.229   Si +0.539
          Cl −0.045 Cl_{eq}−0.139   Cl −0.423
                         Cl_{ax}−0.430

Figure 2.27. The increase in silicon positive charge upon formation of hypervalent species with a Lewis base.

the achiral species will cause a predominantly racemic process to occur (see Chapter 1). The basic premise behind chiral Lewis base activation of an achiral Lewis acid is outlined in **Figure 2.26**.[35,36] Addition of a neutral donor to a Lewis acid with polarizable M–X bonds will cause greater partial positive charge to accumulate on the metal center, which is offset by greater negative charge moving to the X groups. At the far end of the spectrum, ionization results in a complete shift of a negative charge to an X group along with concomitant formation of a cationic Lewis acid. For silicon species, calculations reveal that even a negatively charged ligand can lead to *greater* partial positive charge on the silicon (**Figure 2.27**).[37] This counterintuitive result provides the basis for the reversible activation of achiral Lewis acids with chiral bases.

An example of noncovalent Lewis base activation of silicon species can be found with the chiral phosphoramide catalysts. In this work, two distinct approaches were employed, utilizing silicon-containing substrates or silicon tetrachloride. Early work focused on the former, where preformed trichlorosilane substrates[38–41] were employed, such as the trichlorosilyl enolate illustrated in **Figure 2.28**.[42–44] Here, the phosphoramide directly activates one of the organic substrates.

An intriguing effect was noted between the catalyst loading and the *anti:syn* ratio.[45] As the catalyst loading decreased, the amount of *anti* diastereomer decreased. This observation is consistent with two molecules of (S,S)-Ph$_2$ cat being needed to form the *anti* diastereomer and one to form the *syn*. As the concentration of (S,S)-Ph$_2$ cat decreases, the second-order pathway is less favorable. At low phosphoramide concentration, a first-order pathway leading to the *syn* diastereomer occurs to a greater extent. A first-order pathway is more favorable with more hindered ligands, such as (S,S)-Ph$_4$ cat. Accordingly, use of (S,S)-Ph$_4$ cat forms the *syn* diastereomer exclusively.

On this basis the mechanism in **Figure 2.29**, involving a highly electrophilic, phosphoramide-bound silyl cation, was proposed.[38] The catalytic cycle is initiated by

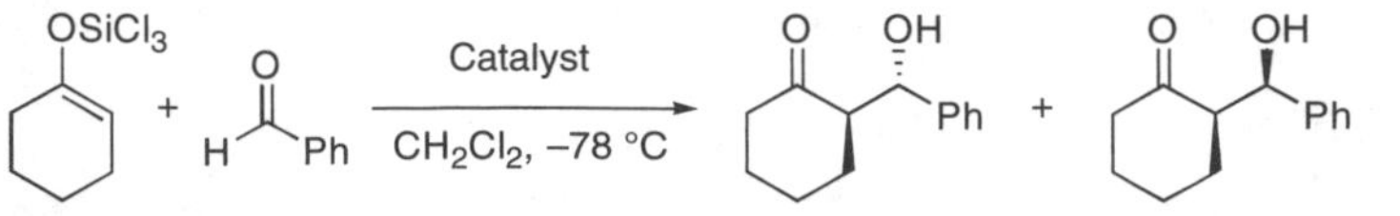

Catalyst	Mol%	Yield (%)	anti:syn	ee major (%)
(S,S)-Ph$_2$ cat	10	95	98:2	92
(S,S)-Ph$_2$ cat	2	96	97:3	91
(S,S)-Ph$_2$ cat	0.5	53	83:17	91
(S,S)-Ph$_4$ cat	10	94	1:99	51

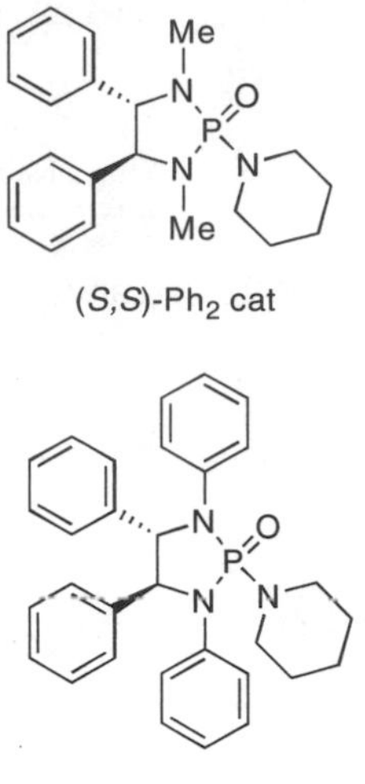

Figure 2.28. Lewis base catalyzed aldol addition with a trichlorosilyl enolate.

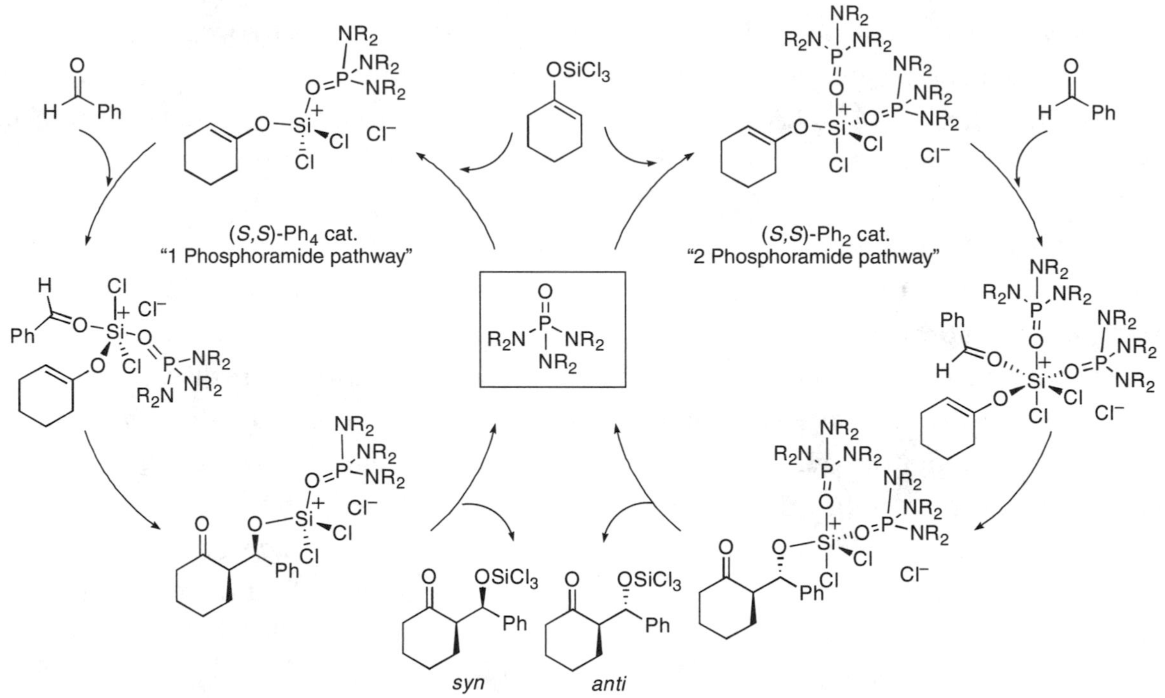

Figure 2.29. The two mechanisms in the phosphoramide-catalyzed aldol addition with a trichlorosilyl enolate.

binding of one or two phosphoramide molecules to the weakly Lewis acidic silicon of the enol trichlorosilane. This binding leads to ionization of a chloride ion and generates two chiral, trichlorosilyl-cation reagents. The adduct with two phosphoramides appears more reactive, which is in accord with the charge trends outlined in **Figure 2.27**. This species can then bind the aldehyde, which initiates the aldol reaction. With release of the trichlorosilyl–aldolate product from the chiral phosphoramide, the catalytic cycle is completed.

Further evidence supporting this reaction pathway included second-order kinetics for the (S,S)-Ph$_2$ cat, indicating the presence of the bisadduct.[46] Dissociation of the chlorides to provide a cationic species was supported by a significant rate suppression in the presence of added n-Bu$_4$NCl.[45]

The culmination of this mechanistic study was the generation of a linked bisphosphoramide catalyst (**Figure 2.30**), which shifted the reaction to the "two phosphoramide pathway" (**Figure 2.29**). This double activation of the substrate provided superior asymmetric induction and alleviated the strong concentration dependence seen in the former second-order process.

In a further development, the organic substrate was activated by means of a *separate* silicon reagent. An example can be seen in the aldol reactions with vinylogous silyl ketene acetals in **Figure 2.30**.[47–49] Under these conditions, silicon tetrachloride is employed as a reagent that, in the presence of catalytic chiral bisphosphoramide, generates the active chiral Lewis acid in situ (**Figure 2.31**).[48] This species then promotes the reaction of the nucleophile and electrophile. Release of the chiral bisphosphoramide generates the product that incorporates the trichlorosilyl group. Thus, the process is not catalytic in the achiral silicon tetrachloride Lewis acid, but is catalytic in the chiral bisphosphoramide Lewis base. This is an extreme example of ligand-accelerated catalysis (see **Figure 2.1**), since silicon tetrachloride without phosphoramide causes no reaction.

The use of silicon tetrachloride, rather than a preformed enol or allyl trichlorosilane, permits utilization of a wider range of substrates. Several reaction classes, including the additions of allylstannanes to aldehydes[50] and the additions of more varied enolic substrates to aldehydes,[48,51] have benefited from this strategy. Furthermore, application of this strategy has been fruitful with other Lewis bases.[52]

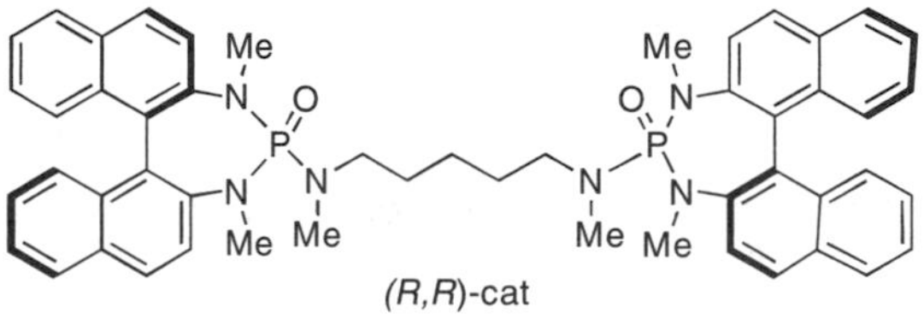

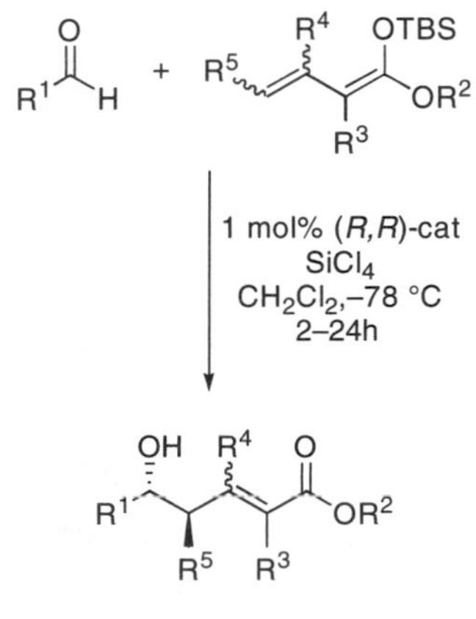

R^1	R^2	R^3	R^4	R^5	Yield (%)	ee (%)
Ph	Et	H	H	H	89	98
PhCH=CH	Et	H	H	H	84	97
PhCH$_2$CH$_2$	Et	H	H	H	68[a]	91
Ph	Me	Me	H	H	93	99
Ph	Et	H	Me	H	91	92
Ph	t-Bu	H	H	Me	92	89

[a]5 mol% (R,R)-cat

Figure 2.30. Bisphosphoramide Lewis base catalyst in the Mukaiyama aldol reaction with vinylogous silyl ketene acetals.

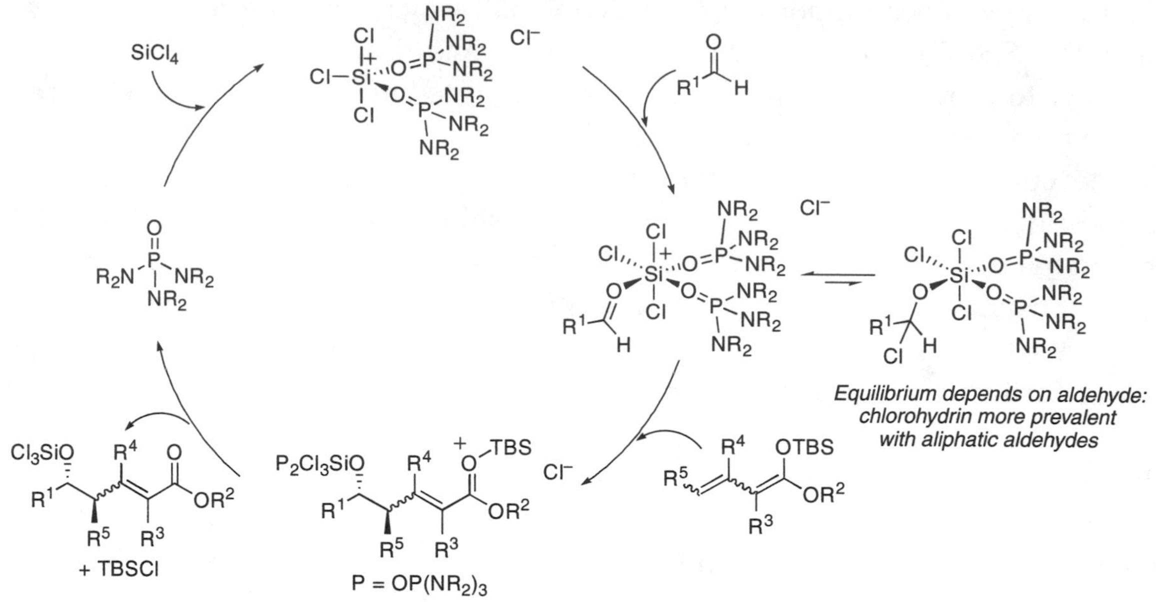

Figure 2.31. Mechanism of the vinylogous silyl ketene acetal Mukaiyama aldol reaction with the bis-phosphoramide catalyst (**Figure 2.30**).

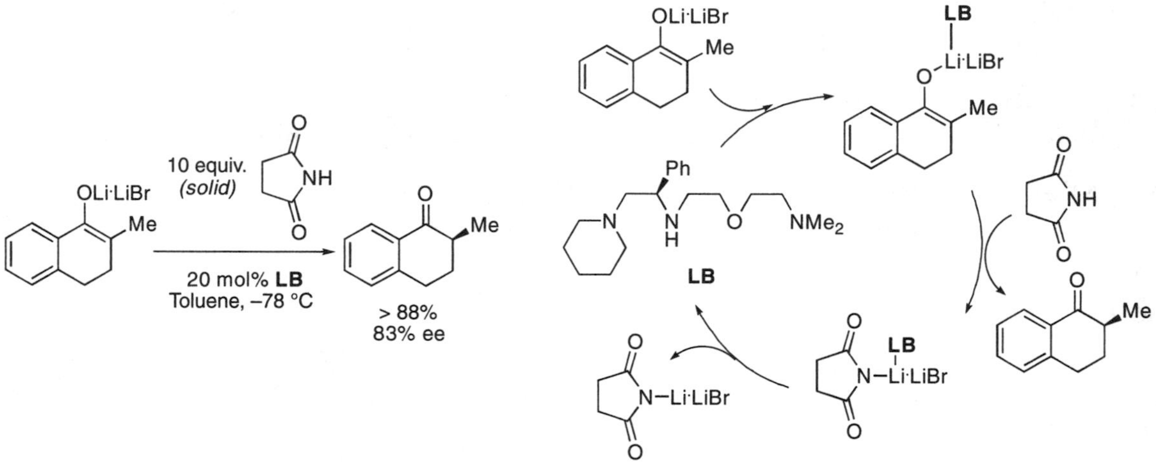

Figure 2.32. Activation of a lithium enolate for protonation with a chiral Lewis base.

In principle, this strategy can be applied to any achiral Lewis acid, so long as the adduct with the chiral Lewis base is more reactive. This holds true whether the Lewis acid is part of an organic substrate or a separate reagent that activates an organic substrate in a reaction. One example where the Lewis acid is part of the substrate can be found in the asymmetric protonation of a lithium enolate in the presence of a catalytic amount of a Lewis base (**Figure 2.32**).[53] Here, the enolate ligated to the Lewis base undergoes protonation much more readily. In addition, the asymmetric environment afforded by the chiral Lewis base causes the protonation to be enantioselective.

An example where the Lewis acid moiety is not part of the reacting organic substrate can be found in the asymmetric hydrosilylation of ketones effected by a copper-hydride species generated in situ (**Equation 2.1**).[54] In this case, both the achiral copper Lewis acid and the chiral phosphine Lewis base are present in substoichiometric amounts with respect to the ketone substrate. The achiral copper Lewis acid and the chiral Lewis base combine to give rise to the catalytic species. Nonetheless, this reaction is classified here as an example of Lewis base catalysis because much less chiral Lewis base is used than achiral Lewis acid (LB*:LA = 1:555).

Equation 2.1

There is a continuum (see **Figure 2.1**) between chiral Lewis base catalysts, used in conjunction with achiral Lewis acids (Section 2.2.1), and chiral Lewis acid catalysts (Section 2.1). This is especially true since most chiral Lewis acid catalysts arise from a combination of an achiral Lewis acid with a chiral Lewis base ligand (see **Figure 2.1**). In this chapter, the distinction lies in the amount of Lewis base employed relative to the Lewis acid. When the LA:LB ratio is 1:1, the reactions have been categorized as Lewis acid catalyzed. Often such Lewis acids are preformed, stable adducts. Such catalysts are particularly desirable when ligand-decelerated catalysis occurs (see the left side of the graph in **Figure 2.1**), because the catalytically active achiral Lewis acid will not be present (see Section 1.2). However, examples are also known that involve ligand-accelerated catalysis (see the right side of the graph in **Figure 2.1**), in which case stable adducts are not necessary since the achiral Lewis acid is not as reactive.

When the ratio of LA:LB exceeds 1:1, the reactions have been categorized as Lewis base catalyzed since the Lewis base is present in the lowest molar ratio. In these reactions, the amount of Lewis acid is typically greater than or equal to that of the organic substrate(s). For these reactions to succeed, ligand-accelerated catalysis (the right side of the graph in **Figure 2.1**) must be operative. Furthermore, ligand exchange is usually facile. The ability to reversibly generate a chiral Lewis acid reagent in situ with substoichiometric amounts of a chiral Lewis base avoids many of the problems with the formulation of preformed chiral Lewis acid–Lewis base adducts (see Section 1.2). The main downside of this strategy is the use of a stoichiometric Lewis acid. Even so, the noncovalent modification of reagents with chiral Lewis bases is a powerful strategy for asymmetric catalysis. Many examples have been reported (see Sections 1.1, 4.5.2, and 11.2), and this strategy will undoubtedly see further use in the future.

2.2.2 Lewis Base Catalysis via Covalent Substrate Modification

While chiral Lewis base catalysts that function via noncovalent, or dative, modification are well-established, there has been a recent resurgence in enantioselective Lewis base catalysis, also referred to as nucleophilic catalysis. As most chiral Lewis bases do not contain metals, these catalysts are often considered a subclass of organocatalysts. For an overview of organocatalysis, including chiral Lewis bases, see the articles in the special issue #8 "Enantioselective Organocatalysis" of *Acc. Chem. Res.* **2004**, *37*, 487–631 and the special issue #12 "Organocatalysis" of *Chem. Rev.* **2007**, *107*, 5413–5883. In contrast to Lewis acids, which form coordination adducts with substrates, the reactions of Lewis base catalysts involve covalent bond formation with one or more substrate molecules to form a transient reactive intermediate. The close proximity afforded by the shorter covalent bonds versus longer coordination bonds allows very efficient transfer of stereochemical information from the organocatalyst to the reacting partners.

Seminal work revolved around the proline-catalyzed cyclization reaction of the triketone substrate in **Figure 2.33**.[55–59] The early mechanistic work on this reaction was vigorously debated with several models proposed involving one to two proline molecules in the transition state. Ultimately, further mechanism[60,61] and computational[62,63] studies have led to the widely accepted mechanism illustrated in **Figure 2.33**. In this mechanism, proline serves two functions (see the transition-state structures[62] in **Figure 2.33**)—as a Lewis base and as a Brønsted acid. As a consequence, this reaction is best considered a result of bifunctional catalysis (see Chapter 12 for further examples). Nonetheless, the precept of a chiral Lewis base was firmly established and has given rise to a new field of asymmetric catalysis.

The key finding in this work is that a reactive enamine forms when the amine portion of proline functions as a Lewis base. The stereochemistry contained within the enamine adduct directs the stereochemistry of the subsequent enamine aldol reaction (compare transition structures in **Figure 2.33**) through a desymmetrization process (see Chapter 10). While advances in understanding of the mechanism have led to a great deal of new reaction development with proline catalysts (see Chapter 12),[64,65] the remainder of this section focuses on asymmetric catalysts that function solely as Lewis bases. The examples provided illustrate the different types of activated intermediates accessible via reaction with chiral Lewis base catalysts. Any re-

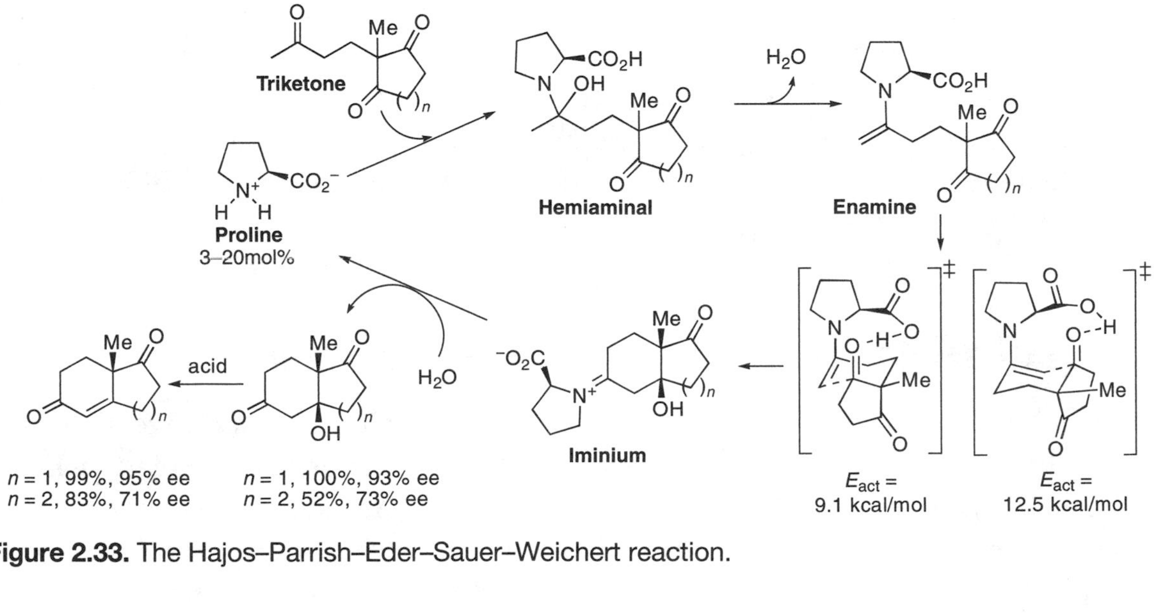

Figure 2.33. The Hajos–Parrish–Eder–Sauer–Weichert reaction.

action that proceeds by transient addition of a nucleophile to generate an activated intermediate is a potential candidate for Lewis base catalysis. To date, most of these reactions involve additions of the Lewis bases to carbonyl electrophiles. However, a broad array of chemistry is possible from this starting point, encompassing direct and indirect activation (c.f. Sections 2.1.1 and 2.1.2). Chiral activated enamines, acyls, imines, and umpolung equivalents can all be generated.

2.2.2.a Nucleophile Activation

Most enamine catalysts in aldol, Michael, and related reactions are comprised of bifunctional proline (see **Figure 2.33**) or chiral amine catalysts combined with a Brønsted acid source. As outlined previously, the Brønsted acid is important in activating the electrophile component (see also Chapter 12). There are, however, exceptions, as illustrated in the amine-catalyzed addition of aldehydes to α,β-unsaturated nitro compounds (**Figure 2.34**).[66] No additional acid is employed and the stereochemical outcome is consistent with the intervention of the illustrated enamine.

Similar enamine activation allows a variety of α-functionalizations of aldehydes and ketones, including amination, hydroxylation, sulfenylation, chlorination, and fluorination. An example in α-chlorination is outlined in **Figure 2.35**.[67] One key to the success of this reaction lies in control of the enamine geometry. By using the substituted imidazolidinone catalyst illustrated in **Figure 2.35**, a discrete adduct is favored. In particular, the dimethyl portion of the catalyst is oriented away from the

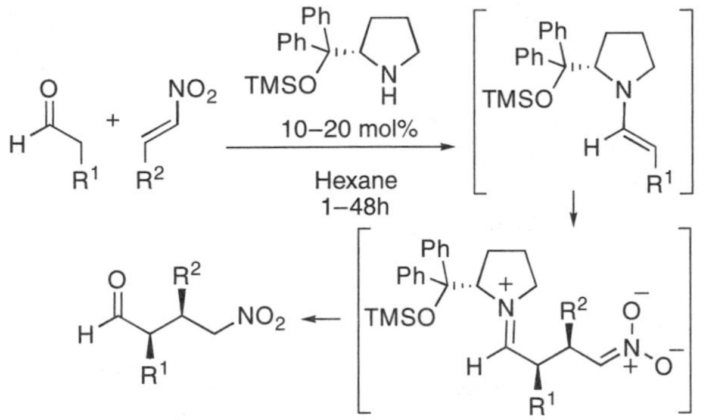

R^1	R^2	Yield (%)	syn:anti	syn ee (%)
Me	Ph	85	94:6	99
Me	n-Bu	52	84:16	99
Me	Cy	56	96:4	99
Et	Ph	66	93:7	99
i-Pr	Ph	72	93:7	99

Figure 2.34. Use of a Lewis base in the enantioselective conjugate addition of an aldehyde to an α,β-unsaturated nitro compound.

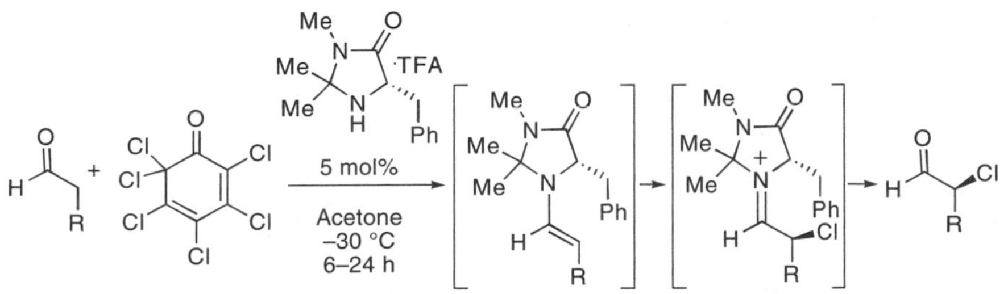

R	Yield (%)	ee (%)
n-Hx	71	92
Cy	87	94
CH$_2$Ph	92	80
(CH$_2$)$_3$COMe	78	87

Figure 2.35. Asymmetric Lewis base catalyzed α-chlorination.

enamine double bond. As a result, the benzyl portion blocks one enamine face very effectively.

Enamine activiation can also be employed in hetero-Diels–Alder[68] chemistry (**Figure 2.36**). The success of enamine catalysis relies upon facile hydrolytic cleavage of the N=C or N–C bond in the catalyst–product adduct. In the cases described above, this occurs to regenerate the carbonyl found in the original substrate. In the case below, this carbonyl product is in equilibrium with the hemiketal form. Oxidation of the mixture affords the lactone product in good overall yield. As a result, the chiral amine catalyst activates an alkene in an overall cycloaddition process.

Enamines have a long history in asymmetric synthesis and can also be obtained from ketenes and acid chlorides. Indeed, one of the earliest asymmetric organocatalytic reactions that proceeded with high selectivity (74% ee) was the methanolysis of methylphenylketene.[69] This reaction used a quinine derivative as a Lewis base catalyst and proceeded via an enamine intermediate. The generation of ketene equivalents in situ has revolutionized this area of research. One reaction series in which a Lewis base activates such an equivalent is illustrated in **Figure 2.37**. A non-nucleophilic

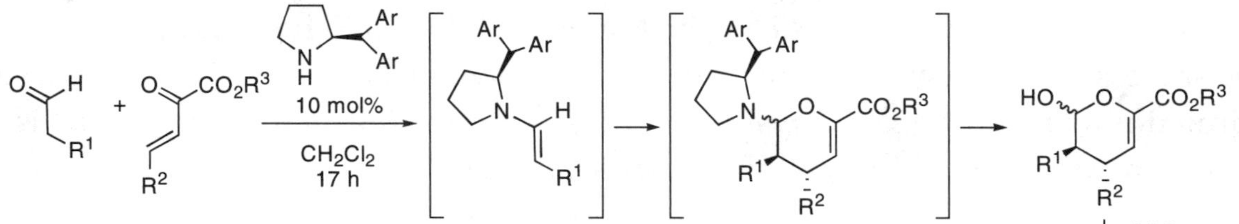

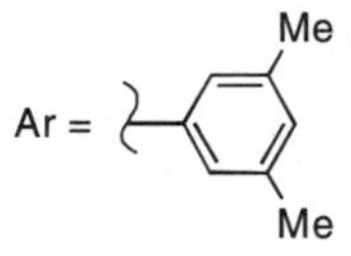

R¹	R²	R³	Yield (%)	ee (%)	R¹	R²	R³	Yield (%)	ee (%)
Et	Ph	Me	69	84	Bn	Ph	Me	65	86
i-Pr	Ph	Me	93	89	i-Pr	Me	Et	75	94

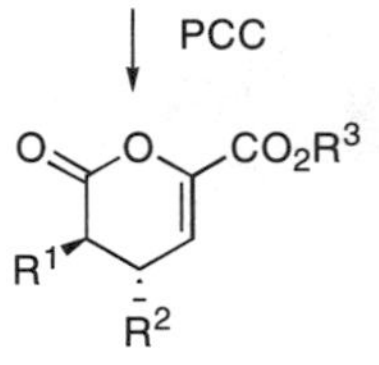

Figure 2.36. Chiral Lewis base catalyzed hetero-Diels–Alder reactions.

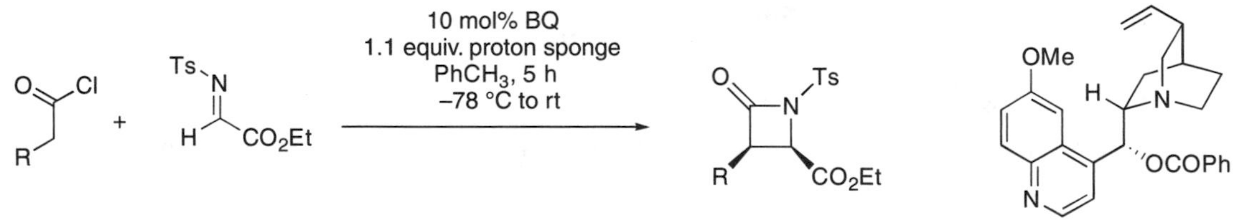

R	Yield (%)	ee (%)	dr	R	Yield (%)	ee (%)	dr
Ph	65	96	99:1	OBn	56	95	99:1
Et	57	99	99:1	CH=CH₂	58	98	99:1
Bn	60	96	33:1	N₃	47	98	25:1
OPh	45	99	99:1	Br	61	96	98:2
OAc	61	98	> 99:1				

Figure 2.37. Use of a Lewis base benzoylquinine (**BQ**) in the enantioselective [2+2] addition of ketene equivalents to imines to form β-lactams.

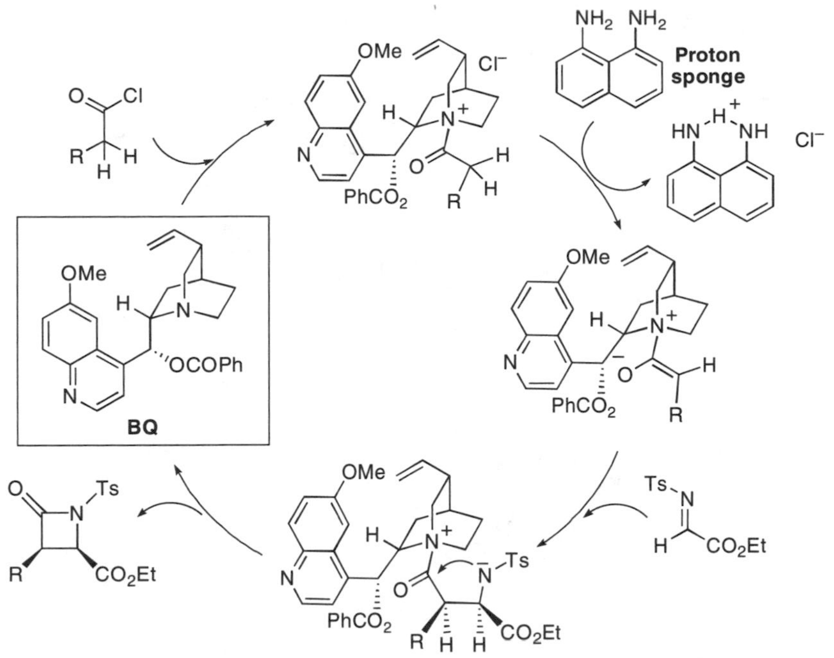

Figure 2.38. Ketene-free pathway in the benzoylquinine- (**BQ**-)catalyzed formation of β-lactams from acid chlorides and imines.

amine base, proton sponge, is combined with the cinchona alkaloid derivative benzoylquinine (**BQ**)[70–72] to catalyze formation of β-lactams. The **BQ** Lewis base also efficiently transmits asymmetry, as judged by the high diastereoselectivity and enantioselectivity of the resultant β-lactams.

The mechanism of the reaction was examined through a series of kinetic studies (**Figure 2.38**).[71,72] Interestingly, acylation of **BQ** by the acid chloride is the turnover-limiting step. The acyl–**BQ** intermediate is then converted to a chiral **BQ**–enolate. As a result, the Lewis base causes nucleophilic activation of the adjacent carbon (rather than the pendant carbon), which then attacks the α-imino ester. Finally, the stereochemistry of the **BQ** Lewis base controls the diastereoselectivity of this attack and hence the product stereochemistry.

2.2.2.b Oxidative Activation in Organocatalysis

Instead of using the well-known propensity for enamines formed under catalytic conditions to act as nucleophiles, oxidation of an enamine via single-electron transfer has set the stage for a whole new series of transformations.[73,74] This strategy takes advantage of the known tendency of enamines to oxidize readily due to their electron-rich character. After formation of an enamine in situ from an aldehyde and a chiral imidazolidinone Lewis base, oxidation with ceric ammonium nitrate (CAN) affords the radical cation, which reacts with an alkene (**Figure 2.39**). Radical cleavage or further oxidation along with hydrolysis of the iminium then provides the observed products. Both unactivated and activated alkenes can participate. However, in the presence of two alkenes, the more nucleophilic alkene reacts selectively (see the first

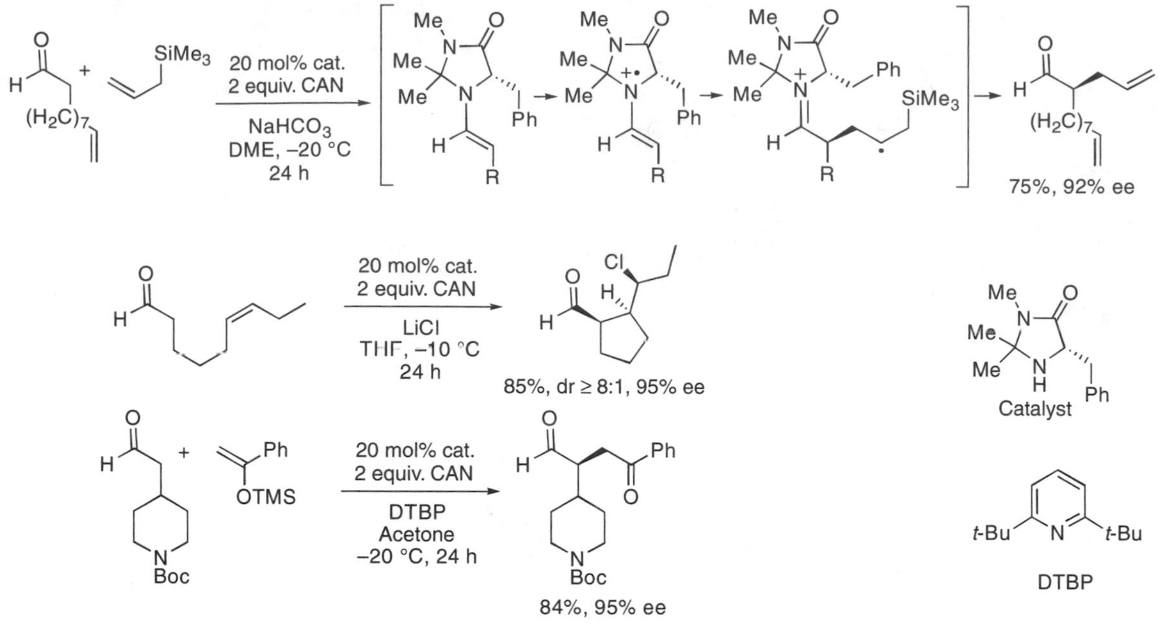

Figure 2.39. Lewis base catalyzed oxidative asymmetric reactions.

example in **Figure 2.39**: the intermolecular vs. intramolecular reaction). When enol silanes are employed, valuable 1,4-dicarbonyl species are produced.

2.2.2.c Electrophile Activation

One familiar activated intermediate is the acyl–DMAP complex, found in DMAP-(4-dimethylaminopyridine) catalyzed acylations of alcohols or amines, commencing with anhydrides or acid chlorides. The development of chiral Lewis bases to substitute for DMAP has received considerable attention[75–77] and a breakthrough was achieved with the axially chiral DMAP–ferrocene analogs[78–80] (**Figure 2.40**). While this catalyst does contain an iron metal center, this center is inert in the subsequent chemistry and serves only as a scaffolding element. Good results have been achieved for the kinetic resolution (see Chapter 7) of several alcohol classes.[78] A representative example is outlined here for the kinetic resolution of the challenging propargylic alcohol substrates (**Figure 2.40**).[81] Good k_{rel} values were obtained, allowing recovery of the starting alcohol with high enantioselectivity at conversions near 50%. An x-ray crystallographic structure of an acyl adduct with the chiral DMAP analog supports the direct activation mechanism illustrated. From this chiral acyl adduct, the two alcohol enantiomers undergo reactions via diastereomeric transition states. Here, the (R)-substrate enantiomer is better accommodated by the asymmetric environment of the acyl adduct and undergoes the more rapid reaction.

Lewis base catalysts have also proved useful in the indirect activation (see Section 2.1.2) of α,β-unsaturated carbonyls. While α,β-unsaturated aldehydes, esters, and amides form well-defined adducts with chiral Lewis acids, the corresponding α,β-unsaturated ketones generate isomeric mixtures (**Figure 2.41**). This activation does

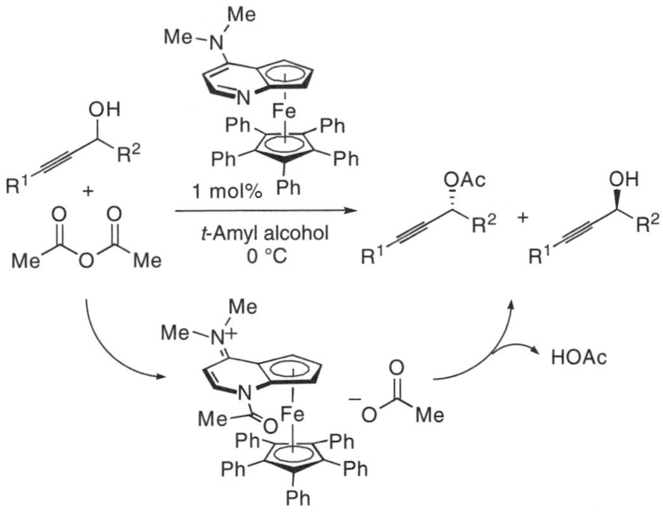

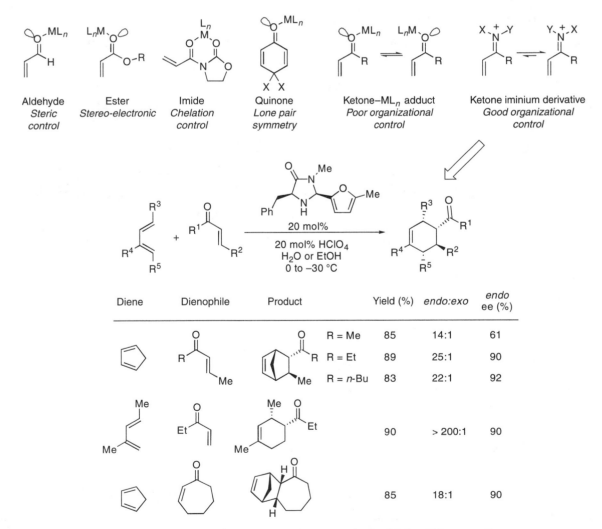

Figure 2.40. Chiral DMAP analog as a Lewis base catalyst in the kinetic resolution of racemic propargylic alcohols.

R^1	R^2	k_{rel}	Conv. (%)	Unreacted alcohol ee (%)
Ph	Me	20	56	96
Ph	Et	18	58	94
Ph	i-Pr	11	63	93
4-MeOC$_6$H$_4$	Me	14	60	94
4-CF$_3$C$_6$H$_4$	Me	10	71	99
Ac	Me	12	64	95
n-BuC≡C	Me	10	66	95
CH$_2$=CMe	Et	8	69	94

Diene	Dienophile	Product		Yield (%)	endo:exo	endo ee (%)
			R = Me	85	14:1	61
			R = Et	89	25:1	90
			R = n-Bu	83	22:1	92
				90	> 200:1	90
				85	18:1	90

Figure 2.41. Activation of various α,β-unsaturated carbonyls for Diels–Alder reaction.

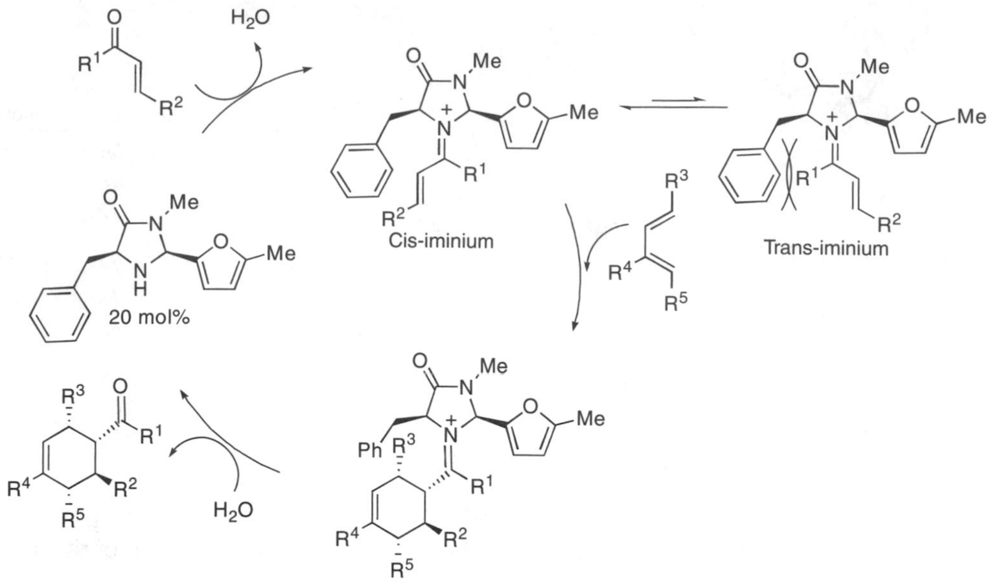

Figure 2.42. Mechanism for the imidazolidione-catalyzed Diels–Alder reaction from **Figure 2.41**.

provide the desired LUMO lowering needed for catalysis, but the resulting asymmetric environment is mixed and does not allow high enantioselection. An elegant solution to this problem utilized imidazolidinone as the Lewis base catalyst.[82] As can be seen from the results in **Figure 2.41**, the *endo:exo* selectivity and the enantioselectivity are excellent. These results hinge on the ability of the Lewis base catalyst to control the iminium geometry.

Figure 2.42 outlines the catalytic cycle for this Diels–Alder reaction. The sense of asymmetric induction observed in all cases is consistent with selective engagement of the diene substrate with the *Si* face of the cis-iminium isomer. Indeed, computational models provide supporting evidence that the trans-iminium isomer will be energetically disfavored on the basis on nonbonding interactions between the benzyl and the R^1 group. Moreover, the cis-iminium isomer selectively exposes the *Si* face to cycloaddition (the bottom-face attack illustrated in **Figure 2.42**).

This type of iminium catalysis is useful for a host of transformations, including Diels–Alder reactions, *endo*-selective Diels–Alder reactions, intramolecular Diels–Alder reactions, [3+2] cycloadditions, [4+3] cycloadditions, transfer hydrogenations, Friedel–Crafts reactions, epoxidations, cyclopropanations, and Michael reactions.[83] Indeed, any reaction that proceeds well via Lewis acid activation of an α,β-unsaturated aldehyde or ketone is certainly a candidate.

2.2.2.d Umpolung Activation

Lewis base catalysts have also proven useful in "umpolung"[84,85] activation. In these reactions, the normal polarity of a substrate is reversed. This is also seen in the oxidative reactions of enamines described in Section 2.2.2.b. Section 2.2.2.d deals with umpolung activations that do not involve oxidation. For example, chiral triazolinylidene carbenes act as Lewis base catalysts for the intramolecular Stetter reaction illustrated in **Figure 2.43**.[86] Here, the polarity of the aldehyde is reversed, generating an acyl anion equivalent, which undergoes conjugate addition to the α,β-unsaturated ester or ketone. As a result, five- and six-membered rings containing a 1,4-dicarbonyl array can be formed efficiently, along with two new stereocenters. In addition, the diastereoselectivity and enantioselectivity are excellent.

Figure 2.44 illustrates the mechanism for the Stetter reaction. After nucleophilic addition of the carbene to the aldehyde, further deprotonation leads to a highly electron-rich double bond. Nucleophilic attack occurs from the phenyl-substituted

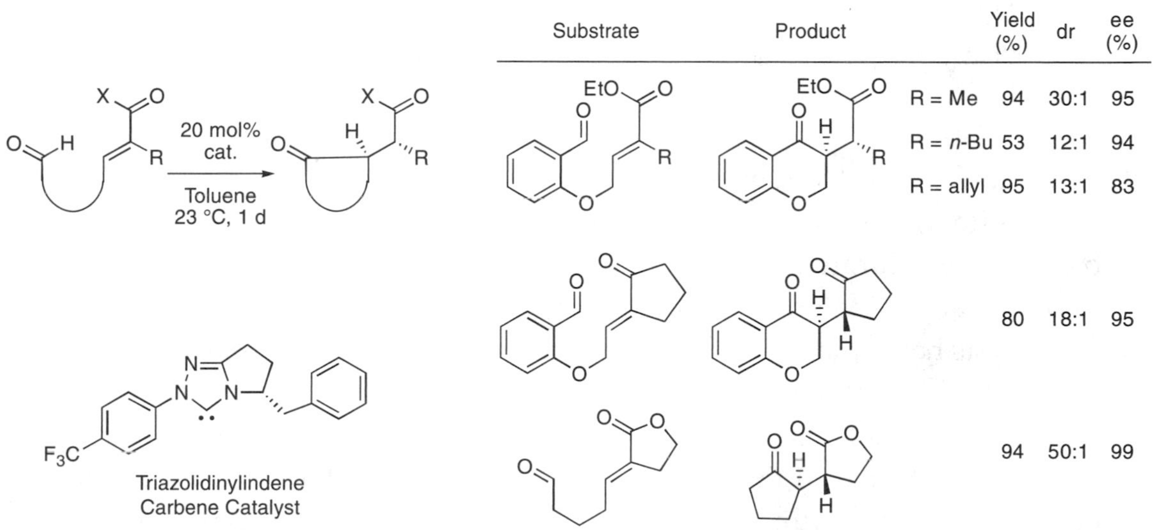

Figure 2.43. Scope of the enantioselective and diastereoselective intramolecular Stetter reaction.

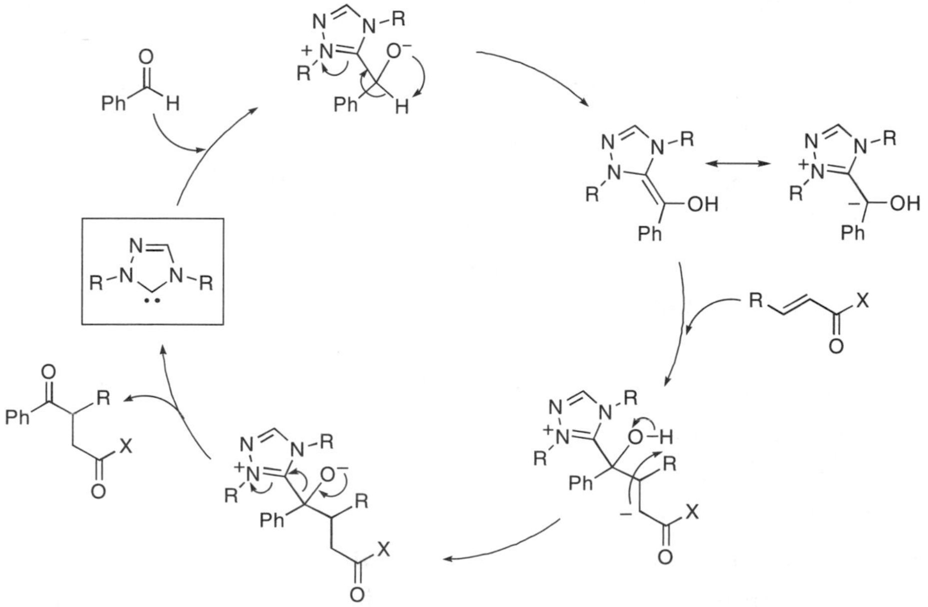

Figure 2.44. Mechanism of the catalytic enantioselective Stetter reaction.

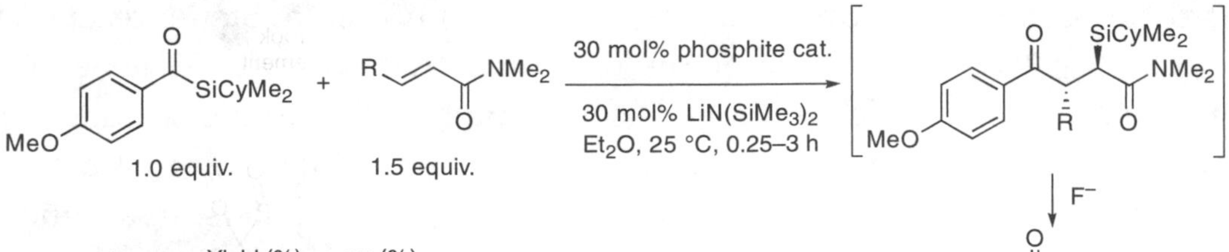

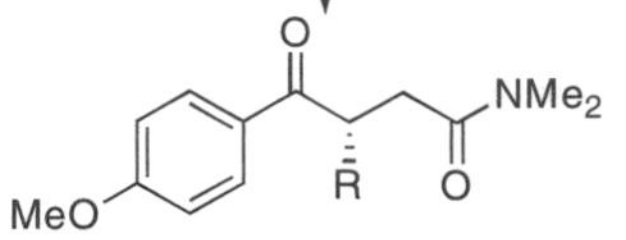

Figure 2.45. Metallophosphite-catalyzed acyl addition to α,β-unsaturated amides.

R	Yield (%)	ee (%)
Ph	68	90
4-MeOC$_6$H$_4$	63	92
4-ClC$_6$H$_4$	66	95
N-tosylindol-3-yl	60	97
2-naphthyl	66	89
Me	56	86
Et	82	71

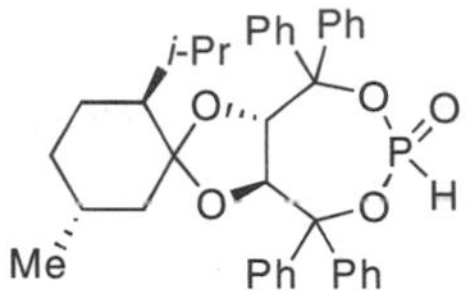

end of this double bond, because the polarization asserted by the "double enamine" (or ketene aminal) dominates the reactivity relative to the enol. After Michael addition to the α,β-unsaturated system, elimination regenerates the carbene catalyst and the carbonyl that originated from the starting aldehyde. The use of carbenes as nucleophilic catalysts has rapidly evolved upon the discovery that heteroatom-substituted carbenes are highly stable.[77,87–90]

Another example of a Lewis base catalyst involving umpolung and an acyl-anion equivalent is illustrated in **Figure 2.45**.[91–93] Here an acyl silane is used as the acyl-anion donor and the lithium anion of a phosphite is employed as the nucleophilic catalyst. Relative to the case above, the more difficult intermolecular version can be accomplished with this catalyst system. Upon reaction with the α,β-unsaturated amide and subsequent silyl removal, an acylic 1,4-dicarbonyl is produced in good yield and high enantioselectivity.

The mechanism of this reaction is distinct from the Stetter reaction (**Figure 2.46**). The first step is deprotonation of the phosphite by the hexamethyldisilazide base, yielding the catalytic species. Addition of this anion to the acyl silane provides a species that undergoes [1,2]-Brook rearrangement to yield a carbanion. This carbanion undergoes conjugate addition to the α,β-unsaturated amide. Catalyst release is triggered from this adduct by an unusual diastereoselective retro-[1,4]-Brook rearrangement.

The variety and scope of Lewis base catalyzed asymmetric reactions is quite astounding. As shown above, Lewis base catalysts are powerful in that new or unusual reaction pathways can be employed. In addition, the absence of metals from most Lewis base catalysts confers advantages with respect to cost and catalyst removal. The biggest disadvantage is the relatively high catalyst loadings employed on average compared to the most efficient asymmetric catalytic processes. It is likely that this problem will be overcome in the near future.

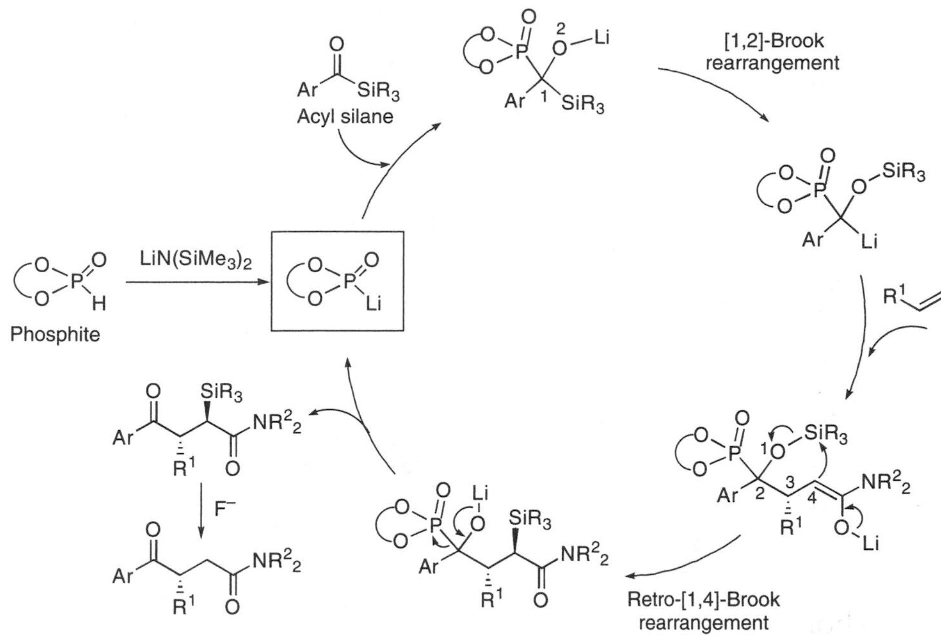

Figure 2.46. Catalytic cycle for metallophosphite catalyzed silylacylation of α,β-unsaturated amides.

Summary and Outlook

In this chapter the basic types of Lewis acid and Lewis base activation have been overviewed. While Lewis acid catalysis is a mature field, new discoveries are continually being made. Due to the strong interactions possible between chiral Lewis acids and a host of functionalized substrates, Lewis acid catalysis will continue to serve as a mainstay of asymmetric catalysis. In Lewis base catalysis, important discoveries are being rapidly advanced. Lewis bases can be used to catalytically activate stoichiometric Lewis acids. As a result, there is a continuum of Lewis base activation: irreversible complexation and modification of a Lewis acid (conventional chiral Lewis acid catalysis) versus reversible complexation and *activation* of a reagent containing a Lewis acid group (chiral Lewis base catalysis). Recently, there has also been a resurgence in organocatalytic Lewis base catalysis. In this type of catalysis, the chiral Lewis bases reversibly form covalent adducts with organic substrates. Examples of Lewis bases activating nucleophilic or electrophilic components are known. In addition, the potential for umpolung activation by Lewis bases provides avenues into many powerful transformations. The true potential of Lewis base catalysis is still be defined, but it will likely become equal to Lewis acid catalysis.

References

(1) Berrisford, D. J.; Bolm, C.; Sharpless, K. B. Ligand-Accelerated Catalysis. *Angew. Chem., Int. Ed. Engl.* **1995**, *34*, 1059–1070.

(2) Mikami, K.; Terada, M.; Sawa, E.; Nakai, T. Asymmetric Catalysis by Chiral Titanium Perchlorate for Carbonyl–Ene Cyclization. *Tetrahedron Lett.* **1991**, *32*, 6571–6574.

(3) Mikami, K.; Yajima, T.; Terada, M.; Uchimaru, T. Asymmetric Catalysis of Ene-type Reaction with Fluoral by Chiral Titanium Complex: A Semiempirical and Ab Initio Analysis of Ene Reactivity. *Tetrahedron Lett.* **1993**, *34*, 7591–7594.

(4) Mikami, K.; Jajima, T.; Terada, M.; Kato, E.; Maruta, M. Diastereoselective and Enantioselective Catalysis of the Carbonyl–Ene Reaction with Fluoral. *Tetrahedron: Asymmetry* **1994**, *5*, 1087–1090.

(5) Mikami, K.; Yajima, T.; Takasaki, T.; Matsukawa, S.; Terada, M.; Uchimaru, T.; Maruta, M. Asymmetric Catalysis of Carbonyl–ene and Aldol Reactions with Fluoral by Chiral Binaphthol-derived Titanium Complex. *Tetrahedron* **1996**, *52*, 85–98.

(6) Gademann, K.; Chavez, D. E.; Jacobsen, E. N. Highly Enantioselective Inverse-Electron-Demand Hetero-Diels–Alder Reactions of α,β-Unsaturated Aldehydes. *Angew. Chem., Int. Ed. Engl.* **2002**, *41*, 3059–3061.

(7) Evans, D. A.; Murry, J. A.; Kozlowski, M. C. C_2-Symmetric Copper(II) Complexes as Chiral Lewis Acids. Catalytic Enantioselective Aldol Additions of Silylketene Acetals to Benzyloxyacetaldehyde. *J. Am. Chem. Soc.* **1996**, *118*, 5814–5815.

(8) Evans, D. A.; Kozlowski, M. C.; Murry, J. A.; Burgey, C. S.; Campos, K. R.; Connell, B. T.; Staples, R. J. C_2-Symmetric Copper(II) Complexes as Chiral Lewis Acids. Scope and Mechanism of Catalytic Enantioselective Aldol Additions of Enolsilanes to (Benzyloxy)acetaldehyde. *J. Am. Chem. Soc.* **1999**, *121*, 669–685.

(9) Mikami, K.; Matsukawa, S. Asymmetric Catalytic Aldol-type Reaction with Ketene Silyl Acetals: Possible Intervention of the Silatropic Ene Pathway. *J. Am. Chem. Soc.* **1994**, *116*, 4077–4078.

(10) Evans, D. A.; Johnson, J. S. Catalytic Enantioselective Hetero Diels–Alder Reactions of α,β-Unsaturated Acyl Phosphonates with Enol Ethers. *J. Am. Chem. Soc.* **1998**, *120*, 4895–4896.

(11) Evans, D. A.; Johnson, J. S.; Olhava, E. J. Enantioselective Synthesis of Dihydropyrans. Catalysis of Hetero Diels–Alder Reactions by Bis(oxazoline) Copper(II) Complexes. *J. Am. Chem. Soc.* **2000**, *122*, 1635–1649.

(12) Abraham, L.; Czerwonka, R.; Hiersemann, M. The Catalytic Enantioselective Claisen Rearrangement of an Allyl Vinyl Ether. *Angew. Chem., Int. Ed. Engl.* **2001**, *40*, 4700–4703.

(13) Bao, J.; Wulff, W. D.; Rheingold, A. L. Vaulted Biaryls as Chiral Ligands for Asymmetric Catalytic Diels–Alder Reactions. *J. Am. Chem. Soc.* **1993**, *115*, 3814–3815.

(14) Hawkins, J. M.; Loren, S. Two-Point-Binding Asymmetric Diels–Alder Catalysts: Aromatic Alkyldichloroboranes. *J. Am. Chem. Soc.* **1991**, *113*, 7794–5.

(15) Jarvo, E. R.; Lawrence, B. M.; Jacobsen, E. N. Highly Enantio- and Regioselective Quinone Diels–Alder Reactions Catalyzed by a Tridentate [(Schiff Base)CrIII] Complex. *Angew. Chem., Int. Ed. Engl.* **2005**, *44*, 6043–6046.

(16) Wang, X.; Adachi, S.; Iwai, H.; Takatsuki, H.; Fujita, K.; Kubo, M.; Oku, A.; Harada, T. Enantioselective Lewis Acid-Catalyzed Mukaiyama–Michael Reactions of Acyclic Enones. Catalysis by *allo*-Threonine-Derived Oxazaborolidinones. *J. Org. Chem.* **2003**, *68*, 10046–10057.

(17) Harada, T.; Adachi, S.; Wang, X. Dimethylsilyl Ketene Acetal as a Nucleophile in Asymmetric Michael Reaction: Enhanced Enantioselectivity in Oxazaborolidinone-Catalyzed Reaction. *Org. Lett.* **2004**, *6*, 4877–4879.

(18) Narasaka, K.; Iwasawa, N.; Inoue, M.; Yamada, T.; Nakashima, M.; Sugimori, J. Asymmetric Diels–Alder Reaction Catalyzed by a Chiral Titanium Reagent. *J. Am. Chem. Soc.* **1989**, *111*, 5340–5345.

(19) Gothelf, K. V.; Hazell, R. G.; Jørgensen, K. A. Crystal Structure of a Chiral Titanium Catalyst–Alkene Complex. The Intermediate in Catalytic Asymmetric Diels–Alder and 1,3-Dipolar Cycloaddition Reactions. *J. Am. Chem. Soc.* **1995**, *117*, 4435–4436.

(20) Narasaka, K.; Hayashi, Y.; Shimadzu, H.; Niihata, S. Asymmetric [2 + 2] Cycloaddition Reaction Catalyzed by a Chiral Titanium Reagent. *J. Am. Chem. Soc.* **1992**, *114*, 8869–8885.

(21) Gothelf, K. V.; Jørgensen, K. A. Transition-Metal Catalyzed Asymmetric 1,3-Dipolar Cycloaddition Reactions Between Alkenes and Nitrones. *J. Org. Chem.* **1994**, *59*, 5687–5691.

(22) Desimoni, G.; Faita, G.; Quadrelli, P. Pyridine-2,6-bis(oxazolines), Helpful Ligands for Asymmetric Catalysts. *Chem. Rev.* **2003**, *103*, 3119–3154.

(23) McManus, H. A.; Guiry, P. J. Recent Developments in the Application of Oxazoline-Containing Ligands in Asymmetric Catalysis. *Chem. Rev.* **2004**, *104*, 4151–4202.

(24) Desimoni, G.; Faita, G.; Jørgensen, K. A. C_2-Symmetric Chiral Bis(oxazoline) Ligands in Asymmetric Catalysis. *Chem. Rev.* **2006**, *106*, 3561–3651.

(25) Evans, D. A.; Miller, S. J.; Lectka, T. Bis(oxazoline) copper(II) Complexes as Chiral Catalysts for the Enantioselective Diels–Alder Reaction. *J. Am. Chem. Soc.* **1993**, *115*, 6460–6461.

(26) Evans, D. A.; Murry, J. A.; von Matt, P.; Norcross, R. D.; Miller, S. J. C_2-Symmetric Cationic Copper(II) Complexes as Chiral Lewis Acids: Counterion Effects in the Enantioselective Diels–Alder Reaction. *Angew. Chem., Int. Ed. Engl.* **1995**, *34*, 798–800.

(27) Evans, D. A.; Miller, S. J.; Lectka, T.; von Matt, P. Chiral Bis(oxazoline)copper(II) Complexes as Lewis Acid Catalysts for the Enantioselective Diels–Alder Reaction. *J. Am. Chem. Soc.* **1999**, *121*, 7559–7573.

(28) Evans, D. A.; Barnes, D. M.; Johnson, J. S.; Lectka, T.; von Matt, P.; Miller, S. J.; Murry, J. A.; Norcross, R. D.; Shaughnessy, E. A.; Campos, K. R. Bis(oxazoline) and Bis(oxazolinyl)pyridine Copper Complexes as Enantioselective Diels–Alder Catalysts: Reaction Scope and Synthetic Applications. *J. Am. Chem. Soc.* **1999**, *121*, 7582–7594.

(29) Johnson, J. S.; Evans, D. A. Chiral Bis(oxazoline) Copper(II) Complexes: Versatile Catalysts for Enantioselective Cycloaddition, Aldol, Michael, and Car-

bonyl Ene Reactions. *Acc. Chem. Res.* **2000**, *33*, 325–335.

(30) Evans, D. A.; Wu, J. Enantioselective Rare-Earth Catalyzed Quinone Diels–Alder Reactions. *J. Am. Chem. Soc.* **2003**, *125*, 10162–10163.

(31) Evans, D. A.; Scheidt, K. A.; Fandrick, K. R.; Lam, H. W.; Wu, J. Enantioselective Indole Friedel–Crafts Alkylations Catalyzed by Bis(oxazolinyl)pyridine–Scandium(III) Triflate Complexes. *J. Am. Chem. Soc.* **2003**, *125*, 10780–10781.

(32) Evans, D. A.; Scheidt, K. A.; Johnston, J. N.; Willis, M. C. Enantioselective and Diastereoselective Mukaiyama–Michael Reactions Catalyzed by Bis(oxazoline) Copper(II) Complexes. *J. Am. Chem. Soc.* **2001**, *123*, 4480–4491.

(33) Evans, D. A.; Johnson, D. S. C_2-Symmetric Copper(II) Complexes as Chiral Lewis Acids. Catalytic Enantioselective Amination of Enolsilanes. *Org. Lett.* **1999**, *1*, 595–598.

(34) Holmes, R. R. Comparison of Phosphorus and Silicon: Hypervalency, Stereochemistry, and Reactivity. *Chem. Rev.* **1996**, *96*, 927–950.

(35) Gutmann, V. *The Donor–Acceptor Approach to Molecular Interactions*; Plenum Press: New York, 1978.

(36) Jensen, W. B. *The Lewis Acid–Base Concept*; Wiley: New York, 1980.

(37) Gordon, M. S.; Carroll, M. T.; Davis, L. P.; Burggraf, L. W. Structure and Stability of Hexacoordinated Sila Dianions. *J. Phys. Chem.* **1990**, *94*, 8125–8128.

(38) Denmark, S. E.; Stavenger, R. A. Asymmetric Catalysis of Aldol Reactions with Chiral Lewis Bases. *Acc. Chem. Res.* **2000**, *33*, 432–440.

(39) Denmark, S. E.; Coe, D. M.; Pratt, N. E.; Griedel, B. D. Asymmetric Allylation of Aldehydes with Chiral Lewis Bases. *J. Org. Chem.* **1994**, *59*, 6161–6163.

(40) Denmark, S. E.; Fu, J. On the Mechanism of Catalytic, Enantioselective Allylation of Aldehydes with Chlorosilanes and Chiral Lewis Bases. *J. Am. Chem. Soc.* **2000**, *122*, 12021–12022.

(41) Denmark, S. E.; Fu, J. Catalytic, Enantioselective Addition of Substituted Allylic Trichlorosilanes Using a Rationally-Designed 1,2′-Bispyrrolidine-Based Phosphoramide. *J. Am. Chem. Soc.* **2001**, *123*, 9488–9489.

(42) Denmark, S. E.; Wong, K.-T.; Stavenger, R. A. The Chemistry of Trichlorosilyl Enolates. 2. Highly-Selective Asymmetric Aldol Additions of Ketone Enolates. *J. Am. Chem. Soc.* **1997**, *119*, 2333–2334.

(43) Denmark, S. E.; Stavenger, R. A.; Wong, K.-T. Lewis Base-Catalyzed, Asymmetric Aldol Additions of Methyl Ketone Enolates. *J. Org. Chem.* **1998**, *63*, 918–919.

(44) Denmark, S. E.; Stavenger, R. A.; Wong, K.-T.; Su, X. Chiral Phosphoramide-Catalyzed Aldol Additions of Ketone Enolates. Preparative Aspects. *J. Am. Chem. Soc.* **1999**, *121*, 4982–4991.

(45) Denmark, S. E.; Su, X.; Nishigaichi, Y. The Chemistry of Trichlorosilyl Enolates. 6. Mechanistic Duality in the Lewis Base-Catalyzed Aldol Addition Reaction. *J. Am. Chem. Soc.* **1998**, *120*, 12990–12991.

(46) Denmark, S. E.; Pham, S. M. Kinetic Analysis of the Divergence of Reaction Pathways in the Chiral Lewis Base Promoted Aldol Additions of Trichlorosilyl Enol Ethers: A Rapid-Injection NMR Study. *Helv. Chim. Acta* **2000**, *83*, 1846–1853.

(47) Denmark, S. E.; Beutner, G. L. Lewis Base Activation of Lewis Acids. Vinylogous Aldol Reactions. *J. Am. Chem. Soc.* **2003**, *125*, 7800–7801.

(48) Denmark, S. E.; Beutner, G. L.; Wynn, T.; Eastgate, M. D. Lewis Base Activation of Lewis Acids: Catalytic, Enantioselective Addition of Silyl Ketene Acetals to Aldehydes. *J. Am. Chem. Soc.* **2005**, *127*, 3774–3789.

(49) Denmark, S. E.; Heemstra, J. R., Jr. Lewis Base Activation of Lewis Acids. Vinylogous Aldol Addition Reactions of Conjugated N,O-Silyl Ketene Acetals to Aldehydes. *J. Am. Chem. Soc.* **2006**, *128*, 1038–1039.

(50) Denmark, S. E.; Wynn, T. Lewis Base Activation of Lewis Acids: Catalytic Enantioselective Allylation and Propargylation of Aldehydes. *J. Am. Chem. Soc.* **2001**, *123*, 6199–6200.

(51) Denmark, S. E.; Heemstra, J. R., Jr. Lewis Base Activation of Lewis Acids. Catalytic Enantioselective Addition of Silyl Enol Ethers of Achiral Methyl Ketones to Aldehydes. *Org. Lett.* **2003**, *5*, 2303–2306.

(52) Malkov, A. V.; Mariani, A.; MacDougall, K. N.; Kocovsky, P. Role of Noncovalent Interactions in the Enantioselective Reduction of Aromatic Ketimines with Trichlorosilane. *Org. Lett.* **2004**, *6*, 2253–2256.

(53) Riviere, P.; Koga, K. An Approach to Catalytic Enantioselective Protonation of Prochiral Lithium Enolates. *Tetrahedron Lett.* **1997**, *38*, 7589–7592.

(54) Lipshutz, B. H.; Noson, K.; Chrisman, W.; Lower, A. Asymmetric Hydrosilylation of Aryl Ketones Catalyzed by Copper Hydride Complexed by Nonracemic Biphenyl Bis-phosphine Ligands. *J. Am. Chem. Soc.* **2003**, *125*, 8779–8789.

(55) Hajos, Z. G.; Parrish, D. R. Asymmetric Synthesis of Optically Active Polycyclic Organic Compounds. **1971**, German Patent: DE 2102623.

(56) Hajos, Z. G.; Parrish, D. R. Stereocontrolled Synthesis of Trans-Hydrindan Steroidal Intermediates. *J. Org. Chem.* **1973**, *38*, 3239–3243.

(57) Hajos, Z. G.; Parrish, D. R. Asymmetric Synthesis of Bicyclic Intermediates of Natural Product Chemistry. *J. Org. Chem.* **1974**, *39*, 1615–1621.

(58) Eder, U.; Sauer, G.; Weichert, R. Optically Active 1,5-Indanone and 1,6-Naphthalenedione. **1971**: DE 2014757.

(59) Eder, U.; Sauer, G.; Weichert, R. Total Synthesis of Optically Active Steroids. 6. New Type of Asymmetric Cyclization to Optically Active Steroid CD Partial Structures. *Angew. Chem., Int. Ed. Engl.* **1971**, *10*, 496–497.

(60) Hoang, L.; Bahmanyar, S.; Houk, K. N.; List, B. Kinetic and Stereochemical Evidence for the Involvement of Only One Proline Molecule in the Transition States of Proline-Catalyzed Intra- and Intermolecular Aldol Reactions. *J. Am. Chem. Soc.* **2003**, *125*, 16–17.

(61) List, B.; Hoang, L.; Martin, H. J. Asymmetric Catalysis Special Feature Part II: New Mechanistic Studies on the Proline-Catalyzed Aldol Reaction. *Proc. Natl. Acad. Sci. U.S.A.* **2004**, *101*, 5839–5842.

(62) Bahmanyar, S.; Houk, K. N. The Origin of Stereoselectivity in Proline-Catalyzed Intramolecular Aldol Reactions. *J. Am. Chem. Soc.* **2001**, *123*, 12911–12912.

(63) Clemente, F. R.; Houk, K. N. Computational Evidence for the Enamine Mechanism of Intramolecular

Aldol Reactions Catalyzed by Proline. *Angew. Chem., Int. Ed. Engl.* **2004**, *43*, 5766–5768.

(64) List, B. Proline-Catalyzed Asymmetric Reactions. *Tetrahedron* **2002**, *58*, 5573–5590.

(65) List, B. Enamine Catalysis Is a Powerful Strategy for the Catalytic Generation and Use of Carbanion Equivalents. *Acc. Chem. Res.* **2004**, *37*, 548–557.

(66) Hayashi, Y.; Gotoh, H.; Hayashi, T.; Shoji, M. Diphenylprolinol Silyl Ethers as Efficient Organocatalysts for the Asymmetric Michael Reaction of Aldehydes and Nitroalkenes. *Angew. Chem., Int. Ed. Engl.* **2005**, *44*, 4212–4215.

(67) Brochu, M. P.; Brown, S. P.; MacMillan, D. W. C. Direct and Enantioselective Organocatalytic α-Chlorination of Aldehydes. *J. Am. Chem. Soc.* **2004**, *126*, 4108–4109.

(68) Juhl, K.; Jørgensen, K. A. The First Organocatalytic Enantioselective Inverse-Electron Demand Hetero-Diels–Alder Reaction. *Angew. Chem., Int. Ed. Engl.* **2003**, *42*, 1498–1501.

(69) Pracejus, H. Asymmetric Syntheses with Ketenes. I. Alkaloid-Catalyzed Asymmetric Syntheses of α-Phenylpropionate Esters. *Liebigs Ann. Chem.* **1960**, *634*, 9–22.

(70) Taggi, A. E.; Hafez, A. M.; Wack, H.; Young, B.; Drury, W. J., III; Lectka, T. Catalytic, Asymmetric Synthesis of β-Lactams. *J. Am. Chem. Soc.* **2000**, *122*, 7831–7832.

(71) Taggi, A. E.; Hafez, A. M.; Wack, H.; Young, B.; Ferraris, D.; Lectka, T. The Development of the First Catalyzed Reaction of Ketenes and Imines: Catalytic, Asymmetric Synthesis of β-Lactams. *J. Am. Chem. Soc.* **2002**, *124*, 6626–6635.

(72) France, S.; Weatherwax, A.; Taggi, A. E.; Lectka, T. Advances in the Catalytic, Asymmetric Synthesis of β-Lactams. *Acc. Chem. Res.* **2004**, *37*, 592–600.

(73) Beeson, T. D.; Mastracchio, A.; Hong, J.-B.; Ashton, K.; MacMillan, D. W. C. Enantioselective Organocatalysis Using SOMO Activation. *Science* **2007**, *316*, 582–585.

(74) Jang, H.-Y.; Hong, J.-B.; MacMillan, D. W. C. Enantioselective Organocatalytic Singly Occupied Molecular Orbital Activation: The Enantioselective α-Enolation of Aldehydes. *J. Am. Chem. Soc.* **2007**, *129*, 7004–7005.

(75) Vedejs, E.; Chen, X. Kinetic Resolution of Secondary Alcohols. Enantioselective Acylation Mediated by a Chiral (Dimethylamino)pyridine Derivative. *J. Am. Chem. Soc.* **1996**, *118*, 1809–1810.

(76) See references cited in Vedejs, E.; MacKay, J. A. Kinetic Resolution of Allylic Alcohols Using a Chiral Phosphine Catalyst. *Org. Lett.* **2001**, *3*, 535–536.

(77) Enders, D.; Balensiefer, T. Nucleophilic Carbenes in Asymmetric Organocatalysis. *Acc. Chem. Res.* **2004**, *37*, 534–541.

(78) Fu, G. C. Enantioselective Nucleophilic Catalysis with "Planar-Chiral" Heterocycles. *Acc. Chem. Res.* **2000**, *33*, 412–420.

(79) Fu, G. C. Asymmetric Catalysis with "Planar-Chiral" Heterocycles. *Pure Appl. Chem.* **2001**, *73*, 347–349.

(80) Fu, G. C. Asymmetric Catalysis with "Planar-Chiral" Derivatives of 4-(Dimethylamino)pyridine. *Acc. Chem. Res.* **2004**, *37*, 542–547.

(81) Tao, B.; Ruble, J. C.; Hoic, D. A.; Fu, G. C. Nonenzymatic Kinetic Resolution of Propargylic Alcohols by a Planar-Chiral DMAP Derivative: Crystallographic Characterization of the Acylated Catalyst. *J. Am. Chem. Soc.* **1999**, *121*, 5091–5092.

(82) Northrup, A. B.; MacMillan, D. W. C. The First General Enantioselective Catalytic Diels–Alder Reaction with Simple α,β-Unsaturated Ketones. *J. Am. Chem. Soc.* **2002**, *124*, 2458–2460.

(83) Lelais, G.; MacMillan, D. W. C. Modern Strategies in Organic Catalysis: The Advent and Development of Iminium Activation. *Aldrichimica Acta* **2006**, *39*, 79–87.

(84) Seebach, D.; Kolb, M. Umpolung (Dipole Inversion) of Carbonyl Reactivity. *Chem. Ind. (London)* **1974**, *7*, 687–692.

(85) Seebach, D. Methods of Reactivity Umpolung. *Angew. Chem., Int. Ed. Engl.* **1979**, *18*, 239–336.

(86) Read de Alaniz, J.; Rovis, T. A Highly Enantio- and Diastereoselective Catalytic Intramolecular Stetter Reaction. *J. Am. Chem. Soc.* **2005**, *127*, 6284–6289.

(87) Arduengo, A. J., III. Looking for Stable Carbenes: The Difficulty in Starting Anew. *Acc. Chem. Res.* **1999**, *32*, 913–921.

(88) Bourissou, D.; Guerret, O.; Gabbai, F. P.; Bertrand, G. Stable Carbenes. *Chem. Rev.* **2000**, *100*, 39–91.

(89) Herrmann, W. A. N-Heterocyclic Carbenes: A New Concept in Organometallic Chemistry. *Angew. Chem., Int. Ed.* **2002**, *41*, 1290–1309.

(90) Perry, M. C.; Burgess, K. Chiral N-Heterocyclic Carbene-Transition Metal Complexes in Asymmetric Catalysis. *Tetrahedron: Asymmetry* **2003**, *14*, 951–961.

(91) Johnson, J. S. Catalyzed Reactions of Acyl Anion Equivalents. *Angew. Chem., Int. Ed. Engl.* **2004**, *43*, 1326–1328.

(92) Nahm, M. R.; Linghu, X.; Potnick, J. R.; Yates, C. M.; White, P. S.; Johnson, J. S. Metallophosphite-Induced Nucleophilic Acylation of α,β-Unsaturated Amides: Facilitated Catalysis by a Diastereoselective Retro [1,4] Brook Rearrangement. *Angew. Chem., Int. Ed. Engl.* **2005**, *44*, 2377–2379.

(93) Nahm, M. R.; Potnick, J. R.; White, P. S.; Johnson, J. S. Metallophosphite-Catalyzed Asymmetric Acylation of α,β-Unsaturated Amides. *J. Am. Chem. Soc.* **2006**, *128*, 2751–2756.

Beyond Lewis Acid and Lewis Base Catalyzed Activation

In the last chapter, Lewis acid and Lewis base catalysts were described. Here, the remaining classes of commonly encountered asymmetric catalysts are outlined—namely, Brønsted acids, Brønsted bases, ionic species, group-transfer agents, cross-coupling agents, and π-activating agents.

3.1 Brønsted Acid and Hydrogen-Bonding Catalysts

As is the case with Lewis base catalysts, there has been a recent resurgence in interest in Brønsted acid catalysis, leading to major new developments. In this section, catalysts that have been referred to as Brønsted acid or hydrogen-bonding catalysts will be discussed.[1] These catalysts can be divided into two functionally distinct categories. In the first, the catalyst activates a substrate by hydrogen-bond donation. In the second, the chiral catalyst causes an enantioselective protonation of a prochiral substrate.

3.1.1 Activation via Hydrogen-Bond-Donating Catalysts

The first of these categories, hydrogen-bonding catalysis, draws inspiration from enzymatic systems,[1] where *general-acid* activation is a common motif. In *general-acid* catalysis, proton transfer occurs in the transition state of the rate-determining step. In *specific-acid* catalysis, a reversible protonation occurs in a pre-equilibrium step.

With small-molecule catalysts, a proton can essentially replace a Lewis acid as an activating element. With Lewis acids (**Figure 3.1**), the nearby ligand sphere provides the ability to tune the metal center and the asymmetric environment. The lesser ability to do so with Brønsted acids (**Figure 3.1**), combined with their weaker and less directional interactions with Lewis bases, would seem to put Brønsted acids at a disadvantage. However, the ability to function in aqueous systems, to provide activation under mild conditions, and to avoid metals offers distinct advantages. After a period of use in primarily bifunctional catalysis, new applications and developments of small-molecule Brønsted acid catalysts are appearing at an increasing rate.[1]

The nature of the hydrogen bond that occurs during general-acid catalysis by small molecules is crucial and depends on the characteristics of the catalyst and the substrate. Hydrogen bonds can vary considerably in strength (< 1–40 kcal/mol) and their dependence on the ideal 180° X–H•••A bond angle (90–180°). For most applications with small-molecule catalysts, the hydrogen bonds arise from electrostatic in-

Lewis acid   **Brønsted acid**

Figure 3.1. Lewis acid and Brønsted acid activation of a representative substrate.

teractions that are of moderate strength (4–15 kcal/mol), and have a relaxed preference for the X–H•••A bond angle (130–180°).[1] Within this paradigm, the terms "Brønsted acid" or "hydrogen-bonding catalysts" are largely semantic. The acidity of the catalytic functional group can span a large range (**Figure 3.2**). The most acidic catalysts, such as phosphoric and carboxylic acids, are commonly referred to as Brønsted acid catalysts.

A number of early reports proposed that hydrogen-bonding catalysis was a crucial control element in catalysts that were ultimately found to be multifunctional (see Chapter 12) in nature (**Figure 3.3**). For example, one early proposal, involving hydrogen bonding[2] as a key element in the proline-catalyzed intramolecular aldol reaction,[3–7] was ultimately supported in later mechanism[8,9] and computational[10,11] studies (see Section 2.2.2 for further discussion of this reaction). In the hydrocyanations of aldehydes catalyzed by cyclic dipeptides,[12] hydrogen bonding from one of the amides

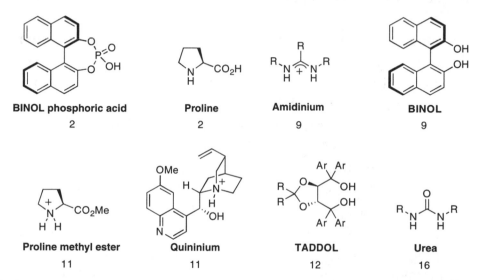

BINOL phosphoric acid
2

Proline
2

Amidinium
9

BINOL
9

Proline methyl ester
11

Quininium
11

TADDOL
12

Urea
16

Figure 3.2. Approximate pK_a values of functional groups found in Brønsted acid and hydrogen-bonding catalysts.

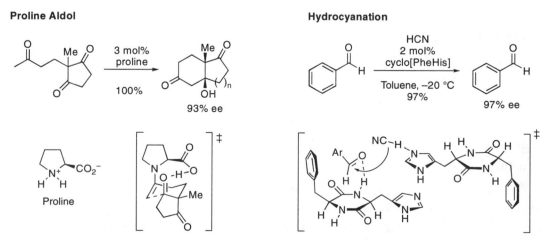

Proline Aldol

3 mol% proline

100%

93% ee

Proline

Hydrocyanation

HCN
2 mol%
cyclo[PheHis]

Toluene, –20 °C
97%

97% ee

Figure 3.3. Early examples of asymmetric catalysts containing Brønsted acids.

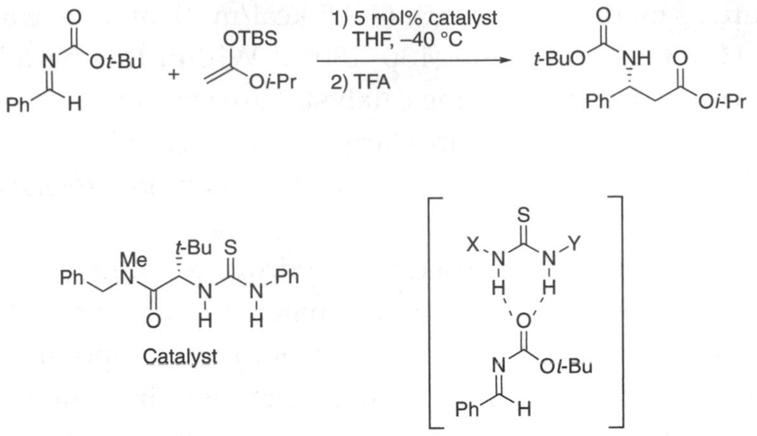

Figure 3.4. Asymmetric addition to an *N*-Boc imine with a urea-based chiral catalyst.

has been invoked as a key control element, although questions about the mechanism remain.[13,14]

With the recognition of hydrogen bonding as an effective control element, chiral species containing an assortment of different catalyst functional groups (see **Figure 3.2**) that can act as hydrogen-bond donors have been investigated in asymmetric catalysis. Amongst the weakest Brønsted acids that have been effective are ureas and thioureas. An example can be seen in the asymmetric addition of silyl ketene acetals to activated imines (**Figure 3.4**).[15] The chiral thiourea catalysts can serve as dual hydrogen-bond donors, and there is good evidence for this motif in a related system.[16a] Based upon these developments, similar catalysts have been applied to a host of reactions, most notably with imine or *N*-acyl–iminium substrates.[1] However, many of these reactions likely involve a different activation mechanism via coordination to the counteranion[16b] (see Section 3.3 for chiral counterion catalysis).

Based upon the well-known capacity of water to undergo hydrogen bonding, the ability of related alcohol derivatives to act as chiral hydrogen bonding catalysts seems a quite reasonable proposition. Even so, catalysts where the alcohol is the sole activation element have only emerged relatively recently. These results have shown that chiral alcohols can be powerful catalysts, as illustrated with the TADDOL-catalyzed Diels–Alder reactions of aldehydes (**Figure 3.5**).[17–19] The mild nature of the catalyst is ideal here where the diene substrates are highly reactive. While TADDOL derivatives could, in principle, contribute two hydrogen bonds in activating the substrates, X-ray crystallographic[20] and computational[21] evidence points to an intramolecular hydrogen bond between the two TADDOL hydroxyls, leaving one free hydrogen-bond donor that interacts with the substrate (**Figure 3.5**).

Based upon the above results, the more acidic hydroxyl groups found in chiral biphenol or BINOL (see **Figure 3.2**) derivatives would be expected to function as slightly stronger Brønsted acid catalysts. Indeed, a chiral biphenol has found utility in Diels–Alder reactions similar to those described in **Figure 3.5** and a crystal struc-

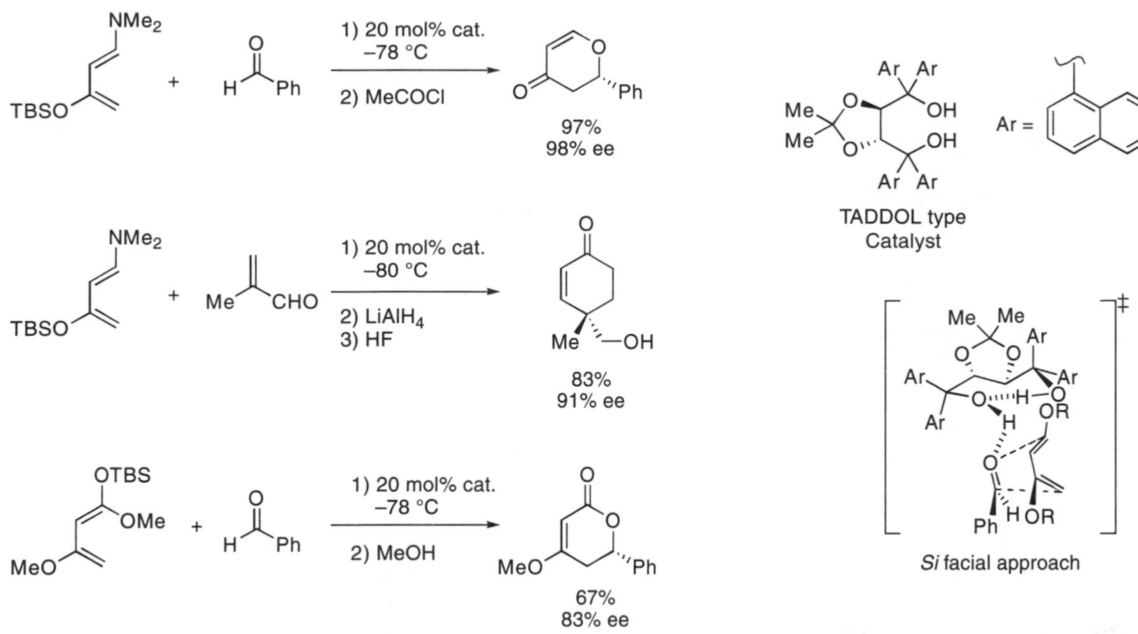

Figure 3.5. Brønsted acid catalyzed asymmetric Diels–Alder reactions using a TADDOL derivative as a single hydrogen-bond donor.

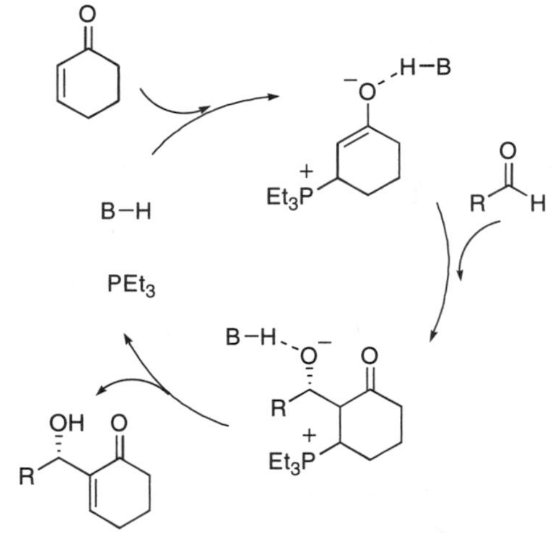

R	X	Yield (%)	ee (%)
PhCH₂CH₂	CF₃	88	90
BnOCH₂CH₂	CF₃	74	82
Cy	CH₃	71	96
i-Pr	CH₃	82	95
Ph	CF₃	40	67
PhCH=CH	CH₃	39	81

Conditions: 10 mol% B–H, 2 equiv. Et₃P, THF, –10 °C, 48 h

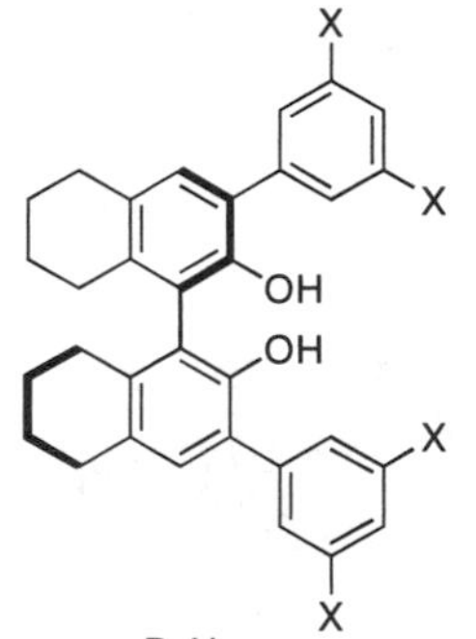

Figure 3.6. Brønsted acid catalyzed asymmetric Morita–Baylis–Hillman reaction using a BINOL derivative as a single hydrogen-bond donor.

ture between the aldehyde and catalyst provides support for the single hydrogen-bond-donor model.[22] The utility of these Brønsted acids has been applied to other classes of reactions, including the Morita–Baylis–Hillman reaction (**Figure 3.6**).[23] In this instance, the chiral biphenol undergoes hydrogen bonding with the enone and catalyzes formation of the zwitterionic adduct with phosphine reagents. The stereochemical environment in the resultant adduct causes stereoselective addition to the aldehyde substrate. Subsequent elimination of the phosphine provides the alcohol proton needed to allow release of the chiral Brønsted acid and turnover.

Amongst the most acidic of the Brønsted acid catalysts are the BINOL phosphoric acids and their derivatives. One application is illustrated with the addition of

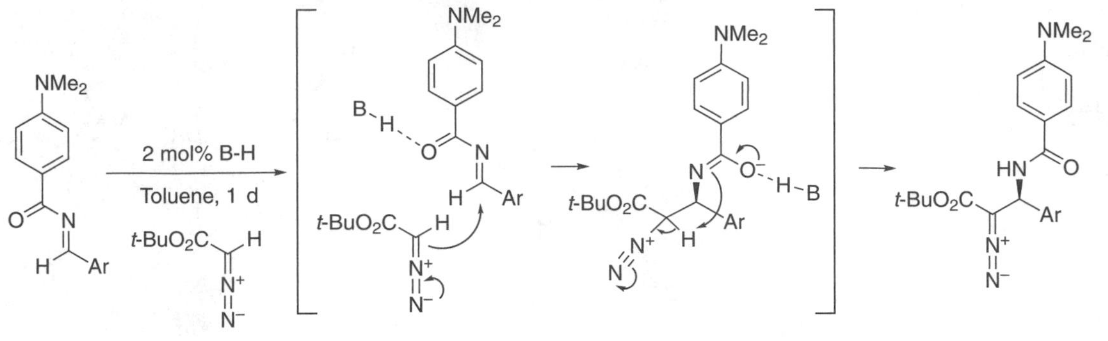

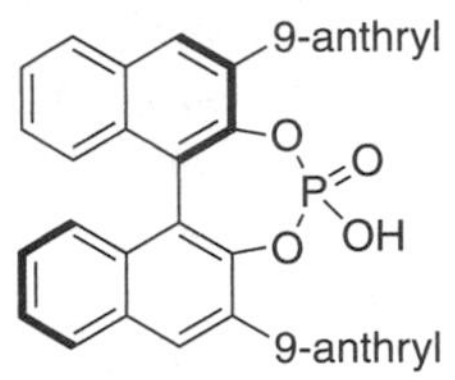

Figure 3.7. A chiral phosphoric-acid catalyst in the addition of diazoesters to *N*-acyl imines.

Ar	Yield (%)	ee (%)
4-FC$_6$H$_4$	74	97
4-PhC$_6$H$_4$	71	97
4-MeOC$_6$H$_4$	62	97
2-FC$_6$H$_4$	89	91
2-MeOC$_6$H$_4$	85	91

α-diazoesters to *N*-acyl imines (**Figure 3.7**).[24] In this reaction, the proton from the phosphoric acid is proposed to interact with the imine to provide electrophilic activation, although the exact mode (N-vs. O-activation) is unclear. Formation of an ion-pair complex (see Section 3.3) by protonation of the imine substrate is an alternate activation mode,[25] but seems unlikely in light of the stronger acidity of *N*-acyl iminiums (p$K_a \approx -2$) relative to phosphoric acids (p$K_a \approx 2$).[26] After bond formation, proton transfer to the nitrogen then allows regeneration of the phosphoric-acid catalyst. The strict requirement for aryl–imine substrates and 3,3'-substituents on the BINOL-derived catalyst suggest that π–π interactions play an important role in orienting the substrate.

The use of hydrogen bonding as an organizational and activating element in catalysis is now thoroughly established. Indeed, the use of Brønsted acids in this way is more widespread than commonly appreciated, because they are often elements of multifunctional catalysts (see Chapter 12). For example, the peptide-derived catalysts used in the kinetic resolution outlined in **Figure A.46** (Section A.2.1.6 of Appendix A) combine Brønsted acid and Lewis base moieties. Specifically, one of the amides is proposed to interact with the substrate via hydrogen bonding.

3.1.2 Catalysts Effecting Substrate Protonation

The above examples involve Brønsted acids acting analogously to Lewis acids, wherein the proton undergoes a transient and reversible interaction with the substrate to provide activation. After the asymmetric step, turnover involves simple release of the hydrogen bond. In a different mode, a Brønsted acid may be used as a

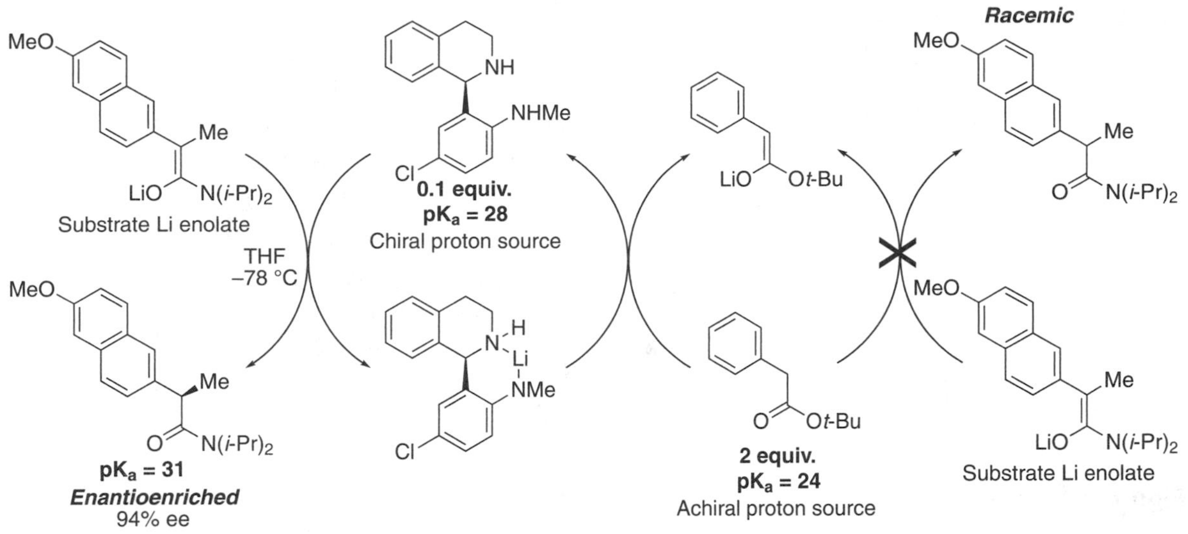

Figure 3.8. Catalytic asymmetric protonation.

chiral proton source to introduce a proton to an achiral substrate.[27] Now, the challenge lies in turnover—namely, in regeneration of the chiral Brønsted acid catalyst.

An early example that illustrates the feasibility of this approach is shown in **Figure 3.8**.[28] Slow addition of the stoichiometric proton source, in combination with a careful pK_a balance and set of relative rates, allows this process to occur. In particular, the reaction of the achiral proton donor, *tert*-butyl phenylacetate, with the substrate lithium enolate (see lower right corner of **Figure 3.8**) must be slower than the reaction of the substrate lithium enolate with the chiral proton source (see upper left corner of **Figure 3.8**). Paradoxically, this circumstance is satisfied even though the achiral proton donor is the most acidic component in the reaction mixture (see the pK_a values in DMSO listed in **Figure 3.8**). Apparently, proton transfer between a heteroatom base (anilide anion) and a carbon acid (*tert*-butyl phenylacetate) is inherently much faster than the proton transfer involving a carbon acid (*tert*-butyl phenylacetate) and a carbon base (substrate lithium enolate).[28]

An alternate approach toward generating potent Brønsted acids, without directly using more acidic species (**Figure 3.2**), centers on the use of an ancillary Lewis acid to boost the acidity of a proton donor (**Figure 3.9**).[29] When tin tetrachloride complexes with a chiral naphthol, binding to the oxygens of the naphthol increases the acidity of the phenolic hydroxyl. The resultant adduct selectively protonates one face of a dimethyl-substituted alkene and triggers a stereoselective polycyclization of a polyprenoid to yield a tricyclic product containing two stereogenic units with 87% ee.[30] Even though this strategy has been applied to very complex substrates (see **Figure 16.22**), the development of proton-transfer catalysts that can be employed with a broader range of substrates and at lower catalyst loading remains a significant challenge.

An interesting paradigm for Brønsted acid catalysis is outlined in **Figure 3.10**. While most reactions of ketenes with the types of basic catalysts illustrated are

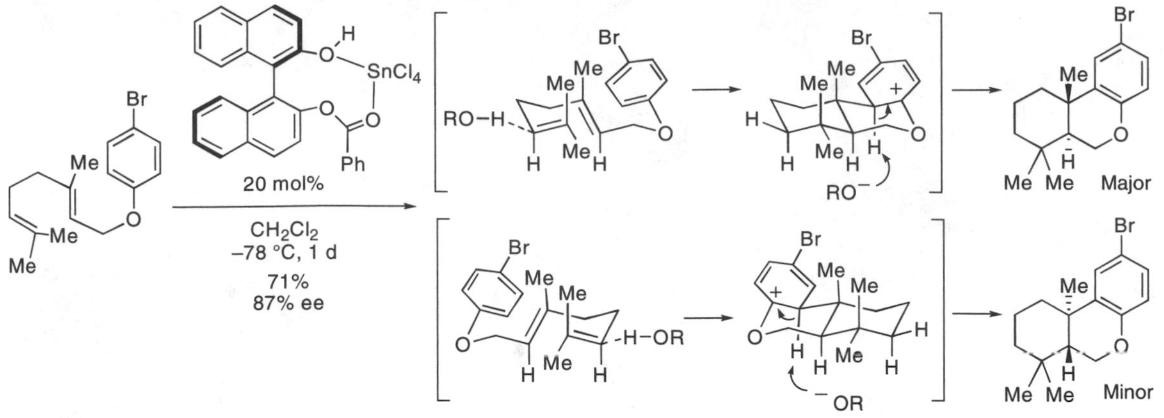

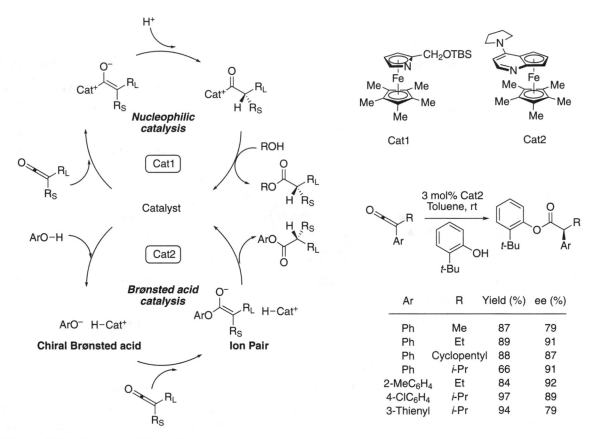

Figure 3.9. Lewis acid assisted Brønsted acid catalysis.

Ar	R	Yield (%)	ee (%)
Ph	Me	87	79
Ph	Et	89	91
Ph	Cyclopentyl	88	87
Ph	i-Pr	66	91
2-MeC$_6$H$_4$	Et	84	92
4-ClC$_6$H$_4$	i-Pr	97	89
3-Thienyl	i-Pr	94	79

Figure 3.10. Two possible pathways for additions to ketenes.

believed to proceed via Lewis base adducts (Cat1 pathway; see Section 2.2.2.c), strong mechanistic evidence indicated that another pathway was viable (Cat2 pathway), depending on the catalyst and substrates.[31] In this alternate pathway, the catalyst first deprotonates the achiral alcohol substrate, generating a chiral Brønsted acid catalyst. Upon reaction of the resultant alkoxide with the ketene, an ion-pair is formed. Since asymmetric induction occurs from within this ion-pair, this scenario may alternately be considered as part of Section 3.3 [Ion-Pairing Catalysts (Electrostatic Catalysts)]. Regardless, stereoselective transfer of a proton from the Brønsted acid of the ion-pair generates the stereochemistry.

To date, Brønsted acid catalysis involving proton transfer has been confined to a small number of systems. However, it is intriguing that asymmetric protonations are the rate-determining steps in a number of other catalytic reactions (e.g., the Nazarov reaction[32]). The use of Brønsted acid catalysts in these contexts is an interesting problem worth further exploration.

3.2 Brønsted Base Catalysts

As was the case for Brønsted acids compared to Lewis acids, Brønsted bases have seen much less development in asymmetric catalysis compared to Lewis bases. While Lewis bases can interact with many different electrophilic functional groups, Brønsted bases are by definition restricted to protons (**Figure 3.11**). This limits Brønsted base catalysts to asymmetric deprotonations and related reactions.

The use of catalysts that are strictly Brønsted bases is relatively rare. However, there are numerous examples of multifunctional asymmetric catalysts in which one of the catalytic components is a Brønsted base (see Chapter 12). An example can be found in the direct aldol reaction catalyzed by the heterobimetallic BINOLate complex illustrated in **Figure 3.12**.[33] This catalyst is proposed to function by a combination of Brønsted base catalyzed deprotonation of the ketone nucleophile and Lewis acid activation of the aldehyde electrophile. Deprotonation by one of the BINOLate oxygens generates a catalyst-bound enolate that undergoes an aldol reaction. Release of the aldolate by proton transfer from the monoprotonated BINOLate regenerates the catalyst.

In the use of Brønsted base catalysts, the most difficult aspect is recycling of the catalyst. Since the first step involves deprotonation of the substrate, conditions must be developed that allow subsequent deprotonation of the protonated inactive catalyst without causing direct deprotonation of the substrate. One of the first solutions to this dilemma is outlined in **Figure 3.13**.[34,35] Here, a chiral lithium amide was used in the catalytic enantioselective deprotonation of *meso*-epoxides.

Crucial to the success of this chemistry is the inability of either the stoichiometric base (LDA) or the lithium-alkoxide product to effect the required deprotonation on

$$LB + S \rightleftharpoons LB-S \qquad S-H + BB^- \rightleftharpoons S^- + BB-H$$

Lewis base | **Brønsted base**

Figure 3.11. Lewis base vs. Brønsted base catalysts.

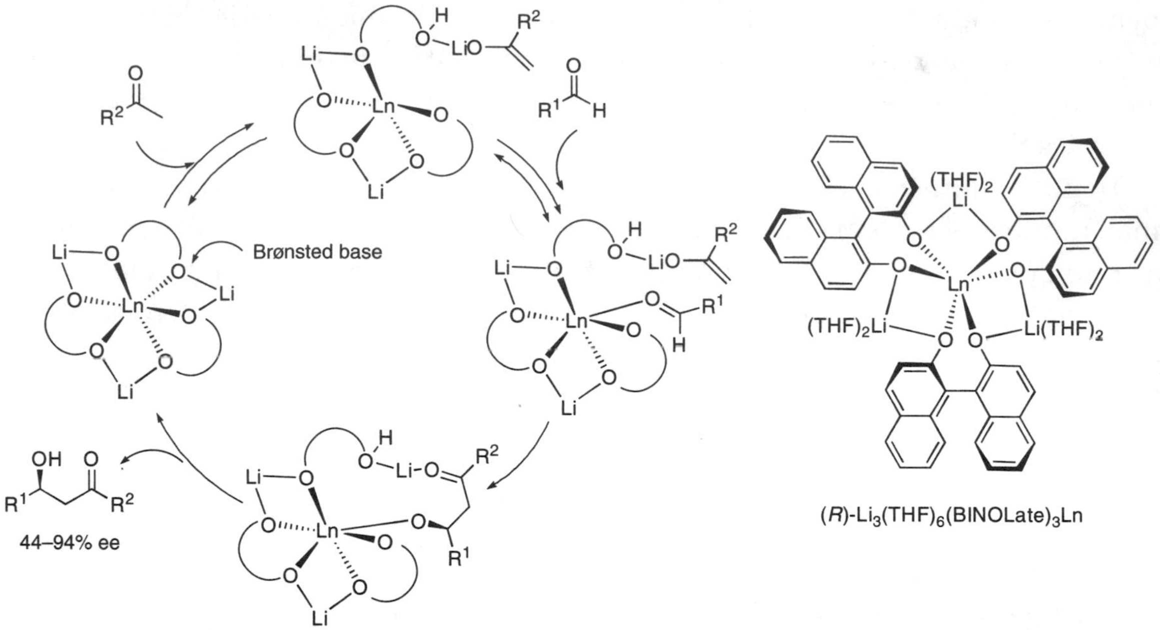

Figure 3.12. A Brønsted base component within a bifunctional catalyst. Ln = lanthanide (III).

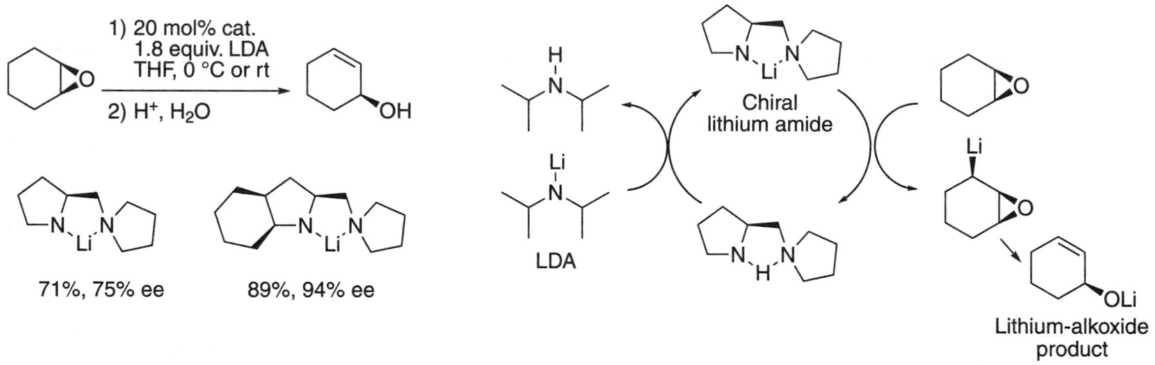

Figure 3.13. Catalytic asymmetric deprotonation of *meso*-expoxides.

the timescale of the catalytic asymmetric reaction (**Equation 3.1**). The former is actually surprising. From a purely thermodynamic treatment of acidity, anything that is sufficiently basic to deprotonate the chiral diamine should also deprotonate the substrate. Here, the more hindered nature of the LDA, relative to the chiral lithium amide, is key to preventing direct reaction of LDA with the substrate.

Equation 3.1

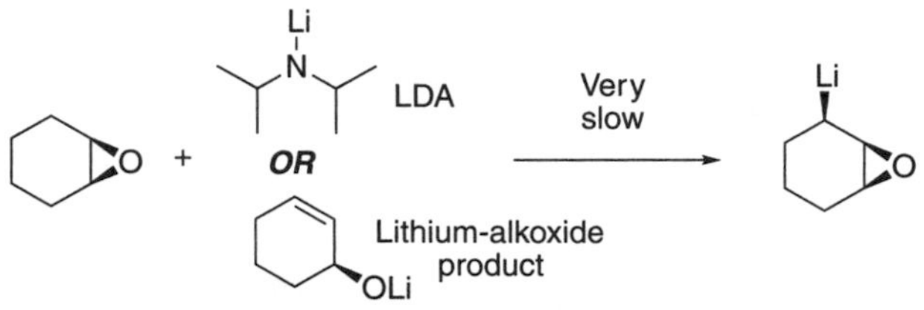

Even in this case, it is not entirely clear that the catalyst functions only as a Brønsted base. The lithium plays a role in directing the catalyst to the reactive site (**Figure**

3.14). The Lewis acid nature of the lithium also likely assists in the opening of the epoxide. Nonetheless, this early work has led to further development of versatile and highly selective lithium-amide catalysts in the catalytic enantioselective deprotonation of *meso*-epoxides.[36,37–39] These catalysts have also found applications in the kinetic resolution of racemic epoxides.[40,41]

Another transformation that takes advantage of differential reactivity is the α-epoxide lithiation illustrated in **Figure 3.15**.[42] Under the reaction conditions employed here, the alkyl lithium alone is insufficient to cause the key deprotonation. Only the diamine–alkyl-lithium adduct can initiate this step. Subsequent exchange of the chiral diamine to a new *i*-PrLi allows turnover.

Conceptually similar is the use of diamine–alkyl-lithium adducts in the asymmetric deprotonation of *N*-Boc pyrrolidine (**Figure 3.16**).[43] Since the electrophile (EX) would react with the alkyl lithium, all of the substrate is deprotonated first. To achieve this end with catalytic chiral diamine, a stoichiometric amount of an achiral diamine

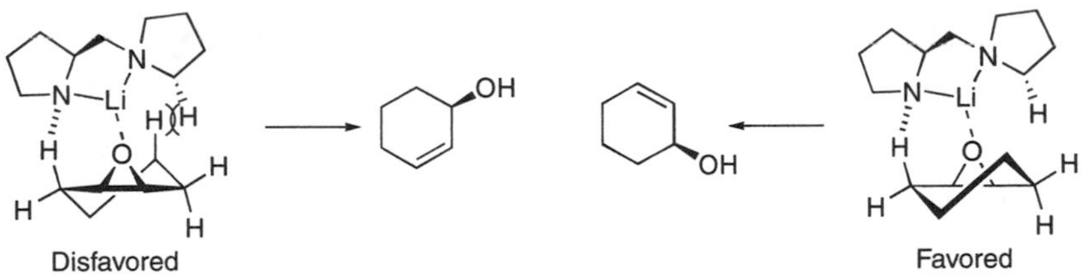

Figure 3.14. Catalytic asymmetric deprotonation: *syn* β-elimination transition structures.

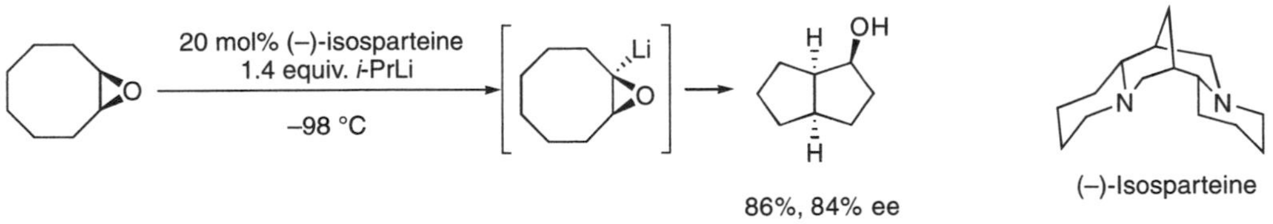

Figure 3.15. Catalytic asymmetric α-deprotonation and rearrangement of *meso*-expoxides.

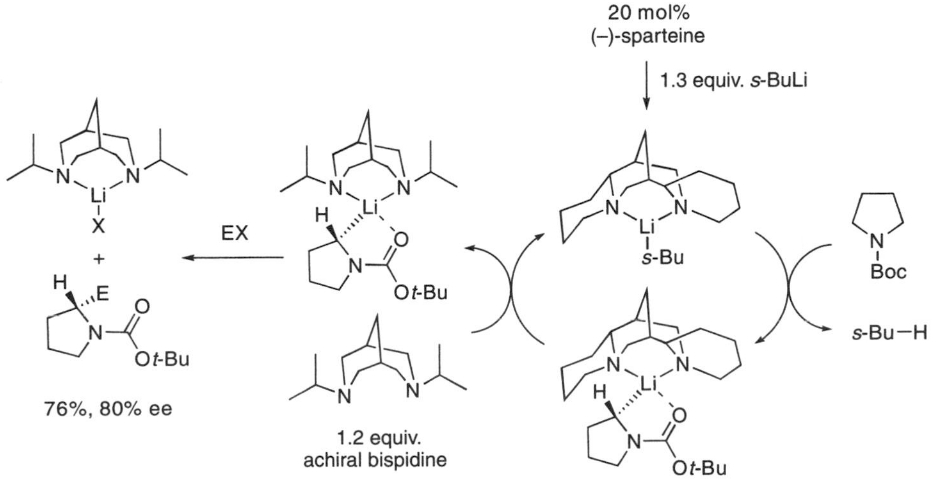

Figure 3.16. Asymmetric catalytic deprotonation with a chiral alkyl lithium.

is used to "trap" the chiral substrate anion after deprotonation. Presumably, the achiral bispidine–alkyl-lithium complex is less reactive in the substrate deprotonation step, due to the large steric bulk of the two *N*-isopropyl groups.

A case with an organocatalyst provides a situation where Brønsted base catalysis occurs without any potential Lewis acid involvement from counterions. Recent efforts have identified the cinchona-alkaloid-derived catalyst (DHQD)$_2$AQN (**Figure 3.17**) as effective in a variety of asymmetric reactions involving anhydride cleavage.[44] In contrast to earlier work with other cinchona catalysts in these same reactions, (DHQD)$_2$AQN lacks any hydroxyl group and cannot act as a Brønsted acid/hydrogen-bond donor (see Section 3.1). In addition, the (DHQD)$_2$AQN displays a broad range of utility. For example, the kinetic resolution (see Chapter 7) of urethane-protected α-amino acid *N*-carboxy anhydrides (UNCAs) proceeds with high selectivity factors (k_{rel} values) for a range of subsitutents and nitrogen-protecting groups (**Figure 3.17**).[45] Study of the mechanism revealed a first-order kinetic dependence in catalyst, alcohol, and UNCA substrate. Together with a kinetic isotope effect ($k_{MeOH}/k_{MeOD} = 1.3$), these results point to the Brønsted base mechanism illustrated in **Figure 3.17**. In this mechanism, an amine of the (DHQD)$_2$AQN acts as a general base, deprotonating the alcohol as it attacks the anhydride moiety. The asymmetric induction within this complex is quite interesting, given the weak hydrogen bond proposed in the transition state. There are likely additional organization elements involved, such as π-stacking between portions of the catalyst and UNCA substrate (see Chapter 5). After asymmetric anhydride cleavage, a thermodynamically favorable proton transfer [pK_a(DMSO)

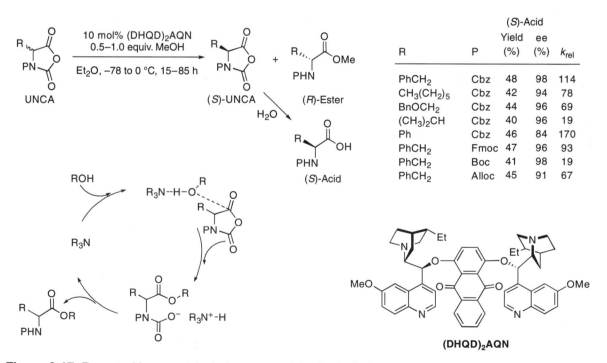

| | | (*S*)-Acid | | |
| | | Yield | ee | |
R	P	(%)	(%)	k_{rel}
PhCH$_2$	Cbz	48	98	114
CH$_3$(CH$_2$)$_5$	Cbz	42	94	78
BnOCH$_2$	Cbz	44	96	69
(CH$_3$)$_2$CH	Cbz	40	96	19
Ph	Cbz	46	84	170
PhCH$_2$	Fmoc	47	96	93
PhCH$_2$	Boc	41	98	19
PhCH$_2$	Alloc	45	91	67

Figure 3.17. Brønsted base catalysis in asymmetric alcoholysis.

quinuclidinium = 9.8; pK_a(DMSO) MeCO$_2$H = 12.3][46,47] within the resultant ion-pair regenerates the catalyst.

Given the recent advances in Brønsted base catalysis, there is a great deal of promise for using catalysts with this type of reactivity in a broader range of transformations.

3.3 Ion-Pairing Catalysts (Electrostatic Catalysts)

So far, all the classic methods of activating organic substrates for chemical reaction have been surveyed, including chiral Lewis acids, Lewis bases, Brønsted acids, and Brønsted bases. What other means are there of activating and/or controlling the approach of achiral substrates to generate new molecules with stereochemical control? In many of the acid–base reactions described above, the role of counterions has not been explicitly addressed. However, the counterions of a reaction can be harnessed in a productive manner by taking advantage of the often close proximity between counterions in a tight ion-pair. If one of the counterions is a chiral species, then stereochemical information can be transferred to the reacting achiral component of the ion-pair. The key to this strategy is the subsequent recycling of the chiral counterion to a fresh substrate to allow turnover, as shown in **Figure 3.18**. In principle, the chiral catalyst component could be the cation or the anion.

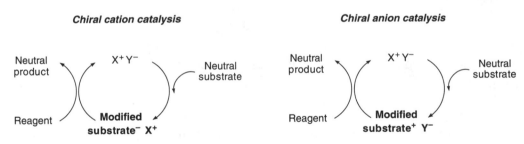

Figure 3.18. General schematic of ion-pair catalysis (either the cation X$^+$ or the anion Y$^-$ is chiral).

3.3.1 Catalysts Containing Chiral Cations

One of the seminal demonstrations of this principle can be found in the asymmetric phase-transfer catalysts for the reactions of enolates (**Figure 3.19**).[48] Under the biphasic reaction conditions, the sodium enolate forms in the aqueous layer or at the interface and is thereby segregated from the organic-soluble alkylating agent (MeCl). This segregation prevents simple racemic alkylation. Rather, upon cation exchange with the ammonium catalyst (**PTC1**), the organic-soluble chiral ion-pair is now available for alkylation, which occurs in the asymmetric environment conferred by the cation. Upon alkylation, the neutral product remains in the organic layer and the catalyst is regenerated.

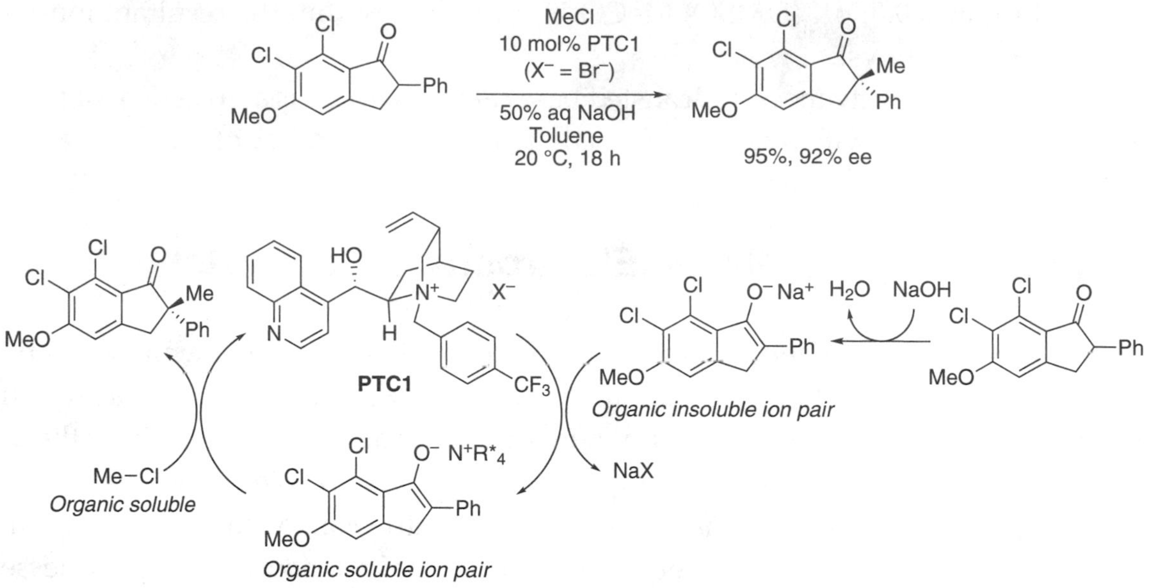

Figure 3.19. Asymmetric enolate alkylation under phase-transfer conditions using an ion-pairing chiral catalyst. R_4N^+ is the chiral phase-transfer catalyst.

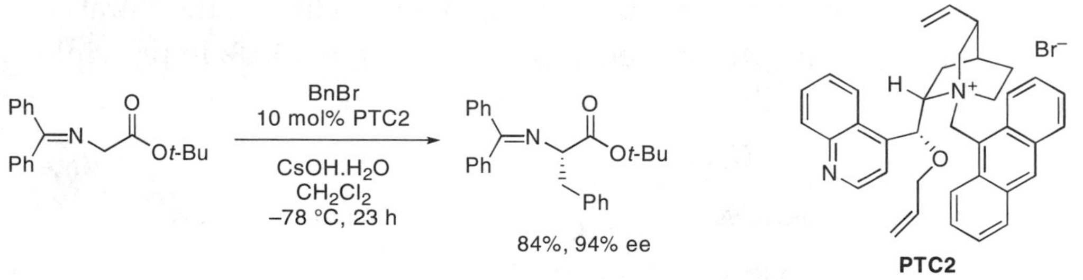

Figure 3.20. α-Amino-acid synthesis with an ion-pairing chiral catalyst.

The utility of these ion-pairing catalysts was advanced considerably by the discovery that glycine-imine enolates[49,50] are suitable substrates and that rational modifications of the catalyst skeleton[51–53] (supported by X-ray crystallographic data)[51] could lead to large improvements in the selectivity. The overall result is a highly efficient α-amino-acid synthesis, as illustrated in **Figure 3.20**. This technology is now broadly applicable to a wide range of enolates with a variety of electrophiles (e.g., alkyl halides, aldehydes, and enones).[54–59]

Even though the above catalysts are all based upon the cinchona skeleton, ion-pairing catalysts are by no means restricted to this structural type.[56] Among the most successful of other ammonium ion-pairing catalysts is the C_2-symmetric binaphthalene-derived type outlined in **Figure 3.21**.[60,61] The ability to tune the Ar groups, along with the accessibility of both enantiomeric forms, has led to numerous applications.[56,59]

While the above cases illustrate the utility of ion-pairing catalysts under phase-transfer conditions, reaction parameters have also been developed that allow their employment under homogenous, nonaqueous conditions. This development has

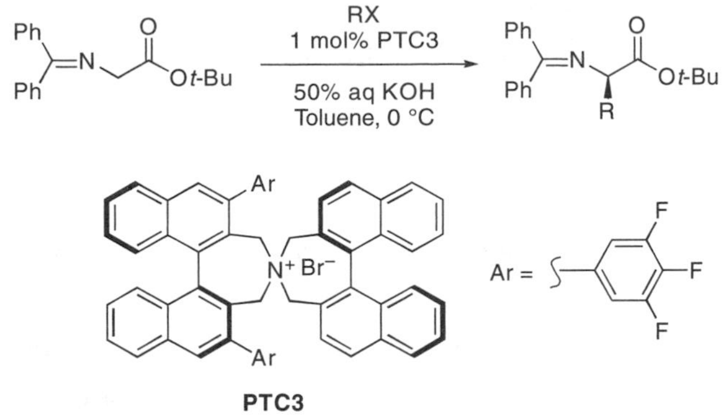

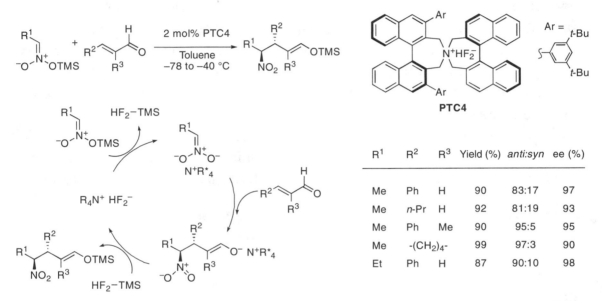

Figure 3.21. α-Amino-acid synthesis with a C_2-symmetric ion-pairing chiral catalyst.

RX	Yield (%)	ee (%)
$H_2C=CHCH_2Br$	80	99
$HCCCH_2Br$	80	99
$4\text{-}MeC_6H_4CH_2Br$	91	99
$4\text{-}FC_6H_4CH_2Br$	92	99
$1\text{-}NaphthylCH_2Br$	90	99
$2,6\text{-}Me_2C_6H_3CH_2Br$	98	99
CH_3CH_2I	89	98
CH_3I	92	96

R^1	R^2	R^3	Yield (%)	anti:syn	ee (%)
Me	Ph	H	90	83:17	97
Me	n-Pr	H	92	81:19	93
Me	Ph	Me	90	95:5	95
Me	-(CH₂)₄-		99	97:3	90
Et	Ph	H	87	90:10	98

Figure 3.22. Use of an ion-pair catalyst under homogenous conditions for addition of silyl nitronate. R_4N^+ is the chiral phase-transfer catalyst **PTC4**.

stimulated the further examination of these catalysts in an array of transformations where ion-pairing is possible. One such situation involves the reactions of charged nitronate species with aldehydes[62] and conjugated systems.[63] An example with the latter is illustrated in **Figure 3.22**. The catalytic cycle is most interesting as the catalyst HF_2^- counterion is also employed to activate the silyl nitronate by removing the TMS group.

3.3.2 Catalysts Containing Chiral Anions

In the preceding cases, the chiral catalytic counterion complements a charged intermediate in the reaction pathway.[64] This strategy can also be applied toward substrates that typically undergo reaction in the neutral form (see the top of **Figure 3.23**). In doing so, the formation of a charged intermediate, such as the iminium at the bottom of **Figure 3.23**, must be viable. A frequent additional advantage of such intermediates is their higher reactivity profile compared to the neutral species. With such a charged intermediate in hand, a chiral counterion can be employed to lead to a chiral intermediate. Since the charged chiral intermediate is more reactive, reaction proceeds only from this point, ensuring limited racemic background reaction. So long as formation and cleavage of the charged intermediate is relatively rapid, turnover can be maintained, leading to efficient and highly selective metal-free catalytic systems.

For the reaction at the bottom of **Figure 3.23**, an iminium is formed from an achiral amine that is part of an ion-pair catalyst. Another departure from the prior examples is that the anion comprises the chiral portion of the catalyst rather than the cation. In the chiral iminium complex, this chiral phosphate anion blocks one face of the alkene. Thus, transfer hydrogenation from the dihydropyridine, an NADH mimic, occurs with high facial selectivity. A key finding revolved around entry 6, in which (*E*)-citral was shown to undergo a much more selective hydrogenation relative to systems employing a chiral amine catalyst (i.e., Lewis base catalysis; see Section 2.2.2.c). The product, (*R*)-citronellal, is an intermediate in the industrial synthesis of menthol and is a perfume ingredient.

The reactivity of many chiral Lewis acid (see Section 2.1) and chiral metal (see Sections 3.4–3.6) catalysts have benefited from cationic activation. Replacement of the conventional counterions (i.e., BF_4^-, SbF_6^-, etc.) with chiral counterions would allow the

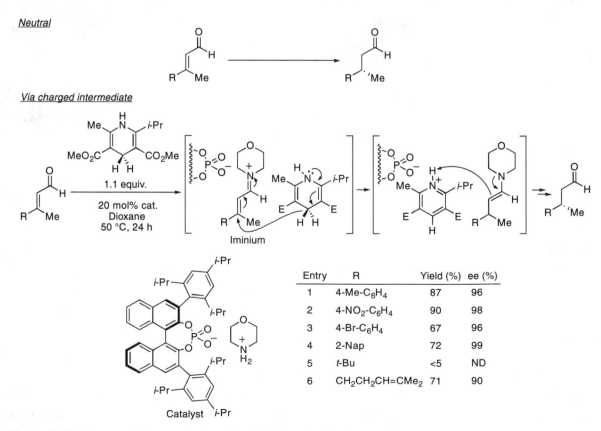

Entry	R	Yield (%)	ee (%)
1	4-Me-C$_6$H$_4$	87	96
2	4-NO$_2$-C$_6$H$_4$	90	98
3	4-Br-C$_6$H$_4$	67	96
4	2-Nap	72	99
5	*t*-Bu	<5	ND
6	CH$_2$CH$_2$CH=CMe$_2$	71	90

Figure 3.23. Asymmetric counterion-directed transfer hydrogenation.

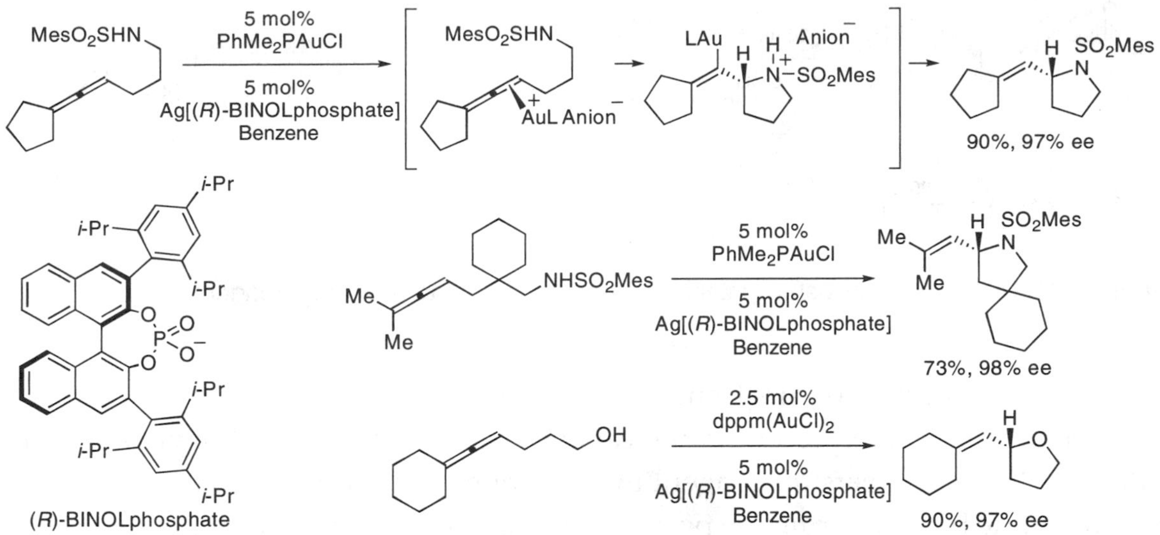

Figure 3.24. Asymmetric induction with a chiral counterion in a metal-catalyzed process.

use of achiral metal species. An elegant example is outlined in **Figure 3.24**.[65] The asymmetric Au(I)-catalyzed hydroalkoxylation of allenes had proven particularly resistant to broad application, which likely reflects the linear coordination geometry of gold placing the chiral components distal from the substrate. However, the use of a chiral counteranion derived from a BINOL compound provided high enantioselectivities in both the hydroamination and hydroalkoxylation reactions. Consistent with the premise that the coordination complex is key to effective transmission of the stereochemical information from the anion, poorer selectivity was seen in polar solvents (e.g., CH_3NO_2, acetone, and THF) compared to nonpolar solvents (e.g., benzene). There is a fine line between chiral anion catalysis, as described in **Figure 3.24**, and chiral Lewis acid catalysis, which would result if the anion were to bind to the gold center.

This approach is likely to have broad applicability and will also allow the combination of the chiral counterions and chiral metal ligands to optimize reactivity/selectivity (see Chapter 6 for examples of optimization with two chiral ligands on one metal center).

3.3.3 Directed Electrostatic Activation

Many structural elements, including steric repulsion, van der Waals attractions, dipole–dipole interactions, π–π interactions, and others (see Chapter 5), have been invoked to explain how catalysts induce asymmetry by control of conformation and steering of substrates. Among these elements, repulsive and attractive electrostatic interactions can be included. The use of electrostatic interactions as an activating element, as well as a directing element, can be found in the above work and is especially clear in the example illustrated in **Figure 3.25**.[66] Here, the catalyst activates the α-β-unsaturated system by forming a charged iminium (see Section 2.2.2.c). Under these

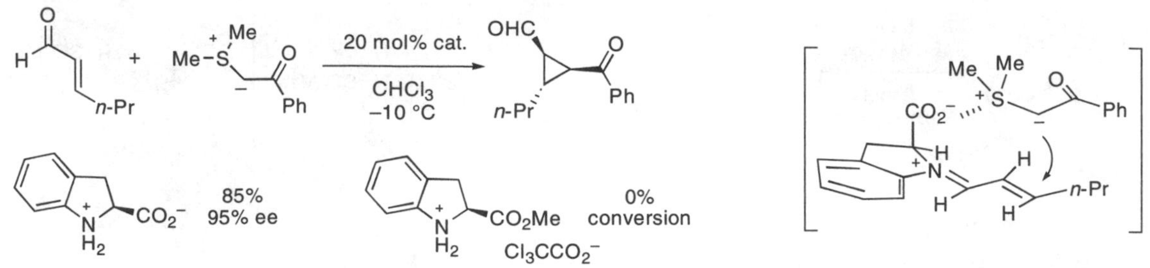

Figure 3.25. Use of an electrostatic interaction from a catalyst to activate a reagent.

conditions, no reaction occurs from the anionic portion of the ylide unless there is an additional carboxylate present in the catalyst. Putatively, this carboxylate interacts with the ylide and steers the reagent into position. In addition, the carboxylate *activates* the ylide by countering the positive charge. This proposal accounts for both the reactivity and selectivity patterns observed with a series of α-β-unsaturated aldehydes and ylides.

3.4 Group-Transfer Catalysts

In this chapter, group-transfer reactions refer to transformations that are comprised of two main phases (**Figure 3.26**). First, a chiral catalyst undergoes reaction with an achiral reagent to generate an activated species containing a new transferable "group." Second, the chiral activated catalyst undergoes reaction with a prochiral substrate, often without coordination, causing stereoselective transfer of that "group" to generate a new chiral product and regenerate the original unactivated catalyst. The group transfer can be a concerted one-step process or it can be stepwise. This mode is distinct from that in which the catalyst remains bound after the key stereochemistry-determining step (e.g., see the Lewis base catalysts in Section 2.2). Many different types of transformation fall into this category, including dihydroxylation, aziridination, cyclopropanation, epoxidation, carbene insertion reactions, olefin metathesis, etc. In some of these transformations, the transferring "group" may even be activated by one of the modes described above (most frequently Lewis acid or Lewis base activation). Often, these transformations involve a change of oxidation state at the catalyst, but not always.

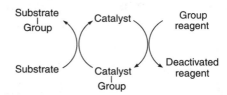

Figure 3.26. A general group-transfer reaction.

3.4.1 Simple Group Transfer

Reactions involving group transfer were among the first to be subjected to asymmetric catalysis. A defining example is the catalytic asymmetric dihydroxylation reaction of alkenes[67] (**Figure 3.27**), which illustrates the concept of ligand-accelerated catalysis (see Chapter 1).[68,69]

The catalytic cycle for the $K_3Fe(CN)_6$ variant[70] in **Figure 3.28** illustrates the group transfer. Mechanistic studies and calculations[71] support a group transfer involving a stereo-determining [3+2] cycloaddition, followed by hydrolysis of the diol from the metal complex. In this reaction mechanism, the metal center cycles between osmium(VIII) and osmium(VI). The use of biphasic reaction conditions allows selective catalyst reoxidation to occur in the absence of any chiral ligands. Under homogeneous conditions, reoxidation of the product-bound osmium gives rise to a separate and less selective catalyst species (see **Figure 1.17**). The ligand (see **Figure 3.27**) not only provides an asymmetric environment for the dihydroxylation upon coordination of an sp^3 hybridized nitrogen to the osmium, but also creates a species that is much

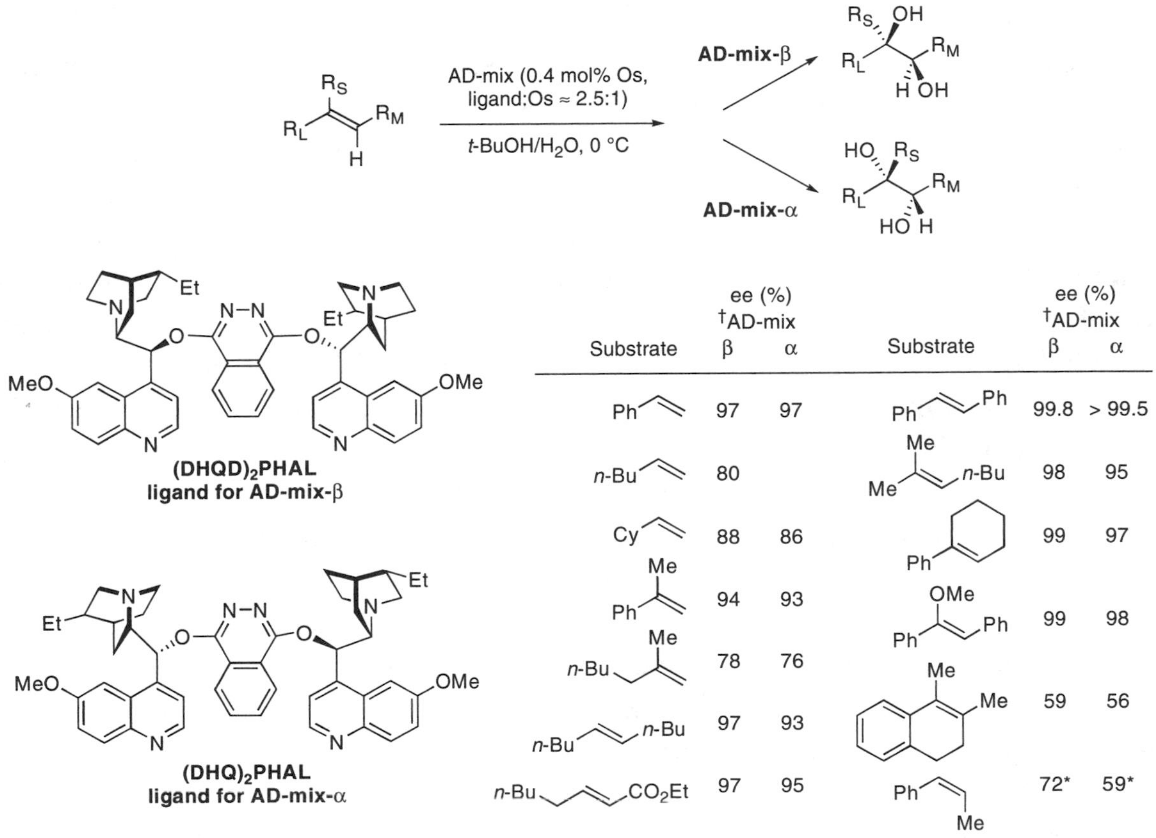

		ee (%) [†]AD-mix			ee (%) [†]AD-mix	
Substrate		β	α	Substrate	β	α
Ph⟋		97	97	Ph⟍⟋Ph	99.8	> 99.5
n-Bu⟋		80		Me₂C⟍⟋n-Bu	98	95
Cy⟋		88	86	Ph-cyclohexene	99	97
Ph(Me)C=		94	93	Ph(OMe)C=Ph	99	98
n-Bu(Me)C=		78	76	dimethyl-dihydronaphthalene	59	56
n-Bu⟍⟋n-Bu		97	93	Ph⟍⟋Me	72*	59*
n-Bu⟍⟋CO₂Et		97	95			

(DHQD)₂PHAL
ligand for AD-mix-β

(DHQ)₂PHAL
ligand for AD-mix-α

[†]Each kg of AD-mix contains: $K_3Fe(CN)_6$, 699.6 g; K_2CO_3, 293.9 g; (DHQD)₂PHAL or (DHQ)₂PHAL, 5.52 g; and $K_2OsO_2(OH)_2$, 1.04 g. The standard procedure calls for 1.4 g of this AD-mix per mmol olefin. In some cases, $MeSO_2NH_2$ was added to accelerate product dissociation.
*Different ligand linker employed.

Figure 3.27. Catalytic asymmetric dihydroxylation.

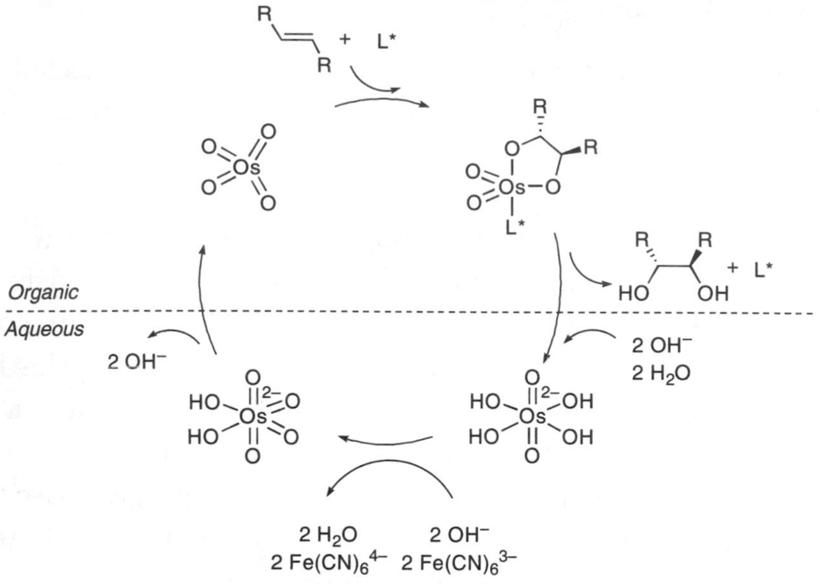

Figure 3.28. The two half-reactions in catalytic asymmetric dihydroxylation. L* = (DHQD$_2$)PHAL or (DHQ)$_2$PHAL.

more reactive than the unbound OsO$_4$. In doing so, the racemic background reaction becomes negligible, even though achiral catalyst is present!

Chiral titanium–tartrate catalysts also display ligand-acclerated catalysis and have been shown to be highly general for the asymmetric epoxidation of prochiral allylic alcohols and related derivatives (**Figure 3.29**).[72–74] The same catalysts are also highly competent in the kinetic resolution of racemic allylic alcohols and the desymmetrization of *meso*-diallylic alcohols (see Chapter 7).[74,75] The titanium–tartrate catalysts are restricted, however, to allylic alcohols, because the alcohol moiety plays a key role in positioning the alkene for reaction in the catalyst coordination sphere (**Figure 3.30**).

The dimeric forms of the titanium adducts are believed to be the active form of the catalyst (**Figure 3.30**).[76–78] In contrast to the dihydroxylation case above, the titanium center serves to assemble the ligands and provide Lewis acid activation of the peroxo group by coordination of both oxygens. The entire redox event occurs between ligands on the catalyst, where the *tert*-butyl hydroperoxide is reduced and the olefin is oxidized upon group transfer of one of the peroxide oxygen atoms.

A catalyst system developed for the epoxidation of alkenes that does not require the allylic alcohol group is the manganese salens (**Figure 3.31**).[79–83] This catalyst class performs best with disubstituted cis-alkenes and trisubstituted alkenes. However, some variants do provide selective epoxidation of certain monosubstituted alkenes, geminally disubstituted alkenes, and disubstituted trans-alkenes.

Many elegant mechanistic studies have been devoted to this reaction, in which an oxygen atom is transferred from the metal center to the alkene. While it is beyond the scope here to review this work, the simplified catalytic cycle outlined in **Figure 3.32** illustrates two of the most likely variants, a concerted process and a stepwise

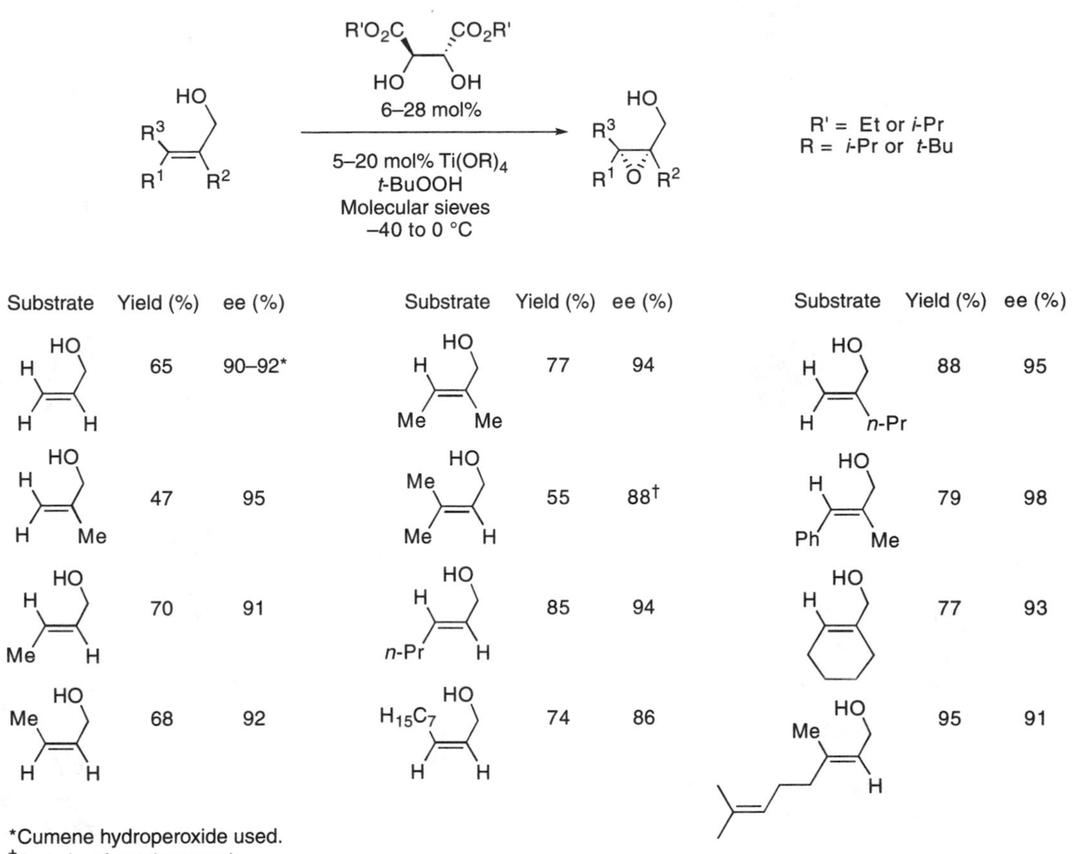

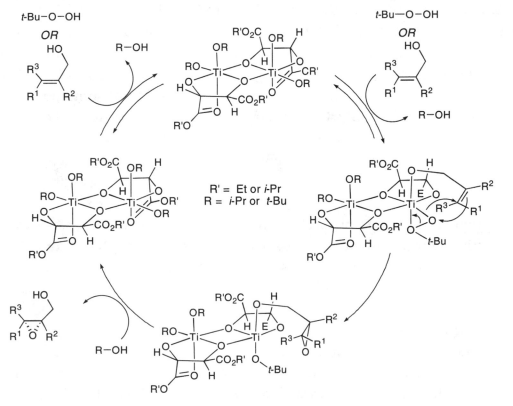

Figure 3.29. Catalytic asymmetric epoxidation of allylic alcohols.

Substrate	Yield (%)	ee (%)	Substrate	Yield (%)	ee (%)	Substrate	Yield (%)	ee (%)
	65	90–92*		77	94		88	95
	47	95		55	88†		79	98
	70	91		85	94		77	93
	68	92		74	86		95	91

*Cumene hydroperoxide used.
†1 equiv. of catalyst used.

Figure 3.30. Catalytic cycle of the asymmetric epoxidation of allylic alcohols.

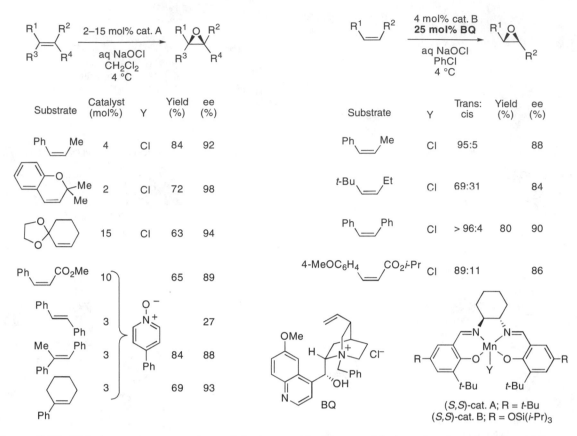

Figure 3.31. Chiral manganese–salen catalysts used in the epoxidation of unfunctionalized alkenes.

process. There is broad agreement for a Mn(V) to Mn(III) redox cycle; however, there has been much debate over whether a concerted oxo-transfer, oxometallacyclic intermediate, or radical intermediate occurs.[82–86] The presence of significant amounts of trans-epoxides from cis-alkenes containing radical stabilizing groups provides strong evidence for the stepwise pathway. For substrates that do not contain radical stabilizing groups, either the concerted pathway or the stepwise pathway with collapse of the radical prior to bond rotation is proposed. The most recent work suggests that multiple pathways can participate in a given reaction (in addition to those outlined below!) and that the participatory pathways depend on the precise catalyst, oxidant, additive, and substrate.[82,83]

The proposed side-on approach of the alkene to the Mn(V) species explains the poor selectivity for trans-alkenes, since one of the alkene substituents undergoes disfavorable steric interactions with the salen face. This can be circumvented by switching the entire catalytic cycle to allow rotation in the radical intermediate, which gives rise to the thermodynamically more stable trans-epoxide (see the right side of **Figure 3.31** and the lower right of **Figure 3.32**).[87]

In a related transformation using an organocatalyst, chiral dioxiranes have been shown to be powerful agents for the asymmetric epoxidation of olefins.[88–91] In particular, the D-fructose-derived catalyst illustrated in **Figure 3.33** is highly selective for a wide variety of trans-disubstituted alkenes (including dienes and enynes) as well as

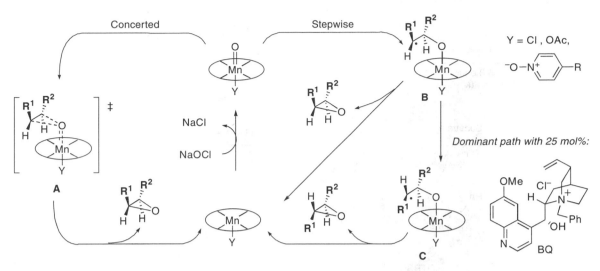

Figure 3.32. Mechanism of the chiral manganese–salen epoxidation. Concerted C–O bond formation (via transition state **A**) and rapid closure of the radical intermediate **B** give the cis-epoxide. Bond rotation in **B** gives intermediate **C**, which closes to the trans-epoxide. This path is favored with addition of BQ.

trisubstituted alkenes.[92–95] Notably, the enantiomeric catalyst is readily available from L-sorbose. Related compounds[89,90] are also proving successful for a number of cis-olefins,[96] with encouraging enantioselectivities for some terminal olefins.[97]

Discovering highly enantioselective chiral ketone catalysts was challenging due to a number of undesired processes during catalyst reoxidation (**Figure 3.34**). It is noteworthy that epoxidation by Oxone® itself under these conditions is negligible. An efficient ketone catalyst embodies a delicate balance of sterics and electronics in the chiral control elements around the carbonyl group. The inductively withdrawing nature of the substituents near the ketone of the carbohydrate-derived catalysts is particularly effective in conferring this balance. Even so, the breakthrough for these reactions came upon recognition of the pH profile.[89,90] In early efforts, nearly neutral reaction conditions were employed as Oxone® undergoes more rapid decomposition at higher pH. Unfortunately, an excess of the chiral ketone was needed because the ketone underwent rapid decomposition, presumably via a Baeyer–Villiger process. Surprisingly, the reactions at higher pH were rapid, selective, and would proceed with substoichiometric ketone. At higher pH, the Baeyer–Villiger process is suppressed, since the precursor is deprotonated more readily. In addition, the key dioxirane forms more rapidly due to the increased nucleophilicity of the Oxone® species. The acceleration of the overall reoxidation process offsets the greater rate of Oxone® decomposition under these conditions. The stereochemical outcome of the oxygen-transfer process itself can be rationalized by a spiro transition-state model, although a planar approach does compete in many cases, accounting for the minor enantiomer (**Figure 3.35**).[97,98]

Group-transfer reactions are not limited to oxygen centers. For example, corresponding catalytic asymmetric aziridination[99] and cyclopropanation[100,101] processes have been developed.

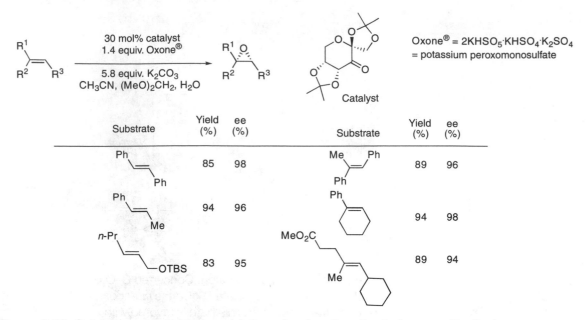

Figure 3.33. Catalytic asymmetric epoxidation of unfunctionalized alkenes with dioxirane.

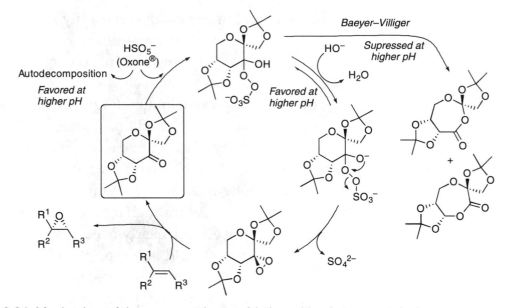

Figure 3.34. Mechanism of the asymmetric epoxidation with a ketone catalyst.

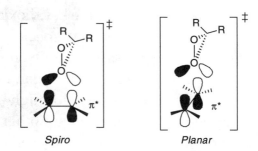

Figure 3.35. Proposed epoxidation transition states.

Based upon the above examples, it might seem that asymmetric group-transfer reactions only occur with alkenes or possibly other π-species. This is not at all the case, as seen in the asymmetric C–H insertion reactions available with rhodium–carbenoid species. The invention of carboxylate-derived chiral catalysts energized this field and early work was largely centered on intramolecular transformations.[102–104] With the discovery of carbene donors containing both electron-withdrawing and electron-donating groups, there has been a great deal of success in the intermolecular processes, especially with very simple starting substrates (**Figure 3.36**).[105–108]

The catalytic cycle for the rhodium-catalyzed insertion is illustrated in **Figure 3.37**. First, the achiral diazo compound reacts with the dirhodium species, expelling nitrogen and generating the carbene group-transfer reagent. The generally accepted

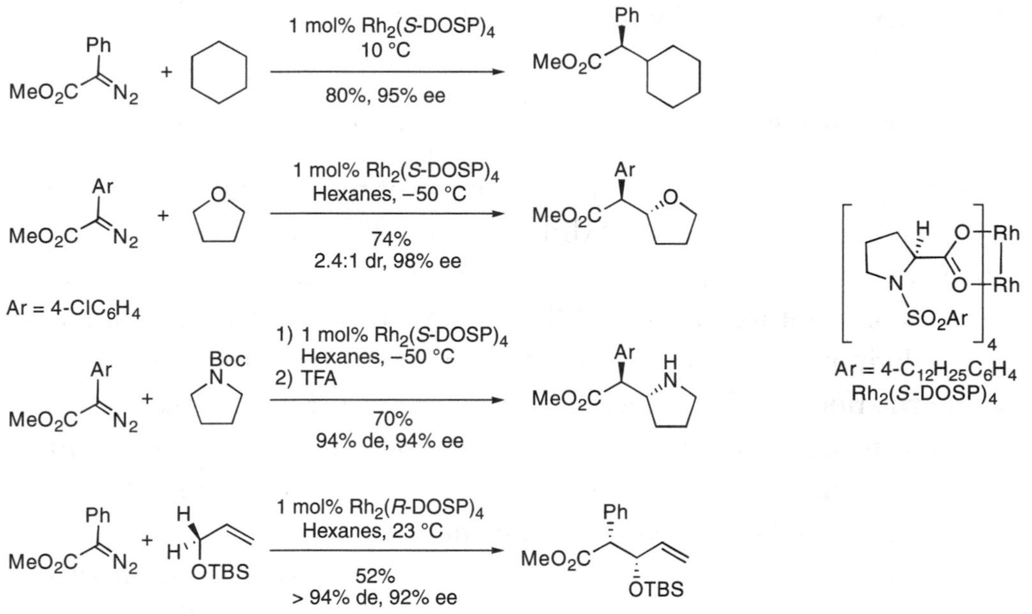

Figure 3.36. Rhodium-catalyzed asymmetric C–H insertions.

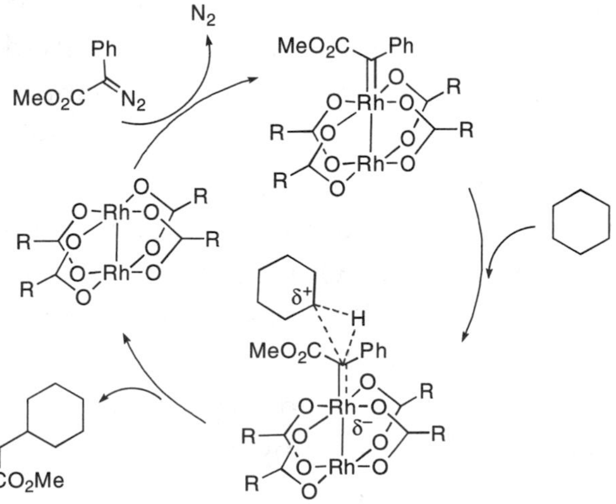

Figure 3.37. Catalytic cycle of rhodium-catalyzed asymmetric C–H insertions.

mechanism then involves direct insertion between the carbenoid and the C–H bond by means of a three-center transition state.[102–104] Kinetic isotope effects support a concerted but nonsynchronous C–H insertion with build-up of positive charge at the carbon of the C–H bond (**Figure 3.37**). The selectivity toward tertiary C–H sites and for the 2-position of tetrahydrofuran are consistent with this mechanism because these are the sites that are best suited to stabilize the positive charge build-up in the transition state.

Group-transfer reactions are among some of the earliest examples of asymmetric catalysis. Initial work focused on the metal-catalyzed transfer of oxygen, nitrogen, and carbon groups to alkenes. More recent work in the field has expanded the scope to include insertions into saturated substrates and has shown that organocatalysts can also be useful. This type of process is a mainstay of asymmetric catalysis and will undoubtedly see further development in the future.

3.4.2 Alkene Metathesis

Alkene metathesis[113] comprises another example of metal-catalyzed group-transfer, although the sequence is somewhat different from those described in the preceding section. In alkene metathesis, the transfer of a metal-bound CR_2 group to an alkene occurs with concomitant loss of one of the original alkene CR_2 groups. Examples of asymmetric olefin metathesis usually hinge on kinetic resolution of a racemic diene or desymmetrization of a diene/triene.[114,115] In each case, the chiral catalyst selectively recognizes and reacts with one enantiotopic alkene. As illustrated in **Figure 3.38** for a series of desymmetrization reactions,[116–118] compounds containing 5-, 6-, 7-, and 8-membered rings can be formed using this method.[114,115]

The mechanism of this group-transfer process involves initial formation of a metallacyclobutane by a [2+2] process from the most reactive alkene (**Figure 3.39**). A reverse [2+2] then loads one-half of the alkene onto the catalyst, generating the alkylidene group-transfer reagent. Intramolecular reaction with one of two available enantiotopic alkenes then constitutes the enantioselectivity-determining step. Unlike the ruthenium catalysts,[119] discrete alkene π-coordination (see Section 3.6) to the molybdenum catalyst does not appear to occur.[120] Ultimately a new alkene is created containing the transferred alkylidene group.

The ability to promote skeletal rearrangements, where simple achiral substrates are transformed into more complex chiral molecules, is a powerful feature of the alkene metathesis catalysts. While all the cases here are intramolecular, intermolecular cross-metathesis examples have also been reported and proceed via similar principles. In addition, many asymmetric tandem processes (see Chapter 14) are possible and comprise a burgeoning area of ongoing development.[114,115,121] While early asymmetric catalysts centered on molybdenum-derived complexes, there are now many examples with tungsten-[115] and ruthenium-derived chiral complexes.[121–124] For the

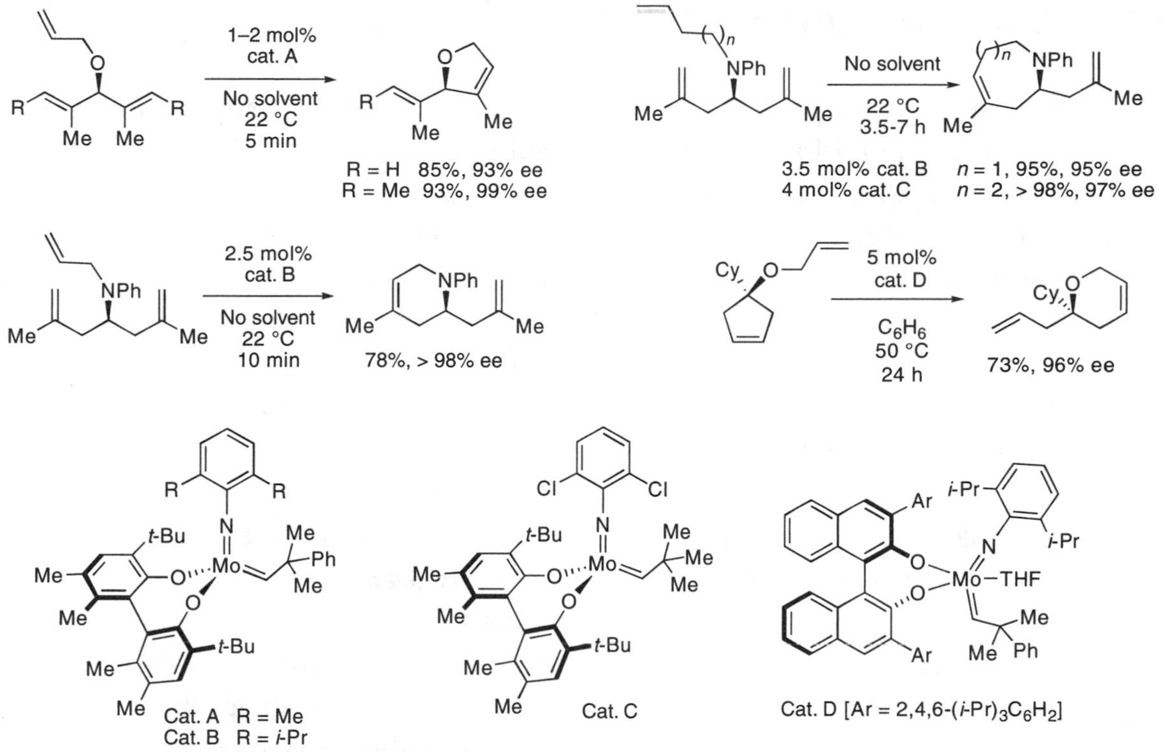

Figure 3.38. Asymmetric alkene metathesis catalysts.

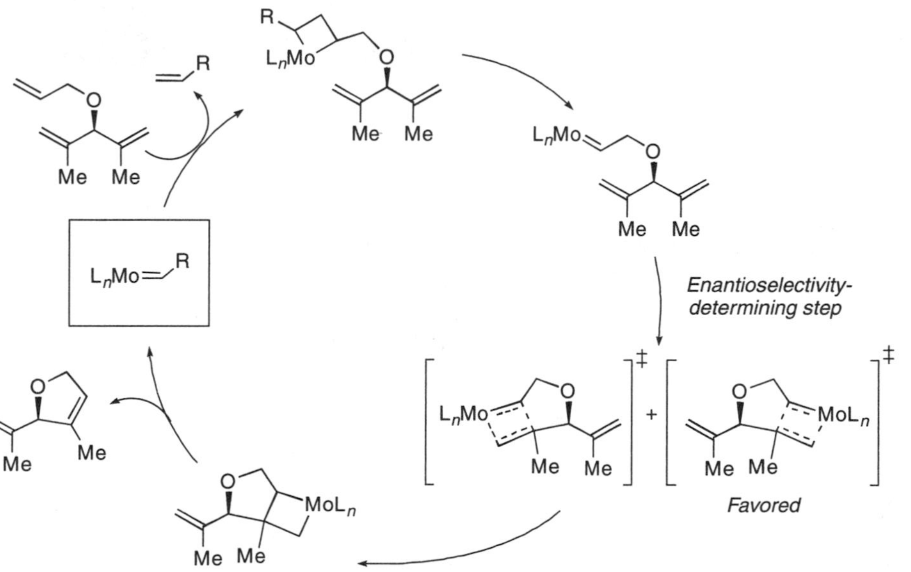

Figure 3.39. A representative catalytic cycle in asymmetric alkene metathesis (L_n = chiral ligand)

ruthenium catalysts, the mechanism is different in that discrete binding of the alkene occurs via η^2-coordination (see Section 3.6.1).[125] This step may be responsible for the stereochemistry alone or in combination with the subsequent metallacyclobutane-forming step.

3.5 Cross-Coupling Catalysts

A process related to group transfer involves the metal-catalyzed coupling of two substrate fragments to yield a new bond. Instead of a metal transferring an attached group to a noncoordinated substrate, the group is transferred to another coordinated substrate (**Figure 3.40**). For this to occur, both fragments must be sequentially incorporated onto the metal catalyst, forming an activated intermediate. The final reaction of the catalytic cycle forms the new bond between the two fragments and releases the metal catalyst.

A classic example of such a cross-coupling reaction is presented in the context of the Suzuki reaction (**Figure 3.41**).[109–112] Here, axial stereochemistry is generated upon coupling of the aryl from the aryl halide to the aryl of the aryl boronic acid.

In this process, multiple steps occur at the metal center and it is unclear exactly which step is enantioselectivity-determining (**Figure 3.42**). Enantioselectivity would be established in the relative conformation of the two aryl groups bound to the palladium prior to reductive elimination. This relationship is initially created upon trans-metallation, but could be altered upon equilibration of an atropodiastereomeric diaryl–palladium species. In addition, reductive elimination of one of the enantiomers from the diastereomeric diaryl–palladium intermediates may be more favorable.

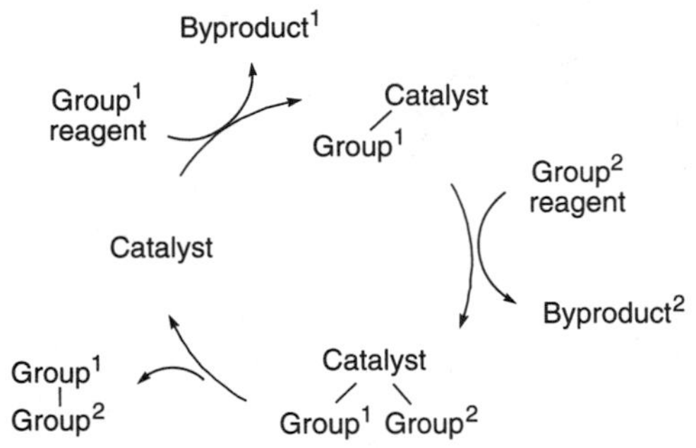

Figure 3.40. A general cross-coupling reaction.

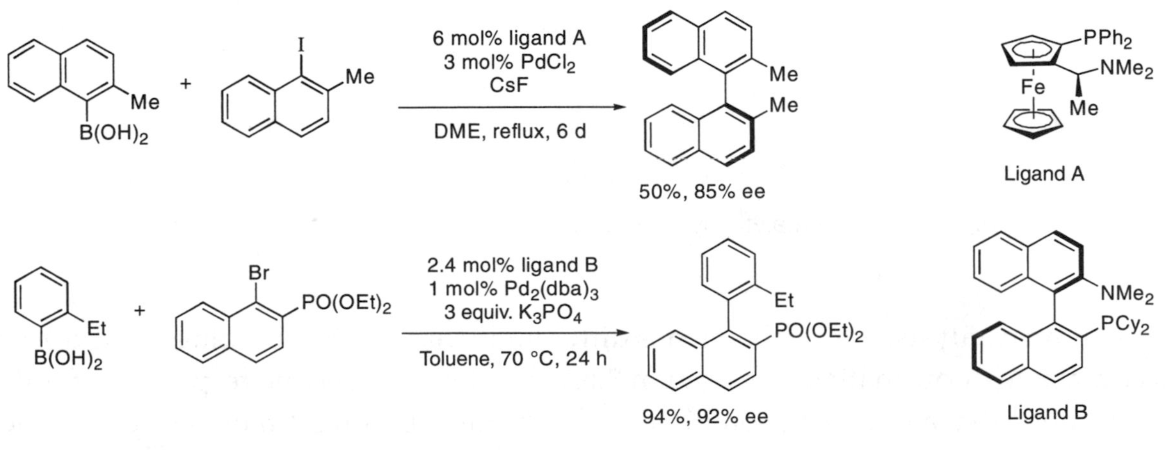

Figure 3.41. Catalytic asymmetric Suzuki cross-coupling reactions.

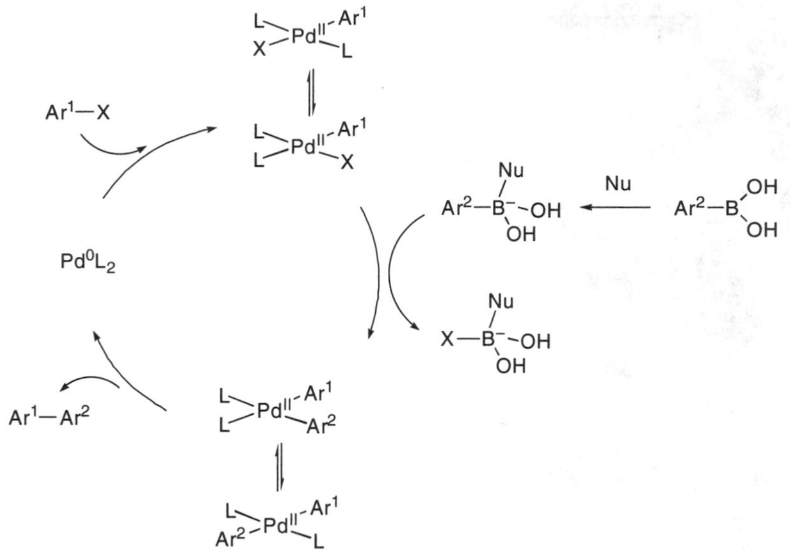

Figure 3.42. Simplified catalytic cycle for the catalytic asymmetric Suzuki cross-coupling.

Aside from the requirements for the enantioselectivity-determining step, several competing reactions, including protodeboration and reductive coupling, have made development of general versions of this reaction challenging.

The number of catalytic *asymmetric* cross-coupling reactions has been limited to date, especially compared to the large number of known catalytic processes. However, the ability of metals to catalyze unusual bond-forming processes will certainly lead to further developments in this area.

3.6 Activation via π-Coordination

The activation of alkenes by metal coordination to the π-orbitals is powerful, because it provides access to a broad range of chemistries distinct from those discussed in the previous sections. The orbital basis for this interaction is shown in **Figure 3.43**. A vacant orbital on the metal center accepts electron donation from the filled π-orbital of the alkene. Simultaneously, two electrons from a filled metal *d*-orbital are donated into the alkene π*-orbital (in an interaction called "backbonding"). In most cases, metals need two or more *d*-electrons to form stable olefin complexes. Further, π-coordination can involve more than two centers and more than two π-electrons at one time, as illustrated with the η^3-, η^4-, and η^6-coordination motifs shown in **Figure 3.43**. Since the π-fragments themselves are achiral, the use of chiral metal centers provides an avenue for introduction of asymmetry in subsequent reactions.[126] Finally, the ability of metals to activate organic π-systems toward nucleophilic and electrophilic additions and insertions, for example, provides entry into a rich array of bond constructions.[127]

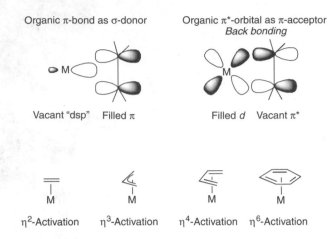

Figure 3.43. Orbital basis and π-coordination modes.

3.6.1 Activation via η² π-Coordination

By far and away the most common π-activation paradigm is η²-coordination. This type of activation is crucial to a host of asymmetric reactions utilizing alkene precursors, including hydrogenation, hydroformylation, hydrocarbonylation, hydroamination, amidocarbonylation, Wacker, Heck, etc. In many of these transformations, there are several distinct steps in the catalytic cycle where the metal may serve very different purposes, including π-activation, group-transfer (see Section 3.4), and others. Any one or even a combination of these steps can be stereochemistry-determining. In this section, examples will first be described where the π-coordination/π-activation is believed to be stereochemistry-determining. Examples at the end will describe the balance of effects that can lead to effective processes, even when π-coordination/π-activation is not the dominant stereochemistry-determining event.

The Wacker process is a classic example of electrophilic alkene activation via π-coordination (**Figure 3.44**, top).[128] When an electron-deficient palladium(II) species coordinates an alkene, different nucleophiles can attack, giving rise to an intermediate alkyl-palladium species. This alkyl-palladium species can undergo a number of different reactions, of which β-hydride elimination is usually most facile. In the classic Wacker process, the final result is generation of a ketone on the more-substituted position of an alkene. However, the nucleophilic attack on the coordinated alkene is usually a trans-addition,[129] which implies that an asymmetric process is viable if the β-hydride elimination is diverted to allow the nascent stereocenter to remain.[130] Examples of this can be found under conditions of high chloride or bromide concentration, giving rise to chlorohydrins[131] and dibromides,[132] respectively (**Figure 3.44**, bottom). When the palladium ligand is homochiral, discrimination of the alkene enantiofaces upon π-coordination can provide the asymmetric induction.

Related intramolecular processes have been exploited with many different nucleophiles. For example, the *ortho*-substituted phenols undergo cyclization to the carbo-

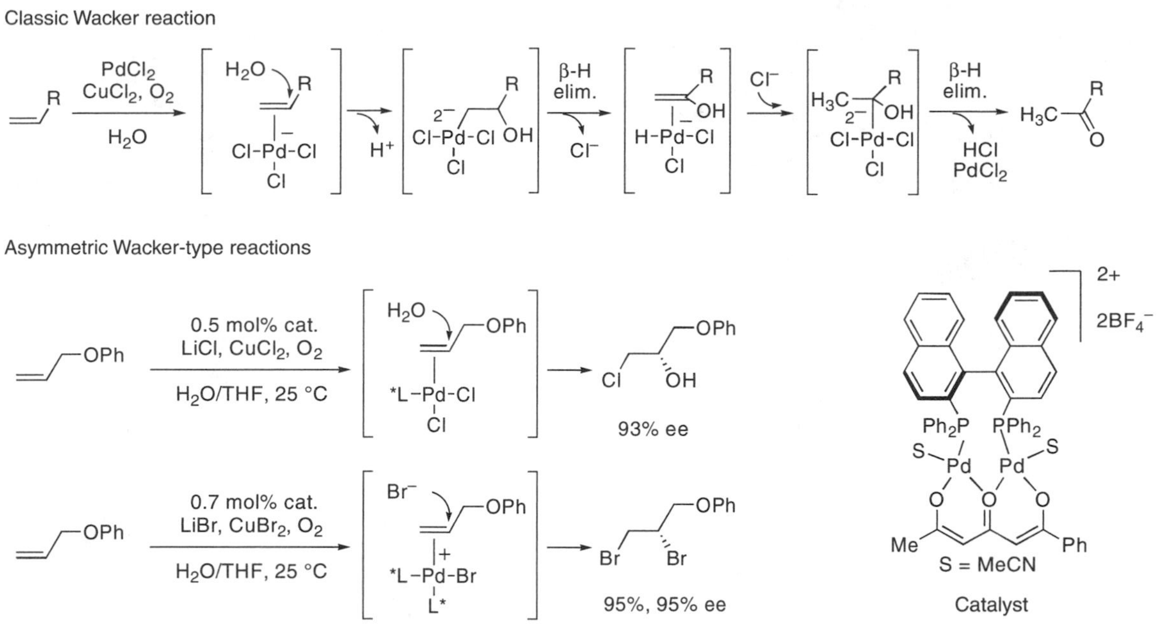

Figure 3.44. Asymmetric Wacker-type intermolecular reactions.

cyles in high enantiomeric excess when treated with a chiral palladium–bisoxazoline catalyst under oxidative conditions **Figure 3.45**.[133] In contrast to the case above, benzoquinone is now employed as the oxidant rather than O_2. Examination of the catalytic cycle is instructive (**Figure 3.45**). In the first step, the alkene forms an η^2-adduct with the palladium(II) catalyst. Electrophilic activation of the alkene by the metal then allows intramolecular nucleophilic attack of the phenol to yield an alkyl-palladium(II) adduct. Notably, there is no β-hydrogen that can eliminate to reform the double bond between the same two centers as are found in the starting material. This clever tactic eliminates the largest problem in asymmetric Wacker reactions—namely, the reformation of the double bond in the original position with concomitant loss of stereochemistry. As a result, β-hydrogen elimination occurs with the trans position to release the product. Notably, this tactic also results in formation of quaternary stereocenters, which are difficult to generate enantioselectively.[134–136] At this point, the palladium catalyst rapidly releases acid (HX), resulting in palladium(0). Regenerating the catalytic palladium(II) is the other significant issue encountered in asymmetric Wacker-type reactions. A mild oxidant that does undergo reaction with the substrate is critical. Typically, O_2 and catalytic copper(II) or benzoquinone is employed in this role.

In the above case, the chiral catalyst differentiates the two enantiotopic faces of the alkene. In the next case (**Figure 3.46**), a symmetric substrate is employed and enantiotopic alkenes are differentiated.[137] This case is also distinct because the initial alkene adduct with the palladium catalyst and the nucleophile is intercepted by the second alkene and undergoes further reaction. After a series of steps, a bicyclic product is produced with high enantioselectivity. This example illustrates a common and powerful characteristic of many transition metal catalyzed reactions—namely, several

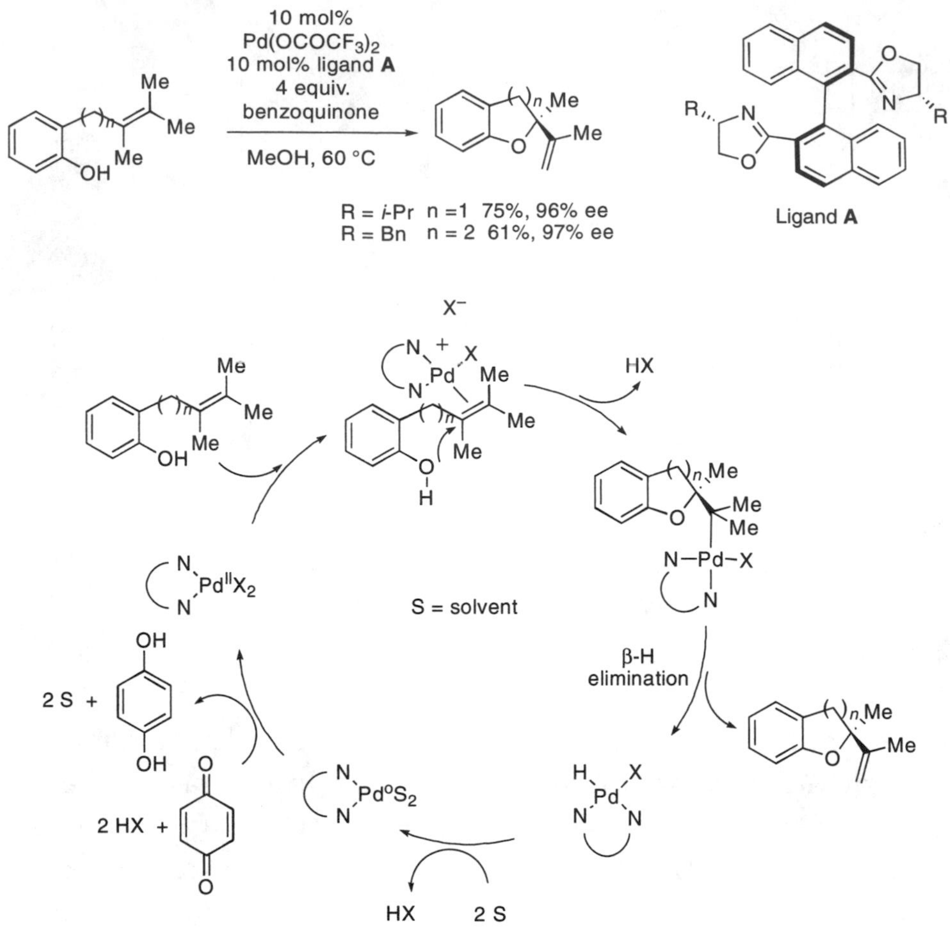

Figure 3.45. Asymmetric Wacker-type intramolecular reaction.

sequential transformations can occur, allowing the rapid construction of complex molecules from simple substrates (see Chapter 14).

Like the Wacker-type reactions, the Heck reaction can also involve a β-hydride elimination to release the palladium species. However, many of the problems are avoided since the catalytic cycle is completed without the need for an external oxidant. Again, the regioselectivity of β-hydride elimination was a key factor in the development of effective asymmetric versions. In spite of this and other significant issues (i.e., the regioselectivity of the initial addition, re-insertion of the palladium hydride, etc.), highly successful asymmetric Heck reactions have been developed.[138] Intermolecular examples have proven the most challenging and limitations still remain. Examples with phosphinooxazoline–palladium complexes are shown in **Figure 3.47**. The mechanism of the asymmetric Heck reaction first involves oxidative addition to a C–X bond (where C is usually aryl or alkenyl) by palladium(0) to yield a palladium(II) species with a reactive Pd–C bond. Depending on the substrates and reaction conditions, subsequent enantiotopic alkene π-coordination can occur via a neutral or cationic (X-dissociation, as illustrated in **Figure 3.47**) pathway.[138] Regard-

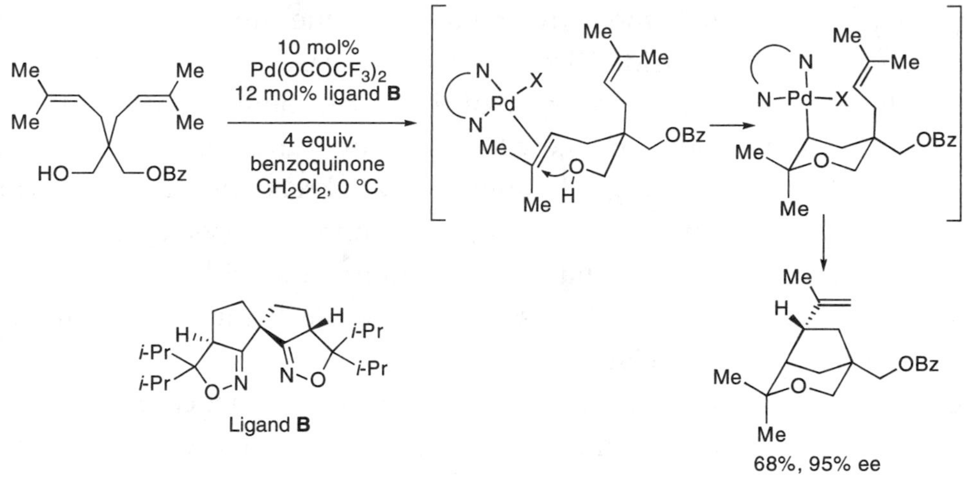

Figure 3.46. Asymmetric Wacker-type desymmetrization.

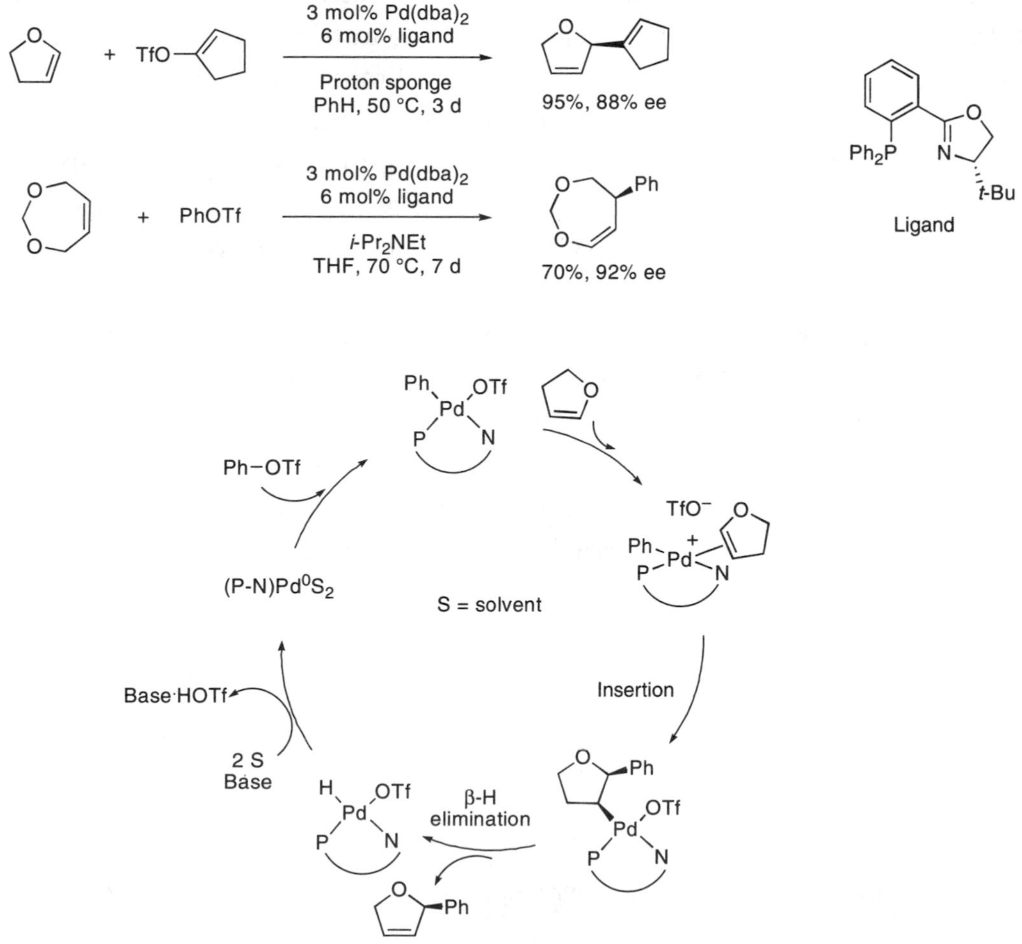

Figure 3.47. Intermolecular catalytic asymmetric Heck reaction.

less, migratory insertion into the activated alkene generates an alkyl-palladium(II) species. Subsequent cis β-hydride elimination in this case cannot regenerate the double bond in the original location because of the stereochemistry of the ring system. Rather, the double bond forms in a new location, retaining the stereogenic center.[139] In contrast to analogous (BINAP)palladium-catalyzed reactions, isomerization of the products by catalyst did not induce double-bond migration (via a Pd–H species).

Intramolecular Heck reactions have proven more amenable to development and are useful for a broad array of substrates.[138] For example, the intramolecular reactions of the (Z)-alkenes illustrated in **Figure 3.48** proceed with high enantioselectivity.[140–142] Again, activation and enantiodifferentiation of the alkene via π-coordination is key to the success of this reaction. Here, there is no β-hydrogen to allow the alkene to reform between the original two carbons. Notably, similar processes have been applied to much more complex substrates, providing concise routes to the complex alkaloids quadrigemine C and psycholeine (see Chapter 16).[143]

The example of enamide hydrogenation[144] is instructive because detailed study of the mechanism has shown that the initial alkene coordination step is not necessarily stereochemistry-determining (**Figure 3.49**).[145,146] Activation does occur upon π-binding, but asymmetric induction occurs both during π-binding and dihydrogen addition. In fact, the selectivity shown by the catalyst in the initial π-coordination of the two enantiotopic alkene faces has little to do with the ultimate course of the asymmetric reaction. Even though π-adduct I_2 forms to a greater extent, due to its higher thermodynamic stability, it undergoes slower reaction in subsequent steps than I_1. Since the two intermediates, I_2 and I_1, can equilibrate via decomplexation/recomplexation, a Curtin–Hammett situation arises (see Chapter 1) and the major product formed arises from I_1.

The hydrometallation (as in hydrogenation) or carbometallation (as in the Heck reaction) of alkenes from metal-hydride or metal-alkyl species is a powerful entry into many different products. The resultant alkyl-metal intermediates can be intercepted by other reagents intramolecularly (see **Figure 3.46** for an example) or intermolecularly (see **Figure 3.47** for an example). With carbon monoxide interception (**Figure 3.50**),[147,148] the result is an asymmetric hydroformylation process. Again, the stereo-

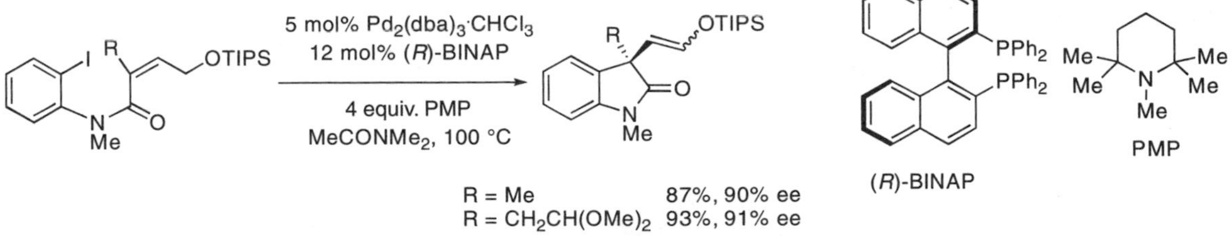

R = Me   87%, 90% ee
R = CH₂CH(OMe)₂ 93%, 91% ee

Figure 3.48. Intramolecular catalytic asymmetric Heck reaction.

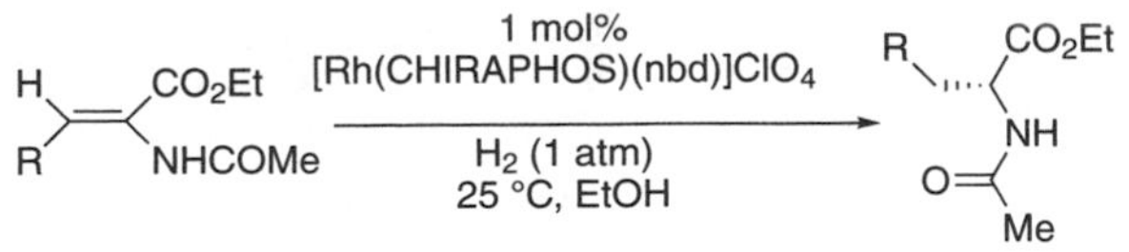

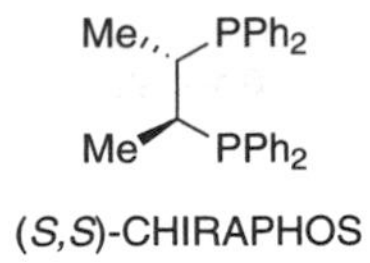

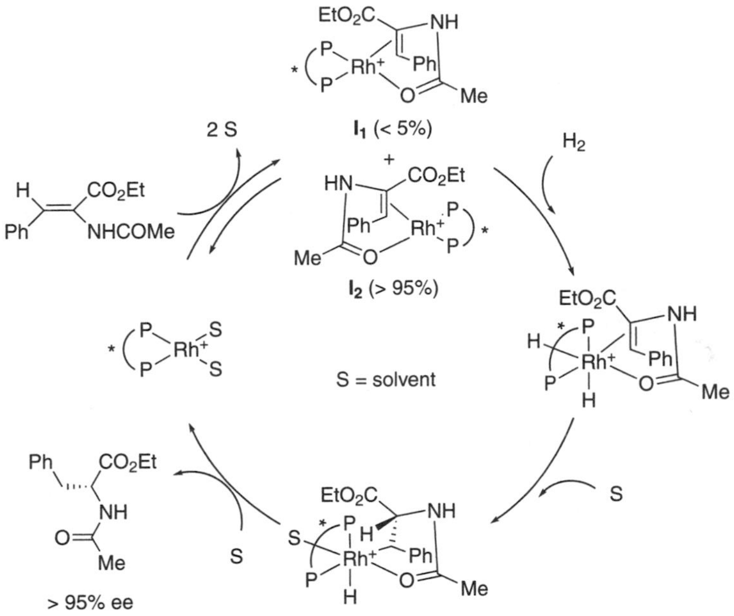

Figure 3.49. π-Activation in catalytic asymmetric hydrogenation.

chemistry-determining step can occur at several stages in the process. Since many of the reactions can be reversible and the precise stereochemistry-determining step may change with the catalyst or substrate formulation, rational modification of these systems can be difficult.

The preceding examples show that several factors can lead to effective asymmetric processes, even when π-coordination/π-activation is not the dominant stereochemistry-determining event. For many processes involving π-activation with transition metals, the mechanisms are complex. Determination of the stereochemistry-determining events is nontrivial and often requires significant study. Nonetheless, the unique activation process afforded by π-activation, compared to those described in the preceding sections, creates an entry into a valuable set of catalytic asymmetric transformations that will undoubtedly see further development. In addition, the durability and high efficiency of these catalysts allow extremely low catalyst loadings. Finally, the orthogonal activation pathway (π-activation) allows the use of many functional groups that would preclude selectivity with alternate activation modes (i.e., Lewis acid, etc.).

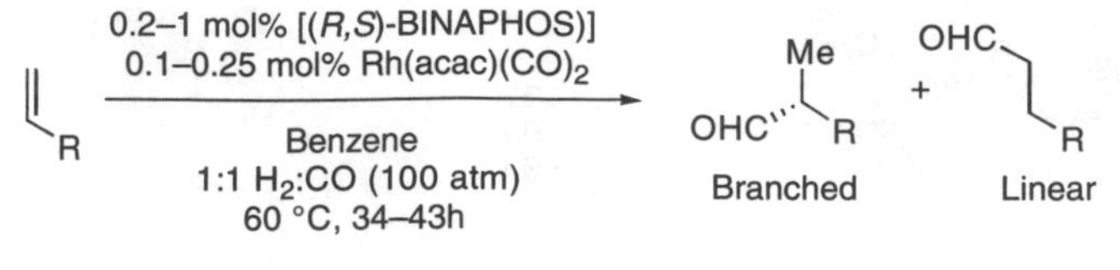

R	Conversion (%)	Branched: linear	ee (%)
OAc	> 99	86:14	92
Ph	> 99	88:12	94
4-MeOPh	> 99	87:13	88
4-ClPh	> 99	87:13	93

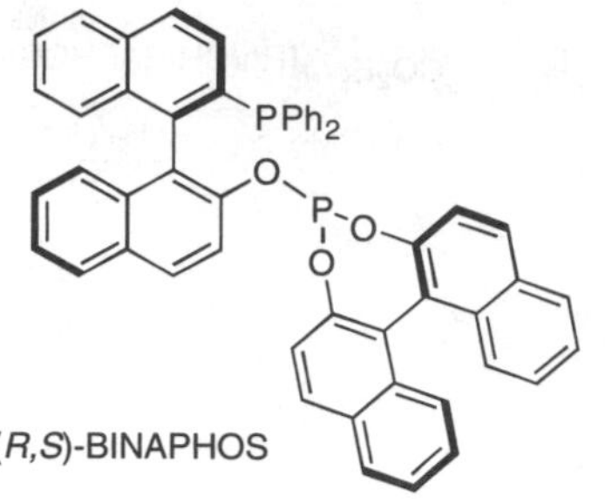

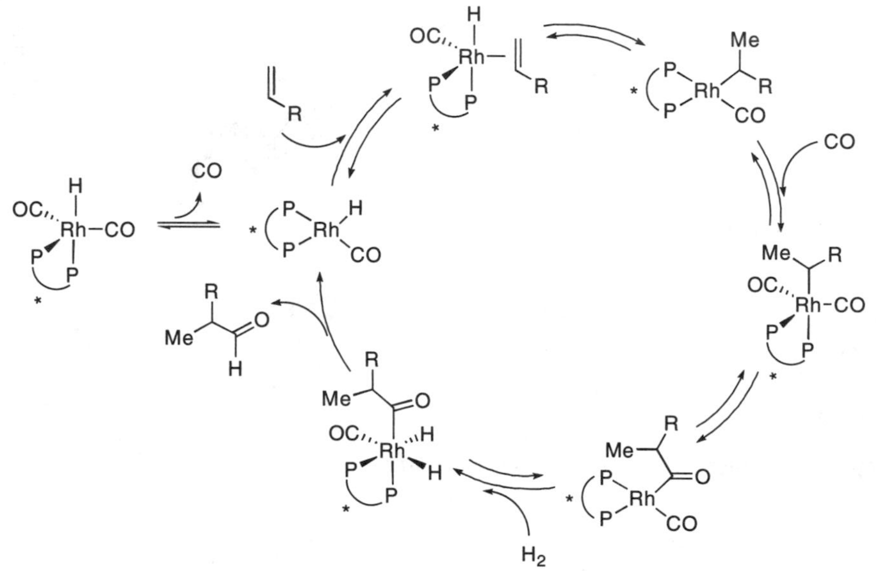

Figure 3.50. π-Activation in catalytic asymmetric hydroformylation.

3.6.2 Activation via η³ π-Coordination

The ability of transition metals to bind π-systems with more than two contiguous unsaturated carbons has also been exploited to good effect in asymmetric catalysis. Coordination over three centers leads to η³-allyl complexes (see **Figure 3.43**). In particular, the chemistry of cationic allyl–palladium complexes has proven extremely fruitful,[149,150] although other metal centers can be used.[149,151,152]

Figure 3.51 illustrates a simplified catalytic cycle for a representative reaction involving a cationic π-allyl. Typically, a palladium(II) catalyst precursor is employed which undergoes reduction in situ to provide the active palladium(0) catalyst. Subsequent η²-alkene coordination brings the substrate into the ligand sphere of the metal, where formal oxidative ionization proceeds with departure of the leaving group (LG). The result is an η³-allyl–palladium(II) complex in which the chiral ligand can influence the subsequent approach of a nucleophile to provide a chiral product. Decomplexation of the η²-bound product then regenerates the palladium(0) catalyst.

Within the above paradigm there are several processes for asymmetric induction, allowing many different substrate types (e.g., achiral, *meso*, racemic mixtures, etc.) to

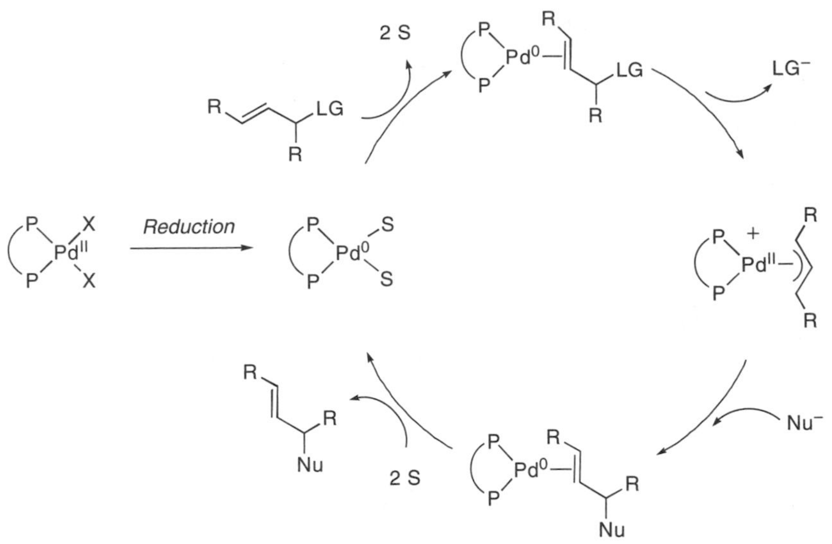

Figure 3.51. Simplified mechanism for asymmetric cationic π-allyl reactions (S = solvent).

be successfully employed.[149,150] Chapter 4 details the origin of the asymmetric induction from the catalyst to the substrate in this reaction. For the specific example in **Figure 3.51**, the stereochemistry from the chiral palladium ligand causes differentiation of the enantiotopic termini.

Examples of the four primary processes are illustrated in **Figure 3.52**. In the first (enantiotopic facial discrimination), an achiral substrate generates an achiral allyl species. The asymmetric induction is exercised via initially selective coordination by the chiral catalyst to one of the two enantiotopic alkene faces[153] or by formation of the more stable η^3-allyl via an η^1- to η^3-isomerization.[154]

In the second (selective reaction of enantiotopic leaving groups), one of the carbonate groups ionizes selectively after the initial alkene coordination.[155] The oxidative ionization and the nucleophilic addition steps may proceed with either retention or inversion with respect to the palladium and depends upon the substrate, catalyst, nucleophile, and reaction parameters. In this case, a double inversion is likely occurring, resulting in an azide substitution reaction with net retention. The use of *meso*-substrates in such reactions allows complexity to be introduced rapidly and, here, four stereocenters are established. The resultant azide serves as a useful precursor to the important antitumor agent (+)-pancratistatin, which is available in 11% overall yield from the dicarbonate.[156]

In the third (selective reaction of enantiotopic allyl termini), both enantiomers of a racemic mixture converge onto a single cationic allyl–palladium adduct, where the faces of the allyl fragment are homotopic (see Appendix A) rather than enantiotopic.[157] As a result, coordination to either allyl face results in the same palladium adduct. However, the termini of the allyl are enantiotopic and asymmetric induction is exercised by controlling the approach of the nucleophile to those termini.

In the fourth [DyKAT (dynamic kinetic asymmetric transformation) of diastereotopic allyl intermediates], a racemic mixture is again employed. However, the

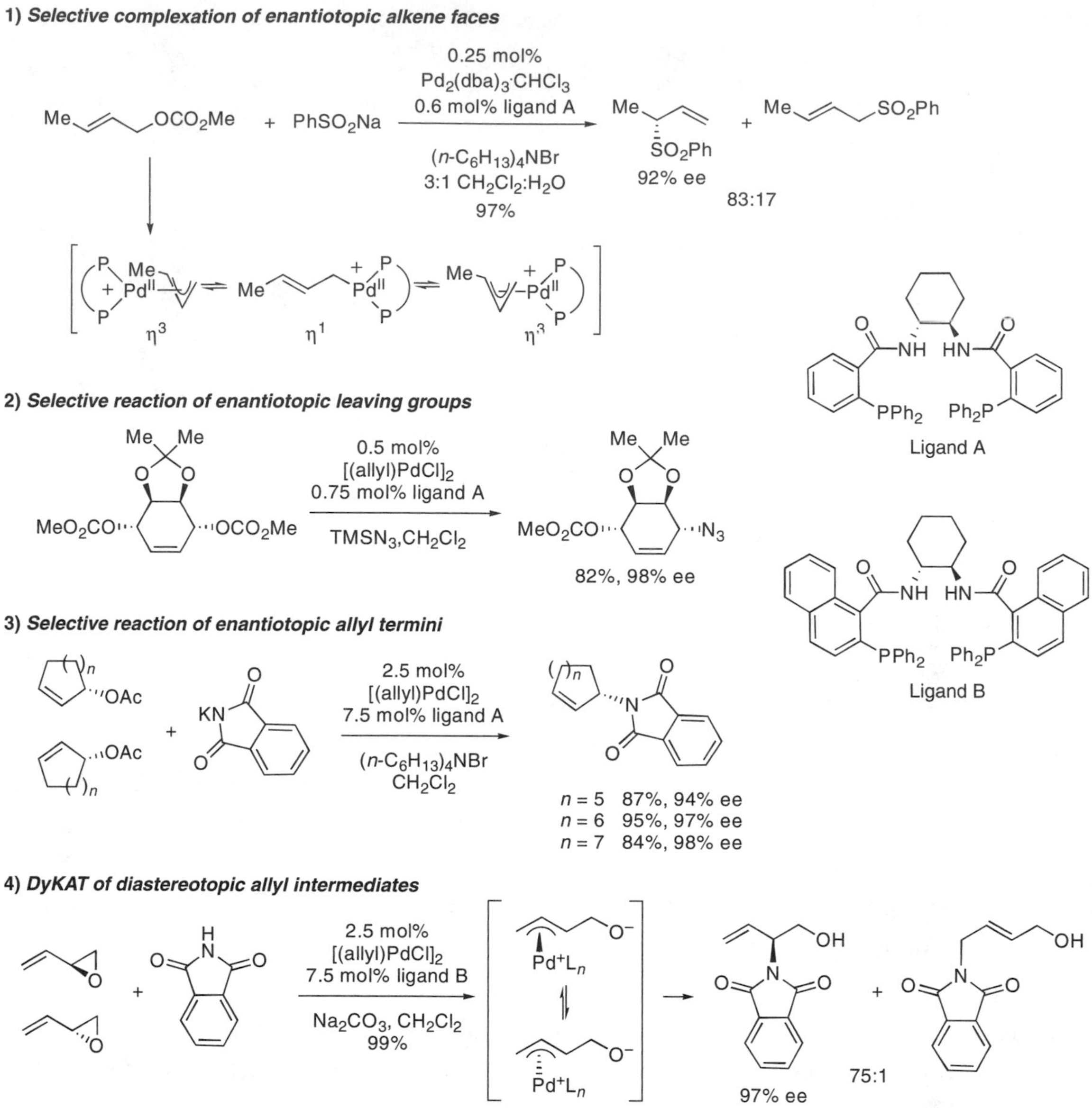

1) *Selective complexation of enantiotopic alkene faces*

2) *Selective reaction of enantiotopic leaving groups*

3) *Selective reaction of enantiotopic allyl termini*

n = 5 87%, 94% ee
n = 6 95%, 97% ee
n = 7 84%, 98% ee

4) *DyKAT of diastereotopic allyl intermediates*

Figure 3.52. Examples of asymmetric π-allylation reactions that proceed with different enantiodiscrimination mechanisms.

allyl fragment now possesses enantiotopic faces that lead to diastereomeric allyl–palladium adducts when a chiral ligand is present. If these two diastereomeric allyl–palladium adducts can equilibrate rapidly, a DyKAT (see Chapter 9) is possible. The stereochemistry arises from the more reactive diastereomeric allyl–palladium adduct. In this example, a key feature is the epoxide that acts as the leaving group. As a result, an alkoxide anion forms, which then directs the nucleophile to the more hindered end of the π-allyl cation.[158]

The above cases all involve intermolecular reactions between an allyl species and a nucleophile. Intramolecular versions are also possible. One example (**Figure 3.53**) cleverly exploits cationic π-allyl chemistry to generate a stabilized cyclopropyl–

methyl carbocation. This intermediate undergoes a Wagner–Meerwein shift to product the ring-expanded product with good enantioselection.[159]

π-Allyl species can be generated in many different ways. The above examples highlight allyl-cation generation via departure of a leaving group from an appropriate substrate. Another entry to η^3-allyl–metal species involves the hydrometallation of a diene, allene, or styrene. For the example in **Figure 3.54**, enantiotopic complexation established the stereochemistry of the π-benzyl adduct. The aniline subsequently reacts with inversion of configuration to provide the product with good enantiomeric excess.[160] Extensive mechanism studies support the catalytic cycle outlined in **Figure 3.54**.[161–163] Even so, optimization of the enantioselectivity remains elusive with a high of 85% ee using a SEGPHOS ligand.[164] Even with well-established

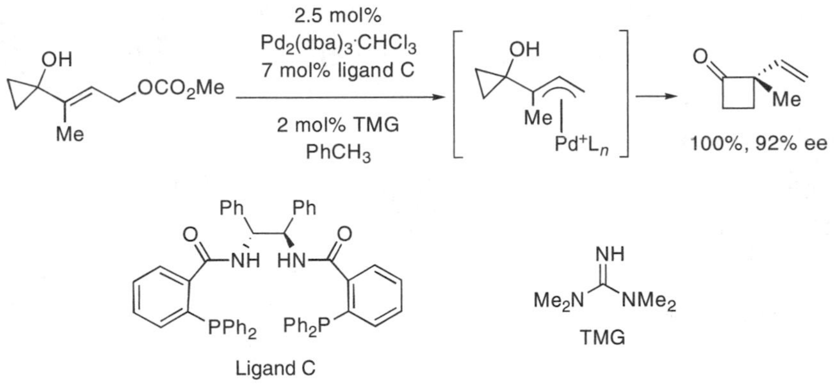

Figure 3.53. Intramolecular asymmetric reaction via a cationic π-allyl intermediate.

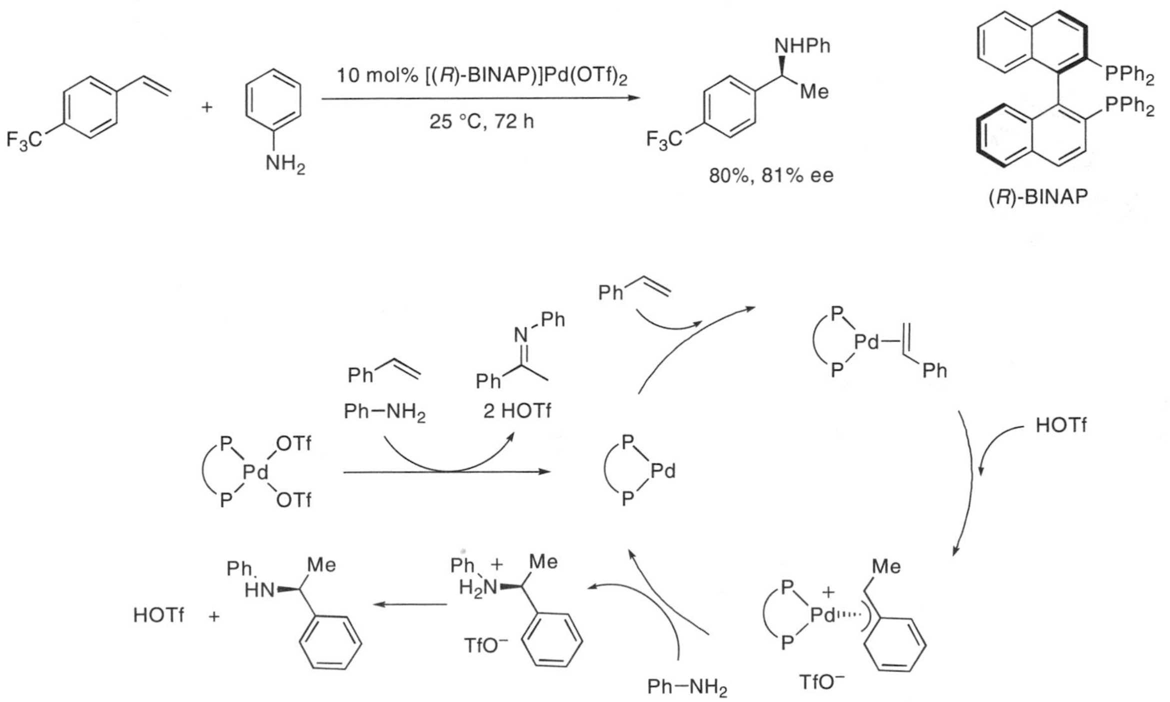

Figure 3.54. An asymmetric cationic π-allyl reaction via a hydrometallation process.

structures, identification of the exact stereoinducing features can be problematic, leading to difficulty in rationally improving catalysts.

3.6.3 Activation via η^4 π-Coordination

There are many additional π-coordination modes possible (see **Figure 3.43**) with η^4-, η^5-, and η^6-metal adducts occupying prominent places in organometallic chemistry.[127] Unfortunately, catalytic turnover is often problematic in the reactions of such complexes, which limits applications in asymmetric catalysis. There are a number of η^4-adducts, however, that are involved in catalytic cycles and play key roles in stereochemical induction.

One example is outlined in **Figure 3.55** for the [4+1] cycloaddition of an unsaturated allene with carbon monoxide. A chiral rhodium catalyst provides asymmetric induction in the process, leading to the cyclopentenone product.[165,166] DFT calculations support the catalytic cycle illustrated.[167] First, the catalyst exchanges the double η^2-coordinated cyclooctadiene (cod) for the η^4-coordinated substrate. In doing so, two different diastereomeric adducts can arise from coordination to the two enantiofaces of the unsaturated allene. Calculations indicate that one of these adducts is more stable. In all further steps, the intermediates corresponding to this diastereomeric adduct constitute the lowest-energy pathway. As a result, the initial η^4-coordination essen-

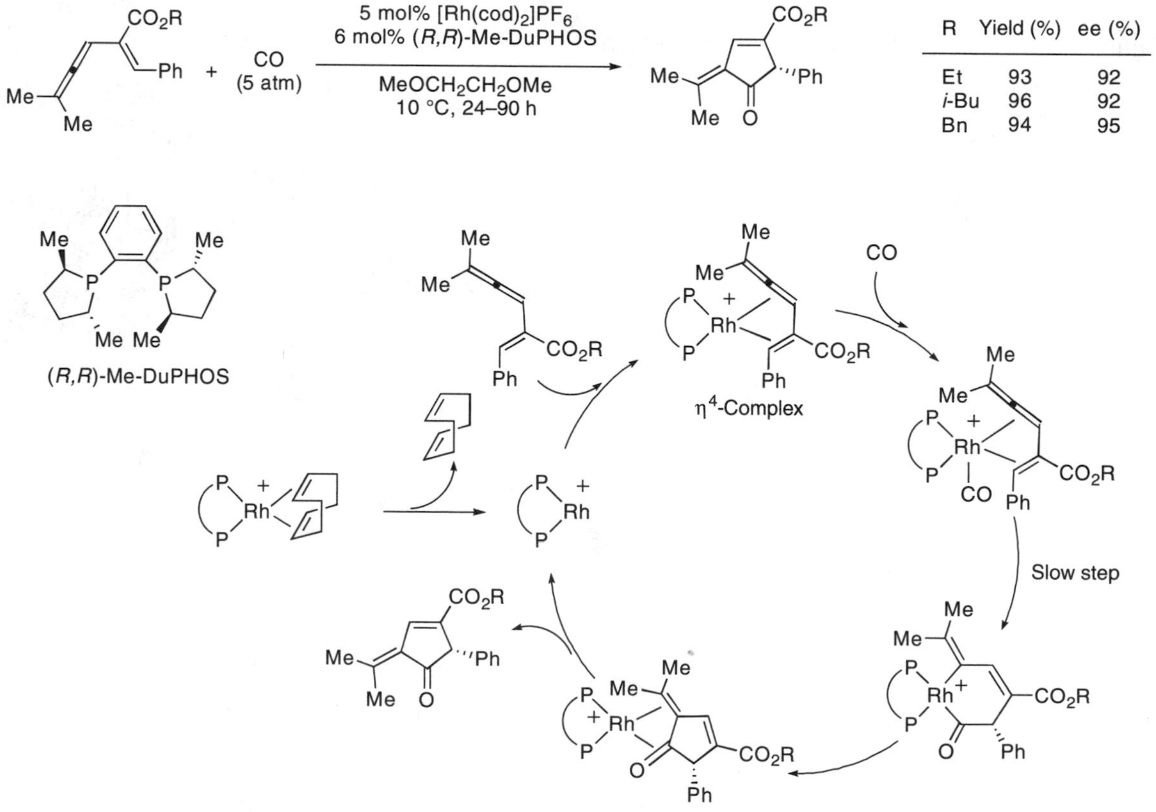

Figure 3.55. η^4-Coordination as part of a [4+1] asymmetric cycloaddition.

tially sets the stereochemistry of the reaction. The turnover-limiting step is the insertion of carbon monoxide and formation of a six-membered metallacycle that also sets the centrochiral stereocenter found in the product. Further reductive elimination and decomplexation provide the product.

3.6.4 Activation via Double η² π-Coordination

Catalytic asymmetric activation can also rely upon two or more unconjugated π-fragments. For example, the [5+2] reaction outlined in **Figure 3.56** commences from a 1,6-dienyl–cyclopropane substrate.[168] In formation of the most stable double η²-complex, the BINAP stereochemistry dictates which enantioface of each alkene undergoes coordination. This sets the first stereocenter that is formed in the vinyl cyclopropane ring-opening and metallacycle formation. The next stereocenter forms in a diastereoselective migratory insertion; it is likely that formation of the cis-substituted ring system in kinetically favored. Reductive elimination then generates the final fused bicyclic product. The ability to make larger ring systems (e.g., seven-, eight-, or nine-membered rings, etc.) with high enantioselectivity is a unique feature that highlights the utility of transition-metal-catalyzed cycloaddition reactions.

Another transformation in which double η²-coordination plays a key role is metal-catalyzed [2+2+2] cycloadditions. These powerful processes bring three π-systems (e.g., alkynes, alkenes, isocyanates, imines, carbon dioxide, ketones, etc.) together to generate an array of interesting six-membered rings.[169,170] Many versions that introduce new centrochiral or axial stereochemistry are possible.[169–171] One example in

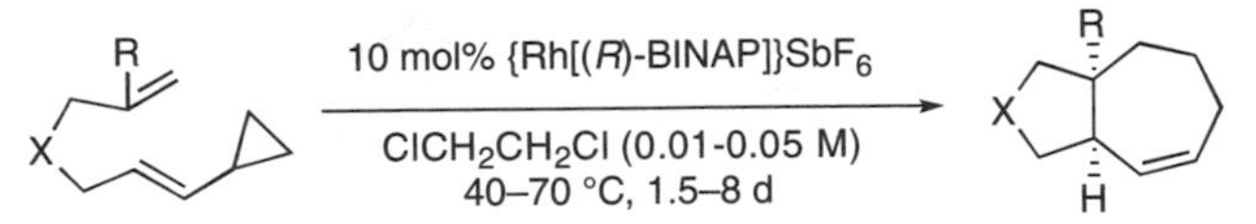

X	R	Yield (%)	ee (%)
C(CO$_2$Me)$_2$	Me	72	> 95
C(CO$_2$Me)$_2$	CH$_2$OBn	80	> 99
C(CO$_2$Me)$_2$	H	73	52
NTs	H	90	96

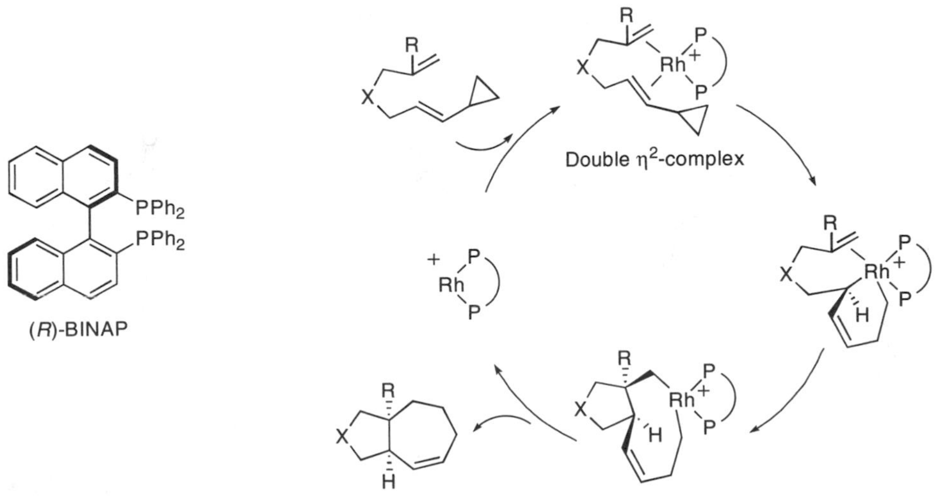

Figure 3.56. Double η²-coordination as part of a [5+2] asymmetric cycloaddition.

which two axial chiral units are generated in the form of a triaryl is illustrated in **Figure 3.57**.[172] High diastereoselectivities and enantioselectivities are noted for these unusual compounds. A catalytic cycle incorporating the generally accepted mechanism of metal-catalyzed acetylene cyclotrimerization[128,169,170] indicates that an early step is formation of a double η^2-diyne adduct. The relative disposition of the naphthyl groups at this stage would be controlled by the chiral DuPHOS ligand and most likely translates into that pictured for the initial metallacycle. At this and all subsequent stages, rotation of the naphthyls is likely hindered, so the axial chirality remains to become a feature of the final product.

The unique activation afforded by π-coordination, compared to the activation processes described in the preceding sections, creates an entry into a valuable set of

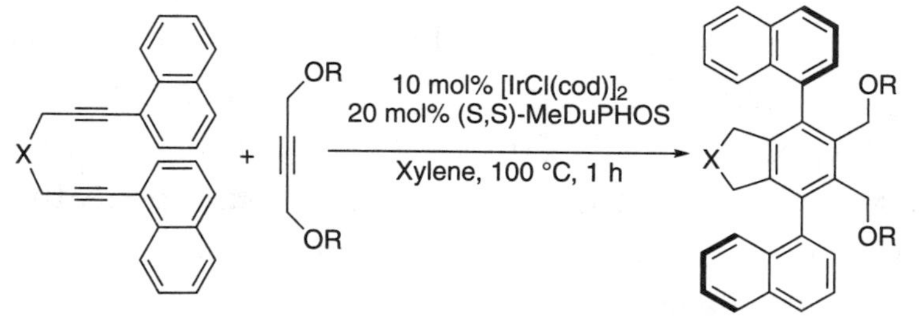

X	R	Yield (%)	dl: meso	ee (%)
O	TBS	74	> 99:1	99.5
O	MOM	76	93:7	98.5
NTs	THP	97	> 99:1	99.1
C(CO$_2$Me)$_2$	Me	77	> 99:1	> 99.8
CH$_2$	TBS	77	91:9	98.6

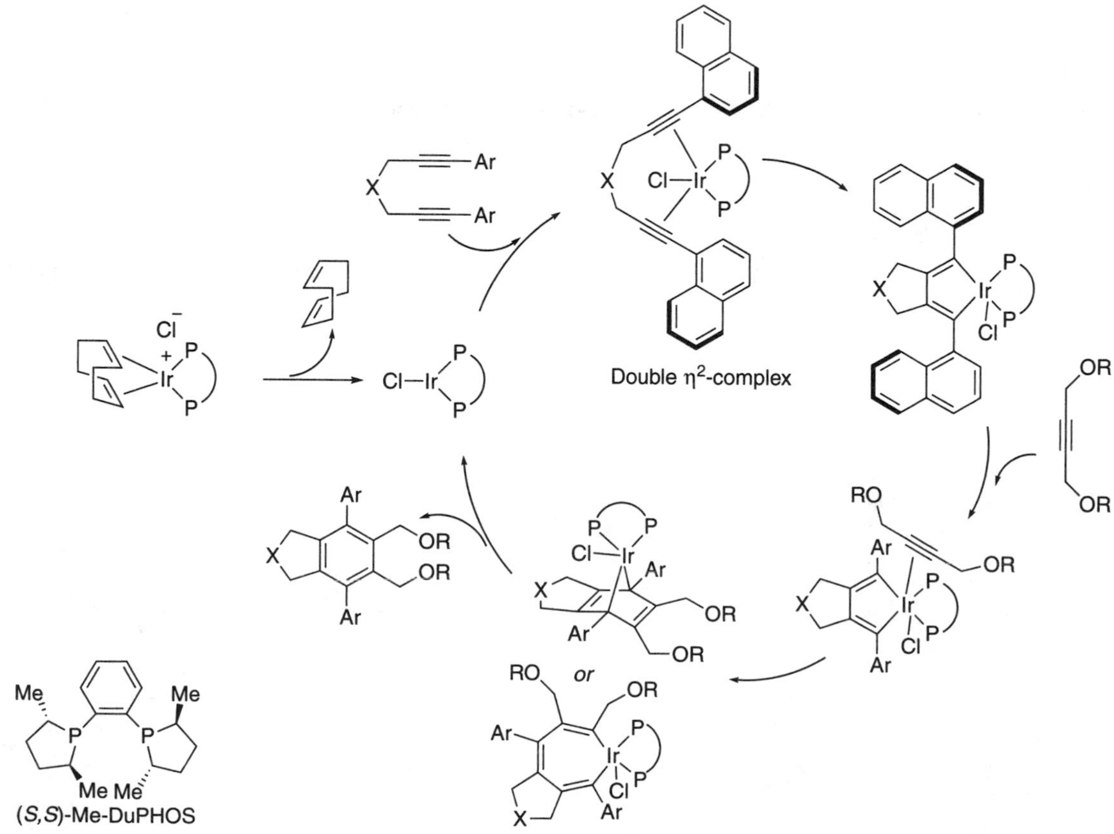

Figure 3.57. Asymmetric double η^2-coordination as part of a [2+2+2] asymmetric cycloaddition.

catalytic asymmetric transformations that will undoubtedly benefit from further investigation.

Summary

The discussion of activation modes exercised by asymmetric catalysts in Chapters 2 and 3 is by no means comprehensive. For example, catalysts that operate by shape exclusion or microenvironment modification (i.e., inclusion complexes and imprinted catalysts[173]) have not been discussed (see Chapter 15). In addition, catalysts can combine modes in dual and multifunctional catalysis (see Chapter 12). Furthermore, some catalytic asymmetric reactions defy classification into the simple categories because the catalyst provides activation for several distinctly different steps (i.e., transition-metal-catalyzed hydrogenation). Even so, the above activation modes provide a framework for understanding the advantages and limitations of many catalytic processes. Within the last 10 years, there has been an explosion in research into some of the less-studied activation modes (i.e., Brønsted acid catalysis) and it is clear that a similar thrust in others of the less-well-explored modes is underway.

References

(1) Taylor, M. S.; Jacobsen, E. N. Asymmetric Catalysis by Chiral Hydrogen-Bond Donors. *Angew. Chem., Int. Ed. Engl.* **2004**, *45*, 1520–1543.

(2) Jung, M. E. A Review of Annulation. *Tetrahedron* **1976**, *32*, 3–31.

(3) Hajos, Z. G.; Parrish, D. R. Asymmetric Synthesis of Optically Active Polycyclic Organic Compounds. **1971**, German Patent: DE 2102623.

(4) Hajos, Z. G.; Parrish, D. R. Stereocontrolled Synthesis of Trans-Hydrindan Steroidal Intermediates. *J. Org. Chem.* **1973**, *38*, 3239–3243.

(5) Hajos, Z. G.; Parrish, D. R. Asymmetric Synthesis of Bicyclic Intermediates of Natural Product Chemistry. *J. Org. Chem.* **1974**, *39*, 1615–1621.

(6) Eder, U.; Sauer, G.; Weichert, R. Optically Active 1,5-Indanone and 1,6-Naphthalenedione. **1971**, German Patent: DE 2014757.

(7) Eder, U.; Sauer, G.; Weichert, R. Total Synthesis of Optically Active Steroids. 6. New Type of Asymmetric Cyclization to Optically Active Steroid CD Partial Structures. *Angew. Chem., Int. Ed. Engl.* **1971**, *10*, 496–497.

(8) Hoang, L.; Bahmanyar, S.; Houk, K. N.; List, B. Kinetic and Stereochemical Evidence for the Involvement of Only One Proline Molecule in the Transition States of Proline-Catalyzed Intra- and Intermolecular Aldol Reactions. *J. Am. Chem. Soc.* **2003**, *125*, 16–17.

(9) List, B.; Hoang, L.; Martin, H. J. Asymmetric Catalysis Special Feature Part II: New Mechanistic Studies on the Proline-Catalyzed Aldol Reaction. *Proc. Natl. Acad. Sci. U.S.A.* **2004**, *101*, 5839–5842.

(10) Bahmanyar, S.; Houk, K. N. The Origin of Stereoselectivity in Proline-Catalyzed Intramolecular Aldol Reactions. *J. Am. Chem. Soc.* **2001**, *123*, 12911–12912.

(11) Clemente, F. R.; Houk, K. N. Computational Evidence for the Enamine Mechanism of Intramolecular Aldol Reactions Catalyzed by Proline. *Angew. Chem., Int. Ed. Engl.* **2004**, *43*, 5766–5768.

(12) Tanaka, K.; Mori, A.; Inoue, S. The Cyclic Dipeptide Cyclo[(*S*)-phenylalanyl-(*S*)-histidyl] as a Catalyst for Asymmetric Addition of Hydrogen Cyanide to Aldehydes. *J. Org. Chem.* **1990**, *55*, 181–185.

(13) Jackson, W. R.; Jayatilake, G. S.; Matthews, B. R.; Wilshire, C. Evaluation of Some Cyclic Dipeptides as Catalysts for the Asymmetric Hydrocyanation of Aldehydes. *Aust. J. Chem.* **1988**, *41*, 201–213.

(14) Shvo, Y.; Gal, M.; Becker, Y.; Elgavi, A. Asymmetric Hydrocyanation of Aldehydes with Cyclodipeptides: A New Mechanistic Approach. *Tetrahedron: Asymmetry* **1996**, *7*, 911–924.

(15) a) Wenzel, A. G.; Jacobsen, E. N. Asymmetric Catalytic Mannich Reactions Catalyzed by Urea Derivatives: Enantioselective Synthesis of β-Aryl-β-Amino Acids. *J. Am. Chem. Soc.* **2002**, *124*, 12964–12965.
b) Wenzel, A. G.; Lalonde, M. P. Jacobsen, E. N. Divergent Stereoinduction Mechanisms in Urea-Catalyzed Additions to Imines. *Synlett* **2003**, 1919–1922.

(16) a) Zuend, S. J.; Jacobsen, E. N. Cooperative Catalysis by Tertiary Amino-Thioureas: Mechanism and Basis for Enantioselectivity of Ketone Cyanosilylation. *J. Am. Chem. Soc.* **2007**, *129*, 15872–15883.
b) Raheem, I. T.; Thiara, P. S.; Peterson, E. A.; Jacobsen, E. N. Enantioselective Pictet-Spengler-Type Cyclizations of Hydroxylactams: H-Bond Donor

Catalysis by Anion Binding. *J. Am. Chem. Soc.* **2007**, *129*, 13404–13405.

(17) Huang, Y.; Unni, A. K.; Thadani, A. N.; Rawal, V. H. Hydrogen Bonding: Single Enantiomers from a Chiral–Alcohol Catalyst. *Nature* **2003**, *424*, 146.

(18) Thadani, A. N.; Stankovic, A. R.; Rawal, V. H. Enantioselective Diels–Alder Reactions Catalyzed by Hydrogen Bonding. *Proc. Natl. Acad. Sci. U.S.A.* **2004**, *101*, 5846–5850.

(19) Du, H.; Zhao, D.; Ding, K. Enantioselective Catalysis of the Hetero-Diels–Alder Reaction Between Brassard's Diene and Aldehydes by Hydrogen-Bonding Activation: A One-Step Synthesis of (*S*)-(+)-Dihydrokawain. *Chem. Eur. J.* **2004**, *10*, 5964–5970.

(20) Seebach, D.; Beck, A. K.; Heckel, A. TADDOLs, Their Derivatives, and TADDOL Analogues: Versatile Chiral Auxiliaries. *Angew. Chem., Int. Ed. Engl.* **2001**, *40*, 92–138.

(21) Zhang, X.; Du, H.; Wang, Z.; Wu, Y.-D.; Ding, K. Experimental and Theoretical Studies on the Hydrogen-Bond-Promoted Enantioselective Hetero-Diels–Alder Reaction of Danishefsky's Diene with Benzaldehyde. *J. Org. Chem.* **2006**, *71*, 2862–2869.

(22) Unni, A. K.; Takenaka, N.; Yamamoto, H.; Rawal, V. H. Axially Chiral Biaryl Diols Catalyze Highly Enantioselective Hetero-Diels–Alder Reactions Through Hydrogen Bonding. *J. Am. Chem. Soc.* **2005**, *127*, 1336–1337.

(23) McDougal, N. T.; Schaus, S. E. Asymmetric Morita-Baylis-Hillman Reactions Catalyzed by Chiral Brønsted Acids. *J. Am. Chem. Soc.* **2003**, *125*, 12094–12095.

(24) Uraguchi, D.; Sorimachi, K.; Terada, M. Organocatalytic Asymmetric Direct Alkylation of α-Diazoester via C–H Bond Cleavage. *J. Am. Chem. Soc.* **2005**, *127*, 9360–9361.

(25) Akiyama, T.; Itoh, J.; Yokota, K.; Fuchibe, K. Enantioselective Mannich-Type Reaction Catalyzed by a Chiral Brønsted Acid. *Angew. Chem., Int. Ed. Engl.* **2004**, *43*, 1566–1568.

(26) Smith, M. B.; March, J. *March's Advanced Organic Chemistry*, 5th ed.; Wiley: New York, 2001.

(27) Duhamel, L.; Duhamel, P.; Plaquevent, J.-C. Enantioselective Protonations: Fundamental Insights and New Concepts. *Tetrahedron: Asymmetry* **2004**, *15*, 3653–3691.

(28) Vedejs, E.; Kruger, A. W. Catalytic Asymmetric Protonation of Amide Enolates: Optimization of Kinetic Acidity in the Catalytic Cycle. *J. Org. Chem.* **1998**, *63*, 2792–2793.

(29) Ishibashi, H.; Ishihara, K.; Yamamoto, H. Chiral Proton Donor Reagents: Tin Tetrachloride-Coordinated Optically Active Binaphthol Derivatives. *Chem. Rev.* **2002**, *2*, 177–188.

(30) Ishihara, K.; Nakamura, S.; Yamamoto, H. The First Enantioselective Biomimetic Cyclization of Polyprenoids. *J. Am. Chem. Soc.* **1999**, *121*, 4906–4907.

(31) Wiskur, S. L.; Fu, G. C. Catalytic Asymmetric Synthesis of Esters from Ketenes. *J. Am. Chem. Soc.* **2006**, *127*, 6176–6177.

(32) Liang, G.; Trauner, D. Enantioselective Nazarov Reactions through Catalytic Asymmetric Proton Transfer. *J. Am. Chem. Soc.* **2004**, *126*, 9544–9545.

(33) Yamada, Y. M. A.; Yoshikawa, N.; Sasai, H.; Shibasaki, M. Direct Catalytic Asymmetric Aldol Re-

actions of Aldehydes with Unmodified Ketones. *Angew. Chem., Int. Ed. Engl.* **1997**, *36*, 871–1873.

(34) Asami, M.; Ishizaki, T.; Inoue, S. Catalytic Enantioselective Deprotonation of *meso*-Epoxides by the Use of Chiral Lithium Amide. *Tetrahedron: Asymmetry* **1994**, *5*, 793–796.

(35) Asami, M.; Suga, T.; Honda, K.; Inoue, S. A Novel Highly Efficient Chiral Lithium Amide for Catalytic Enantioselective Deprotonation of *meso*-Epoxides. *Tetrahedron Lett.* **1997**, *38*, 6425–6428.

(36) Södergren, M. J.; Andersson, P. G. New and Highly Enantioselective Catalysts for the Rearrangement of *meso*-Epoxides into Chiral Allylic Alcohols. *J. Am. Chem. Soc.* **1998**, *120*, 10760–10761.

(37) Södergren, M. J.; Bertilsson, S. K.; Andersson, P. G. Allylic Alcohols via Catalytic Asymmetric Epoxide Rearrangement. *J. Am. Chem. Soc.* **2000**, *122*, 6610–6618.

(38) Bertilsson, S.; Södergren, M. J.; Andersson, P. G. New Catalysts for the Base-Promoted Isomerization of Epoxides to Allylic Alcohols. Broadened Scope and Near-Perfect Asymmetric Induction. *J. Org. Chem.* **2002**, *67*, 1567–1573.

(39) Magnus, A.; Bertilsson, S.; Andersson, P. G. Asymmetric Base-Mediated Epoxide Isomerization. *Chem. Soc. Rev.* **2002**, *31*, 223–229.

(40) Gayet, A.; Bertilsson, S.; Andersson, P. G. Novel Catalytic Kinetic Resolution of Racemic Epoxides to Allylic Alcohols. *Org. Lett.* **2002**, *4*, 3777–3779.

(41) Gayet, A.; Andersson, P. G. Kinetic Resolution of Racemic Epoxides Using a Chiral Diamine Catalyst. *Tetrahedron Lett.* **2005**, *46*, 4805–4807.

(42) Hodgson, D. M.; Lee, G. P.; Marriott, R. E.; Thompson, A. J.; Wisedale; Witherington, R. Isomerisations of Cycloalkene- and Bicycloalkene-Derived Achiral Epoxides by Enantioselective α-Deprotonation. *J. Chem. Soc., Perkin Trans. 1* **1998**, 2151.

(43) O'Brien, P.; McGrath, M. J. Catalytic Asymmetric Deprotonation Using a Ligand Exchange Approach. *J. Am. Chem. Soc.* **2005**, *127*, 16378–16379.

(44) Tian, S.-K.; Chen, Y.; Hang, J.; Tang, L.; McDaid, P.; Deng, L. Asymmetric Organic Catalysis with Modified Cinchona Alkaloids. *Acc. Chem. Res.* **2004**, *37*, 621–631.

(45) Hang, J.; Tian, S.-K.; Tang, L.; Deng, L. Asymmetric Synthesis of α-Amino Acids via Cinchona Alkaloid-Catalyzed Kinetic Resolution of Urethane-Protected α-Amino Acid N-Carboxyanhydrides. *J. Am. Chem. Soc.* **2001**, *123*, 12696–12697.

(46) Spivey, A. C.; Andrews, B. I. Catalysis of the Asymmetric Desymmetrization of Cyclic Anhydrides by Nucleophilic Ring-Opening with Alcohols. *Angew. Chem., Int. Ed. Engl.* **2001**, *40*, 3131–3134.

(47) Bordwell, F. G. Equilibrium Acidities in Dimethyl Sulfoxide Solution. *Acc. Chem. Res.* **1988**, *21*, 456–463.

(48) Dolling, U. H.; Davis, P.; Grabowski, E. J. J. Efficient Catalytic Asymmetric Alkylations. 1. Enantioselective Synthesis of (+)-Indacrinone via Chiral Phase-Transfer Catalysis. *J. Am. Chem. Soc.* **1984**, *106*, 446–447.

(49) O'Donnell, M. J.; Bennett, W. D.; Wu, S. The Stereoselective Synthesis of α-Amino Acids by Phase-Transfer Catalysis. *J. Am. Chem. Soc.* **1989**, *111*, 2353–2355.

(50) Lipkowitz, K. B.; Cavanaugh, M. W.; Baker, B.; O'Donnell, M. J. Theoretical Studies in Molecular Recognition: Asymmetric Induction of Benzophe-

none Imine Ester Enolates by the Benzylcinchoninium Ion. *J. Org. Chem.* **1991**, *56*, 5181–5192.

(51) Corey, E. J.; Xu, F.; Noe, M. C. A Rational Approach to Catalytic Enantioselective Enolate Alkylation Using a Structurally Rigidified and Defined Chiral Quaternary Ammonium Salt Under Phase Transfer Conditions. *J. Am. Chem. Soc.* **1997**, *119*, 12414–12415.

(52) Lygo, B.; Wainwright, P. G. A New Class of Asymmetric Phase-Transfer Catalysts Derived from Cinchona Alkaloids—Application in the Enantioselective Synthesis of α-Amino Acids. *Tetrahedron Lett.* **1997**, *38*, 8595–8598.

(53) Lygo, B.; Crosby, J.; Lowdon, T. R.; Peterson, J. A.; Wainwright, P. G. Studies on the Enantioselective Synthesis of α-Amino Acids via Asymmetric Phase-Transfer Catalysis. *Tetrahedron* **2001**, *57*, 2403–2409.

(54) Shiori, T. In *Handbook of Phase-Transfer Catalysis*; Sasson, Y., Neumann, R., Eds.; Blackie Academic & Professional: London, 1997.

(55) Kacprzak, K.; Gawronski, J. Cinchona Alkaloids and Their Derivatives: Versatile Catalysts and Ligands in Asymmetric Synthesis. *Synthesis* **2001**, 961–998.

(56) Maruoka, K.; Ooi, T. Enantioselective Amino Acid Synthesis by Chiral Phase-Transfer Catalysis. *Chem. Rev.* **2003**, *103*, 3013–3028.

(57) O'Donnell, M. J. The Enantioselective Synthesis of α-Amino Acids by Phase-Transfer Catalysis with Achiral Schiff Base Esters. *Acc. Chem. Res.* **2004**, *37*, 506–517.

(58) Lygo, B.; Andrews, B. I. Asymmetric Phase-Transfer Catalysis Utilizing Chiral Quaternary Ammonium Salts: Asymmetric Alkylation of Glycine Imines. *Acc. Chem. Res.* **2004**, *37*, 518–525.

(59) Ooi, T.; Maruoka, K. Asymmetric Organocatalysis of Structurally Well-Defined Chiral Quaternary Ammonium Fluorides. *Acc. Chem. Res.* **2004**, *37*, 526–533.

(60) Ooi, T.; Kameda, M.; Maruoka, K. Molecular Design of a C_2-Symmetric Chiral Phase-Transfer Catalyst for Practical Asymmetric Synthesis of α-Amino Acids. *J. Am. Chem. Soc.* **1999**, *121*, 6519–6520.

(61) Ooi, T.; Kameda, M.; Maruoka, K. Design of N-Spiro C_2-Symmetric Chiral Quaternary Ammonium Bromides as Novel Chiral Phase-Transfer Catalysts: Synthesis and Application to Practical Asymmetric Synthesis of α-Amino Acids. *J. Am. Chem. Soc.* **2003**, *125*, 5139–5151.

(62) Ooi, T.; Doda, K.; Maruoka, K. Designer Chiral Quaternary Ammonium Bifluorides as an Efficient Catalyst for Asymmetric Nitroaldol Reaction of Silyl Nitronates with Aromatic Aldehydes. *J. Am. Chem. Soc.* **2003**, *125*, 2054–2055.

(63) Ooi, T.; Doda, K.; Maruoka, K. Highly Enantioselective Michael Addition of Silyl Nitronates to α,β-Unsaturated Aldehydes Catalyzed by Designer Chiral Ammonium Bifluorides: Efficient Access to Optically Active γ-Nitro Aldehydes and Their Enol Silyl Ethers. *J. Am. Chem. Soc.* **2003**, *125*, 9022–9023.

(64) Mayer, S.; List, B. Asymmetric Counterion-Directed Catalysis. *Angew. Chem., Int. Ed. Engl.* **2006**, *45*, 4193–4195.

(65) Hamilton, G. L.; Kang, E. J.; Mba, M.; Toste, F. D. A Powerful Chiral Counterion Strategy for Asymmetric Transition Metal Catalysis. *Science* **2007**, *317*, 496–499.

(66) Kunz, R. K.; MacMillan, D. W. C. Enantioselective Organocatalytic Cyclopropanations. The Identification

of a New Class of Iminium Catalyst Based upon Directed Electrostatic Activation. *J. Am. Chem. Soc.* **2005**, *127*, 3240–3241.

(67) Kolb, H. C.; VanNieuwenhze, M. S.; Sharpless, K. B. Catalytic Asymmetric Dihydroxylation. *Chem. Rev.* **1994**, *94*, 2483–2547.

(68) Jacobsen, E. N.; Marko, I.; Mungall, W. S.; Schroeder, G.; Sharpless, K. B. Asymmetric Dihydroxylation via Ligand-Accelerated Catalysis. *J. Am. Chem. Soc.* **1988**, *110*, 1968–1970.

(69) Berrisford, D. J.; Bolm, C.; Sharpless, K. B. Ligand-Accelerated Catalysis. *Angew. Chem., Int. Ed. Engl.* **1995**, *34*, 1059–1070.

(70) Kwong, H.-L.; Sorato, C.; Ogino, Y.; Chen, H.; Sharpless, K. B. Preclusion of the Second Cycle. in the Osmium-Catalyzed Asymmetric Dihydroxylation of Olefins Leads to a Superior Process. *Tetrahedron Lett.* **1990**, *31*, 2999–3002.

(71) DelMonte, A. J.; Haller, J.; Houk, K. N.; Sharpless, K. B.; Singleton, D. A.; Strassner, T.; Thomas, A. A. Experimental and Theoretical Kinetic Isotope Effects for Asymmetric Dihydroxylation. Evidence Supporting a Rate-Limiting "(3 + 2)" Cycloaddition. *J. Am. Chem. Soc.* **1997**, *119*, 9907–9908.

(72) Katsuki, T.; Sharpless, K. B. The First Practical Method for Asymmetric Epoxidation. *J. Am. Chem. Soc.* **1980**, *102*, 5974–5976.

(73) Gao, Y.; Klunder, J. M.; Hanson, R. M.; Masamune, H.; Ko, S. Y.; Sharpless, K. B. Catalytic Asymmetric Epoxidation and Kinetic Resolution: Modified Procedures Including in situ Derivatization. *J. Am. Chem. Soc.* **1987**, *109*, 5765–5780.

(74) Katsuki, T. In *Comprehensive Asymmetric Catalysis*; Jacobsen, E., N., Pfaltz, A., Yamamoto, H., Eds.; Springer-Verlag: Berlin, 1999; Vol. II, pp 621–648.

(75) Martin, V. S.; Woodard, S. S.; Katsuki, T.; Yamada, Y.; Ikeda, M.; Sharpless, K. B. Kinetic Resolution of Racemic Allylic Alcohols by Enantioselective Epoxidation. A Route to Substances of Absolute Enantiomeric Purity?. *J. Am. Chem. Soc.* **1981**, *103*, 6237–6240.

(76) Williams, I. D.; Pedersen, S. F.; Sharpless, K. B.; Lippard, S. J. Crystal Structures of Two Titanium Tartrate Asymmetric Epoxidation Catalysts. *J. Am. Chem. Soc.* **1984**, *106*, 6430–6431.

(77) Woodard, S. S.; Finn, M. G.; Sharpless, K. B. Mechanism of Asymmetric Epoxidation. 1. Kinetics. *J. Am. Chem. Soc.* **1991**, *113*, 106–113.

(78) Finn, M. G.; Sharpless, K. B. Mechanism of Asymmetric Epoxidation. 2. Catalyst Structure. *J. Am. Chem. Soc.* **1991**, *113*, 113–126.

(79) a) Zhang, W.; Loebach, J. L.; Wilson, S. R.; Jacobsen, E. N. Enantioselective Epoxidation of Unfunctionalized Olefins Catalyzed by Salen Manganese Complexes. *J. Am. Chem. Soc.* **1990**, *112*, 2801–2803. b) Jacobsen, E. N.; Zhang, W.; Muci, A. R.; Ecker, J. R.; Deng, L. Highly Enantioselective Epoxidation Catalysts Derived from 1,2-Diaminocyclohexane. *J. Am. Chem. Soc.* **1991**, *113*, 7063–7064. c) Brandes, B. D.; Jacobsen, E. N. Highly Enantioselective, Catalytic Epoxidation of Trisubstituted Olefins. *J. Org. Chem.* **1994**, *59*, 4378–4380.

(80) a) Irie, R.; Noda, K.; Ito, Y.; Matsumoto, N.; Katsuki, T. Catalytic Asymmetric Epoxidation of Unfunctionalized Olefins Using Chiral (Salen)manganese(III) Complexes. *Tetrahedron: Asymm.* **1991**, *2*, 481–494.

(81) b) Hatayama, A.; Hosoya, N.; Irie, R.; Ito, Y.; Katsuki, T. Highly Enantioselective Epoxidation of 2,2-Dimethylchromenes. *Synlett* **1992**, *5*, 407–409.

(81) a) Ito, Y. N.; Katsuki, T. Asymmetric Catalysis of New Generation Chiral Metallolsalen Complexes. *Bull. Chem. Soc. Jpn.* **1999**, *72*, 603–619. b) Katsuki, T. Chiral Metallosalen Complexes: Structures and Catalyst Tuning for Asymmetric Epoxidation and Cyclopropanation. *Adv. Synth. Cat.* **2002**, *344*, 131–147.

(82) Jacobsen, E. N.; Wu, M. H. In *Comprehensive Asymmetric Catalysis*; Jacobsen, E. N., Pfaltz, A., Yamamoto, H., Eds.; Springer-Verlag: Berlin, 1999; Vol. II, pp 649–677.

(83) McGarrigle, E. M.; Gilheany, D. G. Chromium– and Manganese–Salen Promoted Epoxidation of Alkenes. *Chem. Rev.* **2005**, *105*, 1563–1602.

(84) Fu, H.; Look, G. C.; Zhang, W.; Jacobsen, E. N.; Wong, C. H. Mechanistic Study of a Synthetically Useful Monooxygenase Model Using the Hypersensitive Probe *trans*-2-Phenyl-1-vinylcyclopropane. *J. Org. Chem.* **1991**, *56*, 6497–6500.

(85) Zhang, W.; Lee, N. H.; Jacobsen, E. N. Nonstereospecific Mechanisms in Asymmetric Addition to Alkenes Result in Enantiodifferentiation After the First Irreversible Step. *J. Am. Chem. Soc.* **1994**, *116*, 425–426.

(86) Finney, N. S.; Pospisil, P. J.; Chang, S.; Palucki, M.; Konsler, R. G.; Hansen, K. B.; Jacobsen, E. N. On the Viability of Oxametallacyclic Intermediates in the (Salen)Mn-catalyzed Asymmetric Epoxidation. *Angew. Chem., Int. Ed. Engl.* **1997**, *36*, 1720–1723.

(87) Chang, S.; Galvin, J. M.; Jacobsen, E. N. Effect of Chiral Quaternary Ammonium Salts on (Salen)Mn-Catalyzed Epoxidation of Cis-Olefins. A Highly Enantioselective, Catalytic Route to trans-Epoxides. *J. Am. Chem. Soc.* **1994**, *116*, 6937–6938.

(88) Denmark, S. E.; Wu, Z. The Development of Chiral, Nonracemic Dioxiranes for the Catalytic, Enantioselective Epoxidation of Alkenes. *Synlett* **1999**, 847–859.

(89) Frohn, M.; Shi, Y. Chiral Ketone-Catalyzed Asymmetric Epoxidation of Olefins. *Synthesis* **2000**, 1979–2000.

(90) Shi, Y. Organocatalytic Asymmetric Epoxidation of Olefins by Chiral Ketones. *Acc. Chem. Res.* **2004**, *37*, 488–496.

(91) Yang, D. Ketone-Catalyzed Asymmetric Epoxidation Reactions. *Acc. Chem. Res.* **2004**, *37*, 497–505.

(92) Tu, Y.; Wang, Z.-X.; Shi, Y. An Efficient Asymmetric Epoxidation Method for *trans*-Olefins Mediated by a Fructose-Derived Ketone. *J. Am. Chem. Soc.* **1996**, *118*, 9806–9807.

(93) Wang, Z.-X.; Tu, Y.; Frohn, M.; Shi, Y. A Dramatic pH Effect Leads to a Catalytic Asymmetric Epoxidation. *J. Org. Chem.* **1997**, *62*, 2328–2329.

(94) Wang, Z.-X.; Tu, Y.; Frohn, M.; Zhang, J.-R.; Shi, Y. An Efficient Catalytic Asymmetric Epoxidation Method. *J. Am. Chem. Soc.* **1997**, *119*, 11224–11235.

(95) Burke, C. P.; Shi, Y. Regio- and Enantioselective Epoxidation of Dienes by a Chiral Dioxirane: Synthesis of Optically Active Vinyl *cis*-Epoxides. *Angew. Chem., Int. Ed. Engl.* **2006**, *45*, 4475–4478.

(96) Tian, H.; She, X.; Shu, L.; Yu, H.; Shi, Y. Highly Enantioselective Epoxidation of Cis-Olefins by Chiral Dioxirane. *J. Am. Chem. Soc.* **2000**, *122*, 11551–11552.

(97) Hickey, M.; Goeddel, D.; Crane, Z.; Shi, Y. Highly Enantioselective Epoxidation of Styrenes: Implication of an Electronic Effect on the Competition Between Spiro and Planar Transition States. *Proc. Natl. Acad. Sci. U.S.A.* **2004**, *101*, 5794–5798.

(98) Lorenz, J. C.; Frohn, M.; Zhou, X.; Zhang, J.-R.; Tang, Y.; Burke, C.; Shi, Y. Transition State Studies on the Dioxirane-Mediated Asymmetric Epoxidation via Kinetic Resolution and Desymmetrization. *J. Org. Chem.* **2005**, *70*, 2904–2911.

(99) Müller, P. Enantioselective Catalytic Aziridinations and Asymmetric Nitrene Insertions into CH Bonds. *Chem. Rev.* **2003**, *103*, 2905–2919.

(100) Li, A.-H.; Dai, L.-X.; Aggarwal, V. K. Asymmetric Ylide Reactions: Epoxidation, Cyclopropanation, Aziridination, Olefination, and Rearrangement. *Chem. Rev.* **1997**, *97*, 2341–2372.

(101) Lebel, H.; Marcoux, J.-F.; Molinaro, C.; Charette, A. B. Stereoselective Cyclopropanation Reactions. *Chem. Rev.* **2003**, *103*, 977–1050.

(102) Doyle, M. P.; Forbes, D. C. Recent Advances in Asymmetric Catalytic Metal Carbene Transformations. *Chem. Rev.* **1998**, *98*, 911–936.

(103) Doyle, M. P.; McKervey, M. A.; Ye, T. *Modern Catalytic Methods for Organic Synthesis with Diazo Compounds*; Wiley: New York, 1998.

(104) Davies, H. M. L.; Beckwith, R. E. J. Catalytic Enantioselective C–H Activation by Means of Metal Carbenoid-Induced C–H Insertion. *Chem. Rev.* **2003**, *103*, 2861–2904.

(105) Davies, H. M. L.; Hansen, T.; Churchill, M. R. Catalytic Asymmetric C–H Activation of Alkanes and Tetrahydrofuran. *J. Am. Chem. Soc.* **2000**, *122*, 3063–3070.

(106) Davies, H. M. L.; Hansen, T. Asymmetric Intermolecular Carbenoid C–H Insertions Catalyzed by Rhodium(II) (*S*)-*N*-(*p*-Dodecylphenyl)sulfonylprolinate. *J. Am. Chem. Soc.* **1997**, *119*, 9075–9076.

(107) Davies, H. M. L.; Hansen, T.; Hopper, D. W.; Panaro, S. A. Highly Regio-, Diastereo-, and Enantioselective C–H Insertions of Methyl Aryldiazoacetates into Cyclic *N*-Boc-Protected Amines. Asymmetric Synthesis of Novel C_2-Symmetric Amines and *threo*-Methylphenidate. *J. Am. Chem. Soc.* **1999**, *121*, 6509–6510.

(108) Davies, H. M. L.; Antoulinakis, E. G.; Hansen, T. Catalytic Asymmetric Synthesis of Syn-Aldol Products from Intermolecular C–H Insertions Between Allyl Silyl Ethers and Methyl Aryldiazoacetates. *Org. Lett.* **1999**, *1*, 383–386.

(109) Cammidge, A. N.; Crépy, K. V. L. The First Asymmetric Suzuki Cross-Coupling Reaction. *J. Chem. Soc., Chem. Commun.* **2000**, 1723–1724.

(110) Cammidge, A. N.; Crépy, K. V. L. Synthesis of Chiral Binaphthalenes Using the Asymmetric Suzuki Reaction. *Tetrahedron* **2004**, *60*, 4377–4386.

(111) Yin, J.; Buchwald, S. L. A Catalytic Asymmetric Suzuki Coupling for the Synthesis of Axially Chiral Biaryl Compounds. *J. Am. Chem. Soc.* **2000**, *122*, 12051–12052.

(112) Baudoin, O. The Asymmetric Suzuki Coupling Route to Axially Chiral Biaryls. *Eur. J. Org. Chem.* **2005**, 4223–4229.

(113) Buchmeiser, M. R. Homogeneous Metathesis Polymerization by Well-Defined Group VI and Group VIII Transition-Metal Alkylidenes: Fundamentals and Applications in the Preparation of Advanced Materials. *Chem. Rev.* **2000**, *100*, 1565–1604.

(114) Hoveyda, A. H.; Schrock, R. R. Catalytic Asymmetric Olefin Metathesis. *Chem. Eur. J.* **2001**, *7*, 945–950.

(115) Schrock, R. R.; Hoveyda, A. H. Molybdenum and Tungsten Imido Alkylidene Complexes as Efficient Olefin-Metathesis Catalysts. *Angew. Chem., Int. Ed. Engl.* **2003**, *42*, 4592–4633.

(116) La, D. S.; Alexander, J. B.; Cefalo, D. R.; Graf, D. D.; Hoveyda, A. H.; Schrock, R. R. Mo-Catalyzed Asymmetric Synthesis of Dihydrofurans. Catalytic Kinetic Resolution and Enantioselective Desymmetrization Through Ring-Closing Metathesis. *J. Am. Chem. Soc.* **1998**, *120*, 9720–9721.

(117) Cefalo, D. R.; Kiely, A. F.; Wuchrer, M.; Jamieson, J. Y.; Schrock, R. R.; Hoveyda, A. H. Enantioselective Synthesis of Unsaturated Cyclic Tertiary Ethers by Mo-Catalyzed Olefin Metathesis. *J. Am. Chem. Soc.* **2001**, *123*, 3139–3140.

(118) Dolman, S. J.; Sattely, E. S.; Hoveyda, A. H.; Schrock, R. R. Efficient Catalytic Enantioselective Synthesis of Unsaturated Amines: Preparation of Small- and Medium-Ring Cyclic Amines Through Mo-Catalyzed Asymmetric Ring-Closing Metathesis in the Absence of Solvent. *J. Am. Chem. Soc.* **2002**, *124*, 6991–6997.

(119) Costabile, C.; Cavallo, L. Origin of Enantioselectivity in the Asymmetric Ru-Catalyzed Metathesis of Olefins. *J. Am. Chem. Soc.* **2004**, *126*, 9592–9600.

(120) Goumans, T. P. M.; Ehlers, A. W.; Lammertsma, K. The Asymmetric Schrock Olefin Metathesis Catalyst. A Computational Study. *Organometallics* **2005**, *24*, 3200–3206.

(121) Gillingham, D. G.; Kataoka, O.; Garber, S. B.; Hoveyda, A. H. Efficient Enantioselective Synthesis of Functionalized Tetrahydropyrans by Ru-Catalyzed Asymmetric Ring-Opening Metathesis/Cross-Metathesis (AROM/CM). *J. Am. Chem. Soc.* **2004**, *126*, 12288–12290.

(122) Seiders, T. J.; Ward, D. W.; Grubbs, R. H. Enantioselective Ruthenium-Catalyzed Ring-Closing Metathesis. *Org. Lett.* **2001**, *3*, 3225–3228.

(123) Hoveyda, A. H.; Gillingham, D. G.; Van Veldhuizen, J. J.; Kataoka, O.; Garber, S. B.; Kingsbury, J. S.; Harrity, J. P. A. Ru Complexes Bearing Bidentate Carbenes: From Innocent Curiosity to Uniquely Effective Catalysts for Olefin Metathesis. *Org. Biomol. Chem.* **2004**, *2*, 8–23.

(124) Funk, T. W.; Berlin, J. M.; Grubbs, R. H. Highly Active Chiral Ruthenium Catalysts for Asymmetric Ring-Closing Olefin Metathesis. *J. Am. Chem. Soc.* **2006**, *128*, 1840–1846.

(125) Costabile, C.; Cavallo, L. Origin of Enantioselectivity in the Asymmetric Ru-Catalyzed Metathesis of Olefins. *J. Am. Chem. Soc.* **2004**, *126*, 9592–9600.

(126) Gladysz, J. A.; Boone, B. J. Chiral Recognition in π-Complexes of Alkenes, Aldehydes, and Ketones with Transition Metal Lewis Acids; Development of a General Model for Enantioface Binding Selectivities. *Angew. Chem., Int. Ed. Engl.* **1997**, *36*, 550–583.

(127) Hegedus, L. S. *Transition Metals in the Synthesis of Complex Organic Molecules*, 2nd ed.; University Science Books: Sausalito, CA, 1999.

(128) Collman, J. P.; Hegedus, L. S.; Norton, J. R.; Finke, R. G. *Principles and Applications of Organotransition*

Metal Chemistry; University Science Books: Sausalito, CA, 1987.

(129) Stille, J. K.; Divakaruni, R. Stereochemistry of the Hydroxypalladation of Ethylene. Evidence for *trans* Addition in the Wacker Process. *J. Am. Chem. Soc.* **1978**, *100*, 1303–1304.

(130) Tietze, L. F.; Ila, H.; Bell, H. P. Enantioselective Palladium-Catalyzed Transformations. *Chem. Rev.* **2004**, *104*, 3453–3516.

(131) El-Qisairi, A.; Hamed, O.; Henry, P. M. A New Palladium(II)-Catalyzed Asymmetric Chlorohydrin Synthesis. *J. Org. Chem.* **1998**, *63*, 2790–2791.

(132) El-Qisairi, A. K.; Qaseer, H. A.; Katsigras, G.; Lorenzi, P.; Trivedi, U.; Tracz, S.; Hartman, A.; Miller, J. A.; Henry, P. M. New Palladium(II)-Catalyzed Asymmetric 1,2-Dibromo Synthesis. *Org. Lett.* **2003**, *5*, 439–441.

(133) Uozumi, Y.; Kato, K.; Hayashi, T. Catalytic Asymmetric Wacker-Type Cyclization. *J. Am. Chem. Soc.* **1997**, *119*, 5063–5064.

(134) Martin, S. F. Methodology for the Construction of Quaternary Carbon Centers. *Tetrahedron* **1980**, *36*, 419–460.

(135) Corey, E. J.; Guzman-Perez, A. The Catalytic Enantioselective Construction of Molecules with Quaternary Carbon Stereocenters. *Angew. Chem., Int. Ed. Engl.* **1998**, *37*, 388–401.

(136) Christoffers, J. M., A. Enantioselective Construction of Quaternary Stereocenters. *Angew. Chem., Int. Ed. Engl.* **2001**, *40*, 4591–4597.

(137) Arai, M. A.; Kuraishi, M.; Arai, T.; Sasai, H. A New Asymmetric Wacker-Type Cyclization and Tandem Cyclization Promoted by Pd(II)-Spiro Bis(isoxazoline) Catalyst. *J. Am. Chem. Soc.* **2001**, *123*, 2907–2908.

(138) Shibasaki, M.; Vogl, E. M.; Ohshima, T. Asymmetric Heck Reaction. *Adv. Synth. Catal.* **2004**, *346*, 1533–1552.

(139) Loiseleur, O.; Hayashi, M.; Schmees, N.; Pfaltz, A. Enantioselective Heck Reactions Catalyzed by Chiral Phosphinooxazoline–Palladium Complexes. *Synthesis* **1997**, 1338–1345.

(140) Ashimori, A.; Overman, L. E. Catalytic Asymmetric Synthesis of Quaternary Carbon Centers. Palladium-Catalyzed Formation of Either Enantiomer of Spirooxindoles and Related Spirocyclics Using a Single Enantiomer of a Chiral Diphosphine Ligand. *J. Org. Chem.* **1992**, *57*, 4571–4572.

(141) Ashimori, A.; Matsuura, T.; Overman, L. E.; Poon, D. J. Catalytic Asymmetric Synthesis of Either Enantiomer of Physostigmine. Formation of Quaternary Carbon Centers with High Enantioselection by Intramolecular Heck Reactions of (Z)-2-Butenanilides. *J. Org. Chem.* **1993**, *58*, 6949–6951.

(142) Ashimori, A.; Bachand, B.; Calter, M. A.; Govek, S. P.; Overman, L. E.; Poon, D. J. Catalytic Asymmetric Synthesis of Quaternary Carbon Centers. Exploratory Studies of Intramolecular Heck Reactions of (Z)-α,β-Unsaturated Anilides and Mechanistic Investigations of Asymmetric Heck Reactions Proceeding via Neutral Intermediates. *J. Am. Chem. Soc.* **1998**, *120*, 6488–6499.

(143) Lebsack, A. D.; Link, J. T.; Overman, L. E.; Stearns, B. A. Enantioselective Total Synthesis of Quadrigemine C and Psycholeine. *J. Am. Chem. Soc.* **2002**, *124*, 9008–9009.

(144) Fryzuk, M. D.; Bosnich, B. Asymmetric Synthesis. Production of Optically Active Amino Acids by Catalytic Hydrogenation. *J. Am. Chem. Soc.* **1977**, *99*, 6262–6267.

(145) Chan, A. S. C.; Pluth, J. J.; Halpern, J. Identification of the Enantioselective Step in the Asymmetric Catalytic Hydrogenation of a Prochiral Olefin. *J. Am. Chem. Soc.* **1980**, *102*, 5952–5954.

(146) Halpern, J. Mechanism and Stereoselectivity of Asymmetric Hydrogenation. *Science* **1982**, *217*, 401–407.

(147) Sakai, N.; Mano, S.; Nozaki, K.; Takaya, H. Highly Enantioselective Hydroformylation of Olefins Catalyzed by New Phosphine Phosphite–Rhodium(I) Complexes. *J. Am. Chem. Soc.* **1993**, *115*, 7033–7034.

(148) Agbossou, F.; Carpentier, J.-F.; Mortreux, A. Asymmetric Hydroformylation. *Chem. Rev.* **1995**, *95*, 2485–2506.

(149) Trost, B. M.; Van Vranken, D. L. Asymmetric Transition Metal-Catalyzed Allylic Alkylations. *Chem. Rev.* **1996**, *96*, 395–422.

(150) Trost, B. M. Designing a Receptor for Molecular Recognition in a Catalytic Synthetic Reaction: Allylic Alkylation. *Acc. Chem. Res.* **1996**, *29*, 355–364.

(151) Moberg, C.; Belda, O. Molybdenum-Catalyzed Asymmetric Allylic Alkylations. *Acc. Chem. Res.* **2004**, *37*, 159–167.

(152) Trost, B. M.; Machacek, M. R.; Aponick, A. Predicting the Stereochemistry of Diphenylphosphino Benzoic Acid (DPPBA)-Based Palladium-Catalyzed Asymmetric Allylic Alkylation Reactions: A Working Model. *Acc. Chem. Res.* **2006**, *39*, 747–760.

(153) Trost, B. M.; Krische, M. J.; Radinov, R.; Zanoni, G. On Asymmetric Induction in Allylic Alkylation via Enantiotopic Facial Discrimination. *J. Am. Chem. Soc.* **1996**, *118*, 6297–6298.

(154) Trost, B. M.; Toste, F. D. Regio- and Enantioselective Allylic Alkylation of an Unsymmetrical Substrate: A Working Model. *J. Am. Chem. Soc.* **1999**, *121*, 4545–4554.

(155) Trost, B. M.; Stenkamp, D.; Pulley, S. R. An Enantioselective Synthesis of *cis*-4-*tert*-Butoxycarbamoyl-1-methoxycarbonyl-2-cyclopentene—A Useful, General Building Block. *Chem. Eur. J.* **1995**, *1*, 568–572.

(156) Trost, B. M.; Pulley, S. R. Asymmetric Total Synthesis of (+)-Pancratistatin. *J. Am. Chem. Soc.* **1995**, *117*, 10143–10144.

(157) Trost, B. M.; Bunt, R. C. Asymmetric Induction in Allylic Alkylations of 3-(Acyloxy)cycloalkenes. *J. Am. Chem. Soc.* **1994**, *116*, 4089–4090.

(158) Trost, B. M.; Bunt, R. C.; Lemoine, R. C.; Calkins, T. L. Dynamic Kinetic Asymmetric Transformation of Diene Monoepoxides: A Practical Asymmetric Synthesis of Vinylglycinol, Vigabatrin, and Ethambutol. *J. Am. Chem. Soc.* **2000**, *122*, 5968–5976.

(159) Trost, B. M.; Yasukata, T. A Catalytic Asymmetric Wagner–Meerwein Shift. *J. Am. Chem. Soc.* **2001**, *123*, 7162–7163.

(160) Kawatsura, M.; Hartwig, J. F. Palladium-Catalyzed Intermolecular Hydroamination of Vinylarenes Using Arylamines. *J. Am. Chem. Soc.* **2000**, *122*, 9546–9547.

(161) Nettekoven, U.; Hartwig, J. F. A New Pathway for Hydroamination. Mechanism of Palladium-

Catalyzed Addition of Anilines to Vinylarenes. *J. Am. Chem. Soc.* **2002**, *124*, 1166–1167.

(162) Hartwig, J. F. Development of Catalysts for the Hydroamination of Olefins. *Pure Appl. Chem.* **2004**, *76*, 507–516.

(163) Johns, A. M.; Utsunomiya, M.; Incarvito, C. D.; Hartwig, J. F. A Highly Active Palladium Catalyst for Intermolecular Hydroamination. Factors that Control Reactivity and Additions of Functionalized Anilines to Dienes and Vinylarenes. *J. Am. Chem. Soc.* **2006**, *128*, 1828–1839.

(164) Hu, A.; Ogasawara, M.; Sakamoto, T.; Okada, A.; Nakajima, K.; Takahashi, T.; Lin, W. Palladium-Catalyzed Intermolecular Asymmetric Hydroamination with 4,4-Disubstituted BINAP and SEGPHOS. *Adv. Synth. Cat.* **2006**, *348*, 2051–2056.

(165) Murakami, M.; Itami, K.; Ito, Y. Rhodium-Catalyzed Asymmetric [4 + 1] Cycloaddition. *J. Am. Chem. Soc.* **1997**, *119*, 2950–2951.

(166) Murakami, M.; Itami, K.; Ito, Y. Catalytic Asymmetric [4 + 1] Cycloaddition of Vinylallenes with Carbon Monoxide: Reversal of the Induced Chirality by the Choice of Metal. *J. Am. Chem. Soc.* **1999**, *121*, 4130–4135.

(167) Meng, Q.; Li, M.; Zhang, J. The Computational Study on the Mechanism of Rhodium(I)-Catalyzed Asymmetric Carbonylative [4 + 1] Cycloaddition with (*R*,*R*)-Me-DuPHOS-Type Ligand. A DFT Study. *J. Mol. Struct.: THEOCHEM* **2005**, *726*, 47–54.

(168) Wender, P. A.; Haustedt, L. O.; Lim, J.; Love, J. A.; Williams, T. J.; Yoon, J.-Y. Asymmetric Catalysis of the [5 + 2] Cycloaddition Reaction of Vinylcyclopropanes and π-Systems. *J. Am. Chem. Soc.* **2006**, *128*, 6302–6303.

(169) Vollhardt, K. P. C. Cobalt-Mediated [2 + 2 + 2]-Cycloadditions: A Maturing Synthetic Strategy. *Angew. Chem., Int. Ed. Engl.* **1984**, *23*, 539–556.

(170) Grigg, R.; Scott, R.; Stevenson, P. Rhodium-Catalyzed [2 + 2 + 2]-Cycloadditions of Acetylenes. *J. Chem. Soc., Perkin Trans. 1* **1988**, 1357–1364.

(171) Lautens, M.; Klute, W.; Tam, W. Transition Metal-Mediated Cycloaddition Reactions. *Chem. Rev.* **1996**, *96*, 49–92.

(172) Shibata, T.; Fujimoto, T.; Yokota, K.; Takagi, K. Iridium Complex-Catalyzed Highly Enantio- and Diastereoselective [2 + 2 + 2] Cycloaddition for the Synthesis of Axially Chiral Teraryl Compounds. *J. Am. Chem. Soc.* **2004**, *126*, 8382–8383.

(173) a) Dalko, P. I.; Moisan, L. In the Golden Age of Organocatalysis. *Angew. Chem., Int. Ed. Engl.* **2004**, *43*, 5138–5175. b) Tada, M. Iwasawa, Y. Advanced Chemical Design with Supported Metal Complexes for Selective Catalysis. *Chem. Commun.* **2006**, 2833–2844. c) Yang, H.; Li, J.; Yange, J.; Liu, Z.; Yang, Q.; Li, C. Asymmetric Reaction on Chiral Catalysts Entrapped within a Mesoporous Cage. *Chem. Commun.* **2007**, 1086–1088.

Asymmetric Induction in Enantioselective Catalysis

One of the most important goals of asymmetric catalysis is the rational design and optimization of new catalysts. Although this seems like a straightforward exercise, it is far from it. Our understanding of the structure and function of catalysts, and the mechanisms of asymmetric reactions, is rudimentary. Although the lack of emphasis on mechanistic studies is certainly a contributing factor, it is often found that the reaction mechanisms are considerably more complex than anticipated. Furthermore, the difference in energy between diastereomeric transition states ($\Delta\Delta G^\ddagger$), which lead to enantiomeric products, is often small, complicating rational catalyst optimization. Given these challenges, it is not surprising that most catalyst optimization is performed with a combination of rational design, chemical intuition, and serendipity.

To understand how asymmetry is transmitted from the catalyst to the substrate, it is necessary to know the three-dimensional structure of the ligated catalyst. This is particularly challenging if the ligand retains a high degree of conformational flexibility when bonded to the metal, because it is more difficult to determine the conformation of the ligand in the transition state. For these reasons, it is easiest to begin studying the transmission of asymmetry in catalysts where the ligand–metal adduct has little rotational freedom.

This chapter examines how different types of chiral catalysts transmit asymmetry in enantioselective reactions. Emphasis is initially placed on ligands and catalysts that are well-defined. As the chapter progresses, and catalysts with additional degrees of conformational flexibility are introduced, the proposed structures and transition states are more speculative. Nonetheless, much can be learned from such systems.

4.1 Transmission of Asymmetry

In many cases asymmetric catalysts often bind and react preferentially with one of the prochiral faces of the substrate. In others, asymmetric catalysts bind the substrate and shield one of the prochiral faces, thus impeding reaction at that face. Despite the simplicity of these strategies, the mechanics of transmission of asymmetry from the catalyst to the substrate are complex and not well-understood in many systems. Furthermore, there are many classes of chiral ligands and catalysts, and the nature of asymmetry transmission from the catalyst to the substrate varies greatly. Thus, this section first introduces some of the means by which the asymmetry of the catalyst is transmitted to the substrate.

The most common method to transfer asymmetry from a catalyst to the substrate relies on steric biasing. Other catalyst–substrate interactions, such as π-interactions between aromatic groups on the catalyst and substrate, or hydrogen bonding between the catalyst and substrate, etc., can also play important roles and may be used in combination with steric biasing (see Chapter 5).

4.1.1 C_2- vs. Non-C_2-Symmetric Catalysts

C_2-Symmetric catalysts have received much attention and merit special consideration in the context of the transmission of asymmetry. In the early days of asymmetric catalysis, it was often observed that catalysts containing C_2-symmetric ligands were, in general, more enantioselective than those with non-C_2-symmetric ligands. It was proposed that this selectivity resulted from the smaller number of metal–substrate adducts and transition states available to these catalysts than are available to catalysts containing less-symmetric ligands.[1-3] This principle is illustrated in the context of the asymmetric allylation reaction in **Figure 4.1**. The most common mode of nucleophilic attack on η³-allyl is on the face opposite the bulky palladium center.[4,5] Recall that interconversion of the diastereomeric intermediates occurs within the coordina-

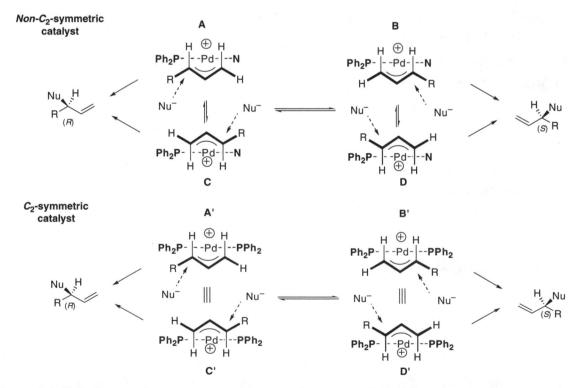

Figure 4.1. Palladium π-allyl complexes with a non-C_2-symmetric P–N ligand and a C_2-symmetric P–P ligand illustrating the number of diastereomeric lowest-energy transition states for nucleophilic attack. The chiral ligand backbone and the Pd–allyl bond are omitted for clarity.

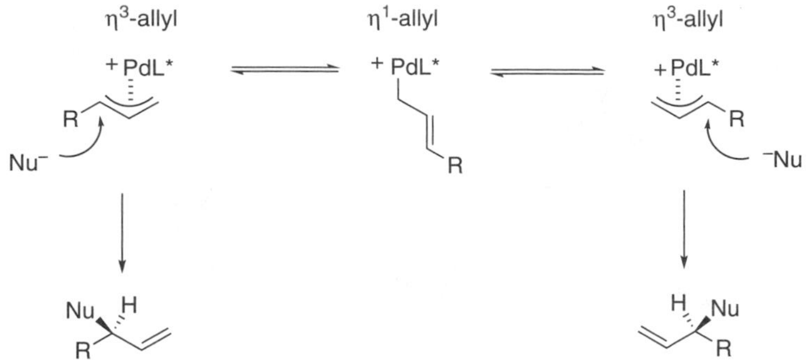

Figure 4.2. Interconversion of diastereomeric palladium–allyl complexes. The interconversion may be assisted by coordination of an additional ligand.

tion sphere of the metal center.[6] This interconversion of diastereomeric intermediates proceeds by reversible generation an η^1-allyl species in a mechanism often described as a π–σ–π interconversion (**Figure 4.2**). The interconversion can be aided by coordination of an additional ligand, as will be seen in subsequent chapters.

Catalysts containing P–N ligands can bind either of the two prochiral faces of an allyl group (**Figure 4.1**, where the chiral ligand backbone and the bond between the allyl and palladium are omitted for clarity). Attack of a nucleophile, such as phenoxide, can occur pseudotrans to phosphorus or pseudotrans to nitrogen in intermediates **A** and **C**, respectively, to generate the (R)-product (**Figure 4.1**, top). These two modes of attack are both sterically inequivalent and electronically distinct (due to the trans effect).[7] Likewise, attack on diastereomers **B** and **D** can give rise to formation of the (S)-product. In contrast, when a C_2-symmetric catalyst is employed, the transition states for attack on intermediates **A'** and **C'** are equivalent, as are **B'** and **D'**. Thus, the number of competing diastereomeric transition states is reduced with C_2-symmetric catalysts, simplifying analysis and catalyst design. The simplification is even greater in five- and six-coordinate intermediates, where the number of diastereomeric metal–ligand adducts is greater (see Figure A.8). Some examples of C_2-symmetric bis(phosphines) and non-C_2-symmetric P–N ligands are illustrated in **Figure 4.3**. Although many catalysts lacking C_2-symmetry exhibit high levels of enantioselectivity, those containing C_2-symmetric ligands comprise one of the most important and selective classes of catalysts.

4.1.2 Quadrant Diagrams

A generic model for steric biasing of chiral metal–ligand adducts has been advanced to facilitate the prediction of the facial stereoselectivity in catalyst–substrate complexes and transition states. In this model, the environment around the metal is divided into quadrants in which the horizontal dividing line is congruent with a plane or pseudoplane in the catalyst. For simplicity, the quadrant diagram is given for a C_2-symmetric catalyst (**Figure 4.4a**). The two shaded diagonal quadrants represent space that is occupied by substituents on the ligand that extend forward, whereas the unshaded rectangles correspond to less-occupied space. Binding of the prochiral faces of an olefin to a metal, for example, would give rise to diastereomers in which the

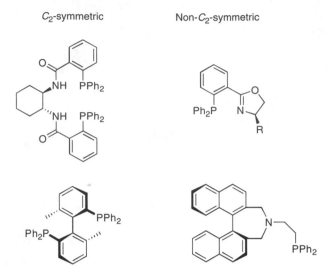

Figure 4.3. Examples of C_2-symmetric bis(phosphines) and non-C_2-symmetric P–N ligands used in catalytic asymmetric allylation reactions.

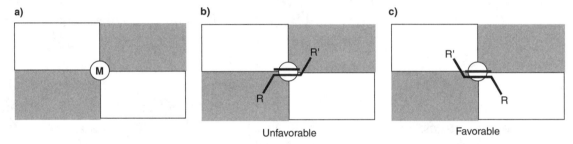

Figure 4.4. a) Quadrant diagrams for a C_2-symmetric catalyst. The shaded regions represent quadrants occupied by substituents on the ligand and the unshaded sections correspond to relatively unoccupied areas. b) Unfavorable binding and c) favorable binding of an olefin in diastereomeric olefin complexes.

more-stable diastereomer contains the R and R′ substituents positioned in the open, unshaded quadrants (**Figure 4.4b** vs. **Figure 4.4c**).

The means by which metal complexes of chiral ligands block quadrants depends on the nature of the ligand and the metal–ligand adduct. In some cases, the ligands possess stereogenic centers in close proximity to the metal. In other cases, the stereogenic centers are located so far from the metal it is not obvious how the effect of these distant stereocenters can be transmitted to the site of reaction. We describe examples of each of these in the following sections.

4.2 Chiral Metal Complexes from Chiral Ligands

4.2.1 Bisoxazolines

Two popular families of ligands that provide rigid chiral environments on coordination are the C_2-symmetric bisoxazoline ligands,[8,9] such as Box and PyBox, and the semicorrin ligands.[10] The structures of these ligands and their coordination complexes

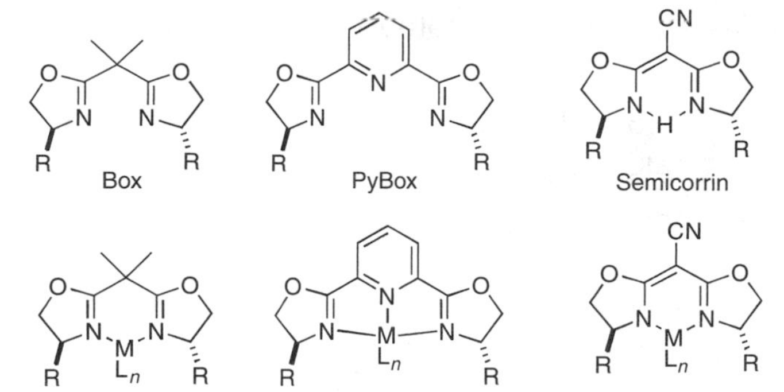

Figure 4.5. Structures of the bis(oxazoline) ligands, Box and PyBox and a semicorrin ligand, along with their metal complexes.

are shown in **Figure 4.5**. In conjunction with a variety of main-group and transition metals, these ligands form highly enantioselective catalysts that have been successfully used in many asymmetric reactions,[8–10] including the aldol, cyclopropanation,[11,12] aziridination,[13] Diels–Alder,[14,15] Michael,[16] and ene reactions.[17] The nearly planar metallacycle formed on binding the bidentate ligand, and the presence of the pendent five-membered rings, limit the flexibility of these ligand systems. The common feature of the ligands in **Figure 4.5** is that the substituents at the stereogenic centers extend forward, positioning them in close proximity to the metal center and the substrate binding sites. As will be illustrated below, the well-defined asymmetric environment of these ligands facilitates use of quadrant diagrams. Electronically, the anionic nature of the semicorrin ligands makes them stronger electron donors than the neutral bisoxazoline ligands. As a result, complexes of the semicorrin ligands are less Lewis acidic and will exhibit stronger binding of the chiral ligand.

4.2.1.a Control of Nucleophile Trajectory by Chiral Ligands

Several successful asymmetric transformations have been developed using chelating substrates in combination with copper(II)–bisoxazoline complexes, creating well-defined substrate–catalyst adducts that allow for more efficient catalyst optimization.[8] The crystal structure[15] of [(S,S)-t-Bu-Box)]Cu(OH$_2$)$_2$(SbF$_6$)$_2$ shown in **Figure 4.6**

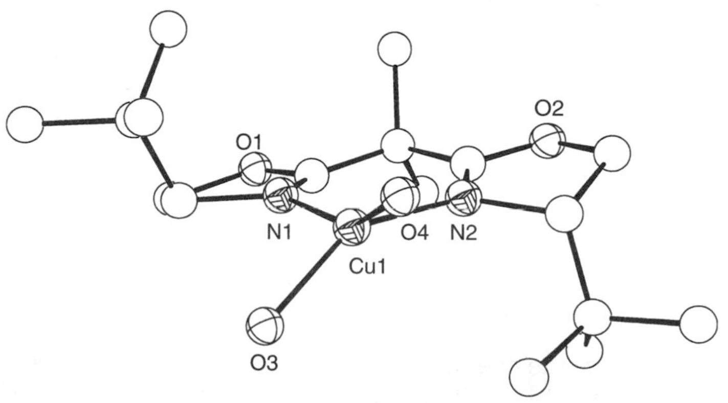

Figure 4.6. Structure of [(S,S)-t-Bu-Box)]Cu(OH$_2$)$_2$(SbF$_6$)$_2$, illustrating the distorted square-planar geometry of the copper and the ligand environment.

illustrates the pseudo-square-planar geometry of the Jahn–Teller-distorted d^9-copper center.[18] The SbF_6 counterions are not interacting with the copper center, and are not shown in the figure. The [(S,S)-t-Bu-Box)Cu^{2+} catalyst has proven particularly effective in the asymmetric Diels–Alder reaction with chelating imide dienophiles (**Equation 4.1**),[15] which prefer the s-cis conformation when coordinated (**Figure 4.7**).

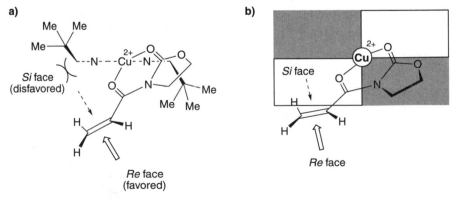

R = H, 98% ee
R = Me, 96% ee
R = Ph, 96% ee
R = Cl, 94% ee

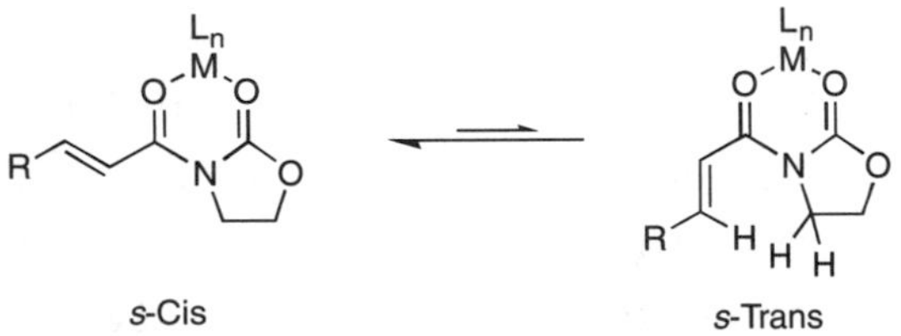

Equation 4.1

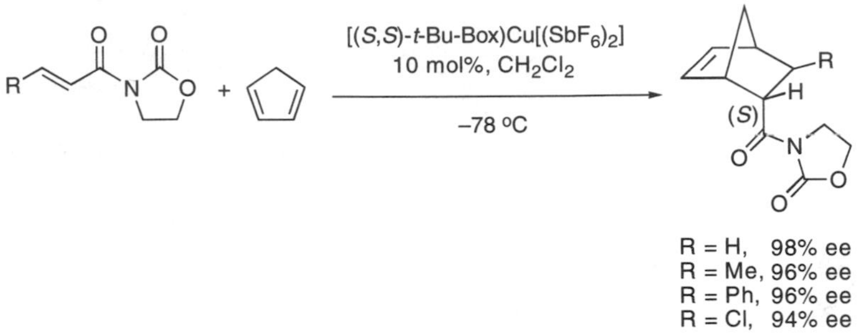

Figure 4.7. The s-cis conformation of the dienophile is strongly favored over the s-trans conformer when bound in a bidentate fashion.

Semiempirical calculations of copper(II) (Box)Cu(sub)$^{2+}$ complexes (where sub = bidentate dicarbonyl substrate) suggest a similar degree of distortion from square-planar geometry to that observed in the structure of [(S,S)-t-Bu-Box)]Cu(H$_2$O)$_2$$^{2+}$ in **Figure 4.6**.[14] To simplify the understanding of the stereochemical outcome of the reaction (**Equation 4.1**), the nearly planar backbone of [(S,S)-t-Bu-Box)]Cu^{2+} is represented by a horizontal line with the protruding substituents extending forward (**Figure 4.8**). Upon coordination of the imide substrate to [(S,S)-t-Bu-Box)]Cu^{2+}, the

Figure 4.8. Proposed models for the stereochemical control in the Diels–Alder reaction. a) The bisoxazoline ligand is represented by the horizontal line and protruding t-Bu groups. b) A quadrant diagram for the same facial selectivity.

prochiral faces of the dienophile are in distinctly different steric environments, with the *Si* face (i.e., the *Si* face of the alkene) blocked by the protruding *t*-Bu group of the *t*-Bu-Box ligand. Attack of the diene, therefore, takes place at the less-hindered *Re* face with excellent enantioselectivities (**Equation 4.1**).[19] This model for facial selectivity with bis(oxazoline) can be applied to several asymmetric reactions involving chelating dicarbonyl substrates and the (Box)Cu^{2+} family of catalysts, provided that the copper geometry is close to square-planar in the catalyst–substrate adducts.

When using such a model for bisoxazoline-based catalysts, caution must be exercised, because changes in metal geometry or coordination number will impact the enantioselectivity. For example, substitution of Cu(II) for its neighbor Zn(II) forming [(*S*,*S*)-*t*-Bu-Box)]Zn(SbF$_6$)$_2$, resulted in formation of the opposite enantiomer of the product in 56% ee (**Equation 4.2**). Likewise, the same facial selection was observed for [(*S*,*S*)-Ph-Box)]Zn(SbF$_6$)$_2$, which gave the product in 92% ee. In changing the metal geometry from distorted square-planar (Cu^{2+}) to tetrahedral (Zn^{2+}), the *Re* face is now shielded (**Figure 4.9**), in contrast to the Cu(II) complex (**Figure 4.8**).[20] Furthermore, switching to the magnesium catalyst, which is also proposed to have the tetrahedral geometry, again gives the (*R*) enantiomer.[21] It is also noteworthy that the Fe(III)–bis(oxazoline) catalysts, which are likely octahedral, gave the same stereochemistry as the Zn(II) and Mg(II) complexes.[21]

Equation 4.2

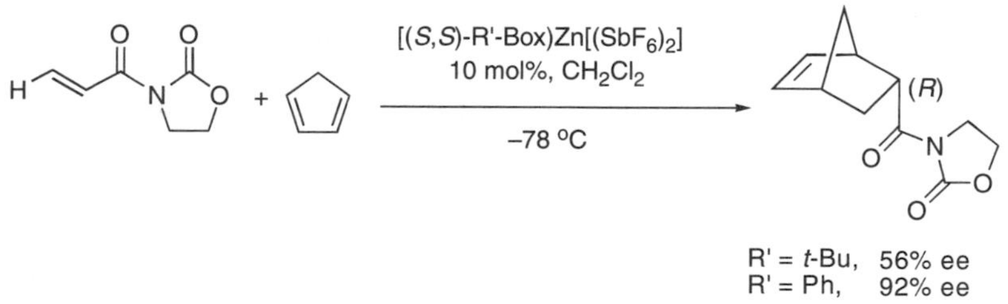

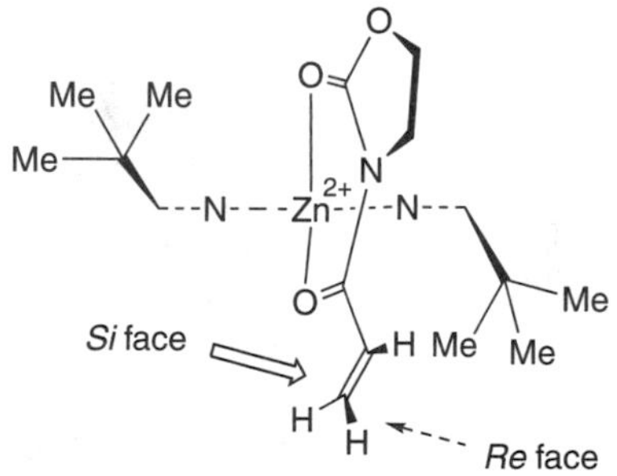

R' = *t*-Bu, 56% ee
R' = Ph,   92% ee

Figure 4.9. Proposed model for the stereochemical control in the Diels–Alder reaction. The bisoxazoline ligand is represented by the horizontal line and protruding *t*-Bu groups.

4.2.1.b Indirect Control of Nucleophile Trajectory via Substrate Relay

Analysis of the stereochemical outcome of a reaction with bis(oxazoline)-based catalysts becomes more difficult when the trajectory of the attacking nucleophile is parallel to the plane defined by the nitrogens of the ligand and the metal center. An interesting example of this behavior arises in the Tsuji–Trost asymmetric allylic-alkylation reaction.[22,23] The key steps and intermediates are shown in **Figure 4.10**. Reaction of either enantiomer of the allylic-acetate with a (Box)Pd(0) intermediate leads to formation of a η^3-allyl complex. The nucleophile, $KCH(CO_2Me)_2$, attacks the η^3-allyl termini from the face opposite the palladium. In the presence of the chiral bisoxazoline ligand, the nucleophile preferentially attacks one of the two termini-leading to the major enantiomer, while the minor enantiomer, is generated by attack at the other terminus.

Insight into the factors that control the regioselectivity of the nucleophilic attack and, therefore, the enantioselectivity, were gleaned from X-ray crystallographic studies of $[(Box)Pd(\eta^3\text{-allyl})]^+$ derivatives. In the structure of the parent cation, $[(Bz\text{-}Box)Pd(allyl)]^+$, the palladium has the expected square-planar geometry and the (Bz-Box)Pd metallacycle is nearly planar (**Figure 4.11**). In contrast, in the structure of the intermediate, $[(Box)Pd(1,3\text{-diphenylallyl})]^+$, there are significant nonbonded interactions between a benzyl substituent of the ligand and the adjacent phenyl group of the η^3-allyl. To minimize the steric repulsion between these groups, the metallacycle has distorted markedly from planarity, as depicted in **Figure 4.11**. Furthermore, there is a lengthening of the Pd–N and Pd–C bond lengths associated with these interacting groups (**Figure 4.11**).

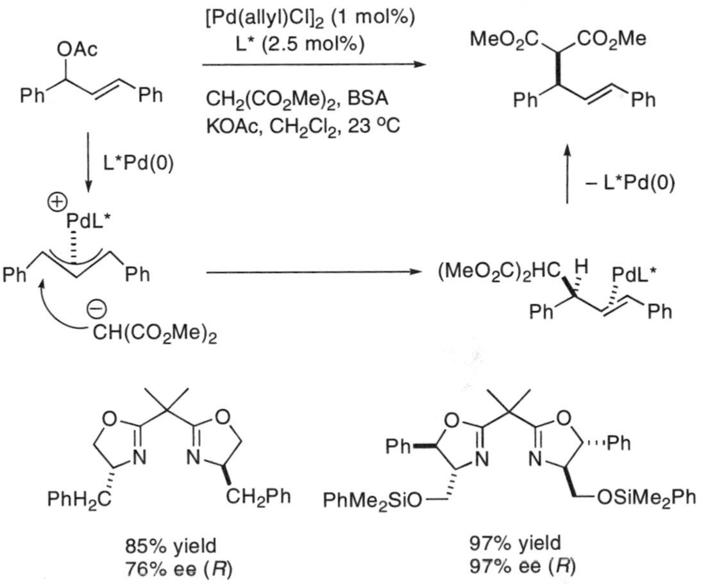

Figure 4.10. The mechanism of the asymmetric allylation reaction (BSA = bissilylacetamide).

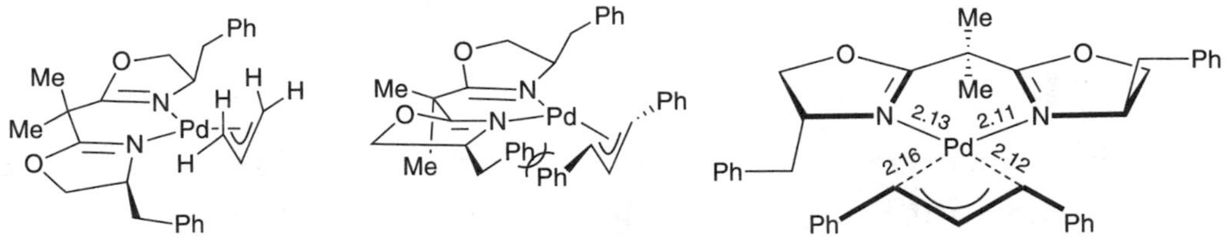

Figure 4.11. Drawings of (Bz-Box)Pd(allyl)⁺ and (Bz-Box)Pd(1,3-diphenylallyl)⁺ based on the X-ray crystal structures. Bond distances are in angstroms (Å).

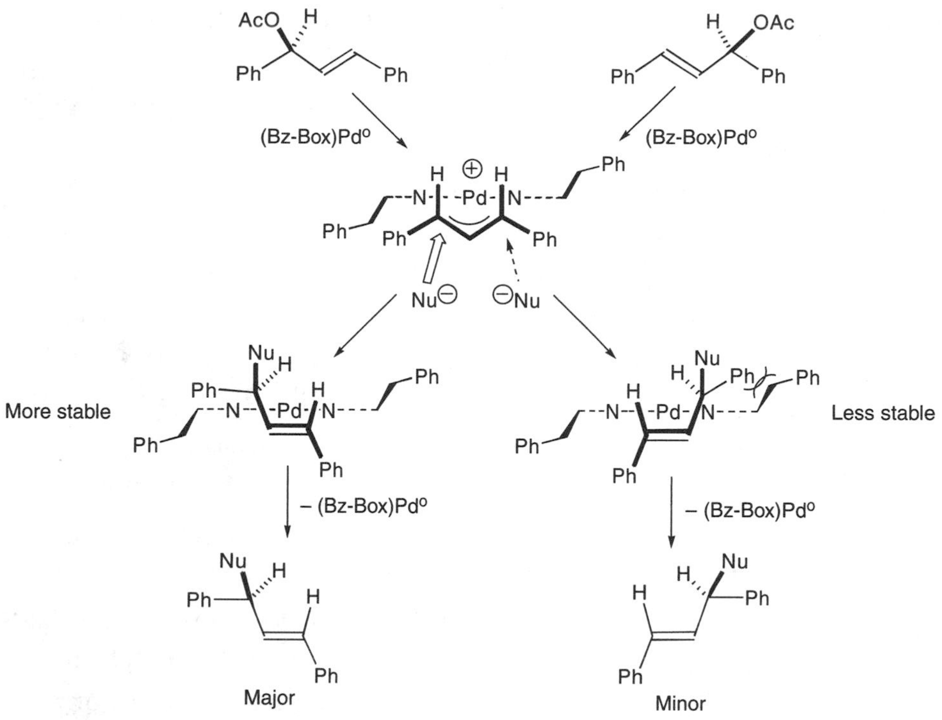

Figure 4.12. Model for stereoselection in the asymmetric allylation reaction. Attack of the nucleophile at the elongated C–Pd bond leads to the less-strained olefin complex.

Based on the configuration of the product formed on reaction of [(Bz-Box)Pd(1,3-diphenylallyl)]$^+$ with the nucleophile, attack has occurred on the allyl terminus with the elongated Pd–C bond. It has been suggested that attack at the more crowded position is favored. As the reaction occurs, the forming carbon–carbon double bond rotates into the plane containing the Pd and nitrogen atoms of the ligand (**Figure 4.12**). Attack at the more-hindered position results in a reduction of nonbonded interactions in the transition state leading to the palladium–olefin complex.[10] Inspection of the proposed geometries of the olefin products highlights the differences in stability of the diastereomeric products (**Figure 4.12**). The idea that the regioselectivity of the nucleophilic attack is controlled by the interaction of the bound substrate with the ligand can be contrasted with the more commonly invoked rational for asymmetric induction, which involves direct interaction of the nucleophile with the chiral ligand.

4.2.2 BINOL-Based Lewis Acids

An important class of "privileged ligands" is based on the parent atropisomeric BINOL (1,1'-bi-2-naphthol) (**Figure 4.13**).[24] BINOL is fairly stable to racemization because of the steric hindrance to rotation about the pivotal 1,1'-bond. Barriers to racemization of BINOL have been determined to be 37–38 kcal/mol.[25,26] No racemization was detected on heating to 100 °C in a dioxane/water mixture for 24 h, although addition of strong acid or base will cause racemization at this temperature.[26] BINOL derivatives bind well to many main-group, transition-metal, and lanthanide compounds to form catalysts that exhibit exceptionally high levels of enantioselectivity.[27–30] On binding to metals, the BINOLate ligand forms a rigid chiral metallacycle. **Figure 4.14** illustrates the skewed conformation of the BINOLate metallacycle viewed from the O–M–O plane.

Numerous highly enantioselective catalysts employ the parent BINOL ligand. In addition, many BINOL derivatives have been introduced and used successfully in asymmetric catalysis.[27–29,31] In understanding how asymmetry is conveyed in BINOLate-based catalysts, it is easiest to examine systems substituted at the 3,3'-positions, because the substituents typically extend in the direction of the substrate binding site (see below). In contrast, the hydrogens at the 3,3'-positions of the parent BINOLate ligand do not protrude into the metal binding site (**Figure 4.14**). Consequently, the transfer of asymmetry with BINOLate-based catalysts is poorly understood. It is possible that the skewed conformation of the BINOLate ligand causes an electronic asymmetry at the metal center that influences the transfer of asymmetry by controlling the orientation of substrate binding (see Section 4.6).[32] Also complicating our understanding of how the BINOLate ligand functions is the fact that many BINOLate-based catalysts involve early transition metals, such as titanium or zirconium, or lanthanides. In these systems, the metal coordination number and aggregation state

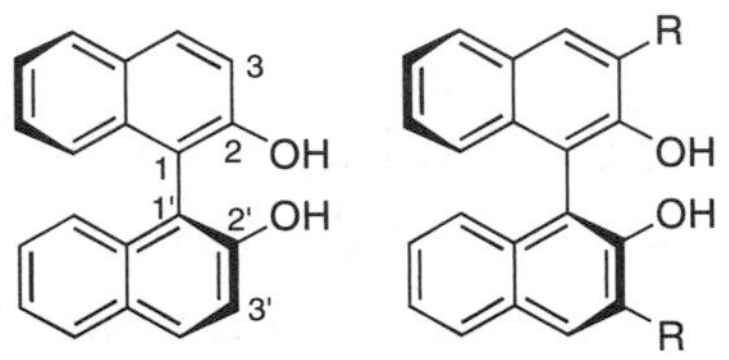

Figure 4.13. Structures of the parent BINOL and common derivatives substituted at the 3,3'-positions.

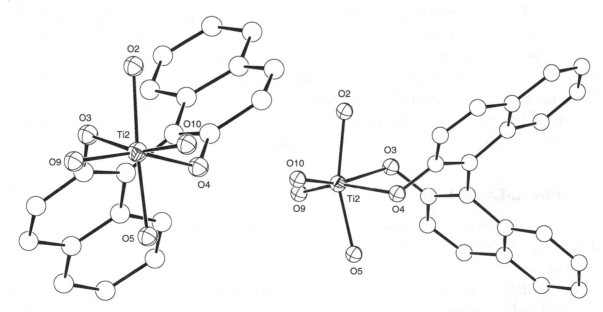

Figure 4.14. Patrial structure illustrating the skewed conformation of the BINOLate metallacycle bound to a pseudo-octahedral titanium center.

are often unknown. Thus, to simplify the discussion of asymmetric induction with BI-NOLate ligands, we will employ a stoichiometric Lewis acid based on boron. Boron compounds of this type are known to be tetrahedral, and as such, are monomeric.

A series of boron Lewis acids bearing ligands derived from BINOL were prepared for the Diels–Alder reaction with peri-hydroxyquinones (**Figure 4.15**).[33,34] The peri-hydroxyquinone substrates chelate to boron, limiting the conformational flexibility of the substrate–Lewis acid adduct. The enantioselectivity in the cycloaddition reaction will be controlled by the ability of the 3,3′-substituents on the BINOL to shield one of the prochiral faces of the dienophile. Thus, when the 3,3′-dimethyl-BINOL ligand was utilized (R = Me, **Figure 4.15**), the enantioselectivity of the anthraquinone product was 70%.[33] Use of the 3,3′-diphenyl BINOL (R = Ph), in which the phenyl substituents project further forward, better shielding one face of the dienophile, gave > 98% ee of the product.[33] The stereochemical outcome can be understood by examination of the drawings in **Figure 4.16**. Although there is no structural information on 3,3′-diphenyl-BINOLate complexes of boron, a partial structure of this ligand bonded to a tetrahedral metal center is shown. From these views it can be seen that the phenyl groups

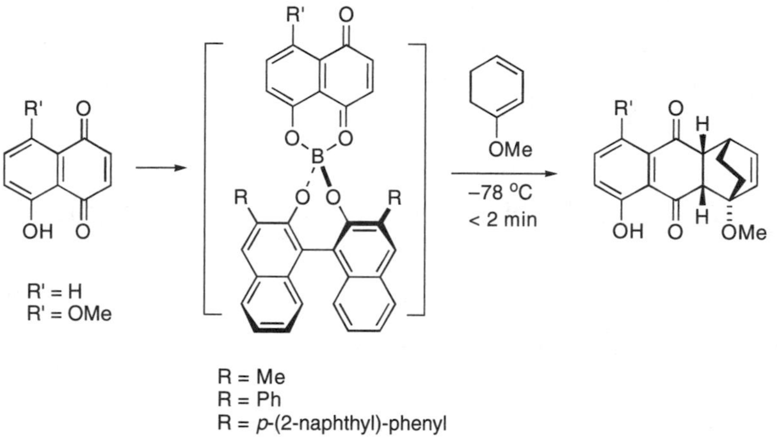

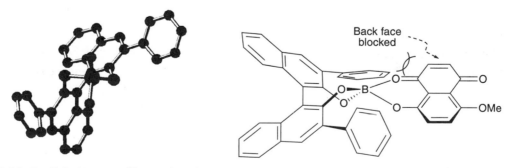

Figure 4.15. Asymmetric Diels–Alder of peri-hydroxyquinone promoted by BINOLate–boron Lewis acids.

Figure 4.16. Partial structure illustrating the skewed conformation of the 3,3′-diphenyl-BINOLate ligand bound to a tetrahedral metal center (left). The same (Ph$_2$-BINOLate)B fragment viewed from a different angle, with the substrate added (right). The back face of the reactive double bond is rendered inaccessible by the distant phenyl group.

block adjacent quadrants, leaving two quadrants open. Upon binding of the substrate, the distant phenyl group shelters the reactive double bond from attack on the back face, while the front face is exposed, allowing the reaction to take place.

This Diels–Alder reaction was employed in the synthesis of (+)-diepoxin σ, which possesses antifungal, antibacterial, and antitumor activity.[34] The dieneophile was the peri-hydroxyquinone, where R' = OMe, and the diene was cyclopentadiene, which is smaller than the 2-methoxy cyclohexadiene used to generate the anthraquinone derivative in **Figure 4.15**. As a result, the 3,3'-diphenyl-BINOL-based system did not exhibit the same degree of steric bias and gave low ee. Further extending the length of the 3,3'-substituents was found to improve enantioselectivity. Thus, when R = *p*-(2-naphthyl)phenyl, the enantioselectivity increased to 93%.

4.3 Asymmetric Induction with Chiral Metallocene Catalysts

Chiral group IV metallocenes have been found to polymerize terminal olefins with a high degree of stereocontrol over the polymer architecture.[35] It is not surprising, then, that resolved versions of these metallocenes have also been successfully applied in asymmetric catalysis.[36] The chiral ligand scaffolding around the metal center in the ansa-metallocene catalysts is conformationally limited by the ethylene tether, and therefore, fairly well-defined. This characteristic facilitates prediction of the sense of enantioselection in the reactions of these C_2-symmetric complexes. The most commonly applied ansa-metallocenes in asymmetric catalysis are based on group IV metals and ethylene-1,2-bis(η^5-4,5,6,7-tetrahydro-1-indenyl), abbreviated EBTHI (**Figure 4.17**).[36,37]

An example of the use of such ansa-metallocene catalysts is the asymmetric reduction of imines to afford the corresponding amines. As little as 0.02 mol% catalyst can be used to generate amine products in high yields and enantioselectivities (**Figure 4.18**).[38-43] The silane employed in this reaction is typically phenylsilane (H_3SiPh) or a soluble polymer, polymethylhydrosiloxane (PMHS), with the repeat unit $[SiH(Me)O]_n$.[43]

The reaction is initiated by activation of (EBTHI)TiF$_2$ (0.5–5 mol%) with phenylsilane, methanol, and pyrrolidine. Under these conditions, an active titanium(III) hydride is likely formed.[38] As illustrated in **Figure 4.19**, the mechanism is postulated to

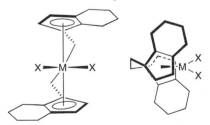

Figure 4.17. Two views of the chiral ansa-metallocene complex (EBTHI)MX$_2$.

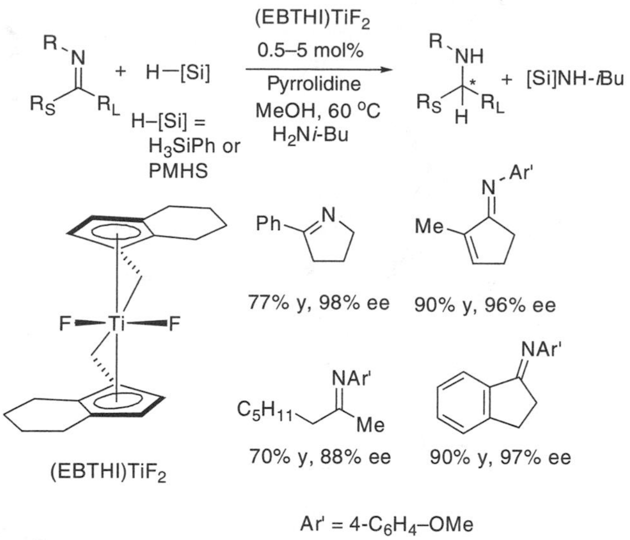

Figure 4.18. Asymmetric hydrosilylation of imines with (EBTHI)TiF$_2$ precatalyst (Ar' = 4-C$_6$H$_4$-OMe).

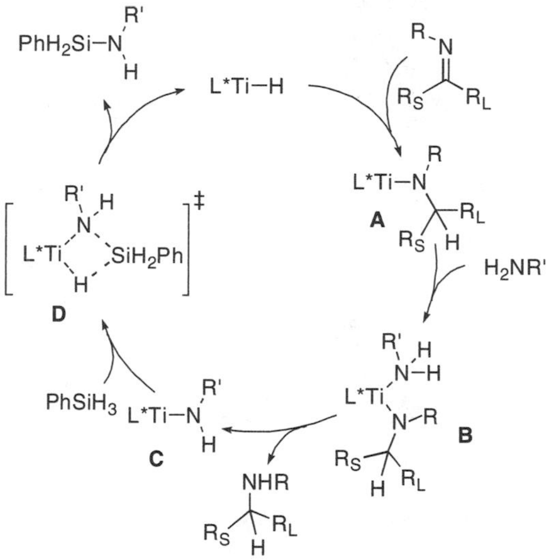

Figure 4.19. Proposed mechanism for the asymmetric hydrosilylation of imines.

proceed by enantioselective insertion of the prochiral ketimine into the titanium(III) hydride to generate a secondary amide (**A**). In the absence of added amine, cleavage of the intermediate amide **A** was the turnover-limiting step.[39] Slow addition of a primary amine, such as isobutylamine resulted in higher TOF, because the isobutylamine exchanges with the bulkier chiral amide to liberate the enantioenriched amine product. The sterically less-encumbering primary amide is more easily cleaved by the silane through a four-centered transition state[44] to regenerate the titanium hydride and form an equivalent of silylisobutylamine.

The model for asymmetric induction for this process is illustrated in **Figure 4.20**, where R$_L$ and R$_S$ are the large and small substituents on the ketimine.[40] Most ketimine substrates can have two isomeric forms, *syn* and *anti*, which give rise to four isomeric

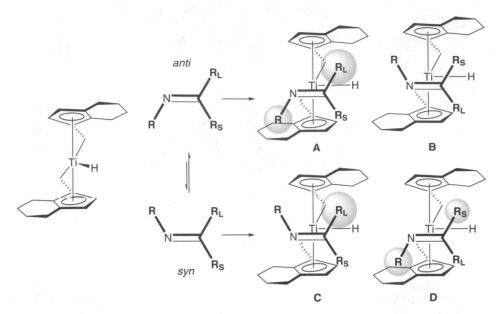

Figure 4.20. Proposed transition states for the asymmetric reduction of *anti* and *syn* ketimines. The spheres highlight steric clashing between the imine substituents and the catalyst.

transition states (**Figure 4.20**). In the transition states **A** and **B** with the *anti* isomer, **A** is destabilized, because interactions of R and R_L with the metallocene ring system. In contrast, in transition state **B** only R_S is directed toward the ligand. In the transition states with the *syn* ketimine, **C** is more favorable than **D**, which has both R_S and R interacting with the ligand. The stereochemistry of the products formed from cyclic ketimine, which proceed through the *syn* transition states, is consistent with this model.

With acyclic ketimines, reaction of the *anti* isomer is predicted to take place through transition state **B** and the *syn* isomer through transition state **C**. This dual-path reaction manifold may result in a decrease in the observed enantioselectivity, because these transition states give opposite configurations of the product. This does not appear to be a serious limitation, however. The aliphatic imine in **Figure 4.18** has an *anti/syn* ratio of 2.6:1, but undergoes reduction with an enantioselectivity of 88%.[43] This result suggests that the *syn* and *anti* ketimines equilibrate under the reduction conditions and the reaction proceeds primarily via transition state **B**.

4.4 Transmission of Asymmetry to Substituents within a Metal–Ligand Adduct

In the early days of asymmetric catalysis, it was thought essential to position the stereogenic center or centers of the ligand as close as possible to the reactive site on the metal.[45] Initial chiral phosphines were therefore prepared that were chiral at phosphorus, as exemplified by the ligand CAMP (**Figure 4.21**).[46] In hindsight, it is surprising that rhodium complexes of this ligand were fairly enantioselective, given the high

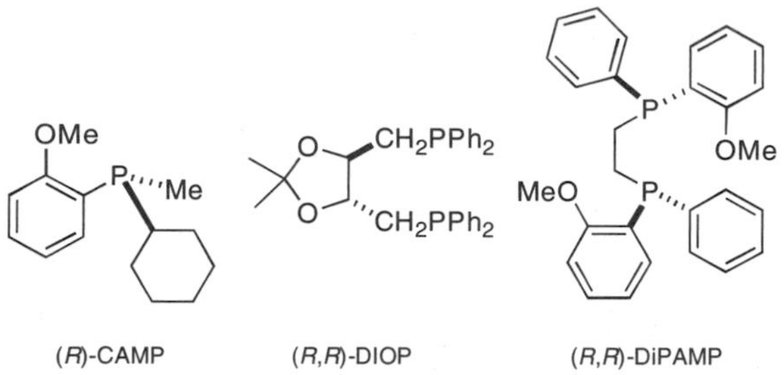

Figure 4.21. Important early chiral phosphine ligands.

degree of rotational freedom about the phosphorus–carbon and the rhodium–phosphorus bonds. When DIOP,[1,47] a chiral bidentate phosphine with two remote stereogenic centers (**Figure 4.21**), was introduced chemists began to appreciate the significant role that ligand conformations would play in enantioselective catalysis. The performance of catalysts based on DIOP inspired the development of the chelating ligand DiPAMP (**Figure 4.21**), which subsequently became the basis of a commercial synthesis of L-DOPA, a medication used in the treatment of Parkinson's disease.[45]

4.4.1 Conformations of Chiral Metallacycles: BINAP and TADDOL

Given the intense interest in asymmetric catalysis, it is not surprising that the synthesis and screening of new chiral ligands remains a very active area of research. Of the myriad enantioenriched ligands that have been employed in asymmetric catalysis, a few ligand classes seem to give consistently high levels of enantioselectivities with an array of metal centers. Because of the effectiveness of these ligands across a broad range of reactions, they have been referred to as "privileged ligands."[24]

4.4.2 BINAP-Based Catalysts

One of the prominent ligands in this class is the commercially available axially chiral BINAP (**Figure 4.22**).[48] This ligand is chiral, because of the skewed conformation of the binaphthyl rings, and it is stable to racemization due to the high barrier to rotation about the central C–C bond. Although best known for its application in catalytic enantioselective hydrogenations of olefins and ketones,[49,50] metal complexes of BINAP have found many applications in asymmetric reactions. BINAP can accommodate metals of different radii by rotating about the central aryl–aryl bond and the 2,2'-P–C

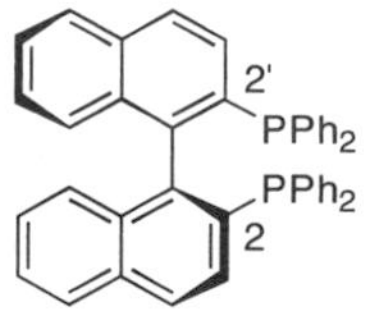

Figure 4.22. Structure of (S)-BINAP.

bonds, but once coordinated, the conformation of the ligand is constrained in the rigid metallacycle.

The efficiency with which BINAP-based catalysts transmit asymmetry to substrates results from transmission of the binaphthyl backbone axial chirality to control the position and orientation of the diastereotopic phenyl substituents. The skewed conformation of the seven-membered metallacycle is chiral and is set by the binaphthyl axial stereochemistry. The phenyl groups assume pseudoaxial and pseudoequatorial positions, as can be seen in the X-ray crystal structure of the BINAP ligand bound to ruthenium (**Figure 4.23**).[51] A stereoview is illustrated in **Figure 4.24**. The pseudoequatorial phenyl rings extend forward past the metal center and the pseudoaxial phenyl groups orient away from the metal. BINAP derivatives, of which there are many, and several chiral bidentate phosphines (see below) display these characteristics, but do not exhibit such pronounced differences in the positions of the pseudoequatorial and pseudoaxial phenyl groups. It is the combination of the protruding equatorial phenyl groups and the degree of orientation of the axial phenyl groups away from the substrate binding site that is thought to be responsible for the exquisite enantiocontrol in this ligand system.[52] The fact that substituents in the *para* positions of the phenyl groups often impact the enantioselectivity of the catalyst is consistent with this hypothesis, although electronic effects may also be important.[53] Additionally, an edge-face (CH-π) interaction between the phosphorus-bound aryl groups contributes to the differentiation of the pseudoaxial and pseudoequatorial

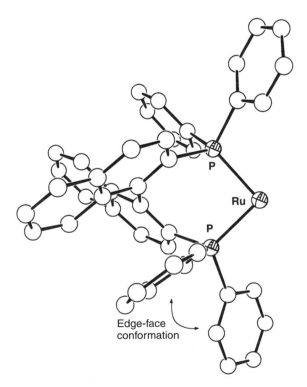

Figure 4.23. Partial structure of [(S)-BINAP)]Ru(O$_2$C-*t*-Bu)$_2$, with carboxylate ligands omitted for clarity. The pseudoequatorial phenyl rings are thrust forward, while the pseudoaxial phenyls are oriented away from the ruthenium center.

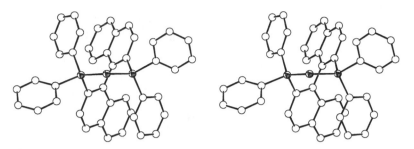

Figure 4.24. Stereoview of the metal–BINAP core from Figure 4.23, illustrating the chiral environment of the ligand.

phenyl groups. As such, the equatorial phenyl groups protrude with the face of the aryl forward. This important interaction can be seen in the structure in **Figure 4.23** and in the stereoview in **Figure 4.24**.[54,55]

To understand the control of asymmetry in catalysts bearing the BINAP ligand, the enantioselective hydrogenation reaction with the precatalyst (BINAP)Ru(O$_2$CR)$_2$ from **Figure 4.23** will be examined. The resultant catalyst exhibits high enantioselectivity with a wide range of substrates.[56] Detailed investigations of this system have led to an understanding of the reaction mechanism,[57] which is believed to proceed through a mono-hydride intermediate, and the proposal of a reliable model for the asymmetric induction.[57,58]

As documented in many examples, the best substrates for the asymmetric hydrogenation with this catalyst possess a pendant Lewis basic site that, along with the functional group undergoing reduction, binds to the ruthenium center, forming a chelate.[56,59] This is illustrated for one of the most important classes of substrates for this catalyst, the β-keto ester (**Equation 4.3**). Reduction of these substrates provides access to products analogous to those formed in aldol condensations. In the model for asymmetric induction, the ketone carbonyl is thought to coordinate in a π-fashion to the ruthenium, while the ester binds through an oxygen lone pair (**Figure 4.25**). Due to these different binding modes, the ketone component places greater steric demands on the ruthenium–BINAP system and occupies the least-hindered site on the catalyst, as shown in **Figure 4.25**. In the structure on the left, the interaction of the ketone with the protruding equatorial phenyl group of the BINAP, drawn in bold, destabilizes this diastereomer. In contrast, the right-handed structure places the ketone carbonyl in the quadrant with the axial P-phenyl group, which is directed away from the substrate. Here, it is assumed that the major diastereomer of the substrate–catalyst adduct will generate the major enantiomer of the product. This assumption, however, is not always valid.[60,61]

$$R \overset{O}{\underset{}{\|}} \overset{O}{\underset{}{\|}} OR' \quad \xrightarrow[\text{H}_2]{\text{[(}R\text{)-BINAP]Ru(OAc)}_2} \quad R \overset{OH}{\underset{}{}} \overset{O}{\underset{}{\|}} OR'$$

Equation 4.3

Another example involves the catalytic intramolecular hydroacylation of 4-pentenals with cationic rhodium catalysts containing bidentate chiral phosphines

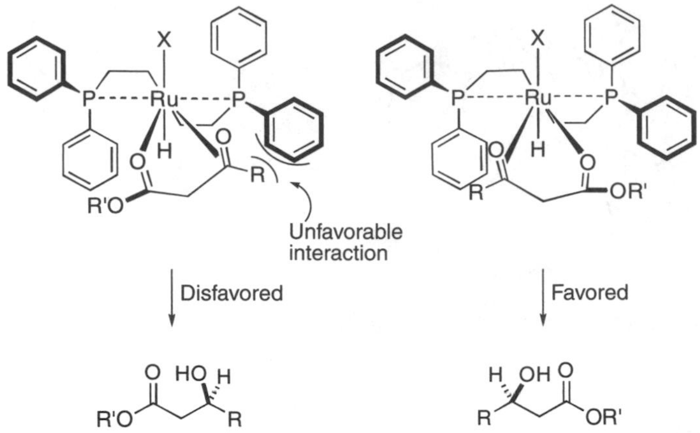

Figure 4.25. Proposed chelation of β-keto esters to the [(R)-BINAP]Ru moiety in the asymmetric hydrogenation (shown in Equation 4.3). The equatorial phenyl groups of the BINAP ligand are shown in bold. The unfavorable interaction of the ketone with the equatorial phenyl on the left destabilizes this diastereomer.

(**Equation 4.4**).[62–65] This reaction takes place at room temperature in acetone or methylene chloride with turnover numbers as high as 500, and isolated yields around 90%. The catalyst precursor, [(S)-BINAP)]Rh(NBD)ClO$_4$[48] (where NBD = norbornadiene), was briefly treated with hydrogen to prepare the catalyst, [(S)-BINAP)]Rh(sol)$_2$ClO$_4$ (where sol = solvent). Based on labeling experiments,[66–68] the mechanism is believed to involve initial oxidative addition of the acyl hydrogen. Like the ruthenium hydrogenation catalyst described above, the coordinatively unsaturated rhodium(III) intermediate possesses open coordination sites after the oxidative-addition step. Coordination of the pendent olefin with the C–C double bond situated roughly parallel to the Rh–H axis gives the proper alignment for the olefin insertion step (**Figure 4.26**). Binding of the olefin to the rhodium will occur to minimize steric interactions between the bound substrate and the equatorial phenyl groups. In the diastereomer on the left in **Figure 4.26**, the bulky *tert*-butyl group lies in the quadrant with the pseudoaxial phenyl group, which is directed away. In contrast, the other diastereomer results in severe steric interactions between the protruding pseudoequatorial phenyl group and the *tert*-butyl group, destabilizing this isomer. The next step in the mechanism is the olefin insertion and subsequent reductive elimination to generate the product with high enantioselectivity.[62]

Equation 4.4

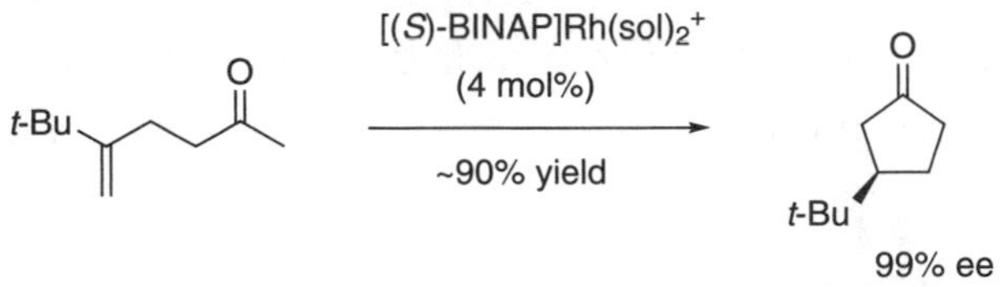

The quadrant model outlined in Section 4.1.2 has been used to facilitate the prediction of the facial stereoselectivity in reactions with catalysts bearing chiral bidentate phosphines (**Figure 4.27**). The protruding equatorial phenyl groups occupy the

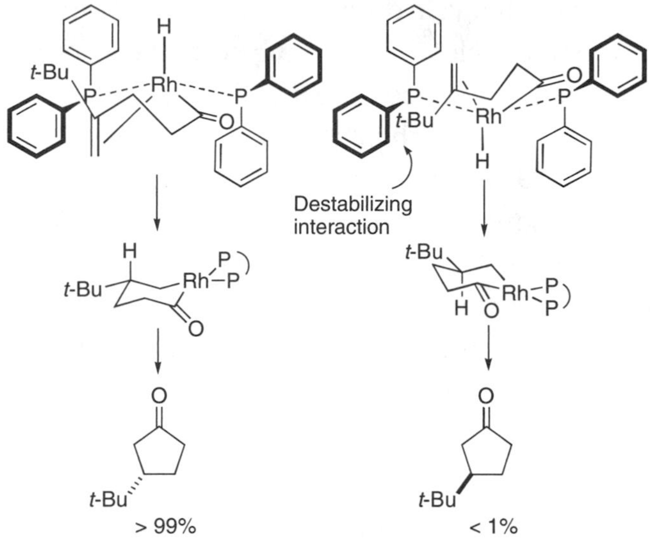

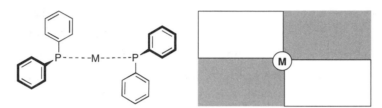

Figure 4.26. Diastereomeric binding modes of the unsaturated acyl substrate to the cationic [(S)-BINAP]Rh(H) moiety and cyclization. The equatorial phenyl groups are shown in bold.

Figure 4.27. Model for the spacial positions of the ligand. The shaded regions are occupied by the pseudoequatorial phenyl groups (bold) that protrude forward.

two shaded diagram quadrants, blocking encroachment into these regions. The quadrants containing the pseudoaxial phenyl groups are unshaded. In order to minimize nonbonded interactions, the substrate avoids contact with the pseudoequatorial phenyl groups (i.e., the shaded regions of the model).

Other types of bidentate phosphine ligands exhibit a similar disposition of the phenyl substituents to BINAP. For example, the five-membered metallacycle formed on binding of (2S,3S)-bis(diphenylphosphino)butane [(S,S)-chiraphos, **Figure 4.28**] to metals adopts a twisted conformation. Although five-membered rings are normally conformationally flexible, the conformation of chiraphos is controlled by the methyl substituents on the ligand backbone, which are pseudoequatorially disposed in the lowest-energy conformation. The conformational twist of the metallacycle causes the phenyl groups to adopt pseudoaxial and pseudoequatorial positions (**Figure 4.28**). Based on structural information,[69] the pseudoequatorial and pseudoaxial phenyl groups of chiraphos are not as dissimilar in their positions when compared to those of BINAP.[70] As a result, the pseudoequatorial phenyl groups do not protrude to the same extent as they do in BINAP complexes, nor are the pseudoaxial phenyl rings of chiraphos directed away from the metal to the same degree. The sp^2 hybridized carbons of the seven-membered metallacycle derived from BINAP induce a conformation that is more skewed than metallacycles with sp^3 hybridized carbons in the backbone, and thus create greater distinction between the axial and equatorial phenyl groups.

Both (S)-BINAP and (S,S)-chiraphos have been employed in the asymmetric acylation reaction in **Figure 4.29**. With simple substrates, where R = Me or i-Pr, both ligands exhibited the same sense of stereoselectivity, as would be expected by the models outlined in **Figures 4.27** and **4.28**. In these models, both ligands have the same twist sense, with the equatorial phenyl groups blocking the same quadrants (**Figure 4.27**). With bulky groups, such as t-Bu and SiMe$_3$, however, the (S)-BINAP- and (S,S)-chiraphos-based catalysts give the opposite stereoselectivity (**Figure 4.29**). Assuming that the mechanisms are the same, the opposite facial selectivity suggests that the more-stable diastereomer of the acyl-hydride complex in the BINAP system becomes the less-stable diastereomer with the chiraphos catalyst.

Although the structures of the relevant intermediates are unknown, based on the ground-state crystal structures and the production of opposite enantiomers of the product with these catalysts, it has been suggested that the position of the axial phenyl groups in chiraphos, which are not directed away from the metal center to the same extent as in BINAP, cause greater steric interactions with the t-Bu group (**Figure 4.26**).[62] Admittedly, however, it is surprising that what is perceived to be a small difference in ligand conformation results in formation of the opposite enantiomer of the product.

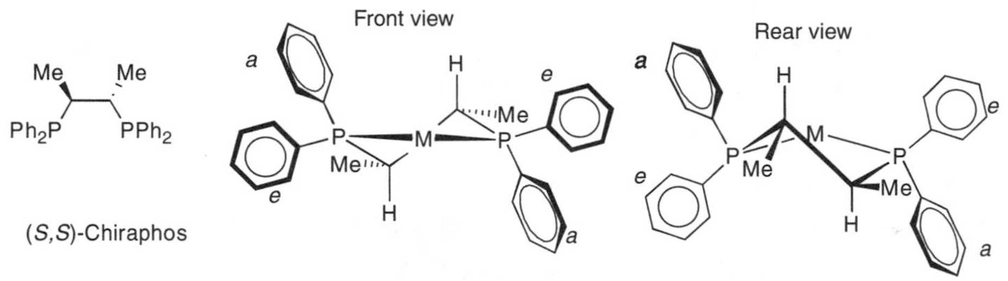

Figure 4.28. Front view of (S,S)-chiraphos with pseudoaxial (a) and pseudoequatorial (e) phenyl groups labeled.

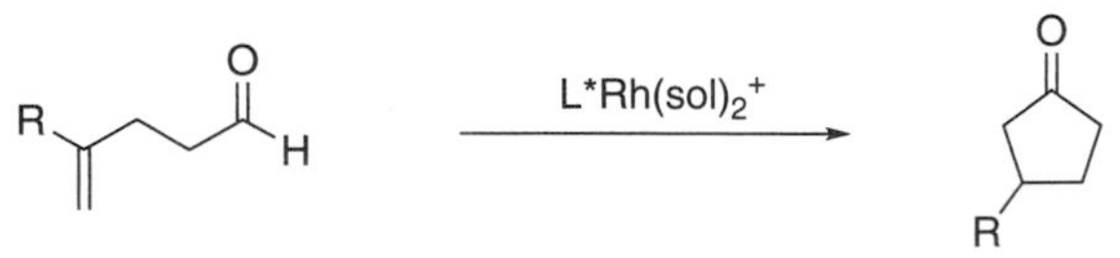

Entry	R	(S)-BINAP % ee (configuration)	(S,S)-chiraphos % ee (configuration)
1	Me	78 (S)	42 (S)
2	i-Pr	60 (S)	45 (S)
3	t-Bu	99 (S)	29 (R)
4	SiMe$_3$	99 (S)	8 (R)
5	CO$_2$i-Pr	99 (S)	11 (R)
6	COt-Bu	94 (S)	63 (R)
7	COPh	94 (S)	64 (R)

Figure 4.29. Comparison of the enantioselectivities between BINAP and chiraphos catalysts in the hydroacylation reaction.

In examples with R = ketone or ester, it is possible that the carbonyl of the R group can undergo different secondary interactions with the (chiraphos)Rh- and (BINAP) Rh-based catalysts, resulting in formation of the opposite enantiomer. Comparisons of entries 1–4 in **Figure 4.29** serve as reminders that subtle differences in conformations of chiral ligands can result in substantial changes in enantioselective reactions.

4.4.3 TADDOL-Based Complexes

A family of ligands that has frequently been employed in asymmetric catalysis with excellent results is the tartaric acid-based TADDOL ligands (**Figure 4.30**). Following the introduction of the parent TADDOL ligand, a large number of analogs with different aromatic substituents and ketal moieties were introduced.[71] TADDOL-based catalysts have found applications in a variety of Lewis acid catalyzed enantioselective reactions.[71] Upon deprotonation, the ligands bind well to early transition metals, forming strong M–O bonds[72] in the resulting TADDOLate complexes (**Figure 4.30**).

The conformational behavior in these ligands is similar to BINAP and chiraphos as outlined above. Based on the X-ray structural studies of unbound TADDOL ligands and of metal-bound TADDOLate complexes (**Figure 4.31**),[73,74] the features that make these ligands exceptional have been surmised. Upon coordination of TADDOL to metals, a trans-fused bicyclo[5.3.0]decane ring system results. The stereocenters of the dioxolane ring are too distant from the metal center in TADDOLate complexes to directly impact the enantioselectivity. Nevertheless, they impart a strong conformational preference on the adjacent metallacycle, positioning the diastereotopic aromatic groups such that they assume a pseudoequatorial/pseudoaxial arrangement. The pseudoaxial phenyl groups are positioned antiperiplanar to the neighboring dioxolane C–H bonds. In this fashion, the asymmetry of the stereogenic centers is relayed forward, toward the metal and the substrate-binding sites. There are some important differences, however, between the TADDOL and BINAP ligand systems. In the case of BINAP, the diastereotopic aryl groups are attached to the metal-bound phosphorus centers, while in TADDOL, these groups are bonded to the adjacent carbons. As a result, the aryl groups of the TADDOLate ligands are further removed from the metal, and the pseudoequatorial phenyl groups do not extend past the metal, as they do in BINAP complexes. Thus, it is the pseudoaxial aryl groups that are believed to be the dominant stereocontrolling element in TADDOL.[75]

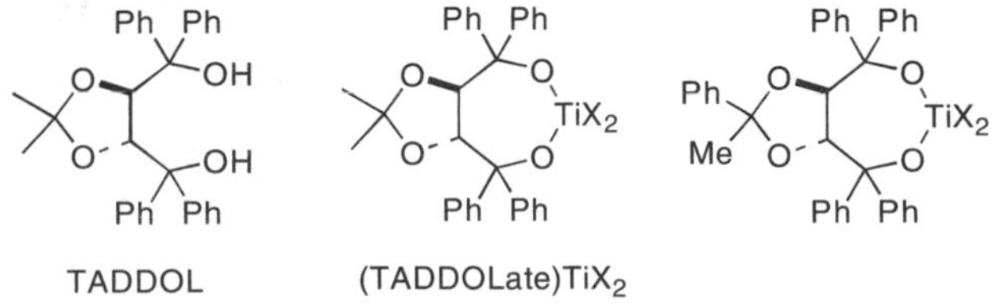

TADDOL   (TADDOLate)TiX$_2$

Figure 4.30. The parent (R,R)-TADDOL ligand and (TADDOLate)Ti complexes (X = halide or isopropoxide).

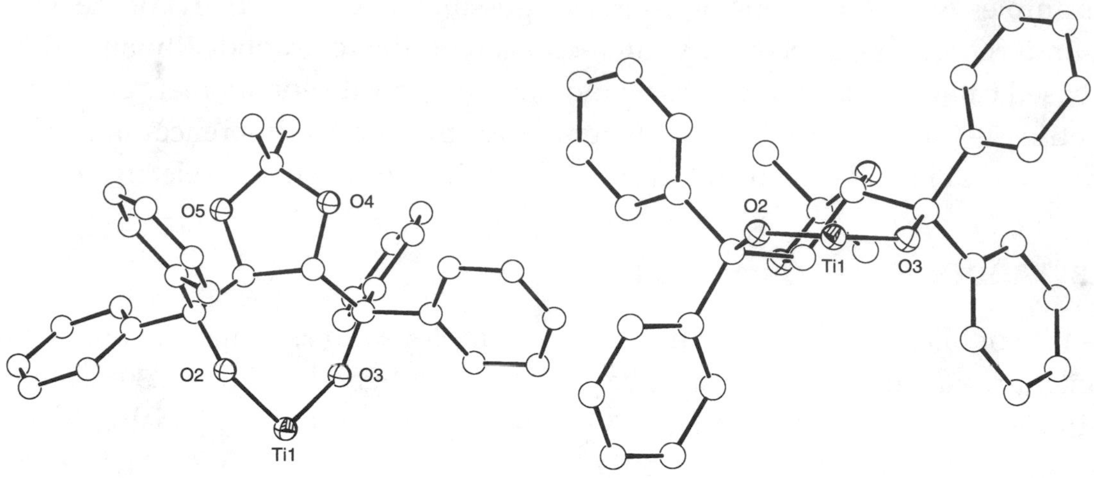

Figure 4.31. Partial structure of Ti(TADDOLate) moiety, illustrating the chiral environment about the ligand. The left view illustrates how the asymmetry in the metallacycle causes two rings to protrude forward and two to orient away from the metal center.

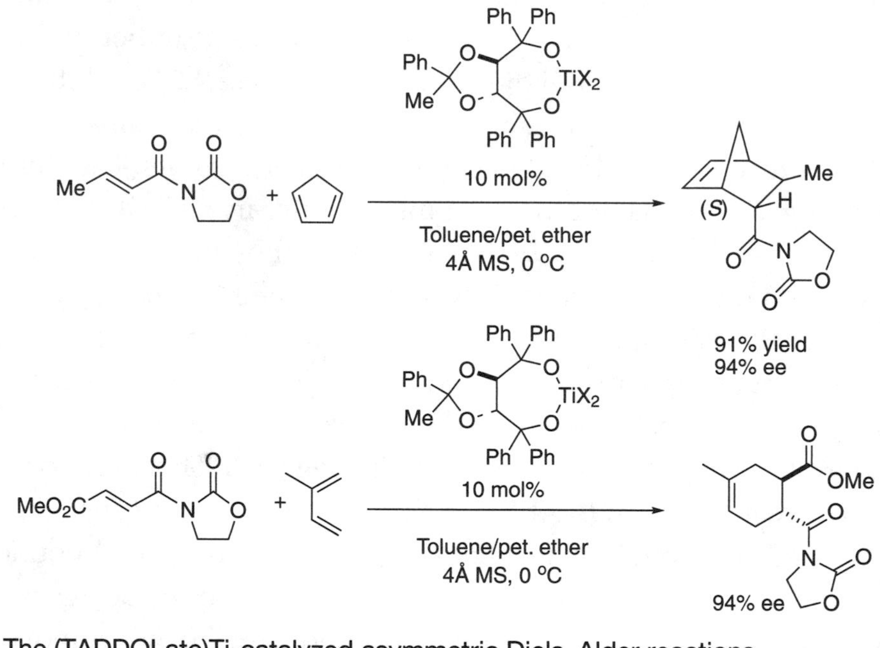

Figure 4.32. The (TADDOLate)Ti-catalyzed asymmetric Diels–Alder reactions.

Early examples of applications of TADDOL ligands in asymmetric catalysis were centered on the catalytic asymmetric addition of alkyl groups to aldehydes[71,76–78] and the asymmetric Diels–Alder reaction (**Figure 4.32**),[79] both using (TADDOLate)Ti-based complexes as the Lewis acids. The rapid exchange of alkoxide ligands and the notorious propensity of early metal alkoxides to aggregate through the formation of bridging alkoxides complicate studies of alkoxide-based catalysts.[72]

The story of the (TADDOLate)Ti-catalyzed Diels–Alder reaction[79] nicely illustrates a number of important points about the challenges of elucidating details of reaction mechanisms and the transmission of asymmetry. Although the mechanism of this Lewis acid catalyzed process has been extensively investigated, some controversy remains about the geometry of the active substrate–catalyst adduct. In the reaction of the *N*-acyloxazolidinone with cyclopentadiene in the presence of (*R*,*R*-TADDOLate)TiCl$_2$, the expected *endo* adduct predominates with the (*S*)-configuration (**Figure 4.32**).[79] In this process, the dienophile is believed be to activated by chelation to titanium, giving rise to an octahedral substrate–catalyst adduct. In support of this binding mode, an X-ray crystallographic study of a substrate–catalyst adduct showed the substrate chelating to titanium. In the structure, the oxygens of the TADDOLate ligand and the coordinated substrate were located in the equatorial plane and the mutually trans chlorides were axial (**Figure 4.33**).[74] It was proposed that this adduct was the reactive intermediate that led directly to product on reaction with the diene.[74,80,81] Because the reactive C–C double bond of the dienophile is distant from the chiral environment of the TADDOLate ligand in this complex, it was unclear how the chiral ligand could bias the facial selectivity in the Diels–Alder cycloaddition. Furthermore, in the geometry of the substrate adduct in the crystal structure, the carbonyl oxygens of the dieneophile are each trans to the basic, electron-donating alkoxides of the TADDOLate ligand. This geometrical arrangement results in a low degree of activation of the substrate. Stronger activation would be expected if the carbonyl oxygen adjacent to the reactive double bond were trans to chloride.

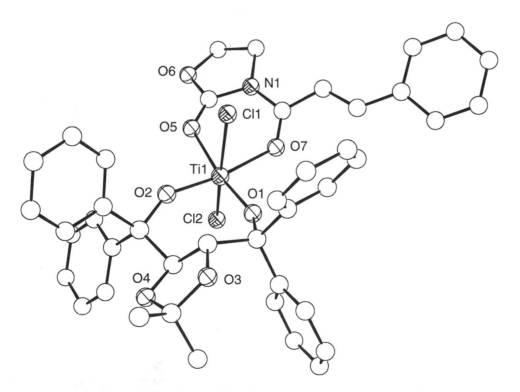

Figure 4.33. Structure of the (TADDOLate)TiCl$_2$ bound to the substrate.

Skepticism about the intermediacy of this adduct inspired further investigations into the solution behavior of the catalyst–substrate adduct in this reaction.[82,83] High-level quantum-chemical calculations were also employed to probe the degree of activation of the substrate with different arrangements of the ligands on the octahedral titanium center.[84] Analysis of the coordination modes of the substrate to the titanium center indicates that there are five possible geometries for this adduct, as illustrated in **Figure 4.34**. The bold vertical ligands in these structures represent the axial phenyl groups of the TADDOLate ligand. In isomer **A**, all of the oxygens lie in the equatorial plane. Diastereomers **B** and **C** have the carbonyl oxygen adjacent to the double bond trans to chloride while in **D** and **E** it is trans to a TADDOLate oxygen. In-depth NMR studies of (TADDOLate)TiCl$_2$ and the substrate indicted that only three of the five possible diastereomers were present in solution in a ratio of 70:24:6.[82] The major diastereomer in solution, **A** (**Figure 4.34**), was that observed in the crystal structure and calculated to be the most stable.[82,84] By positioning the TADDOLate ligand and the substrate in the equatorial plane, nonbonded interactions are minimized.

Experimental evidence into the precise geometry of the two less-abundant substrate–catalyst adducts was inconclusive. Perhaps more important than the geometries of the ground-state structure of these adducts is the degree of activation of the

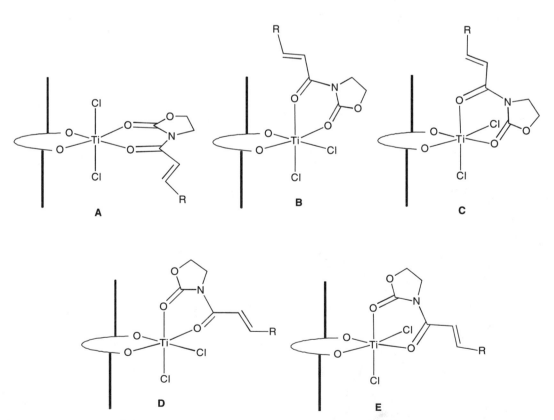

Figure 4.34. Five possible binding modes of the *N*-acyloxazolidinone to [(*R*,*R*)-TADDOLate]TiCl$_2$. The bold lines represent the pseudoaxial aryl groups of the TADDOLate ligand.

substrate by the titanium in geometries **A–E** in **Figure 4.34**, and hence their relative reactivity. High-level calculations on model systems predict that the bound substrate exhibits the greatest degree of activation in geometries **B** and **C**, because the carbonyl flanking the olefin is trans to the weakly donating chloride. The substrate in complexes **D** and **E** is predicted to be less reactive than in **B** and **C**, but more reactive than **A**.[84] Thus, although **A** is the most-stable diastereomer, as reflected in the high solution concentration and in the theoretical studies, it is believed to be the least reactive.

A plausible scenario for the reaction would involve reversible binding of the substrate to the titanium center. It has been found that exchange between bound and free substrate with (TADDOLate)TiCl$_2$ has a low barrier (15 kcal/mol) and proceeds via a dissociative pathway. This exchange is much faster than the cycloaddition, which is the stereoselectivity- and rate-determining step.[82] Based on the data outlined above, it has been proposed that the reaction falls under Curtin–Hammett conditions (see Section 2.5).[85] The reaction is funneled through a less-stable substrate–catalyst adduct, because of the greater reactivity of this intermediate and the fast equilibration of the diastereomeric substrate adducts.

In the substrate adducts **B** and **C** in **Figure 4.34**, the *Si* face is shielded by the pseudoaxial TADDOLate aryl group and the attack of the diene takes place from the *Re* face, as shown in **Figure 4.35**

This example illustrates several important points concerning the study of catalytic asymmetric reactions, including: 1) the way in which different arrangements of the ligands on the metal generate diastereomeric complexes that are chiral at the metal, 2) how different catalyst–substrate geometries exhibit unequal reactivities, and 3) how reactivity can be controlled by the Curtin–Hammett principle. It also demonstrates the application of a variety of experimental and computational tools to gain insight into various aspects of a reaction mechanism.

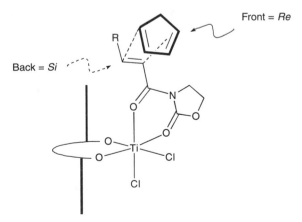

Figure 4.35. A possible transition state for the asymmetric Diels–Alder reaction where attack of the cyclopentadiene takes place on the α-carbon *Re* face of the dienophile, along the Ti–Cl axis.

4.5 Environments from Ligands That Form Diastereotopic Complexes

One of the challenges in the synthesis of chiral ligands is often the resolution of the components from which the ligand is assembled. It is frequently the case that resolutions of enantiomers is the most time-consuming step in the iterative cycle of catalyst screening, which involves the synthesis of new ligands, the preparation of catalysts, and the screening and evaluation of new catalysts. Therefore, if ligands can be designed such that a single stereocenter in the unbound ligand controls the generation of subsequent ligand stereocenters on binding to the metal, the number of stereocenters in the ligand–metal adduct can be increased, without additional resolutions.[86] Examples of this include selective coordination of diastereotopic lone pairs of the ligand to the metal, coordination of diastereotopic groups, and binding of the ligand in a diastereoselective fashion, such that the metal center becomes chiral. Furthermore, as outlined below, this interaction can serve to reduce the degrees of rotational freedom within the metal–ligand scaffold, thus extending and rigidifying the asymmetric environment of the catalyst.

4.5.1 Coordination of Diastereotopic Lone Pairs

The development of new enantioselective catalysts is, in part, dependent on the discovery and identification of innovative methods to transfer chirality from the catalyst to the substrate. A powerful method for catalyst development that has rarely been employed is diastereoselective binding of ligands to metals. In the process of binding the ligand to the metal, new stereogenic centers are formed either on the ligand or at the metal center. The diastereoselective coordination of an sp^3 hybridized lone pair on nitrogen or sulfur, for example, can increase the number of stereogenic centers upon coordination of the ligand.

Chiral diamines have been used in combination with zinc as catalysts for the asymmetric hydrosilylation of ketones with polymethylhydrosiloxane, $[-SiMe(H)O-]_n$. The precatalysts for this reaction were readily formed on combining diamines and dimethylzinc (**Figure 4.36**). In this reaction, the catalyst, which is proposed to be (diamine)ZnH$_2$, delivers hydride to the carbonyl carbon, resulting in a zinc alkoxide. The alkoxide reacts with the Si–H bonds of the polymer to generate a silyl ether, which undergoes hydrolysis on isolation to form the alcohol.[87]

When the reduction of acetophenone was performed using catalyst derived from diamines **A–C** (**Table 4.1**) and dimethylzinc, the (R)-alcohol product was observed.[88] Coordination of **A** to dimethylzinc results in formation of a five-membered metallacycle with fewer degrees of freedom than the six- or seven-membered metallacycles formed from **B** and **C**. When the short-chain ligand N,N'-ethylenebis(1-phenylethylamine) (**A**) was bound to dimethylzinc, the resulting complex showed

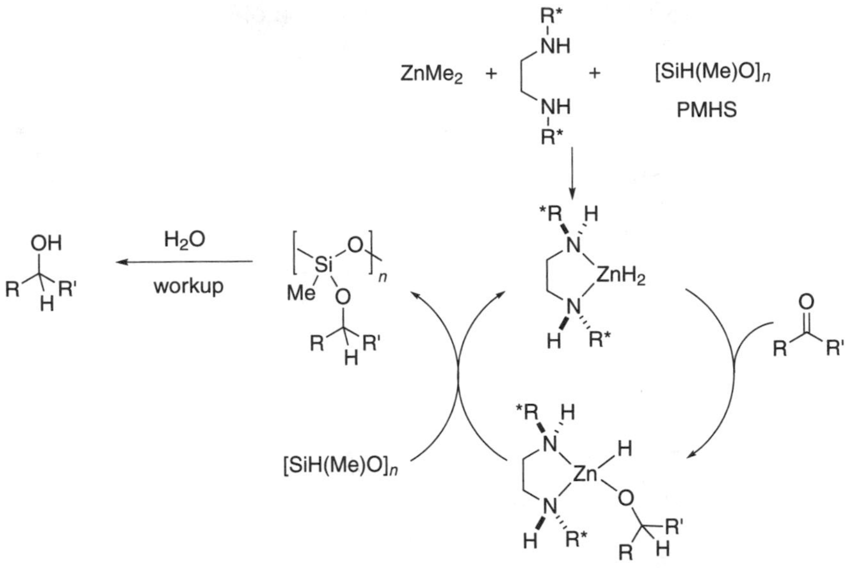

Figure 4.36. Proposed catalytic cycle in the zinc-diamine-catalyzed asymmetric reduction of ketones.

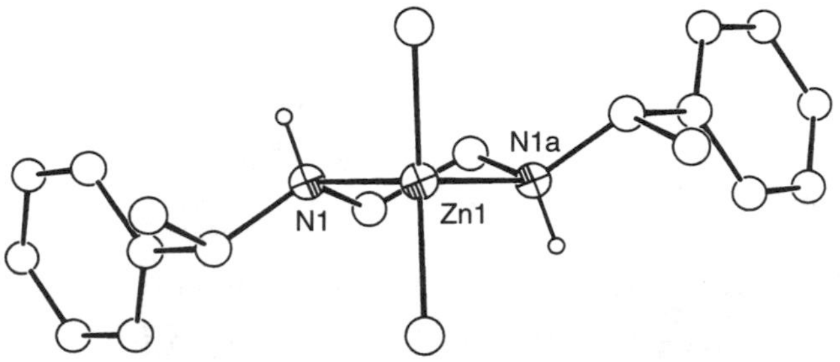

Figure 4.37. Structure of diamine **A** (see Table 4.1) coordinated to ZnMe$_2$, illustrating the configurations of the stereogenic nitrogen centers.

that stereochemical information in the chiral N-phenylethyl groups was effectively relayed to the nitrogen centers, which are rendered configurationally stable upon coordination. The X-ray crystal structure of this complex is illustrated in **Figure 4.37**, where the configuration of the nitrogens are S.[87] The stereogenic nitrogen centers are held in close proximity to the zinc, and have a significant impact on the enantioselectivity of the catalyst (79% enantioselectivity, **Table 4.1**). Increasing the tether length of the diamine backbone results in poor control of the configuration of the coordinating nitrogens, giving rise to diastereomeric zinc complexes. The mixture of diastereomers exhibits low levels of enantioselectivity in the asymmetric reaction (< 17% ee). Interestingly, the dimethylzinc complex of chiral diimine **D** was also shown to catalyze the reaction, but it gave the reduced product with the *opposite absolute configuration* in modest enantioselectivity (48% ee, **Table 4.1**). In this case, the nitrogen atoms in diimine **D** are sp^2 hybridized and the side chains are the only groups responsible for controlling the transfer of asymmetry in the reaction.

Table 4.1. Enantioselective reduction of ketones by polymethylhydrosiloxane in the presence of chiral zinc catalysts

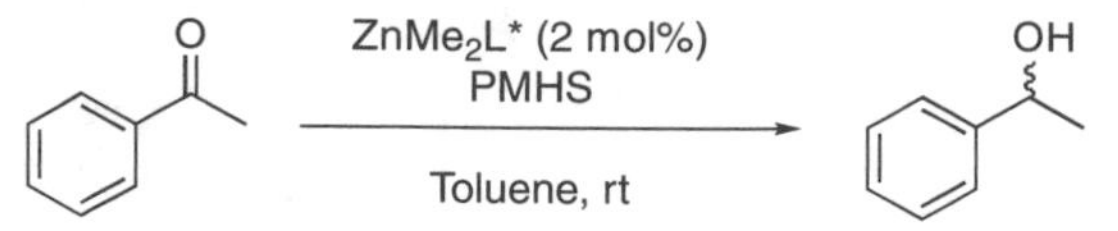

L*		Yield (%)	ee % (configuration)
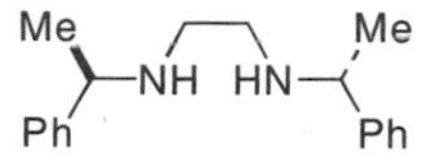	A	100	75 (R)
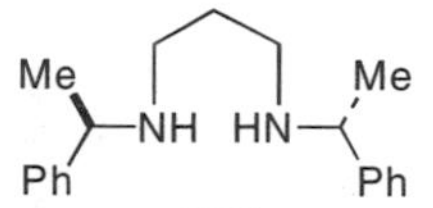	B	100	5 (R)
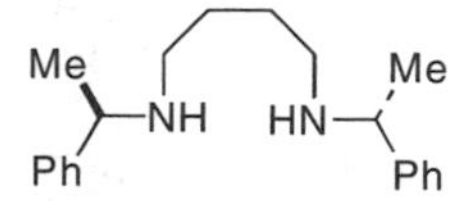	C	100	17 (R)
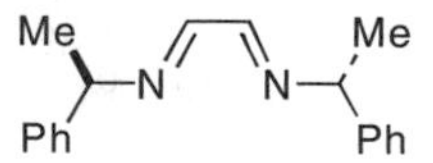	D	98	48 (S)

4.5.2 Binding of Diastereotopic Groups

Coordination of a diastereotopic atom to a coordinatively unsaturated metal can be important in enantioselective catalysis with many ligands. For example, bis(sulfon-amide)-based catalysts can be highly active and enantioselective in a number of reactions, including the asymmetric alkylation of aldehydes (**Equation 4.5**).[89,90] This reaction allows preparation of functionalized secondary alcohols with excellent levels of enantioselectivity.[91]

Equation 4.5

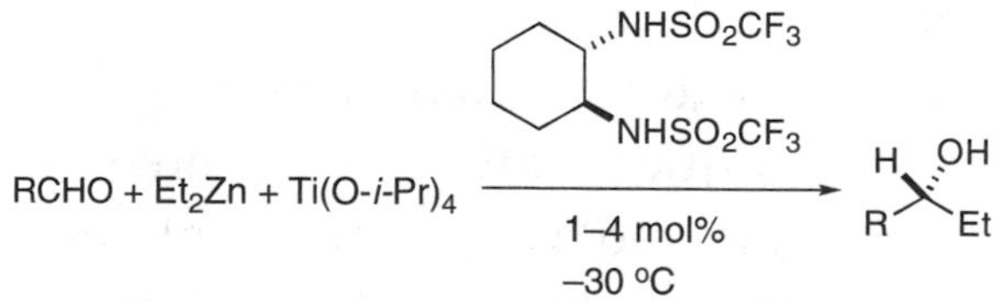

R	ee % (yield %)
Ph	98 (99)
PhCH=CH	85 (99)
PhCH$_2$CH$_2$	99 (92)
n-C$_5$H$_{11}$	78 (99)

Titanium complexes of the bis(sulfonamide) ligands have been proposed to be the active species in this reaction.[89,90,92,93] X-ray structure studies show long dative bond-

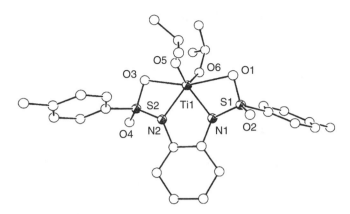

Figure 4.38. Illustration of the bonding in bis(sulfonamido)Ti(O *i*-Pr)$_2$ based on the crystal structure. The dative Ti–O(sulfonyl) bond distances are between 2.2 and 2.4 Å, compared to the covalent Ti–O(alkoxide) distances, which are 1.7–1.8 Å.

ing interactions between one of the diastereotopic oxygen atoms on each sulfonamide group and titanium.[93] This Ti–O(sulfonyl) interaction causes the sulfur atoms to be stereogenic and presumably rigidifies the C_2-symmetric ligand scaffold, as seen in **Figure 4.38**. In a sense, coordination of one of the diastereotopic sulfonyl oxygens to titanium extends the chiral environment of the ligand.[89,93–97]

Bis(sulfonamide) ligands are also important catalyst components in other reactions, including the asymmetric amination of *N*-acyloxazolidinones with magnesium,[98] the asymmetric cyclopropanation of allylic alcohols with zinc,[99–102] and the asymmetric Diels–Alder reaction with aluminum.[103] Magnesium bis(sulfonamido) complexes are Lewis acidic and presumably also coordinate the sulfonyl oxygens. Evidence for such interactions is found in related systems.[104] It is unlikely, however, that aluminum[103] and especially zinc[105] complexes of the bis(sulfonamide) ligands will bind the sulfonyl oxygens. In these systems, conformational gearing of the sulfonyl group with the chiral ligand backbone may extend the asymmetric environment, causing the high levels of enantioselectivity observed with these catalysts.

4.6 Electronic Asymmetry of Coordination Sites

Another example of a family of ligands that form an additional stereocenter upon coordination is the phosphorus-sulfur-based ligands in **Figure 4.39**. These ligands have been successfully employed in the asymmetric allylation with palladium[106,107] and the asymmetric hydrogenation (**Equation 4.6**) and hydrosilylation reactions with rhodium.[108]

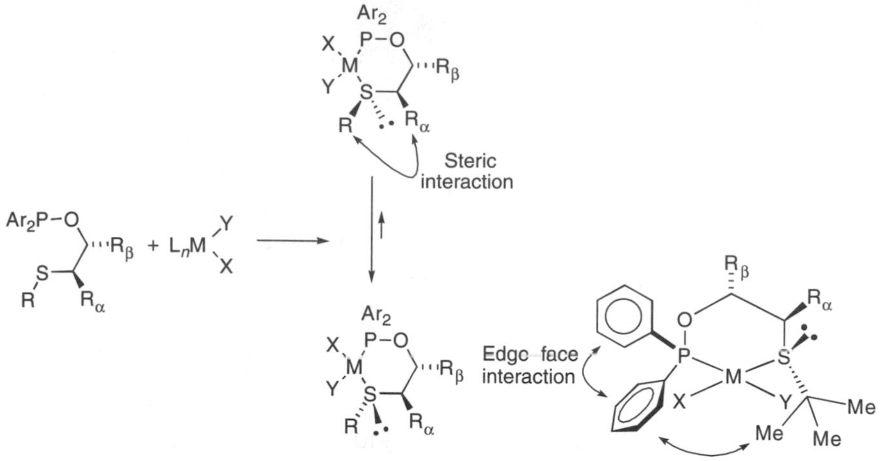

Figure 4.39. Binding of the P-S-ligand to metals forms two diastereomers that can interconvert by sulfur inversion or detachment of sulfur and recoordination.

Equation 4.6

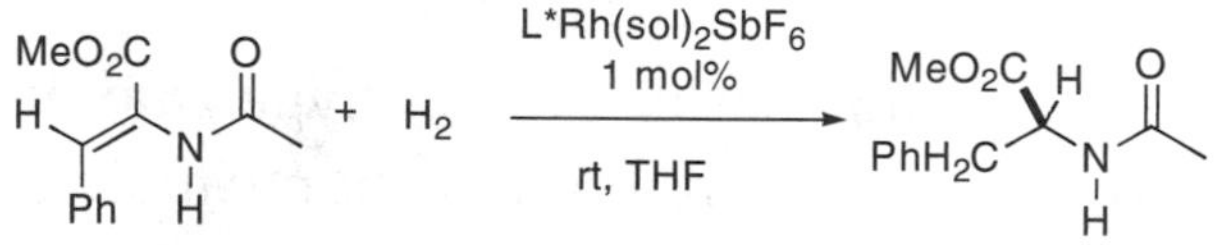

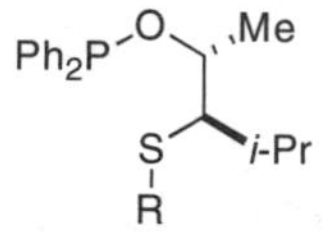

R	ee (%)
Ph	95
$3,5\text{-}C_6H_3\text{-}Me_2$	97
CH_2Ph	90
Cy	NR
t-Bu	68

Metals bind to these ligands in a bidentate fashion, coordinating phosphorus and one of the diastereotopic lone pairs on sulfur (**Figure 4.39**). The resultant diastereomers (with the S-R oriented up and S-R oriented down) can readily interconvert by inversion of the coordinated sulfur lone pair, which has a low energy barrier, or detachment of the sulfur and re-coordination of the other lone pair. The degree of steric interaction between R_α and the sulfur-substituent R controls the extent to which the metal preferentially binds one of the diastereotopic lone pairs over the other and, therefore, the orientation of the S-R. The direct attachment of the stereogenic sulfur center to the metal, and its proximity to the substrate-binding site, make control of the orientation of the S-R substituent crucial to the effective stereochemical communication between the catalyst and the substrate. It has been found that the sulfur substituent is disposed in a pseudoaxial position to avoid nonbonded interactions with the R_α substituent, although electronic effects cannot be ruled out.[108,109] For bulky sulfur substituents, such as t-Bu, this group then influences the orientation of the P-phenyl groups. To minimize interactions between the sulfur substituent and the neighboring P-phenyl, the cis P-phenyl adopts an edge-on conformation, causing the remaining P-phenyl group to adopt a face-on orientation.[109,110]

An additional feature of mixed heteroatom donor ligands, such as the P-S-ligand in **Figure 4.39**, is that they electronically differentiate their respective trans binding sites. In this case, the phosphorus ligand is a stronger trans donor than sulfur, making ligand Y more labile than ligand X (**Figure 4.39**). The trans effect will impact how chelating substrates bind, with the weakest trans donor opposite the strongest.

Using this P-S-ligand system, a detailed study of the mechanism of the asymmetric hydrogenation was undertaken. The catalyst–substrate adduct was independently synthesized and characterized by X-ray crystallography. A drawing of the structure is illustrated in **Figure 4.40**. By ^{1}H and ^{31}P NMR spectroscopy only a single diastereomer was observed in solution out of the four possibilities.

Beginning with the diastereomer in the crystal structure (**A, Figure 4.41**), the unobserved diastereomers can be drawn by reversing the positions of the carbonyl and double bond (**B**) and coordination of the opposite face of the olefin (**C** and **D**).

The observed diastereomer is favored by a combination of the steric constraints of the bound ligand and the electronic factors caused by the unequal trans influence of the phosphorus and sulfur donors. The bulky sulfur substituent and the neighboring edge-on aryl group block the lower two quadrants around the metal center, leaving the upper quadrants sterically unencumbered (**Figure 4.42**). The left half and right half of the diagram are electronically inequivalent by virtue of the differing trans influence of the phosphorus and sulfur, which favor binding the weaker trans donor of the substrate opposite the strongly donating phosphorus center. The trans influence is represented by the checkered regions of the quadrant diagram. In the substrate, the C–C double bond has a stronger trans influence than the carbonyl lone pair and

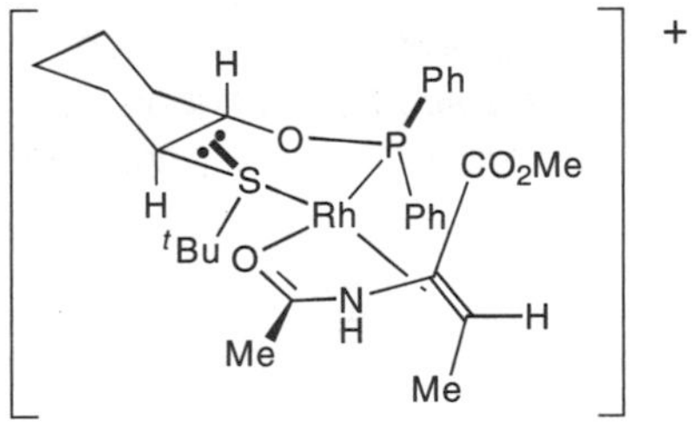

Figure 4.40. Drawing of the intermediate formed on coordination of the substrate to rhodium (based on the an X-ray crystal structure of this complex).

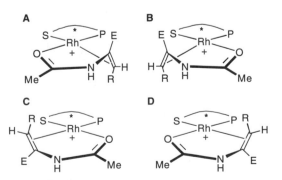

Figure 4.41. Possible diastereomers of the substrate coordinated to (P-S)Rh$^+$ moiety (**Figure 4.40**). Diastereomer **A** is predicted to be the most stable and is observed in the crystal structure.

will prefer to bind opposite the weaker sulfur donor (**Figure 4.42**) With this system, the stereochemistry of the catalyst–substrate complex correlates with the observed stereochemistry of the reduced product. As discussed in Section 2.5, this is not always the case in asymmetric hydrogenation reactions.

Like the (BINAP)Ru-catalyzed hydrogenation of β-keto esters, the mechanism is proposed to involve initial loss of solvent ligands, chelation of the substrate as described above, followed by oxidative addition of H_2 (**Figure 4.43**). Note that in the oxidative-addition product, the ligand with the strongest trans influence (phosphorus and the hydrides) are positioned trans to the three weakest donors. In intermediate **B**, the olefin is correctly aligned for the migratory-insertion step, which establishes the product stereochemistry. The resultant alkyl hydride (**C**) undergoes reductive

Figure 4.42. Model for the occupied space and electronic effects in the rhodium complex of a P-S-chelating ligand. The darker areas are occupied by the ligand. The checkered regions lines indicate the binding site trans to the strong trans influence of the phosphine ligand.

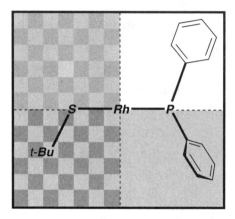

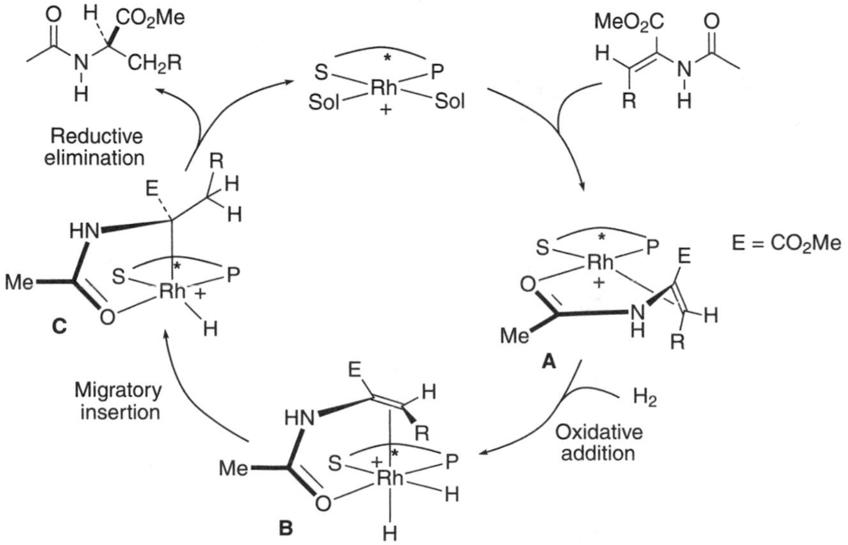

Figure 4.43. Proposed mechanism of the (P-S)Rh⁺-catalyzed hydrogenation of α-acylaminoacrylates.

elimination to form the stereocenter. Subsequent dissociation of the product, which does not bind well to the rhodium, then occurs.

To recap, the transmission of asymmetry in this system is dominated by the orientation of the S-R substituent, which is, in turn, dictated by the adjacent stereocenter. The orientation of the S-R group further biases the neighboring cis P-phenyl substituent. To minimize the transannular interaction, the cis P-phenyl orients an edge toward the substrate binding site, sterically blocking this quadrant. Additionally, the trans influence dictates diastereoselectivity in substrate binding and the oxidative addition of dihydrogen.

4.7 Steric Bias of Configurationally Dynamic Ligand Stereocenters

4.7.1 Ligands with Chiral Relay Groups Not Directly Connected to the Metal

A clever catalyst design based on stereochemically dynamic functional groups employs the dihydropyrazole moiety, where a fixed stereocenter controls the configuration of two fluxional nitrogen stereocenters (**Figure 4.44**).[111] The stereogenic nitrogen bearing the CH_2R substituent is remote, but exhibits a marked impact on the catalyst enantioselectivity. The stereochemistry of the N-methyl group of the amino alcohol is *anti* to the adjacent stereocenter to avoid an unfavorable eclipsing interaction. Data from reactivity studies suggest that the ligand is bound in a tridentate fashion and likely adopts a meridinal geometry as shown in **Figure 4.44**.

Application of these complexes in the asymmetric Diels–Alder reaction was explored to evaluate the influence of the substituent R on the enantioselectivity of the catalyst (**Figure 4.45**). The results of this study suggest that the size of the R group plays a crucial role in the enantioselectivity of the catalyst (**Figure 4.45**). As the size

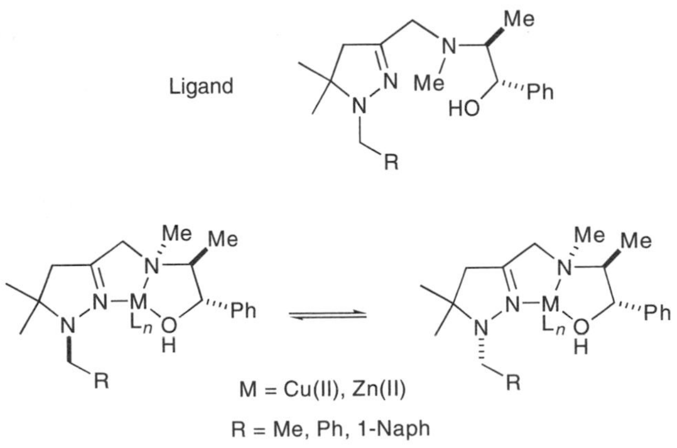

Figure 4.44. A ligand with a chiral relay is shown. Coordination to metals establishes two additional stereocenters. Interconversion of two possible diastereomeric complexes through nitrogen inversion is shown.

of the remote substituent R is increased from methyl to 1-naphthyl, the enantioselectivity of the product rises from 50% to > 90% ee for both diastereomeric products. From these data, it is clear that the size of the stereochemically labile NCH$_2$R group is the primary determinant of the degree of facial selectivity.

A proposed model to rationalize the role of the stereolabile NCH$_2$R group and the enantioselectivity is illustrated in **Figure 4.46**. The dieneophile, shown in bold, is bound in a bidentate fashion. A reasonable assumption is that the N-acyl moiety of the substrate will bind in the axial position on the face opposite the pseudoequatorial phenyl. As a result, the CH$_2$R group will likely be directed downward to reduce nonbonded interactions between the relay substituent and the substrate. To minimize steric strain between one of the geminal dimethyl groups and the chiral relay, the CH$_2$R group is expected to be oriented toward the bound substrate (**Figure 4.46**). In this conformation the naphthyl group shields the *Re* face of the substrate, consistent with the observed sense of stereoselection. One advantage of the catalyst possessing stereochemically dynamic groups is that several ligands can often be constructed

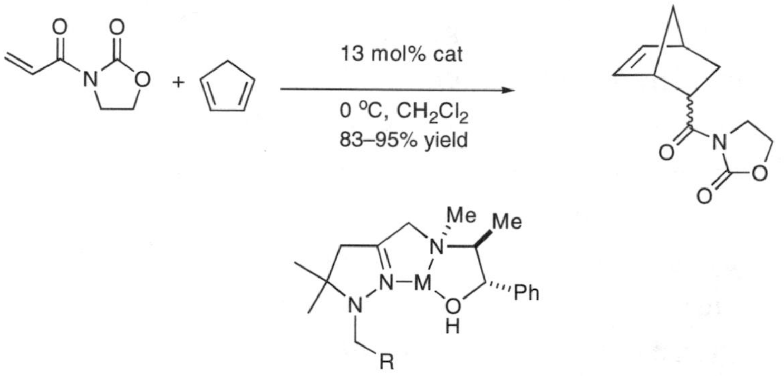

M = Cu(OTf)$_2$			M = Zn(OTf)$_2$		
R	*endo/exo*	ee *endo* [*exo*]	R	*endo/exo*	ee *endo* [*exo*]
Me	4.6	59 [60]	Me	3.9	54 [41]
Ph	3.2	82 [92]	Ph	3.4	89 [87]
1-Naph	2.4	92 [97]	1-Naph	3.8	96 [95]

Figure 4.45. Asymmetric Diels–Alder reaction using stereochemically dynamic ligands.

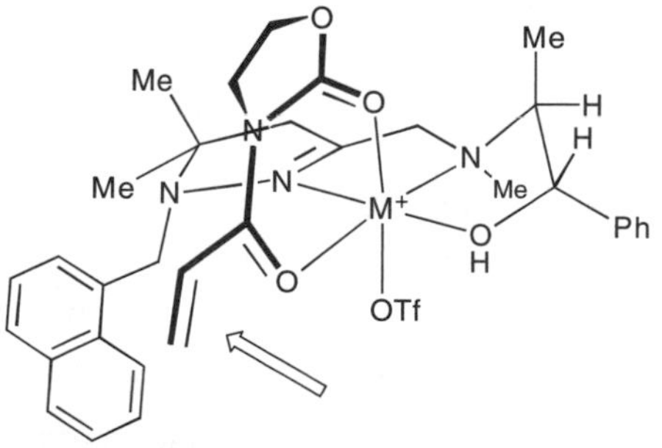

Figure 4.46. Proposed transition state for the asymmetric Diels–Alder reaction. The dienophile is shown in bold, and one of the triflate anions is proposed to occupy an axial position.

from a single resolved starting material, thus permitting rapid assembly of a family of ligands.[111]

4.7.2 Catalysts That Are Chiral at the Metal Center

One of the guiding principles in the design of asymmetric catalysts is that the metal is in a chiral environment. As we have seen, this can be accomplished in several ways, including projection of stereochemical information from remote centers toward the metal by the conformational preferences of the ligand–metal adduct or, at the other extreme, positioning the stereocenters in close proximity to the metal center. This latter strategy can be taken one step further by designing catalysts such that the metal center is a stereogenic center.[112,113] The synthesis of compounds where the sole chiral element is the metal center is very difficult, and is an area that remains to be developed (see Section A.1.1.b). Less difficult, but still tricky, is the synthesis of catalysts in which a chiral ligand binds to the metal center, causing the metal to become a stereocenter. The challenge is maintaining the stereochemical integrity of the catalyst over the course of the reaction, because the coordination number of the metal inevitably changes, providing an opportunity for the stereochemistry of the metal to scramble. As we will see in Chapter 6, use of diastereomeric catalysts can complicate optimization of asymmetric processes. However, if the energy difference between the diastereomeric catalysts is sufficiently large, thermodynamics may insure that only a single diastereomer is generated.

Reaction of dimeric $[(\eta^6\text{-arene})MCl_2]_2$ (M = Ru, Os) with the P/O ligand BINPO and $NaSbF_6$ resulted in formation of $[(\eta^6\text{-arene})MCl(BINAPO)][SbF_6]$, an 18-electron, coordinatively saturated complex (**Figure 4.47**). To facilitate access to an open

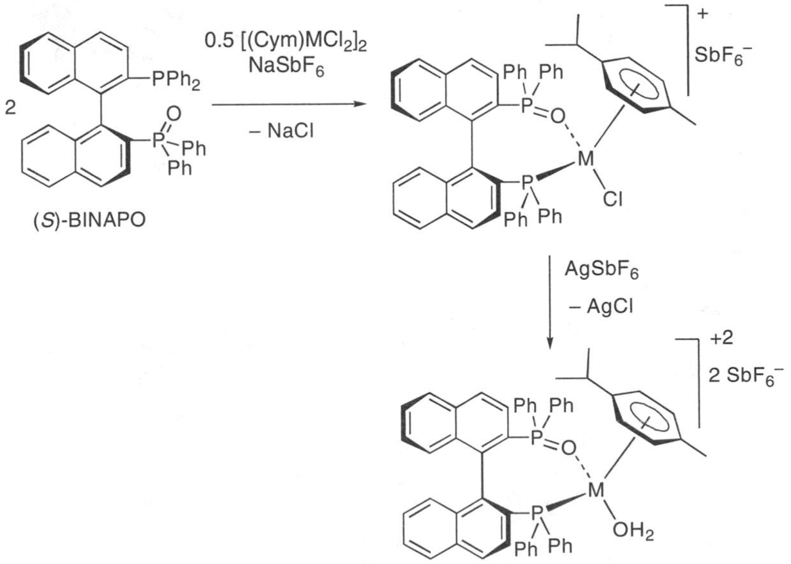

Figure 4.47. The synthesis of diastereomeric BINAPO complexes of ruthenium and osmium. The metal can be considered pseudotetrahedral.

coordination site, the remaining chloride was removed by treatment with $AgSbF_6$ to give the dicationic aqua complex $[(\eta^6\text{-arene})M(OH_2)(BINAPO)][SbF_6]_2$. Analysis of the ^{31}P and ^1H NMR spectra of the ruthenium and osmium complexes indicated formation of a single diastereomer in each case.

These compounds were then used in the Lewis acid catalyzed asymmetric Diels–Alder reaction of cyclopentadiene with methacrolein, a monodentate dienophile (**Figure 4.48**). The enantioselectivities with both the osmium and ruthenium systems were excellent.[114,115]

The X-ray crystal structure of $[(\eta^6\text{-arene})OsCl(BINAPO)]^+$ is illustrated in **Figure 4.49**. Based on this structure, a catalyst–substrate adduct has been proposed, and a partial structure showing the arrangements of ligands around the metal center is illustrated in **Figure 4.50**. A transition state to explain the sense of enantioselectivity is illustrated on the bottom of **Figure 4.50**.

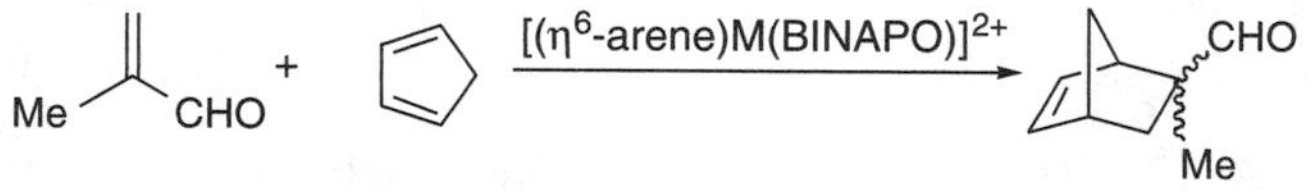

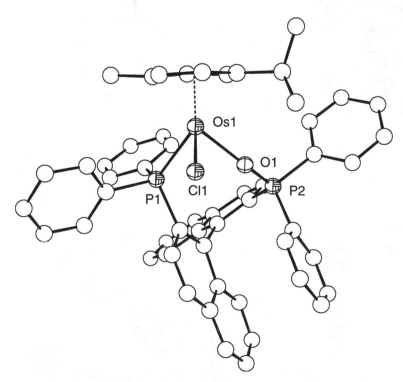

Catalyst (mol%)	Configuration	exo:endo ratio	ee (%)
Os (4)	R_{Os},S_{BINPO}	98:2	93 (S)
Ru (10)	R_{Ru},S_{BINPO}	96:4	99 (S)

Figure 4.48. Asymmetric Diels–Alder reactions catalyzed by osmium and ruthenium complexes of BINAPO.

Figure 4.49. Structure of $[(\eta^6\text{-arene})OsCl(BINAPO)]^+$. The P–Os–O bond angle is 161.5°.

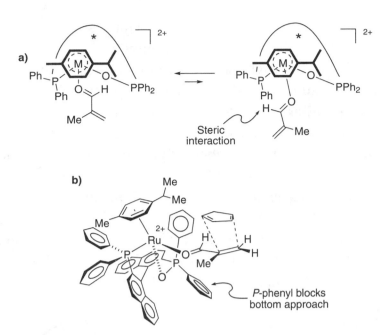

Figure 4.50. a) View from above the η^6-arene of the dienophile coordination (recall that the P–Os–O bond angle of BINAPO is 161.5° in Figure 4.49). Two proposed aldehyde conformations are shown. b) Proposed transition state for the Diels–Alder reaction.

4.7.3 Ligands That Can Adopt Atropisomeric Conformations

Ligands with axial chirality, such as BINAP and BINOL, have been shown to form highly enantioselective catalysts with a variety of metals in many asymmetric transformations. Not only are these ligands extremely adept at inducing asymmetry in prochiral substrates, they also efficiently transfer asymmetry to stereochemically flexible groups within a ligand.[86,116–118] The preparation of substituted derivatives of these popular ligands in enantiomerically pure form can be a laborious process, inspiring investigators to circumvent resolution of atropisomeric moieties, when possible. Below are some methods that have been successfully used in this context.

4.7.3.a Axial Chirality Induced in Backbone Biphenyl-Based Ligands

The binaphthyl ligand in **Figure 4.51** possesses central chirality and a fixed chiral axis,[119] whereas the biphenyl derivative has a conformationally mobile axis.[120] ^{1}H NMR spectra of the biphenyl ligand exhibit two diastereomeric complexes in approximately a 1:1 ratio that interconvert via rotation about the biphenyl axis.[120] Addition of $[(\eta^3\text{-}1,3\text{-diphenylallyl})\text{PdCl}]_2$ (0.5 equivalent) could give up to four diastereomers—namely, two diastereomeric configurations of the chiral axis and two orientations of the 1,3-diphenylallyl ligand (**Figure 4.52**). Only two diastereomers were observed—one W- and one M-type allyl, both with the (*S*)-configuration of the chiral axis. The configuration of the chiral axis was assigned by observation of a negative Cotton effect at 250 nm in the CD spectrum.[121,122] Thus, the central chirality causes the diastereomeric ligand conformations to have significantly different energies once coordinated to the (η^3-1,3-Ph$_2$allyl)Pd moiety.

Comparison of the binaphthyl and biphenyl ligands in **Figure 4.51** was undertaken in the asymmetric allylation of 1,3-dipehnyl-2-propenyl acetate with dimethyl malonate (**Figure 4.53**) under conditions almost identical to those in **Figure 4.10**. Use of the binaphthyl ligand with the fixed (S)-chiral axis (R = *i*-Pr) gave 85% ee of the (S) product (**Figure 4.53**). The conformationally dynamic biphenyl ligand (R = *i*-Pr) exhibited very similar enantioselection [83% ee of the (S)-product], while the *t*-Bu derivative gave slightly higher enantioselectivity. Application of the diastereomeric binaphthyl ligand with the fixed (R)-chiral axis resulted in formation of the opposite enantiomer of the product with 90% ee. Several important conclusions can be drawn from these results. The advantage of the use of the dynamic biphenyl-derived ligand is that it gives slightly higher enantioselectivity than the more synthetically challeng-

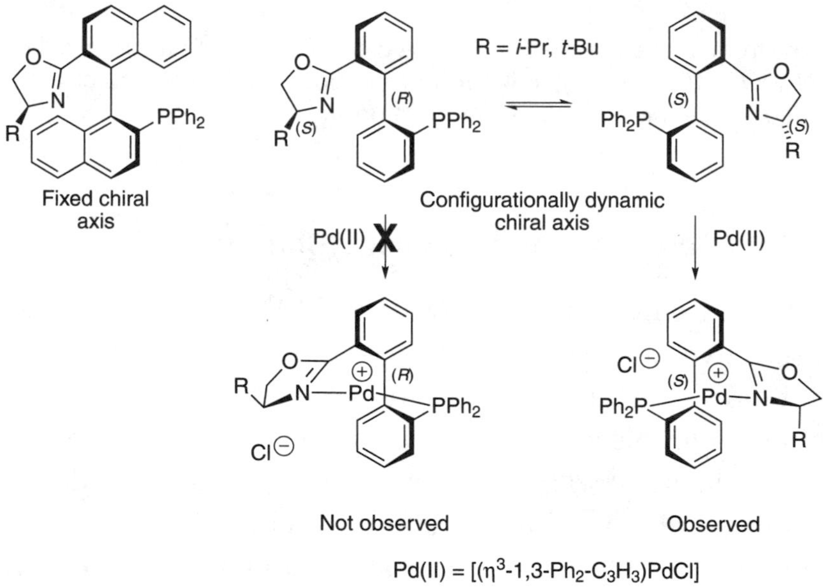

Figure 4.51. A binaphthyl ligand with central chirality and a fixed chiral axis (top left) and a biphenyl ligand with central chirality and a conformationally dynamic axis (top right). Reaction with [(η^3-1,3-Ph$_2$allyl)PdCl] affords the complex with the (S)-chiral axis (also see **Figure 4.52**). The allyl moiety is removed for clarity.

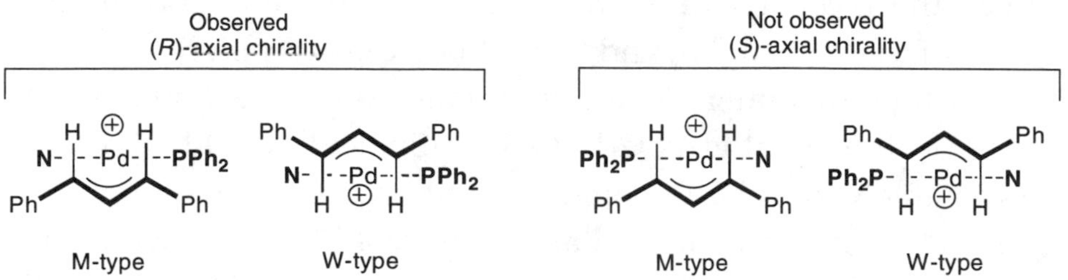

Figure 4.52. The four possible diastereomeric allyl complexes. The allyl can have the M- or W-type orientation with respect to the Ph$_2$P---N ligand plane. The biphenyl-derived ligand can have the (S)- or (R)-axial chirality.

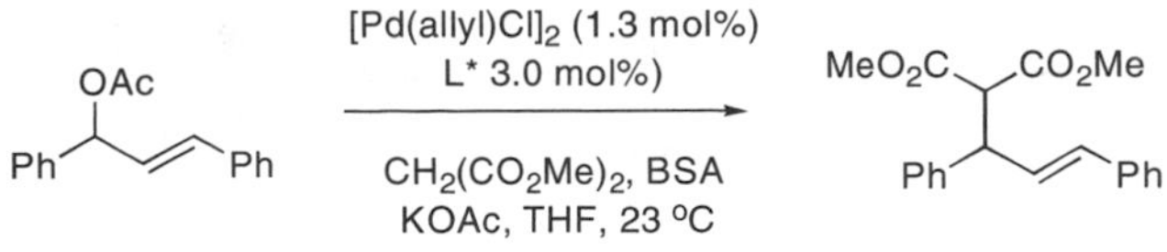

Figure 4.53. Palladium-catalyzed allylic alkylation with ligands from **Figure 4.51**. In each case, the central chirality has the (S)-configuration (BSA = bissilylacetamide).

ing fixed binaphthyl ligand. The disadvantage of the dynamic chiral axis, however, is that only one of the diastereomeric combinations can be access [the (S)-central chirality induces only the (S)-axial chirality, **Figure 4.51**]. In this particular case, the ligand with (S)-central chirality and (R)-axial chirality (**Figure 4.51**) gave slightly higher enantioselectivity (90% ee, entry 4, **Figure 4.53**).

This approach has been applied successfully to other catalytic asymmetric reactions[123] and will likely attract more attention in an effort to streamline ligand syntheses.

4.7.3.b Axial Chirality Induced in Pendant Biphenyl-Based Ligands

Diphosphite ligands **A–D** (**Figure 4.54**) have been used in the rhodium-catalyzed reduction of dimethylitaconate to afford chiral diesters. The phosphite ligands employed are composed of two groups, a nonracemic sugar-based backbone linker and BINOL-, biphenol-, or 2-naphthol-derived pendent groups. BINOL-derived ligands A_S and A_R provide diastereomers that differ in the relative configurations of the backbone and the binaphthyl groups. Ligands **B** and **C** (**Figure 4.54**) contain conformationally flexible groups that can undergo atropisomerization. These groups relay the stereochemistry of the ligand backbone toward the metal center. This process has a low barrier, insuring rapid interconversion of the diastereomeric ligands and, therefore, catalysts.[124] Although three diastereomers of ligands **B** and **C** are possible, it is unlikely they will be present in equal amounts, because the stereochemistry of the biphenol moieties depends on their interaction with the chiral backbone. Ligand **D** contains 2-naphthoxy groups that cannot adopt atropisomeric conformations.

When (S)- or (R)-BINOL-based ligands (A_S and A_R) were employed, the enantioselectivities were very high, but the sense of enantioselection was opposite (**Table 4.2**). Axial chirality of the pendent groups is dominant in determining the configuration of the product. The magnitude of the enantioselectivities from A_S and A_R is similar, suggesting that the chiral linker has almost no impact on the catalyst enantioselectivity.

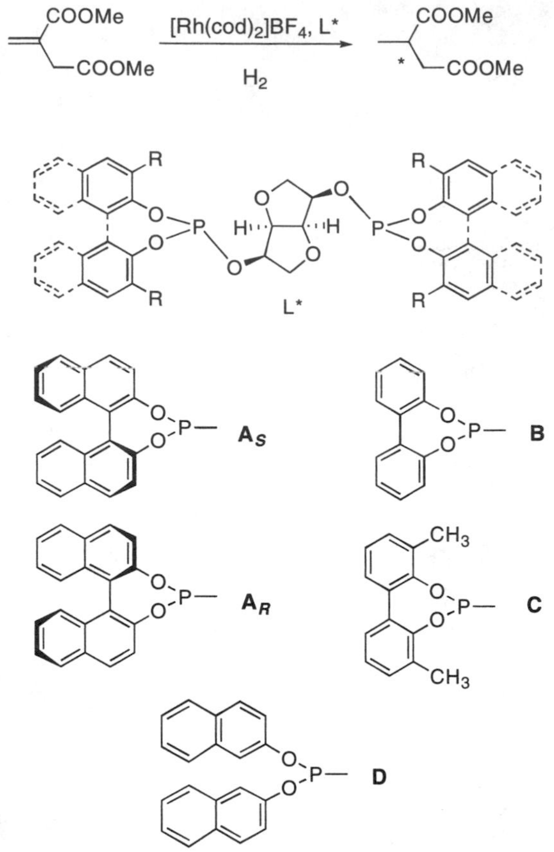

Figure 4.54. Asymmetric hydrogenation of dimethylitaconate with chiral diphosphite ligands.

Employing biphenol-based ligand **B**, which possesses the conformationally flexible pendent moieties, gave intermediate enantioselectivity (39%) favoring the (*S*) product. It is possible that this ligand gives rise to diastereomeric catalysts that operate simultaneously in the reduction reaction. Conclusions about the proportions of the diastereomeric catalyst based on the enantioselectivity, however, is not possible, because the catalysts are likely to have different relative rates.

The highest enantioselectivities (97% ee) were obtained with ligand **C** substituted with methyl groups at the 3- and 3′-postions. The enantioselectivity was slightly higher than with ligands A_S and A_R, based on the stereochemically fixed BINOL (**Table 4.2**). Direct comparison between ligands **C**, A_S, and A_R is complicated due to the absence of the 3,3′-substituents in A_S and A_R. It is interesting to note that the order of activity of the catalysts parallels their enantioselectivity: $A_S < A_R < C$. In the case of ligand **D**, which cannot adopt atropisomeric conformations, the product has low ee (21%).

The advantage of incorporating stereochemically dynamic groups into ligands allows one to create a ligand library based on a single chiral backbone. The drawback is that the optimal catalyst must be generated by screening a series of chiral ligands. As outlined in subsequent chapters, a more efficient method to develop asymmetric catalysts is based on the combination of chiral and achiral ligands.[125] In this fashion, a single chiral ligand can be used to generate a family of catalysts.[126]

Table 4.2. Enantioselectivities in the asymmetric hydrogenaiton of dimethylitaconate with catalysts based on ligands **A–D** (Figure 4.54).

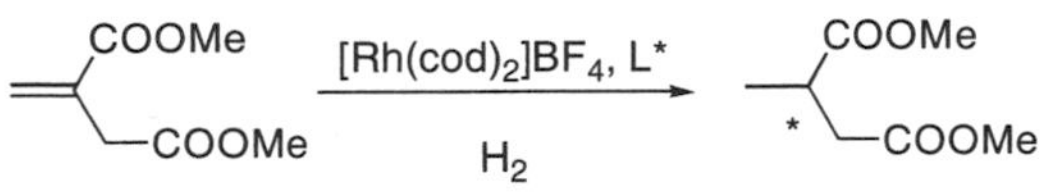

L*	Conversion at 20 h (%)	ee (%)
A$_S$	> 99	88 (S)
A$_R$	> 99	95 (R)
B	74	39 (S)
C	> 99	97 (R)
D	65	21 (S)

4.7.3.c Axial Chirality Induced in Atropisomeric Amides

Almost all of the atropisomeric ligands (i.e., ligands that are chiral by virtue of hindered rotation) are biaryls. Examples that we have seen are BINAP and BINOL. These ligands are stable to racemization under typical reaction conditions. The use of other types of atropisomeric ligands is uncommon, in part due to difficulties in the synthesis of such ligands. Nonetheless, once this obstacle is overcome, the use of other types of atropisomeric ligands will likely be more common. An interesting strategy that takes advantage of atropisomeric conformations has been employed in the asymmetric allylation with palladium. In this system, the central chirality of a fixed stereocenter biases the conformation of a stereochemically dynamic atropisomeric amide.[127,128]

As illustrated in **Figure 4.55**, use of the atropisomeric amide-based phosphine ligand in the palladium-catalyzed allylic alkylation with dimethyl malonate gave 85% ee. Although no structural information has been reported concerning the (π-allyl)Pd complex in the reaction, NMR data suggest the ligand is bidentate, coordinating to the palladium through the phosphorus and amide-carbonyl oxygen. In this adduct, the extant carbon stereochemistry will influence the disposition of the axial chirality of the amide, as well as the orientation of the two *P*-phenyl groups (see Section 4.3.2).

The central chirality of this ligand is sufficiently removed from the metal center that it is unlikely to be directly responsible for the excellent control of enantioselectivity in the allylation reaction. It is more plausible that the transmission of asymmetric induction is relayed through the axial conformation of the amide and the orientation of the diastereotopic *P*-phenyl groups.[110]

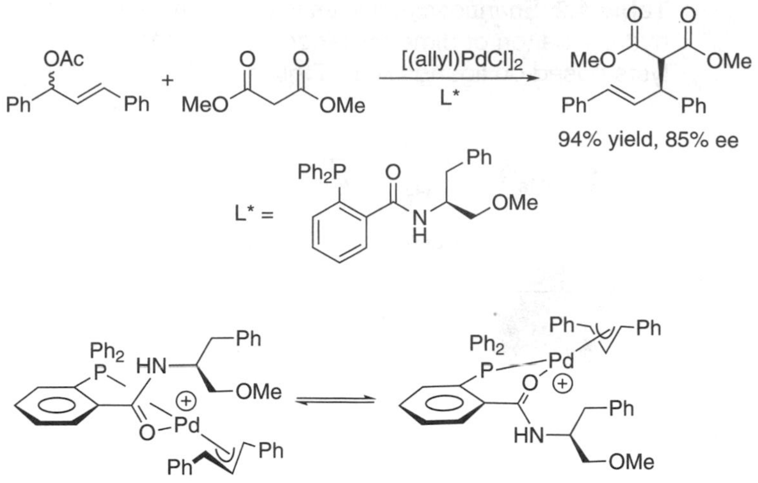

Figure 4.55. Asymmetric allylic alkylation using atropisomeric amides as chiral ligands (top) and the possible diastereomeric allyl intermediates (bottom).

4.8 Induced Asymmetry in the Substrate

4.8.1 Diastereomeric Complexes with Prochiral Substrates

A similar chelation of a sulfonyl oxygen to that in Section 4.5.2 was proposed based on density functional theory calculations in a study of the copper-catalyzed aziridination of alkenes using [N-(p-toluenesulfonyl)imino]phenyliodinane (PhI=NSO$_2$Tol) as the nitrene source (**Figure 4.56**).[129] In systems employing the bisoxazoline[13,130] and diimine[131,132] ligands, the reaction is believed to proceed through a Cu(I)/Cu(III) couple, and evidence for an intermediate copper nitrene has been presented.[129,132]

As illustrated in **Figure 4.57**, coordination of the sulfonyl oxygen gives rise to diastereomeric intermediates differing in the configuration at sulfur. The calculations suggest that the sulfonyl oxygen remains bound to copper during the formation of the N–C bonds.[129] Although the impact of the stereogenic sulfur center has not been in-

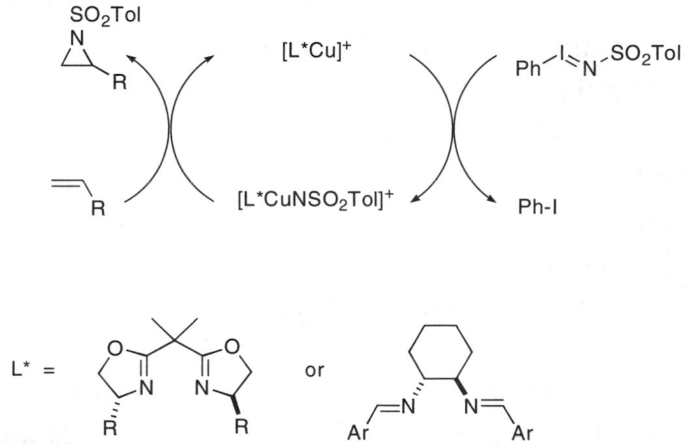

Figure 4.56. Catalytic aziridination of olefins with copper catalysts.

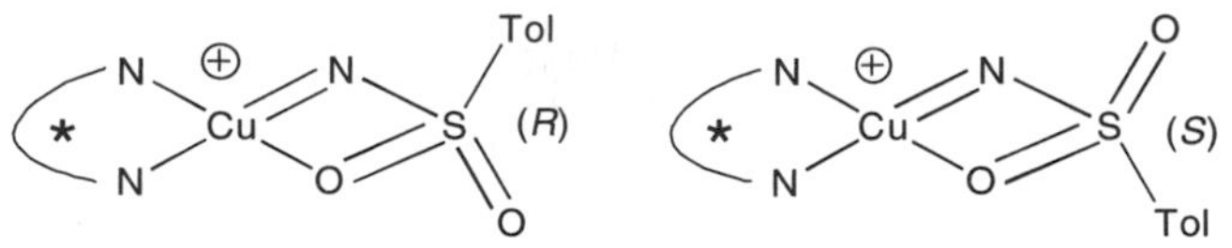

Figure 4.57. Diastereomers formed on coordination of the sulfonyl oxygen in the intermediate nitrene proposed for the aziridination of olefins.

vestigated, it is likely to be significant, based on its proximity to the reactive nitrogen of the nitrene.

4.8.2 Substrates with Chiral Relays

We have seen examples where the chiral metal–ligand adduct controls the attack of a reagent on a bound substrate, and examples in which the catalyst controls the binding of a substrate prochiral face in the enantioselectivity-determining step. A distinct and ingenious approach involves transmission of stereochemical information from a chiral catalyst to a stereolabile center in the substrate, termed a chiral relay, which then controls the stereochemistry of attack at that substrate by an external reagent.[133]

An example of a chiral relay built into the substrate is illustrated in **Figure 4.58**. In this approach, a rapidly racemizing amino group is incorporated into the substrate. The role of the asymmetric catalyst is to bind the substrate and temporarily transform the stereochemically labile nitrogen into a chiral auxiliary. In turn, this stereocenter directs the approach of the incoming reagent.

These substrates were used in asymmetric Diels–Alder reaction catalyzed by the now familiar bisoxazoline–copper(II) complexes. The substrates for this reaction are butenoylpyrazolidinones having different N-alkyl relays, R (**Figure 4.59**). Two routes to coordination and activation of the substrate can be envisioned. The bound substrate can undergo epimerization and equilibrate to the lower-energy diastereomer or

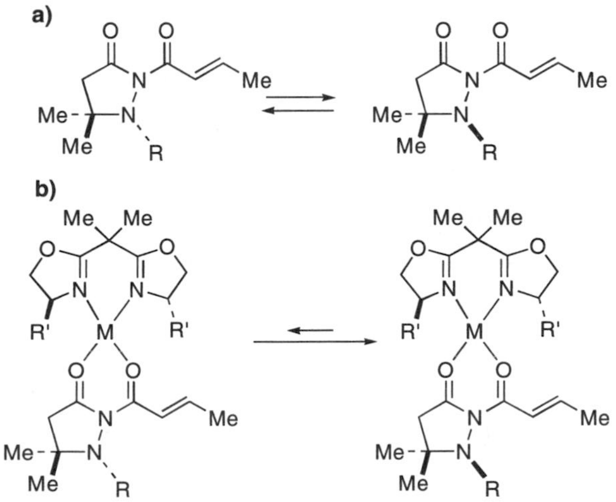

Figure 4.58. a) Nitrogen inversion racemizes the substrate. b) Diastereomeric substrate–catalyst adducts can epimerize through nitrogen inversion. These diastereomers may be very different in energy.

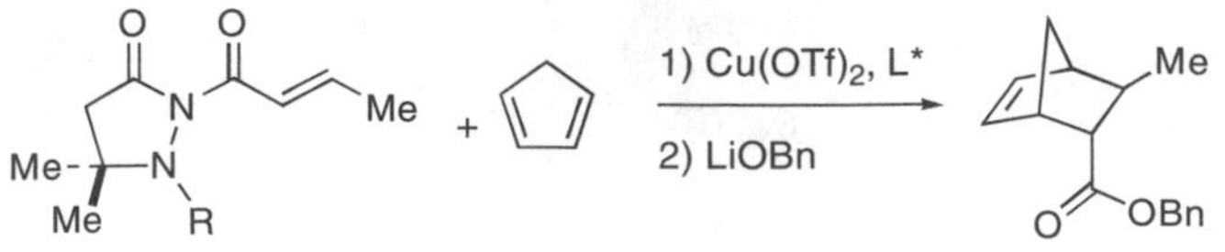

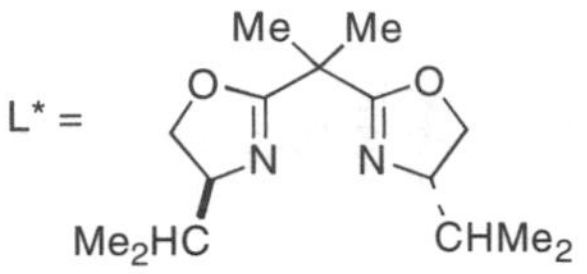

Substrates	ee (%)
R = H	08
R = Et	56
R = Bn	71
R = 2-CH$_2$-Naph	65
R = 1-CH$_2$-Naph	92

Figure 4.59. Impact of chiral relay groups, R, on the enantioselectivity of the asymmetric Diels–Alder reaction. The second step cleaves the relay portion of the substrate.

one enantiomer of the racemizing substrate can selectively bind to the chiral catalyst before activatation (**Figure 4.58**).

In this study, a comparison was made between the relay group on the substrate and the enantioselectivity (**Figure 4.59**). As progressively larger chiral relay groups were used, the enantioselectivity increased. The observed ee range (8–92%) clearly indicates that the enantioselectivity is highly dependent on the size of the chiral relay. Furthermore, the large variation in enantioselectivity suggests that the bisoxazoline ligand is not solely responsible for the enantioselectivity. It was also demonstrated that the enantioselectivities with the most efficient chiral relays were independent of the substituents on C_2-symmetric bisoxazoline ligands. This evidence also indicates that the chiral relay is responsible for control of the facial attack of the diene.[133,134] Although the concept of chiral relays should be applicable to other types of substrates, it is not limited to incorporation into the substrate (see Section 4.7.1).

Concluding Remarks

The basic concepts describing the transmission of asymmetry in chiral catalysts have been outlined in this chapter, with a focus on ligand classes most commonly encountered in asymmetric catalysis. This foundation will be expanded in Chapter 5, which describes nonclassical two-point catalyst-substrate interactions. Some of the systems presented in this chapter have been extensively researched and the proposed models advanced are widely accepted. Other examples have been more recently introduced, yet hold considerable potential for both expansion and applications.

Understanding the fundamental concepts outlined herein is necessary, although not sufficient in and of itself, to rationally design new enantioselective catalysts. It is also valuable to know the basic characteristics of the reaction of interest. For example, how do the ligand and substrate bind to the metal, what is the metal coordination number, and what is the metal geometry? While determination of these features may appear trivial, it is not always the case. Furthermore, to increase the likelihood of success, a detailed knowledge of the reaction mechanism is helpful. Even after achieving the level of understanding outlined above, the design of asymmetric catalysts is challenging, because the energy differences between diastereomeric transition states are inherently small and currently beyond our ability to reliably predict or calculate. In the end, there is no substitute for the iterative cycle of catalyst optimization, consisting of catalyst design, screening, analysis of the results, and catalyst modification. This process can be dramatically streamlined, however, with knowledge of the reaction mechanism and the mode of transmission of asymmetry from the catalyst to the substrate. As is so often true in chemistry, the age-old saying that "luck (success) favors the prepared mind" holds true.

References

(1) Kagan, H. B.; Phat, D.-T. Asymmetric Catalytic Reduction with Transition Metal Complexes. I. Catalytic System of Rhodium(I) with (–)-2,3-O-Isopropylidene-2,3-dihydroxy-1,4-bis(diphenylphosphino)butane, a New Chiral Diphosphine. *J. Am. Chem. Soc.* **1972**, *94*, 6429–6433.

(2) Kagan, H. B. In *Asymmetric Catalysis*; Morrison, J. D., Ed.; Academic Press: New York, 1985; Vol. 5, pp 1–339.

(3) Whitesell, J. K. C_2-Symmetry and Asymmetric Induction. *Chem. Rev.* **1989**, *89*, 1581–1615.

(4) Trost, B. M. Designing a Receptor for Molecular Recognition in a Catalytic Synthetic Reaction: Allylic Alkylation. *Acc. Chem. Res.* **1996**, *29*, 355–364.

(5) Trost, B. M.; Machacek, M. R.; Aponick, A. Predicting the Stereochemistry of Diphenylphosphino Benzoic Acid (DPPBA)-Based Palladium-Catalyzed Asymmetric Allylic Alkylation Reactions: A Working Model. *Acc. Chem. Res.* **2006**, *39*, 747–760.

(6) Hegedus, L. S. *Transition Metals in the Synthesis of Complex Organic Molecules*, 2nd ed.; University Science Books: Sausalito, CA, 1999.

(7) Blochl, P. E.; Togni, A. First-Principles Investigation of Enantioselective Catalysis: Asymmetric Allylic Amination with Pd Complexes Bearing P,N-Ligands. *Organometallics* **1996**, *15*, 4125–4132.

(8) Johnson, J. S.; Evans, D. A. Chiral Bis(oxazoline) Copper(II) Complexes: Versatile Catalysts for Enantioselective Cycloaddition, Aldol, Michael, and Carbonyl Ene Reactions. *Acc. Chem. Res.* **2000**, *33*, 325–335.

(9) Jørgensen, K. A.; Johannsen, M.; Yao, S.; Audrain, H.; Thorhauge, J. Catalytic Asymmetric Addition Reactions of Carbonyls. A Common Catalytic Approach. *Acc. Chem. Res.* **1999**, *32*, 605–613.

(10) Pfaltz, A. Chiral Semicorrins and Related Nitrogen Heterocycles as Ligands in Asymmetric Catalysis. *Acc. Chem. Res.* **1993**, *26*, 339–345.

(11) Evans, D. A.; Woerpel, K. A.; Hinman, M. M.; Faul, M. M. Bis(oxazolines) as Chiral Ligands in Metal-Catalyzed Asymmetric Reactions: Catalytic Asymmetric Cyclopropanation of Olefins. *J. Am. Chem. Soc.* **1991**, *113*, 726–728.

(12) Fritschi, H.; Leutenegger, U.; Pfaltz, A. Semicorrin Metal-Complexes as Enantioselective Catalysts. 2. Enantioselective Cyclopropane Formation from Olefins with Diazo-Compounds Catalyzed by Chiral (Semicorrinato) Copper Complexes. *Helv. Chim. Acta* **1988**, *71*, 1553–1565.

(13) Evans, D. A.; Faul, M. M.; Bilodeau, M. T.; Anderson, B. A.; Barnes, D. M. Bis(oxazoline) Copper-Complexes as Chiral Catalysts for the Enantioselective Aziridination of Olefins. *J. Am. Chem. Soc.* **1993**, *115*, 5328–5329.

(14) Evans, D. A.; Miller, S. J.; Lectka, T. Bis(oxazoline) Copper(II) Complexes as Chiral Catalysts for the Enantioselective Diels–Alder Reaction. *J. Am. Chem. Soc.* **1993**, *115*, 6460–6461.

(15) Evans, D. A.; Miller, S. J.; Lectka, T.; von Matt, P. Chiral Bis(oxazoline)copper(II) Complexes as Lewis Acid Catalysts for the Enantioselective Diels–Alder Reaction. *J. Am. Chem. Soc.* **1999**, *121*, 7559–7573.

(16) Evans, D. A.; Rovis, T.; Kozlowski, M. C.; Tedrow, J. S. C_2-Symmetric Cu(II) Complexes as Chiral Lewis Acids. Catalytic Enantioselective Michael Addition of Silylketene Acetals to Alkylidene Malonates. *J. Am. Chem. Soc.* **1999**, *121*, 1994–1995.

(17) Evans, D. A.; Burgey, C. S.; Paras, N. A.; Vojkovsky, T.; Tregay, S. W. C_2-Symmetric Copper(II) Complexes as Chiral Lewis Acids. Enantioselective Catalysis of the Glyoxylate–Ene Reaction. *J. Am. Chem. Soc.* **1998**, *120*, 5824–5825.

(18) Hathaway, B. J. In *Comprehensive Coordination Chemistry*, Wilkinson, G., Ed.; Pergamon: New York, 1987; Vol. 5, pp 533–774.

(19) Evans, D. A.; Barnes, D. M.; Johnson, J. S.; Lectka, T.; von Matt, P.; Miller, S. J.; Murry, J. A.; Norcross, R. D.; Shaughnessy, E. A.; Campos, K. R. Bis(oxazoline) and Bis(oxazolinyl)pyridine Copper Complexes as Enantioselective Diels–Alder Catalysts: Reaction Scope and Synthetic Applications. *J. Am. Chem. Soc.* **1999**, *121*, 7582–7594.

(20) Evans, D. A.; Kozlowski, M. C.; Tedrow, J. S. Cationic Bis(oxazoline) and Pyridyl-bis(oxazoline)Cu(II) and Zn(II) Lewis Acid Catalysts. A Comparative Study in Catalysis of Diels–Alder and Aldol Reactions. *Tetrahedron Lett.* **1996**, *37*, 7481–7484.

(21) Corey, E. J.; Imai, N.; Zhang, H. Y. Designed Catalyst for Enantioselective Diels–Alder Addition from a C_2-Symmetric Chiral Bis(oxazoline)-Iron(III) Complex. *J. Am. Chem. Soc.* **1991**, *113*, 728–729.

(22) Tsuji, J.; Minami, I. New Synthetic Reactions of Allylalkyl Carbonates, Allyl Beta-Keto Carboxylates, and Allyl Vinylic Carbonates Catalyzed by Palladium Complexes. *Acc. Chem. Res.* **1987**, *20*, 140–145.

(23) Trost, B. M. Cyclizations via Palladium-Catalyzed Allylic Alkylations. *Angew. Chem., Int. Ed. Engl.* **1989**, *28*, 1173–1192.

(24) Yoon, T. P.; Jacobsen, E. N. Privileged Chiral Catalysts. *Science* **2003**, *299*, 1691–1693.

(25) Meca, L.; Reha, D.; Havlas, Z. Racemization Barriers of 1,1'-Binaphthyl and 1,1'-Binaphthalene-2,2'-diol: A DFT Study. *J. Org. Chem.* **2003**, *68*, 5677–5680.

(26) Kyba, E. P.; Gokel, G. W.; De Jong, F.; Koga, K.; Sousa, L. R.; Siegel, M. G.; Kaplan, L.; Sogah, G. D. Y.; Cram, D. J. Host–guest Complexation. 7. The Binaphthyl Structural Unit in Host Compounds. *J. Org. Chem.* **1977**, *42*, 4173–4184.

(27) Rosini, C.; Franzini, L.; Raffaelli, A.; Salvadori, P. Synthesis and Applications of Binaphthylic C_2-Symmetric Derivatives as Chiral Auxiliaries in Enantioselective Reactions. *Synthesis* **1992**, 503–505.

(28) Pu, L. 1,1'-Binaphthyl Dimers, Oligomers, and Polymers: Molecular Recognition, Asymmetric Catalysis, and New Materials. *Chem. Rev.* **1998**, *98*, 2405–2494.

(29) Shibasaki, M.; Yoshikawa, N. Lanthanide Complexes in Multifunctional Asymmetric Catalysis. *Chem. Rev.* **2002**, *102*, 2187–2219.

(30) Aspinall, H. C.; Greeves, N. Defining Effective Chiral Binding Sites at Lanthanides: Highly Enantioselective Reagents and Catalysts from Binaphtholate and PyBox Ligands. *J. Organomet. Chem.* **2002**, *647*, 151–157.

(31) Chen, Y.; Yekta, S.; Yudin, A. K. Modified BINOL Ligands in Asymmetric Catalysis. *Chem. Rev.* **2003**, *103*, 3155–3212.

(32) Brown, S. N.; Chu, E. T.; Hull, M. W.; Noll, B. C. Electronic Dissymmetry in Chiral Recognition. *J. Am. Chem. Soc.* **2005**, *127*, 16010–16011.

(33) Kelly, R. T.; Whiting, A.; Chandrakumar, N. S. A Rationally Designed, Chiral Lewis Acid for the Asymmetric Induction of Some Diels–Alder Reactions. *J. Am. Chem. Soc.* **1986**, *108*, 3510–3512.

(34) Wipf, P.; Jung, J.-K. Formal Total Synthesis of (+)-Diepoxin σ. *J. Org. Chem.* **2000**, *65*, 6319–6337.

(35) Bochmann, M. Cationic Group 4 Metallocene Complexes and Their Role in Polymerisation Catalysis: The Chemistry of Well Defined Ziegler Catalysts. *J. Chem. Soc., Dalton Trans.* **1996**, 255–270.

(36) Hoveyda, A. H.; Morken, J. P. Enantioselective C–C and C–H Bond Formation Mediated or Catalyzed by Chiral EBTHI Complexes of Titanium and Zirconium. *Angew. Chem., Int. Ed. Engl.* **1996**, *35*, 1262–1284.

(37) Wild, F. R. W. P.; Zsolnai, L.; Huttner, G.; Brintzinger, H.-H. *ansa*-Metallocene Derivatives IV. Synthesis and Molecular Structures of Chiral *ansa*-Titanocene Derivatives with Bridged Tetrahydroindenyl Ligands. *J. Organomet. Chem.* **1982**, *232*, 233–247.

(38) Verdaguer, X.; Lange, U. E. W.; Reding, M. T.; Buchwald, S. L. Highly Enantioselective Imine Hydrosilylation Using (*S,S*)-Ethylenebis(5-tetrahydroindenyl)titanium Difluoride. *J. Am. Chem. Soc.* **1996**, *118*, 6784–6785.

(39) Verdaguer, X.; Lange, U. E. W.; Buchwald, S. L. Amine Additives Greatly Expand the Scope of Asymmetric Hydrosilylation of Imines. *Angew. Chem., Int. Ed. Engl.* **1998**, *37*, 1103–1107.

(40) Willoughby, C. A.; Buchwald, S. L. Asymmetric Titanocene-Catalyzed Hydrogenation of Imines. *J. Am. Chem. Soc.* **1992**, *114*, 7562–7564.

(41) Willoughby, C. A.; Buchwald, S. L. Synthesis of Highly Enantiomerically Enriched Cyclic Amines by the Catalytic Asymmetric Hydrogenation of Cyclic Imines. *J. Org. Chem.* **1993**, *58*, 7627–7629.

(42) Yun, J.; Buchwald, S. L. Efficient Kinetic Resolution in the Asymmetric Hydrosilylation of Imines of 3-Substituted Indanones and 4-Substituted Tetralones. *J. Org. Chem.* **2000**, *65*, 767–774.

(43) Hansen, M. C.; Buchwald, S. L. A Method for the Asymmetric Hydrosilylation of *N*-Aryl Imines. *Org. Lett.* **2000**, *2*, 713–715.

(44) Woo, H. G.; Tilley, T. D. Dehydrogenative Polymerization of Silanes to Polysilanes by Zirconocene and Hafnocene Catalysts. A New Polymerization Mechanism. *J. Am. Chem. Soc.* **1989**, *111*, 8043–8044.

(45) Knowles, W. S. Asymmetric Hydrogenation. *Acc. Chem. Res.* **1983**, *16*, 106–112.

(46) Knowles, W. S.; Sabacky, M. J.; Vineyard, B. D. Catalytic Asymmetric Hydrogenation. *J. Chem. Soc., Chem. Commun.* **1972**, 10–11.

(47) Dang, T. P.; Kagan, H. B. The Asymmetric Synthesis of Hydratropic Acid and Amino-Acids by Homogeneous Catalytic Hydrogenation. *J. Chem. Soc., Chem. Commun.* **1971**, 481–482.

(48) Miyashita, A.; Yasuda, A.; Takaya, H.; Toriumi, T.; Ito, K.; Souchi, T.; Noyori, R. Synthesis of 2,2'-Bis(diphenylphosphino)-1,1'-Binaphthyl (BINAP), an Atropisomeric Chiral Bis(triaryl)phosphine, and its use in the Rhodium(I)-Catalyzed Asymmetric Hydrogenation of α-(Acylamino)acrylic Acids. *J. Am. Chem. Soc.* **1980**, *102*, 7932–7934.

(49) Noyori, R.; Ohkuma, T. Asymmetric Catalysis by Architectural and Functional Molecular Engineering: Practical Chemo- and Stereoselective Hydrogenation of Ketones. *Angew. Chem., Int. Ed. Engl.* **2001**, *40*, 40–73.

(50) Noyori, R.; Yamakawa, M.; Hashiguchi, S. Metal–Ligand Bifunctional Catalysis: A Nonclassical Mechanism for Asymmetric Hydrogen Transfer Between Alcohols and Carbonyl Compounds. *J. Org. Chem.* **2001**, *66*, 7931–7944.

(51) Ohta, T.; Takaya, H.; Noyori, R. Bis(diarylphosphino)-1,1'-Binaphthyl BINAP-Ruthenium(II) Dicarboxylate Complexes: New, Highly Efficient Catalysts for Asymmetric Hydrogenations. *Inorg. Chem.* **1988**, *27*, 566–569.

(52) Noyori, R. *Asymmetric Catalysis in Organic Synthesis*; Wiley: New York, 1994.

(53) Hao, J.; Hatano, M.; Mikami, K. Chiral Palladium(II)-Catalyzed Asymmetric Glyoxylate–Ene Reaction: Alternative Approach to the Enantioselective Synthesis of α-Hydroxy Esters. *Org. Lett.* **2000**, *3*, 4059–4062.

(54) Morton, D. A. V.; Orpen, A. G. Structural Systematics 4. Conformations of the Diphosphine Ligands in $M_2(Ph_2PCH_2PPh_2)$ and $M(Ph_2PCH_2CH_2PPh_2)$ Complexes. *J. Chem. Soc., Dalton Trans.* **1992**, 641–653.

(55) Brunner, H.; Winter, A.; Breu, J. Enantioselective Catalysis Part 114. Chirostructural Analysis of Bis(diphenylphosphanyl)ethane Transition Metal Chelates. *J. Organomet. Chem.* **1998**, *553*, 285–306.

(56) Noyori, R.; Takaya, H. BINAP: An Efficient Chiral Element for Asymmetric Catalysis. *Acc. Chem. Res.* **1990**, *23*, 345–350.

(57) Ashby, M. T.; Halpern, J. Kinetics and Mechanism of Catalysis of the Asymmetric Hydrogenation of α,β-Unsaturated Carboxylic Acids by Bis(carboxylato) [2,2'-bis(diphenylphosphino)-1,1'-binaphthyl] ruthenium(II), [RuII(BINAP) $(O_2CR)_2$]. *J. Am. Chem. Soc.* **1991**, *113*, 589–594.

(58) Noyori, R.; Tokunaga, M.; Kitamura, M. Stereoselective Organic Synthesis via Dynamic Kinetic Resolution. *Bull. Chem. Soc. Jpn.* **1995**, *68*, 36–56.

(59) Kitamura, M.; Ohkuma, T.; Inoue, S.; Sayo, N.; Kumobayashi, H.; Akutagawa, S.; Ohta, T.; Takaya, H.; Noyori, R. Homogeneous Asymmetric Hydrogenation of Functionalized Ketones. *J. Am. Chem. Soc.* **1988**, *110*, 629–631.

(60) Halpern, J. Mechanism and Stereoselectivity of Asymmetric Hydrogenation. *Science* **1982**, *217*, 401–407.

(61) Landis, C. R.; Halpern, J. Asymmetric Hydrogenation of Methyl (Z)-α-Acetamidocinnamate Catalyzed by [1,2-Bis(phenyl-o-anisoyl)phosphinoethane] rhodium(I): Kinetics, Mechanism and Origin of Enantioselection. *J. Am. Chem. Soc.* **1987**, *109*, 1746–1754.

(62) Barnhart, R. W.; Wang, X.; Noheda, P.; Bergens, S. H.; Whelan, J.; Bosnich, B. Asymmetric Catalysis. Asymmetric Catalytic Intramolecular Hydroacylation of 4-Pentenals Using Chiral Rhodium

Diphosphine Catalysts. *J. Am. Chem. Soc.* **1994**, *116*, 1821–1830.

(63) Taura, Y.; Tanaka, M.; Funakoshi, K.; Sakai, K. Asymmetric Cyclization Reactions by Rh(I) with Chiral Ligands. *Tetrahedron Lett.* **1989**, *30*, 6349–6352.

(64) Taura, Y.; Tanaka, M.; Wu, X.-M.; Funakoshi, K.; Sakai, K. Asymmetric Cyclization Reactions. Cyclization of Substituted 4-Pentenals into Cyclopentanone Derivatives by Rhodium(I) with Chiral Ligands. *Tetrahedron* **1991**, *47*, 4879–4888.

(65) Wu, X.-M.; Funakoshi, K.; Sakai, K. Highly Enantioselective Cyclization Using Cationic Rh(I) with Chiral Ligands. *Tetrahedron Lett.* **1992**, *33*, 6331–6334.

(66) Bosnich, B. In *Encyclopedia of Inorganic Chemistry*; King, R. B., Ed.; Wiley: New York, 1994; 219–236.

(67) Fairlie, D. P.; Bosnich, B. Homogeneous Catalysis. Mechanism of Catalytic Hydroacylation: The Conversion of 4-Pentenals to Cyclopentanones. *Organometallics* **1988**, *7*, 946–954.

(68) Fairlie, D. P.; Bosnich, B. Homogeneous Catalysis. Conversion of 4-Pentenals to Cyclopentanones by Efficient Rhodium-Catalyzed Hydroacylation. *Organometallics* **1988**, *7*, 936–945.

(69) Ball, R. G.; Payne, N. C. Chiral Phosphine Ligands in Asymmetric Synthesis. Molecular Structure and Absolute Configuration of (1,5-Cyclooctadiene)-(2S,3S)-2,3-bis(diphenylphosphino)butanerhodium(I) Perchlorate Tetrahedrofuran Solvate. *Inorg. Chem.* **1977**, *16*, 1187–1191.

(70) Fryzuk, M. D.; Bosnich, B. Asymmetric Synthesis. Production of Optically Active Amino Acids by Catalytic Hydrogenation. *J. Am. Chem. Soc.* **1977**, *99*, 6262–6267.

(71) Seebach, D.; Beck, A. K.; Heckel, A. TADDOLs, Their Derivatives, and TADDOL Analogues: Versatile Chiral Auxiliaries. *Angew. Chem., Int. Ed. Engl.* **2001**, *40*, 92–138.

(72) Bradley, D. C.; Mehrotra, R. C.; Rothwell, I. P.; Singh, A. *Alkoxo and Aryloxo Derivatives of Metals*; Academic Press: New York, 2001.

(73) Seebach, D.; Plattner, D. A.; Beck, A. K.; Wang, Y. M.; Hunziker, D.; Petter, W. On The Mechanism of Enantioselective Reactions Using α, α, α', α'-Tetraaryl-1,3-dioxolane-4,5-dimethanol(TADDOL)-Derived Titanates—Differences Between C_2-Symmetrical and C_1-Symmetrical TADDOLs: Facts, Implications and Generalizations. *Helv. Chim. Acta* **1992**, *75*, 2171–2209.

(74) Gothelf, K. V.; Hazzel, R., G; Jørgensen, K. A. Crystal Structure of a Chiral Titanium-Catalyst–Alkene Complex. The Intermediate in the Catalytic Asymmetric Diels–Alder and 1,3-Dipolar Cycloaddition Reactions. *J. Am. Chem. Soc.* **1995**, *117*, 4435–4436.

(75) Braun, M. The "Magic" Diarylhydroxylmethyl Group. *Angew. Chem., Int. Ed. Engl.* **1996**, *35*, 519–522.

(76) Weber, B.; Seebach, D. Ti-TADDOLate-Catalyzed, Highly Enantioselective Addition of Alkyl- and Aryl-Titanium Derivatives to Aldehydes. *Tetrahedron* **1994**, *50*, 7473–7484.

(77) Seebach, D.; Beck, A. K.; Schmidt, B.; Wang, Y. M. Enantio- and Diastereoselective Titanium-TADDOLate Catalyzed Addition of Diethyl and Bis(3-buten-1-yl) Zinc to Aldehydes. A Full Account with Preparative Details. *Tetrahedron* **1994**, *50*, 4363–4384.

(78) Seebach, D.; Beck, A. K.; Roggo, S.; Wonnacott, A. Chiral Alkoxytitanium(IV) Complexes for Enantioselective Nucleophilic Additions to Aldehydes and as Lewis Acids in Diels–Alder Reactions. *Chem. Ber.* **1985**, *118*, 3673–3682.

(79) Narasaka, K.; Iwasawa, N.; Inoue, M.; Yamada, T.; Kakashima, M.; Sugimori, J. Asymmetric Diels–Alder Reaction Catalyzed by a Chiral Titanium Reagent. *J. Am. Chem. Soc.* **1989**, *111*, 5340–5345.

(80) Gothelf, K. V.; Jørgensen, K. A. On the Mechanism of Ti-TADDOLate-Catalyzed Asymmetric Diels–Alder Reactions. *J. Org. Chem.* **1995**, *60*, 6847–6851.

(81) Gothelf, K. V.; Jørgensen, K. A. On the *trans-cis* Controversy in Ti-TADDOLate-Catalysed Cycloadditions. Experimental Indications for the Structure of the Reactive Catalyst–Substrate Intermediate. *J. Chem. Soc., Perkin Trans.1* **1997**, 111–116.

(82) Haase, C.; Sarko, C. R.; DiMare, M. TADDOL-Based Titanium Catalysts and Their Adducts: Understanding Asymmetric Catalysis of Diels–Alder Reactions. *J. Org. Chem.* **1995**, *60*, 1777–1787.

(83) Seebach, D.; Dahinden, R.; Marti, R. E.; Beck, A. K.; Plattner, D. A.; Kühnle, F. N. M. On the Ti-TADDOLate-Catalyzed Diels–Alder Addition of 3-Butenoyl-1,3-oxazolidine-2-one to Cyclopentadiene. General Features of Ti-BINOLate and Ti-TADDOLate-Mediated Reactions. *J. Org. Chem.* **1995**, *60*, 1788–1799.

(84) García, J. I.; Martinez-Merino, V.; Mayoral, J. A. Quantum Chemical Insights into the Mechanism of the TADDOL-TiCl$_2$ Catalyzed Diels–Alder Reactions. *J. Org. Chem.* **1998**, *63*, 2321–2324.

(85) Seeman, J. I. Effect of Conformational Change on Reactivity in Organic Chemistry. Evaluations, Applications, and Extensions of Curtin–Hammett–Winstein-Holness Kinetics. *Chem. Rev.* **1983**, *83*, 83–134.

(86) Muñiz, K.; Bolm, C. Configurational Control in Stereochemically Pure Ligands and Metal Complexes for Asymmetric Catalysis. *Chem. Eur. J.* **2000**, *6*, 2309–2316.

(87) Mimoun, H.; Yves de Saint Laumer, J.; Giannini, L.; Scopelliti, R.; Floriani, C. Enantioselective Reduction of Ketones by Polymethylhydrosiloxane in the Presence of Chiral Zinc Catalysts. *J. Am. Chem. Soc.* **1999**, *121*, 6158–6166.

(88) Mastranzo, V. M.; Quintero, L.; Anaya de Parrodi, C.; Juaristi, E.; Walsh, P. J. Use of Diamines Containing the α-Phenylethyl Group as Chiral Ligands in the Asymmetric Hydrosilylation of Prochiral Ketones. *Tetrahedron* **2004**, *60*, 1781–1789.

(89) Takahashi, H.; Kawakita, T.; Ohno, M.; Yoshioka, M.; Kobayashi, S. A Catalytic Enantioselective Reaction Using a C_2-Symmetric Disulfonamide as a Chiral Ligand: Alkylation of Aldehydes Catalyzed by Disulfonamide-Ti(O-*i*-Pr)$_4$-Dialkylzinc Reagents. *Tetrahedron* **1992**, *48*, 5691–5700.

(90) Takahashi, H.; Kawakita, T.; Yoshioka, M.; Kobayashi, S.; Ohno, M. Enantioselective Alkylation of Aldehyde Catalyzed by Bissulfonamide-Ti(O-*i*-Pr)$_4$-Dialkylzinc System. *Tetrahedron Lett.* **1989**, *30*, 7095–7098.

(91) Knochel, P.; Jones, P. *Organozinc Reagents*; Oxford University Press: New York, 1999.

(92) Ostwald, R.; Chavant, P.-Y.; Stadtmuller, H.; Knochel, P. Catalytic Asymmetric Addition of

Polyfunctional Dialkylzincs to β-Stannylated and β-Silylated Unsaturated Aldehydes. *J. Org. Chem.* **1994**, *59*, 4143–4153.

(93) Pritchett, S.; Woodmansee, D. H.; Gantzel, P.; Walsh, P. J. Synthesis and Crystal Structures of Bis(sulfonamide) Titanium Bis(alkoxide) Complexes: Mechanistic Implications in the Bis(sulfonamide) Catalyzed Asymmetric Addition of Dialkylzinc Reagents to Aldehydes. *J. Am. Chem. Soc.* **1998**, *120*, 6423–6424.

(94) Pritchett, S.; Gantzel, P.; Walsh, P. J. Synthesis and Structural Study of Titanium Bis(sulfonamido) Bis(amide) Complexes. *Organometallics* **1999**, *18*, 823–831.

(95) Royo, E.; Betancort, J. M.; Davis, T. J.; Walsh, P. J. Synthesis, Structure and Catalytic Properties of Bis[bis(sulfonamido)] Titanium Complexes. *Organometallics* **2000**, *19*, 4840–4851.

(96) Armistead, L. T.; White, P. S.; Gagné, M. R. Synthesis and Structure of Titanium(IV) Amido Complexes Containing C_2-Symmetric Bis(sulfonamide) Ligands. *Organometallics* **1998**, *17*, 216–220.

(97) Pritchett, S.; Woodmansee, D. H.; Davis, T. J.; Walsh, P. J. Improved Methodology for the Asymmetric Alkylation of Aldehydes Employing Bis(sulfonamide) Complexes. *Tetrahedron Lett.* **1998**, *39*, 5941–5944.

(98) Evans, D. A.; Nelson, S. G. Chiral Magnesium Bis(sulfonamide) Complexes as Catalysts for the Merged Enolization and Enantioselective Amination of N-Acyloxazolidinones. A Catalytic Approach to the Synthesis of Arylglycines. *J. Am. Chem. Soc.* **1997**, *119*, 6452–6453.

(99) Takahashi, H.; Yoshioka, M.; Ohno, M.; Kobayashi, S. A Catalytic Enantioselective Reaction Using a C_2-symmetric Disulfonamide as a Chiral Ligand: Cyclopropanation of Allylic Alcohols by the Et_2Zn–CH_2I_2–Disulfonamide System. *Tetrahedron Lett.* **1992**, *33*, 2575–2578.

(100) Takahashi, H.; Yoshioka, M.; Shibasaki, M.; Ohno, M.; Imai, N.; Kobayashi, S. A Chiral Enantioselective Reaction Using a C_2-Symmetric Disulfonamide as a Chiral Ligand: Simmons–Smith Cyclopropanation of Allylic Alcohols by the Et_2Zn–CH_2I_2–Disulfonamide System. *Tetrahedron* **1995**, *51*, 12013–12026.

(101) Denmark, S. E.; O'Connor, S. P. Enantioselective Cyclopropanation of Allylic Alcohols. The Effect of Zinc Iodide. *J. Org. Chem.* **1997**, *62*, 3390–3401.

(102) Denmark, S. E.; O'Connor, S. P. Catalytic, Enantioselective Cyclopropanation of Allylic Alcohols. Substrate Generality. *J. Org. Chem.* **1997**, *62*, 584–594.

(103) Corey, E. J.; Sarshar, S.; Lee, D.-H. First Example of a Highly Enantioselective Catalytic Diels–Alder Reaction of an Achiral C_{2v}-Symmetric Dienophile and an Achiral Diene. *J. Am. Chem. Soc.* **1994**, *116*, 12089–12090.

(104) Ichiyanagi, T.; Shimizu, M.; Fujisawa, T. Enantioselectie Diels–Alder Reaction Using Chiral Mg Complexes Derived from Chiral 2-[2-[(Alkyl- or 2-[2-[(Arylsulfonyl)amino]phenyl]-4-phenyl-1,3-oxazoline. *J. Org. Chem.* **1997**, *62*, 7937–7941.

(105) Denmark, S. E.; O'Connor, S. P.; Wilson, S. R. Solution and Solid-State Studies of a Chiral Zinc-Sulfonamide Relevant to an Enantioselective Cyclopropanation Reaction. *Angew. Chem., Int. Ed. Engl.* **1998**, *37*, 1149–1151.

(106) Evans, D. A.; Campos, K. R.; Tedrow, J. S.; Michael, F. E.; Gagné, M. R. Application of Chiral Mixed Phosphorus/Sulfur Ligands to Palladium-Catalyzed Allylic Substitutions. *J. Am. Chem. Soc.* **2000**, *122*, 7905–7920.

(107) Evans, D. A.; Campos, K. R.; Tedrow, J. S.; Michael, F. E.; Gagné, M. R. Chiral Mixed Phosphorus/Sulfur Ligands for Palladium-Catalyzed Allylic Alkylations and Aminations. *J. Org. Chem.* **1999**, *64*, 2994–2995.

(108) Evans, D. A.; Michael, F. E.; Tedrow, J. S.; Campos, K. R. Application of Chiral Mixed Phosphorus/Sulfur Ligands to Enantioselective Rhodium-Catalyzed Dehydroamino Acid Hydrogenation and Ketone Hydrosilylation Processes. *J. Am. Chem. Soc.* **2003**, *125*, 3534–3543.

(109) Barbaro, P.; Currao, A.; Herrmann, J.; Nesper, R.; Pregosin, P. S.; Salzmann, R. Chiral P,S-Ligands Based on D-Thioglucose Tetraacetate. Palladium(II) Complexes and Allylic Alkylation. *Organometallics* **1996**, *15*, 1879–1888.

(110) Brown, J. M.; Evans, P. Structure and Reactivity in Asymmetric Hydrogenation; a Molecular Graphics Analysis. *Tetrahedron* **1988**, *4*, 4905–4916.

(111) Sibi, M. P.; Zhang, R.; Manyem, S. A New Class of Modular Chiral Ligands with Fluxional Groups. *J. Am. Chem. Soc.* **2003**, *125*, 9306–9307.

(112) von Zelewsky, A. *Stereochemistry of Coordination Compounds*; Wiley: New York, 1996.

(113) Brunner, H. Optically Active Organometallic Compounds of Transition Elements with Chiral Metal Atoms. *Angew. Chem. Int., Ed. Engl.* **1999**, *38*, 1194–1208.

(114) Faller, J. W.; Parr, J. Utility of Osmium(II) in the Catalysis of Asymmetric Diels–Alder Reactions. *Organometallics* **2001**, *20*, 697–699.

(115) Faller, J. W.; Grimmond, B. J.; D'Alliessi, D. G. An Application of Electronic Asymmetry to Highly Enantioselective Catalytic Diels–Alder Reactions. *J. Am. Chem. Soc.* **2001**, *123*, 2525–2529.

(116) Babin, J. E.; Whiteker, G. T. U.S. Patent 5,360,938, *Chem. Abstr.* **1994**, *122*, 18660.

(117) Nozaki, K.; Sakai, N.; Nanno, T.; Higashijima, T.; Mano, S.; Horiuchi, T.; Takaya, H. Highly Enantioselective Hydroformylation of Olefins Catalyzed by Rhodium(I) Complexes of New Chiral Phosphine-Phosphite Ligands. *J. Am. Chem. Soc.* **1997**, *119*, 4413–4423.

(118) Reetz, M. T.; Neugebauer, T. New Diphosphite Ligands for Catalytic Asymmetric Hydrogenation: The Crucial Role of Conformationally Enantiomeric Diols. *Angew. Chem., Int. Ed. Engl.* **1999**, *38*, 179–181.

(119) Imai, Y.; Zhang, W.; Kida, T.; Nakatsuji, Y.; Ikeda, I. Diphenylphosphinooxazoline Ligands with a Chiral Binaphthyl Backbone for Pd-Catalyzed Allylic Alkylation. *Tetrahedron Lett.* **1998**, *39*, 4343–4346.

(120) Zhang, W.; Xie, F.; Yoshinaga, H.; Kida, T.; Nakatsuji, Y.; Ikeda, I. A Novel Axially Chiral Phosphine-Oxazoline Ligand with an Axis-Unfixed Biphenyl Backbone: Preparation, Complexation, and Application in an Asymmetric Catalytic Reaction. *Synlett* **2006**, 1185–1188.

(121) Superchi, S.; Casarini, D.; Laurita, A.; Bavoso, A.; Rosini, C. Induction of a Preferred Twist in a Biphenyl Core by Stereogenic Centers: A Novel Approach to the Absolute Configuration of 1,2- and

1,3-Diols. *Angew. Chem., Int. Ed. Engl.* **2001**, *40*, 451–454.

(122) Isaksson, R.; Rashidi-Ranjbar, P.; Sandstrom, J. Synthesis and Chromatographic Resolution of Some Chiral Four-Carbon 2,2'-Bridged Biphenyls. Some Unusually High Selectivity Factors. *J. Chem. Soc., Perkin Trans. 1* **1991**, 1147–1152.

(123) Imai, Y.; Zhang, W.; Kida, T.; Nakatsuji, Y.; Ikeda, I. Novel Chiral Bisoxazoline Ligands with a Biphenyl Backbone: Preparation, Complexation, and Application in Asymmetric Catalytic Reactions. *J. Org. Chem.* **2000**, *65*, 3326–3333.

(124) Pastor, S. D.; Shum, S. P.; Rodebaugh, R. K.; Debellis, A. D.; Clarke, F. H. Sterically Congested Phosphite Ligands: Synthesis, Crystallographic Characterization, and Observation of Unprecedented Eight-Bond ^{31}P, ^{31}P Coupling in the ^{31}P-NMR Spectra. *Helv. Chim. Acta* **1993**, *76*, 900–915.

(125) Hartwig, J. Synthetic Chemistry—Recipes for Excess *Nature* **2005**, *437*, 487–488.

(126) Costa, A. M.; Jimeno, C.; Gavenonis, J.; Carroll, P. J.; Walsh, P. J. Optimization of Catalyst Enantioselectivity and Activity Using Achiral and Meso Ligands. *J. Am. Chem. Soc.* **2002**, *124*, 6929–6941.

(127) Mino, T.; Kashihara, K.; Yamashita, M. New Chiral Phosphine-Amide Ligands in Palladium-Catalyzed Asymmetric Allylic Alkylations. *Tetrahedron: Asymmetry* **2001**, *12*, 287–291.

(128) Clayden, J.; Lai, L. W.; Helliwell, M. Using Amide Conformation to "Project" the Stereochemistry of an (–)-Ephedrine-Derived Oxazolidine: A Pair of Pseudoenantiomeric Chiral Amido-Phosphine Ligands. *Tetrahedron: Asymmetry* **2001**, *12*, 695–698.

(129) Brandt, P.; Södergren, M. J.; Andersson, P. G.; Norrby, P.-O. Mechanistic Studies of Copper-Catalyzed Alkene Aziridination. *J. Am. Chem. Soc.* **2000**, *122*, 8013–8020.

(130) Evans, D. A.; Bilodeau, M. T.; Faul, M. M. Development of the Copper-Catalyzed Olefin Aziridination Reaction. *J. Am. Chem. Soc.* **1994**, *116*, 2742–2753.

(131) Li, Z.; Conser, K. R.; Jacobsen, E. N. Asymmetric Alkene Aziridination with Readily Available Chiral Diimine-Based Catalysts. *J. Am. Chem. Soc.* **1993**, *115*, 5326–5327.

(132) Li, Z.; Quan, R. W.; Jacobsen, E. N. Mechanism of the (Diimine)copper-Catalyzed Asymmetric Aziridination of Alkenes. Nitrene Transfer via Ligand-Accelerated Catalysis. *J. Am. Chem. Soc.* **1995**, *117*, 5889–5890.

(133) Sibi, M. P.; Venkatraman, L.; Liu, M.; Jasperse, C. P. A New Approach to Enantiocontrol and Enantioselectivity Amplification: Chiral Relay in Diels–Alder Reactions. *J. Am. Chem. Soc.* **2001**, *123*, 8444–8445.

(134) Sibi, M. P.; Chen, J.; Stanley, L. Enantioselective Diels–Alder Reactions: Effect of the Achiral Template on Reactivity and Selectivity. *Synlett* **2007**, 298–302.

5

Nonclassical Catalyst–Substrate Interactions

We have seen several examples of substrates that are activated by catalysts through two-point binding (see Chapter 2). The chelation of the substrate to the catalyst increases the binding constant while decreasing the degrees of freedom in the catalyst–substrate adduct. Not only does chelation reduce the number of competing transition states, it also simplifies the interpretation of the stereochemical outcome of the reaction. In general, bidentate substrates form two dative bonds to the catalyst. Other types of attractive substrate–catalyst interactions include those involving polarizable π-systems. These weak forces (secondary interactions) have rarely been applied in catalyst design, but they have been shown to exert a significant influence on catalyst enantioselectivity levels.

5.1 π–π Interactions in Substrate Binding

An important example of the utility of secondary interactions was demonstrated using the Diels–Alder reaction of α,β-unsaturated esters,[1,2] acid chlorides, and ketones.[3] As shown in **Figure 5.1**, α,β-unsaturated esters can coordinate to Lewis acids with four possible binding modes. The Lewis acid can bind *syn* or *anti* with respect to the –OR group, and with the s-trans and s-cis conformations of the ester.[4] Uncoordinated esters preferentially adopt a conformation with the R group *syn* to the carbonyl oxygen, disfavoring the *syn* binding modes in **Figure 5.1**. The *anti* s-cis conformation is disfavored on steric grounds. Thus, the *anti* s-trans conformation is favored.

   To investigate the possible role of secondary interactions, a Lewis acid catalyst was designed with a well-defined ligand framework around a three-coordinate boron center (**Figure 5.2**). Boron-based catalysts exhibit a strong tendency to go from three-coordinate to four-coordinate on binding the substrate. Unlike heavier main-group elements and transition metals, boron does not exceed a coordination number of four. A key feature of this system is the close proximity of the boron to an electron-rich and polarizable naphthyl group. Upon coordination and activation of the car-

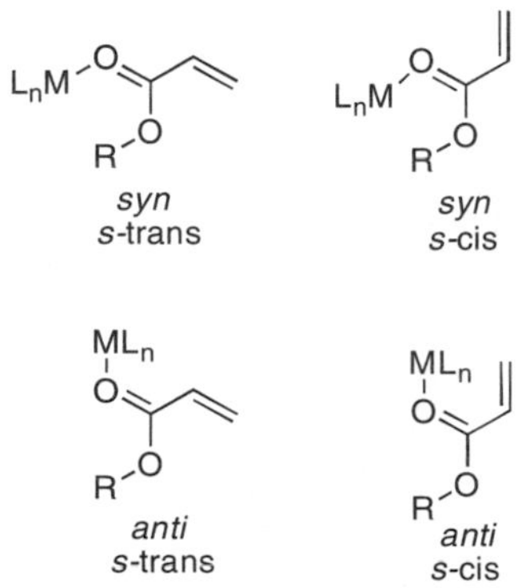

Figure 5.1. Possible geometries of Lewis acid–ester adducts.

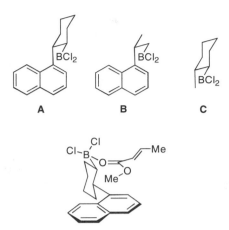

Figure 5.2. Catalyst **A** is illustrated along with controls **B** and **C** (top). Below is shown a drawing of the solid-state conformation of **A** with methyl crotonate.

bonyl group by the boron, electrostatic and dipole–induced-dipole interactions create an attraction between the polarizable aromatic group and the electron-deficient carbonyl of the dienophile. These favorable secondary interactions serve to limit the conformational freedom of the bound substrate and simultaneously shield one of the prochiral faces from reaction (**Figure 5.2**).

Support for the two-point substrate-coordination model was obtained from X-ray crystallographic analysis of a series of substrate–catalyst adducts, variable-temperature NMR studies, and correlation of the predicted facial selectivity in the cycloaddition reaction.[1,2] The solid-state structure of **A** bound to methyl crotonate shows boron binding the carbonyl oxygen *anti* to the –OR group (**Figure 5.2**).[4] The ester is found over the naphthalene ring system and is roughly parallel to it. The dienophile is within the van der Waals radii of the naphthyl group, with the closest distance between the carbonyl carbon and the naphthyl ring being 3.17 Å. A series of variable-temperature ^{1}H NMR experiments with Lewis acids **A**, **B**, and **C**, in combination with one equivalent of methyl crotonate, were performed to evaluate the secondary interactions in solution. Groups positioned over the aromatic ring will exhibit an upfield shift in the ^{1}H NMR spectrum due to ring anisotropy. With Lewis acids **A** and **B**, the methoxy resonance of the ester shifted upfield as the temperature was lowered, whereas in **C** a downfield shift was observed. Lowering the temperature favors the formation of the Lewis acid-substrate adduct, which increases its concentration. At –55 °C, the difference in chemical shift of the methoxy resonances in substrate bound to **A** versus **C** was 1.2 ppm, indicating a time-averaged preferential orientation of the methoxy group above the naphthyl ring in **A**, consistent with the solid-state structure (**Figure 5.2**). Additionally, the Diels–Alder reaction with catalyst **A** exhibited high levels of enantioselectivity with cyclopentadiene and cyclohexadiene, as illustrated in **Equation 5.1**. The configurations of the Diels–Alder products were consistent with attack of the diene on the exposed face of the dienophile, as shown in the proposed transition state (**Figure 5.3**).

Equation 5.1

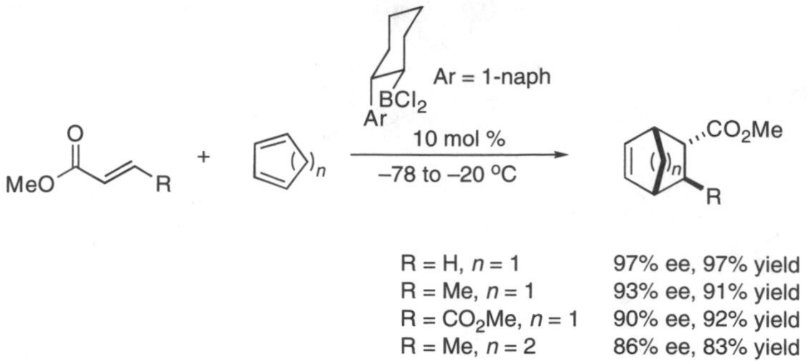

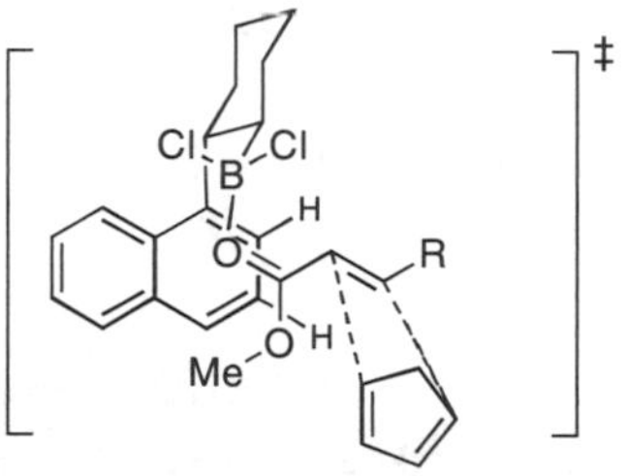

R = H, n = 1   97% ee, 97% yield
R = Me, n = 1   93% ee, 91% yield
R = CO_2Me, n = 1   90% ee, 92% yield
R = Me, n = 2   86% ee, 83% yield

Figure 5.3. Proposed transition state for the Lewis acid catalyzed asymmetric Diels–Alder reaction. A secondary interaction between the naphthyl group and the substrate carbonyl favors the orientation shown.

The two-point bonding model of the type outlined above is an important design concept that will likely be applicable to other asymmetric catalysts and reactions.

5.2 C–H•••π Interactions in Substrate Binding

Although C–H•••π interactions have been well-characterized outside the realm of asymmetric catalysis,[5,6] evidence for the influence of such interactions in enantioselection is rare. It is likely, however, that C–H•••π interactions are actually more common, but have not been documented due to their subtle nature. Two examples are provided below, one identified by X-ray crystallographic characterization of likely intermediates in a catalytic asymmetric reaction and the other characterized by high-level computational studies.

5.2.1 C–H•••π Interactions in 1,3-Dipolar Cycloadditions

A fascinating example of a catalytic asymmetric reaction bringing together several important concepts is found in the rhodium- and iridium-catalyzed 1,3-dipolar cycloaddition of nitrones with methacrolein.[7] The reaction and representative results are illustrated in **Figure 5.4**. Despite the simplicity of the chiral diphosphine (*R*)-prophos (**Figure 5.4**), excellent control of regiochemistry as well as diastereoselectivity and enantioselectivity were observed. Solution and solid-state structural studies elucidated the origin of stereochemical control in this system.

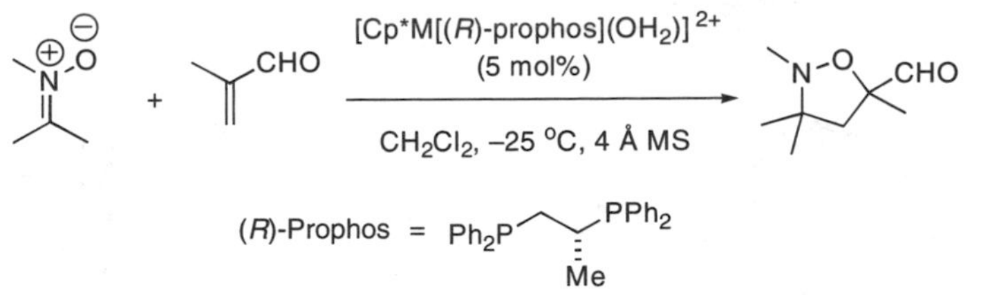

Substrate	Product	M	Yield (%)	ee (%)
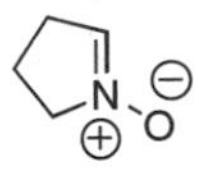	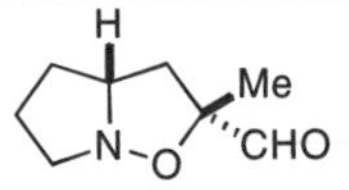	Rh Ir	100 100	86 86
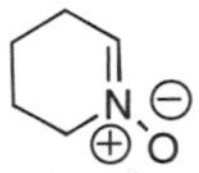	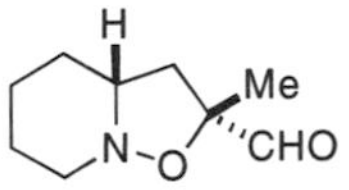	Rh Ir	99 100	91 91
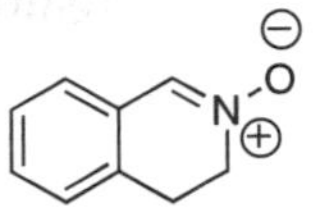	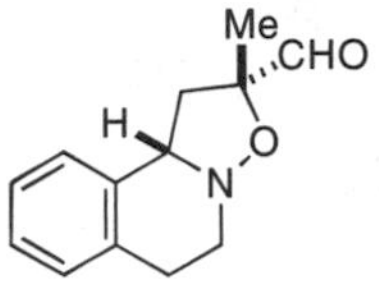	Rh Ir	58 100	81 93

Figure 5.4. The rhodium- and iridium-catalyzed 1,3-dipolar cycloaddition with methacrolein.

The precatalyst for this reaction was prepared by combining Cp*M(acetone)$_3^{2+}$ (M = Rh, Ir) with (*R*)-prophos in the presence of a trace amount of water (**Equation 5.2**). Only the (*S*)-configuration at the metal center was detected by variable-temperature ^{1}H and ^{31}P NMR spectroscopy. The (*R*)-configuration would result in steric clashing between the bulky Cp*M group and the methyl of the phosphine back-bone. The aquo complexes reversibly bind methacrolein with displacement of water (**Equation 5.2**). Addition of molecular sieves drives the equilibrium to the metha-crolein adduct. The ^{1}H NMR of the Rh– and Ir–methacrolein adducts exhibited resonances for the aldehyde CH at 7.06 and 7.23 ppm, a shift to higher field compared to the free aldehyde by over 2.5 ppm. X-ray structure determination of the rhodium and iridium derivatives revealed that the positioning of the aldehyde CH over the *pro*-(*S*) *P*-phenyl ring (**Figure 5.5**) was the origin of the large upfield shift. The CH-π interactions are characterized by C–H•••C_{Ph} distances under 3.05 Å. It is proposed that this interaction stabilizes the conformation about the metal–methacrolein bond.

Equation 5.2

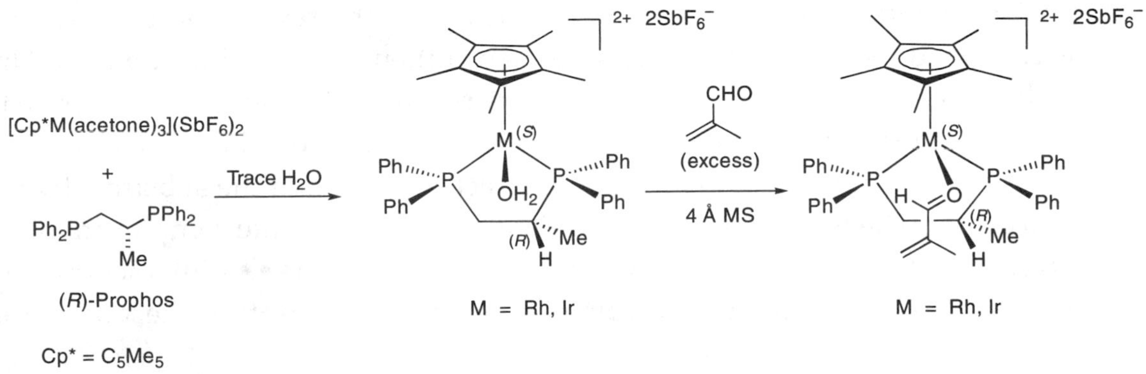

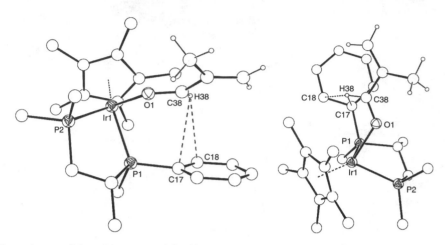

Figure 5.5. Two views of the structure of Cp*(prophos)Ir(methacrolein)$^{2+}$, illustrating the CH–π interaction with H38. The anions and three of the *P*-phenyl groups have been omitted for clarity.

On the basis of NMR and X-ray crystallographic characterization of the intermediates in the 1,3-dipolar cycloaddition, the catalytic cycle in **Figure 5.6** was proposed. This cycle includes the spectroscopically observed reversible coordination of the nitrone to the catalyst, which inhibits the reaction. Slow addition of the nitrone minimizes the impact of this inhibition. A transition state is illustrated in which the nitrone attacks the *Re* face of the bound methacrolein in the *s*-trans conformation to give the observed stereochemistry of the product. Key to the transmission of asymmetry in this reaction is the stereochemistry of the metal center. It appears that the most important role of the methyl group on prophos is to control the diastereoselectivity in coordination of prophos to the metal precursor in **Equation 5.2**, rather than to directly impact the enantioselection in the cycloaddition transition state.

5.2.2 Computational Identification of C–H•••π Interactions in Asymmetric Transfer Hydrogenations

An interesting example of two-point binding is found in the asymmetric transfer hydrogenation of ketones (**Figure 5.7**).[8–10] Isopropanol serves as both reagent and solvent in this reaction. Furthermore, the high concentration of isopropanol prevents erosion of the enantioselectivity by reducing the rate of the reverse process. The reaction is catalyzed by an electronically saturated ruthenium–hydride complex. Unlike most hydrogenation catalysts, which require precoordination of the substrate to the metal, the reduction occurs in this system in an outer coordination sphere. In other words, the reaction occurs without the direct activation of the substrate by the ruthenium center. The formation of the transient catalyst–substrate complex involves a hydrogen bond from the catalyst to the substrate and a C–H•••π interaction with an η^6-arene as the C–H donor and an aromatic ring on the substrate as the π-acceptor.

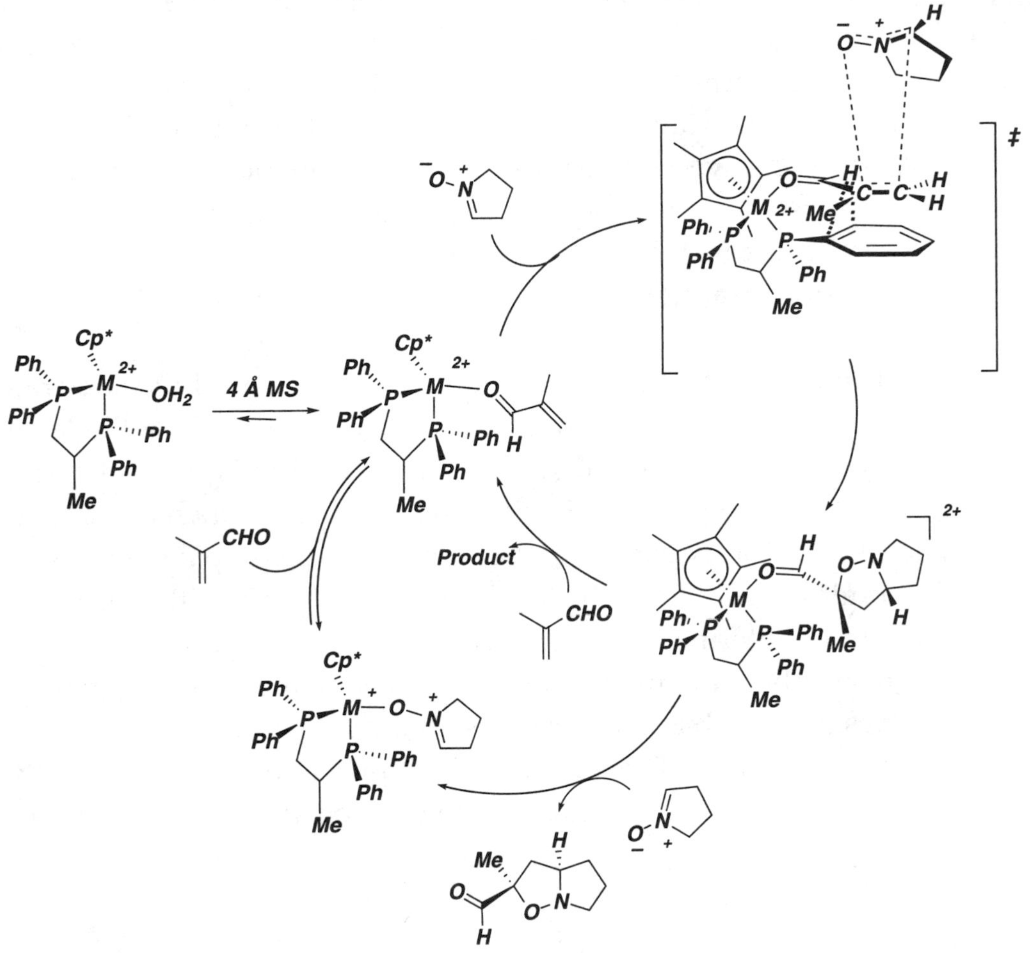
4 Å MS
CHO
Product
Cp*
Ph
Me

Figure 5.6. Possible transition state for the asymmetric 1,3-dipolar cycloaddition.

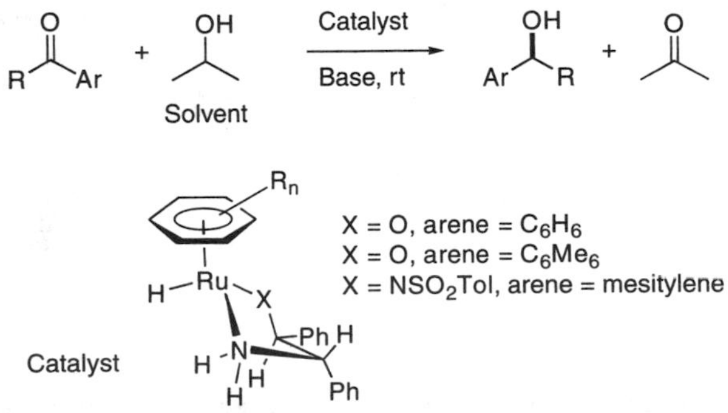
Catalyst
Base, rt
Solvent
X = O, arene = C6H6
X = O, arene = C6Me6
X = NSO2Tol, arene = mesitylene
Catalyst

Figure 5.7. The asymmetric transfer hydrogenation reaction proceeds with high levels of enantioselectivity and excellent yields.

As shown in **Figure 5.7**, catalysts with different combinations of η^6-arenes and either amino alcohols or an *N*-sulfonylated diamine can be used.[9] The pseudoequatorial phenyl substituents on the backbone of the chiral ligand bias the conformation of the skewed five-membered ring metallacycle such that it adopts the δ-configuration (see Section A.1.1.b). As we will see, the ring conformation is important in the diastereoselective regeneration of the ruthenium catalyst with the (*R*)-configuration at the metal center (**Figure 5.8**). The stereochemistry at the ruthenium has been determined by X-ray crystallography[11] and is consistent with the results of computational studies.[12]

The mechanism of the reduction entails initial association of the substrate with the catalyst via hydrogen bonding between the pseudoaxial N–H of the ligand and the carbonyl oxygen to give intermediate **A**, as illustrated in **Figure 5.8**.[10] This transient intermediate is proposed to directly precede the six-membered pericyclic transition state (**B**). Hydride delivery to the carbonyl is accompanied by transfer of hydrogen from nitrogen to the incipient alkoxide during this processes. A 16-electron ruthenium intermediate, initially associated with the product alcohol by a hydrogen bond to the ruthenium amide, dissociates the product, generating a 16-electron species (**C**) that has been characterized by X-ray crystallography.[11] In a reversal of the above process, **C** reacts with the 2-propanol diastereoselectively to produce acetone and reform the saturated ruthenium catalyst that is chiral at the metal.

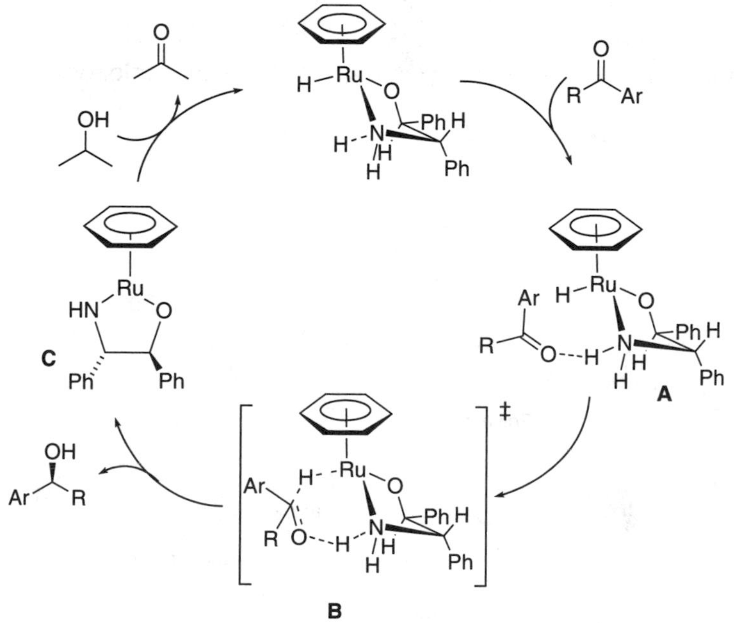

Figure 5.8. Proposed catalytic cycle for the transfer hydrogenation reaction. The catalyst exhibits a high turnover frequency and excellent enantioselectivities.

In the six-membered transition state in **Figure 5.8**, two possible orientations of the ketone are envisioned that lead to enantiomeric products (**Figure 5.9**). Surprisingly, analysis of the configuration of the alcohol product implies that the reaction proceeds via the sterically more encumbered transition state, with delivery of the hydride to the *Re* face of the carbonyl. After further scrutiny employing high-level theoretical calculations (R = H), it was found that the stereochemical induction is not only a result of the chiral chelating ligand and the configuration at ruthenium, but also an attractive C–H•••π interaction between a C–H of the η^6-arene ligand and the substrate (**Figure 5.9**).[13]

In the model system with benzaldehyde as substrate (**Figure 5.9**, R = H), the calculated (arene)C–H•••Ph distance to the *ortho* carbon (2.86 Å) is close to the sum of the van der Waals radii (2.9 Å). The C–H•••π attraction arises from both electrostatic and charge-transfer interactions.[6] This favorable interaction is accentuated by two factors: 1) the increased partial positive charge (and acidity[14]) imparted on the η^6-arene hydrogens upon coordination to the ruthenium and 2) the increased partial negative charge at the *ortho* carbon, because of the alkoxide character developing on the substrate in the transition state.[13] This interaction stabilizes the *Re* transition state.

Experimental support for this hypothesis is found in the reduction of a series of substituted benzaldehyde-d_1 derivatives (**Figure 5.10**).[15] Compounds with electron-donating substituents were reduced with greater enantioselectivity due to the increased acceptor character of the electron-rich aromatic ring. Conversely, electron-deficient substrates gave lower enantioselectivities. Consistent with the proposed transition-state models (**Figure 5.9**), reaction of an α,β-unsaturated aldehyde (cinnamaldehyde-d_1) and a saturated aldehyde (dihydrocinnamaldehyde-d_1) under these conditions gave nearly racemic alcohol product, illustrative of the importance of the C–H•••π interactions.

Another probe of the C–H•••π attraction is 4-cyano-4'-methoxybenzophenone, a ketone that is essentially sterically symmetrical, but electronically unsymmetrical

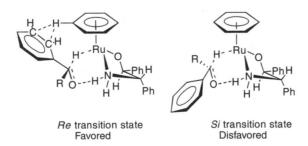

Re transition state
Favored

Si transition state
Disfavored

Figure 5.9. *Re* and *Si* transition states for the asymmetric transfer hydrogenation. Although the *Re* transition state is more sterically hindered, it is stabilized by a C–H•••π interaction between the η^6-arene C–H and the substrate *ortho* and *meta* carbons.

due to the different electronic donor abilities of the cyano and methoxy groups. As illustrated in **Figure 5.11**, the reduction preferentially generates the (S)-alcohol, consistent with a stronger C–H•••π interaction involving the electron-rich 4-methoxy aryl group (34% ee).[13]

Based on the transition states outlined above with the (η^6-C_6H_6)Ru-based catalyst, it might be anticipated that changing the η^6-arene to C_6Me_6 would result in a change from the *Re* to the *Si* transition state. Based on the configuration of the alcohol products, this was not the case. A computational analysis of the *Re* and *Si* transition states suggested that a C(sp^3)H•••π attraction stabilized the *Re* transition state sufficiently to override the greater steric congestion between the substrate and the η^6-C_6Me_6 (**Figure 5.12**). The distances between the interacting carbon of the methyl group and the *ortho* and *meta* carbons of the substrate in the *Re* transition state were calculated to be 2.94 Å and 3.18 Å, respectively. As in the parent C_6H_6 system, the increased positive character of the benzylic hydrogens in coordinated η^6-arenes relative to the uncoordinated arenes increases the strength of the C(sp^3)H•••π attraction.

The attractive C–H•••π interactions outlined here are very important in the stabilization of one of the diastereomeric transition states. Nonetheless, other factors are

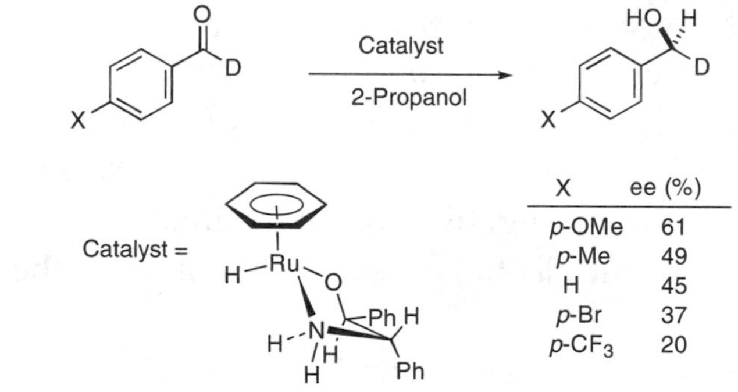

X	ee (%)
p-OMe	61
p-Me	49
H	45
p-Br	37
p-CF$_3$	20

Figure 5.10. Reductions of substituted benzaldehydes-d_1. The hydride catalyst was generated in situ from the precursor chloride by addition of base.

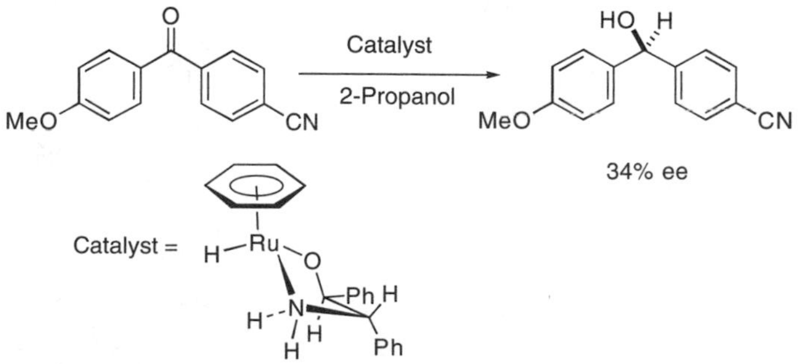

Figure 5.11. The lone pairs of the benzophenone derivative are in sterically similar environments, but they are electronically distinct. The configuration of the product is consistent with the more electron-rich π-system proximal to the η^6-arene because of the more favorable C–H•••π interaction.

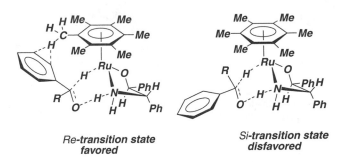

Re-transition state
favored

Si-transition state
disfavored

Figure 5.12. Re and Si transition states for the asymmetric transfer hydrogenation. The Re transition state is stabilized by a $C(sp^3)H•••\pi$ interaction.

also crucial to the enantioselectivity of the catalyst, such as the ligand stereochemistry and the configuration of the ruthenium center. An analogy has been drawn between the C–H•••π interaction described above and the stabilization of the *endo* pathway in the Diels–Alder reaction through secondary orbital interactions in the absence of a catalyst or promoter.[10]

5.3 Formyl C–H•••O Interactions in Lewis Acid Catalysis

To rationalize the stereochemical induction in Lewis acid catalyzed asymmetric reactions of aldehydes, an interaction between the formyl C–H bond of the coordinated aldehyde and a basic site (usually oxygen) on the catalyst has been proposed to stabilize one of the diastereomeric transition states.[16–19] This interaction, although likely weak, is not the overriding factor in determining stereoselectivity, but is an important consideration along with the more classical structural elements. This organizing factor serves to restrict rotation about the dative bond between the carbonyl oxygen and the Lewis acid catalyst.

The formyl hydrogen of an aldehyde or formamide is not usually considered highly electron deficient or a participant in hydrogen bonds. There is a large body of X-ray crystallographic evidence, however, that formyl C–H bonds interact with neighboring carbonyl oxygens, as judged by the occurrence of close C•••O distances in C–H•••O interactions (within 3–4 Å).[20–22] The frequency with which short C–H•••O contacts occur in the solid-state suggests an influence on the crystal packing. Furthermore, their geometrical characteristics, in this case the orientation of the formyl C–H bond toward the sp^2 hybridized lone pair of the carbonyl oxygen, are similar to those of O–H•••O hydrogen bonds.[20] These attractive C–H•••O interactions are electrostatic in nature. Whether or not these interactions should be called hydrogen bonds is still under debate.[21] Nonetheless, the term hydrogen bonding will suffice in this context.

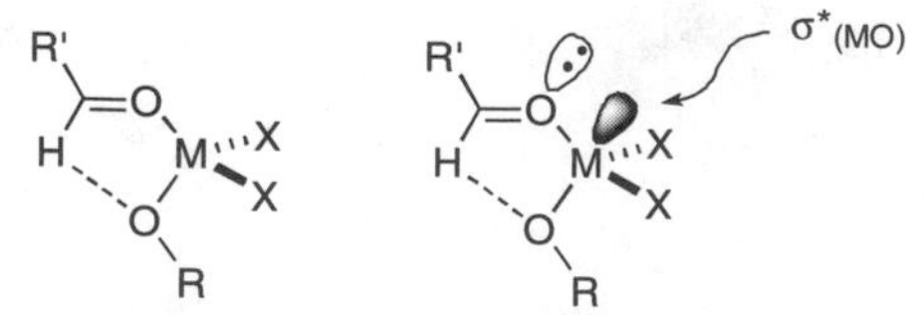

Figure 5.13. On the left is shown an interaction between the formyl C–H of a coordinated aldehyde and the basic oxygen of an alkoxide ligand on the Lewis acid M. On the right is shown the alignment of the M–O σ* orbital of the ligand and the lone pair of the aldehyde.

The short formyl C–H•••O interactions described above are mostly intermolecular, solid-state phenomena with unactivated aldehydes. Several significant differences arise when aldehydes are activated by Lewis acids. Coordination of the carbonyl oxygen to a Lewis acid increases the positive character of the carbonyl carbon and the formyl hydrogen. It also increases the electron density on the Lewis acidic site and the Lewis bacisity of heteroatoms attached to the Lewis acidic atom. Both of these factors lead to more favorable interactions in these intramolecular hydrogen bonds (**Figure 5.13**) relative to those in unactivated aldehydes. The conformation shown, with the carbonyl eclipsing the M–O bond, is that often found in the solid-state structures. Additionally, the uncoordinated lone pair of the aldehyde is *syn peri-planar* to the M–O σ* orbital. Delocalization of the carbonyl oxygen lone pair into the M–O σ* orbital, in an analogous fashion to the anomeric effect,[23,24] may also contribute to the stabilization of this conformation. The existence of the C–H•••O hydrogen bond[23-28] and the n to σ* orbital interaction[23,24] are in agreement with molecular-orbital calculations.

An example of how the effects of a C–H•••O hydrogen bond, electrostatic and dipole–induced-dipole attractions, and Lewis acid coordination can all combine and result in very high enantioselectivity is illustrated with a boron-based Lewis acid in **Equation 5.3**. The process of interest is the Diels–Alder reaction between 2-substituted acrolein derivatives and cyclopentadiene in the presence of a catalyst derived from *n*-butylboronic acid and a tryptophan derivative (**Equation 5.3**).

Equation 5.3

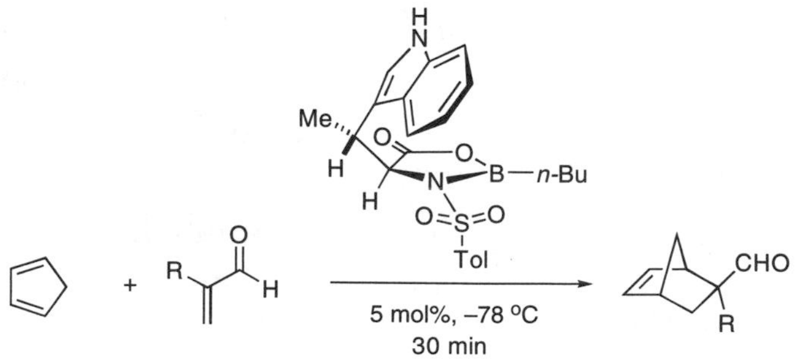

R	ee (%)
Br	> 99
Cl	> 99
Me	92
Et	92

An in-depth study of the catalyst in the absence and presence of a substrate provided vital insight into the interactions that control enantioselectivity in this system. Variable-temperature nOe studies of the catalyst indicate that it is conformationally dynamic. In contrast, upon addition of an equivalent of 2-methylacrolein and by cooling to low temperature (210 K), nOe effects were observed between nonvicinal hydrogens. The results of this experiment led to the conclusion that the catalyst–substrate adduct is conformationally rigid and that the conformation is close to that shown in **Figure 5.14**. Further evidence for this conformation is the detection of a charge-transfer complex between the indole and the bound substrate, indicating a separation of about 3 Å between these groups. Additionally, coordination of the aldehyde is reversible, even at low temperature. As the temperature is lowered, the methyl resonance of the aldehyde moves to higher field, indicating that the aldehyde is located above the indole moiety. In contrast, in the complex between 2-methylacrolein and BF$_3$, the methyl resonance shifts downfield as the temperature is lowered.[29]

The substrate can bind to boron with either the s-cis or s-trans geometries, and these conformers will lead to enantiomeric products (**Figure 5.15**). Interconversion of the conformers can take place by bond rotation or exchange of the substrate, both of which are faster than the Diels–Alder reaction. Thus, the system is believed to be under Curtin–Hammett control. Product formation from these two conformers will not depend on their prevalence, but on their free energies of activation. Based on the sense of enantioselection in the product, the s-cis conformation is more reactive.

In the transition state for the Diels–Alder cycloaddition, the reacting double bond of the dienophile undergoes a change in hybridization from sp^2 to sp^3, increasing its

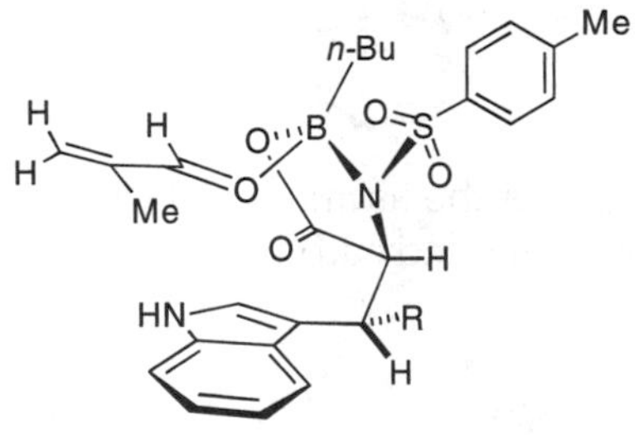

Figure 5.14. Solution structure based on observed nOe's.

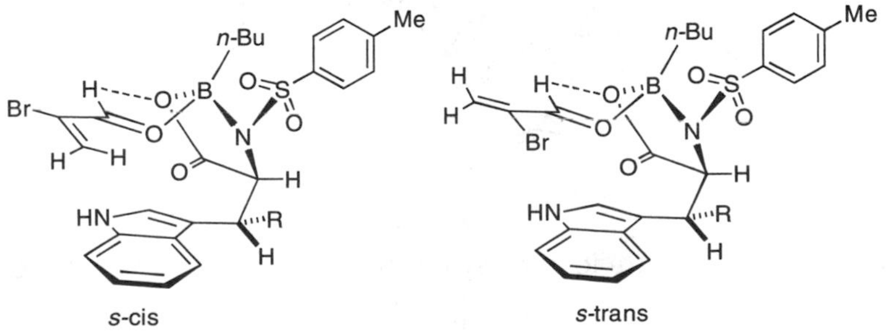

Figure 5.15. Proposed structures of the s-cis (left) and s-trans conformers.

steric demands. It was proposed that in the *s*-trans conformer, rehybridization of these carbons during the Diels–Alder reaction would drive the bromide toward the indole ring, causing an increase in nonbonded interactions in the transition state. In the *s*-cis conformation, the effect is greatly reduced. A proposed transition state is shown in **Figure 5.16**.

A closely related catalyst has been employed in the Mukaiyama aldol reaction and exhibited excellent enantioselectivities (**Equation 5.4**).[30,31] The electron-withdrawing 3,5-bis(trifluoromethyl)phenyl group (see **Figure 5.17**) increases the Lewis acidity and the turnover frequency of the catalysts, relative to the *n*-Bu derivative used in the Diels–Alder reaction above.

Equation 5.4

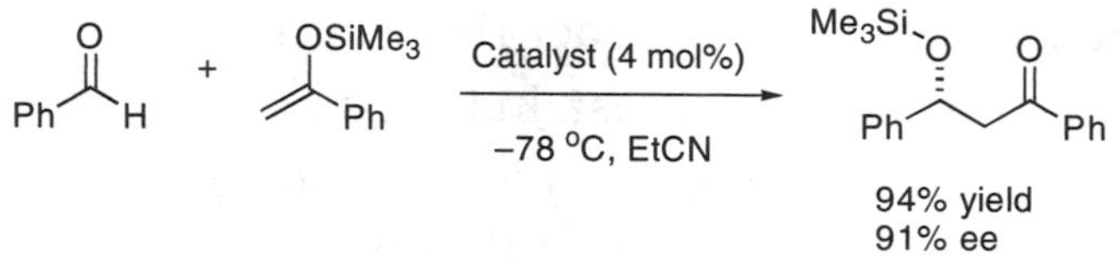

A similar model for the activation of the aldehyde, involving coordination to boron, π-stacking between the indole and aldehyde, and a C–H•••O hydrogen bond between the substrate and the carboxylate, was proposed (**Figure 5.17**).

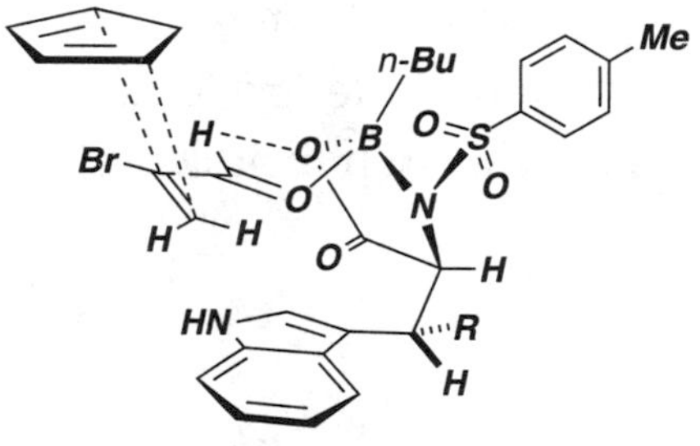

Figure 5.16. Proposed transition state for the asymmetric Diels–Alder reaction, illustrating the interaction of the formyl C–H with the boron-bound carboxylate.

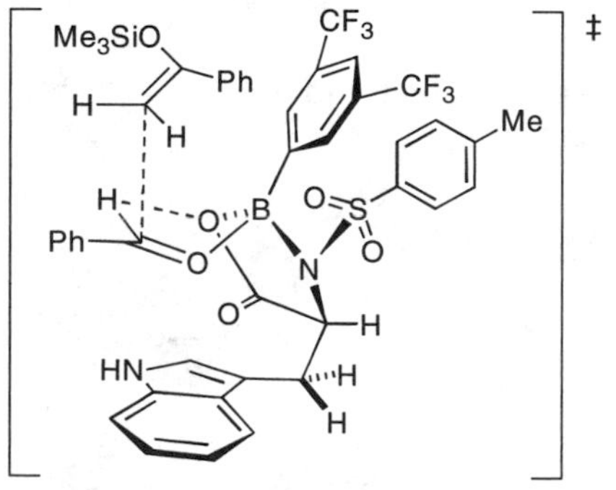

Figure 5.17. Proposed transition state for the asymmetric Mukaiyama aldol reaction.

5.4 Charge–Charge Interactions

Many catalytic asymmetric reactions involve charged intermediates. If the catalyst or catalyst–substrate adduct is charged, it will associate with a counterion due to electrostatic forces. In this way, the structure of the counterion will likely impact the reactivity and enantioselectivity of the catalyst. If the counter ion is in the chiral catalyst–substrate adduct, its nature and position may be likened to a secondary interaction. As such, it can be a significant factor in controlling the reaction enantioselectivity. In the case of asymmetric phase-transfer catalysis, where there are no covalent bonds between the substrate and the catalyst, the attractive electrostatic interactions are responsible for maintaining the substrate in the chiral environment of the catalyst, and may orient the substrate and control the sense of asymmetric induction (see Section 3.3).

5.4.1 Asymmetric Silver Catalysts with an Internal Counterion

One of the first and most significant contributions in which charge–charge interactions were proposed to play an important role in the enantioselectivity of an asymmetric catalyst was the asymmetric aldol reaction of isocyanoacetates with aldehydes to give 5-alkyl-2-oxazoline-4-carboxylates (**Figure 5.18**).[32–39] The catalyst for this reaction is a gold(I) complex, prepared in situ from a chiral phosphine with both central and planar chirality. The ligand is proposed to coordinate to the gold center through the two phosphorus centers,[34,39] while the amino group remains unattached to the metal center, as has been observed in Pd complexes of this ligand.[40] There is evidence, however, that the gold is coordinating to only one phosphorus at a time, rapidly moving between the two phosphorus centers.[39] Whatever the ligand binding

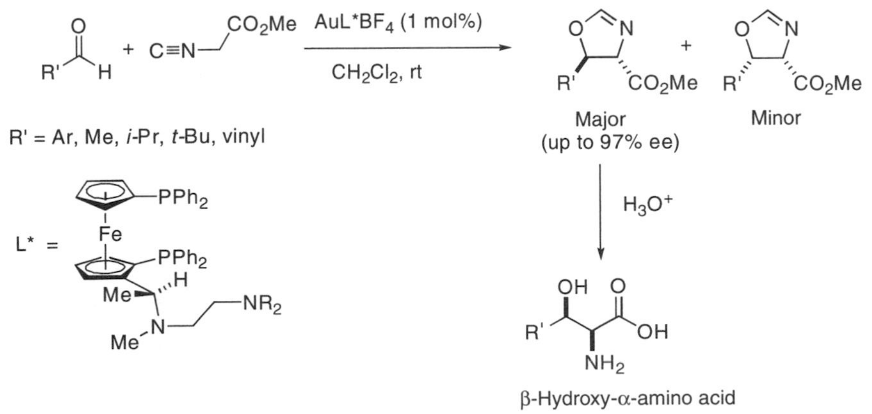

Figure 5.18. The asymmetric aldol reaction of isocyanoacetates with aldehydes catalized by cationic gold(I) bis(phosphine) catalysts.

mode, the resultant gold catalyst is highly efficient [turnover numer (TON) up to 10,000], giving excellent enantioselectivities and good diastereoselectivities with a variety of aldehyde substrates (**Figure 5.18**).[36] The heterocyclic products of this aldol condensation can be hydrolyzed to provide β-hydroxy-α-amino acids, which have been used in natural product syntheses.[41]

The mode of operation of the gold catalyst has been probed through both analysis of the substrate scope[33,38,42] and rigorous mechanistic studies.[34,39] An examination of a series of ferrocene bis(phosphine) ligands with different amino tethers indicated that the enantioselectivity, and to a lesser extent the diastereoselectivity, were strongly dependent on the nature of the amino side chain.[33,38,42] As outlined in **Figure 5.19**, high enantioselectivity was observed when two methylene units connected the amino groups, as in ligand **A** and a number of closely related ligands with different alkyl groups on the terminal amine. In contrast, when the number of methylene groups was increased to three, as in ligand **B**, the catalyst enantioselectivity dropped dramatically. Replacement of the terminal dimethylamino group in **A** with a hydroxyl group again resulted in a significant decrease in the observed enantioselectivity. Truncating the side chain after the first amino group resulted in production of racemic product. Interestingly, use of **E**, which has the opposite central chirality, but the same sense planar chirality as **A**, resulted in production of the opposite enantiomer of the trans product with low enantioselectivity.[37] These results clearly indicate that both the nature and length of the amino side chain affect the catalyst enantioselectivity.

The reaction works with a variety of aldehydes, including formaldehyde (**Figure 5.20**).[43] Based on the results in **Figures 5.19** and **5.20**, and mechanistic studies that in-

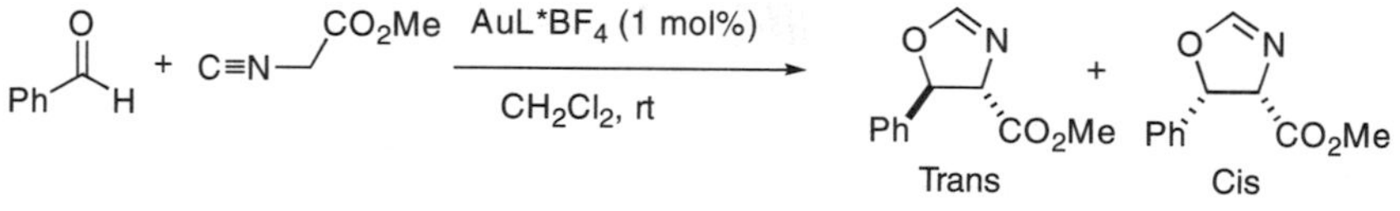

L*	Yield (%)	Trans/cis	ee (%) of trans (configuration)
A	91	90/10	91 (4S,5R)
B	99	89/11	23 (4S,5R)
C	91	69/31	37 (4S,5R)
D	80	68/32	0 —
E	—	84/16	41 (4R,5S)

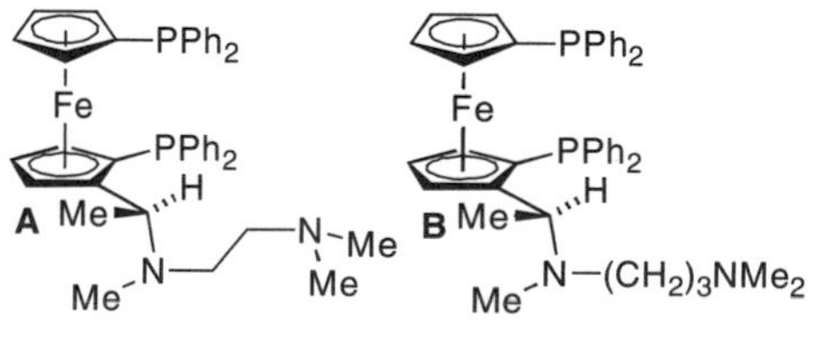

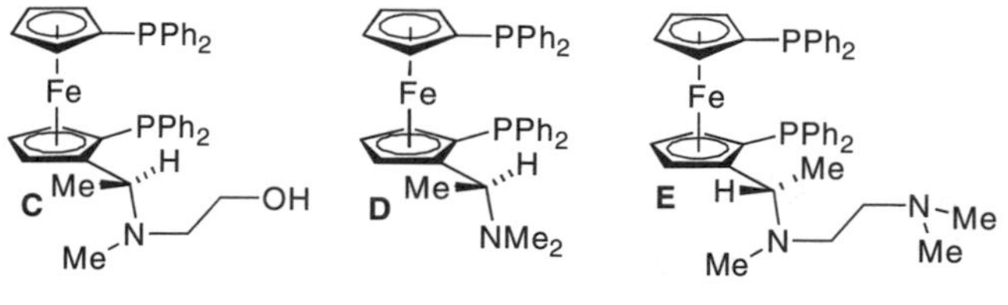

Figure 5.19. Impact of ligand structure on the enantioselectivity and diastereoselectivity of the isocyanoacetate aldol reaction with benzaldehyde.

dicate the rate-determining step is attack of the enolate on the aldehyde,[39] models for the enolate intermediate have been proposed, as illustrated in **Figure 5.21**. The key features of this proposal include the coordination of the enolate through the iso-cyanide by the Lewis acidic gold center and deprotonation of the α-carbon of the ester by the terminal amino group of the ligand (Bifunctional catalysis, see Chapter 12). The proposed electrostatic interaction between the positively charged ammonium ion and the enolate anion presumably orients the enolate. As a consequence, one face of the enolate is shielded from attack of the electrophile by the ammonium ion side chain. This proposal explains the strong dependency of the enantioselectivity on the nature

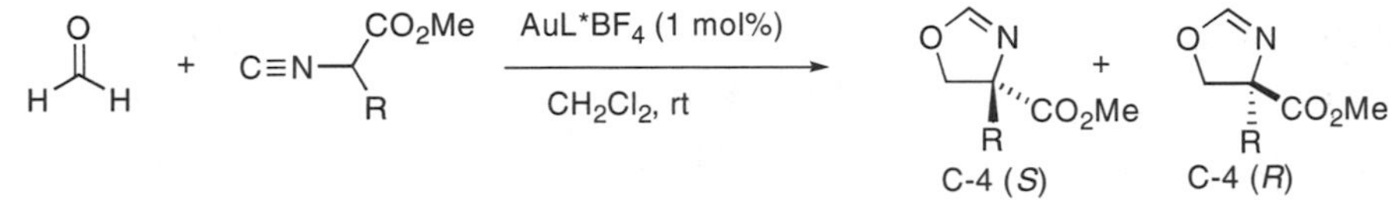

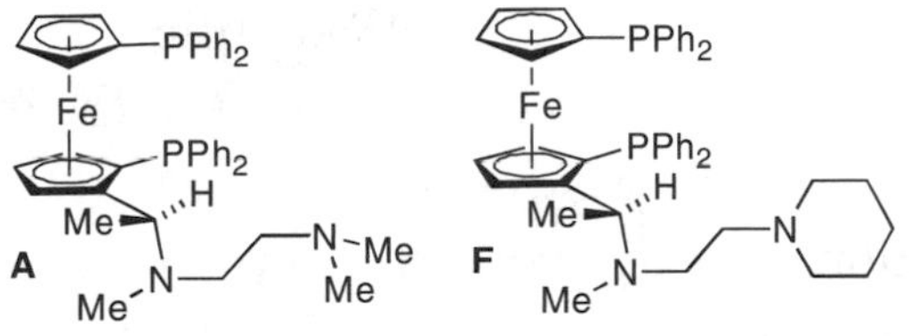

R	Ligand	Yield (%)	ee (%)	Configuration
H	A	99	52	(S)
H	F	89	44	(S)
Me	A	100	64	(S)
Me	F	95	63	(S)
i-Pr	A	99	71	(S)
i-Pr	F	96	81	(S)

Figure 5.20. The isocyanoacetate aldol reaction with formaldehyde generates a single stereocenter.

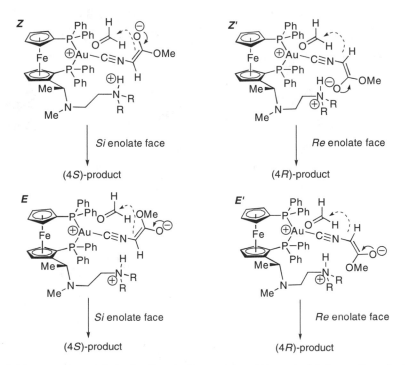

Figure 5.21. Model for asymmetric induction in the gold-catalyzed aldol reaction. In transition states **Z** and **Z′**, the enolates have the (**Z**)-geometry, while in transition states **E** and **E′**, they have the (**E**)-geometry. The enolate facial selectivity is independent of the geometry of the enolate. Note that formaldehyde is behind the gold catalyst.

and length of the pendent amine chain. Apparently, the extended chain of ligand **B** (**Figure 5.19**) positions the ammonium group such that the enolate cannot achieve the orientation necessary for high enantioselectivity in the aldol condensation. It should be noted that formation of a hydrogen bond between the ammonium and enolate ions cannot be ruled out.

As illustrated in **Figure 5.21**, the four transition states differ in the enolate geometry and the face of the enolate undergoing reaction (*Re* vs. *Si*) with the electrophile.[39] To simplify the discussion, we will initially consider reaction with formaldehyde as the electrophile, because it does not give a mixture of diastereomers (**Figure 5.20**).

As illustrated in **Figure 5.21**, the front face of the enolate is blocked by the pendant ammonium side chain and formaldehyde is positioned behind the plane of the page. **Z** and **Z′**, the (Z) enolates, undergo reaction at the *Si* and *Re* faces of the enolate, giving rise to the (4S)- and (4R)-products, respectively. Likewise, **E** and **E′** are the (E)-enolates, which react at the *Si* and *Re* faces, also giving the (4S)- and (4R)-configurations. In these proposed intermediates, the configuration at C-4 is independent of the enolate geometry, because the (S)-configuration at C-4 of the product can be formed from intermediates **Z** and **E** and the (R)-configuration can be generated from both **Z′** and **E′**. The stereoselectivity is determined by the enolate facial selectivity on attack of formaldehyde with pathways **Z** and **E** being more favorable.[39]

When prochiral aldehyde substrates are employed, diastereomers can be formed, because the intermediate enolates that attack the aldehyde in **Figure 5.21** can react with either the *Re* or *Si* face of the aldehyde. From analysis of the stereochemistry of the products from formaldehyde, it is clear that the major diastereomer is formed through either intermediate **Z** or **E** (**Figure 5.21**). The Curtin–Hammett principle (see

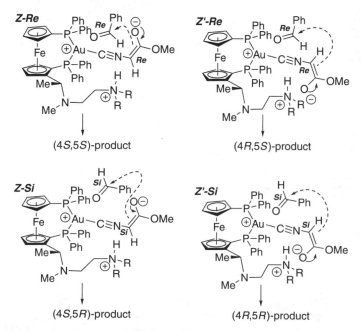

Figure 5.22. Model for the enantioselectivity and diastereoselectivity in the gold-catalyzed aldol reaction, leading to the trans and cis products. In all of the transition states, the enolate has the (Z)-geometry.

Section 1.5), however, prevents us from drawing conclusions concerning the reactivity of these intermediates, even if we knew their relative concentrations. We will, therefore, consider only intermediates **Z** and **Z′** in the reaction with prochiral aldehyde substrates, keeping in mind that the reaction may well occur through **E** and **E′** (**Figure 5.22**). In the reaction with formaldehyde, the *Si* face of the enolate reacts with the electrophile. With a prochiral substrate, such as benzaldehyde (**Figure 5.22**), we again expect the *Si* face of the enolate to exhibit greater reactivity. The reaction then occurs through the (Z)-enolate and not the (Z′)-enolate. The *Si* face of the enolate can react with either the *Re* or *Si* face of the aldehyde (which are labeled in **Figure 5.22**), leading to the (4S,5R)-trans or the (4S,5S)-cis products, respectively. Note that attack on the *Si* face of the aldehyde gives the (5R)-product, while attack at the *Re* face gives the (5S)-configuration. Addition to the *Si* face of benzaldehyde is observed, generating the (4S,5R)-product, presumably through transition state **Z-Si**. Note that this model does not take into account the orientations of the *P*-phenyl groups, which play an important role in asymmetric catalysis (see Chapter 4). In this case, attack at the *Re* face is likely disfavored, because the phenyl of the benzaldehyde would clash with the *P*-phenyl groups of the ligand.

It is noteworthy that the stereochemistry at C-4 depends on facial selectivity in the attack of the enolate on the electrophile. The stereochemistry at C-5 depends on the approach of the aldehyde to the enolate. It is likely independent of the facial selectivity at the nucleophile, which leads to the stereochemistry at C-4 (however, see Section A.2.2). An important concept in asymmetric catalysis is that the enantioselectivity and diastereoselectivity of a reaction are not necessarily related. Nonetheless, in reactions in which diastereomers are formed, catalysts that give high enantioselectivity frequently give high diastereoselectivity, too.

Another noteworthy observation in this system is that both the central and planar chirality are acting in a cooperative fashion in ligand **A**, **Figure 5.19**, whereas in the diastereomeric ligand **E**, the two stereochemical elements are opposed.[37] This phenomenon is related to the matched and mismatched terminology in asymmetric synthesis (see Chapter 13). In these ligands, the central chirality dominates the control of stereochemistry. Changing the configuration of the stereocenter results in a reversal in the stereoselectivity. Interestingly, the dominant element of a ligand can vary from catalyst to catalyst. In the catalytic asymmetric cross-coupling of a secondary Grignard reagent and an aryl halide (Kumada coupling) using bis(phosphine) ligands similar to **A** and **E** (**Figure 5.19**), the planar chirality played a larger role in the enantioselectivity of the reaction.[44,45]

In the previous example, experimental evidence indicated that the position of a positively charge ammonium ion of the chiral ligand influences the orientation of the intermediate enolate. As a result, small changes in the length of the amino side chain or amino group substituents have a significant impact on catalyst enantioselectivity. In this system, the catalyst assumes a dual role: it activates the isocyanate by coordination to the gold center and positions the enolate through ion paring.

5.4.2 Asymmetric Phase-Transfer Catalysts

The gold system outlined above has some features in common with ion-pair mediated reactions under phase-transfer conditions. In phase-transfer catalysis (PTC),[46,47] the primary interactions that hold the substrate in the chiral environment of the catalyst are a combination of electrostatic and van der Waals forces, because there are no covalent or dative bonds formed between the catalyst and substrate (also see Section 3.3). Reactions conducted under PTC are becoming increasingly important in asymmetric catalysis[48] and have found applications in enantioselective alkylations,[49,50] Michael additions,[51,52] aldol and related condensations,[53,54] 1,2-addition reactions,[55] epoxide formation,[56,57] and reductions.[58] As an example, the phase-transfer-catalyzed alkylation of an enolate is presented (**Equation 5.5**). This is an excellent method for the synthesis of unnatural amino acid derivatives with high levels of enantioselectivity.[59–63]

Equation 5.5

A simplified view of the mechanics of the PTC process is outlined in **Figure 5.23**, where Q^+ is the chiral cation. In this procedure a biphasic mixture, typically consisting of organic and aqueous phases, is stirred rapidly to increase the contact between the phases. In the example in **Equation 5.5**, the ester and the alkyl halide reside in the organic layer, where they do not react in the absence of base. The ionic inorganic base remains in the aqueous phase. In step **A** of the reaction (**Figure 5.23**), the ester is deprotonated at the aqueous organic interface. The resultant sodium enolate undergoes ion exchange with the chiral cationic phase-transfer catalyst in step **B**, forming the lipophilic contact ion-pair. Extraction of the lipophilic ion-pair into the organic phase follows, where the catalyst–enolate adduct and the alkyl halide react in step **C**, generating the neutral alkylated product with concomitant association of the cationic phase-transfer catalyst with the halide. To understand the stereochemical induction in the alkylation, it is necessary to know how the enolate interacts with the chiral template. Thus, questions such as, "What is the stereochemistry of the enolate (E or Z)?" "Where does the enolate bind to the catalyst?" "What other interactions exist between the catalyst and the enolate?" and "What is the origin of facial selectivity in the nucleophilic displacement reaction?" must be answered to expedite rational catalyst design. These questions are, however, difficult to answer definitively.

One of the most-studied classes of chiral nonracemic phase-transfer catalysts is derived from the *cinchona alkaloids*, such as cinchonine (**A, Figure 5.24**).[48–50] Alkylation of the tertiary amine gives the quaternary ammonium salt. Such catalysts have been

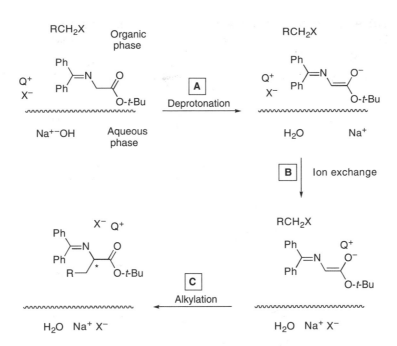

Figure 5.23. Mechanism of phase-transfer catalysis in the asymmetric enolate alkylation.

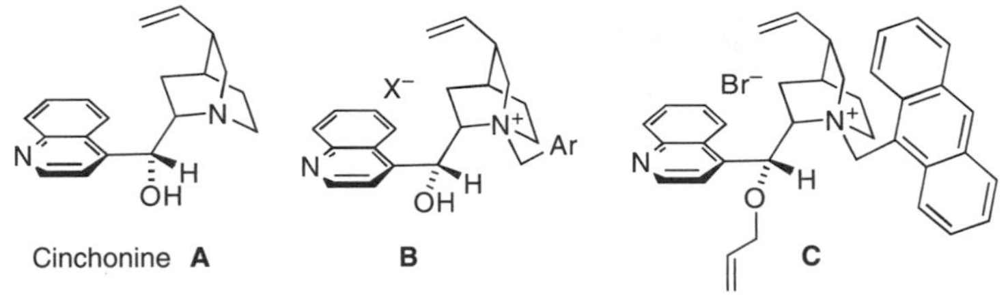

Figure 5.24. Structure of cinchonine (A), a class of phase-transfer catalysts (B), and *O*(9)-allyl-*N*-(9-anthracenylmethyl)cinchonidinium bromide (C).

used successfully in many reactions, including the alkylation of *tert*-butylglycinate–benzophenone Schiff base shown in **Equation 5.5**.[48]

An in-depth study of this family of catalysts led to the rational design of *O*(9)-allyl-*N*-(9-anthracenylmethyl)cinchonidinium bromide (**C**, **Figure 5.24**), a highly enantioselective catalyst for the asymmetric alkylation reaction in **Equation 5.5**.[61,62] When used at low temperature with CsOH•H$_2$O, the enantioselectivities in the alkylation reaction were uniformly very high (**Table 5.1**).[61] The use of CsOH•H$_2$O reduces the amount of the aqueous phase and allows the reaction to be performed at temperatures far below those accessible with aqueous solutions.

Conformational analysis of the anthracenyl derivative **C** indicates that the cation has a well-defined, rigid geometry with the anthracenylmethyl substituent staggered with respect to the *N*-CH$_2$ groups of the bridgehead nitrogen. Additionally, the *O*-allyl group helps to orient the quinoline ring. An X-ray crystal structure of **C** with the 4-nitrophenoxide counterion was obtained that provides crucial insight into how the enolate is likely to associate with the chiral cation.[61] The phenoxide behaves as a

Table 5.1. Enantioselective phase-transfer catalysis with *O*(9)-allyl-*N*-(9-anthracenylmethyl)cinchonidinium bromide (**C**, Figure 5.24).

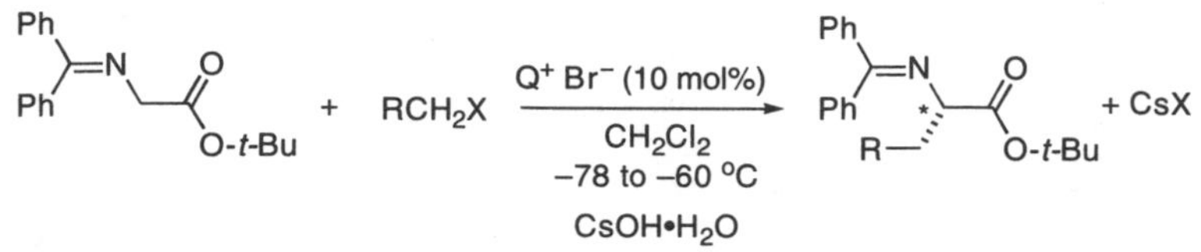

RCH₂X	ee (%)	Isolated yield (%)
EtI	98	82
(cyclopropylmethyl bromide)	99	75
(allyl bromide)	97	89
(allyloxy-methyl-benzodioxole-CH₂Br)	96	81
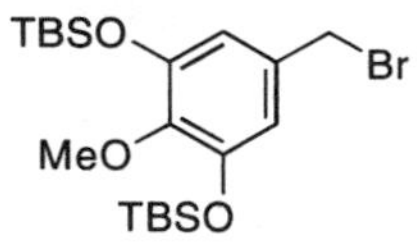 (TBSO/MeO-substituted benzyl bromide)	97	67

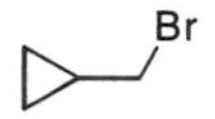

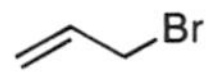

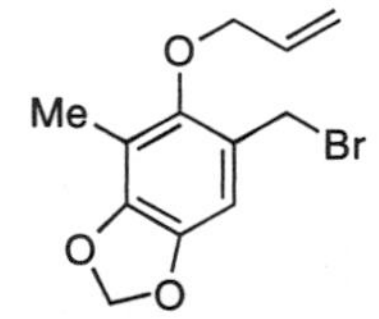

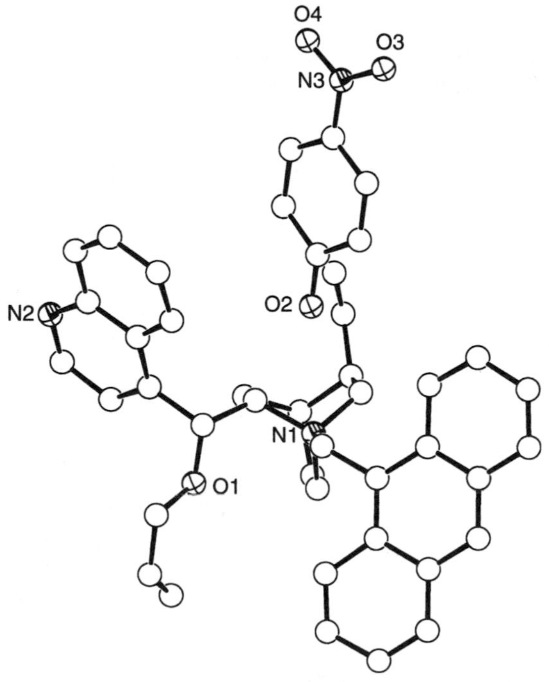

Figure 5.25. Structure of *O*(9)-allyl-*N*-(9-anthracenylmethyl)cinchonidinium 4-nitrophenolate.

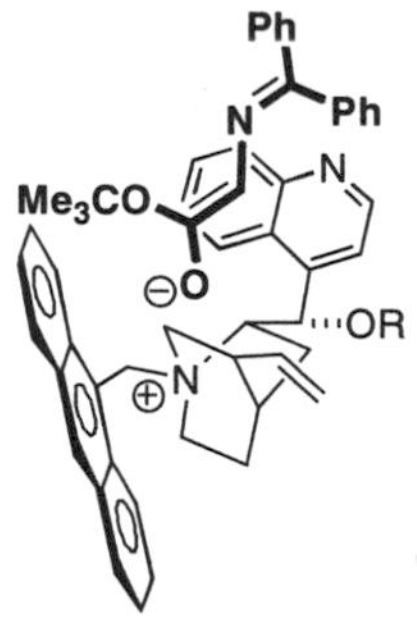

Figure 5.26. Proposed orientation of the ion-pair formed in the asymmetric enolate alkylation

substrate analog in this model for the phase-transfer-catalyst–enolate ion pair. As illustrated in the structure of **Figure 5.25**, the oxygen of the 4-nitrophenoxide is positioned near to the charged nitrogen at a distance of 3.46 Å and is located opposite the unsubstituted CH_2–CH_2 bridge of the quinuclidinium bicycle. This is the sterically least-hindered approach to the cationic nitrogen. Based on this and related structures, it is predicted that the enolate will occupy the same position as the phenoxide in the structure.[61] A model for the position of the enolate (drawn in bold) in the cleft of the chiral ammonium ion is shown in **Figure 5.26**. Attractive interactions between the enolate and the quinoline ring are likely to shield one face of the enolate from attack of the electrophile. The electrophile, which is not shown, would attack the unshielded face of the enolate from the top, giving the observed stereochemistry of the product.

Concluding Remarks

Under the proper conditions, nonclassical interactions between the catalyst and the substrate can contribute to, or even control, the enantioselectivity of catalytic asymmetric reactions. As can be judged by the examples outlined above, some secondary interactions are often difficult to identify and it would be challenging to rationally incorporate them into the design of asymmetric catalysts. Nonetheless, the greater our understanding of these interactions, the more likely we can achieve this goal.

References

(1) Hawkins, J. M.; Loren, S. Two-Point-Binding Asymmetric Diels–Alder Catalysts: Aromatic Alkyldichloroboranes. *J. Am. Chem. Soc.* **1991**, *113*, 7794–7795.

(2) Hawkins, J. M.; Loren, S.; Nambu, M. Asymmetric Lewis Acid-Dienophile Complexation: Secondary Attraction Versus Catalyst Polarizability. *J. Am. Chem. Soc.* **1994**, *116*, 1657–1660.

(3) Hawkins, J. M.; Nambu, M.; Loren, S. Asymmetric Lewis Acid-Catalyzed Diels–Alder Reactions of α,β-Unsaturated Ketones and α,β-Unsaturated Acid Chlorides. *Org. Lett.* **2003**, *5*, 4293–4295.

(4) Shambayati, S.; Crowe, W. E.; Schreiber, S. L. On the Conformation and Structure of Organometal Complexes in the Solid State: Two Studies Relevant to Chemical Synthesis. *Angew. Chem., Int. Ed. Engl.* **1990**, *29*, 256–272.

(5) Nishio, M. CH/p Hydrogen Bonds in Crystals. *Cryst. Eng. Comm.* **2004**, *6*, 130–158.

(6) Nishio, M.; Hirota, M.; Umezawa, Y.; In the CH/π Interaction: Evidence, Nature and Consequences, Marchand, A. P., Ed.; Wiley: New York, 1998; Chapter 2, pp 11–45.

(7) Carmona, D.; Lamata, M. P.; Viguri, F.; Rodriguez, R.; Oro, L. A.; Lahoz, F. J.; Balana, A. I.; Tejero, T.; Merino, P. Enantioselective 1,3-Dipolar Cycloaddition of Nitrones to Methacrolein Catalyzed by (η⁵-C₅Me₅)M[(R)-Prophos] Containing Complexes [M = Rh, Ir; (R)-Prophos = 1,2-bis(Diphenylphosphino)propane]: On the Origin of the Enantioselectivity. *J. Am. Chem. Soc.* **2005**, *127*, 13386–13398.

(8) Noyori, R.; Ohkuma, T. Asymmetric Catalysis by Architectural and Functional Molecular Engineering: Practical Chemo- and Stereoselective Hydrogenation of Ketones. *Angew. Chem., Int. Ed. Engl.* **2001**, *40*, 40–73.

(9) Noyori, R.; Hashiguchi, S. Asymmetric Transfer Hydrogenation Catalyzed by Chiral Ruthenium Complexes. *Acc. Chem. Res.* **1997**, *30*, 97–102.

(10) Noyori, R.; Yamakawa, M.; Hashiguchi, S. Metal–Ligand Bifunctional Catalysis: A Nonclassical Mechanism for Asymmetric Hydrogen Transfer Between Alcohols and Carbonyl Compounds. *J. Org. Chem.* **2001**, *66*, 7931–7944.

(11) Haack, K.-J.; Hasiguchi, S.; Fujii, A.; Ikariya, T.; Noyori, R. The Catalyst Precursor, Catalyst and Intermediate in the Ru(II)-Promoted Asymmetric Hydrogenation Transfer Between Alcohols and Ketones. *Angew. Chem., Int. Ed. Engl. Engl.* **1997**, *36*, 285–288.

(12) Yamakawa, M.; Ito, H.; Noyori, R. The Metal–Ligand Bifunctional Catalysis: A Theoretical Study on the Ruthenium(II)-Catalyzed Hydrogen Transfer Between Alcohols and Carbonyl Compounds. *J. Am. Chem. Soc.* **2000**, *122*, 1466–1478.

(13) Yamakawa, M.; Yamada, I.; Noyori, R. CH/π Attraction: The Origin of Enantioselectivity in Transfer Hydrogenation of Aromatic Carbonyl Compounds Catalyzed by Chiral η⁶-Arene-Ruthenium(II) Complexes. *Angew. Chem., Int. Ed. Engl.* **2001**, *40*, 2818–2821.

(14) Crabtree, R. H. *The Organometallic Chemistry of the Transition Metals*, 3rd ed.; Wiley: New York, 2001.

(15) Yamada, I.; Noyori, R. Asymmetric Transfer Hydrogenation of Benzaldehydes. *Org. Lett.* **2000**, *2*, 3425–3427.

(16) Corey, E. J.; Rohde, J. J.; Fisher, A.; Azimioara, M. D. A Hypothesis for Conformation Restriction in Complexes of Formyl Compounds with Boron Lewis Acids. Experimental Evidence for Formyl CH••O and CH••F Hydrogen Bonds. *Tetrahedron Lett.* **1997**, *38*, 33–36.

(17) Corey, E. J.; Barnes-Seeman, D.; Lee, T. W. The Formyl C–H•••O Hydrogen Bond as a Key to Transition-State Organization in Enantioselective Allylation, Aldol and Diels–Alder Reactions Catalyzed by Chiral Lewis Acids. *Tetrahedron Lett.* **1997**, *38*, 1699–1702.

(18) Corey, E. J.; Rohde, J. J. The Application of the Formyl C–H•••O Hydrogen Bond Postulate to the Understanding of Enantioselective Reactions Involving Chiral Boron Lewis Acids and Aldehydes. *Tetrahedron Lett.* **1997**, *38*, 37–40.

(19) Corey, E. J.; Lee, T. W. The Formyl C–H•••O Hydrogen Bond as a Critical Factor in Enantioselective Lewis-Acid Catalyzed Reactions of Aldehydes. *J. Chem. Soc., Chem. Commun.* **2001**, 1321–1329.

(20) Taylor, R.; Kennard, O. Crystallographic Evidence for the Existence of CH•••O, CH•••N and CH•••Cl Hydrogen Bonds. *J. Am. Chem. Soc.* **1982**, *104*, 5063–5070.

(21) Desiraju, G. R. The C–H•••O Hydrogen Bond in Crystals: What Is It? *Acc. Chem. Res.* **1991**, *24*, 290–296.

(22) Chaney, J. D.; Goss, C. R.; Folting, K.; Santarsiero, B. Formyl C–H•••O Hydrogen Bonding in Crystalline Bis-Formamides? *J. Am. Chem. Soc.* **1996**, *118*, 9432–9433.

(23) Mackey, M. D.; Goodman, J. M. Conformational Preferences of R¹R²C=O•H₂BF Complexes. *J. Chem. Soc., Chem. Commun.* **1997**, 2383–2384.

(24) Goodman, J. M. Molecular Orbital Calculations on R¹R²C=O•H₂BF Complexes: Anomeric Stabilization and Conformational Preferences. *Tetrahedron Lett.* **1992**, *33*, 7219–7222.

(25) Bernardi, N.; Bottoni, A.; Casolari, S.; Tagliavini, E. Zirconium Tetrachloride-Formaldehyde Complexes: A Computational and Spectroscopic Investigation. *J. Org. Chem.* **2000**, *65*, 4783–4790.

(26) Gung, B. W. The Difference in Strength of BF₃-Complexes of Aromatic and Aliphatic Aldehydes. *Tetrahedron Lett.* **1991**, *32*, 2867–2870.

(27) Gung, B. W.; Wolf, M. A. A Study of the Electronic Structures of Boron Trifluoride Complexes with Carbonyl Compounds by Ab Initio MO Methods. *J. Org. Chem.* **1992**, *57*, 1370–1375.

(28) Salvatella, L.; Mokrane, A.; Cartier, A.; Ruiz-López, M. F. The Role of Menthyl Group in Catalyzed Asymmetric Diels–Alder Reactions. A Combined Quantum Mechanics/Molecular Mechanics Study. *J. Org. Chem.* **1998**, *63*, 4664–4670.

(29) Corey, E. J.; Loh, T.-P.; Roper, T. D. The Origin of Greater Than 200:1 Enantioselectivity in a Catalytic Diels–Alder Reaction as Revealed by Physical and Chemical Studies. *J. Am. Chem. Soc.* **1992**, *114*, 8290–8292.

(30) Corey, E. J.; Cywin, C. L.; Roper, T. D. Enantioselective Mukaiyama-Aldol and Aldoldihydropyrone Annulation Reactions Catalyzed by a Tryptophan-Derived Oxazaborolidine. *Tetrahedron Lett.* **1992**, *33*, 6907–6910.

(31) Ishihara, K.; Kondo, S.; Yamamoto, H. Scope and Limitations of Chiral *B*-[3,5-Bis(trifluoromethyl)phenyl]oxazaborolidine Catalyst for Use in the Mukaiyama Aldol Reaction. *J. Org. Chem.* **2000**, *65*, 9125–9128.

(32) Ito, Y.; Sawamura, M.; Hayashi, T. Catalytic Asymmetric Aldol Reaction: Reaction of Aldehydes with Isocyanoacetate Catalyzed by a Chiral Ferrocenylphosphine-Gold(I) Complex. *J. Am. Chem. Soc.* **1986**, *108*, 6405–6406.

(33) Ito, Y.; Sawamura, M.; Hayashi, T. Asymmetric Aldol Reaction of an Isocyanoacetate with Aldehydes by Chiral Ferrocenylphosphine-gold(I) Complexes: Design and Preparation of New Efficient Ferrocenylphosphine Ligands. *Tetrahedron Lett.* **1987**, *28*, 6215–6218.

(34) Sawamura, M.; Ito, Y.; Hayashi, T. NMR Studies of the Gold(I)-Catalyzed Asymmetric Aldol Reaction of Isocyanoacetate. *Tetrahedron Lett.* **1990**, *31*, 2723–2726.

(35) Kuwano, R.; Ito, Y. Addition of Isocyanocaroxylates to Aldehydes in *Comprehensive Asymmetric Catalysis*. Jacobsen, E. N.; Pfaltz, A.; Yamoto, H., Eds.; Springer–Verlag: Berlin, 1999; Vol. 2, pp 1067–1074.

(36) Sawamura, M.; Ito, Y. Catalytic Asymmetric Synthesis by Means of Secondary Interaction Between Chiral Ligands and Substrates. *Chem. Rev.* **1992**, *92*, 857–871.

(37) Pastor, S. D.; Togni, A. Asymmetric Synthesis with Chiral Ferrocenylamine Ligands: The Importance of Central Chirality. *J. Am. Chem. Soc.* **1989**, *111*, 2333–2334.

(38) Pastor, S. D.; Togni, A. Chiral Cooperativity: The Effect of Distant Chiral Centers in Ferrocenylamine Ligands upon Enantioselectivity in the Gold(I)-Catalyzed Aldol Reaction. *Helv. Chim. Acta* **1991**, *74*, 905–933.

(39) Togni, A.; Pastor, S. D. Chiral Cooperativity: The Nature of the Diastereoselective and Enantioselective Step in the Gold(I)-Catalyzed Aldol Reaction Utilizing Chiral Ferrocenylamine Ligands. *J. Org. Chem.* **1990**, *55*, 1649–1664.

(40) Hayashi, T.; Kumada, M.; Higuchi, T.; Hirotsu, K. Asymmetric Synthesis Catalyzed by Chiral Ferrocenylpheophine Transition Metal Complexes. 4. Crystal Structure of Dichloro{*N,N*-dimethyl-1[1′,2-bis(diphenylphosphino)ferrocenyl]ethylamine} palladium(II) [PdCl$_2$(BPPFA)]. *J. Organomet. Chem.* **1987**, *334*, 195–203.

(41) Ito, Y.; Sawamura, M.; Hayashi, T. Asymmetric Synthesis of *threo*- and *erythro*-Sphingosines by Asymmetric Aldol Reaction of α-Isocyanoacetate Catalyzed by a Chiral Ferrocenylphosphine-Gold(I) Complex. *Tetrahedron Lett.* **1988**, *29*, 239–240.

(42) Hayashi, T.; Sawamura, M.; Ito, Y. Asymmetric Synthesis Catalyzed by Chiral Ferrocenylphosphine Transition Metal Complexes. 10. Gold(I)-Catalyzed Asymmetric Aldol Reaction of Isocyanoacetate. *Tetrahedron* **1992**, *48*, 1999–2012.

(43) Ito, Y.; Sawamura, M.; Shirakawa, E.; Hayashizaki, K.; Hayashi, T. Asymmetric Aldol Reaction of α-Isocyanocarboxylates with Paraformaldehyde Catalyzed by Chiral Ferrocenylphosphine-Gold(I) Complexes: Catalytic Asymmetric Synthesis of α-Alkylserines. *Tetrahedron Lett.* **1988**, *29*, 235–238.

(44) Hayashi, T.; Tajika, M.; Tamao, K.; Kumada, M. High Stereoselectivity in Asymmetric Grignard Cross-Coupling Catalyzed by Nickel Complexes of Chiral (Aminoalkylferrocenyl)phosphines. *J. Am. Chem. Soc.* **1976**, *98*, 3718–3719.

(45) Hayashi, T.; Konishi, K.; Fukushima, M.; Mise, T.; Kagotani, M.; Tajika, M.; Kumada, M. Asymmetric Synthesis Catalyzed by Chiral Ferrocenylphosphine-Transition Metal Complexes. 2. Nickel- and Palladium-Catalyzed Asymmetric Grignard Cross-Coupling. *J. Am. Chem. Soc.* **1982**, *104*, 180–186.

(46) Rabinovitz, M.; Cohen, Y.; Halpern, M. Hydroxide Ion Initiated Reactions Under Phase Transfer Catalysis Conditions: Mechanism and Implications. *Angew. Chem., Int. Ed. Engl.* **1986**, *25*, 960–970.

(47) Halpern, M., Ed. In *A.C.S. Symposium 659*; American Chemical Society: Washington, DC, 1997.

(48) O'Donnell, M. J. In *Catalytic Asymmetric Synthesis*, 2nd ed.; Ojima, I., Ed.; Wiley: New York, 2000; pp 727–755.

(49) Dolling, U. H.; Davis, P.; Grabowski, E. J. J. Efficient Catalytic Asymmetric Alkylations. 1. Enantioselective Synthesis of (+)-Indacrinone via Chiral Phase-Transfer Catalysis. *J. Am. Chem. Soc.* **1984**, *106*, 446–447.

(50) Hughes, D. L.; Dolling, U. H.; Ryan, K. M.; Schoenewaldt, E. F.; Grabowski, E. J. J. Efficient Catalytic Asymmetric Alkylations. 3. A Kinetic and Mechanistic Study of the Enantioselective Phase-Transfer Methylation of 6,7–Dichloro-5-methoxy-2-phenyl-1-indanone. *J. Org. Chem.* **1987**, *52*, 4745–4752.

(51) Corey, E. J.; Zhang, F.-Y. Enantioselective Michael Addition of Nitromethane to α,β-Enones Catalyzed by Chiral Quaternary Ammonium Salts. A Simple Synthesis of (*R*)-Baclofen. *Org. Lett.* **2000**, *2*, 4257–4259.

(52) Zhang, F.-Y.; Corey, E. J. Enantio- and Diastereoselective Michael Reactions of Silyl Enol Ethers and Chalcones by Catalysis Using a Chiral Quaternary Ammonium Salt. *Org. Lett.* **2001**, *3*, 639–641.

(53) Corey, E. J.; Zhang, F.-Y. *Re*- and *Si*-Face-Selective Nitroaldol Reactions Catalyzed by a Rigid Chiral Quaternary Ammonium Salt: A Highly Stereoselective Synthesis of the HIV Protease Inhibitor Amprenavir (Vertex 478). *Angew. Chem., Int. Ed. Engl.* **1999**, *38*, 1931–1934.

(54) Ando, A.; Miura, T.; Tatematsu, T.; Shioiri, T. Chiral Quarternary Ammonium Fluoride. A New Rreagent for Catalytic Asymmetric Aldol Reactions. *Tetrahedron Lett.* **1993**, *34*, 1507–1510.

(55) Arai, S.; S, H.; Shioiri, T. Catalytic Asymmetric Horner–Wadsworth–Emmons Reaction Under Phase-Transfer-Catalyzed Conditions. *Tetrahedron Lett.* **1998**, *39*, 2997–3000.

(56) Wynberg, H.; Marsman, B. Synthesis of Optically Active 2,3-Epoxycyclohexanone and the Determination of its Absolute Configuration. *J. Org. Chem.* **1980**, *45*, 158–161.

(57) Alcaraz, L.; Macdonald, G.; Ragot, J. P.; Lewis, N. Manumycin A: Synthesis of the (+)-Enantiomer and Revision of Stereochemical Assignment. *J. Org. Chem.* **1998**, *63*, 3526–3527.

(58) Drew, M. D.; Lawrence, N. J.; Watson, W.; Bowles, S. A. The Asymmetric Reduction of Ketones Using Chiral Ammonium Fluoride Salts and Silanes. *Tetrahedron Lett.* **1997**, *38*, 5857–5860.

(59) O'Donnell, M. J.; Wu, S.; Huffman, J. C. A New Active Catalyst Species for Enantioselective Alkylation by Phase-Transfer Catalysis. *Tetrahedron* **1994**, *50*, 4507–4618.

(60) O'Donnell, M. J.; Bennett, W. D.; Wu, S. The Stereoselective Synthesis of α-Amino Acids by Phase-Transfer Catalysis. *J. Am. Chem. Soc.* **1989**, *111*, 2353–2355.

(61) Corey, E. J.; Noe, M. C. A Rational Approach to Catalytic Enantioselective Enolate Alkylation Using a Structurally Rigidified and Defined Chiral Quaternary Ammonium Salt Under Phase Transfer Conditions. *J. Am. Chem. Soc.* **1997**, *119*, 12414–12415.

(62) Lygo, B.; Wainwright, P. G. A New Class of Asymmetric Phase-Transfer Catalysts Derived from Cinchona Alkaloids—Application in the Enantioselective Synthesis of Alpha-Amino Acids. *Tetrahedron Lett.* **1997**, *38*, 8595–8598.

(63) Maruoka, K.; Ooi, T. Enantioselective Amino Acid Synthesis by Chiral Phase-Transfer Catalysis *Chem. Rev.* **2003**, *103*, 3013–3028.

6

Chiral Poisoning, Chiral Activation, and Screening Achiral Ligands

The traditional approach to enantioselective catalysis with metal-based catalysts is an iterative operation dependent on the synthesis of highly enantioenriched chiral ligands.[1–3] After formation of the metal–ligand adducts, the newly formed catalysts or precatalysts are screened to determine their enantioselectivity and activity. The results of the screening process are analyzed and the next generation of ligands and catalysts are designed, synthesized, and screened. Despite our increasing ability to rationally design ligands and catalyst, the optimization process remains largely intuitive and the path to discovering the best catalysts depends upon the ability of the research team and good fortune. In this optimization process, therefore, it is crucial to rapidly generate numerous catalysts with diverse chiral environments to increase the odds of success. Unfortunately, the synthesis of enantiopure ligands can be an arduous task, severely hampering the optimization of asymmetric processes.[4] As a result, alternative approaches to catalyst optimization have been investigated to increase the efficiency with which asymmetric catalysts can be prepared, screened, and successfully optimized. In particular, catalysts that employ two separate ligands (both chiral or one chiral) provide further opportunities for catalyst development.

6.1 Optimization of Asymmetric Catalysts with Enantioenriched Chiral Ligands

6.1.1 Asymmetric Deactivation of Racemic Catalysts

In the synthesis of enantiopure ligands for asymmetric catalysis, chemists often turn to the pool of Nature's chiral products for ligand building blocks. Ligand precursors come in a variety of structures, such as amino acids, and can provide diversity in ligand architecture and expedite catalyst synthesis and optimization. Naturally derived chiral building blocks are also advantageous, because they are available in nearly optically pure form. Unfortunately, Nature is rarely so generous as to provide both antipodes of the desired ligand precursors in resolved form. Furthermore, many of the common and most useful ligand motifs are based on non-natural structures, most of which must be synthesized as racemates and the enantiomers separated. The resolution of racemic ligands is often tricky and resolving agents that are successful in the resolution of a particular ligand precursor are unlikely to exhibit similar levels of success with closely related analogs. As a result, development of resolution conditions is time-consuming and expensive. Racemic ligands and ligand precursors, on the other hand, are often much less costly. These factors have inspired new approaches to asymmetric catalysis based on the use of racemic chiral ligands.[5] A technique that involves the use of racemic ligands in asymmetric catalysis has been termed "chiral poisoning" and is outlined below.[6–8]

In this method, a racemic catalyst is combined with a resolved ligand (the chiral poison). Binding of the chiral poison (P*) to the racemic catalyst would give rise to diastereomers, which would then have different association constants (**Figure 6.1**). Ideally, the chiral ligand binds specifically to one of the enantiomeric catalysts, deactivating it. There are several mechanisms for deactivation, the most common being tight binding of the chiral poison to a coordination site needed for catalysis. The unreacted or unpoisoned enantiomer of the catalyst or catalyst precursor is then left to promote the asymmetric reaction.

Ideal systems of the type in **Figure 6.1** are rare, because in most cases the chiral poison will not interact specifically with one catalyst enantiomer. More likely, one enantiomer of the catalyst will be inhibited to a greater extent than the other (**Figure 6.2**). The product ee and the rate relative to the enantiopure catalyst will reflect the degree of inhibition of each catalyst enantiomer. For example, if $K_{(S)\text{-cat}\bullet P*}/K_{(R)\text{-cat}\bullet P*}$ = 20, and the enantiopure catalyst exhibited 100% enantioselectivity, the products would be predicted to be 90% ee. For $K_{(S)\text{-cat}\bullet P*}/K_{(R)\text{-cat}\bullet P*}$ = 10, the product would be generated with 82% ee.

Other issues that must be considered when implementing chiral poisoning techniques are the relative binding constants of the substrate and poison. The poison, which is generally used in catalytic amounts, must exhibit much greater affinity for the catalyst than the substrate, otherwise the substrate will displace the chiral

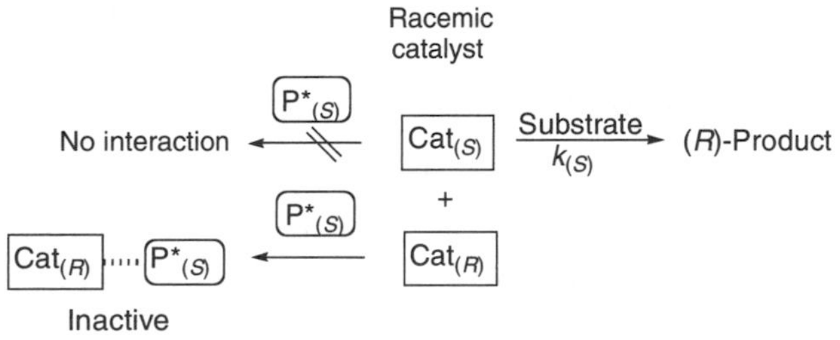

Figure 6.1. An example of chiral poisoning under optimal circumstances, where the chiral poison [P*(S)] only binds to one enantiomer of the catalyst [Cat(R)].

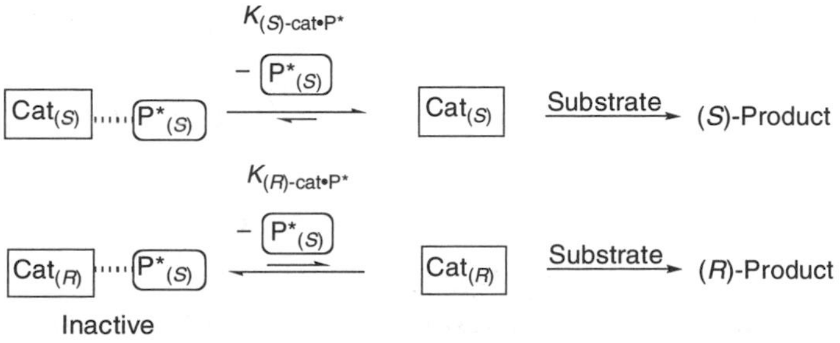

Figure 6.2. An example of chiral poisoning in which the poison P*(S) exhibits higher affinity for Cat(R) than Cat(S), resulting in product predominantly with the (S)-configuration.

poison. Complications can also arise, however, if the poison binds so well to the metal center that it displaces the chiral racemic ligand L*,[9] or if the chiral poison causes ligand redistribution, perhaps to give ML*$_2$ and MP*$_2$, either of which might also catalyze the reaction. Finally, nonlinear effects (Chapter 11) can be of great benefit in catalytic asymmetric reactions,[10] but implementing the chiral poisoning strategy in a reaction that displays nonlinear behavior will add another layer of complexity to the mechanistic analysis.

The application of the chiral poisoning method entails selection of a reaction for which a highly enantioselective catalyst has been developed. The racemic version of the catalyst is then prepared and a series of enantioenriched chiral poisons is employed to evaluate their ability to selectively deactivate one enantiomer of the racemic catalyst. As mentioned, the efficiency with which the enantioenriched poison inhibits each enantiomer of the racemic catalyst is judged by the product ee. Complete and selective inhibition of one enantiomer of the catalyst would result in formation of the product in the same enantioselectivity observed with the enantiopure catalyst. In contrast to approaches outlined in later sections, one of the limitations of chiral poisoning is that the product ee cannot exceed that of the enantiopure catalyst, and in several cases, it falls short by a significant margin.

A rare example of a system that exhibits near perfect deactivation is the ruthenium–Xyl-BINAP catalyst (where Xyl = 3,5-dimethylphenyl). Enantiopure (Xyl-BINAP)RuCl$_2$(dmf)$_n$ precatalyst, after activation, is known to efficiently catalyze the asymmetric hydrogenation of β-keto esters with very high levels of enantioselectivity.[11,12] In the asymmetric deactivation, 0.5 equivalents of (S)-3,3'-dimethyl-1,1'-binaphthyl-2,2'diamine, (S)-DM-DABN, was used as chiral poison. It selectively binds to the [(S)-Xyl-BINAP]RuCl$_2$ center (**Figure 6.3**). The [(R)-Xyl-BINAP]RuCl$_2$ (dmf)$_n$ does not bind the (S)-DM-DABN, due to an unfavorable steric interaction of the 3,3'-dimethyl substituents with the pseudoequatorial Xyl groups on the chelating phosphine.[13] This is more clearly seen in the illustration in **Figure 6.4**.

This system has been applied to the asymmetric hydrogenation of methyl 3-oxobutanoate, as illustrated in **Figure 6.5**. Employing racemic (Xyl-BINAP)RuCl$_2$ (dmf)$_n$ with (S)-DM-DABN gave essentially identical enantioselectivity compared to the enantioenriched (Xyl-BINAP)RuCl$_2$(dmf)$_n$ precatalyst.[13] This example illustrates the great potential of chiral poisoning in catalytic asymmetric systems.

The same strategy has been used in the kinetic resolution of racemic 2-cyclohexenol (**Figure 6.6**).[14] Use of racemic (Xyl-BINAP)RuCl$_2$(dmf)$_n$ and 0.5 equivalents of the (S)-DM-DABN poison resulted in reduction of the allylic alcohol with $k_{rel} < 100$. In this kinetic resolution, essentially enantiopure starting material remained after 53% conversion.[13] To measure k_{rel}, the reaction was sampled after 48% conversion (see Chapter 7 on kinetic resolution for details).

The (Xyl-BINAP)Ru-based catalysts above are unique in that one of the enantiomeric catalysts can be almost completely deactivated. More frequently, the chiral

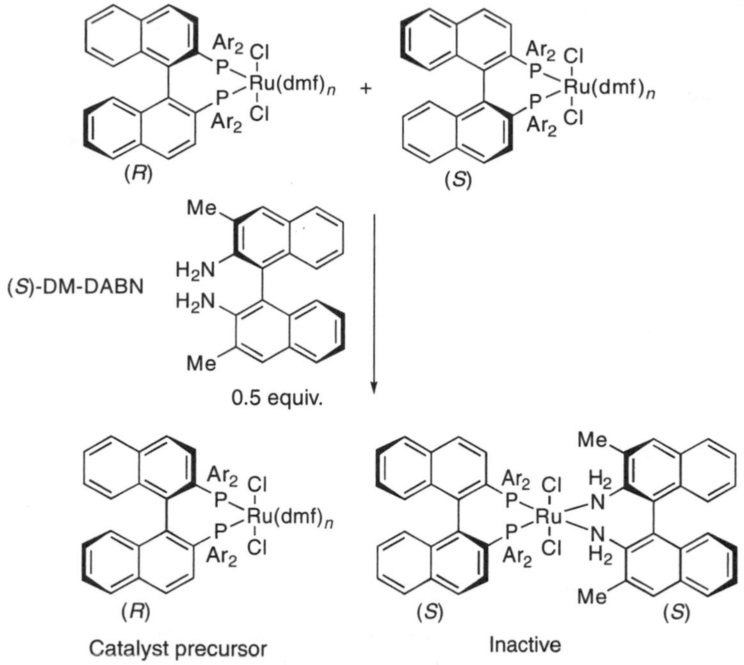

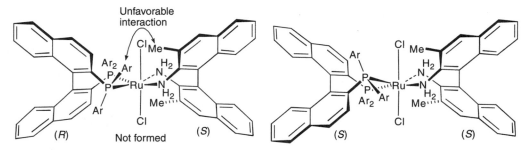

Figure 6.3. The DM-DABN selectively binds to the (S)-enantiomer of the ruthenium precatalyst, resulting in an inactive compound.

Figure 6.4. The diastereomers that would be formed on coordination of DM-DABN to both enantiomers of the (Xyl-BINAP)RuCl₂ moiety. The nonbonded repulsion between the pseudoequatorial P-Xyl groups and DM-DABN methyl groups in the diastereomer on the left is sufficiently severe as to prevent formation of substantial amounts of this complex.

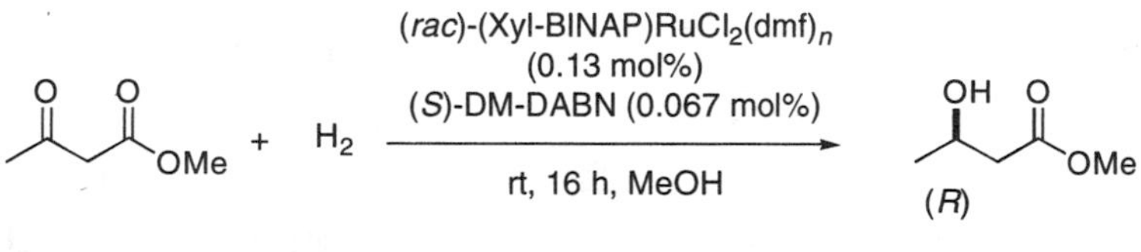

Catalyst	(S)-DM-DABN	Product ee (%)	Yield (%)
Racemic[a]	0.5 equiv.	99.3 (R)	100
(R)[b]	None	99.9 (R)	100

[a] Substrate:catalyst = 750.
[b] Substrate:catalyst = 1500.

Figure 6.5. Carbonyl reduction using racemic (Xyl-BINAP)RuCl₂(dmf)ₙ precatalyst and (S)-DM-DABN poison.

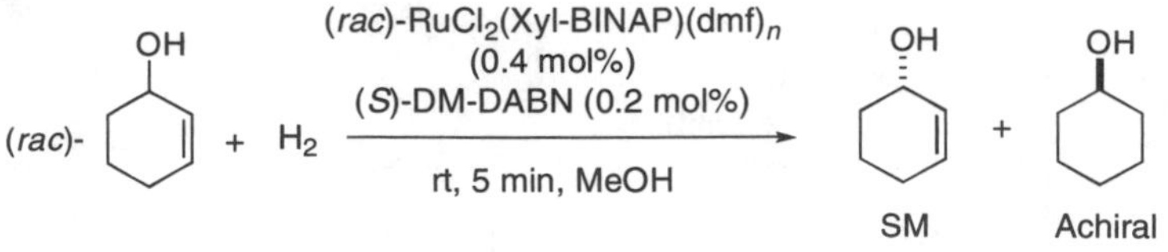

Catalyst	(S)-DM-DABN	% Conversion	% ee of SM	k_{rel}
Racemic[a]	0.5 equiv.	53	100 (S)	
(R)[b]	None	53	100 (S)	
Racemic[a]	0.5 equiv.	48	88 (S)	102

[a] Substrate:catalyst = 250.
[b] Substrate:catalyst = 500.

Figure 6.6. Kinetic resolution of 2-cyclohexenol using racemic catalyst and (S)-DM-DABN.

poison reversibly inhibits one enantiomer of the catalyst to a greater extent than the other. This scenario has been observed in the aluminum–catalyzed hetero-Diels–Alder reaction with the Danishefsky diene and benzaldehyde. When enantiopure 3,3'-di(triphenylsilyl)BINOL, $(Ph_3Si)_2$-BINOL, was employed, the enantioselectivity of the cis product was 95% ee (**Equation 6.1**).[15] The catalyst was found to be monomeric in solution, suggesting nonlinear effects are likely not involved.

Equation 6.1

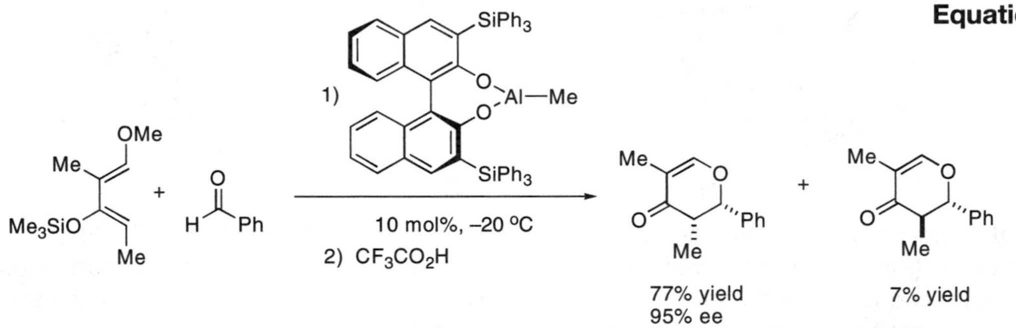

Use of the catalyst prepared from racemic $(Ph_3Si)_2$-BINOL and various chiral ketone poisons was then examined in the hetero-Diels–Alder reaction with the goal of selectively inhibiting one enantiomer of the catalyst. Some of the results of this study are illustrated in **Figure 6.7**, and indicate that when resolved bromocamphor was employed with the racemic catalyst, the cis diastereomer was produced in 82% ee (75% yield). Application of camphor or iodocamphor in this reaction exhibited significantly reduced enantioselectivities.[16]

Several aspects of this system are noteworthy. First, it is surprising that the ketone binds to the catalyst better than the less sterically hindered aldehyde substrate. Initially, investigations employed a 2:1 ratio of racemic catalyst to camphor derivative (**Figure 6.7**) and indicated that bromocamphor was the most efficient inhibitor. It was found, however, that the highest enantioselectivities were observed when the ratio of racemic catalyst to bromocamphor was 1:1—in other words, there is a two-fold excess of bromocamphor with respect to the enantiomer of the BINOL-based catalyst being inhibited. It is also surprising that there is such a large difference in enantioselectivities between the bromocamphor and iodiocamphor inhibitors. At this point, it is unclear what interactions are responsible for the significant difference in affinity of bromocamphor for the enantiomeric catalysts.[16]

A fascinating example of selective removal of one enantiomer of a racemic catalyst involves asymmetric autocatalysis,[17] a topic that is covered in detail in Chapter 11.[18,19] The reaction is the asymmetric addition of diisopropylzinc to pyrimidine carboxaldehydes in **Figure 6.8**. The product of the reaction is a catalyst that promotes the asymmetric addition with very high enantioselectivity. With no enantioenriched material in the reaction mixture, the initial product formed is racemic. When enantiomorphic quartz crystals (i.e., crystals composed of achiral molecules that pack in a chiral space group in the solid state) are present, however, one enantiomer of the product (catalysts) is selectively absorbed.[20] The imbalance created by this selective absorption renders the solution enantioentriched, albeit only slightly. The enormous positive nonlinear effect[21] in this system then generates the product with very high ee (93–97%!). As expected, using the opposite quartz enantiomorph generates the antipode. In this example, the chiral crystal is playing the part of a chiral poison by selectively deactivating one enantiomer of the product/catalyst.

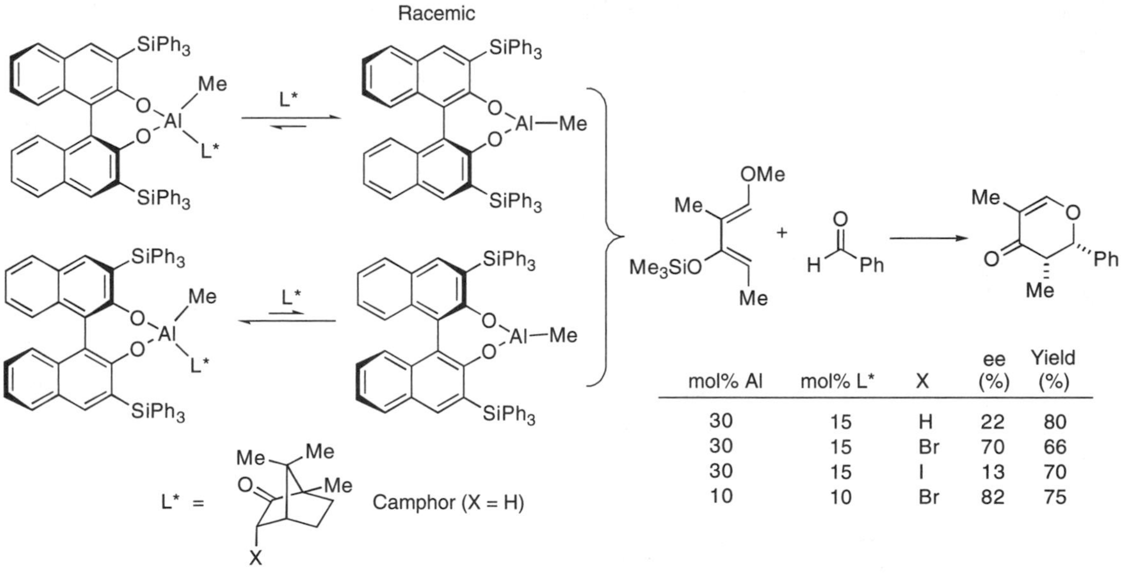

mol% Al	mol% L*	X	ee (%)	Yield (%)
30	15	H	22	80
30	15	Br	70	66
30	15	I	13	70
10	10	Br	82	75

Figure 6.7. Chiral poisoning of a racemic aluminum catalyst with camphor-based ligands.

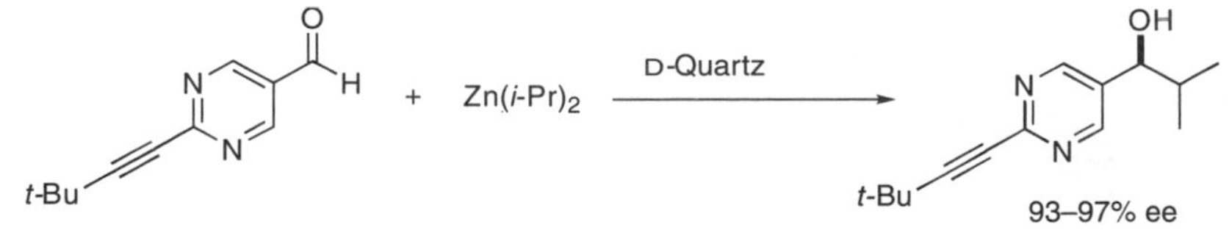

Figure 6.8. Selective absorption of one enantiomer of the product/catalyst onto chiral crystals of D-quartz in this autocatalytic system creates an imbalance that is multiplied as a result of the strong positive nonlinear effect in this asymmetric addition.

6.1.2 Asymmetric Activation of a Racemic Catalyst

The strategy of deactivation of one enantiomer of a racemic catalyst with an enantioenriched poison is somewhat limited. In the absence of a positive nonlinear effect [(+)-NLE, see Chapter 11 for a detailed discussion], the ee of the product will not exceed that generated by the analogous enantiopure catalyst.[5] Another approach to the use of racemic catalysts in asymmetric catalysis, which is related to chiral poisoning, also involves the addition of an enantioenriched ligand to a racemic catalyst. Rather than *deactivate* one enantiomer, however, the idea is to *activate* one enantiomer of the racemate.

In this approach, a racemic catalyst is treated with resolved activator (L*) that selectively activates one enantiomer of the racemic catalyst precursor. The activated catalyst is more active and enantioselective than the catalyst precursors.

An ideal version of asymmetric activation is illustrated in **Figure 6.9**, where the racemic catalyst is completely converted to two diastereomeric catalysts. These diastereomers must display very different activities, with the more active catalyst exhibiting very high enantioselectivity. The rare exception to this would be when both catalysts are highly enantioselective and give the same configuration of the product. The advantage of this method is that the activated catalyst can exhibit increased activity and enantioselectivity with respect to the enantioenriched catalysts or precatalysts. This approach to asymmetric catalysis has been called asymmetric activation.[22,23]

The picture can be more complicated than the simplified scheme in **Figure 6.9**, with equilibria between the catalyst and the activated catalyst (**Figure 6.10**). Nonetheless, if one diastereomer of the catalyst is significantly more active than the other species in solution, and displays high enantioselectivity, the net result may be the same—namely, a highly enantioselective and efficient process.

There are some pitfalls to the asymmetric activation strategy, however. If the diastereomeric catalysts have similar activities, enantioselectivity will generally be reduced. Furthermore, the product ee is a function of the relative concentrations of the diastereomeric catalysts and their activity, complicating rational catalyst optimization. Like most catalyst systems, the rate of reaction depends on the structure of the

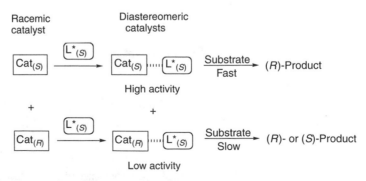

Figure 6.9. A simplified model for asymmetric activation.

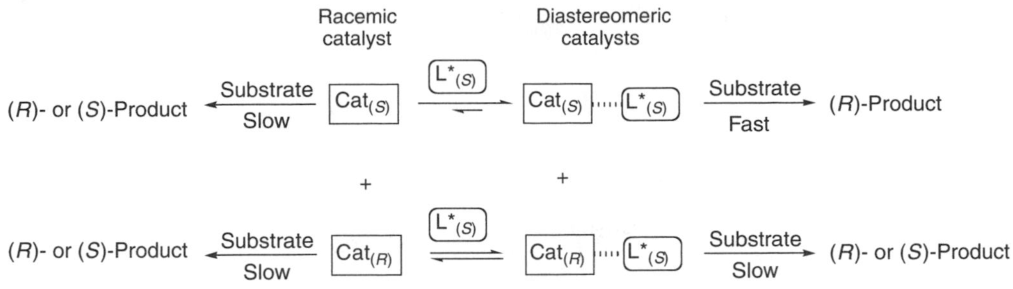

Figure 6.10. Asymmetric activation scheme involving several equilibria.

substrate. With diastereomeric catalysts operating in the same reaction mixture, there can be a dramatic dependency of the product ee on the substrate structure, because the product enantioselectivity depends on both the relative rate *and* the enantioselectivity of each catalyst. This can be contrasted with traditional single-catalyst systems, where the rate of the catalyst does not impact the enantioselectivity (unless the background reaction is competitive with the ligand-accelerated reaction; see Chapter 1). Despite these drawbacks, the asymmetric activation strategy can be a very effective method.

6.1.2.a Asymmetric Activation of Racemic Carbonyl–Ene Catalysts

Asymmetric activation has been examined in the context of the carbonyl–ene reaction,[24] for which a generic mechanism is shown in **Equation 6.2**. The reaction proceeds with transposition of the double bond to provide a functionalized secondary alcohol.

Equation 6.2

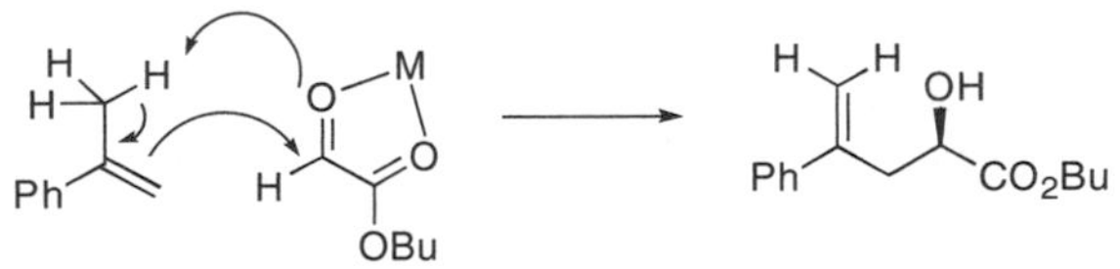

The reaction is catalyzed by (BINOLate)Ti-based catalysts.[25] Use of (BINOLate)Ti(O*i*-Pr)$_2$, which is a trimer in the solid state[26,27] and a dimer in solution,[28] gives excellent enantioselectivity but poor yield (**Figure 6.11**). Activation of the *racemic* catalyst (10 mol%) with 5 mol% (*R*)-BINOL resulted in an increase in the yield with only a slight decrease in the enantioselectivity with respect to the use of enantiopure (BINOLate)Ti(O*i*-Pr)$_2$ (**Figure 6.11**). This can be compared to activation of (*R*)-(BINOLate)Ti(O*i*-Pr)$_2$ with (*R*)-BINOL, which resulted in the highest yield and 96.8% enantioselectivity.

Since the individual enantioselectivities are known (**Figure 6.11**), the relative rates of the activated and inactivated catalysts can be calculated. It is assumed that the (*R*)-BINOL only activates the (*R*)-enantiomer of the racemic catalyst and does so completely. Nonetheless, a lower limit for the relative rates of the activated versus the inactivated catalyst (k_{rel}) is 26.3 (**Figure 6.12**). These results also provide insight into the different roles of the (BINOLate)Ti and added (*R*)-BINOL. If rapid exchange

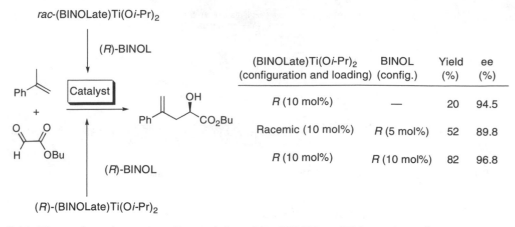

(BINOLate)Ti(O*i*-Pr)$_2$ (configuration and loading)	BINOL (config.)	Yield (%)	ee (%)
R (10 mol%)	—	20	94.5
Racemic (10 mol%)	*R* (5 mol%)	52	89.8
R (10 mol%)	*R* (10 mol%)	82	96.8

Figure 6.11. The carbonyl–ene reaction catalyzed by (BINOLate)Ti-based catalysts.

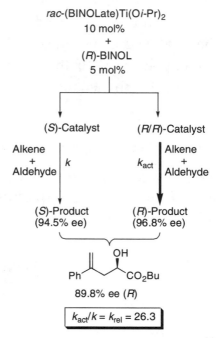

Figure 6.12. Determination of the relative rates of the activated and inactivated catalysts in the asymmetric carbonyl–ene reaction.

of the BINOL ligands occurs on the timescale of the reaction, as might be expected based on exchange rates of titanium alkoxides,[29,30] then the catalyst ee would be 33%. In the absence of nonlinear effects, low product ee would be anticipated. This is not the case. These observations suggest that the original titanium-bound BINOLate ligand and the added (*R*)-BINOL interact differently with the titanium center. Further investigations are necessary to delineate the catalyst structure.

6.1.2.b Asymmetric Activation of Racemic Hydrogenation Catalysts

Interesting and informative results were found when racemic hydrogenation-catalyst precursors (related to those introduced in Chapter 5) were activated with resolved diamines. This detailed study provides insightful observations concerning diastereomeric catalysts and the importance of their relative rates.[35]

Combination of diphosphine–ruthenium complexes with diamines, such as [(R)-Tol-BINAP]RuCl$_2$(dmf)$_n$ and (S,S)-DPEN, gives precatalysts of the type [(R)-Tol-BINAP]RuCl$_2$[(S,S)-DPEN] (**Equation 6.3**). Catalyst prepared from this precursor exhibited 96% ee in the reduction of 2,4,4-trimethyl-2-cyclohexenone to (S)-allylic alcohol (**Equation 6.4, Table 6.1**).

Equation 6.3

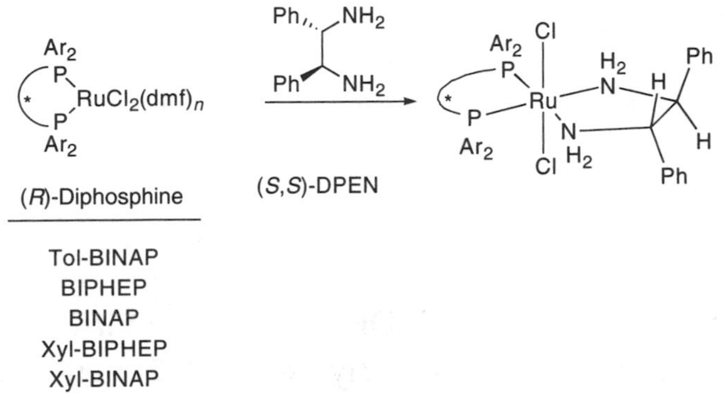

(R)-Diphosphine   (S,S)-DPEN

Tol-BINAP
BIPHEP
BINAP
Xyl-BIPHEP
Xyl-BINAP

Equation 6.4

(S)-Product

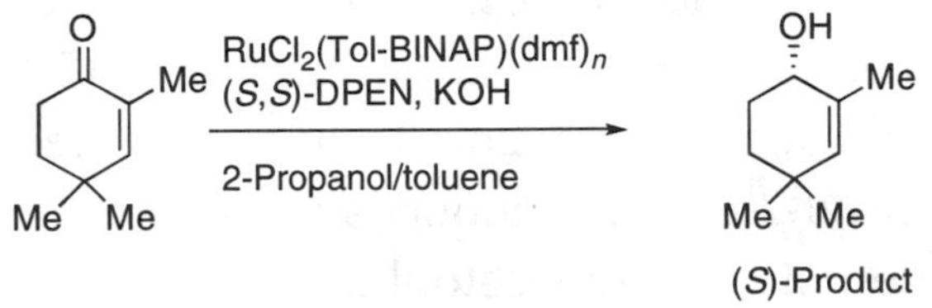

The catalyst comprised of the mismatched ligands, (R)-Tol-BINAP and (R,R)-DPEN, was much slower and furnished the reduced (S)-alcohol with only 26% ee (**Table 6.1**). Employing racemic Tol-BINAP and (S,S)-DPEN gave a 50:50 mixture of diastereomeric ruthenium complexes that reduced 2,4,4-trimethyl-2-cyclohexenone to (S)-allylic alcohol with a remarkable level of enantioselectivity (95% ee). This high enantioselectivity indicates that diastereomeric catalysts have not only different enantioselectivities, but also very different TOFs. As shown in **Table 6.1**, the faster catalyst reduced the substrate 121 times faster than its diastereomer! This huge rate difference accentuates an important point: when diastereomeric catalysts are applied together in the same reaction, the product ee is controlled by the enantioselectivity of each of the diastereomeric catalysts *and* their relative rates.[31–34] Thus, *enantioselectivities and TOFs both depend on the nature of the substrate.*

This fact complicates reaction optimization in processes that employ diastereomeric catalysts in the same reactor. An illustration of this point is the reduction of 2-methylacetophenone (**Equation 6.5**). Use of racemic Tol-BINAP and (S,S)-DPEN afforded the (R)-enantiomer of the alcohol with 90% ee (**Table 6.1**). When

Table 6.1. Comparison of the reduction enantioselectivities and efficiencies of diastereomeric catalysts with ketone substrates (Equations 6.4 and 6.5).

Ketone	Tol-BINAP configuration	DPEN configuration	Alcohol ee % (configuration)	Relative rate
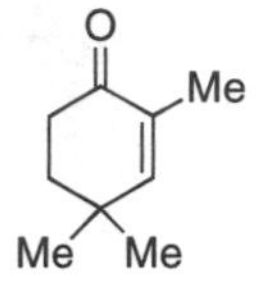	R	S,S	96 (S)	121
	R	R,R	26 (S)	1
	Racemic	S,S	95 (S)	
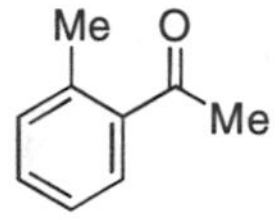	S	S,S	97.5 (R)	13
	S	R,R	8 (R)	1
	Racemic	S,S	90 (R)	

enantiopure (S)-Tol-BINAP and (S,S)-DPEN were employed in the precatalyst preparation, the resulting catalyst rapidly reduced 2-methylacetophenone with 97.5% ee, giving the (R)-product. The diastereomer incorporating (S)-Tol-BINAP and (R,R)-DPEN provided (R)-alcohol of only 8% ee (**Table 6.1**). In this dual catalyst system with 2-methylacetophenone, the faster catalyst is 13 times faster than its diastereomer.[35] Note that the faster, more enantioselective catalyst diastereomer in the reduction of 2,4,4-trimethyl-2-cyclohexenone is the slower, less enantioselective diastereomer in the reaction of 2-methylacetophenone.

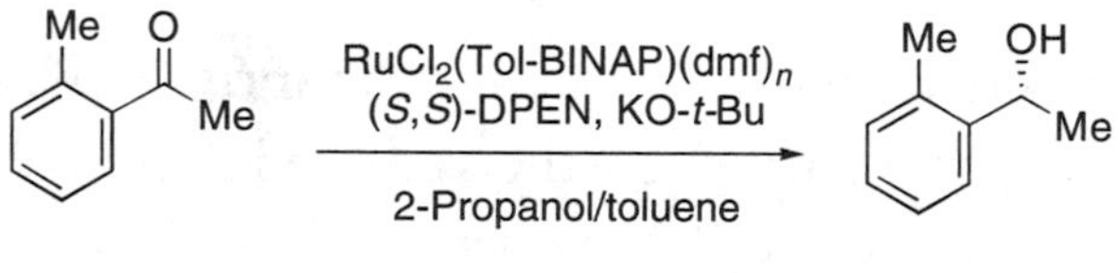

Equation 6.5

$$\text{RuCl}_2(\text{Tol-BINAP})(\text{dmf})_n$$
$$(S,S)\text{-DPEN, KO-}t\text{-Bu}$$
$$\text{2-Propanol/toluene}$$

(R)-Product

The results outlined in **Table 6.1** illustrate how well the chiral activation strategy can work with the correct combination of chiral ligands *and* substrate.

6.1.3 Combined Asymmetric Activation and Deactivation of Racemic Catalysts

Both asymmetric activation[22] and asymmetric deactivation[5] of racemic catalyst precursors have proven to be effective methods in enantioselective catalysis. By simultaneously activating one enantiomer of a racemic catalyst or precatalyst and deactivating the other, the difference in rate of the two catalysts can be increased over the catalyst generated through either of these approaches alone. The combined activation/deactivation strategy is illustrated in **Figure 6.13**.

The system employed for the proof of concept of this strategy was the *rac*-(Xyl-BINAP)RuCl$_2$(dmf)$_n$.[36] As outlined in **Figure 6.3**, *rac*-(Xyl-BINAP)RuCl$_2$(dmf)$_n$ reacts

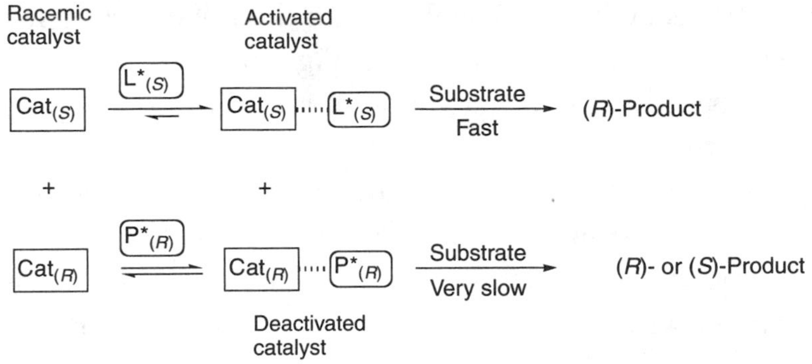

Figure 6.13. Use of the asymmetric activation and deactivation strategies together.

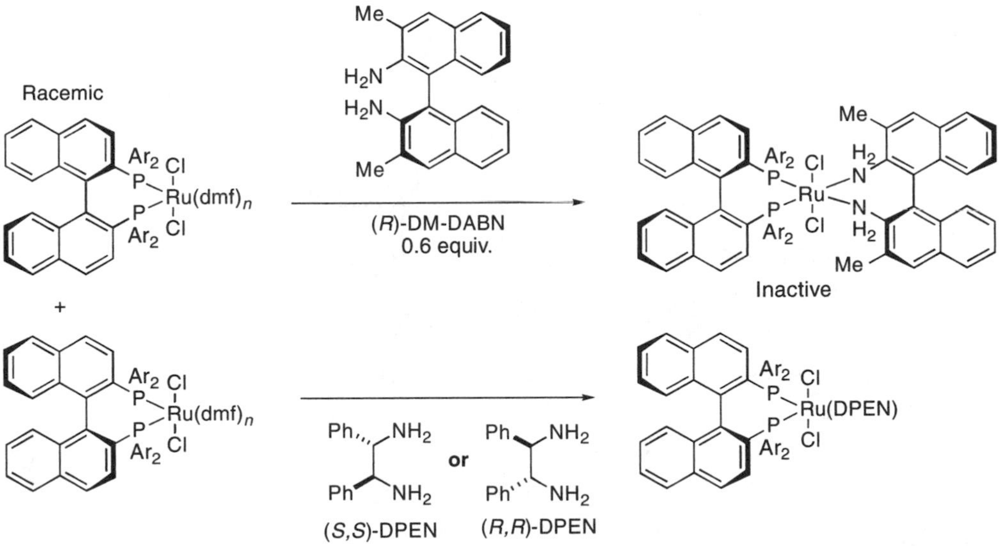

Figure 6.14. Simultaneous deactivation/activation applied to *rac*-(Xyl-BINAP)RuCl$_2$(dmf)$_n$.

with (R)-DM-DABN to give exclusively [(R)-Xyl-BINAP]RuCl$_2$[(R)-DM-DABN] by ^{1}H NMR spectroscopy. No formation of the diastereomer [(S)-Xyl-BINAP]RuCl$_2$[(R)-DM-DABN] was detected, consistent with the model outlined in **Figure 6.4**. The stereochemistry of [(R)-Xyl-BINAP]RuCl$_2$[(R)-DM-DABN] was determined by X-ray diffraction. The remaining [(S)-Xyl-BINAP]RuCl$_2$(dmf)$_n$ was then treated with either (S,S)-DPEN or (R,R)-DPEN to give the diastereomers [(S)-Xyl-BINAP]RuCl$_2$[(S,S)-DPEN] or [(S)-Xyl-BINAP]RuCl$_2$[(R,R)-DPEN] (**Figure 6.14**).

To evaluate separately the activation and deactivation, [(R)-Xyl-BINAP]RuCl$_2$ (dmf)$_n$ was combined with a series of diamine additives and the resultant catalyst mixtures used in the reduction of 1-acetonaphthone (**Figure 6.15**). When no diamine was added, the reaction was very slow and the product nearly racemic. Likewise, when the poison (R)-DM-DABN, which is the matched deactivator, was added, similar results were obtained. Addition of (R,R)-DPEN or (S,S)-DPEN to [(R)-Xyl-BINAP]RuCl$_2$(dmf)$_n$ provided catalysts with matched or mismatched ligand (see Chapter 13) combinations, respectively. The catalyst with the matched ligands gave the (S)-alcohol in 99% ee, while the catalyst with the mismatched ligands generated the same product configuration with 56% ee (**Figure 6.15**).

The simultaneous activation/deactivation of racemic catalyst precursor was then performed to compare the efficiency of asymmetric activation against the combined asymmetric activation/deactivation procedure. In these experiments (*rac*-Xyl-BINAP)RuCl$_2$(dmf)$_n$ was combined with slightly over 0.5 equivalents of (R)-DM-DABN as the chiral poison and 0.5 equivalents of either (R,R)-DPEN or (S,S)-DPEN. The results of these studies with 1-acetonaphthone and 2,4,4-trimethyl-2-cyclohexenone are illustrated in **Figures 6.16** and **6.17**.

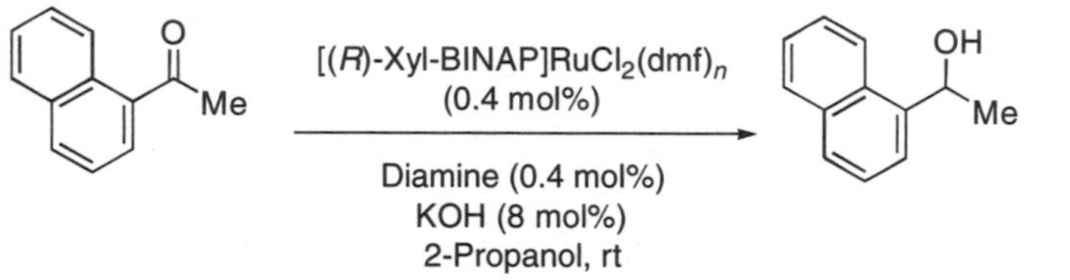

Diamine	Time (h)	Alcohol ee % (config.)	Yield (%)
None	14	4 (R)	4
(R)-DM-DABN	14	7 (R)	6
(S,S)-DPEN	4	56 (S)	> 99
(R,R)-DPEN	4	99 (S)	> 99

Figure 6.15. Deactivation and activation of [(R)-Xyl-BINAP]RuCl$_2$(dmf)$_n$ with diamines in the reduction of 1-acetonaphthone.

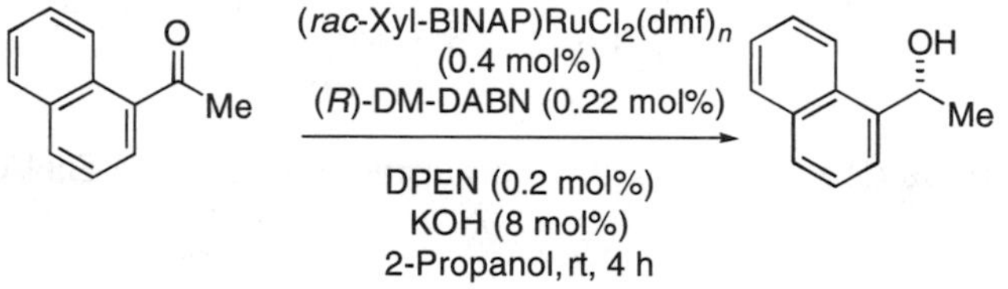

DM-DABN	DPEN	Alcohol ee % (config.)	Yield (%)
None	(S,S)	80 (R)	> 99
(R)	(S,S)	96 (R)	> 99

Figure 6.16. Comparison of the asymmetric activation method with the asymmetric activation/deactivation strategy in the hydrogenation of 1-acetonaphthone (c.f. **Table 6.1**).

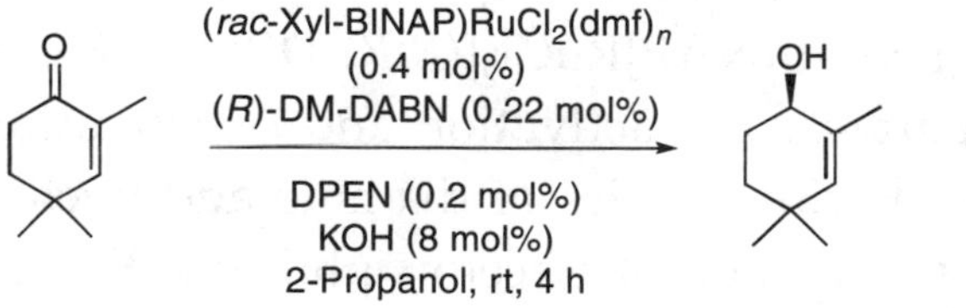

DM-DABN	DPEN	Alcohol ee % (config.)	Yield (%)
None	(R,R)	84 (R)	> 99
(R)	(R,R)	92 (R)	> 99

Figure 6.17. Comparison of the asymmetric activation method with the asymmetric activation/deactivation strategy in the hydrogenation of 2,4,4-trimethyl-2-cyclohexenone (c.f. **Table 6.1**).

As illustrated in **Figures 6.14, 6.16,** and **6.17,** the results employing the asymmetric activation/deactivation of *rac*-(Xyl-BINAP)RuCl$_2$(dmf)$_n$ proved to be significantly better than asymmetric activation alone.[36] Again, the required stereochemistry of the activating agent was substrate dependent (c.f. **Table 6.1**)

6.1.4 Asymmetric Activation of Enantioenriched Catalysts

As alluded to by some of the results above, the technique of chiral activation can also be used to increase the efficiency and enantioselectivity of resolved asymmetric catalysts. An example of this strategy[22] is the asymmetric addition of alkyl groups to aldehydes using a zinc-based catalyst.[37,38]

The catalyst generated from 3,3'-diphenyl BINOL (Ph$_2$-BINOL) and diethylzinc was sluggish and not very enantioselective (8% ee under these conditions) (**Figure 6.18**). In contrast, treatment of this same catalyst with an enantiopure diimine ligand generated a new activated catalyst that exhibited a high degree of ligand acceleration[39] and increased enantioselectivity. The researchers then optimized the catalyst enantioselectivity by screening enantioenriched BINOL derivatives and diimine ligands. The optimal diimine ligand for this reaction was found to be a stilbene diamine derivative (**Figure 6.18**). Combination of ZnEt$_2$, Ph$_2$-BINOL, and a diimine generated a highly enantioselective catalyst that produced 1-phenyl-1-propanol with 90% ee at 0 °C and 99% ee at –78 °C. This catalyst-optimization process illustrates the strength of the chiral activation strategy to maximize catalyst efficiency and enantioselectivity: generation of many catalysts quickly in a modular fashion.

The chiral activation strategy represents a useful technique in asymmetric catalysis. Racemic catalysts can be selectively activated such that they give high enantioselectivity. Resolved catalysts can also be activated with chiral activators, which can increase their efficiency and enantioselectivity.

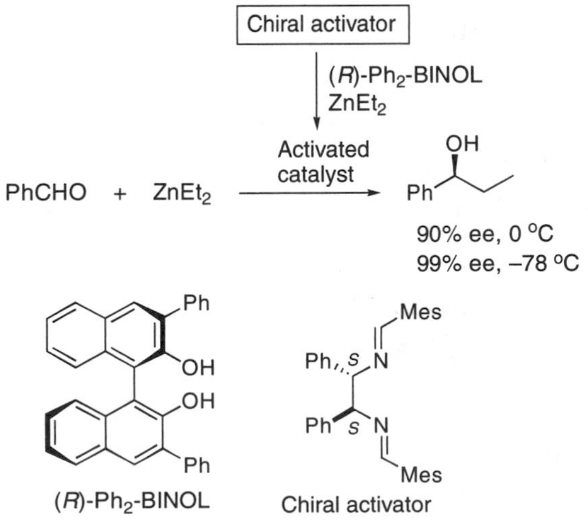

Figure 6.18. The asymmetric activation strategy applied to resolved catalysts results in enhanced activity and enantioselectivity (Mes = 2,4,6–C$_6$H$_2$–Me$_3$).

6.2 Optimization of Asymmetric Catalysts with Achiral Ligands

The chiral ligands used in asymmetric catalysis are generally acquired in enantioenriched form and are stable to racemization, as exemplified by the axially chiral ligand BINAP. As we have seen in Chapter 4, however, the dominant factor in controlling enantioselectivity with BINAP-based catalysts is not the fixed axial chirality, but the positions and orientations of the phenyl rings in the coordinated ligand. The axial chirality causes a twist in the metallacycle, orienting the phenyl rings in equatorial and axial positions, which define the chiral environment of the substrate binding site. In a similar fashion, the asymmetry of one ligand can enforce an asymmetric conformation in an adjacent ligand if the neighboring ligand is stereochemically dynamic.

While the focus of this section is the use of achiral ligands that have chiral conformation or chiral binding modes, it is well known that achiral ligands without chiral conformations can impact both enantioselectivity and activity of a catalyst, as outlined below.[40]

6.2.1 Application of Achiral Additives Void of Chiral Conformations

Most achiral additives are Lewis bases and can influence catalyst efficiency and enantioselectivity through a variety of mechanisms. These include 1) dissociating inactive catalyst aggregates, leading to more active monomeric species; 2) coordinating to metal centers, resulting in a change in the metal coordination number, metal geometry, electronic character of the metal, or even the reaction mechanism; 3) facilitating removal of the product from the catalyst to regenerate the active species and increase the TOF; 4) acting as a poison for a less-enantioselective, more-active catalyst; and 5) associating with the catalyst by means other than covalent and dative bonding.[40] It is also possible that the additive can operate on the catalyst through more than one of these modes of action. The nature of the interaction between the asymmetric catalyst and the achiral additive, however, is often challenging to determine and even more difficult to predict.

6.2.1.a Coordination of Achiral Additives

The dramatic impact of simple achiral ligands on enantiofacial selectivity is illustrated in a lanthanide Lewis acid catalyzed asymmetric Diels–Alder reaction (**Equation 6.6**).[41,42] The catalyst for the this reaction was generated in situ by combining Yb(OTf)$_3$ and (R)-BINOL, followed by addition of two equivalents of meso-1,2,6-trimethylpiperidine. While the catalyst nature remains elusive, a speculative structure is illustrated in **Equation 6.7**. The (BINOL)Yb-based catalyst promotes the asymmetric Diels–Alder reaction of cyclopentadiene with 3-acyl-1,3-oxazolidin-2-ones with high enantioselectivity with up to 95% ee.

Equation 6.6

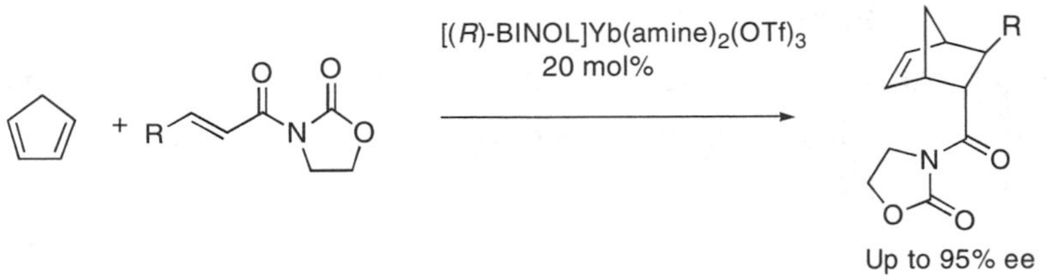

Up to 95% ee

Equation 6.7

Proposed structure

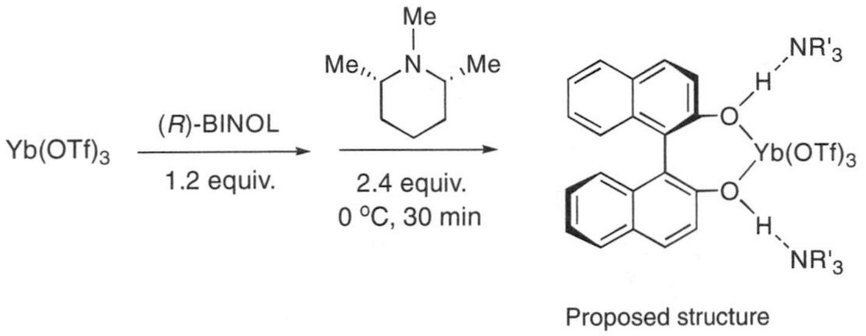

In the course of these investigations, it was found that aging the catalyst without substrate led to lower enantioselectivity and increasing aging periods resulted in decreasing enantioselectivities. In contrast, when the catalyst was aged in the presence of the substrate, higher enantioselectivities were obtained. These observations suggested that the dienophile was stabilizing the catalyst and prompted examination of the impact of substrate-like additives on the catalyst enantioselectivity. Use of 3-acetyl-1,3-oxazolidin-2-one (**A, Figure 6.19**) as additive resulted in generation of the *endo* (2*S*,3*R*)-adduct with 93% ee. In sharp contrast, addition of acetylacetone derivatives resulted in a reversal in the facial selectivity. For example, addition of 3-phenylacetylacetone (**B, Figure 6.19**) led to the generation of the (2*R*,3*S*)-adduct in 81% ee. In these reactions, the same enantiomer of BINOL was employed.

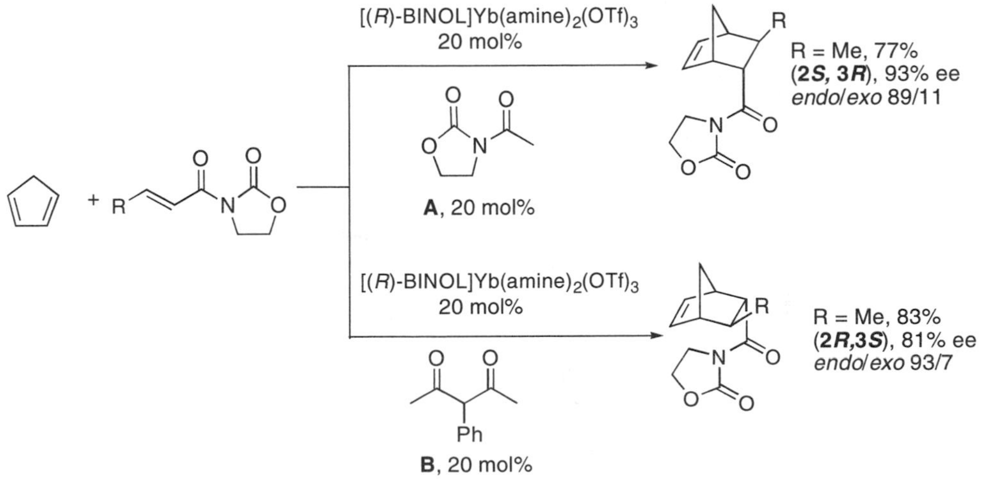

Figure 6.19. Asymmetric Diels–Alder reaction with [(*R*)-BINOL]Yb-based catalyst. Use of additives **A** and **B** results in formation of the opposite product enantiomers.

Lanthanides like ytterbium can achieve coordination numbers of eight or greater. In this example, the additives probably chelate to the ytterbium center through the carbonyl oxygens (the diketone additive may undergo deprotonation to make an acac derivative). Chelation of these additives will alter the geometry about the lanthanide center, changing the chiral substrate-binding pocket. While an astute series of observations led the investigators to explore the use of various additives, discovering an additive with such striking impact often requires extensive experimentation. As outlined in Section 6.2.2, application of ligands that have chiral conformations increases the likelihood of discovering a successful additive.

6.2.1.b Achiral Hydrogen-Bond Donors and Acceptors

Although the addition of additives to increase catalyst efficiency and enantioselectivity is most frequently observed with metal-containing catalysts that can bind an additive, applications in the realm of organocatalysis are also possible. An interesting approach to the use of achiral additives in organocatalytic systems is based on complementary hydrogen bonding between a proline-based precatalyst and a series of additives.[43]

Proline has been shown to be an excellent catalyst for a wide range of reactions (see Chapter 12).[44,45] The enantioselective Michael addition of ketones to nitroalkenes, however, has proven challenging for this versatile catalyst. By preparing proline-based precatalysts tethered to amidonaphthyridine heterocycles via a linker, the environment around the proline can be altered by association with a complementary hydrogen-bonding partner. This general approach is illustrated **Figure 6.20**.

Pyridinones are complementary hydrogen-bonding partners for amidonaphthyridines, as shown in **Figure 6.20**. Two proline-linked amidonaphthyridines and a series of pyridinones employed in these investigations are illustrated in **Figure 6.21**. Solution NMR studies of proline amidonaphthyridines with the pyridinones indi-

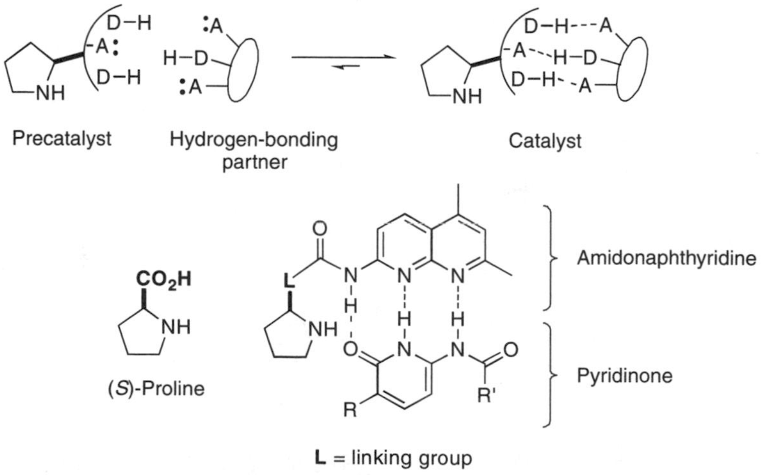

Figure 6.20. Complementary hydrogen-bonding precatalyst and additive partners and their proposed self-assembly. D is a hydrogen-bond donor and A is a hydrogen-bond acceptor.

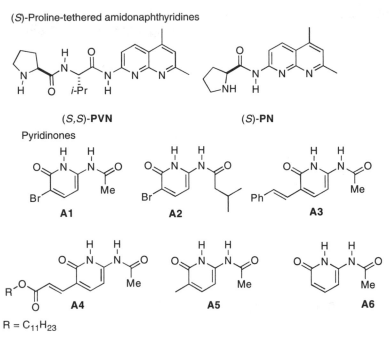

Figure 6.21. (S)-Proline-tethered amidonaphthyridines [(S,S)-**PVN** and (S)-**PN**] and a series of pyridinones (**A1–A6**).

cated that the self-assembled partners were favored over the individual components by about 50:1.

The asymmetric nitro-Michael reaction is illustrated in **Table 6.2**. The mechanism involves reaction of the proline moiety with the cyclohexanone to generate an enamine. The enamine then undergoes conjugate addition with the nitroalkene and the resulting iminium ion is hydrolyzed to regenerate the proline-based catalyst (see Section 2.2.2.a and 12.2.3). When (S)-proline was used as catalyst, the reaction proceeded to only 1% conversion and the product was formed with 27% ee (**Table 6.2**, entry 1). Use of the proline–valine–amidonaphthyridine [(S,S)-**PVN**] with and without pyridinones **A1–A6** gave similarly low enantioselectivities, presumably because the amidonaphthyridine was too far from the proline moiety. Catalytic amounts of truncated (S)-**PN** in the absence of pyridinone additive also resulted in formation of nearly racemic product (7% ee, entry 2). When pyridinones **A1–A6** were used in combination with (S)-**PN**, however, the diastereoselectivity and enantioselectivity of the product were impacted (**Table 6.2**, entries 3–9). Pyridinone **A6** afforded the product with the highest enantioselectivity (72% at 15 °C and 79% at 3–5 °C). Enantioselectivities as high as 94% were measured with related substrates. It was also noted that reactions were significantly faster when (S)-**PN** was used with pyridinones than when it was used alone.[43]

This chemistry demonstrates that catalyst efficiency and enantioselectivity can be altered through formation of intermolecular hydrogen bonds. It is easy to imagine optimizing a catalytic asymmetric reaction with a library of amidonaphthyridine- and pyridinone-based precatalysts and additives. Optimization of catalysts with this approach will certainly generate more attention in the future.

Table 6.2. Results from the asymmetric nitro-Michael reaction with (S)-**PN** precatalyst and hydrogen-bonding additives from **Figure 6.21**.

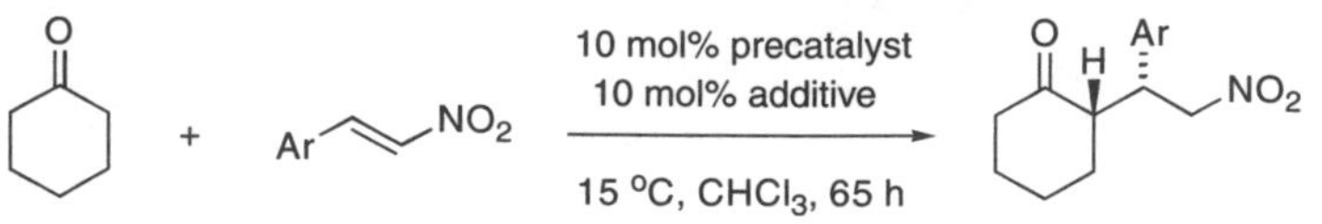

Entry	Catalyst	Additive	NMR yield (%)	dr	ee (%)
1	(S)-Proline	None	1	—	27
2	(S)-**PN**	None	70	41:1	7
3	(S)-**PN**	**A1**	82	31:1	47
4	(S)-**PN**	**A2**	74	33:1	34
5	(S)-**PN**	**A3**	91	42:2	32
6	(S)-**PN**	**A4**	69	77:1	16
7	(S)-**PN**	**A5**	59	41:1	35
8	(S)-**PN**	**A6**	98	59:1	72
9	(S)-**PN**	**A6**	63	58:1	79[a]

[a] Reaction at 3–5 °C.

6.2.2 Conformational Chirality in Atropisomeric Systems

Important recent contributions highlight the utility of ligands that have chiral conformations, yet are not configurationally stable at room temperature. In particular, enantioselective reactions with 2,2′-bis(diarylphosphino)biphenyl derivatives (**Figure 6.22**) coordinated to late transition metals have proven ideal for such studies. Even though the individual skew conformations of compounds derived from BIPHEP (**Figure 6.22**, Ar = Ph) are chiral, the barrier to racemization of the unbound parent ligand is 22 kcal/mol.[46] A barrier of this magnitude does not allow resolution of the enantiomeric conformations at room temperature, and these ligands are considered configurationally dynamic.

Disubstituted biphenyl derivatives can be resolvable with two very large ortho substituents or with certain tethers linking the 2- and 2′- positions,[47] as illustrated by the constrained biphenyl derivatives in **Figure 6.23**. The barrier to racemization of these tethered biphenyls strongly depends on the characteristics of the tether.[48–51] Coordination of BIPHEP to a metal center forms a metallacycle that is related to the constrained biphenyl derivatives and will have a higher barrier to rotation about the central C–C bond than the free ligand.

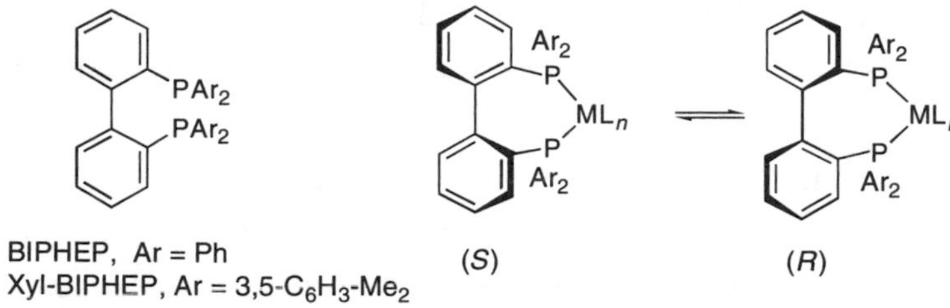

BIPHEP, Ar = Ph
Xyl-BIPHEP, Ar = 3,5-C_6H_3-Me_2

(S)   (R)

Figure 6.22. Configurationally dynamic 2,2′-bis(diarylphosphino)biphenyl derivatives and their enantiomeric complexes.

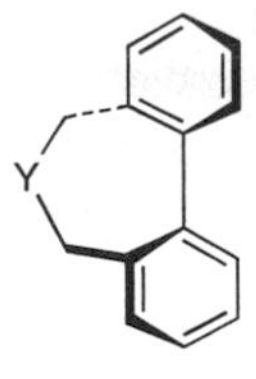

Y	$\Delta G^{\ddagger}$ (kcal/mol)
O	9
S	17
CH_2	12
$C(CO_2H)_2$	23
CH_2CH_2	24

Figure 6.23. Racemization barriers of constrained biphenyl derivatives.

6.2.2.a Asymmetric Hydrogenation Reactions Using BIPHEP Ligands

Based on the results from the asymmetric ketone reductions with racemic Tol-BINAP outlined above[35] (**Equation 6.4, Table 6.1**), it was envisioned that stereochemically dynamic ligands could be substituted for the racemic Tol-BINAP ligand. Synthesis of the precatalyst with the conformationally flexible diphosphine Xyl-BIPHEP (**Figure 6.24**) led to rac-(Xyl-BIPHEP)RuCl$_2$(dmf)$_n$. Upon combining the (Xyl-BIPHEP)-RuCl$_2$(dmf)$_n$ with (S,S)-DPEN, two diastereomers were initially observed by NMR spectroscopy in a 1:1 ratio, as expected (**Figure 6.24, Table 6.3**). Over three hours at room temperature, the (S/S,S) and (R/S,S) diastereomers were observed to slowly equilibrate to a 1:3 mixture.

The enantioselective reduction of 1′-acetonaphthone was studied with catalysts derived from atropisomeric BIPHEN ligands and (S,S)-DPEN. The initial diastereomeric ratio of the catalysts influenced the product ee. Diastereomeric ratios of the precatalyst of 1:1, 2:1, and 3:1 produced the alcohol product with 63, 73, and 84% ee, respectively, indicating that the major diastereomer is more enantioselective (**Table 6.4**, entries 1–3). Notably, the diastereomeric ratios in entries 1–3 corresponded to the dichloride precatalysts. The dihydride complex, however, is believed to be the catalytically active species.[52–54] The diastereomeric ratio of the dihydride species was not investigated.

The benefits of using the stereochemically dynamic Xyl-BIPHEP ligands are apparent on comparison of product ee when a 3:1 mixture of diasteromeric catalysts was used versus the racemic Xyl-BINAP-based catalyst in **Table 6.4**. The 1:1 mixture of diastereomers formed on addition of (S,S)-DPEN to [(±)-Xyl-BINAP]RuCl$_2$(dmf)$_n$

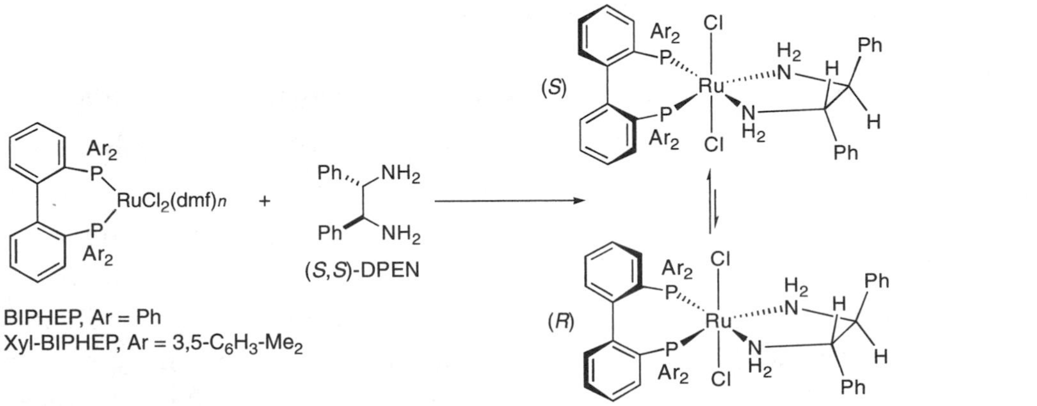

BIPHEP, Ar = Ph
Xyl-BIPHEP, Ar = 3,5-C$_6$H$_3$-Me$_2$

Figure 6.24. Reaction of (BIPHEP)RuCl$_2$(dmf)$_n$ with (S,S)-DPEN generates diastereomeric complexes.

Table 6.3. Kinetic and thermodynamic diastereomeric ratios from the reaction of (BIPHEP)RuCl$_2$(dmf)$_n$ derivatives with (S,S)-DPEN.

	RuCl$_2$(diphosphine)(dmf)$_n$	Diamine	S/S,S : R/S,S
Kinetic	Xyl-BIPHEP	(S,S)-DPEN	1 : 1
Thermodynamic			1 : 3
Kinetic	BIPHEP	(S,S)-DPEN	1 : 1
Thermodynamic			1 : 2

Table 6.4. Initial ratios of diastereomeric catalysts (**Figure 6.24**) and their enantioselectivities in the reduction of 1′-acetonaphthone.

Entry	Diphosphine	S/S,S : R/S,S	T (°C)	ee
1	Xyl-BIPHEP	1:1	28	63
2	Xyl-BIPHEP	2:1	28	73
3	Xyl-BIPHEP	3:1	28	84
4	(±)-Xyl-BINAP	1:1	28	80
5	Xyl-BIPHEP	3:1	-35	92
6	(±)-Xyl-BINAP	1:1	-35	89

gave product of slightly lower ee than the diastereomeric catalysts generated from Xyl-BIPHEP (i.e., entries 3 vs. 4 and entries 5 vs. 6). These important investigations laid the groundwork for related studies involving enantioselective reactions with normally unresovable phosphine ligands, such as BIPHEP, and examination of the interactions responsible for the relative populations of the observed diastereomers. An understanding of the factors that control both the kinetics and thermodynamics of equilibration of the diastereomeric catalysts is crucial to developing catalysts based on the use of conformationally flexible ligands.

6.2.2.b Asymmetric Catalysis with Normally Unresovable Ligands

In the *rac*-(Xyl-BIPHEP)RuCl$_2$[(S,S)-DPEN]-catalyzed hydrogenation outlined in **Table 6.4**, the stereochemically dynamic BIPHEP derivative was an essential element of the catalyst asymmetric environment. Subsequent investigators have taken advantage of the barrier associated with atropisomerization in the (BIPHEP)Pt system[55] to prepare an asymmetric catalyst in which the only chiral ligand is the normally unresolvable BIPHEP frozen in a skewed conformation.[56]

Reaction of the *rac*-(BIPHEP)PtCl$_2$ with *rac*-(BINOLate)Na$_2$ furnished a 1:1 kinetic ratio of the (BINOLate)Pt(BIPHEP) diastereomers (**Equation 6.8**).

Alkoxide exchange reactions can often be catalyzed by trace amounts of the alcohol.[29] With this in mind, the equilibrium constant between the (BINOLate)Pt (BIPHEP) diastereomers in **Equation 6.8** was measured by treatment of these compounds with catalytic *rac*-BINOL and found to be > 17 (**Equation 6.9**).

In order to understand the molecular interactions that cause the strong thermodynamic preference in the (BINOLate)Pt(BIPHEP) diastereomers, the structure of the minor diastereomer was determined and is shown in **Figure 6.25.**[57] In this compound, the BIPHEP ligand is markedly distorted in order to reduce nonbonded interactions between the BINOLate 3,3′-hydrogens, which are directed toward the protruding equatorial *P*-phenyl groups. In the more stable diastereomer, the 3,3′-hydrogens would point toward the pseudoaxial *P*-phenyl groups, which are oriented away from the platinum center, thereby greatly reducing the interligand steric interactions.

Equation 6.8

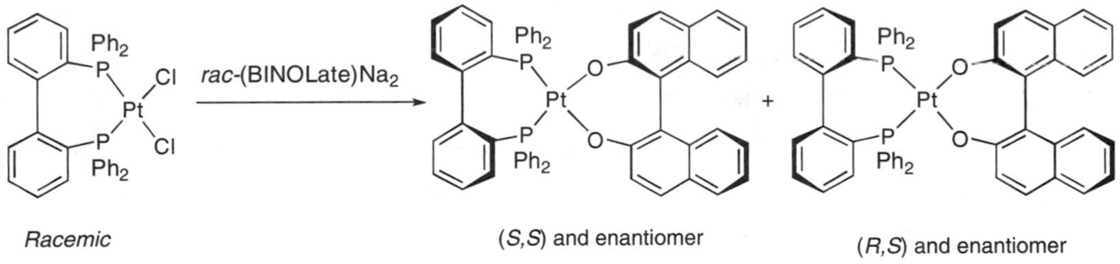
Ph2
P
Cl
Pt
P
Cl
Ph2
rac-(BINOLate)Na2
O
+
Racemic
(S,S) and enantiomer
(R,S) and enantiomer

Equation 6.9

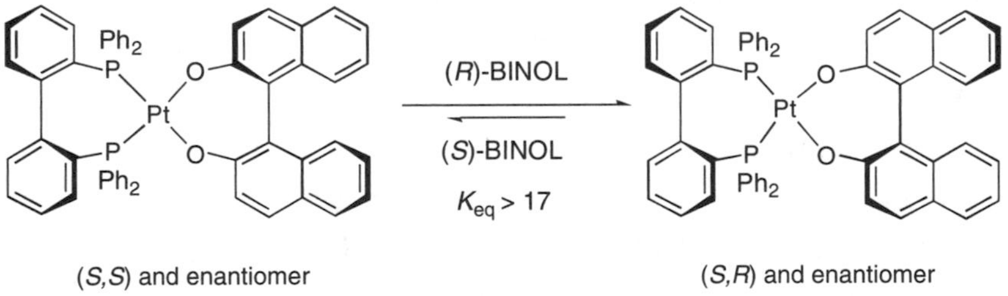
Ph2
P
Pt
O
(R)-BINOL
(S)-BINOL
Keq > 17
(S,S) and enantiomer
(S,R) and enantiomer

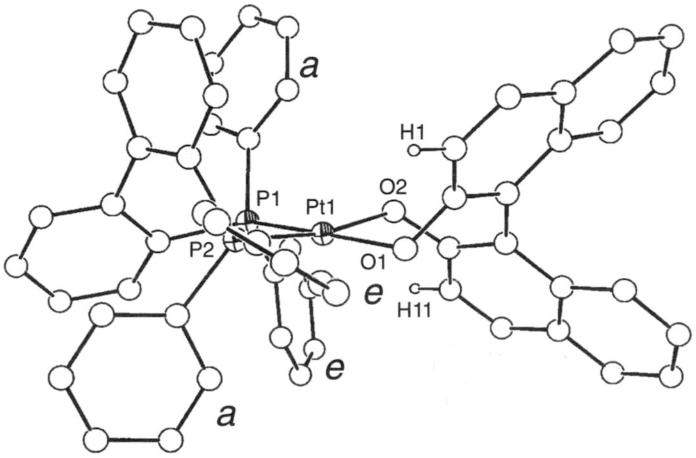
a
H1
P1
Pt1
O2
P2
O1
e
H11
e
a

Figure 6.25. Structure of the less stable diastereomer (BINOLate)Pt(BIPHEP) diastereomer. The pseudoaxial phenyl groups are labeled ***a*** and the pseudoequatorial groups are labeled ***e***.

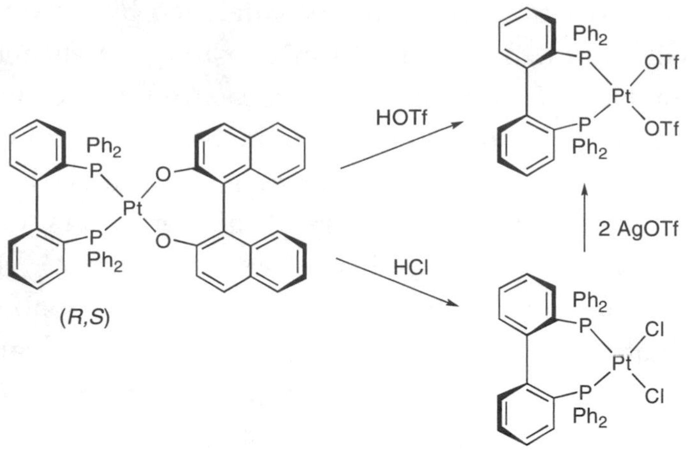

Figure 6.26. Protonation of [(R)-BIPHEP]Pt[(S)-BINOLate] with acids to give (BIPHEP)PtX$_2$ derivatives of high enantiopurity.

The next step in this work was to remove the BINOLate ligand, leaving the BIPHEN ligand as the only chiral element in the platinum complex. In this case, (S)-BINOLateNa$_2$ was used in the synthesis of the platinum complex in place of *rac*-BINOLateNa$_2$ shown in **Equation 6.8**. As illustrated in **Figure 6.26**, reaction of [(R)-BIPHEP]Pt[(S)-BINOLate] of high diastereopurity with triflic acid resulted in liberation of (S)-BINOL and formation of [(R)-BIPHEP]Pt(OTf)$_2$. Similarly, reaction of the [(R)-BIPHEP]Pt[(S)-BINOLate] with hydrochloric acid gave [(R)-BIPHEP]PtCl$_2$. The dichloride, [(R)-BIPHEP]PtCl$_2$, could be converted to the triflate, [(R)-BIPHEP]Pt(OTf)$_2$, by treatment with two equivalents of silver triflate.

To determine if any racemization occurred during the protonation process, samples of the bis(triflate) were treated with (S,S)-DPEN (**Equation 6.10**), and the diastereomeric excess of the cationic diamine adducts was determined by ^{31}P NMR spectroscopy. The [(R)-BIPHEP]Pt(OTf)$_2$ formed in **Figure 6.26** was found to have 98% ee. An aliquot of [(R)-BIPHEP]Pt(OTf)$_2$ that was allowed to stand for eight hours at room temperature was similarly measured and found to have 96% ee. Racemization of [BIPHEP]Pt[(S,S)-DPEN]$^{2+}$ in solution at room temperature was very slow.

Equation 6.10

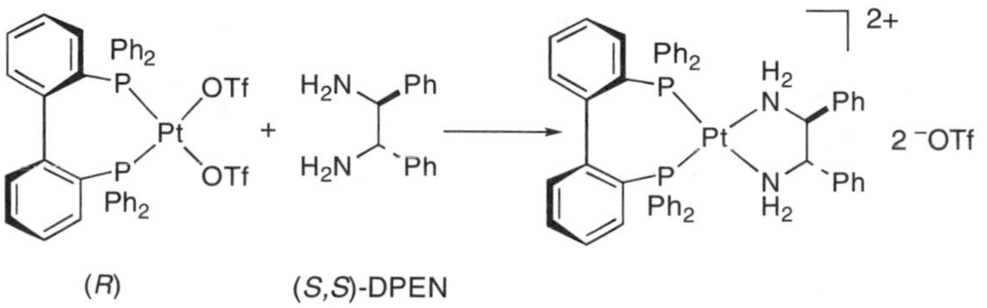

The configurational stability of (BIPHEP)PtX$_2$ compounds to racemization over several hours at room temperature indicated that these compounds could be employed in catalytic asymmetric reactions that proceeded readily at room temperature or below. The potential of these BIPHEN-based complexes was examined in the asymmetric Diels–Alder reaction outlined in **Equation 6.11**. Application of freshly generated [(R)-BIPHEP]Pt(OTf)$_2$ in **Equation 6.11** resulted in formation of the cyclo-adduct in 92–94% ee (94:6 *endo:exo* ratio).[56] The enantiomeric excess of the platinum catalyst [(R)-BIPHEP]Pt(OTf)$_2$ was not eroded over the course of the reaction, as de-termined by quenching the reaction at > 90% conversion with (S,S)-DPEN and sub-sequent ^{31}P NMR analysis of the resulting mixture. The stereochemistry of the Diels–Alder adduct was consistent with previous reports[58] employing [(R)-BIPHEP]Pt(OTf)$_2$ with the same substrates.

Equation 6.11

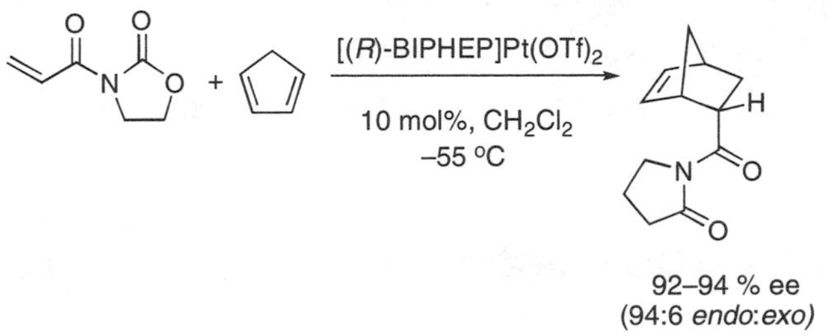

92–94 % ee
(94:6 *endo:exo*)

A related catalyst has been generated that catalyzes a highly enantioselective ene-type reaction. In this example, *rac*-[(BIPHEP)Rh(NBD)]$^+$ (where NBD = norbor-nadiene) reacts with hydrogen (to hydrogenate the NBD) and (R)-DABN, giving a single diastereomer of [(BIPHEP)Rh$^+$[(R)-DABN] after equilibration. The diamine is removed from the coordination sphere of the metal by treatment with two equiva-lents triflic acid, which generates a very active catalyst for the enantioselective cy-clization outlined in **Equation 6.12**.[59]

Equation 6.12

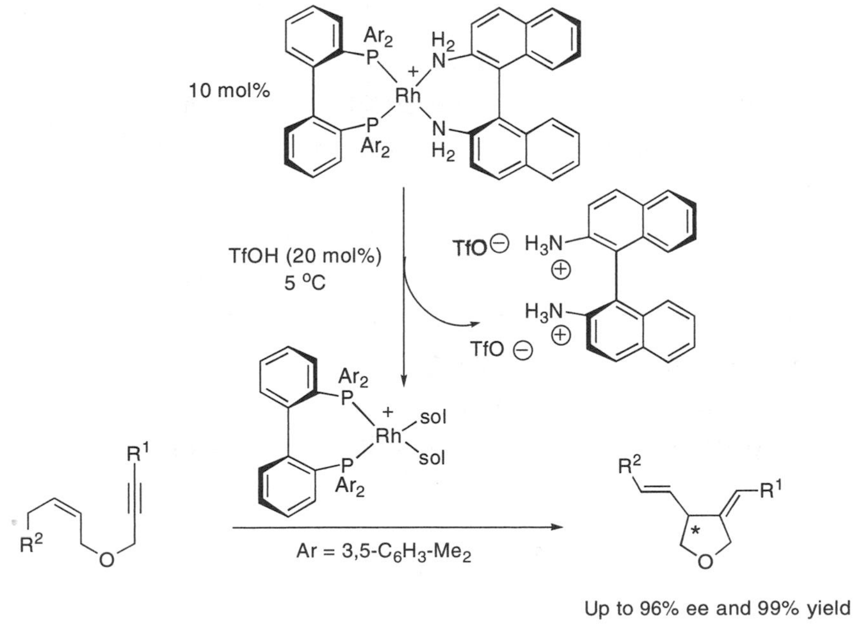

Ar = 3,5-C$_6$H$_3$-Me$_2$

Up to 96% ee and 99% yield

These, and related systems,[60,61] demonstrate that diphosphine ligands that are normally not resolvable can adopt stereochemically robust conformations when coordinated to substitutionally inert metals. The ubiquitous nature of chiral chelating diphosphines should render this strategy useful in further applications in asymmetric catalysis.

6.2.3 Conformational Chirality in Nonatropisomeric Systems

6.2.3.a Optimization of Asymmetric Catalysts with Achiral and Meso Ligands

Optimization of asymmetric catalysts has traditionally been performed by the synthesis and screening of chiral ligands.[2] In the asymmetric activation of enantiopure catalysts, as outlined above, two chiral ligands are needed to generate the catalysts. The results outlined below demonstrate that catalyst enantioselectivity and activity can also be optimized by modification of achiral and meso ligands.[62]

6.2.3.b Achiral and Meso Diamine and Diimine Ligands with Zinc(BINOLate)

A more advanced strategy than chiral activation would be *achiral* activation (**Figure 6.27**), in which achiral or meso ligands with chiral conformations are attached to a metal coordinated to a resolved chiral ligand.[63,64] Catalyst optimization using this approach is more effective, because numerous catalysts can be synthesized from each resolved ligand by combination with a series of achiral and meso ligands. Many more achiral and meso ligand precursors and ligands are commercially available than enantioenriched ligands. Furthermore, catalysts with a wider variety of shapes can be synthesized, because there is more diversity among achiral ligands than chiral ligands. Finally, the price of achiral and meso components is generally much less than the cost of related enantiopure materials.

The application of achiral ligands in asymmetric catalysis has been demonstrated and is outlined below. The catalyst system employed in this proof-of-concept study was outlined in Section 6.1.4 (**Figure 6.18**).[37,38] In reactions using $ZnEt_2$ and 10 mol% (S)-Ph$_2$-BINOL with no additional ligands, benzaldehyde was slowly converted to (S)-1-phenyl-1-propanol with 44% ee (**Table 6.5**, entry 1). After 28 h, the reaction was only 83% complete. Several classes of achiral and meso diimine and diamine additives were then examined, with representative examples outlined in **Table 6.5**. Both the Ph$_2$-BINOL and achiral or meso ligands were employed at 10 mol%, and three equivalents of diethylzinc were used relative to aldehyde substrate. In the asymmetric activation study (**Figure 6.18**), resolved diimines derived from 2,4,6-

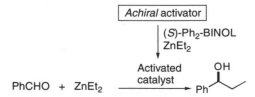

Figure 6.27. Activation of resolved catalysts can result in enhanced enantioselectivity and efficiency.

Table 6.5. Results of screening achiral diimine and diamine activators with (S)-Ph$_2$-BINOL in **Figure 6.27.**[a]

Entry	Achiral ligand	ee (%) at 0 °C (configuration)	ee (%) at −45 °C (configuration)
1	No achiral activator	44 (S)	—
2	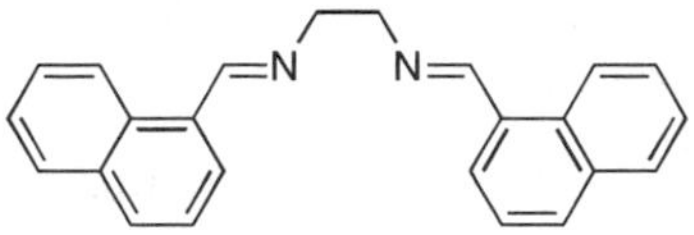	52 (S)	—
3	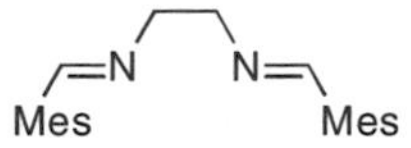 (Mes ... Mes)	75 (R)	87 (R)
4	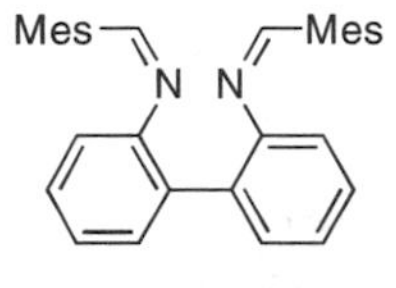 (Mes ... Mes)	89 (R)	96 (R)
5	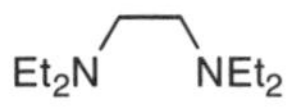 Et$_2$N NEt$_2$	36 (R)	72 (R)
6	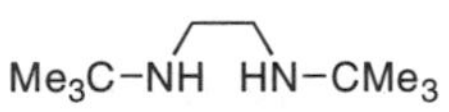 Me$_3$C−NH HN−CMe$_3$	73 (S)	76 (S)
7	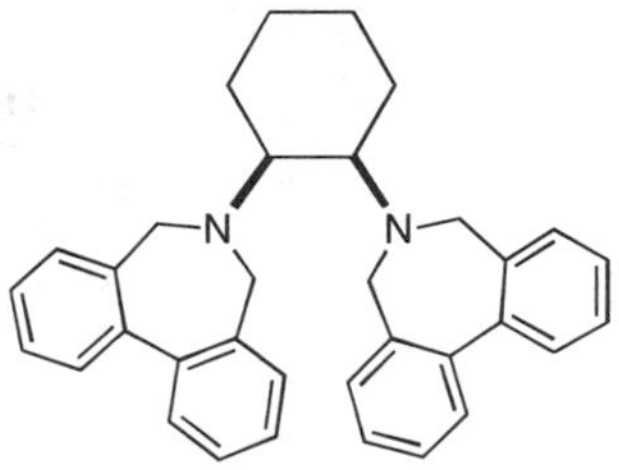	83 (R)	92 (R)

[a]Reactions employed 10 mol% Ph$_2$-BINOL, 10 mol% diimine ligand, and 300 mol% diethylzinc (Mes = 2,4,6−C$_6$H$_2$−Me$_3$).

trimethylbenzaldehyde were found to provide the most enantioselective catalysts. Likewise, achiral diimines derived from this aldehyde formed the most enantioselective catalysts. Use of the simple diimines derived from ethylenediamine and 1-naphthaldehyde (**Table 6.5**, entry 2), in combination with (S)-Ph$_2$-BINOL, resulted in a slight increase in the enantioselectivity over the reaction with only (S)-Ph$_2$-BINOL. In contrast, use of the mesityl-diimine derivative resulted in enantioselectivities as high as 87%, with the opposite enantiomer predominating (entry 3) relative to the catalyst lacking the achiral additive. This dramatic change in selectivity is a testament to the catalyst modification possible with even achiral ligands.

Several of the ligands examined were imines prepared from the 2,2′-diaminobiphenyl backbone. Coordination of these ligands increases the barrier to interconversion of their atropisomeric conformations. As shown in **Figure 6.28**, the two enantiomeric conformations of the zinc-bound diiminobiphenyl ligand become diastereomeric in the presence of the (S)-Ph$_2$-BINOLate (chiral relay). If these diastereomers have sufficiently different energies, the equilibrium will lie to the side

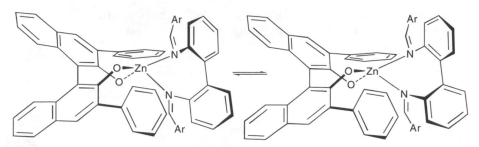

Figure 6.28. Diastereomeric catalysts derived from different conformations of the diaminobiphenyl ligand (entry 4, **Table 6.5**).

of the more stable diastereomer, in which case coordination is similar to a dynamic resolution of the diaminobiphenyl ligand (see Section 6.2.2). The chiral conformation of the biphenyl-diimine ligand extends the chiral space around the catalyst (**Figure 6.28**). Use of the biphenyl-diamine derivative (entry 4) in combination with (*S*)-Ph$_2$-BINOL in **Figure 6.27** resulted in very high enantioselectivities (96% ee at –45 °C).

Diamines also had a significant impact on the catalyst enantioselectivities. The simple diamine, *N,N,N′,N′*-tetraethylethylenediamine caused formation of the (*R*)-enantiomer in 72% ee (**Table 6.5**, entry 5). Other diamines that were particularly interesting are the secondary derivatives, such as *N,N′*-di-*tert*-butylethylenediamine (entry 6). Coordination of the nitrogens to zinc freezes nitrogen inversion, creating two new stereogenic centers directly bonded to the metal center (see Section 4.6.2).[47,65] The coordination is likely reversible, allowing equilibration of the stereochemistry at nitrogen through a three-coordinate zinc species. It was expected that the *C*$_2$-symmetric (Ph$_2$-BINOLate)Zn would favor binding of the diamine in a *C*$_2$-symmetric fashion (see Figure 6.30). The chiral nitrogen centers would then influence the stereochemistry of the asymmetric C–C bond-forming process. Examination of this diamine for the reaction in **Figure 6.27** resulted in product with 76% ee, the highest enantioselectivity achieved for the (*S*)-configuration of the alcohol in this study.

Meso diamines with pendant biphenyl groups were also examined (**Table 6.5**, entry 7). Atropisomerization of the chiral biphenyl moiety is anticipated to have a low barrier. Interaction of the stereochemically dynamic biphenyl groups with the (Ph$_2$-BINOLate)Zn center is expected to influence the biphenyl stereochemistry and extend the chiral space around the Lewis acidic zinc center. The resultant catalyst was efficient and gave the product with the (*R*)-configuration as high as 92% ee at –45 °C (entry 7). Notably, this is significantly higher than the enantioselectivity measured when *N,N,N′,N′*-tetraethylethylenediamine was employed (entry 5). It is possible that this difference in enantioselectivity is due to the chiral conformations of the biphenyl-diamine derivative.

Experiments to probe the effect of the achiral and meso ligands on the catalyst TOF were also performed as outlined in **Figure 6.27**. In the absence of other ligands, the reaction of diethylzinc with benzaldehyde is 1% complete after 8 h at 0 °C (**Fig-**

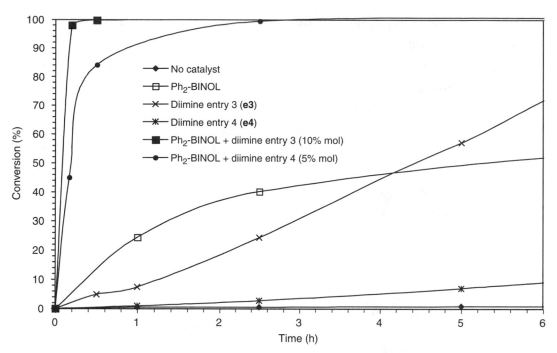

Figure 6.29. Plot of conversion vs. time for the addition with no catalyst, Ph$_2$-BINOL alone, diimine ligands in entries 3 and 4 in **Table 6.5**, and the diimine ligands in combination with Ph$_2$-BINOL **(Figure 6.27)**.

ure 6.29). Use of Ph$_2$-BINOL in this reaction resulted in an increase in the TOF, giving almost 25% conversion after 1 h. Combination of 10 mol% diimine in entry 3 of **Table 6.5** (denoted **e3** in **Figure 6.29**) or the biphenyl diimine (entry 4), diethylzinc, and benzaldehyde (in the absence of Ph$_2$-BINOL) resulted in < 10% conversion with the diimine after 1 h. When Ph$_2$-BINOL was used in combination with diimine ligands **e3** and **e4**, the reactions both showed dramatic TOF enhancements. In the case of ligand **e4**, the reaction was complete in about 10 min. The increase in TOF is even more dramatic with the diamines.[64]

The catalysts in these reactions are proposed to be the four-coordinate (BINOLate)Zn(diimine) and (BINOLate)Zn(diamine) complexes. An example of the latter is shown in the X-ray crystal structure in **Figure 6.30**. Coordination of the aldehyde substrate, followed by addition of the ethyl group from ZnEt$_2$, results in product formation.

In this detailed study of asymmetric catalyst optimization, the only chiral ligand employed was (S)-Ph$_2$-BINOL and enantioselectivities between 96% (R) and 76% (S) could be obtained.[64] Clearly asymmetric catalysts can be efficiently optimized by screening achiral and meso ligands. This approach has been successfully applied to other reactions[66,67] and holds great promise for future catalyst development.

6.2.3.c Metal Geometry-Induced Ligand Asymmetry

An alternative strategy to optimizing asymmetric catalysts involves ligands that are symmetric in certain metal geometries, but become asymmetric when the metal binds an additional ligand, such as the substrate in an asymmetric reaction. Under these circumstances it is possible that the achiral ligand can contribute to the chiral environment of the catalyst and, as a result, impact the enantioselectivity of the process. This concept is outlined below.

The four-coordinate compounds (MBP)TiCl₂ [**Figure 6.31**, where MBP = methylene bis(phenoxide)] are achiral because the X–Ti–X plane is a mirror plane that bisects the boat-shaped conformation of the MBP ligand. On recrystallization of this species from THF, the Lewis acidic titanium coordinates a THF solvent molecule, thus forming the five-coordinate (MBP)TiCl₂(THF). In this five-coordinate compound, the apical (*a*) and equatorial (*e*) oxygens of the MBP metallacycle are inequiv-

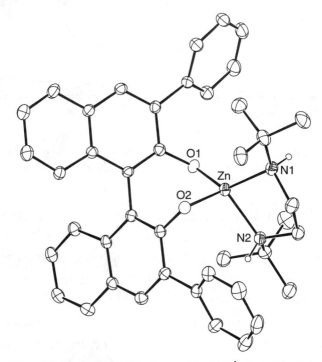

Figure 6.30. Structure of the (Ph₂-BINOLate)Zn bound to *N,N'*-di-*tert*-butylethylenediamine. The ethylenediamine nitrogens have the expected (*R,R*)-configurations.

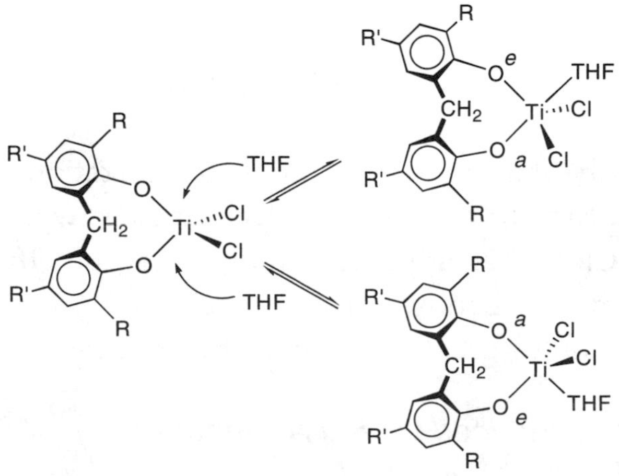

Figure 6.31. Monomeric (MBP)TiCl₂ is achiral because the Cl–Ti–Cl plane is contained in a mirror plane. Coordination of THF in the apical position causes the complex to become chiral if one MBP oxygen is apical (*a*) and the other is equatorial (*e*).

alent by virtue of their binding sites (**Figure 6.31**). The (MBP)Ti metallacycle is then asymmetric and (MBP)TiCl$_2$(THF) exists as a racemic mixture in the solid-state crystal structure.[68]

Inversion of the boat-shaped (MBP)Ti metallacycle (**Figure 6.32**) is slow on the NMR timescale, because the strain that is incurred inverting the methylene through the metallacycle is severe. H$_a$ and H$_b$ are inequivalent and couple with each other in the ^{1}H NMR spectrum, because one is above the (MBP)Ti metallacycle and the other is outside.

The goal of this work was to use the asymmetric (MBP)Ti metallacycle in the trigonal bipyramidal geometry as an integral part of the chiral environment of an asymmetric catalyst. Conceptually, if it were possible to substitute a substrate for the THF molecule in one of the enantiomers of (MBP)TiCl$_2$(THF) without stereochemical scrambling (**Figure 6.31**), the (MBP)Ti metallacycle would form a chiral environment for the substrate. The action of a reagent on this bound substrate would give nonracemic product. By incorporating enantioenriched alkoxide ligands (OR*) into the titanium coordination sphere, the configuration of the (MBP)Ti metallacycle can be biased (**Figure 6.33**). Now the two diastereomers in **Figure 6.33** are diastereomeric and will likely be present in different amounts.

Figure 6.32. Inversion of the methylene through the metallacycle is one method by which the enantiomeric (MBP)TiCl$_2$(THF) complexes can interconvert (and racemize). In general, the methylene hydrogens, H$_a$ and H$_b$, are inequivalent on the NMR timescale.

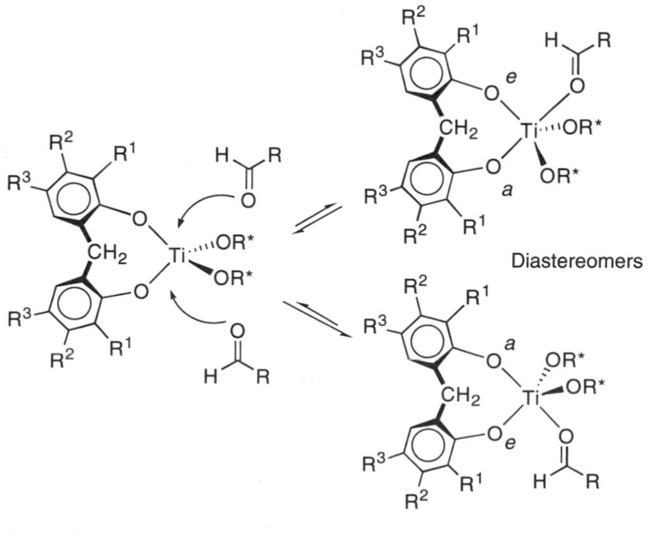

Figure 6.33. Incorporation of chiral alkoxides into the (MBP)Ti complex causes the aldehyde adducts to become diastereomeric, because the MBP ligand can adopt both enantiomeric forms. These diastereomers are unlikely to have the same energy.

A series of achiral derivatives of MBP-H$_2$ (20 mol%) were combined with Ti(OR*)$_4$ in the asymmetric addition to aldehydes (**Equation 6.13**). Examination of the results in **Table 6.6** indicates that changing the achiral MBP ligand can have a striking effect on the ee of the product [9% (R) to 83% (S)]. MBP-H$_2$ ligands with small R^1 substituents (entries 1–4) gave lower enantioselectivities. Note that catalysts with R^1 = H (entries 1 and 2) exhibited a significant enantioselectivity difference with respect to the background reaction [9% ee (R) vs. 39% ee (S)] with the chiral alkoxide complex Ti(OR*)$_4$ alone. Increasing the bulk of the R^1 substituent on the MBP ligand to t-Bu resulted in an increase in enantioselectivities (entries 6–8). A further increase in ee was observed when R^1 was adamantyl (entry 9, 83% ee). Thus, modification of the achiral MBP ligand resulted in an ee change of over 90%.

Equation 6.13

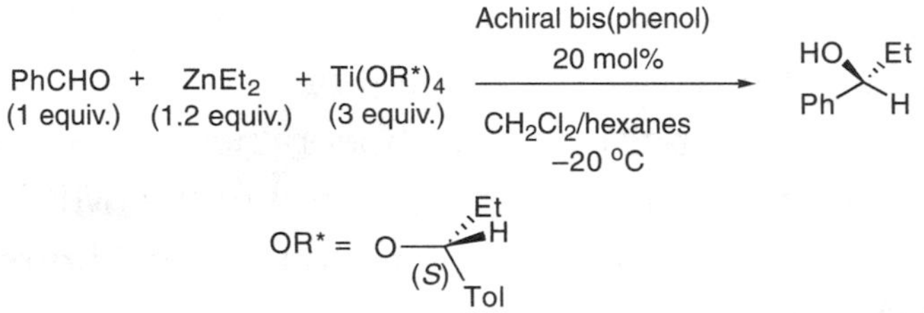

Substrate coordination can temporarily increase the number of stereogenic units in a catalyst. In this system, it is proposed that a change in metal geometry from tetrahedral to trigonal bipyramidal on coordination of a substrate can induce asymmetry in the (MBP)Ti metallacycle (**Figure 6.31**) by creating a chiral conformation. Once in an asymmetric geometry, the (MBP)Ti moiety can participate in, or even control, the relay of asymmetry to the substrate. The asymmetry of the bound MBP ligand is illustrated in the X-ray crystal structure in **Figure 6.34**. A dimethylamine serves as a

Table 6.6. MBP-H$_2$ ligands used in the asymmetric addition of alkyl groups to aldehydes (**Equation 6.13**).

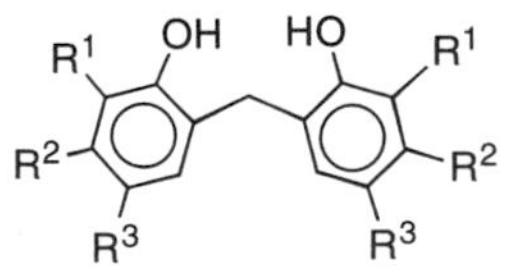

Entry	R^1	R^2	R^3	ee %[a] (configuration)
1	H	H	H	1 (S)
2	H	H	Cl	**9 (R)**
3	Cl	Cl	Cl	24 (S)
4	Me	H	Me	16 (S)
5	Ph	H	H	36 (S)
6	t-Bu	H	t-Bu	68 (S)
7	t-Bu	H	Me	79 (S)
8	t-Bu	H	H	73 (S)
9	Adamantyl	H	Me	**83 (S)**
10	No MBP-H$_2$ ligand added			39 (S)

[a] ee after 1 h.

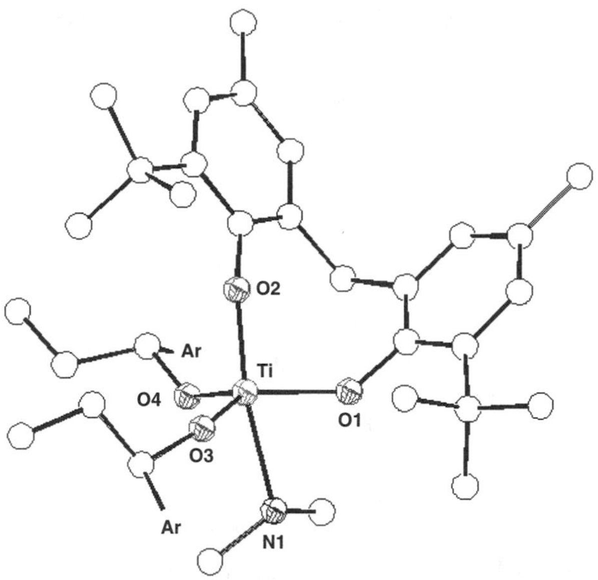

Figure 6.34. Solid-state structure of $(MBP)Ti(OR^*)_2(NHMe_2)$. The alkoxide aryl groups (**Equation 6.13**) have been removed for clarity.

substrate analog for the aldehyde in this example. The principle of geometry-induced ligand asymmetry can be applied to the optimization of other asymmetric catalysts.

6.3 Asymmetric Activation of Achiral Catalysts with Enantioenriched Ligands

6.3.1 Application of Achiral Salen Complexes to Asymmetric Cyclopropanation

One of the most useful classes of asymmetric catalysts is based on chiral salen complexes. Despite the widespread applications of salen-based catalysts, however, questions remain about how asymmetry is transmitted from the catalyst to the substrate. One proposal is that metal–salen complexes exist in nonplanar, or stepped, conformations and that this nonplanar characteristic is an important element in the chiral environment of the catalyst (**Figure 6.35**).[69,70] The stepped conformation has been observed in some, but not all, X-ray crystallographically characterized (salen)M complexes.[71,72] The hypothesis of a stepped conformation in the active catalyst species inspired researchers to employ achiral (salen)M complexes in combination with chiral ligands (L*) that would coordinate to the metal center.[73,74] As shown in **Figure**

Figure 6.35. Stepped conformation of an achiral (salen)M moiety proposed to be important in the transfer of asymmetry. If L* is achiral, the equilibrium constant will be 1. If L* is chiral, the complexes are diastereomers and one will likely be lower in energy.

6.35, coordination of a chiral ligand L* to the metal gives rise to diastereomeric complexes, which are unlikely to have an equilibrium constant of 1. This concept has been applied to the asymmetric cyclopropanation of olefins.

The asymmetric cyclopropanation of styrene was examined with an achiral (salen)Ru(PPh$_3$)$_2$ precatalyst (1 mol%), a series of chiral sulfoxides (10 mol%), and ethyl diazoacetate (EDA).[75] Reaction of the EDA with the precatalyst in the presence of the sulfoxides results in formation of the active catalysts, which then promote the cyclopropanation reaction (**Figure 6.36**). In this reaction it was found that the enantioselectivities depended on the structure of the sulfoxide, although the diastereoselectivities did not. Coordination of the sulfoxide to the ruthenium will impact the equilibrium between the stepped conformations of the salen ligand, as illustrated in **Figure 6.35**. While the complexes are diastereomeric, the chiral sulfoxide is distant from the reactive carbene. As such, this model predicts that the sulfoxides will change the enantioselectivity of the catalysts and not the diastereoselectivity.

The most promising sulfoxide, methyl tolyl sulfoxide, was then employed in further catalyst-optimization experiments. Increasing the sulfoxide to 50 mol%, or increasing the reaction concentration, resulted in an increase in the enantioselectivity. The largest increase in enantioselectivity was found when the reaction was per-

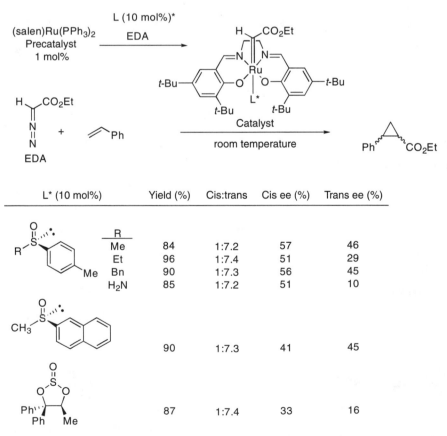

L* (10 mol%)		Yield (%)	Cis:trans	Cis ee (%)	Trans ee (%)
	R				
	Me	84	1:7.2	57	46
	Et	96	1:7.4	51	29
	Bn	90	1:7.3	56	45
	H$_2$N	85	1:7.2	51	10
		90	1:7.3	41	45
		87	1:7.4	33	16

Figure 6.36. Optimization of an achiral (salen)Ru(PPh$_3$)$_2$ precatalyst and chiral sulfoxides in the asymmetric cyclopropanation of styrene.

formed at −78 °C, in which case the cis and trans cyclopropanes were formed in 93 and 87% ee, respectively. Interestingly, NMR binding studies with the methyl tolyl sulfoxide and carbene complex, (salen)Ru(=CHCO$_2$Et), indicated that the chiral ligand binds through the sulfoxide oxygen rather than the sulfur, and that K_{eq} for binding of the sulfoxide was 129.[75]

One can imagine expanding this catalyst screening by using 10 achiral salen ligands with different substitution patterns in combination with 10 chiral axial ligands to generate a library of 100 catalysts. This study illustrates the potential of employing achiral salen complexes with chiral axial ligands in asymmetric catalysis.

6.3.2 Optimization of Achiral Diels–Alder Catalysts Using DNA

Outside of being an icon of modern science, DNA, with its double-stranded, right-handed helical structure, is an attractive scaffold for the construction of asymmetric catalysts. One of the challenges to developing DNA-based catalysts is incorporating catalytically active metal sites into the backbone of DNA to exploit its extended chiral environment. A clever approach to the assembly of a DNA-based asymmetric catalyst involves the use of achiral complexes that associate with DNA in a non-covalent fashion. In binding to the DNA, the achiral complex will be situated in the chiral environment of the double strand, allowing transmission of the asymmetry of the DNA to the product. By adjusting the properties of the achiral complex, the location or locations of the catalyst on the DNA and the distance of the metal from the double helix can be modulated. This strategy has been employed in the development of catalysts for the copper-based asymmetric Diels–Alder reaction and is illustrated in **Figure 6.37**.[76,77]

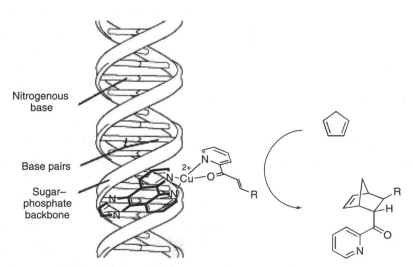

Figure 6.37. Binding of an achiral copper(II) center to DNA positions the catalytically active copper in the chiral environment of the DNA, allowing for efficient transmission of asymmetry from the double helix to the substrate.

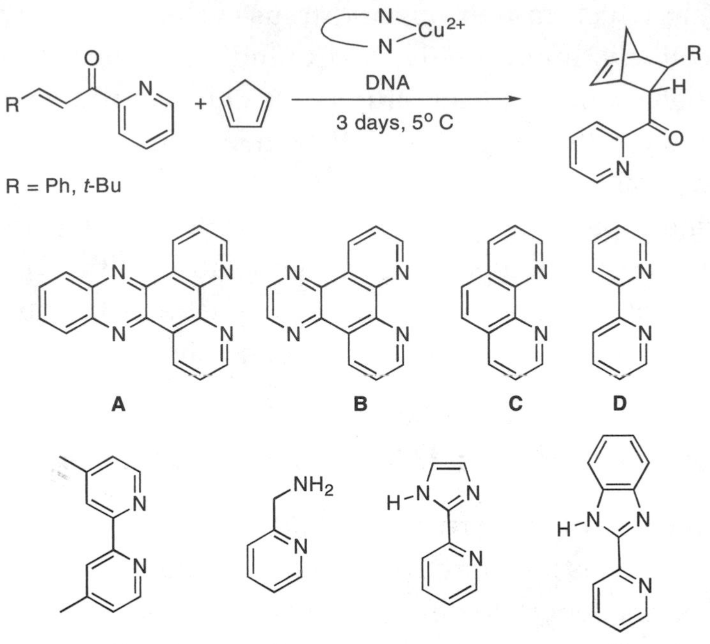

Figure 6.38. Copper(II) complexes of achiral ligands **A–H** were used to form DNA-based catalysts for the asymmetric Diels–Alder reaction.

Achiral ligands **A–H** were used in this study (**Figure 6.38**). Each contains a bidentate metal binding site and a hydrophobic aromatic π-system. The DNA binding constants of the copper(II) ligand adducts were determined to be between $9.3 \times 10^3\ M^{-1}$ and $8 \times 10^5\ M^{-1}$ (**Table 6.7**), indicating moderate to strong binding to DNA. As seen in **Table 6.7**, as high as 99% ee was obtained in the Diels–Alder reaction (**Figure 6.38**). The ligand binding sites of **A–D** contain similar coordination environments, allowing comparison of the DNA binding strength versus enantioselectivity. Intriguingly, an inverse relationship between binding strength and enantioselectivity was observed, with (bipy)Cu^{2+} exhibiting the highest enantioselectivity (90%) and lowest affinity for the DNA. It is noteworthy that when DNA and Cu(NO$_3$)$_2$ were used in the absence of ligand, the Diels–Alder adduct of opposite configuration was formed with a mere 10% ee.

Based on the initial results from screening ligands **A–D**, additional bipy-type ligands were examined. It was found that the Cu(II) complex of 4,4′-dimethyl-2,2′-bipyridine (**E**) exhibited complete *endo*-selectivity and 97–99% ee with both substrates examined (**Table 6.7**).

In this DNA-based catalyst system, the asymmetry is transmitted from the DNA. It is expected that the enantioselectivity of the catalyst will strongly depend on the nature of the interaction between the DNA and copper complex. It is known that the adducts of ligands **A** and **B**, (dppz)Cu(NO$_3$)$_2$ and (dpq)Cu(NO$_3$)$_2$, bind via intercalation into the DNA base pairs.[76] The nature of the interaction between the other ligand adducts and DNA is currently unknown, but will shed light on this important

Table 6.7. Binding constants and selectivities for DNA-based catalysts derived from ligands **A–H** in **Figure 6.38.**[a]

Achiral spacer	$K_{b\ (DNA)}$ M^{-1}	Dienophile R	endo:exo	ee (%)
A	$8 \pm 3 \times 10^5$	Ph	96:4	49
B	$7.2 \pm 1.2 \times 10^4$	Ph	95:5	61
C	$1.3 \pm 0.1 \times 10^4$	Ph	96:4	73
D	$9.4 \pm 0.3 \times 10^4$	Ph	98:2	90
D		t-Bu	94:6	83
E	$1.12 \pm 0.02 \times 10^4$	Ph	99:1	99
E		t-Bu	99:1	97
F	—	Ph	97:3	< 5
G	$5.2 \pm 0.3 \times 10^3$	Ph	99:1	91
H	$1.5 \pm 0.1 \times 10^4$	Ph	98:2	92
H		t-Bu	98:2	90
No ligand	—	Ph	95:5	−10

[a] Reaction conditions: 1.3 mg/L DNA, 0.3 mM Cu(NO$_3$)$_2$, and 15 mM cyclopentadiene, 3 d, 5 °C.

approach to catalyst development. Nonetheless, the assembly and optimization of a modular DNA-derived catalyst from readily available achiral ligands is a powerful technique that will likely be employed in additional reactions in the future.

Summary

In this chapter, nontraditional approaches to asymmetric catalysis have been outlined. Techniques such as chiral poisoning allow use of less-expensive racemic ligands in combination with resolved ligands. The resolved poison inhibits one enantiomer of the racemic catalyst, leaving the other to perform the catalytic process. Although chiral poisons have exhibited excellent efficiency in some cases, most examples employing this approach are not highly effective. A related, but more advanced strategy is the use of chiral activators, wherein one enantiomer of a racemic catalyst is selectively activated, generating a new catalytic species that exhibits increased activity, and often increased enantioselectivity, over the initial catalyst. This method has been extended to include the activation of resolved catalysts with resolved activators. The drawback to this latter approach is that two enantioenriched ligands are needed. Nonetheless, the benefits of a highly efficient and enantioselective catalyst often outweigh the additional effort of ligand resolution. Finally, the use of achiral activators with chiral conformations enables efficient extension of the chiral environment about the catalyst. As we have seen, for each resolved catalyst, many new catalysts can be generated by addition of achiral ligands. Combinatorial catalyst screening using these methods promises to be efficient and cost effective for the optimization of catalyst enantioselectivity and activity.

References

(1) Ojima, I., Ed. *Catalytic Asymmetric Synthesis*, 2nd ed.; Wiley: New York, 2000.

(2) Jacobsen, E. N.; Pfaltz, A.; Yamamoto, H. *Comprehensive Asymmetric Catalysis*; Springer-Verlag: Berlin, 1999; Vol. 1–3.

(3) Noyori, R. *Asymmetric Catalysis in Organic Synthesis*; Wiley: New York, 1994.

(4) Seyden-Penne, J. *Chiral Auxiliaries and Ligands in Asymmetric Synthesis*; Wiley: New York, 1995.

(5) Faller, J. W.; Lavoie, A. R.; Parr, J. Chiral Poisoning and Asymmetric Activation. *Chem. Rev.* **2003**, *102*, 3345–3367.

(6) Faller, J. W.; Parr, J. Chiral Poisoning: A Novel Strategy for Asymmetric Catalysis. *J. Am. Chem. Soc.* **1993**, *115*, 804–805.

(7) Alcock, N. W.; Brown, J. M.; Maddox, P. J. Substrate-Induced Kinetic Resolution of Racemic Biphosphines in situ for Homogeneous Catalysis. *J. Chem. Soc., Chem. Commun.* **1986**, 1532–1534.

(8) Brown, J. M.; Maddox, P. J. Iridium Complexes of Dehydroamino Acids: The Kinetic Resolution of Racemic Diphosphines and Their Application in Catalytic Asymmetric Hydrogenation. *Chirality* **1991**, *3*, 345–354.

(9) Faller, J. W.; Lavoie, A. R.; Grimmond, B. J. Application of the Chiral Poisoning Strategy: Enantioselective Diels–Alder Catalysis with a Racemic Ru/BINAP-Monoxide Lewis Acid. *Organometallics* **2002**, *21*, 1662–1666.

(10) Girard, C.; Kagan, H. B. Nonlinear Effects in Asymmetric Synthesis and Stereoselective Reactions: Ten Years of Investigation. *Angew. Chem., Int. Ed. Engl.* **1998**, *37*, 2922–2959.

(11) Noyori, R.; Ohkuma, T. Asymmetric Catalysis by Architectural and Functional Molecular Engineering: Practical Chemo- and Stereoselective Hydrogenation of Ketones. *Angew. Chem., Int. Ed. Engl.* **2001**, *40*, 40–73.

(12) Agera, D. J.; Lanemana, S. A. Reduction of 1,3-Dicarbonyl Systems with Ruthenium-Biarylbisphosphine Catalysts. *Tetrahedron: Asymmetry* **1997**, *8*, 3327–3355.

(13) Mikami, K.; Yusa, Y.; Korenaga, T. Asymmetric Deactivation of Racemic BINAP-Ru (II) Catalysts Through Complete Enantiomer Discrimination by Dimethylbinaphthylamine: Highly Enantioselective Hydrogenation of Olefin and β-Keto Ester. *Org. Lett.* **2002**, *4*, 1643–1645.

(14) Faller, J. W.; Tokunaga, M. Chiral Poisoning in the Kinetic Resolution of Allylic Alcohols. *Tetrahedron Lett.* **1993**, *34*, 7359–7362.

(15) Maruoka, K.; Itoh, T.; Shirasaka, T.; Yamamoto, H. Asymmetric Hetero-Diels–Alder Reaction Catalyzed by Chiral Organoaluminum Reagent. *J. Am. Chem. Soc.* **1988**, *110*, 310–312.

(16) Maruoka, K.; Yamamoto, H. Generation of Chiral Organoaluminum Reagent by Discrimination of the Racemates with Chiral Ketone. *J. Am. Chem. Soc.* **1989**, *111*, 789–790.

(17) Todd, M. H. Asymmetric Autocatalysis: Product Recruitment for the Increase in the Chiral Environment (PRICE). *Chem. Soc. Rev.* **2002**, *31*, 211–222.

(18) Soai, K.; Shibata, T.; Morloka, H.; Choji, K. Asymmetric Autocatalysis and Amplification of Enantiomeric Excess of a Chiral Molecule. *Nature* **1995**, *378*, 767–768.

(19) Shibata, T.; Yonekubo, S.; Soai, K. Practically Perfect Asymmetric Autocatalysis with (2-Alkynyl-5-pyrimidyl)alkanols. *Angew. Chem., Int. Ed. Engl.* **1999**, *38*, 659–661.

(20) Soai, K.; Osanai, S.; Kadowaki, K.; Yonekubo, S.; Shibata, T.; Sato, I. D- and L-Quartz-Promoted Highly Enantioselective Synthesis of a Chiral Organic Compound. *J. Am. Chem. Soc.* **1999**, *121*, 11235–11236.

(21) Sato, I.; Urabe, H.; Ishiguro, S.; Shibata, T.; Soai, H. Amplification of Chirality from Extremely Low to Greater than 99.5% ee by Asymmetric Autocatalysis. *Angew. Chem., Int. Ed. Engl.* **2003**, *42*, 315–317.

(22) Mikami, K.; Terada, M.; Korenaga, T.; Matsumoto, Y.; Ueki, M.; Angelaud, R. Asymmetric Activation. *Angew. Chem., Int. Ed. Engl.* **2000**, *39*, 3532–3556.

(23) Mikami, K.; Terada, M.; Korenaga, T.; Matsumoto, Y.; Matsukawa, S. Enantiomer-Selective Activation of Racemic Catalysts. *Acc. Chem. Res.* **2000**, *33*, 391–401.

(24) Mikami, K.; Shimizu, M. Asymmetric Ene Reactions in Organic Synthesis. *Chem. Rev.* **1992**, *92*, 1021–1050.

(25) Mikami, K.; Matsukawa, S. Asymmetric Synthesis by Enantiomer-Selective Activation of Racemic Catalysts. *Nature* **1997**, *385*, 613–615.

(26) Davis, T. J.; Balsells, J.; Carroll, P. J.; Walsh, P. J. Snapshots of Titanium BINOLate Complexes: Diverse Structures with Implications in Asymmetric Catalysis. *Org. Lett.* **2001**, *3*, 699–702.

(27) Balsells, J.; Davis, T. J.; Carroll, P. J.; Walsh, P. J. Insight into the Mechanism of the Asymmetric Addition of Alkyl Groups to Aldehydes Catalyzed by Titanium-BINOLate Species. *J. Am. Chem. Soc.* **2002**, *124*, 10336–10348.

(28) Martin, C. Design and Application of Chiral Ligands for the Improvement of Metal-Catalyzed Asymmetric Transformations. Ph.D. Thesis, Massachusetts Institute of Technology, Cambridge, MA, 1989.

(29) Bradley, D. C.; Mehrotra, R. C.; Rothwell, I. P.; Singh, A. *Alkoxo and Aryloxo Derivatives of Metals*; Academic Press: New York, 2001.

(30) Katsuki, T. In *Comprehensive Asymmetric Catalysis*; Jacobsen, E. N., Pfaltz, A., Yamamoto, H., Eds.; Springer-Verlag: Berlin, 1999; Vol. 2, pp 621–648.

(31) Blackmond, D. G.; Rosner, T.; Neugebauer, T.; Reetz, M. T. Kinetic Influences on Enantioselectivity for Non-Diastereopure Catalysts Mixtures. *Angew. Chem., Int. Ed. Engl.* **1999**, *38*, 2196–2199.

(32) Bolm, C.; Muniz, K.; Hildebrand, J. P. Planar-Chiral Ferrocenes in Asymmetric Catalysis: The Impact of Stereochemically Inhomogeneous Ligands. *Org. Lett.* **1999**, *1*, 491–494.

(33) Kitamura, M.; Suga, S.; Niwa, M.; Noyori, R. Self and Nonself Recognition of Asymmetric Catalysts. Nonlinear Effects in the Amino Alcohol-Promoted Enantioselective Addition of Dialkylzincs to Aldehydes. *J. Am. Chem. Soc.* **1995**, *117*, 4832–4842.

(34) Balsells, J.; Walsh, P. J. Design of Diastereomeric Self-Inhibiting Catalysts for Control of Turnover Frequency and Enantioselectivity. *J. Am. Chem. Soc.* **2000**, *122*, 3250–3251.

(35) Ohkuma, T.; Doucet, H.; Pham, T.; Mikami, K.; Korenaga, T.; Terada, M.; Noyori, R. Asymmetric Activation of Racemic Ruthenium (II) Complexes for Enantioselective Hydrogenation. *J. Am. Chem. Soc.* **1998**, *120*, 1086–1087.

(36) Mikami, K.; Korenaga, T.; Ohkuma, T.; Noyori, R. Asymmetric Activation/Deactivation of Racemic Ru Catalysts for Highly Enantioselective Hydrogenation of Ketonic Substrates. *Angew. Chem., Int. Ed. Engl.* **2000**, *39*, 3707–3710.

(37) Mikami, K.; Angelaud, R.; Ding, K. L.; Ishii, A.; Tanaka, A.; Sawada, N.; Kudo, K.; Senda, M. Asymmetric Activation of Chiral Alkoxyzinc Catalysts by Chiral Nitrogen Activators for Dialkylzinc Addition to Aldehydes: Super High-Throughput Screening of Combinatorial Libraries of Chiral Ligands and Activators by HPLC-CD/UV and HPLC-OR/RIU Systems. *Chem. Eur. J.* **2001**, *7*, 730–737.

(38) Ding, K.; Ishii, A.; Mikami, K. Super High Throughput Screening (SHTS) of Chiral Ligands and Activators: Asymmetric Activation of Chiral Diol-Zinc Catalysts by Chiral Nitrogen Activators for the Enantioselective Addition of Diethylzinc to Aldehydes. *Angew. Chem., Int. Ed. Engl.* **1999**, *38*, 497–501.

(39) Berrisford, D. J.; Bolm, C.; Sharpless, K. B. Ligand-Accelerated Catalysis. *Angew. Chem., Int. Ed. Engl.* **1995**, *34*, 1059–1070.

(40) Vogl, E. M.; Groger, H.; Shibasaki, M. Towards Perfect Asymmetric Catalysis: Additives and Cocatalysts. *Angew. Chem., Int. Ed. Engl.* **1999**, *38*, 1570–1577.

(41) Kobayashi, S.; Ishitani, H. Lanthanide (III)-Catalyzed Enantioselective Diels–Alder Reactions. Stereoselective Synthesis of Both Enantiomers by Using a Single Chiral Source and a Choice of Achiral Ligands. *J. Am. Chem. Soc.* **1994**, *116*, 4083–4084.

(42) Kobayashi, S.; Hachiya, I.; Ishitani, H.; Araki, M. Asymmetric Diels–Alder Reaction Catalyzed by a Chiral Ytterbium Trifluoromethanesulfonate. *Tetrahedron Lett.* **1993**, *34*, 4535–4538.

(43) Clarke, M. L.; Fuentes, J. A. Self-Assembly of Organocatalysts: Fine-Tuning Organocatalytic Reactions. *Angew. Chem., Int. Ed. Engl.* **2007**, *46*, 930–933.

(44) Berkessel, A.; Gröger, H. *Asymmetric Organocatalysis*; Wiley: New York, 2005.

(45) Houk, K. N.; List, B. Eds. Asymmetric Organocatalysis (special issue). *Acc. Chem. Res.* **2004**, *37*, 487–631.

(46) Desponds, O.; Schlosser, M. The Activation Barrier to Axial Torsion in 2,2'-Bis(diphenylphosphino) biphenyl. *Tetrahedron Lett.* **1996**, *37*, 47–48.

(47) Eliel, E. L.; Wilen, S. H. *Stereochemistry of Organic Compounds*; Wiley: New York, 1994.

(48) Kurland, R. J.; Rubin, M. B.; Wise, W. B. Inversion Barrier in Singly Bridged Biphenyls. *J. Chem. Phys.* **1964**, *40*, 2426–2431.

(49) Müllen, K.; Heinz, W.; Klärner, F.-G.; Roth, W. R.; Kindermann, I.; Adamczak, O.; Wette, M.; Lex, J. Inversion Barriers of ortho,ortho'-Bridged Biphenyls. *Chem. Ber.* **1990**, *23*, 2349–2371.

(50) Iffland, D. C.; Siegel, H. A Biphenyl Whose Optical Activity is Due to a Three-carbon Bridge Across the 2,2'-Positions. *J. Am. Chem. Soc.* **1958**, *80*, 1947–1950.

(51) Mislow, K.; Hyden, S.; Schaefer, H. Stereochemistry of the 1,2,3,4-Dibenzcyclonona-1,3-diene System. A Note on the Racemization Barrier in Bridged Biphenyls. *J. Am. Chem. Soc.* **1962**, *84*, 1449–1455.

(52) Noyori, R.; Yamakawa, M.; Hashiguchi, S. Metal-Ligand Bifunctional Catalysis: A Nonclassical Mechanism for Asymmetric Hydrogen Transfer Between Alcohols and Carbonyl Compounds. *J. Org. Chem.* **2001**, *66*, 7931–7944.

(53) Abdur-Rashid, K.; Faatz, M.; Lough, A. J.; Morris, R. H. Catalytic Cycle for the Asymmetric Hydrogenation of Prochiral Ketones to Chiral Alcohols: Direct Hydride and Proton Transfer from Chiral Catalysts *trans*-Ru(H)$_2$(diphosphine)(diamine) to Ketones and Direct Addition of Dihydrogen to the Resulting Hydridoamido Complexes. *J. Am. Chem. Soc.* **2001**, *123*, 7473–7474.

(54) Abdur-Rashid, K.; Clapham, S. E.; Hadzovic, A.; Harvey, J. N.; Lough, A. J.; Morris, R. H. Mechanism of the Hydrogenation of Ketones Catalyzed by *trans*-Dihydrido(diamine)ruthenium(II) Complexes. *J. Am. Chem. Soc.* **2002**, *124*, 15104–15118.

(55) Becker, J. J.; White, P. S.; Gagné, M. R. Synthesis and Characterization of Chiral Platinum(II) Sulfonamides: (dppe)Pt(NN) and (dppe)Pt(NO) Complexes. *Inorg. Chem.* **1999**, *38*, 798–801.

(56) Becker, J. J.; White, P. S.; Gagné, M. R. Asymmetric Catalysis with the Normally Unresolvable, Conformationally Dynamic 2,2'-Bis(diphenylphosphino)-1,1'-biphenyl (BIPHEP). *J. Am. Chem. Soc.* **2001**, *123*, 9478–9479.

(57) Tudor, M. D.; Becker, J. J.; White, P. S.; Gagné, M. R. Diastereoisomer Interconversion in Chiral Biphep-PtX$_2$ Complexes. *Organometallics* **2000**, *19*, 4376–4384.

(58) Ghosh, A. K.; Matsuda, H. Counterions of BINAP-Pt(II) and -Pd(II) Complexes: Novel Catalysts for Highly Enantioselective Diels–Alder Reaction. *Org. Lett.* **1999**, *1*, 2157–2159.

(59) Mikami, K.; Kataoka, S.; Yusa, Y.; Aikawa, K. Racemic but Tropos (Chirally Flexible) BIPHEP Ligands for Rh(I)-Complexes: Highly Enantioselective Ene-Type Cyclization of 1,6-Enynes. *Org. Lett.* **2004**, *6*, 3699–3701.

(60) Mikami, K.; Aikawa, K.; Yusa, Y. Asymmetric Activation of the Pd Catalyst Bearing the Tropos Biphenylphosphine (BIPHEP) Ligand with the Chiral Diaminobinaphthyl (DABN) Activator. *Org. Lett.* **2002**, *4*, 95–97.

(61) Mikami, K.; Aikawa, K.; Yusa, Y.; Hatano, M. Resolution of Pd Catalyst with tropos Biphenylphosphine (BIPHEP) Ligand by DM-DABN: Asymmetric Catalysis by an Enantiopure BIPHEP-Pd Complex. *Org. Lett.* **2002**, *4*, 91–94.

(62) Hartwig, J. Synthetic Chemistry–Recipes for Excess. *Nature* **2005**, *437*, 487–488.

(63) Balsells, J.; Walsh, P. J. The Use of Achiral Ligands to Convey Asymmetry: Chiral Environment Amplification. *J. Am. Chem. Soc.* **2000**, *122*, 1802–1803.

(64) Costa, A. M.; Jimeno, C.; Gavenonis, J.; Carroll, P. J.; Walsh, P. J. Optimization of Catalyst Enantioselectivity and Activity Using Achiral and Meso Ligands. *J. Am. Chem. Soc.* **2002**, *124*, 6929–6941.

(65) von Zelewsky, A. *Stereochemistry of Coordination Compounds*; Wiley: New York, 1996.

(66) Reetz, M. T.; Mehler, G. Mixtures of Chiral and Achiral Monodentate Ligands in Asymmetric Rh-Catalyzed Olefin Hydrogenation: Reversal of Enantioselectivity. *Tetrahedron Lett.* **2003**, *44*, 4593–4596.

(67) Reetz, M. T.; Li, X. Mixtures of Configurationally Stable and Fluxional Atropisometric Monodentate P Ligands in Asymmetric Rh-Catalyzed Olefin Hydrogenation. *Angew. Chem., Int. Ed. Engl.* **2005**, *44*, 2959–2962.

(68) Okuda, J.; Fokken, S.; Kang, H.-C.; Massa, W. Synthesis and Characterization of Mononuclear Titanium Complexes Containing a Bis(phenoxy) Ligand Derived from 2,2′-Methylene-bis(6-*tert*-butyl-4-methylphenol). *Chem. Ber.* **1995**, *128*, 221–227.

(69) Norrby, P.-O.; Linde, C.; Åkermark, B. On the Chirality Transfer in the Epoxidation of Alkenes Catalyzed by Mn(salen) Complexes. *J. Am. Chem. Soc.* **1995**, *119*, 11035–11036.

(70) Hamada, T.; Fukuda, T.; Imanishi, H.; Katsuki, T. Mechanism of One Oxygen Atom Transfer from Oxo (Salen)manganese(V) Complex to Olefins. *Tetrahedron* **1996**, *52*, 515–530.

(71) Samsel, E. G.; Srinivasan, K.; Kochi, J. K. Mechanism of the Chromium-Catalyzed Epoxidation of Olefins. Role of Oxochromium(V) Cations. *J. Am. Chem. Soc.* **1985**, *107*, 7606–7617.

(72) Pospisil, P. J.; Carsten, D. H.; Jacobsen, E. N. X-Ray Structural Studies of Highly Enantioselective Mn(salen) Epoxidation Catalysts. *Chem. Eur. J.* **1996**, *2*, 974–980.

(73) Hashihayata, T.; Ito, Y.; Katsuki, T. Enantioselective Epoxidation of 2,2-Dimethylchromenes Using Achiral Mn–Salen Complexes as a Catalyst in the Presence of Chiral Amine. *Synlett* **1996**, 1079–1081.

(74) Hashihayata, T.; Ito, Y.; Katsuki, T. The First Asymmetric Epoxidation Using a Combination of Achiral (Salen)manganese(III) Complex and Chiral Amine. *Tetrahedron* **1997**, *53*, 9541–9552.

(75) Miller, J. A.; Gross, B. A.; Zhuravel, M. A.; Jin, W.; Nguyen, S. T. Axial Ligand Effects: Utilization of Chiral Sulfoxide Additives for the Induction of Asymmetry in (Salen)ruthenium(II) Olefin Cyclopropanation Catalysts. *Angew. Chem., Int. Ed. Engl.* **2005**, *44*, 3885–3889.

(76) Roelfes, G.; Feringa, B. L. DNA-Based Asymmetric Catalysis. *Angew. Chem., Int. Ed. Engl.* **2005**, *44*, 3230–3232.

(77) Roelfes, G.; Boersma, A. J.; Feringa, B. L. Highly Enantioselective DNA-Based Catalysis. *J. Chem. Soc., Chem. Commun.* **2006**, 635–637.

7

Kinetic Resolution

With increasing emphasis on enantioselective catalysis, kinetic resolutions have been perceived to be less elegant. One reason for this is that the maximum yield of a standard resolution is 50%. The continued use of resolutions in academic and industrial settings, however, certainly attests to their utility. Kinetic resolutions are often the best option when the racemate is inexpensive, when no reasonable enantioselective method is available, or when classical resolution does not provide the desired material with high ee.[1-4]

While classical resolutions entail the use of stoichiometric resolving agents, the kinetic resolutions outlined here use asymmetric catalysts that react much faster with one enantiomer of a racemate to furnish enantioenriched unreacted substrate and enantioenriched product. To be useful, therefore, the resolved starting material and product *must be easily separable*, preferably with minimal effort. Another feature of a successful kinetic resolution is that starting material (or product) of high ee (> 98%) is obtained near 50% conversion. Desirable characteristics of all catalytic asymmetric reactions, such as high yields, short reaction times, scalability (with high volumetric through-put), low catalyst loading, inexpensive catalysts, minimal generation of waste, reproducibility, broad substrate scope, and functional group compatibility, are also important considerations in kinetic resolutions.

7.1 Basic Concepts of Kinetic Resolution

The efficiency of a kinetic resolution is given by the relative rates of reaction of the substrate enantiomers (S_R and S_S) with the chiral catalyst (Cat_R in this example) to generate the products (P_R and P_S). The relative rate, $k_{rel} = k_{fast}/k_{slow}$, and the selectivity factor, s, are used interchangeably (**Figure 7.1**). The ideal situation is when k_{rel} is very large and the reaction can be quenched at 50% conversion to yield nearly enantiopure starting material or product. The selectivity factor is related to $\Delta\Delta G^{\ddagger}$, the difference in free energy between the selectivity-determining diastereomeric transition states (**Figure 7.1**). The k_{rel} is easily calculated using **Equation 7.1**, by experimentally determining the conversion and the ee of the starting material at a given time. The ee of the starting material increases as the reaction progresses, as shown graphically in **Figure 7.2** for various values of k_{rel}. Thus, high substrate ee can be obtained, even when k_{rel} is moderate, by running the reaction to higher conversion. The increased substrate ee, however, comes at the expense of the yield of recovered resolved substrate. In general, systems with $k_{rel} > 10$ are useful.

$$s = k_{rel} = \frac{\ln[(1 - c)(1 - ee)]}{\ln[(1 - c)(1 + ee)]}$$

Equation 7.1

c = Conversion of starting material
ee = ee of starting material

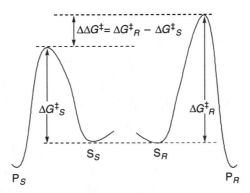

$$s = k_{rel} = k_{fast}/k_{slow} = e^{\Delta\Delta G^{\ddagger}/RT}$$

Figure 7.1. Kinetic resolution involving the reaction of a racemic substrate, achiral reagent, and a resolved catalyst. The relative rates of reaction of the enantiomers, shown above in an energy diagram, determines the efficiency of the kinetic resolution.

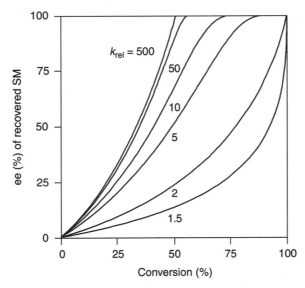

Figure 7.2. Graph representing the different k_{rel} values as a function of conversion (%) and recovered starting material (SM) ee (%).

7.2 Kinetic Resolutions That Create No Additional Stereocenters

7.2.1 Kinetic Resolution of Alcohols with Planar Chiral Catalysts

Racemic alcohols are readily available and are excellent substrates for enzymatic kinetic resolutions. A recent focus has been to develop nonenzymatic acyl transfer methods to accomplish kinetic resolutions of alcohols, due to a lack of generality and the high dilution that most enzymatic catalysts require. Kinetic resolutions of racemic alcohols catalyzed by chiral small-molecule acylation catalysts, however, were unsuccessful until recently.

An innovative concept for nucleophilic asymmetric catalysis using organometallic ferrocene derivatives was introduced in 1996.[5] It is well known that nucleophiles, such as DMAP [4-(dimethylamino)pyridine], catalyze the acylation of alcohols to form esters. Chiral DMAP analogs in which a 2-substituted heterocycle is π-bonded to a transition metal were prepared (**Figure 7.3**). The nucleophilic nitrogen is in an asymmetric environment created by the annulated cyclopentadienyl group, which differentiates the right and left sides of the heterocycle, and the $Fe(\eta^5\text{-}C_5Ph_5)$ ensemble, which differentiates the top and bottom faces. Each quadrant of the diagram is filled to a different extent, with darker shading indicating greater steric hindrance. These catalysts are highly efficient and stereoselective in the acylation of secondary aryl alcohols, as illustrated in **Table 7.1**.[6] The best catalyst for the kinetic resolution is based on the robust ferrocene unit (**Figure 7.3**). It can be used at low catalyst loading (0.5 mol%) and exhibited excellent selectivity factors with a variety of substrates (**Table 7.1**).

The ferrocene-derived catalyst has also been employed in the kinetic resolution of allylic alcohols.[7] One exciting example is the efficient kinetic resolution of the β-hydroxyketone in **Figure 7.4**, in which the enantiomeric planar chiral catalyst was employed. The selectivity factor in the kinetic resolution was 107, allowing the isolation of the desired alcohol in 47% yield with an ee of 98%.[7] The product alcohol has been used in the synthesis of epothilone A, a potent anticancer agent.

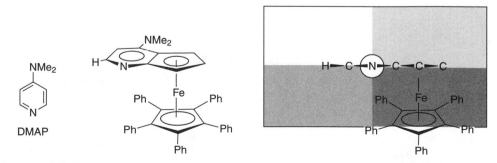

Figure 7.3. DMAP and the planar chiral DMAP analog based on ferrocene. The quadrant diagram on the right illustrates the degree of blocking in each quadrant, with darker shading indicating greater steric hindrance. Only the relevant atoms of the heterocycle are included.

The proposed mechanism involves rapid reaction of the catalyst with the acylating agent to form an ion-pair (**Figure 7.5**). The structure of the cation has been determined crystallographically and is shown in **Figure 7.6**. The acyl group lies in the plane of the ring system with the smaller carbonyl oxygen near the cyclopentadienyl group and the larger methyl in the less sterically hindered position. Attack of the alcohol to the top face of this intermediate proceeds through a tetrahedral intermediate that collapses to give product and regenerates the catalyst. The large C_5Ph_5 ring system shields the acyl group from attack on the bottom face (**Figure 7.5**).

The X-ray crystal structure of the catalyst in **Figure 7.6** clearly shows how the C_5Ph_5 phenyl groups block the bottom face of the sp^2 hybridized nitrogen. When an analogous catalyst with a less sterically demanding C_5Me_5 group was employed, the catalyst was not as enantioselective, because the smaller C_5Me_5 group does not shield the bottom face of the acyl as well as the larger C_5Ph_5.

Table 7.1. Kinetic resolution of aryl alcohols with the planar chiral ferrocene-derived catalyst in **Figure 7.3**.

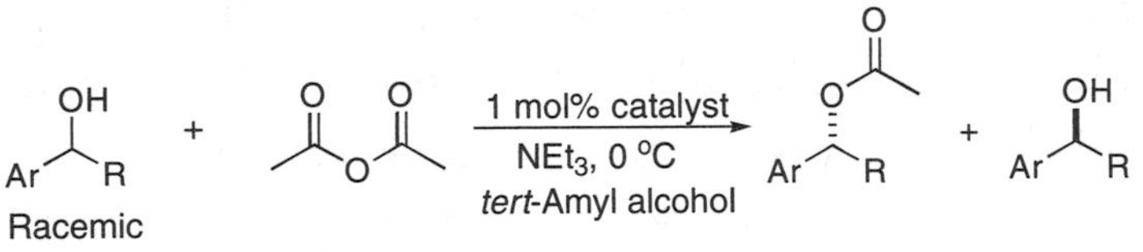

Unreacted alcohol (major enantiomer)		k_{rel}	ee (%)	Conversion (%)
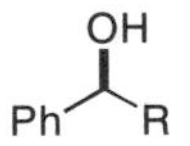 Ph–CH(OH)–R	R = Me	43		55
	Et	59	99	54
	i-Pr	87	99	52
	t-Bu	95	97	51
			96	
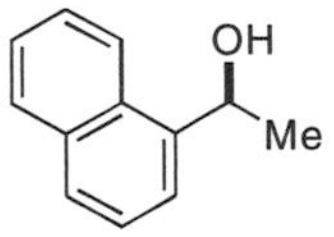 naphthyl–CH(OH)–Me		65	95	52
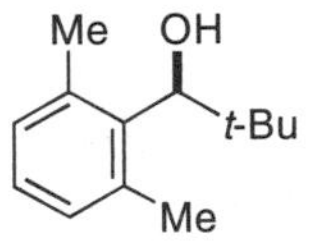 (2,6-Me$_2$-phenyl)–CH(OH)–t-Bu		> 200	99	51

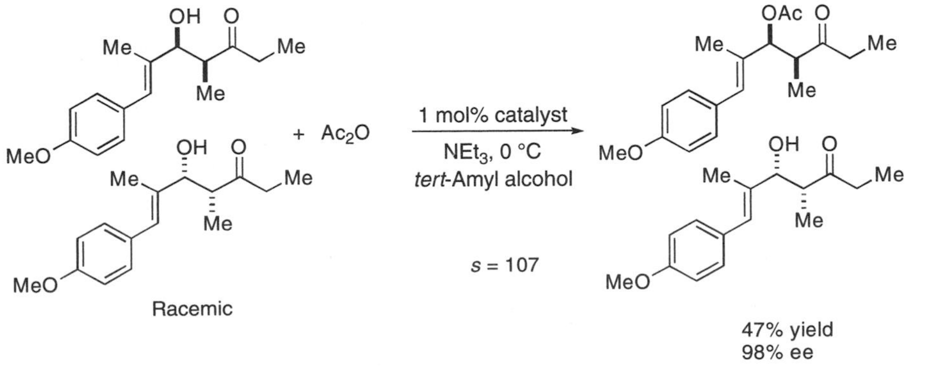

Figure 7.4. Kinetic resolution of a precursor to epothilone A using the antipode of the planarchiral catalyst in **Figure 7.3**.

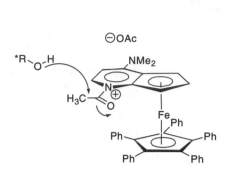

Figure 7.5. Proposed attack of the alcohol on the cationic acylated intermediate.

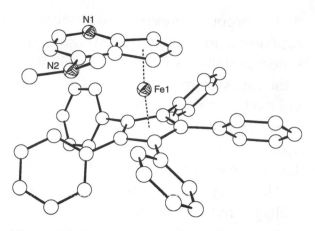

Figure 7.6. X-ray structure of the enantiomeric ferrocene-based nucleophilic catalyst from **Figure 7.5.**

7.2.2 Kinetic Resolution of Amines

The planar chiral ferrocene-based catalyst of **Figure 7.3** also performs well in the kinetic resolution of racemic primary amines,[8] a reaction that has proved difficult to accomplish with nonenzymatic catalysts (**Table 7.2**). Due to the increased nucleophilicity of the amine substrates, the background reaction with common acylating reagents (i.e., the reaction that takes place without the participation of the catalyst, see Section 1.2) is very fast. The background reaction leads to racemic product, thus eroding selectivity. In contrast, an O-acylated azlactone acylating agent (where Ar = β-naphthyl) was found to react much faster with the ferrocene-based catalyst than with primary amines (**Table 7.2**).[8] Good selectivity factors were observed for this challeng-

Table 7.2. Kinetic resolution of amines with the planar chiral ferrocene-based catalyst from **Figure 7.3**.

Unreacted amine (major enantiomer)	k_{rel}
X = H	12
2-Me	16
4-OMe	13
3-OMe	22
4-CF₃	13
	11

ing kinetic resolution. A closely related catalyst has also been successfully applied to the kinetic resolution of indolines.[9]

The reactions presented above are excellent examples of kinetic resolutions. More importantly, however, they illustrate the concept of the planar chiral design feature of nucleophilic catalysts, a concept that has widespread applicability in catalytic asymmetric systems.[10,11]

7.2.3 Hydrolytic Kinetic Resolution (HKR) of Epoxides with (Salen)CoX Catalysts

Certain classes of catalysts exhibit excellent enantioselectivities over a broad range of mechanistically distinct reactions. For this reason, they have become known as "privileged structures."[12] Transition-metal–salen complexes are one such class of catalysts that continues to find exciting applications. These catalysts have been especially useful in the generation of chiral epoxides. Early investigations into the challenging enantioselective epoxidation of unfunctionalized olefins with manganese–salen complexes provided access to a range of epoxides with high levels of enantioselectivity.[13,14] Despite considerable effort, however, no general method for the direct synthesis of highly enantioenriched terminal epoxides has been advanced. Terminal epoxides of high ee are among the most useful chiral building blocks in organic synthesis.

Racemic terminal epoxides are readily synthesized from inexpensive alkenes and some are even commodity chemicals. Chiral salen complexes of Cr and Co will catalyze the opening of epoxides with a variety of nucleophiles, such as carboxylic acids, phenols, azide, and water.[15] The Co(III)–salen complexes are extremely efficient at promoting the hydrolytic kinetic resolution (HKR) of epoxides (**Figure 7.7**).[16,17] With these catalysts, epoxides with > 99% ee can be obtained in yields approaching the theoretical maximum of 50% (**Table 7.3**). Selectivity factors for the HKR are usually outstanding, most being over 50 and several exceeding 200. In some cases reactions are conducted under solvent-free conditions and have been used to produce hundreds of kilograms of resolved epoxide.[17]

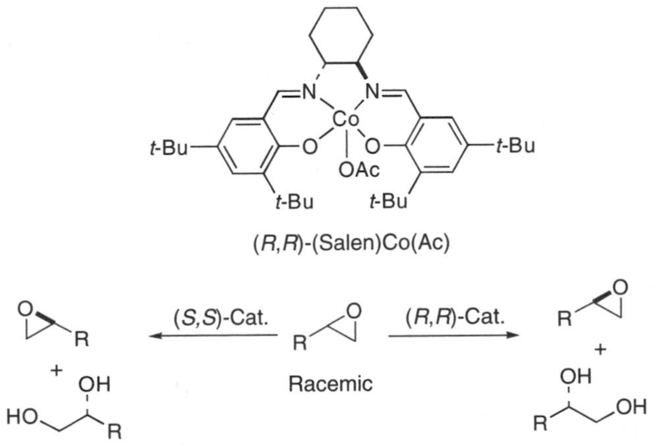

Figure 7.7. Cobalt–salen-catalyzed HKR.

Table 7.3. Kinetic resolution of racemic epoxides with (salen)Co(OAc) catalyst.

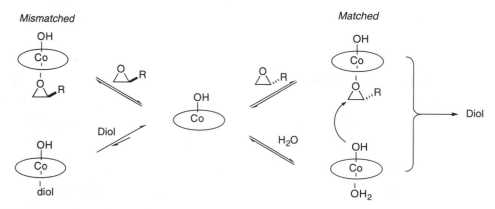

R	k_{rel}	Epoxide ee (%)	Yield (%)
Me	> 400	> 98	44
CH_2Cl	50	98	44
$(CH_2)_3CH_3$	290	98	46
$(CH_2)_5CH_3$	260	99	45
$CH=CH_2$	30	99	29

Figure 7.8. Relevant equilibria in the (salen)Co(OH)-catalyzed HKR.

The most commonly employed catalyst is the acetate derivative, although a dependency on the counterion has been noted.[17,18] The active form of the catalyst is (salen)Co(OH), which can be generated from either (salen)Co(OAc) or (salen)Co(Cl) by addition of the counterion to the epoxide, as shown in **Equation 7.2**.

Equation 7.2

X = Cl, OAc

A detailed kinetic study of the HKR has provided insight into the working of this exceptional catalyst system. The HKR has been found to be second-order in catalyst,[16,19] proceeding via a cooperative spelling bimetallic mechanism with simultaneous activation of both the epoxide and nucleophile by different monomeric catalysts, as illustrated in **Figure 7.8** (see Chapter 12 for further discussion). Surprisingly, it was found that both enantiomers of the epoxide coordinate to the (salen)Co(OH) complex with comparable binding constants. The high selectivity in the ring-opening, therefore, must arise from the preferential reaction of one of the diastereomeric catalyst–epoxide adducts. It was also found that water and epoxide bind with similar affinities to (salen)Co(OH), but the diol product has a much lower affinity.

When (salen)Co(X) compounds with less nucleophilic counterions, such as tosylate, were present with (salen)Co(OH), the activity increased by up to 30-fold. Under these conditions, the (salen)Co(X) and (salen)Co(OH) complexes are believed to play different roles in the catalytic cycle (**Figure 7.9**). The (salen)Co(X) complex is more Lewis acidic and is expected to bind the epoxide more tightly, better activating it toward nucleophilic attack than the less Lewis acidic (salen)Co(OH). The (salen)Co(X) complex alone, however, does not promote the addition of water to the epoxide.[19] Rather, the (H$_2$O)(salen)Co(OH) species, which activates the nucleophile, is also required (**Figure 7.9**).

The proposed dual catalyst mechanism in **Figure 7.9** accounts for the observation that when the (salen)Co(OAc) is employed, the HKR slows considerably in the late stages. As the reaction progresses, the concentration of the more Lewis acidic

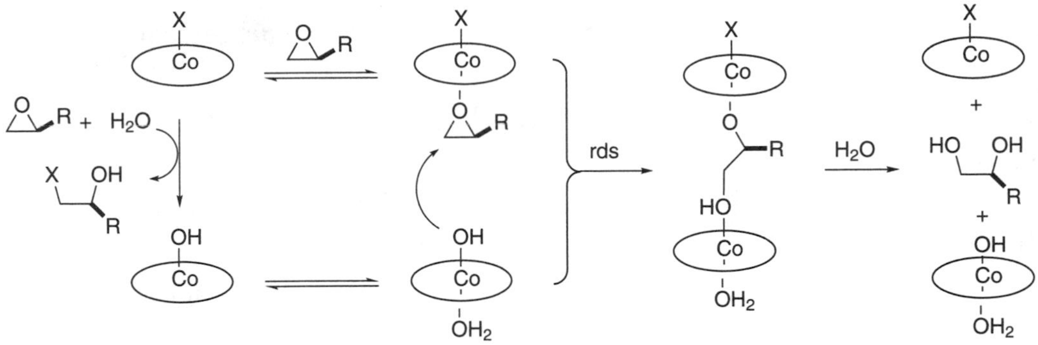

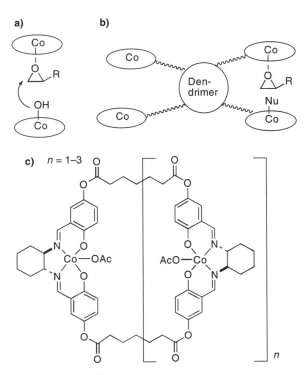

Figure 7.9. Proposed mechanistic scheme for the dual catalyst role in the HKR.

Figure 7.10. Cooperative interactions in the HKR of epoxides by a) monomeric and b) dendrimeric catalysts. c) Structure of oligomeric HKR catalysts.

(salen)Co(OAc) is depleted, because the acetate opens the epoxide to generate (salen) Co(OH). By the time the HKR approaches completion, all the (salen)Co(OAc) has been converted into (salen)Co(OH), which is less Lewis acidic and less active. This example is also unique in that the (salen)Co(X) compound is fulfilling the roles of pre-catalyst and cocatalyst.

The dual role of the catalyst has inspired the design of catalysts with dendrimeric[20] and oligomeric[21,22] architectures to hold Co–salen units in close proximity (**Figure 7.10**). The resulting catalysts exhibit intramolecular cooperative interactions between two metals, increasing catalyst efficiency and allowing reduced catalyst loadings.

7.2.4 Kinetic Resolution with Ring-Closing Metathesis Catalysts

Carbon–carbon bond-forming reactions are central to the synthesis of natural and synthetic products. As such, kinetic resolutions that form carbon–carbon bonds can be used to construct complex structures.

Ring-closing metathesis (RCM) reactions have emerged as an important method in the assembly of ring systems and are frequently employed as the key step in a number of total syntheses.[23,24] Chiral, nonracemic metathesis catalysts have been introduced for the asymmetric ring-closing metathesis of prochiral substrates and have been applied to kinetic resolutions of a variety of racemic dienes. In the kinetic resolution of dienes with asymmetric ring-closing metathesis catalysts, the relative rates of the fast- and slow-reacting enantiomers have been found to depend highly on the position of the stereocenter with respect to the alkenes and their substitution patterns.

Ring-closing metathesis catalysts are compatible with a variety of functional groups, thus allowing the synthesis of heterocycles.[24] An example is the application of the chiral molybdenum catalyst in **Figure 7.11** to the kinetic resolution of acyclic amino dienes.[25] Initial investigation found low k_{rel} values of 3–4 (**Figure 7.11**). In this kinetic resolution, the catalyst likely reacts with the less-hindered terminal alkene ini-

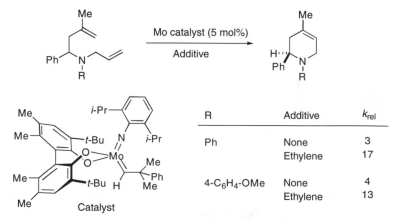

R	Additive	k_{rel}
Ph	None	3
	Ethylene	17
4-C$_6$H$_4$-OMe	None	4
	Ethylene	13

Figure 7.11. Kinetic resolution of amino dienes with an asymmetric ring-closing metathesis catalyst.

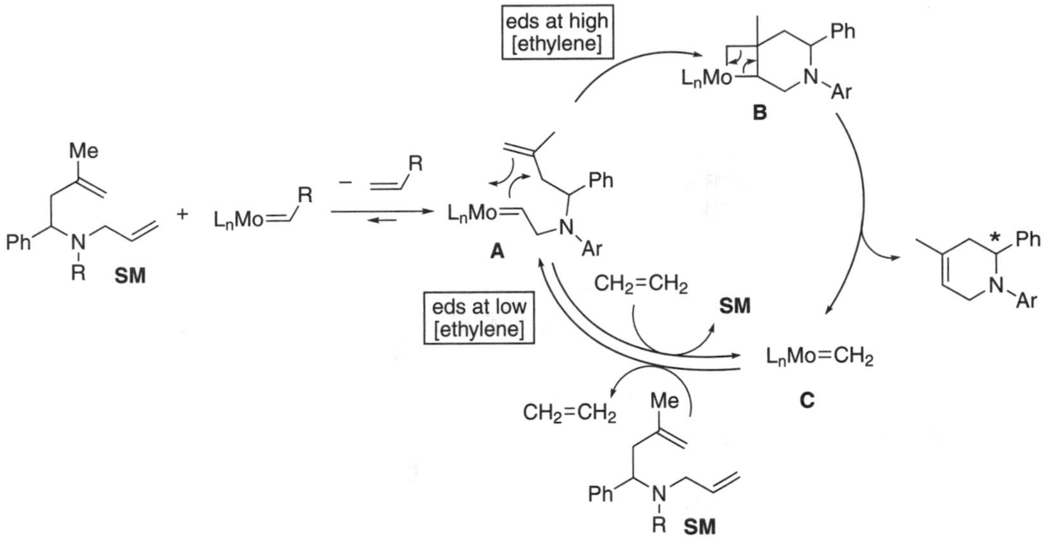

Figure 7.12. Mechanistic scheme for the kinetic resolution of amino dienes. In the presence of ethylene, the initial metathesis is readily reversible (eds = enantiodetermining step).

tially to form a new alkylidene (**Figure 7.12**). The barriers for reaction of the catalyst with the different **SM** enantiomers to form **A** in this step are likely similar, because the benzylic stereocenter is distant from the chiral environment of the catalyst (**Figure 7.12**). If this intermediate were to proceed directly to product (via **B**), we would expect the relative rates of the fast- and slow-reacting enantiomers to be comparable, as observed (k_{rel} = 3–4).

If, however, the initial alkylidene formation was fast and reversible (i.e., **A** + ethylene $\rightleftarrows$ **C** + **SM**), the ring-closing step to form **B** would then become the stereochemistry-determining step and, therefore, have a significant impact on the the relative rates of the reaction of the enantiomers (**Figure 7.12**). By conducting the reaction under an atmosphere of ethylene,[26,27] the initial metathesis step is more readily reversible, resulting in higher observed k_{rel} values of 13–17 (**Figure 7.11**).[25] Another benefit of the addition of ethylene is that it suppresses intermolecular cross-metathesis that leads to the coupling of two substrate molecules through the unsubstituted double bonds.

Other substrates bearing different functional groups have been used with varying degrees of success in the kinetic resolution with the same molybdenum catalyst from **Figure 7.11**. Silyl ethers are very useful in this chemistry, because the size of the protecting group can be altered to maximize the selectivity factor (**Equation 7.3**). Silyl ethers can also be used as temporary tethers that can be cleaved at a later stage (**Equation 7.4**). After cleavage, the products are equivalent to a cross-metathesis reaction that has proceeded to give only the (Z)-olefin.[28]

Equation 7.3

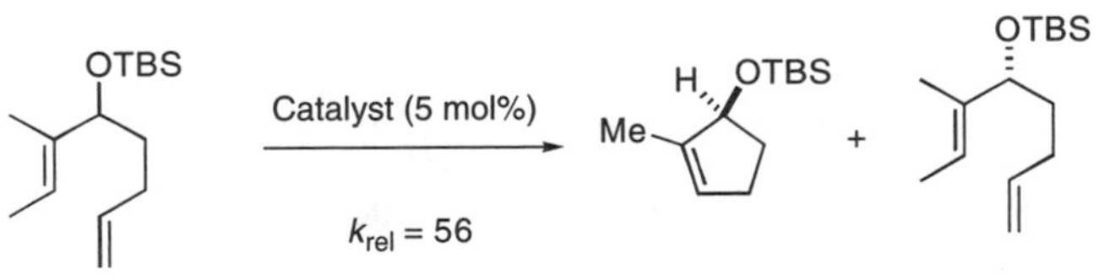

Equation 7.4

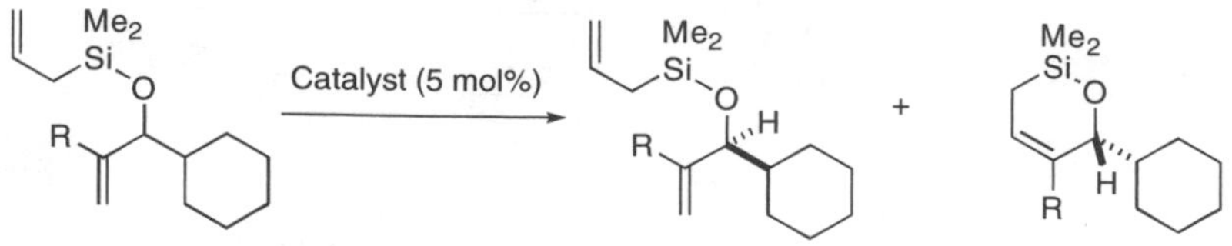

R = H, k_{rel} = 56
R = Me, k_{rel} = 1.1

7.3 Kinetic Resolutions That Create Further Stereocenters

7.3.1 Kinetic Resolution of Allylic Alcohols with the Sharpless–Katsuki Epoxidation Catalyst

Two of the most significant developments in the history of asymmetric catalysis were the introduction of the Sharpless–Katsuki asymmetric epoxidation of allylic alcohols[29] and the use of this reaction in the kinetic resolution of racemic allylic alcohols.[30,31] The Sharpless–Katsuki asymmetric epoxidation was the first highly enantioselective oxidation reaction, exhibiting > 90% enantioselectivity for a wide range of allylic alcohols. It has been used in numerous enantioselective syntheses and is still viewed as one of the most useful reactions in asymmetric catalysis. Before the use of this reaction in the kinetic resolution of allylic alcohols can be discussed, background information concerning the asymmetric epoxidation of prochiral allylic alcohols is warranted.

The Sharpless–Katsuki asymmetric epoxidation (AE) is illustrated in **Figure 7.13**. The reaction protocol is straightforward, typically conducted with 5–20 mol% titanium tetraisopropoxide and a 10–20% excess of the chiral dialkyl tartrate ligand over titanium tetraisopropoxide. Comparable enantioselectivities are observed with DMT, DET, and DIPT ligands (**Figure 7.13**) in the asymmetric epoxidation, although DIPT and DCHT give better results in the kinetic resolution. To preserve catalyst lifetimes,

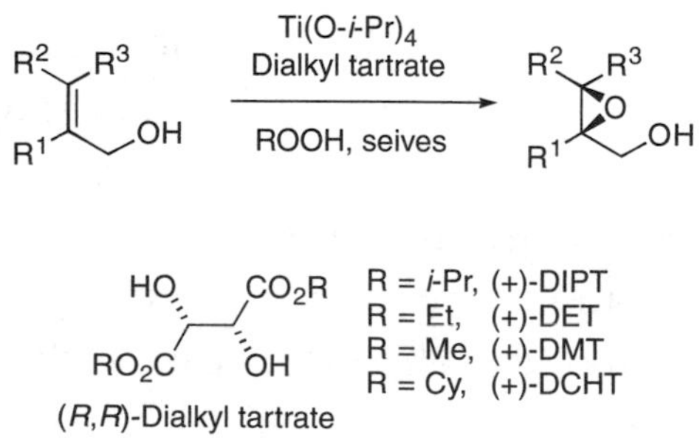

R = i-Pr, (+)-DIPT
R = Et,   (+)-DET
R = Me,   (+)-DMT
R = Cy,   (+)-DCHT

(R,R)-Dialkyl tartrate

R^1 = Me, R^2 = H, R^3 = H   92% ee

R^1 = H, R^2 = Me, R^3 = H   91% ee

R^1 = H, R^2 = H, R^3 = Me   88% ee

R^1 = Me, R^2 = Me, R^3 = H   94% ee

Figure 7.13. The scope of the Sharpless–Katsuki asymmetric epoxidation reaction.

3 or 4 Å molecular sieves are added to the reaction mixture, because the catalyst is very sensitive to traces of water. Several stoichiometric anhydrous hydroperoxide oxidants (ROOH) can be used, with *tert*-butyl hydroperoxide (TBHP) being the most common. Isolated yields of the epoxy alcohols typically range from 65–90% and enantioselectivities are high (**Figure 7.13**). The only class of allylic alcohols that exhibit reduced enantioselectivities is (Z)-disubstituted allylic alcohols. Good enantioselectivities are obtained with these substrates when R^3 does not contain branching near the double bond. When R^3 is more sterically demanding, however, the levels of enantioselectivity are reduced.

The structure of the catalyst and its mechanism of operation have been extensively studied.[32] The dialkyl tartrate ligands chelate the titanium center, liberating two equivalents of isopropanol. Strong binding of the chiral ligand to the metal is important, because titanium tetraisopropoxide itself will catalyze the epoxidation of allylic alcohols to give racemic epoxy alcohols. Fortunately, the titanium tartrate complexes exhibit much higher TOF than titanium tetraisopropoxide, a phenomenon termed "ligand acceleration."[33] Based on X-ray crystal structures of closely related titanium tartrate complexes[34] and solution studies,[32] the catalyst is believed to be dimeric in solution, with alkoxide bridging between the chelating ligand and the neighboring titanium center (**Figure 7.14**). The Lewis basic ester carbonyl groups can act as dative ligands to the electron-deficient titanium centers, an interaction that is believed to be important in maintaining an octahedral geometry at the titanium center that is not directly involved in the oxygen-transfer process. Binding of the TBHP to titanium liberates another equivalent of isopropanol. The oxidant is then activated by chelation to titanium, which lowers the energy of the O–O σ^* orbital. It is this orbital that is attacked by the olefin in the transition state. The intermediate formed just prior to oxygen transfer has been proposed to contain the chiral ligand, allylic alcohol, and oxidant assembled on a single titanium center, as shown in **Figure 7.14**. Based on this model, it can be understood why (Z)-substituted allylic alcohols give lower enantioselectivities when R^3 is large: interaction of R^3 with the ester of the neighboring titanium destabilizes this conformation and the subsequent transition state. To alleviate this nonbonded interaction, the olefin can rotate 180° to expose the opposite face to the activated peroxy group.

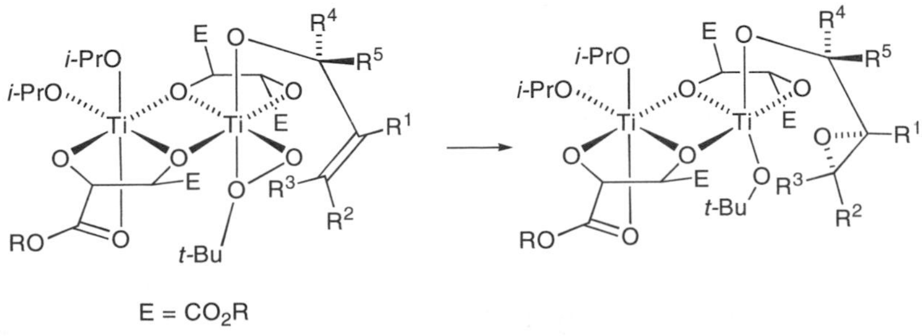

Figure 7.14. Proposed structure of the activated complex in the Sharpless–Katsuki asymmetrical epoxidation reaction. The oxygen atom is transferred to the bottom face of the double bond.

The kinetic resolution of allylic alcohols with the Sharpless–Katsuki AE was the first example of a nonenzymatic catalyst to exhibit high selectivity factors with a range of racemic substrates. Following this seminal work, many catalysts were successfully employed in kinetic resolutions.[1,2]

When using the Sharpless–Katsuki catalyst in the kinetic resolution of secondary allylic alcohols, there are two stereochemical issues that must be considered: 1) the kinetic resolution of the racemic allylic alcohol and 2) the diastereoselection in the epoxidation. If the goal is to isolate the starting allylic alcohol with high enantioselectivity, the product diastereoselectivity is irrelevant. On the other hand, if the epoxy alcohol is the desired product, both the efficiency of the resolution and the diastereoselectivity will be crucial (see also Section 16.2.3).

An important factor in the kinetic resolution is the choice of the dialkyl tartrate, with the selectivity factor improving with increasing the size of the ester (DCHT > DIPT > DET > DMT). When (+)-DIPT is employed the (S)-enantiomer reacts quickly to give the *anti*-epoxide product with high diastereoselectivity in most cases, while the (R)-enantiomer undergoes epoxidation slowly, usually affording a mixture of the *syn* and *anti* epoxides (**Figure 7.15**). One of the most remarkable features of this system is that the fast-reacting enantiomer generally undergoes epoxidation over 30 times faster than the slow-reacting enantiomer, and k_{rel} values as high as 700 have been reported.[35] It should be noted, however, as the relative rates of reaction increase, greater error is introduced when measuring k_{rel}. With relative rates over 50, reactions that are halted at 55% conversion will allow recovery of starting material of > 99% ee. **Table 7.4** lists some representative kinetic resolution results with the Sharpless–Katsuki catalyst that were conducted under stoichiometric and catalytic conditions using 0.6–0.7 equivalents of TBHP.[30,35]

Kinetic resolutions of (Z)-allylic alcohols exhibit significantly lower k_{rel} values (**Table 7.4**, entry 6). This behavior is consistent with the reduced stereoselectivity of the catalyst with prochiral (Z)-allylic alcohols outlined earlier.

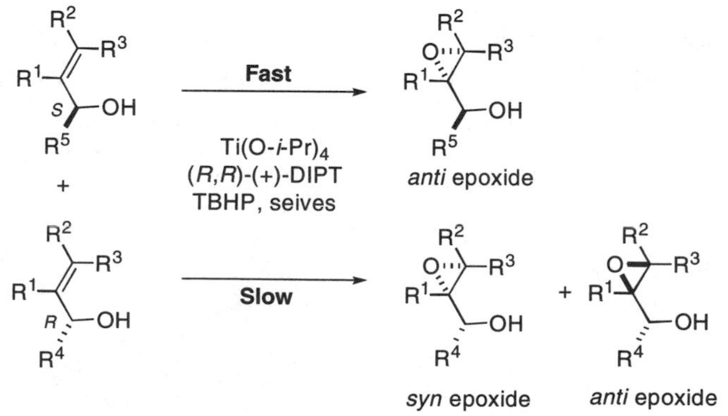

Figure 7.15. Kinetic resolution of racemic secondary allylic alcohols, showing the reaction of both enantiomers of the racemate. The fast-reacting enantiomer exhibits high diastereoselectivity but the slow-reacting enantiomer does not. See **Figure 7.14** for the proposed stereochemical model for the matched catalyst–substrate combination (R^4 = H).

Table 7.4. Kinetic resolution of secondary allylic alcohols with the Sharpless–Katsuki asymmetric epoxidation reaction.

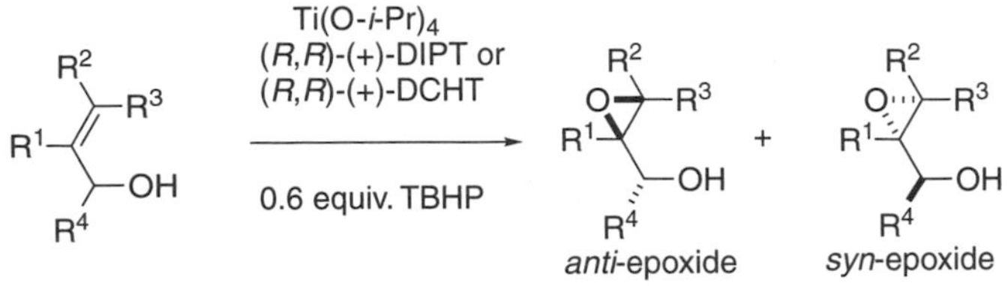

Entry		Ligand	mol%	Allylic alcohol Config.	Allylic alcohol ee (%)	k_{rel}	Epoxy alcohol anti:syn
1	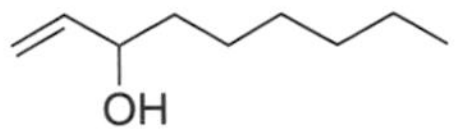	DIPT	120	R	> 96	83	99:1
		DIPT	15	R	86	29	-
		DCHT	15	R	> 98	-	-
2	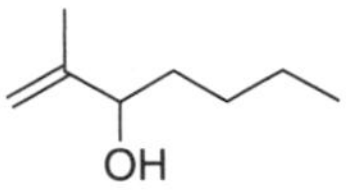	DIPT	120	R	> 96	138	98:2
		DIPT	15	R	94	33	-
		DCHT	15	R	97	76	-
3	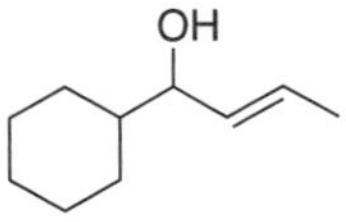	DIPT	120	R	> 96	104	97:3
		DIPT	15	R	94	32	-
		DCHT	15	R	95	56	-
4	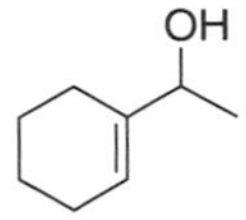	DIPT	120	R	> 96	83	98:2
		DIPT	15	R	> 98	-	-
		DCHT	15	R	> 98	-	-
5	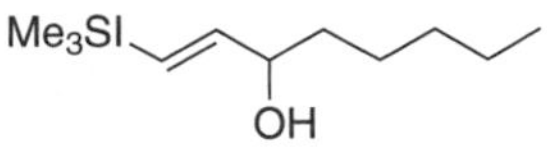	DIPT	120	R	> 99	700	> 99:1[a]
6	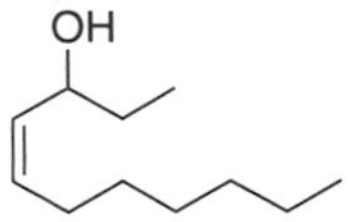	DIPT	120	R	95	16	40:60

[a] The anti-epoxy alcohol formed with >99% ee.

If the aim of the kinetic resolution is to isolate the epoxy alcohol product, the reaction must be quenched early, before the less-reactive starting material enantiomer undergoes significant reaction, typically after 40–45% conversion. It is often found that the epoxy alcohol can be obtained with higher enantioselectivity if the kinetic resolution is performed and the enantioenriched allylic alcohol is isolated. Subsequent diastereoselective-directed epoxidation in a separate step then provides the desired epoxy alcohol.

The model for stereoinduction in the asymmetric epoxidation of prochiral allylic alcohols also serves well for the prediction of the fast- and slow-reacting secondary allylic alcohols. As shown in **Figure 7.14**, when R^5 = alkyl it is oriented away from the ligand system and does not interfere with the reactivity (matched substrate and catalyst). In contrast, the R^4 group is directed toward the metal–ligand framework, destabilizing the intermediate shown and impeding the epoxidation. This yields the mismatched catalyst–substrate combination (see Chapter 13 for a further discussion of matched and mismatched systems). Even if the k_{rel} values are acceptable with certain (Z)-allylic alcohols, the epoxide products are typically formed with low diastereoselectivity (**Table 7.4**, entry 6).[30]

7.3.2 Kinetic Resolution via the Hydrosilation of Imines

In Section 7.2.2, the kinetic resolution of amines by selective acylation of one enantiomer does not generate additional stereocenters. Chiral amines with multiple stereocenters can be generated by the kinetic resolution of racemic imines through reduction of the C=N bond and formation of a second stereocenter. In this instance, the formation of diastereomeric products is possible and may cause difficulties in product isolation. In contrast to kinetic resolutions where no new chiral centers are formed, the mathematical treatment of such processes that generate an additional stereocenter is more complicated.[36]

The asymmetric hydrosilylation of imines to yield amines with catalysts derived from (EBTHI)TiF$_2$ [**Figure 7.16**, where EBTHI = ethylene-1,2-bis(η^5-4,5,6,7-tetrahydro-1-indenyl)] was introduced in Chapter 4.[37–41] This catalyst has also been employed in the kinetic resolution of imines. Here, the unreacted imine is isolated as the ketone and the hydrosilylated amine as the free amine.[42] The catalyst exhibits high selectivity in the kinetic resolution, allowing for the isolation of the amine with high enantioselectivity and diastereoselectivity at conversions around 50%.

The control factors behind the facial selectivity in the reduction can be seen in **Figure 7.17**. There are four possible approaches of the imine substrates to the (R,R)-(EBTHI)Ti-H catalyst. In all of the approaches, the benzo group is positioned in front of the metallocene to minimize interaction with the ligand system. In structure **A** (**Figure 7.17**) the N-methyl is positioned in a quadrant that is not occupied by the chiral

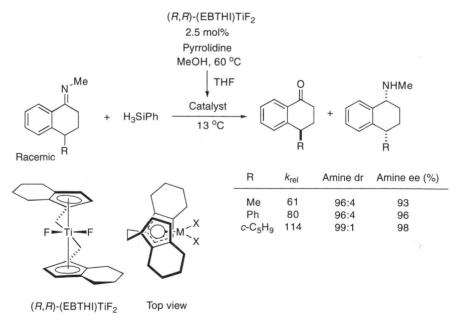

R	k_{rel}	Amine dr	Amine ee (%)
Me	61	96:4	93
Ph	80	96:4	96
c-C$_5$H$_9$	114	99:1	98

Figure 7.16. Kinetic resolution of *N*-methyl imines of 4-substituted tetralones.

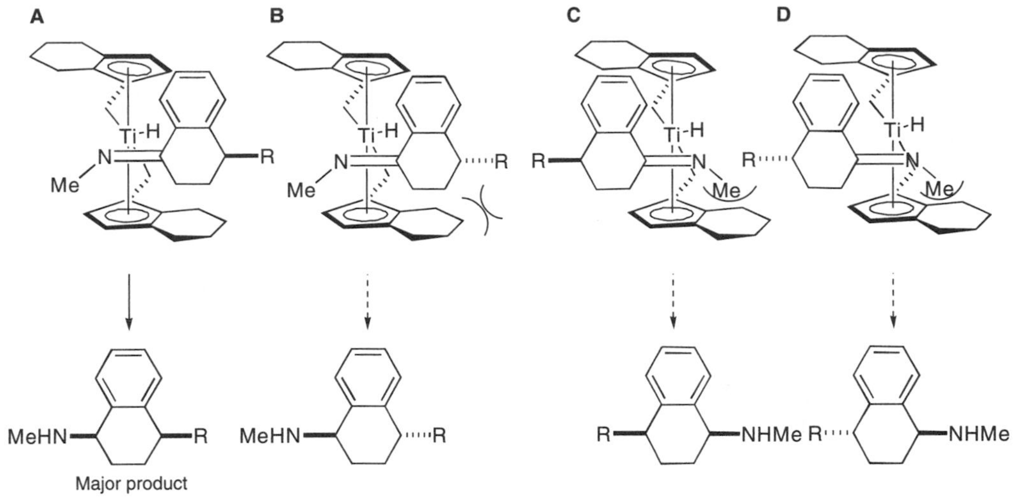

Figure 7.17. Approaches of 4-substituted *N*-methyl imines to the chiral catalyst, (*R,R*)-(EBTHI)TiH. Pathway **A** is the lowest in energy and accounts for the relative and absolute stereochemistry of the predominant product.

ligand system. Furthermore, the R group is pointing away from the catalyst. This orientation is expected to be the lowest in energy and lead to rapid reduction of the imine to generate the amine with the observed relative and absolute stereochemistry. Structure **B** is higher in energy than **A** due to the projection of the substituent R toward the metal. In both **C** and **D**, the *N*-methyl group is directed into a quadrant containing the chiral ligand, rendering these pathways less favorable. Consistent with this model, *N*-methyl imines of tetralone with larger 4-substituents result in higher selectivity factors and increased diastereoselectivity (**Figure 7.16**).

The enantiomeric catalyst has been applied to the synthesis of (1*S*,4*S*)-sertraline, an antidepressant marketed under the trade name Zoloft (**Equation 7.5**). Commercially, a classical resolution is performed with D-mandelic acid.[43]

Equation 7.5

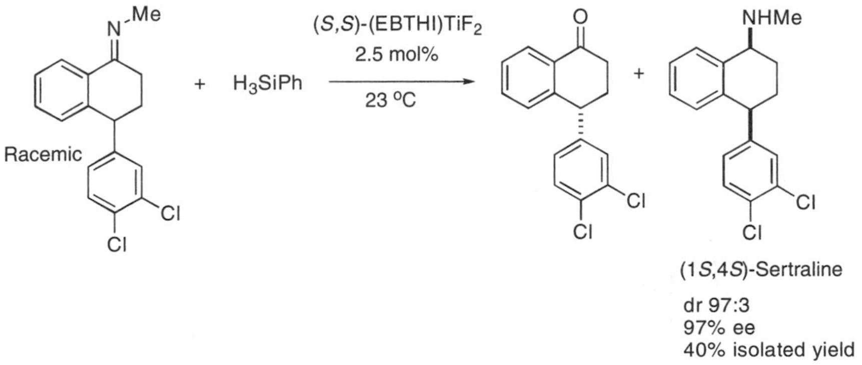

(1*S*,4*S*)-Sertraline

dr 97:3
97% ee
40% isolated yield

7.3.3 Kinetic Resolution of Atropisomeric Amides

Kinetic resolutions often allow access to highly enantioenriched chiral compounds that are not otherwise accessible. One such example is the resolution of atropisomeric amides with the Sharpless–Katsuki asymmetric dihydroxylation (AD, **Figure 7.18**).[44,45] The AD of prochiral olefins exhibits excellent enantioselectivities with all olefin

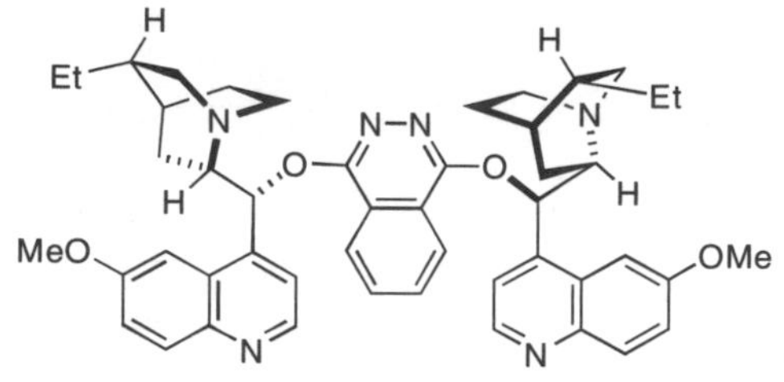

(DHQ)₂PHAL

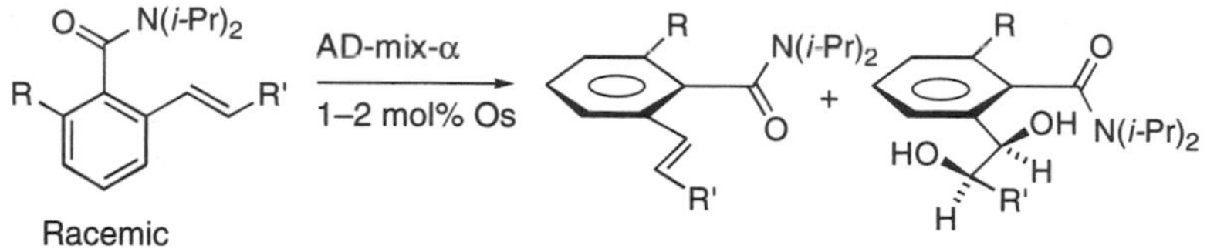

Racemic

Figure 7.18. Use of the Sharpless asymmetric dihydroxylation (AD) of atropisomeric amides with pendant olefins.

classes, with the exception of cis olefins.[46] The AD is experimentally simple, reliable, and easily reproducible. All these features have led to its extensive use in enantiose-lective synthesis.

Aryl amides, such as those in **Figure 7.18**, are chiral due to the orthogonal nature of the aryl ring and the amide carbonyl. Racemization occurs by rotation about the aryl–amide bond. The barrier to rotation is controlled by the size of the substituents at the 2- and 6-positions of the aryl ring and the size of the N,N'-dialkyl groups.[47]

The kinetic resolution of these atropisomers was carried out with commercially avaliable AD-mix-α, which contains 1 mol% $K_2OsO_2(OH)_2$, 1 mol% of the chiral ligand (DHQ)₂PHAL, the stoichiometric oxidant $[K_3Fe(CN)_3]$, and base (K_2CO_3) (**Figure 7.18**).[45] Under the reaction conditions, the osmium is oxidized to osmium tetroxide, which binds the chiral ligand at the sp^3 hybridized nitrogen in a monodentate fashion. Oxidation of the pendant olefin of the atropisomers takes place with fair to excellent selectivity factors (**Table 7.5**).[48]

Isolation of the highly enantioenriched atropisomers allowed determination of their barriers to rotation and half-lives to racemization, which ranged from 7–135 h at room temperature.

Table 7.5. Kinetic resolution of atropisomeric amides with the Sharpless asymmetric dihydroxylation.

Substrate	mol% Os	T (°C)	k_{rel}
	1	25	6.2
	1	0	16
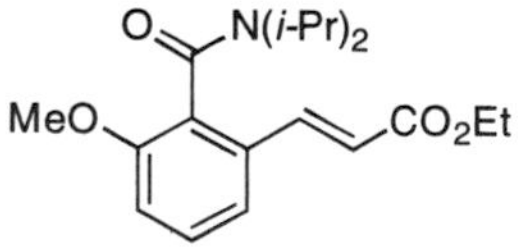	1	25	32
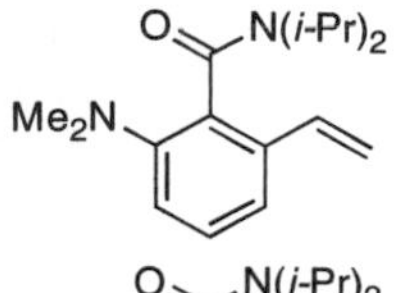	2	25	26
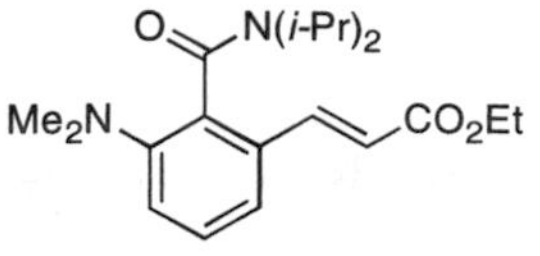	2	25	19

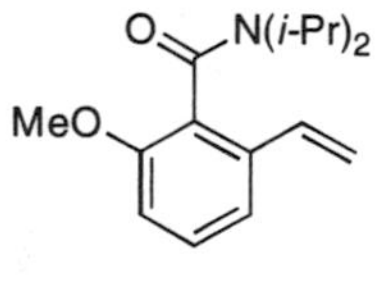

7.4 Conversion of a Racemate to One Enantioenriched Product: Combined Kinetic Resolution and Stereospecific Reaction

The most significant drawback of the kinetic resolution is the maximum yield of 50%. To circumvent this issue, elaborate resolutions that rely on the racemization of the substrate have been introduced, as outlined in Chapter 9 on dynamic kinetic resolutions. A clever application that achieves yields > 50% is the sequential kinetic resolution of enol-ester epoxides, followed by acid-catalyzed rearrangement of the remaining starting material.[49] This process allows conversion of each enantiomer of a racemate into the same enantiomer of the product. The procedure is based on a mechanistic duality in the acid-catalyzed rearrangement illustrated in **Figure 7.19** for Lewis acids (LA). The rearrangement favors retention of configuration when strong acids are employed (path A) and inversion with weak acids (path B).[50,51]

Treatment of the enol-ester epoxides with 5 mol% of a (BINOLate)Ti-based catalyst results in kinetic resolution of the epoxide via the weak-acid mechanism (path B) and generation of the 2-benzyloxy ketones as shown in **Table 7.6**. Either the starting material (SM) or the product can be isolated with good to excellent enantioselectivity, depending on conversion. With cyclic enol-ester epoxide derivatives, the relative rates of the fast- versus slow-reacting enantiomer are usually high.

As indicated in **Table 7.6**, the unreacted epoxide and the rearranged product have the same configuration at C-2, because the rearrangement proceeds with inversion of configuration (**Figure 7.19**, path B). Advantage can now be taken of the dual reaction pathway by proper choice of a catalyst to promote the rearrangement of the remaining epoxide *with retention of configuration* (**Figure 7.19**, path A). Thus, on completion of the kinetic resolution and removal of the catalyst by filtration, treatment of the resulting mixture with toluenesulfonic acid causes reaction via path A, providing 2-(benzoyloxy)cyclohexanone in 78% yield with 93% ee (**Figure 7.20**).[49]

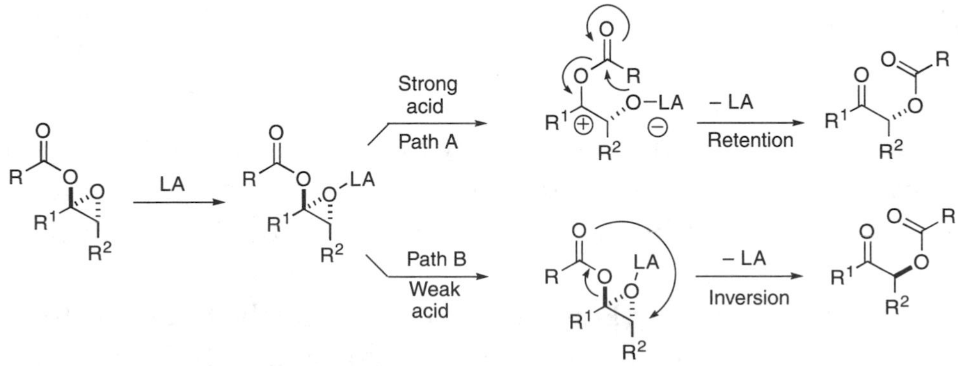

Figure 7.19. Dual reaction manifold for the Lewis acid catalyzed rearrangement of enol ester epoxides.

Table 7.6. Kinetic resolution of enol-ester epoxides.

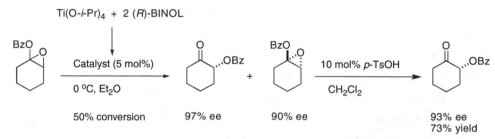

Substrate	Conversion (%)	SM ee (%)	Product ee (%)	k_{rel}
$n = 1$	55	99	89	49
$n = 2$	50	97	90	> 100
$n = 3$	54	98	80	50
$n = 4$	63	97	71	14

Figure 7.20. Sequential kinetic resolution of an enol-ester epoxide and acid-catalyzed rearrangement.

The chiral catalyst in this kinetic resolution is generated by combining titanium tetraisopropoxide and BINOL in a 1:2 ratio. Removal of the solvent and the liberated isopropanol affords the catalyst.[49] It is believed that the BINOL ligands displace all four isopropoxy groups from titanium, but the structure of the catalyst remains elusive.

7.5 Kinetic Resolution of Mixtures of Racemic Diastereomers

The cycloisomerization of enynes is another useful C–C bond-forming reaction[52,53] that has been performed enantioselectivily with prochiral substrates[54,55] and applied to kinetic resolutions with racemic substrates.[56] In the case of the kinetic resolution, the reaction generates an additional stereocenter, thus providing products with multiple stereocenters.

Reaction of racemic enyne (**Equation 7.6**) with a catalyst derived from [(COD) RhCl]$_2$, (COD = 1,5-cyclooctadiene) racemic BINAP, and AgSbF$_6$ at room temperature for 2 min provided the racemic trans tetrahydrofuran in 89% yield as the sole product of the cycloisomerization. Surprisingly, the racemic cis tetrahydrofuran was not observed, indicating that the reaction is highly diastereoselective. Since the chiral catalyst is responsible for the newly formed stereocenter, the cis and trans diastereomers should form in equal amounts if a statistical distribution occurs [i.e., half of each substrate enantiomer undergoes reaction with the (S)-BINAP-based catalyst and the other half with the (R)-BINAP-derived catalyst]. To explain the observations, each enantiomer of the catalyst must react specifically with one enantiomer of the enyne with excellent diastereoselectivity, as shown in **Figure 7.21**. This is an extreme example of double diastereoselection (see Chapter 13) and indicates that the use of an enantioenriched catalyst will result in a kinetic resolution.[57]

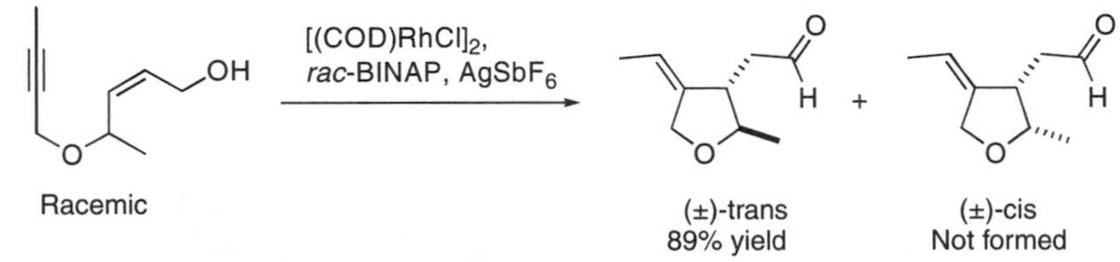

Equation 7.6

A proposed mechanism is illustrated in **Figure 7.22**. Coordination of the enyne is followed by cyclization with oxidation of the metal. Regioselective β-hydride elimination and reductive elimination give the trans tetrahydrofuran enol that readily isomerizes to the observed aldehyde.

A clever application of this kinetic resolution takes advantage of the loss of stereochemistry at the carbons undergoing β-hydride elimination. A racemic mixture of *syn* and *anti* diastereomers (four compounds) was subjected to the cycloisomerization reaction with catalyst prepared from (S)-BINAP. Only the (2R)-starting material reacts, yielding the trans isomer with the (2R,3S)-configuration in 49% yield and > 99% ee (**Equation 7.7**). The (2S,5R)- and (2S,5S)-enynes, both with > 99% ee, remained and were isolated in a combined 48% yield.[57]

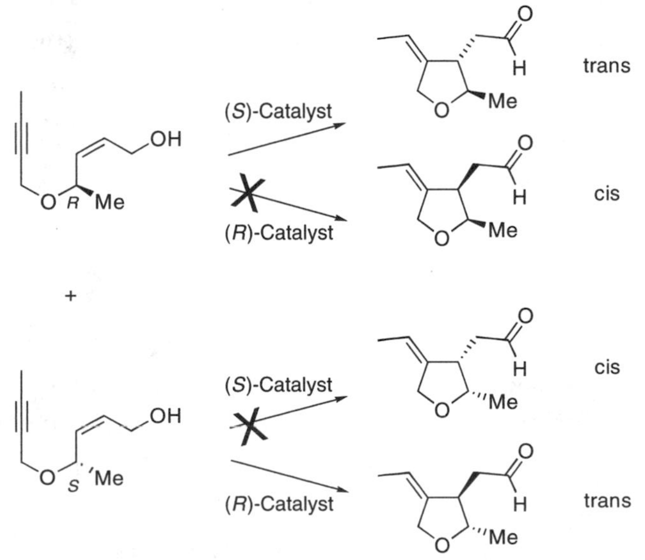

Figure 7.21. Reaction of each enantiomer of a racemic enyne with the racemic [(BINAP)-Rh]⁺-based catalyst in **Equation 7.6**.

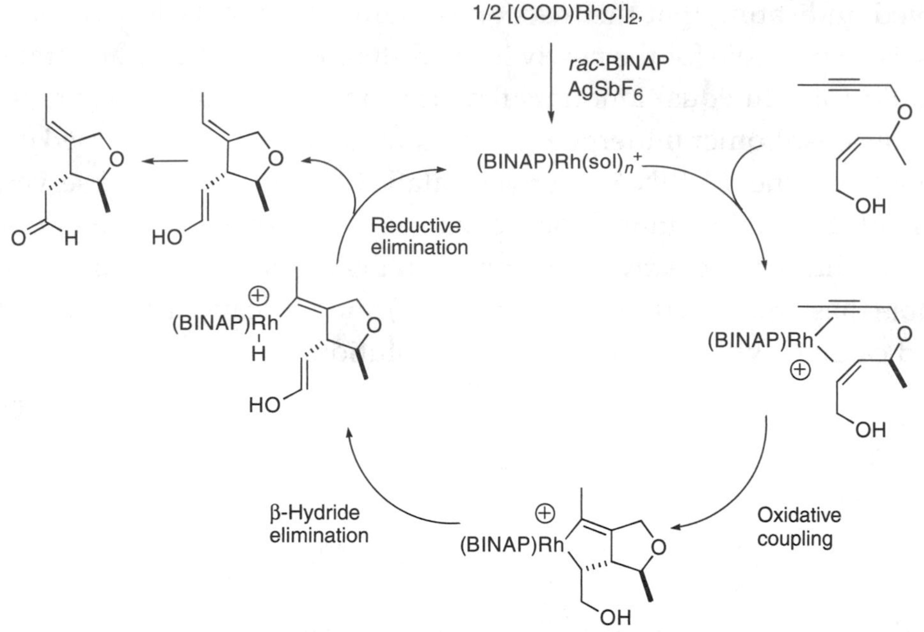

Figure 7.22. Proposed mechanism for the cycloisomerization of enynes.

Equation 7.7

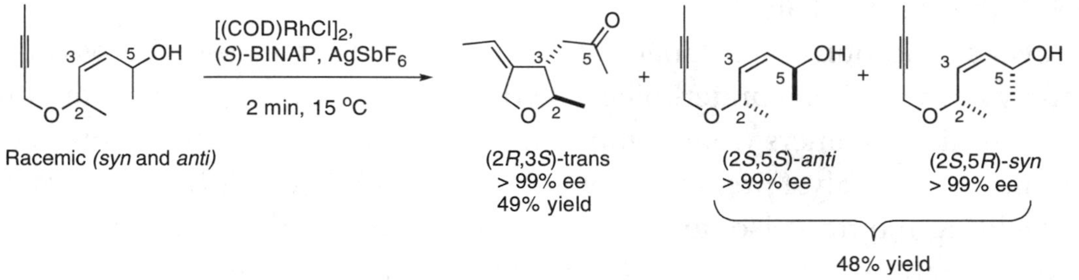

Both enantiomers of each diastereomer of the starting material can be isolated when the kinetic resolution is carried out separately with the (±)-*syn*- and (±)-*anti* enynes with catalysts derived from (R)- and (S)-BINAP. It is also impressive that the catalyst exhibits such exquisite selectivity in the kinetic resolution.

Summary

Kinetic resolutions are not only intriguing and challenging processes to design, they can be synthetically very useful. Oftentimes other methods are unavailable for the isolation of the desired enantioenriched compounds or both the enantioenriched starting material and enantioenriched product are valuable. Kinetic resolutions can be performed on a preparative scale for the synthesis of large quantities of material. If the racemate and catalyst are inexpensive or easily prepared, kinetic resolution may be the method of choice to access highly enantioenriched materials.

References

(1) Kagan, H. B.; Fiaud, J. C. Kinetic Resolution. *Top. Stereochem.* **1988**, *18*, 249–330.

(2) Keith, J. M.; Larrow, J. F.; Jacobsen, E. N. Practical Considerations in Kinetic Resolution Reactions. *Adv. Synth. Catal.* **2001**, *1*, 5–26.

(3) Cook, G. R. Transition Metal-Mediated Kinetic Resolution. *Curr. Org. Chem.* **2000**, *4*, 869–885.

(4) Hoveyda, A. H.; Didiuk, M. T. Metal-Catalyzed Kinetic Resolution Processes. *Curr. Org. Chem.* **1998**, *2*, 489–526.

(5) Ruble, J. C.; Fu, G. C. Chiral π-Complexes of Heterocycles with Transition Metals: A Versatile New Family of Nucleophilic Catalysts. *J. Org. Chem.* **1996**, *61*, 7230–7231.

(6) Ruble, J. C.; Tweddell, J.; Fu, G. C. Kinetic Resolution of Arylalkylcarbinols Catalyzed by a Planar-Chiral Derivative of DMAP: A New Benchmark for Nonenzymatic Acylation. *J. Org. Chem.* **1998**, *63*, 2794–2795.

(7) Bellemin-Laponnaz, S.; Tweddell, J.; Ruble, J. C.; Breitling, F. M.; Fu, G. C. The Kinetic Resolution of Allylic Alcohols by a Non-enzymatic Acylation Catalyst; Application to Natural Product Synthesis. *J. Chem. Soc., Chem. Commun.* **2000**, 1009–1010.

(8) Arai, S.; Bellemin-Laponnaz, S.; Fu, G. C. Kinetic Resolution of Amines by a Nonenzymatic Acylation Catalyst. *Angew. Chem., Int. Ed. Engl.* **2001**, *40*, 234–236.

(9) Arp, F. O.; Fu, G. C. Kinetic Resolutions of Indolines by a Nonenzymatic Acylation Catalyst. *J. Am. Chem. Soc.* **2006**, *128*, 14264–14265.

(10) Fu, G. F. Enantioselective Nucleophilic Catalysis with "Planar-Chiral" Heterocycles. *Acc. Chem. Res.* **2000**, *33*, 412–420.

(11) Fu, G. C. Asymmetric Catalysis with "Planar-Chiral" Derivatives of 4-(Dimethylamino)pyridine. *Acc. Chem. Res.* **2004**, *37*, 542–547.

(12) Yoon, T. P.; Jacobsen, E. N. Privileged Chiral Catalysts. *Science* **2003**, *299*, 1691–1693.

(13) Jacobsen, E. N.; Wu, M. H. In *Comprehensive Asymmetric Catalysis*; Jacobsen, E. N., Pfaltz, A., Yamamoto, H., Eds.; Springer-Verlag: Berlin, 1999; Vol. 2, pp 649–678.

(14) Bosnich, B. In *Encyclopedia of Inorganic Chemistry*; King, R. B., Ed.; John Wiley & Sons: New York, 1994; pp 219–236.

(15) Jacobsen, E. N. Asymmetric Catalysis of Epoxide Ring-Opening Reactions. *Acc. Chem. Res.* **2000**, *33*, 421–431.

(16) Tokunaga, M.; Larrow, J. F.; Kakuichi, F.; Jacobsen, E. N. Asymmetric Catalysis with Water: Efficient Kinetic Resolution of Terminal Epoxides by Means of Catalytic Hydrolysis. *Science* **1997**, *277*, 936–938.

(17) Schaus, S. E.; Brandes, B. D.; Larrow, J. F.; Tokunaga, M.; Hansen, K. B.; Gould, A. E.; Furrow, M. E.; Jacobsen, E. N. Highly Selective Hydrolytic Kinetic Resolution of Terminal Epoxides Catalyzed by Chiral (Salen)Co³⁺ Complexes. Practical Synthesis of Enantioenriched Terminal Epoxides and 1,2-Diols. *J. Am. Chem. Soc.* **2002**, *124*, 1307–1315.

(18) Kim, G.-J.; Leeb, H.; Kim, S.-J. Catalytic Activity and Recyclability of New Enantioselective Chiral Co–Salen Complexes in the Hydrolytic Kinetic Resolution of Epichlorohydrine. *Tetrahedron Lett.* **2003**, *44*, 5005–5008.

(19) Nielsen, L. P. C.; Stevenson, C. P.; Blackmond, G. D.; Jacobsen, E. N. Mechanistic Investigation Leads to a Synthetic Improvement in the Hydrolytic Kinetic Resolution of Terminal Epoxides. *J. Am. Chem. Soc.* **2004**, *126*, 1360–1362.

(20) Breinbauer, R.; Jacobsen, E. N. Cooperative Asymmetric Catalysis with Dendrimeric [Co(Salen)] Complexes. *Angew. Chem., Int. Ed. Engl.* **2000**, *39*, 3604–3607.

(21) Ready, J. M.; Jacobsen, E. N. Highly Active Oligomeric (Salen)Co Catalysts for Asymmetric Epoxide Ring-Opening Reactions. *J. Am. Chem. Soc.* **2001**, *123*, 2687–2688.

(22) Ready, J. M.; Jacobsen, E. N. A Practical Oligomeric [(Salen)Co] Catalyst for Asymmetric Epoxide Ring-Opening Reactions. *Angew. Chem., Int. Ed. Engl.* **2002**, *41*, 1374–1377.

(23) Grubbs, R. H.; Chang, S. Recent Advances in Olefin Metathesis and its Application in Organic Synthesis. *Tetrahedron* **1998**, *54*, 4413–4450.

(24) Deiters, A.; Martin, S. F. Synthesis of Oxygen- and Nitrogen-Containing Heterocycles by Ring-Closing Metathesis. *Chem. Rev.* **2004**, *104*, 2199–2238.

(25) Dolman, S. J.; Sattely, E. S.; Hoveyda, A. H.; Schrock, R. R. Efficient Catalytic Enantioselective Synthesis of Unsaturated Amines: Preparation of Small- and Medium-Ring Cyclic Amines Through Mo-Catalyzed Asymmetric Ring-Closing Metathesis in the Absence of Solvent. *J. Am. Chem. Soc.* **2002**, *124*, 6991–6997.

(26) Harrity, J. P. A.; Visser, M. S.; Gleason, J. D.; Hoveyda, A. M. Ru-Catalyzed Rearrangement of Styrenyl Ethers. Enantioselective Synthesis of Chromenes Through Zr- and Ru-Catalyzed Processes. *J. Am. Chem. Soc.* **1997**, *119*, 1488–1489.

(27) Weatherhead, G. S.; Ford, J. G.; Alexanian, E. J.; Schrock, R. R.; Hoveyda, A. H. Tandem Catalytic Asymmetric Ring-Opening Metathesis/Ring-Closing Metathesis. *J. Am. Chem. Soc.* **2000**, *122*, 1828–1829.

(28) Zhu, S. S.; Cefalo, D. R.; La, D. S.; Jamieson, J. Y.; Davis, W. M.; Hoveyda, A. H.; Schrock, R. R. Chiral Mo-BINOL Complexes: Activity, Synthesis, and Structure. Efficient Enantioselective Six-Membered Ring Synthesis Through Catalytic Metathesis. *J. Am. Chem. Soc.* **1999**, *121*, 8251–8259.

(29) Katsuki, T.; Sharpless, K. B. The First Practical Method for Asymmetric Expoxidation. *J. Am. Chem. Soc.* **1980**, *102*, 5974–5776.

(30) Martín, V. S.; Woodard, S. S.; Katsuki, T.; Yamada, Y.; Ikeda, M.; Sharpless, K. B. Kinetic Resolution of Racemic Allylic Alcohols by Enantioselective Epoxidation. A Route to Substances of Absolute Enantiomeric Purity? *J. Am. Chem. Soc.* **1981**, *103*, 6237–6240.

(31) Gao, Y.; Klunder, J. M.; Hanson, R. M.; Masamune, H.; Ko, S. Y.; Sharpless, K. B. Catalytic Asymmetric Epoxidation and Kinetic Resolution: Modified Procedures Including in situ Derivatization. *J. Am. Chem. Soc.* **1987**, *109*, 5765–5780.

(32) Finn, M. G.; Sharpless, K. B. Mechanism of Asymmetric Epoxidation. 2. Catalyst Structure. *J. Am. Chem. Soc.* **1991**, *113*, 113–126.

(33) Berrisford, D. J.; Bolm, C.; Sharpless, K. B. Ligand-Accelerated Catalysis. *Angew. Chem., Int. Ed. Engl.* **1995**, *34*, 1059–1070.

(34) Williams, I. D.; Pedersen, S. F.; Sharpless, K. B.; Lippard, S. L. Crystal Structures of Two Titanium Tartrate Asymmetric Epoxidation Catalysts. *J. Am. Chem. Soc.* **1984**, *106*, 6430–6431.

(35) Kitano, Y.; Matsumoto, T.; Sato, F. A Highly Efficient Kinetic Resolution of γ-, and β-Trimethylsilyl Secondary Allylic Alcohols by the Sharpless Asymmetric Epoxidation. *Tetrahedron* **1988**, *44*, 4073–4086.

(36) Guette, J.-P.; Horeau, A. Asymmetric Synthesis by Action of an Optically Active Reagent on a Substrate Having At Least 1 Asymmetric Center. Relation Between Quantities of 4 Enantiomers Formed. *Bull. Soc. Chim. Fr.* **1967**, 1747–1752.

(37) Verdaguer, X.; Lange, U. E. W.; Reding, M. T.; Buchwald, S. L. Highly Enantioselective Imine Hydrosilylation Using (*S,S*)-Ethylenebis(5-tetrahydroindenyl) titanium Difluoride. *J. Am. Chem. Soc.* **1996**, *118*, 6784–6785.

(38) Verdaguer, X.; Lange, U. E. W.; Buchwald, S. L. Amine Additives Greatly Expand the Scope of Asymmetric Hydrosilylation of Imines. *Angew. Chem., Int. Ed. Engl.* **1998**, *37*, 1103–1107.

(39) Willoughby, C. A.; Buchwald, S. L. Asymmetric Titanocene-Catalyzed Hydrogenation of Imines. *J. Am. Chem. Soc.* **1992**, *114*, 7562–7564.

(40) Willoughby, C. A.; Buchwald, S. L. Synthesis of Highly Enantiomerically Enriched Cyclic Amines by the Catalytic Asymmetric Hydrogenation of Cyclic Imines. *J. Org. Chem.* **1993**, *58*, 7627–7629.

(41) Hansen, M. C.; Buchwald, S. L. A Method for the Asymmetric Hydrosilylation of *N*-Aryl Imines. *Org. Lett.* **2000**, *2*, 713–715.

(42) Yun, J.; Buchwald, S. L. Efficient Kinetic Resolution in the Asymmetric Hydrosilylation of Imines of 3-Substituted Indanones and 4-Substituted Tetralones. *J. Org. Chem.* **2000**, *65*, 767–774.

(43) Williams, M.; Quallich, G. Sertraline-Development of a Chiral Inhibitor of Serotonin Uptake. *Chem. Ind. (London)* **1990**, 315–319.

(44) Jacobsen, E. N.; Marko, I.; Mungall, W. S.; Schroeder, G.; Sharpless, K. B. Asymmetric Dihydroxylation via Ligand-Accelerated Catalysis. *J. Am. Chem. Soc.* **1988**, *110*, 1968–1970.

(45) Kolb, H. C.; VanNieuwenhze, M. S.; Sharpless, K. B. Catalytic Asymmetric Dihydroxylation. *Chem. Rev.* **1994**, *94*, 2483–2547.

(46) Wang, L.; Sharpless, K. B. Catalytic Asymmetric Dihydroxylation of *cis*-Disubstituted Olefins. *J. Am. Chem. Soc.* **1992**, *114*, 7568–7570.

(47) Clayden, J. Stereocontrol with Rotationally Restricted Amides. *Synlett* **1998**, 810–816.

(48) Rios, R.; Jimeno, C.; Carroll, P. J.; Walsh, P. J. Kinetic Resolution of Atropisomeric Amides. *J. Am. Chem. Soc.* **2002**, *124*, 10272–10273.

(49) Feng, X.; Shu, L.; Shi, Y. Chiral Lewis Acid Catalyzed Resolution of Racemic Enol Ester Epoxides. Conversion of Both Enantiomers of an Enol Ester Epoxide to the Same Enantiomer of Acyloxy Ketone. *J. Org. Chem.* **2002**, *67*, 2831–2836.

(50) Zhu, Y.; Manske, K. J.; Shi, Y. Dual Mechanisms of Acid-Catalyzed Rearrangement of Enol Ester Epoxides: Enantioselective Formation of α-Acyloxy Ketones. *J. Am. Chem. Soc.* **1999**, *121*, 4080–4081.

(51) Zhu, Y.; Shu, L.; Tu, Y.; Shi, Y. Enantioselective Synthesis and Stereoselective Rearrangements of Enol Ester Epoxides. *J. Org. Chem.* **2001**, *66*, 1818–1826.

(52) Trost, B. M.; Krische, M. J. Transition Metal-Catalyzed Cycloisomerizations. *Synlett* **1998**, 1–16.

(53) Aubert, C.; Buisine, O.; Malacria, M. The Behavior of 1,*n*-Enynes in the Presence of Transition Metals. *Chem. Rev.* **2002**, *102*, 813–834.

(54) Lei, A.; He, M.; Zhang, X. Highly Enantioselective Syntheses of Functionalized α-Methylene-γ-butyrolactones via Rh(I)-catalyzed Intramolecular Alder Ene Reaction: Application to Formal Synthesis of (+)-Pilocarpine. *J. Am. Chem. Soc.* **2002**, *124*, 8198–8199.

(55) Lei, A.; Waldkirch, J. P.; He, M.; Zhang, X. Highly Enantioselective Cycloisomerization of Enynes Catalyzed by Rhodium for the Preparation of Functionalized Lactams. *Angew. Chem., Int. Ed. Engl.* **2002**, *41*, 4526–4529.

(56) Lei, A.; Wu, S.; He, M.; Zhang, X. Highly Enantioselective Asymmetric Hydrogenation of α-Phthalimide Ketone: An Efficient Entry to Enantiomerically Pure Amino Alcohols. *J. Am. Chem. Soc.* **2004**, *126*, 1626–1627.

(57) Lei, A.; He, M.; Zhang, X. Rh-Catalyzed Kinetic Resolution of Enynes and Highly Enantioselective Formation of 4-Alkenyl-2,3-disubstituted Tetrahydrofurans. *J. Am. Chem. Soc.* **2003**, *125*, 11472–11473.

8

Parallel Kinetic Resolution

The success of a kinetic resolution depends on the relative rates of reaction of a chiral nonracemic catalyst with each enantiomer of a racemic substrate (Chapter 7). In performing a kinetic resolution, the reaction must be quenched at some point that strikes a balance between enantiomeric excess and yield. Imagine a case where the goal of a kinetic resolution is to isolate the (R)-enantiomer of the substrate (S_R) with high enantiomeric excess. As the reaction progresses, the fast-reacting enantiomer, S_S, is consumed and the ee of the starting material increases. When the ee of the starting material is high, the relative amount of the slow-reacting enantiomer is also high. As the resolution nears completion, the enantiomers can undergo reaction at equal rates, because the rate depends on both the relative rate constants (k_R and k_S) and the concentrations of S_R and S_S. For this reason, very high relative rate values (k_{rel}) are necessary to recover the starting material with high enantiomeric excess and in a yield approaching 50%.

8.1 Parallel Kinetic Resolution Concepts

A better resolution strategy than the kinetic resolution is one that minimizes the build-up of the less-reactive substrate enantiomer as the kinetic resolution approaches completion. This can be achieved by simultaneously converting each enantiomer of

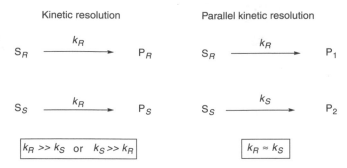

Figure 8.1. Comparison between a kinetic resolution and a parallel kinetic resolution. In an ideal parallel kinetic resolution, the starting material remains at 0% ee over the course of the reaction.

the starting material into distinct products, P_1 and P_2, at similar rates. Under these conditions, the ratio of concentrations of the substrate enantiomers is near one and the starting material remains close to 0% ee (**Figure 8.1**). When the two resolutions work synergistically, the efficiency of the resolution is greatly increased. This strategy has been termed "parallel kinetic resolution" (PKR).[1–3]

Parallel kinetic resolution is powerful because the selectivity factor ($s = k_{rel}$) can be considerably lower than that needed to achieve the same result in a standard kinetic resolution. For example, a PKR (at 100% conversion) employing two simultaneous reactions with the same moderate selectivity factor ($s_1 = s_2 = 49$) would yield the same result as a standard kinetic resolution (at 50% conversion) with a much higher selectivity factor ($s = 200$). The theoretical yield of each resolved enantiomer in both of these experiments is 50%, with an enantiomeric excess of 96%.[1] For a successful PKR, the reactions should 1) proceed without mutual interference of the catalyst or reagents, 2) have rates that are similar, 3) have opposite enantiocontrol with respect to the substrate, and 4) produce different products that can be easily separated.[1] Subclasses of PKRs have been proposed based on the relationship of the products. Chemodivergent PKRs afford different, nonisomeric products from the substrate enantiomers, regiodivergent PKRs give isomeric products that are not diastereomers, and stereodivergent PKRs give diastereomeric products.[3] Examples of each of these cases are introduced below.

8.2 Chemodivergent PKR

In a chemodivergent PKR, the products of the reactions are not isomers and can, in fact, have very different structures. An early example of a catalytic chemodivergent PKR was found during a study of a classical kinetic resolution in the intramolecular cyclopropanation of racemic secondary allylic diazoacetates.[4] The catalyst for this reaction is the dirhodium(II) carboxamidate $Rh_2(MEOX)_4$, the structure of which is illustrated in **Figure 8.2**. The catalyst reacts with the diazo group to form a transient carbene and dinitrogen. It was found that the enantiomers of the starting diazo com-

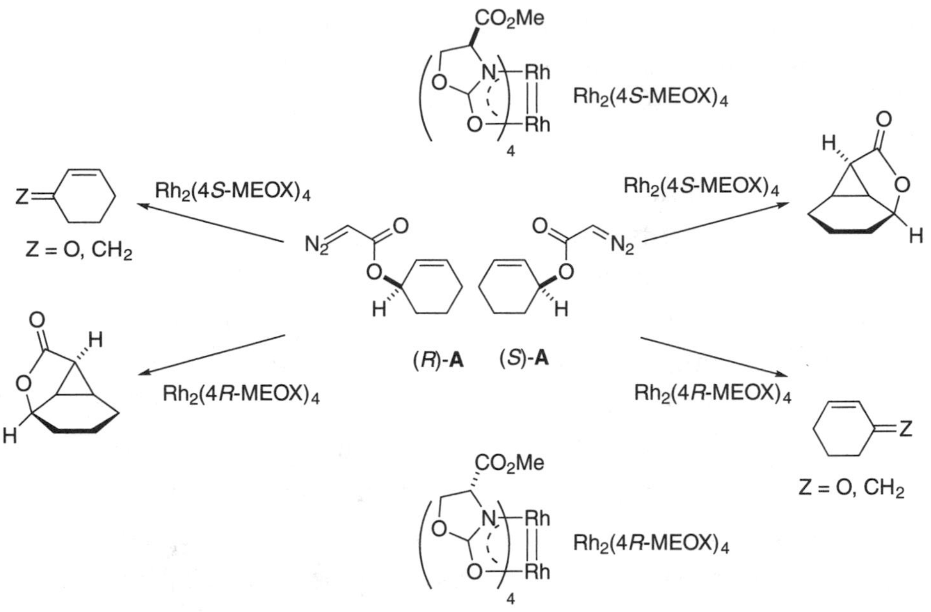

Figure 8.2. Regiodivergent reaction of racemic **A** with $Rh_2(4S\text{-MEOX})_4$ affords the cyclopropane derivative from (S)-**A** and conjugated products (X = O and CH_2) from (R)-**A** (top half).

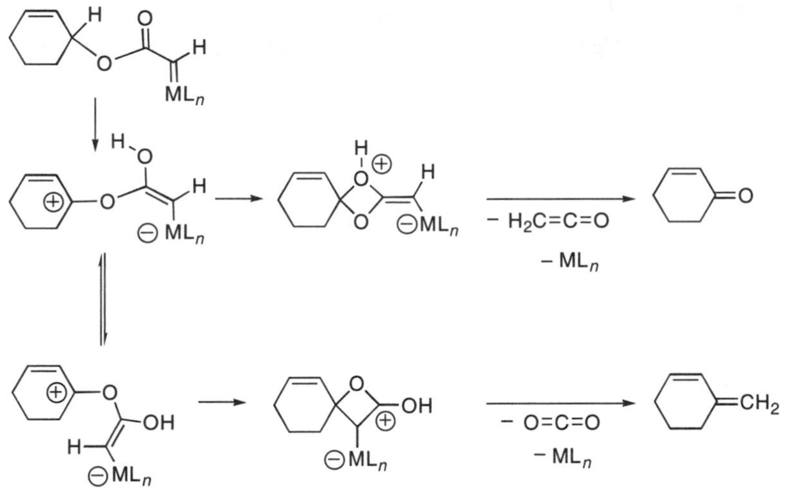

Figure 8.3. Proposed mechanism of formation of the ketone and diene products generated in **Figure 8.2**.

pound, 1-cyclohexen-1-yl diazoacetate, led to entirely different products, as shown in **Figure 8.2**. Using $Rh_2(4S\text{-MEOX})_4$ the (S)-enantiomer of **A**, (S)-**A**, underwent the expected intramolecular cyclopropanation with an enantioselectivity of 94% (40% yield).[4] Surprisingly, the products formed from (R)-**A** with $Rh_2(4S\text{-MEOX})_4$ were the diene (Z = CH_2) and cyclohexenone (Z = O). In this case, the propensity of the chiral catalyst to react with the *Re* olefin face is countered by the stereochemistry of the tethered diazo, which directs addition to the *Si* face. Thus, only one diastereomeric combination has a barrier to intramolecular cyclopropanation lower in energy than the competing extrusion pathways (**Figure 8.3**). The proposed mechanism for formation of the achiral products involves an intramolecular hydride abstraction followed by elimination of either ketene or carbon dioxide (**Figure 8.3**).[5]

A beautiful application of a chemodivergent PKR is found in the key step of the total synthesis of (–)-colombiasin A.[6] In the presence of the prolinate-based catalyst $Rh_2(R\text{-DOSP})_4$, one enantiomer of the starting 1-methyl-1,2-dihydronaphthylene derivative undergoes what amounts to a C–H activation/Cope rearrangement,[7–10] while the other enantiomer is converted to a cyclopropane (**Figure 8.4**). The mechanism of the C–H activation/Cope rearrangement has not been elucidated, but does not likely occur via a stepwise pathway involving the C–H activation product undergoing a Cope rearrangement. The PKR products were not easily separated, so the mixture was hydrogenated and reduced to provide the desired product in 34% yield (68% theoretical yield) over three steps as a single diastereomer with > 95% ee. Conversion of this intermediate to (–)-colombiasin A was readily accomplished using standard procedures.

The dramatic difference in reaction pathways experienced by the two enantiomers in **Figure 8.4** can be understood by using the predictive models that have been proposed for the C–H activation/Cope rearrangement[8] and cyclopropanation[11] with this catalyst (**Figure 8.5**). It is believed that the catalyst adopts a D_2-symmetric shape with the SO_2Ar substituents acting as blocking groups oriented up and down in an alternating fashion. The methyl groups attached to the stereogenic center in the 1-methyl-

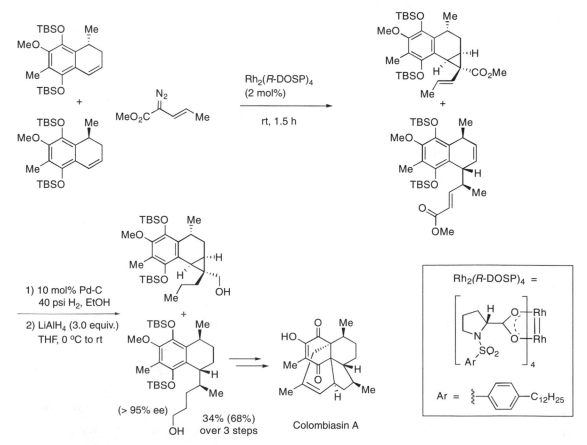

Figure 8.4. Rhodium-prolinate-catalyzed PKR in the total synthesis of (–)-colombiasin.

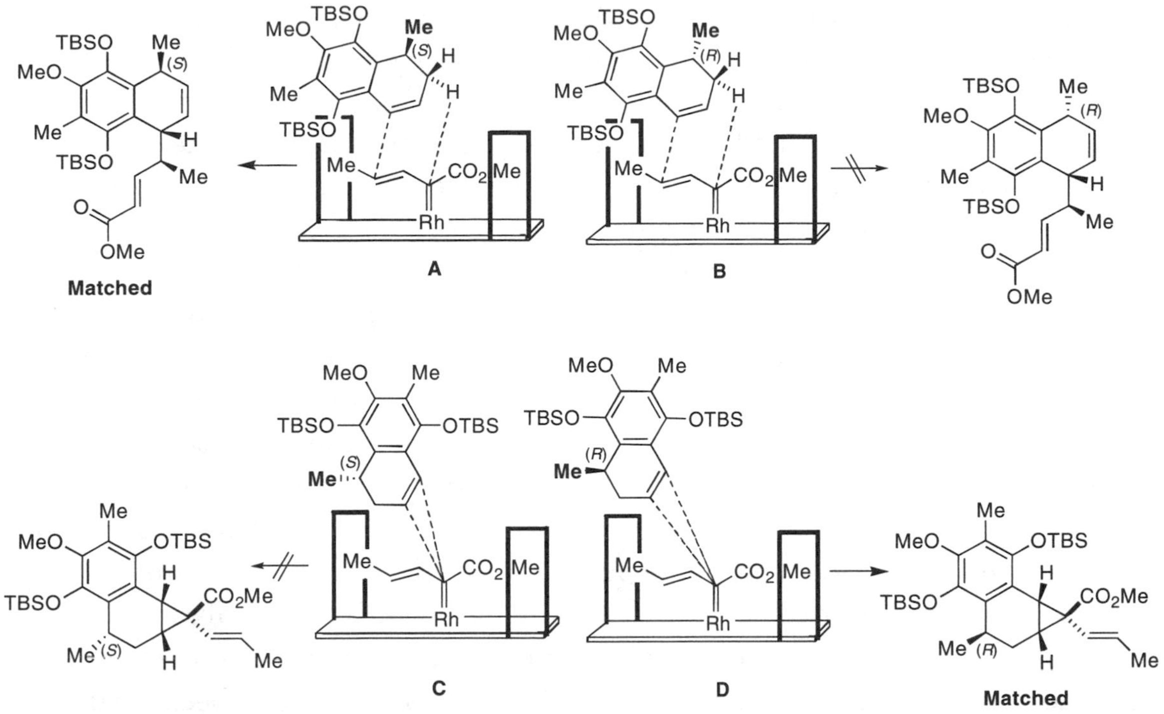

Figure 8.5. In these models of the Rh$_2$(R-DOSP)$_4$ catalyst, only the rhodium bearing the carbene is illustrated. The ligand aryl groups are represented by the bold rectangles. The methyl group of the stereogenic center in the (R)-substrate experiences steric interactions with the ligand in the C–H activation/Cope rearrangement (model **B**), while the methyl of the (S)-substrate interacts unfavorably with the ligand in the cyclopropanation transition state (model **C**).

1,2-dihydronaphthylene derivatives are bold to facilitate the discussion (**Figure 8.5**). In model **A**, the (S)-substrate allows the methyl group to orient away from the blocking aryl group of the ligand (depicted with bold lines), while the (R)-substrate would experience a steric clash in **B** between the methyl and the ligand, disfavoring the C–H activation/Cope rearrangement pathway. The situation is reversed in the cyclopropanation manifold. Here, reaction of the (S)-substrate is unfavorable due to interaction of the methyl group with the ligand system. The (R)-substrate can orient the methyl group away from the ligand scaffolding, resulting in facile cyclopropanation. It is remarkable that pathways that are so different can have such similar barriers that allow both reactions to take place simultaneously.

In each example outlined above, a single asymmetric catalyst promotes distinct reactions with each enantiomer of the starting material. It is also possible to perform a PKR in which two different catalysts are employed, each reacting selectively with one enantiomer of the substrate and generating dissimilar products. In this scenario, each catalyst must be selectively activated by only one of the two derivatizing reagents. Selective activation would be straightforward if the catalysts promote different types of reactions. If they catalyze similar reactions, the derivatizing reagents can be separated by phase isolation—that is, by constraining them to different phases during the reaction.

As an example of a dual-catalyst PKR, two acylation catalysts with selectivities for opposite enantiomers of the starting material were employed. One catalyst is an insoluble cross-linked lipase sold as ChiroCLEC-PC,[12] an enzymatic catalyst that is used in combination with the lipase-specific acyl donor, vinyl pivalate (**Figure 8.6**). The second catalyst is a chiral phosphine-based nucleophilic acylation catalyst that is used with an anhydride activating agent. The vinyl pivalate does not react with chiral phosphine, but the anhydride would activate the ChiroCLEC-PC. These reagents therefore, must be prevented from intermixing by constraining each to a different phase. To accomplish the phase isolation, an insoluble polystyrene- (PS-)bound mixed anhydride activator was employed (**Figure 8.6**). No chemical reaction can take place between the supported enzyme and the supported anhydride, because they are both insoluble solids. The carbonyl group distal to the polymer is shielded with a bulky mesityl group (Mes = $2,4,6\text{-}C_6H_2\text{-}Me_3$) to ensure that the phosphine catalyst reacts with the mixed anhydride to form the polymer-bound acylphosphonium intermediate and not the unsupported acylphosphonium by attack at the distal carbonyl of the anhydride. Formation of the polymer-bound acylphosphonium intermediate is crucial to the successful separation of the quasi-enantiomeric products, because the solid polystyrene bearing one enantiomer of the derivatized starting alcohol can be separated from the organic soluble material simply by filtration.

A preliminary kinetic resolution was performed with the PS-bound anhydride activator, as illustrated in **Figure 8.7**. After isolation of the polymer-bound product and liberation of the alcohol, the selectivity factor was determined to be 23. The high se-

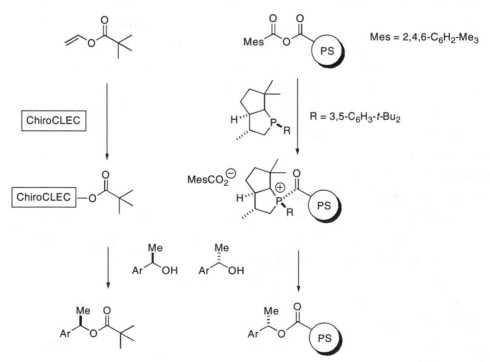

Figure 8.6. A dual catalyst PKR with phase separation of the cross-linked enzyme ChiroCLEC, the polymer-bound anhydride, and the organic solution phase.

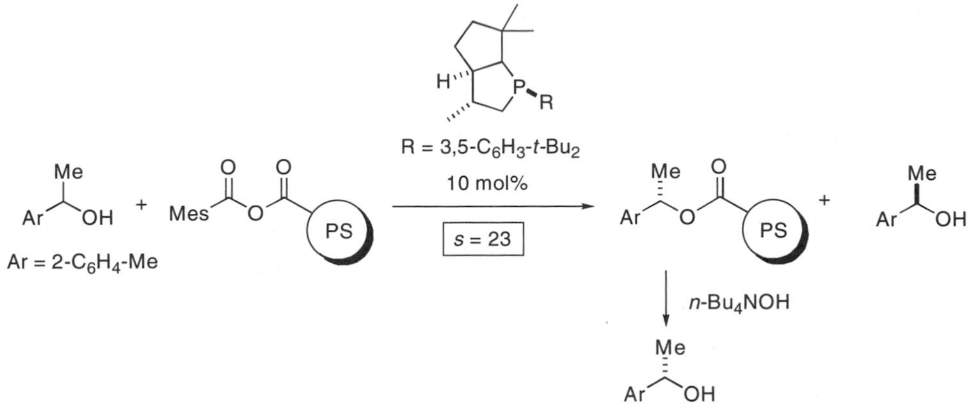

Figure 8.7. Kinetic resolution with phosphine catalyst, illustrating the modest selectivity factor (s = 23) and the ee and theoretical yield as a function of conversion.

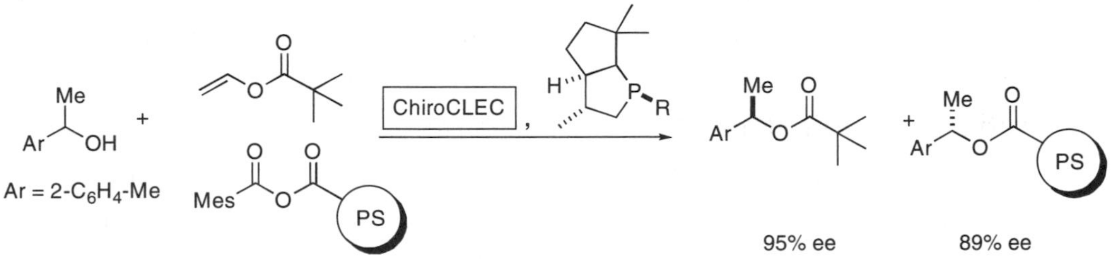

Conversion (%)	SM ee (%)	Maximum yield (%)	Product ee (%)	Maximum yield (%)
50	81	50	81	50
54	89	46	77	54
56	94	44	74	56

Figure 8.8. PKR of a benzylic alcohol with the dual catalyst system. Comparison of the PKR above with the kinetic resolution illustrated in **Figure 8.7** illustrates the advantage of the PKR in this system.

lectivity factor indicates that diffusion of substrate into the polymer phase is rapid on the timescale of the acylation reaction.

The PKR experiment consisted of three phases,[13] the organic insoluble cross-linked ChiroCLEC-PC, the polystyrene-bound anhydride, and the organic-soluble components: the chiral phosphine catalyst and vinyl pivalate (**Figure 8.6**).

A direct comparison of the kinetic resolution with the PKR highlights the potential benefits of the multicatalyst system. As mentioned above, the selectivity factor in the classical kinetic resolution outlined in **Figure 8.7**, was moderate (s = 23). As the conversion increased from 50 to 56%, the recovered starting material ee rose at the expense of the theoretical yield and the product ester ee. In contrast, in the PKR shown in **Figure 8.8**, the polymer-bound product had 89% ee and the pivalate had 95% ee after complete consumption of the starting material. The total recovery was > 80% in the PKR.[14] Under optimal conditions in the PKR, the starting alcohol would remain racemic as the reaction progressed, indicating that both enantiomers are being consumed at equal rates. In these experiments the ee of the starting alcohol was measured to be 6–13% with 40–70% conversion.

The phase isolation used in this system keeps the two catalytic reactions from interfering with each other and it greatly facilitates separation of the quasi-enantiomeric esters formed in the PKR. In principle, two catalysts that are both soluble could be used together if they exhibit orthogonal reactivity.

These examples demonstrate why chemodivergent PKRs are rare. First, it is challenging to devise a two-catalyst system with the characteristics necessary for a successful PKR. Second, with single asymmetric catalyst systems, it is rare that two *nonisomeric* products form from enantiomeric starting materials. More common, however, are asymmetric catalysts that give *isomeric* products in the PKR.

8.3 Regiodivergent PKR

In a regiodivergent PKR with a single enantioenriched catalyst, the enantiomeric substrates react via different pathways, affording regioisomeric products. Reactions of this type are fascinating, in that the course of the reaction depends highly on both the structure of the ligand and the substrate. Under very similar reaction conditions, closely related catalyst and substrate combinations may exhibit dissimilar reactivities.

The kinetic resolution of a 4-alkynal with chiral cationic rhodium bis(phosphine) catalysts is an interesting case in point. It was found that the cationic [Rh(*i*-Pr-DUPHOS)]-based catalyst exhibits a selectivity factor of 21 for the reaction in **Equation 8.1**. This kinetic resolution is an effective method to generate either the 2-cyclopentenones with high ee at lower conversion or the starting 4-alkynals at slightly higher conversion (93% ee at 56% conversion).[15]

Equation 8.1

In a study of the impact of ligand structure on the efficiency of the kinetic resolution, it was found that the cationic (Tol-BINAP)Rh⁺-based catalyst led to simultaneous formation of two different enones, each with high enantioselectivity (**Equation 8.2**). Thus, the fate of the enantiomeric 4-alkynals depends on the ligand structure, with *i*-Pr-DUPHOS- and Tol-BINAP-based catalysts promoting a kinetic resolution (**Equation 8.1**) and a PKR (**Equation 8.2**), respectively.

Equation 8.2

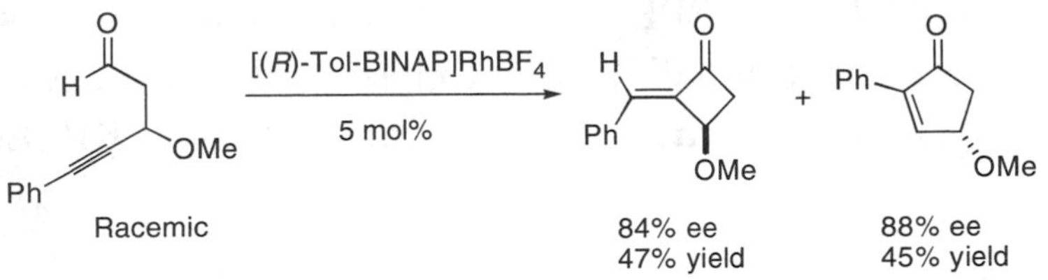

Two reaction pathways for the formation of the enones from **Equation 8.2** are proposed in **Figure 8.9**. Insertion of the Rh(I) into the aldehyde C–H bond generates the Rh(III)–acyl-hydride intermediate **A**. Addition of the rhodium hydride to the alkyne can occur with a net trans addition to form the six-membered rhodacycle **B,** or in a cis fashion, leading to the five-membered metallacycle **C**. Reductive elimination of intermediates **B** and **C** leads to the isomeric enones.

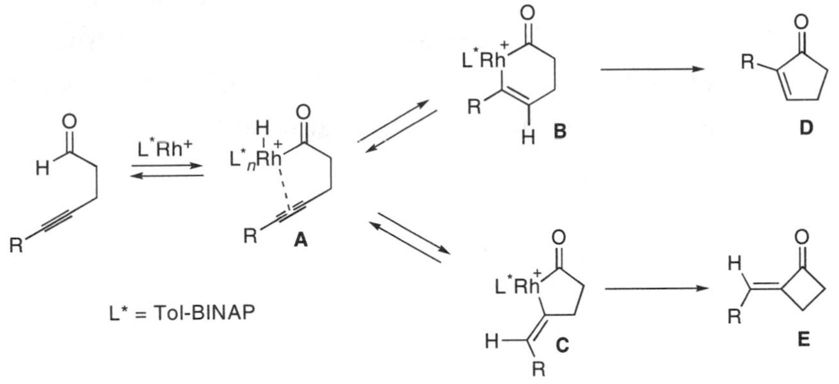

Figure 8.9. Proposed divergent pathways for the (Tol-BINAP)rhodium-catalyzed formation of cyclopentenone and cyclobutanone from 4-alkynals.

To illustrate the degree of catalyst configurational control over the fate of the substrate, enantioenriched 4-alkynal was subjected to the reaction with enantiomeric catalysts (**Equation 8.3**). When (R)-Tol-BINAP was employed, the catalyst preferentially generated the strained cyclobutanone product, whereas catalyst derived from (S)-Tol-BINAP produced the cyclopentenone almost exclusively. In this system, it is possible to generate either product by selecting the appropriate enantiomer of the catalyst.

Equation 8.3

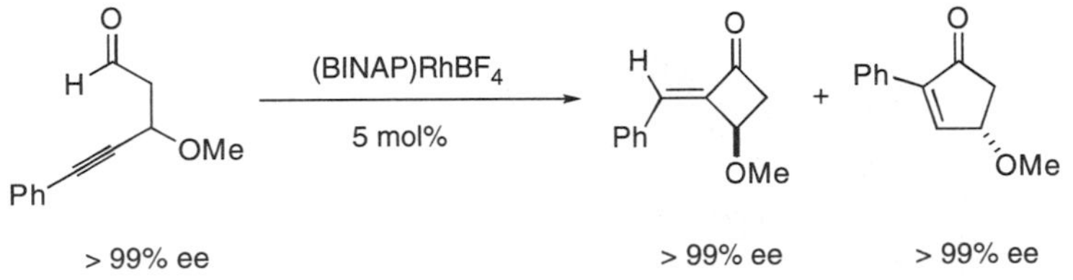

(R)-Tol-BINAP	89 :	11
(S)-Tol-BINAP	2 :	98

Reactions of this type are rare and usually require a specific combination of catalyst and substrate to succeed. In this system, for example, use of other phosphines (e.g., CHIRAPHOS, DUPHOS, BPE, and JOSIPHOS) resulted in essentially no cyclobutanone formation. Additionally, use of a closely related substrate with the same catalyst resulted in a kinetic resolution ($s = 22$) rather than a PKR (**Equation 8.4**).[15,16]

Equation 8.4

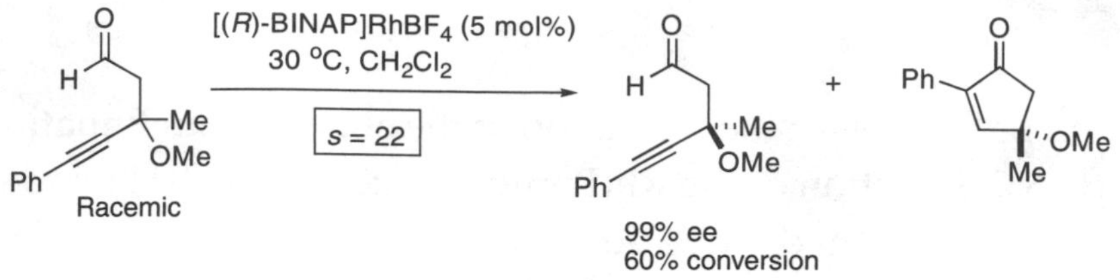

Modified cinchona alkaloids, such as (DHQD)$_2$AQN (**Figure 8.10**), have been shown to be effective for a variety of processes, including desymmetrization of cyclic anhydrides,[17] as well as the kinetic[18] and dynamic kinetic resolution[19] of N-carboxyanhydrides. During a study of the kinetic resolution of monosubstituted succinic anhydrides with these alkaloid-based catalysts (**Equation 8.5**), unexpected results were obtained that were inconsistent with a kinetic resolution.[20]

Equation 8.5

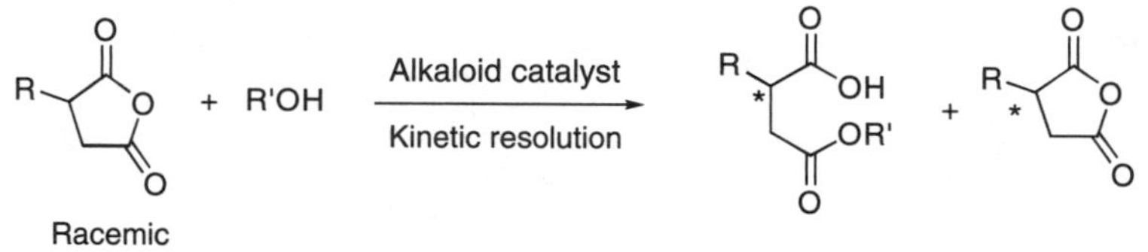

In the reaction of racemic 2-methylsuccinic anhydride (**A**) with an excess of methanol in the presence of (DHQD)$_2$AQN (10 mol%), two mono-ester products (**B** and **C**) were produced at similar rates over the course of the reaction (**Figure 8.11**). Analysis of the products after completion of the reaction indicated that the ratio of **B** to **C** was 39 to 61, respectively, and that each was produced with moderate ee (< 65%). These results were consistent with a regiodivergent PKR in which the enantiomers of the racemic substrate were converted into isomeric products in the anhydride alcoholysis. As expected, the PKR depends on the nature of the alcohol used in ring-opening. While ethanol afforded similar results to methanol, trifluoroethanol resulted in generation of the ester with as high as 91% ee at low temperature (**Figure 8.11**).

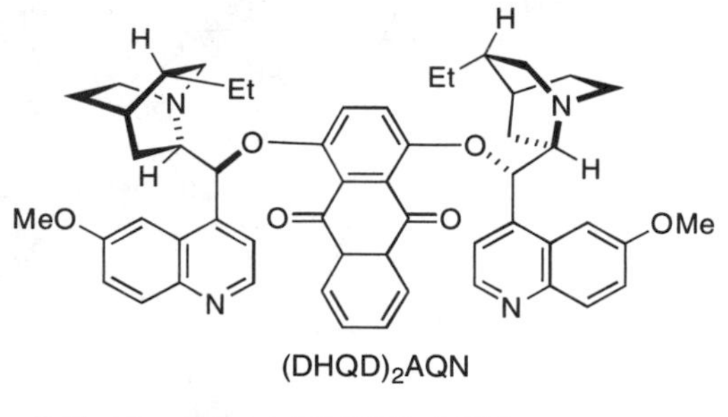

(DHQD)$_2$AQN

Figure 8.10. Structure of the alkaloid catalyst (DHQD)$_2$AQN.

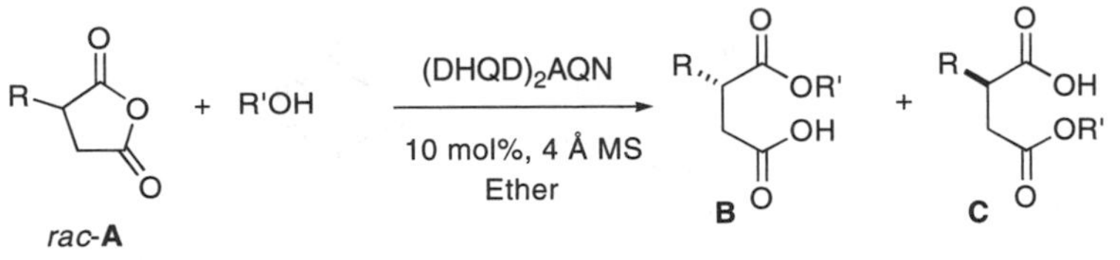

Figure 8.11. Results of the regiodivergent PKR in the opening of succinic anhydrides catalyzed by the modified alkaloid (DHQD)$_2$AQN.

R'OH	T (°C)	B/C	% ee B	% ee C
MeOH	25	39/61	74	67
EtOH	25	49/51	82	67
CF$_3$CF$_2$OH	25	49/51	85	72
CF$_3$CF$_2$OH	−25	44/56	91	80

The regiodivergent alcoholysis of the enantiomeric anhydrides (S)-**A** and (R)-**A** is illustrated in **Figure 8.12**. Reaction of anhydride (S)-**A** of high enantiopurity with tri-fluoroethanol, catalyzed by (DHQD)$_2$AQN, afforded (S)-**B** and (S)-**C** in a ratio of 92:8. Under identical conditions, (R)-**A** underwent alcoholysis to afford (R)-**B** and (R)-**C** in a ratio of 3:97. These results clearly indicate that the regioselectivity of the reaction is catalyst-controlled. The substrate scope in the PKR with 2-alkyl- and 2-arylsuccinic anhydrides was investigated and found to be broad.[20]

Based on related studies,[18] the mechanism probably involves a termolecular reaction, as shown in **Figure 8.13**. In this process, an alcohol substrate that is hydrogen bonded to the bridgehead nitrogen of the chiral alkaloid-based catalyst attacks one of the anhydride carbonyl groups. Ring-opening is rate-determining in the general-base-catalyzed mechanism.

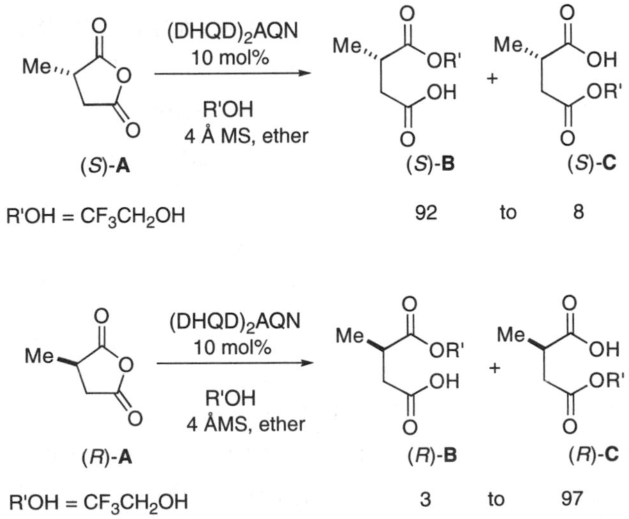

Figure 8.12. Catalyst-controlled regioselectivity of (S)- and (R)-methylsuccinic anhydride with trifluoroethanol and catalytic (DHQD)$_2$AQN.

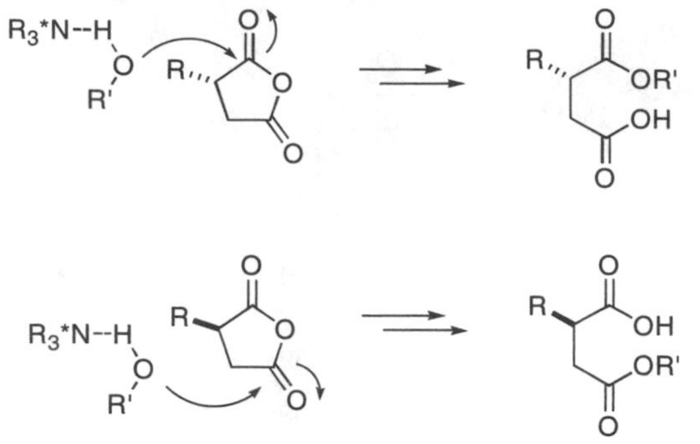

Figure 8.13. Regiodivergent pathways in the PKR of anhydrides.

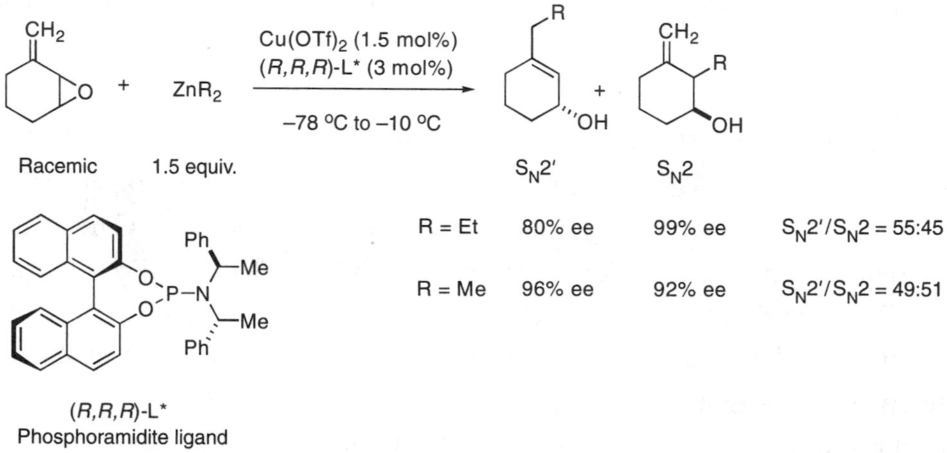

	S_N2'	S_N2	
R = Et	80% ee	99% ee	$S_N2'/S_N2 = 55{:}45$
R = Me	96% ee	92% ee	$S_N2'/S_N2 = 49{:}51$

(R,R,R)-L*
Phosphoramidite ligand

Figure 8.14. Regiodivergent asymmetric ring-opening of allylic epoxides.

A rare example of a regiodivergent reaction that entails C–C bond formation is the ring-opening of allylic epoxides catalyzed by $Cu(OTf)_2$ and the phosphoramidite ligand (R,R,R)-L* (**Figure 8.14**).[21] The chiral catalyst formed in this reaction is exceptional at discriminating between enantiomers of some allylic epoxides. As shown in **Figure 8.14**, reaction of the organozinc reagents with the (1R,2S)-allylic epoxide leads to the allylic alcohol through addition to the terminus of the double bond (S_N2'), while the enantiomeric (1S,2R)-allylic epoxide is transformed into the homoallylic alcohol via an S_N2 epoxide-opening. Analysis of the products after completion of the reaction indicated that both regioisomers were formed with high enantioselectivity, confirming that the process is a regioisomeric PKR.[21] When racemic catalyst generated from (R,R,R)- and (S,S,S)-L* was employed, the ratio of the $S_N2'{:}S_N2$ products was 92:2, indicating that it is the configuration of the catalyst that gives rise to the regiodivergent reactivity.

Insight into this process was gained by monitoring the relative concentrations and enantioselectivities of the allylic epoxide starting material and alcohol products during the reaction with dimethylzinc. The data from these experiments indicated that the reaction proceeded in two distinct phases. The fast-reacting enantiomer of the allylic epoxide underwent ring-opening at –78 °C in 15 min via an S_N2' pathway, generating the allylic alcohol with high regioselectivity. At this point in the reaction, the starting allylic epoxide was nearly enantiopure. The remaining epoxy alcohol slowly reacted, reaching completion after the reaction temperature had been increased to $-10\ °C$. This later reaction proceeded via an S_N2 pathway, leading to the homoallylic alcohol, also in a highly regioselective fashion. These results indicated that the first phase of this reaction would be better characterized as a kinetic resolution than a parallel kinetic resolution, because the regioisomeric products are not formed simultaneously. The difference in the reaction rates of the enantiomeric allylic epoxides in this transformation is substrate dependent. If the substrate enantiomers exhibit relative rates that are similar, the reaction would meet the criteria of a PKR.[22]

A proposed mechanism to rationalize the divergent reactivity in the allylic epoxide-opening chemistry is illustrated in **Figure 8.15**.[21] It is likely that the Cu(II) is reduced to Cu(I) by the dialkylzinc reagent. Initial formation of diastereomeric π-complexes (**A**) is followed by an oxidative ring-opening reaction to give Cu(III)[23] alkyl complexes **B** and **B′**. The regioselectivity of the product is proposed to depend on the relative rates of reductive elimination and isomerization of **B** and **B′**. Direct reductive elimination from **B** and **B′** affords **C** and **C′**, leading to the allylic alcohols on hydrolysis. On the other hand, isomerization of **B** and **B′** can take place with generation of **D** and **D′**. This reversible isomerization may proceed via a π-allyl intermediate or 1,3-sigmatropic shift. Reductive elimination of intermediates **D** and **D′** gives **E** and **E′**, resulting in the formation of the homoallylic alcohol products on work-up. Use of

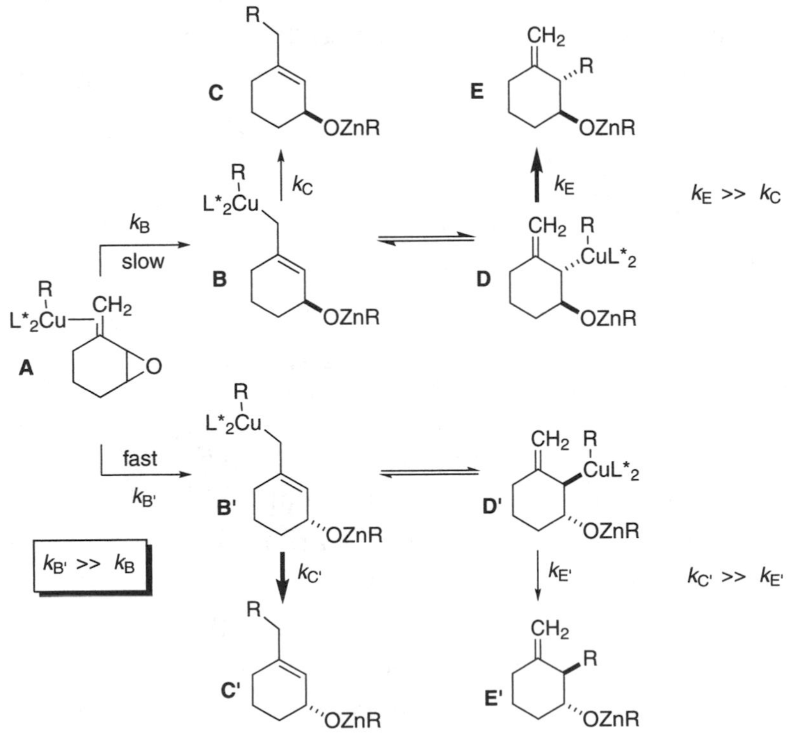

Figure 8.15. Proposed mechanism of the regiodivergent addition of dialkylzinc reagents to allylic epoxides with (R,R,R)-L*.

(R,R,R)-L* in the copper-catalyzed reaction results in rapid ring-opening of the (1R,2S)-allylic epoxide to give intermediate **B′**, which quickly undergoes reductive elimination to give **C′** without isomerization via **D′** and reductive elimination to form **E′**. After the (1R,2S)-allylic epoxide is consumed and the temperature is raised, oxidative ring-opening of the enantiomeric allylic epoxide gives intermediate **B**. Complex **B** does not readily undergo direct reductive elimination to give **C**, but isomerizes via **D**. Reductive elimination from **D** results in the formation of **E**, the S_N2 product. The influence of the L* configuration results in significantly different rates of reductive elimination from intermediates **B** and **D** on the top manifold and from **B′** and **D′** on the bottom. An alternative mechanism involves regioselective reductive elimination from a common π-allyl intermediate.[22]

In this process, the matched and mismatched catalyst–substrate combinations lead to discrimination in binding of the enantiomeric epoxides by the chiral catalysts. Furthermore, the diastereomeric interactions between the chiral phosphoramidate ligand and the bound substrate determined the regioselectivity of the product by selective acceleration of one of the reductive elimination pathways.

8.4 Stereodivergent PKR

Stereodivergent PKR reactions transform racemic substrates into diastereomeric products. This can be accomplished, for instance, by introduction of a new stereocenter into a racemic substrate, as exemplified in **Equation 8.6**.[24] Here the catalyst dictates the stereochemistry of the newly formed stereocenter, regardless of the extant stereochemistry in the substrate (see Chapter 13).

Equation 8.6

73% ee   99% ee

dr = 58 : 42

Although several examples of stereodivergent PKR reactions catalyzed by enzymes are known,[3] cases with small-molecule catalysts are uncommon. The scarcity of such stereodivergent PKRs may be due, in part, to the difficulties associated with the separation of diastereomers. This was likely the case in an attempted kinetic resolution of the racemic alcohol in **Equation 8.7** using the Sharpless–Katsuki kinetic resolution.[25,26] Allylic alcohols with trans-disubstituted double bonds are excellent substrates for the asymmetric epoxidation. Based on the transition-state model of the asymmetric epoxidation reaction (Chapter 7), the stereocenter at C-4 is directed away from the catalyst and not expected to exhibit a significant impact on the relative rates of epoxidation. In the kinetic resolution with 0.6 equivalents of TBHP, the alcohol was recovered with only 6% ee.[27] Based on the high enantioselectivity of the Sharpless–Katsuki asymmetric epoxidation, it is anticipated that each of the diastereomeric epoxy alcohols was generated with high enantioselectivity.

Equation 8.7

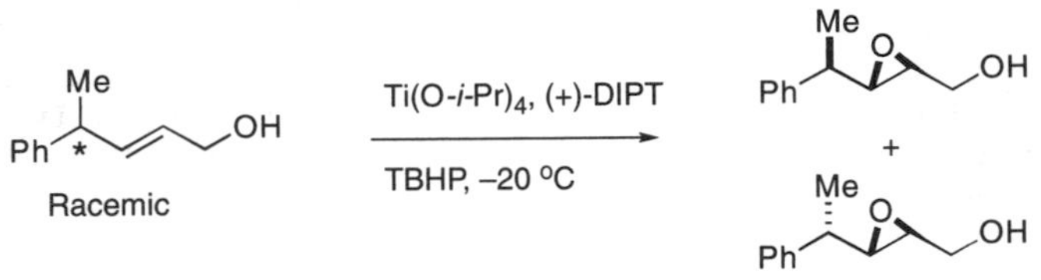

Summary

The build-up in concentration of the slow-reacting enantiomer near completion of a kinetic resolution can lead to the two enantiomers reacting at equal rates, since rate = k_{obs}[SM]. To prevent the participation of the slow-reacting enantiomer as the kinetic resolution nears completion, a parallel kinetic resolution can be performed instead. Under optimal conditions, the two enantiomers are removed at equal rates and converted to distinct, easily separable products. Under these circumstances, the selectivity factor can be significantly less than the selectivity factor in a kinetic resolution, while achieving comparable results. PKR reactions, however, involve more variables and require more reaction engineering to be successful. The greater challenges and benefits of the PKR will certainly continue to attract the attention of practitioners of asymmetric catalysis for many years.

References

(1) Vedejs, E.; Chen, X. Parallel Kinetic Resolution. *J. Am. Chem. Soc.* **1997**, *119*, 2584–2585.

(2) Eames, J. Parallel Kinetic Resolutions. *Angew. Chem., Int. Ed. Engl.* **2000**, *39*, 885–888.

(3) Dehli, J. R.; Gotor, V. Parallel Kinetic Resolution of Racemic Mixtures: A New Strategy for the Preparation of Enantiopure Compounds? *Chem. Soc. Rev.* **2002**, *31*, 365–370.

(4) Doyle, M. P.; Dyatkin, A. B.; Kalinin, A. V.; Ruppar, D. A.; Martin, S. F.; Spaller, M. R.; Liras, S. Highly Selective Enantiomer Differentiation in Intramolecular Cyclopropanation Reactions of Racemic Secondary Allylic Diazoacetates. *J. Am. Chem. Soc.* **1995**, *117*, 11021–11022.

(5) Doyle, M. P.; Dyatkin, A. B.; Autry, C. L. A New Catalytic Transformation of Diazo Esters: Hydride Abstraction in Dirhodium(II)-Catalyzed Reactions. *J. Chem. Soc., Perkin Trans. 1* **1995**, 619–621.

(6) Davies, H. M. L.; Dai, X.; Long, M. S. Combined C–H Activation/Cope Rearrangement as a Strategic Reaction in Organic Synthesis: Total Synthesis of (–)-Colombiasin A and (–)-Elisapterosin B. *J. Am. Chem. Soc* **2006**, *128*, 2485–2490.

(7) Davies, H. M. L.; Q, J. Catalytic Asymmetric Reactions for Organic Synthesis: the Combined C–H activation/Cope Rearrangement. *Proc. Natl. Acad. Sci. U.S.A.* **2004**, *101*, 5472–5475.

(8) Davies, H. M. L.; Jin, Q. Highly Diastereoselective and Enantioselective C–H Functionalization of 1,2-Dihydronaphthalenes: A Combined C–H Activation/Cope Rearrangement Followed by a Retro-Cope Rearrangement. *J. Am. Chem. Soc.* **2004**, *126*, 10862–10863.

(9) Davies, H. M. L.; Oystein, L. Intermolecular C–H Insertions of Donor/Acceptor-Substituted Rhodium Carbenoids: A Practical Solution for Catalytic Enantioselective C–H Activation. *Synthesis* **2004**, *16*, 2595–2608.

(10) Davies, H. M. L.; Beckwith, R. E. J. Catalytic Enantioselective C–H Activation by Means of Metal-Carbenoid-Induced C–H Insertion. *Chem. Rev.* **2003**, *103*, 2861–2903.

(11) Nowlan, D. T.; Gregg, T. M.; Davies, H. M. L.; Singleton, D. A. Isotope Effects and the Nature of Selectivity in Rhodium-Catalyzed Cyclopropanations. *J. Am. Chem. Soc* **2003**, *125*, 15902–15911.

(12) Khalaf, N.; Govardhan, C. P.; Lalonde, J. J.; Persichetti, R. A.; Wang, Y.-F.; Margolin, A. L. Cross-Linked Enzyme Crystals as Highly Active Catalysts in Organic Solvents. *J. Am. Chem. Soc.* **1996**, *118*, 5494–5495.

(13) Rebek, J.; Brown, D.; Zimmerman, S. Three-Phase Test for Reaction Intermediates. Nucleophilic Catalysis and Elimination Reactions. *J. Am. Chem. Soc.* **1975**, *97*, 454–455.

(14) Vedejs, E.; Rozners, E. Parallel Kinetic Resolution Under Catalytic Conditions: A Three-Phase System Allows Selective Reagent Activation Using Two Catalysts. *J. Am. Chem. Soc.* **2001**, *123*, 2428–2429.

(15) Tanaka, K.; Fu, G. C. Enantioselective Synthesis of Cyclopentenones via Rhodium-Catalyzed Kinetic Resolution and Desymmetrization of 4-Alkynals. *J. Am. Chem. Soc.* **2002**, *124*, 10296–10297.

(16) Tanaka, K.; Fu, G. C. Parallel Kinetic Resolution of 4-Alkynals Catalyzed by Rh(I)/Tol-BINAP: Synthesis of Enantioenriched Cyclobutanones and Cyclopentenones. *J. Am. Chem. Soc.* **2003**, *125*, 8078–8079.

(17) Chen, Y.; Tian, S.-K.; Deng, L. A Highly Enantioselective Catalytic Desymmetrization of Cyclic Anhydrides with Modified Cinchona Alkaloids. *J. Am. Chem. Soc.* **2000**, *122*, 9542–9543.

(18) Hang, J.; Tian, S.-K.; Tang, L.; Deng, L. Asymmetric Synthesis of β-Amino Acids via Cinchona Alkaloid-Catalyzed Kinetic Resolution of Urethane-Protected α-Amino Acid N-Carboxyanhydrides. *J. Am. Chem. Soc.* **2001**, *123*, 12696–12697.

(19) Hang, J.; Li, H.; Deng, L. Development of a Rapid, Room-Temperature Dynamic Kinetic Resolution for Efficient Asymmetric Synthesis of α-Aryl Amino Acids. *Org. Lett.* **2002**, *4*, 3321–3324.

(20) Chen, Y.; Deng, L. Parallel Kinetic Resolutions of Monosubstituted Succinic Anhydrides Catalyzed by a Modified Cinchona Alkaloid. *J. Am. Chem. Soc.* **2001**, *123*, 11302–11303.

(21) Bertozzi, F.; Crotti, P.; Macchia, F.; Pineschi, M.; Feringa, B. L. Highly Enantioselective Regiodivergent and Catalytic Parallel Kinetic Resolution. *Angew. Chem., Int. Ed. Engl.* **2001**, *40*, 930–932.

(22) Pineschi, M.; Del Moro, F.; Crotti, P.; Di Bussolo, V.; Macchia, F. Catalytic Regiodivergent Kinetic Resolution of Allylic Epoxides: A New Entry to Allylic and Homoallylic Alcohols with High Optical Purity. *J. Org. Chem.* **2004**, *69*, 2099–2105.

(23) Sofia, A.; Karlström, E.; Bäckvall, J.-E. Experimental Evidence Supporting a Cu(III) Intermediate in Cross-Coupling Reactions of Allylic Esters with Diallylcuprate Species. *Chem. Eur. J.* **2001**, *7*, 1981–1989.

(24) Dehli, J. R.; Gotor, V. Preparation of Enantiopure Ketones and Alcohols Containing a Quaternary Stereocenter Through Parallel Kinetic Resolution of α-Keto Nitriles. *J. Org. Chem.* **2002**, *67*, 1716–1718.

(25) Martín, V. S.; Woodard, S. S.; Katsuki, T.; Yamada, Y.; Ikeda, M.; Sharpless, K. B. Kinetic Resolution of Racemic Allylic Alcohols by Enantioselective Epoxidation. A Route to Substances of Absolute Enantiomeric Purity? *J. Am. Chem. Soc.* **1981**, *103*, 6237–6240.

(26) Gao, Y.; Klunder, J. M.; Hanson, R. M.; Masamune, H.; Ko, S. Y.; Sharpless, K. B. Catalytic Asymmetric Epoxidation and Kinetic Resolution: Modified Procedures Including in situ Derivatization. *J. Am. Chem. Soc.* **1987**, *109*, 5765–5780.

(27) Sharpless, K. B.; Behrens, C. H.; Katsuki, T.; Lee, A. W. M.; Martin, V. S.; Takatani, M.; Viti, S. M.; Walker, F. J.; Woodard, S. S. Stereo and Regioselective Openings of Chiral 2,3-Epoxy Alcohols. Versatile Routes to Optically Pure Natural Products and Drugs. Unusual Kinetic Resolutions. *Pure Appl. Chem.* **1983**, *55*, 589–604.

Dynamic Kinetic Resolution and Dynamic Kinetic Asymmetric Transformations

9.1 Dynamic Kinetic Resolution

The maximum theoretical yield of a kinetic resolution (50%, see Chapter 7) has inspired chemists to explore methods for converting both enantiomers of a racemic mixture into a single stereoisomeric product. Although this can be accomplished in a two-step reaction sequence as outlined in Section 7.4 in the rearrangement of enol–ester epoxides,[1] a more general and practical method would be to realize these processes concurrently.[2-7] One method to achieve such results is a dynamic kinetic resolution (DKR), which is compared to a kinetic resolution in **Figure 9.1** (see Chapter 1 for energy diagrams).

The DKR couples a kinetic resolution with a rapid in situ racemization of the chiral substrate through an achiral intermediate (I, **Figure 9.1**) or transition state. As the fast-reacting enantiomer of the substrate is converted to product in an enantioselective resolution, the equilibrium is constantly adjusted through a dynamic racemization process. The racemization is thermodynamically favorable because an increase in entropy is realized upon mixing the two enantiomers.[5] Like the kinetic resolution, the rate of the fast-reacting enantiomer (k_{fast}) must be significantly faster than the slow-reacting enantiomer (k_{slow}). Very good results are obtained when $k_{fast}/k_{slow} > 20$. A general condition for a successful dynamic kinetic resolution is that the racemization process k_{rac} should proceed at an equal or greater rate than the catalytic asymmetric processes (i.e. $k_{rac} \geq k_{fast}$). If k_{fast}/k_{slow} is very high, it is possible for $k_{rac} < k_{fast}$ and still obtain product of high ee, however. Furthermore, if $k_{rac} \geq k_{fast}$, the DKR, will afford higher enantiomeric excesses of the product than the classical kinetic resolution, because the continuous racemization will then prevent the build-up of one enantiomer of the starting material that plagues the kinetic resolution. The mathematical treatment of DKR processes is complicated, but has been reported.[8-10] In a DKR the racemization process does not involve the chiral catalyst. As outlined below, when the interconversion of the substrate enantiomers takes place on the chiral catalyst, the process is called a dynamic kinetic asymmetric transformation (DyKAT).

To devise a successful DKR, an effective kinetic resolution must be identified, a method for the racemization must be found, and the racemization and resolution must be compatible. Furthermore, the product of the dynamic kinetic resolution must

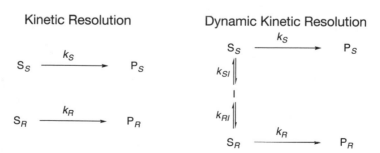

Figure 9.1. Comparison of a classical kinetic resolution with a dynamic kinetic resolution. In the dynamic kinetic resolution I is an achiral intermediate or transition state.

be stable to racemization under the reaction conditions. Perhaps the most challenging aspect of a DKR is optimizing conditions to promote facile racemization of the substrate that are chemically and kinetically well matched with the resolution step. As will be discussed, racemization can occur by several mechanisms, including base- or acid-catalyzed racemizations, thermal processes, addition/elimination sequences, oxidation–reduction couples, and racemization through a degenerative nucleophilic substitution reaction.[4] The DKR process will be contrasted with a related method called a dynamic kinetic asymmetric transformation (DyKAT). At first glance these processes appear to be the same, but there are some important differences.[11]

9.1.1 Racemization by Base Catalysis

Substrates that possess a stereogenic center with an acidic hydrogen can undergo base-catalyzed racemization and have proven ideal for many DKR processes.

9.1.1.a *Dynamic Kinetic Resolution of 1,3-Dicarbonyl Compounds Through Asymmetric Hydrogenation*

A common entry into the DKR manifold involves racemization of stereocenters alpha to carbonyl groups.[3,12] The achiral intermediate formed through base-catalyzed enolization is in equilibrium with the keto counterpart, allowing for facile racemization of the substrate. An ideal class of substrates for the DKR is α-substituted β-dicarbonyl compounds, because of the increased acidity of these compounds and the mild conditions under which they undergo racemization. When this rapid racemization is coupled with a hydrogenation catalyst that reduces one enantiomer much faster than the other, the resulting DKR generates a variety of useful α-substituted β-hydroxy carbonyl compounds with high enantioselectivities and diastereoselectivities (**Figure 9.2**). Although reduction of α-substituted β-keto esters can give rise to four stereoisomers, conditions can often be found that give a single stereoisomer with high selectivity.[3]

As outlined in Chapter 4, (BINAP)Ru-based catalysts reduce prochiral β-keto esters with excellent enantioselectivities.[12] These catalysts can also be used in the enantioselective and diastereoselective hydrogenation of α-substituted β-keto esters.

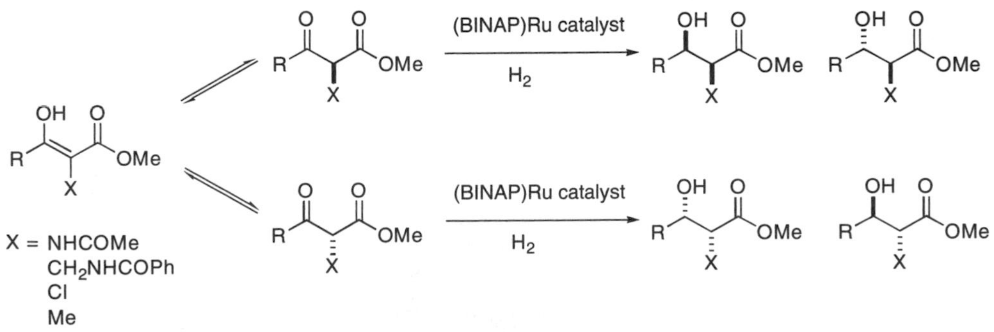

Figure 9.2. DKR in the asymmetric reduction of keto esters, illustrating the substrate racemization and the four possible diastereomeric products.

In reactions with these substrates, the configuration of the catalyst determines which face of the ketone is reduced, while the stereochemistry at the β-carbon is based on the substrate structure (**Figure 9.2**). Thus, use of (R)-BINAP gives product with the (R)-configuration at the β-carbon, generally with excellent selectivity. The reduction is chemoselective for the carbonyl and reduction of the enol is not observed.

The DKR of a variety of acyclic β-dicarbonyl derivatives has been successfully accomplished with excellent control over enantioselectivity and diastereoselectivity, as exemplified in **Figure 9.3**. The *syn* β-hydroxy ester formed in this process is used to prepare azetidinone on an industrial scale (120 ton/year),[3] which is a key intermediate in the synthesis of carbapenem antibiotics.

The proposed model for the stereoinduction in the DKR is similar to that introduced in Chapter 4. In this example, the (S)-enantiomer of the substrate reacts faster with the [(R)-BINAP]Ru-based catalyst to minimize steric interactions with the pseudoequatorial phenyl group nearest the η^1-coordinated ester (see Chapter 4 for a more in-depth discussion), as shown in **Figure 9.4**.[3]

An interesting example of how the diastereoselectivity of the DKR process can be reversed involves the synthesis of threonine derivatives.[13,14] In the (BINAP)Ru(II)-catalyzed reduction of the β-keto ester with α-NHAc or NHCOPh groups, the reduction gives the *syn* diastereomer with excellent enantioselectivity and diastereoselectivity. The impressive level of *syn* selectivity in this DKR is not consistent with an amide-directed hydrogenation. Instead, it is believed that the amide group participates in an intramolecular hydrogen bond, as illustrated in the transition-state model in **Figure 9.5**.[13]

In the DKR of the α-phthalimido β-keto ester using the BINAP analog C3-Tunephos,[15] the reaction favors the *anti* diastereomer with high enantioselectivity and diastereoselectivity (**Figure 9.6**).[14] Unlike the amide in **Figure 9.5**, which is proposed to stabilize the transition state leading to the *syn* diastereomer via a hydrogen bond, no such interaction is possible with the α-phthalimido substrate. Based on the

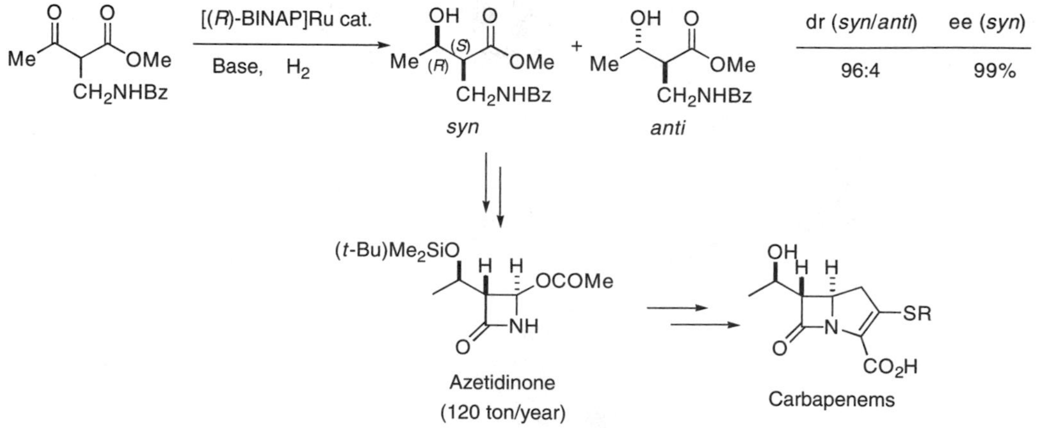

Figure 9.3. Synthesis of β-hydroxy ester via DKR with a (BINAP)-Ru-based catalyst. The *syn* product has been prepared on an industrial scale using this method.

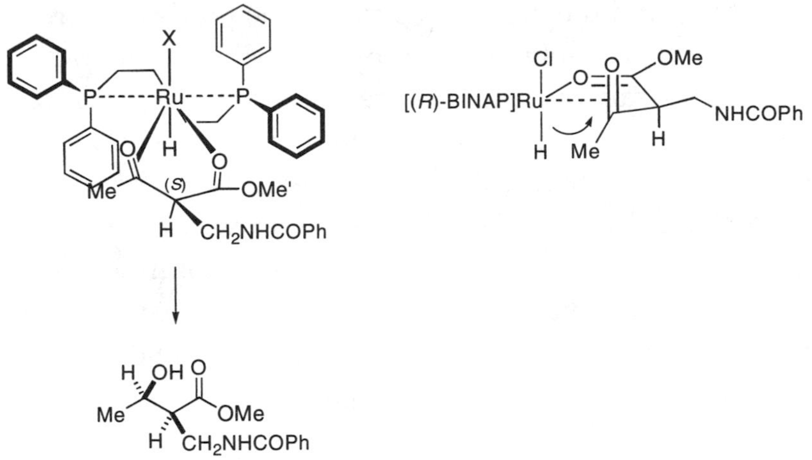

Figure 9.4. Face-on (BINAP backbone shown schematically) and side view of the stereochemical model for the [(R)-BINAP]Ru-based DKR of α-substituted β-keto esters.

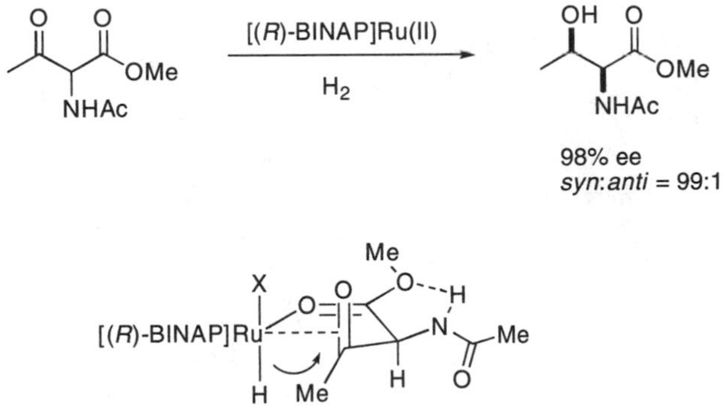

98% ee
syn:anti = 99:1

Figure 9.5. DKR of a β-keto ester in the synthesis of a threonine derivative. A hydrogen bond between the amide and ester is thought to control the substrate binding mode, giving rise to the high *syn:anti* selectivity observed.

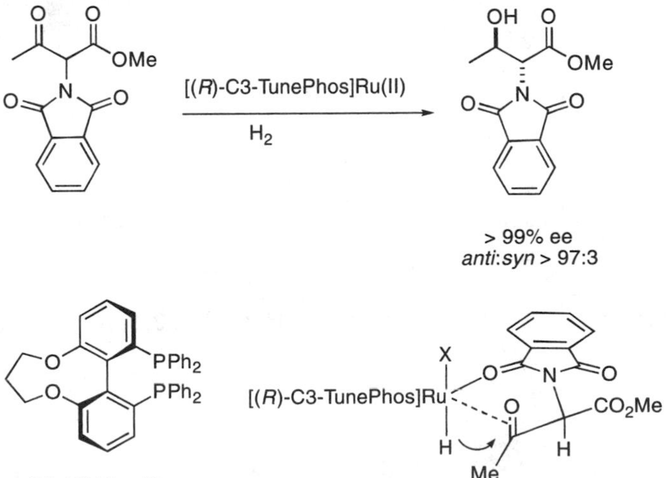

> 99% ee
anti:syn > 97:3

(R)-C3-TunePhos

Figure 9.6. DKR of the α-phthalimido-protected β-keto ester in the synthesis of a threonine derivative using (C3-Tunephos)Ru-based catalyst. The imide, rather than the keto group, is proposed to coordinate to ruthenium, resulting in a reversal of configuration at the α-carbon.

transition-state models and the greater basicity of the imide carbonyl group relative to the ester, this reaction probably proceeds by way of chelation to the imide, as illustrated in **Figure 9.6**. Thus, all four diastereomers of threonine can be obtained with excellent enantioselectivity and diastereoselectivity by manipulating the *N*-protecting group and catalyst configuration.

When the [(*R*)-BINAP]Ru-based catalyst is employed in the DKR of the keto lactone in **Equation 9.1**, the major product is the (*R*)-alcohol in 94% ee with a *syn:anti* ratio of 98:2. In contrast, racemic keto esters in **Equation 9.2** are reduced with *syn:anti* ratios ranging from 1:99 to 3:97 with enantioselectivities of 90–95% ee. The diastereoselectivity in these reactions can be explained by a similar stereochemical model (**Figure 9.7**).[3]

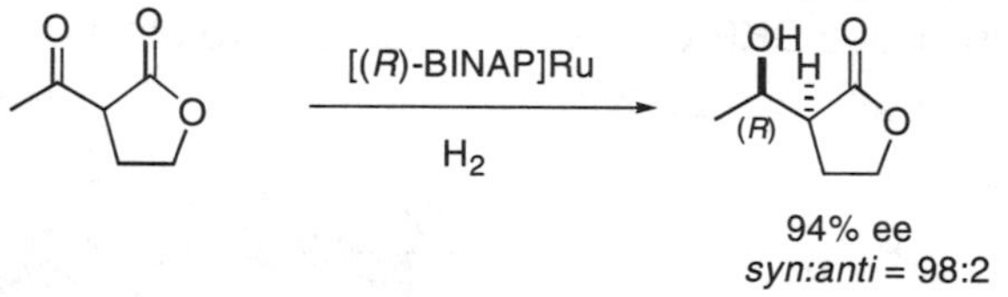

Equation 9.1

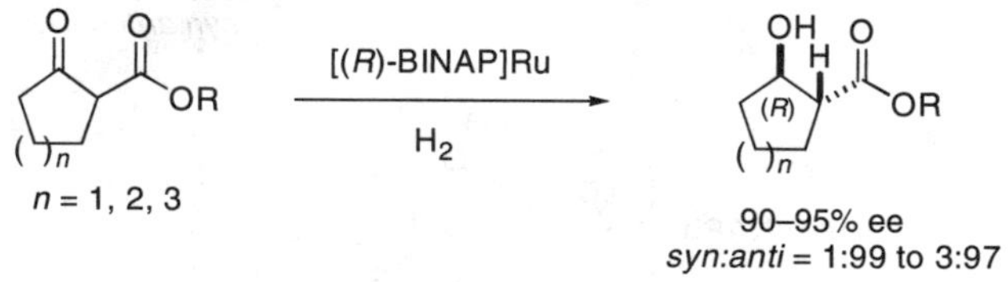

94% ee
syn:anti = 98:2

Equation 9.2

n = 1, 2, 3

90–95% ee
syn:anti = 1:99 to 3:97

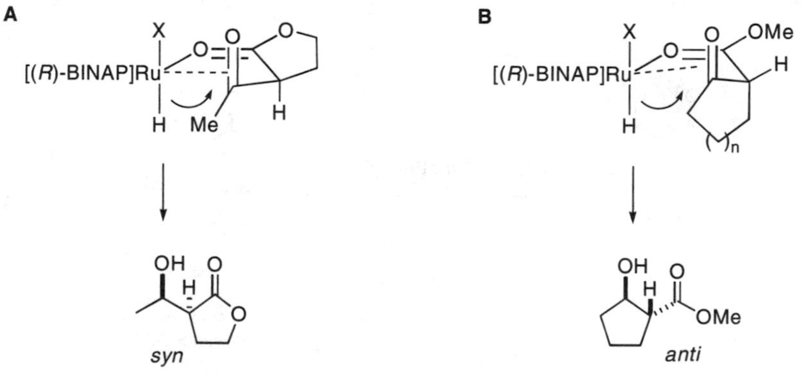

Figure 9.7. Models for the stereochemical outcome in **Equation 9.1 (A)** and **Equation 9.2 (B)**.

DKR via asymmetric hydrogenation is not limited to the highly acidic β-dicarbonyl compounds. In the hydrogenation of racemic 2-isopropylcyclohexanone with [(*S*)-BINAP]RuCl$_2$[(*R,R*)-DPEN] in the presence of base, the alcohol products were obtained with a 99.8:0.2 diastereomeric ratio. The cis product had 93% ee (**Equation 9.3**).[16]

Equation 9.3

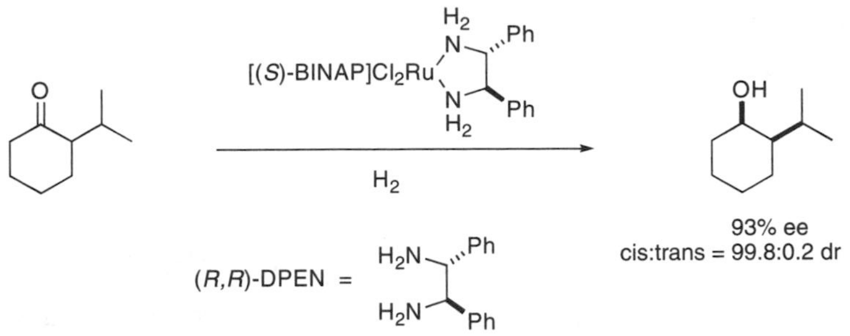

Further analysis of the DKR parameters[9] in **Equation 9.3** indicated that the (R)-ketone was hydrogenated 36 times faster than its enantiomer and that the slow-reacting (S)-ketone undergoes racemization 47 times faster than it hydrogenates (**Figure 9.8**).[16] Based on these values, k_{rac} is 1.3-fold faster than the fast reaction (k_R), fulfilling the optimal requirement for a DKR (see Section 9.1).

A similar method has been applied to the synthesis of (1R,2S)-2-methoxycyclohexanol (**Figure 9.9**). [17] The product of this reaction is a key intermediate in the synthesis of sanfetrinem, an antibiotic with a potent and broad spectrum of antibacterial activity.[18] A diastereomeric catalyst was required to yield the cis diastereomer in **Figure 9.9**, compared to the DKR of the structurally related 2-isopropylcyclohexanone (**Equation 9.3**).

Although use of α-substituted β-keto esters in the DKR is the most common, this method has also been used successfully with other substrates, including α-substituted β-keto phosphonates.[19] The application of (BINAP)Ru-based catalysts to the synthe-

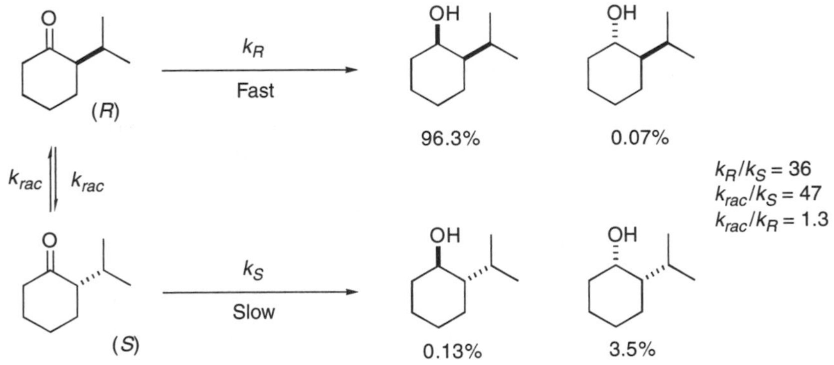

Figure 9.8. DKR of 2-isopropylcyclohexanone (**Equation 9.3**).

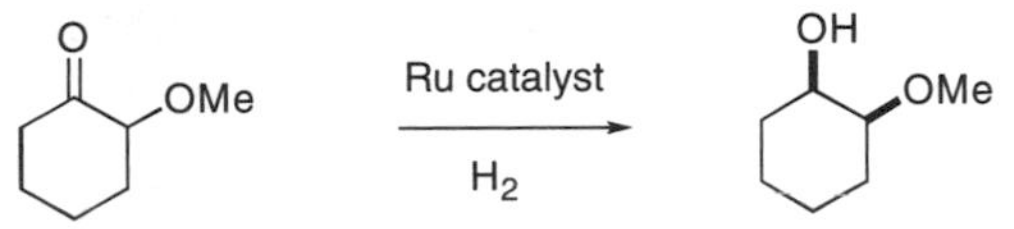

Catalyst	T (°C)	cis:trans	ee (%)
[(S)-BINAP]RuCl$_2$[(S,S)-DPEN]	50	96.5:2.5	87
[(S)-Xyl-BINAP]RuCl$_2$[(S,S)-DPEN]	50	98:2	96
[(S)-Xyl-BINAP]RuCl$_2$[(S,S)-DPEN]	5	> 99:1	99

Figure 9.9. DKR of 2-methoxycyclohexanone.

sis of precursors to several bioactive compounds via DKR has made this method one of the most useful for the generation of chiral building blocks with multiple stereocenters.

9.1.1.b DKR of N-Carboxyanhydrides

The synthesis of optically active non-natural α-amino acid derivatives has attracted great attention, because of their utility in peptide and protein synthesis. Initial studies in the kinetic resolution of N-carboxyanhydrides, which are easily prepared from racemic amino acids, indicated that modified cinchona alkaloids were efficient catalysts for this process (**Figure 9.10**).[20,21]

Kinetic studies of this kinetic resolution with closely related substrates showed that the reaction followed a first-order dependence on the substrate, alcohol, and amine catalyst.[20] Under pseudo-first-order reaction conditions, a kinetic isotope effect (k_{ROH}/k_{ROD}) of 1.3 was measured. These observations are most likely due to a general-base catalysis mechanism (**Equation 9.4**). Alcoholysis is the turnover-limiting step that is mediated by the amine catalyst. An alternative mechanism involving nucleophilic attack on the anhydride by the amine catalyst could also be imagined. If the rate-limiting step were formation of the acylammonium salt, the reaction would not depend on the alcohol concentration. On the other hand, if the turnover-limiting step were reaction of the acylammonium salt with alcohol, it would be expected that more nucleophilic alcohols would react faster. It was found, however, that trifluoroethanol reacted faster than ethanol, a result that is inconsistent with the nucleophilic catalyst mechanism.[22–24]

Termolecular ring-opening   **Equation 9.4**

In principle, the basic cinchona alkaloid catalysts could play the dual role of reversibly deprotonating the substrate, resulting in racemization as well as catalyzing the alcoholysis (**Equation 9.5**). Under the low-temperature conditions of the kinetic resolution (**Figure 9.10**), however, the rate of racemization of the substrate is insignificant compared to the rate of alcoholysis of the N-carboxyanhydrides ($k_{rac} \ll k_{fast}$, k_{slow}).

Bimolecular racemization   **Equation 9.5**

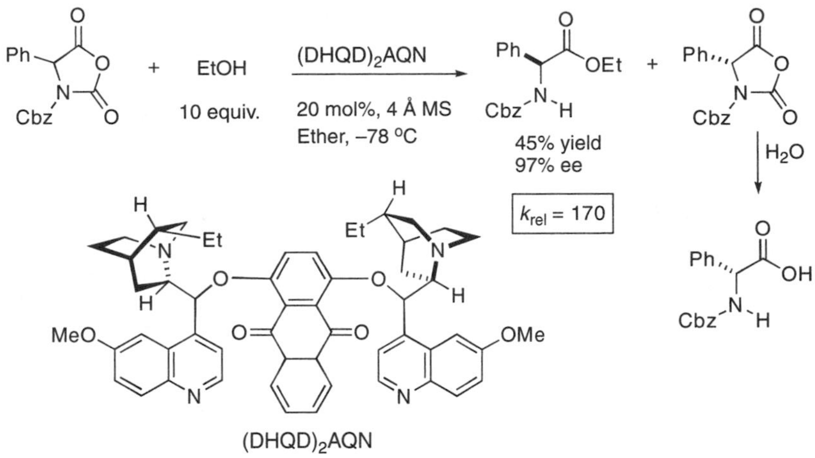

Figure 9.10. Kinetic resolution of *N*-carboxyanhydrides.

Table 9.1. Variation of conditions in the alcoholysis of *N*-carboxyanhydrides results in transition from a kinetic resolution at low temperature to an efficient DKR at higher temperatures.

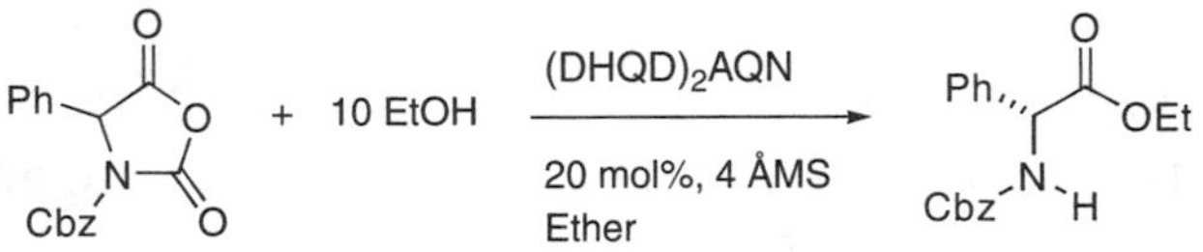

	Temperature (°C)	Time (h)	Conversion (%)	ee (%) of product
1	−78	2.0	48	94
2	−78	336	100	12
3	−40	22	100	22
4	−20	3.0	100	32
5	0	1.0	100	44
6	23	0.3	100	56
7	34	0.2	100	58
8	34[a]	2.0	100	86

[a]Slow addition of only 1.2 equiv. EtOH.

The difference in rate laws for the termolecular alcoholysis and the bimolecular racemization enabled the researchers to modify the reaction conditions such that a dynamic kinetic resolution could be performed.[21] The entropy of activation for the bimolecular racemization is generally less negative than the termolecular alcoholysis. Thus, based on the different temperature behaviors of the alcoholysis and the racemization, it was rationalized that raising the temperature would result in a greater acceleration of the racemization with respect to the alcoholysis. As shown in **Table 9.1**, increasing the temperature from −78 °C to 34 °C allowed transition from the kinetic resolution regime (entry 1) to the DKR with the enantiomeric excess of the alcoholysis product reaching 58% (entries 2–7). Racemization was further favored over alcoholysis by performing a slow addition of 1.2 equivalents of ethanol at 34 °C, resulting in product of 86% ee (entry 8).

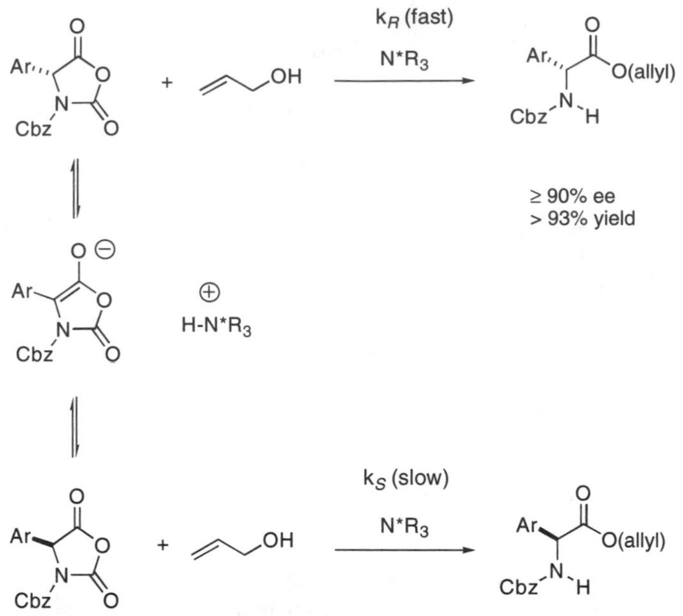

Figure 9.11. DKR of *N*-carboxyanhydrides with (DHQD)$_2$AQN and allyl alcohol.

Changing the alcohol to allyl alcohol also resulted in an increase in the product ee's to ≥ 90% and yields ≥ 93% with a wide variety of aromatic and heteroaromatic substrates (**Figure 9.11**).[21]

9.1.2 Dual-Catalyst Oxidation–Reduction Processes for DKR of Secondary Alcohols

Many examples of the kinetic resolution of alcohols have been reported, with enzymatic catalysts among the most efficient at this process. To effect an enzymatic DKR of racemic secondary alcohols would be challenging, because it is difficult to perform the racemization step with biocatalysts.[25,26] However, certain transition-metal complexes can catalyze the racemization of secondary alcohols under mild conditions. Combining the ability of a transition-metal catalyst to racemize secondary alcohols with the efficiency of enzyme-catalyzed acylation reactions, one of the most impressive classes of a dynamic kinetic resolution has been developed.

The transition-metal catalyst chosen for the racemization of the alcohol substrate is a dimeric ruthenium complex with a bridging hydride (**Equation 9.6**).[27,28] Unlike other ruthenium complexes that will promote hydrogen transfer,[29,30] this ruthenium catalyst does not require an external base to racemize the secondary alcohol, thereby avoiding interference with enzyme activity and problems due to base-catalyzed transesterification. As illustrated in **Equation 9.6**, the dimeric precatalyst dissociates into a saturated ruthenium–hydride complex and an unsaturated ruthenium–dicarbonyl complex depicted as two resonance forms. **Equation 9.7** shows a partial mechanism with key intermediates for the racemization of a secondary alcohol. Concerted hydrogen transfer mediated by the cyclopentadienyl hydroxyl group is supported by mechanistic studies and computations.[31-34] The ruthenium-catalyzed racemization of the alcohol was found to be slow at ambient temperature and required higher temperature to proceed at reasonable rates.

Equation 9.6

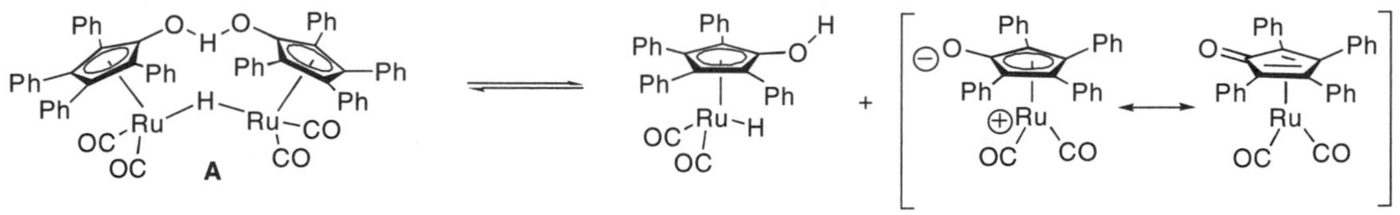

Equation 9.7

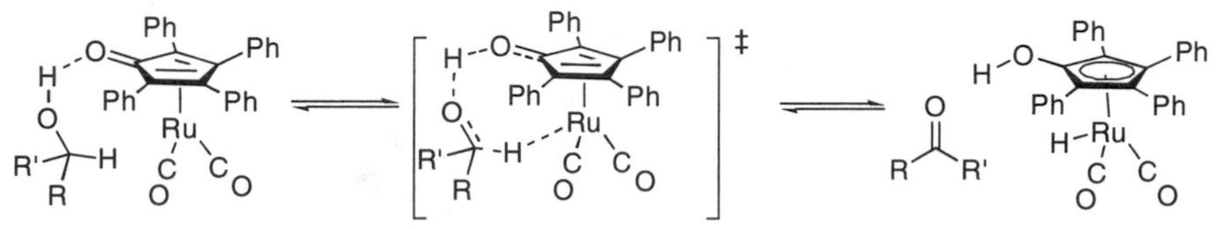

Once the parameters of the racemization were determined, the ruthenium-catalyzed racemization was coupled with the biocatalytic kinetic resolution. An enzyme that exhibited high thermostability was required to be compatible with the conditions required for high TOF of the racemization catalyst. The enzyme employed was a lipase from *Candida Antarctica* supported on acrylic resin (Novozym 435®). The acyl donor 4-chlorophenyl acetate was chosen, because it was found to be compatible with the enzyme and because the byproduct of the transesterification, 4-chlorophenol, did not interfere with the ruthenium catalyst. Comparison of the TOF of the two catalytic cycles indicated that racemization was significantly slower than acylation by the enzyme. In this system, however, k_{fast}/k_{slow} in the enzymatic resolution step is very high, on the order of 200. Thus, acylated alcohol products with excellent enantioselectivity and high yields can be obtained (**Table 9.2**).

Similar methods have also been successfully applied to diols (**Equation 9.8**)[35] and α-hydroxy esters (**Equation 9.9**).[36,37]

Equation 9.8

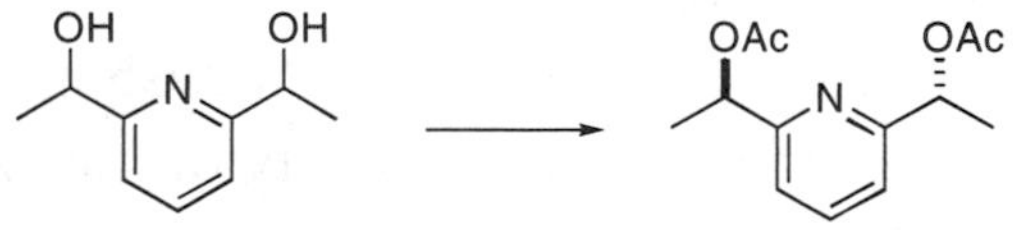

78% yield, 99% ee

Equation 9.9

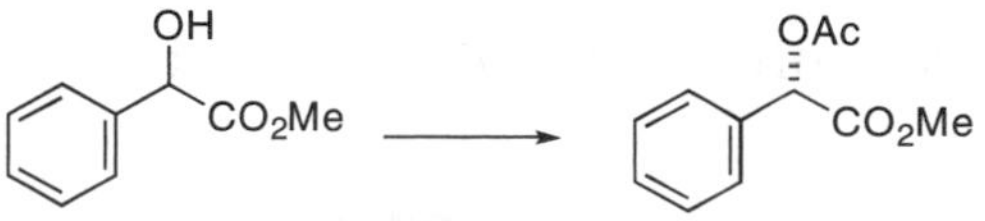

74% yield, 94% ee

The ruthenium- and enzyme-catalyzed DKR has also been applied to β-hydroxy acid derivatives in a one-pot aldol/DKR process (Figure 9.12).[37] Initial formation of the

Table 9.2. Dual-catalyst DKR of secondary alcohols with the ruthenium dimer (from **Equation 9.6**) and the biocatalyst Novozym 435.

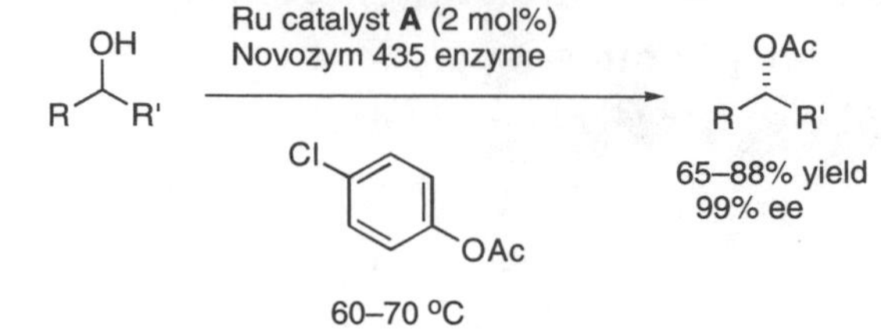

Substrate	Product	Yield (%)	ee (%)
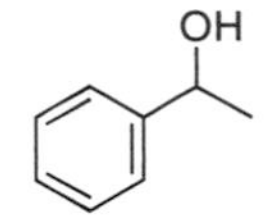 OH (1-phenylethanol)	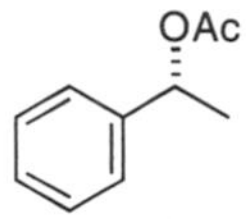 OAc	80	99
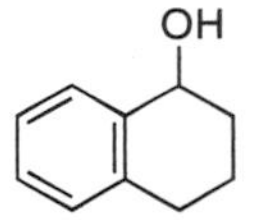 OH (1-tetralol)	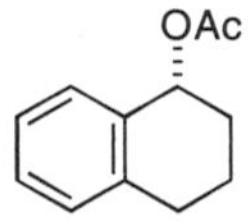 OAc	65	99
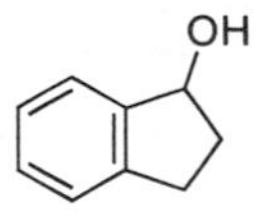 OH (1-indanol)	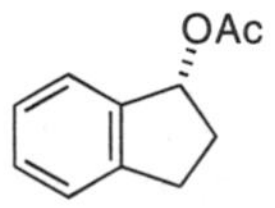 OAc	77	99
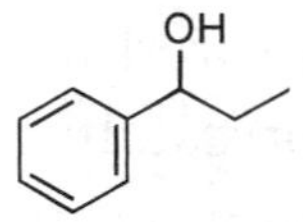 OH (1-phenylpropanol)	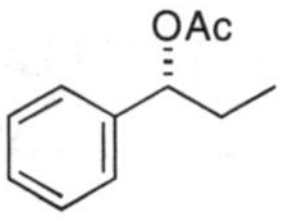 OAc	80	98
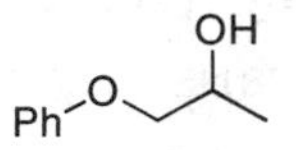 OH (PhO-CH2-CH(OH)-CH3)	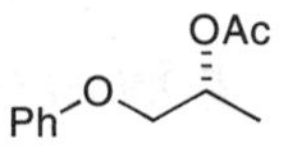 OAc	88	99
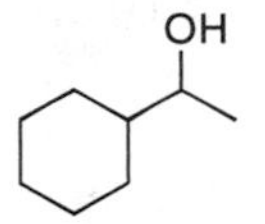 OH (1-cyclohexylethanol)	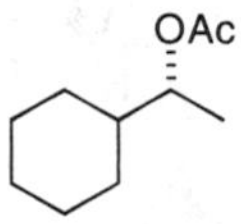 OAc	79	99

enolate with LDA and coupling with an aldehyde resulted in formation of the racemic aldol adduct. Removal of the THF and addition of *tert*-butyl methyl ether (TBME), enzyme, acyl donor, and the dimeric ruthenium catalyst from **Equation 9.6** allowed isolation of the aldol products with very high enantioselectivities (**Figure 9.12**). For more on tandem reactions, see Chapter 14.

A significant drawback to employing enzyme catalysts is that only one enantiomer of the enzyme is naturally available. In the majority of the examples of dual-catalyst metal–enzyme systems studied, a lipase catalyst has been employed. As a result, only products with the (R)-configuration were accessible.[38] More recently, however, (S)-selective enzymes have been introduced, allowing direct access to both enantiomers of the resolved alcohol.[39,40] In one case,[39] the limited thermal stability of the enzyme required that a metal catalyst with increased activity be developed for the alcohol racemization. Therefore, the aminocyclopentadienylruthenium catalyst (**Figure 9.13**), activated by addition of potassium *tert*-butoxide, was employed. Excellent enantioselectivities and yields were observed in this system at lower temperatures.

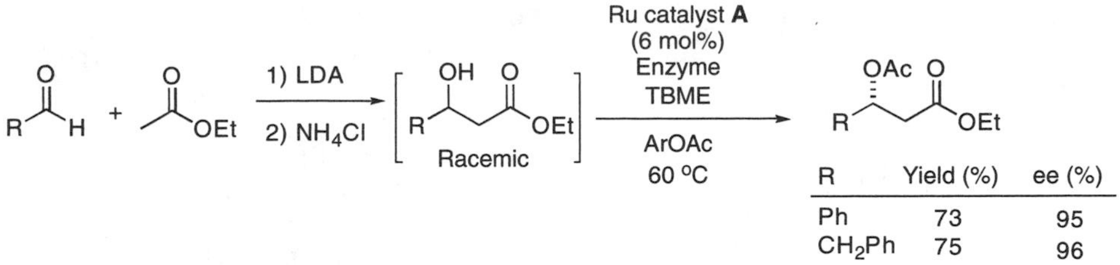

R	Yield (%)	ee (%)
Ph	73	95
CH_2Ph	75	96

Figure 9.12. Tandem aldol/DKR process for the enantioselective synthesis of β-hydroxy ester derivatives.

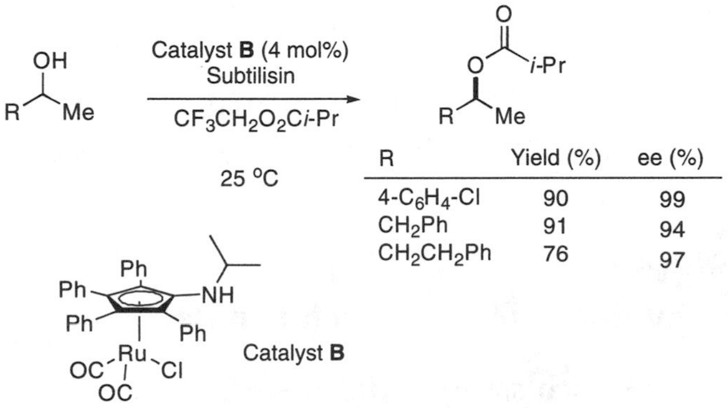

R	Yield (%)	ee (%)
4-C_6H_4-Cl	90	99
CH_2Ph	91	94
CH_2CH_2Ph	76	97

Figure 9.13. DKR with subtilisin and an aminocyclopentadienylruthenium catalyst.

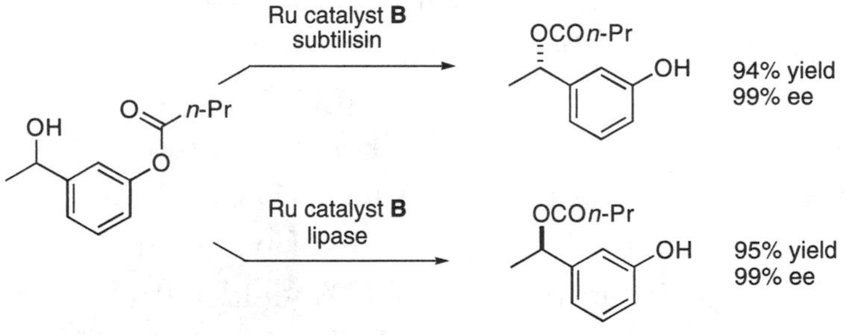

Figure 9.14. DKR catalyzed with aminocyclopentadienylruthenium catalyst (from **Figure 9.13**) and subtilisin or lipase affords either enantiomer of the product in this isomerization.

As illustrated in the DKR in **Figure 9.14**, with this more active catalyst, racemic alcohol can be transformed into either enantiomer of the acylated product with excellent enantioselectivities and yields.

A great deal of progress has been made with metal–enzyme dual catalysts in the DKR of alcohols and the method has been extended to diols.[41,42] An analogous process beginning with imines leads to resolved amines.[43–45] Other functional groups also participate in this class of reactions but have been less explored. An example is the DKR of allylic acetates using a palladium(II) source for the racemization catalyst and a lipase as the biocatalyst.[46] In the presence of $Cl_2Pd(NCMe)_2$, the allylic acetate undergoes racemization through an intramolecular 1,3-acetate migration (**Figure 9.15**).[47] In this system, the lipase hydrolyzes the one enantiomer of the allylic acetate,

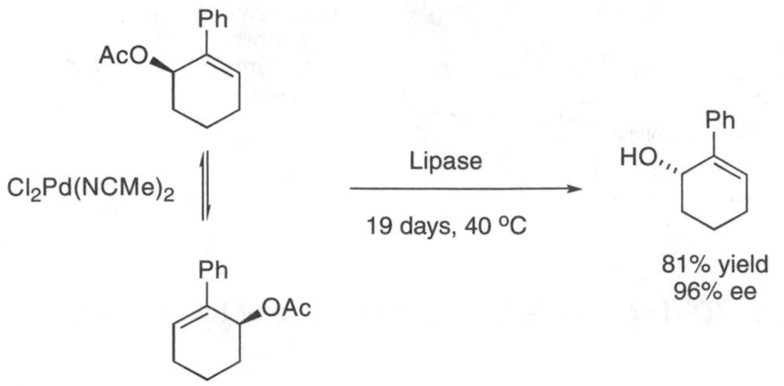

Figure 9.15. Palladium-catalyzed racemization and lipase-catalyzed ester hydrolysis are coupled for the DKR of racemic allylic acetates.

yielding the allylic alcohol, which does not participate in further reaction under these conditions. The racemization process, however, is slow, and the DKR requires several days to achieve high yields.

9.1.3 Racemization by Nucleophilic Substitution Reactions

Another mechanism by which sp^3 hybridized stereocenters can undergo racemization is by degenerative nucleophilic displacements, and DKRs based on such reactions have appeared.[5] In a study directed toward the hydrolytic kinetic resolution of functionalized epoxides with (salen)Co complexes it was observed that epibromohydrin underwent rapid racemization during the kinetic resolution, indicating that it would be a good candidate for a DKR (**Equation 9.10**).[48]

The racemization of epibromohydrin is likely catalyzed by trace amounts of bromide present in the reaction mixture (**Equation 9.11**). As shown in **Equation 9.10**, reaction of epibromohydrin with 1.5 equivalents of water in the presence of 2 mol% (R,R)-(salen)Co(OAc) furnished the diol in 93% yield with 96% enantioselectivity. This diol undergoes closure to the epoxide glycidol, a very useful C-3 building block in enantioselective synthesis, by treatment with potassium carbonate.

Equation 9.10

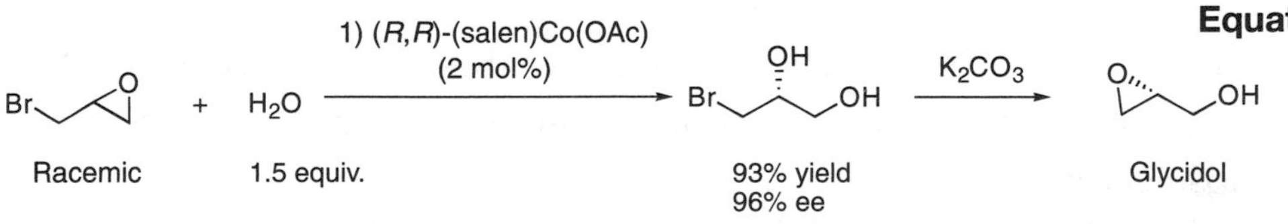

Equation 9.11

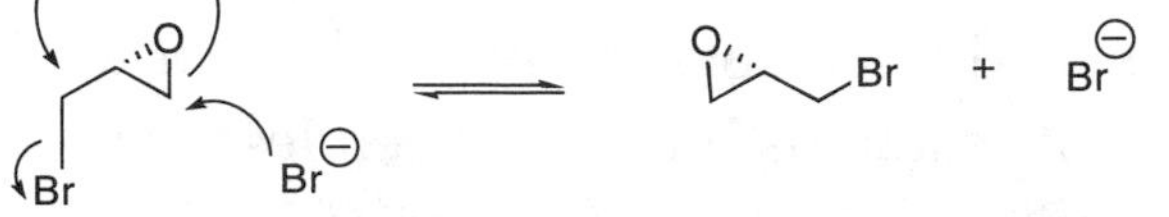

9.1.4 DKR in Asymmetric Cross-Coupling Reactions

Because facial racemization of reagents in the absence of a catalyst is uncommon, DKRs incorporating spontaneous racemization of the substrate at room temperature or below are rare. One important class of C–C bond-forming reactions that does fall into this category is asymmetric cross-coupling reactions.[49,50] Asymmetric cross-coupling reactions generally employ secondary Grignard or organozinc reagents with stereogenic carbons attached directly to the metal center (**Figure 9.16**). These organometallic reagents racemize faster that they undergo cross-coupling, making them suitable partners for DKR processes.

The asymmetric cross-coupling reaction of vinyl bromides with Grignard and organozinc reagents has been successfully catalyzed by nickel and palladium complexes with bidentate *P,N*-ligands based on ferrocene (**Figure 9.17**).[51,52] The Grignard reagent [(α-trimethylsilyl)benzyl]magnesium bromide has proven to be one of the best organometallic partners for asymmetric cross-coupling reactions, providing chiral allylsilanes. When disubstituted vinyl bromides are employed, no E–Z isomerization of the double bond is detected.

The details of the mechanism and stereoinduction in the asymmetric cross-coupling reaction are not that well understood. Nonetheless, a proposed catalytic cycle is illustrated in **Figure 9.18** with a model for stereoinduction. After reduction of the palladium(II) to palladium(0) by the Grignard reagent, oxidative addition generates a chiral palladium vinyl bromide. Selective transmetallation of one enantiomer of the Grignard reagent fixes the configuration of the palladium alkyl that is carried through to the final allylsilane. It is likely that reductive elimination from palladium(II) proceeds with retention of configuration at carbon.[53–56]

In the model for asymmetric induction, the hydrogen of the newly formed palladium alkyl bisects the *N,N*-dimethyl groups of the ligand to minimize steric repulsion. The trimethylsilyl group, being larger than the phenyl substituent, will preferentially orient above the ferrocene moiety. This stereochemical model also predicts correctly the sense of stereoinduction in asymmetric cross-coupling with 1-phenylethyl and 2-butyl Grignard reagents.[51]

Enantioenriched allylsilanes, such as those in **Figure 9.17**, are useful reagents in organic synthesis for the Lewis acid promoted allylation of aldehydes.

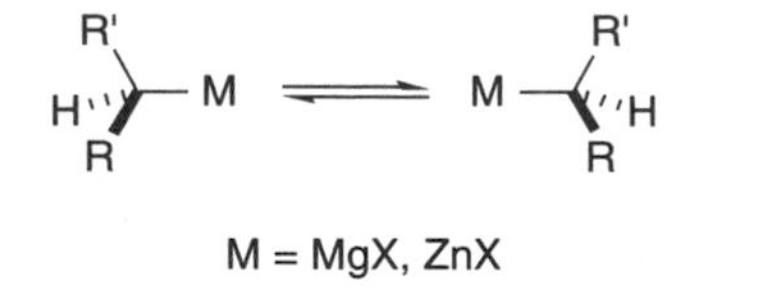

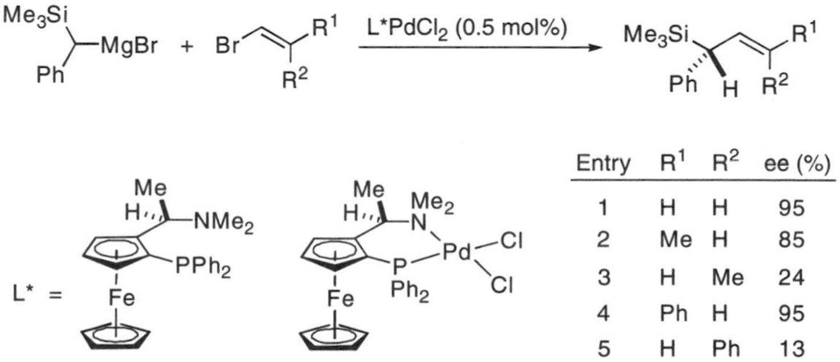

Entry	R[1]	R[2]	ee (%)
1	H	H	95
2	Me	H	85
3	H	Me	24
4	Ph	H	95
5	H	Ph	13

M = MgX, ZnX

Figure 9.16. Racemization of chiral organometallic reagents.

Figure 9.17. Catalytic asymmetric cross-coupling reactions catalyzed by palladium *P,N*-complexes.

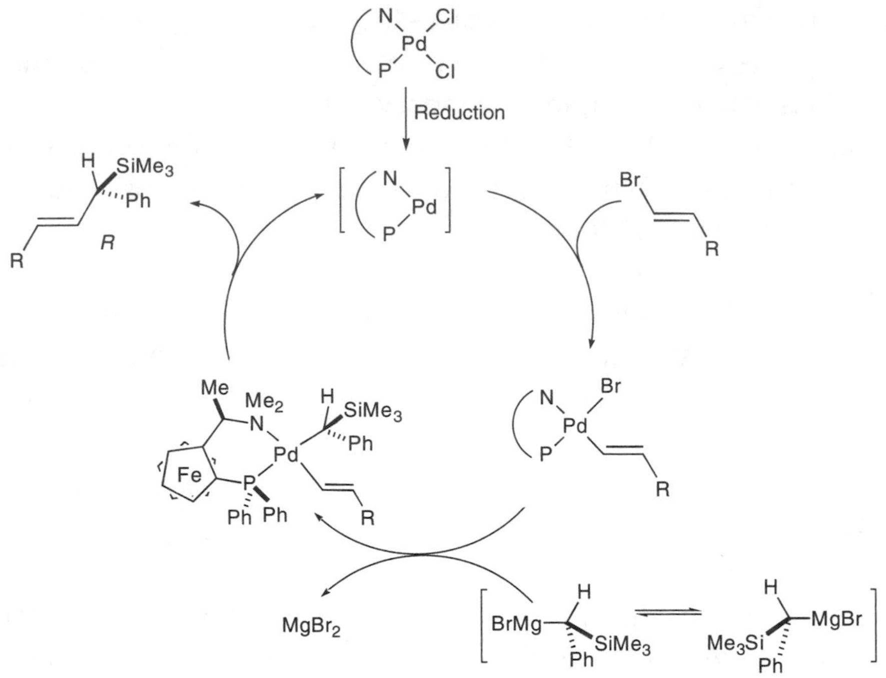

Figure 9.18. Proposed mechanism and model for stereochemical induction in the asymmetric cross-coupling reaction.

9.2 Dynamic Kinetic Asymmetric Transformations

As mentioned in the introduction to this chapter, DKRs and dynamic kinetic asymmetric transformations (DyKATs) are similar and have, in fact, led to some confusion.[11] The primary difference between a DKR and a DyKAT concerns the mechanism of inversion of the pre-existing substrate stereocenter. As described in the DKR in **Figure 9.1**, racemization of the substrate occurs via an achiral intermediate or transition state. In general, the catalyst that promotes the racemization is achiral and not related to the catalyst that performs the resolution step. In a DyKAT the interconversion of the substrate stereochemistry occurs on the chiral catalyst. Since the catalyst is chiral, inversion of the stereochemistry of the substrate gives a diastereomeric catalyst–substrate adduct and is, therefore, an epimerization.[57] There are two subclasses of DyKAT, both of which are illustrated in **Figure 9.19** (see Chapter 1 for energy diagrams).

In class **A** the substrate enantiomers S_R and S_S bind to the catalyst (cat*) to give diastereomeric catalyst–substrate adducts (Cat*S_R and Cat*S_S). Epimerization occurs on the catalyst via an intermediate or transition state (Cat*S) that is chiral. The outcome of this class of DyKAT depends on several factors, including the relative rates of product formation ($k_{S'cat*}/k_{R''cat*}$) and the relative concentrations of the catalyst–substrate adducts (Cat*S_R and Cat*S_S). The concentrations of the catalyst–substrate adducts also depend on their rates of formation and loss.[11]

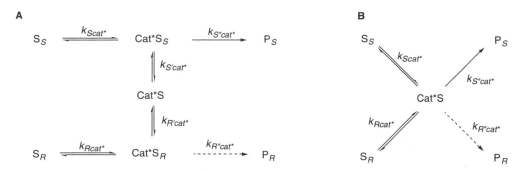

Figure 9.19. DyKAT reaction pathways involving epimerization (**A**, left) and desymmetrization (**B**, right). The (R)- and (S)-substrates and products are S_R, S_S, P_R, and P_S. The chiral catalyst is cat*.

In a type **B** DyKAT, a single catalyst–substrate adduct is formed from both enantiomers, followed by diastereomeric reaction pathways to generate the product. As such, the selectivity depends only on the relative rates of the product-forming steps, $k_{S''cat*}/k_{R''cat*}$. The rates of formation of the intermediate Cat*S governs only the rate at which product is generated.

9.2.1 DyKAT via Epimerization of Type A

Two examples of a type **A** DyKAT are illustrated below—namely, asymmetric π-allylation and asymmetric phosphination.

9.2.1.a DyKAT in Asymmetric π-Allylation

An interesting example of a DyKAT with readily available butenolides has been studied in detail (**Equation 9.12**). Butenolides are useful chiral building blocks that are easily functionalized to provide rapid access to a variety of structures.

Equation 9.12

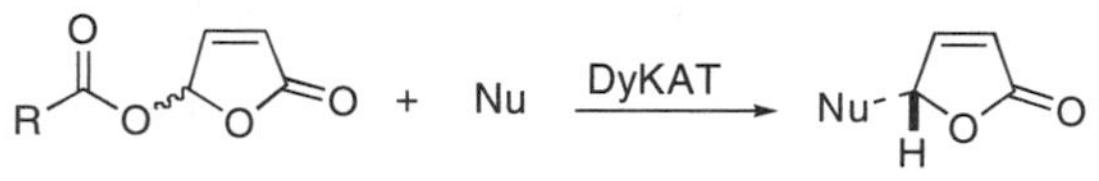

In the study highlighted here, initial attempts to perform kinetic resolutions of a butenolide with a chiral palladium catalyst resulted in recovery of starting material

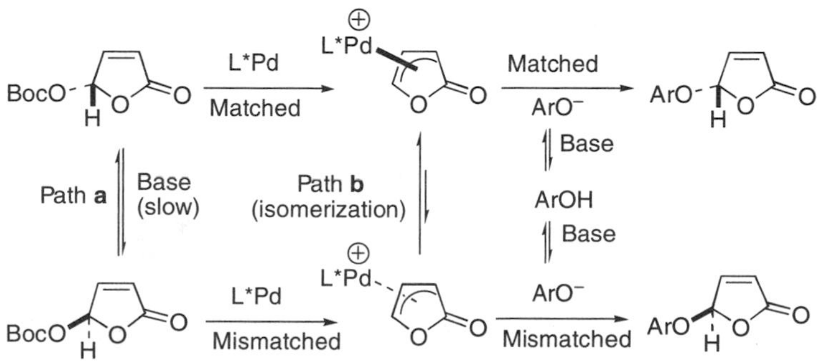

Figure 9.20. Possible reaction pathways in the DyKAT of butenolides.

with low enantiomeric excess at conversions near 50%. It was hypothesized that racemization of the starting butenolide was responsible, suggesting that DyKAT might prove a better option.

Attempts to perform a DKR and promote the racemization of the starting butenolide by addition of base (path **a**, **Figure 9.20**, Boc = *t*-BuOCO) resulted in reduced enantioselectivities. An alternative strategy involved isomerization of the diastereomeric π-allyl palladium complexes (path **b**). Two possible mechanisms for isomerization of the π-allyl diastereomers were considered most probable and are illustrated in **Figure 9.21**.

The bimetallic mechanism involves attack of a palladium(0) species on the face of the π-allyl opposite the coordinated PdL* cation. Such a pathway leads to interconversion of diastereomeric π-allyl palladium intermediates.[58] If the bimetallic mechanism were in effect, increasing the concentration of the catalyst would be expected to increase the isomerization rate. It was found, however, that *decreasing* the catalyst concentration resulted in an *increase* in product ee, inconsistent with the bimetallic pathway. The second mechanism is an intramolecular isomerization in which a sigma-bound alkoxide intermediate is formed (**Figure 9.21**). Reformation of either diastereomer of the π-allyl complex can take place from this intermediate. It is also

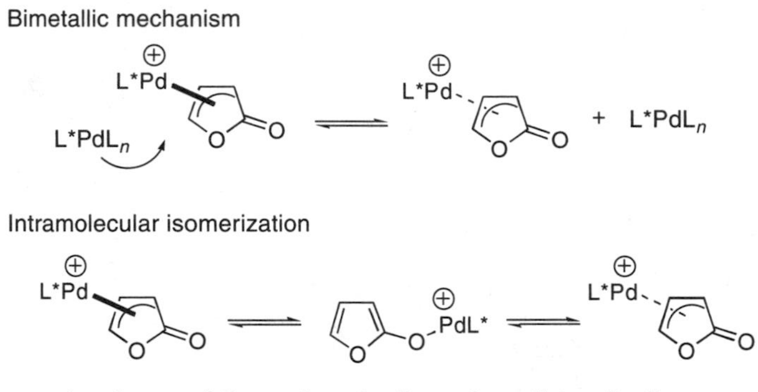

Figure 9.21. Possible mechanisms of the epimerization of π-allyl palladium complexes.

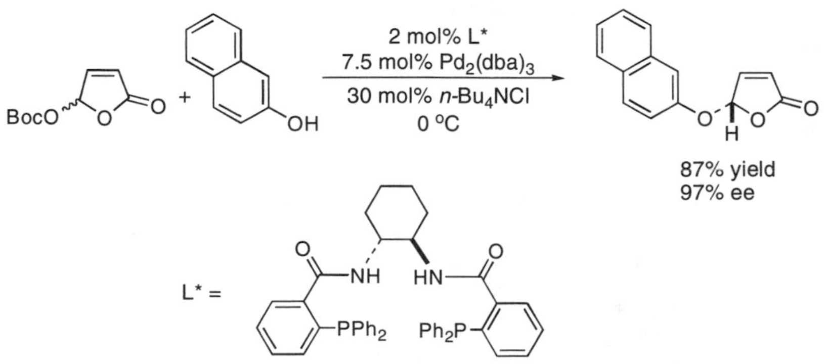

Figure 9.22. The optimized DyKAT of butenolides provides useful chiral building blocks for enantioselective synthesis.

known that halide can facilitate the interconversion of η^3-allyl and η^1-allyl complexes.[59] Addition of 30 mol% n-Bu$_4$NCl resulted in an increase in ee from 48 to 75%, which was attributed to the acceleration of the isomerization of the π-allyl intermediates. It was also found that reducing the amount of base resulted in an increase in the product ee. Decreasing the amount of base will reduce the concentration of ArO$^-$ and, therefore, slow the nucleophilic attack on the allyl with respect to the isomerization. After significant experimentation, optimal conditions were found that provided high yields of the butenolides with excellent enantioselectivities, as exemplified in **Figure 9.22**.

9.2.1.b DyKAT with Asymmetric Phosphination

One of the most useful classes of ligands in transition-metal-catalyzed asymmetric synthesis is chiral phosphines. The synthesis of these important ligands usually involves classical resolutions or the use of stoichiometric chiral auxiliaries. A catalytic asymmetric route to chiral phosphines would be very attractive. Furthermore, such methods could potentially lead to *P*-chirogenic phosphine ligands, which are not only historically important but have been shown to be highly effective and enantioselective on an industrial scale, as in the synthesis of L-dopa.[60]

An interesting approach to this challenging synthesis involves a DyKAT strategy employing an enantioselective phosphination reaction.[61–69] The reaction transforms a racemic secondary phosphine into a configurationally stable tertiary phosphine with excellent conversion and enantioselectivities up to 88% (**Figure 9.23**). The precatalyst for this reaction is [(*R,R*)-Me-DUPHOS)Pd(*trans*-stilbene)], which reacts with the aryl iodide via oxidative addition to generate [(*R,R*)-Me-DUPHOS)Pd(Ar)I].

A detailed study of the reaction provided important insight into the factors that control enantioselectivity and provided evidence for the mechanism illustrated in **Figure 9.24** [Ar = 2,4,6-C$_6$H$_2$(*i*-Pr)$_3$]. Combination of L*Pd(Ph)I with the racemic secondary phosphine HPArMe resulted in reversible displacement of iodide. The resulting

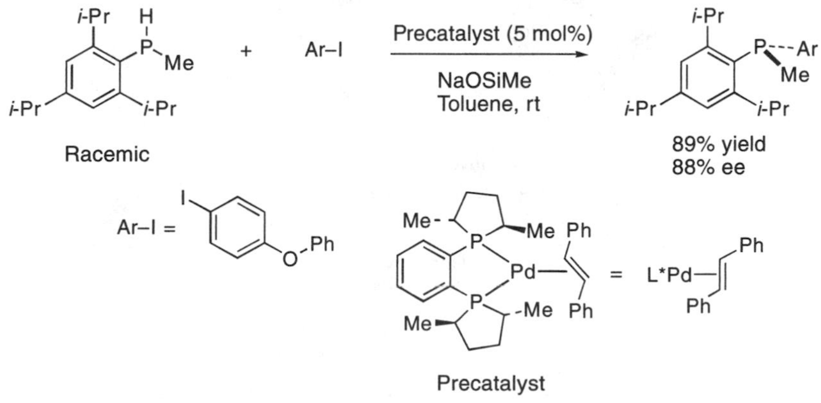

Figure 9.23. DyKAT of racemic phosphine to generate a tertiary *P*-chirogenic phosphine.

cation was observed by NMR spectroscopy as a mixture of diastereomers that differ in their configuration at the pyramidal phosphorus center. Deprotonation of the P–H of the diastereomeric cations gives the phosphide L*Pd(Ph)P(Ar)Me. Epimerization via inversion at phosphorus interconverts the two diastereomers (**Figure 9.25**). Reductive elimination results in P–C bond formation and sets the configuration at the tertiary phosphorus center. The resulting [(Me-DUPHOS)Pd] can undergo oxidative addition (the RE/OA pathway) or can coordinate two secondary phosphines, which dissociate before or during oxidative addition to regenerate L*Pd(Ph)I (**Figure 9.24**).

Understanding the factors that control enantioselection is vital to catalyst optimization. In this case the relative rates of interconversion vs. reductive elimination of the diastereomers **B** and **B'** (**Figure 9.25**) can control enantioselection. If reductive elimination of **B** and **B'** is faster than their interconversion via phosphorus inversion,

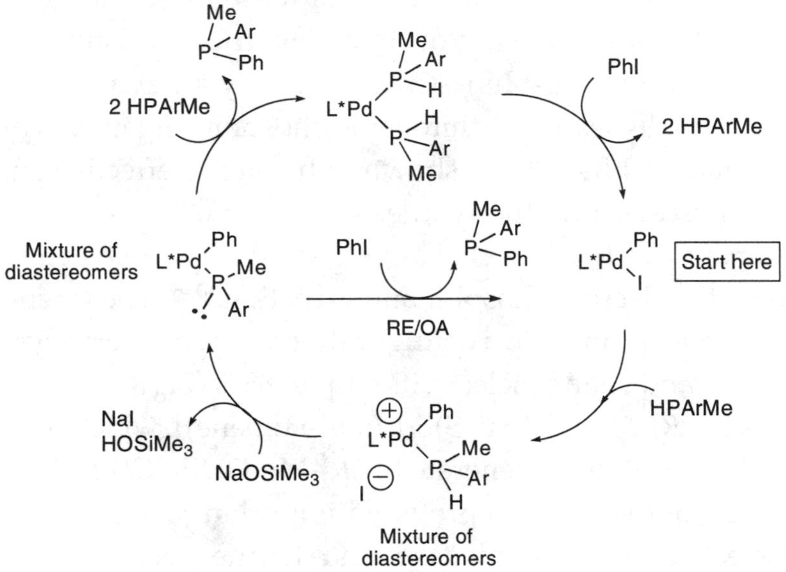

Figure 9.24. Proposed mechanism for the phosphination reaction.

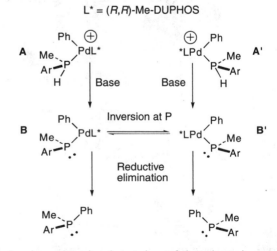

Figure 9.25. Proposed mechanism showing inversion of the phosphorus being faster than reductive elimination.

the enantioselectivity will reflect the ratio of the diastereomers **A** and **A'**. On the other hand, if **B** and **B'** interconvert faster than reductive elimination, the ratio of the enantiomeric products will most likely be different than the ratio of **A** and **A'**. To probe the enantioselectivity-determining step, a 1:1 and 1.4:1 ratio of **A** and **A'** were treated with base, yielding the enantioenriched phosphine products. In both experiments, the ratios of the enantiomers were identical, indicating that the rate of inversion of **B** and **B'** was faster than reductive elimination. Upon further experimentation it was found that interconversion of **B** and **B'** was much faster than reductive elimination, indicating that the enantiomeric ratio of phosphines depended on both the equilibrium ratio of diastereomers **B** and **B'** and their relative rates of reductive elimination (i.e. Curtin–Hammett behavior, see Chapter 1). Interestingly, it was found that the equilibrium constant between **B** and **B'** was large (ca. 50), but the minor diastereomer underwent reductive elimination three times faster than the major, eroding the product enantiomeric ratio. These studies indicate that the key to success in such systems is designing catalysts where the more prevalent diastereomer undergoes reductive elimination more rapidly.[63]

Due to the importance of chiral phosphines in asymmetric catalysis, it is certain that catalytic asymmetric methods for the generation of these compounds will be a focus of significant future investigations.

9.2.2 DyKAT via Desymmetrization (Type B)

An example of a type B DyKAT is the asymmetric allylation with the allylic acetate and a chiral palladium catalyst in **Figure 9.26**. Both enantiomers of the substrate react with L*Pd(0) to give a common palladium π-allyl complex where the π-allyl ligand is achiral. Attack at the two termini proceeds via diastereomeric pathways, which in this example occur at different rates. For a related example see **Figure 1.21**. A wide variety of chelating ligands have been employed in this reaction and many give excellent levels of enantioselectivity. New ligands are often examined in this reaction to gauge their potential utility in asymmetric catalysis.[70]

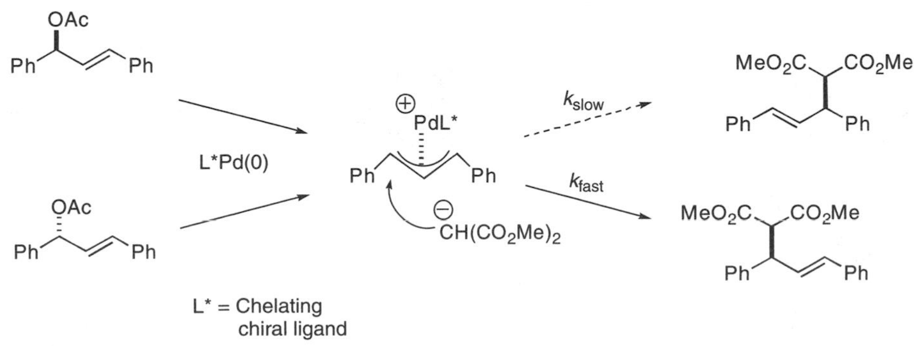

Figure 9.26. The palladium-catalyzed asymmetric allylation reaction above is an example of a type **B** DyKAT.

9.3 Reactions Between Two Racemic Substrates

An example of a reaction employing two racemic reaction partners following the above principles is illustrated in **Figure 9.27**. If no simple diastereoselection (see Chapter 13) occurs, then the four isomers shown would be formed with no greater than 25% yield each. In this asymmetric π-allylation reaction, the racemic allylic acetate undergoes a type B DyKAT and the racemic azlactone undergoes a DKR. The dynamic kinetic resolution of the nucleophile is interesting, because both enantiomers of the azlactone converge onto the same achiral enolate (**Figure 1.20d**) and may formally be regarded as a simple asymmetric reaction of an achiral substrate (differentiation of the prochiral faces). However, if the azlactone enolate is associated with the chiral palladium counterion, then a type A DyKAT is occurring with this component. Regardless of the precise mechanism, one of the four potential product isomers forms with high yield and high selectivity in this unusual process. A model for the asymmetric induction is illustrated in **Figure 9.28** with the ligand coordinated by both phosphorus atoms in a C_2-symmetric fashion.[71] Spectroscopic data, however, suggest that ligand coordination is more complicated, with monomeric allylpalladium complexes in equilibrium with oligomeric species.[72,73] The model represents a time-averaged monomer with a C_2-symmetric ligand that is useful in rationalizing stereochemical outcomes.[74] The model illustrates the position of the P-phenyl groups by the flaps protruding toward and away from the palladium center. For a further discussion of how P-phenyl groups convey asymmetry via edge-face interactions, see Chapter 4.

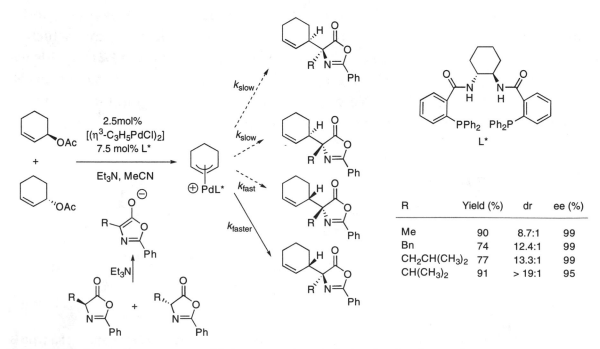

R	Yield (%)	dr	ee (%)
Me	90	8.7:1	99
Bn	74	12.4:1	99
CH$_2$CH(CH$_3$)$_2$	77	13.3:1	99
CH(CH$_3$)$_2$	91	> 19:1	95

Figure 9.27. Dynamic processes with two racemic substrates leading to one isomer.

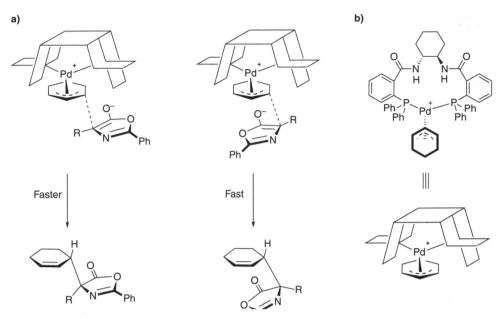

Figure 9.28. a) Stereochemical model for the asymmetric induction. b) Structure of the intermediate and model for the chiral environment around the palladium. The flaps represent the phenyl groups protruding toward and away from the palladium center.

Summary

Chapter 9 describes the application of catalytic kinetic resolutions to a variety of important processes, demonstrating that this is an effective route to enantioenriched compounds. In kinetic resolutions, like conventional resolutions, the maximum yield is 50%. Furthermore, kinetic resolutions must exhibit enormous differences in the relative rates of the fast- and slow-reacting enantiomers to achieve high enantioselectivities and maximize yields. The high relative rates are necessary to offset the concentration dilemma caused by build-up of the slow-reacting enantiomer toward the end of the resolution process. In contrast, DKR offers the advantage of an in situ racemization of the substrate, thus allowing complete conversion of the starting material to a single product with high enantioselectivity. Furthermore, because the substrate is undergoing racemization, the relative concentrations of the substrate enantiomers will remain close to the ideal value of unity.

What makes the DKR more difficult, and more limited in its applications, than kinetic resolution, is the added complexity of the process. Many times, however, the increased challenge and greater rewards of the DKR offset these negative aspects.

References

(1) Zhu, Y.; Shu, L.; Tu, Y.; Shi, Y. Enantioselective Synthesis and Stereoselective Rearrangements of Enol Ester Epoxides. *J. Org. Chem.* **2001**, *66*, 1818–1826.

(2) Ward, R. S. Dynamic Kinetic Resolution. *Tetrahedron: Asymmetry* **1995**, *6*, 1475–1490.

(3) Noyori, R.; Tokunaga, M.; Kitamura, M. Stereoselective Organic Synthesis via Dynamic Kinetic Resolution. *Bull. Chem. Soc. Jpn.* **1995**, *68*, 36–56.

(4) El Gihani, M. T.; Williams, J. M. J. Dynamic Kinetic Resolution. *Curr. Opin. Chem. Bio.* **1999**, *3*, 11–15.

(5) Huerta, F. F.; Minidis, A. B. E.; Bäckvall, J.-E. Racemization in Asymmetric Synthesis. Dynamic Kinetic Resolution and Related Processes in Enzyme and Metal Catalysis. *Chem. Soc. Rev.* **2001**, *30*, 321–331.

(6) Caddick, S.; Jenkins, K. Dynamic Resolutions in Asymmetric Synthesis. *Chem. Soc. Rev.* **1996**, *25*, 447–456.

(7) Pamies, O.; Backvall, J.-E. Combination of Enzymes and Metal Catalysts. A Powerful Approach in Asymmetric Catalysis. *Chem. Rev.* **2003**, *103*, 3247–3262.

(8) Kitamura, M.; Tokunaga, M.; Noyori, R. Mathematical Treatment of Kinetic Resolution of Chirally Labile Substrates. *Tetrahedron* **1993**, *49*, 1853–1860.

(9) Kitamura, M.; Tokunaga, M.; Noyori, R. Quantitative Expression of Dynamic Kinetic Resolution of Chirally Labile Enantiomers: Stereoselective Hydrogenation of 2-Substituted 3-Oxo Carboxylic Esters Catalyzed by BINAP-Ruthenium(II) Complexes. *J. Am. Chem. Soc.* **1993**, *115*, 144–152.

(10) Andraos, J. Quantification and Optimization of Dynamic Kinetic Resolution. *J. Phys. Chem. A* **2003**, *107*, 2374–2387.

(11) Faber, K. Non-sequential Processes for the Transformation of a Racemate into a Single Stereoisomeric Product: Proposal for Stereochemical Classification. *Chem. Eur. J.* **2001**, *7*, 5004–5010.

(12) Noyori, R.; Takaya, H. BINAP: An Efficient Chiral Element for Asymmetric Catalysis. *Acc. Chem. Res.* **1990**, *23*, 345–350.

(13) Noyori, R.; Ikeda, T.; Ohkuma, T.; Widhalm, M.; Kitamura, M.; Takaya, H.; Akutagawa, S.; Sayo, N.; Saito, T.; Taketomi, T.; Kumobayashi, H. Stereoselective Hydrogenation via Dynamic Kinetic Resolution. *J. Am. Chem. Soc.* **1989**, *111*, 9134–9135.

(14) Lei, A.; Wu, S.; He, M.; Zhang, X. Highly Enantioselective Asymmetric Hydrogenation of α-Phthalimide Ketone: An Efficient Entry to Enantiomerically Pure Amino Alcohols. *J. Am. Chem. Soc.* **2004**, *126*, 1626–1627.

(15) Zhang, Z.; Qian, H.; Longmire, J.; Zhang, X. Synthesis of Chiral Bisphosphines with Tunable Bite Angles and Their Applications in Asymmetric Hydrogenation of β-Ketoesters. *J. Org. Chem.* **2000**, *65*, 6223–6226.

(16) Ohkuma, T.; Ooka, H.; Yamakawa, M.; Ikariya, T.; Noyori, R. Stereoselective Hydrogenation of Simple Ketones Catalyzed by Ruthenium(II) Complexes. *J. Org. Chem.* **1996**, *61*, 4872–4873.

(17) Matsumoto, T.; Murayama, T.; Mitsuhashi, S.; Miura, T. Diastereoselective Synthesis of a Key Intermediate for the Preparation of Tricyclic β-Lactam Antibiotics. *Tetrahedron Lett.* **1999**, *40*, 5043–5046.

(18) Di Modugno, E.; Erbetti, I.; Ferrari, L.; Galassi, G.; Hammond, S. M.; Xerri, L. In-Vitro Activity of the Tribactam GV104326 Against Gram-Positive, Gram-Negative, and Anaerobic-Bacteria. *Antimicrob. Agents Chemother.* **1994**, *38*, 2362–2368.

(19) Kitamura, M.; Tokunaga, M.; Noyori, R. Asymmetric Hydrogenation of β-Keto Phosphonates: A Practical Way to Fosfomycin. *J. Am. Chem. Soc.* **1995**, *117*, 2931–2932.

(20) Hang, J.; Tian, S.-K.; Tang, L.; Deng, L. Asymmetric Synthesis of β-Amino Acids via Cinchona Alkaloid-Catalyzed Kinetic Resolution of Urethane-Protected α-Amino Acid N-Carboxyanhydrides. *J. Am. Chem. Soc.* **2001**, *123*, 12696–12697.

(21) Hang, J.; Li, H.; Deng, L. Development of a Rapid, Room-Temperature Dynamic Kinetic Resolution for Efficient Asymmetric Synthesis of α-Aryl Amino Acids. *Org. Lett.* **2002**, *4*, 3321–3324.

(22) Hiratake, J.; Yamamoto, Y.; Oda, J. Catalytic Asymmetric Induction from Prochiral Cyclic Acid Anhydrides Using Cinchona Alkaloids. *J. Chem. Soc., Chem. Commun.* **1985**, 1717–1719.

(23) Hiratake, J.; M., I.; Yamamoto, Y.; Oda, J. Enantiotopic-Group Differentiation. Catalytic Asymmetric Ring-Opening of Prochiral Cyclic Anhydrides with Methanol, Using Cinchona Alkaloids. *J. Chem. Soc., Perkin Trans. 1* **1987**, 1053–1058.

(24) Chen, Y.; McDaid, P.; Deng, L. Asymmetric Alcoholysis of Cyclic Anhydrides. *Chem. Rev.* **2003**, *103*, 2965–2984.

(25) Ebbers, E. J.; Ariaans, G. J. A.; Houbiers, J. P. M.; Bruggink, A.; Zwanenburg, B. Controlled Racemization of Optically Active Organic Compounds: Prospects for Asymmetric Transformation. *Tetrahedron* **1997**, *53*, 9417–9476.

(26) Strauss, U. T.; Felfer, U.; Faber, K. Biocatalytic Transformation of Racemates into Chiral Building Blocks in 100% Chemical Yield and 100% Enantiomeric Excess. *Tetrahedron: Asymmetry* **1999**, *10*, 107–117.

(27) Blum, Y.; Czarkie, D.; Rahamim, Y.; Shvo, Y. (Cyclopentadienone)ruthenium Carbonyl Complexes—A New Class of Homogeneous Hydrogenation Catalysts. *Organometallics* **1985**, *4*, 1459–1461.

(28) Shvo, Y.; Czarkie, D.; Rahamim, Y.; Chodosh, D. F. A New Group of Ruthenium Complexes: Structure and Catalysis. *J. Am. Chem. Soc.* **1986**, *108*, 7400–742.

(29) Chowdhury, R. L.; Bäckvall, J.-E. Efficient Ruthenium-Catalyzed Transfer Hydrogenation of Ketones by Propan-2-ol. *J. Chem. Soc., Chem. Commun.* **1991**, 1063–1064.

(30) Samec, J. S. M.; Bäckvall, J.-E.; Andersson, P. G.; Brandt, P. Mechanistic Aspects of Transition Metal-Catalyzed Hydrogen Transfer Reactions. *Chem. Soc. Rev.* **2006**, *35*, 237–248.

(31) Casey, C. P.; Singer, S. W.; Powell, D. R.; Hayashi, R. K.; Kavana, M. Hydrogen Transfer to Carbonyls and Imines from α-Hydroxycyclopentadienyl Ruthenium Hydride: Evidence for Concerted Hydride and Proton Transfer. *J. Am. Chem. Soc.* **2001**, *123*, 1090–1100.

(32) Pàmics, O.; Bäckvall, J.-E. Studies on the Mechanism of Metal-Catalyzed Hydrogen Transfer from Alcohols to Ketones. *Chem. Eur. J.* **2001**, *7*, 5052–5058.

(33) Martín-Matute, B.; Edin, M.; Bogár, K.; Kaynak, F. B.; Bäckvall, J.-E. Combined Ruthenium(II) and Lipase Catalysis for Efficient Dynamic Kinetic Resolution of Secondary Alcohols. Insight into the Racemization Mechanism. *J. Am. Chem. Soc.* **2005**, *127*, 8817–8825.

(34) Casey, C. P.; Johnson, J. B. Kinetic Isotope Effect Evidence for the Concerted Transfer of Hydride and Proton from Hydroxycyclopentadienyl Ruthenium Hydride in Solvents of Different Polarities and Hydrogen Bonding Ability. *Can. J. Chem.* **2005**, *83*, 1339–1346.

(35) Persson, B. A.; Huerta, F. F.; Bäckvall, J.-E. Dynamic Kinetic Resolution of Secondary Diols via Coupled Ruthenium and Enzyme Catalysis. *J. Org. Chem.* **1999**, *64*, 5237–5240.

(36) Huerta, F. F.; Laxmi, Y. R. S.; Bäckvall, J.-E. Dynamic Kinetic Resolution of α-Hydroxy Acid Esters. *Org. Lett.* **2000**, *2*, 1037–1040.

(37) Huerta, F. F.; Bäckvall, J.-E. Enantioselective Synthesis of β-Hydroxy Acid Derivatives via a One-Pot Aldol Reaction-Dynamic Kinetic Resolution. *Org. Lett.* **2001**, *3*, 1209–1212.

(38) Pamies, O.; Bäckvall, J.-E. Combination of Enzymes and Metal Catalysts. A Powerful Approach in Asymmetric Catalysis. *Chem. Rev.* **2003**, *103*, 3247–3262.

(39) Kim, M.-J.; Chung, Y. I.; Choi, Y. K.; Lee, H. K.; Kim, D.; Park, J. (*S*)-Selective Dynamic Kinetic Resolution of Secondary Alcohols by the Combination of Subtilisin and an Aminocyclopentadienylruthenium Complex as the Catalysts. *J. Am. Chem. Soc.* **2003**, *125*, 11494–11495.

(40) Borén, L.; Martín-Matute, B.; Xu, Y.; Córdova, A.; Bäckvall, J.-E. (*S*)-Selective Kinetic Resolution and Chemoenzymatic Dynamic Kinetic Resolution of Secondary Alcohols. *Chem. Eur. J.* **2006**, *12*, 225–232.

(41) Martín-Matute, B.; Edin, M.; Bäckvall, J.-E. Highly Efficient Synthesis of Enantiopure Diacetylated C_2-Symmetric Diols by Ruthenium- and Enzyme-Catalyzed Dynamic Kinetic Asymmetric Transformation (DyKAT). *Chem. Eur. J.* **2006**, *12*, 6053–6061.

(42) Fransson, A.-B. L.; Xu, Y.; Leijondahl, K.; Bäckvall, J.-E. Enzymatic Resolution, Desymmetrization, and Dynamic Kinetic Asymmetric Transformation of 1,3-Cycloalkanediols. *J. Org. Chem.* **2006**, *71*, 6309–6316.

(43) Paetzold, J.; Bäckvall, J.-E. Chemoenzymatic Dynamic Kinetic Resolution of Primary Amines. *J. Am. Chem. Soc.* **2005**, *127*, 17620–17621.

(44) Privalov, T.; Samec, J. S. M.; Bäckvall, J.-E. DFT Study of an Inner-Sphere Mechanism in the Hydrogen Transfer from a Hydroxycyclopentadienyl Ruthenium Hydride to Imines. *Organometallics* **2007**, *26*, 2840–2848.

(45) Samec, J. S. M.; Ell, A. H.; Aberg, J. B.; Privalov, T.; Eriksson, L.; Bäckvall, J.-E. Mechanistic Study of Hydrogen Transfer to Imines from a Hydroxycyclopentadienyl Ruthenium Hydride. Experimental Support for a Mechanism Involving Coordination of Imine to Ruthenium Prior to Hydrogen Transfer. *J. Am. Chem. Soc.* **2006**, *128*, 14293–14305.

(46) Allen, J. V.; Williams, J. M. J. Dynamic Kinetic Resolution with Enzyme and Palladium Combinations. *Tetrahedron Lett.* **1996**, *37*, 1859–1962.

(47) Overman, L. E. Mercury(II)- and Palladium(II)-Catalyzed [3,3]-Sigmatropic Rearrangements. *Angew. Chem., Int. Ed. Engl.* **1984**, *23*, 579–586.

(48) Furrow, M. E.; Schaus, S. E.; Jacobsen, E. N. Practical Access to Highly Enantioenriched C-3 Building Blocks via Hydrolytic Kinetic Resolution. *J. Org. Chem.* **1998**, *63*, 6776–6777.

(49) Hayashi, T.; Sawamura, M.; Ito, Y. Asymmetric Synthesis Catalyzed by Chiral Ferrocenylphosphine Transition Metal Complexes. 10. Gold(I)-Catalyzed Asymmetric Aldol Reaction of Isocyanoacetate. *Tetrahedron* **1992**, *48*, 1999–2012.

(50) Ogasawara, M.; Hayashi, T. In *Catalytic Asymmetric Synthesis*, 2nd ed.; Ojima, I., Ed.; Wiley: New York, 2000; pp 651–674.

(51) Hayashi, T.; Konishi, K.; Fukushima, M.; Mise, T.; Kagotani, M.; Tajika, M.; Kumada, M. Asymmetric Synthesis Catalyzed by Chiral Ferrocenylphosphine-Transition Metal Complexes. 2. Nickel- and Palladium-Catalyzed Asymmetric Grignard Cross-Coupling. *J. Am. Chem. Soc.* **1982**, *104*, 180–186.

(52) Hayashi, T.; Konishi, M.; Okamoto, Y.; Kabeta, K.; Kumada, M. Asymmetric Synthesis Catalyzed by Chiral Ferrocenylphosphine-Transition-Metal Complexes. 3. Preparation of Optically Active Allylsilanes by Palladium-Catalyzed Asymmetric Grignard Cross-Coupling. *J. Org. Chem.* **1986**, *51*, 3772–3781.

(53) Netherton, M. R.; Fu, G. C. Suzuki Cross-Couplings of Alkyl Tosylates that Possess β Hydrogen Atoms: Synthetic and Mechanistic Studies. *Angew. Chem., Int. Ed. Engl.* **2002**, *41*, 3910–3912.

(54) Milstein, D.; Stille, J. K. Mechanism of Reductive Elimination. Reaction of Alkylpalladium(II) Complexes with Tetraorganotin, Organolithium, and Grignard Reagents. Evidence for Palladium(IV) Intermediacy. *J. Am. Chem. Soc.* **1979**, *101*, 4981–4991.

(55) Moncarz, J. R.; Brunker, T. J.; Glueck, D. S.; Sommer, R. D.; Rheingold, A. L. Stereochemistry of Palladium-Mediated Synthesis of PAMP-BH$_3$: Retention of Configuration at P in Formation of Pd–P and P–C Bonds. *J. Am. Chem. Soc.* **2003**, *125*, 1180–1181.

(56) Moncarz, J. R.; Brunker, T. J.; Jewett, J. C.; Orchowski, M.; Glueck, D. S.; Sommer, R. D.; Lam, K.-C.; Incarvito, C. D.; Concolino, T. E.; Ceccarelli, C.; Zakharov, L. N.; Rheingold, A. L. Palladium-Catalyzed Asymmetric Phosphination. Enantioselective Synthesis of PAMP-BH$_3$, Ligand Effects on Catalysis, and Direct Observation of the Stereochemistry of Transmetalation and Reductive Elimination. *Organometallics* **2003**, *22*, 3205–3221.

(57) Eliel, E. L.; Wilen, S. H. *Stereochemistry of Organic Compounds*; Wiley: New York, 1994.

(58) Granberg, K. L.; Bäckvall, J. E. Isomerization of (π-Allyl)palladium Complexes via Nucleophilic Displacement by Palladium(0). A Common Mechanism in Palladium(0)-Catalyzed Allylic Substitution. *J. Am. Chem. Soc.* **1992**, *114*, 6858–6863.

(59) Faller, J. W.; Tully, M. T. Organometallic Conformational Equilibriums. XV. Preparation and Resolution of 1,2,3-η^3-(1-Acetyl-2,3-dimethylallyl)[(*S*)-α-phenethylamine]chloropalladium. *J. Am. Chem. Soc.* **1972**, *94*, 2676–2679.

(60) Knowles, W. S. Asymmetric Hydrogenation. *Acc. Chem. Res.* **1983**, *16*, 106–112.

(61) Scriban, C.; Glueck, D. S.; Zakharov, L. N.; Kassel, W. S.; DiPasquale, A. G.; Golen, J. A.; Rheingold, A. L. P–C and C–C Bond Formation by Michael Addition in Platinum-Catalyzed Hydrophosphination and in the Stoichiometric Reactions of Platinum Phosphido Complexes with Activated Alkenes. *Organometallics* **2006**, *25*, 5757–5767.

(62) Moncarz, J. R.; Laritcheva, N. F.; Glueck, D. S. Palladium-Catalyzed Asymmetric Phosphination: Enantioselective Synthesis of a *P*-Chirogenic Phosphine. *J. Am. Chem. Soc.* **2002**, *124*, 13356–13357.

(63) Blank, N. F.; Moncarz, J. R.; Brunker, T. J.; Scriban, C.; Anderson, B. J.; Amir, O.; Glueck, D. S.; Zakharov, L. N.; Golen, J. A.; Incarvito, C. D.; Rheingold, A. L. Palladium-Catalyzed Asymmetric Phosphination. Scope, Mechanism, and Origin of Enantioselectivity. *J. Am. Chem. Soc.* **2007**, *129*, 6847–6858.

(64) Chan, V. S.; Stewart, I. C.; Bergman, R. G.; Toste, F. D. Asymmetric Catalytic Synthesis of P-Stereogenic Phosphines via a Nucleophilic Ruthenium Phosphido Complex. *J. Am. Chem. Soc.* **2006**, *128*, 2786–2787.

(65) Kovacik, I.; Wicht, D. K.; Grewal, N. S.; Glueck, D. S.; Incarvito, C. D.; Guzei, I. A.; Rheingold, A. L. Pt(Me-Duphos)-Catalyzed Asymmetric Hydrophosphination of Activated Olefins: Enantioselective Synthesis of Chiral Phosphines. *Organometallics* **2000**, *19*, 950–953.

(66) Scriban, C.; Glueck, D. S. Platinum-Catalyzed Asymmetric Alkylation of Secondary Phosphines: Enantioselective Synthesis of P-Stereogenic Phosphines. *J. Am. Chem. Soc.* **2006**, *128*, 2788–2789.

(67) Scriban, C.; Glueck, D. S.; Golen, J. A.; Rheingold, A. L. Platinum-Catalyzed Asymmetric Alkylation of a Secondary Phosphine: Mechanism and Origin of Enantioselectivity. *Organometallics* **2007**, *26*, 1788–1800.

(68) Korff, C.; Helmchen, G. Preparation of Chiral Triarylphosphines by Pd-catalysed Asymmetric P–C Cross-Coupling. *J. Chem. Soc., Chem. Commun.* **2004**, 530–531.

(69) Brunker, T. J.; Anderson, B. J.; Blank, N. F.; Glueck, D. S.; Rheingold, A. L. Enantioselective Synthesis of P-Stereogenic Benzophospholanes via Palladium-Catalyzed Intramolecular Cyclization. *Org. Lett.* **2007**, *9*, 1109–1112.

(70) Pfaltz, A.; Lautens, M. In *In Comprehensive Asymmetric Catalysis*; Jacobsen, E. N., Pfaltz, A., Yamamoto, H., Eds.; Springer-Verlag: Berlin, 1999; Vol. 2; 833–884.

(71) Trost, B. M.; Ariza, X. Catalytic Asymmetric Alkylation of Nucleophiles: Asymmetric Synthesis of α-Alkylated Amino Acids. *Angew. Chem., Int. Ed. Engl.* **1997**, *36*, 2635–2637.

(72) Lloyd-Jones, G. C.; Stephen, S. C.; Fairlamb, I. J. S.; Martorell, A.; Dominguez, B.; Tomlin, P. M.; Murray, M.; Fernandez, J. M.; Jeffery, J. C.; Riis-Johannessen, T.; Guerziz, T. Coordination of the Trost Modular Ligand to Palladium Allyl Fragments: Oligomers, Monomers, and Memory Effects in Catalysis. *Pure Appl. Chem.* **2004**, *76*, 589–601.

(73) Amatore, C.; Jutand, A.; Mensah, L.; Ricard, L. On the Formation of Pd(II) Complexes of Trost Modular Ligand Involving N–H Activation or P,O Coordination in Pd-Catalyzed Allylic Alkylations. *J. Organomet. Chem.* **2007**, *692*, 1457–1464.

(74) Trost, B. M.; Machacek, M. R.; Aponick, A. Predicting the Stereochemistry of Diphenylphosphino Benzoic Acid (DPPBA)-Based Palladium-Catalyzed Asymmetric Allylic Alkylation Reactions: A Working Model. *Acc. Chem. Res.* **2006**, *39*, 747–760.

10

Desymmetrization Reactions

Desymmetrization of meso and centrosymmetric substrates by asymmetric catalysts is one of the most powerful approaches for the construction of enantioenriched organic compounds.[1] Such symmetry-breaking reactions are unique because they can establish stereochemistry at several centers simultaneously. Furthermore, these stereocenters may be distant from the site of reaction and would be difficult to install by other methods. This strategy generally involves the differential reactivity of enantiotopic atoms or functional groups of a substrate with a chiral reagent or catalyst. As will be illustrated below, in some cases the desymmetrization is very similar to a kinetic resolution; instead of choosing between different enantiomers of a substrate, the catalyst chooses between enantiotopic groups on a single substrate. As a consequence, catalysts that perform well in kinetic resolutions can also be efficient at the desymmetrization of meso substrates. In other desymmetrizations, the reactive functional group on the molecule is capable of undergoing reaction only once, such as the nucleophilic opening of an epoxide. The distinct advantage of desymmetrization over kinetic resolution is that the theoretical yield in desymmetrization is 100%.

Desymmetrization substrates can be classified into those that can undergo a single transformation (such as the opening of meso epoxides and aziridines) and those that can undergo two reactions (such as the acylation of diols).

10.1 Compounds with One Reactive Functional Group Containing Enantiotopic Centers

10.1.1 Opening of Meso Epoxides

In the desymmetrization of simple meso epoxides, such as cyclohexene oxide derivatives, the substrate contains enantiotopic carbon atoms (**Figure 10.1**). Once the epoxide has been opened, it is inert to further reaction. Thus, the product ee remains constant over the course of the reaction, provided that catalyst remains unchanged. Nucleophilic ring-opening of this reactive heterocycle establishes two or more stereogenic centers.

A successful catalyst for the desymmetrization of meso epoxides with silyl azides was developed based on C_3-symmetric trialkanolamine ligands that bind to zirconium, forming dimeric precatalysts (**Equation 10.1**).[2] These ligands bind tightly to zirconium(IV) and exchange very slowly, unlike simple monodentate alkoxides which often exchange rapidly.[3,4]

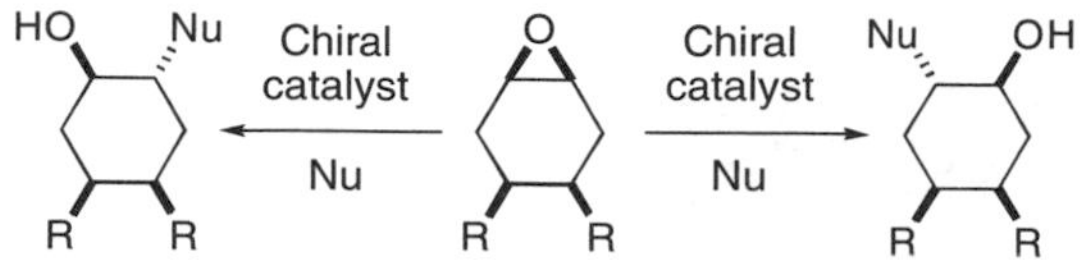

Figure 10.1. Desymmetrization of meso epoxides.

Equation 10.1

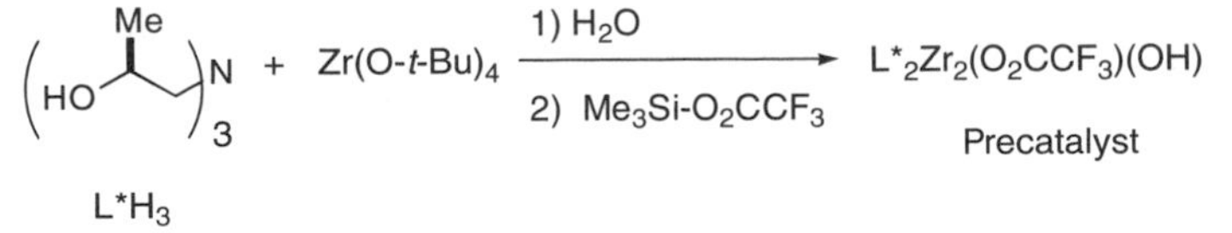

Mechanistic studies on the asymmetric epoxide-opening by these catalysts with trimethylsilyl azide support a catalyst with two zirconium centers that act in a cooperative, bifunctional manner.[5] As shown in **Figure 10.2**, it is proposed that activation of the epoxide occurs at one zirconium center and binding of azide nucleophile to the other. The initial azido alkoxide is silylated to form the organic product and regenerate the zirconium-azide catalyst. In the reaction with cyclohexene oxide illustrated in Figure 10.2, the azide product is formed with 93% ee.

As exemplified in **Figure 10.2**, the mechanism of epoxide-opening requires that the nucleophile attacks the epoxide from the backside, resulting in trans stereochemistry of the ring-opened product. This constraint limits the utility of the reaction in that the cis-disubstituted products cannot be prepared. One approach to circumvent this limitation would be to open the epoxide with a halide that could be displaced by a nucleophile in a separate step. Initial ring-opening would result in formation of the trans product and subsequent substitution of the halide would give a net double inversion with formation of the cis ring-opened product.

The enantioselective addition of trialkylsilyl halides to meso epoxides would provide access to the desired β-halohydrins, but unfortunately has proven difficult. An ingenious method to address this challenging problem involved a "bait-and-switch" technique (**Equation 10.2**).[6] The catalyst system for the azide opening of epoxides was used, because it efficiently activates both the epoxide and the nucleophile. The idea was to replace the azide with a halide, which would then be delivered to the epoxide. This was accomplished using the reactive allyl halide as a trap for the azide. Once the N_3^- is generated, it reacts more rapidly with the allyl iodide to produce allyl azide and liberate the iodide nucleophile. The iodide then adds to the epoxide to give the observed β-iodohydrin. The reaction of cyclopentene oxide with trimethylsilyl azide, allyl iodide, and the dimeric precatalyst (**Equation 10.2**) resulted in formation of only about 4% of the azide-opened product and 96% formation of the protected β-iodohydrin (**Equation 10.2**). As expected, the coproduct, allyl azide, was also formed. No reaction was observed in the absence of the zirconium catalyst.

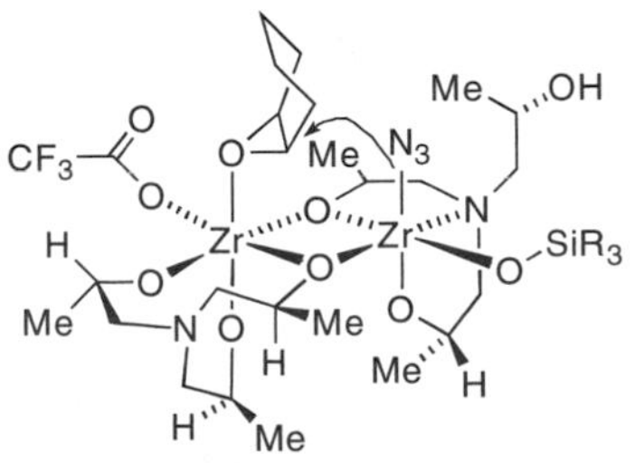

Figure 10.2. Proposed transition state for the desymmetrization of cyclohexene oxide.

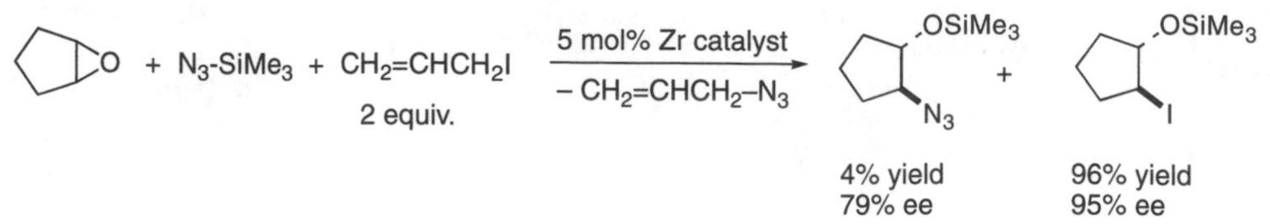

Equation 10.2

It was also demonstrated that bromide was a suitable nucleophile, although an excess of the less reactive allyl bromide (up to 20 equivalents) was needed to suppress epoxide-opening by azide. The reaction shows good tolerance of functional groups and excellent enantioselectivities with a variety of substrates (**Table 10.1**).

The (salen)Co catalysts that exhibit high efficiency and enantioselectivity in the hydrolytic kinetic resolution of terminal epoxides do not readily promote ring-opening reactions of 1,2-disubstituted epoxides (see Chapter 7).[7,8] The (salen)Cr-based catalysts, however, do promote ring-opening of internal epoxides with nucleophiles like trimethylsilyl azide with high levels of enantioselectivity.[9–11]

Table 10.1. Catalytic formation of β-bromohydrins.

Substrate	Product	Yield (%)	ee (%)
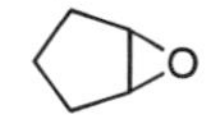	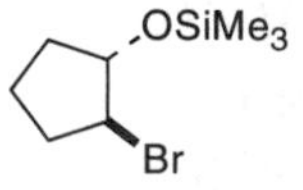	81	95
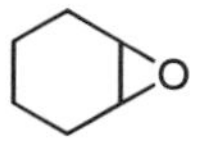	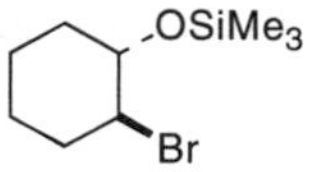	86	91
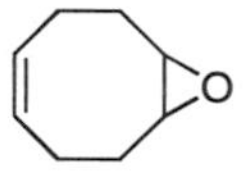	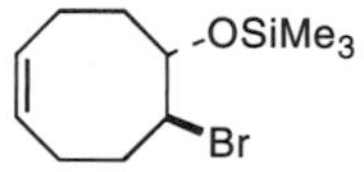	92	84
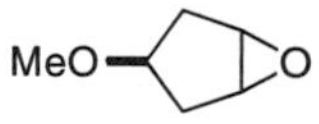	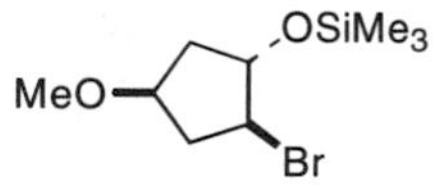	89	95
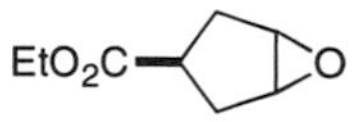	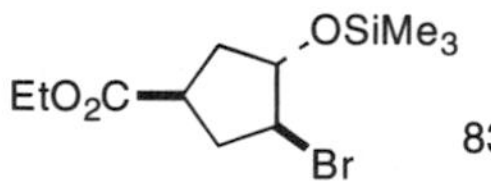	83	95

10.1.2 Desymmetrization via C–H Activation with Achiral and Meso Compounds

10.1.2.a C–H Oxidation: Advantage of Desymmetrization with Two Enantioselectivity-Determining Steps

Catalytic enantioselective oxidation of unactivated C–H bonds is another area that remains very challenging. The unreactive nature of these bonds requires the generation of highly active oxidants able to abstract H•, yet sufficiently selective to discriminate among the various types of C–H bonds in a chemoselective and enantioselective fashion. Nature has developed biological catalysts that perform C–H oxidations with high stereoselectivities and regioselectivities. Examples include the iron–porphyrin-based cytochrome P450 and the nonheme dinuclear iron center in methane monooxygenase. The remarkable efficiency and selectivity of these enzymes have inspired the design of related small-molecule systems. Initial successes have recently been reported for the asymmetric hydroxylation of more reactive C–H bonds, such as benzylic, allylic, and those neighboring oxygen and nitrogen.[12–14]

The mechanism of hydrocarbon hydroxylation is proposed to involve hydrogen abstraction by a reactive metal–oxo intermediate to generate a radical. Attack of the radical on the newly formed metal hydroxyl gives the hydroxylated product and the reduced metal complex (**Figure 10.3**). The extent of asymmetric induction in this stepwise hydroxylation depends on capture of the radical before it reorients, exposing the opposite face to attack, or before it strays from the chiral metal complex. The further the dissociation of the radical from the chiral catalyst, the greater the loss of stereochemistry (**Figure 10.3**). [12,15] An increase in the solvent viscosity would be expected to hinder dissociation of the radical from the chiral catalyst. Consistent with this hypothesis, a more viscous solvent results in higher enantioselectivities in the asymmetric hydroxylation reaction.[16]

Unlike most asymmetric processes, where either the substrate or reagent is bound to the catalyst, the carbon-based radical intermediate in the asymmetric hydroxylation is not. For a small-molecule catalyst, limiting radical dissociation and decay is a challenging task. To circumvent this problem, desymmetrization reactions were examined, as shown in **Figure 10.4**. The enantioselectivity is established in the hydrogen abstraction step by breaking the symmetry of the substrate. Radical decay in this example does not affect the enantioselectivity of the reaction but may impact the diastereoselectivity.

As illustrated in **Figure 10.5**, the chiral (salen)Mn$^+$ is able to hydroxylate cyclic ethers with very high levels of enantioselectivity, albeit in low to moderate yields.[17,18]

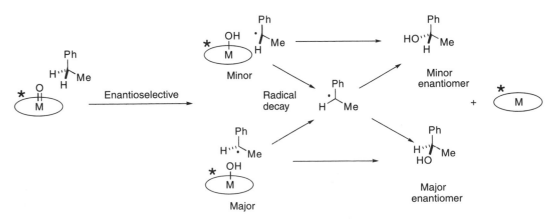

Figure 10.3. Proposed mechanism of C–H oxidation and decay of stereochemistry in the intermediate radical.

The tetrahydropyran derivative underwent asymmetric hydroxylation with reduced rate and enantioselectivity. Additionally, the product was a mixture of the lactol and lactone, and was isolated in low yield. Given the challenging nature of this reaction, these results are impressive and represent an important step in the development of asymmetric hydroxylation processes.

Two possible mechanisms have been proposed for the hydroxylation reaction. A hydrogen abstraction mechanism is illustrated in **Figure 10.3** for the hydroxylation of ethylbenzene. An alternative mechanism for the cyclic ether in **Figure 10.5** involves single electron transfer from oxygen to the metal–oxo intermediate, followed by enantioselective deprotonation (**Figure 10.6**).[15] Kinetic-isotope-effect measurements with deuterium labels α to oxygen were inconsistent with mechanisms involving direct H• abstraction.

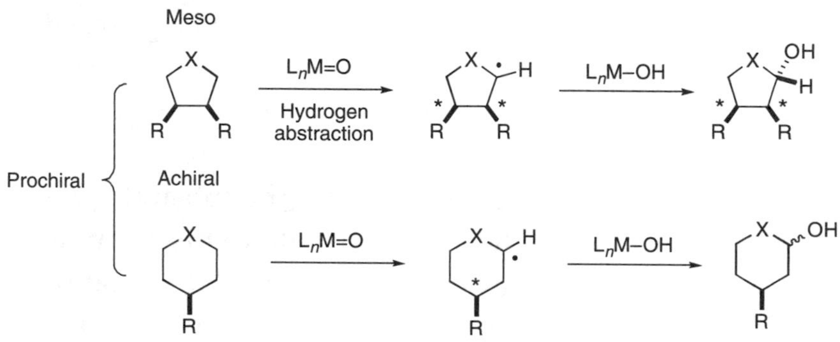

Figure 10.4. Hydrogen abstraction from meso and achiral substrates (X = O, NCO$_2$Ph) results in desymmetrization and establishes the enantioselectivity. Radical decay can result in formation of diastereomers.

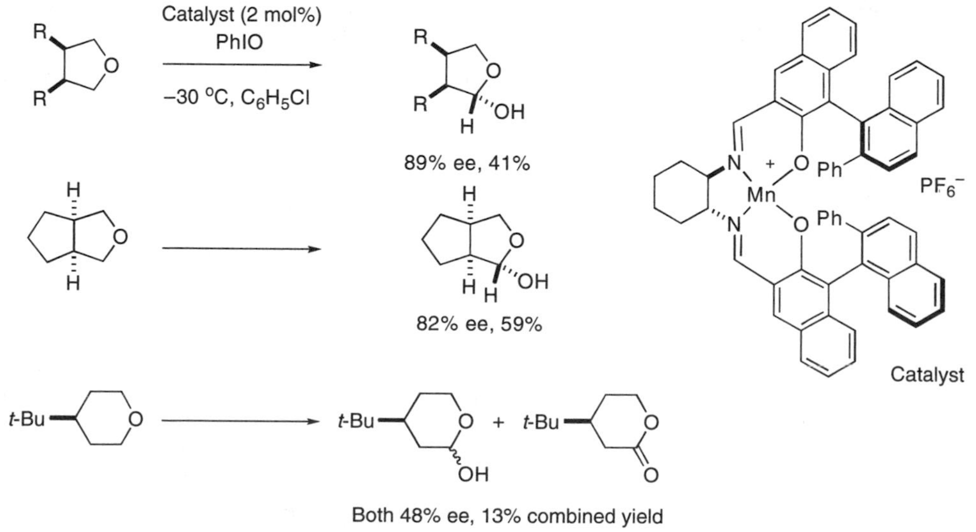

Figure 10.5. Asymmetric hydroxylation of meso and prochiral substrates with cationic (salen)Mn-based catalyst.

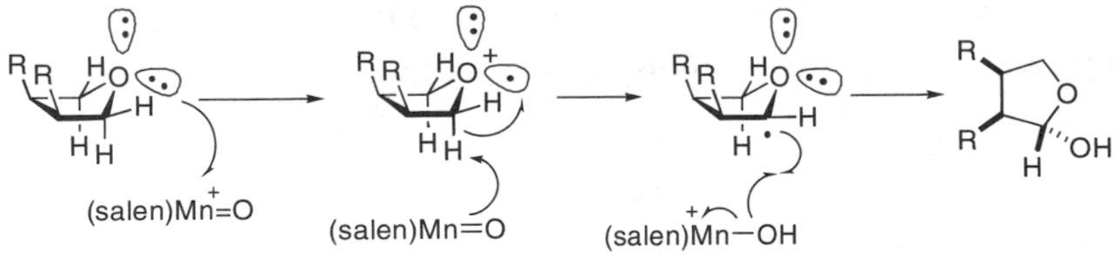

Figure 10.6. Electron-transfer mechanism for the asymmetric hydroxylation (Figure 10.5).

10.1.2.b C–H Insertions with Metal Carbenoids

An exciting C–C bond-forming reaction that shows great promise in enantioselective synthesis is insertion of a metal carbenoid into an unactivated C–H bond.[19,20] In the context of desymmetrization reactions, most examples involve intramolecular C–H insertions, but intermolecular versions are known, too. One of the best catalyst systems for the intramolecular process is based on dirhodium complexes with four monoanionic ligands, each of which spans both metal centers, as exemplified in the structure of $Rh_2(5S\text{-MEPY})_4$, an important member of this family of catalysts (**Figure 10.7**). These catalysts have the paddlewheel structural motif, which is common in such dirhodium complexes. Each rhodium of these C_2-symmetric complexes has cis nitrogens and cis oxygens with an open coordination site opposite the Rh–Rh axis.[21]

The mechanism of the rhodium-catalyzed C–H insertion is believed to proceed through metal-catalyzed extrusion of dinitrogen from the diazoester to form a transient metallocarbenoid. Reaction of the carbenoid with a C–H bond results in C–C bond formation (**Figure 10.8**). A simplified transition state is shown in **Figure 10.8**, although the details remain a topic of discussion.[22–26] The C–H bond undergoing insertion is believed to be parallel to the rhodium–carbon bond, as shown in the transition state. Only one rhodium is directly involved in the carbene insertion and it has been proposed that the second rhodium center acts cooperatively by accepting electron density and increasing the electrophilicity of the carbene moiety.[26]

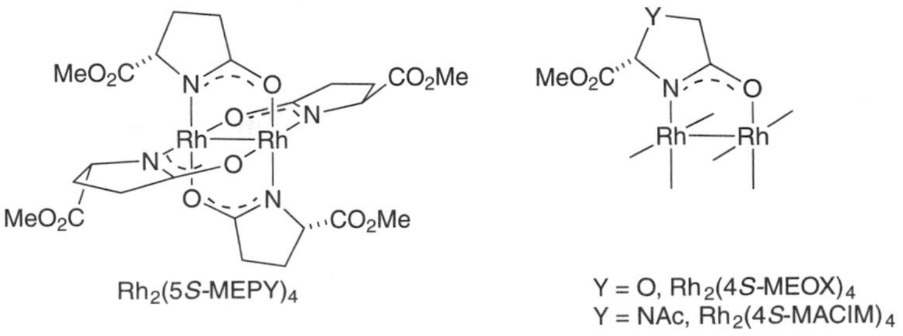

$Rh_2(5S\text{-MEPY})_4$

Y = O, $Rh_2(4S\text{-MEOX})_4$
Y = NAc, $Rh_2(4S\text{-MACIM})_4$

Figure 10.7. Structure of $Rh_2(5S\text{-MEPY})_4$ and related dirhodium catalysts.

Catalysts such as $Rh_2(5R\text{-MEPY})_4$ (**Figure 10.7**) have proven to be well-suited for desymmetrization of cyclic and acyclic diazoacetates via C–H insertion reactions.[27] As illustrated in **Figure 10.9**, insertion occurs into a C–H bond, giving a five-membered lactone with very good control over both the enantioselectivity and diastereoselectivity. The catalyst loading for these reactions can be as low as 0.1 mol%, allowing isolation of the product in 65–70% yield. In the case of the dibenzyl derivative, removal of the protecting groups is easily accomplished, leading to 2-deoxyxylolactones of high ee.

A simplified model for the chiral environment around the rhodium centers is illustrated in **Figure 10.10**. The model of $Rh_2(5S\text{-MEPY})_4$ is shown viewed down the Rh–Rh axis, where the ester groups (E) protrude forward, blocking the top two quadrants and defining the orientation of the carbene moiety. Also illustrated is a proposed intermediate and transition state leading to the observed product.

Cyclic *cis*-4-methylcyclohexyl diazoacetate also proved to be a very good substrate for desymmetrization via an intramolecular C–H insertion reaction. The product was almost entirely the cis fused bicyclic lactone (**Equation 10.3**). Use of the

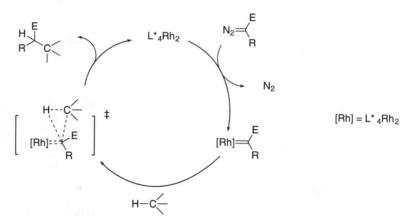

Figure 10.8. Proposed mechanism for the insertion of a metallocarbenoid into a C–H bond with a three-centered transition state.

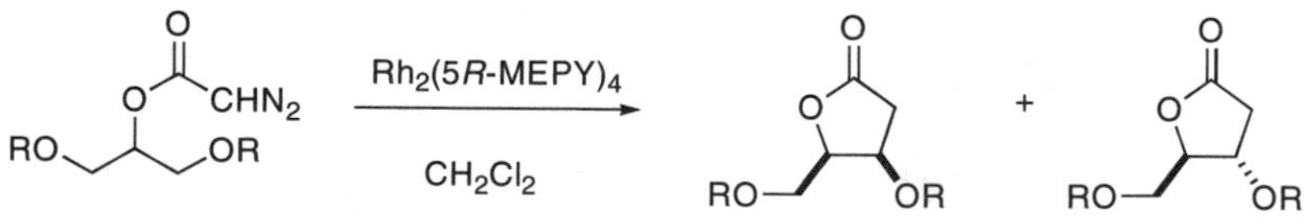

R	cis:trans	cis ee (%)	trans ee (%)
Me	93:7	97	50
Et	93:7	89	50
CH₂Ph	93:7	94	45

Figure 10.9. Highly enantioselective C–H insertion reactions catalyzed by $Rh_2(5R\text{-MEPY})_4$ under mild conditions.

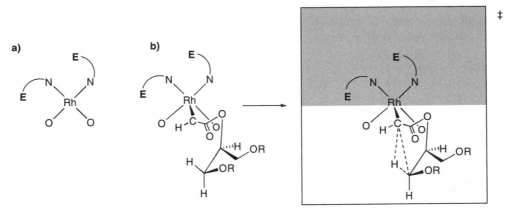

Figure 10.10. a) Stereochemical model for the chiral environment around the rhodium centers of $Rh_2(5S\text{-MEPY})_4$ (**Figure 10.7**) viewed down the Rh–Rh linkage. **E** represents the forward-protruding esters. b) The carbene intermediate and a possible transition state in the intramolecular C–H insertion reaction. The top face of the catalyst is blocked by the esters and the reaction takes place on the bottom face.

diastereomeric substrate *trans*-4-methylcyclohexyl diazoacetate resulted in predominant formation of the trans-fused lactone (**Equation 10.4**). When the newly formed carbene is in the axial position, as in **Equation 10.3**, insertion into the equatorial C–H predominates. In the case of *trans*-4-methylcyclohexyl diazoacetate, the carbene is equatorial and can attack either equatorial or axial C–H bonds, with preferential insertion into the equatorial C–H (**Equation 10.4**).[28]

Equation 10.3

$$Me \cdots \text{(cyclohexyl)} \; CHN_2, \; O\text{-}C(=O) \quad \xrightarrow[\substack{0.5 \text{ mol\%} \\ CH_2Cl_2, \text{ reflux}}]{Rh_2(5S\text{-MEPY})_4} \quad 98 : 2$$

98
(98% ee)

2
(84% ee)

Equation 10.4

$$Me \cdots \text{(cyclohexyl)} \; CHN_2, \; O\text{-}C(=O) \quad \xrightarrow[\substack{0.5 \text{ mol\%} \\ CH_2Cl_2, \text{ reflux}}]{Rh_2(5S\text{-MEPY})_4} \quad 15 : 85$$

15
(82% ee)

85
(91% ee)

These results collectively demonstrate the power of metallocarbenoid C–H insertions to functionalize unreactive C–H bonds with high enantioselectivity and diastereoselectivity under mild conditions. Related reactions are being used in the synthesis of a variety of natural and non-natural products that would be difficult to prepare efficiently with other methods.[19,20]

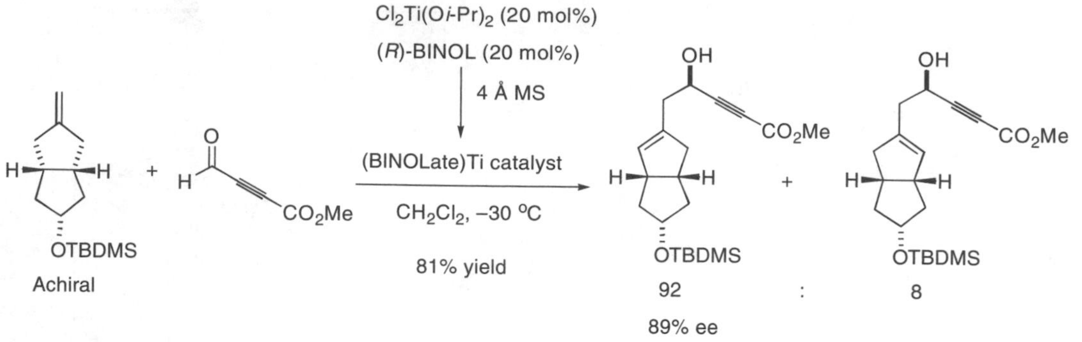

Figure 10.11. Desymmetrization by (BINOLate)Ti-based catalysts in the carbonyl–ene reaction.

10.1.3 Desymmetrization with the Carbonyl–Ene Reaction

One of the advantages of the desymmetrization strategy is that stereochemistry can be established at several centers simultaneously, frequently even when these centers are remote to the reactive site of the substrate. The carbonyl–ene reaction has been employed in the desymmetrization of olefins, furnishing homoallylic alcohols with good control over enantioselectivity and diastereoselectivity. The bicyclic prochiral olefin (**Figure 10.11**) participated in the Lewis acid catalyzed carbonyl-ene reaction with 3-formylpropiolate in the presence of a (BINOLate)Ti-based catalyst. Two products were formed in 81% yield with a 92:8 diastereomeric ratio and the major product in 89% ee.[29] The net result is efficient generation of four stereocenters in one step.

As shown in **Figure 10.11**, the catalyst is prepared by combining $Cl_2Ti(O\text{-}i\text{-}Pr)_2$ with BINOL in the presence of molecular sieves.[30] It is often extremely difficult to determine the structure of early-transition-metal catalysts due to their coordinative unsaturation[3] and their propensity to react with even traces of water. The structure of the catalyst in this reaction remains elusive.

10.1.4 Ring-Opening of Meso Oxabicyclic Alkenes

A class of meso substrates that has attracted significant attention is the oxabicyclic compounds illustrated in **Figure 10.12**.[31,32] These strained unsaturated ring systems open with formation of C–H,[33–35] C–C,[36] C–N,[37] and C–O[37] bonds. Depending on the catalyst and reaction conditions, different stereoisomers can be formed from the same starting material.[38]

A specific example of ring-opening with dimethylzinc catalyzed by palladium(II) is shown in **Figure 10.13**.[36] This reaction can be envisioned to proceed by two mechanisms. Coordination of palladium to the C–C double bond of the substrate followed by ring-opening with cleavage of the C–O bond would give an intermediate palladium allyl. Mechanistic evidence, however, points to a carbopalladation mechanism,

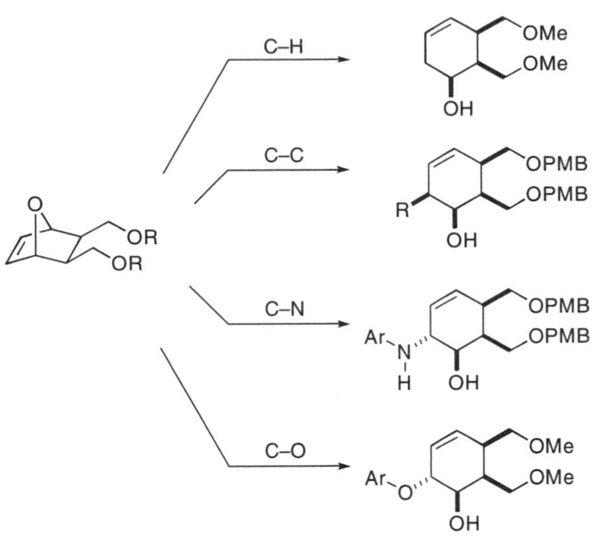

Figure 10.12. Ring-opening of oxabicyclo[2.2.1]heptenes with a variety of nucleophiles gives products with multiple stereocenters in > 90% ee.

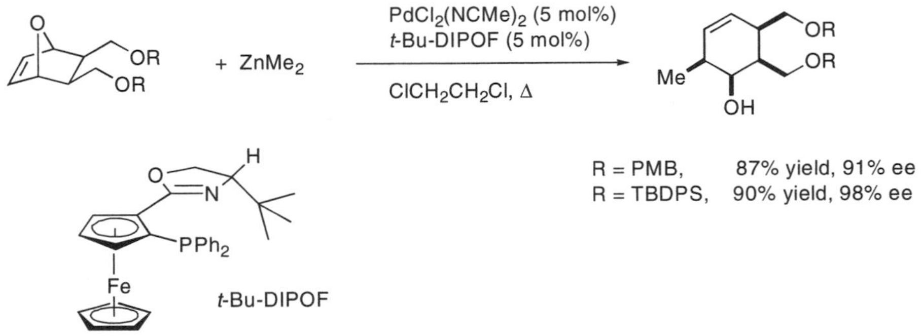

R = PMB,   87% yield, 91% ee
R = TBDPS,   90% yield, 98% ee

Figure 10.13. Enantioselective desymmetrization/ring-opening of an oxabicyclic system can be used in the synthesis of densely functionalized building blocks.

as shown in **Figure 10.14**.[39] Reversible activation of the palladium catalyst by Lewis acidic ZnMe$_2$ generates a cationic intermediate. Addition of the Pd–Me bond across the alkene is followed by β-oxygen elimination to give a ring-opened palladium alkoxide that undergoes subsequent transmetallation with zinc to liberate the palladium catalyst.

The utility of these ring-opening reactions was demonstrated in the total synthesis of the polyether antibiotic ionomycin (**Figure 10.15**).[40] In this synthesis, two of the four fragments were prepared by ring-opening of the [3.2.1]-oxabicyclic alkene. The C$_{17}$–C$_{23}$ fragment was synthesized by Ni(COD)$_2$/(S)-BINAP-catalyzed desymmetrization with ring-opening, employing DIBAL-H as the hydride source, in 95% yield and 93–95% ee. The synthesis of the C$_2$–C$_{10}$ fragment began with the desymmetrization of the oxabicyclic alkene catalyzed by Cl$_2$Pd[(R)-DIPOF] with ZnMe$_2$ and catalytic Zn(OTf)$_2$ Lewis acid to aid formation of the cationic palladium catalyst. The product was isolated in 80% yield and 94% ee.

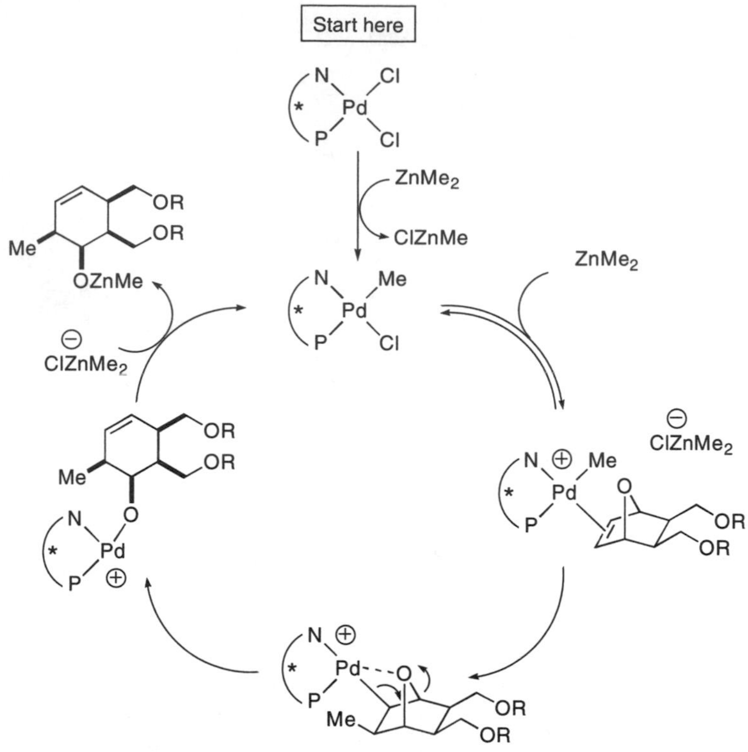

Figure 10.14. Proposed mechanism for the asymmetric ring-opening of oxabicyclic alkenes catalyzed by palladium.

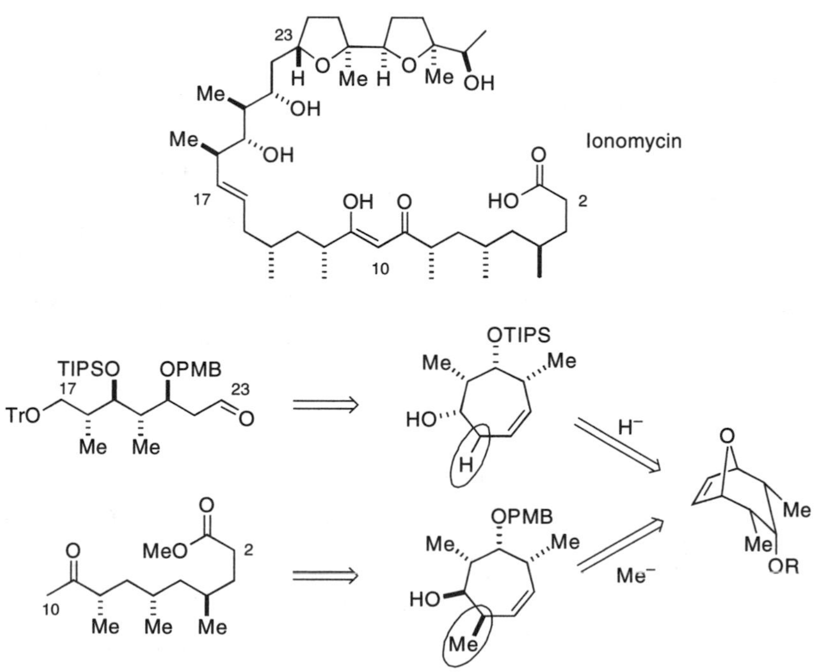

Figure 10.15. Disconnection of two fragments utilized in the synthesis of the polyether antibiotic ionomycin. The bonds formed in the forward sense are indicated.

As shown above, these desymmetrization reactions can be performed with a variety of nucleophiles in high yields and with high enantioselectivities (> 90% in most cases). Ring-opening occurs with the establishment of multiple stereocenters, making such methods attractive for applications in organic synthesis.

10.2 Compounds with Two Reactive Enantiotopic Groups

10.2.1 Reaction with Only One Enantiotopic Functional Group

10.2.1.a Combining Desymmetrization with Kinetic Resolution: Meso Diols

Meso diols are an example of a class of substrates that contain symmetrically disposed enantiotopic functional groups. Significant effort has been expended on the development of nonenzymatic methods for the desymmetrization of these important substrates and planar chiral complexes have been shown to be excellent catalysts for this process (**Figure 10.16**).[41] The mechanism of operation of nucleophilic planar catalysts was discussed in Chapters 2 and 7.

The desymmetrization of meso substrates with enantiotopic functional groups is synthetically very attractive, because the reactions involve an initial desymmetrization that can be coupled with a kinetic-resolution step, leading to enhancement of the product ee. The idea is illustrated in the context of the desymmetrization of diols outlined above (**Figure 10.16**), but the concept is general and has been applied to a variety of desymmetrization processes to prepare material of extremely high ee (see Section 16.2.3).[42]

In the reaction of the meso diol with the planar chiral catalyst and acetic anhydride, the initial desymmetrization results in formation of unequal amounts of the enantiomeric monoesters (**Figure 10.17**). The enantiomeric carbinols of the starting diol constitute the matched and mismatched combinations with the catalyst. The matched pair will lead to rapid formation of the major enantiomer of the monoester, while reaction of the mismatched combination will be slower and generate the minor enantiomer. Since both the monoesters contain one remaining reactive carbinol, they can undergo subsequent acylation, which comprises a kinetic resolution. It is now the minor enantiomer of the monoester that forms the matched pair with the chiral

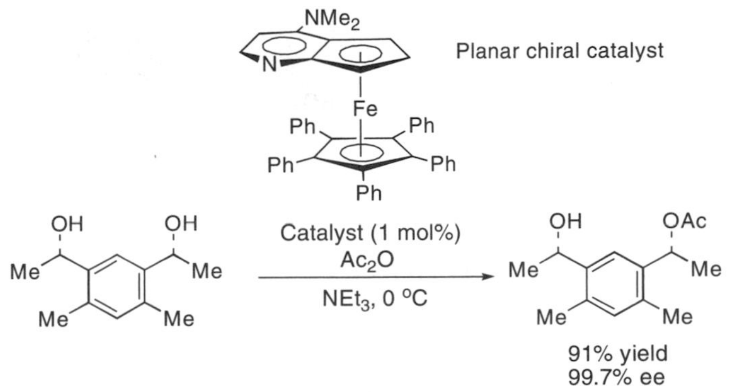

Figure 10.16. Efficient desymmetrization of a meso diol with a planar chiral catalyst.

catalyst and reacts more quickly. As the minor enantiomer is converted to the meso diester, the enantiomeric excess of the monoester will increase. The longer the reaction is allowed to continue, the greater the ee of the monoester. Exceptionally high enantioselectivities can be obtained with a selective catalyst. As with all kinetic resolutions, however, increased product ee comes at the expense of the yield of the desired monoester.

As exemplified by the case above, small-molecule catalysts are particularly adept at desymmetrizations where the stereogenic center is in close proximity to the reactive site. As the distance between the reactive site and the stereogenic or prochiral center increases, the desymmetrization becomes more challenging. A remarkable example of desymmetrization of a substrate with a distance of over 5.7 Å between the reactive hydroxyl groups and the prochiral center is illustrated in **Figure 10.18**.[43] It is possible that subtle molecular recognition interactions are the key to success in this system. Understanding how peptide-based catalysts of this type discriminate be-

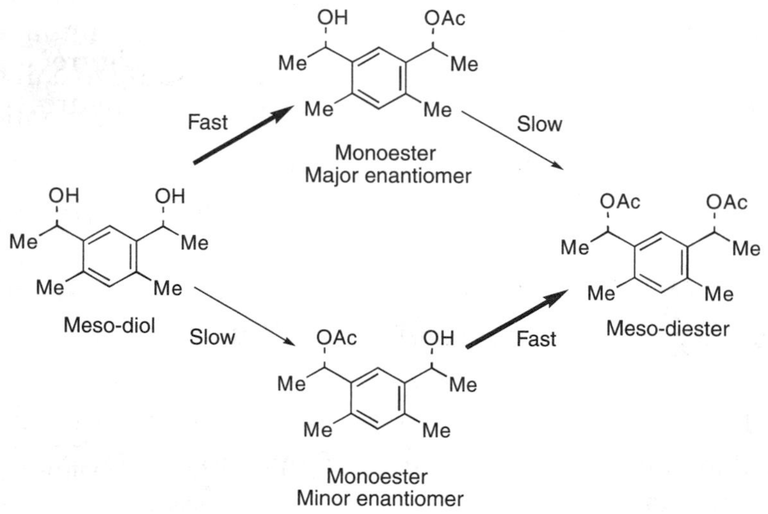

Figure 10.17. Desymmetrization/kinetic resolution of a meso diol.

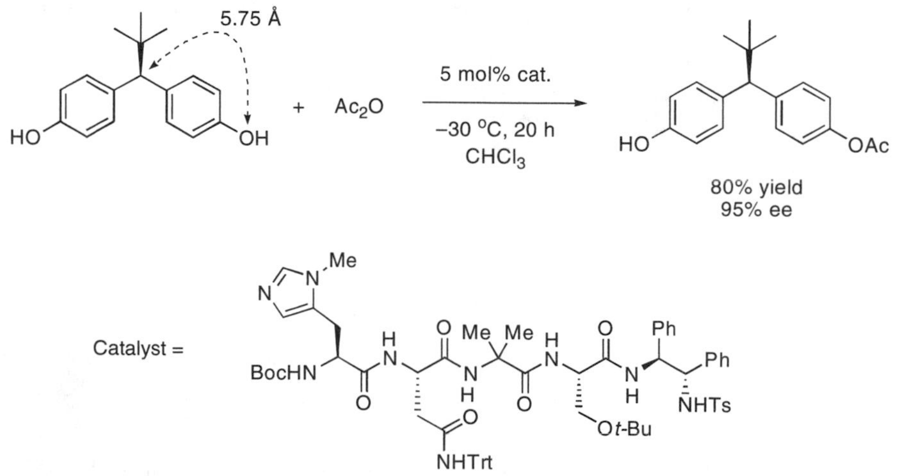

Figure 10.18. Remote desymmetrization of a bis(phenol).

tween enantiotopic groups and the advancement of catalytic systems to differentiate remote sites will be areas of activity in the future.

10.2.1.b Ring-Opening of Meso Centrosymmetric Bis(epoxides)

The beauty of desymmetrization reactions is that they can be applied to complex meso and centrosymmetric substrates to establish multiple stereocenters. Several natural products contain hidden symmetry. Strategies that target the synthesis of symmetrical intermediates and then break the symmetry, often lead to more expedient routes.[44,45] Polycyclic ethers are a class of natural products that contain pseudo-centrosymmetric motifs buried in the complex molecular architecture and hemibrevetoxin B is one member of this group. The pseudo-centrosymmetric portion of the structure is highlighted in **Figure 10.19**. The synthesis of hemibrevetoxin B consisted of the retrosynthetic disconnection to a key monoepoxide intermediate. In the forward direction, an elegant double tetrahydropyran-to-oxepane ring-expansion gives the required skeleton.[46,47] Retrosynthetic disconnection of the monoepoxide leads to the achiral centrosymmetric bis(epoxide).

Desymmetrization of the centrosymmetric bis(epoxide) (**Figure 10.20**) for the synthesis of hemibrevetoxin B was accomplished by the selective hydrolysis of one of the

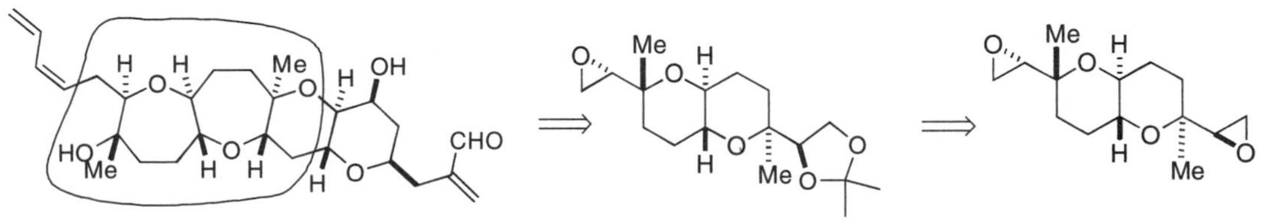

Hemibrevetoxin B   Key intermediate   Meso

Figure 10.19. Structure of hemibrevetoxin B, highlighting a fragment that was disconnected to a epoxide key intermediate. This epoxide can be further disconnected to a centrosymmetric bis(epoxide).

(salen)Co(OAc)
20 mol%
1.1 equiv. H₂O

Achiral
centrosymmetric
bis(epoxide)

98% yield
> 95% ee

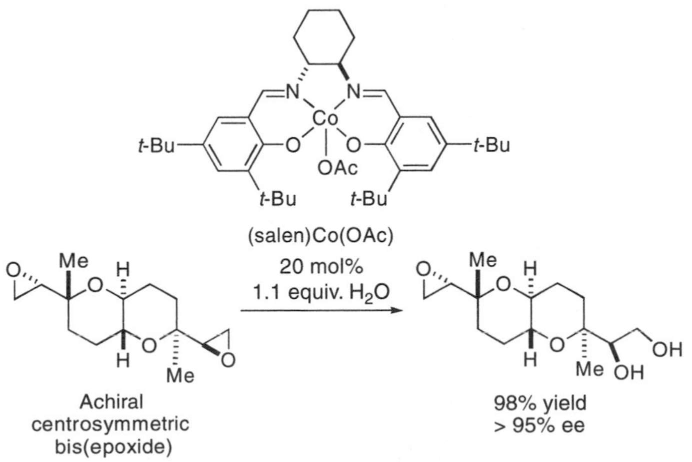

Figure 10.20. Hydrolytic epoxide-opening of the centrosymmetric bis(epoxide). Desymmetrization of the bis(epoxide) provides access to a key intermediate in the synthesis of hemibrevetoxin B (**Figure 10.19**).

enantiotopic epoxides, which have opposite configurations. The catalyst chosen for this application was (salen)Co(OAc), because it exhibits exquisite selectivity in the hydrolytic kinetic resolution of terminal epoxides (Chapter 7).[7,50] The (salen)Co (OAc) complex promoted the epoxide-opening with water to furnish the diol precursor to the key intermediate for the synthesis of hemibrevetoxin B with excellent selectivity.[48,49]

10.2.1.c Epoxidation of Dienes

10.2.1.c.1 Achiral Divinyl Carbinols

The desymmetrization of symmetric divinyl carbinols with the Sharpless–Katsuki epoxidation catalyst provides access to densely functionalized epoxy alcohols with very high selectivities. The high intrinsic ability of the Sharpless–Katsuki catalyst to discriminate between the diastereotopic *and* enantiotopic olefin faces makes the diisopropyl tartrate-based system particularly attractive. These same features are responsible for the high levels of efficiency in the kinetic resolution of secondary allylic alcohols,[51,52] as detailed in Chapter 7. The desymmetrization combines an asymmetric synthesis with a subsequent kinetic resolution to furnish products with extraordinary levels of enantiopurity.[53] For a further discussion, see Figure 16.18 and the accompanying text. This method has been used to install the stereochemistry in a variety of natural product syntheses, some of which are highlighted below.

Examination of the asymmetric epoxidation of the divinylcarbinol in **Figure 10.21** clearly indicated that the ee of the product increased with conversion. This observation is consistent with further epoxidation of the minor epoxy alcohol enantiomer in the kinetic-resolution process.[42] The epoxy alcohol product of this reaction was subsequently converted to riboflavin. It should be noted that these desymmetrization reactions are usually conducted using stoichiometric diisopropyl tartrate (DIPT) and titanium tetraisopropoxide.

The model for the differential reactivity of the enantiotopic groups with the Sharpless–Katsuki catalyst (**Figure 10.22**) is similar to the model outlined in Chapter 7. Orientation of the mismatched double bond away from the catalyst lowers the activation energy to epoxidation (left structure in **Figure 10.22**).

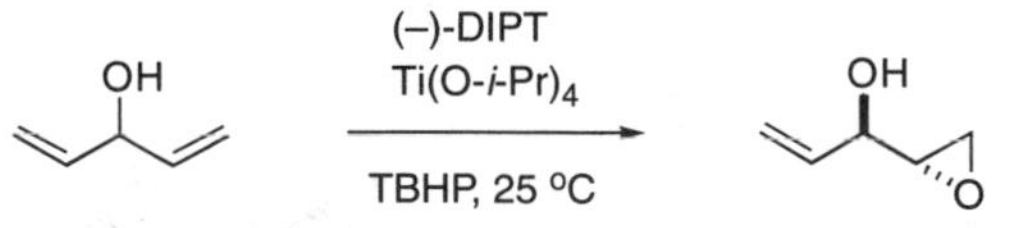

Time (h)	ee (%)
3	84
24	93
140	≥97

Figure 10.21. Desymmetrization of the divinyl carbinol with the Sharpless–Katsuki asymmetric epoxidation exhibits an increase in the ee of the epoxy alcohol due to more rapid second epoxidation of the minor enantiomer.

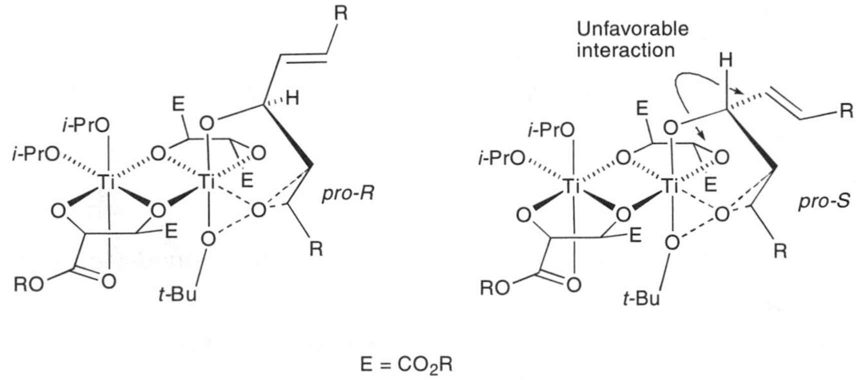

E = CO₂R

Figure 10.22. Models for the enantiotopic group differentiation in desymmetrizations of divinyl carbinols with the Sharpless–Katsuki catalyst.

10.2.1.c.2 Meso Divinyl Carbinols

A remarkable example of the utility of desymmetrization reactions with allylic alcohols and the Sharpless–Katsuki asymmetric epoxidation reaction is shown in **Figure 10.23**.[54] The meso diene is transformed into an epoxide product that has seven chiral centers and was produced in high yield with excellent enantioselectivity.[44] Similar desymmetrizations have been applied to the synthesis of other natural products.[55-57]

Another key desymmetrization of an advanced intermediate is in the synthetic route to a secondary metabolites of crude plant extracts of *Celastraceae*. The polyhydroxylated 4β-hydroxyalatol is one such example and is illustrated in **Figure 10.24**. Disconnection leads to an epoxy alcohol that could arise from desymmetrization of the divinyl carbinol.[58]

Use of stoichiometric titanium tetraisopropoxide and (+)-DIPT to desymmetrize the divinyl carbinol gave the monoepoxide in low yield with 14% ee. Using a

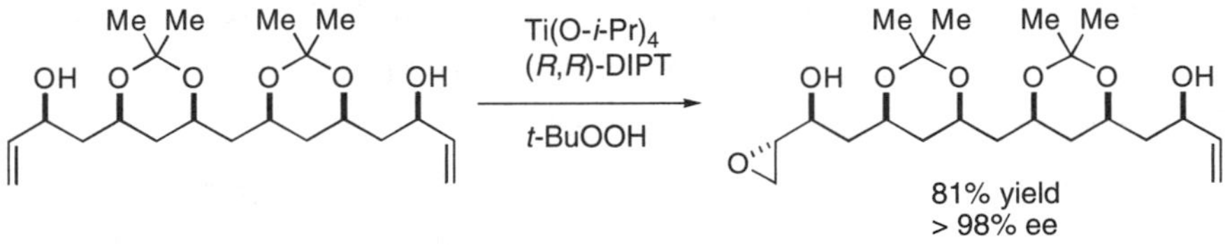

Figure 10.23. Desymmetrization of a meso diol with the Sharpless–Katsuki asymmetric epoxidation.

4β-Hydroxyalatol

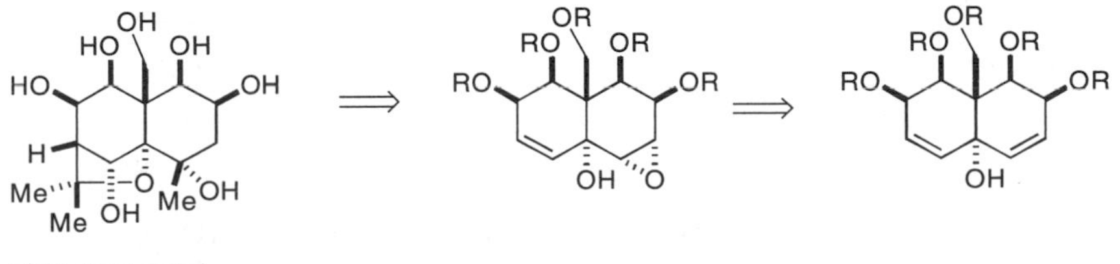

Figure 10.24. Retrosynthetic analysis of 4β-hydroxyalatol leads to a meso-divinyl carbinol.

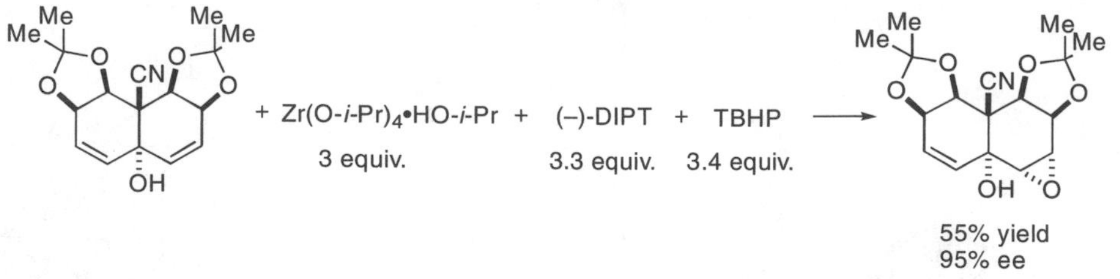

Figure 10.25. Desymmetrization of a divinyl carbonyl with a zirconium-modified Sharpless reagent.

zirconium-modified reagent gave the desired product, which was isolated in 55% yield based on recovered starting material (**Figure 10.25**).

The above case illustrates that there is room for improvement in the efficiency of some of these desymmetrization processes with complex substrates. However, they do allow rapid access to advanced intermediates that are of demonstrated utility in natural product synthesis. Furthermore, these substrates may be prepared via a two-directional synthesis, such that the two chain ends are elaborated simultaneously with subsequent chain-terminus differentiation, as exemplified in **Figure 10.23**.[45]

10.2.2 Reaction with Both Enantiotopic Functional Groups

10.2.2.a Desymmetrization of Achiral Dienes via Catalytic Asymmetric Hydrosilylation

Hydrosilylation of olefins results in the addition of an Si–H bond across the C–C double bond[59] and allows access not only to organosilanes but also alcohols and alkyl halides upon oxidation of the newly formed Si–C bond.[60,61] Intramolecular variants have proven more successful, because it is easier to control the regioselectivity in the intramolecular hydrosilylation than in the intermolecular counterpart. Additionally, the cyclic intermediates in the intramolecular hydrosilylation allow for better control over enantioselectivities. Like the hydroacylation reaction introduced in Chapter 4,[62] the intermediate in the catalytic reaction is likely to have two new sigma bonds (M–H, M–Si bonds formed after oxidative addition of the silane), and a C–C double bond coordinated before the cyclization ensues. Therefore, the catalyst should have three coordination sites available, making (diphosphine)Rh(solvent)$_2^+$ catalysts attractive.

Desymmetrization of divinyl carbinols has been examined as an entry into enantioenriched 1,3-diols,[63] which are useful synthetic intermediates in the synthesis of polypropionate-derived natural products. In the presence of an (R,R)-DIOP-based rhodium catalyst, intramolecular hydrosilylation proceeded over the course of several days. It was found that the enantioselectivities depended on the silicon substituents, with aryl groups giving the highest enantioselectivities and diastereoselectivities (**Figure 10.26**). Enantioselectivities in this process reached 93% when a 3,5-dimethylphenyl derivative was employed.[63]

The mechanism of the hydrosilylation reaction is shown in **Figure 10.27** and likely proceeds via initial Si–H oxidative addition, followed by olefin coordination and silyl olefin insertion, which is believed to be the enantioselective and turnover-limiting step.[64-66] Detailed deuterium isotopic labeling experiments have also demonstrated that there are several other unproductive steps that have been excluded from the simplified mechanism in **Figure 10.27**.

The model for asymmetric induction is similar to other diphosphine-based catalysts discussed in Chapter 4. The enantioselection is thought to be controlled by the position of the pseudoequatorial phenyl groups of the diphosphine and the olefin face selection, as shown in **Figure 10.28**. In the olefin complex, interaction of the coordinated olefin with the pseudoequatorial phenyl groups is minimized.

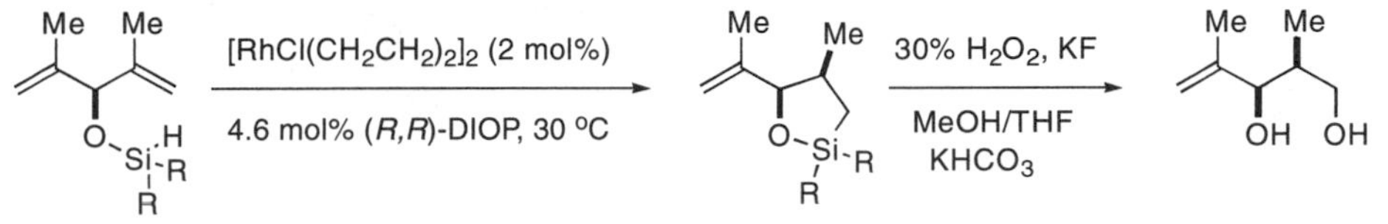

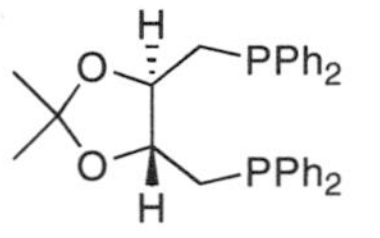

R	Yield (%)	syn:anti	syn ee (%)
Me	60	86:14	18
Ph	94	98:2	84
3-C$_6$H$_4$-Me	80	99:1	87
3,5-C$_6$H$_3$-Me$_2$	66	99:1	93

Figure 10.26. Desymmetrization of dienes employing catalytic asymmetric hydrosilylation chemistry. Oxidation of the product provides a valuable 1,3-diol.

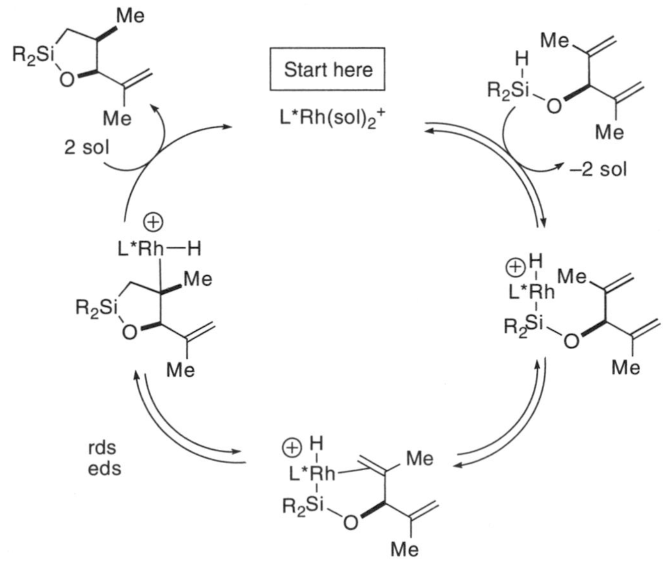

Figure 10.27. Proposed mechanism for the asymmetric hydrosilylation.

A fascinating example of a desymmetrization reaction involving hydrosilylation to form chiral spiranes that have axial chirality has been reported (**Figure 10.29**).[67] Oxidative addition of one of the homotopic Si–H bonds gives an intermediate silyl–hydride complex. In the intramolecular hydrosilylation that follows, the catalyst differentiates between the diastereotopic alkenyl groups as well as their diastereotopic faces in the silyl olefin insertion (**Figure 10.29**). Stereocenters are generated at carbon and silicon with only the (S)-configuration shown at the silicon center. The two diastereomers formed in this process are the (S,S_{Si})- and (R,S_{Si})-configurations. The next step is a diastereoselective hydrosilylation in which the two faces of the remaining alkene are differentiated, generating the third stereocenter with three sets of diastereomers possible (the C_1-symmetric diastereomers are equivalent). Employing an analog of (R,R)-DIOP, namely (R,R)-TBDMS-SILOP, the diastereomeric ratio was 98:2 with the (S,R_{Si},S)-diastereomer predominating.

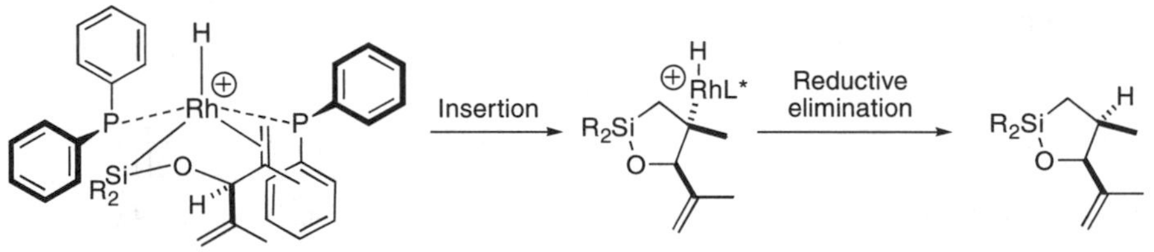

Figure 10.28. Model for the stereochemical induction in the desymmetrization via asymmetric hydrosilylation using (S,S)-DIOP. The pseudoequatorial P-phenyl rings are drawn in bold.

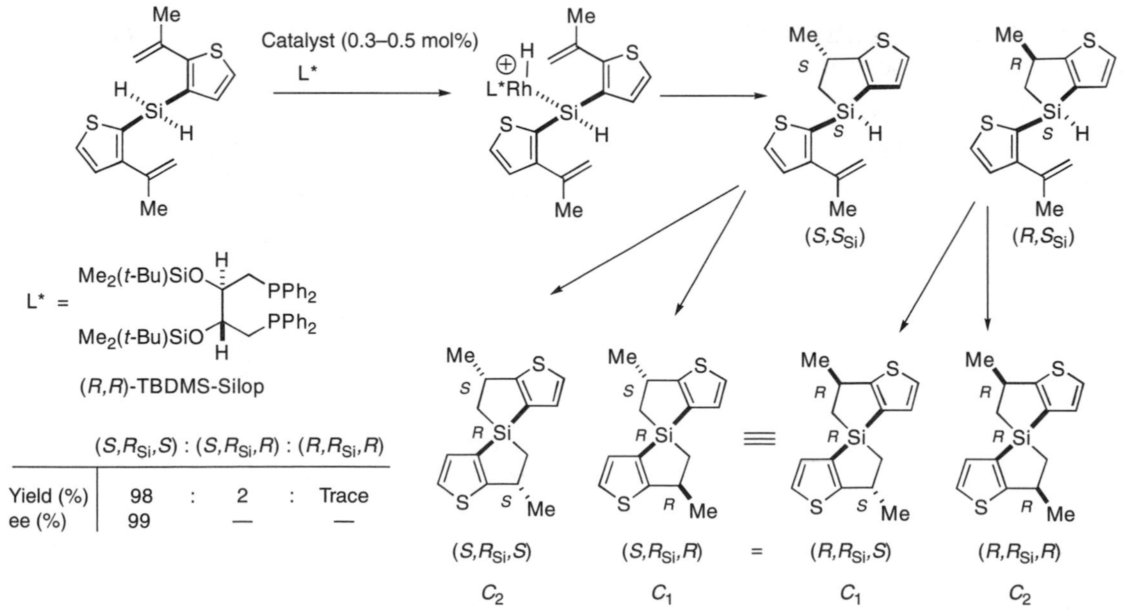

	(S,R_{Si},S) : (S,R_{Si},R) : (R,R_{Si},R)		
Yield (%)	98 :	2 :	Trace
ee (%)	99	—	—

Figure 10.29. Hydrosilylation begins with homotopic Si–H insertion, followed by olefin insertion and reductive elimination [only the (S)-configuration at Si is shown]. A diastereoselective hydrosilylation provides the spirocycles with high selectivity.

10.2.2.b Desymmetrization of Achiral and Meso Dienes with the Heck Reaction

Despite the large number of catalytic asymmetric reactions, examples of C–C coupling reactions that are general and highly enantioselective remain small. In this regard, one reaction that has proven to be very powerful in synthesis is the asymmetric Heck reaction.[68] The Heck reaction typically entails the addition of an aryl or vinyl group across an olefin, followed by reformation of the C–C double bond. A proposed mechanism for this reaction is illustrated in **Figure 10.30**.[69–72] Palladium(II) acetate, a common starting material for the Heck reaction, is reduced by tertiary amines to palladium(0). Oxidative addition of the aryl or vinyl halide or triflate forms a palladium(II) intermediate possessing a Pd–C bond. Coordination of the olefin and migratory insertion of the vinyl or aryl group results in formation of C–C and Pd–C bonds. If the newly formed Pd–C bond has β-hydrogens that can be oriented *syn* to the Pd–C bond, β-hydride elimination will occur to provide an olefin and a Pd(II) hydride that undergoes deprotonation to reform palladium(0).

The enantioselectivity-determining steps in the mechanism are coordination of the olefin and the migratory insertion of the Pd–R substituent. Under conditions where the group X coordinates strongly to palladium, the reaction is believed to involve loss of phosphine to open a coordination site for the olefin. When a chiral bidentate ligand is employed, the consequences of dissociation of one arm of the chelating ligand may be detrimental to the enantioselectivity (**Figure 10.30**, path **A**). In contrast, in the cationic route (**Figure 10.30**, path **B**) the X⁻ group dissociates while

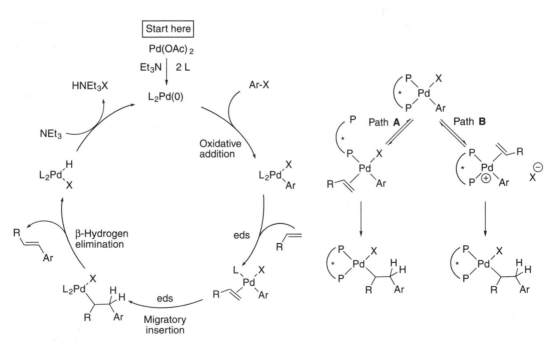

Figure 10.30. Left: Proposed mechanism of the Heck reaction, where the group X does not dissociate. Right: Comparison of the neutral and cationic mechanisms.

the bis(phosphine) ligand remains chelating to the palladium center. Path **B** can be favored by addition of silver salts with weakly coordinating anions to sequester the halide or by employing vinyl or aryl triflates. Triflates are weakly coordinating ligands that readily dissociate in the polar solvents employed in the asymmetric Heck reaction. Reactions proceeding through the cationic pathway generally exhibit higher enantioselectivities.[68,72,73]

The asymmetric Heck reaction has been applied to a number of elegant syntheses for the construction of complex ring systems. Illustrated in **Figure 10.31** is the desymmetrization of an achiral diene via a related process to give a 5,5-ring system that is

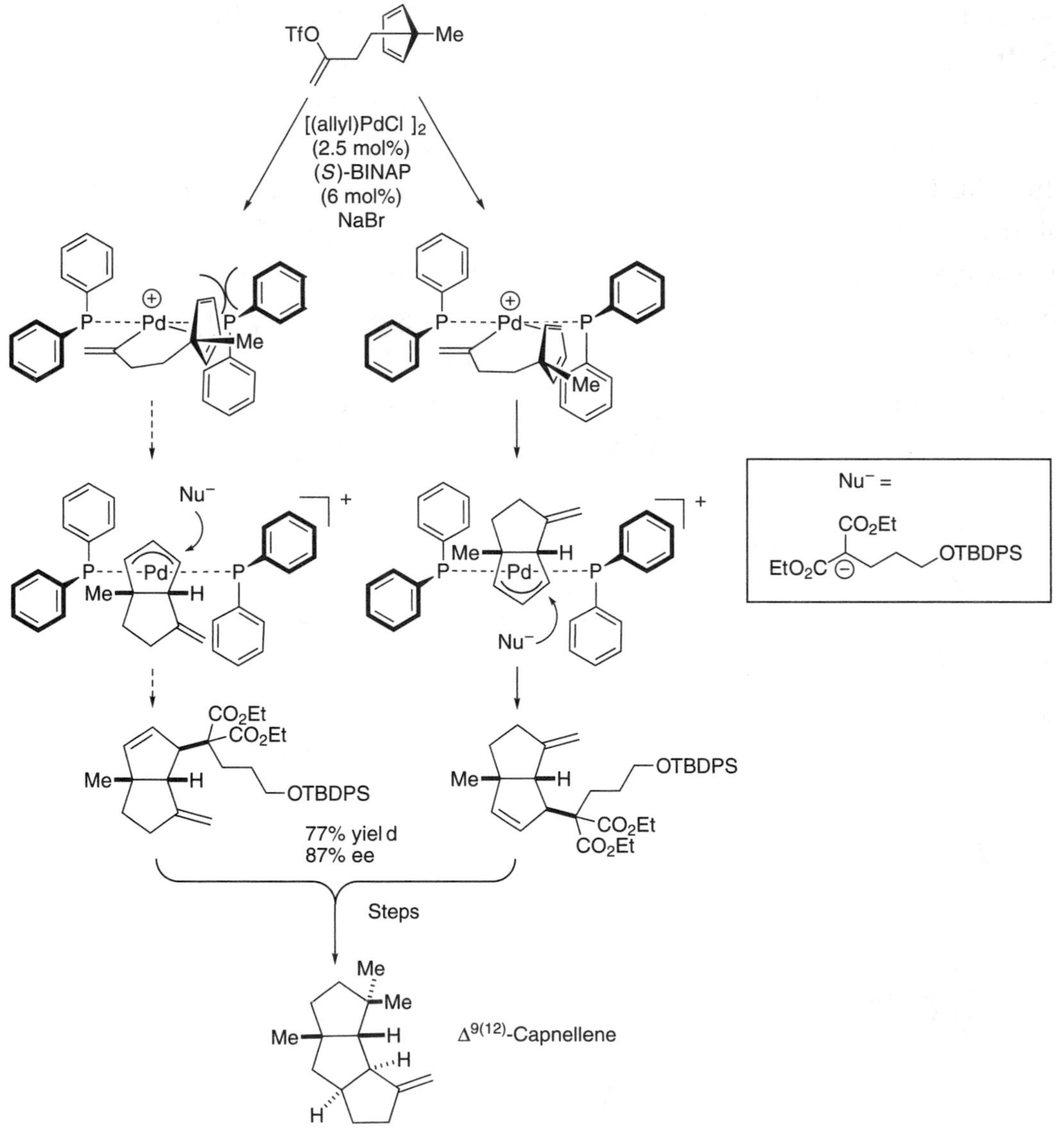

Figure 10.31. Proposed model for the stereoinduction in the asymmetric palladium cascade reaction to form the 5,5-ring system. The left pathway leading to the minor enantiomer suffers from an unfavorable steric interaction between the bound substrate and pseudoequaltorial phenyl of (S)-BINAP.

found in the backbone of numerous natural products. The reaction is promoted by a (BINAP)Pd-based catalyst formed by the combination of [Pd(allyl)Cl]$_2$ with (S)-BINAP, NaBr, and the nucleophile. This example is peculiar, because NaBr is added to the reaction mixture. Nonetheless, it is proposed that the reaction proceeds via the cationic mechanism in **Figure 10.30**. Oxidative addition of the vinyl triflate is followed by coordination of a double bond, giving two possible diastereomeric intermediates. In the model for stereochemical induction (**Figure 10.31**), coordination of the lower double bond results in encroachment of the cyclopentadiene group into the upper-right quadrant, which is occupied by the protruding pseudoequatorial phenyl of the (S)-BINAP. In contrast, when the upper double bond is coordinated, the cyclopentadienyl group lies in the open lower-right quadrant, where nonbonded interactions are minimized. Recall that the pseudoaxial phenyl groups of BINAP complexes are directed away from the metal binding sites (Chapter 4). Insertion results in formation of a palladium π-allyl intermediate for which β-hydride elimination is not viable, because it would result in formation of an allene in a five-membered ring. Rather, the π-allyl palladium intermediate is trapped by a β-dicarbonyl anion in a diastereoselective fashion. The product, formed in 77% yield and 87% ee, is an intermediate in the synthesis of $\Delta^{9(12)}$-capnellene.[74,75]

Equation 10.5 shows the application of the asymmetric Heck reaction to the desymmetrization of an enantiotopic diene to afford a bicyclic enone with very good enantioselectivity and yield. After the insertion step in the mechanism, the β-hydrogen elimination generates an enol that undergoes isomerization to the ketone product. This ketone, formed in 76% yield and 86% enantioselectivity, was taken on to a key intermediate in the synthesis of vernolepin.[76,77]

Equation 10.5

An impressive example of the power of the asymmetric Heck reaction and the ingenuity of chemists is the late-stage desymmetrization of a meso intermediate in the synthesis of the polypyrrolidinoindoline alkaloids quadrigemine C and psycholeine (**Figure 10.32**). Quadrigemine C represents a formidable challenge due to the contiguous stereogenic quarternary centers connecting the two halves of the molecule and the stereogenic diaryl quaternary stereocenters. The palladium-controlled double Heck cyclization was performed at 80 °C with 1,2,2,6,6-pentamethylpiperidine (PMP) as base and stoichiometric palladium and (R)-Tol-BINAP. The major product was the

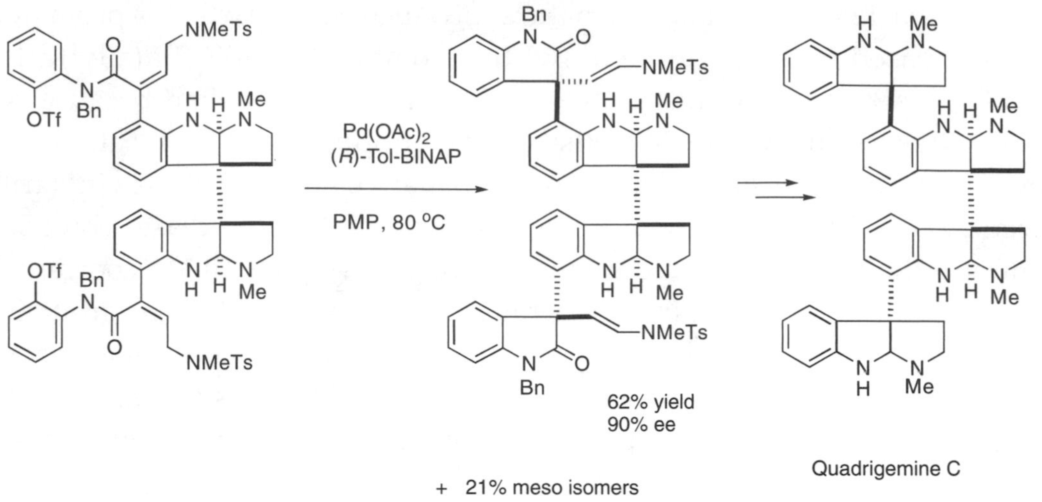

Figure 10.32. Desymmetrization of a late-stage intermediate with the asymmetric Heck reaction in the synthesis of quadrigemine C.

C_1-symmetric dioxindole, isolated in 62% yield with 90% ee.[78] See **Figure 16.21** and the accompanying text for further discussion.

10.2.2.c Desymmetrization of Biaryls Through Enantioselective Coupling

Enantioenriched 1,1'-binaphthyls, such as BINOL, BINAP, and related derivatives, have had an enormous impact on asymmetric catalysis. The enantioselective synthesis of enantioenriched ligand precursors for the construction of these ligands has focused on the formation of the central C–C biaryl bond. A unique approach to atropisomeric biaryl ligand precursors involves desymmetrization of preformed achiral biaryls through enantioselective cross-coupling reactions.[79,80]

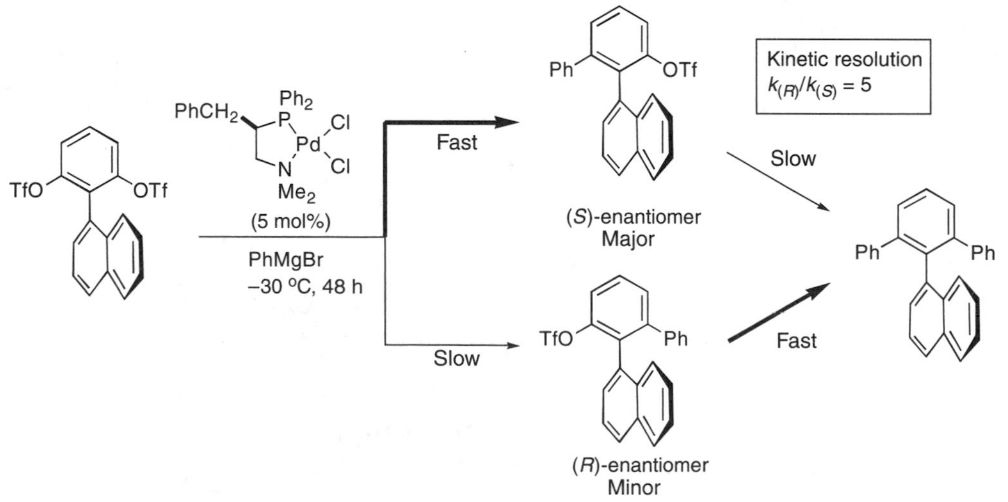

Figure 10.33. Enantioselective desymmetrization of the bis(triflate) provides the monophenyl product with high enantioselectivity. Further reaction of the enantioenriched monophenyl triflate results in a kinetic resolution of the product with a selectivity factor of 5.

Kumada coupling with phenylmagnesium bromide and a bis(triflate) (**Figure 10.33**) catalyzed by $Cl_2Pd([(S)$-PhePhos] at low temperature resulted in formation of the monophenylated product in 87% yield and 93% ee. The monophenylated product ee depended on the conversion, suggesting that a kinetic resolution in the second step of the reaction had a significant impact. In a separate control experiment with the racemic monophenylated product, the monophenyl product has 17% ee after 20% conversion to the diphenyl product, corresponding to $k_{rel} = 5$ (**Figure 10.33**). In this experiment it was the minor (R)-enantiomer of the monophenylation product that underwent the second phenylation faster.

The monotriflate products formed in the desymmetrization are useful in further coupling reactions for the synthesis of chiral ligands. As outlined in **Equation 10.6**, diphenylphosphinylation of the monotriflate takes place without racemization. Reduction with trichlorosilane and triethylamine provides the atropisomeric monophosphine in good overall yield. This ligand has been demonstrated to be useful in the asymmetric hydrosilylation–oxidation of styrene to provide 1-phenylethanol in 91% ee.[79]

Equation 10.6

10.2.2.d π-Allyl Palladium Reactions with Meso Substrates

Transition-metal-catalyzed allylic substitution reactions have emerged as one of the most powerful methods in asymmetric catalysis, because they lend themselves to the formation of both carbon–carbon and carbon–heteroatom bonds.[81–83] This class of reactions involves the generation of a metal π-allyl intermediate that can participate in several types of transformations with excellent control over chemoselectivity, regioselectvity and enantioselectivity. Numerous investigations have been performed with late-transition-metal complexes, particularly of palladium and rhodium. Extensive effort in ligand design has resulted in the development of many chiral ligands. As a result, high levels of enantioselectivity can be obtained with a variety of nucleophiles and metal allyls.

An important subclass of asymmetric allylic substitution reactions is desymmetrization of meso enediol diesters and related substrates. These substrates have been exploited in a wide array of enantioselective total syntheses of biologically - active compounds. This general class of reaction is illustrated in **Figure 10.34** in the

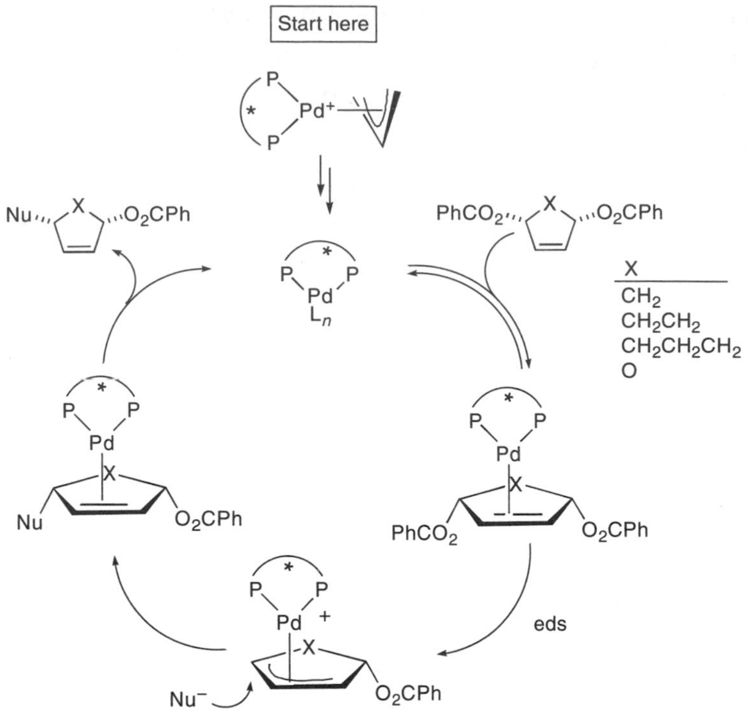

Figure 10.34. Proposed mechanism for the desymmetrization of meso substrates.

context of the mechanism of the desymmetrization of a generic *meso*-1,4-bis(acyloxy)-2-ene substrate. The mechanism is believed to proceed by generation of a palladium(0) species that complexes the substrate opposite to the leaving groups. The next step is ionization of one of the enantiotopic leaving groups to generate a π-allyl intermediate that subsequently undergoes nucleophilic addition.[83] Dissociation of the product regenerates the catalyst. The enantioselectivity-determining step in this process is the ionization of the leaving group. Despite the fact that the product still posseses an allylic benzoate, displacement of the remaining leaving group by the catalyst represents the mismatched combination and is significantly slower than the first ionization. This difference in reactivity has been exploited in the synthesis of complex natural products as outlined below.

One of the key factors in the phenomenal success of this process was the development of diphosphine ligands based on easily resolvable chiral diamines.[84] The ligand that has received the most attention is illustrated in **Figure 10.35**. It is believed to bind to palladium through the phosphorus centers. The situation is more complicated, however, with non-C_2-symmetric palladium complexes formed that are in equilibrium with oligomeric compounds, wherein the phosphorus centers of the ligand bind to different palladiums.[85] In the monomeric adducts, the ligand enforces a large P–Pd–P bite angle, θ. Opening of the bite angle in diphosphine ligands pushes the *P*-bound aryl groups that transmit the asymmetry forward to envelope the π-allyl moiety.[86,87] A related ligand that was characterized crystallographically exhibited a P–Pd–P bite angle of 110.5°,[88] significantly larger than the idealized 90° bite angle preferred by square-planar compounds. The ligand and catalyst shown in **Figure 10.35** will be used throughout the remainder of this section and denoted L* (ent-L* for the antipode).

Asymmetric desymmetrization of 2,5-diacyloxy-2,5-dihydrofurans with the palladium–diphosphine catalyst from **Figure 10.35** allowed construction of the key building blocks for the synthesis of a variety of natural products. In the syn-thesis of C-2-*epi*-hygromycin, the desymmetrization with the sodium salt of 1-phenylsulfonylnitroethane, an acyl anion equivalent, proceeded in excellent yield and enantioselectivity to give the monosubstitution product (**Figure 10.36**).[89–91] Although inconsequential in the synthesis, the monobenzoate product is formed as a 5:1 mixture of diastereomers, indicating that the allyl intermediate is imparting stereochemical control on the attacking nucleophile.[91] The monobenzoate product has a mismatched stereochemical relation with the catalyst employed in the first step and, therefore, reacts only very slowly. Conversely, the enantiomeric catalyst has a matched relationship and reacts rapidly. Thus, treatment of the monobenzoate

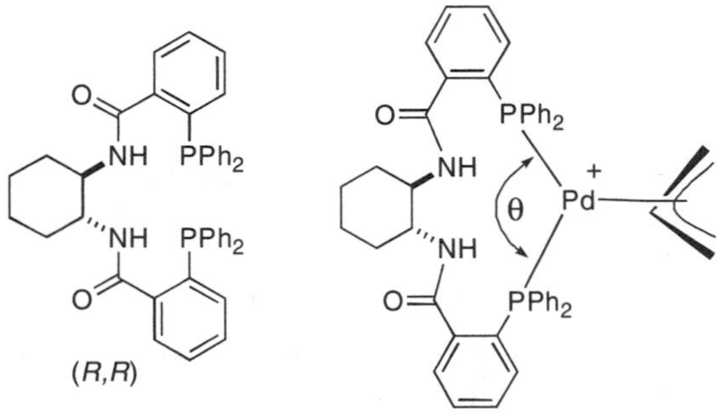

Figure 10.35. Chelating diphosphine for asymmetric allylic substitutions.

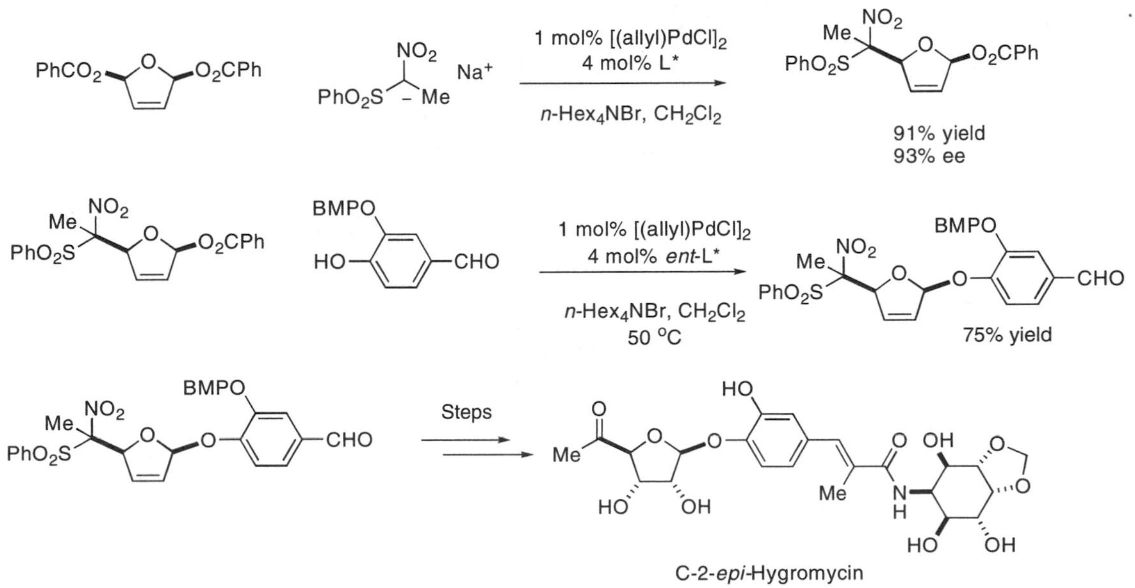

Figure 10.36. The crucial desymmetrization of the diester in the synthesis of C-2-*epi*-hygromycin.

product with catalyst from the enantiomeric ligand (ent-L*), and with a phenol as the nucleophile resulted in facile substitution to provide the aldehyde product. In the second substitution, a chiral catalyst was not required to achieve the desired regioselectivity and diastereoselectivity (see below). Nonetheless, the chiral catalyst was chosen because of its higher reactivity with respect to alternative achiral catalysts examined for this reaction. The aldehyde intermediate was subsequently transformed into C-2-*epi*-hygromycin.

A similar approach was successful in the enantioselective synthesis of (+)-valienamine (**Figure 10.37**).[92] Desymmetrization of the dibenzoate, employing the palladium catalyst in **Figure 10.35** and (phenylsulfonyl)nitromethane, resulted in efficient generation of the monobenzoate product. As discussed above, cyclization of this intermediate to the isoxazoline-N-oxide via ionization of the remaining benzoate was slow under the reaction conditions. In this case, however, addition of an achiral catalyst, Pd(PPh$_3$)$_4$, completed the cyclization to give the bicyclic product in 87% yield and 99% enantioselectivity. This desymmetrization is very attractive, especially considering that the starting dibenzoate is available in one step from cyclohexadiene.

In the two preceding examples, the nucleophilic attack in the desymmetrization results in formation of a C–C bond. Desymmetrization can also be accomplished with nitrogen nucleophiles in an intramolecular fashion (**Figure 10.38**) or an intermolecular manner (**Figure 10.39**). Desymmetrization of the meso bis(imide) derivative (**Figure 10.38**)[93,94] resulted in cyclization to provide the cyclic carbamate in high yield and enantioselectivity. This important intermediate, or its antipode, has been used in several syntheses, including (–)-swainsonine[95] and mannostatin.[93]

Desymmetrization of meso diesters has also been realized with azide as nucleophile (**Figure 10.39**).[96,97] The synthesis of (+)-pancratistatin, a biologically active compound of low natural abundance, was initiated with the meso diester by desymmetrization using the palladium catalyst in **Figure 10.35** and NaN$_3$ in 83% isolated yield with 95% enantioselectivity. This valuable intermediate was carried on in the synthesis of (+)-pancratistatin.[96]

The palladium-catalyzed desymmetrization of meso substrates has been employed in the asymmetric synthesis of a variety of natural products and demonstrated the exceptional utility of this process. In this section, we have focused on a

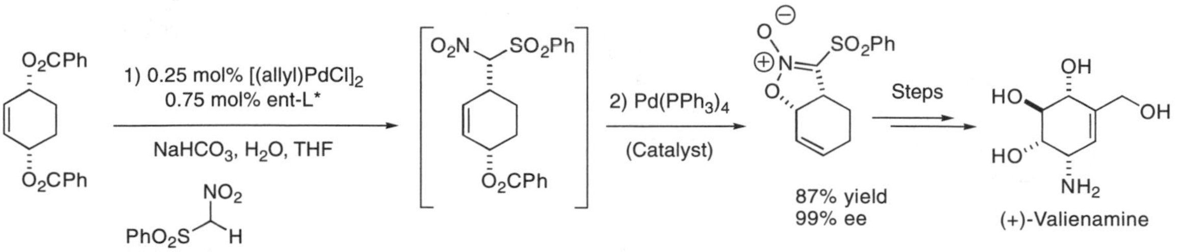

Figure 10.37. Synthesis of (+)-valienamine based on the desymmetrization of a readily available *meso* diester.

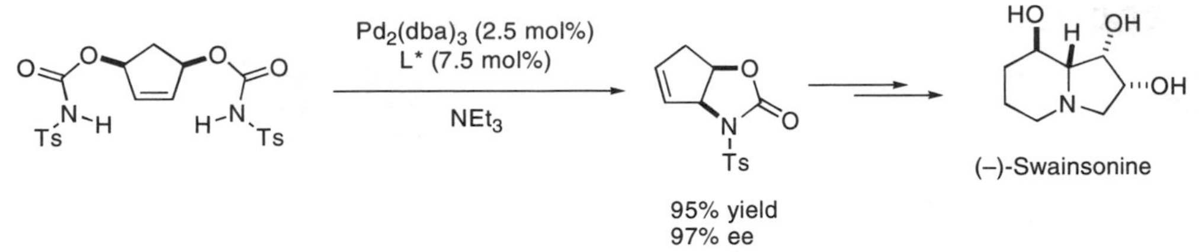

Figure 10.38. Desymmetrization of the meso diimide to furnish a cyclic carbamate that is an important intermediate in the synthesis of (–)-swainsonine.

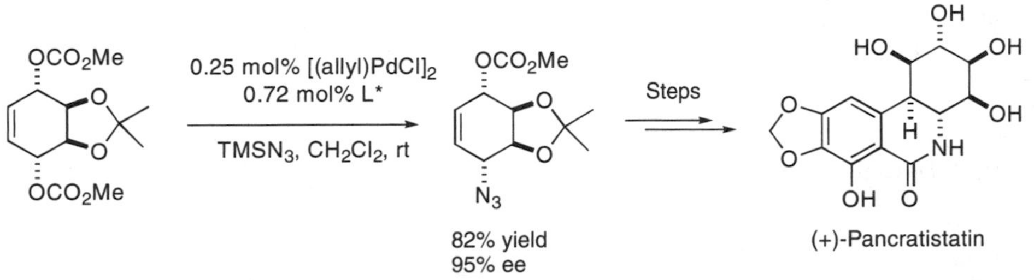

Figure 10.39. Palladium-catalyzed desymmetrization with sodium azide exhibits excellent enantioselectivity and establishes stereochemistry at four stereogenic centers.

single palladium catalyst, but the reaction is also effective with other ligands and metals.[83] Asymmetric allylic substitution reactions can lead to C–C, C–N, C–O and C–S bond formation with excellent control over enantioselectivity and diastereoselectivity and will likely witness considerable development and utility in the future.

Summary

Synthetic strategies involving desymmetrization of complex meso and centrosymmetric substrates are not only elegant, but often represent the most efficient routes to key chiral building blocks and synthetic intermediates. Compounds with multiple stereocenters can be generated rapidly (see Chapter 16). The desymmetrization of such compounds is an area of asymmetric catalysis that has not been greatly explored, but will be more fully developed in the coming years.

References

(1) Willis, M. C. Enantioselective Desymmetrisation. *J. Chem. Soc., Perkin Trans. 1* **1999**, 1765–1784.

(2) Nugent, W. A. Chiral Lewis Acid Catalysis. Enantioselective Addition of Azide to *Meso* Epoxides. *J. Am. Chem. Soc.* **1992**, *114*, 2768–2769.

(3) Bradley, D. C.; Mehrotra, R. C.; Rothwell, I. P.; Singh, A. *Alkoxo and Aryloxo Derivatives of Metals*; Academic Press: New York, 2001.

(4) Balsells, J.; Costa, A. M.; Walsh, P. J. Temperature-Dependent Nonlinear Effects and Catalyst Evolution in the Asymmetric Addition of Diethylzinc to Benzaldehyde. *Isr. J. Chem.* **2001**, *41*, 251–261.

(5) McCleland, B. W.; Nugent, W. A.; Finn, M. G. Mechanistic Studies of the Zirconium-Triisopropanolamine-Catalyzed Enantioselective Addition of Azide to Cyclohexene Oxide. *J. Org. Chem.* **1998**, *63*, 6656–6666.

(6) Nugent, W. A. Desymmetrization of *Meso* Epoxides with Halides: A New Catalytic Reaction Based on Mechanistic Insight. *J. Am. Chem. Soc.* **1998**, *120*, 7139–7140.

(7) Schaus, S. E.; Brandes, B. D.; Larrow, J. F.; Tokunaga, M.; Hansen, K. B.; Gould, A. E.; Furrow, M. E.; Jacobsen, E. N. Highly Selective Hydrolytic Kinetic Resolution of Terminal Epoxides Catalyzed by Chiral (Salen)Co³⁺ Complexes. Practical Synthesis of Enantioenriched Terminal Epoxides and 1,2-Diols. *J. Am. Chem. Soc.* **2002**, *124*, 1307–1315.

(8) Jacobsen, E. N. Asymmetric Catalysis of Epoxide Ring-Opening Reactions. *Acc. Chem. Res.* **2000**, *33*, 421–431.

(9) Schaus, S. E.; Larrow, J. F.; Jacobsen, E. N. Practical Synthesis of Enantiopure Cyclic 1,2-Amino Alcohols via Catalytic Asymmetric Ring Opening of *Meso* Epoxides. *J. Org. Chem.* **1997**, *62*, 4197–4199.

(10) Martínez, L. E.; Leighton, J. L.; Carsten, D. H.; Jacobsen, E. N. Highly Enantioselective Ring Opening of Epoxides Catalyzed by (Salen)Cr(III) Complexes. *J. Am. Chem. Soc.* **1995**, *117*, 5897–5898.

(11) Hansen, K. B.; Leighton, J. L.; Jacobsen, E. N. On the Mechanism of Asymmetric Nucleophilic Ring-Opening of Epoxides Catalyzed by (Salen)Cr(III) Complexes. *J. Am. Chem. Soc.* **1996**, *118*, 10924–10925.

(12) Groves, J. T.; Viski, P. Asymmetric Hydroxylation by a Chiral Iron Porphyrin. *J. Am. Chem. Soc.* **1989**, *111*, 8537–8538.

(13) Katsuki, T. Some Recent Advances in Metallosalen Chemistry. *Synlett* **2003**, 281–297.

(14) Larrow, J. F.; Jacobsen, E. N. Kinetic Resolution of 1,2-Dihydronaphthalene Oxide and Related Epoxides via Asymmetric C–H Hydroxylation. *J. Am. Chem. Soc.* **1994**, *116*, 12129–12130.

(15) Nishida, T.; Miyafuji, A.; Ito, Y. N.; Katsuki, T. Enthalpy- and/or Entropy-Controlled Asymmetric Oxidation: Stereocontrolling Factors in Mn–Salen-Catalyzed Oxidation. *Tetrahedron Lett.* **2000**, *41*, 7053–7058.

(16) Hamachi, K.; Irie, R.; Katsuki, T. Asymmetric Benzylic Oxidation Using a Mn–Salen Complex as Catalyst. *Tetrahedron Lett.* **1996**, *37*, 4979–4982.

(17) Miyafuji, A.; Katsuki, T. Enantiotopic-Place Selective C–H Oxidation Using a (Salen)manganese(III) Complex as a Catalyst. *Synlett* **1997**, 836–838.

(18) Miyafuji, A.; Katsuki, T. Asymmetric Desymmetrization of *Meso*-Tetrahydrofuran Derivatives by Highly Enantiotopic Selective C–H Oxidation. *Tetrahedron* **1998**, *54*, 10339–10348.

(19) Doyle, M. P.; McKervey, A. M.; Ye, T. *Modern Catalytic Methods for Organic Synthesis with Diazo Compounds*; Wiley: New York, 1998.

(20) Davies, H. M. L.; Beckwith, R. E. J. Catalytic Enantioselective C–H Activation by Means of Metal-Carbenoid-Induced C–H Insertion. *Chem. Rev.* **2003**, *103*, 2861–2903.

(21) Timmons, D. J.; Doyle, M. P. Catalyst Selection for Metal Carbene Transformations. *J. Organomet. Chem.* **2001**, *617–618*, 98–104.

(22) Doyle, M. P.; Westrum, L. J.; Wolthuis, W. N. E.; See, M. M.; Boone, W. P.; Bagheri, V.; Pearson, M. M. Electronic and Steric Control in Carbon–Hydrogen Insertion Reactions of Diazoacetoacetates Catalyzed by Dirhodium(II) Carboxylates and Carboxamides. *J. Am. Chem. Soc.* **1993**, *115*, 958–964.

(23) Davies, H. M. L.; Hansen, T.; Churchill, M. R. Catalytic Asymmetric C–H Activation of Alkanes and Tetrahydrofuran. *J. Am. Chem. Soc.* **2000**, *122*, 3063–3070.

(24) Pirrung, M. C.; Morehead, A. T., Jr. Electronic Effects in Dirhodium(II) Carboxylates. Linear Free Energy Relationships in Catalyzed Decompositions of Diazo Compounds and CO and Isonitrile Complexation. *J. Am. Chem. Soc.* **1994**, *116*, 8991–9000.

(25) Taber, D. F.; Malcolm, S. C. Rhodium-Mediated Intramolecular C–H Insertion: Probing the Geometry of the Transition State. *J. Am. Chem. Soc.* **1998**, *63*, 3717–3721.

(26) Nakamura, E.; Yoshikai, N.; Yamanaka, M. Mechanism of C–H Bond Activation/C–C Bond Formation Reaction Between Diazo Compound and Alkane Catalyzed by Dirhodium Tetracarboxylate. *J. Am. Chem. Soc.* **2002**, *124*, 7181–7192.

(27) Doyle, M. P.; Dyatkin, A. B.; Tedrow, J. S. Synthesis of 2-Deoxyxylolactone from Glycerol Derivatives via Highly Enantioselective Carbon–Hydrogen Insertion Reactions. *Tetrahedron Lett.* **1994**, *35*, 3853–3856.

(28) Doyle, M. P.; Dyatkin, A. B.; Roos, G. H. P.; Canas, F.; Pierson, D.; van Basten, A.; Mueller, P.; Polleux, P. Diastereocontrol for Highly Enantioselective Carbon–Hydrogen Insertion Reactions of Cycloalkyl Diazoacetates. *J. Am. Chem. Soc.* **1994**, *116*, 4507–4508.

(29) Mikami, K.; Yoshida, A.; Matsumoto, Y. Catalytic Asymmetric Carbonyl–Ene Reactions with Alkynylogous and Vinylogous Glyoxylates: Application to Controlled Synthesis of Chiral Isocarbacyclin Analogues. *Tetrahedron Lett.* **1996**, *37*, 8515–8518.

(30) Mikami, K.; Terada, M.; Nakai, T. Asymmetric Glyoxylate–Ene Reaction Catalyzed by Chiral Titanium Complexes: A Practical Access to α-Hydroxy Esters in High Enantiomeric Purity. *J. Am. Chem. Soc.* **1989**, *111*, 1940–1941.

(31) Woo, S.; Keay, B. A. "S$_N$2'" and "S$_N$2' Like" Ring Openings of Oxa-*N*-Cyclo Systems. *Synthesis* **1996**, 669–686.

(32) Lautens, M.; Fagnou, K.; Hiebert, S. Transition Metal-Catalyzed Enantioselective Ring-Opening Reactions of Oxabicyclic Alkenes. *Acc. Chem. Res.* **2003**, *36*, 48–58.

(33) Lautens, M.; Chiu, P.; Ma, S.; Rovis, T. Nickel-Catalyzed Hydroalumination of Oxabicyclic Alkenes Ligand Effects on the Regio- and Enantioselectivity. *J. Am. Chem. Soc.* **1995**, *117*, 532–533.

(34) Lautens, M.; Ma, S.; Chiu, P. Synthesis of Cyclohexenols and Cycloheptenols via the Regioselective Reductive Ring Opening of Oxabicyclic Compounds. *J. Am. Chem. Soc.* **1997**, *119*, 6478–6487.

(35) Lautens, M.; Dockendorff, C.; Fagnou, K.; Malicki, A. Rhodium-Catalyzed Asymmetric Ring Opening of Oxabicyclic Alkenes with Organoboronic Acids. *Org. Lett.* **2002**, *4*, 1311–1314.

(36) Lautens, M.; Hiebert, S.; Renaud, J.-L. Enantioselective Ring Opening of Aza and Oxabicyclic Alkenes with Dimethylzinc. *Org. Lett.* **2000**, *2*, 1971–1973.

(37) Lautens, M.; Fagnou, K. Effects of Halide Ligands and Protic Additives on Enantioselectivity and Reactivity in Rhodium-Catalyzed Asymmetric Ring-Opening Reactions. *J. Am. Chem. Soc.* **2001**, *123*, 7170–7171.

(38) Bertozzi, F.; Pineschi, M.; Macchia, F.; Arnold, L. A.; Minnaard, A. J.; Feringa, B. L. Copper Phosphoramidite Catalyzed Enantioselective Ring-Opening of Oxabicyclic Alkenes: Remarkable Reversal of Stereocontrol. *Org. Lett.* **2002**, *4*, 2703–2705.

(39) Lautens, M.; Hiebert, S.; Renaud, J.-L. Mechanistic Studies of the Palladium-Catalyzed Ring Opening of Oxabicyclic Alkenes with Dialkylzinc. *J. Am. Chem. Soc.* **2001**, *123*, 6834–6839.

(40) Lautens, M.; Colucci, J. T.; Hiebert, S.; Smith, N. D.; Bouchain, G. Total Synthesis of Ionomycin Using Ring-Opening Strategies. *Org. Lett.* **2002**, *4*, 1879–1882.

(41) Ruble, J. C.; Tweddell, J.; Fu, G. C. Kinetic Resolution of Arylalkylcarbinols Catalyzed by a Planar-Chiral Derivative of DMAP: A New Benchmark for Nonenzymatic Acylation. *J. Org. Chem.* **1998**, *63*, 2794–2795.

(42) Schreiber, S. L.; Schreiber, T. S.; Smith, D. B. Reactions That Proceed with a Combination of Enantiotopic Group and Diastereotopic Face Selectivity Can Deliver Products with Very High Enantiomeric Excess: Experimental Support of a Mathematical Model. *J. Am. Chem. Soc.* **1987**, *109*, 1525–1529.

(43) Lewis, C. A.; Chiu, A.; Kubryk, M.; Balsells, J.; Pollard, D.; Esser, C. K.; Murry, J.; Reamer, R. A.; Hansen, K. B.; Miller, S. J. Remote Desymmetrization at Near-Nanometer Group Separation Catalyzed by a Miniaturized Enzyme Mimic. *J. Am. Chem. Soc.* **2006**, *128*, 16454–16455.

(44) Schreiber, S. L.; Goulet, M. T.; Schulte, G. Two Directional Chain Synthesis. The Enantioselective Preparation of *syn*-Skipped Polyol Chains from *meso* Precursors. *J. Am. Chem. Soc.* **1987**, *109*, 4718–4720.

(45) Poss, C. S.; Schreiber, S. L. Two-Directional Chain Synthesis and Terminus Differentiation. *Acc. Chem. Res.* **1994**, *27*, 9–17.

(46) Nakata, T. Synthetic Study of Marine Polycyclic Ethers. Total Synthesis of Hemibrevetoxin B. *Synth. Org. Chem. Jpn.* **1998**, *56*, 940–951.

(47) Morimoto, M.; Matsukura, H.; Nakata, T. Total Synthesis of Hemibrevetoxin B. *Tetrahedron Lett.* **1996**, *37*, 6365–6368.

(48) Holland, J. M.; Lewis, M.; Nelson, A. First Desymmetrization of a Centrosymmetric Molecule in Natural Product Synthesis: Preparation of a Key Fragment in the Synthesis of Hemibrevetoxin B. *Angew. Chem., Int. Ed. Engl.* **2001**, *40*, 4082–4084.

(49) Holland, J. M.; Lewis, M.; Nelson, A. Desymmetrization of a Centrosymmetric Diepoxide: Efficient Synthesis of a Key Intermediate in a Total Synthesis of Hemibrevetoxin B. *J. Org. Chem.* **2003**, *68*, 747–753.

(50) Tokunaga, M.; Larrow, J. F.; Kakuichi, F.; Jacobsen, E. N. Asymmetric Catalysis with Water: Efficient Kinetic Resolution of Terminal Epoxides by Means of Catalytic Hydrolysis. *Science* **1997**, *277*, 936–938.

(51) Martín, V. S.; Woodard, S. S.; Katsuki, T.; Yamada, Y.; Ikeda, M.; Sharpless, K. B. Kinetic Resolution of Racemic Allylic Alcohols by Enantioselective Epoxidation. A Route to Substances of Absolute Enantiomeric Purity? *J. Am. Chem. Soc.* **1981**, *103*, 6237–6240.

(52) Gao, Y.; Klunder, J. M.; Hanson, R. M.; Masamune, H.; Ko, S. Y.; Sharpless, K. B. Catalytic Asymmetric Epoxidation and Kinetic Resolution: Modified Procedures Including in situ Derivatization. *J. Am. Chem. Soc.* **1987**, *109*, 5765–5780.

(53) Smith, D. B.; Wang, D.; Schreiber, S. L. The Asymmetric Epoxidation of Divinyl Carbinols: Theory and Applications. *Tetrahedron* **1990**, *46*, 4793–4808.

(54) Katsuki, T. In *Comprehensive Asymmetric Catalysis*; Jacobsen, E. N., Pfaltz, A., Yamamoto, H., Eds.; Springer-Verlag: Berlin, 1999; Vol. 2, pp 621–648.

(55) Hatakeyama, S.; Sakurai, K.; Numata, H.; Ochi, N.; Takano, S. A Novel Chiral Route to Substituted Tetrahydrofurans. Total Synthesis of (+)-Verrucosidin and Formal Synthesis of (–)-Citreoviridin. *J. Am. Chem. Soc.* **1988**, *110*, 5201–5203.

(56) Nakatsuka, M.; Ragan, J. A.; Sammakia, T.; Smith, D. B.; Uehling, D. E.; Schreiber, S. L. Total Synthesis of FK506 and an FKBP Probe Reagent, (C$_8$,C$_{9-13}$C$_2$)-FK506. *J. Am. Chem. Soc.* **1990**, *112*, 5583–5601.

(57) Okamoto, S.; Kobayashi, Y.; Kato, H.; Hori, K.; Takahashi, T.; Tsuji, J.; Sato, F. Prostaglandin Synthsis via Two-Component Coupling. Highly Efficient Synthesis of Chiral Prostaglandin Intermediates 4-Alkoxy-2-alkyl-2-cyclopenten-1-one and 4-Alkoxy-3–alkenyl-2 methylenecyclopentan-1-one. *J. Org. Chem.* **1988**, *53*, 5590–5592.

(58) Spivey, A. C.; Woodhead, S. J.; Weston, M.; Andrews, B. I. Enantioselective Desymmetrization of *meso*-Decalin Diallylic Alcohols by a New Zr-Based Sharpless AE Process: A Novel Approach to the Asymmetric Synthesis of Polyhydroxylated Celas-

traceae Sesquiterpene Cores. *Angew. Chem., Int. Ed. Engl.* **2001**, *40*, 769–771.

(59) Hishiyama, H.; Itoh, K. In *Catalytic Asymmetric Synthesis*; Ojima, I., Ed.; Wiley: New York, 2000; pp 111–143.

(60) Tamao, K.; Ishida, N.; Tanaka, T.; Kumada, M. Sila-functional Compounds in Organic Synthesis. Part 20. Hydrogen Peroxide Oxidation of the Silicon–Carbon Bond in Organoalkoxysilanes. *Organometallics* **1983**, *2*, 1694–1696.

(61) Tamao, K.; Kakui, T.; Kumada, M. Organofluorosilicates in Organic Synthesis. 2. A Convenient Procedure for Preparing Primary Alcohols from Olefins. A Novel Facile Oxidative Cleavage of Carbon–Silicon Bonds by *m*-Chloroperoxybenzoic Acid. *J. Am. Chem. Soc.* **1978**, *100*, 2268–2269.

(62) Barnhart, R. W.; Wang, X.; Noheda, P.; Bergens, S. H.; Whelan, J.; Bosnich, B. Asymmetric Catalysis. Asymmetric Catalytic Intramolocular Hydrosilation and Hydroacylation. *Tetrahedron* **1994**, *50*, 4335–4346.

(63) Tamao, K.; Tohma, T.; Inui, N.; Nakayama, O.; Ito, Y. Catalytic Asymmetric Intramolecular Hydrosilation. *Tetrahedron Lett.* **1990**, *31*, 7333–7336.

(64) Tamao, K.; Ito, Y. Deuterium-Labeling Studies on the Regioselective and Stereoselective Intramolecular Hydrosilation of Allyl Alcohols and Allylamines Catalyzed by Platinum and Rhodium Complexes. *Organometallics* **1993**, *12*, 2297–2308.

(65) Bergens, S. H.; Noheda, P.; Whelan, J.; Bosnich, B. Asymmetric Catalysis. Production of Chiral Diols by Enantioselective Catalytic Intramolecular Hydrosilation of Olefins. *J. Am. Chem. Soc.* **1992**, *114*, 2121–2128.

(66) Bergens, S. H.; Noheda, P.; Whelan, J.; Bosnich, B. Asymmetric Catalysis. Mechanism of Asymmetric Catalytic Intramolecular Hydrosilylation. *J. Am. Chem. Soc.* **1992**, *114*, 2128–2135.

(67) Tamao, K.; Nakamura, K.; Ishii, H.; Yamaguchi, S.; Shiro, M. Axially Chiral Spirosilanes via Catalytic Asymmetric Intramolecular Hydrosilation. *J. Am. Chem. Soc.* **1996**, *118*, 12469–12470.

(68) Shibasaki, M.; Vogl, E. M. The Palladium-Catalysed Arylation and Vinylation of Alkenes—Enantioselective Fashion. *J. Organomet. Chem.* **1999**, *576*, 1–15.

(69) de Meijere, A.; Meyer, F. E. Clothes Make the People: The Heck Reaction in New Clothing. *Angew. Chem., Int. Ed. Engl.* **1994**, *33*, 2379–2411.

(70) Brown, J. M.; Perez-Torrente, J. J.; Alcock, N. W.; Clase, H. J. Stable Arylpalladium Iodides and Reactive Arylpalladium Trifluoromethanesulfonates in the Intramolecular Heck Reaction. *Organometallics* **1995**, *14*, 207–213.

(71) Brown, J. M.; Hii, K. K. Characterization of Reactive Intermediates in Palladium-Catalyzed Arylation of Methyl Acrylate (Heck Reaction). *Angew. Chem., Int. Ed. Engl.* **1996**, *35*, 657–659.

(72) Cabri, W.; Candiani, I. Recent Developments and New Perspectives in the Heck Reaction. *Acc. Chem. Res.* **1995**, *28*, 2–7.

(73) Ashimori, A.; Overman, L. E. Catalytic Asymmetric Synthesis of Quarternary Carbon Centers. Palladium-Catalyzed Formation of Either Enantiomer of Spirooxindoles and Related Spirocyclics Using a Single Enantiomer of a Chiral Diphosphine Ligand. *J. Org. Chem.* **1992**, *57*, 4571–4572.

(74) Kagechika, K.; Shibasaki, M. Asymmetric Heck Reaction: A Catalytic Asymmetric Synthesis of the Key Intermediate for Δ-9(12)-Capnellene-3β,8β,10α-triol and Δ-9(12)-Capnellene-3β,8β,10α,14–tetrol. *J. Org. Chem.* **1991**, *56*, 4093–4094.

(75) Ohshima, T.; Kagechika, K.; Adachi, M.; Sodeoka, M.; Shibasaki, M. Asymmetric Heck Reaction—Carbanion Capture Process. Catalytic Asymmetric Total Synthesis of (–)-Δ9(12)-Capnellene. *J. Am. Chem. Soc.* **1996**, *118*, 7108–7116.

(76) Kondo, K.; Sodeoka, M.; Mori, M.; Shibasaki, M. Asymmetric Heck Reaction. A Catalytic Asymmetric Synthesis of the Key Intermediate for Vernolepin. *Tetrahedron Lett.* **1993**, *34*, 4219–4222.

(77) Danishefsky, S.; Schuda, P. F.; Kitahara, T.; Etheredge, S. J. The Total Synthesis of *dl*-Vernolepin and *dl*-Vernomenin. *J. Am. Chem. Soc.* **1977**, *99*, 6066–6075.

(78) Lebsack, A. D.; Link, J. T.; Overman, L. E.; Stearns, B. A. Enantioselective Total Synthesis of Quadrigemine C and Psycholeine. *J. Am. Chem. Soc.* **2002**, *124*, 9008–9009.

(79) Hayashi, T.; Niisuma, S.; Kamikawa, T.; Suzuki, N.; Uozumi, Y. Catalytic Asymmetric Synthesis of Axially Chiral Biaryls by Palladium-Catalyzed Enantioposition-Selective Cross-Coupling. *J. Am. Chem. Soc.* **1995**, *117*, 9101–9102.

(80) Kamikawa, T.; Uozumi, Y.; Hayashi, T. Enantioposition-Selective Alkynylation of Biaryl Ditriflates by Palladium-Catalyzed Asymmetric Cross-Coupling. *Tetrahedron Lett.* **1996**, *37*, 3161–3164.

(81) Trost, B. M.; Van Vranken, D. L. Asymmetric Transition-Metal-Catalyzed Allylic Alkylations. *Chem. Rev.* **1996**, *96*, 395–422.

(82) Trost, B. M. Designing a Receptor for Molecular Recognition in a Catalytic Synthetic Reaction: Allylic Alkylation. *Acc. Chem. Res.* **1996**, *29*, 355–364.

(83) Trost, B. M.; Crawley, M. L. Asymmetric Transition-Metal-Catalyzed Allylic Alkylations: Applications in Total Synthesis. *Chem. Rev.* **2003**, *103*, 2921–2944.

(84) Trost, B. M.; Vranken, D. L. V.; Bingel, C. A Modular Approach for Ligand Design for Asymmetric Allylic Alkylations via Enantioselective Palladium-Catalyzed Ionizations. *J. Am. Chem. Soc.* **1992**, *114*, 9327–9343.

(85) Lloyd-Jones, G. C.; Stephen, S. C.; Fairlamb, I. J. S.; Martorell, A.; Dominguez, B.; Tomlin, P. M.; Murray, M.; Fernandez, J. M.; Jeffery, J. C.; Riis-Johannessen, T.; Guerziz, T. Coordination of the Trost Modular Ligand to Palladium Allyl Fragments: Oligomers, Monomers, and Memory Effects in Catalysis. *Pure Appl. Chem.* **2004**, *76*, 589–601.

(86) Trost, B. M.; Murphy, D. J. A Model for Metal-Templated Catalytic Asymmetric Induction via π-Allyl Fragments. *Organometallics* **1985**, *4*, 1143–1145.

(87) Hayashi, T.; Ohno, A.; Lu, S.-J.; Matsumoto, Y.; Fukuyo, E.; Yanagi, K. Optically Active Ruthenocenylbis(phosphines): New Efficient Chiral Phosphine Ligands for Catalytic Asymmetric Reactions. *J. Am. Chem. Soc.* **1994**, *116*, 4221–4226.

(88) Trost, B. M.; Breit, B.; Peukert, S.; Zambrano, J.; Ziller, J. W. A New Platform for Designing Ligands for Asymmetric Induction in Allylic Alkylations. *Angew. Chem., Int. Ed. Engl.* **1995**, *34*, 2386–2388.

(89) Trost, B. M.; Dirat, O.; Dudash, J.; Hembre, E. J. An Asymmetric Synthesis of C-2-*epi*-Hygromycin A. *Angew. Chem., Int. Ed. Engl.* **2001**, *40*, 3658–3660.

(90) Trost, B. M.; Dudash, J.; Hembre, E. J. Asymmetric Induction of Conduritols via AAA Reactions: Synthesis of the Aminocyclohexitol of Hygromycin A. *Chem. Eur. J.* **2001**, *7*, 1691–1629.

(91) Trost, B. M.; Dudash, J.; Dirat, O. Application of the AAA Reaction to the Aynthesis of the Furanoside of C-2-*epi*-Hygromycin Synthesis: A Total Synthesis of C-2-*epi*-Hygromycin A. *Chem. Eur. J.* **2002**, *8*, 259–268.

(92) Trost, B. M.; Chupak, L. S.; Lubbers, T. Total Synthesis of (±)- and (+)-Valienamine via a Strategy Derived from New Palladium-Catalyzed Reactions. *J. Am. Chem. Soc.* **1998**, *120*, 1732–1740.

(93) Trost, B. M.; Van Vranken, D. L. A General Synthetic Strategy Toward Aminocyclopentitol Glycosidase Inhibitors. Application of Palladium Catalysis to the Synthesis of Allosamizoline and Mannostatin A. *J. Am. Chem. Soc.* **1993**, *115*, 444–458.

(94) Trost, B. M.; Patterson, D. E. Enhanced Enantioselectivity in the Desymmetrization of *Meso*-Biscarbamates. *J. Org. Chem.* **1998**, *63*, 1339–1341.

(95) Buschmann, N.; Ruckert, A.; Blechert, S. A New Approach to (–)-Swainsonine by Ruthenium-Catalyzed Ring Rearrangement. *J. Org. Chem.* **2002**, *67*, 4325–4329.

(96) Trost, B. M.; Pulley, S. R. Asymmetric Total Synthesis of (+)-Pancratistatin. *J. Am. Chem. Soc.* **1995**, *117*, 10143–10144.

(97) Trost, B. M.; Cook, G. R. An Asymmetric Synthesis of (–)-Epibatidine. *Tetrahedron Lett.* **1996**, *37*, 7485–7488.

Nonlinear Effects, Autocatalysis, and Autoinduction

11.1 Diastereomeric Interactions in Chemistry

The concept of nonlinear effects in asymmetric catalysis falls under the broader heading of diastereomeric interactions. Recall that racemates and their enantiopure counterparts can have different scalar physical properties, such as melting point, boiling point, solubility, vapor pressure, etc.[1-3] For example, in the solid state each crystal of the racemate usually (but not always!) contains equal amounts of both enantiomers. The same crystalline form is not possible, however, with the enantiopure compound, because the packing of the molecules in the crystal must be different. For this reason, crystals of the racemic and enantiopure compounds will have different melting points, densities, and solubilities. In more extreme cases, the racemate and resolved compound may exhibit differences in structure.[4]

A dramatic example of the different properties of racemic and enantiopure compounds is found in α-(trifluoromethyl)lactic acid (**Figure 11.1**). The racemate melts at 88 °C while enantiopure material melts at 110 °C. As a consequence, the racemate sublimes *faster than the enantiopure material*. Starting with crystalline α-(trifluoromethyl)lactic acid of 80% ee, the ee increases to > 99% on standing at room temperature over 60 h. Insight into this remarkable phenomenon can be gained by inspection of intermolecular interactions in the solid state structures of the racemic and enantiopure compounds. Crystals of the enantiopure acid contain more intermolecular hydrogen bonds than crystals of the racemate.[5] Increased hydrogen bonding in the solid state can lead to higher density and a more stable crystalline form.

In the neat liquid phase, the different intramolecular interactions can also result in different physical properties for racemic versus enantiopure compounds. For example, racemic isopropyl trifluorolactate, $CF_3CH(OH)CO_2(i\text{-}Pr)$, boils at 93 °C and the pure enantiomer boils at 136 °C.[6] Partial distillation of enantioenriched isopropyl

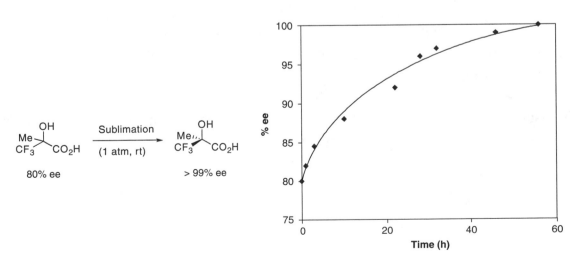

Figure 11.1. Sublimation of (S)-α-(trifluoromethyl)lactic acid (initially of 80% ee) results in an increase due to the faster rate of sublimation of the racemate.

trifluorolactate results in a change in the ee of the distillate and the residue relative to the initial ee.

When dissolved, the different interactions present in crystals of the racemate and enantiopure compounds are usually replaced by interactions with the solvent. It is often observed, therefore, that the solution spectra (NMR, IR, UV/vis, etc.) of racemic and enantioenriched compound appear identical. In some cases, however, intermolecular associations persist in solution.[7–10] In such instances, preferential formation of either the homochiral or heterochiral versions of dimers, oligomers, or polymeric aggregates can occur. For example, the ^{1}H NMR spectra of nearly enantiopure phosphinic amide, PhMeP(O)NHPh, which contains a stereogenic phosphorus center, exhibits a doublet for the P–Me group (because of coupling to ^{31}P, spin $1/2$) at 1.74 ppm. In contrast, the P–Me group of the racemate resonates at 1.68 ppm.[11] A PhMeP(O)NHPh solution of 80% ee was found to exhibit two doublets in a ratio of 90:10.

On the basis of IR studies, it has been shown that phosphinic amides such as PhMeP(O)NHPh associate by hydrogen bonding. One possibility in the present case is the formation of diastereomeric homochiral and heterochiral dimers, as illustrated in **Figure 11.2**. If the diastereomeric heterochiral and homochiral adducts have similar energies, the major enantiomer is most likely to interact with a molecule of the same configuration. In contrast, the minor enantiomer will likely interact with a molecule of the major enantiomer, because the major enantiomer is present at much higher concentration. Thus, even if the phosphinic amide molecules rapidly exchange positions in the diastereomeric dimers on the NMR timescale, different signals for the enantiomers will be observed. These signals will coalesce as the PhMeP(O)NHPh approaches 0% ee. Association of this type can lead to a nonlinear relationship between the solution ee and the measured optical rotation.[12,13] It is also important to note that aggregation states will be influenced by the nature of the solvent and the concentration. Less-polar solvents devoid of hydrogen-bond donors and acceptors will favor aggregation, whereas those with hydrogen-bond donors and acceptors will disfavor association. For this reason, the use of polar protic solvents when recording optical rotations will minimize the likelihood of errors due to solute-solute interactions.

The diastereomeric associations in solution that can give rise to ee-dependent NMR spectra can also have surprising and serious consequences on the ratio of enantiomers during purification of nonracemic compounds. For simplicity, this discus-

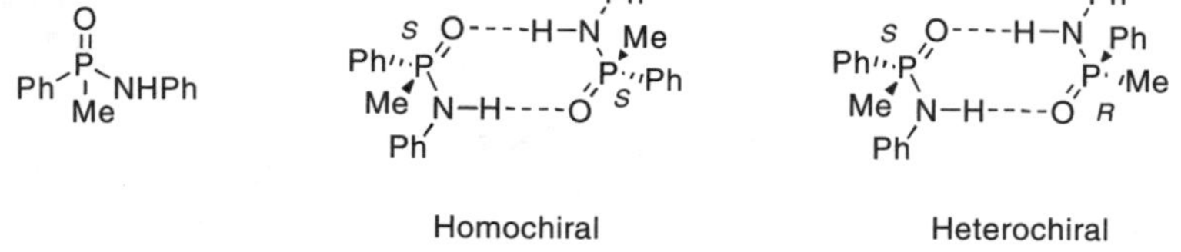

Homochiral   Heterochiral

Figure 11.2. The structure of a P-chiral phosphinic amide and possible association of monomers into homochiral and heterochiral dimers.

sion will be limited to diastereomeric dimers, although it also applies to oligomers and higher aggregates. Consider a case of monomer association in enantioenriched substances where the heterochiral dimers are significantly more stable than their homochiral counterparts. In a racemic solution, the majority of the compound will associate to form heterochiral dimers. In a solution enriched in the (R)-enantiomer, heterochiral dimers will form until all the minor (S)-enantiomer has been incorporated into the heterochiral dimers. The remaining (R)-enantiomer cannot associate in the same fashion, and may form homochiral dimers (or remain unassociated). The homochiral and heterochiral dimers are diastereomeric, and diastereomers can be separated using achiral chromatography methods.

Consider the β-amino acid of 66.6% ee in **Figure 11.3**. When it was subjected to column chromatography with regular, achiral silica gel with hexanes and ethyl acetate (5:1) as the mobile phase, 20 fractions containing the compound were collected. The first fraction to elute contained β-amino acid of only 8% ee. The ee of the β-amino acid increased dramatically in the later fractions, which contained β-amino acid derivative of > 99.9% ee![10] The magnitude of the ee change is over 90% in this example. This phenomenon has been of called "enantiomer self-disproportionation" or "self-disproportionation of enantiomers."[10] Importantly, the ee of the purified material will depend on which of the fractions are combined. In contrast, when the same β-amino acid derivative was subjected to chromatography under identical conditions, with the exception of a change in eluent from hexanes and ethyl acetate to chloroform, the fraction of lowest ee was 65.8% ee and the highest was 68.8% ee.[10] While such phenomena are believed to be rare, this example cautions us that care must be used even when enantioenriched samples are subjected to standard chromatography (especially those containing fluorinated alkyl groups near the stereogenic center[8]). Other compounds which undergo self-disproportionation are illustrated in **Figure**

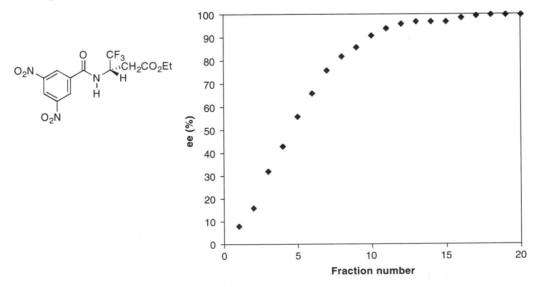

Figure 11.3. Chromatography of the depicted β-amino acid derivative of 66.6% ee on standard silica gel. Twenty fractions containing the acid were collected and the ee of each fraction plotted.

11.4.[7,8] These compounds have axial, central, and helical chirality, suggesting that the self-disproportionation of enantiomers may be general.

In summary, crystallization of nonenantiopure compounds can result in an increase in the ee of the recovered crystals. This valuable technique is widely used in the pharmaceutical industry to "upgrade" the ee in the preparation of medications as single enantiomers. It is not widely appreciated, however, that standard organic purification techniques, including sublimation, distillation, and chromatography with achiral stationary phases can lead to changes in the ee of enantioenriched materials. Thus, care must be used handling enantioenriched compounds. The intermolecular interactions that give rise to such behavior can be attributed to hydrogen bonding, van der Waals forces, and dipole–dipole interactions. Examples of the interactions commonly encountered with enantioenriched samples are depicted in **Figure 11.5** with a sample of 60% ee.[8] Some of these situations fall outside of diastereomeric interactions, because they give rise to entirely different chemical species. If there are no intermolecular interactions or the interactions consist only of the formation of homochiral dimers, the molecules are enantiomers and no increase in ee will be observed employing the purification methods outlined above (Group A, **Figure 11.5**). In

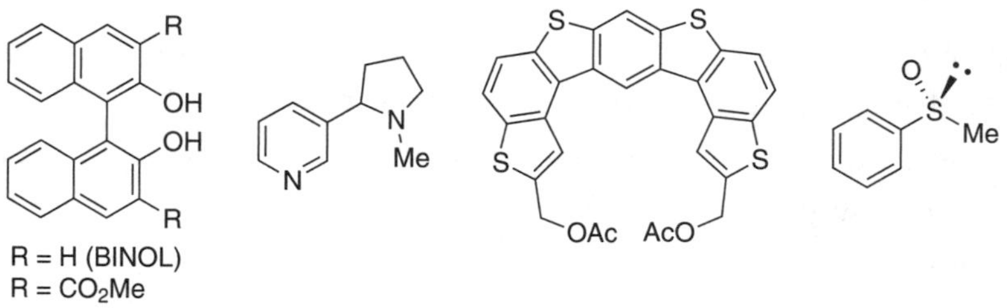

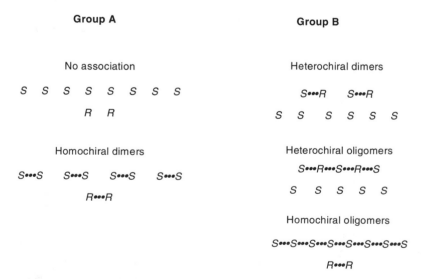

R = H (BINOL)
R = CO$_2$Me

Figure 11.4. Compounds that undergo self-disproportionation under certain conditions when chromatographed on achiral silica gel.

Group A

No association

S S S S S S S S
R R

Homochiral dimers

S•••S   S•••S   S•••S   S•••S
R•••R

Group B

Heterochiral dimers

S•••R   S•••R
S   S   S   S   S   S

Heterochiral oligomers

S•••R•••S•••R•••S
S   S   S   S   S

Homochiral oligomers

S•••S•••S•••S•••S•••S•••S•••S
R•••R

Figure 11.5. Enantioenriched materials with no interactions or that form only homochiral dimers (group A) cannot be separated by sublimation, distillation, or chromatography on achiral silica gel. Interactions in group B, however, can lead to changes in ee during purification.

contrast, heterochiral dimers, oligomers, or homochiral oligomers with different chain lengths and, therefore, molecular weights can, in principle, be separated by standard achiral techniques.

11.2 Nonlinear Effects in Asymmetric Catalysis

Progress in asymmetric catalysis and in the separation of enantiomers via chiral chromatography has increased the number and types of highly enriched organic compounds that are readily available. As such, the demand for compounds approaching enantiopurity by pharmaceutical, agrochemical, and materials industries continues to increase. Efforts to maximize product enantioselectivity to meet the standards of these industries inevitably lead to questions concerning catalyst enantiopurity. Although it is often assumed that catalysts are very close to enantiopure, this is not always the case. What happens to the product ee when an asymmetric catalyst is enantiomerically *impure*? As an example, the asymmetric addition of diethylzinc to aldehydes is presented in **Figure 11.6**.[14,15] In this reaction a catalytic amount of BINOL is employed with 1.2 equivalents titanium tetraisopropoxide, 3 equivalents diethylzinc and one equivalent benzaldehyde. The BINOL reacts with the titanium tetraisopropoxide to generate (BINOLate)Ti(O-*i*-Pr)$_2$, an asymmetric Lewis acid. When enantiopure (R)-BINOL is employed, the product alcohol is formed with 89% ee. Use of 20 mol% BINOL of 80, 60, 40, 20% ee resulted in a linear decrease in the product ee, as represented by the diamonds on the graph of BINOL ee versus product ee in **Figure 11.7**.[14,16] Many reactions in asymmetric catalysis show this type of linear relationship between catalyst ee and product ee.

Interestingly, when the reaction was conducted with 1.2 equivalents preformed (BINOLate)Ti(O-*i*-Pr)$_2$ of 100, 80, 60, 40, and 20% ee and *no* titanium tetraisopropoxide was added, the product ee was less than that predicted from a linear relationship. For example, when ligand of 60% ee was used, the product ee was 46% (see the dashed curve in **Figure 11.7**) rather than the 53% ee expected from a linear relationship.[16] How can these results be rationalized?

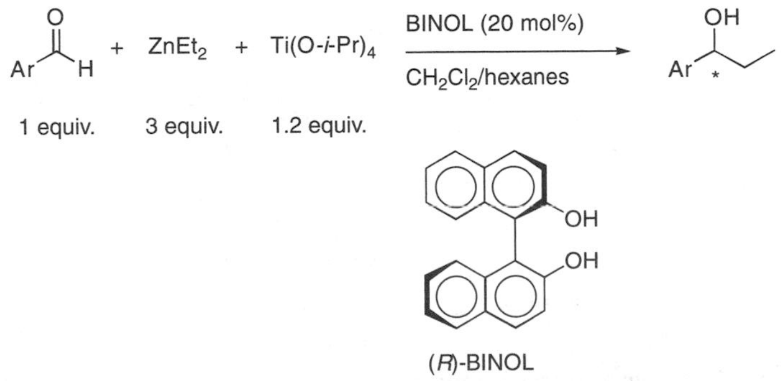

Figure 11.6. Asymmetric addition of diethylzinc to benzaldehyde.

Titanium alkoxide complexes are usually electron deficient and coordinatively unsaturated and often form dimers or oligomers.[17] Mechanistic studies on the asymmetric addition in **Figure 11.6** indicate that when catalytic BINOL is used to form (BINOLate)Ti(O-*i*-Pr)$_2$, this Lewis acid reacts with an equivalent of titanium tetraisopropoxide to form a binuclear species (**Figure 11.8**). The (BINOLate)Ti(O-*i*-Pr)$_2$ interacts with titanium tetraisopropoxide rather than dimerizing because the isopropoxy oxygens on titanium tetraisopropoxide are more basic than those in the stronger Lewis acid (BINOLate)Ti(O-*i*-Pr)$_2$. When there is no titanium tetraisopropoxide present, however, (BINOLate)Ti(O-*i*-Pr)$_2$ dimerizes. With nonenantiopure (BINOLate)Ti(O-*i*-Pr)$_2$, the dimers can be either heterochiral or homochiral. Like the diastereomeric interactions described in the previous section, the diastereomeric [(BINOLate)Ti(O-*i*-Pr)$_2$]$_2$ dimers will exhibit different reactivity in the asymmetric addition, giving rise to the observed nonlinear behavior. In contrast, the binuclear compound contains only a single BINOLate ligand.

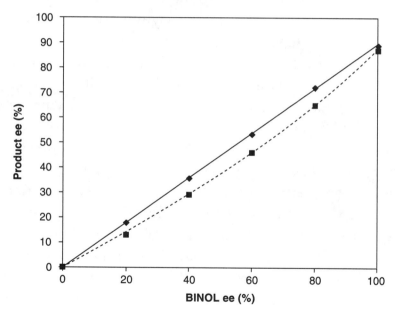

Figure 11.7. Relation between BINOL ee and product ee for the asymmetric addition of diethylzinc to benzaldehyde. The diamonds represent points when the reaction is conducted as in **Figure 11.6**. The squares are points for the stoichiometric reaction using (BINOLate)Ti(O-*i*-Pr$_2$) in the absence of additional titanium tetraisopropoxide.

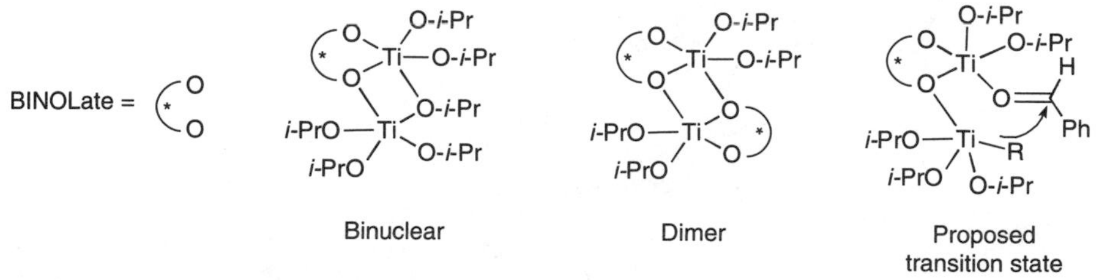

Figure 11.8. Structures of binuclear and dimeric species in the asymmetric addition of diethylzinc to benzaldehyde and the proposed transition state.

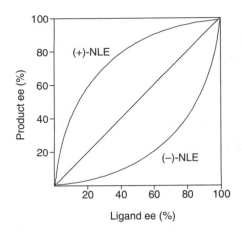

Figure 11.9. Plot of catalyst ee vs. product ee. In this example, the product ee is 100% when enantiopure catalyst is used. Convex and concave curves are (+)-NLE and (–)-NLE, respectively.

Perhaps surprisingly, there are many examples of such nonlinear relationships (NLE = nonlinear effects) between the catalyst ee and product ee. For ease of discussion, in the generic plot in **Figure 11.9**, the enantioselectivity of the catalyst is assigned as 100% when enantiopure ligands (or catalyst) are used. The straight line illustrates a linear relationship. Here, when ligand of 50% ee is used, the product ee will be 50%. When the ee of the product is lower than predicted it is termed "asymmetric diminution" and the reaction exhibits a negative nonlinear effect [(–)-NLE], as shown in **Figure 11.9**. In this hypothetical example, ligand of 50% ee results in product of 15% ee. Conversely, if the product ee is higher than expected, it is called "asymmetric amplification" and a positive nonlinear effect [(+)-NLE] is observed. In this case, ligand of 50% ee will give product of > 85% ee. The benefits of positive nonlinear effects are significant.

In addition to being of practical importance in asymmetric catalysis, nonlinear effects have been used extensively to obtain information on reaction mechanisms. The concept of nonlinear effects in asymmetric catalysis, however, was only recently introduced,[18] and continues to evolve into a more powerful tool to probe catalyst structure and function.[19,20] Like the complexity of reaction mechanisms, the theoretical and mathematical treatment[19,20] of nonlinear behavior can be quite involved and is beyond the scope of this work. Nonetheless, coverage here will include the most common types of molecular interactions that give rise to nonlinear behavior.

11.2.1 The Origins of Nonlinear Effects

Catalyst and catalyst precursors can associate or aggregate in solution in various ways, depending on the nature of the complexes and even on the reaction they promote. When chiral ligands of reduced ee are employed, these interactions can lead to nonlinear behavior. The interactions that are responsible for the nonlinear behavior can become quite complex as the number of chiral ligands bound to a central metal increases or the number of ligands contained in an aggregate increases.[19] Fortunately, the majority of reactions that exhibit nonlinear behavior can be understood based on relatively simple models that contain one or two chiral ligands. We will focus, therefore, on the molecular interactions that give rise to nonlinear effects in these systems. This section will provide examples of how nonlinear effects occur in reactions that are second-order in catalyst and result in diastereomeric transition states. Reactions that are first-order in catalyst, but that still display a nonlinear effect (a reservoir effect) will also be outlined.

At this point, a cautionary note is necessary. Although nonlinear effects are easily determined simply by employing ligand or catalyst of various ee and plotting the ligand ee's against the product ee's, experimental errors can lead to false positive nonlinear behavior. It is important to visually examine reaction mixtures to be sure that they are homogeneous. Recall that racemates and enantiopure compounds often have significantly different solubilities and it is frequently found that the racemate is less soluble than the enantiopure compound. It is not uncommon that the use of enantiopure catalyst results in a homogeneous reaction mixture, but catalyst of low ee leads to heterogeneous reaction mixtures. This is often due to the precipitation of racemic catalyst, effectively increasing the ee of the soluble catalyst and resulting in a false positive nonlinear effect.

11.2.2 Processes That Are Second-Order in Catalyst

11.2.2.a Ring-Opening of Meso Epoxides by (Salen)Cr(III) Complexes

Catalytic asymmetric processes that are second-order in catalyst can give rise to nonlinear behavior. One such example is the nucleophilic ring opening of meso epoxides by azide with chiral Cr–salen catalysts (**Equation 11.1**). As outlined in earlier chapters on desymmetrization and kinetic resolution, several catalysts will promote the enantioselective ring-opening of meso epoxides.

Equation 11.1

The mechanism of the epoxide-opening with azide is similar to that of the hydrolytic kinetic resolution (HKR) of epoxides in that it is second-order in catalyst, and proposed to involve activation of the epoxide and the azide by two Cr–salen complexes.[21,22] When nonenantiopure catalyst was employed, the two chiral (salen)Cr[III] catalysts participating in the ring-opening can either have the same or opposite configurations (**Figure 11.10**). The resultant transition states, therefore, have a diastereomeric relationship with different activation energies. Furthermore, the enantioselectivity of the two processes will differ with both possible heterochiral

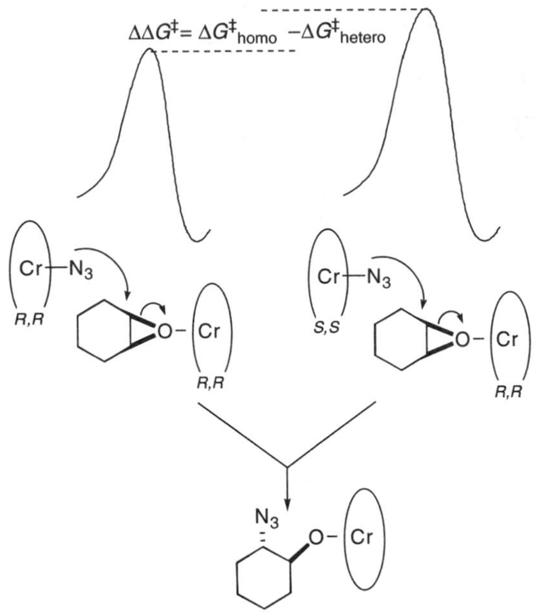

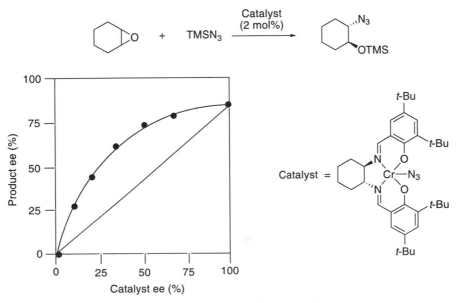

Figure 11.10. Proposed mechanism for the epoxide-opening catalyst by Cr(salen) complexes. In this example, reaction of the homochiral catalyst combination has a lower barrier than the reaction of the heterochiral pair. The reaction, therefore, exhibits a (+)-NLE.

combinations (cat$_{RR}$•epoxide + cat$_{SS}$•Nu and cat$_{SS}$•epoxide + cat$_{RR}$•Nu) canceling each other out and giving no net enantioselectivity. If the homochiral combination then has a lower barrier, as shown in **Figure 11.10**, there will be a (+)-NLE. A positive nonlinear effect is observed experimentally in this system (**Figure 11.11**), providing additional support for the dual-substrate activation of the proposed mechanism.

Figure 11.11. Nonlinear effect observed in the Cr(salen)-catalyzed opening of epoxides with TMSN$_3$.

11.2.3 Processes That Are First-Order in Catalyst: Reservoir Effects

11.2.3.a The ML*$_2$ Reservoir Effect: Asymmetric Mukaiyama Aldol Reactions with Chiral Copper Catalysts

In the previous example, there are two possible diastereomeric catalyst combinations in the transition state, each involving two (salen)Cr moieties, that lead to the observed nonlinear behavior. There are several mechanisms, however, in which a single metal is responsible for promoting the asymmetric reaction that also exhibit nonlinear behavior. One such class of catalysts is comprised of metals with a single chiral ligand, L*M, where the nonlinear behavior originates outside the catalytic cycle by a disproportionation of the catalyst (**Equation 11.2**). In this example, the major enantiomer of the ligand is L_R^*. Two diastereomeric complexes, $(L_R^*)_2M$ [plus $(L_S^*)_2M$] and $(L_R^*)(L_S^*)M$, can be formed. The diastereomers $(L_R^*)_2M$ and $(L_R^*)(L_S^*)M$ are likely to have different energies, and it is usually observed that the heterochiral diastereomer, $(L_R^*)(L_S^*)M$, is more stable, as illustrated in the energy diagram in **Figure 11.12**. Furthermore, it is often the case that these L_2^*M complexes are catalytically inactive. When enantiomerically impure ligand with known enantiomeric excess (ee_{ligand}) is used in the preparation of the catalyst, a portion of this ligand forms the catalytically inactive L_2^*M diastereomers. The enantiomeric excess of the ligand in the L_2^*M complexes will be lower than the enantiomeric excess of the ligand originally added (ee_{ligand}), because the heterochiral complex $(L_R^*)(L_S^*)M$ is formed to a greater extent than the homochiral complexes $(L_R^*)_2M$ and $(L_S^*)_2M$. The enantiomeric excess of the active catalyst L*M (ee_{act}) will, therefore, be higher than that of ee_{ligand} and a positive nonlinear effect will be observed. This model is often called the reservoir effect, because much of the minor enantiomer of the ligand is in the unreactive L_2^*M complexes. Many reactions that exhibit nonlinear effects can be explained with this model.

$$2\ L^*ML_n \rightleftharpoons L^*_2M\ +\ ML_m \qquad \text{\textbf{Equation 11.2}}$$

A well-studied system that exhibits nonlinear behavior attributed to the reservoir effect is the asymmetric aldol reaction catalyzed by C_2-symmetric copper(II) bis(oxa-

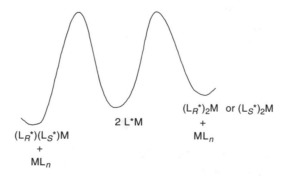

Figure 11.12. Energy diagram for the equilibrium between the catalytically active L*M, the homochiral $(L_R^*)_2M$ (plus its enantiomer), and the heterochiral $(L_R^*)(L_S^*)M$. Also formed in this equilibrium is L_nM, where L is an achiral donor ligand, such as the solvent or water.

zolinyl)pyridine (PyBox) complexes (**Equation 11.3**).[23] In this reaction a *tert*-butyl thioacetate-derived trimethylsilylketene acetal adds to (benzyloxy)acetaldehyde to afford the protected β-hydroxy ester. Subsequent silyl ether cleavage yields the β-hydroxy esters.

Equation 11.3

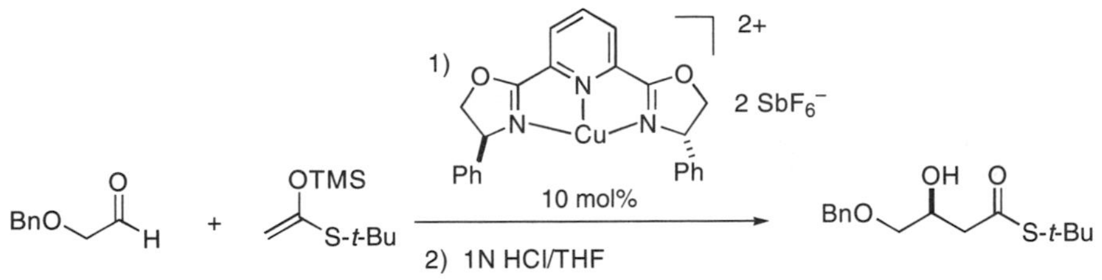

The proposed mechanism for this reaction involves chelation of the (benzyloxy) acetaldehyde to the copper center as shown in **Figure 11.13**. The d^9 Cu(II) copper complex is Jahn–Teller-distorted and such complexes are known to coordinate and activate substrates in the distorted square plane most strongly.[24] The *Re* face of the activated substrate is blocked by the phenyl group of the PyBox ligand, so attack on the activated carbonyl occurs on the *Si* face of the aldehyde with high enantioselectivity.

To examine the nonlinear behavior in the (PyBox)Cu(II)-catalyzed aldol reaction, the catalyst was generated from PyBox ligand of reduced ee. A strong positive nonlinear effect was observed under these conditions and is illustrated graphically in **Figure 11.14**.

This nonlinear effect can be rationalized by the generation of stable (PyBox)$_2$ Cu(II)$^{2+}$ diastereomers formed through disproportionation of (PyBox)Cu(II)$^{2+}$, as outlined in **Figure 11.15**. The (PyBox)$_2$Cu^{2+} complexes formed are catalytically inactive, as they have no coordination site available to bind the substrate. The structural aspects that give rise to the greater stability of the *meso*-(PyBox)$_2$Cu^{2+} complex, and the nonlinear effect in this system can be seen by consideration of the spatial arrangement of the ligand substituents in the heterochiral and homochiral complexes. Representations are

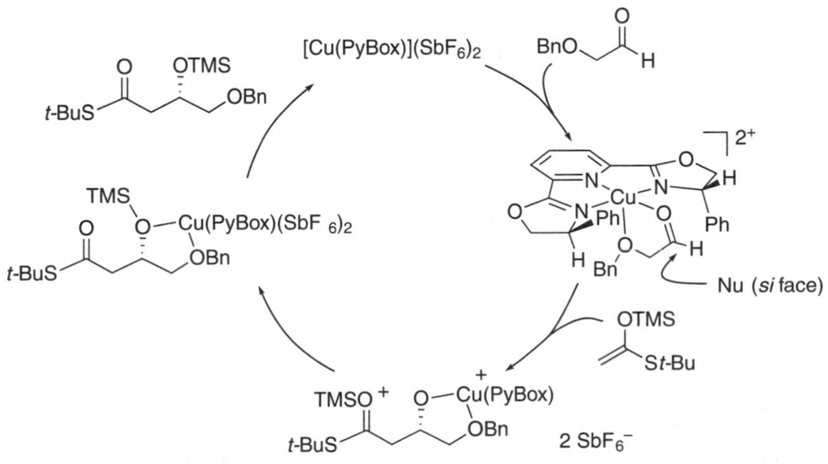

Figure 11.13. Proposed mechanism of the (PyBox)Cu(II)-catalyzed asymmetric aldol reaction.

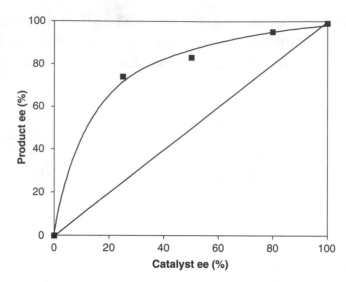

Figure 11.14. Nonlinear effect in the (benzyloxy)acetaldehyde aldol reaction catalyzed by a (PyBox)Cu(II)-based catalyst.

shown in **Figure 11.16**, where the ligand backbone is represented by a straight line and only the oxazoline nitrogens are shown. Based on this simplified model, the heterochiral $(PyBox)_2Cu^{2+}$ is the more stable of the two diastereomers, as each phenyl group is located in a different quadrant. In contrast, the homochiral diastereomer possess two destabilizing phenyl–phenyl interactions, because the four phenyl groups occupy only two quadrants. Support for these models was gained through characterization of both the homochiral and heterochiral compounds by X-ray crystallography. Semiempirical calculations corroborated the greater stability of the heterochiral dimer, which was approximately 3 kcal/mol lower in energy than the homochiral complex.

The strong positive nonlinear effect observed is evidence that the copper byproduct formed in the disproportionation, which may be independently generated from $Cu(SbF_6)_2$, does not promote the aldol reaction to an appreciable extent. The fate of this copper species is unknown, but reasonable possibilities include formation of an

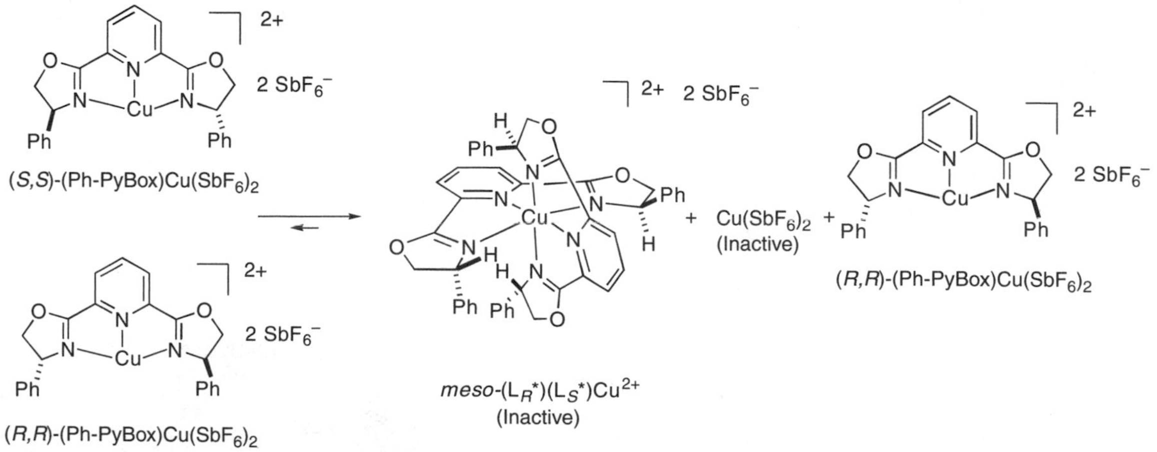

Figure 11.15. Equilibration between monomeric catalysts and catalytically inactive heterochiral $(L_R^*)(L_S^*)Cu^{2+}$. The major enantiomer of the catalyst in this example is derived from (R,R)-Phy-PyBox).

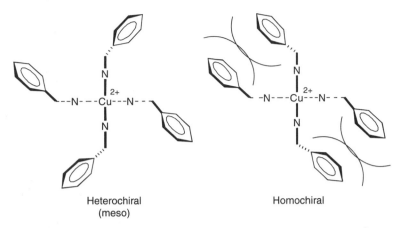

Heterochiral
(meso)

Homochiral

Figure 11.16. A representation of the heterochiral and homochiral (PyBox)$_2$Cu^{2+} diastereomers. In these drawings the ligand backbones are represented by horizontal and vertical straight lines and only the oxazoline nitrogens are shown for clarity (the pyridine nitrogens are omitted). In the actual structure, the angles between the backbones will not be exactly 90°, because of the Jahn–Teller distortion of Cu(II).

insoluble salt or oxide. The stable [(*S,S*)-Ph-PyBox]Cu[(*R,R*)-Ph-PyBox]$^{2+}$ complex acts as a reservoir for the minor enantiomer of the ligand, sequestering it into the inactive heterochiral adduct. As a result, the enantiomeric excess of the catalytically active (PyBox)Cu^{2+} is effectively increased, resulting in the observed positive nonlinear behavior.

11.2.3.b The (ML*)$_2$ Reservoir Effect: The Asymmetric Addition of Alkylzinc Reagents to Aldehydes

Of all the reactions studied that exhibit nonlinear behavior, none have contributed as much to our fundamental understanding of asymmetric catalysis as the asymmetric addition of alkylzinc reagents to aldehydes catalyzed by amino-alcohol-derived Lewis acids.[25–27] This highly enantioselective class of reactions[28,29] also show the strongest positive nonlinear effects observed.[19] A system that has been extensively studied is based on the (–)-3-exo-(dimethylamino)isoborneol, (–)-DAIB,[26,27,30] and the morpholino derivative (–)-MIB[31] (**Equation 11.4**). Enantiopure (–)-DAIB promotes the addition of ethyl groups to benzaldehyde with an enantioselectivity of 98%. Using catalytic (–)-DAIB of 15% ee resulted in production of 1-phenyl-1-propanol of 95% ee, almost the same as enantiopure (–)-DAIB!

PhCHO + ZnEt$_2$

(1.0 equiv.)   (2.0 equiv.)

Chiral ligand

4 mol%
Toluene/hexanes
0 °C

Equation 11.4

(–)-DAIB

(–)-MIB

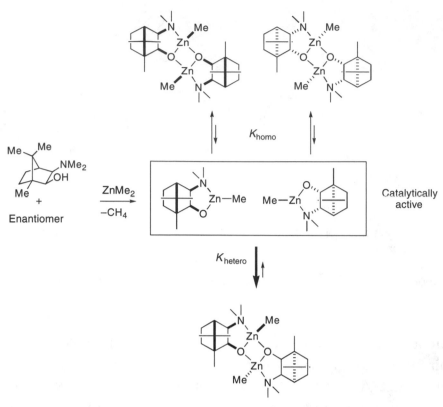

Figure 11.17. Reaction of DAIB with ZnMe$_2$ and the equilibria that are established.

Understanding the mechanism of this unique process is essential to the discussion of autocatalysis, which is the subject of the next section. In this process, the background reaction, which is the direct addition of dialkylzinc reagents to aldehydes in the absence of Lewis acids, is very slow. Nonetheless, dimethylzinc readily undergoes protonolysis with amino alcohols, generating methane and the ligated zinc-alkoxide intermediate, L*ZnMe (**Figure 11.17**). This unsaturated three-coordinate species undergoes reversible dimerization, with the equilibrium lying far toward the dimers. When nonenantiopure ligand is used, two diastereomeric dimers form; the C_2-symmetric homochiral dimer and the meso heterochiral dimer, which is achiral. The structures of these complexes are drawn in **Figure 11.17**. Their three-dimensional structures have been determined by X-ray crystallography and are shown in **Figure 11.18**.[27]

The dimers in **Figure 11.17** are catalytically inactive and do not alkylate benzaldehyde under stoichiometric conditions.[27] Addition of excess dimethylzinc, however, is required for the transfer of the methyl group to the aldehyde carbonyl, suggesting that the zinc adducts of the chiral amino-alkoxide ligand act as Lewis acids and activate the carbonyl, allowing dimethylzinc to deliver the methyl group. This proposal has been supported by calculations, and two calculated transition states are shown in **Figure 11.19**.[32] Binding of the zinc *syn* to the aldehydic hydrogen, in addition to the *Si* face, is favored. In contrast, binding of the zinc *syn* to the phenyl group and *Re*-face

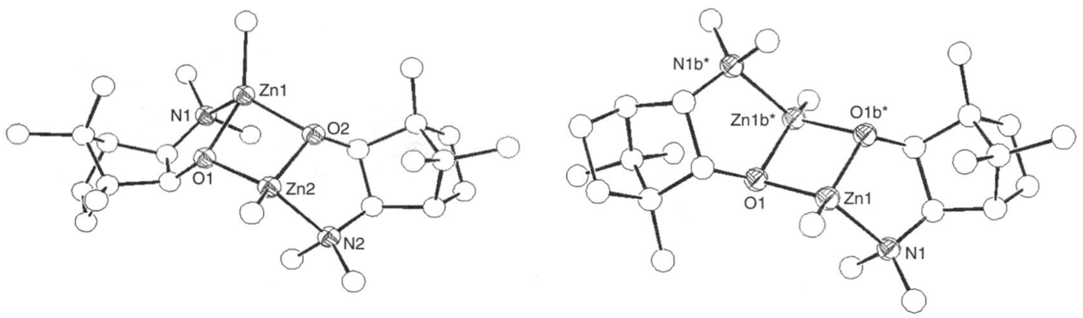

Figure 11.18. X-ray structures of the homochiral dimer (left) and the meso-heterochiral dimer (right).

addition is unfavorable due to steric interactions between the spectator Zn–Me and phenyl group. Alternative transition states have also been calculated.[33]

Solution studies indicate that the homochiral and heterochiral dimers in **Figure 11.17** have very different reactivities in the presence of additional dimethylzinc and aldehyde substrate. The ^{1}H NMR spectrum of the heterochiral dimer is unchanged upon addition of an equivalent of dimethylzinc and benzaldehyde. In contrast, addition of dimethylzinc and benzaldehyde to the homochiral dimer results in coalescence of the methyl resonances of dimethylzinc and L*ZnMe. Additionally, broadening of the aldehydic C–H resonance is detected. These observations suggest that the homochiral dimer dissociates and the dimethylzinc, benzaldehyde, and ligated zinc assembles, poised for the asymmetric addition to occur as illustrated **Figure 11.20**.[27] On the basis of experimental data, it has been estimated that the turnover efficiency of the catalyst derived from the homochiral dimer is 1200 times that of the heterochiral complex, giving rise to the remarkable nonlinear behavior observed in this system. Note that the catalyst and the product are both zinc alkoxides. The product, however, forms a tetrameric [MeZn(OCHPhMe)]$_4$ species with unreactive four-coordinate zinc species.

In this example, the heterochiral dimer is thermodynamically very stable. As a result, almost all of the minor enantiomer of the ligand is trapped in the heterochiral dimer where it is effectively removed from the catalytic cycle. The racemic portion of

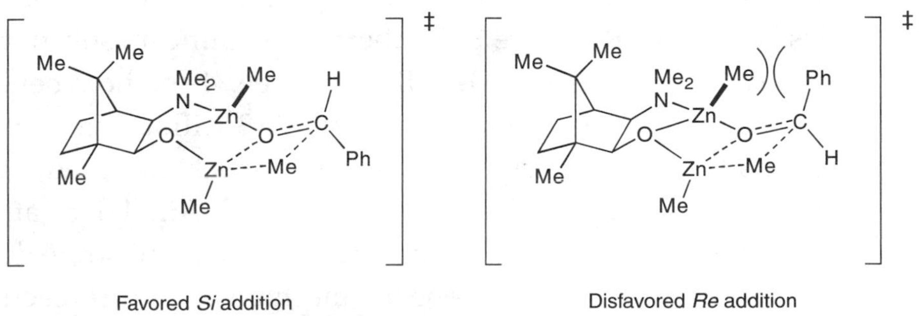

Favored *Si* addition   Disfavored *Re* addition

Figure 11.19. Proposed transition state for the asymmetric addition of methyl groups from dimethylzinc to benzaldehyde.

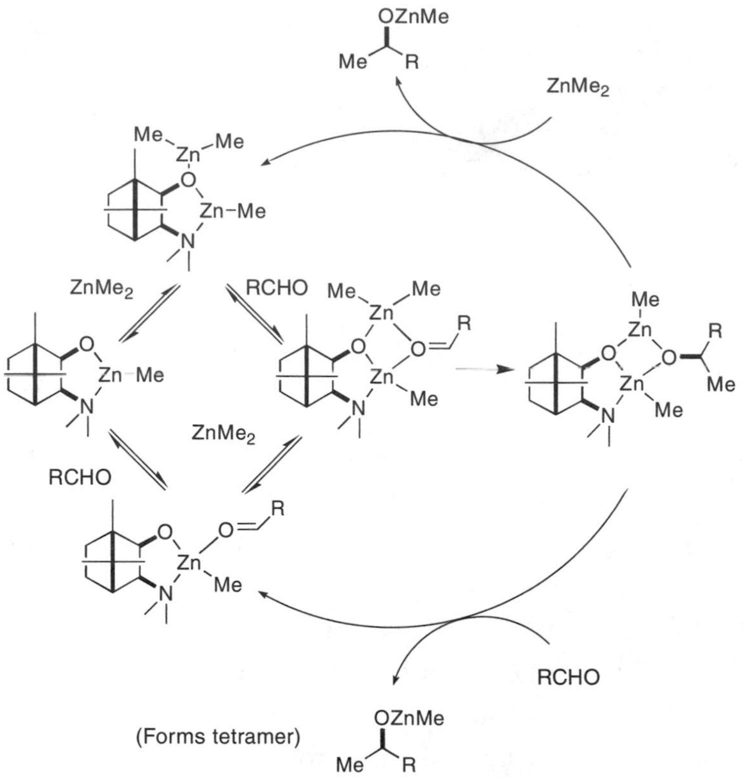

Figure 11.20. Proposed catalytic cycle for the asymmetric addition of alkyl groups to aldehydes, showing the active monomeric intermediates.

the ligand is retained as an unreactive reservoir, hence the term reservoir effect. In essence, the system can be thought of as an in situ resolution of nonenantiopure ligand.

An important consequence of the use of catalysts that exhibit strong (+)-NLEs with low catalyst ee is that the reactions are considerably slower than those with enantiopure ligand. This is not surprising, because the majority of the catalyst is tied up in the unreactive heterochiral dimer when ligand of low ee is employed.

The asymmetric addition of alkyl groups to aldehydes is a well-behaved reaction and has been frequently employed in subsequent studies of nonlinear behavior.[34] One such case is the substrate dependency of nonlinear effects.[35] Using the closely related ligand (–)-MIB (**Equation 11.4**) of 10 and 20% ee, a series of benzaldehyde derivatives was examined to determine the impact of steric and electronic effects on enantioselectivities. As shown in **Figure 11.21**, there is a significant substrate dependency on the nonlinear effects. Using (–)-MIB of 10% ee, 4-methoxybenzaldehyde underwent addition with 94.5% enantioselectivity, while 3-triflouromethyl benzaldehyde reacted with 62.1% enantioselectivity. Both substrates led to product with 98% enantioselectivity when enantiopure ligand was employed. Alkyl derivatives exhibited a lower (+)-NLE than the benzaldehyde derivatives.[35] The substrate dependency of nonlinear effects has been applied to probe the mechanism of this reaction and, in principle, can be used to gain mechanistic insight into other catalytic asymmetric reactions.[35,36]

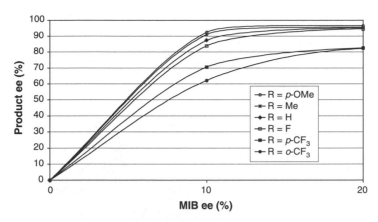

Figure 11.21. Substrate dependency of nonlinear effects in the addition of $ZnEt_2$ to substituted benzaldehydes in the presence of (–)-MIB. The substituents on the benzaldehyde derivatives are shown in the legend.

11.2.4 Dependency of Nonlinear Effects on Catalyst Preparation

In the generation of nonenantiopure catalysts for nonlinear studies, one can use enantioimpure ligand to prepare the catalyst or one can prepare both enantiomers of the enantiopure catalyst separately and mix them in the desired ratio to generate the nonenantiopure catalyst. These two methods generally lead to catalyst mixtures that exhibit the same reactivity and enantioselectivity, even if dimeric intermediates are involved, because rapid equilibration of the diastereomeric dimeric catalysts results in formation of the thermodynamically most favorable mixture of complexes. In rare cases, however, the nonlinear behavior depends on the method of catalyst preparation, as illustrated in the following example.

The asymmetric Diels–Alder reaction between 1-acetoxy-1,3-butadiene and methacrolein is catalyzed by (BINOLate)Ti-based catalysts.[37] These catalysts are prepared by combining $Cl_2Ti(O\text{-}i\text{-}Pr)_2$ with BINOL in dichloromethane in the presence of activated molecular sieves. The newly formed titanium complex was separated from the molecular sieves by centrifugation followed by transfer of the dichloromethane solution of the titanium complex by cannula. When this (BINOLate)Ti species was used in the asymmetric Diels–Alder reaction illustrated in **Equation 11.5**, the product was almost entirely the *endo* adduct (> 98%) formed with 94% enantioselectivity.

Equation 11.5

To examine the nonlinear behavior of this reaction, catalyst with varying degrees of enantiopurity was prepared by mixing the preformed (*R*)-catalyst with the preformed (*S*)-catalyst. Catalyst mixture prepared in this fashion exhibited a linear correlation between the catalyst ee and product ee (**Figure 11.22**, line A). In contrast, when the nonenantiopure catalyst mixture was prepared by combining the preformed (*R*)-catalyst and the appropriate amount of preformed catalyst prepared from *racemic* BINOL, a positive nonlinear effect was observed (line B, **Figure 11.22**). These results can be explained if kinetically stable dimeric (BINOLate)Ti-based catalysts are involved. In the first catalyst preparation, the (*R,R*)- and (*S,S*)-catalyst dimers do not exchange BINOLate ligands during the reaction process to generate the meso (*S,R*)-dimer. Thus, only homochiral dimers are involved and these are stable under the reaction conditions. When racemic BINOL is used in the catalyst preparation, however, the initially formed (BINOLate)Ti species can undergo condensation with species containing either enantiomer of the ligand. In this fashion both the homochiral and heterochiral dimers are generated. On combination of catalyst formed from racemic ligand with enantiopure catalyst to prepare catalyst of variable ee, asymmetric amplification is observed (line B, **Figure 11.22**). Nonlinear behavior in this system can arise if the heterochiral dimer is more stable and/or less reactive.

Despite the level of interest in this system, the structure of the catalyst remains elusive. Generation of titanium complexes in the presence of molecular sieves can result in the formation of stable titanium–oxo complexes, which are almost always polynuclear. Therefore, it is likely that the active species in this reaction are oxo-bridged dimers.[38–42] It is curious that the BINOLate ligands do not undergo exchange under the reaction conditions, because most titanium alkoxide and aryloxide complexes undergo facile ligand exchange.[17]

Systems that display linear and nonlinear behavior that depends on the method of catalyst preparation are uncommon.[43] It is possible that such behavior would be more common if researchers tested catalysts for nonlinear effects using these two different methods of catalyst preparation.

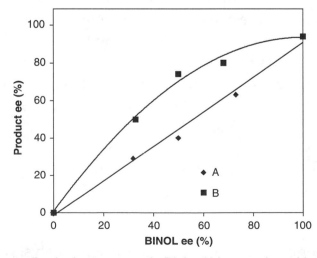

Figure 11.22. Nonlinear studies in the asymmetric Diels–Alder reaction with (BINOLate)Ti-based catalysts. In A, catalyst of various ee was prepared by mixing enantiopure catalysts. In B, the catalyst was made by combining an enantiopure catalyst with catalyst from racemic BINOL.

11.2.5 Applications of NLEs to Mechanistic Questions and Catalyst Optimization

Nonlinear effects have become an important diagnostic tool in mechanistic studies of catalytic asymmetric processes. They are easily measured and there are few techniques that can provide so much information about a catalytic asymmetric reaction with such a minimal investment of effort. However, attention to experimental detail is crucial in obtaining accurate data in these experiments. Errors in weighing enantiomeric catalyst for the preparation of catalyst of reduced ee can result in the observation of false nonlinear behavior. Likewise, formation of a precipitate during the course of the reaction can also lead to anomalous behavior. In this context, it is important to recall that the homochiral and heterochiral complexes are diastereomers and will have different solubility properties. Finally, nonlinear behavior can arise because of catalyst decomposition or evolution over the course of the reaction. As will be outlined in subsequent sections, the product of an asymmetric reaction can become incorporated into the catalyst, generating new catalysts. If these new catalysts exhibit different reactivities, nonlinear behavior may be observed. For this reason, it is important to examine the product ee as a function of conversion to insure that the ee is not changing over the course of the reaction.

In the next section some informative examples are presented that demonstrate how nonlinear behavior can be used as a starting point to unravel the mechanism of catalytic asymmetric reactions. Knowledge of the reaction mechanisms is vital for the rational design of new catalyst generations. Caution must also be exercised in interpreting nonlinear effects. For example, the absence of a nonlinear effect does not rule out a second-order catalyst mechanism. The presence of a nonlinear effect, however, does indicate a more complex scenario that could arise from several sources.

11.2.5.a Asymmetric Isomerization of Meso Epoxides

The asymmetric isomerization of meso epoxides provides a convenient route to allylic alcohols (**Equation 11.6**) and has been applied to natural-product synthesis.[44–46] Excellent results have been obtained with catalysts derived from diamine ligands possessing both secondary and tertiary amino groups.[46,47] The crucial elimination step is believed to occur by coordination of the epoxide to the lithium amide, which activates the hydrogens of the β-carbon and facilitates a *syn* elimination pathway. A simplified picture of the epoxide-opening that does not include solvation or aggregation is illustrated in **Equation 11.6**.

Equation 11.6

One catalyst system that exhibits high levels of enantioselectivity with a variety of meso epoxides, and has also been used in the kinetic resolution of racemic epoxides with excellent results, is illustrated in **Equation 11.7**.[48–50] Lithium diisopropylamide (LDA) is often chosen as the stoichiometric base in these reactions, because it is more efficient at deprotonating the diamine ligand to generate the catalyst than deprotonating the epoxide. With slower catalysts, however, the background reaction of LDA with the epoxide can erode the enantioselectivity of the process. In this and related systems, it is found that addition of DBU results in increased levels of enantioselectivity. The results using five equivalents of DBU with selected meso epoxides are compiled in **Table 11.1**.

Equation 11.7

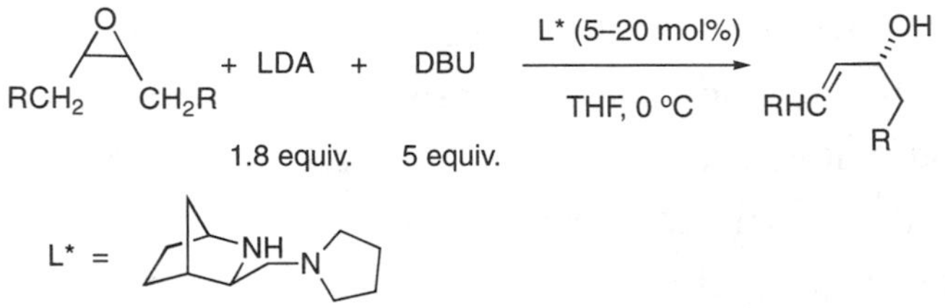

Aggregation and solvation in lithium-amide-mediated reactions can impact both reactivity and selectivity.[46,51] Originally, it was speculated that the role of the DBU was to act as a dative ligand and break up the less-enantioselective catalyst aggregates $(LiL^*)_n$ by coordination to lithium. In the hope of exploring aggregation state as a function of DBU concentration, a series of studies examining the nonlinear behavior of the diamine-based system in **Equation 11.7** with cyclohexene oxide were undertaken. The catalyst ee was varied from 25–100% in THF containing 0–6.0 equivalents of DBU with respect to epoxide.[48,49] A plot of the ee of the ring-opened

Table 11.1. Enantioselective opening of meso epoxides as outlined in **Equation 11.7**.

Epoxide	Product	L* (mol%)	Yield (%)	ee (%)
	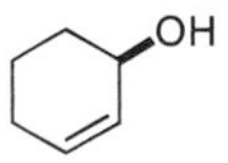	5	91	96
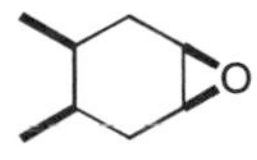	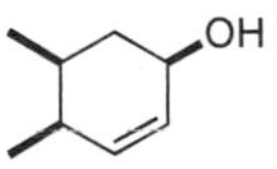	5	95	97
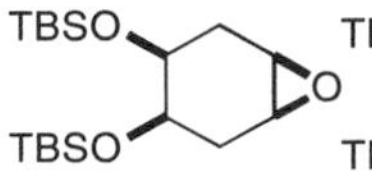	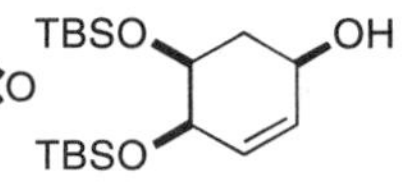	5	60	97
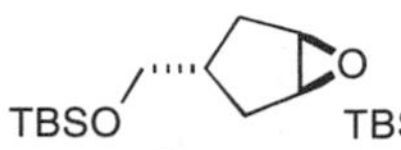	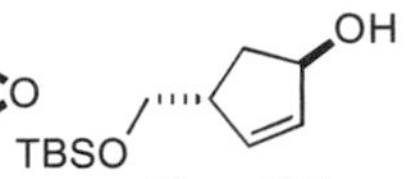	20	42	95
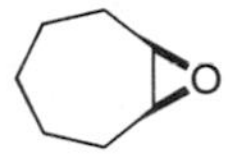	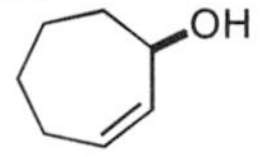	5	89	96

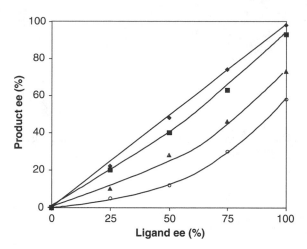

Figure 11.23. Nonlinear studies performed with different amounts of DBU: (♦) 6 equiv., (■) 3 equiv., (▲) 1 equiv., and (o) no added DBU.

product, (R)-cyclohex-2-ol, versus catalyst ee for selected DBU equivalents is shown in **Figure 11.23**. At higher concentrations of DBU (six equivalents), there is a linear correlation between catalyst ee and product ee. With less DBU, however, there is a negative nonlinear effect, in addition to a reduction of the catalyst enantioselectivity. Initially these data were interpreted to suggest the formation of a monomeric catalyst of the type L*Li(DBU) at higher DBU concentrations. The nonlinear behavior at lower concentrations of DBU was attributed to the higher propensity of DBU to disrupt the heterochiral dimer over the homochiral dimer.

Subsequent solution studies into the origin of the DBU effect on enantioselectivity with ^{6}Li and ^{15}N NMR spectroscopy revealed a more intricate role for the DBU. In these investigations, the LDA was found to react with the DBU to establish the equilibrium in **Equation 11.8**, which has an equilibrium constant greater than unity. DBU also reacted with n-BuLi to cleanly generate Li$^+$(DBU$^-$), allowing study of Li$^+$(DBU$^-$) in the absence of LDA and diisopropylamine (DIPA).

Equation 11.8

DBU　　　　LDA　　　　Li$^+$(DBU$^-$)　　　DIPA

Detailed spectroscopic studies have been performed with the norephederine-derived chiral lithium amide, which also promotes the enantioselective elimination of meso epoxides.

Solution ^{6}Li NMR spectroscopy of the ^{15}N-labeled norephederine-derived dimer was comprised of two triplets of equal intensity, indicating that the homochiral

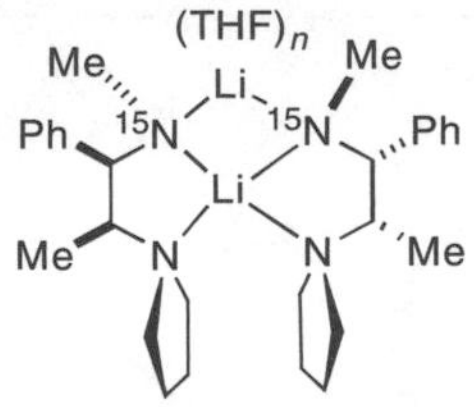

Figure 11.24. Structure of the homochiral dimer determined by solution NMR experiments based on ^6Li–^{15}N coupling and molecular modeling.

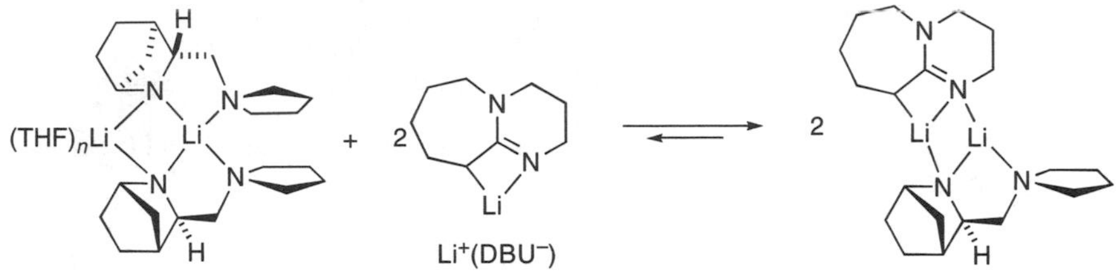

Figure 11.25. Possible equilibrium resulting between the bicyclic dimer and Li(DBU).

dimer had two inequivalent lithium centers coordinated to the labeled nitrogen (**Figure 11.24**).

An analogous homochiral dimer is depicted in **Figure 11.25** that may be responsible for the observed negative nonlinear effect in **Figure 11.23**. $Li^+(DBU^-)$ reacts with the homochiral dimer to form dinuclear species.[52] It is these dinuclear species that are believed to be responsible for the enantioselective epoxide-opening reactions. At low DBU concentration, the equilibrium lies toward the dimers and nonlinear behavior is observed. As the concentration of $Li^+(DBU^-)$ is increased, the equilibrium shifts toward the right. Because the dinuclear species only contains one chiral ligand, is it not expected to exhibit nonlinear behavior.

In these investigations, the analysis of nonlinear effects, in combination with solution-phase NMR studies, provided insight into the nature of the catalytically active species.

11.2.5.b Mechanistic Studies of Lewis Base Catalyzed Allylation of Aldehydes

As we have seen, chiral Lewis acids promote a wide variety of asymmetric reactions. Most of these processes involve coordination of unsaturated substrates, which increases the electrophilicity of the substrate and facilitates its reaction with nucleophiles. In these systems it is generally true that coordination of a Lewis base to a Lewis acid with multiple open coordination sites reduces the propensity of the Lewis acid to bind additional ligands or substrates. This is not always the case, however. A counterintuitive behavior of Lewis acids involves activation of the Lewis acid with a chiral Lewis basic ligand. Upon binding of the Lewis basic ligand, the ability of the resulting Lewis acid to bind the substrate *increases* rather than decreases. Such is the

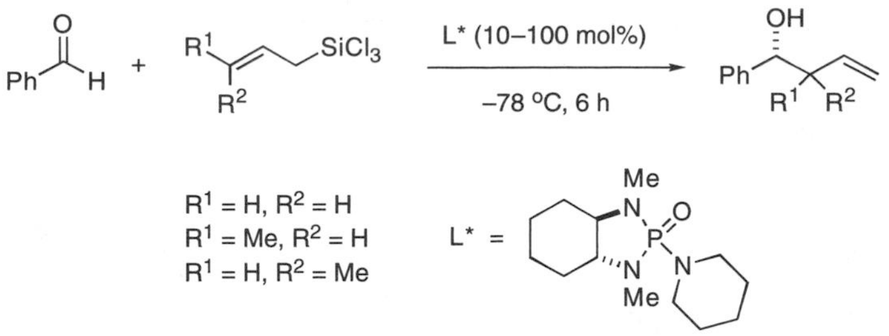

Figure 11.26. Lewis base catalyzed allylation of aldehydes.

case with some organosilanes that contain electronegative substituents. These compounds can bind Lewis bases to achieve hypercoordinate species that may exhibit enhanced reactivity toward Lewis basic substrates relative to four-coordinate precursors.[53] See Chapter 2 for a further discussion.

Hypervalent silanes have been applied to the asymmetric aldol[54] and the allylation of aldehydes[55,56] (**Figure 11.26**).[57] As outlined below, examination of the nonlinear effects in the allylation reaction have been crucial in unraveling some of the mechanistic aspects of this process.

Initial studies concerning the asymmetric allylation of benzaldehyde indicated that the reaction was first-order in both benzaldehyde and allylsilane. The order in phosphoramide ligand was also examined by variation of its concentration and found to be 1.77. An order between 1 and 2 suggests that the reaction may consist of two competing pathways involving coordination of one or two equivalents of the chiral phosphoramide groups to silicon. Consistent with this possibility, determination of the ee of the homoallylic alcohol product as a function of the phosphoramide ligand ee indicated that there is a small positive nonlinear effect, as shown in **Figure 11.27**. The nonlinearity in **Figure 11.27** explains another initially puzzling observation made in the course of these studies—namely, as the phosphoramide loading was decreased from 100 mol% to 10 mol%, the product ee dropped from 81% to 40%, *despite the fact that there is no competitive background reaction between the allylsilane and benzaldehyde*.[55,56] The dependency of the product ee on phosphoramide concentration lends additional support to the existence of two competing pathways that are different in their order in phosphoramide. At higher phosphoramide concentration, allylation proceeds primarily with two phosphoramides bound to silicon. Decreasing the concentration of the chiral Lewis base activator results in a shift from coordination of two phosphoramide ligands toward a less enantioselective pathway with a single phorphoramide bound to silicon.

A reaction pathway based on binding of two phosphoramide ligands to the allyl silane is only possible if one of the chloride ligands dissociates from silicon, forming a cationic intermediate. Such is the case in aldol reactions of trichlorosilyl enolates,[54] epoxide-opening reactions with silicon tetrachloride,[58] and allylation of aldehydes with trichlorosilanes,[59] all of which are promoted by chiral Lewis bases. Shown in

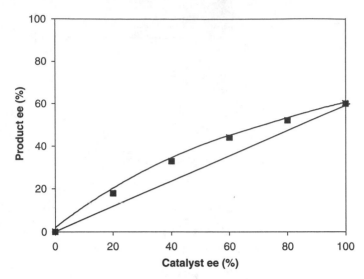

Figure 11.27. Plot of the ee dependence of the allylic alcohol as a function of the phosphoramide ee for the reaction in **Figure 11.26**.

Figure 11.28 are proposed trigional-bipyramidal and octahedral intermediates that precede the transition state.[55]

To explore the possibility of decreasing the catalyst loading without erosion of the enantioselectivity due to competition from the less selective monophosphoramide pathway, a series of chelating bis(phosphoramide) ligands were prepared with various tether lengths (**Figure 11.29**). These ligands were evaluated and compared to a monophosphoramide, which serves as a model for binding of the bis(phosphoramide) ligands in a monodentate fashion. The enantioselectivities in the allylation of benzaldehyde with this family of bis(phosphoramide) ligands and the monodentate ligands are also given in **Figure 11.29**. The ee of the homoallylic alcohol products depends on the number of methylene groups of the bis(phosphoramide) ligands. When $n = 4$, the enantioselectivity is below 20%. Extension of the tether to $n = 5$ causes the enantioselectivity to increase abruptly to as high as 72%. In contrast, addition of another methylene unit ($n = 6$) results in a decrease in the product ee to 46%. These sharp changes are at odds with a mechanism involving monodentate coordination of the ligands, in which case all of the chelating and monodentate ligands would be expected to exhibit similar enantioselectivities. The

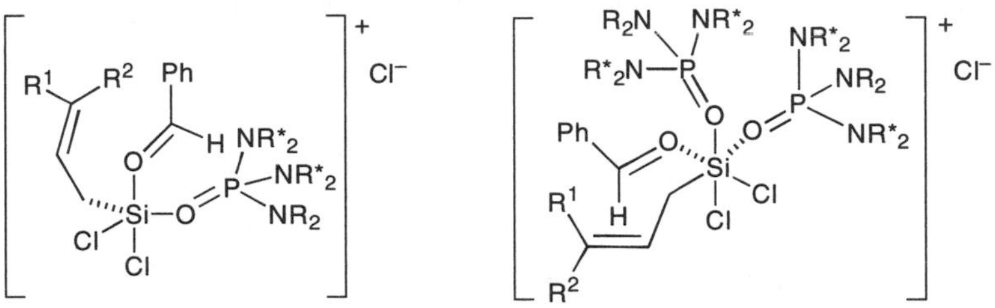

Figure 11.28. Proposed trigonal-bipyramidal and octahedral complexes preceding the transition states.

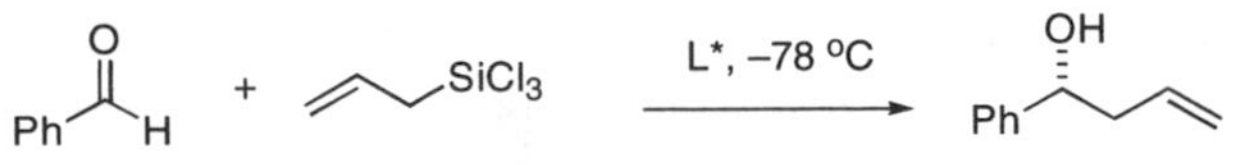

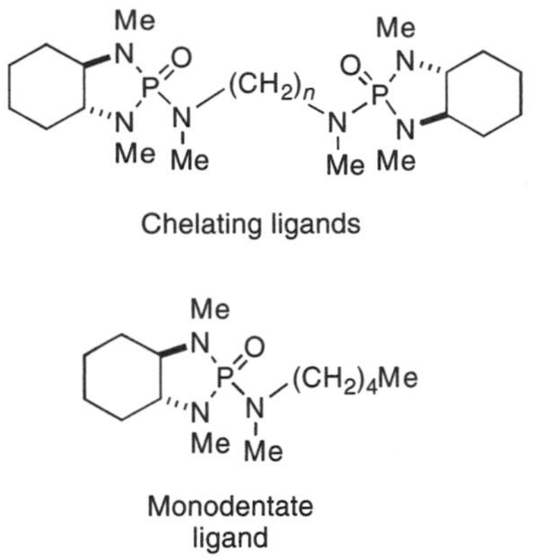

n	mol%	% ee
2	50	0
3	50	35
4	50	17
4	10	10
5	50	65
5	10	72
6	50	46
Mono-dentate	100	51

Figure 11.29. Dependence of ee on the tether length with a chelating and monodentate phosphoramide ligand.

strong dependency of the product ee on the tether length is consistent with a mechanism involving chelation of the bis(phosphoramide). A subsequent study with related bis(phosphoramide) ligands and substituted allyl trichlorosilane resulted in a similar dependency of enantioselectivity on tether length.[60]

The two competing allylation pathways with the monodentate phosphoramides are first- and second-order in the chiral ligand. As such, the allylation reaction will exhibit enantioselectivities that depend on the phosphoramide concentration. If the bis(phosphoramide) ligands chelate to the silicon, they should not exhibit a strong concentration dependency. Comparison of the enantioselectivities with 50 mol% and 10 mol% of the bis(phosphoramide) ligands with $n = 4$ and $n = 5$ showed only minor differences. Particularly interesting is the increase in enantioselectivity to 72% observed on decreasing the ligand loading to 10 mol% with $n = 5$. These results lend further support to the chelation of the bis(phosphoramide) ligands to silicone in the allylation transition state.

The initial breakthrough in this research into the mechanism of the Lewis base catalyzed asymmetric allylation reaction can be traced back to the studies of the nonlinear behavior in **Figure 11.27**. The combination of the nonlinear effects and kinetic analysis enabled the researchers to develop improved reaction conditions and catalysts for this fascinating reaction.

11.2.6 Impact of Phase Behavior on Nonlinear Effects

In order to evaluate the possibility of nonlinear behavior for a catalyst system, one must take great care to ensure that the reaction mixture is homogeneous. To demonstrate this point, researchers performed two in situ preparations of the *same catalyst with ligand of low enantiopurity*. In the first case, a strong positive nonlinear effect was observed, while in the second there was no nonlinear behavior. How can this be? In the first experiment, the reaction mixture was heterogeneous, while in the second case it was homogeneous. Filtration of the heterogeneous solution indicated it contained ligand in the filtrate of very low ee and ligand in the solution of high ee.[61]

Recall that highly enantioenriched and racemic compounds generally have different physical properties, such as solubility (see Section 11.1). Thus, if the reaction mixture is heterogeneous, because some of the precatalyst or catalyst has precipitated from solution, the ee of the catalyst in solution will likely be different from the ee of the catalyst added to the solution (initial = soluble + precipitated). The measurement of nonlinear behavior under heterogeneous conditions can lead to the observation of nonlinear effects that are erroneously attributed to interactions between homochiral and heterochiral catalysts. What was initially viewed as an experimental technicality, however, has powerful implications, as outlined below. To fully understand the origins of nonlinear effects in heterogeneous systems, it would be necessary to review phase diagrams, which is outside the scope of this text. A more detailed discussion can be found in the original manuscripts[61-65] or in a concise summary.[66] Nonetheless, an example of this behavior is given below to provide the reader with a basic understanding of this phenomenon.

Enantioenriched and racemic proline are nearly insoluble in chloroform. In the presence of a trace of ethanol (chloroform:ethanol = 99:1), highly enantioenriched proline exhibits a significant increase in solubility relative to racemic proline. In this solvent system, enantiopure proline is *over 100 times* more soluble than the racemate. Thus, solid proline of 1% ee was prepared by combining enantiopure (R)- and (S)-proline. The solid of 1% ee was stirred for 24 h with a solution of chloroform/ethanol at 0 °C, during which time the solution remained heterogeneous. The insoluble proline was filtered and the filtrate was collected. The ee of the proline in the filtrate was found to be 97–99%! The insoluble proline was determined to be the crystalline racemate. As the enantiopure materials dissolve, the less soluble racemate forms and crystallizes, leaving a highly enriched solution of proline.

As will be discussed in Chapter 12 on bifunctional catalysis, proline is a highly enantioselective catalyst for a number of reactions, one of which is the α-aminoxylation of aldehydes. Taking advantage of the crystallization behavior of proline, solid proline of 10% ee was stirred with a chloroform/ethanol (99:1) solution. After sufficient time for equilibration between the phases, the solution was filtered and the filtrate used in the α-aminoxylation. Because the filtrate contained nearly enantiopure proline, the product in the α-aminoxylation–reduction sequence was formed with very high enantioselectivity (**Figure 11.30**).[67-69]

The origin of the large difference in solubility between enantiomerically pure proline and its racemate was proposed. It is found that racemic proline crystallized from chloroform with a chloroform molecule of solvation in the crystal. The chloroform forms a hydrogen bond with the racemic proline. It is thought that this interaction re-

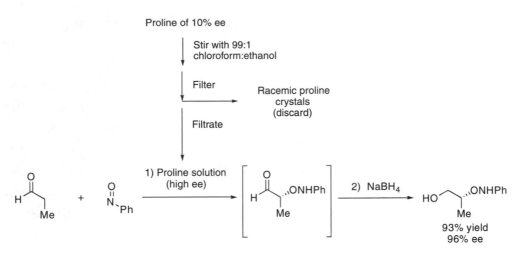

Figure 11.30. α-Amination beginning with a proline solution of very low ee.

duces the solubility of racemic proline relative to enantiopure proline. No chloroform was incorporated into crystals of enantiopure proline formed in chloroform.[63] The observation that small achiral molecules, such as chloroform, can increase the solution ee has important implications.

11.3 Autocatalytic Reactions

Autocatalysis occurs when a catalyst self-replicates—that is, promotes its own formation (**Figure 11.31**). This should not be confused with autoinduction, wherein the product of an asymmetric reaction becomes incorporated into the catalyst, generating new catalysts with different enantioselectivities and TOFs (see Section 11.4).

Autocatalysis was proposed as a process for the propagation of homochirality in the prebiotic era.[70] For the successful propagation of homochirality, a reaction must be perfectly enantioselective, which is highly unlikely. A small decrease in the enantioselectivity from 100% will lead to a decrease in the catalyst ee. Over time, the catalyst ee will erode, ultimately generating catalyst that is close to racemic. For example, an autocatalyst with an initial ee of 99% that generates product of 99% ee will exhibit a drop to less than 94% ee after 200 turnovers. In the same number of turnovers, a catalyst of 88% ee that generates itself with 88% ee will drop to under 50% ee.[71] To circumvent this problem of decreasing ee, a subsequent model was introduced[72] based on an association of the product enantiomers to form, for example, a thermodynamically stable heterochiral dimer that exhibits little or no catalytic

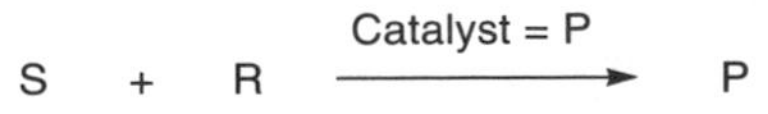

$$S \ + \ R \ \xrightarrow{\text{Catalyst = P}} \ P$$

Figure 11.31. An autocatalytic reaction between the substrate (S) and the reagent (R) generates the product (P), which has the same structure as the catalyst.

activity. Such a dimerization results in the continuous sequestering of the minor enantiomer of the catalyst (i.e., the reservoir effect, see Section 11.2.3), increasing the ee of the active catalyst.

Examples of autocatalysis in asymmetric catalysis are very rare,[73–75] but represent a challenging area for future research. One of the most extraordinary discoveries in asymmetric catalysis is an autocatalytic reaction that exhibits asymmetric amplification, as described above.[73,76,77] The remarkable combination of autocatalysis and asymmetric amplification means that catalyst of very low ee can reproduce itself with very high enantioselectivity. A system that exhibits autocatalysis and a strong positive nonlinear effect employs pyrimidinyl alcohols as the precatalysts for the asymmetric addition of isopropyl groups to pyrimidinyl aldehydes (**Equation 11.9**). The reaction is specific for 2-substituted pyrimidine-based aldehydes and diisopropylzinc.[78]

Equation 11.9

Initial studies on this autocatalytic process focused on the use of nearly enantiopure catalysts and the optimization of conditions to maximize enantioselectivities. After screening several catalyst–substrate systems, it was found that when the pyrimidyl catalyst in **Equation 11.9** was used with the corresponding pyrimidyl aldehyde substrate, the catalyst replicated itself with near perfect enantioselectivity. Thus, using catalyst of 99.5% ee, the product was generated with > 99.5% enantioselectivity. When the reaction was carried out sequentially, with the product of each reaction serving as the catalyst for the next cycle, the catalyst replicated itself, giving > 99.5% ee in > 99% yield *after 10 cycles*!

Like the asymmetric additions of alkyl groups to aldehydes introduced earlier in this chapter (**Equation 11.4**), the reaction in **Equation 11.9** exhibits a strong positive nonlinear effect. Employing nearly racemic ligand in this autocatalytic reaction resulted in an incredible *increase* in the ee of the catalyst. For example, beginning with 5.5% ee of pyrimidyl alcohol (R = *t*-Bu) resulted in production of the addition product with 69.6% ee after one cycle (taking into consideration the initial catalyst ee, **Equation 11.10**). Likewise, use of the derivative with R = SiMe$_3$ of 8.4% ee resulted in production of catalyst of 74.2% ee. In an effort to determine the limit of this system, pyrimidyl alcohol with R = *t*-Bu of 0.00005% ee was used in the autocatalytic reaction with slow addition of the substrate aldehyde. After the first cycle, product of 57% ee was obtained. Utilization of this material as catalyst in the second cycle resulted in product generation of 99% ee. Repeating this procedure for a third cycle gave the product with > 99.5% ee. These results attest to the remarkable power of autocatalysis when combined with nonlinear behavior.[79,80]

Equation 11.10

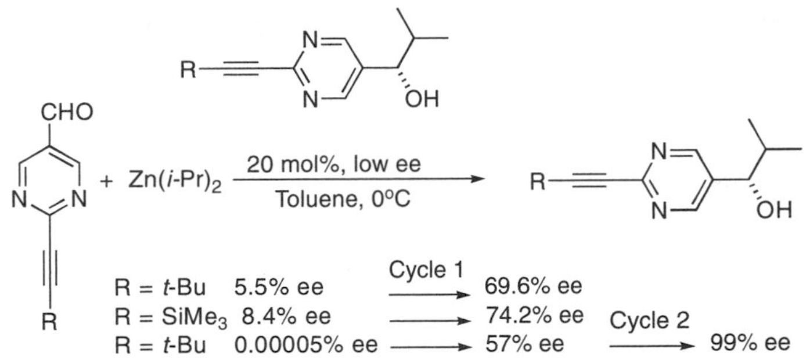

R = t-Bu	5.5% ee		69.6% ee	
R = SiMe$_3$	8.4% ee	Cycle 1	74.2% ee	Cycle 2
R = t-Bu	0.00005% ee		57% ee	99% ee

Despite detailed kinetic and solution studies,[71,78,81,82] a full explanation of the reaction mechanism and the nature of the catalyst remain elusive. Homochiral and heterochiral aggregates are in equilibrium with an equilibrium constant near 1. Spectroscopic evidence supports zinc-alkoxide dimers as the resting state of the catalyst (**Figure 11.32**).[82] Also shown is a tetramer that has two three-coordinate Lewis acidic zinc centers.[78,83,84] The reactions of four-coordinate zinc dimers based on the amino alcohol ligands, introduced earlier in this chapter (**Figure 11.17**), involve dissociation into catalytically active three-coordinate monomers.[29] In contrast, the

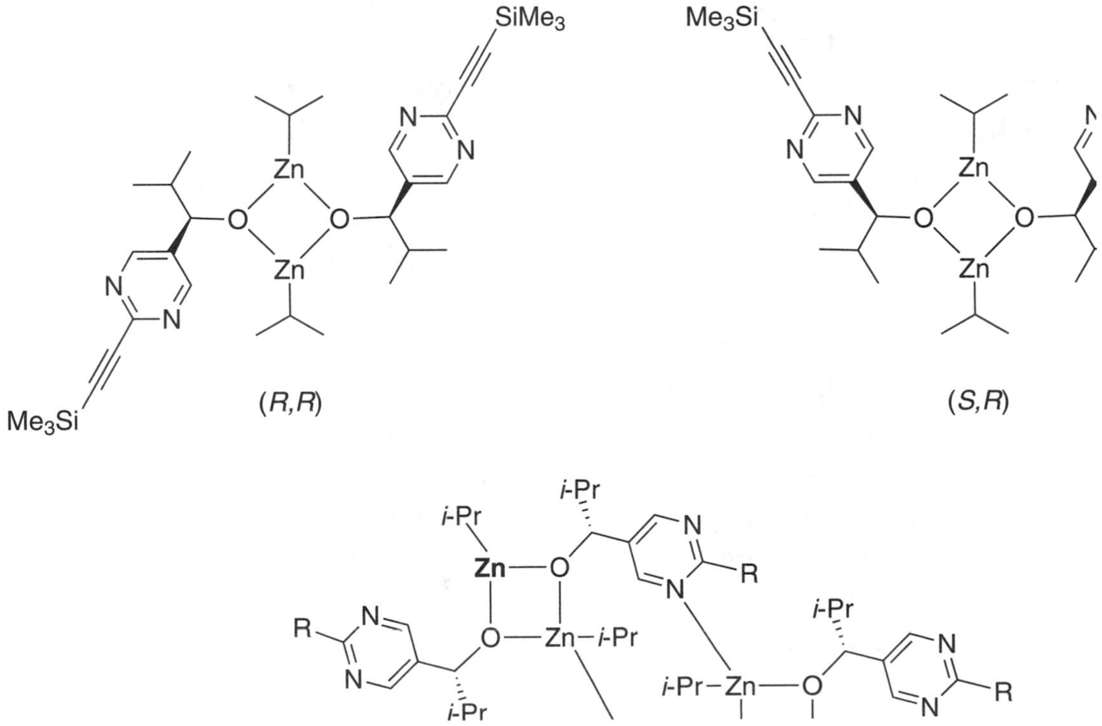

Figure 11.32. Proposed homochiral and heterochiral dimers in the autocatalytic reactions (**Equations 11.9** and **11.10**). Also shown is a dimer of dimers with the bold zinc three-coordinate and potentially active.

aggregates in the Soai reaction contain catalytically active three-coordinate zinc centers and are unlikely to form monomers. Kinetic studies suggest that the homochiral dimer is catalytically active but that the heterochiral dimer is inactive.[81] Analysis of the kinetic data from this reaction suggests that a tetranuclear zinc species is involved.[83] Clearly, further work will be forthcoming on the mechanism of this fascinating reaction.

11.3.1 Absolute Asymmetric Synthesis

In the context of this autocatalytic process, it is appropriate to introduce the topic of absolute asymmetric synthesis. Absolute asymmetric synthesis has been defined as "the formation of enantiomerically enriched products from achiral precursors without the intervention of chiral reagents or catalysts."[85] When the synthesis of molecules possessing chiral centers is performed in the absence of enantioenriched reagents or catalysts, the product obtained generally does not have a measurable ee and is said to be "racemic." It has been understood for many years, however, that the product is not truly racemic but will have some very small imbalance in the ratio of enantiomers. Enantioenriched products are formed from achiral precursors as a result of statistical fluctuations in the ee in the "racemic" starting material. In other words, a small imbalance in the product enantiomers will exist, although the ee will usually be below detection limits. If this imbalance could be amplified in an autocatalytic reaction, it would be possible to spontaneously produce highly enantioenriched product. Caution must be used, because a chiral nonracemic impurity could act as a seed to bias the chirality of the reactions.[86] This nonracemic impurity could come from any one of several sources, including the solvent, the glass reaction vessel, the atmosphere, etc.

In research that set the stage for the investigation of absolute asymmetric synthesis, the Soai reaction (**Equation 11.9**) was performed in the absence of added zinc catalyst, but in the presence of trace amounts of enantiopure materials. The alcohol product of these experiments displayed a high level of ee through autocatalytic asymmetric amplification.[75] For example, quartz, which crystallizes in a chiral space group, can be obtained with very high enantiopurity. When the reaction was run in the presence of enantioenriched quartz crystals, the quartz served to initiate the autocatalytic reaction to provide pyrimidinyl alcohol of high ee.[87] This is particularly impressive, because the generation of the catalyst in the earliest stages of the reaction must occur at the liquid–solid interface. Remarkably, even a trace of (S)-α-dueterobenzyl alcohol of 95% ee resulted in production of the (R)-product with 95% ee.[88]

Several experimental reports document the use of the Soai reaction in absolute asymmetric synthesis.[80,84,89] In one study, the Soai reaction was conducted without the addition of chiral substances, yet still yielded enantioenriched pyrimidyl alcohol.

In these investigations, the reaction was conducted 37 times under identical reaction conditions. The product pyrimidyl alcohol ranged from 15% ee to 91% ee! Of the 37 experiments, the (S)-configuration of the pyrimidyl alcohol was obtained 19 times and the (R)-configuration was found 18 times.[89] Similar results were obtained by other researchers under different reaction conditions.[84] It was proposed that these results were indicative of random fluctuations in the ee of the initially formed pyrimidyl alkoxide catalyst, which is amplified in the autocatalytic process. In these experiments the researchers were extremely careful to prevent the introduction of trace chiral contaminants. These results indicate that the Soai reaction can be used to produce enantioenriched products in absolute asymmetric synthesis employing nominally achiral starting materials in the absence of chiral substances.[85]

This system is an extraordinary illustration of the potential of autocatalytic reactions that exhibit strong (+)-NLE. Furthermore, the ramifications of the Soai reaction provide insight into a possible chemical origin of homochirality on Earth from a small imbalance in enantiomeric precursors that is amplified through autocatalysis and asymmetric amplification. Although it is inconceivable that the addition of alkylzinc reagents to aldehydes could have played a role in an aqueous prebiotic soup, the Soai reaction provides food for thought. Considerable challenges remain ahead as chemists focus on developing autocatalytic processes that involve other reactions.

11.4 Autoinduction in Asymmetric Catalysis

The preceding section on autocatalysis illustrated how the product of a reaction can have the same structure as the catalyst. Examples of autocatalytic reactions are rare in asymmetric catalysis. More common is the situation in which the product associates with the catalyst, altering the catalyst reactivity and enantioselectivity. This interaction can take several forms, ranging from weak association through hydrogen bonds to direct incorporation of the product into the catalyst via strong covalent bonds. Behavior of this type has been termed "autoinduction"[90,91] or "asymmetric autoinduction" and is illustrated in **Figure 11.33**, where S is the substrate, R is the reagent, and P is the product.

The association of the product with the catalyst can impact the catalyst enantioselectivity and/or turnover frequency. Autoinduction can often be detected simply by

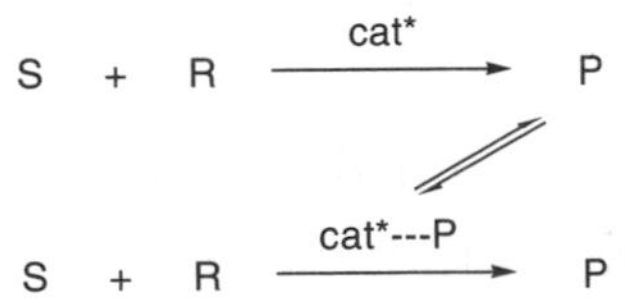

$$S + R \xrightarrow{\text{cat*}} P$$

$$S + R \xrightarrow{\text{cat*---P}} P$$

Figure 11.33. Generic mechanism for autoinduction. The initial catalyst (cat*) and the newly formed catalyst (cat*---P) typically exhibit different reactivity and enantioselectivity.

monitoring the enantiomeric excess of the product at various conversions. At early conversions, where the concentration of the product is low, the impact of the product on the initial catalyst enantioselectivity is typically small. As the reaction progresses, however, the ratio of product to catalyst increases, shifting the equilibrium toward cat*---P and increasing the impact of the product on the catalyst.

11.4.1 The Influence of Catalyst–Substrate vs. Catalyst–Product Interactions

An interesting example of how autoinduction was used to gain mechanistic information, which ultimately led to rational development of an improved catalyst, involves the asymmetric Diels–Alder reaction catalyzed by aluminum complexes of axially chiral vaulted biphenanthrol ligands (**Equation 11.11**).[92] Reaction of diethylaluminum chloride with (S)-VAPOL results in liberation two equivalents of ethane to generate an aluminum Lewis acid catalyst. Combining 10 mol% of this catalyst with methyl acrylate and cyclopentadiene led to formation of the product with very high *endo* selectivity (99:1).

Equation 11.11

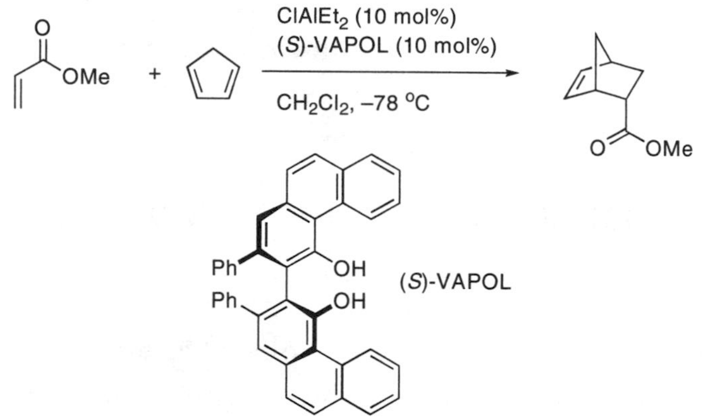

When the progress of the reaction was monitored by removal of aliquots and analysis by chiral stationary-phase gas chromatography, the Diels–Alder adduct was determined to be 48% ee at 20% conversion. As shown in **Figure 11.34**, the product rose to over 80% ee by 60% conversion.

To rationalize the increasing enantioselectivity with conversion, it was proposed that the aluminum catalyst is five-coordinate with two ester carbonyl groups bound to the metal and that there are competing dual catalytic cycles (**Figure 11.35**). Hypervalent aluminum complexes of the type proposed are well known.[93] The aluminum species present at low conversion binds two equivalents of methyl acrylate (**Figure 11.35**). As the reaction progresses and the ratio of product to methyl acrylate increases, more of the product binds to the aluminum catalyst. Incorporation of the product into the catalyst generates a more enantioselective catalyst and is responsible for the rise in enantioselectivity. Support of this hypothesis was obtained when

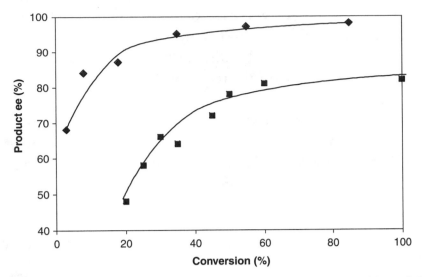

Figure 11.34. Dependence of product ee on conversion in **Equation 11.11** without (■) and with (◆) pivaldehyde additive.

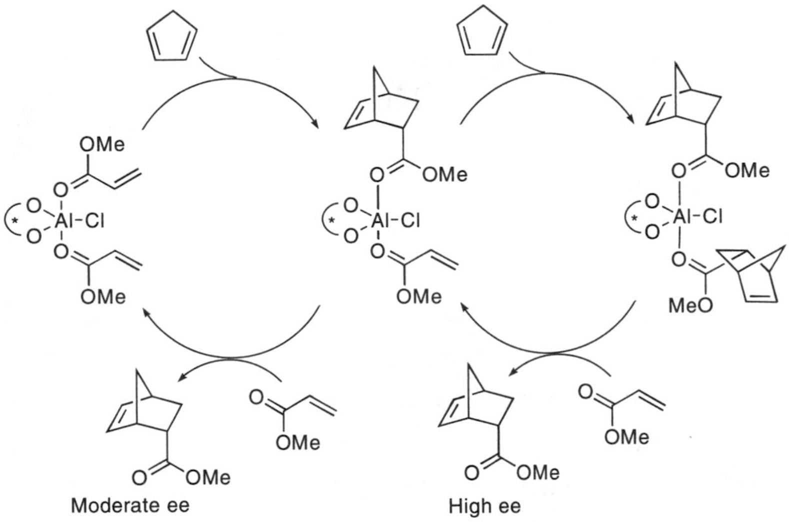

Figure 11.35. The proposed dual pathway in the catalytic asymmetric Diels–Alder reaction. At low conversion the cycle on the left is dominant. At higher conversions the more enantioselective cycle on the right predominates.

the reaction was conducted in the presence of 0.5 equivalents of the enantioenriched product (> 99% ee). Removal of an aliquot at 40% conversion indicated that the product was 93% ee after correction for the 0.5 equivalents of enantioenriched product initially added to the reaction mixture. When the reaction was performed without addition of enantioenriched product under otherwise identical conditions, the product was only 70% ee at 40% conversion. These results indicate that the product impacts the catalyst enantioselectivity.

Once it was discovered that spectator carbonyl groups could bind to the aluminum center and impact the enantioselectivity of the catalyst, a series of carbonyl compounds were screened as additives to optimize the ee of the Diels–Alder adduct.

One additive that was employed was pivaldehyde. A comparison of the product ee with and without pivaldehyde is illustrated in **Figure 11.34**. Examination of the enantioselective behavior indicates that binding of pivaldehyde leads to formation of a better catalyst in the early stages of the reaction. As the reaction proceeds, however, the ee with pivaldehyde increases, indicating the participation of the product through autoinduction. Enantioselectivities over 99% were observed with the proper choice of carbonyl additive.[92] This study not only shows how mechanistic insight can be gained by monitoring the level of enantioselectivity as the reaction progresses, but provides hints as to how the catalyst can be efficiently optimized.

11.4.2. Autoinduction in Reactions Catalyzed by Alkoxide-Based Catalysts That Form Alkoxide Products

Many asymmetric reactions generate chiral alkoxides as the initially formed products, which are then protonated on isolation to yield the alcohols. Examples are abundant and include the asymmetric addition of alkyl, aryl, allyl, and vinyl groups to aldehydes and ketones, and some asymmetric ketone reductions, to name a few. These reactions are frequently catalyzed by Lewis acid catalysts that contain alkoxide ligands. Given that alkoxide-exchange processes are generally fast,[17] it is not surprising that the chiral alkoxide product often becomes incorporated into the catalyst, resulting in catalyst evolution and changes in the enantioselectivity over the course of the reaction.

Development of methods for the synthesis of enantiomerically enriched β-hydroxy carbonyl compounds has been a long-standing goal in organic synthesis. Earlier efforts toward enantioselective aldol-type reactions employed preformed enolate equivalents that must be prepared prior to use. Methods for direct aldol condensations of un-derivatized carbonyl partners are more complex but have enjoyed recent success.[94–101] One system that has proven efficient and highly enantioselective at catalyzing the direct aldol is based on a dinuclear-zinc catalyst prepared from proline (**Equation 11.12**).[96]

Equation 11.12

This dinuclear catalyst has been used in the aldol reaction between ynones and pyruvate-derived aldehydes (**Equation 11.13**). The reaction provides functionalized β-hydroxy ketones in good yields with excellent enantioselectivities.[102]

Equation 11.13

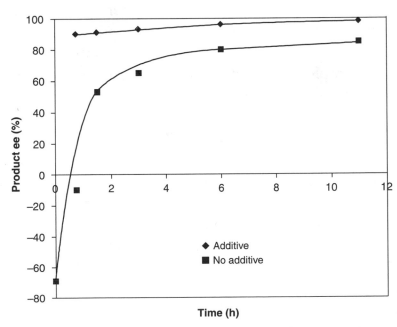

R = CH₂CH=CH₂   76% yield, > 98% ee
R = CH₂C(Me)=CH₂   79% yield, > 98% ee
R = CH₂OTBS   84% yield, > 95% ee

When the enantioselectivity of the product was monitored, it was found that the product ee varied dramatically over the course of the reaction. After 5 min the (*S*)-product was formed with 69% enantioselectivity. Upon further reaction, the product ee decreased and the enantiofacial selectivity switched. At the completion of the reaction, the opposite enantiomer predominated, as illustrated graphically in **Figure 11.36**. A slow kinetic resolution of the product, in which the minor enantiomer was selectively destroyed, was observed. Given that the product yield was significantly greater than 50%, the kinetic resolution can only account for a small amount of the increase in product ee shown in **Figure 11.36**. To rationalize this peculiar behavior, a second catalyst, which exhibited the opposite enantiofacial selectivity with respect to the initial catalyst, was hypothesized to form early in the reaction. The new catalyst was thought to arise from autoinduction. To examine this possibility, the catalyst in **Equation 11.12** was incubated with an equivalent of high-ee β-hydroxy-ketone product. The resulting catalyst led to consistently high product ee over the course of the reaction (**Figure 11.36**, ◆ points). These results suggest that the product is

Figure 11.36. Plot of the ee vs. time in the asymmetric aldol reaction (**Equation 11.13**) (■) in the absence of added product and (◆) in the presence of added product.

incorporated into the catalyst to generate a new catalyst that is significantly more enantioselective. It was also found that incubation of *achiral* β-hydroxy-carbonyl additives with the initial catalyst resulted in generation of the predominant enantiomer of the product in the early stages of the reaction, but with lower enantioselectivity than when the enantioenriched β-hydroxy ketone was used as the additive. Addition of simple alcohols, such as isopropanol, gave higher enantioselectivities at low conversion than reactions without additive, but were not as enantioselective as reactions with enantioenriched β-hydroxy ketone added (see Chapter 6 for additional examples of achiral additives in asymmetric catalysis). These results suggest that the ketone carbonyl is important and that it probably associates with one of the zinc centers.

Many asymmetric catalysts result in the formation of alkoxide products, but experiments to probe autoinduction are frequently either not reported or performed. Yet the presence of autoinduction can provide important mechanistic insight that can facilitate catalyst development and optimization. Like the detection of nonlinear effects, experiments to evaluate reactions for the characteristics of autoinduction are easily performed. Ideally, scientists will measure the product ee not only at the end of the reaction, but also over the course of the reaction.

11.4.3 Autoinduction with Nonenantiopure Catalysts

In the examples of autoinduction above, the catalysts are believed to be monomeric and, therefore, they are expected to exhibit a linear relation between the ee of the initial catalyst and the product ee *at low conversion*. Recall that the ee of the product in reactions that exhibit autoinduction will change as a function of conversion. To minimize the impact of the autoinduction on the ee, the product ee must be measured at early conversion. As we have seen in the aldol reaction outlined above, when enantiopure catalyst was employed, the incorporation of enantiomeric products into the catalyst will lead to formation of two new catalysts, cat*---P_R and cat*---P_S. These catalysts are diastereomeric and, therefore, are expected to have different reactivities, enantioselectivities, and association constants. Now imagine using a catalyst that exhibits autoinduction and is not enantiopure. With enantio*im*pure catalyst, autoinduction can lead to the generation of four new catalysts, cat_R---P_R, cat_R---P_S, cat_S---P_R, and cat_S---P_S. In such a system, it is possible that the ee of the product exceeds the ee of the starting catalyst. This situation could arise if the chiral product of a reaction selectively activates one enantiomer of the initial catalyst, forming a new catalyst, such as cat_R---P_R, that is more active and enantioselective than the original catalyst and the diastereomeric catalyst, cat_S---P_R. In this situation, the product ee would increase with increasing product concentration.

If the system also exhibits positive nonlinear behavior, it is anticipated that the initial product ee will be greater than the catalyst ee. Under the influence of both autoinduction and a strong positive nonlinear effect, a mediocre catalyst of low ee could produce product of high ee. A system that employs catalyst of low ee and product of high ee to generate product with ee even higher than the initial catalyst, or the initially added product, is described below.

In the previous two sections, the autoinduction described was based on the formation of strong bonds—namely, covalent bonds in the asymmetric aldol reaction and dative bonds in the asymmetric Diels–Alder reaction. Weaker interactions, such as hydrogen bonds between the catalyst and product, can also have a profound impact on catalyst activity and enantioselectivity (see Chapter 5). A fascinating example of autoinduction based on hydrogen bonds arose during the asymmetric hydrocyanation of 3-phenoxybenzaldehyde[103] catalyzed by a cyclic dipepeptide (**Equation 11.14**).[104] The (S)-enantiomer of the cyanohydrin product of this reaction is an important precursor to a pyrethroid insecticide.

Equation 11.14

The enantioselectivity of the hydrocyanation of 3-phenoxybenzaldehyde, catalyzed by the cyclic dipeptide (2.2 mol%), was determined at several points over the course of the reaction. As shown in the plot in **Figure 11.37**, the ee of the product was not constant over the course of the reaction, but exhibited a dramatic increase from 34% ee at 21% conversion to 92% ee at the end of the reaction (◆).[103] When the reaction with the (R,R)-cyclic dipeptide was spiked with 8.8 mol% of the (S)-product of 92% ee, the product was > 95% ee (■). In contrast, upon addition of the (R)-enantiomer of the product (8.8 mol%, 85% ee) to the (R,R)-cyclic dipeptide, which is the minor enantiomer produced from this configuration of the catalyst, the ee of the product (▲) was similar to the reaction without added product (◆). These results indicate that only the (S)-enantiomer of the product interacts with the (R,R)-configuration of the catalyst to generate a new catalyst that is more enantioselective and exhibits higher turnover frequency than the (R,R)-catalyst alone.

Clearly from the change in ee as a function of time in the absence of additive, the (S)-product has a dramatic impact on the enantioselectivity in this autocatalytic reaction. To evaluate the specificity of this interaction, nearly racemic dipeptide was used in combination with either enantiomer of the product. From these experiments it was found that the configuration of the *spiked product determined which enantiomer of the product predominated*. The cyanohydrin product does not catalyze the reaction by itself, ruling out the possibility of an autocatalytic process. It does, however, control the stereochemical outcome of the reaction. This situation arises because only the

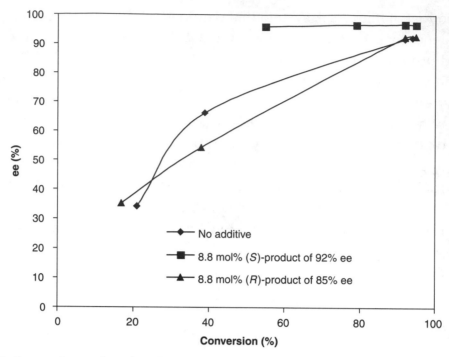

Figure 11.37. Conversion and ee data for the reaction in **Equation 11.14** catalyzed by the (R,R)-dipeptide catalyst (♦) with no additive, (■) with 8.8 mol% added (S)-product of 92% ee, and (▲) with 8.8 mol% added (R)-product of 85% ee. Product ee has been corrected for the addition of the initially added product.

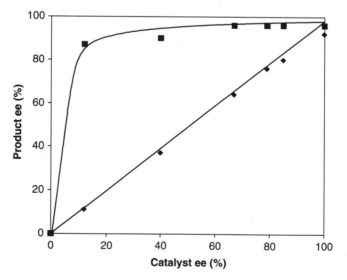

Figure 11.38. Plot of catalyst ee vs. product ee (♦) without added product and (■) with 8.8 mol% added (S)-product of 92% ee (**Equation 11.14**).

matched catalyst-product combination [(R,R)-catalyst with (S)-product or (S,S)-catalyst with (R)-product] is highly reactive.

The variation of the ee of this reaction as a function of catalyst ee was then investigated under several conditions. Initially, the catalyst ee was varied and the product ee was observed to vary linearly (**Figure 11.38**). The reaction was then conducted in

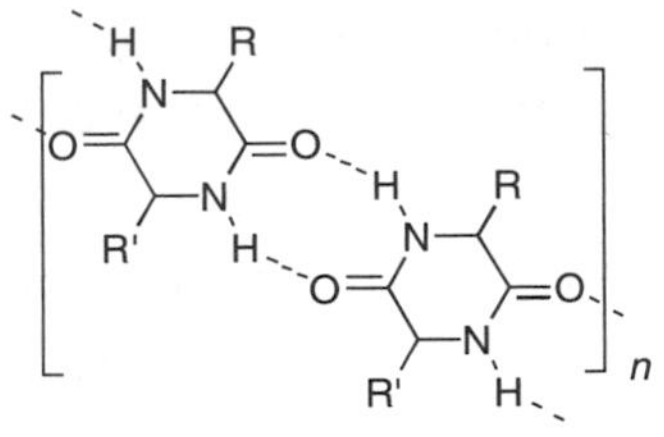

Figure 11.39. Proposed structure of the dipeptide polymer.

a similar fashion, with the addition of 8.8 mol% of 92% ee (*S*)-product. In the presence of added product, a strong positive nonlinear effect was observed, indicating a substantially different catalyst species.

Mechanistic studies on this intriguing system have been complicated by the heterogeneous nature of the reaction system. Nonetheless, studies suggest that the dipeptide forms a catalytically inactive polymer through hydrogen bonding (**Figure 11.39**). The (*S*)-cyanohydrin product is thought to cause dissociation of a portion of the polymeric dipeptide, leading to a catalyst incorporating both the (*R,R*)-dipeptide and the (*S*)-product associated by hydrogen bonds. Kinetic studies performed in the presence of the (*S*)-product suggest that the reaction is second-order in the (*R,R*)-dipeptide.[105] Consistent with this proposal, when the cyclic dipeptide was bound to a polymer support and the (*S*)-product was added, the enantioselectivity under these conditions was low.[106] By linking the cyclic dipeptide to the polymer, self-association through hydrogen bonding is largely prevented. Conducting the reaction in methanol results in low enantioselectivity. Methanol solutions of the dipeptide are homogeneous and the methanol likely disrupts the crucial hydrogen bonds between dipeptide monomers.

Despite all the unanswered questions concerning the mechanism of this reaction, it represents a unique and remarkable example of how autoinduction can be coupled with nonlinear effects to generate a highly enantioselective process. Like the Soai reaction outlined earlier, a process akin to the hydrocyanation above could serve as a mechanistic model for early events on Earth in a primordial soup.

Summary

The importance of nonlinear effects in asymmetric catalysis cannot be understated. Outside of the obvious practical significance in the production of highly enantioenriched product from catalyst of lower ee, nonlinear effects are extremely useful in understanding reaction mechanisms and can also guide the catalyst-optimization process.

A fascinating and thought-provoking class of reactions are those that combine autocatalytic reactions with a strong (+)-NLE. In this way, the catalyst can self-replicate and maintain very high levels of enantioselectivity. Significantly more common than autocatalytic processes are reactions in which the enantioenriched product becomes incorporated into the catalyst, generating a new catalyst. This new catalyst may have very different activity and enantioselectivity, and can greatly change the product ee and even the enantiofacial selectivity. While the use of nonenantioenriched catalysts in reactions that display autoinduction adds another layer of complexity to the analysis, the resulting data can be useful in understanding reaction mechanisms.[107]

One of the most common oversights in the catalyst-optimization process is failure to determine if a catalyst exhibits nonlinear behavior. The potential wealth of insight into the catalytic system is well worth the minimal effort required to evaluate nonlinear behavior.

References

(1) Eliel, E. L.; Wilen, S. H. *Stereochemistry of Organic Compounds*; Wiley: New York, 1994.

(2) Heller, D.; Drexler, H.-J.; Fischer, C.; Buschmann, H.; Baumann, W.; Heller, B. How Long Have Nonlinear Effects Been Known in the Field of Catalysis? *Angew. Chem., Int. Ed. Engl.* **2000**, *39*, 495–499.

(3) Jacques, J.; Collet, A.; Wilen, S. H. *Enantiomers, Racemates, and Resolutions*; Wiley: New York, 1981.

(4) Pritchett, S.; Gantzel, P.; Walsh, P. J. Synthesis and Crystal Structures of Chiral Titanium Bis(sulfonamido) Bis(amide) Complexes: Differences in Ligand Hapticity Caused by Crystal Packing Forces. *Organometallics* **1997**, *16*, 5130–5132.

(5) Soloshonok, V. A.; Ueki, H.; Yasumoto, M.; Mekala, S.; Hirschi, J. S.; Singleton, D. A. Phenomenon of Optical Self-Purification of Chiral Non-Racemic Compounds *J. Am. Chem. Soc.* **2007**, *129*, 12112–12113.

(6) Katagiri, T.; Yoda, C.; Furuhashi, K.; Ueki, K.; Kubota, T. Separation of an Enantiomorph and its Racemate by Distillation: Strong Chiral Recognizing Ability of Trifluorolactates. *Chem. Lett.* **1996**, 115–116.

(7) Noyori, R.; Kitamura, M. Enantioselective Addition of Organometallic Reagents to Carbonyl Compounds: Chirality Transfer, Multiplication, and Amplification. *Angew. Chem., Int. Ed. Engl.* **1991**, *30*, 49–69.

(8) Soloshonok, V. A.; Berbasov, D. O. Self-disproportionation of Enantiomers on Achiral Phase Chromatography. One More Example of Fluorine's Magic Powers. *Chim. Oggi/Chemistry Today* **2006**, *24*, 44–47.

(9) Soloshonok, V. A.; Berbasov, D. O. Self-disproportionation of Enantiomers of (*R*)-Ethyl 3-(3,5-dinitrobenzamido)-4,4,4-trifluorobutanoate on Achiral Silica Gel Stationary Phase. *J. Fluro. Chem* **2006**, *127*, 597–603.

(10) Soloshonok, V. A. Remarkable Amplification of the Self-disproportionation of Enantiomers on Achiral-Phase Chromatography Columns. *Angew. Chem., Int. Ed. Engl.* **2006**, *45*, 766–769.

(11) Harger, M. J. P. Proton Magnetic Resonance Nonequivalence of the Enantiomers of Alkylphenylphosphinic Amides. *J. Chem. Soc,. Perkin. Trans. 2* **1977**, 1882–1887.

(12) Horeau, A.; Guette, J. P. Diastereoisomeric Interactions of Antipodes in the Liquid Phase. *Tetrahedron* **1974**, *30*, 1923–1931.

(13) Horeau, A. Interactions Between Enantiomers in Solution; Effect on the Rotatory Power. Optical Purity and Enantiomeric Purity. *Tetrahedron Lett.* **1969**, *10*, 3121–3124.

(14) Mori, M.; Nakai, T. Asymmetric Catalytic Alkylation of Aldehydes with Diethylzinc Using Chiral Binaphthol-Titanium Complex. *Tetrahedron Lett.* **1997**, *38*, 6233–6236.

(15) Zhang, F.-Y.; Yip, C. W.; Cao, R.; Chan, A. S. C. Enantioselective Addition of Diethylzinc to Aromatic Aldehydes Catalyzed by a Ti(BINOL) Complex. *Tetrahedron: Asymmetry* **1997**, *8*, 585–589.

(16) Balsells, J.; Davis, T. J.; Carroll, P. J.; Walsh, P. J. Insight into the Mechanism of the Asymmetric Addition of Alkyl Groups to Aldehydes Catalyzed by Titanium-BINOLate Species. *J. Am. Chem. Soc.* **2002**, *124*, 10336–10348.

(17) Bradley, D. C.; Mehrotra, R. C.; Rothwell, I. P.; Singh, A. *Alkoxo and Aryloxo Derivatives of Metals*; Academic Press: New York, 2001.

(18) Puchot, C.; Samuel, O.; Dunach, E.; Zhao, S. H.; Agami, C.; Kagan, H. B. Nonlinear Effects in Asymmetric Synthesis. Examples in Asymmetric Oxidations and Aldolization Reactions. *J. Am. Chem. Soc.* **1986**, *108*, 2353–2357.

(19) Girard, C.; Kagan, H. B. Nonlinear Effects in Asymmetric Synthesis and Stereoselective Reactions: Ten Years of Investigation. *Angew. Chem., Int. Ed. Engl.* **1998**, *37*, 2922–2959.

(20) Blackmond, D. G. Kinetic Aspects of Nonlinear Effects in Asymmetric Catalysis. *Acc. Chem. Res.* **2000**, *33*, 402–411.

(21) Martínez, L. E.; Leighton, J. L.; Carsten, D. H.; Jacobsen, E. N. Highly Enantioselective Ring Opening of Epoxides Catalyzed by (Salen)Cr(III) Complexes. *J. Am. Chem. Soc.* **1995**, *117*, 5897–5898.

(22) Jacobsen, E. N. Asymmetric Catalysis of Epoxide Ring-Opening Reactions. *Acc. Chem. Res.* **2000**, *33*, 421–431.

(23) Evans, D. A.; Kozlowski, M. C.; Murry, J. A.; Burgey, C. S.; Campos, K. R.; Connell, B. T.; Staples, R. J. C$_2$-Symmetric Copper(II) Complexes as Chiral Lewis Acids. Scope and Mechanism of Catalytic Enantioselective Aldol Additions of Enolsilanes to (Benzyloxy)acetaldehyde. *J. Am. Chem. Soc.* **1999**, *121*, 669–685.

(24) Hathaway, B. J. In *Comprehensive Coordination Chemistry*; Wilkinson, G., Ed.; Pergamon: New York, 1987; Vol. 5, pp 533–774.

(25) Oguni, N.; Matsuda, Y.; Kaneko, T. Asymmetric Amplification Phenomena in Enantioselective Addition of Diethylzinc to Benzaldehyde. *J. Am. Chem. Soc.* **1988**, *110*, 7877–7878.

(26) Kitamura, M.; Suga, S.; Oka, H.; Noyori, R. Quantitative Analysis of the Chiral Amplification in the Amino Alcohol-Promoted Asymmetric Alkylation of Aldehydes with Dialkylzincs. *J. Am. Chem. Soc.* **1998**, *120*, 9800–9809.

(27) Kitamura, M.; Okada, S.; Suga, S.; Noyori, R. Enantioselective Addition of Dialkylzincs to Aldehydes Promoted by Chiral Amino Alcohols. Mechanism and Nonlinear Effect. *J. Am. Chem. Soc.* **1989**, *111*, 4028–4036.

(28) Soai, K.; Niwa, S. Enantioselective Additions of Organozinc Reagents to Aldehydes. *Chem. Rev.* **1992**, *92*, 833–856.

(29) Pu, L.; Yu, H.-B. Catalytic Asymmetric Organozinc Additions to Carbonyl Compounds. *Chem. Rev.* **2001**, *101*, 757–824.

(30) Noyori, R. *Asymmetric Catalysis in Organic Synthesis*; Wiley: New York, 1994.

(31) Nugent, W. A. MIB: An Advantageous Alternative to DAIB for the Addition of Organozinc Reagents. *J. Chem. Soc., Chem. Commun.* **1999**, 1369–1370.

(32) Yamakawa, M.; Noyori, R. Asymmetric Addition of Dimethylzinc to Benzaldehyde Catalyzed by (2S)-3-exo-(Dimethylamino)isobornenol. A Theoretical Study on the Origin of Enantioselection. *Organometallics* **1999**, *18*, 128–133.

(33) Rasmussen, T.; Norrby, P.-O. Characterization of New Six-Membered Transition States of the Amino-Alcohol Promoted Addition of Dialkyl Zinc to Aldehydes. *J. Am. Chem. Soc.* **2001**, *123*, 2464–2465.

(34) Ding, K.; Du, H.; Yuan, Y.; Long, J. Combinatorial Chemistry Approach to Chiral Catalyst Engineering and Screening: Rational Design and Serendipity. *Chem. Eur. J.* **2004**, *10*, 2872–2884.

(35) Chen, Y. K.; Costa, A. M.; Walsh, P. J. Substrate Dependence of Nonlinear Effects: Mechanistic Probe and Practical Applications. *J. Am. Chem. Soc.* **2001**, *123*, 5378–5379.

(36) Buono, F.; Walsh, P. J.; Blackmond, D. G. Rationalization of Anomalous Nonlinear Effects in the Alkylation of Substituted Benzaldehydes. *J. Am. Chem. Soc.* **2002**, *124*, 13652–13653.

(37) Mikami, K.; Motoyama, Y.; Terada, M. Asymmetric Catalysis of Diels–Alder Cycloadditions by an MS-Free Binaphthol-Titanium Complex: Dramatic Effect of MS, Linear vs. Positive Nonlinear Relationship, and Synthetic Applications. *J. Am. Chem. Soc.* **1994**, *116*, 2812–2820.

(38) Terada, M.; Matsumoto, Y.; Nakamura, Y.; Mikami, K. Molecular Assembly of BINOL-Ti Complexes into an Active μ_3-Oxo Titanium Catalyst. *Inorg. Chim. Acta* **1999**, *296*, 267–272.

(39) Terada, M.; Matsumot, Y.; Nakamura, Y.; Mikami, K. Anomalous Role of Molecular Sieves 4Å in the Preparation of a Binaphthol-Derived Active μ^3-Oxo Titanium Catalyst. *J. Chem. Soc., Chem. Commun.* **1997**, 281–282.

(40) Pandiaraju, S.; Chen, G.; Lough, A.; Yudin, A. K. Generation of Highly Enantioselective Catalysts from the Pseudoenatiomeric Assembly of BINOL, F_8–BINOL, and Ti(Oi–Pr)$_4$. *J. Am. Chem. Soc.* **2001**, *123*, 3850–3851.

(41) Mikami, K.; Ueki, M.; Matsumoto, Y.; Terada, M. Tetranuclear Titanium 7,7-Modified Binapholate Cluster as a Novel Chiral Lewis Acid Catalyst. *Chirality* **2001**, *13*, 541–544.

(42) Hanawa, H. J.; Hashimoto, T.; Maruoka, K. Bis{[(S)-binaphthoxy](isopropoxy)titanium} Oxide as a μ-Oxo-Type Chiral Lewis Acid: Application to Catalytic Asymmetric Allylation of Aldehydes. *J. Am. Chem. Soc.* **2003**, *125*, 1708–1709.

(43) Luukas, T. O.; Fenwick, D. R.; Kagan, H. B. Presence or Absence of a Nonlinear Effect According to the Asymmetric Catalyst Preparation in the Alkylation of Benzaldehyde. *C. R. Chim.* **2002**, *5*, 487–491.

(44) Hodgson, D. M.; Gibbs, A. R.; Lee, G. P. Enantioselective Desymmetrization of Achiral Epoxides. *Tetrahedron* **1996**, *52*, 14361–14384.

(45) O'Brien, P. Recent Advances in Asymmetric Synthesis Using Chiral Lithium Amide Bases. *J. Chem. Soc., Perkin. Trans. 1* **1998**, 439–1458.

(46) Magnus, A.; Bertilsson, S. K.; Andersson, P. G. Asymmetric Base-Mediated Epoxide Isomerisation. *Chem. Soc. Rev.* **2002**, *31*, 223–229.

(47) Asami, M.; Suga, T.; Honda, K.; Inoue, S. A Novel Highly Effective Chiral Lithium Amide for Catalytic Enantioselective Deprotonation of *meso* Epoxides. *Tetrahedron Lett.* **1997**, *38*, 6425–6428.

(48) Södergren, M. J.; Andersson, P. G. New and Highly Enantioselective Catalysts for the Rearrangment of *meso* Epoxides into Chiral Allylic Alcohols. *J. Am. Chem. Soc.* **1998**, *120*, 10760–10761.

(49) Södergren, M. J.; Bertilsson, S. K.; Andersson, P. G. Allylic Alcohols via Catalytic Asymmetric Epoxide Rearrangement. *J. Am. Chem. Soc.* **2000**, *122*, 6610–6618.

(50) Gayet, A.; Bertilsson, S.; Andersson, P. G. Novel Catalytic Kinetic Resolution of Racemic Epoxides to Allylic Alcohols. *Org. Lett.* **2002**, *4*, 3777–3779.

(51) Collum, D. B. Solution Structures of Lithium Dialkylamides and Related *N*-Lithiated Species: Results from Lithium-6–Nitrogen-15 Double Labeling Experiments. *Acc. Chem. Res.* **1993**, *26*, 227–234.

(52) Pettersen, D.; Amedjkouh, M.; Lill, S. O. N.; Ahlberg, P. On the Novel Function of the Additive DBU. Catalytic Stereoselective Deprotonation by a Mixed Dimer of Lithiated DBU and a Chiral Lithium Amide. *J. Chem. Soc., Perkin. Trans. 2* **2002**, 1397–1405.

(53) Chuit, C.; Corriu, R. J. P.; Reye, C.; Young, Y. C. Reactivity of Penta- and Hexacoordinate Silicon Compounds and Their Role as Reaction Intermediates. *Chem. Rev.* **1993**, *93*, 1371–1448.

(54) Denmark, S. E.; Stavenger, R. A. Asymmetric Catalysis of Aldol Reactions with Chiral Lewis Bases. *Acc. Chem. Res.* **2000**, *33*, 432–440.

(55) Denmark, S. E.; Fu, J. On the Mechanism of Catalytic, Enantioselective Allylation of Aldehydes with Chlorosilanes and Chiral Lewis Bases. *J. Am. Chem. Soc.* **2000**, *122*, 12021–12022.

(56) Denmark, S. E.; Coe, D. M.; Pratt, N. E.; Griedel, B. D. Asymmetric Allylation of Aldehydes with Chiral Lewis Bases. *J. Org. Chem.* **1994**, *59*, 6161–6163.

(57) Denmark, S. E.; Fu, J. Catalytic Enantioselective Addition of Allylic Organometallic Reagents to Aldehydes and Ketones. *Chem. Rev.* **2003**, *103*, 2763–2794.

(58) Denmark, S. E.; Barsanti, P. A.; Wong, K.-T.; Stavenger, R. A. Enantioselective Ring Opening of Epoxides with Silicon Tetrachloride in the Presence of a Chiral Lewis Base. *J. Org. Chem.* **1998**, *63*, 2428–2429.

(59) Nakajima, M.; Saito, M.; Shiro, M.; Hashimoto, S.-I. (S)-3,3'-Dimethyl-2,2'-biquinoline *N,N'*-Dioxide as an Efficient Catalyst for Enantioselective Addition of Allytrichlorosilanes to Aldehydes. *J. Am. Chem. Soc.* **1998**, *120*, 6419–6420.

(60) Denmark, S. E.; Fu, J. Catalytic, Enantioselective Addition of Substituted Allylic Trichlorosilanes Using a Rationally-Designed 2,2'-Bispyrrolidine-Based Bisphosphoramide. *J. Am. Chem. Soc.* **2001**, *123*, 9488–9489.

(61) Satyanarayana, T.; Ferber, B.; Kagan, H. B. Asymmetric Amplification in Catalysis by *trans*-1,2-Diaminocyclohexane Bistriflamide. *Org. Lett.* **2007**, *9*, 251–253.

(62) Klussmann, M.; Iwamura, H.; Mathew, S. P.; Wells, D. H. J.; Pandya, U.; Armstrong, A.; Blackmond, D. G. Thermodynamic Control of Asymmetric Amplification in Amino Acid Catalysis. *Nature* **2006**, *441*, 621–623.

(63) Klussmann, M.; White, A. J. P.; Armstrong, A.; Blackmond, D. G. Rationalization and Prediction of Solution Enantiomeric Excess in Ternary Phase Systems. *Angew. Chem., Int. Ed. Engl.* **2006**, *45*, 7985–7989.

(64) Klussmann, M.; Mathew, S. P.; Iwamura, H.; Wells, D. H. J.; Armstrong, A.; Blackmond, D. G. Kinetic Rationalization of Nonlinear Effects in Asymmetric Catalysis Based on Phase Behavior. *Angew. Chem., Int. Ed. Engl.* **2006**, *45*, 7989–7992.

(65) Hayashi, Y.; Matsuzawa, M.; Yamaguchi, J.; Yonehara, S.; Matsumoto, Y.; Shoji, M.; Hashizume, D.; Koshino, H. Large Nonlinear Effect Observed in the Enantiomeric Excess of Proline in Solution and That in the Solid State. *Angew. Chem., Int. Ed. Engl.* **2006**, *45*, 4593–4597.

(66) Kellogg, R. M. The Crystallization Behavior of Proline and its Role in Asymmetric Organocatalysis. *Angew. Chem., Int. Ed. Engl.* **2007**, *46*, 494–497.

(67) Zhong, G. A Facile and Rapid Route to Highly Enantiopure 1,2-Diols by Novel Catalytic Asymmetric α-Aminoxylation of Aldehydes. *Angew. Chem., Int. Ed. Engl.* **2003**, *42*, 4247–4250.

(68) Brown, S. P.; Brochu, M. P.; Sinz, C. J.; MacMillan, D. W. C. The Direct and Enantioselective Organocatalytic α-Oxidation of Aldehydes. *J. Am. Chem. Soc.* **2003**, *125*, 10808–10809.

(69) Hayashi, Y.; Yamaguchi, J.; Hibino, K.; Shoji, M. Direct Proline Catalyzed Asymmetric α-Aminooxylation of Aldehydes. *Tetrahedron Lett.* **2003**, *44*, 8293–8296.

(70) Calvin, M. *Chemical Evolution*; Clarendon: London, 1969; Chapter 7.

(71) Blackmond, D. G. Asymmetric Autocatalysis and its Implications for the Origin of Homochirality. *Proc. Natl. Acad. Sci. U.S.A.* **2004**, *101*, 5732–5736.

(72) Frank, F. C. Spontaneous Asymmetric Synthesis. *Biochem. Biophys. Acta* **1953**, *11*, 459–463.

(73) Soai, K.; Shibata, T.; Morloka, H.; Choji, K. Asymmetric Autocatalysis and Amplification of Enantiomeric Excess of a Chiral Molecule. *Nature* **1995**, *378*, 767–768.

(74) Mauksch, M.; Tsogoeva, S. B.; Martynova, I. M.; Wei, S. Evidence of Asymmetric Autocatalysis in Organocatalytic Reactions. *Angew. Chem., Int. Ed. Engl.* **2006**, *46*, 393–396.

(75) Soai, K.; Shibata, T.; Sato, I. Enantioselective Automultiplication of Chiral Molecules by Asymmetric Autocatalysis. *Acc. Chem. Res.* **2000**, *33*, 382–390.

(76) Shibata, T.; Yonekubo, S.; Soai, K. Practically Perfect Asymmetric Autocatalysis with (2-Alkynyl-5-pyrimidyl)alkanols. *Angew. Chem., Int. Ed. Engl.* **1999**, *38*, 659–661.

(77) Shibata, T.; Yamamoto, J.; Matsumoto, N.; Yonekubo, S.; Osanai, S.; Soai, K. Amplification of a Slight Enantiomeric Imbalance in Molecules Based on Asymmetric Autocatalysis: The First Correlation Between High Enantiomeric Enrichment in a Chiral Molecule and Circularly Polarized Light. *J. Am. Chem. Soc.* **1998**, *120*, 12157–12158.

(78) Klankermayer, J.; Gridnev, I. D.; Brown, J. M. Role of the Isopropyl Group in Asymmetric Autocatalytic Zinc Alkylations. *J. Chem. Soc., Chem. Commun* **2007**, 3151–3153.

(79) Sato, I.; Urabe, H.; Ishiguro, S.; Shibata, T.; Soai, H. Amplification of Chirality from Extremely Low to Greater than 99.5% ee by Asymmetric Autocatalysis. *Angew. Chem., Int. Ed. Engl.* **2003**, *42*, 315–317.

(80) Singleton, D. A.; Vo, L. K. Enantioselective Synthesis Without Discrete Optically Active Additives. *J. Am. Chem. Soc.* **2002**, *124*, 10010–10011.

(81) Blackmond, D. G.; McMillan, C. R.; Ramdeehul, S.; Schorm, A.; Brown, J. M. Origins of Asymmetric Amplification in Autocatalytic Alkylzinc Additions. *J. Am. Chem. Soc.* **2001**, *123*, 10103–10104.

(82) Gridnev, I. D.; Serafimov, J. M.; Brown, J. M. Solution Structure and Reagent Binding of the Zinc Alkoxide Catalyst in the Soai Asymmetric Autocatalytic Reaction. *Angew. Chem., Int. Ed. Engl.* **2004**, *43*, 4884–4887.

(83) Buono, F. G.; Blackmond, D. G. Kinetic Evidence for a Tetrameric Transition State in the Asymmetric Autocatalytic Alkylation of Pyrimidyl Aldehydes. *J. Am. Chem. Soc.* **2003**, *125*, 8978–8979.

(84) Gridnev, I. D.; Serafimov, J. M.; Quiney, H.; Brown, J. M. Reflections on Spontaneous Asymmetric Synthesis by Amplifying Autocatalysis. *Org. Biomol Chem.* **2003**, *1*, 3811–3819.

(85) Mislow, K. Absolute Asymmetric Synthesis: A Commentary. *Collect. Czech. Chem. Commun.* **2003**, *68*, 849–864.

(86) Podlech, J.; Gehring, T. New Aspects of Soai's Asymmetric Autocatalysis. *Angew. Chem., Int. Ed. Engl.* **2005**, *44*, 5776–5777.

(87) Soai, K.; Osanai, S.; Kadowaki, K.; Yonekubo, S.; Shibata, T.; Sato, I. D- and L-Quartz-Promoted Highly Enantioselective Synthesis of a Chiral Organic Compound. *J. Am. Chem. Soc.* **1999**, *121*, 11235–11236.

(88) Sato, I.; Omiya, D.; Saito, T.; Soai, K. Highly Enantioselective Synthesis Induced by Chiral Primary Alcohols Due to Deuterium Substitution. *J. Am. Chem. Soc.* **2000**, *122*, 11739–11740.

(89) Soai, K.; Sato, I.; Shibata, T.; Komiya, S.; Hayashi, M.; Matsueda, Y.; Imamura, H.; Hayase, T.; Morioka, H.; Tabira, H.; Yamamoto, J.; Kowata, Y. Asymmetric Synthesis of Pyrimidyl Alkanol Without Adding Chiral Substances by the Addition of Diisopropylzinc to Pyrimidine-5-carbaldehyde in Conjunction with Asymmetric Autocatalysis. *Tetrahedron: Asymmetry* **2003**, *14*, 185–188.

(90) Alberts, A. H.; Wynberg, H. The Role of the Product in Asymmetric C–C Bond Formation: Stoiciometric and Catalytic Enantioselective Autoinduction. *J. Am. Chem. Soc.* **1989**, *111*, 7265–7266.

(91) Bolm, C.; Bienewald, F.; Seger, A. Asymmetric Autocatalysis with Amplification of Chirality. *Angew. Chem., Int. Ed. Engl.* **1996**, *35*, 1657–1659.

(92) Heller, D. P.; Goldberg, D. R.; Wulff, W. D. Positive Cooperativity of Product Mimics in the Asymmetric Diels–Alder Reaction Catalyzed by a VAPOL-Aluminum Catalyst. *J. Am. Chem. Soc.* **1997**, *119*, 10551–10552.

(93) Atwood, D. A.; Hutchison, A. R.; Zhang, Y. Z. Compounds Containing Five-Coordinate Group 13 Elements. *Struct. Bonding* **2003**, *105*, 167–201.

(94) Yamada, Y. M. A.; Yoshikawa, N.; Sasai, H.; Shibasaki, M. Direct Catalytic Asymmetric Aldol Reactions of Aldehydes with Unmodified Ketones. *Angew. Chem., Int. Ed. Engl.* **1997**, *36*, 1871–1873.

(95) List, B.; Lerner, R. A.; Barbas, C. F., III. Proline-Catalyzed Direct Asymmetric Aldol Reactions. *J. Am. Chem. Soc.* **2000**, *122*, 2395–2396.

(96) Trost, B. M.; Ito, H. A Direct Catalytic Enantioselective Aldol Reaction via a Novel Catalyst Design. *J. Am. Chem. Soc.* **2000**, *122*, 12003–12004.

(97) Northrup, A. B.; Mangion, I. K.; Hettche, F.; MacMillan, D. W. C. Enantioselective Organocatalytic Direct Aldol Reactions of α-Oxyaldehydes: Step One in a Two-Step Synthesis of Carbohydrates. *Angew. Chem., Int. Ed. Engl.* **2004**, *43*, 2152–2154.

(98) Palomo, C.; Oiarbide, M.; García, J. M. Current Progress in the Asymmetric Aldol Addition Reaction. *Chem. Soc. Rev.* **2004**, *33*, 65–75.

(99) Tang, Z.; Jiang, F.; Yu, L.-T.; Cui, X.; Gong, L.-Z.; Mi, A.-Q.; Jiang, Y.-Z.; Wu, Y.-D. Novel Small Organic Molecules for a Highly Enantioselective Direct Aldol Reaction. *J. Am. Chem. Soc.* **2003**, *125*, 5262–5263.

(100) Yao, W.; Wang, J. Direct Catalytic Asymmetric Aldol-Type Reaction of Aldehydes with Ethyl Diazoacetate. *Org. Lett.* **2003**, *5*, 1527–1530.

(101) Yoshikawa, N.; Kumagai, N.; Matsunaga, S.; Moll, G.; Ohshima, T.; Suzuki, T.; Shibasaki, M. Direct Catalytic Asymmetric Aldol Reaction: Synthesis of Either syn- or anti-alpha,beta-Dihydroxy Ketones. *J. Am. Chem. Soc.* **2001**, *123*, 2466–2467.

(102) Trost, B. M.; Fettes, A.; Shireman, B. T. Direct Catalytic Asymmetric Aldol Additions of Methyl Ynones. Spontaneous Reversal in the Sense of Enantioinduction. *J. Am. Chem. Soc.* **2004**, *126*, 2660–2661.

(103) Danda, H.; Nishikawa, H.; Otaka, K. Enantioselective Auto Induction in the Asymmetric Hydrocyanation of 3-Phenoxybenzaldehyde Catalyzed by Cyclo(R)-Phenylalanyl-(R)-Histidyl. *J. Org. Chem.* **1991**, *56*, 6740–6741.

(104) Tanaka, K.; Mori, A.; Inoue, I. The Cyclic Dipeptide Cyclo[(S)-phenylalanyl-(S)-histidyl] as a Catalyst for Asymmetric Addition of Hydrogen Cyanide to Aldehydes. *J. Org. Chem.* **1990**, *55*, 181–185.

(105) Kogut, E. F.; Thoen, J. C.; Lipton, M. A. Examination and Enhancement of Enantioselective Autoinduction in Cyanohydrin Formation by Cyclo[(R)-His-(R)-Phe]. *J. Org. Chem.* **1998**, *63*, 4604–4610.

(106) Shvo, Y.; Gal, M.; Becker, Y.; Elgavi, A. Asymmetric Hydrocyanation of Aldehydes with Cyclo-Dipeptides: A New Mechanistic Approach. *Tetrahedron: Asymmetry* **1996**, *7*, 911–924.

(107) Balsells, J.; Costa, A. M.; Walsh, P. J. Temperature-Dependent Nonlinear Effects and Catalyst Evolution in the Asymmetric Addition of Diethylzinc to Benzaldehyde. *Isr. J. Chem.* **2001**, *41*, 251–261.

Bifunctional, Dual, and Multifunctional Catalyst Systems

For any catalytic asymmetric transformation, at least one activation mode (see Chapters 2 and 3) is necessary to confer the activation energy lowering (via transition-state stabilization or ground-state destabilization) required for catalysis to occur. This is exemplified by coordination of a carbonyl oxygen to a Lewis acid catalyst, thereby reducing the barrier to nucleophilic addition relative to attack on the unactivated carbonyl compound. Alternatively, binding of a nucleophile to a catalyst can serve to increase its nucleophilic character and, therefore, its reactivity toward electrophiles. Not only are many activation modes possible, often more than one activation mode is incorporated in a single transformation.   Such combinations may occur sequentially at the same metal (hydrogenation, alkene coordination, oxidative addition of hydrogen, then migratory insertion, etc.).

There are also examples where activation of the substrate or reagent does not sufficiently lower the activation energy, such that the reaction can proceed under the desired conditions. In these cases, a dual activation strategy, wherein both electrophile and nucleophile are simultaneously activated may suffice. The combination of simultaneous activation modes is the subject of this chapter with a focus on those systems in which more than one of the activating components is chiral and plays a role in the ultimate stereoselection conferred to the product.[1-8]

Dual catalyst activation arises from the activation of chemically distinct reagents by two separate catalysts in an intermolecular sense. While this activation mode is attractive, in that complex bifunctional or multifunctional catalyst assemblies are not required, downsides include possible interactions between two or more catalytic entities, the need for selective recognition of each substrate by a given catalyst in a mixture, and the high molecularity of the reaction (i.e., entropy limitations). *Bifunctional catalysts* can solve many of these issues by containing two catalytic moieties in a defined relationship. Dependent bifunctional catalysts still suffer the first of these problems since the catalytic sites can influence each other. Even so, many useful catalysts of this type have been found. Independent bifunctional catalysts possess two catalytic moieties that can be modulated independently and, in principle, can be more easily fine-tuned.   Bifunctional catalysts can be obtained by linking dual catalytic moieties or by design from first principles. Multifunctional catalysts are also possible that contain more than two catalytic entities. The characterization of catalyst systems as dual, bifunctional, or multifunctional is often not recognized, however, until mechanism studies are undertaken. The principles unearthed during such studies are powerful in the improvement of catalysts and in the conception of new catalyst systems.

Nature has made ample use of multifunctional catalysts to achieve high turnover frequencies and turnover numbers in reactions that would otherwise be difficult to promote. Many enzymatic catalysts involve cooperation between two or more reactive sites that activate both reaction partners. In such systems, the catalyst also maintains the activated components at a favorable distance with the optimal relative orientation for maximum reactivity and selectivity.

Two examples of multifunctional enzymes from Nature that catalyze the asymmetric aldol reaction of ketone donors to aldehyde acceptors are illustrated in **Figure 12.1.**[9] Aldolase enzymes are classified as type I, which operate without a metal ion

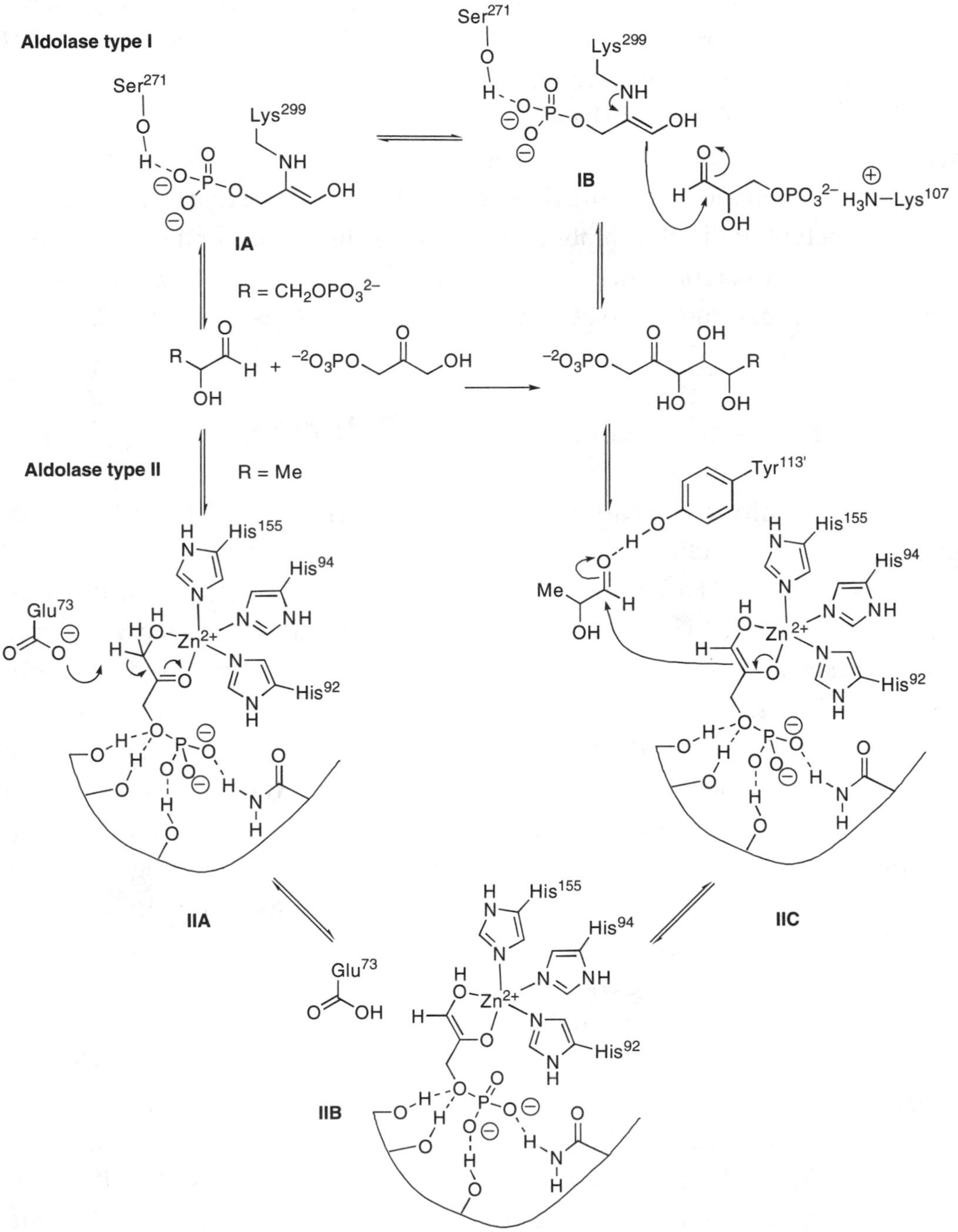

Figure 12.1. Mechanisms of type I (top) and II (bottom) aldolase reactions. These enzymes catalyze the asymmetric addition of dihydroxyacetone phosphate donors to aldehydes.

cofactor, or type II, which involve a zinc cofactor. In the type I aldolase at the top of **Figure 12.1**, a lysine residue adds to the ketone carbonyl to form an imine that tautomerizes to enamine **IA**. Several hydrogen bonds are believed to aid in the orientation of the reaction partners during stereoselective addition of the enamine donor to the acceptor aldehyde as shown in **IB**. On the bottom of **Figure 12.1**, dihydroxyacetone phosphate is activated by coordination to zinc while the phosphate group is held in a pocket containing several hydrogen-bond donors. Binding of dihydroxyacetone phosphate to zinc increases the acidity of the α-hydrogens, one of which is then deprotonated by a neighboring glutamate residue, as shown in **IIA**, to generate zinc enolate **IIB**. A tyrosine residue hydrogen bonds to the aldehyde substrate in **IIC**, activating it and stabilizing the developing charge in the addition transition state.

Like the organocatalytic type I and zinc-based type II aldolases described above, the following sections also show how both purely organic and metal-based small molecules can act as bifunctional and multifunctional catalysts. The topics that follow are organized based on characteristics of the catalytic systems. Modifications of a catalyst system can give rise to catalysts that operate by the same mechanism but belong to a different subset of catalysts. In such cases, the catalysts are most easily discussed together.

12.1 Intramolecular Bifunctional Catalysis

Catalysts that contain two catalytically active sites in the same complex are defined as intramolecular bifunctional catalysts. This class can be subdivided into catalysts with two active sites that are either electronically coupled or electronically independent. In interdependent bifunctional catalysts, the two active sites are electronically coupled. In this way, changes at one site impact the reactivity at the neighboring site. The sites in these catalysts are typically connected by a common atom or functional group. An example is a metal alkoxide, $L_nM–OR$. The metal can act as a Lewis acid, while the oxygen can behave as a Brønsted base, deprotonating the substrate, or as a Lewis base by donating a lone pair of electrons to a second metal. The advantages of interdependent bifunctional catalysts often stem from their high degree of organization. Additionally, these bifunctional catalysts often self-assemble to form the active species. The drawback to interdependent bifunctional catalysts is that one site is not easily tuned without affecting the other.

Intramolecular bifunctional catalysts that have separated, noninteracting sites are termed independent bifunctional catalysts. In these systems, the two sites are usually separated by several intervening atoms. Independent bifunctional catalysts have the benefit of greater tunability, because each site can be modulated without impacting the other. Often independent bifunctional catalysts are less rigid, with a higher degree of rotational freedom. While increased flexibility in the transition state may lead to greater tunability and broader substrate generality, and it can complicate rational catalyst design.

Some of the examples outlined in this chapter have been previously discussed in other chapters with a different focus. These cases are presented in less detail here and the reader is referred to prior sections for further discussion.

12.1.1 Interdependent Lewis Acid and Lewis Base Components

Some of the earliest bifunctional catalysts developed combined Lewis acidic and Lewis basic sites that were directly connected. Modification of either site changes the reactivity of the other, so they are classified as interdependent bifunctional catalysts. The bifunctional nature of these catalysts, their intriguing reaction mechanisms, and their excellent control over stereoinduction has raised significant interest and inspired the development of related catalysts based on these concepts.

12.1.1.a Bifunctional Oxazaborolidine-Catalyzed Asymmetric Reduction of Ketones (CBS Reduction)

An excellent example of interdependent bifunctional catalysis is the oxazaborolidine-catalyzed asymmetric reduction of ketones.[10–16] This system is often referred to as the "CBS reduction" after the researchers who proposed the accepted mechanism and developed the catalyst outlined below.[13] As shown in **Equation 12.1**, the oxazaborolidine catalyst is generated by reaction of an amino alcohol with BH_3. The oxazaborolidine alone does not react with ketones. Addition of BH_3 to the mixture of oxazaborolidine and ketone, however, results in the rapid reduction of a variety of ketone substrates with high levels of enantioselectivity. The B-Me derivative is the catalyst of choice, because it is less air-sensitive and gives slightly higher enantioselectivities.

Equation 12.1

The basic steps of the proposed reaction mechanism are illustrated in **Figure 12.2**. Coordination of BH_3 to nitrogen is followed by binding of the ketone substrate. Subsequent hydride transfer to the carbonyl carbon forms a boron alkoxide that dissociates to regenerate the catalyst.

Consideration of each step of the mechanism illustrates the interdependent nature of this classic bifunctional catalyst. The nitrogen lone pair of the oxazaborolidine is delocalized into the empty p-orbital on boron, giving the B–N bond partial double bond character. Binding of BH_3 to the nitrogen lone pair ruptures this interaction, increasing the Lewis acidity of the endocyclic boron. Coordination of the BH_3 to the oxazaborolidine also serves to enhance the nucleophilicity of the boron hydrides. Note that the BH_3 binds to give a cis-fused ring juncture. The alternative, coordination of the BH$_3$ *syn* to the carbinol results in a trans-fused 5,5-ring system that is estimated to be 10–15 kcal/mol higher in energy by calculations (**Figure 12.3**).[17]

In the next step, binding of the ketone to the unsaturated boron center could give rise to diastereomeric substrate adducts. Coordination of the ketone to the top face of the oxazaborolidine is strongly disfavored by the proline backbone. As a result, the bottom face of the unsaturated boron center binds the ketone, coordinating its sterically least encumbered lone pair. Not only does binding of the ketone to the Lewis acidic boron activate the carbonyl, it orients the substrate exposing one face to the neighboring BH$_3$ group. The hydride transfer then proceeds through a six-membered transition state.

Earlier generations of oxazaborolidine catalysts were derived from acyclic amino alcohols (**Figure 12.4**). In this system, coordination of BH$_3$ and ketone can give rise to diastereomeric adducts that are closer in energy than the analogous intermediates in the proline-based system. These isomeric adducts lead to enantiomeric products, resulting in slightly lower enantioselectivity relative to the proline derivative.

Although not included in the catalytic cycle in **Figure 12.2**, two of the three boron hydrides can be transferred to ketone substrates. After reduction of two equivalents

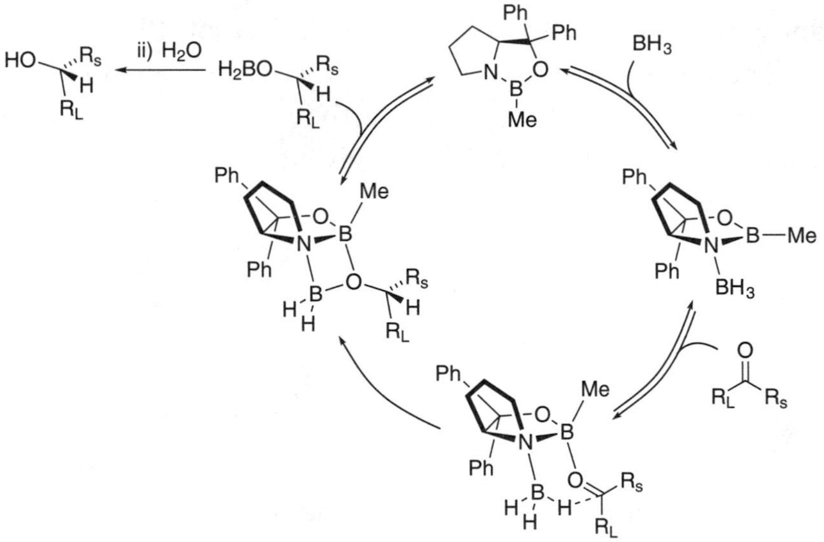

Figure 12.2. Proposed mechanism of the catalytic asymmetric oxazaborolidine-catalyzed reduction of ketones.

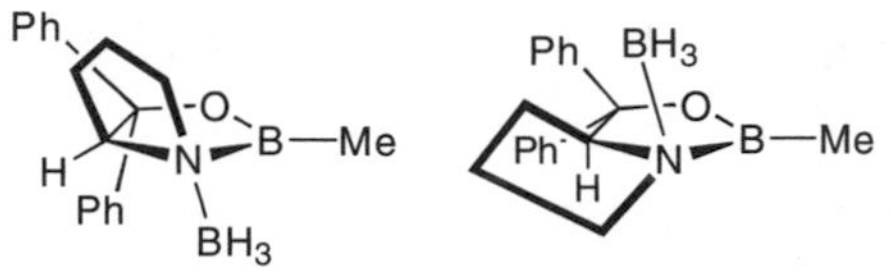

Figure 12.3. Coordination of the BH$_3$ to the top face of the oxazaborolidine (right) is disfavored due to formation of a strained bicyclic ring system.

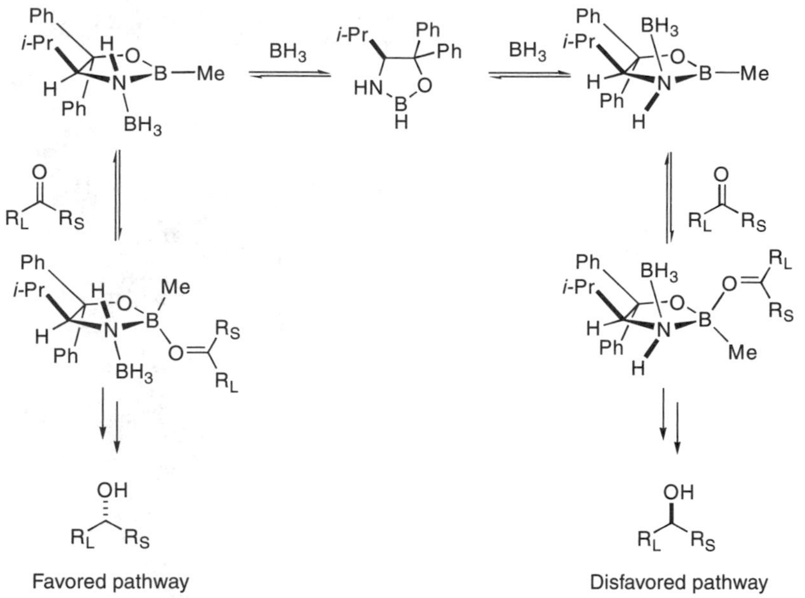

Figure 12.4. Diastereomeric oxazaborolidines formed on coordination of BH_3 lead to opposite enantiofacial selectivity. Coordination of the BH_3 to the top face of the ring system leads to the minor enantiomer.

of ketone, the third hydride is inactive, probably due to the increased steric interactions between the dialkoxyborane and the oxazaborolidine. It is also interesting to note that oxazaborolidine catalysts have been applied to the asymmetric reduction of oxime ethers and β-keto oxime ether to provide enantioenriched amines and amino alcohols, respectively.[10,11,18,19]

The ease of generation of the CBS catalyst, combined with the high enantioselectivity in the reduction of ketones, make this system a popular choice for the generation of enantioenriched secondary alcohols.

12.1.1.b Amino Alcohol-Based Catalysts for the Asymmetric Addition of Alkyl Groups to Aldehydes

Like the CBS ketone-reduction catalysts described above, the amino alcohol-based catalysts for the asymmetric addition of alkyl groups to aldehydes are also interdependent bifunctional catalysts that operate through a dual activation mechanism.[20–24] The reaction mechanism and nonlinear behavior of these catalysts is described in Chapter 7, so only the bifunctional nature of these catalysts is highlighted here. Reaction of the amino alcohol DAIB with the dialkylzinc reagent generates the three-coordinate amino alkoxide, which has a dimeric resting state (**Figure 12.5**). Binding of a dialkylzinc to the Lewis basic alkoxide increases the Lewis acidity of the central zinc, which can then better activate the aldehyde. Not only is the O-bound ZnR'_2 in a more reactive geometry, the alkyl groups are more nucleophilic due to the increased electron density on the bound ZnR'_2 relative to the free ZnR'_2.

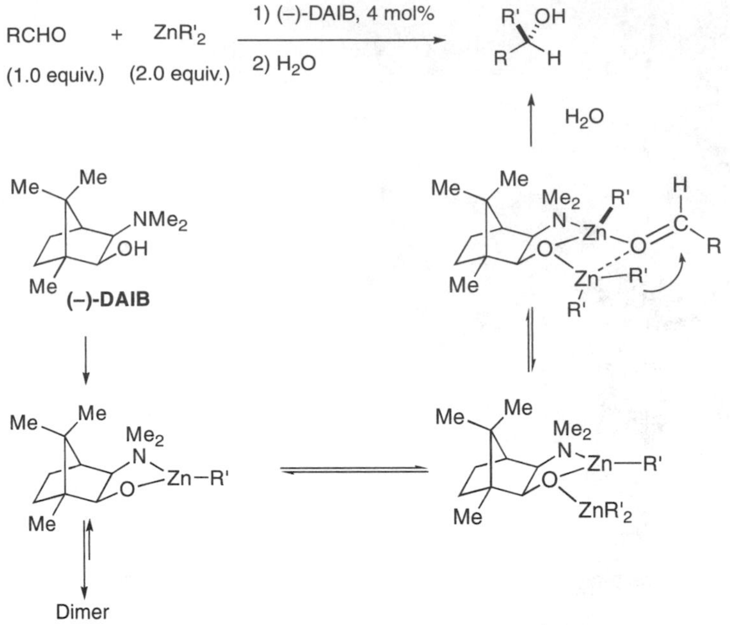

Figure 12.5. Basic steps in the amino alcohol-based catalytic asymmetric addition of alkylzinc reagents to aldehydes. See Chapter 11 for additional details.

Many efficient and highly enantioselective catalysts were subsequently developed for this reaction,[23–25] based on the concepts derived from this historically important system.

12.1.2 Interdependent Brønsted Acid–Metal-Hydride Activation in the Asymmetric Hydrogenation of Ketones

Asymmetric hydrogenation of prochiral ketones is an important method for the preparation of chiral secondary alcohols. These enantioenriched secondary alcohols are important intermediates in the pharmaceutical, agrochemical, and flavor and fragrance industries. Until recently, however, such reactions were limited to substrates with pendent binding sites that could chelate to the catalyst, like β-keto esters.[3] Many of the catalysts that efficiently hydrogenate C–C double bonds exhibit little or no reactivity with isolated ketones. This discrepancy may be ascribed to the different binding modes of alkenes and ketones. While substrates with C–C double bonds easily form metal π-complexes, this binding mode is uncommon for ketones, which predominantly bind through their oxygen lone pairs. The distinct binding mode of olefins versus carbonyls leads to the proposal that successful hydrogenation catalysts for these substrates will likely operate by dissimilar mechanisms. This has indeed been found to be the case, and mechanistic investigations indicate that the ketone reduction proceeds by an unexpected pathway that does not involve coordination of the ketone substrate to the metal center.

A key discovery in the development of ketone hydrogenation catalysts was the beneficial effect of diamines with at least one NH_2 group on the efficiency of ruthenium phosphine-based catalysts. Thus, ruthenium complexes bearing chiral diamine and bidentate phosphine ligands, such as the precatalysts illustrated in **Figure 12.6**, exhibit high enantioselectivity and efficiency in ketone reductions in the presence of base.[3,4,26,27] Base and isopropanol are necessary to convert the precatalyst into the dihydride catalyst. Additional attractive features of these catalysts include low catalyst loadings, functional group tolerance, and high selectivity for C=O bonds over C=C bonds. Furthermore, they can be used in the asymmetric hydrogenation of C=N bonds to furnish enantioenriched amines, although comparable enantioselectivities to those observed with ketones have yet to be achieved.[28,29]

As illustrated in **Table 12.1**, catalyst derived from *trans*-RuCl$_2$[(*S*)-Xyl-BINAP)][(*S*)-DAIPEN] will reduce aromatic, heteroaromatic, α,β-unsaturated, and cyclopropyl ketones with extremely high control over enantioselectivity.[26,30]

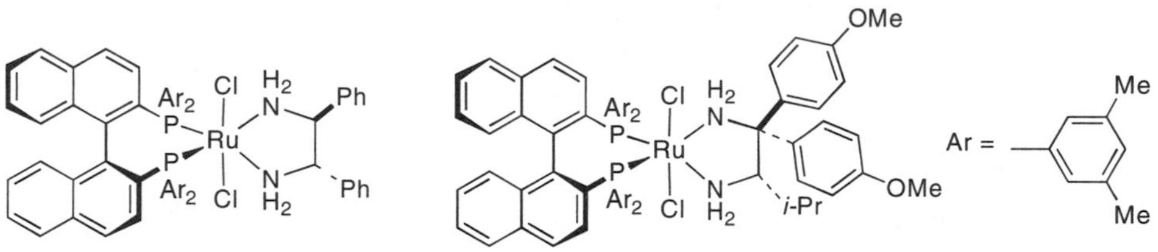

trans-RuCl$_2$[(*S*)-Xyl-BINAP)][(*S*,*S*)-DPEN]   *trans*-RuCl$_2$[(*S*)-Xyl-BINAP)][(*S*)-DAIPEN]

Figure 12.6. *trans*-RuCl$_2$[(*S*)-Xyl-BINAP)][diamine] precatalysts.

Table 12.1. Enantioselective reduction of ketones with *trans*-RuCl$_2$[(*S*)-Xyl-BINAP)][(*S*)-DAIPEN].

Substrate		% ee	Substrate		% ee
(phenyl ketone)	R = Me	99	(Ph enone)		97
	Et	99			
	i-Pr	99			
	Cyclopropyl	96	(C$_5$H$_{12}$ enone)		97
(ortho-X acetophenone)	X = Me	99			
	Br	97		R^1 R^2	
	CF$_3$	96	(cyclopropyl ketone)	H Me	96
	OMe	92		H Ph	96
(furyl ketone)		99		Me Me	98
(thiazolyl ketone)		96	(cyclohexyl methyl ketone)		85
(pyridyl ketone)		94	(aliphatic ketone)		01

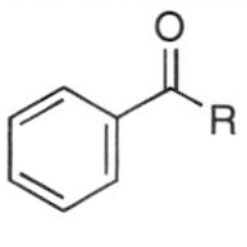

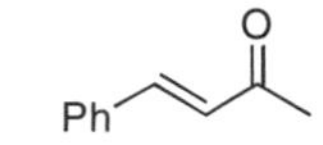

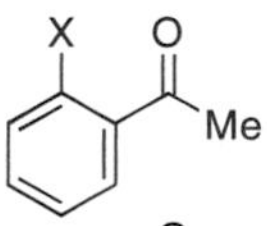

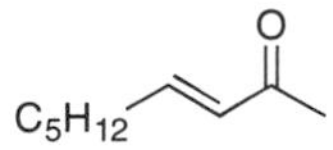

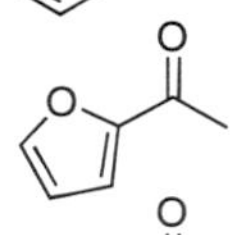

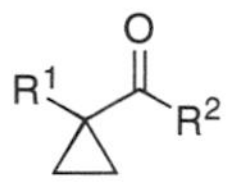

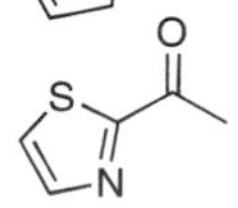

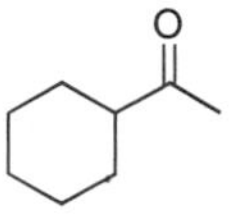

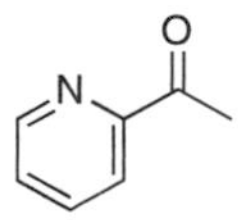

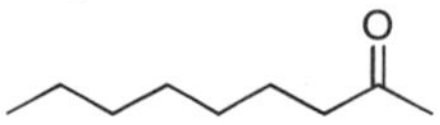

Particularly impressive is the chemoselectivity exhibited by these catalysts for the reduction of carbonyl groups in the presence of alkenes. Dialkyl ketones remain a difficult class of substrates for these catalysts,[3] although related systems bearing novel phosphines exhibit enantioselectivities as high at 75% for 2-hexanone.[31]

This catalyst system has been applied to the synthesis of various pharmaceutical agents,[3] as illustrated in the synthesis of the antidepressant (R)-fluoxetine (**Equation 12.2**). Using 7.3 mg of the precatalyst, activation with base in 2-propanol (30 mL) and addition of 3-dimethylaminopropiophenone substrate (10.6 g, S/C = 10 000), under 8 atm H_2 at 25 °C for 5 h gave the precursor to (R)-fluoxetine in 97.5% ee (96% yield).[32]

Equation 12.2

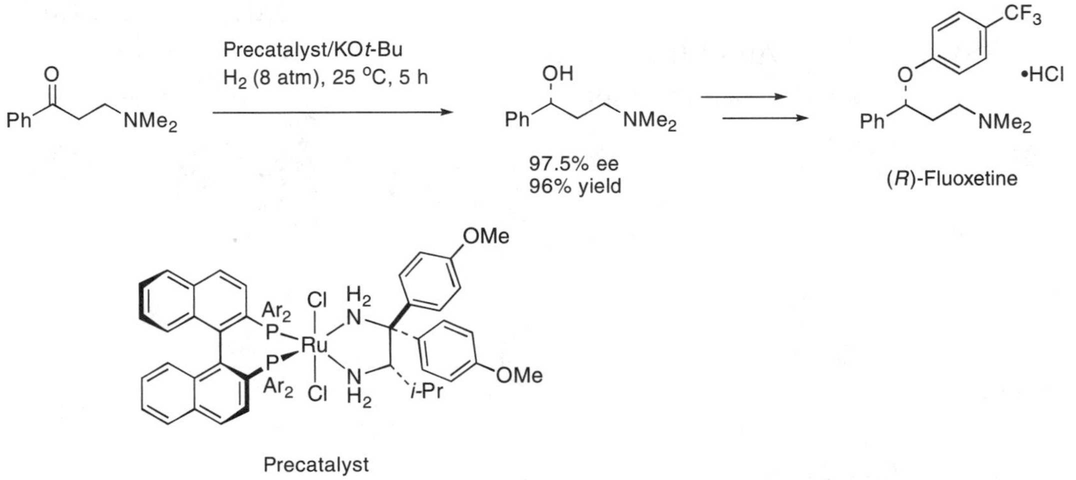

It is notable that the correct combination of ligand configurations of the BINAP derivative and the diamine are essential for good enantioselectivity (**Figure 12.7**). The mismatched combination gives lower enantioselectivity than the catalyst derived from substitution of either the chiral diamine or the chiral bis(phosphine) by achiral ligands.

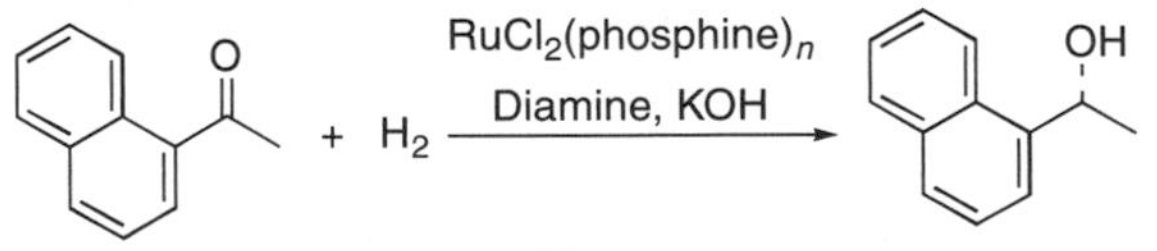

Phosphine	Diamine	% ee
(S)-BINAP	(S,S)-DPEN	97
(S)-BINAP	(R,R)-DPEN	14
(S)-BINAP	$NH_2CH_2CH_2NH_2$	57
PPh₃	(S,S)-DPEN	75

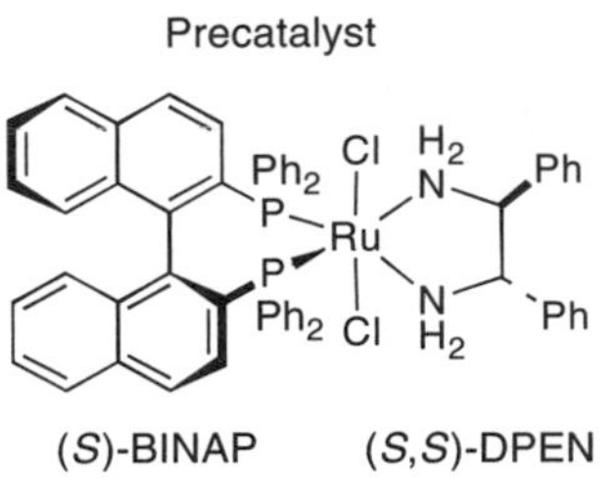

Figure 12.7. The complementary interaction of (S)-BINAP with (S,S)-DPEN can be compared to the noncomplementary combination, which gives lower enantioselectivity than when each chiral ligand is used in combination with an achiral ligand.

Detailed experimental and theoretical investigations have provided a comprehensive picture of the unique mechanism of these remarkable catalysts.[4,33-35] The active catalyst is believed to be the dihydride (**Figure 12.8**), which is generated by combining the dichloride precatalyst (**Figure 12.6**), alkoxide base, and dihydrogen in isopropanol. Analogous dihydride compounds have been prepared and characterized crystallographically.[34] The first step in the catalytic cycle is believed to be formation of a weak hydrogen bond between the ketone substrate and the catalyst N–H, as shown in **Figure 12.8**. From this hydrogen-bonded intermediate, facile reduction takes place through a six-membered transition state, resulting in formation of the chiral alcohol and the Ru-amide. In this case, the catalyst is acting in a bifunctional capacity; the N–H is a Brønsted acid, while the metal center is a hydride donor.

Late-transition-metal–amide complexes of this type are very reactive,[36-38] and have been characterized in model systems.[34] The heterolytic cleavage of dihydrogen is the turnover-limiting step in the catalytic cycle (**Figure 12.9**). It likely occurs via initial formation of a dihydrogen complex, which has been observed in model systems.[39] Cleavage of the dihydrogen complex has been proposed to involve proton abstraction by the amide nitrogen to give the dihydride complex, deprotonation of the dihydrogen ligand by alkoxide base,[40,41] or under neutral conditions, an alcohol-assisted pathway.[35] Consistent with these hypotheses, a first-order dependency on the ruthenium catalyst and the dihydrogen, along with zero-order dependency on ketone

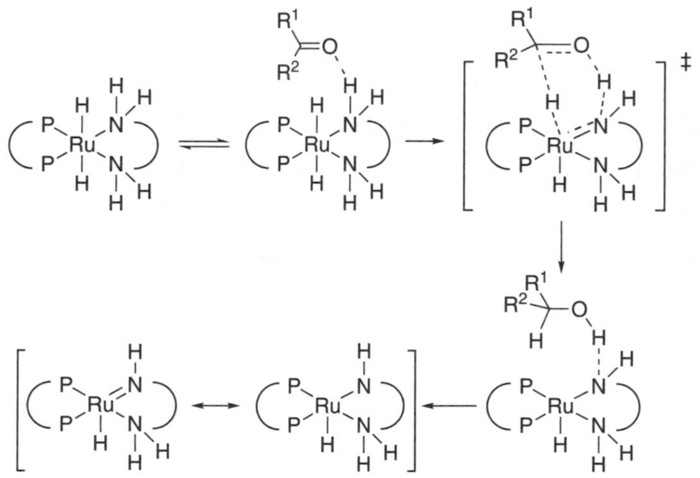

Figure 12.8. Proposed mechanism for the hydrogenation of ketones with (BINAP)RuH$_2$(diamine) derivatives.

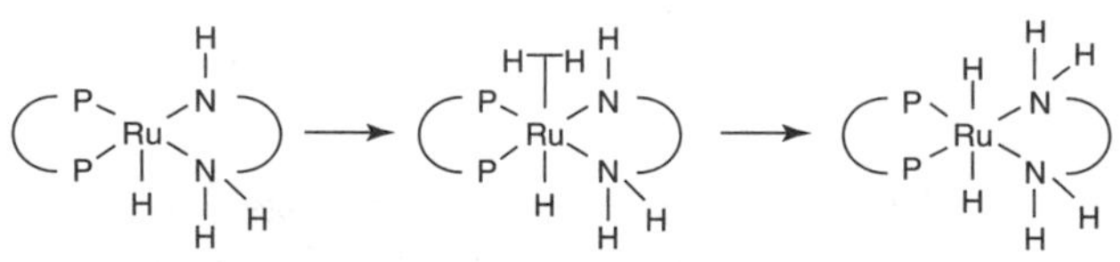

Figure 12.9. Proposed mechanism for the heterolytic cleavage of dihydrogen.

concentration, were observed. Additionally, $\Delta S^{\ddagger}$ is on the order of -25 cal/mol-K, suggesting a highly ordered transition state.

It is interesting to note that these hydrogenation catalysts are 18-electron species and need not open up a coordination site to generate a 16-electron intermediate, as observed with many alkene hydrogenation catalysts.[42] The bifunctional nature of these catalysts allows them to access an unconventional second coordination sphere hydrogen-transfer mechanism. Activation of the ketone by a hydrogen bond from the ligand alludes to the power of such subtle interactions in enantioselective catalysis.

12.2 Independent Bifunctional Catalysis

In the class of independent bifunctional catalysts, the active sites are separated by several intervening atoms. This allow for greater flexibility in tuning catalyst activity and enantioselectivity. As will be seen, a variety of different functionalities can be combined in the development of these fascinating catalysts.

12.2.1 Independent Lewis Acid–Brønsted Base Components

The asymmetric reaction of isocyanoacetate with aldehydes catalyzed by gold complexes bearing chiral phosphines with pendent amino groups is a classic example of bifunctional catalysis. The reaction was covered in Chapter 5 and will be outlined briefly here. Recall that the proposed reaction mechanism involves deprotonation of the gold-bound isocyanide by the pendant amino group, generating an ion pair (**Figure 5.21**). Coordination of the aldehyde, followed by attack by the enolate in an aldol-like reaction, forms the product. The enantioselectivity of this reaction strongly depends on the structure of the side chain and substituents on the amino group. This is particularly striking with acetaldehyde as substrate, as can be seen in **Figure 12.10**.

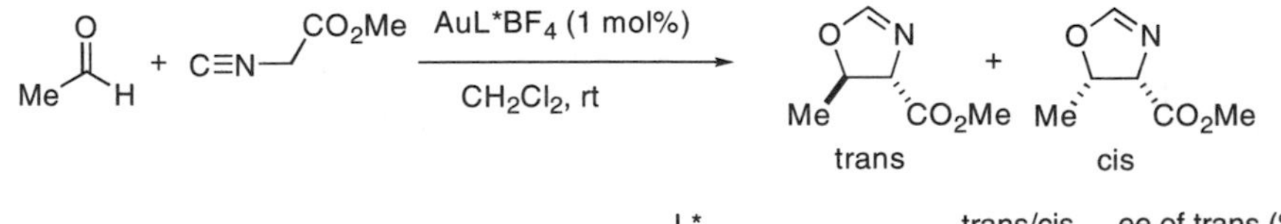

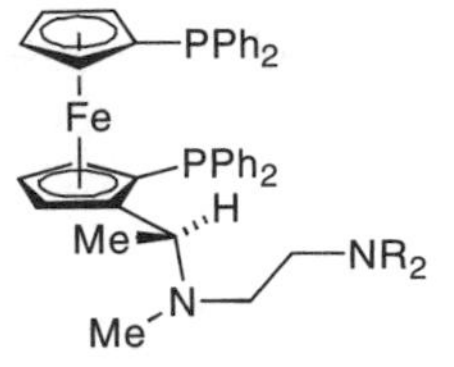

L*	trans/cis	ee of trans (%) (configuration)
$NR_2 = NMe_2$	78/22	37 (4S,5R)
$NR_2 = NEt_2$	84/16	72 (4S,5R)
$NR_2 = N\!\!\bigcirc\!\!O$ 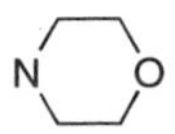	89/11	89 (4S,5R)

Figure 12.10. Bifunctional cationic gold(I) bis(phosphine) catalysts can promote the asymmetric aldol-type reaction. Acetaldehyde is a very sensitive substrate in this reaction. Seemingly minor changes in the catalyst structure have tremendous impact on enantioselectivity.

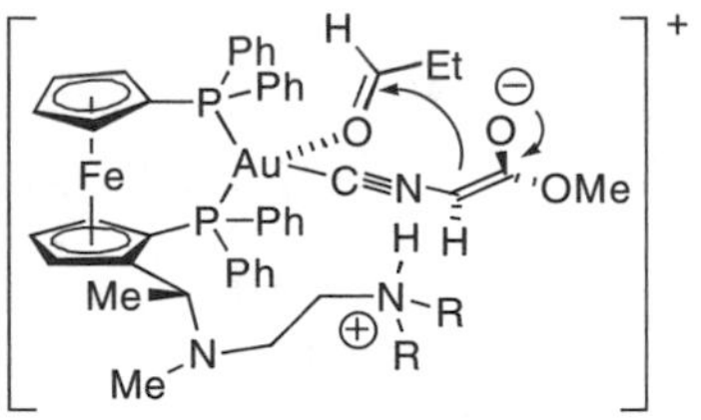

Figure 12.11. Transition state for the formation of the major diastereomer in the asymmetric aldol reaction.

The proposed transition state is illustrated in **Figure 12.11**. In this ensemble, the enolate is held in place by the gold center and the charge–charge interaction between the enolate and the ammonium ions. Because of the close proximity of the ammonium group to the enolate, the substituents on the pendent nitrogen will influence the enolate conformation and impact the enantioselectivity of the reaction.

12.2.2 Independent Lewis Acid–Lewis Base Containing Catalysts

Catalysts that have both Lewis acidic and Lewis basic sites must be carefully designed to avoid internal or intermolecular complexation of these moieties, resulting in catalyst deactivation. The addition of trimethylsilyl cyanide to carbonyl groups is a reaction that is known to be catalyzed by both Lewis acids and Lewis bases. The Lewis acid activates the carbonyl group and the Lewis base activates the trimethylsilyl cyanide. With these characteristics, this reaction has received significant attention from researchers working with Lewis acid–Lewis base bifunctional catalysts.

Bifunctional catalysts based on a modified BINOL skeleton have been designed to bind a central Lewis acidic metal and to activate trimethylsilyl cyanide via coordination to pendent Lewis basic sites at the 3,3′-positions (**Figure 12.12**).[43,44] It was found that aluminum-based catalysts performed well in this reaction, especially in the presence of phosphine oxide additives (**Figure 12.13**). The phosphine oxide additive is proposed to moderate the Lewis acidity of the aluminum center through coordination.[43] In its absence, the high Lewis acidity of the aluminum caused nonselective addition of cyanide to the aldehyde substrate. Binding the phosphine oxide additive also increases the coordination number of the aluminum, and changes the geometry of the metal center and the chiral environment.[43]

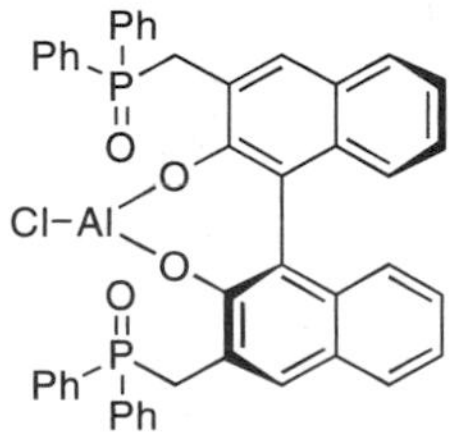

Figure 12.12. Proposed structure of a bifunctional catalyst for the addition of cyanide to aldehydes. The Lewis basic phosphine-oxide moiety is proposed to activate trimethylsilyl cyanide.

It is often difficult to determine if a catalyst is promoting a reaction through a bifunctional pathway. One way to probe the mechanism is with catalysts in which one of the reactive sites has been deleted. When the 3,3'-positions of the ligand were substituted with a sterically bulky noncoordinating group (CH_2CHPh_2), the opposite enantiomer of the product was obtained in low enantioselectivity. Additionally, a shift in the IR spectrum of trimethylsilyl cyanide indicates Lewis basic activation of the trimethylsilyl cyanide upon addition of phosphine oxides. Although further work is necessary to clarify the mechanism of this process, a working transition state is proposed in **Figure 12.14**.

This catalyst system also performs well in a modified asymmetric Strecker reaction. The Strecker reaction is important in the synthesis of α-amino acid precursors and involves the addition of cyanide to imines (**Figure 12.15**). In this reaction, optimal conditions include 20 mol% TMS–CN and 120 mol% HCN.[45] It was found that the enantioselectivities were strongly dependent on the nature of the N-alkyl substituent, with fluorenyl giving the best results. With most imine substrates, the yields were excellent and enantioselectivities were > 90%.

It is well known in asymmetric catalysis that reactions that give excellent enantioselectivities with aldehydes often do not give satisfactory results with ketones. There are two issues that must be overcome to develop successful catalysts for ketone substrates: reactivity and enantioselectivity. Aldehydes usually bind to metal centers

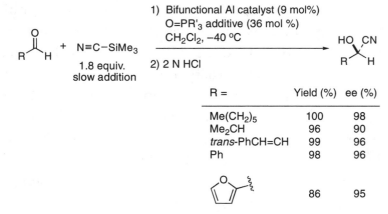

R =	Yield (%)	ee (%)
$Me(CH_2)_5$	100	98
Me_2CH	96	90
trans-PhCH=CH	99	96
Ph	98	96

Figure 12.13. Bifunctional Lewis acid–Lewis base catalyzed addition of TMS–CN to aldehydes.

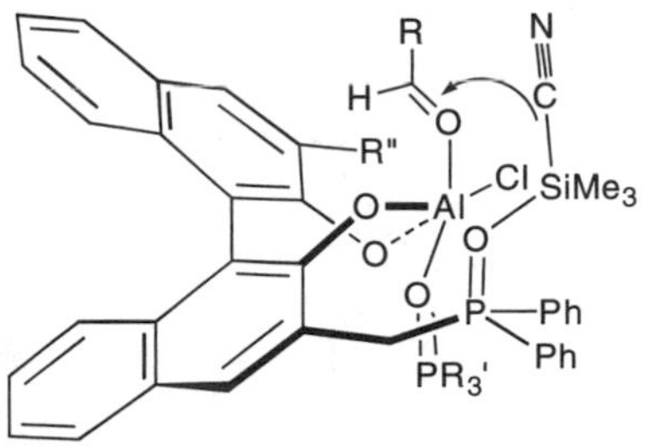

Figure 12.14. Proposed transition state for the addition of cyanide to aldehydes with a bifunctional catalyst.

with the lone pair that is *syn* to the aldehydic hydrogen (**Figure 12.16**). Ketones, in contrast, must coordinate to the metal *syn* to a sterically more encumbering alkyl substituent. As a result, ketones will not bind as tightly as aldehydes, resulting in decreased reactivity. Second, it is generally accepted that, in order to achieve high enantioselectivity, the catalyst must be able to differentiate between the two lone pairs on the carbonyl oxygen of the ketone. This is often a difficult task if the lone pairs are in sterically similar environments. Given these constraints, it is not surprising that reactions with ketones often require more active catalysts than aldehydes. This is the case with the asymmetric addition of trimethylsilyl cyanide to ketones.

Catalysts for the asymmetric addition of cyanide to ketones were designed with a similar dual activation strategy, but employed a carbohydrate scaffold and an early-transition-metal Lewis acid. The proposed catalyst structure is illustrated in **Figure 12.17**. This catalyst promotes the addition of cyanide to ketones at low catalyst loading

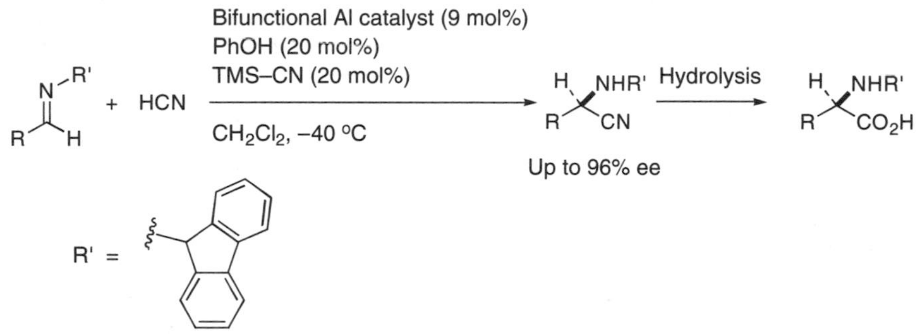

Figure 12.15. Asymmetric Strecker reaction catalyzed by the bifunctional catalyst from **Figure 12.12**.

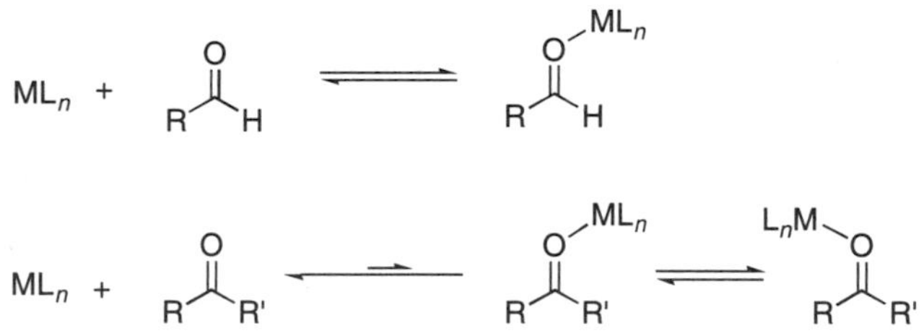

Figure 12.16. Comparison of aldehyde vs. ketone binding to Lewis acids.

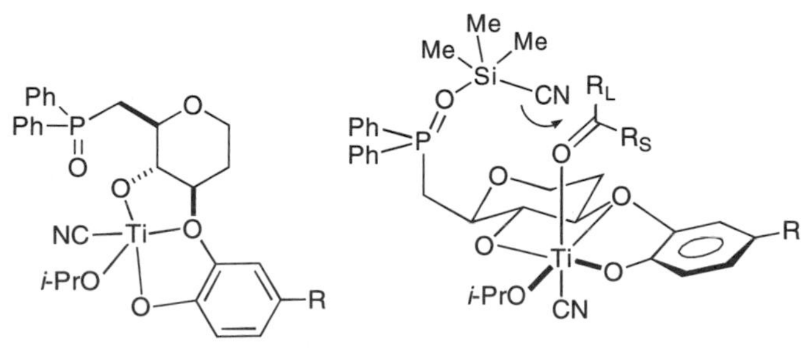

Proposed transition state

Figure 12.17. Proposed catalyst structure and mode of dual activation in the trimethylsilyl cyanide addition to ketones.

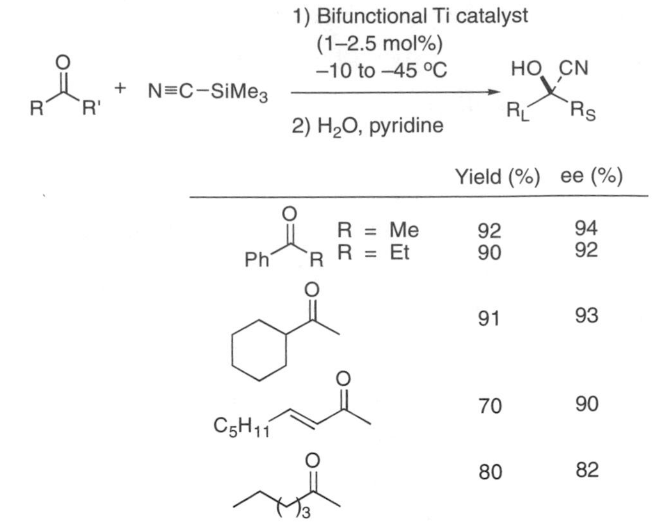

Figure 12.18. Trimethylsilyl cyanide addition to ketones catalyzed by a bifunctional titanium-based catalyst (**Figure 12.17**).

(1–2.5 mol%) with excellent enantioselectivities for a variety of ketone substrates (**Figure 12.18**).

Interdependent bifunctional catalysts based on chiral amino alcohol ligands for the asymmetric addition of alkylzinc reagents to aldehydes were introduced earlier in this chapter. Application of related amino alcohol-based catalysts to the asymmetric addition of alkylzinc reagents to α-ketoesters proved to be a more challenging problem. Initial studies indicated that the major product was derived from reduction of the keto group and that the addition product that was isolated was nearly racemic (**Equation 12.3**). Alkyl additions to α-ketoesters are considerably more difficult than the additions to aldehydes because the direct reaction of diethylzinc with α-ketoesters is rapid, in contrast to aldehydes, and the catalyst must accelerate the addition pathway over the ketone reduction path. The addition products are precursors to α-keto acids, which are valuable synthetic materials.[46] Efforts to develop bifunctional catalysts for this important reaction were undertaken.[47–50]

Equation 12.3

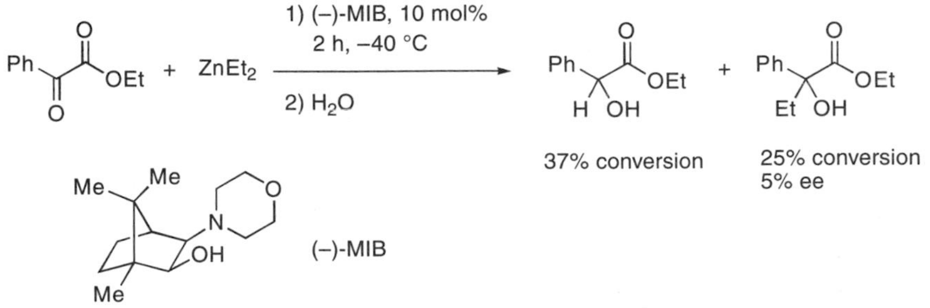

A novel set of modular bifunctional catalysts based on the salen scaffold has recently been shown to exhibit exceptional control over chemoselectivity in the alkylzinc addition to α-ketoesters (**Figure 12.19**).[47,48] Enantioselectivities as high as 88% were achieved with these bifunctional catalysts.[49]

Evidence for a bifunctional mechanism in the addition to α-keto esters included a comparison between structurally similar catalysts, one of which lacked the pendent amine activator. Under identical conditions, the standard salen-based catalyst exhibited increased reduction product relative to the bifunctional catalyst. Furthermore, the standard salen-based catalyst produced addition product that was nearly racemic (**Figure 12.20**).

The proposed mechanism involves alkoxide ionization to generate a five-coordinate titanium cation that binds the keto group of the substrate (**Figure 12.21**). Support for this hypothesis includes the same enantioselectivities when catalyst derived from Ti(O-i-Pr)$_4$ and Ti(O-t-Bu)$_4$ were used. The bifunctional catalysts outlined here are not only highly efficient, but positively impact the product distribution such that the desired addition product is almost exclusively obtained.

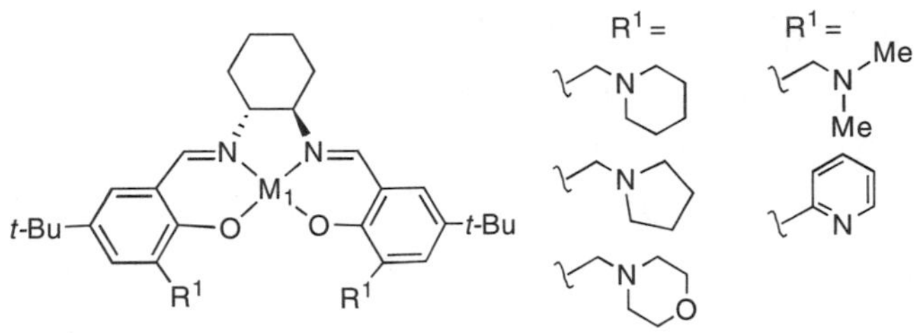

Figure 12.19. Bifunctional salen-based catalysts for the asymmetric addition of alkyl groups to α-keto esters.

Bifunctional salen: 0%   99% [62% ee (*R*)]
Standard salen: 20%   56% [4% ee (*S*)]

Bifunctional salen
M = Ti(O-i-Pr)$_2$

Standard salen
M = Ti(O-i-Pr)2

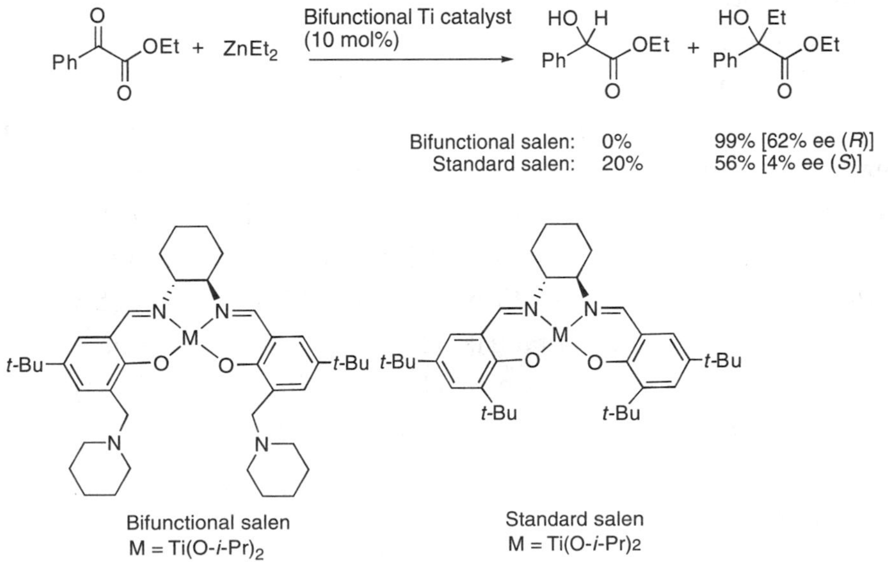

Figure 12.20. Comparison of the bifunctional salen with a standard salen in the asymmetric addition of ethyl groups to α-keto esters.

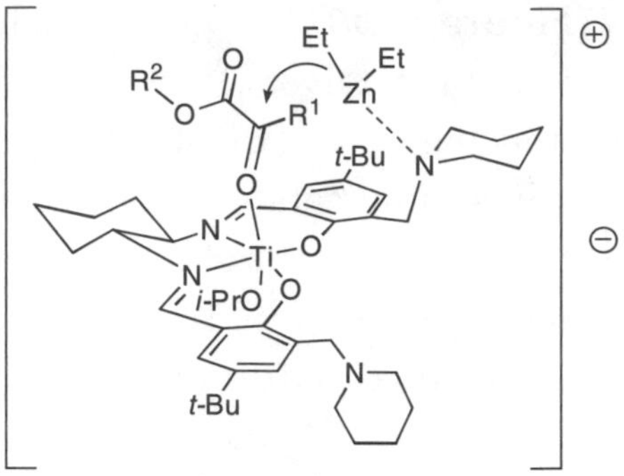

Figure 12.21. Proposed transition state for the additions to α-keto esters catalyzed by a bifunctional salen catalyst.

12.2.3 Independent Lewis Base and Brønsted Base Catalysts

Lewis and Brønsted basicity are common characteristics of organic molecules and are often found in the active sites of enzymes. It is not surprising, then, that purely organic small molecules, called "organocatalysts," also serve as asymmetric catalysts. Although asymmetric catalysis is still dominated by metal-based catalysts, drawbacks such as air- and moisture-sensitivity, high costs of some metal precursors, potential metal toxicity, and the removal of trace metal-containing products, are circumvented by most organocatalysts. As a result, the development and investigation of asymmetric organocatalytic reactions has become an active research area.[51–56] In the context of bifunctional catalysts, our discussion will highlight important organocatalysts.

The ability of enantioenriched organic compounds to catalyze asymmetric reactions had received little attention until recently.[53,57] The 1970s witnessed a milestone in this emerging field with the discovery that simple enantioenriched organic compounds could catalyze aldol reactions with high enantioselectivity. This seminal work, reported from two industrial laboratories,[58,59] employed the natural amino acid proline. Organocatalysis was largely dormant (outside the use of cinchona alkaloids, see Chapter 5, for example) for many years until recent developments in asymmetric epoxidation reactions, phase-transfer reactions, and aldol chemistry, to name a few.

The reaction credited for the initiation of organocatalysis is the Hajos–Parrish–Ender–Sauer–Wiechert reaction (**Equation 12.4**). The reaction involves an intramolecular aldol condensation of a triketone with proline and proceeds through an enamine intermediate. Depending on reaction conditions, the initially formed aldol product can undergo elimination to the conjugated enone.[58,59]

Equation 12.4

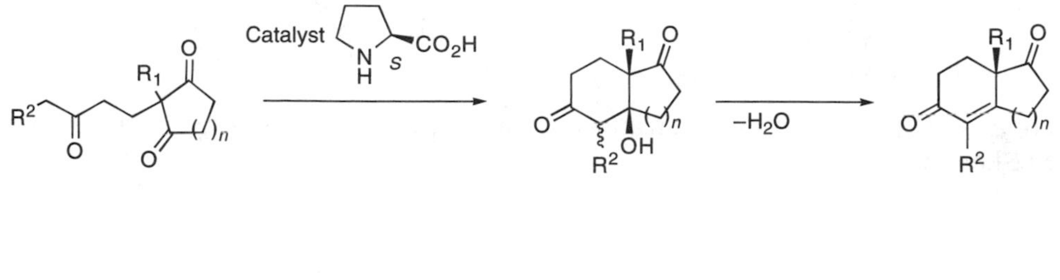

Kinetic studies indicate that there is a single proline in the transition state. Additionally, no nonlinear effects were observed with proline between 10 and 100% ee.[60] As shown in **Figure 12.22**, proline reacts with the methyl ketone to generate a nucleophilic enamine intermediate. Based on quantum-mechanical calculations and experimental studies, the cyclization is believed to proceed by nucleophilic addition of the enamine to the carbonyl carbon with simultaneous hydrogen transfer from the carboxylic moiety of the proline to the developing alkoxide. The combination of nucleophilic enamine addition and ketone activation by the carboxylic acid group are believed to be responsible for the high enantioselectivity of this bifunctional catalyst.[61,62]

Application of the (*S*)-proline-catalyzed intramolecular aldol has been successfully applied to a variety of other substrates with high enantioselectivities and moderate to high diastereoselectivities.[63] Representative products are illustrated in **Figure 12.23.**

Proline also catalyzes the asymmetric intermolecular aldol reaction between ketones[51,64,65] and aldehydes and even the aldehyde cross-aldol reaction[66] (**Figure 12.24**). Enantioselective and diastereoselective cross-aldol reactions represent a formidable challenge, especially when both aldehyde components have enolizable hydrogens. The two aldehydes must play distinct roles in the reaction—namely, one must be the nucleophilic donor and its partner must be the electrophilic acceptor. Furthermore, the cross-aldol product must be inert to enolization or carbonyl addition under the reaction conditions. In the case of the aldehyde cross-aldol in **Figure 12.24**, the propionaldehyde is added slowly to suppress dimerization and favor the cross-aldol reaction. Under these conditions, only one regioisomer of the product was observed by ^1H NMR spectroscopy.[66]

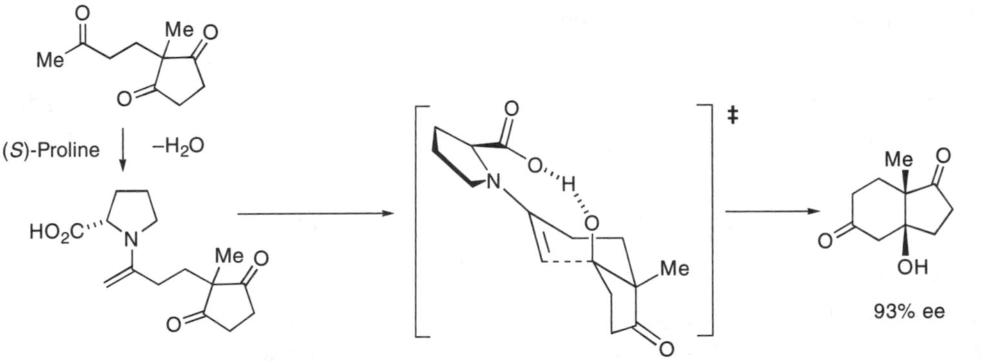
Me
Me O
O
(S)-Proline
$-H_2O$
HO_2C
N
Me O
O
O
N
O
H
O
Me
O
‡
Me O
O
OH
93% ee

Figure 12.22. Proposed mechanism of the Hajos–Parrish–Ender–Sauer–Wiechert reaction based on quantum-mechanical models.

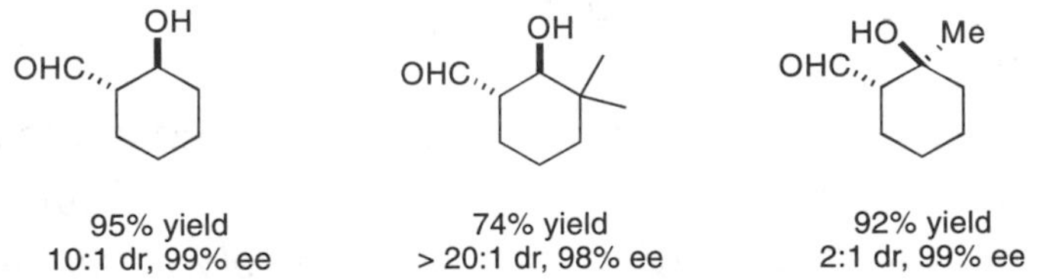
OHC
OH
95% yield
10:1 dr, 99% ee
OHC
OH
74% yield
> 20:1 dr, 98% ee
OHC
HO
Me
92% yield
2:1 dr, 99% ee

Figure 12.23. Selected results from the intramolecular aldol reaction catalyzed by (*S*)-proline.

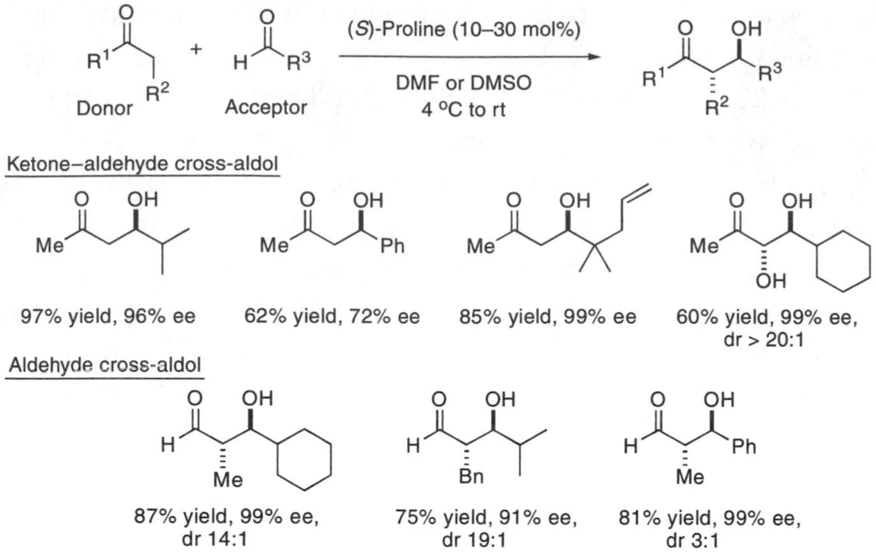

Figure 12.24. Proline-catalyzed aldol reactions between ketones and aldehydes and between different aldehydes.

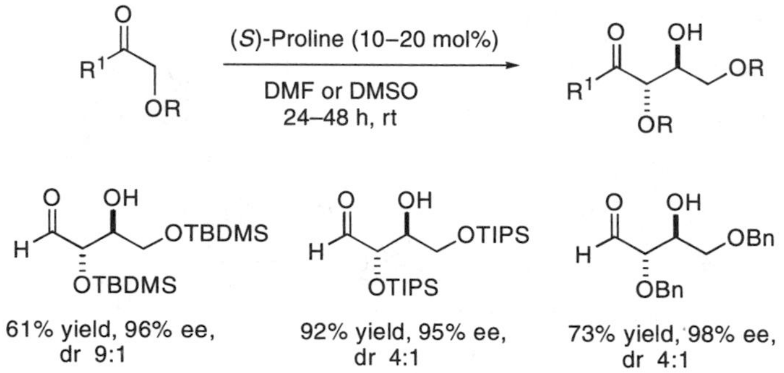

Figure 12.25. Proline-catalyzed aldol dimerization of α-oxyaldehydes.

Proline will also catalyze the enantioselective dimerization of α-oxyaldehydes to form protected erythrose derivatives, which are useful chiral building blocks (**Figure 12.25**).[67]

The dimerization of α-oxyaldehydes is the first step in a two-step synthesis of carbohydrates, as outlined in **Figure 12.26**.[68] The second step in the synthesis is a Mukaiyama aldol reaction followed by a cyclization to provide rapid access to a variety of protected sugars. Proper choice of Lewis acid for the Mukaiyama aldol allows synthesis of isomeric hexose derivatives with very good stereochemical control.[68] This two-step method for the synthesis of differentially protected hexoses enables entry into a variety of pharmaceutically important building blocks that were previously quite difficult to prepare.

The proline-catalyzed cross-aldol reaction has been used in the efficient synthesis of the natural product (–)-prelactone B (**Figure 12.27**), a metabolite in the biosynthesis of polyketide antibiotics.[69] The cross-aldol reaction between propionaldehyde and isobutyraldehyde was conducted under slow addition of propionaldehyde to four

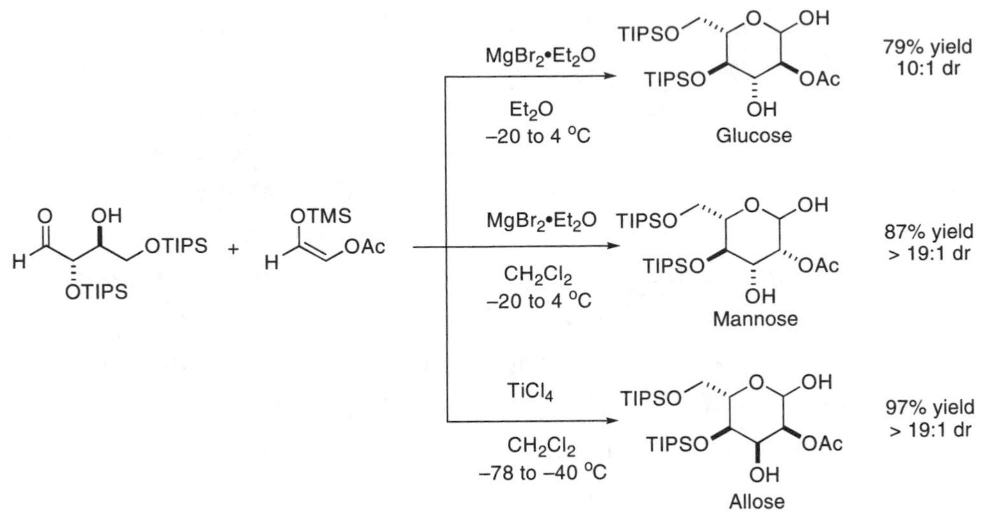

Figure 12.26. Diastereoselective Mukaiyama aldol reaction promoted by Lewis acids provides access to differentially protected sugars.

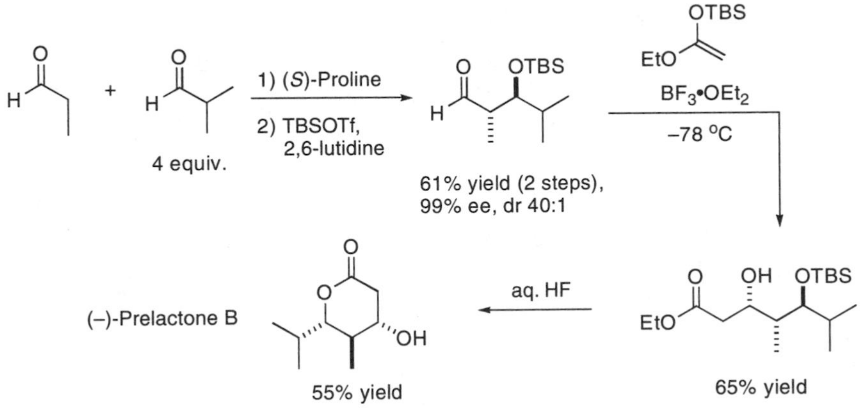

Figure 12.27. Concise synthesis of prelactone B, highlighting the proline-catalyzed aldehyde cross-aldol reaction.

equivalents of isobutyraldehyde and (S)-proline to minimize the dimerization of propionaldehyde. The protected aldehyde was subject to a Felkin-selective Mukaiyama-type aldol, followed by deprotection/lactonization initiated by aqueous HF to furnish (–)-prelactone B.[69]

Stereochemical models to account for the proline-catalyzed aldol reaction have been developed based on calculations and are shown in **Figure 12.28**.[61,62,70,71] A model derived from a Zimmerman–Traxler-like transition state with chair-like disposition of the carbonyl and enamine groups can be used to explain and predict the stereoselectivity in these reactions. The diastereoselectivity arises from reaction through the *anti*-enamine, whereas the *syn*-enamine transition state (not shown) is estimated to be 2–10 kcal/mol higher in energy.[62]

The success of proline in the aldol transformations has stimulated many investigations and proline has been found to catalyze a variety of reactions with high enantioselectivity, including the Mannich reaction,[72–75] α-amination of aldehydes,[76,77] and α-oxyamination of aldehydes[78–80] (**Figure 12.29**). A common thread in these reactions is the existence of a lone pair on the electrophile that can accept a hydrogen bond from the proline carboxylic acid group in the transition state. The stereoselectivity models proposed for the aldol reaction above can be adapted to rationalize the stereochemical outcome in these reactions. In the (S)-proline-catalyzed Mannich reaction, the opposite enantiofacial selectivity relative to the aldol reaction is observed.[72–75] The N-aryl substituent is *anti* to the proline carboxylic acid group and the imine is in the more favorable (E)-geometry (**Figure 12.29a**).[73]

Dialkyl azodicarboxylates are electrophilic reagents that undergo addition to give α-amination reactions with high levels of enantioselectivities (**Figure 12.29b**).[76,77] The initially formed α-hydrazino aldehydes can racemize on standing and were, therefore, reduced to the α-hydrazino alcohols. In the proline-catalyzed asymmetric α-oxamination reaction,[78–80] calculations suggest a proton transfer from the proline carboxylic acid and *anti* attack of the enamine on the nitroso oxygen (**Figure 12.29c**).[81]

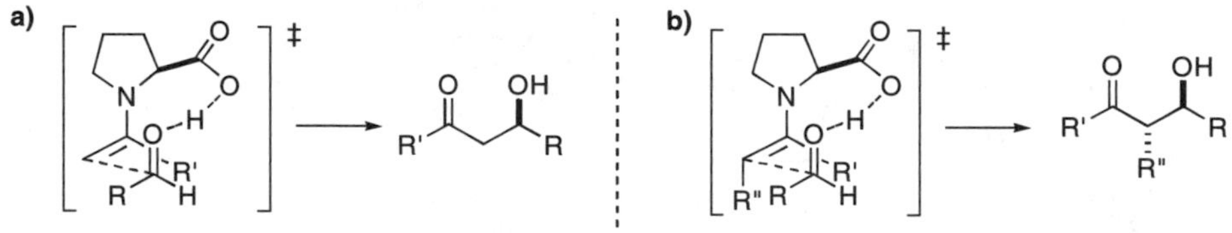

Figure 12.28. Stereoselectivity models for the (a) enantioselective and (b) diastereoselective (S)-proline-catalyzed aldol reactions.

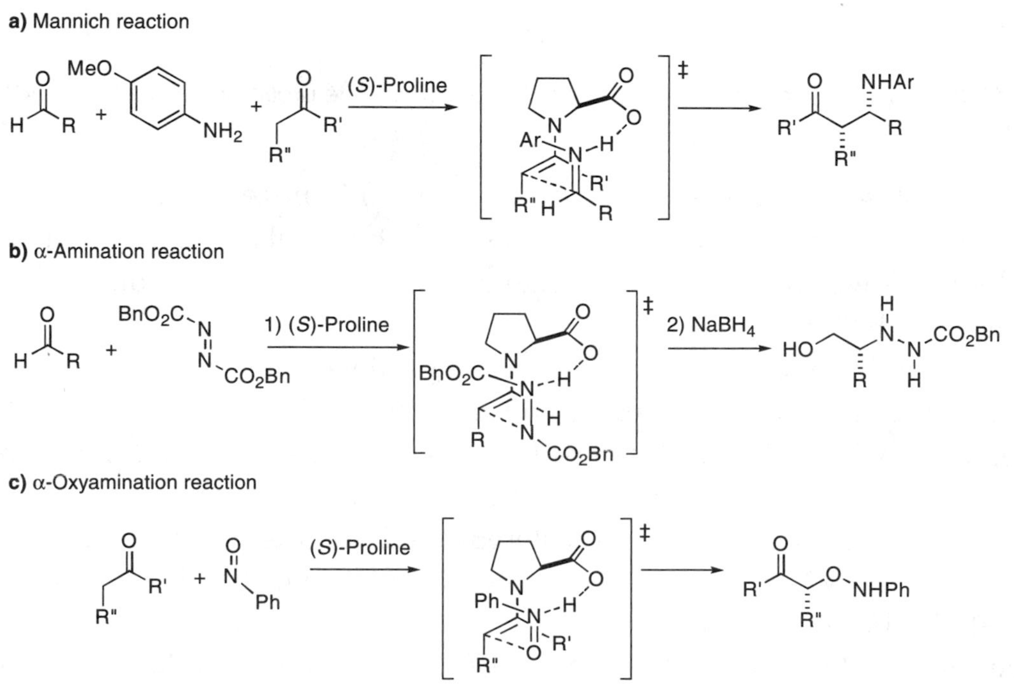

Figure 12.29. Proposed transition states for the Mannich reaction (a), α-amination of aldehydes (b), and α-oxyamination of aldehydes (c), illustrating the common features of these reactions.

A nice example of a proline-catalyzed tandem reaction that highlights the remarkable ability of this simple catalyst to generate stereochemically complex molecules is the generation of the hexose carbohydrate allose (**Equation 12.5**).[82] Treatment of α-benzyloxy acetaldehyde with 10 mol% proline in DMF resulted in the formation of a mixture of erythrose (95–99% ee) and allose (> 99% ee) derivatives.

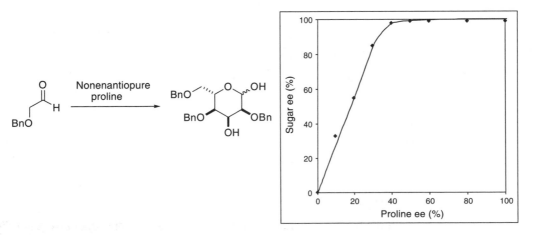

Equation 12.5

Erythrose
95–99% ee
45% yield

Allose
> 99% ee
40% yield

In this reaction, the erythrose derivative formed from the highly enantioselective and diastereoselective proline-catalyzed dimerization of α-benzyloxy acetaldehyde undergoes a subsequent diastereoselective aldol reaction and cyclization to afford the protected allose (**Equation 12.6**). The condensation of the erythrose derivative with the proline-derived enamine of α-benzyloxy acetaldehyde is the rate-determining step in this process.

Equation 12.6

Erythrose

Allose

Mechanistic studies with proline of reduced enantiopurity exhibited a strong positive nonlinear effect in the formation of the hexose derivative, as shown in **Figure 12.30**. No change in the hexose ee was observed until the proline ee was dropped

Figure 12.30. Nonlinear effect in the proline-catalyzed hexose formation.

below 40%.[82] It is likely that the nonlinear behavior in this system is due to the greater insolubility of the racemate catalyst, which could precipitate from solution. The ee of the dissolved proline is therefore likely to be much higher than the initial proline ee (see Chapter 11 on nonlinear effects).[83]

The observation of a nonlinear effect in this process is seemingly in contrast to the proline-catalyzed Hajos–Parrish–Ender–Sauer–Wiechert reaction (**Figure 12.22**), which exhibits a linear relationship between the enantiopurity of proline and the aldol product ee.[60] Investigations into the origin of the nonlinear effect observed in the hexose formation focused on the second, diastereoselective aldol condensation. In a closely related model system (**Equation 12.7**), treatment of racemic *anti*-β-hydroxy aldehyde with catalytic (*R*)-proline produced the polyketide product in 22% yield with 95% ee and > 19:1 dr.[82] The starting cross-aldol product was isolated in 72% yield with < 5% ee, ruling out the possibility of a simple kinetic resolution (see Chapter 7).

Equation 12.7

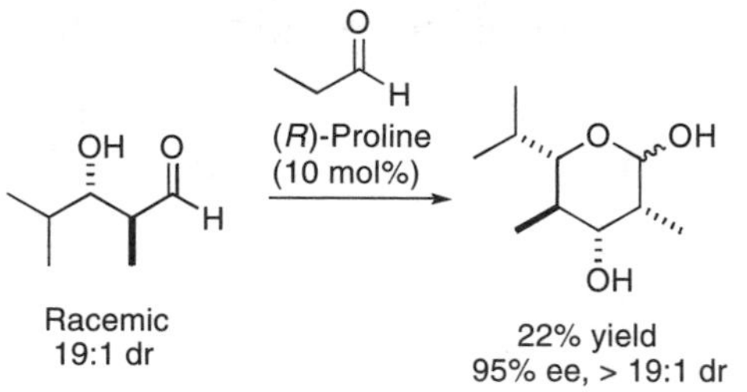

Racemic
19:1 dr

(R)-Proline
(10 mol%)

22% yield
95% ee, > 19:1 dr

The nonlinear behavior of the hexose formation and the recovery of racemic *anti*-β-hydroxy aldehyde lead one to conclude that proline catalyzes the retro-aldol of the *anti*-β-hydroxy aldehyde, thus serving to equilibrate the *anti*-β-hydroxy aldehyde enantiomers and maintain the near 0% ee (**Figure 12.31**). It is possible that the (*R*)-proline-derived enamine reacts significantly faster with (2*S*,3*S*)-*anti*-β-hydroxy aldehyde than with its enantiomer and the system undergoes a dynamic kinetic resolution (see Chapter 9).

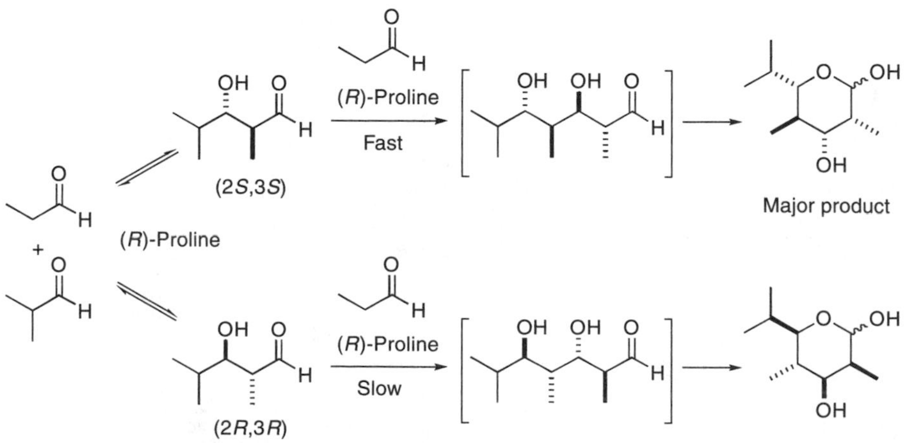

Figure 12.31. Proposed dynamic kinetic resolution in the enantioselective and diastereoselective formation of hexose catalyzed by (*R*)-proline.

12.2.4 Independent Brønsted Acid and Lewis Base Components

The Morita–Baylis–Hillman reaction[84] (**Figure 12.32**) typically involves attack of a nitrogen- or phosphorus-based nucleophile on an unsaturated carbonyl compound to generate an intermediate enolate. This enolate can then undergo 1,2-addition to an aldehyde, or in the case of the aza-Morita–Baylis–Hillman reaction, an imine. Subsequent elimination (retro-Michael) expels the Lewis base catalyst and re-establishes the conjugation of the carbonyl group. The elimination can proceed with the participation of a second equivalent of aldehyde (not shown).[85] Several studies have investigated methods to accelerate the Morita–Baylis–Hillman reaction. Lewis acids accelerate the reaction either by stabilizing the intermediate enolate, activating the aldehyde to addition, or both.[86] The introduction of Lewis acids into the reaction can cause complications, however, because they can coordinate the nucleophilic catalyst and inhibit the reaction.[86,87] Modest increases in the Morita–Baylis–Hillman reaction rate have been observed in the presence of hydrogen-bond donors such as methanol[88] or substrates containing pendent hydroxyl groups.[89] Like the Lewis acids, hydrogen-bond donors can stabilize the intermediate enolate and increase its concentration or can activate the aldehyde.

The beneficial effects of Lewis acids or Brønsted acids in the Lewis base catalyzed Morita–Baylis–Hillman and aza-Morita–Baylis–Hillman reactions have inspired the design of enantioselective bifunctional catalysts to both accelerate the reaction and position the two reaction partners to maximize enantioselectivity. One such class of bifunctional catalysts is based on BINOL with appended pyridyl Lewis bases (**Figure 12.33**).[90] Initial examination of the 2- and 3-pyridyl derivatives in the aza-Morita–Baylis–Hillman reaction indicated that the 2-pyridyl derivative (**A, Figure 12.33**) was ineffective at promoting the reaction and the 3-pyridyl analogue (**B**) gave the product in 73% enantioselectivity (41% yield). These results can be compared with the control experiment employing (S)-BINOL (10 mol%) together with DMAP (10 mol%), which afforded the product in 3% ee (48% yield).

To probe the role of each of the components of the bifunctional catalyst **B**, the mono O-methylated derivatives **C** and **D** were screened in the reaction. Catalyst **C**

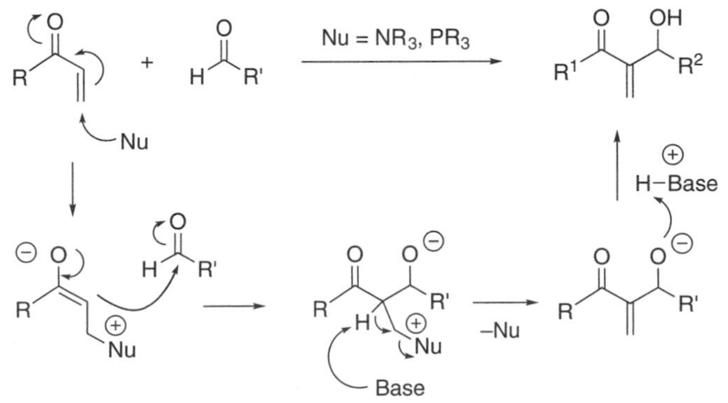

Figure 12.32. Proposed mechanism for the Morita–Baylis–Hillman reaction.

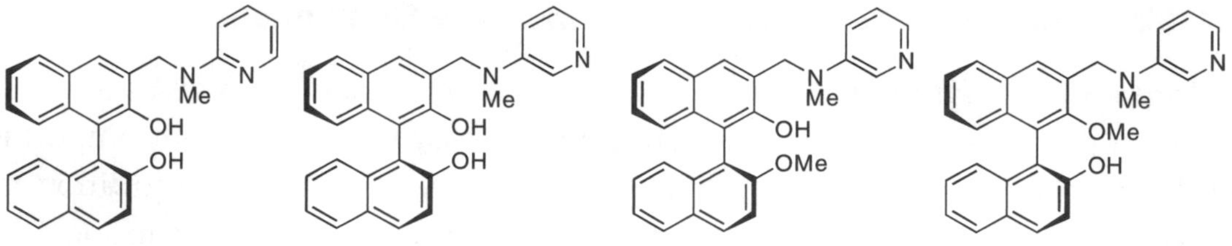

Figure 12.33. Chiral bifunctional catalysts containing Lewis basic pyridyl groups and Brønsted acidic hydrogen-bond donors.

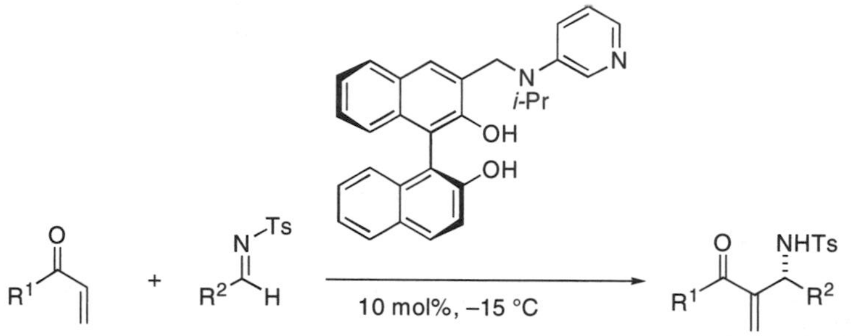

R¹	R²	ee (%)	Yield (%)
Me	Ph	87	93
Me	2-furyl	88	100
Et	4-C₆H₄-NO₂	88	96
H	4-C₆H₄-NO₂	94	36

Figure 12.34. Selected results in the enantioselective aza-Morita–Baylis–Hillman reaction.

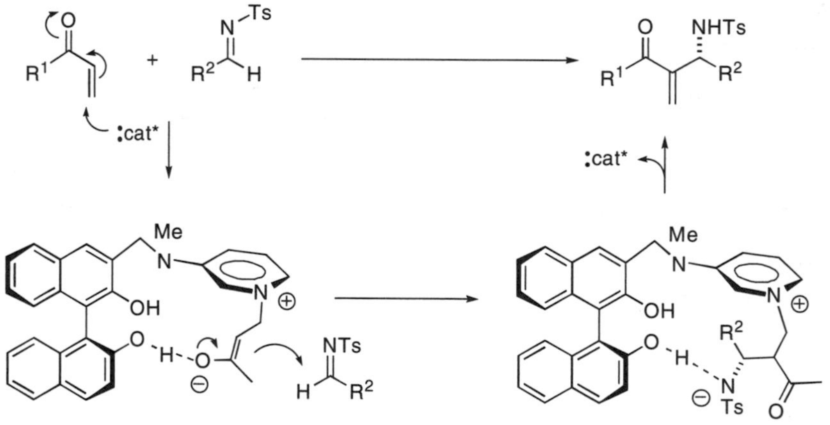

Figure 12.35. Possible mechanism of a Lewis base–Brønsted acid catalyst for the aza-Morita–Baylis–Hillman reaction.

exhibited reduced reactivity (5% yield), but catalyst **D** exhibited good enantioselectivity and yield (79% ee, 85% yield). Further optimization of the catalyst structure led to the *N-i*-Pr catalyst illustrated in **Figure 12.34**, along with selected results.

Based on the mechanism of the aza-Morita–Baylis–Hillman reaction and the structure-activity relationships outlined above, a possible mechanism is illustrated in **Figure 12.35**.[90]

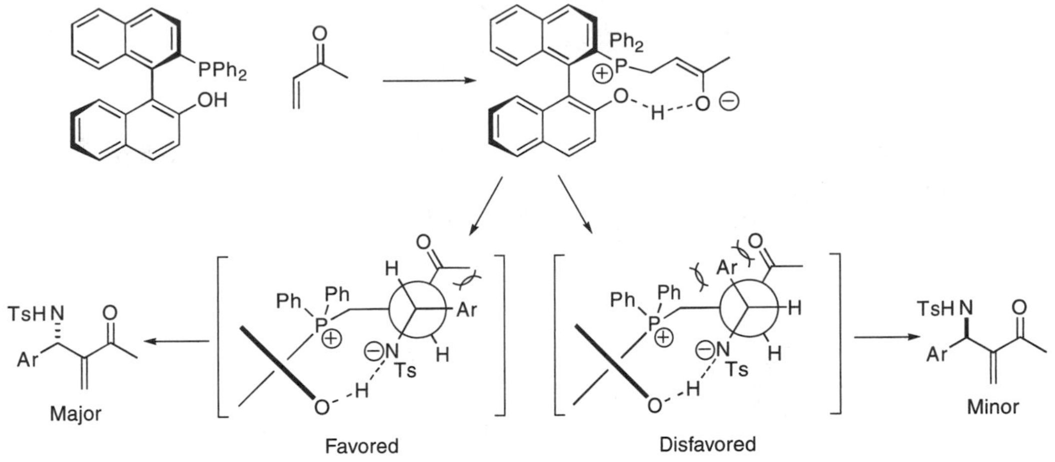

Figure 12.36. Stereochemical model to rationalize the enantioselectivity observed in the aza-Morita–Baylis–Hillman reaction.

Related bifunctional catalysts have been developed and are believed to operate in a similar fashion. One such highly enantioselective catalyst is shown in **Figure 12.36**, in which a stereochemical model is proposed to explain the observed facial selectivity.[87,91]

12.2.5 Independent π-Complexation and Lewis Base Components

In the study of reaction mechanisms of asymmetric catalysts, researchers have occasionally prepared bifunctional catalysts in the hopes of understanding monofunctional catalysts. One such example is in the asymmetric allylic allylation (**Figure 12.37**), where several proposals have been advanced to explain the sense of asymmetric induction.[92] These include steric interactions with the chiral environment of the palladium catalyst and secondary interactions between the cationic counterion of the nucleophile with Lewis basic sites on the catalyst.[1,93] Evidence for the latter hypothesis is the strong counterion effect on enantioselectivity in the asymmetric allylation, increasing with $Na^+ < K^+ < Rb^+ < Cs^+$.[92] Increasing enantioselectivity with ionic radius is also observed with tetraalkylammonium counterions ($Me_4N^+ < Et_4N^+ < n\text{-}Bu_4N^+ < n\text{-}Hex_4N^+$).[94] In the asymmetric allylation, the nature of the cation has a greater impact on the enantioselectivity than the anion. This is counterintuitive, because the role of the metal counterion is to escort the anionic nucleophile, although it is the nucleophile that attacks the activated allyl in the enantioselectivity-determining step.

To probe the role of the cation in this reaction, a series of chiral ligands were designed that incorporated cation-binding sites based on glyme-like subunits (**Figure 12.38**). It was envisioned that these metal-binding tentacles would coordinate the cation and guide the ion pair to the bound substrate.

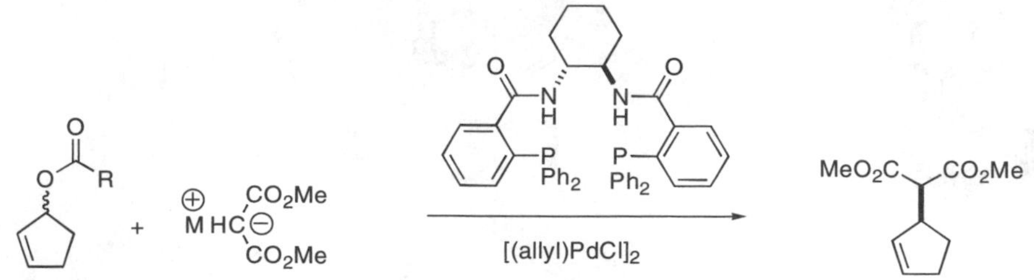

Figure 12.37. Asymmetric allylation reaction with malonate nucleophiles.

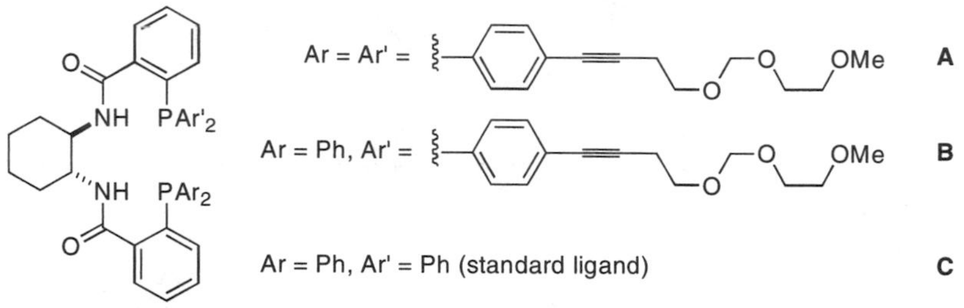

Figure 12.38. C_2-symmetric tetrapodal ligand and C_1-symmetric dipodal ligands.

Examination of the asymmetric allylation reaction in **Equation 12.8** with 0.6 mol% of the tetrapodal ligand (**A, Figure 12.38**) and 0.2 mol% of dimeric [(allyl)PdCl]$_2$, resulted in generation of the allylation product with 99% enantioselectivity. In a comparison, the reaction with the catalyst prepared from the standard ligand **C** required 16 h, while reaction with the tetrapodal ligand reached completion in only 2 h (99% yield).

Equation 12.8

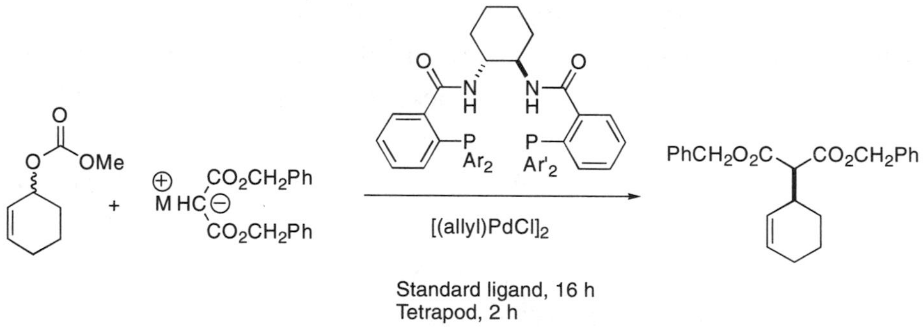

In the reaction of sodium benzenesulfinate as nucleophile and the cyclohexenyl substrate (**Figure 12.39**), the standard ligand was again significantly slower. While the dipodal ligand achieved 50% conversion in under 10 min, the standard ligand reached only 15% conversion in 180 min.

If the reaction mechanism with the dipodal- and tetrapodal-based catalysts does involve bifuctional activation of the allyl and the cation, a dependency on the identity of the cation is expected. This dependency could manifest itself in the reaction enantioselectivity, the rate, or both. To determine the magnitude of this effect, the

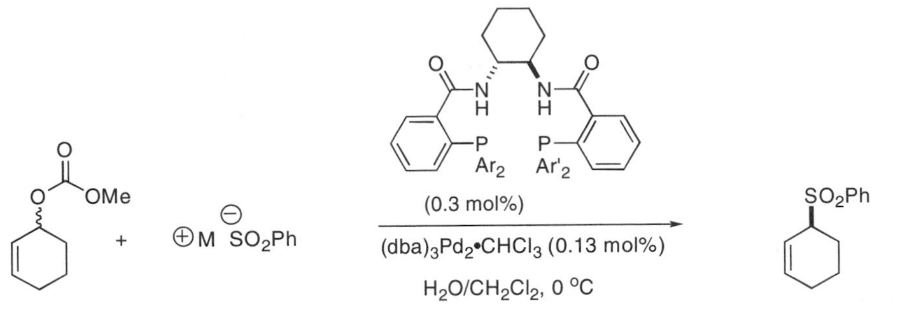

Figure 12.39. A comparison of the cation nature on the half-life of the allylation reaction with dipod ligand **B** from **Figure 12.38**.

benzenesulfinate substitution reaction was monitored and the time to achieve 50% conversion recorded. As outlined in **Figure 12.39**, the reaction rate exhibits a strong dependence on the counterion, with the relative rates being Na > Li > K >> NH_4. The rate enhancement can be attributed to binding of the cations to the polyether tentacles, which will decrease the association of the ion pair, thus increasing the nucleophilicity of the anion (nucleophile activation). Despite the dependency of reaction rate on the cation, the enantioselectivity was independent of the cation, indicating that the catalyst–cation interaction does not *impact* the enantioselectivity. In this case, the interface between the nucleophilic anion and the chiral environment of the catalyst is responsible for controlling asymmetric induction.[92]

12.3 Dual Catalyst Systems

As the name implies, dual catalyst systems involve two chiral catalysts that activate the reaction partners simultaneously to promote an asymmetric transformation. In some cases, a reaction can be second-order in catalyst, with one molecule of catalyst activating the reagent and a second molecule simultaneously activating the substrate. Alternatively, the catalysts can have completely different structures, with each designed to activate a specific reagent. By employing distinct structures to activate the reacting components, greater control over activity and enantioselectivity is possible through independent optimization of the individual catalysts for their specific roles. If the catalysts have different structures, they can easily be screened in a combinatorial fashion to facilitate reaction optimization. Of course, the catalysts must be compatible to avoid catalyst decomposition. Additionally, if the reaction is first-order in each catalyst, low catalyst loadings may lead to long reaction times. Given the challenges and potential of dual catalyst systems, it is anticipated that interest in exploring this area will increase in the coming years.

12.3.1 Identical Catalysts

In Chapter 7 we outlined the application of the cobalt(salen)-based catalysts for the hydrolytic kinetic resolution (HKR) of terminal epoxides. Recall that detailed kinetic studies indicated that there was second-order dependency on the cobalt catalyst, leading the authors to propose a mechanism involving activation of the epoxide and nucleophile (hydroxide) by the cobalt(salen).[95] A similar bifunctional catalyst mechanism has been found in the (salen)Co(III)-catalyzed kinetic resolution of terminal epoxides with phenols[96] and the asymmetric ring-opening of epoxides with (salen)Cr-based catalysts.[97,98] In its most simple form, the cobalt precursors that activate the epoxide and the hydroxide are identical, as illustrated in **Equation 12.9** for the HKR. It is not necessary that the cobalt species be the same, however, because they play different roles by activating either the substrate or the nucleophile.

Equation 12.9

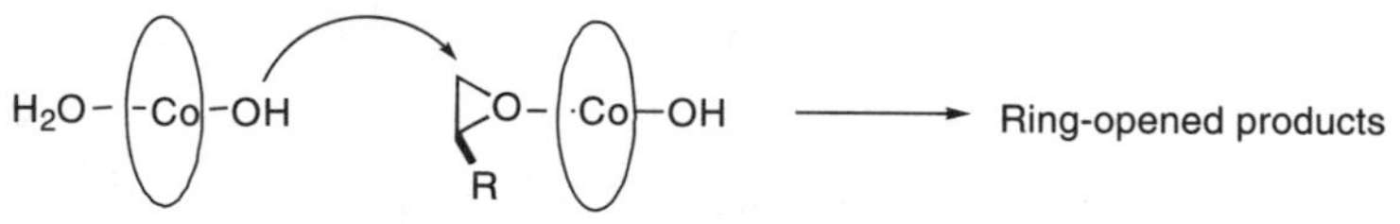

$$H_2O - \text{-Co-} OH \quad \overset{R}{\triangleright}O - \text{-Co-} OH \longrightarrow \text{Ring-opened products}$$

Increasing the Lewis acidity of the epoxide-activating catalyst would likely result in stronger binding of the epoxide and enhanced electrophilicity of the bound epoxide. Both of these effects would be predicted to result in increased reaction rates. Relative to the HKR carried out with the (salen)Co–OH catalyst depicted in **Equation 12.9**, use of a mixed catalyst system employing (salen)Co–OH as the nucleophilic component and (salen)Co–OTs as the epoxide-activating component was up to 30-fold faster (**Equation 12.10**). The maximum rate was achieved when the concentrations of (salen)Co–OH and (salen)Co–OTs were identical. Interestingly, the selectivity factors were similar when the pure (salen)Co–OH catalyst (**Equation 12.9**) and the mixed (salen)Co–OH/(salen)Co–OTs systems were used. This is likely due to the binding of the OTs group opposite the coordinated epoxide, and hence a minimal effect on the asymmetric induction.

Equation 12.10

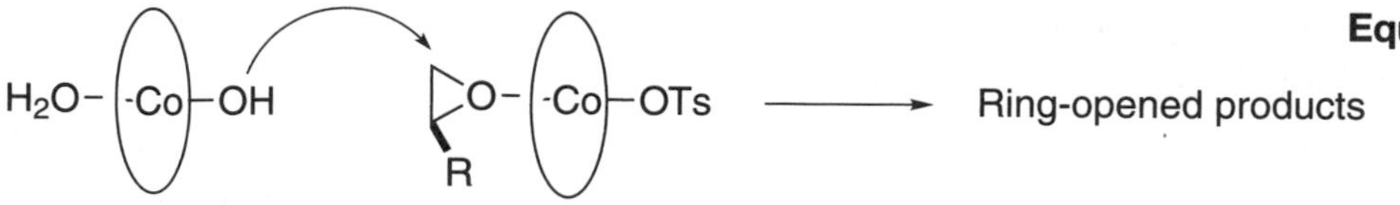

$$H_2O - \text{-Co-} OH \quad \overset{R}{\triangleright}O - \text{-Co-} OTs \longrightarrow \text{Ring-opened products}$$

In the examples above, the catalysts for activation of the nucleophile and the electrophile were either identical or closely related. As illustrated below, the catalyst components need not be related, allowing greater flexibility in reaction optimization.

12.3.2 Structurally Different Dual Catalyst Systems

12.3.2.a Lewis Acid–Lewis Acid

It was reported that (salen)AlCl complexes catalyzed the conjugate addition of hydrogen cyanide to unsaturated imides to generate β-cyano imides with high enantioselectivity (**Figure 12.40**).[99] The efficiency of this (salen)Al-based catalyst for conjugate addition to imides was significantly less for cyanide additions than for the addition of other nucleophiles, such as azide[100] and di- and trisubstituted nitriles.[101] It was also found that the oxo dimer [(salen)Al]₂O did not catalyze the conjugate addition of cyanide to the imide, which is relevant to the discussion below. Based on these observations, and mechanistic studies that indicate that the reaction catalyzed by (salen)AlCl proceeded by dual activation of the imide and cyanide, it was hypothesized that the low reactivity in this system originated from ineffective activation of the cyanide by the (salen)Al-based catalyst. To improve the reaction rates, catalysts known to efficiently activate cyanide were screened in combination with (salen)AlCl.

Lanthanide complexes such as (PyBox)YbCl₃ have been reported to promote nucleophilic addition of cyanide to epoxides via a cooperative dual catalyst mechanism.[102] Despite the ability of the lanthanide complexes to activate both cyanide and epoxides in this reaction, they did not readily catalyze the cyanide addition to unsaturated imides, possibly because the lanthanide-based catalyst did not activate the imide substrate.[103] Given the efficiency of the lanthanide-based catalysts at the activation of cyanide, and the ability of (salen)Al-based complexes to activate imides, it was envisioned that these two catalysts could form a dual catalyst heterobimetallic system for the conjugated cyanation of unsaturated imides. The catalysts employed in this study are outlined in **Figure 12.41**.[103]

As outlined in **Table 12.2**, use of either the dimeric [(salen)Al]₂O complex (2 mol%, entry 1) or the (PyBox)EuCl₃ (3 mol%, entry 2) alone resulted in > 3% conversion.

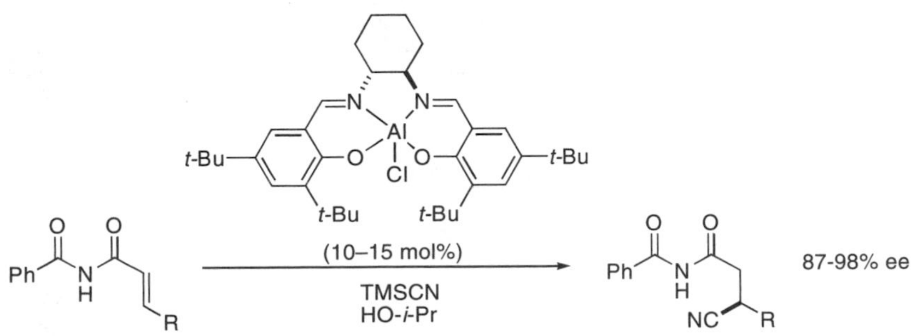

Figure 12.40. Conjugate addition of hydrogen cyanide to unsaturated imides.

R = –(CH₂)₄–, X = Cl
R = –(CH₂)₄–, X = OAl(salen)
R = H, X = Cl

R = i-Pr, R' = H
R = Me, R' = Me

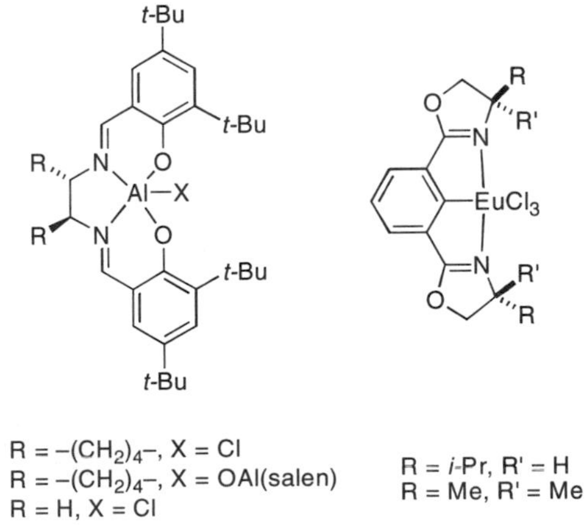

Figure 12.41. Complexes screened in the heterobimetallic dual catalyst system for the asymmetric addition of cyanide to imides.

Combination of these two compounds, however, resulted in 99% conversion and 96% enantioselectivity. Kinetic analysis of the dual catalyst system indicated that there was a first-order dependence on both dimeric [(salen)Al]$_2$O and i-Pr$_2$(PyBox)EuCl$_3$, consistent with both complexes participating in the rate-determining transition state.

Further insight into the importance of the chiral environments of each complex was gained by employing combinations of chiral and achiral catalysts in the conjugate addition reaction (**Table 12.3**). While the combination of (S,S)-[(salen)Al]$_2$O and (S,S)-i-Pr$_2$(PyBox)EuCl$_3$ exhibited high efficiency and enantioselectivity, the diastereomeric combination of (S,S)-[(salen)Al]$_2$O and (R,R)-i-Pr$_2$(PyBox)EuCl$_3$ resulted in a decrease in enantioselectivity and TOF (not shown in the **Table 12.3**). Substitution of achiral Me$_4$(PyBox)EuCl$_3$ (**Figure 12.41**, R = R′ = Me), for (S,S)-i-Pr$_2$(PyBox)EuCl$_3$ resulted in a decrease in enantioselectivity (entry 3). Use of (S,S)-i-Pr$_2$(PyBox)EuCl$_3$ alone exhibited low enantioselectivity and TOF. The reactivity and much of the enantioselectivity were restored when (S,S)-i-Pr$_2$(PyBox)EuCl$_3$ was used with the achiral [(salen)Al]$_2$O, the structure of which is illustrated in **Figure 12.41**. With this combination of catalysts, conversion was high (98%) and enantioselectivity was good (78%, entry 5). The results of the mechanistic and stereochemical studies indicate that the

Table 12.2. Conjugate cyanide additions with a dimeric aluminium complex (entry 1), with a europium complex (entry 2), and with both (entry 3).

Entry	(Salen)AlX	(PyBox)ErCl$_3$	Conversion (%)	ee (%)
1	(S,S)-[(Salen)Al]$_2$O	—	< 3	—
2	—	(S,S)-i-Pr(PyBox)EuCl$_3$	< 3	—
3	(S,S)-[(Salen)Al]$_2$O	(S,S)-i-Pr(PyBox)EuCl$_3$	99	96

Table 12.3. Influence of ligand stereochemistry on the efficiency and enantioselectivity in the asymmetric conjugate addition of cyanide to imides. (See **Figure 12.41** for catalyst structures.)

Entry	(Salen)AlX	(PyBox)ErCl$_3$	Conversion after 24 h (%)	ee (%)
1	(S,S)-[(Salen)Al]$_2$O	(S,S)-i-Pr$_2$(PyBox)EuCl$_3$	99	96
2	(S,S)-[(Salen)Al]$_2$O	(R,R)-i-Pr$_2$(PyBox)EuCl$_3$	99	72
3	(S,S)-[(Salen)Al]$_2$O	Me$_4$(PyBox)EuCl$_3$	45	84
4	—	(S,S)-i-Pr$_2$(PyBox)EuCl$_3$	< 3	16
5	Achiral [(Salen)Al]$_2$O	(S,S)-i-Pr$_2$(PyBox)EuCl$_3$	98	78

catalysts operate in a cooperative fashion in the transition state to impart asymmetry on the reactants.[103]

12.3.2.b Lewis Base–Lewis Base Catalysis

A key feature of many enzyme and peptide catalysts is the ability to form noncovalent interactions in both an intramolecular fashion, to confer secondary structure, and in an intermolecular fashion to facilitate reactions with substrates. In the context of dual chiral catalyst systems, peptide-based catalysts that form noncovalent interactions with a co-catalyst or co-catalyst–substrate complexes represents a unique approach to asymmetric catalysis. This amino-acid-based dual catalyst strategy was applied to the Morita–Baylis–Hillman reaction,[84] because the accepted multistep reaction involves charged intermediates that could be stabilized by hydrogen bonds, as described in Section 12.2.4.[85]

Initial investigations found that 10 mol% *N*-methylimidazole (NMI) catalyzed the Morita–Baylis–Hillman reaction between methyl vinyl ketone and 2-nitrobenzaldehyde (**Figure 12.42**), albeit with low TOF (40% conversion after 24 h). With proline, no reaction was observed, while a combination of proline and NMI was the fastest (75% conversion after 24 h). Despite the spectacular success of proline in a variety of transformations,[51] the product here was produced with only 10% enantioselectivity, indicating that the proline exhibited little control over the enantioselectivity in this process.[104,105]

Substitution of NMI with Me-His (Pmh) (**Table 12.4**) introduces chirality into the co-catalyst and serves as a platform on which additional amino acid groups can be appended. When proline and the Pmh derivative were used together to promote the Morita–Baylis–Hillman reaction, the enantioselectivity increased to 19% (**Table 12.4**, entry 1). It was found that increasing the number of amino acids in the peptide resulted in increasing enantioselectivities until the eight amino acids had been added. Peptides longer than the octamer resulted in slight decreases in enantioselectivity. The peptides used in this study were not fully optimized by variation of the individual amino acids, however, and are unlikely to represent the most enantioselective catalysts.[104]

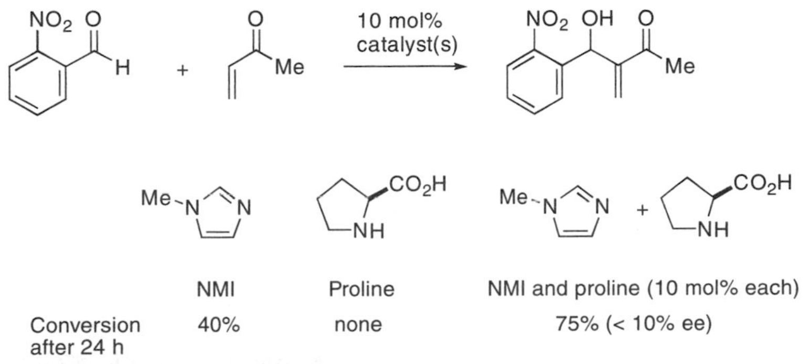

Figure 12.42. Control experiments in the asymmetric Morita–Baylis–Hillman reaction.

Control experiments conducted with the peptides shown in **Table 12.4** in the absence of proline co-catalyst indicated that the peptides alone were poor catalysts, affording product with low yields and enantioselectivities (< 10%). This observation suggests that the proline plays an important role in both efficiency and enantioselectivity. To evaluate the impact of the proline stereochemistry on the reaction enantioselectivity, the reactions were conducted with the octamer and each enantiomer of proline (**Figure 12.43**). Combination of (S)-proline and the octamer catalyst generated the (R)-product with 78% ee. In sharp contrast, use of (R)-proline and the octamer furnished the (S)-product with 39% ee. The change in enantiofacial selectivity with enantiomeric proline catalysts suggests that the proline stereochemistry exhibits greater control in the enantioselectivity-determining step than the octamer.

Table 12.4. Pmh-based catalysts (10 mol%) used with proline (10 mol%) in the asymmetric Morita–Baylis–Hillman reaction of **Figure 12.42**.

Entry	Catalyst	Peptide length	ee (%)
1	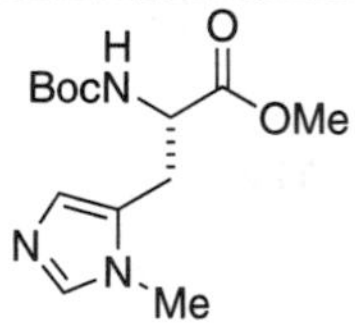 (Boc-Pmh-OMe structure) = Boc-Pmh-OMe		19
2	Boc-Pmh-Aib-OMe	Dipeptide	33
3	Boc-Pmh-Aib-Phe-OMe	Tripeptide	33
4	Boc-Pmh-Aib-Phe-DPhe-OMe	Tetrapeptide	40
5	Boc-Pmh-Aib-Cha-hPhe-DPhe-OMe	Pentamer	47
6	Boc-Pmh-Aib-Gln(trt)-DPhe-Phe-OMe	Hexamer	61
7	Boc-Pmh-Aib-Chg-Gln(trt)-DPhe-(Boc)Trp-Phe-OMe	Heptamer	73
8	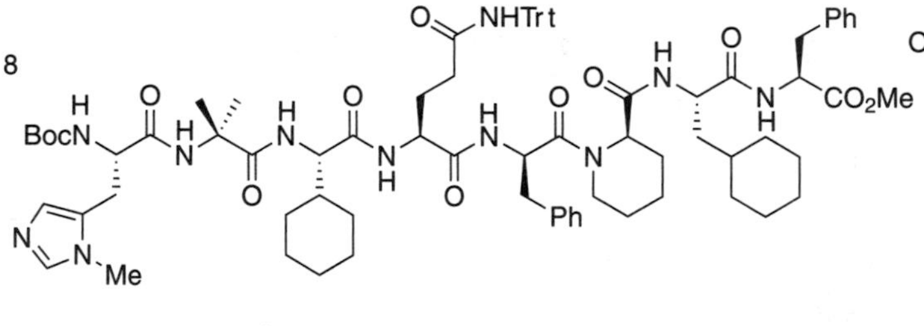 (octamer structure)	Octamer	78

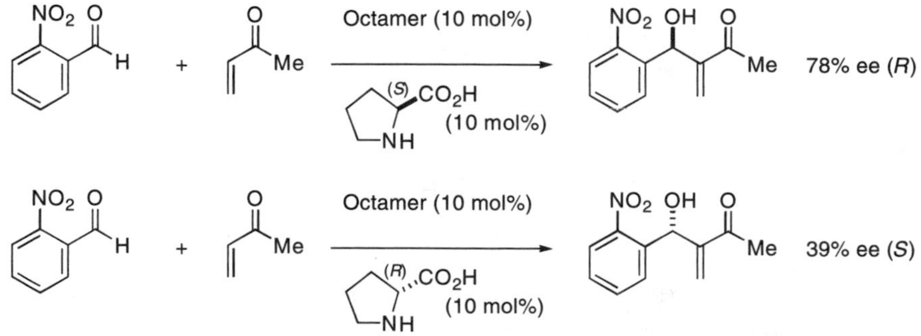

Figure 12.43. Combination of proline enantiomers with octamer from **Table 12.4** led to opposite stereoselectivity.

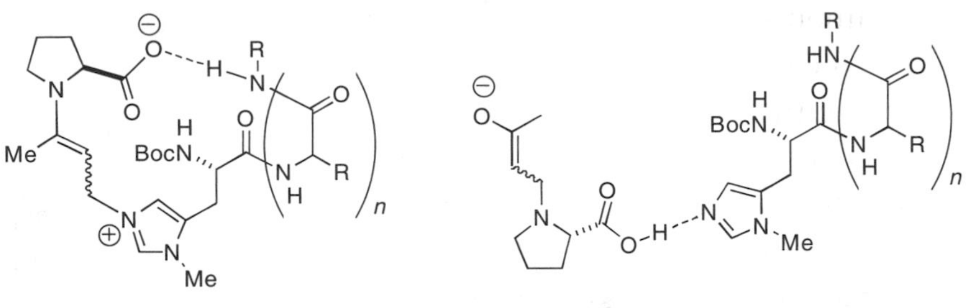

Proline enamine   Proline enolate

Figure 12.44. Possible proline–peptide–substrate interactions in the Morita–Baylis–Hillman reaction

The importance of the relative stereochemistries of proline and the octamer, the poor enantioselectivity and efficiency of proline or the octamer when operating alone, and the dependency of the enantioselectivity on the peptide length in the dual catalyst system suggest a higher-order transition state. Two possible intermediates are shown in **Figure 12.44**, in which the proline forms an enamine or undergoes conjugate addition to generate an enolate.[104] Presumably the enantioselectivity in this dual catalyst system could be improved by optimization of the individual amino acid residues in the peptide-based catalyst.

12.3.2.c Interconversion of Enantioselective, Dual, and Bifunctional Catalyst Systems

At this point, we have studied enantioselective catalysts that activate only a single substrate, catalysts that are bifunctional, and dual catalyst systems. In the development and optimization of an enantioselective process, it is possible that each class of catalyst can be examined. In this section, an example of catalyst optimization is outlined that illustrates a rational approach to the development of dual and bifunctional catalysts from a single point activation catalyst.

β-Lactams are one of the most important pharmacophores and are found in the penam antibiotics. Since the discovery of penicillin in 1929, the synthesis of β-lactams has received considerable attention and enantioselective variants have now been developed. Certain enantioenriched β-lactams can be accessed through a Lewis base catalyzed enantioselective [2+2] cycloaddition. The reaction proceeds by deprotonation of an acid chloride to generate a reactive ketene intermediate (**Figure 12.45**). Attack of a Lewis base on the ketene forms a zwitterionic enolate that undergoes addition to the imino ester, followed by ring-closure with expulsion of the nucleophilic catalyst. When this reaction was catalyzed by the chiral Lewis base benzoylquinine (BQ*), the resulting yields were low (45–65%), possibly due to polymerization of the acid chloride and imino ester.[106]

It was envisioned that introduction of a Lewis acid co-catalyst to coordinate the imino ester might facilitate attack of the zwitterionic enolate and increase the yield of β-lactams (**Figure 12.46**).[107,108] A series of achiral Lewis acid co-catalysts therefore,

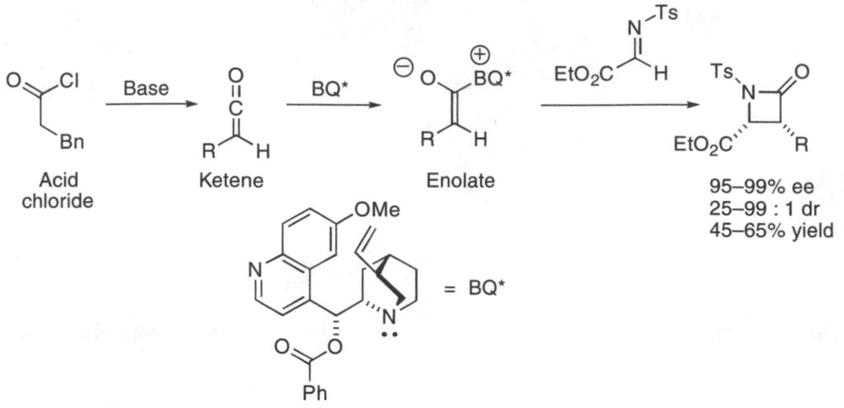

Figure 12.45. The Lewis basic benzoylquinine reacts with a ketene to form a zwitterionic enolate that reacts with imines to give β-lactams.

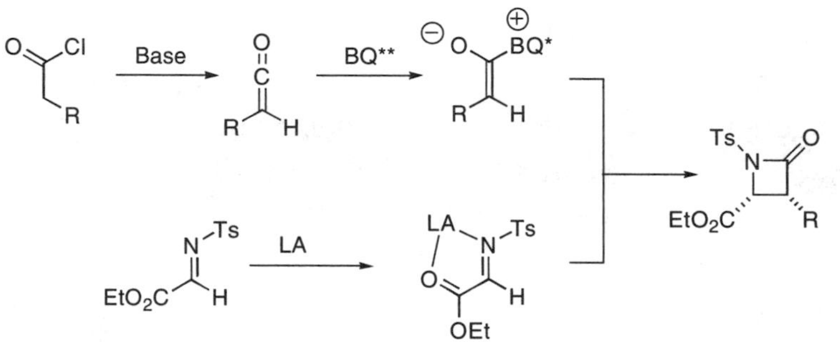

Figure 12.46. Chiral Lewis base–Lewis acid catalyzed generation of β-lactams.

were screened at 10 mol% loading and the reactions evaluated based on yield (**Table 12.5**).

Lewis acidic phosphine-ligated late-transition-metal catalysts closely related to $Rh(PPh_3)_3(OTf)$ and $Cu(PPh_3)_2(OTf)$ have been shown to catalyze the reaction of imino esters[107] and were examined in the β-lactam synthesis in **Figure 12.46**. In the presence of these complexes, however, the yields decreased (entries 1–6). This observation can be attributed to the coordination of the azaphilic Lewis acids to the benzoylquinine tertiary amine, a hypothesis subsequently supported spectroscopically in this system.[109] Pearson's[110] observation that soft, polarizable metals preferentially bind to soft Lewis bases like amines is consistent with the experimental finding in **Table 12.5**. In contrast, hard Lewis acids preferentially bind to hard Lewis bases, such as those containing oxygen. Thus, the more oxophilic $Al(OTf)_3$, $Zn(OTf)_2$, and $In(OTf)_3$ are less likely to coordinate and deactivate the BQ* than softer Lewis acids. Co-catalysts $Al(OTf)_3$, $Zn(OTf)_2$, and $In(OTf)_3$ (entries 8–10) significantly improved yields over the reaction conducted in the absence of metal co-catalysts (entry 1).

Consistent with binding of the imino ester to the Lewis acid co-catalyst, combination of either $In(OTf)_3$ or $Zn(OTf)_2$ with the imino ester resulted in lower C=O and C=N stretching frequencies and small shifts in the NMR spectra relative to the unbound substrate.

Although reactions catalyzed by the combination of BQ* and In(OTf)$_3$ generated β-lactam of > 95% ee, the diastereoselectivities decreased slightly in some cases [from 50:1 with BQ* only to 10:1 with BQ* and In(OTf)$_3$].[109] High selectivity is particularly important in the synthesis and study of biologically active compounds such as the β-lactam in **Figure 12.47**, which is a potent inhibitor of human leukocyte elastase.[111] To address this selectivity issue, as well as study the impact of employing *chiral* Lewis acid co-catalysts on the enantioselectivity and diastereoselectivity, selected chiral Lewis acids were screened with BQ* (**Figure 12.47**).[112] Initially, (*R*)-(BINOLate)Al(OTf) was chosen as the chiral Lewis acid. The dual catalyst system comprised of 10 mol% each of BQ* and (*R*)-(BINOLate)Al(OTf) resulted in generation of β-lactam of 80% ee with 8:1 dr. This selectivity can be compared to use of the simple achiral Lewis acid Al(OTf)$_3$, which furnished product of 99% ee and 11:1 dr

Table 12.5. Lewis acids screened as co-catalysts in **Figure 12.46** (R = Ph) and isolated yields of β-lactam product.

Entry	Lewis acid (10 mol%)	Isolated yield (%)	
1	None	65	
2	Rh(PPh$_3$)$_3$(OTf)	27	
3	Pd(PPh$_3$)$_2$Cl$_2$	30	Soft, azaphilic Lewis acids
4	Cu(OTf)$_2$	35	
5	Cu(PPh$_3$)$_2$ClO$_4$	49	
6	Sn(OTf)$_2$	44	
7	Sm(OTf)$_3$	45	
8	Al(OTf)$_3$	67	Harder, more oxophilic Lewis acids
9	Zn(OTf)$_2$	85	
10	In(OTf)$_3$	95	

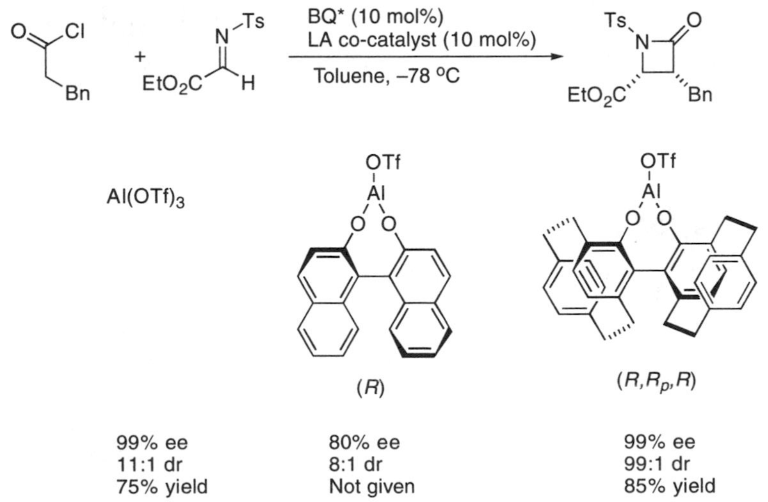

Figure 12.47. Combination of achiral or chiral Lewis acid co-catalysts with BQ* in the synthesis of a β-lactam that inhibits human leukocyte elastase.

under the same conditions. The aluminum complex of chiral (R,R_p,R)-cyclophane-based ligand (**Figure 12.47**) was also examined. Like BINOL, this ligand has axial chirality that projects away from the metal center. In addition, the cyclophane moieties project up and down, increasing the steric differentiation in adjacent quadrants about the metal center. Application of aluminum cyclophane-based catalyst afforded product of 99% ee and exceptional dr (99:1). These results demonstrate that the dual chiral catalyst approach can provide increased tunability in catalytic asymmetric processes.

As outlined in the introduction, one of the possible drawbacks in dual catalyst systems is higher molecularity (i.e., the rate can depend on the concentrations of both catalysts). In such cases, increased catalyst loading may be necessary to achieve the desired TOF. This issue can be circumvented by incorporation of both catalytically active sites into a single bifunctional catalyst. Of course, the two sites would need to operate in a cooperative, intramolecular fashion to benefit from the potential rate enhancement of the bifunctional design.

By adding In(OTf)$_3$ to BQ* in the β-lactam-forming cycloaddition reactions, the yields were found to increase over reactions conducted in the absence of Lewis acids (**Table 12.5**). The difference in enantioselectivity between these two reactions, however, was found to be < 2%, indicating that the In(OTf)$_3$ had little impact on the stereoinduction. Transformation of this chiral Lewis base–achiral Lewis acid combination into a single bifunctional catalyst was accomplished by addition a metal-binding site to the quinine-based catalyst. Since BQ* proved to be a highly enantioselective catalyst, the idea was to conserve as much of the BQ* structure as possible. With this in mind, a salicylic-acid-based binding site was installed (**Figure 12.48**). Deprotonation of the new Lewis basic BQ* derivative and addition of In(OTf)$_3$ gave the bifunctional In–BQ*, which catalyzed the β-lactam formation with excellent enantioselectivity (99%) and yield (90%) and good dr (10:1).

The β-lactam-forming reaction with BQ*–In was determined to be first-order in BQ*–In,[109] ruling out an intermolecular process involving generation of the zwitterionic enolate on one BQ*–In and activation of the imino ester by indium on another BQ*–In. Based on these kinetic data, as well as other experimental observations, a

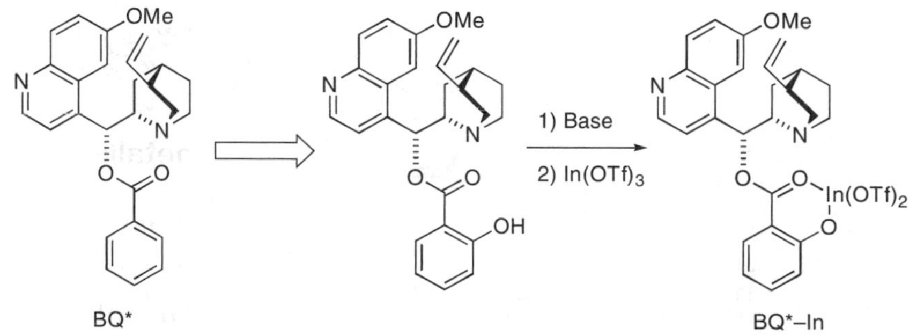

Figure 12.48. Replacement of the benzoyl group with a salicylate provides a metal-binding site. Deprotonation and addition of In(OTf)$_3$ generates the bifunctional catalyst.

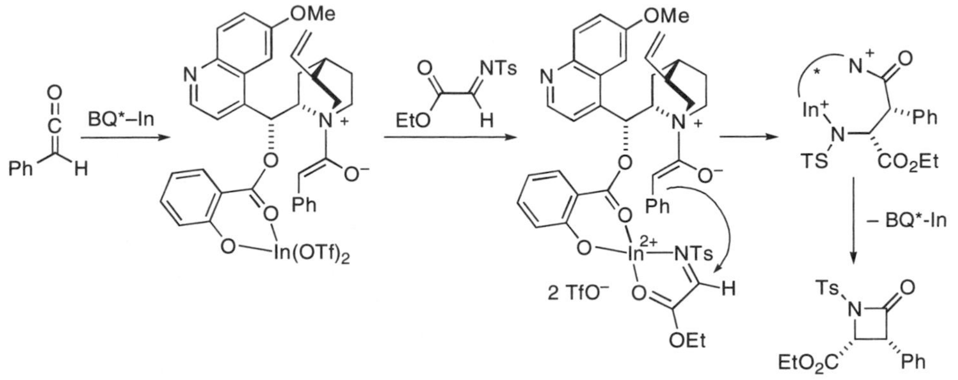

Figure 12.49. Proposed mechanism for the β-lactam-forming reaction with bifunctional catalyst BQ*–In.

cooperative intramolecular mechanism was proposed. As illustrated in **Figure 12.49**, formation of the zwitterionic enolate at the tertiary nitrogen and coordination of the imino ester to the indium center gives a ternary intermediate. C–C bond formation occurs next in the rate-determing step, followed by transacylation, which liberates the catalyst.[109]

12.4 Multifunctional Catalysis

Catalysts that possess more than two sites to activate substrates and reagents are classified as multifunctional catalysts. Nature has developed many enzymes that contain multiple functional groups in the active site for substrate activation, as exemplified in the discussion of aldolase enzymes in the introduction to this chapter (**Figure 12.1**). The multiple activation modes of enzymes act synergistically by controlling the positions and orientations of the reaction partners. Such attributes allow enzymes to operate under very mild conditions with extremely high enantioselectivities and TOFs. These same characteristics usually give rise to a high degree of substrate specificity. Unlike enzymes, however, small-molecule catalysts often promote mechanistically distinct transformations. The close proximity of the active sites of multifunctional catalysts appears to be a key feature that enables these catalysts to achieve high enantioselectivities in unrelated reactions.

12.4.1 Multifunctional Lanthanide–Main-Group Heterobimetallic Catalysts

One of the most versatile and successful catalyst systems in asymmetric synthesis is the heterobimetallic lanthanide BINOL derivatives (Ln = lanthanide, **Figure 12.50**).[7,113–115] These multifunctional catalysts consist of a central lanthanide with coordination numbers of 6–8.[113,116] In the lanthanide series, the preferred oxidation state is +3 and the coordination numbers of 8–9 are most common.[117] The three BINOLate

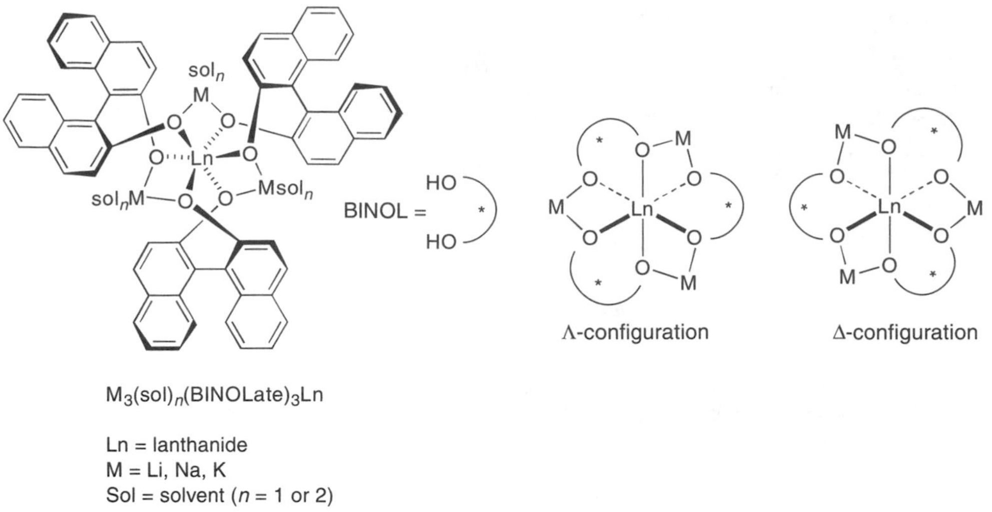

$M_3(sol)_n(BINOLate)_3Ln$

Ln = lanthanide
M = Li, Na, K
Sol = solvent (n = 1 or 2)

Figure 12.50. Drawing of heterobimetallic $M_3(sol)_n(BINOLate)_3Ln$ (M = Li, Na, or K; Ln = lanthanide^{3+}; sol = solvent, typically THF or ether). Use of (S)-BINOL results in formation of the Δ-configuration at the lanthanide center.

ligands surrounding the lanthanide center create the chiral environment and carry a total charge of –6. To balance the excess charge at the lanthanide metal, three group I counterions (Li, Na, or K) are necessary. As shown in **Figure 12.50**, the group I metals bind to two naphthylate oxygens on adjacent BINOLate ligands. The bridging interactions stabilize the heterobimetallic structures while increasing the Lewis acidity of the lanthanide center through removal of electron density from the BINOLate oxygens.[7,115]

Further contributing to the chiral environment of these complexes, the lanthanide center is chiral due to the arrangement of the BINOLate ligands (See Appendix A).[118] Only one diastereomer is observed in solution (NMR) and the solid state (X-ray), indicating that the chirality of the BINOLate ligands strongly influences the configuration at the metal center (**Figure 12.50**).

These catalysts can be prepared by combining three equivalents of (BINOLate)M$_2$ (M = Li, Na, or K) with the LnX$_3$ (X = Cl, OTf).[7,119] Alternatively, they can be generated by reaction of La(O-i-Pr)$_3$ or Ln[N(SiMe$_3$)$_2$]$_3$ with three equivalents of the mono-lithio salt of BINOL, liberating three equivalents of isopropanol or HN(SiMe$_3$)$_2$, respectively.[7,119] Several derivatives have been characterized in the solid state, such as the six-coordinate complex Li$_3$(THF)$_6$(BINOLate)$_3$Yb and seven-coordinate Li$_3$(THF)$_4$ (BINOLate)$_3$La(THF)[116,120] (**Figure 12.51**).[119,121,122] From these structures it can be seen that both the lanthanide and the main-group metals bind dative ligands and are, therefore, Lewis acids. The basic naphthylate oxygens can act as Brønsted bases, deprotonating acidic substrates to generate nucleophiles. The nature of the Ln–O–M(I) linkage in the heterobimetallic complexes makes them interdependent multifunctional catalysts. The multifunctional properties of the M$_3$(THF)$_n$(BINOLate)$_3$Ln complexes allow them

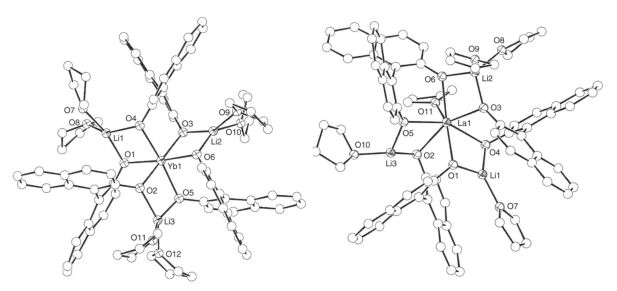

Figure 12.51. Structures of six-coordinate $Li_3(THF)_6(BINOLate)_3Yb$ (left) and seven-coordinate $Li_3(THF)_4(BINOLate)_3La(THF)$.

to exhibit various modes of reactivity, contributing to their successful application across a broad spectrum of asymmetric reactions. At the same time, the multifaceted reactivity complicates mechanistic studies.

One of the unique features of lanthanide-based catalysts, such as the heterobimetallic $M_3(THF)_n(BINOLate)_3Ln$ complexes, is the approach to catalyst optimization. Traditionally, asymmetric catalysts are optimized by the synthesis and screening of chiral ligands, which can be a time-consuming process if the ligands are difficult to synthesize. In contrast, $M_3(THF)_n(BINOLate)_3Ln$-based catalysts have been optimized primarily by using different lanthanide and main-group combinations.[7,115] The optimization process depends on a distinctive feature of the lanthanide elements, the lanthanide contraction, which is a steady decrease in ionic radius with increasing atomic number. The ionic radius of eight-coordinate trivalent lanthanide elements range from 1.17 Å for La to 1.00 Å for Lu. As discussed below, the lanthanide radius can have a dramatic impact on the enantioselectivity of these isostructural catalysts.

12.4.2 Application of Heterobimetallic Catalysts to the Nitroaldol Reaction

Multifunctional $M_3(THF)_n(BINOLate)_3Ln$ catalysts have been successfully applied to a range of Lewis acid catalyzed processes.[7] One of these is the nitroaldol or Henry reaction (**Equation 12.11**), which is quite useful in organic synthesis, because the nitro group of the product is readily reduced, thus providing access to valuable β-amino alcohols.

Equation 12.11

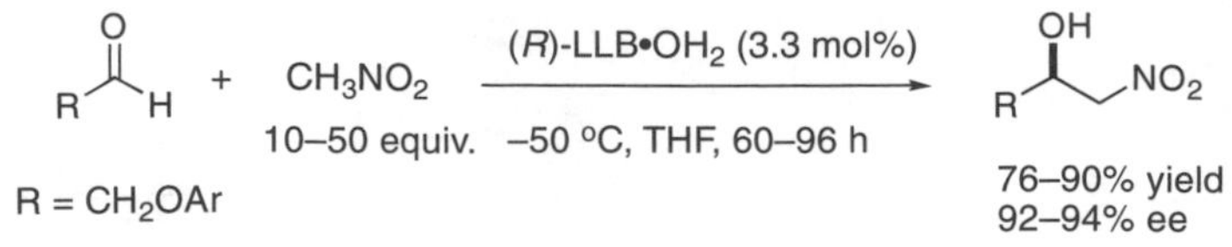

$$R = CH_2OAr$$

$$LLB \cdot OH_2 = Li_3(THF)_n(BINOLate)_3La \cdot OH_2$$

Although the details of the mechanism are not completely understood, a working model is presented in **Figure 12.52**. It is postulated that a basic aryloxide oxygen acts as a Brønsted base to deprotonate the nitromethane ($pK_a = 10$), generating the lithium nitronate. Coordination of the aldehyde to the Lewis acidic lanthanide center activates it to attack by the nitronate. The resulting alkoxide product then abstracts the naphthol hydrogen to form the alcohol and regenerate the catalyst. The role of the water molecule is not known, but it may increase the coordination number of the lanthanide. Based on X-ray crystallographic studies, it is known that in seven-coordinate $M_3(THF)_{n-}$(BINOLate)$_3$Ln(L) compounds the lanthanide center is displaced out of the plane of the main-group metals, increasing the distance between the chiral BINOLate ligands

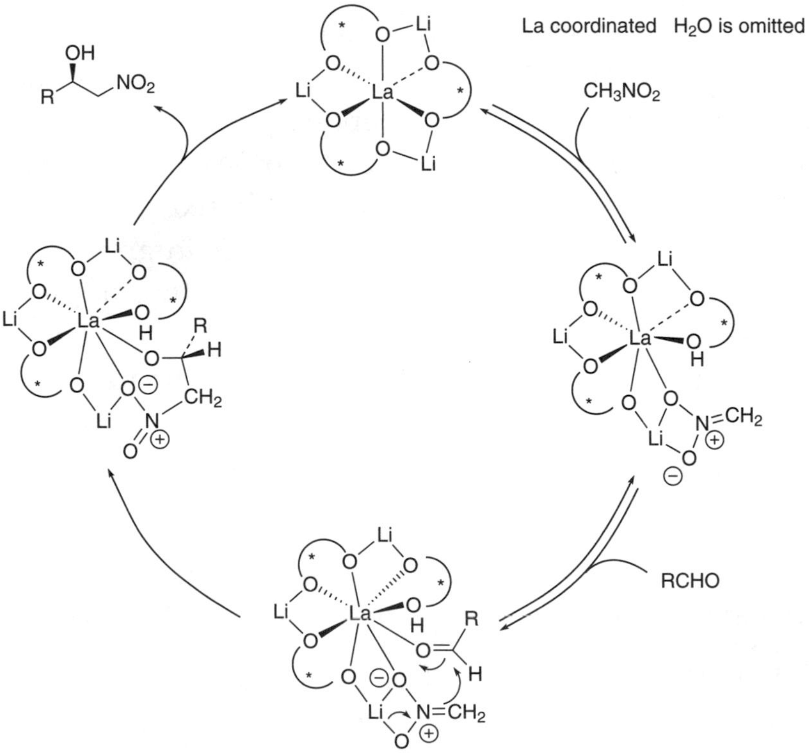

Figure 12.52. Possible mechanism for the nitroaldol reaction catalyzed by Li$_3$(THF)$_n$(BINOLate)$_3$ La(OH$_2$). The THF solvent molecules bound to lithium and the La-bound H$_2$O have been omitted for clarity.

and the bound substrate. In eight-coordinate complexes $M_3(THF)_n(BINOLate)_3LnL_2$ with trans ligands L, the lanthanide moves back into the plane of the main-group metals.[116]

Although the bifunctional catalysts gave excellent enantioselectivities, the reactions were sluggish. In an effort to increase the efficiency of the catalyst, a catalytic amount of base was generated from water and BuLi to facilitate the deprotonation of the nitroalkane. Application of this modified catalyst is shown in **Equation 12.12**, where high enantioselectivity and diastereoselectivity were realized. In the absence of LiOH, only trace products were formed.

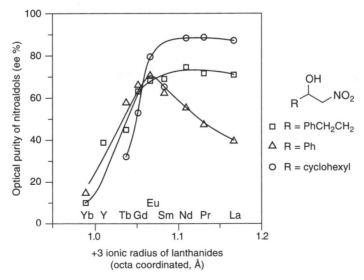

As alluded to above, the enantioselectivity of these reactions depend on the ionic radii of the lanthanide-based catalyst. This is illustrated in a plot of eight-coordinate lanthanide(III) radius versus enantioselectivity in the nitroaldol for three substrates in **Figure 12.53**.[7] Note that different substrates exhibit distinct dependencies of enantioselectivity on ionic radius.

Figure 12.53. Effects of the ionic radii of lanthanides on the product ee in the nitroaldol reaction. Used with permission of the American Chemical Society.

12.4.2.a Application of Heterobimetallic Catalysts to the Direct Aldol Reaction

A reaction related to the nitro aldol that demonstrates the tunability of the $M_3(THF)_n$ $(BINOLate)_3Ln$ catalyst system is the direct aldol. Early developments in the asymmetric aldol reactions involved the use of preformed silyl enol ethers or silyl ketene acetals (Mukaiyama-type donors), which circumvented problems of competitive aldehyde and ketone enolization and limited formation of undesired aldol products.[9,122] While the importance of this chemistry cannot be understated, the drawbacks of this approach are that the donor is prepared separately and the aldol product is obtained after deprotection (**Figure 12.54**). The direct application of unactivated ketones has proven to be more challenging (**Figure 12.54**). One reason for this is that the deprotonation ketones (pK_a around 17) requires more basic conditions than the deprotonation of nitroalkanes (pK_a around 10 in water).

In the search for catalysts for the direct aldol reaction, a series of $M_3(THF)_n(BINO-Late)_3Ln$ complexes was screened. This study revealed that the $Li_3(THF)_n(BINO-Late)_3La$ generated the aldol adducts with the highest enantioselectivity.[123] The catalyst TOF was low, however, and 20 mol% catalyst loading was needed. It was later found that addition of a catalytic amount of base, generated from $KN(SiMe_3)_2$, enhanced the activity of the catalyst, most likely by facilitating the formation of the enolate. In support of this proposal, the use of deuterated acetophenone, $PhCOCD_3$, resulted in $k_H/k_D = 5$. Additionally, the rate is independent of the concentration of aldehyde. These results suggest that the rate-determining step is enolate formation. Some results from this study are illustrated in **Figure 12.55**.

This catalyst system has been applied to the synthesis of important natural product precursors (**Figure 12.56**). Conversion of the enantioenriched aldol product from **Figure 12.55** to the corresponding ester via Baeyer–Villiger oxidation was followed by a four-step sequence to give an intermediate in the epothilone synthesis.[124] The Baeyer–Villiger oxidation product was also protected as the TBS ether and converted to the β-hydroxy aldehyde. Diastereoselective addition of the enolate formed from acetophenone and LDA gave a 3:1 mixture of the *anti:syn* diastereomers. In contrast, the same aldol condensation with acetophenone catalyzed by (*S*)-LLB/KOH afforded

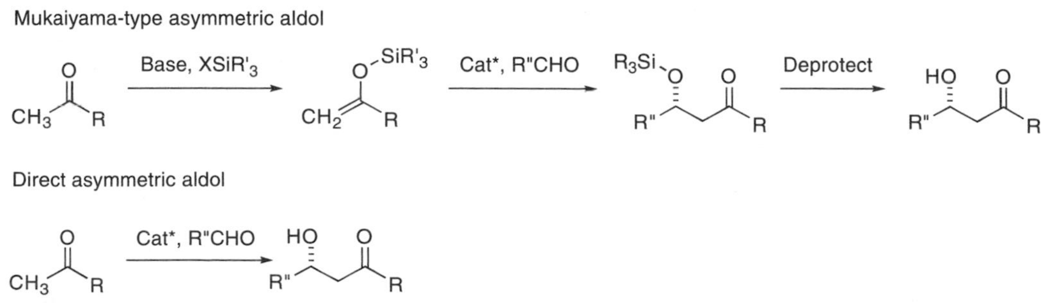

Mukaiyama-type asymmetric aldol

Direct asymmetric aldol

Figure 12.54. Comparison of the Mukaiyama-type aldol with the direct aldol.

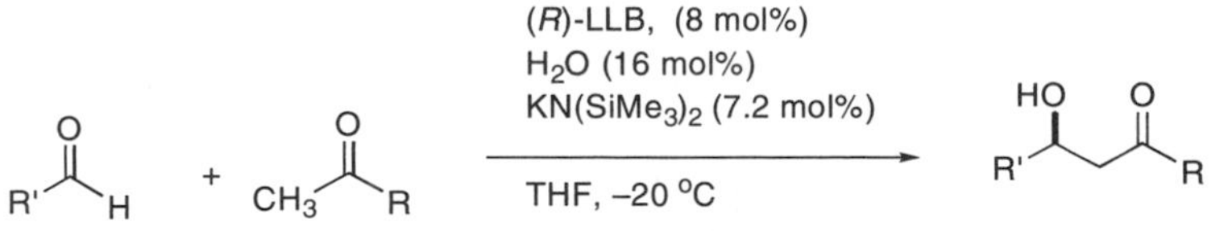

LLB = $Li_3(THF)_n(BINOLate)_3La$

R'	R	ee (%)	Yield (%)
t-Bu	Ph	89	75
t-Bu	Me	76	62
$BnOCH_2CMe_2$	Ph	93	70
$PhCH_2CH_2$	$3\text{-}C_6H_4\text{-}NO_2$	80	60

Figure 12.55. Selected results from the direct aldol reaction catalyzed by the heterobimetallic catalyst composed of lanthanide, lithium, BINOLate (LLB), and base.

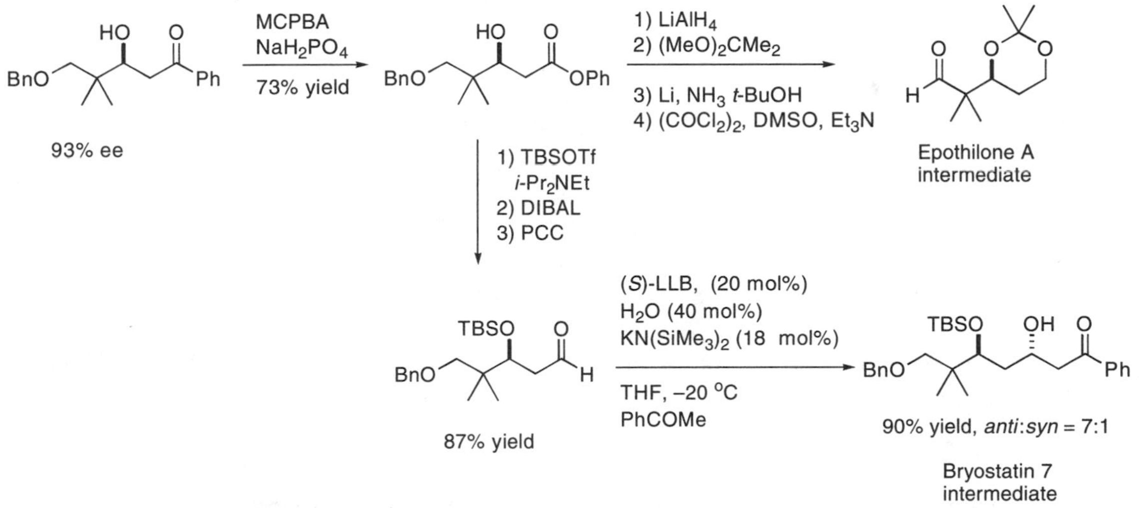

Figure 12.56. Applications of the aldol products, from heterobimetallic chiral catalysis, to the synthesis of intermediates in the synthesis of epothilone and bryostatin 7.

the *anti*-aldol product (dr = 7:1) with 95% ee and 90% combined yield. This *anti* product was an intermediate in the synthesis of bryostatin 7.[125] This is an example of a double diastereoselective reaction (Chapter 13).

The base additive that was found to give the highest enantioselectivities and TOF in the direct aldol was KOH, generated from $KN(SiMe_3)_2$ or KO-*t*-Bu and water. Interestingly, the catalyst derived from $Li_3(THF)_n(BINOLate)_3La$ incorporating KOH gave higher enantioselectivity and yield than when LiOH was used. Furthermore, the combination of $K_3(THF)_n(BINOLate)_3La$ with KOH was less effective. The authors proposed that the core structure of the catalyst contains the three lithiums with a coordinated KOH to the central lanthanide. A proposed mechanism based on the observations outlined above is illustrated in **Figure 12.57**. It is noteworthy that the nature of the main group metal M has a dramatic impact on substrate binding ability of $M_3(THF)_n(BINOLate)_3Ln$ complexes.[126]

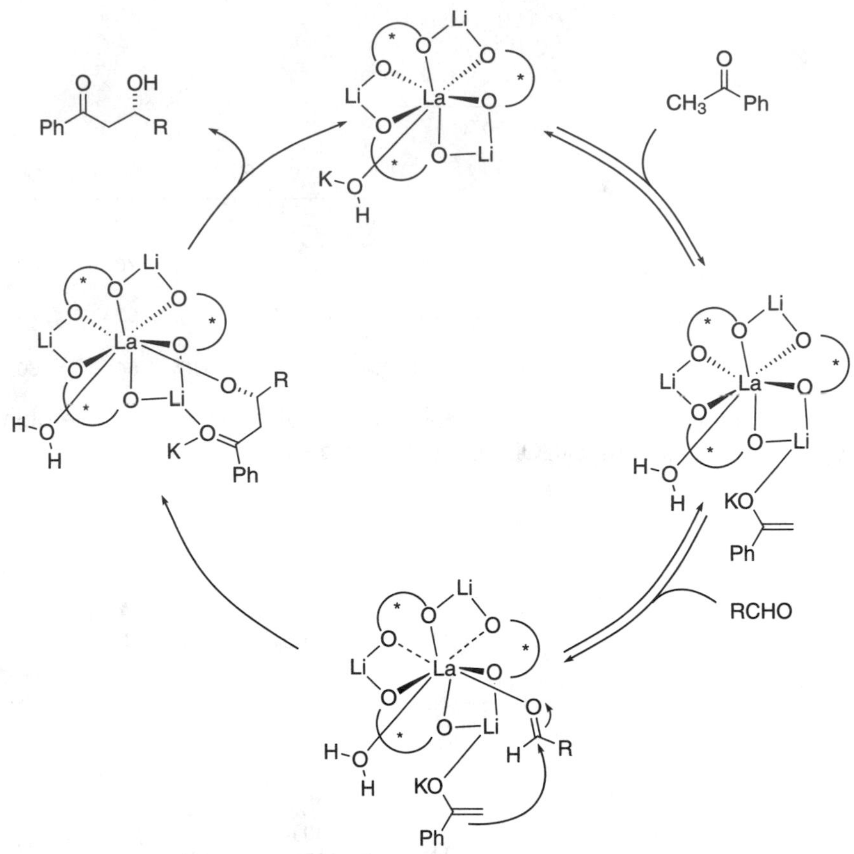

Figure 12.57. Proposed mechanism for the enantioselective direct aldol reaction. The THF solvates on the lithium and potassium centers have been omitted for clarity.

12.4.3 Asymmetric Conjugate Addition of Amines to Enones: Mechanistic Observations

The $M_3(THF)_n(BINOLate)_3Ln$ class of catalysts has also been used to promote the 1,4-addition of amines to α,β-unsaturated carbonyl compounds to provide β-amino carbonyl compounds (**Figure 12.58**), which are of biological interest.[127]

Mechanistic studies in this system have provided important insight into the operation of $M_3(THF)_n(BINOLate)_3Ln$ catalysts. Kinetic analysis indicates that the reaction is first-order in both the $Li_3(THF)_n(BINOLate)_3Y$ and chalcone substrate, but zero-order in amine.[128] In studies of the related ytterbium complex, $Na_3(THF)_n$-$(BINOLate)_3Yb$, it has been proposed that the Yb center is too sterically crowded to bind an additional ligand and that Yb must dissociate $(BINOLate)Na_2$ before it can coordinate a substrate.[129] It should be noted that the ionic radius of Yb is one of the smallest of the rare earth elements (see **Figure 12.53**). In the conjugate addition of amines (**Figure 12.58**), use of 1 mol% $Li_3(THF)_n(BINOLate)_3Y$ with 0–3% $(BINOLate)Li_2$ did not result in a change in enantioselectivity (96%) or TOF. Use of only $(BINOLate)Li_2$, however, led to much lower TOF and formation of the product with only 17% enantioselectivity. The absence of inhibition suggests that $Li_3(THF)_n(BI$-

NOLate)$_3$Y does not lose (BINOLate)Li$_2$ to generate a Li(BINOLate)$_2$Y species in the conjugate addition of amines to enones.[128] It was also observed that the reaction is inhibited by water and there is a strong positive nonlinear effect (see Chapter 11).[128] In this case, the water likely competes with the amine or ketone binding to the central lanthanide. A possible mechanism for the conjugate addition is illustrated in **Figure 12.59**.

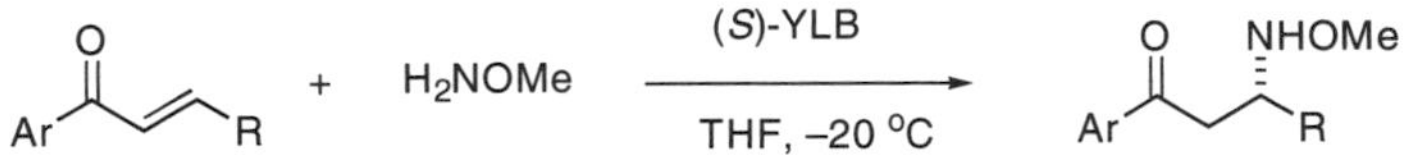

Ar	R	ee (%)	Yield (%)
Ph	Ph	95	97
2-Furyl	Ph	94	95
Ph	4-Pyridyl	85	91
Ph	CH$_2$i-Pr	93	93

YLB = Li$_3$(THF)$_n$(BINOLate)$_3$Y

Figure 12.58. Selected results in the asymmetric addition of amines to enones with heterobimetallic Li$_3$(THF)$_n$(BINOLate)$_3$Y.

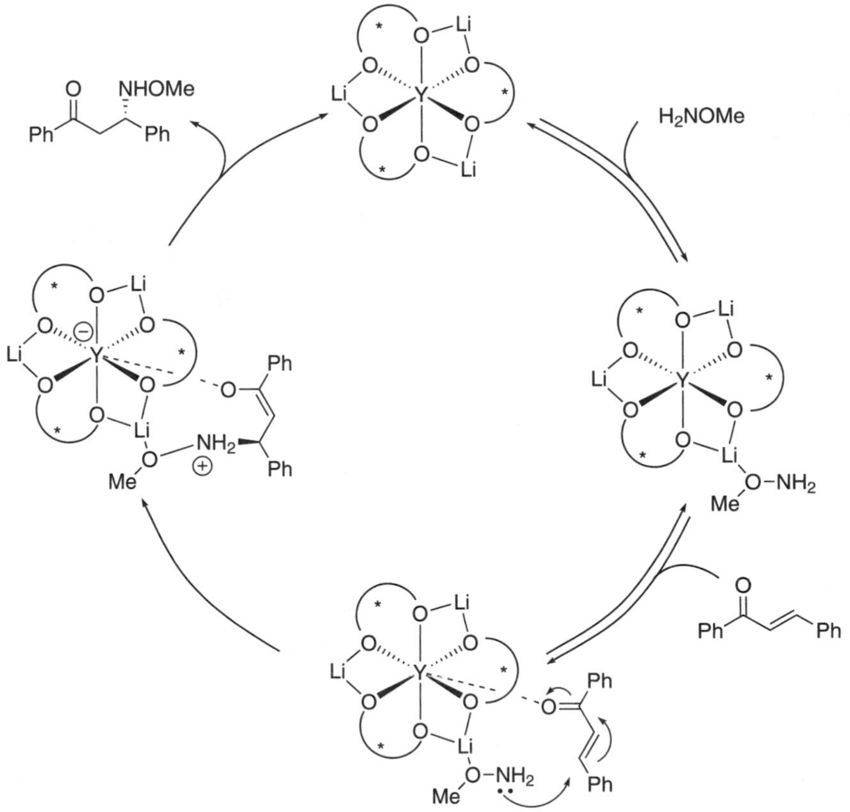

Figure 12.59. Possible mechanism for the 1,4-addition of *O*-methylhydroxylamine to chalcone. The THF solvates on the lithium centers have been omitted for clarity.

The multifunctional $M_3(THF)_n(BINOLate)_3Ln$ catalysts outlined in this section have several modes of reactivity. Both the main-group and lanthanide centers are Lewis acidic and have open coordination sites. The BINOLate oxygens can act as Brønsted bases, deprotonating substrates and activating them for reaction. Finally, additives such as water and hydroxide may also increase the reactivity and enantioselectivity of these catalysts. The ability of lanthanides to simultaneously bind various reaction partners and hold them in close contacts is a key feature of the $M_3(THF)_n(BINOLate)_3Ln$ family of catalysts.

Conclusions

Chiral catalysts that provide multiple avenues for activation, such as those described in this chapter, hold great promise in asymmetric synthesis. In contrast to many conventional asymmetric catalysts, dual, bifunctional, and multifunctional catalysts can promote reactions that cannot be catalyzed through single-site activation. As outlined here, various permutations allow activation modes to be combined to provide a diverse range of efficient and highly enantioselective dual, bifunctional, and multifunctional catalysts. Given these attributes, research into catalysts with multiple activation modes will expand and is expected to lead to new catalytic enantioselective processes.

References

(1) Sawamura, M.; Ito, Y. Catalytic Asymmetric Synthesis by Means of Secondary Interaction Between Chiral Ligands and Substrates. *Chem. Rev.* **1992**, *92*, 857–871.

(2) van den Beuken, E. K.; Feringa, B. L. Bimetallic Catalysis by Late Transition Metal Complexes. *Tetrahedron* **1998**, *54*, 12985–13011.

(3) Noyori, R.; Ohkuma, T. Asymmetric Catalysis by Architectural and Functional Molecular Engineering: Practical Chemo- and Stereoselective Hydrogenation of Ketones. *Angew. Chem., Int. Ed. Engl.* **2001**, *40*, 40–73.

(4) Noyori, R.; Yamakawa, M.; Hashiguchi, S. Metal-Ligand Bifunctional Catalysis: A Nonclassical Mechanism for Asymmetric Hydrogen Transfer Between Alcohols and Carbonyl Compounds. *J. Org. Chem.* **2001**, *66*, 7931–7944.

(5) Rowlands, G. J. Ambifunctional Cooperative Catalysts. *Tetrahedron* **2001**, *57*, 1865–1882.

(6) Shibasaki, M.; Kanai, M.; Funabashi, K. Recent Progress in Asymmetric Two-Center Catalysis. *J. Chem. Soc., Chem. Commun.* **2002**, 1989–1999.

(7) Shibasaki, M.; Yoshikawa, N. Lanthanide Complexes in Multifunctional Asymmetric Catalysis. *Chem. Rev.* **2002**, *102*, 2187–2219.

(8) Ma, J.-A.; Cahard, D. Towards Perfect Catalytic Asymmetric Synthesis: Dual Activation of the Electrophile and the Nucleophile. *Angew. Chem., Int. Ed. Engl.* **2004**, *43*, 4566–4583.

(9) Machajewski, T. D.; Wong, C.-H.; Lerner, R. A. The Catalytic Asymmetric Aldol Reaction. *Angew. Chem., Int. Ed. Engl.* **2000**, *39*, 1352–1374.

(10) Itsuno, S.; Nakano, M.; Miyazaki, K.; Masuda, H.; Ito, K. Asymmetric Synthesis Using Chirally Modified Borohydrides. Part 3. Enantioselective Reduction of Ketones and Oxime Ethers with Reagents Prepared from Borane and Chiral Amino Alcohols. *J. Chem. Soc., Perkin. Trans.* **1985**, 2039–2044.

(11) Itsuno, S.; Sakurai, Y.; Ito, K.; Hirao, A. Catalytic Behavior of Optically Active Amino Alcohol-Borane Complex in the Enantioselective Reduction of Acetophenone Oxime O-Alkyl Ethers. *Bull. Chem. Soc. Jpn.* **1987**, *60*, 395–396.

(12) Hirao, A.; Itsuno, S.; Nakahama, S.; Yamazaki, N. Asymmetric Reduction of Aromatic Ketones with Chiral Alkoxy-Amineborane Complexes. *J. Chem. Soc., Chem. Commun.* **1981**, 315–317.

(13) Corey, E. J.; Bakshi, B. K.; Shibata, S. Highly Enantioselective Borane Reduction of Ketones Catalyzed by Chiral Oxazaborolidines. Mechanism and Synthetic Implications. *J. Am. Chem. Soc.* **1987**, *109*, 5551–5553.

(14) Corey, E. J.; Bakshi, R. K.; Shibata, S.; Chen, C. P.; Singh, V. A Stable and Easily Prepared Catalyst for the Enantioselective Reduction of Ketones. Applications to Multistep Syntheses. *J. Am. Chem. Soc.* **1987**, *109*, 7925–7926.

(15) Corey, E. J.; Shibata, S.; Bakshi, R. K. An Efficient and Catalytically Enantioselective Route to (S)-(–)-Phenyloxirane. *J. Org. Chem.* **1988**, *53*, 2861–2863.

(16) Corey, E. J.; Helal, C. J. Reduction of Carbonyl Compounds with Chiral Oxazaborolidine Catalysts: A New Paradigm for Enantioselective Catalysis and a Powerful New Synthetic Method. *Angew. Chem., Int. Ed. Engl.* **1998**, *37*, 1986–2012.

(17) Alagona, G.; Ghio, C.; Persico, M.; Tomasi, S. Quantum Mechanical Study of Stereoselectivity in the Oxazaborolidine-Catalyzed Reduction of Acetophenone. *J. Am. Chem. Soc.* **2003**, *125*, 10027–10039.

(18) Sakito, Y.; Yoneyoshi, Y.; Suzukamo, G. Asymmetric Reduction of Oxime Ethers. Distinction of *anti* and *syn* Isomers Leading to Enantiomeric Amines. *Tetrahedron Lett.* **1988**, *29*, 223–224.

(19) Tillyer, R. D.; Boudreau, C.; Tschaen, D.; Dolling, U.-H.; Reider, P. J. Asymmetric Reduction of Keto Oxime Ethers Using Oxazaborolidine Reagents. The Enantioselective Synthesis of Cyclic Amino Alcohols. *Tetrahedron Lett.* **1995**, *36*, 4337–4340.

(20) Kitamura, M.; Suga, S.; Kawai, K.; Noyori, R. Catalytic Asymmetric Induction. Highly Enantioselective Addition of Dialkylzincs to Aldehydes. *J. Am. Chem. Soc.* **1986**, *108*, 6071–6072.

(21) Kitamura, M.; Suga, S.; Oka, H.; Noyori, R. Quantitative Analysis of the Chiral Amplification in the Amino Alcohol-Promoted Asymmetric Alkylation of Aldehydes with Dialkylzincs. *J. Am. Chem. Soc.* **1998**, *120*, 9800–9809.

(22) Kitamura, M.; Oka, H.; Noyori, R. Asymmetric Addition of Dialkylzincs to Benzaldehyde Derivatives Catalyzed by Chiral β-Almino Alcohols. Evidence for the Monomeric Alkylzinc Aminoalkoxide as Catalyst. *Tetrahedron* **1999**, *55*, 3605–3614.

(23) Pu, L.; Yu, H.-B. Catalytic Asymmetric Organozinc Additions to Carbonyl Compounds. *Chem. Rev.* **2001**, *101*, 757–824.

(24) Soai, K.; Niwa, S. Enantioselective Additions of Organozinc Reagents to Aldehydes. *Chem. Rev.* **1992**, *92*, 833–856.

(25) Pu, L. Asymmetric Alkynylzinc Additions to Aldehydes and Ketones. *Tetrahedron* **2003**, *59*, 9873–9886.

(26) Ohkuma, T.; Ooka, H.; Hashiguchi, S.; Ikariya, T.; Noyori, R. Practical Enantioselective Hydrogenation of Aromatic Ketones. *J. Am. Chem. Soc.* **1995**, *117*, 2675–2676.

(27) Xie, J.-H.; Wang, L.-X.; Fu, Y.; Zhu, S.-F.; Fan, B.-M.; Duan, H.-F.; Zhou, Q.-L. Synthesis of Spiro Diphosphines and Their Application in Asymmetric Hydrogenation of Ketones. *J. Am. Chem. Soc.* **2003**, *125*, 4404–4405.

(28) Cobley, C. J.; Henschke, J. P. Enantioselective Hydrogenation of Imines Using a Diverse Library of Ruthenium Dichloride(diphosphine)(diamine) Precatalysts. *Adv. Synth. Catal.* **2003**, *345*, 195–201.

(29) Abdur-Rashid, K.; Lough, A. J.; Morris, R. H. RuHCl(diphosphine)(diamine): Catalyst Precursors for the Stereoselective Hydrogenation of Ketones and Imines. *Organometallics* **2001**, *20*, 1047–1049.

(30) Ohkuma, T.; Koizumi, M.; Doucet, H.; Pham, T.; Kozawa, M.; Murata, K.; Katayama, E.; Yokozawa, T.; Ikariya, T.; Noyori, R. Asymmetric Hydrogenation of Alkenyl, Cyclopropyl, and Aryl Ketones. RuCl$_2$(Xyl-BINAP)(1,2-diamine) as a Precatalyst Exhibiting a Wide Scope. *J. Am. Chem. Soc.* **1998**, *120*, 13529–13530.

(31) Jiang, Q.; Jiang, Y.; Xiao, D.; Cao, P.; Zhang, X. Highly Enantioselective Hydrogenation of Simple Ketones Catalyzed by a Rh-PennPhos Complex. *Angew. Chem., Int. Ed. Engl.* **1998**, *37*, 1100–1103.

(32) Ohkuma, T.; Ishii, D.; Takeno, H.; Noyori, R. Asymmetric Hydrogenation of Amino Ketones Using Chiral RuCl₂(diphophine)(1,2-diamine) Complexes. *J. Am. Chem. Soc.* **2000**, *122*, 6510–6511.

(33) Noyori, R. Asymmetric Catalysis: Science and Opportunities. *Angew. Chem., Int. Ed. Engl.* **2002**, *41*, 2008–2022.

(34) Abdur-Rashid, K.; Clapham, S. E.; Hadzovic, A.; Harvey, J. N.; Lough, A. J.; Morris, R. H. Mechanism of the Hydrogenation of Ketones Catalyzed by *trans*-Dihydrido(diamine)ruthenium(II) Complexes. *J. Am. Chem. Soc.* **2002**, *124*, 15104–15118.

(35) Sandoval, C. A.; Ohkuma, T.; Muñiz, K.; Noyori, R. Mechanism of Asymmetric Hydrogenation of Ketones Catalyzed by BINAP/1,2-Diamine-Ruthenium(II) Complexes. *J. Am. Chem. Soc.* **2003**, *125*, 13490–13503.

(36) Fulton, J. R.; Bouwkamp, M. W.; Bergman, R. G. Reactivity of a Parent Amidoruthenium Complex: A Transition Metal Amide of Exceptionally High Basicity. *J. Am. Chem. Soc.* **2000**, *122*, 8799–8800.

(37) Bryndza, H. E.; Tam, W. Monomeric Metal Hydroxides, Alkoxides, and Amides of the Late Transition Metals: Synthesis, Reactions, and Thermochemistry. *Chem. Rev.* **1988**, *88*, 1163–1188.

(38) Fulton, J. R.; Holland, A. W.; Fox, D. J.; Bergman, R. G. Formation, Reactivity, and Properties of Nondative Late Transition Metal–Oxygen and –Nitrogen Bonds. *Acc. Chem. Res.* **2002**, *35*, 44–56.

(39) Hamilton, R. J.; Leong, C. G.; Bigam, G.; Miskolzie, M.; Bergens, S. H. A Ruthenium-Dihydrogen Putative Intermediate in Ketone Hydrogenation. *J. Am. Chem. Soc.* **2005**, *127*, 4152–4153.

(40) Hartmann, R.; Chen, P. Noyori's Hydrogenation Catalyst Needs a Lewis Acid Cocatalyst for High Activity. *Angew. Chem., Int. Ed. Engl.* **2001**, *40*, 3581–3585.

(41) Hartmann, R.; Chen, P. Numerical Modeling of Differential Kinetics in the Asymmetric Hydrogenation of Acetophenone by Noyori's Catalyst. *Adv. Synth. Catal.* **2003**, *345*, 1353–1359.

(42) Hegedus, L. S. *Transition Metals in the Synthesis of Complex Organic Molecules*, 2nd ed.; University Science Books: Sausalito, CA, 1999.

(43) Hamashima, Y.; Sawada, D.; Kanai, M.; Shibasaki, M. A New Bifunctional Asymmetric Catalyst: An Efficient Catalytic Asymmetric Cyanosilylation of Aldehydes. *J. Am. Chem. Soc.* **1999**, *121*, 2641–2642.

(44) North, M. Synthesis and Applications of Non-Racemic Cyanohydrins. *Tetrahedron: Asymmetry* **2003**, *14*, 147–176.

(45) Takamura, M.; Hamashima, Y.; Usuda, H.; Kanai, M.; Shibasaki, M. A Catalytic Asymmetric Strecker-Type Reaction: Interesting Reactivity Difference Between TMSCN and HCN. *Angew. Chem., Int. Ed. Engl.* **2000**, *39*, 1650–1652.

(46) Coppola, G. M.; Schuster, H. F. *α-Hydroxy Acids in Enantioselective Synthesis*; VCH Publishers: New York, 1997.

(47) DiMauro, E. F.; Kozlowski, M. C. The First Catalytic Asymmetric Addition of Dialkylzincs to α-Ketoesters. *Org. Lett.* **2002**, *4*, 3781–3784.

(48) DiMauro, E. F.; Kozlowski, M. C. Development of Bifunctional Salen Catalysts: Rapid, Chemoselective Alkylations of α-Ketoesters. *J. Am. Chem. Soc.* **2002**, *124*, 12668–12669.

(49) Fennie, M. W.; DiMauro, E. F.; O'Brien, E. M.; Annamalai, V.; Kozlowski, M. C. Mechanism and Scope of Salen Bifunctional Catalysts in Asymmetric Aldehyde and α-Ketoester Alkylation. *Tetrahedron* **2005**, *61*, 6249–6265.

(50) Funabashi, K.; Jachmann, M.; Kanai, M.; Shibasaki, M. Multicenter Strategy for the Development of Catalytic Enantioselective Nucleophilic Alkylation of Ketones: Me₂Zn Addition to α-Ketoesters. *Angew. Chem., Int. Ed. Engl.* **2003**, *42*, 5489–5492.

(51) List, B. Proline-Catalyzed Asymmetric Reactions. *Tetrahedron* **2002**, *58*, 5573–5590.

(52) Seayad, J.; List, B. Asymmetric Organocatalysis. *Org. Biomol. Chem.* **2005**, *3*, 719–724.

(53) Berkessel, A.; Gröger, H. *Asymmetric Organocatalysis*; Wiley: New York, 2005.

(54) Lelais, G.; MacMillan, D. W. C. Modern Strategies in Organic Catalysis: The Advent and Development of Iminium Activation. *Aldrichimica Acta* **2006**, *39*, 79–87.

(55) Dalko, P. I.; Moisan, L. In the Golden Age of Organocatalysis. *Angew. Chem., Int. Ed. Engl.* **2004**, *43*, 5138–5175.

(56) Enders, D.; Grondal, C.; Hüttl, M. R. M. Asymmetric Organocatalytic Domino Reactions. *Angew. Chem., Int. Ed. Engl.* **2007**, *46*, 1570–1581.

(57) a) Houk, K. N.; List, B. Asymmetric Organocatalysis (special issue). *Acc. Chem. Res.* **2004**, *37*, 487–631.

(58) Eder, U.; Sauer, G.; Wiechert, R. New Type of Asymmetric Cyclization to Optically Active Steroid CD Partial Structures. *Angew. Chem., Int. Ed. Engl.* **1971**, *10*, 496–497, b) list, B (Ed.) Organocatalysis (special issue) *Chem. Rev.* **2007**, *107*, 5413–5883.

(59) Hajos, Z. G.; Parrish, D. R. Asymmetric Synthesis of Bicyclic Intermediates of Natural Product Chemistry. *J. Org. Chem.* **1974**, *39*, 1615–1621.

(60) Hoang, L.; Bahmanyar, S.; Houk, K. N.; List, B. Kinetic and Stereochemical Evidence for the Involvement of Only One Proline Molecule in the Transition States of Proline-Catalyzed Intra- and Intermolecular Aldol Reactions. *J. Am. Chem. Soc.* **2003**, *125*, 16–17.

(61) Clemente, F. R.; Houk, K. N. Computational Evidence for the Enamine Mechanism of Intramolecular Aldol Reactions Catalyzed by Proline. *Angew. Chem., Int. Ed. Engl.* **2004**, *43*, 5766–5768.

(62) Allemann, C.; Gordillo, R.; Clemente, F. R.; Cheong, P. H.-Y.; Houk, K. N. Theory of Asymmetric Organocatalysis of Aldol and Related Reactions: Rationalizations and Predictions. *Acc. Chem. Res.* **2004**, *37*, 558–569.

(63) Pidathala, C.; Hoang, L.; Vignola, N.; List, B. Direct Catalytic Asymmetric Enolexo Aldolizations. *Angew. Chem., Int. Ed. Engl.* **2003**, *42*, 2785–2788.

(64) List, B.; Lerner, R. A.; Barbas, C. F., III. Proline-Catalyzed Direct Asymmetric Aldol Reactions. *J. Am. Chem. Soc.* **2000**, *122*, 2395–2396.

(65) Notz, W.; List, B. Catalytic Asymmetric Synthesis of *anti*-1,2-Diols. *J. Am. Chem. Soc.* **2000**, *122*, 7386–7387.

(66) Northrup, A. B.; MacMillan, D. W. C. The First Direct and Enantioselective Cross-Aldol Reaction of Aldehydes. *J. Am. Chem. Soc.* **2002**, *124*, 6798–6799.

(67) Northrup, A. B.; Mangion, I. K.; Hettche, F.; MacMillan, D. W. C. Enantioselective Organocatalytic Direct Aldol Reactions of α-Oxyaldehydes: Step One in a Two-Step Synthesis of Carbohydrates. *Angew. Chem., Int. Ed. Engl.* **2004**, *43*, 2152–2154.

(68) Northrup, A. B.; MacMillan, D. W. C. Two-Step Synthesis of Carbohydrates by Selective Aldol Reactions. *Science* **2004**, *305*, 1752–1755.

(69) Pihko, P. M.; Erkkilä, A. Enantioselective Synthesis of Prelactone B Using a Proline-Catalyzed Crossed-Aldol Reaction. *Tetrahedron Lett.* **2003**, *44*, 7607–7609.

(70) Bahmanyar, S.; Houk, K. N. The Origin of Stereoselectivity in Proline-Catalyzed Intramolecular Aldol Reactions. *J. Am. Chem. Soc.* **2001**, *123*, 12911–12912.

(71) Bahmanyar, S.; Houk, K. N.; Martin, H. J.; List, B. Quantum Mechanical Predictions of the Stereoselectivities of Proline-Catalyzed Asymmetric Intermolecular Aldol Reactions. *J. Am. Chem. Soc.* **2003**, *125*, 2475–2479.

(72) List, B. The Direct Catalytic Asymmetric Three-Component Mannich Reaction. *J. Am. Chem. Soc.* **2000**, *122*, 9336–9337.

(73) Bahmanyar, S.; Houk, K. N. Origins of Opposite Absolute Stereoselectivities in Proline-Catalyzed Direct Mannich and Aldol Reactions. *Org. Lett.* **2003**, *5*, 1249–1251.

(74) Notz, W.; Tanaka, F.; Watanabe, S.-i.; Chowdari, N. S.; Turner, J. M.; Thayumanavan, R.; Barbas, F. C. The Direct Organocatalytic Asymmetric Mannich Reaction: Unmodified Aldehydes as Nucleophiles. *J. Org. Chem.* **2003**, *68*, 9624–9634.

(75) List, B.; Pojarliev, P.; Biller, W. T.; Martin, H. J. The Proline-Catalyzed Direct Asymmetric Three-Component Mannich Reaction: Scope, Optimization, and Application to the Highly Enantioselective Synthesis of 1,2-Amino Alcohols. *J. Am. Chem. Soc.* **2002**, *124*, 827–833.

(76) List, B. Direct Catalytic Asymmetric α-Amination of Aldehydes. *J. Am. Chem. Soc.* **2002**, *124*, 5656–5657.

(77) Bøgevig, A.; Juhl, K.; Kumaragurubaran, N.; Zhuang, W.; Jørgensen, K. A. Direct Organo-Catalytic Asymmetric α-Amination of Aldehydes—A Simple Approach to Optically Active α-Amino Aldehydes, α-Amino Alcohols, and α-Amino Acids. *Angew. Chem., Int. Ed. Engl.* **2002**, *41*, 1790–1793.

(78) Brown, S. P.; Brochu, M. P.; Sinz, C. J.; MacMillan, D. W. C. The Direct and Enantioselective Organocatalytic α-Oxidation of Aldehydes. *J. Am. Chem. Soc.* **2003**, *125*, 10808–10809.

(79) Hayashi, Y.; Yamaguchi, J.; Hibino, K.; Shoji, M. Direct Proline Catalyzed Asymmetric α-Aminooxylation of Aldehydes. *Tetrahedron Lett.* **2003**, *44*, 8293–8296.

(80) Zhong, G. A Facile and Rapid Route to Highly Enantiopure 1,2-Diols by Novel Catalytic Asymmetric α-Aminoxylation of Aldehydes. *Angew. Chem., Int. Ed. Engl.* **2003**, *42*, 4247–4250.

(81) Cheong, P. H.-Y.; Houk, K. N. Origins of Selectivities in Proline-Catalyzed α-Aminoxylations. *J. Am. Chem. Soc.* **2004**, *126*, 13912–13913.

(82) Córdova, A.; Engqvist, M.; Ibrahem, I.; Casas, J.; Sundén, H. Plausible Origins of Homochirality in the Amino Acid Catalyzed Neogenesis of Carbohydrates. *J. Chem. Soc., Chem. Commun.* **2005**, 2047–2049.

(83) Klussmann, M.; Mathew, S. P.; Iwamura, H.; Wells, D. H. J.; Armstrong, A.; Blackmond, D. G. Kinetic Rationalization of Nonlinear Effects in Asymmetric Catalysis Based on Phase Behavior. *Angew. Chem., Int. Ed. Engl.* **2006**, *45*, 7989–7992.

(84) Basavaiah, D.; Rao, A. J.; Satyanarayana, T. Recent Advances in the Baylis–Hillman Reaction and Applications. *Chem. Rev.* **2003**, *103*, 811–892.

(85) Price, K. E.; Broadwater, S. J.; Walker, B. J.; McQuade, D. T. A New Interpretation of the Baylis–Hillman Mechanism. *J. Org. Chem.* **2005**, *70*, 3980–3987.

(86) Aggarwal, V. K.; Mereu, A.; Tarver, G. J.; McCague, R. Metal- and Ligand-Accelerated Catalysis of the Baylis–Hillman Reaction. *J. Org. Chem.* **1998**, *63*, 7183–7189.

(87) Shi, M.; Chen, L.-H.; Li, C.-Q. Chiral Phosphine Lewis Bases Catalyzed Asymmetric aza-Baylis–Hillman Reaction of N-Sulfonated Imines with Activated Olefins. *J. Am. Chem. Soc.* **2005**, *127*, 3790–3800.

(88) Ameer, F.; Drewes, S. E.; Freese, S.; Kaye, P. T. Rate Enhancement Effects in the DABCO-Catalyzed Synthesis of Hydroxyalkenoate Esters. *Synth. Commun.* **1988**, *18*, 495–500.

(89) Basavaiah, D.; Sarma, P. K. S. Terminal Hydroxyalkyl Acrylates as Substrates for Baylis–Hillman Reaction. *Synth. Commun.* **1990**, *20*, 1611–1615.

(90) Matsui, K.; Takizawa, S.; Sasai, H. Bifunctional Organocatalysts for Enantioselective aza-Morita—Baylis–Hillman Reaction. *J. Am. Chem. Soc.* **2005**, *127*, 3680–3681.

(91) Shi, M.; Chen, L.-H. Chiral Phosphine Lewis Base Catalyzed Asymmetric aza-Baylis–Hillman Reaction of N-Sulfonated Imines with Methyl Vinyl Ketone and Phenyl Acrylate. *J. Chem. Soc., Chem. Commun.* **2003**, 1310–1311.

(92) Trost, B. M.; Radinov, R. On the Effect of a Cation Binding Site in an Asymmetric Ligand for a Catalyzed Nucleophilic Substitution Reaction. *J. Am. Chem. Soc.* **1997**, *119*, 5962–5963.

(93) Sawamura, M.; Nakayama, Y.; Tang, W.-M.; Ito, Y. Enantioselective Allylation of Nitro Group-Stabilized Carbanions Catalyzed by Chiral Crown Ether Phosphine–Palladium Complexes. *J. Org. Chem.* **1996**, *61*, 9090–9096.

(94) Trost, B. M.; Bunt, R. C. Asymmetric Induction in Allylic Alkylations of 3-(Acyloxy)cycloalkenes. *J. Am. Chem. Soc.* **1994**, *116*, 4089–4090.

(95) Nielsen, L. P. C.; Stevenson, C. P.; Blackmond, G. D.; Jacobsen, E. N. Mechanistic Investigation Leads to a Synthetic Improvement in the Hydrolytic Kinetic Resolution of Terminal Epoxides. *J. Am. Chem. Soc.* **2004**, *126*, 1360–1362.

(96) Ready, J. M.; Jacobsen, E. N. Asymmetric Catalytic Synthesis of α-Aryloxy Alcohols: Kinetic Resolution of Terminal Epoxides via Highly Enantioselective Ring-Opening with Phenols. *J. Am. Chem. Soc.* **1999**, *121*, 6086–6087.

(97) Hansen, K. B.; Leighton, J. L.; Jacobsen, E. N. On the Mechanism of Asymmetric Nucleophilic Ring-Opening of Epoxides Catalyzed by (Salen)Cr(III) Complexes. *J. Am. Chem. Soc.* **1996**, *118*, 10924–10925.

(98) Jacobsen, E. N. Asymmetric Catalysis of Epoxide Ring-Opening Reactions. *Acc. Chem. Res.* **2000**, *33*, 421–431.

(99) Sammis, G. M.; Jacobsen, E. N. Highly Enantioselective, Catalytic Conjugate Addition of Cyanide to α,β-Unsaturated Imides. *J. Am. Chem. Soc.* **2003**, *125*, 4442–4443.

(100) Myers, J. K.; Jacobsen, E. N. Asymmetric Synthesis of β-Amino Acid Derivatives via Catalytic Conjugate Addition of Hydrazoic Acid to Unsaturated Imides. *J. Am. Chem. Soc.* **1999**, *121*, 8959–8960.

(101) Taylor, M. S.; Jacobsen, E. N. Enantioselective Michael Additions to α,β-Unsaturated Imides Catalyzed by a Salen–Al Complex. *J. Am. Chem. Soc.* **2003**, *125*, 11204–11205.

(102) Schaus, S. E.; Jacobsen, E. N. Asymmetric Ring Opening of Meso Epoxides with TMSCN Catalyzed by (PyBox)lanthanide Complexes. *Org. Lett.* **2000**, *2*, 1001–1004.

(103) Sammis, G. M.; Danjo, H.; Jacobsen, E. N. Cooperative Dual Catalysis: Application to the Highly Enantioselective Conjugate Cyanation of Unsaturated Imides. *J. Am. Chem. Soc.* **2004**, *126*, 9928–9929.

(104) Imbriglio, J. E.; Vasbinder, M. M.; Miller, S. J. Dual Catalyst Control in the Amino Acid-Peptide-Catalyzed Enantioselective Baylis–Hillman Reaction. *Org. Lett.* **2003**, *5*, 3741–3743.

(105) Shi, M.; Jiang, J. K.; Li, C. Q. Lewis Base and Image-Proline Co-Catalyzed Baylis–Hillman Reaction of Arylaldehydes with Methyl Vinyl Ketone. *Tetrahedron Lett.* **2002**, *43*, 127–130.

(106) France, S.; Wack, H.; Hafez, A. M.; Taggi, A. E.; Witsil, D. R.; Lectka, T. Bifunctional Asymmetric Catalysis: A Tandem Nucleophile/Lewis Acid Promoted Synthesis of β-Lactams. *Org. Lett.* **2002**, *4*, 1603–1605.

(107) Taggi, A. E.; Hafez, A. M.; Lectka, T. α-Imino Esters: Versatile Substrates for the Catalytic, Asymmetric Synthesis of α- and β-Amino Acids and β-Lactams. *Acc. Chem. Res.* **2003**, *36*, 10–19.

(108) France, S.; Weatherwax, A.; Taggi, A. E.; Lectka, T. Advances in the Catalytic, Asymmetric Synthesis of β-Lactams. *Acc. Chem. Res.* **2004**, *37*, 592–600.

(109) France, S.; Shah, M. H.; Weatherwax, A.; Wack, H.; Roth, J. P.; Lectka, T. Bifunctional Lewis Acid-Nucleophile-Based Asymmetric Catalysis: Mechanistic Evidence for Imine Activation Working in Tandem with Chiral Enolate Formation in the Synthesis of β-Lactams. *J. Am. Chem. Soc.* **2005**, *127*, 1206–1215.

(110) Pearson, R. G. Hard and Soft Acids and Bases. *J. Am. Chem. Soc.* **1963**, *85*, 3533–3539.

(111) Firestone, R. A.; Barker, P. L.; Pisano, J. M.; Ashe, B. M.; Dahlgren, M. E. Monocyclic β-Lactam Inhibitors of Human Leukocyte Elastase. *Tetrahedron* **1990**, *46*, 2255–2262.

(112) Wack, H.; France, S.; Hafez, A. M.; Drury, W. J., III; Weatherwax, A.; Lectka, T. Development of a New Dimeric Cyclophane Ligand: Application to Enhanced Diastereo- and Enantioselectivity in the Catalytic Synthesis of β-Lactams. *J. Org. Chem.* **2004**, *69*, 4531–4533.

(113) Aspinall, H. C. Chiral Lanthanide Complexes: Coordination Chemistry and Applications. *Chem. Rev.* **2002**, *102*, 1807–1850.

(114) Shibasaki, M.; Gröger, H. Chiral Heterobimetallic Lanthanoid Complexes: Highly Efficient Multifunctional Catalysts for the Asymmetric Formation of C–C, C–O and C–P Bonds. *Top. Organometal. Chem.* **1999**, *2*, 200–232.

(115) Shibasaki, M.; Sasai, H.; Arai, T. Asymmetric Catalysis with Heterobimetallic Compounds. *Angew. Chem., Int. Ed. Engl.* **1997**, *36*, 1236–1256.

(116) Wooten, A. J.; Carroll, P. J.; Walsh, P. J. Evidence for Substrate Binding by the Lanthanide Centers in Li$_3$(THF)$_n$(BINOLate)$_3$Ln: Solution and Solid-State Characterization of 7- and 8-Coordinate Li$_3$(sol)$_n$(BINOLate)$_3$Ln(S)$_m$ Adducts. *Angew. Chem., Int. Ed. Engl.* **2006**, *45*, 2549–2552.

(117) Cotton, F. A.; Wilkinson, G. *Advanced Inorganic Chemistry*, 5th ed.; Wiley: New York, 1988.

(118) von Zelewsky, A. *Stereochemistry of Coordination Compounds*; Wiley: New York, 1996.

(119) Aspinall, H. C.; Bickley, J. F.; Dwyer, J. L. M.; Greeves, N.; Kelly, R. V.; Steiner, A. Pinwheel-Shaped Heterobimetallic Lanthanide Alkali Metal Binaphtholates: Ionic Size Matters! *Organometallics* **2000**, *19*, 5416–5423.

(120) Aspinall, H. C.; Greeves, N. Defining Effective Chiral Binding Sites at Lanthanides: Highly Enantioselective Reagents and Catalysts from Binaphtholate and PyBox Ligands. *J. Organomet. Chem.* **2002**, *647*, 151–157.

(121) Sasai, H.; Suzuki, T.; Itoh, N.; Tanaka, K.; Date, T.; Okamura, K.; Shibasaki, M. Catalytic Asymmetric Nitroaldol Reaction Using Optically Active Rare Earth BINOL Complexes: Investigation of the Catalyst Structure. *J. Am. Chem. Soc.* **1993**, *115*, 10372–10373.

(122) Gröger, H.; Vogl, E. M.; Shibasaki, M. New Catalytic Concepts for the Asymmetric Aldol Reaction. *Chem. Eur. J.* **1998**, *4*, 1137–1333.

(123) Yamada, Y. M. A.; Yoshikawa, N.; Sasai, H.; Shibasaki, M. Direct Catalytic Asymmetric Aldol Reactions of Aldehydes with Unmodified Ketones. *Angew. Chem., Int. Ed. Engl.* **1997**, *36*, 1871–1873.

(124) Schinzer, D.; Limberg, A.; Bauer, A.; Böhm, O. M.; Cordes, M. Total Synthesis of (–)-Epothilone A. *Angew. Chem., Int. Ed. Engl.* **1997**, *36*, 523–524.

(125) Kageyama, M.; Tamura, T.; Nantz, M. H.; Roberts, J. C.; Somfai, P.; Whritenour, D. C.; Masamune, S. Synthesis of Bryostatin 7. *J. Am. Chem. Soc.* **1990**, *112*, 7407–7408.

(126) Wooten, A. J.; Carroll, P. J.; Walsh, P. J. Impact of Na- and K-C π-Interactions on the Structure and Binding of M$_3$(sol)$_n$(BINOLate)$_3$Ln Complexes. *Org. Lett.* **2007**, *9*, 3359–3362.

(127) Yamagiwa, N.; Matsunaga, S.; Shibasaki, M. Heterobimetallic Catalysis in Asymmetric 1,4-Addition of O-Alkylhydroxylamine to Enones. *J. Am. Chem. Soc.* **2003**, *125*, 16178–16179.

(128) Yamagiwa, N.; Matsunaga, S.; Shibasaki, M. Mechanistic Studies of a Reaction Promoted by the {YLi$_3$[tris(binaphthoxide)]} Complex: Are Three 1,1'-Bi-2-naphthol Units in a Rare-Earth-Alkali-Metal Heterobimetallic Complex Necessary? *Angew. Chem., Int. Ed. Engl.* **2004**, *43*, 4493–4497.

(129) Di Bari, L.; Lelli, M.; Pintacuda, G.; Pescitelli, G.; Marchetti, F.; Salvadori, P. Solution versus Solid-State Structure of Ytterbium Heterobimetallic Catalysts. *J. Am. Chem. Soc.* **2003**, *125*, 5549–5558.

Asymmetric Catalysis with Enantiopure Substrates: Double Diastereoselection

Double diastereoselection occurs when an enantiopure substrate is employed in a reaction with an enantiopure reagent or catalyst *and* at least one new stereogenic unit is generated in the product. Under these circumstances there are two separate factors that can affect the forming stereocenter(s): 1) the extant substrate stereochemistry (internal stereocontrol) and 2) the asymmetric features of the chiral reagent/catalyst (external stereocontrol). These two factors can either act in concert, leading to a highly stereoselective process (matched), or be mutually opposed, leading to a poorly stereoselective process (mismatched). In some cases, the mismatch is so strong that the expected reaction pathway is kinetically unfavorable and entirely different products form instead. Under conditions where the substrate does not exert strong stereochemical control, complete external stereocontrol may occur. In such cases, the use of enantiomeric catalysts with the same substrate would then provide different diastereomers with high selectivity. This chapter begins with an introduction into internal stereoselection to provide an overview of the features that cause matched/mismatched interactions. The remainder describes double diastereocontrol in asymmetric catalysis, which is quite common, as well as the less common triple diastereocontrol.

13.1 Simple Diastereoselection

The concept of diastereoselective synthesis was first recognized by Emil Fischer,[1] wherein one stereogenic unit in a structure can be used to influence the creation of subsequent stereogenic units. With 2^n stereoisomers possible for molecules with n stereogenic units, this concept has profoundly effected organic synthesis.[2] There are two main types of diastereoselection, that arising from the reaction between achiral substrates (simple diastereoselection, **Figure 13.1**) and that arising from the reaction of a chiral substrate that gives rise to new stereogenic centers (substrate-controlled or internal diastereoselection, **Figure 13.2**). For both types of processes, a chiral catalyst can influence the resulting diastereoselectivity. The effect of an asymmetric catalyst on the first of these processes has been discussed (see Section A.3 in the Appendix and Section 2.1.2). In this chapter, the latter type of diastereoselection, between enantiopure substrates and asymmetric catalysts, will be outlined. To understand the challenges faced by catalysts (external stereocontrol) in such diastereoselective processes, a brief overview of substrate diastereoselection (internal stereocontrol) is provided below.

13.1.1 Diastereoselection with Chiral Substrates

In cyclic systems, diastereoselective processes are controlled by relatively straightforward steric, torsional, and stereoelectronic elements. For example, in reactions of

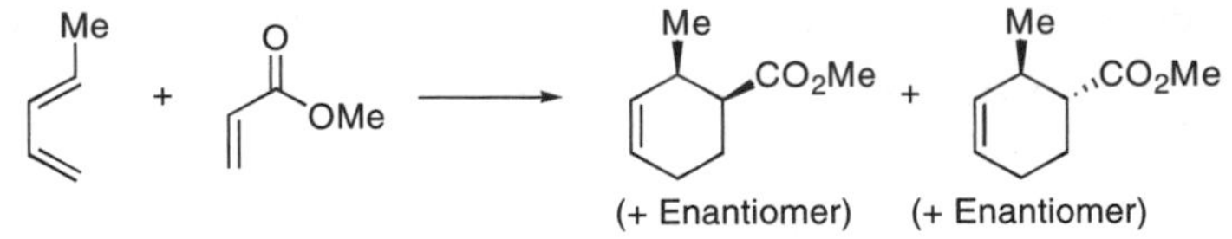

Conditions	Yield	*endo*	*exo*
120 °C (7 h)	50%	95	5
$AlCl_3$, 10–20 °C (3 h)	52%	64	36

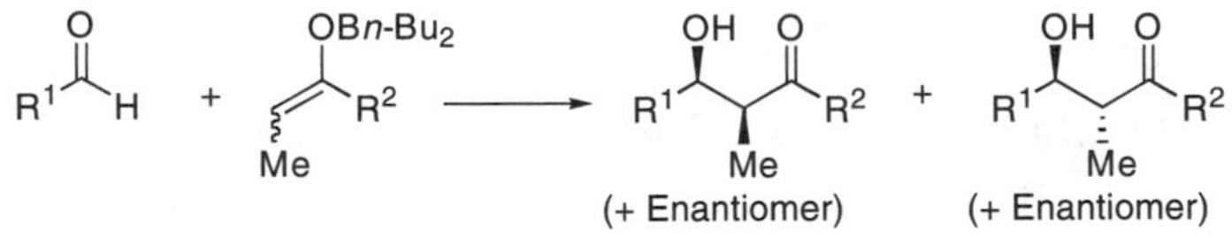

R^1	R^2	*Z:E*	Yield	*syn*	*anti*
Ph	Et	> 97:3	77%	> 97	< 3
i-Pr	S-*t*-Bu	< 5:95	63%	9	91

Figure 13.1. Simple diastereoselection with achiral substrates in Diels–Alder[3] and aldol[4] reactions.

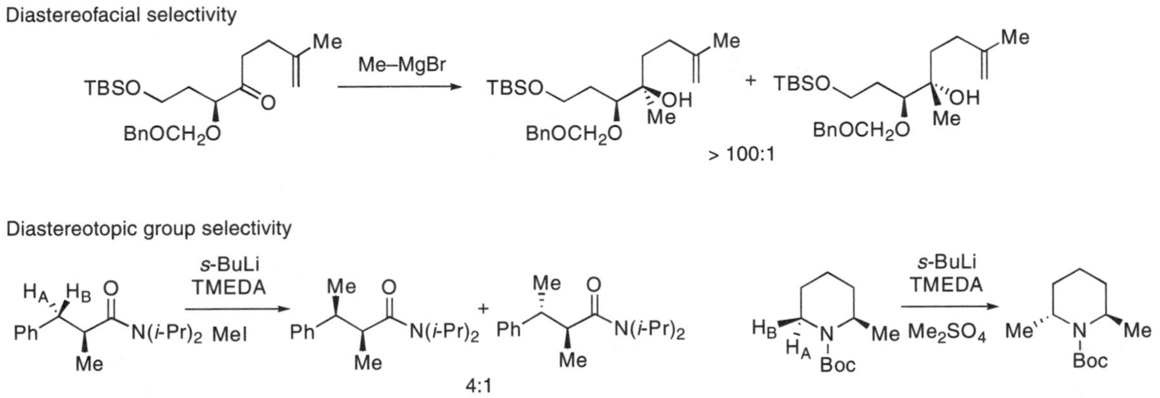

Figure 13.2. Diastereofacial[5] and diastereotopic group[6,7,8] selectivity with chiral substrates (internal diastereoselection).

4-*tert*-butylcylcohexanone, torsional effects control the addition of sterically small nucleophiles (**Equation 13.1**). [9]

Equation 13.1

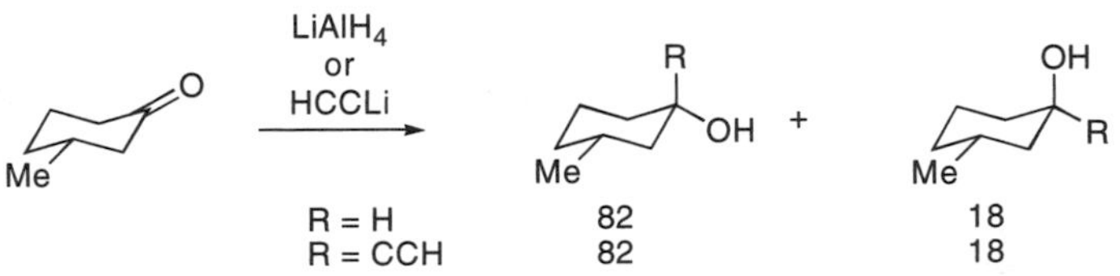

While the control elements in acyclic diastereoselective synthesis are less obvious, many selective processes have been discovered and have succumbed to analysis.[10,11] For example, nucleophilic additions to α-chiral aldehydes often occur with a high level of stereoselection, as in the case in **Equation 13.2.**[12] This type of 1,2-stereocontrol arises due to stereoelectronic factors that have been widely debated.[9,13]

For example, the Felkin-Anh model, which hinges on orbital overlap between σ^* $(R_L–C)$ and $\pi(C=O)$, is highly predictive and intuitively reasonable (**Figure 13.3**).[14,15]

Equation 13.2

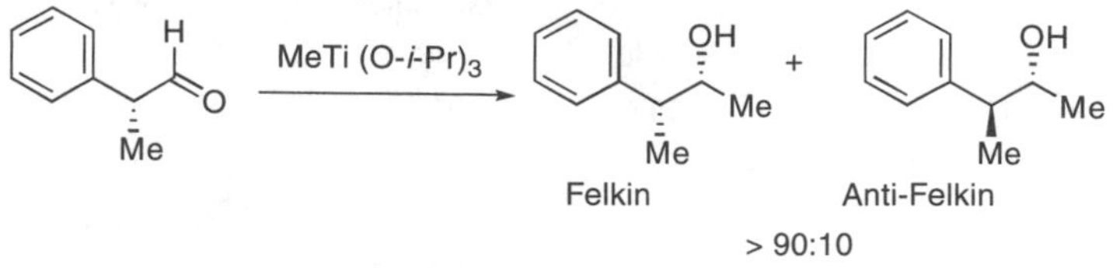

Felkin   Anti-Felkin

> 90:10

With substrates containing nearby coordinating groups, such as those in **Equation 13.3**,[16] metallic reagents allow another mode of 1,2-stereocontrol via chelation (**Figure 13.4**)[17] In essence, the metal center creates a cyclic system that allows effective transmission of stereochemical information. Since many chiral catalysts contain metal centers, chelation needs to be considered with such substrates. Chelation can be disfavored by adjusting the subsitutents on potential coordinating moieties; for example, bulky silyl ethers rarely undergo chelation, whereas alkyl ethers and smaller silyl ethers chelate readily.

Equation 13.3

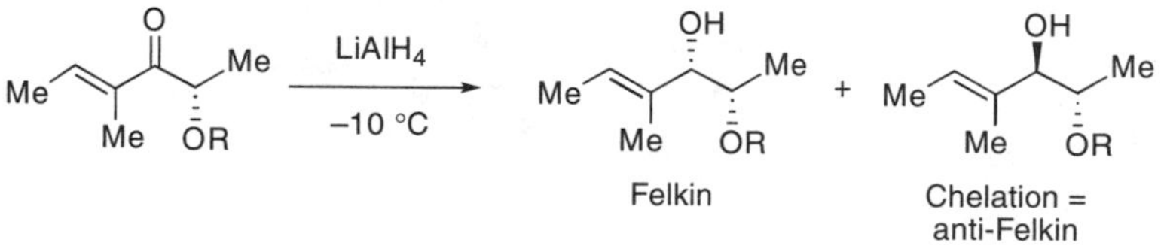

Felkin   Chelation = anti-Felkin

R = CH₂OBn	THF	30:70	Chelation
R = CH₂OBn	Et₂O	2:98	Chelation
R = SiPh₂-t-Bu	THF	95:5	Felkin: R_L=OR

With chelation, 1,3-stereocontrol (**Figure 13.5**) is also very reliable, as for the case in **Equation 13.4**.[18] Again, chelation needs to be considered with such substrates when using metal-derived chiral catalysts.

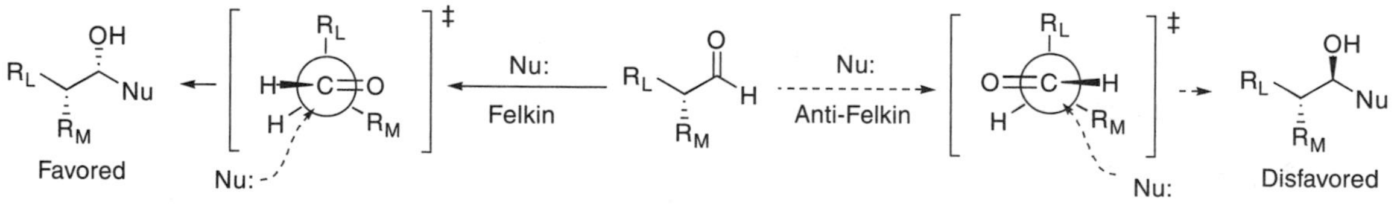

Figure 13.3. Felkin–Anh–Eisenstein model for 1,2-stereocontrol in the addition of nucleophiles to carbonyl compounds. R_L and R_M are the largest- and medium-sized substituents, respectively.

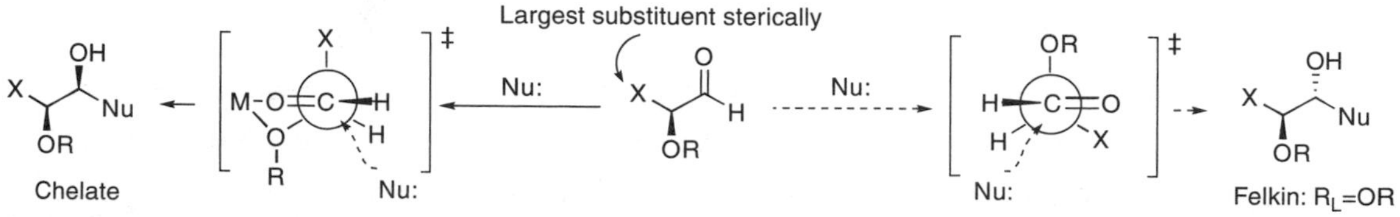

Figure 13.4. α-Chelate vs. Felkin–Anh–Eisenstein model for 1,2-stereocontrol in the addition of nucleophiles to carbonyl compounds.

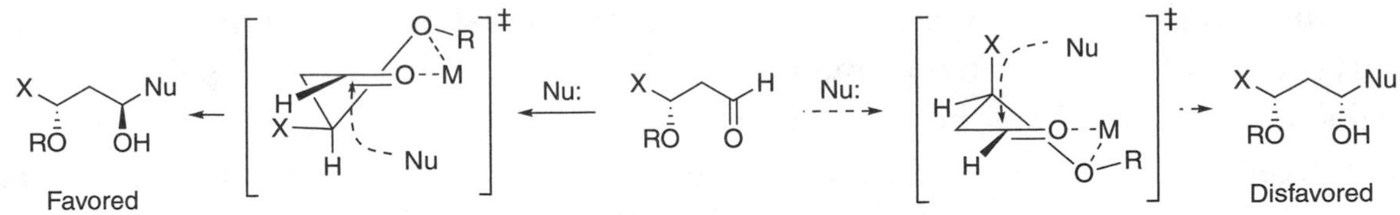

Figure 13.5. β-Chelation model for 1,3-stereocontrol in the addition of nucleophiles to carbonyl compounds.

Equation 13.4

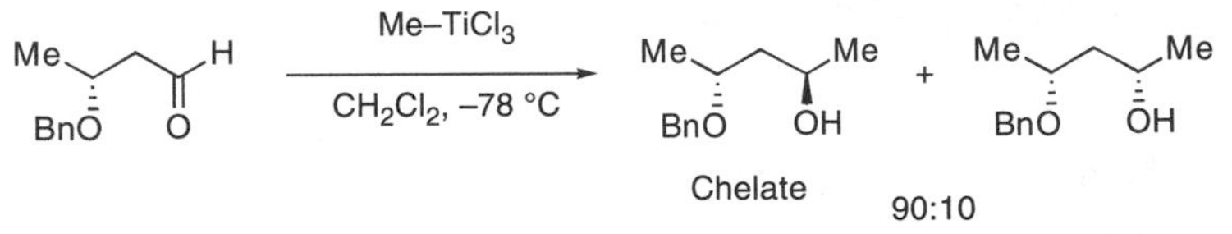

1,3-Diastereocontrol in the absence of chelation is also viable, as illustrated in **Equation 13.5**. Here the substituents at the stereogenic unit are not capable of undergoing chelation, yet reasonable levels of diastereoselectivity can be obtained.[19] Several models have been put forth to explain the outcome in this type of system.[19-26] In general, 1,3-diastereocontrol is typically weaker than 1,2-diastereocontrol and can be overcome by an asymmetric catalyst.

Equation 13.5

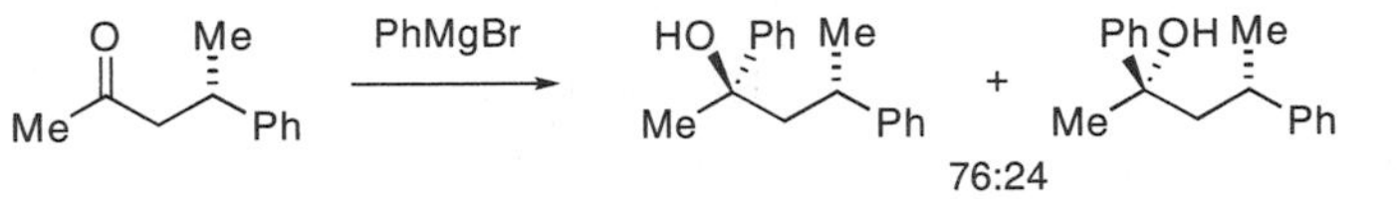

Short-range stereocontrol is not limited to carbonyl groups and has been observed in many systems. Examples that depend upon allylic strain to organize the conformation in a Diels–Alder reaction[27] and a dihydroxylation reaction,[28] respectively, are illustrated in **Figure 13.6**.

Asymmetric catalysts that can overcome the strong effects of short-range diastereocontrol are highly desirable. Unfortunately, internal 1,2- and 1,3-diastereocontrol

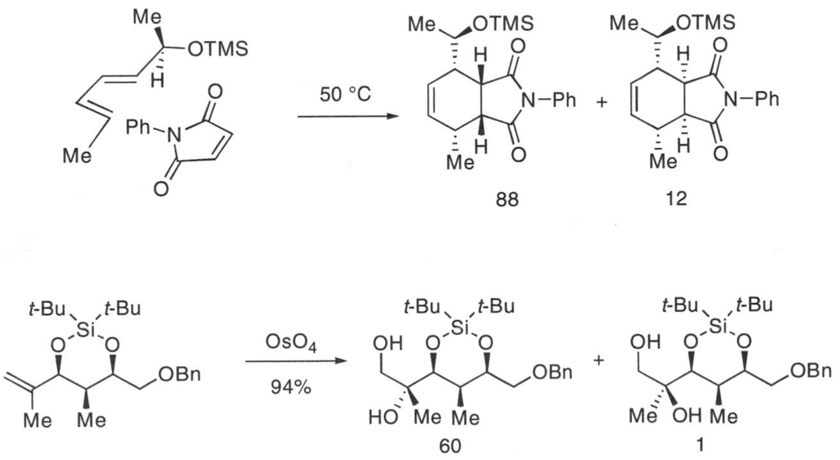

Figure 13.6. Short-range diastereocontrol in noncarbonyl additions.

elements can play a dominant role in determining diastereoselectivity, overriding catalyst control. The internal elements are covalently attached to the substrates by means of short and strong bonds, whereas most metal catalysts function via weaker and longer coordination bonds. When a steric clash arises between substituents at a stereogenic substrate unit and portions of a catalyst, the catalyst coordination is often compromised, resulting in low levels of catalyst stereocontrol. Chiral organocatalysts, many of which form covalent adducts (see Section 2.2) with one substrate, may exert greater levels of diastereocontrol.

13.1.2 Substrate-Directed Reactions

Directed reactions are powerful, in that a functional group on a substrate is used to position a reagent or catalyst to subsequently direct some aspect of the regiochemistry or stereochemistry of a reaction. Examples in which both of these aspects are controlled are illustrated in the epoxidation[29] and hydrogenation[30] reactions in **Figure 13.7**. More distal functional groups can also provide highly effective stereocontrol and regiocontrol.[31]

Directed reactions are not limited to oxidation and reduction. An example, involving C–C bond formation, is illustrated with the cyclopropanation in **Figure 13.8**.[32] Here, an in situ formed zinc alkoxy group is directing the approach of the reagent. Directing groups are not restricted to hydroxyl moieties, as a host of Lewis basic functional groups, including amides, esters, ethers, etc., have been demonstrated as useful.[31]

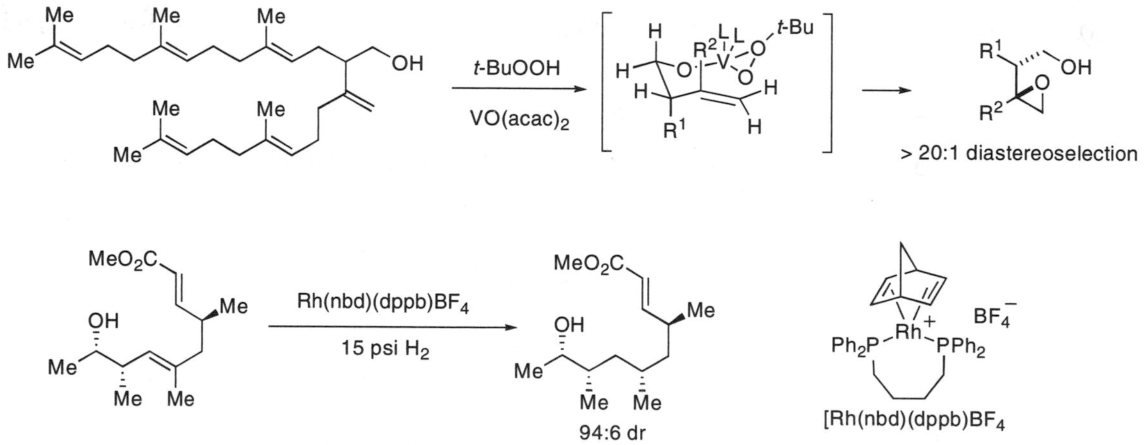

Figure 13.7. Regioselection and diastereoselection in substrate-directed oxidation and reduction.

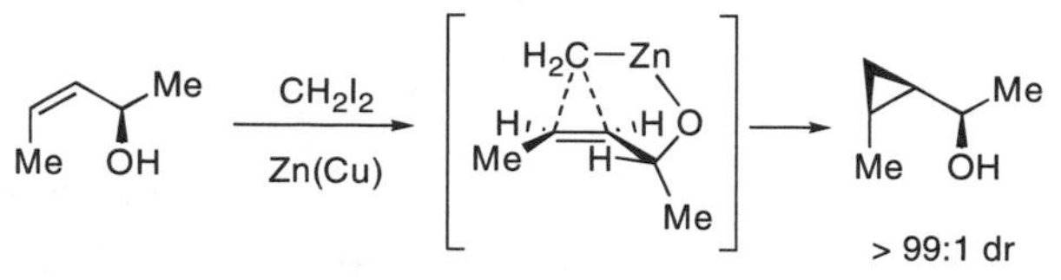

Figure 13.8. Diastereoselective substrate-directed reaction involving C–C bond formation.

Directing groups can also be crucial components in asymmetric catalytic reactions. Many examples are known in hydrogenation, a few of which have been discussed in prior chapters (see Sections 1.5 and 4.4.2). A well-known case outside of hydrogenation can be found in the asymmetric epoxidation[33,34] shown in **Equation 13.6**. Here, the hydroxyl group coordinates the titanium catalyst, bringing it into proximity to only one of the two alkenes. This proximity effect is crucial, as alkene substrates without a hydroxyl-directing group undergo little or no reaction. Furthermore, the chiral catalyst imparts a large enantiofacial bias, leading to the illustrated epoxide in high yield and selectivity.

Equation 13.6

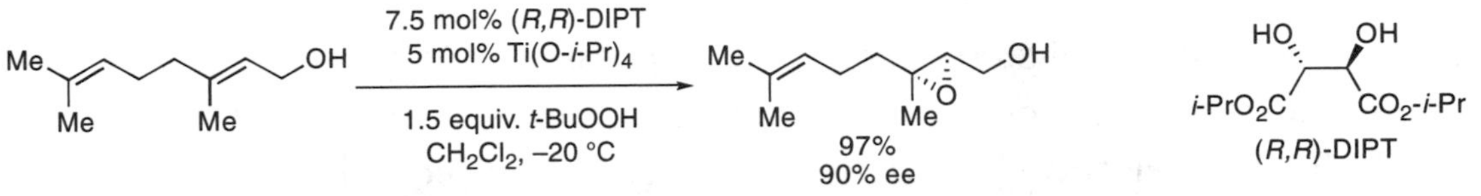

A classic example of a directed reaction can be found in the asymmetric isomerization of *N,N*-diethylgeranylamine[35] outlined in **Figure 13.9**. In the isomerization reaction, the amine directs hydrogen transfer to one face of the proximal alkene, causing regioselective and enantioselective isomerization to form citronellal (*E*)-enamine. This highly efficient process was implemented on a seven-ton scale for the commercial production of (–)-menthol and serves as a seminal effective application of a catalytic asymmetric reaction.

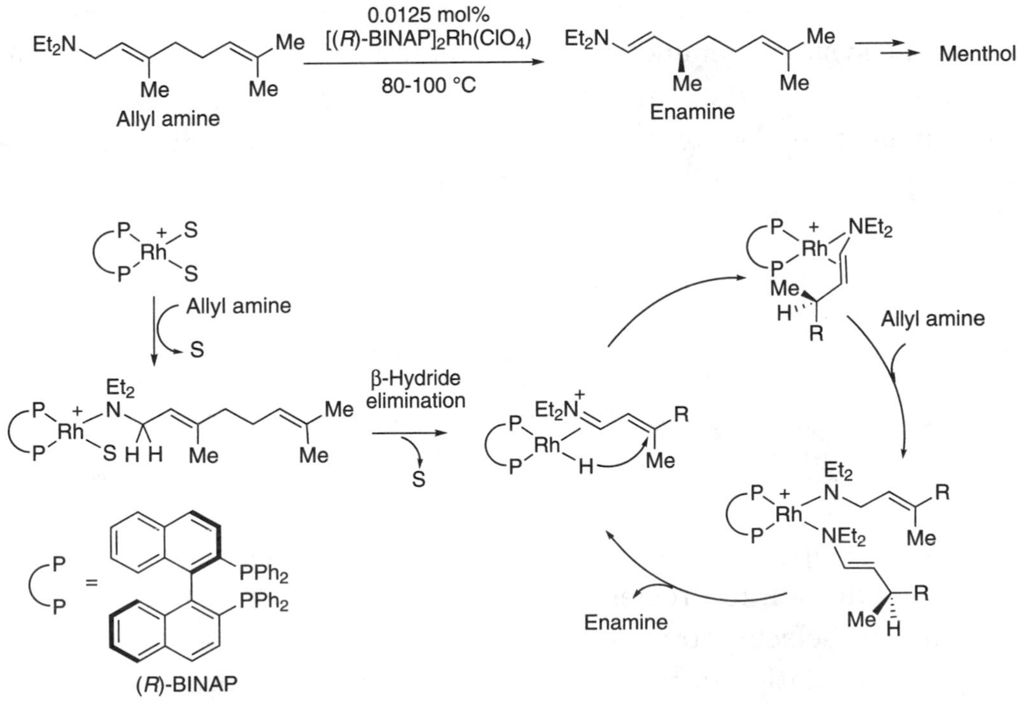

Figure 13.9. Regioselective substrate-directed isomerization with a chiral catalyst.

Substrate direction occurs in many reaction types. Directing groups exert strong and regular stereochemical effects that often predominate when enantiopure catalysts are employed with enantiopure substrates. Such strong substrate control can lead to dramatic matched versus mismatched cases. An undesirable consequence is that some stereochemical arrays may not be accessible (the mismatched cases). On the other hand, when this difference in reaction rate between the matched and mismatched is particularly large, kinetic resolutions become viable with the corresponding racemic mixture (see Chapters 7–9).

13.1.3 Remote Stereocontrol

Stereocontrol from distal stereogenic units is possible, but becomes less reliable the greater the distance between the stereodirecting group and the reactive center.[36] Thus, 1,2- and 1,3-stereoselection (see Section 13.1.1) can be employed with a fairly high degree of reliability, but 1,4-stereoselection and higher cannot. Remote stereoselection typically occurs when some type of long-range organization is possible and is most frequently utilized in cyclic systems where stereodirecting groups are held in proximity to the reacting center via the molecular framework. For these reasons, remote stereocontrol in acyclic systems is consequently more difficult.[36] Successful examples of 1,4-,[37] 1,5-,[38,39] 1,6-,[38,39] and 1,7-stereoselection[40] in acyclic systems[36] are known. In general, fewer examples are found the longer the distance between the reacting center and the stereodirecting groups. For example, 1,7- or further acyclic diastereocontrol is relatively rare. One such case is illustrated in **Equation 13.7**.[40] In these systems, the long-range organization is not imposed by a cyclic framework but by transient cyclic systems generated via chelation with the reagents. Here, internal coordination between the carbonyl and the boronate gives rise to a six-membered twist-boat chelate. External hydride delivery is then controlled by the substitution pattern on the twist boat, allowing high levels of diastereoselection (98.5:1.5 = 97% de).

Equation 13.7

The above examples of 1,4-, 1,5-, 1,6-, and 1,7-stereoselection all involve diastereoselective addition to prostereogenic carbonyls. Any prostereogenic unit can undergo diastereoselective transformations if the appropriate conditions prevail. For example, epoxidation of the alkene in **Equation 13.8** proceeds with 90:10 1,4-diastereoselection due to direction from the hydroxyl stereocenter.[41]

Equation 13.8

90 : 10

Diastereoselective control can also occur intermolecularly (**Equation 13.9**[42]) rather than intramolecularly (**Equations 13.1–13.8**). Here, the newly created stereogenic unit originates from a second substrate, the aldehyde, rather than from the substrate, the allylsilane, which contains the stereochemistry-controlling subunit.

Equation 13.9

1,4-*syn* (Racemic)   1,4-*anti* (Racemic)

TiCl$_4$, Et$_2$O	100 : 0
BF$_3$·OEt$_2$	13 : 87

While the examples in **Equations 13.1–13.9** involve diastereoselection from and to centrochiral units, other types of stereogenic units can act as diastereoselective-directing groups or can be created. **Equation 13.10** illustrates 1,5-diastereoselection from an axial chiral unit to a centrochiral unit.[43]

Equation 13.10

1) *s*-BuLi
2) PhCH=NMe

Single diastereomer

With certain systems even more remote diastereoselection is possible with 1,9-diastereoselection[44] and 1,23-diastereoselection having been demonstrated.[45] The latter case is particularly remarkable in that stereocontrol is exercised over 20 bond lengths via conformational changes relayed through the molecule. Similar concepts have been employed in the design of asymmetric catalysis, allowing the use of simpler and/or less expensive chiral catalysts that convey stereochemical information from distal catalyst sites to the reactive center (see Chapters 4 and 6).

Chiral catalysts can disrupt the weak organizing elements described above, such as chelation or geared interactions, allowing independent-catalyst stereocontrol with a high degree of fidelity in many cases (see Section 13.2.6). Independent-catalyst stereocontrol is readily achieved when additional stereocenters are isolated from the center of reaction by an intervening structural feature. For example, the arene unit in the asymmetric hydrogenation of bis(styrene) in **Figure 13.10** effectively insulates the stereochemical result in the enantioselective hydrogenation of the left-most alkene from influencing the hydrogenation of the right-most alkene.[46] As a consequence,

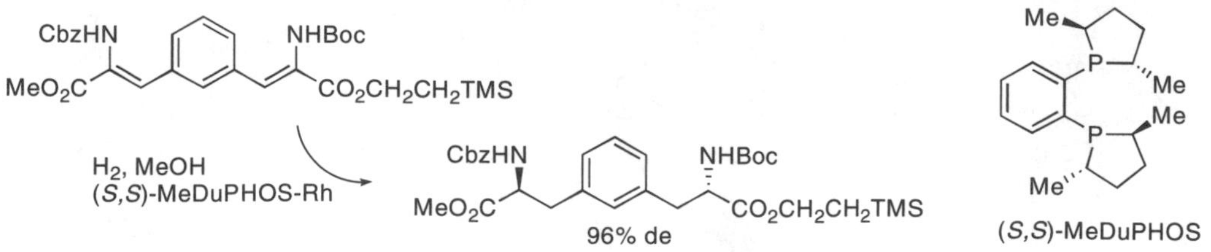

Figure 13.10. Example of stereocontrol independent of distal stereocenters.

both hydrogenations are effectively dictated by the chiral rhodium catalysts, with the first being highly enantioselective and the second highly diastereoselective. For further examples, see Section 16.2.3.

13.2 Double Diastereocontrol

In this chapter, the discussion of double diastereocontrol will be confined to enantiomerically pure substrates. With racemates and a chiral catalyst, a substantial difference in matched and mismatched cases provides the underpinnings for kinetic resolutions (see Chapters 7–9).

When a new stereogenic unit is introduced in a reaction employing an enantiopure catalyst and an enantiopure substrate, diastereomeric products are produced. In such double stereodifferentiating reactions, stereochemical control elements from the substrate (internal diastereocontrol) and from the catalyst (external diastereocontrol) act either constructively or destructively. Depending on the situation, internal or external diastereocontrol may predominate and account for the observed product ratio. External catalyst stereocontrol with enantiopure substrates is easiest to achieve when additional stereocenters are remote. Substrate stereogenic units closely neighboring the reactive centers may strongly influence the stereochemical course of a reaction and override the ability of a catalyst to control stereochemistry (see Section 13.2.2). A highly desirable feature of a chiral catalyst is the ability to completely or nearly completely control the stereochemistry in such situations. Occasionally this goal is met, but more often the results are mixed, where both the substrate stereochemistry and the chiral catalyst contribute to the outcome. This is due to the additive nature of substrate control and catalyst control effects. Aside from the synthetic utility that arises from double stereodifferentiating reactions, the results can also be used to glean insight into the mechanism of a catalytic reaction, especially with respect to the means of stereochemical induction.

13.2.1 Additive Effects

The outcome in double diastereocontrol cases arises from the stereocontrol of the substrate and the catalyst being additive[47] (**Figure 13.11**). For a reaction with a *chiral* substrate and an *achiral* catalyst that produces the major product in a 10:1 diastereomeric

Reference reactions

$$A^* + B \xrightarrow{\text{Catalyst}} C^* + C^{*\prime}$$
$$10:1 \text{ dr}$$

$\Delta\Delta G^{\ddagger}$ (chiral substrate) ≈ 1.4 kcal/mol

$$A + B \xrightarrow{\text{Cat}^*} C + C'$$
$$10:1 \text{ er}$$

$\Delta\Delta G^{\ddagger}$ (chiral substrate) ≈ 1.4 kcal/mol

$$A + B \xrightarrow{\text{Ent-cat}^*} C + C'$$
$$1:10 \text{ er}$$

$\Delta\Delta G^{\ddagger}$ (ent chiral catalyst) ≈ 1.4 kcal/mol

Double stereodifferentiation

Matched reactant pair: Stereoinduction from both partners consonant

$$A^* + B \xrightarrow{\text{Cat}^*} C^* + C^{*\prime}$$
$$100:1 \text{ dr}$$

$\Delta\Delta G^{\ddagger}$ (rxn) $\approx \Delta\Delta G^{\ddagger}$ (chiral substrate) + $\Delta\Delta G^{\ddagger}$ (chiral catalyst) ≈ 2.8 kcal/mol

log [Product ratio] $\approx$ log [substrate ratio] + log [catalyst ratio]

[Product ratio] $\approx$ [substrate ratio] x [catalyst ratio]

Hence: [product ratio] $\approx$ [10] x [10] $\approx$ 100

Mismatched reactant pair: Stereoinduction from partners dissonant

$$A^* + B \xrightarrow{\text{Ent-cat}^*} C^* + C^{*\prime}$$
$$1:1 \text{ dr}$$

$\Delta\Delta G^{\ddagger}$ (rxn) $\approx \Delta\Delta G^{\ddagger}$ (chiral substrate) $-$ $\Delta\Delta G^{\ddagger}$ (chiral catalyst) ≈ 0 kcal/mol

Hence: [product ratio] $\approx$ [10] x [0.1] $\approx$ 1

Figure 13.11. Additive effects in double diastereoselective reactions.

ratio at room temperature, the energy difference between the two reaction pathways corresponds to 1.4 kcal/mol. This analysis requires kinetic control, as is the case with asymmetric catalysts, rather than thermodynamic control. Furthermore, suppose that a closely analogous reaction with an *achiral* substrate and *chiral* catalyst gives rise to a product with a 10:1 enantiomeric ratio. When the *chiral* catalyst is then combined with the *chiral* substrate, two different outcomes can arise. A *matched* case occurs when the inherent substrate control and the inherent catalyst selectivity both favor formation of the same configuration at the newly generated stereogenic unit(s). If the separate effects are strictly additive (1.4 kcal/mol + 1.4 kcal/mol = 2.8 kcal/mol), then the observed diastereoselection should be $(10 \times 10){:}(1 \times 1) = 100{:}1$. If the substrate and the catalyst generate the opposite configurations at the newly generated stereogenic unit, then they are *mismatched*. For the case under discussion, the diastereoselection should be $(10 \times 1){:}(1 \times 10) = 1{:}1$.

This additivity of the energies and the subsequent results are easily seen in reaction coordinate diagrams (**Figure 13.12**). In **Figure 13.12a**, the reaction of a reference *chiral* substrate is shown with an *achiral* catalyst. In **Figures 13.12b** and **13.12c**, the reactions of an *achiral* substrate are shown with the two enantiomers of a *chiral* catalyst; it is important that the substrate and catalysts in **Figures 13.12a–13.12c** be closely analogous. If the *chiral* substrate from **Figure 13.12a** and the *chiral* catalyst from **Figure 13.12b** are combined, the reaction diagram in **Figure 13.12d** results, leading to a larger energy difference between the two paths and a *matched* case. If the *chiral* substrate from **Figure 13.12a** and the *enantiomeric chiral* catalyst from **Figure 13.12c** are combined, the reaction diagram in **Figure 13.12e** results, leading to a smaller energy difference between the two paths and a *mismatched* case.

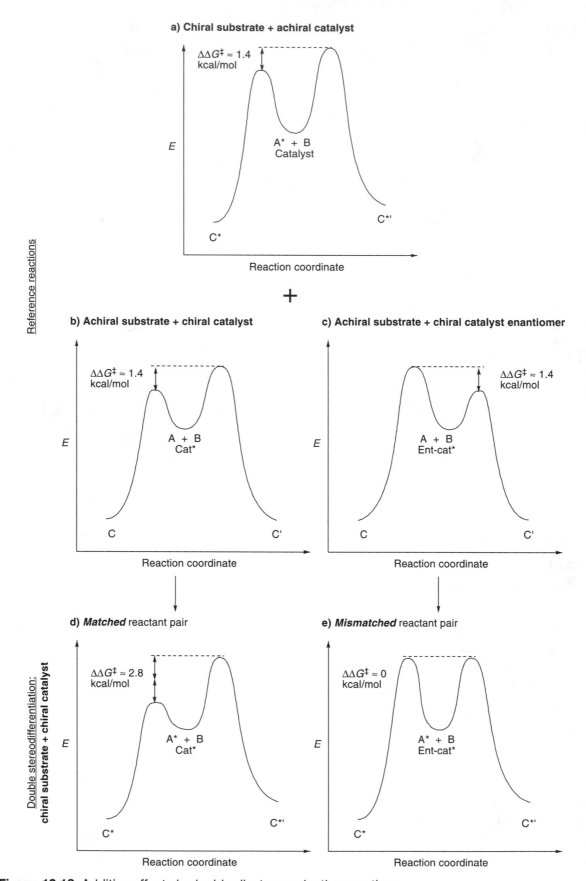

Figure 13.12. Additive effects in double diastereoselective reactions.

For the enantiomeric chiral substrate (A*′), the same reactivity patterns will be observed, except all of the stereochemistries and ratios will be mirrored. Finally, the reaction in **Figure 13.12d** shows that the major product is also the more thermodynamically stable compound (C*). Since these reactions are under kinetic control, it is also possible for situations to arise in which the barrier to the less thermodynamically stable compound (C*′) is smaller, leading to a predominance of C*′.

An example that follows the above precepts is illustrated in the asymmetric Diels–Alder reaction of α,β-unsaturated imides in **Figure 13.13**.[48] Here, the catalyst stereochemistry is held constant and the two substrates enantiomers are employed (c.f. **Figures 13.11** and **13.12**); nonetheless, the analysis is the same. The two reference

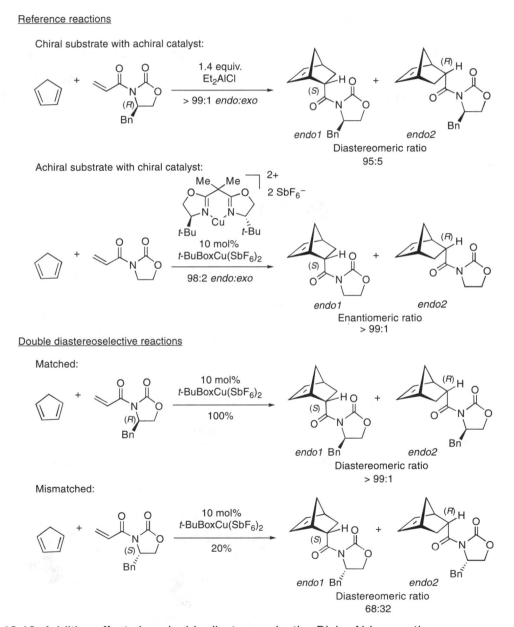

Figure 13.13. Additive effects in a double diastereoselective Diels–Alder reaction.

reactions illustrate the levels of chiral substrate control (95:5)[49] and chiral catalyst control (> 99:1) in closely related systems. The reaction of the (R)-substrate with the (S,S)-t-BuBox–copper catalyst gives rise to a *matched* case where a diastereomeric ratio of (95 × > 99):(1 × 1) = > 99:1 is estimated from the above data and is indeed observed. The reaction of the (S)-substrate with the (S,S)-t-BuBox–copper catalyst gives rise to a *mismatched* case where a diastereomeric ratio of (1 × > 99):(95 × 1) = > 51:49 is estimated from the above data. Qualitatively, the above result is in line with the observed 68:32 ratio. In both cases, catalyst control dominates, giving rise to (S)-*endo1* as the major adduct. Unfortunately, the selectivity levels in the *mismatched* case would not be synthetically useful.

13.2.2 Predominant Substrate Control

When chiral-catalyst stereocontrol is weak, a chiral substrate may exert a net controlling influence in an asymmetric transformation. In the example in **Figure 13.14**,[48] low enantioselection (65:35 er) is observed in the Diels–Alder reaction using a slightly dif-

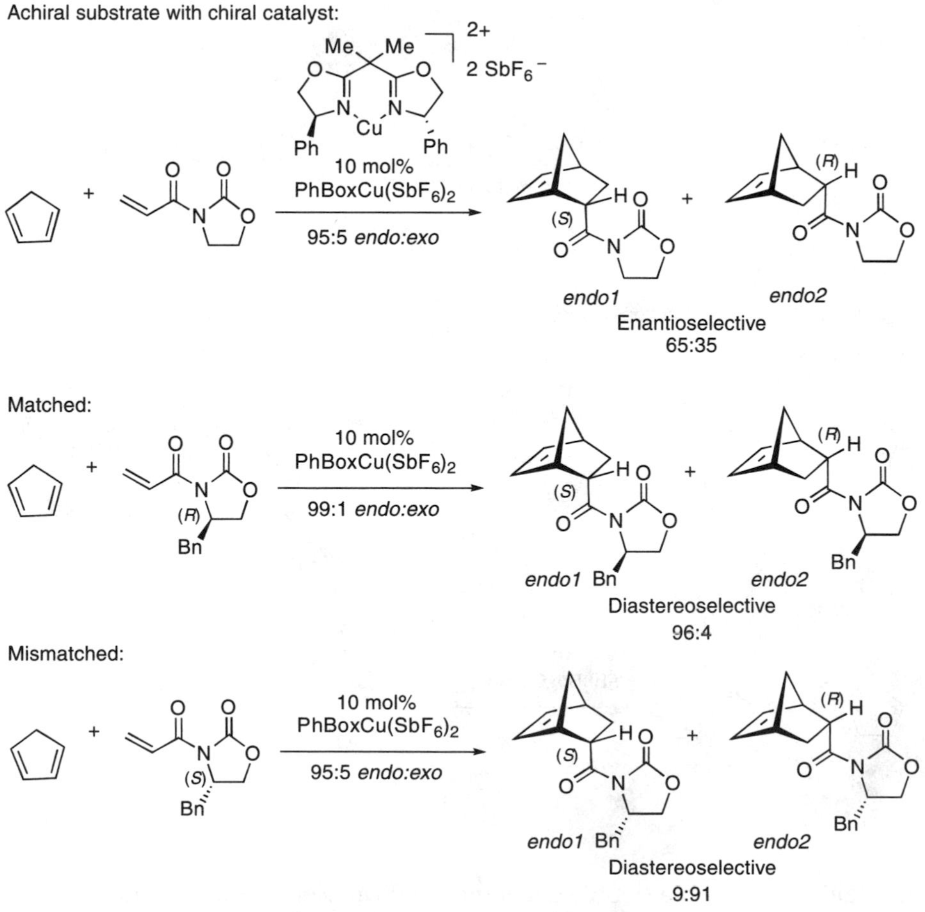

Figure 13.14. Substrate control in a double diastereoselective Diels–Alder reaction.

ferent chiral copper catalyst with the (*S,S*)-PhBox ligand. Recalling that the chiral substrate gives 95:5 diastereoselection (see **Figure 13.13**), it is expected that the substrate stereochemistry should control the stereochemical course of the reaction with a 97:2 ratio in the *matched* case and a 9:91 ratio in the *mismatched* case. The observed ratios of 96:4 and 9:91, respectively, are in close accord with these estimates. Such reactions with enantiopure substrates are typically not synthetically useful, since high selectivity arising solely from substrate control can be achieved without a costly chiral catalyst. An additional disadvantage of reactions with weak catalyst diastereocontrol is the inability to access diastereomeric arrays not available via substrate diastereocontrol.

13.2.3 Substrate Control Resulting in Kinetic Resolution

While the above case of substrate control in a diastereoselective process is typically not useful with enantiopure chiral catalysts and substrates, the utility with racemic substrates can be high. If the stereocontrol elements combine appropriately, the barriers to reaction in the mismatched case may be sufficiently high to allow a kinetic resolution (see Chapter 7). An example is illustrated in the asymmetric C–H insertion reactions in **Figure 13.15**.[50] In the reference reaction with achiral *N*-Boc pyrrolidine,

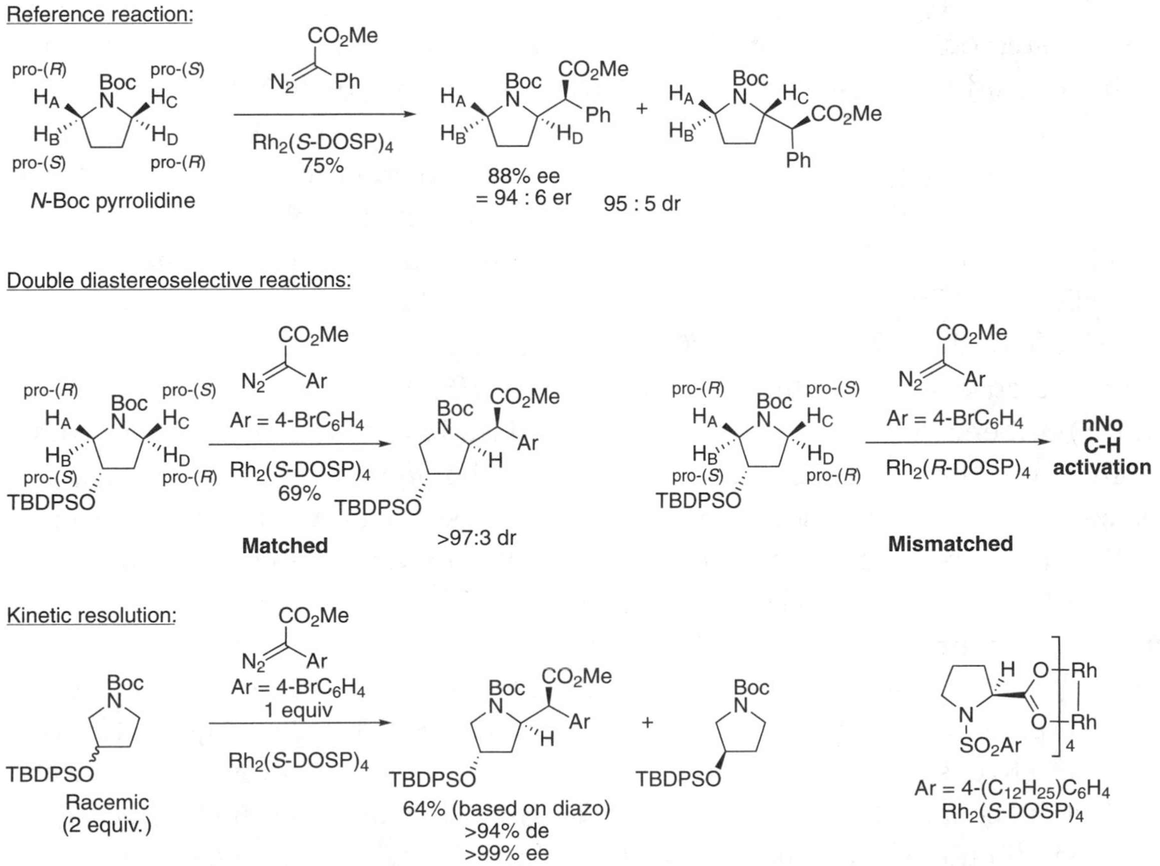

Figure 13.15. Kinetic resolution as a consequence of double diastereoselection.

the (S)-rhodium catalyst favors insertion into the pro-(S) hydrogen to yield the major diastereomer. With a chiral 3-*tert*-butyldiphenylsiloxy-substituted N-Boc pyrrolidine, the four C–H groups available for insertion (H_A, H_B, H_C, H_D) are all unique. The two C–H groups (H_A and H_B) flanking the 3-*tert*-butyldiphenylsiloxy (OTBDPS) group are sterically blocked from undergoing insertion, leaving only the two C–H groups (H_C and H_D) on the right-hand side of the molecule available for reaction. With the (S)-rhodium catalyst, pro-(S) insertion proceeds well (> 97:3 dr) as H_C is on the less sterically congested face of the ring away from the large OTBDPS substituent. In contrast, no reaction occurs with the (R)-rhodium catalyst because the pro-(R) insertion into H_D requires reaction on the same face of the ring as the large OTBDPS substituent. Here, the reaction barrier for one diastereomer in the *matched* case is sufficiently low that it can form. In contrast, the reaction barriers for both diastereomers in the *mismatched* case are high, such that no product is observed under the same conditions. These results are an exaggerated form of typical double diastereoselective processes, where the *matched* reactions are more rapid than the *mismatched* reactions (see **Figures 13.11–13.13**). Based on the above observations, a racemic mixture of the chiral N-Boc pyrrolidine was subjected to the chiral (S)-rhodium catalyst and a highly selective kinetic resolution was achieved.

13.2.4 Mixed Substrate and Catalyst Control

Double stereodifferentiating Mukaiyama aldol reactions have been carried out with (R)- and (S)-α-(benzyloxy)propionaldehyde (**Figure 13.16**) using the chiral PhPyBox–copper catalysts (**Figure 13.17**).[51] In the first reference reaction, $SnCl_4$-mediated addition of the silylketene acetal derived from *tert*-butyl thioacetate to α-(benzyloxy)propionaldehyde provides the chelation-controlled adduct with high selectivity (98:2).[52] In the second reference reaction, the chiral catalyst [(S,S)-(PhPyBox)Cu]$(SbF_6)_2$ forms a chelated substrate adduct and directs addition to the top *Si* face of the achiral α-(benzyloxy)acetaldehyde to yield the (S)-product. From these reference reactions, the combination of the [(S,S)-(PhPybox)Cu]$(SbF_6)_2$ catalyst and (S)-α-(benzyloxy)propionaldehyde should lead to a *matched* case where both the catalyst and the substrate dispose addition to the *Si* aldehyde face. In fact, (S)-α-(benzyloxy)propionaldehyde undergoes a rapid reaction, providing a 98.5:1.5 mixture of diastereomers, favoring the chelation-controlled product. This case highlights how the effects may not be perfectly additive, as a ratio of > 99.9:0.1 is anticipated using the analysis in **Figures 13.11** and **13.12**. Such a result arises because the reference reactions are not perfectly analogous to the double diastereoselective reactions. Finally, reaction of (R)-α-(benzyloxy)propionaldehyde catalyzed by [(S,S)-(PhPyBox)Cu]$(SbF_6)_2$ afforded an unselective, slow reaction (*mismatched*). It is noteworthy, however, that the chiral catalyst did change the diastereoselectivity significantly in the mismatched case, even though it did not provide the product in synthetically useful diastereoselectivity.

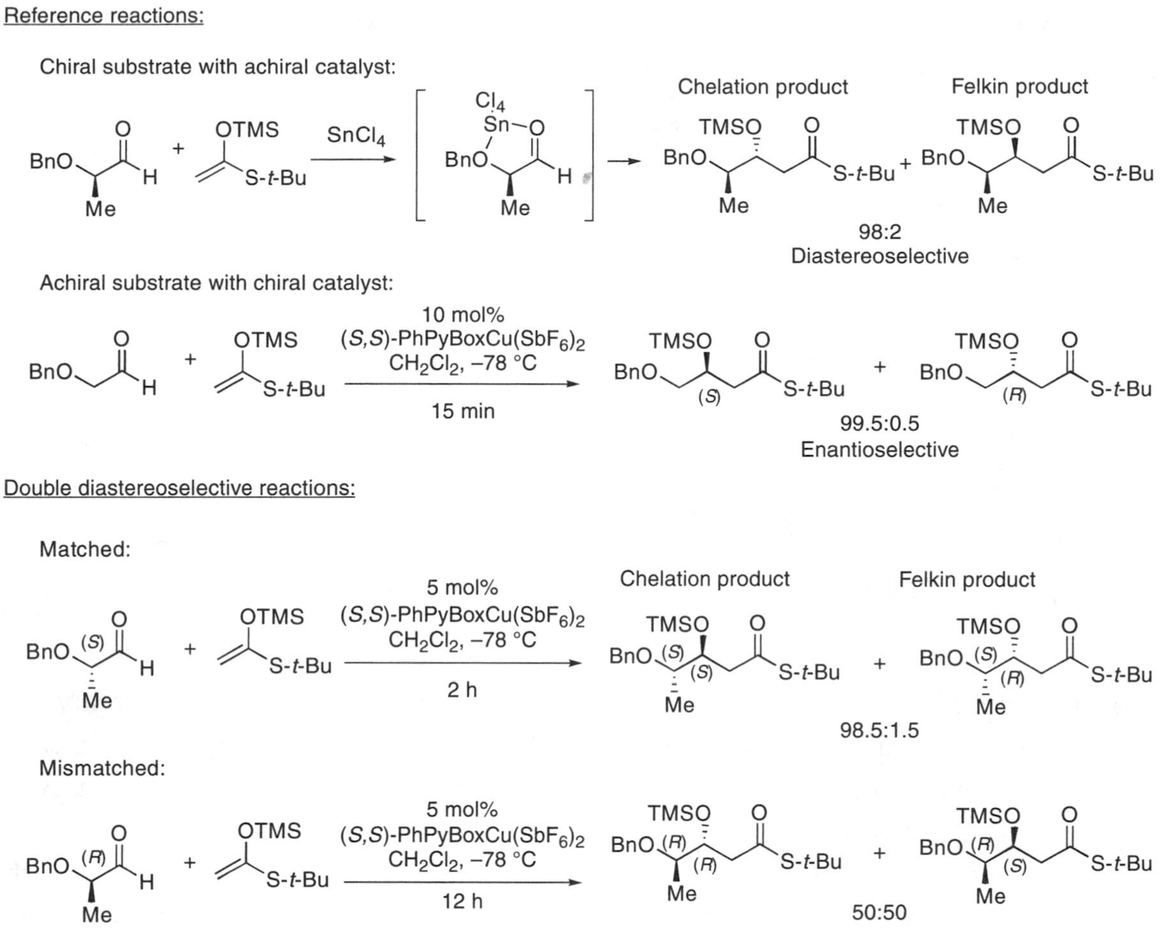

Reference reactions:

Chiral substrate with achiral catalyst:

Achiral substrate with chiral catalyst:

Double diastereoselective reactions:

Matched:

Mismatched:

Figure 13.16. Double diastereoselection in copper-catalyzed Mukaiyama aldol reactions.

13.2.5 Insight into Mechanism

The double stereodifferentiating experiments described above for the asymmetric Mukaiyama aldol reactions (**Figure 13.16**) also provide support for stereochemical models.[51] Specifically, a square-pyramidal catalyst–substrate model with equatorial carbonyl coordination and axial benzyloxy coordination is invoked to account for the observed stereoselection with achiral aldehydes (**Figure 13.17**, upper left). In accord with this model, reaction of (R)-α-(benzyloxy)propionaldehyde catalyzed by $[(S,S)$-(PhPyBox)Cu](SbF$_6$)$_2$ afforded an unselective, slow reaction (**Figure 13.16**, mismatched). This result is consistent with catalyst–substrate square-pyramidal coordination, where the substrate (Me) and ligand (Ph) substituents mask opposite aldehyde carbonyl enantiofaces (**Figure 13.17**). In the matched case, (S)-α-(benzyloxy) propionaldehyde underwent a rapid reaction, providing a 98.5:1.5 mixture of diastereomers, favoring the chelation-controlled product (**Figure 13.16**). In the square-pyramidal complex (**Figure 13.17**), the α-methyl substituent of (S)-α-(benzyloxy) propionaldehyde reinforces the facial bias imposed by the catalyst. A corollary to these experiments is that (R)-α-(benzyloxy)propionaldehyde would be anticipated to

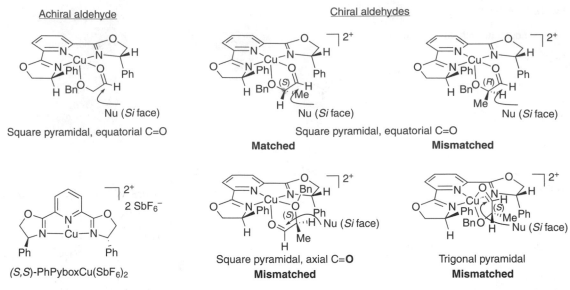

Figure 13.17. Mechanistic understanding of chiral copper catalysts from double diastereoselective reactions.

act as a catalyst inhibitor, since this enantiomer ideally complements the catalyst by orienting a Me group in the only open quadrant available in the square-pyramidal complex. Indeed, this has been shown to be the case, as demonstrated by the low reactivity of (R)-α-(benzyloxy)propionaldehyde. The manifestation of matched and mismatched reaction partners is consistent with this square pyramidal model. On the other hand, the diastereomeric square-pyramidal model with axial carbonyl coordination and the trigonal-bipyramidal model predict the opposite matched and mismatched relationships relative to those observed.

13.2.6 Predominant Catalyst Control

There is great synthetic utility when catalyst control predominates in the reaction of a chiral substrate with a chiral catalyst, as diastereomeric arrays can be established by the simple selection of a catalyst. Furthermore, there is no conjecture about the stereochemical outcome if catalyst control is certain and stereoregular. These features have allowed asymmetric catalysts to enjoy considerable utility with advanced chiral intermediates (see Section 16.3).

With the great progress in the development of asymmetric reduction and oxidation catalysts, numerous examples of highly diastereoselective processes have been reported using chiral substrates in these transformations. For example, the hydrogenation of the tetrasubstituted alkene in **Equation 13.11** was accomplished with a JOSIPHOS2–rhodium catalyst.[53,54] Here the substrate diastereocontrol from the α-methylbenzyl group is weak, such that a more costly chiral catalyst is more efficient on a large scale.

Equation 13.11

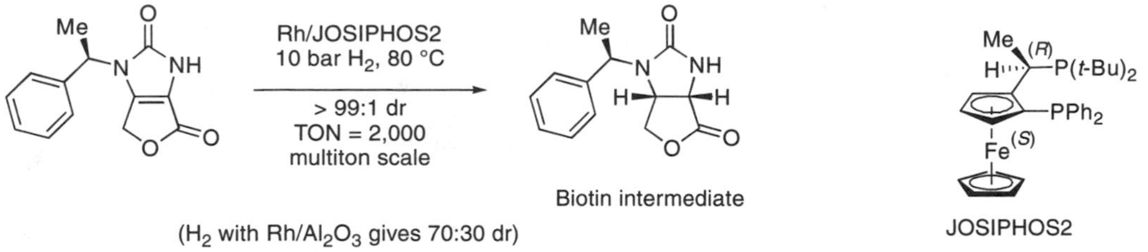

(H₂ with Rh/Al₂O₃ gives 70:30 dr)

The asymmetric transfer hydrogenation of acetylenic ketones with a pre-existing stereogenic center leads to diastereomeric propargylic alcohols. When the chiral ketone in **Figure 13.18** was reduced with isopropanol containing the (R,R)-ruthenium catalyst, the *anti* diastereomeric product was produced in > 97% yield with > 99% ee, together with small amounts of other stereoisomers.[55] In a similar manner, reduction with the enantiomeric (S,S)-ruthenium catalyst gave the *syn* diastereomeric product in > 97% yield with > 99% ee. Thus, the carbonyl diastereofaces in the chiral α,β-unsaturated ketone are efficiently differentiated by the chirality of the Ru template, while the adjacent nitrogen-substituted stereogenic center does not play any significant role. The reaction of the racemic ketone with the (R,R)-ruthenium catalyst indicated that the (S)-enantiomer is four times more reactive than the (R)-isomer. Thus, the matched case is the (R,R)-ruthenium catalyst with the (S)-ketone and the mismatched case is the (R,R)-ruthenium catalyst with the (R)-ketone. Here, both the matched and mismatched cases give rise to high stereoselectivity due to the dominant influence of the catalyst on the course of the reaction.

The concept of reagent control is also well-established in oxidation chemistry. For example, the complex chiral alkene in **Figure 13.19** undergoes asymmetric dihydroxylation with high stereoselectivity using the AD-mix catalysts.[56] With the AD-mix-β using (DHQD)₂PHAL, a ratio of greater than 20:1 in favor of the *anti* product is observed, while the opposite *syn* diastereomer dominates in a 10:1 ratio with the

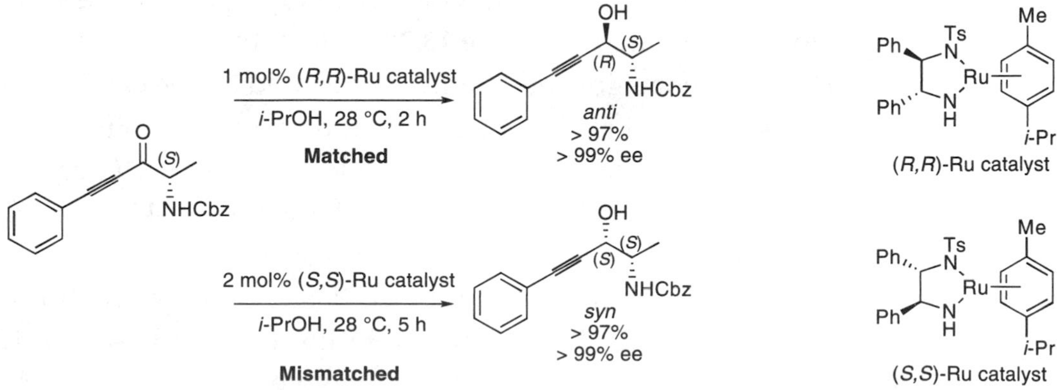

Figure 13.18. Double diastereocontrol in asymmetric catalytic reduction.

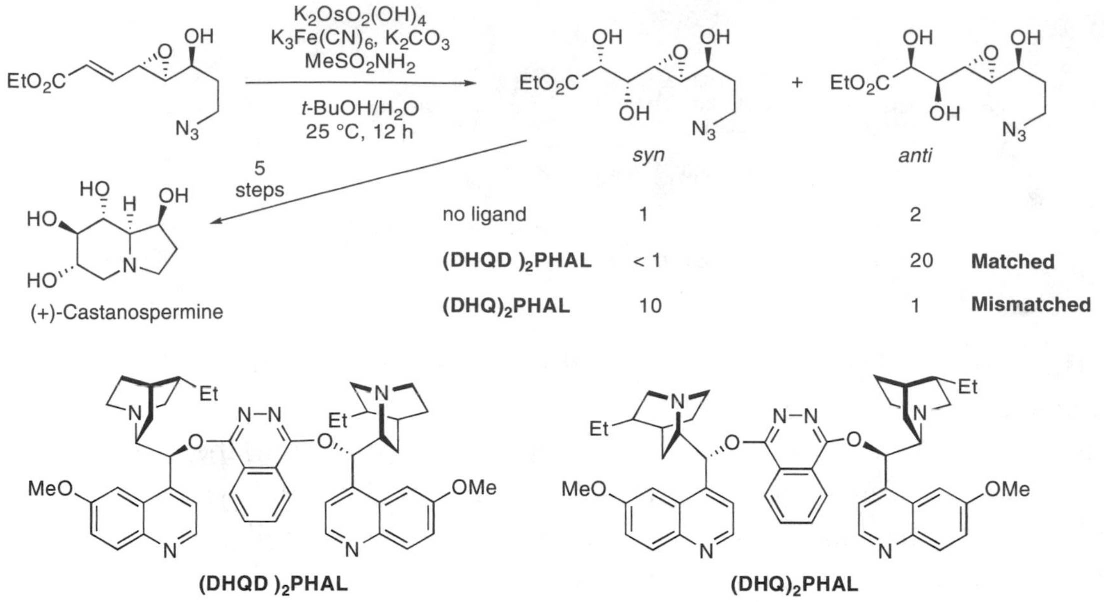

Figure 13.19. Double diastereocontrol in asymmetric catalytic oxidation.

AD-mix-β catalyst using (DHQ)₂PHAL. The *syn* diastereomer was successfully converted to the natural product (+)-castanospermine.

In this example, the substrate diastereocontrol is weak, favoring the *anti* diastereomer in only a 2:1 ratio when dihydroxylation is undertaken in the absence of a chiral ligand. On the other hand, the catalyst stereocontrol is substantially stronger; even in the mismatched case, useful product ratios (10:1 *syn:anti*) can be obtained. The difference in the stereoselectivity of the matched and mismatched cases is determined by the energetic differences exerted from the substrate and by the catalyst in approach to the two alkene diastereofaces. The energy difference from the substrate is small (~0.4 kcal/mol) while the energy differences with the catalysts are substantial (~1.4–1.8 kcal/mol).

Routine use of asymmetric catalysts in double diastereoselective reactions involving C–C bond formation has lagged these other processes. Nevertheless, many examples have been reported (see **Figure 13.13**). One involving a catalytic asymmetric phase-transfer alkylation is illustrated in **Figure 13.20**.[57] Here, the stereochemical outcome at the newly formed stereogenic center is clearly dictated by the catalyst stereochemistry and high stereoselection is observed with both substrates. However, the second reaction is the mismatched case, as the stereoselection is lower (11:1 vs. only one isomer) and, more importantly, the yield is substantially lower (34% vs. 89% over the same time period). This observation indicates a slower net reaction for the mismatched substrate and catalyst, indicating that both diastereomeric reaction pathways are higher in energy for the (R)-substrate relative to the (S)-substrate. As mentioned before, such a bias can allow for highly effective kinetic resolutions (see Chapters 7–9).

A C–C bond-forming reaction in which comparable yields were obtained in the matched and mismatched cases is illustrated in **Figure 13.21**.[58] In this hetero-

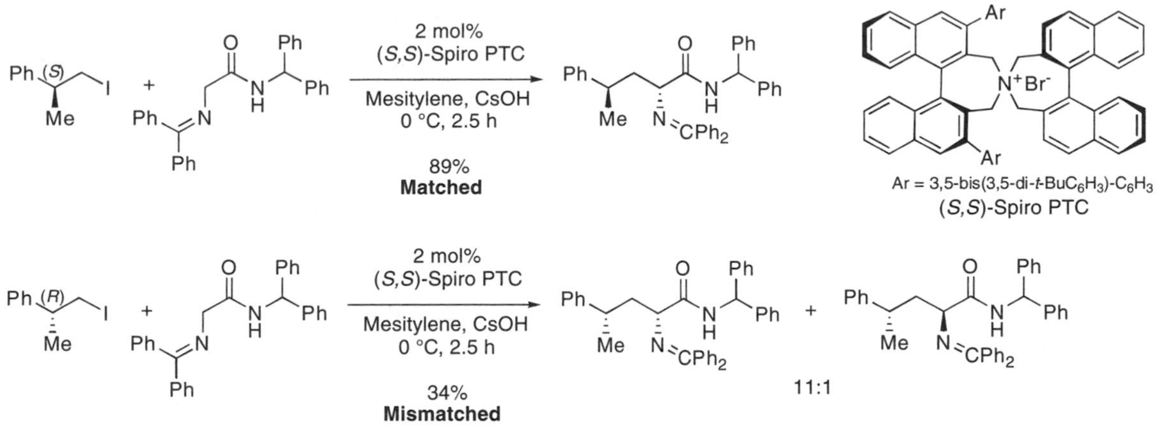

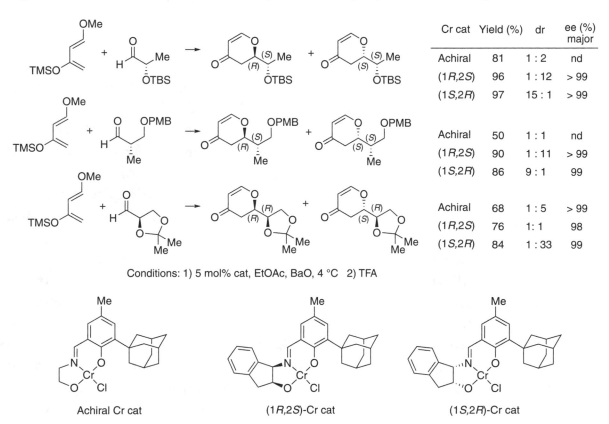

Figure 13.20. Double diastereocontrol in asymmetric catalytic phase-transfer alkylation.

Cr cat	Yield (%)	dr	ee (%) major
Achiral	81	1 : 2	nd
(1*R*,2*S*)	96	1 : 12	> 99
(1*S*,2*R*)	97	15 : 1	> 99
Achiral	50	1 : 1	nd
(1*R*,2*S*)	90	1 : 11	> 99
(1*S*,2*R*)	86	9 : 1	99
Achiral	68	1 : 5	> 99
(1*R*,2*S*)	76	1 : 1	98
(1*S*,2*R*)	84	1 : 33	99

Conditions: 1) 5 mol% cat, EtOAc, BaO, 4 °C   2) TFA

Achiral Cr cat   (1*R*,2*S*)-Cr cat   (1*S*,2*R*)-Cr cat

Figure 13.21. Double diastereocontrol in hetero-Diels–Alder reactions.

Diels–Alder reaction, results were obtained using a chiral aldehyde substrate with achiral and chiral chromium Lewis acid catalysts. For the first two examples, the substrate diastereocontrol is nominal, as illustrated by the results with the achiral catalyst. As such, the chiral catalysts allow generation of the product with high external stereocontrol of the newly formed stereogenic center (9–15:1 diastereomeric ratios). In addition, little racemization of the aldehyde substrate is observed under these conditions, as judged by the uniformly high enantiomeric excess (≥ 99%) of the major diastereomers. If the enantiomeric substrates were employed with the same catalysts, it would be possible to generate *any of the four product diastereomers selectively*. For the third example, 5:1 substrate (internal) diastereocontrol is observed, leading to mismatched (1:1 dr) and matched cases (1:33 dr). This case illustrates that the precise nature of the substrate has a profound effect on substrate diastereocontrol (see Section 13.1) and the outcome of a double diastereoselective process.

Reliability is a key feature when employing asymmetric catalysts in preparative double diastereoselective processes. Ideally, catalyst control is strong (external control dominates) and substrate control is weak. In other words, the catalyst should cause a larger $\Delta\Delta G^{\ddagger}$ between the diastereomeric pathways relative to that from the substrate. In addition, stereoregular catalysts are the most desirable, as the product stereochemistry can be estimated with a high degree of confidence. Further development of catalyst systems that can achieve these goals is needed.

13.2.7 Catalyst Regiocontrol vs. Diastereocontrol

An example in which an asymmetric catalyst influences regioselection rather than diastereoselection or enantioselection is shown in **Figure 13.22**.[59] Here, the substrate is chiral and the stereochemical course of azide attack is controlled by an S_N2 opening of the epoxide. However, attack of azide at either terminus of the epoxide leads to different regioisomeric products. For the first substrate, no regioselection is observed (1:1 ratio) when an achiral chromium–salen catalyst is employed, indicating that there is no strong bias toward attack of one terminus or the other. When chiral catalysts are employed, different regioisomers predominate (2:1 vs. 1:4) with the (R,R)- and (S,S)-salen catalysts, respectively.

The basis of the selectivity is readily understood if the same reactions of the meso epoxides are considered. For these reactions, the chiral salen complexes catalyze a selective cleavage of one of the enantiotopic C–O bonds. The enantiomeric salen complex catalyzes the cleavage of the other C–O bond, leading to the enantiomeric product. With enantioenriched chiral epoxides (**Figure 13.22**), the heterotopic C–O bonds may be considered pseudo-enantiotopic; the same selectivity principles then lead to enantiomeric catalysts affording stereocomplementary products. Because the C–O bonds are only pseudoenantiotopic, the substituents at the centers undergoing cleavage will also contribute to the reaction pattern. As a result, mismatched and matched reaction patterns are observed with the chiral catalysts with the (S,S)-salen catalyst reinforcing the bias observed with the achiral salen catalyst. In general, as the regioselection with the achiral catalyst increases, the disparity between the matched and mismatched cases increases.

In the above case, only regioisomers could be formed upon S_N2 opening with azide. More complicated situations can arise if multiple diastereomers as well as regioisomers are possible when using an enantioenriched substrate. For the case illustrated in **Figure 13.23**,[50] there are two possible regioisomers, each of which can exist as four diastereomers. Thus, as many as eight compounds are possible in the reaction

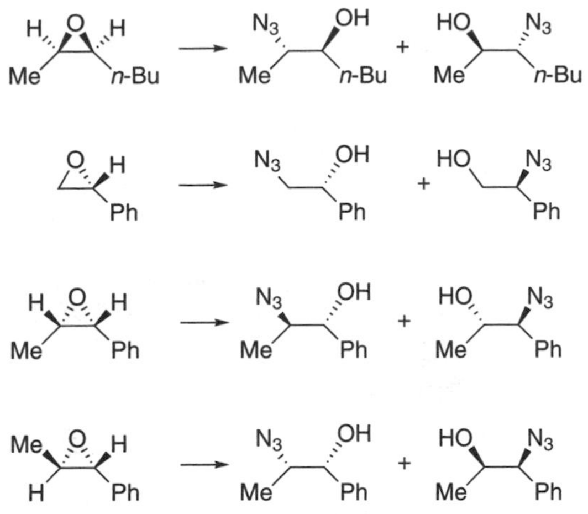

Cr cat	Yield (%)	Regioselection
achiral	81	1 : 1
(R,R)	79	2 : 1
(S,S)	77	1 : 4
achiral	88	4 : 1
(R,R)	87	1 : 7
(S,S)	87	18 : 1
achiral	73	3 : 1
(R,R)	85	1 : 4
(S,S)	70	45 : 1
achiral	89	1 : 9
(R,R)	85	1 : 1
(S,S)	88	1 : 84

Conditions: 5 mol% (salen)Cr, *t*-BuOMe, TMSN$_3$

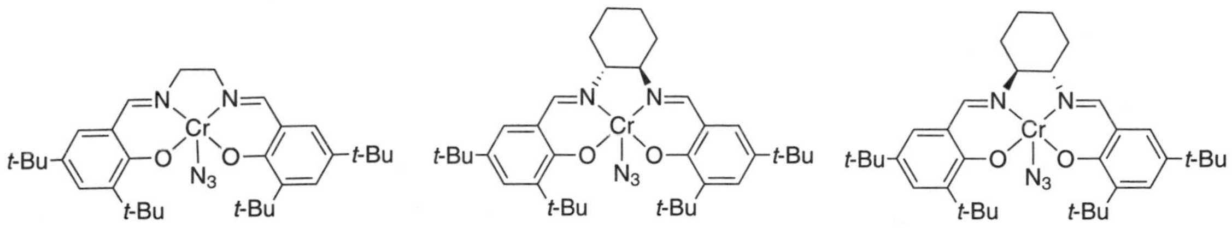

Achiral (salen)Cr   (R,R)-(salen)Cr   (S,S)-(salen)Cr

Figure 13.22. Regiocontrol dictated by an asymmetric catalyst in the azide ring-opening of chiral epoxides.

of the chiral material with the chiral catalyst. With this substrate, insertion into H$_D$ or H$_B$ is blocked sterically by the methoxy group. This leaves two pseudoenantiotopic hydrogens, pro-(R) H$_A$ and pro-(S) H$_C$. The Rh$_2$(R-DOSP)$_4$ catalyst is selective for pro-(R) hydrogens, leading to complete control of regiochemistry by insertion into H$_A$. Furthermore, there is a high degree of stereocontrol in formation of the second stereogenic center adjacent to the aryl group, enabling isolation of a single compound in 72% yield. With the Rh$_2$(S-DOSP)$_4$ catalyst, insertion into the pseudoenantiotopic pro-(S) H$_C$ predominates, leading to the illustrated product in 63% yield; however, other insertion products are also observed. The methoxy group adjacent to H$_C$ creates steric congestion, leading to less selective insertion in this mismatched case. Even so, the diastereoselection remains high.

This type of double asymmetric synthesis defines a unique strategy for effecting regioselective reactions of sterically and electronically unbiased substrates. The ability to generate regioisomers in organic transformations at will is as important as the generation of pure enantiomers and diastereomers. These transformations highlight the utility of asymmetric catalysts outside of conventional enantioselective processes. Application of the selectivity principles outlined above to racemic mixtures leads to parallel kinetic resolutions, as discussed further in Chapter 8.

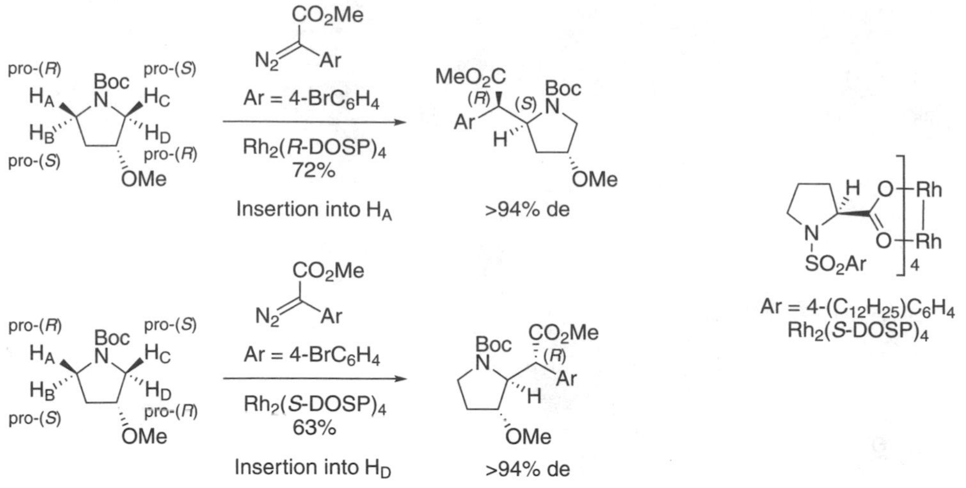

Figure 13.23. Chiral catalyst regiocontrol as well as double diastereocontrol in asymmetric C–H insertions.

13.3 Triple Diastereocontrol

While examples of double diastereocontrol in asymmetric catalysis are relatively common, triple diastereocontrol is quite rare. In triple diastereocontrol, three enantiopure components participate in a reaction. Since each component can exert some control on stereochemical events, the permutations of matched and mismatched cases are more complex. It is possible to have completely matched cases (highly stereoselective), partially matched cases (moderately stereoselective), and completely mismatched cases (poorly stereoselective).

As was the case for double diastereocontrol, one of the first examples of triple diastereocontrol can be found in the context of the aldol reaction (**Figure 13.24**). In this reaction utilizing a stoichiometric boron reagent,[60] a chiral ketone is converted to an enolate. When an achiral boron reagent is used to create this enolate, a fairly stereoselective addition (88:12 *anti:syn*) is observed with the illustrated chiral aldehyde. When chiral boron enolizing agents were employed, the results varied dramatically. With the (*S,S*)-dimethylborolane, the internal diastereoselection conferred by the substrates was eliminated. On the other hand, the (*R,R*)-dimethylborolane reinforced the internal diastereoselection and provided the product with a 96:4 *anti:syn* ratio. Additional examples of triple diastereoselection with stoichiometric asymmetric reagents have been reviewed.[61]

Triple diastereoselection with asymmetric catalysts is even less explored. A notable exception can be found in the Nozaki–Hiyama–Kishi couplings, for which catalytic enantioselective,[62] double diastereoselective,[62] and triple diastereoselective[63] reactions have been developed (**Figure 13.25**) in addition to triple diastereoselective[64] versions with stoichiometric chiral chromium reagents. The catalytic enantioselective reactions with achiral substrates proceed quite well (~90% ee), as illustrated in the first reaction equation of **Figure 13.25**). The ability to modulate conditions to successfully

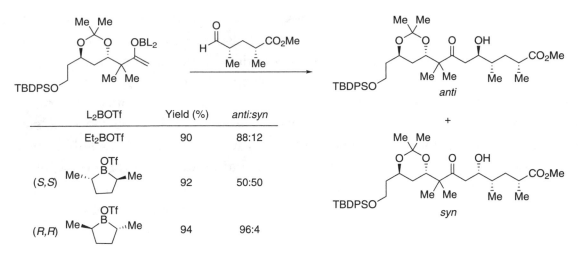

Figure 13.24. Triple diastereocontrol in asymmetric aldol chemistry.

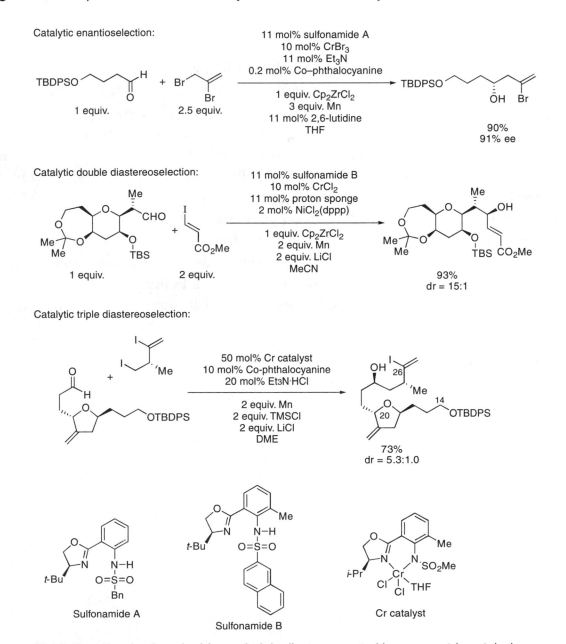

Figure 13.25. Enantioselective, double, and triple diastereocontrol in asymmetric catalysis.

add allyl, vinyl, and alkyl halides provides a broad scope to these Cr-catalyzed reactions. With a chiral aldehyde, the chiral catalyst also allows a highly diastereoselective reaction (15:1 dr; **Figure 13.25**, second equation). When a chiral aldehyde, chiral halide, *and* chiral catalyst were employed, good triple diastereoselection was observed (**Figure 13.25**, third equation). The stereochemical contribution from each individual component can not be assessed at this time, as results for all of the possible stereochemical permutations have not been reported with these complex substrates. As such, an understanding of the pertinent factors that would permit rational selection of substrates/catalysts awaits further study. Even so, this methodology is powerful, as demonstrated in the efficient synthesis of the C14–C26 segment of the halichondrins (**Figure 13.25**, third equation).[63]

Outlook

Asymmetric catalysts can be powerful, even with enantiomerically enriched substrates that are subject to diastereoselection. This double asymmetric synthesis is especially useful when stereochemical elements are distant from the site of reaction and provide nominal control otherwise. Double diastereoselective asymmetric reactions can also provide information about mechanism. The few instances of double asymmetric regioselection and triple asymmetric synthesis that have been reported highlight the developing areas in asymmetric catalysis with enantiopure substrates. The development of low-loading catalyst systems that dominate diastereoselection in double asymmetric synthesis would expand the utility of many transformations, as all possible diastereomers could be generated via the same process. To achieve this goal and rationally select substrate–catalyst combinations, a better understanding of the sterochemical control factors from substrate and catalyst components and their additivity, is needed.

References

(1) Fischer, E. Synthesen in der Zuckergruppe II. *Chem. Ber.* **1894**, *27*, 3189–3232.

(2) Eliel, E. L.; Wilen, S. H.; Mander, L. N. *Stereochemistry of Organic Compounds*; Wiley: New York, 1994.

(3) Inukai, T.; Kojima, T. Aluminum Chloride Catalyzed Diene Condensation. III. Reaction of *trans*-Piperylene with Methyl Acrylate. *J. Org. Chem.* **1967**, *32*, 869–871.

(4) Evans, D. A.; Nelson, J. V.; Vogel, E.; Taber, T. R. Stereoselective Aldol Condensations via Boron Enolates. *J. Am. Chem. Soc.* **1981**, *103*, 3099–3111.

(5) Still, W. C.; McDonald, J. H., III. Chelation-Controlled Nucleophilic Additions. 1. A Highly Effective System for Asymmetric Induction in the Reaction of Organometallics with α-Alkoxy Ketones. *Tetrahedron Lett.* **1980**, *21*, 1031–1034.

(6) Lutz, G. P.; Wallin, A. P.; Kerrick, S. T.; Beak, P. Complex Induced Proximity Effects: β-Lithiations of Carboxamides. *J. Org. Chem.* **1991**, *56*, 4938–4943.

(7) Beak, P.; Lee, W. K. α-Lithioamine Synthetic Equivalents: Syntheses of Diastereoisomers from the Boc Piperidines. *J. Org. Chem.* **1990**, *55*, 2578–2580.

(8) Beak, P.; Basu, A.; Gallagher, D. J.; Park, Y. S.; Thayumanavan, S. Regioselective, Diastereoselective, and Enantioselective Lithiation-Substitution Sequences: Reaction Pathways and Synthetic Applications. *Acc. Chem. Res.* **1996**, *29*, 552–560.

(9) Gung, B. W. Diastereofacial Selection in Nucleophilic Additions to Unsymmetrically Substituted Trigonal Carbons. *Tetrahedron* **1996**, *52*, 5263–5301.

(10) Nógrádi, M. *Stereoselective Synthesis*; VCH Publishers: New York, 1995.

(11) Bartlett, P. A. Stereocontrol in the Synthesis of Acyclic Systems—Applications to Natural Product Synthesis. *Tetrahedron* **1980**, *36*, 3–72.

(12) Reetz, M. T.; Steinbach, R.; Westermann, J.; Urz, R.; Wenderoth, B.; Peter, R. Stereoselectivity and Relative Reactivity in the Reaction of Organotitanium and Organozirconium Reagents with Carbonyl-Compounds. *Angew. Chem., Int. Ed. Engl.* **1982**, *21*, 135.

(13) Mengel, A.; Reiser, O. Around and Beyond Cram's Rule. *Chem. Rev.* **1999**, *99*, 1191–1223.

(14) Anh, N. T.; Eisenstein, O. Theoretical Interpretation of 1,2 Asymmetric Induction—Importance of Anti-Periplanarity. *Nouv. J. Chim.* **1977**, *1*, 61–70.

(15) Anh, N. T. Regio- and Stereo-Selectivities in Some Nucleophilic Reactions. *Top. Curr. Chem.* **1980**, *88*, 146–162.

(16) Overman, L. E.; McCready, R. J. Marine Natural-Products from the Atlantic Zone. 30. Highly Stereocontrolled Reduction of Alpha'-Alkoxyenones to Give Either the Threo or Erythro Allylic 1,2-Diol—Assignment of the Threo Configuration to the C-15, C-16 Diol of Pumiliotoxin-B. *Tetrahedron Lett.* **1982**, *23*, 2355–2358.

(17) Reetz, M. T. Structural, Mechanistic, and Theoretical Aspects of Chelation-Controlled Carbonyl Addition Reactions. *Acc. Chem. Res.* **1993**, *26*, 462–468.

(18) Reetz, M. T.; Jung, A. 1,3-Asymmetric Induction in Addition-Reactions of Chiral β-Alkoxy Aldehydes—Efficient Chelation Control via Lewis Acidic Titanium Reagents. *J. Am. Chem. Soc.* **1983**, *105*, 4833–4835.

(19) Leitereg, T. J.; Cram, D. J. Studies in Stereochemistry. XXXVII. Open-Chain Models for 1,3-Asymmetric Induction in Stereospecific Addition Polymerization. *J. Am. Chem. Soc.* **1968**, *90*, 4011–4018.

(20) Leitereg, T. J.; Cram, D. J. Studies in Stereochemistry. XXXVIII. Open-Chain vs. Cyclic Models for 1,3-Asymmetric Induction in Addition Reactions. *J. Am. Chem. Soc.* **1968**, *90*, 4019–4026.

(21) Brienne, M. J.; Ouannes, C.; Jacques, J. 1,3-Asymmetric Induction. 2. Lithium Aluminium Hydride Reduction of Ketones Having a β-Asymmetric Center. *Bull. Soc. Chim. Fr.* **1968**, 1036–1047.

(22) Evans, D. A.; Bartroli, J.; Godel, T. Acyclic Diastereoselection in the Hydroboration Process—Documented Cases of 1,3-Asymmetric Induction. *Tetrahedron Lett.* **1982**, *23*, 4577–4580.

(23) Reetz, M. T.; Kesseler, K.; Jung, A. Concerning the Role of Lewis-Acids in Chelation Controlled Addition to Chiral Alkoxy Aldehydes. *Tetrahedron Lett.* **1984**, *25*, 729–732.

(24) Fleming, I.; Kilburn, J. D. The Diastereoselectivity of Electrophilic Attack on Trigonal Carbon Adjacent to a Stereogenic Center—Diastereoselective Aldol Reactions of Open-Chain Enolates Having a Stereogenic Center Carrying a Silyl Group at the Beta Position. *J. Chem. Soc., Perkin Trans. 1* **1992**, 3295–3302.

(25) Nakada, M.; Urano, Y.; Kobayashi, S.; Ohno, M. Nonchelation Controlled 1,3-Asymmetric Induction in β-Chiral Acylsilanes. *Tetrahedron Lett.* **1994**, *35*, 741–744.

(26) Evans, D. A.; Dart, M. J.; Duffy, J. L.; Yang, M. G. A Stereochemical Model for Merged 1,2- and 1,3-Asymmetric Induction in Diastereoselective Mukaiyama Aldol Addition Reactions and Related Processes. *J. Am. Chem. Soc.* **1996**, *118*, 4322–4343.

(27) Tripathy, R.; Franck, R. W.; Onan, K. D. Diels–Alder Reaction of Dienes Having Stereogenic Allylic Substituents: Control of Diastereoface Selectivity by the Dienophile. *J. Am. Chem. Soc.* **1988**, *110*, 3257–3262.

(28) Evans, D. A.; Kaldor, S. W. Stereoselective Osmylation of 1,1-Disubstituted Olefins: Effect of Allylic Substituents on Reaction Diastereoselectivity. *J. Org. Chem.* **1990**, *55*, 1698–1700.

(29) Van Tamelen, E. E.; Leopold, E. J. Mechanism of Presqualene Pyrophosphate-Squalene Biosynthesis. 2. Synthesis of Bifarnesol. *Tetrahedron Lett.* **1985**, *26*, 3303–3306.

(30) Evans, D. A.; Morrissey, M. M.; Dow, R. L. Hydroxyl-Directed Hydrogenation of Homoallylic Alcohols. Effects of Achiral and Chiral Rhodium Catalysts on 1,3-Stereocontrol. *Tetrahedron Lett.* **1985**, *26*, 6005–6008.

(31) Hoveyda, A. H.; Evans, D. A.; Fu, G. C. Substrate-Directable Chemical Reactions. *Chem. Rev.* **1993**, *93*, 1307–1370.

(32) Ratier, M.; Castaing, M.; Godet, J.-Y.; Pereyre, M. Stereochemistry in the Simmons-Smith Reaction on Acyclic Allylic Alcohols. *J. Chem. Res., Synop.* **1978**, 179.

(33) Hanson, R. M.; Sharpless, K. B. Procedure for the Catalytic Asymmetric Epoxidation of Allylic Alcohols in the Presence of Molecular Sieves. *J. Org. Chem.* **1986**, *51*, 1922–1925.

(34) Hashimoto, M.; Yanagiya, M.; Shirahama, H. Total Synthesis of *meso*-Triterpene Ether, Teurilene. *Chem. Lett.* **1988**, 645–646.

(35) Inoue, S.-i.; Takaya, H.; Tani, K.; Otsuka, S.; Sate, T.; Noyori, R. Mechanism of the Asymmetric Isomerization of Allylamines to Enamines Catalyzed by 2,2′-Bis(diphenylphosphino)-1,1′-Binaphthyl-Rhodium Complexes. *J. Am. Chem. Soc.* **1990**, *112*, 4897–4905.

(36) Mikami, K.; Shimizu, M.; Zhang, H.-C.; Maryanoff, B. E. Acyclic Stereocontrol Between Remote Atom Centers via Intramolecular and Intermolecular Stereo-communication. *Tetrahedron* **2001**, *57*, 2917–2951.

(37) Reetz, M. T.; Kesseler, K.; Schmidtberger, S.; Wenderoth, B.; Steinbach, R. Chelation or Non-chelation Control in Stereoselective Reactions of Titanium Reagents with Chiral Alkoxycarbonyl Compounds. *Angew. Chem., Int. Ed. Engl.* **1983**, *22*, 989–990.

(38) Lawson, E. C.; Zhang, H. C.; Maryanoff, B. E. Remote Stereocontrol in Acyclic Systems. Hydride Addition to 1,5- and 1,6-Hydroxy Ketones Mediated by Metal Chelation. *Tetrahedron Lett.* **1999**, *40*, 593–596.

(39) Zhang, H.-C.; Harris, B. D.; Costanzo, M. J.; Lawson, E. C.; Maryanoff, C. A.; Maryanoff, B. E. Stereocontrol Between Remote Atom Centers in Acyclic Substrates. *Anti* Addition of Hydride to 1,5-, 1,6-, and 1,7-Hydroxy Ketones. *J. Org. Chem.* **1998**, *63*, 7964–7981.

(40) Molander, G. A.; Bobbitt, K. L. Keto Boronate Reduction: 1,7-Asymmetric Induction. *J. Am. Chem. Soc.* **1993**, *115*, 7517–7518.

(41) Fukuyama, T.; Vranesic, B.; Negri, D. P.; Kishi, Y. Synthetic Studies on Polyether Antibiotics. 2. Stereocontrolled Syntheses of Epoxides of Bis-Homoallylic Alcohols. *Tetrahedron Lett.* **1978**, 2741–2744.

(42) Nishigaichi, Y.; Takuwa, A.; Jodai, A. Divergently Stereocontrolled Reaction of an Allylic Silane Bearing an Asymmetric Ethereal Carbon Toward Aldehydes. *Tetrahedron Lett.* **1991**, *32*, 2383–2386.

(43) Clayden, J.; Darbyshire, M.; Pink, J. H.; Westlund, N.; Wilson, F. X. Remote Stereocontrol Using Rotationally Restricted Amides: (1,5)-Asymmetric Induction. *Tetrahedron Lett.* **1997**, *38*, 8587–8590.

(44) Linnane, P.; Magnus, N.; Magnus, P. Induction of Molecular Asymmetry by a Remote Chiral Group. *Nature (London)* **1997**, *385*, 799–801.

(45) Clayden, J.; Lund, A.; Vallverdú, L.; Helliwell, M. Ultra-remote Stereocontrol by Conformational Communication of Information Along a Carbon Chain. *Nature (London)* **2004**, *431*, 966–971.

(46) Travins, J. M.; Etzkorn, F. A. Design and Enantioselective Synthesis of a Peptidomimetic of the Turn in the Helix-Turn-Helix DNA-Binding Protein Motif. *J. Org. Chem.* **1997**, *62*, 8387–8393.

(47) Masamune, S.; Choy, W.; Petersen, J. S.; Sita, L. R. Double Asymmetric Synthesis and a New Strategy for Stereochemical Control in Organic Synthesis. *Angew. Chem., Int. Ed. Engl.* **1985**, *24*, 1–76.

(48) Evans, D. A.; Miller, S. J.; Lectka, T.; von Matt, P. Chiral Bis(oxazoline)copper(II) Complexes as Lewis Acid Catalysts for the Enantioselective Diels–Alder Reaction. *J. Am. Chem. Soc.* **1999**, *121*, 7559–7573.

(49) Evans, D. A.; Chapman, K. T.; Bisaha, J. Asymmetric Diels–Alder Cycloaddition Reactions with Chiral α,β-Unsaturated *N*-Acyloxazolidinones. *J. Am. Chem. Soc.* **1988**, *110*, 1238–1256.

(50) Davies, H. M. L.; Venkataramani, C.; Hansen, T.; Hopper, D. W. New Strategic Reactions for Organic Synthesis: Catalytic Asymmetric C–H Activation α to Nitrogen as a Surrogate for the Mannich Reaction. *J. Am. Chem. Soc.* **2003**, *125*, 6462–6468.

(51) Evans, D. A.; Kozlowski, M. C.; Murry, J. A.; Burgey, C. S.; Campos, K. R.; Connell, B. T.; Staples, R. J. C₂-Symmetric Copper(II) Complexes as Chiral Lewis Acids. Scope and Mechanism of Catalytic Enantioselective Aldol Additions of Enolsilanes to (Benzyloxy)acetaldehyde. *J. Am. Chem. Soc.* **1999**, *121*, 669–685.

(52) Gennari, C.; Cozzi, P. G. Chelation Controlled Aldol Additions of the Enolsilane Derived from *tert*-Butyl Thioacetate : A Stereosetective Approach to 1β-Methylthienamycin. *Tetrahedron* **1988**, *44*, 5965–5974.

(53) Blaser, H. U.; Spindler, F.; Studer, M. Enantioselective Catalysis in Fine Chemicals Production. *Appl. Catal., A* **2001**, *221*, 119–143.

(54) Imwinkelried, R. Catalytic Asymmetric Hydrogenation in the Manufacture of d-Biotin and Dextromethorphan. *Chimia* **1997**, *51*, 300–302.

(55) Matsumura, K.; Hashiguchi, S.; Ikariya, T.; Noyori, R. Asymmetric Transfer Hydrogenation of α,β-Acetylenic Ketone. *J. Am. Chem. Soc.* **1997**, *119*, 8738–8739.

(56) Kim, N. S.; Choi, J. R.; Cha, J. K. A Concise, Enantioselective Synthesis of Castanospermine. *J. Org. Chem.* **1993**, *58*, 7096–7099.

(57) Ooi, T.; Takeuchi, M.; Kato, D.; Uematsu, Y.; Tayama, E.; Sakai, D.; Maruoka, K. Highly Enantioselective Phase-Transfer-Catalyzed Alkylation of Protected α-Amino Acid Amides Toward Practical Asymmetric Synthesis of Vicinal Diamines, α-Amino Ketones, and α-Amino Alcohols. *J. Am. Chem. Soc.* **2005**, *127*, 5073–5083.

(58) Joly, G. D.; Jacobsen, E. N. Catalyst-Controlled Diastereoselective Hetero-Diels–Alder Reactions. *Org. Lett.* **2002**, *4*, 1795–1798.

(59) Brandes, B. D.; Jacobsen, E. N. Regioselective Ring Opening of Enantiomerically Enriched Epoxides via Catalysis with Chiral (Salen)Cr(III) Complexes. *Synlett* **2001**, 1013–1015.

(60) Duplantier, A. J.; Nantz, M. H.; Roberts, J. C.; Short, R. P.; Somfai, P.; Masamune, S. Triple Asymmetric Synthesis for Fragment Assembly: Validity of Approximate Multiplicativity of the Three Diastereofacial Selectivities. *Tetrahedron Lett.* **1989**, *30*, 7357–7360.

(61) Cowden, C. J.; Paterson, I. Asymmetric Aldol Reactions Using Boron Enolates. *Org. React.* **1997**, *51*, 1–200.

(62) Namba, K.; Kishi, Y. New Catalytic Cycle for Couplings of Aldehydes with Organochromium Reagents. *Org. Lett.* **2004**, *6*, 5031–5033.

(63) Choi, H.-w.; Nakajima, K.; Demeke, D.; Kang, F.-A.; Jun, H.-S.; Wan, Z.-K.; Kishi, Y. Asymmetric Ni(II)/Cr(II)-Mediated Coupling Reaction: Catalytic Process. *Org. Lett.* **2002**, *4*, 4435–4438.

(64) Wan, Z.-K.; Choi, H.-w.; Kang, F.-A.; Nakajima, K.; Demeke, D.; Kishi, Y. Asymmetric Ni(II)/Cr(II)-Mediated Coupling Reaction: Stoichiometric Process. *Org. Lett.* **2002**, *4*, 4431–4434.

14

Multistep Asymmetric Catalysis

Aside from discovering new transformations or developing efficient enantioselective versions of known processes, one of the most important future contributions to synthesis will be the development of asymmetric catalysts for processes in which several steps can be combined. Such processes fulfill one of the great challenges in modern organic synthetic chemistry—namely, enabling the creation of significant complexity with high efficiency and without the need to purify intermediates. Processes in which multiple chemical transformations occur in one reaction vessel are known by several names, including tandem, domino, cascade, sequential, and multistep processes.[1] In this chapter, all such transformations will be referred to as multistep processes.

Multistep processes are defined herein as a sequence of reactions that are performed in one process without isolation or purification of any intermediates. There are three distinct genres of multistep processes (**Figure 14.1**). In sequential reactions, reagents are added and/or the conditions are altered to initiate subsequent reactions. In terms of development, *sequential, consecutive, or iterative processes* are the most straightforward, as incompatible reagents can be employed in the different stages without cross-reactivity problems. In a *cascade or domino process*, one reaction initiates an entire sequence. For example, the formation of the product in the first transformation reveals functionality that then allows a second transformation to occur. The product of this transformation reveals functionality that then allows a third transformation to occur, and so on. In sequential processes, much the same occurs except that additional reagents or catalysts need to be added at specific junctures to allow subsequent steps to occur. Operationally, cascade processes are highly desirable, as only one set of reagents is added for an entire sequence to occur. On the other hand, fewer reagent combinations are viable in a cascade process due to compatibility issues. The third class of multistep processes are *concurrent processes* in which all the reagents are present at once and in which the first reaction is not necessary for the second to occur. This last type is the most challenging, as the compatibility of the reagents and their high selectivity for the different reactive groups is crucial. It is also possible that a multistep processes may combine aspects of these three genres (i.e., a sequential-cascade process; see **Figure 14.1**).

Many different reaction types are amenable to multistep processes, including reactions that form C–C bonds through neutral (e.g., pericyclic), anionic (e.g., aldol),

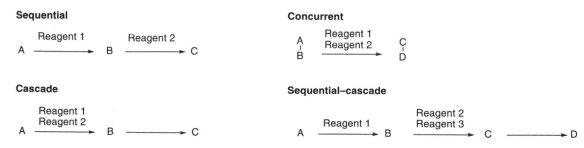

Figure 14.1. Schematic representations of various multistep reaction processes.

cationic, or organometallic (e.g., hydroformylation) processes, reactions involving redox (e.g, alkene dihydroxylation or alkene reduction), and reactions involving functional group transformations (e.g., acylation). A number of comprehensive reviews are available on various aspects of multistep processes dealing primarily with systems in which no stereochemistry occurs or only diastereoselection is considered.[2–17] Additional reviews focus on multistep process involving the rapid assembly of multiple components, but such processes utilizing catalytic enantioselective reactions are rare.[18–27] In this chapter, only processes which commence from achiral or racemic starting materials and involve asymmetric catalysis in at least one step are discussed.

14.1 Catalytic Asymmetric Induction as One Part of a Multistep Process

The introduction of stereocenters with control can occur at many different stages in multistep processes. Such asymmetric reactions may comprise the first step, last step, or some intervening step. Importantly, multistep processes allow the generation of multiple stereocenters by relative diastereoselection in subsequent stages after an enantiomerically pure material is formed.

14.1.1 Catalytic Asymmetric Induction as the First Part of a Multistep Process

Catalytic asymmetric reactions are typically more difficult than nonselective or diastereoselective processes since many issues of chemoselectivity, regioselectivity, diastereoselectivity, and enantioselectivity need to be addressed and chiral catalysts are not always highly robust. As a consequence, a majority of the multistep asymmetric reactions reported to date employ a chiral catalyst in the first step.

For example, the sequential asymmetric enamide hydrogenation and Suzuki coupling to yield a range of novel non-natural α-amino acid derivatives have been described (**Figure 14.2**).[28] In this case, different transition-metal catalysts are used for each transformation of this two-step process. For the first step, a chiral rhodium–DuPHOS catalyst is used to accomplish the asymmetric hydrogenation. For the second step, an achiral palladium catalyst is used to accomplish the Suzuki cross-coupling reaction. A number of different enamides and boronic acids can be used, allowing the synthesis of diverse α-amino acids with high yield and enantioselectivity. Since either DuPHOS enantiomer is available, both enantiomeric series of these α-amino acid derivatives are readily available. Formally, this sequential set of reactions does not fall under the multistep processes as defined in the introduction since the catalyst from the first step was removed by filtration and the reaction mixture was transferred to a second vessel for the subsequent step. Likely, the soluble forms of the asymmetric hydrogenation and Suzuki cross-coupling catalysts were not

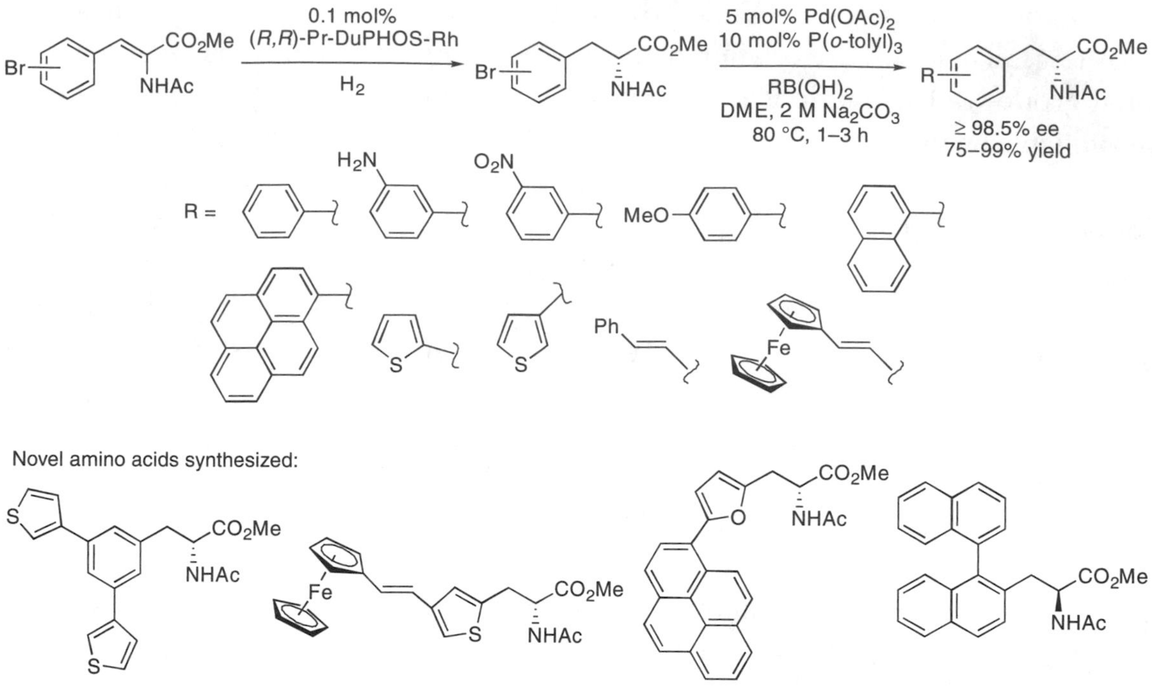

Figure 14.2. Sequential asymmetric enamide hydrogenation and Suzuki coupling to provide novel α-amino acid derivatives.

compatible, highlighting the main difficulty in the development of multistep processes. Nonetheless, this example demonstrates the utility of sequenced methods for generating diverse structures in a rapid and efficient manner.

In the next example (**Figure 14.3**), two catalysts were initially identified that were compatible. Furthermore, the entire issue of catalyst compatibility was eventually circumvented by finding conditions under which a single complex could catalyze multiple reactions. A Rh–DuPHOS-catalyzed asymmetric hydrogenation of the diene in **Figure 14.3** was followed by a Rh–PPh$_3$- or Rh–BIPHEPHOS-catalyzed hydroformylation/dehydrative cyclization.[29] Under mild reaction pressures, only the Rh–Et-DuPHOS-based catalyst promotes the enantioselective enamide hydrogenation in the presence of the achiral hydroformylation catalysts (Rh–BIPHEPHOS or Rh–PPh$_3$) as evidenced by the high enantioselection (≥ 95% ee). Further experiments revealed that the role of the hydroformylation/dehydrative cyclization catalysts could be performed by the Rh–Et-DuPHOS-based catalyst, albeit under more forcing conditions, because the Rh–Et-DuPHOS-based catalyst was not as efficient in these steps. Thus, use of the Rh–Et-DuPHOS-based catalyst, first with H$_2$ at room temperature and then with 80–400 psi 1:1 H$_2$:CO, provided the products directly. The cyclic amino acid derivatives could be obtained in good yield, high regioselectivity, and high enantiomeric excess in this efficient process. Since different reaction conditions (H$_2$ vs. pressurized H$_2$/CO and heat) were needed for each step, this multistep transformation is formally a sequential process, even though only one catalyst was employed. Sequential processes with different gas mixtures are especially appealing, as one catalyst can be

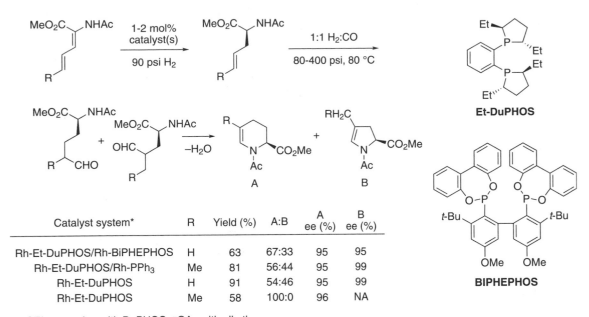

Figure 14.3. Sequential one-pot asymmetric hydrogenation, hydroformylation, and condensation to yield chiral cyclic amino acids.

employed over several steps and the chemoselectivity is readily controlled by manipulation of the reaction atmosphere.

While olefin metathesis catalysts have been shown to be highly versatile, allowing the development of multistep olefin metathesis/hydrogenation,[30] olefin metathesis/olefin isomerization,[31] multiple olefin metatheses,[32] and ring-opening metathesis polymerization (ROMP)/atom-transfer radical polymerization (ATRP)/hydrogenation processes,[33] the use of such catalysts in asymmetric multistep processes is less established. In one example, an asymmetric allylic substitution reaction was combined with a ring-closing metathesis (RCM) reaction (**Figure 14.4**). Initially, the enantioselectivities in this sequential process were low.[34] However, optimization of the phosphoramidite ligand in the first step provided high enantioselectivity (> 90% ee).[35] In this transformation, the alkenyl Grignard reagent presumably forms a cuprate with the catalyst, which then undergoes an asymmetric S_N2' reaction with the allylic chloride to provide a diene adduct. Further addition of 5 mol% Grubbs ruthenium complex at room temperature catalyzed the intramolecular RCM reaction to provide chiral cyclopentenes and cyclohexenes in good overall yield. Notably, the Grubbs ruthenium catalyst was compatible with excess Grignard reagent and trace copper salts. The intermolecular version of this process was also viable wherein a non-alkenyl Grignard reagent was employed. Subsequent addition of the Grubbs ruthenium catalyst and an excess of a second olefin, ethyl acrylate, yielded the cross-metathesis product, albeit in somewhat diminished yield. Unlike many multistep processes that utilize simple functional group transformations, both steps of this process implement catalyzed carbon–carbon bond-forming reactions; two different organometallic

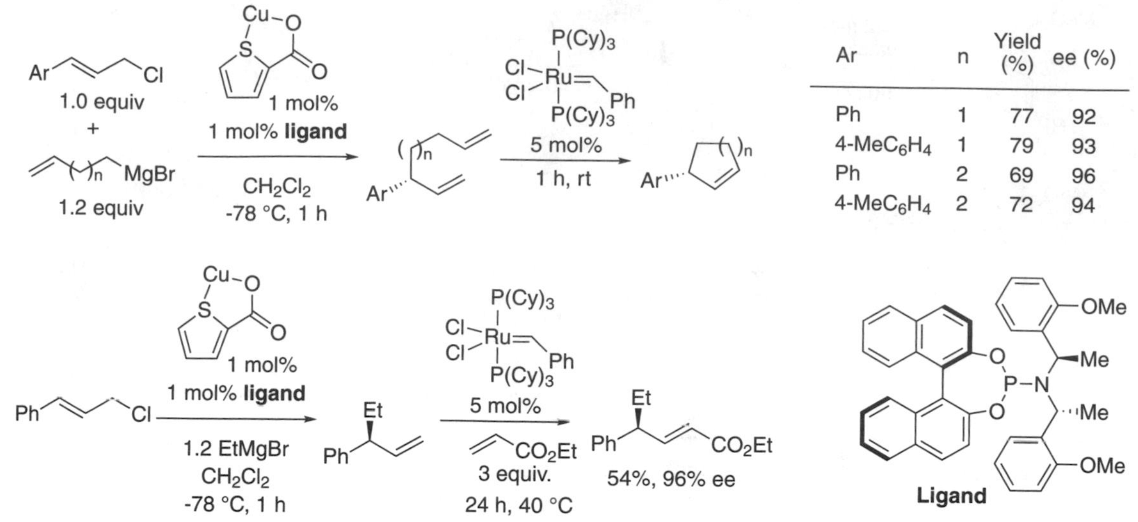

Ar	n	Yield (%)	ee (%)
Ph	1	77	92
4-MeC$_6$H$_4$	1	79	93
Ph	2	69	96
4-MeC$_6$H$_4$	2	72	94

Figure 14.4. Sequential asymmetric allylic substitution and ring-closing metathesis.

reagents are utilized (metal alkyl and metal alkylidene) and two different types of carbon–carbon bonds (σ and π) are formed.

In the above example, an activated carbon coupling partner, the Grignard reagent, was preformed and used. In the multistep process illustrated in **Figure 14.5**, the activated carbon coupling partner, a 1,2-diboronate ester compound, is generated in the first step of the process, allowing the use of stable alkene and aryl halide starting materials.[36] In this sequential process, the chiral rhodium-catalyzed asymmetric diboration proceeds with high enantioselectivity. Further addition of base, an aryl halide, and a palladium catalyst allows a direct Suzuki coupling with the less-hindered boron moiety. Importantly, the rhodium and palladium catalysts are compatible under these conditions. After Suzuki coupling, peroxide and aqueous sodium hydroxide are added to initiate oxidation of the remaining boronate ester. Ultimately, the three-step sequential process yields the product of carbohydroxylation of the alkene in a stereoselective and regioselective manner.

In this chapter, most of the tandem processes discussed are comprised of reactions that each provide distinct isolable intermediates (e.g., the diboronate in **Figure 14.5**). However, there are an enormous number of cascade processes in which an organometallic intermediate undergoes a further transformation. This class of transformation will not be discussed extensively here, but an example is instructive (**Figure 14.6**).[37] In this reaction, an initial enantioselective oxypalladation gives rise to a chiral alkyl–palladium intermediate that would not normally be isolated. In this particular case, the alkyl–palladium species is trapped via a Heck reaction to provide the final product with very good yield and high selectivity. Due to the ability of organopalladium[38] and other organometal intermediates to participate in many different types of transformations, numerous cascade processes of this type have been designed.

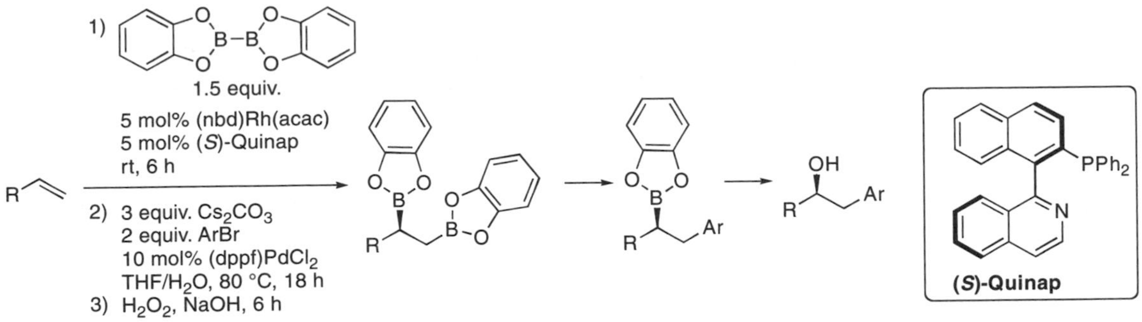

Product	Yield (%)	ee (%)	Product	Yield (%)	ee (%)
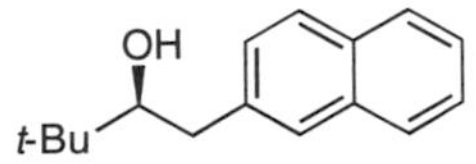	69	92	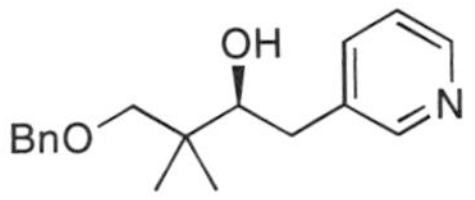	58	93
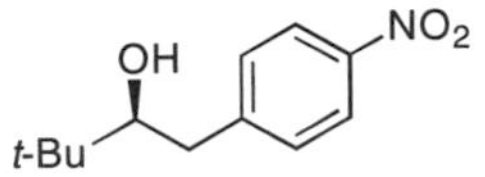	62	94	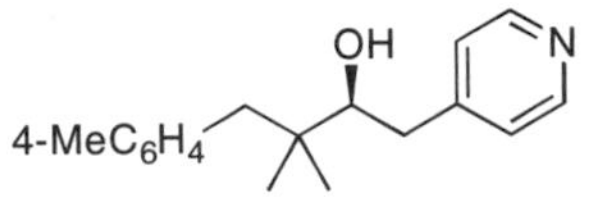	58	96
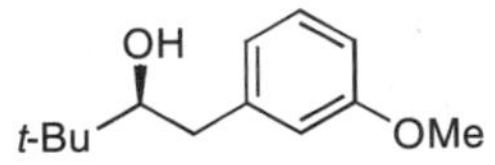	77	95	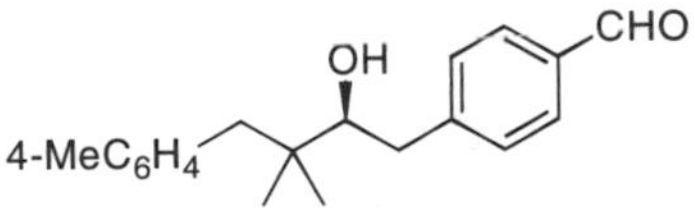	48	96

Figure 14.5. Alkene carbohydroxylation via sequential asymmetric alkene diboration, regioselective Suzuki cross-coupling, and alkyl borane oxidation.

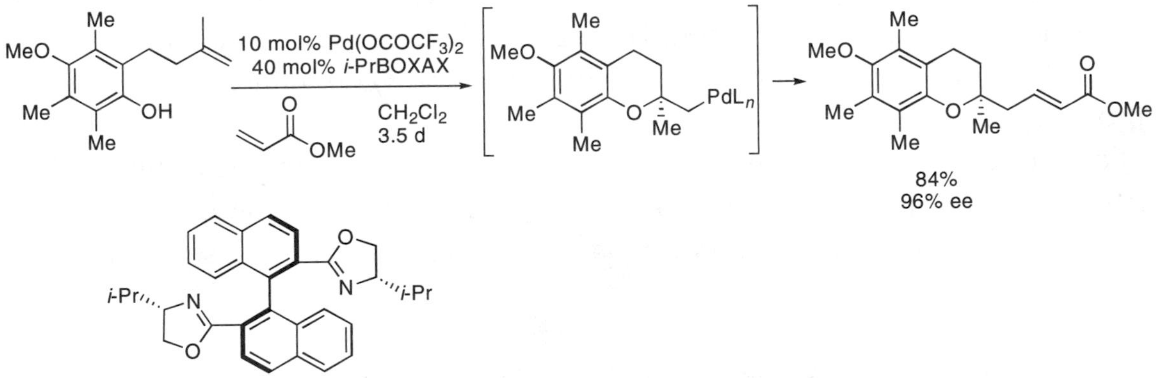

i-PrBOXAX

Figure 14.6. Domino enantioselective oxypalladation and Heck reaction.

14.1.2 Catalytic Asymmetric Induction in a Later Part of a Multistep Process

In the preceding section, examples where catalytic asymmetric reactions were employed in the first step of a multistep process were outlined. Since chiral catalysts are often the most sensitive component in a multistep process, it is not surprising that the majority of catalytic asymmetric multistep processes follow this pattern. However, under the right circumstances, catalytic asymmetric reactions can occur at any juncture in a multistep process.

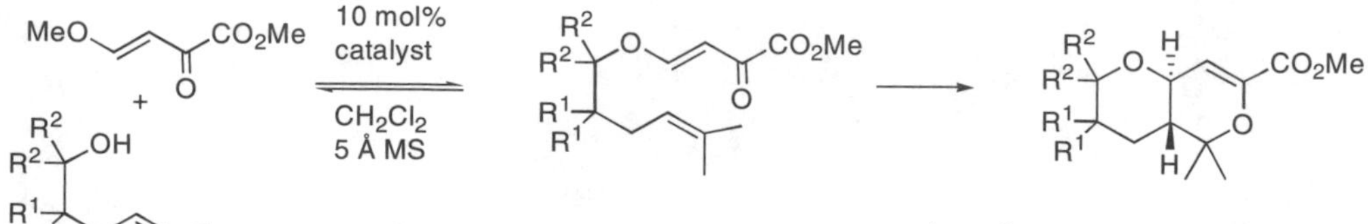

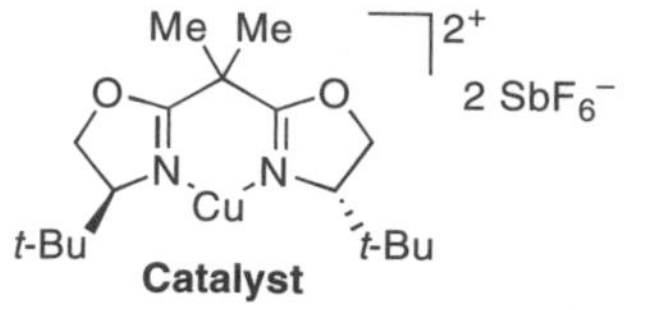

Figure 14.7. Cascade transetherification and intramolecular hetero–Diels–Alder reaction, where both steps are catalyzed by the same copper bis(oxazoline) complex.

For example, the multistep process in **Figure 14.7** implements a catalytic asymmetric reaction in the second step.[39] In this cascade process, the chiral bis(oxazoline)–copper catalyst serves two distinct roles (see Section 14.4 for further discussion of multiple catalyst roles). In the first step, the copper complex catalyzes the transetherification of the enone substrate with the alcohol substrate. This first step involves no asymmetric induction, and in principle an achiral catalyst could have been employed. However, the ability of the copper complex to serve in two capacities by catalyzing this first transformation simplifies and streamlines the process. After the two substrates are joined, the chiral copper complex next catalyzes an intramolecular hetero–Diels–Alder reaction, which proceeds with high enantioselectivity. This multistep process has several merits including 1) operational convenience (only one catalyst is required), 2) convergent union of two components, and 3) rapid increase in structural complexity.

For further examples of multistep processes involving catalytic asymmetric induction in later steps, see Section 14.4. For examples of multistep processes involving separate catalytic enantioselection versus internal diastereocontrol, see Section 14.3.

14.2 Sequential Stereoselective Reactions in a Multistep Process

In contrast to the examples from the preceding section, most catalytic asymmetric multistep processes with stereoselection in later steps involve diastereoselective processes. Again, the most sensitive component, the chiral catalyst, is used in the first step to generate chiral structures. Subsequently, diastereoselective reactions can occur with further chemical transformation.

One of the most common paradigms encountered in such sequential stereoselective processes involves an enantioselective reaction, such as a conjugate addition to an enone, to form a chiral enolate followed by a diastereoselective trapping step with an intermolecular or intramolecular electrophile. For example, an aluminum–lithium bis(BINOLate) catalyst (ALB) has been used to initiate the Michael addition to an enone (**Figure 14.8**).[40] Mechanistic studies indicate that ALB causes deprotonation of the diethyl methylmalonate to form an adduct with a lithium-centered enolate. The aluminum of this adduct coordinates the enone carbonyl (see **Figure 14.8**). From this heterobimetallic complex, the Michael reaction occurs, yielding an aluminum-centered chiral enolate. Subsequent aldol addition to the aldehyde present in the reaction mixture can occur with high (i.e., for dihydrocinnamaldehyde) or low (i.e., for benzaldehyde) π-facial selectivity. Regardless, the thermodynamically more stable trans-substituted cyclopentanone is observed due to the facile epimerization of the acidic center adjacent to the carbonyl. Notably, two electrophiles, the aldehyde and the enone, are present from the inception of the reaction. Addition of the malonate anion to the aldehyde is reversible, whereas the illustrated cascade is irreversible. Furthermore, the intermediate enolate is hindered and adds to the more reactive aldehyde rather than the enone. As such, the order of the reaction sequence is self-determined and a sequential addition of the two electrophiles is not necessary.

This type of three-component coupling with cyclopentenone[41] is particularly attractive for the synthesis of prostaglandin compounds because of the limited number of steps and its convergent nature. Prostaglandins are widely distributed in mammalian systems and their pharmacological effects are so diverse that they have been the subjects of intensive research.[42] Several synthetic prostaglandin derivatives are currently used as drugs, but their synthesis is often still the subject of improvement and innovation.[43] The cascade Michael–aldol asymmetric process has been applied to a stereoselective synthesis of an 11-deoxyprostagladin $F_{1\alpha}$ precursor (**Figure 14.9**).[44] Use of a more complex aldehyde substrate proved to be straightforward. Dehydration of the product of the multistep ALB-catalyzed reaction yields only the (E)-enone

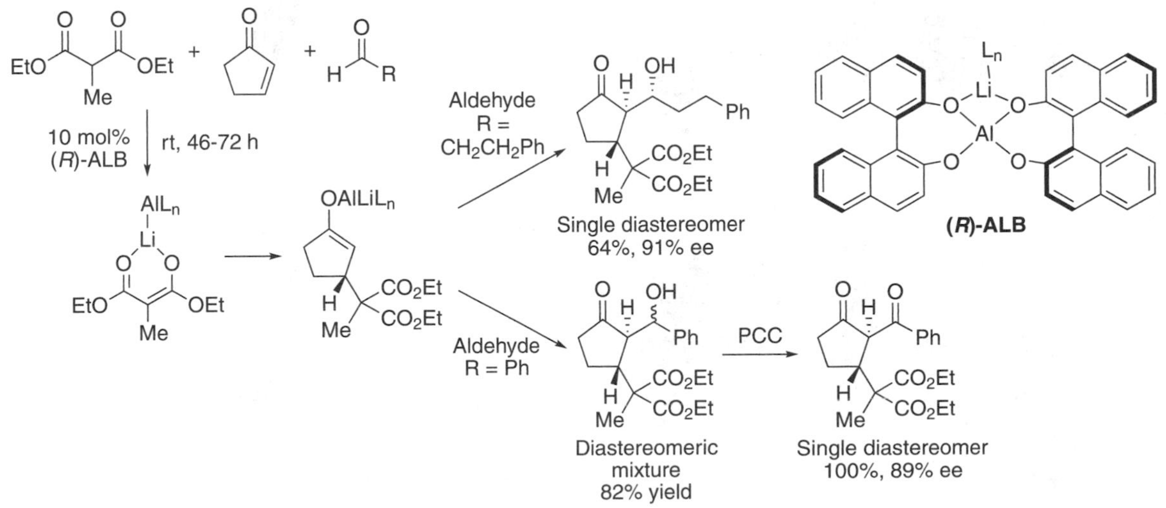

Figure 14.8. Cascade enantioselective Michael addition and diastereoselective intermolecular aldol reaction.

diastereomer in 92% ee. 11-Deoxyprostagladin $F_{1\alpha}$ was subsequently formed from the illustrated (*E*)-enone in 14 further steps.

Related chiral enolates could be produced via catalytic enantioselective conjugate addition (**Figure 14.10**).[45,46] In this case, a more functionalized cyclopentenone was employed and a chiral nucleophilic reagent was formed by action of the chiral copper phosphoramidite catalyst on a dialkylzinc reagent. Asymmetric addition of this species to the enone yields a chiral copper or zinc enolate that can undergo further aldol reaction with the aldehyde present. Both stereocenters formed in the aldol reaction were generated with high selectivity. As expected, the aldehyde attacks from the less hindered face of the enolate, which is opposite of the R^2 group introduced in the conjugate addition. Here again, both the enone and aldehyde electrophiles are present from the inception of the reaction. High chemoselectivity in each step drives

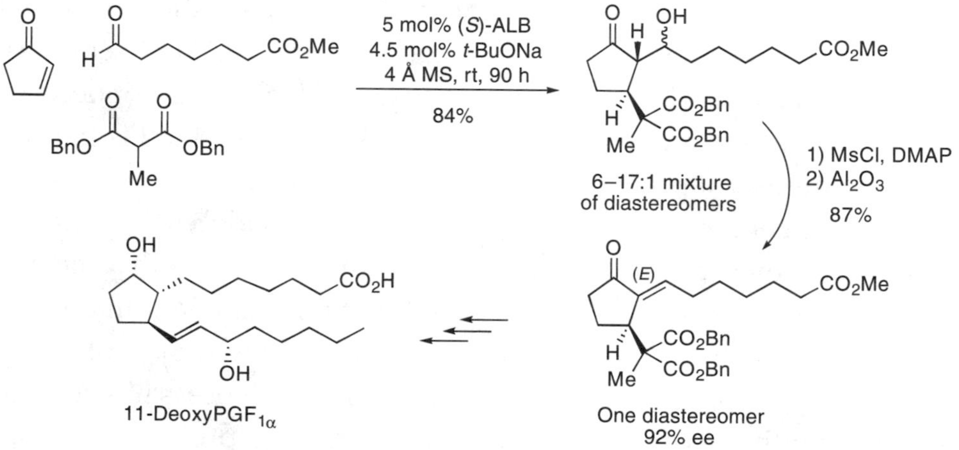

Figure 14.9. Application of sequential enantioselective Michael addition and intermolecular aldol reaction to the synthesis of 11-deoxyprostagladin $F_{1\alpha}$.

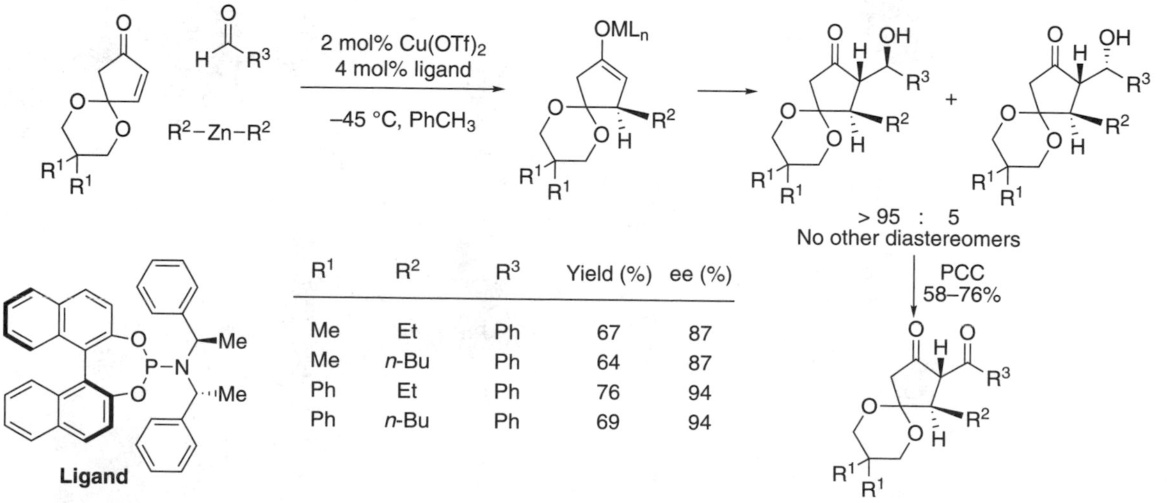

R^1	R^2	R^3	Yield (%)	ee (%)
Me	Et	Ph	67	87
Me	*n*-Bu	Ph	64	87
Ph	Et	Ph	76	94
Ph	*n*-Bu	Ph	69	94

Figure 14.10. Cascade enantioselective conjugate addition and diastereoselective intermolecular aldol reaction.

this cascade process. Namely, the dialkylzinc reagent cannot add to either electrophile without the intervention of a catalyst. The reaction likely proceeds through a copper alkyl formed via transmetallation with the dialkylzinc. The resulting copper alkyl is soft and adds chemoselectively to the soft enone C=C bond more readily than to the harder carbonyl of the aldehyde. The metal enolate that is formed then reacts with the more reactive, harder aldehyde versus the softer enone β-carbon or the less electrophilic carbonyl of the enone.

This type of three-component multistep process is highly applicable to the preparation of prostaglandins, as seen in the highly efficient synthesis of prostaglandin E_1 methyl ester (**Figure 14.11**).[45,46] This synthesis commences with an enone, aldehyde (ω-chain precursor), and functionalized zinc reagent (α-chain precursor). To differentiate between the unsaturated carbonyl moieties of the two substrates, the unsaturated aldehyde was equipped with a removable silyl group, exploiting the fact that β,β-disubstituted enones are not reactive in 1,4-additions under these conditions. In the presence of 3 mol% of the copper phosphoramidite catalyst, the conjugate addition/aldol adduct was obtained in 60% yield with an 83:17 mixture of diastereomers at the hydroxylic center. Reduction of the major diastereomer with $Zn(BH_4)_2$ provided a single isomer in 63% yield and 94% ee, indicating that the stereochemical fidelity in the first step was high. With five further steps, a concise synthesis of prostaglandin E_1 methyl ester was achieved.

Intramolecular versions of the conjugate addition/aldol multistep process are also viable and permit the use of a less reactive second electrophile. For instance, the catalytic enantioselective carbometalation of ketoenone structures has been described using a chiral rhodium catalyst and aryl boronic acids as aryl donors (**Figure 14.12**).[47] Selective reaction at the enone yields a carbon-bound rhodium species which readily isomerizes to the oxygen-bound enolate. Diastereoselective intramolecular addition

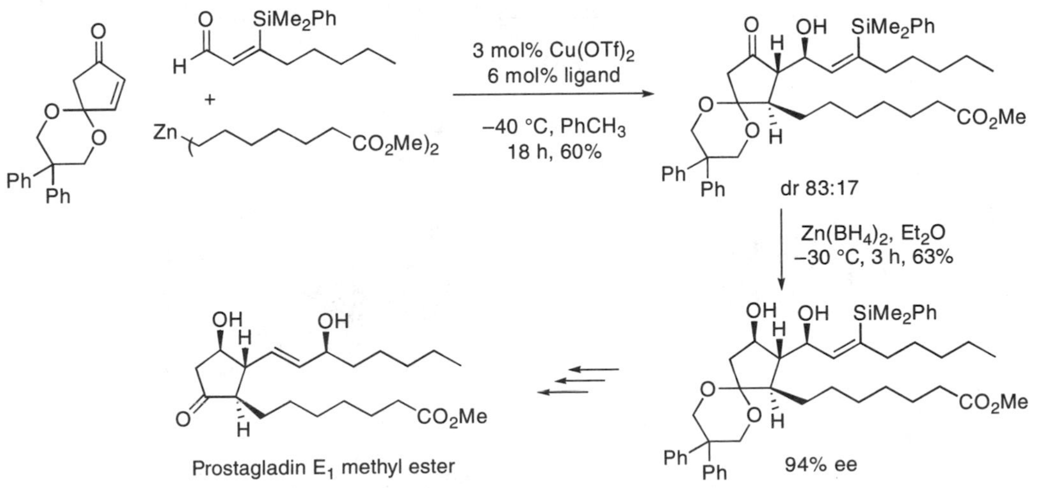

Figure 14.11. Application of sequential conjugate addition and intermolecular aldol reaction to the synthesis of prostaglandin E_1 methyl ester. See **Figure 14.10** for ligand structure.

of this chiral enolate to ketone then provides the product with high diastereo-selectivity and enantioselectivity. In the design of this cascade process, the facility of rhodium–catalyzed addition of aryl boronic acids to aldehyde carbonyls was taken into account. Thus, ketones instead of aldehydes were employed as the intramolecu-lar electrophilic trap. Again, the reaction sequence is dictated by the relative reactiv-ities. Aryl transfer occurs more readily to the carbon–carbon double bond rather than either ketone carbonyl. From the enolate, intramolecular addition to the ketone is much more facile than intermolecular reaction with either carbonyl of the original substrate. In contrast, in the conjugate addition/intermolecular aldol reaction dis-cussed above (**Figure 14.10**), use of ketones as the second electrophile is problematic due to the competing reaction of the enolate with the original enone substrate.[46]

In further work, the above premise has been expanded by changing the second diastereoselective trapping step from prochiral addition to a single carbonyl group to desymmetrization of a symmetric dicarbonyl group (**Figure 14.13**).[48] This transfor-

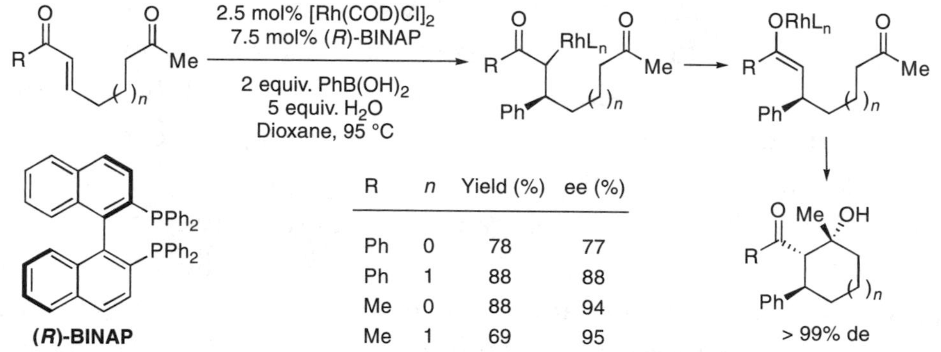

Figure 14.12. Cascade enantioselective conjugate addition and diastereoselective cycloaldolization.

R	n	Yield (%)	ee (%)
Ph	0	78	77
Ph	1	88	88
Me	0	88	94
Me	1	69	95

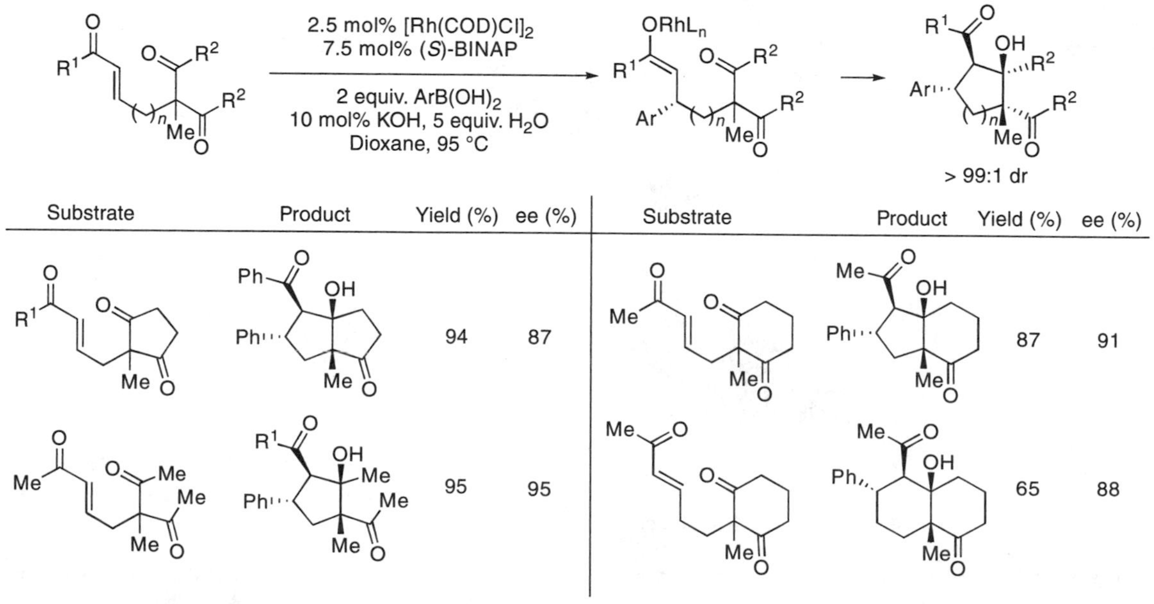

Figure 14.13. Cascade enantioselective conjugate addition and diastereoselective cycloaldolization to form bicyclic structures.

Substrate	Product	Yield (%)	ee (%)	Substrate	Product	Yield (%)	ee (%)
		94	87			87	91
		95	95			65	88

mation enables the rapid assembly of a range of complex fused [3.3.0], [4.3.0], and [4.4.0] bicyclic ring structures from simple precursors. In this cascade process, two carbon–carbon bonds are formed in high chemical yield and four contiguous asymmetric centers, two of which are quaternary, are established with high diastereoselectivity and enantioselectivity.

Rather than carbonyls as electrophilic traps for chiral enolates formed via conjugate addition, alkyl tosylates and halides have also been found to be competent in this role (**Figure 14.14**). [49] First, olefin cross-metathesis is used to unite enone and tosylate reaction partners in a separate step. Subsequent asymmetric conjugate addition to the enone with a dialkylzinc reagent and a chiral copper catalyst provides a chiral zinc enolate that undergoes diastereoselective intramolecular alkylation. For cyclopentyl ($n = 1$) and cyclohexyl ($n = 2$) products, the overall process is efficient. For $n = 3$, the asymmetric conjugate addition proceeds well but cyclization to the larger seven-membered ring does not occur; instead, the tosylate product was isolated in high yield (91%) and selectivity (95% ee). This cascade process proceeds well because dialkylzinc reagents do not react efficiently with tosylate electrophiles. An intermolecular version of this process has also been achieved via asymmetric conjugate addition followed by addition of benzyl bromide to the resultant chiral enolate (bottom of **Figure 14.14**). In this latter case, a sequential process must be used because benzyl bromide will react with the dialkylzinc reagent.

Many further uses are possible for the chiral enolates formed in catalytic asymmetric transformations. For example, a sequential process has been developed in which chiral zinc enolates, formed by analogous asymmetric enone conjugate additions, are trapped as the silyl enol ethers using trimethylsilyl triflate (**Figure 14.15**).[50] Further

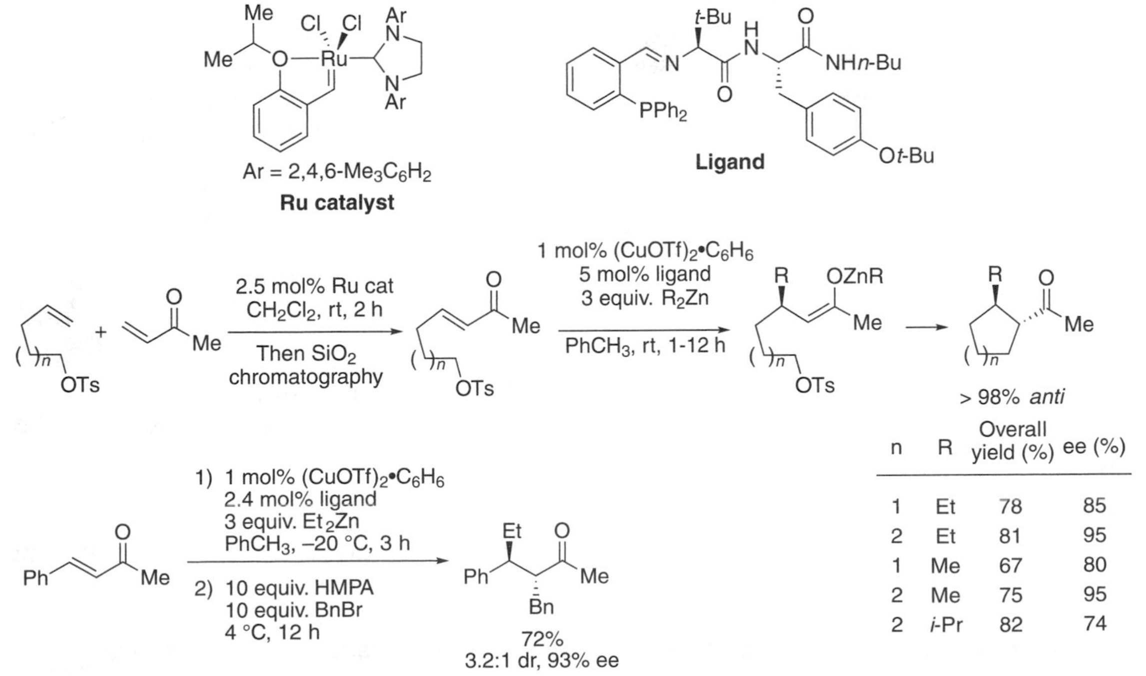

n	R	Overall yield (%)	ee (%)
1	Et	78	85
2	Et	81	95
1	Me	67	80
2	Me	75	95
2	i-Pr	82	74

Figure 14.14. Sequential olefin cross-metathesis, asymmetric enone conjugate addition, and diastereoselective enolate alkylation

addition of diiodomethane, in the presence of the dialkylzinc reagent remaining from the conjugate addition step, initiated a Simmons–Smith-type cyclopropanation of the enol ether double bond. While the diastereoselection in the third step of the process is moderate, the enantioselection from the first step is very high. The product of this sequential process is a silylated cyclopropanol that can undergo several different types of rearrangement reactions to form chiral saturated cycloalkanones, exocyclic enones, and ring-expanded cycloalkenones. Overall a range of six-, seven-, and eight-membered chiral ring systems can be generated. The reaction is not limited to cyclic precursors, although selectivity suffered slightly when an acyclic enone was employed.

While the above examples have illustrated the versatility of chiral enolate intermediates formed from catalytic asymmetric additions in a variety of diastereoselective transformations, many other chiral intermediates can also participate in further stereoselective multistep processes. For example, a chiral bis(oxazoline)–copper catalyst has been used to effect the asymmetric amination of an allenyl β-ketoester with an azodicarboxylate (**Figure 14.16**).[51] The resulting allenyl hydrazide is poised to undergo a carbopalladation reaction with an achiral palladium catalyst to yield a π-allyl palladium intermediate. Further intramolecular trapping of this intermediate with the hydrazide yields the pyrazolidine derivative. Diastereoselection may be established either during the carbopalladation reaction or in the reaction with the π-allyl palladium intermediate. In either case, the two stereofacial approaches are very similar (the keto and ester groups are of like size) and hence poor diastereoselection is ob-

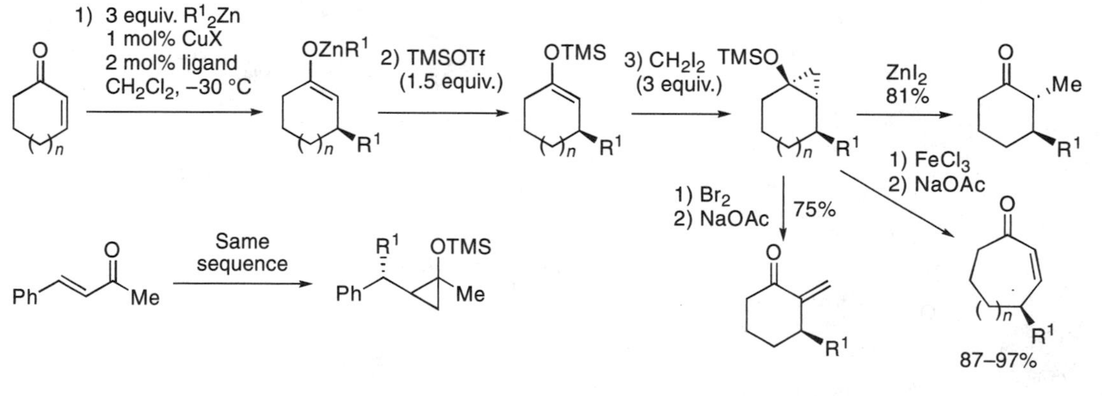

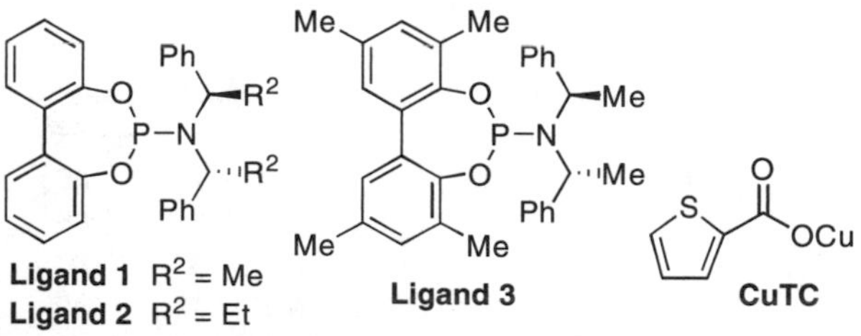

n	R^1	CuX	Ligand	Yield (%)	ee (%)	de (%)
1	Et	CuTC	2	95	98	77
2	Et	CuTC	1	97	93	84
2	Me	Cu(OTf)$_2$	2	91	97	71
Acyclic	Et	Cu(OAc)$_2$	3	97	89	60

Figure 14.15. Sequential asymmetric conjugate addition, silylation, and diastereoselective cyclopropanation.

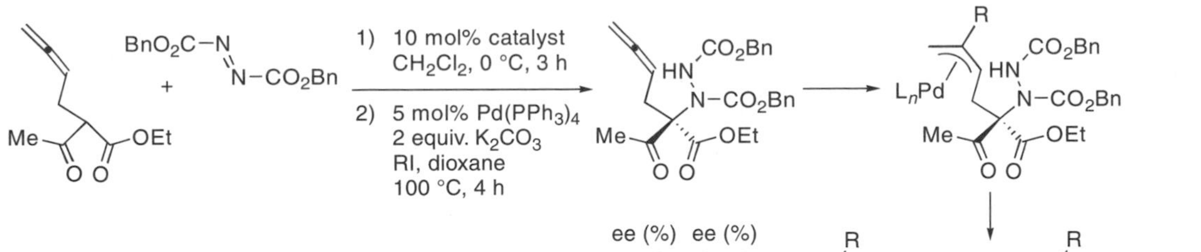

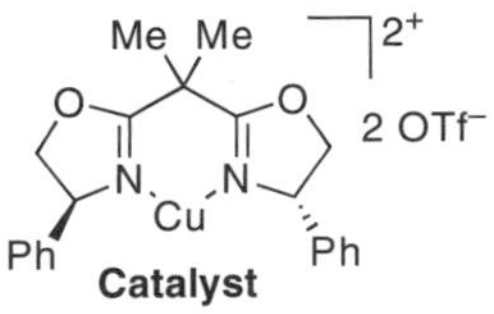

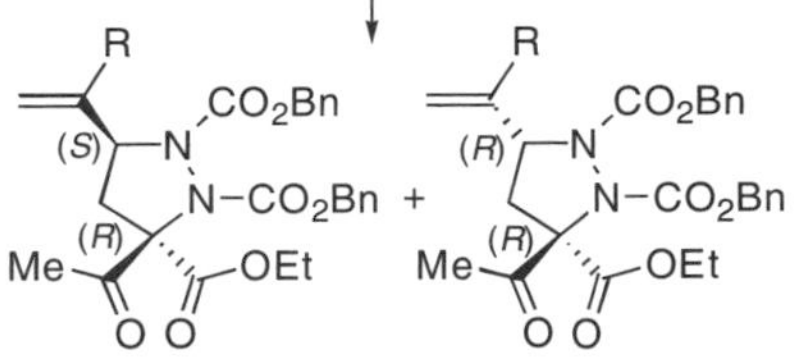

Figure 14.16. Formation of pyrazoline derivatives via sequential asymmetric enolate amination, carbopalladation, and π-allyl palladium substitution.

R	Yield (%)	dr	ee (%) (R,S)	ee (%) (R,R)
Ph	85	67:33	98	98
4-MeOC$_6$H$_4$	92	67:33	98	98
4-MeO$_2$CC$_6$H$_4$	96	62:38	98	97
trans-PhCH=CH	86	63:37	99	99
trans-n-BuCH=CH	75	66:34	98	97
2-thienyl	71	64:36	99	91

served. Even so, these diastereomers have markedly different polarities and are readily separable.

While the diastereoselection in the above example is low, this reaction highlights that it is possible to combine the features available from chiral Lewis acid catalysts for conventional electrophilic activation and late-transition-metal catalysts for organometallic transformations. Importantly, these two catalyst types can be compatible.

Some of the most useful chiral building blocks in enantioselective synthesis are epoxy alcohols, which are readily available via the Sharpless–Katsuki asymmetric epoxidation of prochiral allylic alcohols.[52] More complex epoxy alcohols containing three contiguous stereogenic centers are accessible using the Sharpless kinetic resolution of chiral allylic alcohols (see Chapter 7 on kinetic resolutions). This process, however, gives a maximum of 50% yield and usually requires stoichiometric titanium tetraisopropoxide and tartrate ligand.[53,54] A one-pot approach to the synthesis of complex epoxy alcohols from achiral precursors is illustrated in **Figure 14.17**.[55–57] It is initiated by hydroboration of a terminal alkyne with diethylborane generating a vinylborane. To this intermediate is added diethylzinc, which undergoes transmetallation to generate a vinylzinc reagent. In the presence of a catalyst derived from an amino alcohol such as (–)-MIB,[58] the vinylzinc reagent adds to the enal to generate a bis(allylic) alkoxide with high enantioselectivity. Dioxygen is then introduced into the reaction vessel, which reacts with the excess dialkylzinc reagent to produce a zinc peroxide. Zinc peroxides are insufficiently electrophilic to epoxidize the allylic double bond. Titanium tetraisopropoxide, however, is well-known to be an excellent catalyst for the diastereoselective epoxidation of allylic alcohols. Addition of titanium tetraisopropoxide to the solution of zinc peroxide and zinc bis(allylic) alkoxide results in directed epoxidation of one of the four alkene faces with high diastereoselectivity. The more electron-rich double bond is oxidized most rapidly. Control over

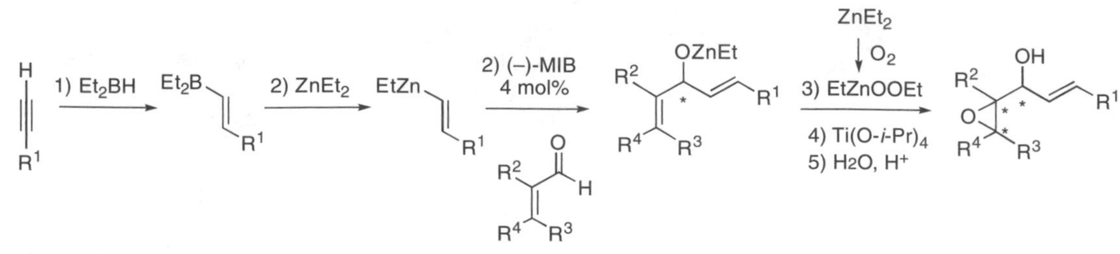

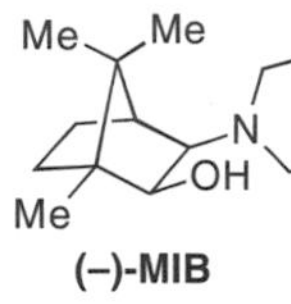

Allylic epoxy alcohol	ee (%)	Yield (%)	dr (*erythro*:*threo*)
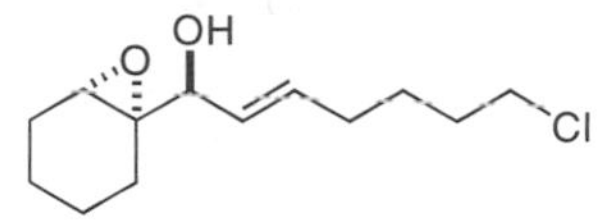	99	78	20:1
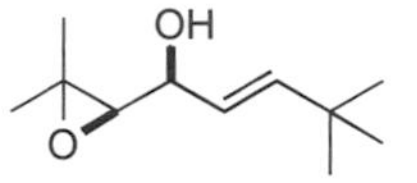	98	76	1:20
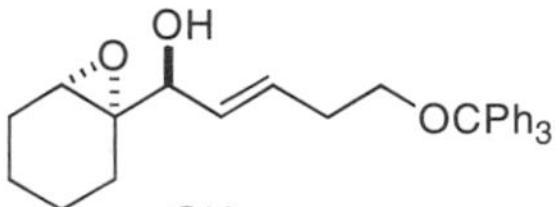	92	60	20:1
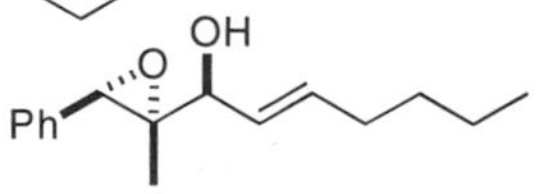	90	80	20:1

Figure 14.17. One-pot synthesis of allylic epoxy alcohols from alkynes and aldehydes.

the facial selectivity in the epoxidation is governed by the presence of $A^{1,2}$- or $A^{1,3}$-strain in one of the diastereomeric transition states.[59] As outlined in **Figure 14.17**, densely functionalized allylic epoxy alcohols can be prepared with high enantioselectivity and diastereoselectivity.

Numerous natural and non-natural products containing cyclopropyl groups exhibit important biological activity. A modular approach to these chiral building blocks that allows further elaboration of the cyclopropane ring is desirable. A flexible method for the synthesis of cyclopropyl alcohols that is related to the allylic epoxy alcohol synthesis outlined in **Figure 14.17** is presented in **Figure 14.18**.[60] In this chemistry, the asymmetric addition of a dialkylzinc reagent to an α,β-unsaturated aldehyde is followed by a directed Simmons–Smith cyclopropanation of the allylic zinc alkoxide with RO–Zn–CH_2–I. Excellent enantioselectivity was observed in the addition step and very high diastereoselectivity was found in the cyclopropanation. Interestingly, when RO–Zn–CHI_2 was employed as the carbenoid precursor, an iodo carbene was transferred, and a fourth stereogenic unit was formed. The iodocyclopropyl alcohols were isolated in 60–78% yield with ≥ 95% ee. The remarkable diastereoselection observed in the generation of the final three stereocenters attests to the ability of zinc to control stereoselectivity in this reaction. The cis disposition of the carbinol and the iodo substituents suggests that the iodo substituent is coordinated to zinc during the cyclopropanation step. These chiral building blocks should enable the synthesis

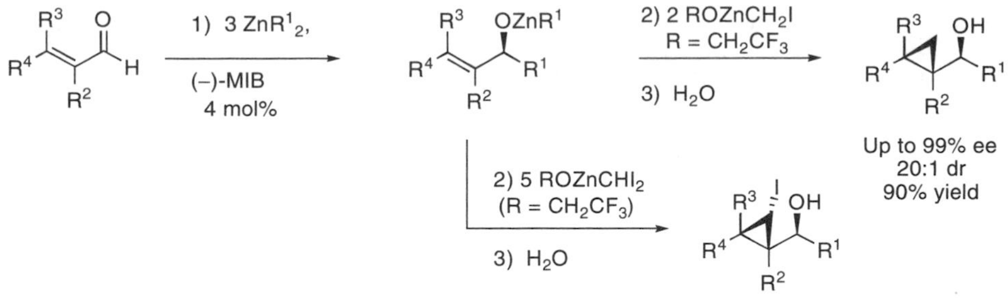

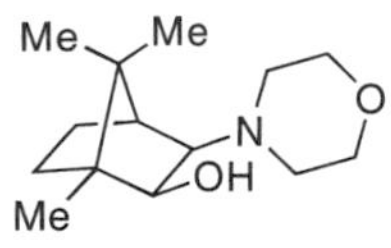

(–)-MIB

R^1	Cyclopropyl alcohol	ee (%)	dr	Yield (%)
Et	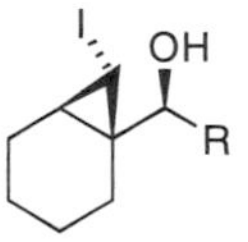	99	> 20:1	68
$(CH_2)_5OTBDPS$		98	> 20:1	70
Me	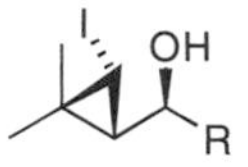	99	> 20:1	78
Et		95	> 20:1	62
$(CH_2)_4$*i*-Pr		96	> 20:1	60
Et	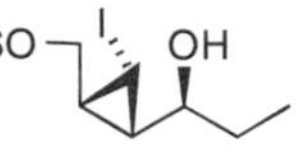	98	> 20:1	74

Figure 14.18. One-pot synthesis of cyclopropyl alcohols and iodocyclopropyl alcohols.

of 1,2,3-substituted cyclopropanes, which are common in nature and biologically active compounds.

There has been a recent resurgence in enantioselective organocatalytic chemistry.[61] Typically, these reactions involve covalent-bond formation between a nonmetallic catalyst species and one or more substrate molecules to form a transient reactive intermediate. The close proximity afforded by the covalent bond often allows very high enantioselection, because the stereochemical information of the organocatalyst is efficiently conveyed to the reacting partners. As is the case for many other types of asymmetric catalysts, organocatalysts are also amenable to incorporation into multistep processes[26] and may, in principle, be utilized at early or later stages in multistep process. To date, however, most multistep processes with chiral organocatalysts employ them in the first stage. Subsequent reactions occur either in sequential or cascade processes and can afford high diastereoselection.

In the first example illustrated below, an intermolecular enantioselective organocatalytic reaction is followed by a diastereoselective intramolecular cyclization (**Figure 14.19**).[62] A chiral imidazolidine catalyzes the enantioselective Michael addition of β-ketoesters to enones and the resultant adduct then undergoes a diastereoselective aldol cyclization to yield the cyclohexanone product. The observed high enantioselectivity is a consequence of the first step in which one face of the iminium intermediate is effectively shielded by the benzyl group of the chiral catalyst. However, a mixture of diastereomers is observed at this stage due to the epimerizable β-ketoester center. High diastereoselection at two further stereocenters, including the β-ketoester center,

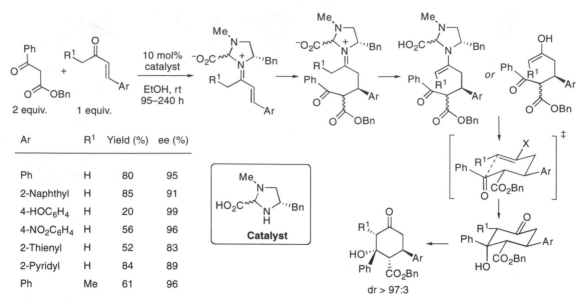

Figure 14.19. Organocatalyzed intermolecular Michael addition followed by intramolecular aldol cyclization.

is established in the last step, where the most stable six-membered transition structure with all equatorial groups intervenes. In this cascade process, the chiral imidazolidine serves three roles. First, it activates the Michael acceptor by iminium-ion formation. Second, it epimerizes the β-ketoester via deprotonation. Third, it acts as a base in the intramolecular aldol reaction. However, the possibility of intramolecular aldol via an enamine mechanism cannot be completely excluded.

Another organocatalytic cascade reaction that utilizes an iminium activation reaction is illustrated in **Figure 14.20**.[63] Here, the second step does not rely on the enamine/enol intermediate formed upon addition to the chiral α,β-unsaturated iminium. Rather, the iminium formed from the indole nucleophile traps the Boc-protected amine in a diastereoselective process that generates the fused pyrroloindoline ring system. The rapid generation of complex systems containing challenging quaternary centers underscores the utility of this approach.

Chiral secondary amine catalysts that form reactive enamines versus iminium ions can also be employed in multistep processes. For example, **Figure 14.21** illustrates a sequential process in which proline catalyzes the asymmetric addition of an enolizable aldehyde to an iminoester via a chiral enamine adduct.[64] After hydrolysis of the resultant iminium, the β-amino aldehyde is treated further with diethyl aluminum cyanide, which results in highly diastereoselective cyanohydrin formation. If the second portion of the sequential process is conducted at low temperature, the cyanohydrin adduct can be isolated directly. Alternatively, reaction at room temperature provides the cyclic adduct. Overall, this sequential process provides a simple and highly stereoselective means for generating γ-cyano-γ-hydroxy-α-amino acids via a three-component coupling.

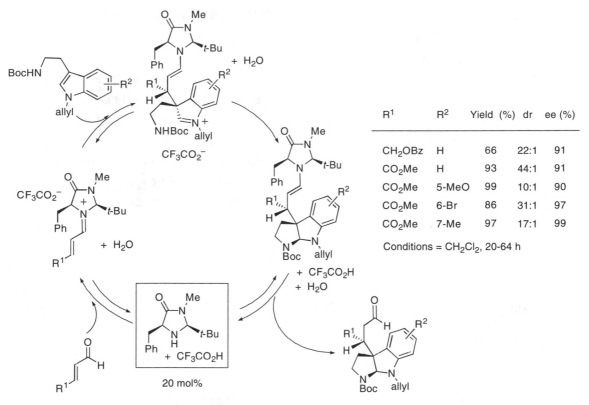

Figure 14.20. Enantioselective pyrroloindoline formation via a catalytic cascade reaction.

R¹	R²	Yield (%)	dr	ee (%)
CH₂OBz	H	66	22:1	91
CO₂Me	H	93	44:1	91
CO₂Me	5-MeO	99	10:1	90
CO₂Me	6-Br	86	31:1	97
CO₂Me	7-Me	97	17:1	99

Conditions = CH₂Cl₂, 20-64 h

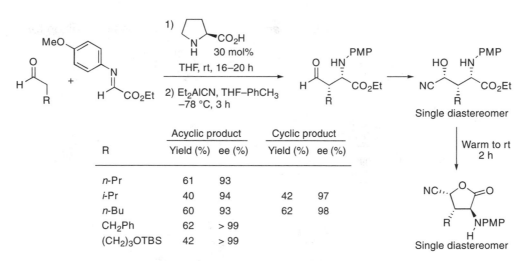

Figure 14.21. Sequential process consisting of an organocatalyzed iminoester addition followed by diastereoselective hydrocyanation reaction.

	Acyclic product		Cyclic product	
R	Yield (%)	ee (%)	Yield (%)	ee (%)
n-Pr	61	93		
i-Pr	40	94	42	97
n-Bu	60	93	62	98
CH₂Ph	62	> 99		
(CH₂)₃OTBS	42	> 99		

Since the aldehyde products of the first portion of the above process can serve as partners in many types of reactions, considerable versatility is possible in these multi-step processes. For example, the aldehyde products described above have been intercepted with an allyl nucleophile that is generated in situ from allyl bromide and indium metal (**Figure 14.22**).[65] While enantioselection in the first step of this sequential process was very high (≥ 93% ee), the diastereoselection in the second step was low, ranging from 1:1 to 2:1. Even so, the fact that a nonbasic nucleophilic species (vs. a Grignard or alkyl lithium nucleophile) could be generated in situ under mild conditions and without removal of the proline catalyst holds promise.

A related report describes a sequential enantioselective/diastereoselective process[66] (**Figure 14.23**) in which the first step is an organocatalytic enantioselective aldehyde hydroxylation catalyzed by proline.[67,68] After completion of the rapid hydroxylation, indium metal and allyl bromide are added to generate an allyl–indium nucleophile. Subsequent diastereoselective addition provides the diol equivalent, which can be readily transformed to the diol with either H_2/PtO_2 or $Cu(OAc)_2$. While enantioselection in the first step was very high ($\geq 97\%$ ee), the diastereoselection in the second step was again low, ranging from 3:2 to 4:1. Clearly, a sufficiently high bias

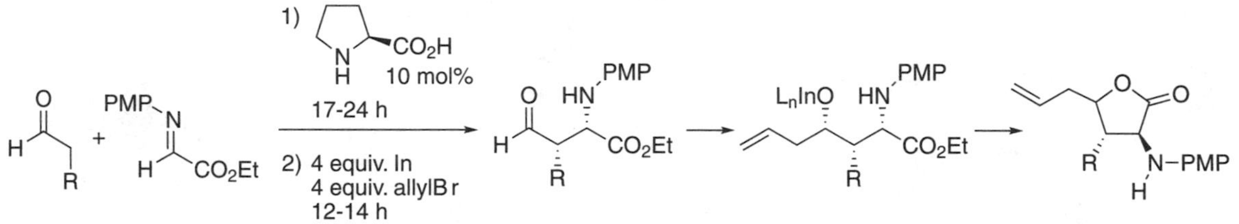

R	Solvent	Yield (%)	dr	ee (%)
i-Pr	THF	63	2:1	93
⌇⌇⌇	H₂O/THF	77	1:1	>99

Figure 14.22. Sequential process consisting of an organocatalyzed iminoester addition followed by allylation.

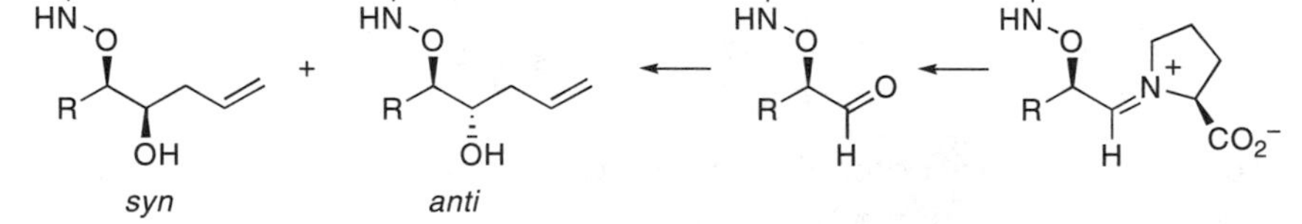

R	Yield (%)	syn:anti	ee (%) syn/anti
Me	80	3:2	98/98
i-Pr	71	5:3	99/97
n-Pr	65	3:2	98/98
n-Bu	82	3:2	98/98
CH₂Ph	74	3:2	99/>99
⌇⌇⌇	70	4:1	97/>99

Figure 14.23. Sequential enantioselective organocatalytic hydroxylation and aldehyde allylation.

between aldehyde stereofaces is absent with the allyl–indium nucleophile. This situation highlights a distinct problem in stereoselective synthesis—namely, the use of internal versus external diastereocontrol (see Chapter 13). In the next section, this problem is discussed in the context of several examples.

14.3 Independent Enantioselective Catalytic Reactions in a Multistep Process

Much effort has been dedicated to use of internal stereocontrol to afford highly diastereoselective processes. However, a need for optimization of conditions and a lack of generality can detract from this approach. The use of multiple chiral catalysts in later steps to effect stereoselective reactions, after formation of one stereocenter or set of stereocenters, is typically viewed as less than ideal. For the synthesis of a given structure with defined stereochemistry, this may be true, as chiral catalysts are often high-value items. With respect to efficient access to all possible diastereomers in pure form, however, the use of additional chiral catalysts offers the advantages of speed and flexibility. This premise requires that catalyst stereocontrol dominates substrate diastereocontrol in such diastereoselective processes (see Chapter 13).

If two (or more) different catalysts were employed for two (or more) stereoselective reactions in a reaction sequence, and each catalyst would give rise to complete stereocontrol, a great deal of diversity could be efficiently assembled. The potential advantages of this approach have been recognized with respect to carbohydrate compounds and a *general* method for the controlled stereoselective synthesis of a stereoisomeric polyol series has been devised. The selective synthesis of the many stereoisomeric versions of such carbohydrate precursors and their derivatives has important applications to problems in carbohydrate recognition.[69] This general method revolves around the used of successive asymmetric dihydroxylation and asymmetric enzymatic aldol reactions (**Figure 14.24**).[70] A key feature of this approach is that the second stereoselective reaction catalyzed by the enzymatic aldolases are subject to complete catalyst stereocontrol. Thus, each stereoisomeric compound in this complex series can be synthesized selectively. In contrast, use of internal substrate diastereocontrol to achieve such levels of stereoselection for different stereoisomeric arrays is notoriously substrate-dependent and can require significant optimization.[71–73]

In the first step of this method, asymmetric osmium-catalyzed dihydroxylation of the unsaturated acetal yields either enantiomeric diol, depending on whether AD-mix-α or AD-mix-β is employed (**Figure 14.25**). Removal of the acetal then provides an aldehyde that is subjected to an enzymatic aldol addition to dihydroxyacetone phosphate, followed by an enzymatic dephosphorylation with acid phosphatase. Several aldolases that are specific for the dihydroxyacetone donor, which provide strong stereochemical induction, and which possess wide substrate specificity, are available,

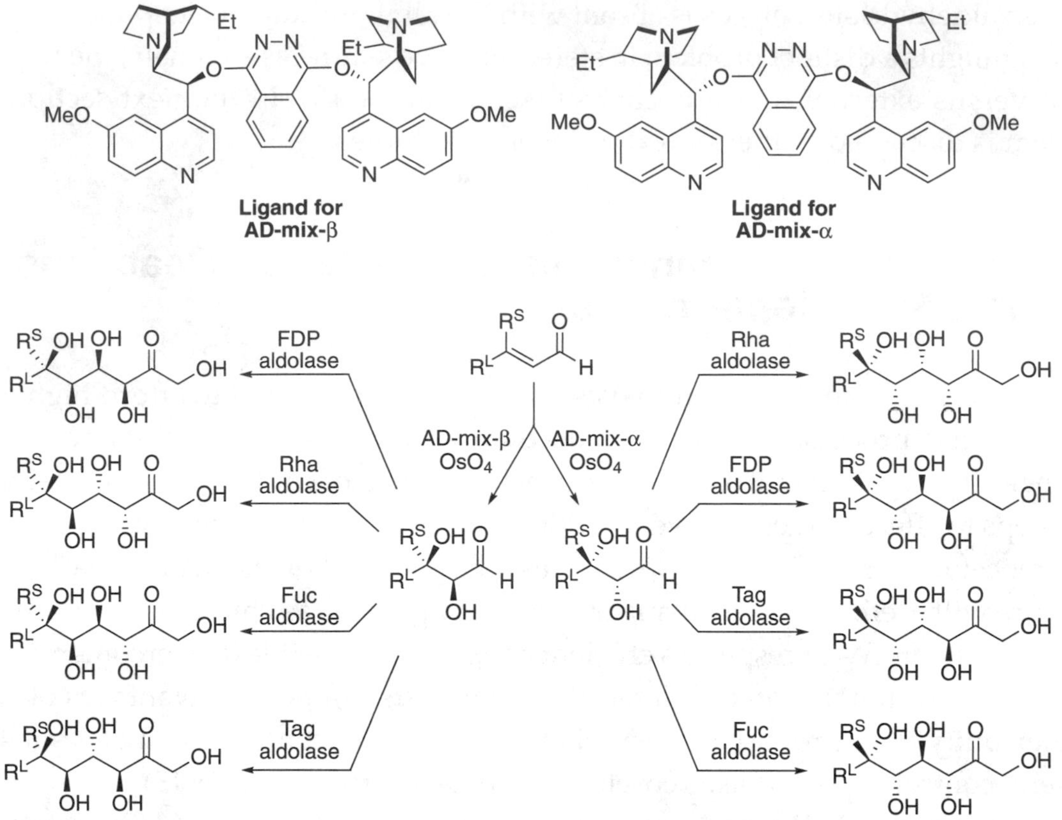

Figure 14.24. Independent stereocontrol in the sequential alkene dihydroxylation and enzymatic aldol reactions to provide novel carbohydrate derivatives.

including D-fructose diphosphate (FDP—provides 3S,4R stereochemistry) and L-rhamnulose 1-phosphate (Rha—provides 3R,4S stereochemistry) aldolase. Complete diastereocontrol was exercised by the aldolase in the aldol addition, allowing the products to be isolated as single diastereomers and with the enantioselectivity afforded by the asymmetric dihydroxylation. In the application shown in **Figure 14.25**, the combination of dihydroxylation catalysts and aldolases allowed the enantiomeric series to be accessed in pure form.

Formally, this sequential set of reactions does not fall under the multistep processes, as defined in the introduction, since not all of the reactions were performed in one operation without purification. However, the last portion of this method, the aldol and dephosphorylation reactions, does represent a sequential multistep process. Regardless, this example demonstrates the power of sequenced methods for generating diverse structures in a rapid and efficient manner. The remainder of this section is devoted to multistep processes in which two or more steps are affected with complete catalyst stereocontrol.

In further studies, a multistep aldol/aldol process was demonstrated as possible with the 2-deoxyribose-5-phosphate (DER) aldolase (**Figure 14.26**).[74] DER aldolase is one of the only aldolases demonstrated to accept two or three aldehyde substrates in a sequential and stereoselective manner. When α-substituted aldehydes are used that

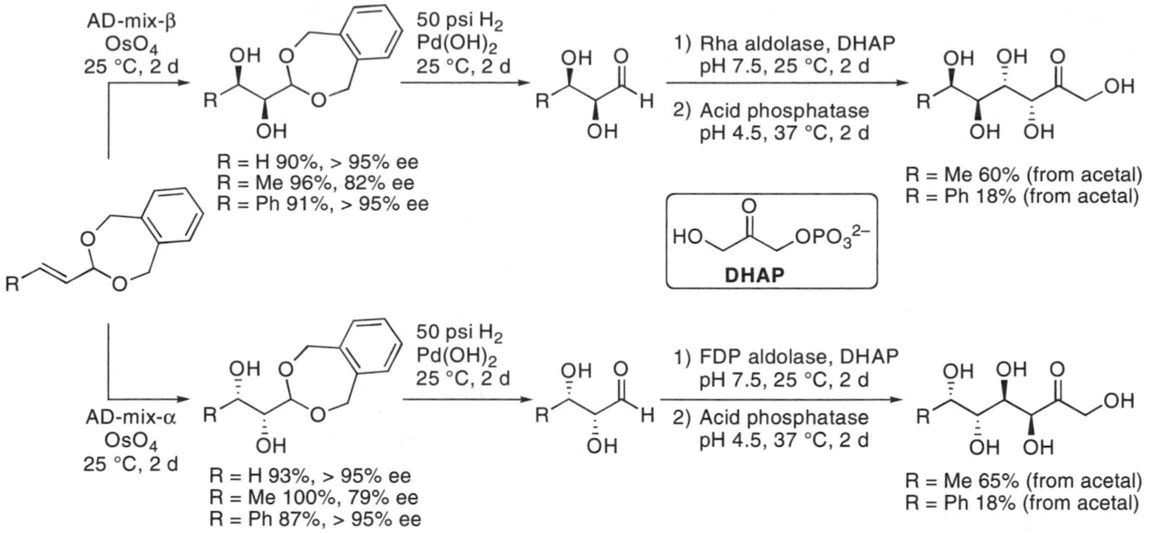

Figure 14.25. Sequential asymmetric alkene dihydroxylation, acetal deprotection, stereoselective aldol, and dephosphorylation to provide novel carbohydrate derivatives.

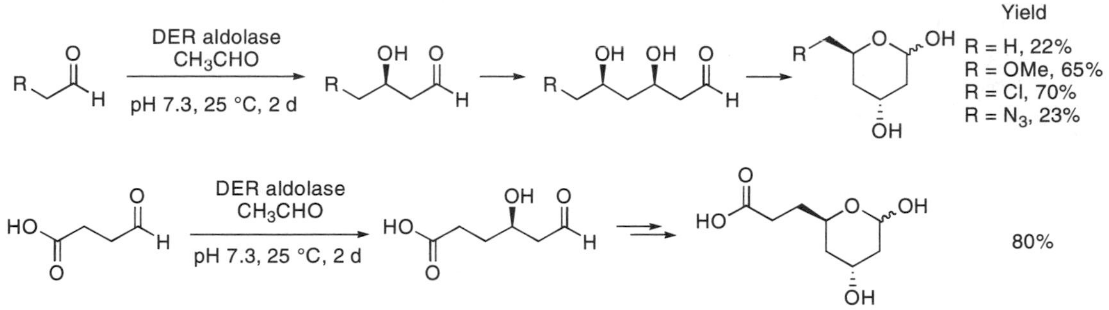

Figure 14.26. Enzyme-catalyzed cascade aldol/aldol/acetalization process.

contain functionality that will not cyclize after the first addition, a second aldol addition can occur, leading to products that cyclize to pyranose forms. In this cascade process, the products are formed with high enantioselectivity and diastereoselectivity.

In the previous process, aldol reactions were performed iteratively by one catalyst building up the structure in one direction. With the appropriate choice of precursor, a similar process should be possible in which aldol reactions are performed on different portions of a precursor. Assembly of new structures by building up two ends is a powerful and efficient method, allowing the rapid generation of complexity.[75] An example of a two-directional asymmetric reaction is illustrated in **Figure 14.27**.[76] Here, an ene reaction of a terminal double bond of a silyl enol ether with an aldehyde provides a new silyl enol ether with a terminal double bond. This intermediate can then undergo a second ene reaction to provide the (R,R)-product, where bonds have been made to each terminus of the original silyl enol ether precursor. This type of cascade process is highly advantageous, as reaction with one functional group reveals the second new functional group. As such, the relative reactivity of the two functional groups does not need to be considered (see **Figure 14.29** below).

For each ene reaction in **Figure 14.27**, the titanium–BINOL catalyst predominates the stereocontrol and provides the (R)-configuration. This premise was tested by subjecting the (R)- and (S)-adducts from the first reaction to the same chiral (BINOLate)Ti-based catalyst (**Figure 14.28**, top). In each instance, the new stereocenter was formed with the (R)-configuration. A slight degree of substrate diastereocontrol was observed as the (R)-compound reacted with higher selectivity than the (S)-compound. The fact that the (R)-stereocenter is formed more rapidly from either the (S)- or (R)-configured monoadduct leads to asymmetric amplification in the two-component reaction (**Figure 14.28**, top). Namely, the (R)-monoadduct forms the (R,R)-diastereomer 34-fold more rapidly than the meso, (R,S)-diastereomer. Furthermore, the chiral catalyst acts on the (S)-monoadduct to also provide the (R)-configuration at the second stereocenter 17-fold more rapidly. As a result, the (S)-monoadduct is siphoned off to the meso bis(adduct) to a greater extent than the (R). As a result, the enantiomeric excess increases for the reaction in **Figure 14.27** from 98.4% in the

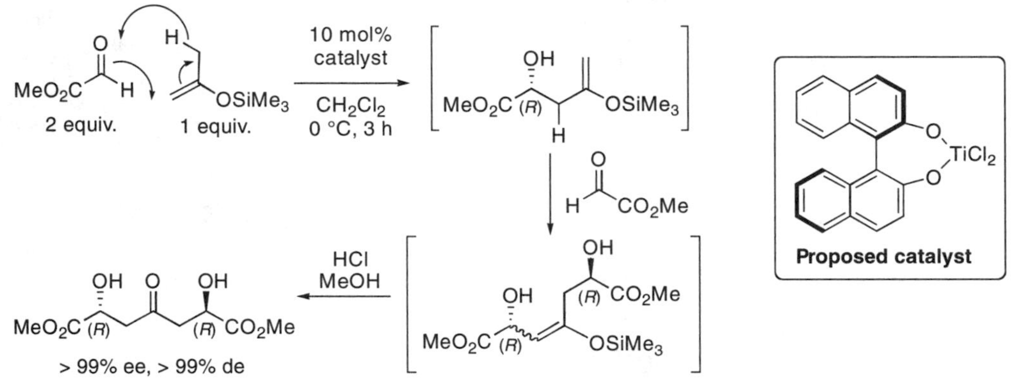

Figure 14.27. Two-directional asymmetric ene reactions.

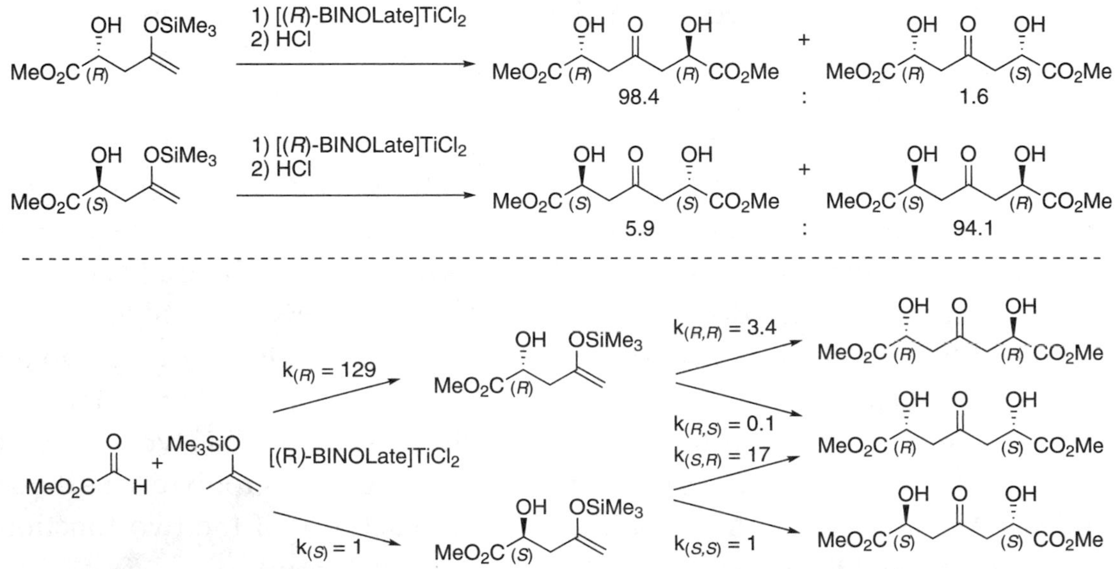

Figure 14.28. Substrate vs. catalyst stereocontrol in the two-directional ene reaction.

monoadduct to 99.6% in the bis(adduct). This increase in enantiomeric excess comes at the expense of formation of a greater quantity of the meso diastereomer. This example is reminiscent of certain desymmetrization reactions presented in Chapter 10.

In the examples of multistep processes with independent enantioselective catalytic reactions discussed above, the two independent reactions are very similar (i.e., two aldol reactions or two ene reactions). The development of multistep processes where the independent reactions are fundamentally different is much more challenging.[17] Such a process has been realized in the asymmetric cyanation and nitroaldol (or Henry) reaction of the bis(aldehyde) illustrated in **Figure 14.29**.[77] In this sequential process, the order of the reactions is controlled by the reagents present. In addition, the two reactive aldehyde moieties are chemically distinct and the more electrophilic alkyl aldehyde undergoes reaction first. Due to the distance between the two reactive sites (seven bonds), substrate diastereocontrol in the second reaction will be weak. Thus, both stereocenters are controlled by the chiral yttrium–lithium–BINOLate-based catalyst (YLB). In the first cyanation reaction, the chiral catalyst generates the acyl cyanohydrin stereocenter with 98% ee. In the second nitroaldol reaction, the chiral catalyst generates the hydroxyl stereocenter with approximately 57% ee, leading to a diastereomeric ratio of 3.7:1. Control experiments with monoaldheydes confirm this assertion: 1) the chiral catalyst provides the hexanal cyanation product in 94% ee and 2) the chiral catalyst provides the benzaldehyde nitroaldol product in 59% ee. This sequential process is even more complex, in that the catalyst identity is different in the two steps. In the first step, the additives water, n-butyl lithium, and a triarylphosphine oxide are crucial to the high enantioselection and presumably alter the active catalyst. Unfortunately, these same additives were found to be deleterious to the enantioselection in the second step (decreases to 11% ee). Thus, lithium tetrafluoroborate was added in the second step to offset the additives introduced in the first step. For discussion on the substrate activation by YLB-type catalysts, see Chapter 12 on bifuctional catalysis.

Another example of two independent enantioselective reactions by one chiral catalyst is illustrated in **Figure 14.30**.[78] First, an asymmetric hetero-Diels–Alder reaction

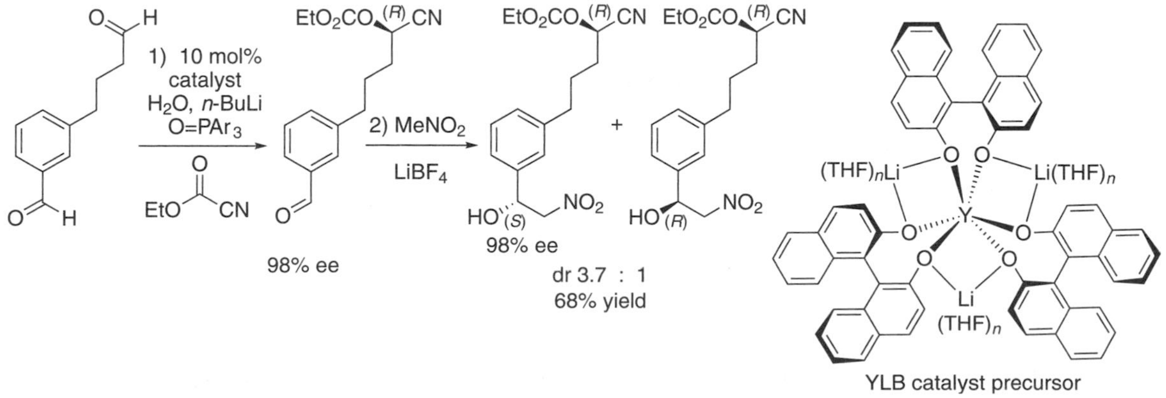

Figure 14.29. Sequential independent enantioselective cyanation and enantioselective Henry reactions of chemically distinct aldehydes catalyzed by the same metal complex.

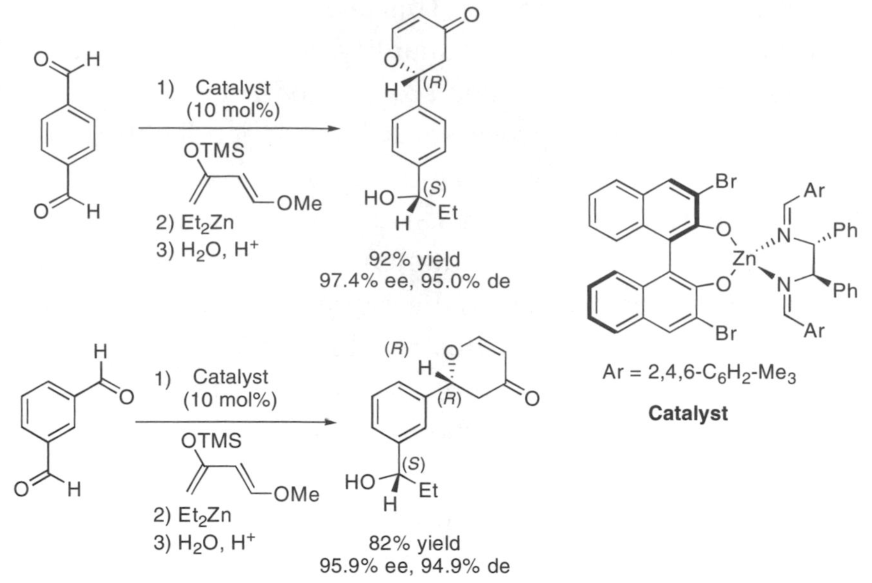

Figure 14.30. Sequential independent enantioselective hetero-Diels–Alder and enantioselective aldehyde alkylation reactions with chemically equivalent aldehydes catalyzed by the same metal complex.

with Danishefsky's diene occurs, followed by an asymmetric aldehyde alkylation with Et$_2$Zn. For the *para*-substituted case, < 3% of the double hetero-Diels–Alder reaction and double aldehyde-alkylation products were seen. For the *meta*-substituted case, 6.4% of the double hetero-Diels–Alder reaction and 10% of the double aldehyde-alkylation products were seen. Again, one metal complex catalyzes two very different transformations sequentially to two reactive functional groups present at the beginning of the reaction. The order of the transformations is controlled by the stepwise addition of the two reagents. Notably, if both aldehydes have equal reactivity in the first step, then a statistical mixture would be expected (1:2:1 starting material:mono-Diels–Alder product:double-Diels–Alder product). However, after the first Diels–Alder reaction, the remaining aldehyde is less electrophilic due to the loss of the other electron-withdrawing aldehyde. As such, double hetero-Diels–Alder products can be avoided in the first step.

In contrast to the prior example, the enantioselectivities exerted by the chiral zinc catalyst in the two reactions are similar, with 97.4% ee in the hetero Diels–Alder reaction and approximately 95% ee in the aldehyde alkylation. Again, these selectivity values are comparable to those observed in the same reactions with the monofunctional substrate, benzaldehyde, indicating that substrate diastereocontrol is overshadowed by catalyst stereocontrol.

In the above reactions (**Figures 14.27–14.30**), the same catalyst is used in two steps. The use of two different catalysts is challenging, as the catalysts must not act upon each other and must only react with their targeted chemical moiety. An elegant example is illustrated in **Figure 14.31.**[79] Both the rhodium and ruthenium catalysts are present during the entire reaction sequence. At a pressure of 10 atm, the ruthenium catalyst does not react and only a highly enantioselective rhodium-catalyzed enamide hydrogenation occurs. Upon raising the pressure, the ruthenium catalyst then reduces the ketone with high diastereoselectivty.

Another interesting strategy toward multistep processes with independent catalyst stereocontrol in two steps is seen in the sequential cyclopropanation and alkene dihydroxylation reactions shown in **Figure 14.32**.[80] Rather than formulating slightly different metal catalysts during the reaction sequence or using the same metal catalyst moiety for two different reactions, a complex containing two separate catalyst species was employed. In generating the linked catalyst, the order of metal addition is crucial, since the different metal binding sites of the bifunctional ligand can compete for

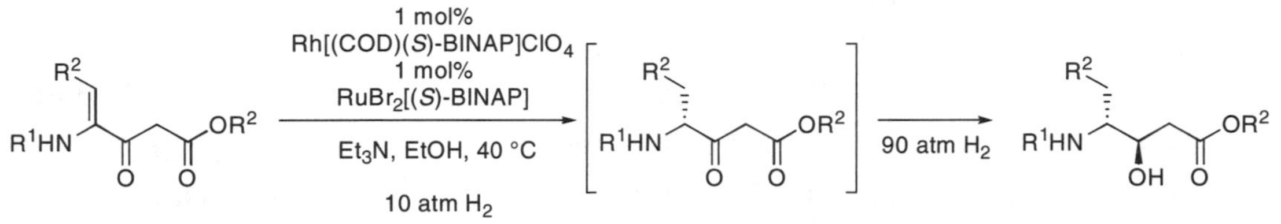

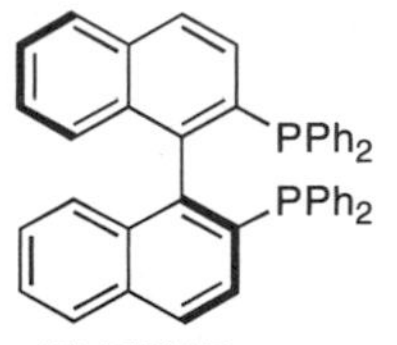

R^1	R^2	R^3	Yield (%)	dr	ee (%)
Ac	Ph	Et	99	> 95:5	> 95
Ac	Ph	Me	99	> 95:5	> 95
Boc	Ph	Et	99	> 95:5	> 95
Ac	4-ClC$_6$H$_4$	Et	90	> 95:5	> 95

Figure 14.31. Sequential independent enantioselective enamide and ketone hydrogenations.

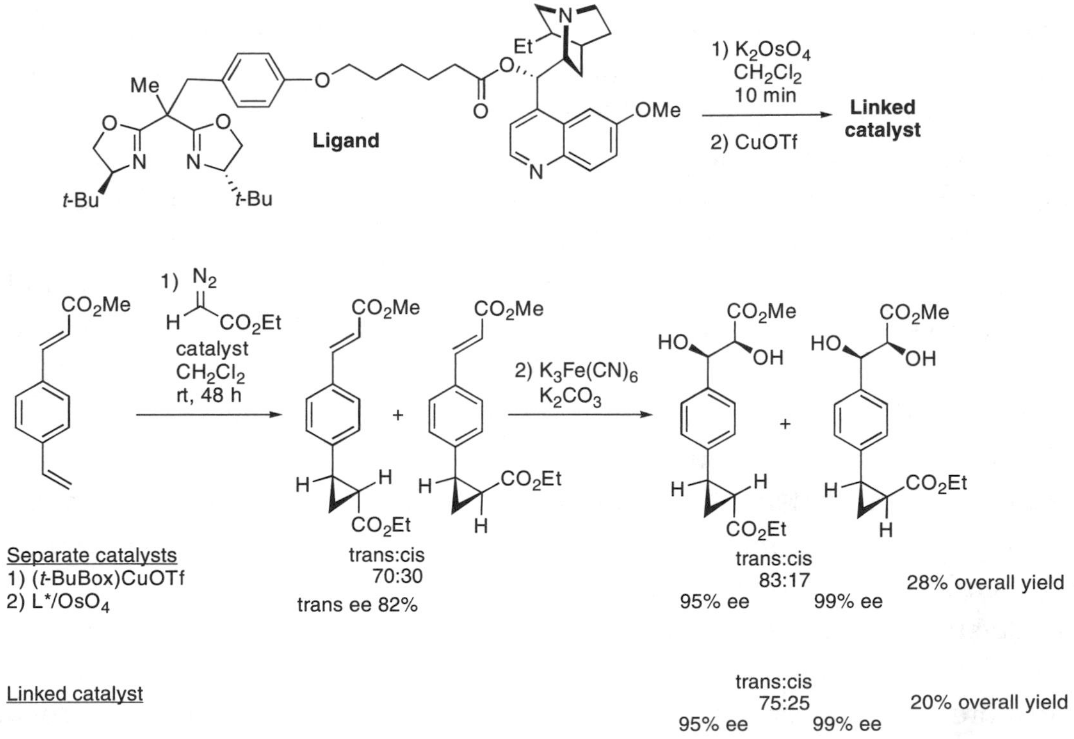

Figure 14.32. A linked catalyst containing two different catalytic subunits that catalyze sequential asymmetric cyclopropanation and dihydroxylation reactions of chemically distinct alkenes within one substrate.

the copper and osmium cations. Adding potassium osmate first and then copper(I) triflate allows selective complexation of copper and osmium ions by the bis(oxazoline) and the dihydroquinidine moieties, respectively.[81] After the reaction, the chiral ligand could be recovered in 91% yield after the chromatographic purification of the products.

The order of reagent addition in this sequential process controls the order of reaction of the functional groups of the substrate. The stereochemistry in each step is established independently by each respective metal center (catalyst control vs. substrate diastereocontrol; see Chapter 13). For the cyclopropanation, 82% ee is observed when mononuclear (t-BuBox)Cu(OTf) is employed, whereas 50% ee was observed with the linked catalyst. The asymmetric dihydroxylation component was more well-behaved and provided 95–99% ee with both the mononuclear and linked dihydroxylation catalysts. Notably, the overall yield for the linked catalyst is comparable to that obtained with the separate catalysts. It should be borne in mind that, although the enantioselectivity in the cyclopropanation is low, the highly selective dihydroxylation converts the minor enantiomer of the cyclopropanation product into a diastereomer, resulting in an increase in the enantiomeric excess of the dihydroxylated product (see Chapter 8, Parallel Kinetic Resolution).

Many of the issues encountered in this archetype of linking two catalysts into one structure, including deleterious interactions between the two catalysts and catalyst recovery, can be resolved by using solid supports to contain two or more different catalysts. The rigidity of the solid support limits the catalyst mobility and thereby prevents different catalyst sites from interacting. In addition, solid-supported catalysts are conveniently recovered by filtration. A further discussion of multiple catalysts on solid supports can be found in Section 15.5.

A rare example of a cascade-type process with two different chiral catalysts can be found in the asymmetric hydrofluorination reaction illustrated in **Figure 14.33**.[82] In this one-pot process, the two different catalysts exercise stereocontrol in two separate catalytic cycles. In the first catalytic cycle (the iminium cycle), an imidazolidinone catalyst forms an iminium species that reacts with a hydride nucleophile. Subsequent hydrolysis of the product enamine gives rise to an aldehyde intermediate with an enolizable hydrogen, which is required in the second catalytic cycle (the enamine cycle). Reaction with the second imidazolidinone catalyst generates an enamine that then reacts with a fluoride electrophile. This transformation is especially striking, given the similarity between the two catalysts and the fact that they are both present during the latter portion of the sequence (the enamine cycle). Nonetheless, cycle-specific catalysis must be occurring, due to the observed stereochemical outcome (the *syn* vs. *anti* selectivity) with the two different catalyst combinations. Furthermore, catalyst stereocontrol predominates over substrate stereocontrol in the second reaction (see Chapter 13). A consequence of this catalyst stereocontrol is that all four diastereomeric products are available by selection of the appropriate catalyst combinations (those illustrated in **Figure 14.33** and their corresponding enantiomeric pairs).

An important benefit of combining asymmetric catalytic reactions in a sequence is the mathematical requirement leading to enantioenrichment in the second cycle. Even if the two individual catalytic cycles are only moderately selective, very high enantiomeric excesses can still be achieved (≥ 99% ee). For example, combining two 86% ee catalytic cycles in a cascade would furnish a 7:1 mixture of diastereomers, with the major diastereomer being formed in 99% ee (see Section 16.2.3 and Chapter 13 for further discussion).

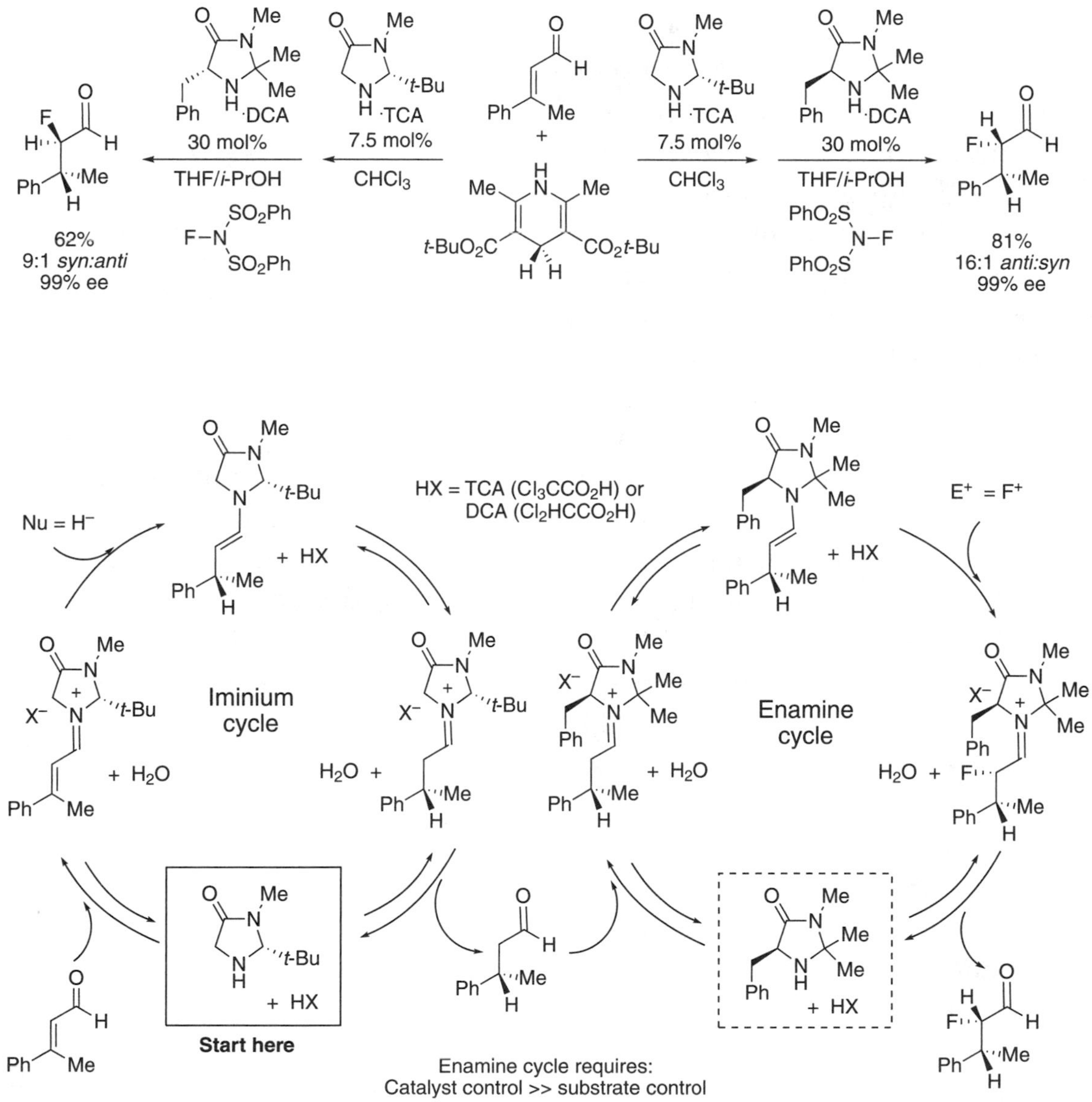

Figure 14.33. One-pot sequential process with two different chiral catalysts operating in separate cycles.

14.4 Chiral Catalysts Participating in Several Stages of a Multistep Process

Many multistep reactions are comprised of reactions activated in different ways. For example, one transformation in a multistep reaction might require a metal catalyst, a second transformation a base, and a third transformation occurs spontaneously after functionality is introduced in a prior transformation. In these cases, the most important criterion is that all the reagents/catalysts be mutually compatible or that sequential addition of reagents/catalysts be feasible. Other multistep processes utilize the same catalyst in more than one transformation

In the preceding sections, we have already seen several examples of a chiral catalyst participating in multiple steps of a process. In one instance, a chiral catalyst participates in a reaction in which no stereochemistry is introduced as well a second enantioselective transformation (Section 14.1.2; **Figure 14.7**). In other cases, the chiral catalyst initiates an asymmetric reaction and in a second step forms a reactive enolate (Section 14.2; **Figures 14.8–14.15** and **14.19**). In the last section, a single chiral catalyst participated in two steps of a process with independent enantioselection exercised at each stage (Section 14.3; **Figures 14.26, 14.27, 14.29,** and **14.30**). In this section, examples are highlighted in which a chiral catalyst catalyzes chiral and achiral transformations in a multistep process.

For example, the chiral copper catalyst utilized in asymmetric oxidative biaryl coupling was also highly effective in the Glaser–Hay coupling of alkynes to yield diynes.[83] For bifunctional substrates containing both an alkyne and 2-naphthol, the copper complex catalyzed both transformations with high efficiency and yielded polymeric materials containing alternating diyne and chiral binaphthyl linkages (**Figure 14.34**). Independent catalyst enantioinduction was found to occur in each successive biaryl bond formation (~73% ee at each biaryl center). The two reactions catalyzed by the copper complex are substantially different. The oxidative asymmetric biaryl coupling involves a chelated copper intermediate and redox chemistry, whereas the Glaser–Hay coupling involves π-complexation to copper and alkynyl copper intermediates. Even so, no crossover is apparent between these two divergent sets of copper adducts, as judged by similar levels of enantioselection observed in the biaryl coupling of the silylated alkyne. The chemoselection in both of these processes is extremely high (alkynes only couple with alkynes, naphthols only couple with naphthols), which eliminates the need for reaction sequences and allows for a concurrent process.

The alkyne portion of the substrate could be generated in several positions (i.e., aryl alkyne vs. alkyl alkyne), but in each case the Glaser–Hay and enantioselective biaryl couplings were highly efficient. Interestingly, these two reactions did occur at different rates, with Glaser–Hay coupling occurring first for the aryl alkyne followed

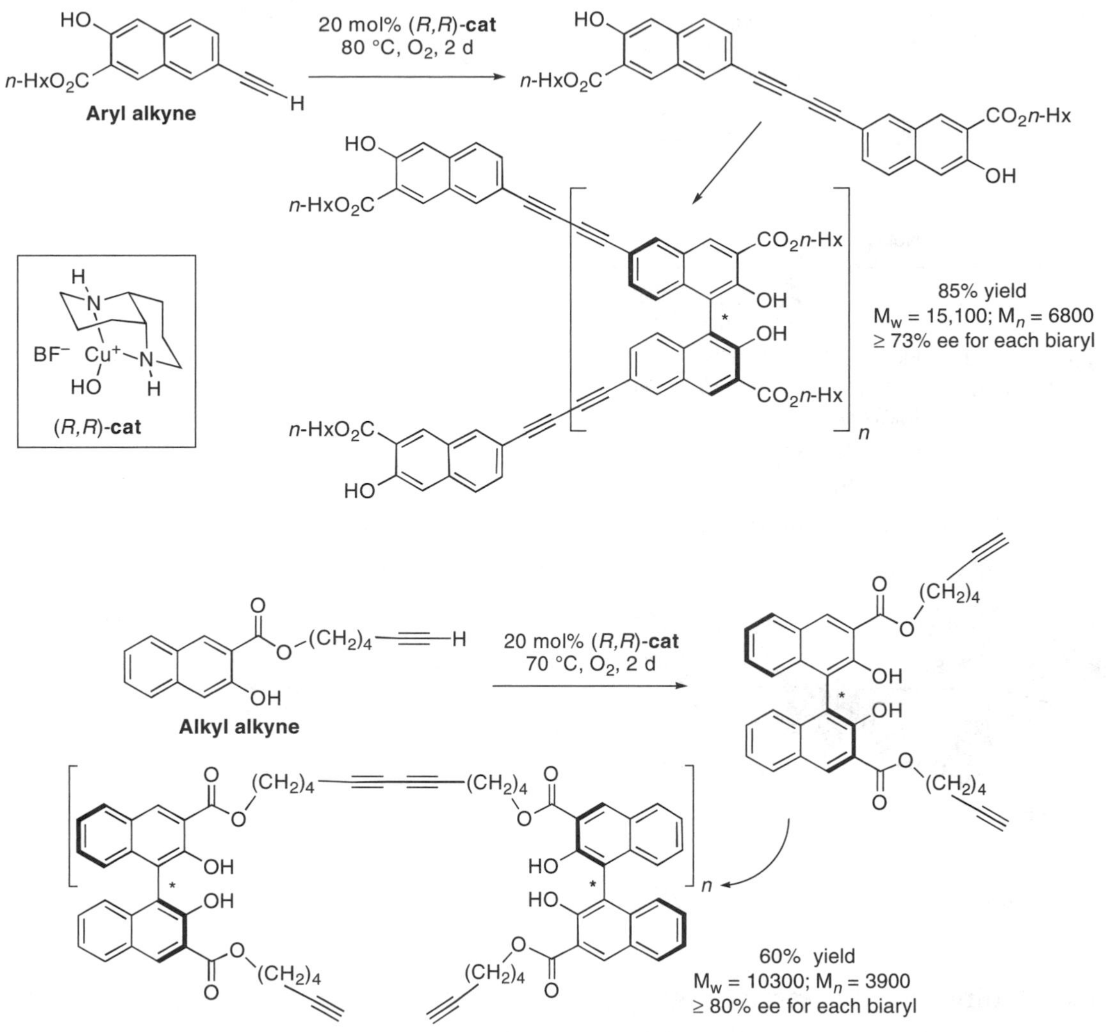

Figure 14.34. Concurrent Glaser–Hay coupling and enantioselective oxidative biaryl coupling catalyzed by one chiral copper complex (M_w = weight of an average polymer molecule, M_n = weight of all polymer molecules/number of polymer molecules).

by the asymmetric biaryl couplings. In contrast, the order is reversed for the second reaction illustrated, which commences with an alkyl-alkyne substrate. Notably, either reaction order is viable.

An elegant example of one organocatalyst performing multiple tasks is illustrated in **Figure 14.35** for the asymmetric synthesis of β-amino acid derivatives with high enantioselectivity and diastereoselectivity from acid chlorides and α-chloroamines.[84,85]

In the process illustrated in **Figure 14.35**, the chiral cinchona alkaloid catalyst benzoyl quinine (BQ) serves to catalyze no less than four separate steps (**Figure 14.36**): 1) dehydrohalogenation to form the ketene, 2) dehydrohalogenation to form the imine, 3) formal [2+2] cycloaddition by means of an enolate adduct, and 4) β-lactam cleavage to a β-amino ester or amide. In addition, the BQ catalyst forms a chiral enolate and is hence responsible for the absolute stereoselection in the [2+2] cycloaddition. With

BQ

R	NuH	yield	ee	dr
Ph	MeOH	62%	95%	12:1
PhO	MeOH	63%	95%	14:1
4-MeOC$_6$H$_4$	MeOH	62%	94%	10:1
4-ClC$_6$H$_4$	MeOH	60%	94%	12:1
Ph	PhCH$_2$NH$_2$	61%	95%	11:1
Ph	MeO$_2$CCH$_2$NH$_2$	43%	95%	11:1
Ph	SerOMe	42%	95%	12:1

SerOMe

proton sponge

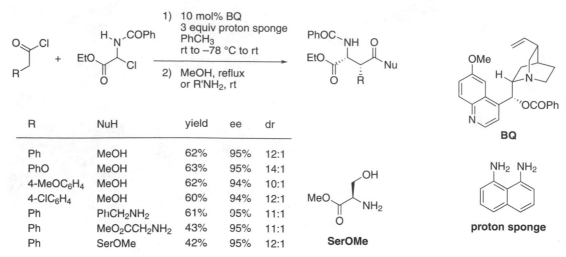

Figure 14.35. Formation of β-amino acid derivatives in a one-pot multistep procedure.

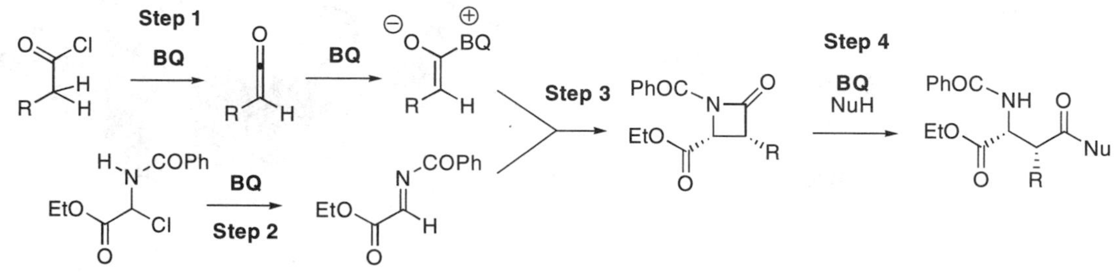

Figure 14.36. Steps of the BQ-catalyzed process.

benzoyl quinidine, the enantiomeric series is accessible in a comparable manner. Several experiments confirmed that the proton sponge is not the base for either dehydrohalogenation reaction. Rather, it is proposed that the BQ initiates the deprotonation and then the proton is shuttled to the stoichiometric proton sponge. Thus, BQ acts as a simple catalyst (no enantioinduction) for three reactions and as a chiral catalyst for one reaction (step 3). The first two of these reactions occur concurrently, while the remainder occur in a cascade fashion. Thus, this is an example of a mixed concurrent-cascade process.

When a trifluoroethyl ester is employed, a fifth role for the BQ is also possible. For the example in **Equation 14.1**, a further reaction can occur after cleavage of the β-lactam with serine to form an amide. BQ catalyzes the displacement of the trifluoroethyl ester with glycine to yield a further amide bond and form a tripeptide.[84,85]

Equation 14.1

Tripeptide
44%, 95% ee, 8:1 dr

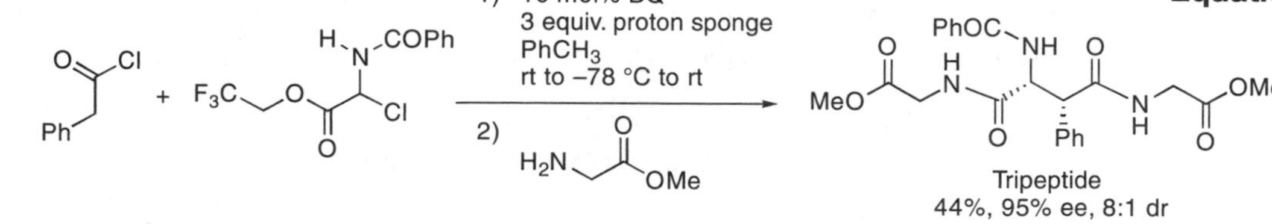

The processes described in this section are highly desirable in that a minimum of different catalysts and reagents are required to initiate complex multistep reactions that assemble multiple components. The advantages from the standpoint of efficiency are obvious, but are offset to some extent by the requirement that all of the reagents and catalysts be mutually compatible and that each component participates in a chemoselective manner.

14.5 Kinetic Resolution as Part of a Multistep Process

Much rarer are multistep processes in which one step is a kinetic resolution or dynamic kinetic resolution. In most of the reported cases, the first step is the kinetic resolution such that the reaction commences from racemic material. In principle, achiral materials could also be employed with the first step(s), generating a racemic intermediate followed by a later resolution step.

14.5.1 Simple Kinetic Resolution as Part of a Multistep Process

In this example, we revisit an unusual cascade transetherification/intramolecular hetero-Diels–Alder reaction (**Figure 14.7**).[39] The first transetherification step with an achiral alcohol and an achiral enol ether was catalyzed by the same chiral bis(oxazoline)–copper catalyst used to achieve the second intramolecular asymmetric Diels–Alder reaction. When a racemic alcohol was employed (**Figure 14.37**), it was found that the chiral bis(oxazoline)–copper catalyst effected a kinetic resolution in the first step, as well as a subsequent stereoselective Diels–Alder reaction.[86] By using two equivalents of the racemic alcohol, the Diels–Alder adduct from the (R)-alcohol could be obtained in 80% yield from the enol-ether precursor. Furthermore, very high stereocontrol was provided (97% ee, 92% de). For additional examples of kinetic resolutions, see Chapter 7.

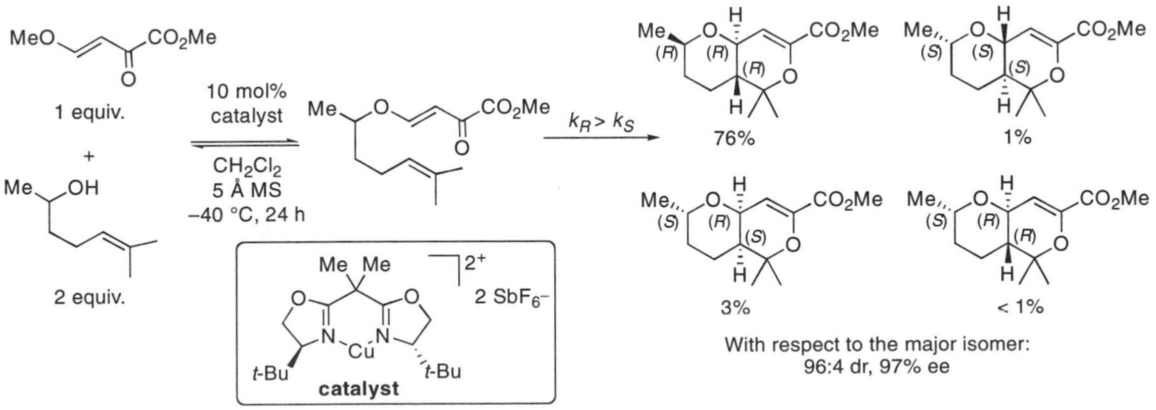

Figure 14.37. Cascade kinetic resolution via transetherification and asymmetric intramolecular hetero-Diels–Alder reactions catalyzed by a one metal complex.

14.5.2 Dynamic Kinetic Resolution as Part of a Multistep Process

In combining kinetic resolution with a diastereoselective intramolecular Diels–Alder reaction, high enantioselectivities were obtained (**Figure 14.37**).[87,88] The maximum yield based on the racemic alcohol precursor in this process, however, is 50%. Further work showed that a dynamic kinetic resolution (see Chapter 9 for further examples) was viable in such a sequence.[89] **Figure 14.38** outlines this strategy and its application to tricyclic enantioentriched intermediates useful for the synthesis of natural products such as compactin[90] and forskolin.[91–93]

The first two steps are catalyzed by a ruthenium complex that racemizes the starting alcohol. This reversible reaction is responsible for the dynamic portion of the kinetic resolution, converting the unreactive (S)-alcohol into the (R)-alcohol, which can undergo facile enzyme-catalyzed acylation. The product of the acylation reaction is poised for an intramolecular Diels–Alder reaction, which creates three additional stereocenters in a highly diastereoselective process (0–8% of other diastereomers observed). Overall, the entire sequence consists of four reactions and creates a single enantiomer containing four stereocenters from a racemic starting material with one stereocenter. This multistep process is a mixed concurrent-cascade process. The racemization steps occur concurrently while the transesterification/Diels–Alder reaction is a cascade process.

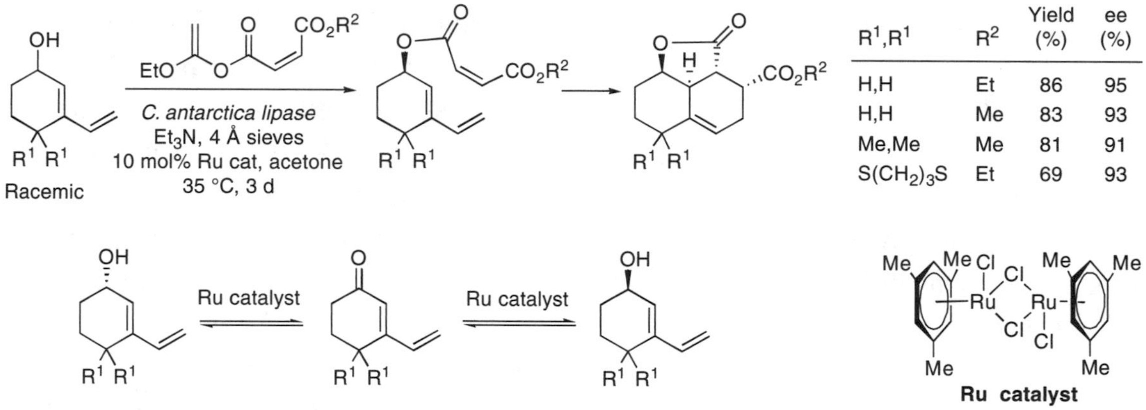

R¹,R¹	R²	Yield (%)	ee (%)
H,H	Et	86	95
H,H	Me	83	93
Me,Me	Me	81	91
S(CH₂)₃S	Et	69	93

Figure 14.38. Dynamic kinetic resolution in transesterification followed by diastereoselective intramolecular Diels–Alder cycloaddition.

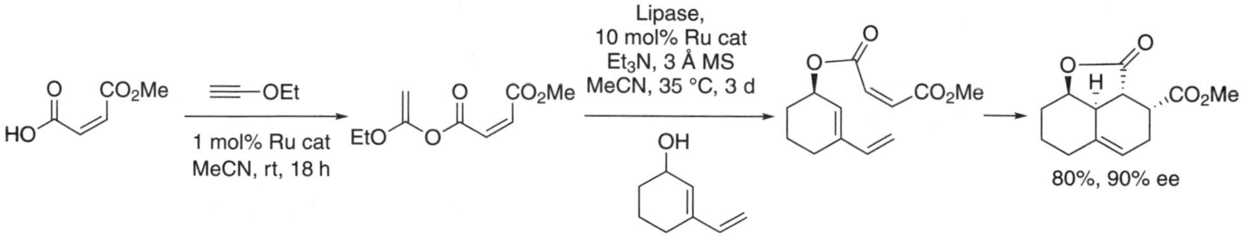

Figure 14.39. Combined sequential–cascade–concurrent multistep process. Active ester formation, dynamic kinetic resolution in transesterification, and diastereoselective intramolecular Diels–Alder cycloaddition.

In a further extension of this work, an additional transformation was added (**Figure 14.39**). Namely, the reaction commences with an acid and the requisite enol ether is made by ruthenium-catalyzed addition. Interestingly, the same ruthenium catalyst is employed in the later racemization process, albeit at a higher loading. Subsequent addition of the remaining components then initiates the same concurrent-cascade process outlined in **Figure 14.38**. In this mixed sequential-concurrent-cascade process, the product can be assembled from three components efficiently (80% yield) and with high selectivity (90% ee).

14.5.3 Parallel Kinetic Resolution as Part of a Multistep Process

Based upon earlier work in enantioselective carbometalation and diastereoselective intramolecular aldol reactions onto symmetric dicarbonyls (**Figure 14.40a**), an opportunity for parallel kinetic resolution was discovered with unsymmetric dicarbonyl substrates in the analogous reaction (**Figure 14.40b**).[48] In the example in **Figure 14.40b**, no chemoselectivity is exercised in the cyclization step, as both of the two different prostereoisomeric carbonyls undergo nucleophilic addition. This results in the formation of two different constitutional isomers. For each of the two prostereoisomeric carbonyls, only one diastereoface is attacked by the chiral enolate, which establishes three new stereocenters. In principle, this cascade reaction could have given rise to two different constitutional isomers, each with eight stereoisomeric compositions (all enantiomers and diastereomers included except for trans ring fusions). In practice, one

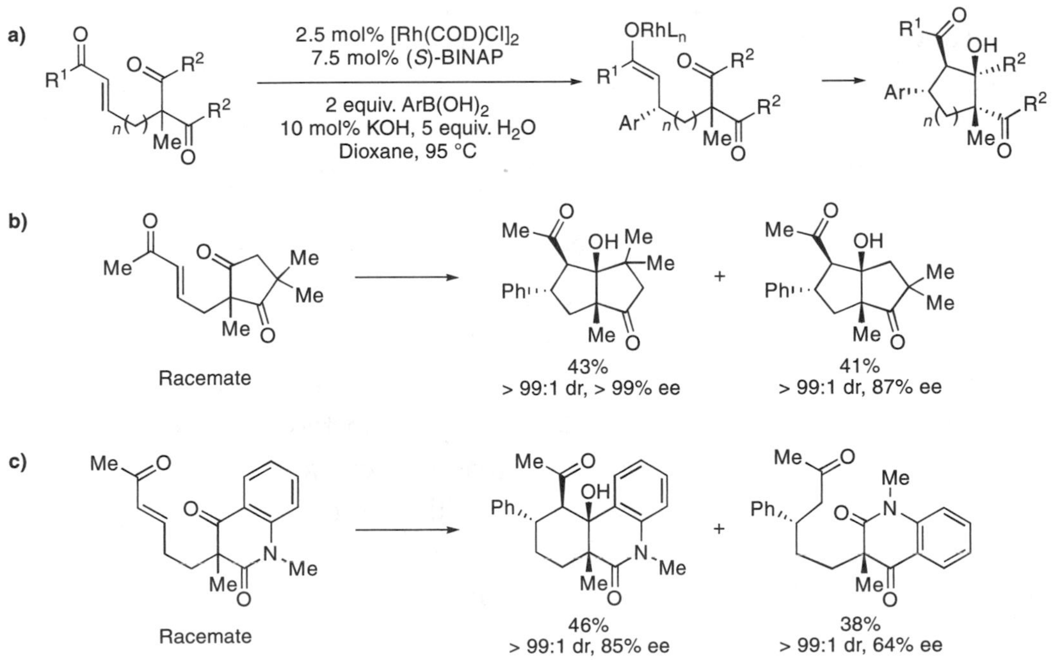

Figure 14.40. Parallel kinetic resolution in cascade enantioselective conjugate addition and diastereoselective cycloaldolization.

stereoisomer is obtained predominantly for each of the two constitutional isomers in **Figure 14.40b**.

In **Figure 14.40c**, the two carbonyls of the β-ketoamide moiety have different reactivities. Thus, the chiral rhodium enolate formed after the carbometalation can only attack the keto carbonyl. Attack of one face of the prostereoisomeric carbonyl by the chiral enolate is facile, resulting in the tricyclic product. On the other hand, attack by the chiral enolate onto the other carbonyl face to provide a diastereomer of the first product is unfavorable, giving rise to an acyclic product.

In both of these cascade reactions, the racemic precursor gives rise to two stereoisomerically pure products with different connectivities. Since the maximum yield of each product is 50%, this process occurs with high efficiency. In essence, the chiral catalyst mediates the transformation of one enantiomer of the precursor into one discrete compound and the transformation of the second enantiomer into a second discrete compound. Thus, the two enantiomers of the precursor are resolved from one another. For more examples of parallel kinetic resolutions, see Chapter 8.

Concluding Remarks

In this day and age, given sufficient time, resources, and determination, even the most complex target molecules can be prepared in high enantiopurity and diastereopurity. With the increasing complexity of targets undertaken by synthetic chemists, however, there is an escalating demand for the development of sophisticated methodologies to enable rapid assembly of simple precursors into complex products in an economical and environmentally conscious fashion. In order to enhance the efficiency with which target molecules can be assembled, new methods for constructing carbon skeletons must be introduced. Ideally, these improved methodologies exhibit high chemoselectivity, regioselectivity, enantioselectivity and diastereoselectivity. It is also desirable that the new reactions are compatible with many functional groups in order to reduce the number of protection and deprotection steps. The development of methodologies that fulfill these demands has been the central focus of this chapter.

One of the obstacles facing chemists developing tandem reactions is catalyst compatibility. As outlined in Chapter 15 on supported catalysts, multistep reactions that employ several catalysts can be performed using supported catalysts, even if the catalysts are not entirely compatible. Multistep reactions can also involve supported catalysts linked in a continuous flow reactor. Given the increasing importance of concise and efficient syntheses, it is anticipated that research into the development of new and more efficient multistep reactions will continue to grow in the coming years.

References

(1) Wender, P. A. Introduction: Frontiers in Organic Synthesis. *Chem. Rev.* **1996**, *96*, 1–2.

(2) Posner, G. H. Multicomponent One-Pot Annulations Forming 3 to 6 Bonds. *Chem. Rev.* **1986**, *86*, 831–844.

(3) Ho, T.-L. *Tandem Organic Reactions*; Wiley: New York, 1992.

(4) Tietze, L. F.; Belfuss, U. Sequential Transformations in Organic Chemistry: A Synthetic Strategy with a Future. *Angew. Chem., Int. Ed. Engl.* **1993**, *32*, 131–163.

(5) Bunce, R. A. Recent Advances in the Use of Tandem Reactions for Organic Synthesis. *Tetrahedron* **1995**, *51*, 13103–13159.

(6) Tietze, L. F. Domino Reactions in Organic Synthesis. *Chem. Rev.* **1996**, *96*, 115–136.

(7) Denmark, S. E.; Thorarensen, A. Tandem [4+2]/[3+2] Cycloadditions of Nitroalkenes. *Chem. Rev.* **1996**, *96*, 137–166.

(8) Winkler, J. D. Tandem Diels–Alder Cycloadditions in Organic Synthesis. *Chem. Rev.* **1996**, *96*, 167–176.

(9) Ryu, I.; Sonoda, N.; Curran, D. P. Tandem Radical Reactions of Carbon Monoxide, Isonitriles, and Other Reagent Equivalents of the Geminal Radical Acceptor/Radical Precursor Synthon. *Chem. Rev.* **1996**, *96*, 177–194.

(10) Parsons, P. J.; Penkett, C. S.; Shell, A. J. Tandem Reactions in Organic Synthesis: Novel Strategies for Natural Product Elaboration and the Development of New Synthetic Methodology. *Chem. Rev.* **1996**, *96*, 195–206.

(11) Wang, K. K. Cascade Radical Cyclizations via Biradicals Generated from Enediynes, Enyne-Allenes, and Enyne-Ketenes. *Chem. Rev.* **1996**, *96*, 207–222.

(12) Padwa, A.; Weingarten, M. D. Cascade Processes of Metallo Carbenoids. *Chem. Rev.* **1996**, *96*, 223–270.

(13) Harvey, D. F.; Sigano, D. M. Carbene–Alkyne–Alkene Cyclization Reactions. *Chem. Rev.* **1996**, *96*, 271–288.

(14) Malacria, M. Selective Preparation of Complex Polycyclic Molecules from Acyclic Precursors via Radical Mediated- or Transition Metal-Catalyzed Cascade Reactions. *Chem. Rev.* **1996**, *96*, 289–306.

(15) Molander, G. A.; Harris, C. R. Sequencing Reactions with Samarium(II) Iodide. *Chem. Rev.* **1996**, *96*, 307–338.

(16) Snider, B. B. Manganese(III)-Based Oxidative Free-Radical Cyclizations. *Chem. Rev.* **1996**, *96*, 339–364.

(17) Ajamian, A.; Gleason, J. L. Two Birds with One Metallic Stone: Single-Pot Catalysis of Fundamentally Different Transformations. *Angew. Chem., Int. Ed. Engl.* **2004**, *43*, 3754–3760.

(18) Dax, S. L.; McNally, J. J.; Youngman, M. A. Multi-Component Methodologies in Solid-Phase Organic Synthesis. *Curr. Med. Chem.* **1999**, *6*, 355–370.

(19) Weber, L.; Illgen, K.; Almstetter, M. Discovery of New Multi Component Reactions with Combinatorial Methods. *Synlett* **1999**, 366–374.

(20) Tietze, L. F.; Modi, A. Multicomponent Domino Reactions for the Synthesis of Biologically Active Natural Products and Drugs. *Med. Res. Rev.* **2000**, *20*, 304–322.

(21) Dömling, A.; Ugi, I. Multicomponent Reactions with Isocyanides. *Angew. Chem., Int. Ed. Engl.* **2000**, *39*, 3168–3210.

(22) Bienaymé, H.; Hulme, C.; Oddon, G.; Schmitt, P. Maximizing Synthetic Efficiency: Multi-Component Transformations Lead the Way. *Chem. Eur. J.* **2000**, *6*, 3321–3329.

(23) Hulme, C.; Gore, V. Multi-Component Reactions: Emerging Chemistry in Drug Discovery "From Xylocain to Crixivan." *Curr. Med. Chem.* **2003**, *10*, 51–80.

(24) Zhu, J. Recent Developments in the Isonitrile-Based Multicomponent Synthesis of Heterocycles. *Eur. J. Org. Chem.* **2003**, 1133–1144.

(25) Orru, R. V. A.; de Greef, M. Recent Advances in Solution-Phase Multicomponent Methodology for the Synthesis of Heterocyclic Compounds. *Synthesis* **2003**, 1471–1499.

(26) Enders, D.; Grondal, C.; Hüttl, M. R. M. Asymmetric Organocatalytic Domino Reactions. *Angew. Chem., Int. Ed. Engl.* **2007**, *46*, 1570–1581.

(27) Chapman, C. J.; Frost, C. G. Tandem and Domino Catalytic Strategies for Enantioselective Synthesis Enantioselective Tandem and Domino Catalytic Strategies. *Synthesis* **2007**, 1–21.

(28) Burk, M. J.; Lee, J. R.; Martinez, J. P. A Versatile Tandem Catalysis Procedure for the Preparation of Novel Amino Acids and Peptides. *J. Am. Chem. Soc.* **1994**, *116*, 10847–10848.

(29) Teoh, E.; Campi, E. M.; Jackson, W. R.; Robinson, A. J. A Highly Enantioselective Synthesis of Cyclic α-Amino Acids Involving a One-Pot, Single Catalyst, Tandem Hydrogenation–Hydroformylation Sequence *Chem. Commun.* **2002**, 978–979.

(30) Louie, J.; Bielawski, C. W.; Grubbs, R. H. Tandem Catalysis: The Sequential Mediation of Olefin Metathesis, Hydrogenation, and Hydrogen Transfer with Single-Component Ru Complexes. *J. Am. Chem. Soc.* **2001**, *123*, 11312–11313.

(31) Sutton, A. E.; Seigal, B. A.; Finnegan, D. F.; Snapper, M. L. New Tandem Catalysis: Preparation of Cyclic Enol Ethers Through a Ruthenium-Catalyzed Ring-Closing Metathesis–Olefin Isomerization Sequence. *J. Am. Chem. Soc.* **2002**, *124*, 13390–13391.

(32) Zuercher, W. J.; Scholl, M.; Grubbs, R. H. Ruthenium-Catalyzed Polycyclization Reactions. *J. Org. Chem.* **1998**, *63*, 4291–4298.

(33) Bielawski, C. W.; Louie, J.; Grubbs, R. H. Tandem Catalysis: Three Mechanistically Distinct Reactions from a Single Ruthenium Complex. *J. Am. Chem. Soc.* **2000**, *122*, 12872–12873.

(34) Alexakis, A.; Croset, K. Tandem Copper-Catalyzed Enantioselective Allylation-Metathesis. *Org. Lett.* **2002**, *4*, 4147–4149.

(35) Tissot-Croset, K.; Polet, D.; Alexakis, A. A Highly Effective Phosphoramidite Ligand for Asymmetric Allylic Substitution. *Angew. Chem., Int. Ed. Engl.* **2004**, *43*, 2426–2428.

(36) Miller, S. P.; Morgan, J. B.; Nepveux, F. J. V.; Morken, J. P. Catalytic Asymmetric Carbohydroxylation of Alkenes by a Tandem Diboration/Suzuki

Cross-Coupling/Oxidation Reaction. *Org. Lett.* **2004,** *6,* 131–133.

(37) Tietze, L. F.; Sommer, K. M.; Zinngrebe, J.; Stecker, F. Palladium-Catalyzed Enantioselective Domino Reaction for the Efficient Synthesis of Vitamin E. *Angew. Chem., Int. Ed. Engl.* **2005,** *44,* 257–259.

(38) Tietze, L. F.; Ila, H.; Bell, H. P. Enantioselective Palladium-Catalyzed Transformations. *Chem. Rev.* **2004,** *104,* 3453–3516.

(39) Wada, E.; Koga, H.; Kumaran, G. A Novel Catalytic Enantioselective Tandem Transetherification-Intramolecular Hetero Diels–Alder Reaction of Methyl (*E*)-4-Methoxy-2-oxo-3-butenoate with δ,ε-Unsaturated Alcohols. *Tetrahedron Lett.* **2002,** *43,* 9397–9400.

(40) Arai, T.; Sasai, H.; Aoe, K.-I.; Okamura, K.; Date, T.; Shibasaki, M. A New Multifunctional Heterobimetallic Asymmetric Catalyst for Michael Additions and Tandem Michael-Aldol Reactions. *Angew. Chem., Int. Ed. Engl.* **1996,** *35,* 104–106.

(41) Suzuki, M.; Yanagisawa, A.; Noyori, R. Prostaglandin Synthesis. 16. The Three-Component Coupling Synthesis of Prostaglandins. *J. Am. Chem. Soc.* **1988,** *110,* 4718–4726.

(42) Marks, F.; Furstenberger, G. *Prostagladins, Leukotrienes, and Other Eicosanoids. From Biogenesis to Clinical Applications;* Wiley: New York, 1999.

(43) Collins, P. W.; Djuric, S. W. Synthesis of Therapeutically Useful Prostaglandin and Prostacyclin Analogs. *Chem. Rev.* **1993,** *93,* 1533–1564.

(44) Yamada, K.-I.; Arai, T.; Sasai, H.; Shibasaki, M. A Catalytic Asymmetric Synthesis of 11-Deoxy-PGF$_{1\alpha}$ Using ALB, a Heterobimetallic Multifunctional Asymmetric Complex. *J. Org. Chem.* **1998,** *63,* 3666–3672.

(45) Arnold, L. A.; Naasz, R.; Minnaard, A. J.; Feringa, B. L. Catalytic Enantioselective Synthesis of Prostaglandin E1 Methyl Ester Using a Tandem 1,4-Addition-Aldol Reaction to a Cyclopenten-3,5-dione Monoacetal. *J. Am. Chem. Soc.* **2001,** *123,* 5841–5842.

(46) Arnold, L. A.; Naasz, R.; Minnaard, A. J.; Feringa, B. L. Catalytic Enantioselective Synthesis of (–)-Prostaglandin E1 Methyl Ester Based on a Tandem 1,4-Addition-Aldol Reaction. *J. Org. Chem.* **2002,** *67,* 7244–7254.

(47) Cauble, D. F.; Gipson, J. D.; Krische, M. J. Diastereo- and Enantioselective Catalytic Carbometallative Aldol Cycloreduction: Tandem Conjugate Addition-Aldol Cyclization. *J. Am. Chem. Soc.* **2003,** *125,* 1110–1111.

(48) Bocknack, B. M.; Wang, L.-C.; Krische, M. J. Desymmetrization of Enone-Diones via Rhodium-Catalyzed Diastereo- and Enantioselective Tandem Conjugate Addition-Aldol Cyclization. *Proc. Natl. Acad. Sci. U.S.A.* **2004,** *101,* 5421–5424.

(49) Mizutani, H.; Degrado, S. J.; Hoveyda, A. H. Cu-Catalyzed Asymmetric Conjugate Additions of Alkylzinc Reagents to Acyclic Aliphatic Enones. *J. Am. Chem. Soc.* **2002,** *124,* 779–781.

(50) Alexakis, A.; March, S. Tandem Enantioselective Conjugate Addition-Cyclopropanation. Application to Natural Products Synthesis. *J. Org. Chem.* **2002,** *67,* 8753–8757.

(51) Ma, S.; Jiao, N.; Zheng, Z.; Ma, Z.; Lu, Z.; Ye, L.; Deng, Y.; Chen, G. Cu- and Pd-Catalyzed Asymmetric One-Pot Tandem Addition-Cyclization Reaction of 2-(2′,3′-Alkadienyl)-β-keto Esters, Organic Halides, and Dibenzyl Azodicarboxylate: An Effective Protocol for the Enantioselective Synthesis of Pyrazolidine Derivatives. *Org. Lett.* **2004,** *6,* 2193–2196.

(52) Katsuki, T. In *Comprehensive Asymmetric Catalysis;* Jacobsen, E. N., Pfaltz, A., Yamamoto, H., Eds.; Springer-Verlag: Berlin, 1999; Vol. II, pp 621–648.

(53) Martín, V. S.; Woodard, S. S.; Katsuki, T.; Yamada, Y.; Ikeda, M.; Sharpless, K. B. Kinetic Resolution of Racemic Allylic Alcohols by Enantioselective Epoxidation. A Route to Substances of Absolute Enantiomeric Purity? *J. Am. Chem. Soc.* **1981,** *103,* 6237–6240.

(54) Gao, Y.; Klunder, J. M.; Hanson, R. M.; Masamune, H.; Ko, S. Y.; Sharpless, K. B. Catalytic Asymmetric Epoxidation and Kinetic Resolution: Modified Procedures Including in situ Derivatization. *J. Am. Chem. Soc.* **1987,** *109,* 5765–5780.

(55) Lurain, A. E.; Maestri, A.; Kelly, A. R.; Carroll, P. J.; Walsh, P. J. Highly Enantio- and Diastereoselective One-Pot Synthesis of Acyclic Epoxy Alcohols with Three Contiguous stereocenters. *J. Am. Chem. Soc.* **2004,** *126,* 13608–13609.

(56) Lurain, A. E.; Carroll, P. J.; Walsh, P. J. One-Pot Asymmetric Synthesis of Chiral Epoxy Alcohols: A Tandem Approach Using Vinylzinc Addition to an Aldehyde and in situ Epoxidation with Molecular Oxygen. *J. Org. Chem.* **2005,** *70,* 1262–1268.

(57) Rowley Kelly, A.; Lurain, A. E.; Walsh, P. J. Highly Enantio- and Diastereoselective One-Pot Synthesis of Acyclic Epoxy Alcohols and Allylic Epoxy Alcohols. *J. Am. Chem. Soc.* **2005,** *127,* 14668–14674.

(58) Nugent, W. A. MIB: An Advantageous Alternative to DAIB for the Addition of Organozinc Reagents. *J. Chem. Soc., Chem. Commun.* **1999,** 1369–1370.

(59) Adam, W.; Wirth, T. Hydroxy Group Directivity in the Epoxidation of Chiral Allylic Alcohols: Control of Diastereoselectivity Through Allylic Strain and Hydrogen Bonding. *Acc. Chem. Res.* **1999,** *32,* 703–710.

(60) Kim, H. Y.; Lurain, A. E.; García-García, P.; Carroll, P. J.; Walsh, P. J. Highly Enantio- and Diastereoselective Tandem Generation of Cyclopropyl Alcohols with up to Four Contiguous stereocenters. *J. Am. Chem. Soc.* **2005,** *127,* 13138–13139.

(61) For an overview, see the articles in a) special issue #8: Enantioselective Organocatalysis. *Acc. Chem. Res.* **2004,** *37,* 487–631. b) special issue #12: Organocatalysis. *Chem Rev.* **2007,** *107,* 5413–5883.

(62) Halland, N.; Aburel, P. S.; Jørgensen, K. A. Highly Enantio- and Diastereoselective Organocatalytic Asymmetric Domino Michael-Aldol Reaction of α-Ketoesters and α,β-Unsaturated Ketones. *Angew. Chem., Int. Ed. Engl.* **2004,** *43,* 1272–1277.

(63) Austin, J. F.; Kim, S.-G.; Sinz, C. J.; Xiao, W.-J.; MacMillan, D. W. C. Enantioselective Organocatalytic Construction of Pyrroloindolines by a Cascade Addition–Cyclization Strategy: Synthesis of (–)-Flustramine B. *Proc. Natl. Acad. Sci. U.S.A.* **2004,** *101,* 5482–5487.

(64) Watanabe, S.-I.; Cordova, A.; Tanaka, F.; Barbas, C. F., III One-Pot Asymmetric Synthesis of α-Cyanohydroxymethyl α-Amino Acid Derivatives: Formation of Three Contiguous Stereogenic Centers. *Org. Lett.* **2002**, *4*, 4519–4522.

(65) Córdova, A.; Barbas, C. F., III. Direct Organocatalytic Asymmetric Mannich-Type Reactions in Aqueous Media: One-Pot Mannich-Allylation Reactions. *Tetrahedron Lett.* **2003**, *44*, 1923–1926.

(66) Zhong, G. A Facile and Rapid Route to Highly Enantiopure 1,2-Diols by Novel Catalytic Asymmetric α-Aminoxylation of Aldehydes. *Angew. Chem., Int. Ed. Engl.* **2003**, *42*, 4247–4250.

(67) Zhong, G. Tandem Aminoxylation-Allylation Reactions: A Rapid, Asymmetric Conversion of Aldehydes to Mono-Substituted 1,2-Diols. *J. Chem. Soc., Chem. Commun.* **2004**, 606–607.

(68) Brown, S. P.; Brochu, M. P.; Sinz, C. J.; MacMillan, D. W. C. The Direct and Enantioselective Organocatalytic α-Oxidation of Aldehydes. *J. Am. Chem. Soc.* **2003**, *125*, 10808–10809.

(69) Wong, C. H.; Halcomb, R. L.; Ichikawa, Y.; Kajimoto, T. Enzymes in Organic Synthesis: Application to the Problems of Carbohydrate Recognition (Part 1). *Angew. Chem., Int. Ed. Engl.* **1995**, *43*, 412–432.

(70) Henderson, I.; Sharpless, K. B.; Wong, C. H. Synthesis of Carbohydrates via Tandem Use of the Osmium-Catalyzed Asymmetric Dihydroxylation and Enzyme-Catalyzed Aldol Addition Reactions. *J. Am. Chem. Soc.* **1994**, *116*, 558–561.

(71) Evans, D. A.; Dart, M. J.; Duffy, J. L.; Yang, M. G.; Livingston, A. B. Diastereoselective Aldol and Allylstannane Addition Reactions. The Merged Stereochemical Impact of α and β Aldehyde Substituents. *J. Am. Chem. Soc.* **1995**, *117*, 6619–6620.

(72) Evans, D. A.; Dart, M. J.; Duffy, J. L.; Yang, M. G. A Stereochemical Model for Merged 1,2- and 1,3-Asymmetric Induction in Diastereoselective Mukaiyama Aldol Addition Reactions and Related Processes. *J. Am. Chem. Soc.* **1996**, *118*, 4322–4343.

(73) Evans, D. A.; Allison, B. D.; Yang, M. G.; Masse, C. E. The Exceptional Chelating Ability of Dimethylaluminum Chloride and Methylaluminum Dichloride. The Merged Stereochemical Impact of α- and β-stereocenters in Chelate-Controlled Carbonyl Addition Reactions with Enol Silane and Hydride Nucleophiles. *J. Am. Chem. Soc.* **2001**, *123*, 10840–10852.

(74) Wong, C.-H.; Garcia-Junceda, E.; Chen, L.; Blanco, O.; Gijsen, H. J. M.; Steensma, D. H. Recombinant 2-Deoxyribose-5-phosphate Aldolase in Organic Synthesis: Use of Sequential Two-Substrate and Three-Substrate Aldol Reactions. *J. Am. Chem. Soc.* **1995**, *117*, 3333–3339.

(75) Poss, C. S.; Schreiber, S. L. Two-Directional Chain Synthesis and Terminus Differentiation. *Acc. Chem. Res.* **1994**, *27*, 9.

(76) Mikami, K.; Matsukawa, S.; Nagashima, M.; Funabashi, H.; Morishima, H. Tandem and Two-Directional Asymmetric Catalysis of the Mukaiyama Aldol Reaction. *Tetrahedron Lett.* **1997**, *38*, 579–582.

(77) Tian, J.; Yamagiwa, N.; Matsunaga, S.; Shibasaki, M. An Asymmetric Cyanation Reaction and Sequential Asymmetric Cyanation-Nitroaldol Reaction Using a {YLi₃[tris(binaphthoxide)]} Single Catalyst Component: Catalyst Tuning with Achiral Additives. *Angew. Chem., Int. Ed. Engl.* **2002**, *41*, 3636–3638.

(78) Du, H.; Ding, K. Enantioselective Catalysis of Hetero Diels–Alder Reaction and Diethylzinc Addition Using a Single Catalyst. *Org. Lett.* **2003**, *5*, 1091–1093.

(79) Doi, T.; Kokubo, M.; Yamamoto, K.; Takahashi, T. One-Pot Sequential Asymmetric Hydrogenation Utilizing Rh(I) and Ru(II) Catalysts. *J. Org. Chem.* **1998**, *63*, 428–429.

(80) Annunziata, R.; Benaglia, M.; Cinquini, M.; Cozzi, F.; Puglisi, A. Sequential Stereoselective Catalysis: Two Single-Flask Reactions of a Substrate in the Presence of a Bifunctional Chiral Ligand and Different Transition Metals. *Eur. J. Org. Chem.* **2003**, 1428–1432.

(81) Annunziata, R.; Benaglia, M.; Cinquini, M.; Cozzi, F. Synthesis of a Bifunctional Ligand for the Sequential Enantioselective Catalysis of Various Reactions. *Eur. J. Org. Chem.* **2001**, 1045–1048.

(82) Huang, Y.; Walji, A. M.; Larsen, C. H.; MacMillan, D. W. C. Enantioselective Organo-Cascade Catalysis. *J. Am. Chem. Soc.* **2005**, *127*, 15051–15053.

(83) Xie, X.; Phuan, P.-W.; Kozlowski, M. C. Novel Pathways for the Formation of Chiral Binaphthyl Polymers: Oxidative Asymmetric Phenolic Coupling Alone and in Tandem with the Glaser–Hay Coupling. *Angew. Chem., Int. Ed. Engl.* **2003**, *42*, 2168–2170.

(84) Dudding, T.; Hafez, A. M.; Taggi, A. E.; Wagerle, T. R.; Lectka, T. A Catalyst that Plays Multiple Roles: Asymmetric Synthesis of β-Substituted Aspartic Acid Derivatives Through a Four-Stage, One-Pot Procedure. *Org. Lett.* **2002**, *4*, 387–390.

(85) Hafez, A. M.; Dudding, T.; Wagerle, T. R.; Shah, M. H.; Taggi, A. E.; Lectka, T. A Multistage, One-Pot Procedure Mediated by a Single Catalyst: A New Approach to the Catalytic Asymmetric Synthesis of β-Amino Acids. *J. Org. Chem.* **2003**, *68*, 5819–5825.

(86) Koga, H.; Wada, E. A New Strategy in Enantioselective Intramolecular Hetero Diels–Alder Reaction: Catalytic Double Asymmetric Induction During the Tandem Transetherification-Intramolecular Hetero Diels–Alder Reaction of Methyl (E)-4-Methoxy-2-oxo-3-butenoate with rac-6-Methyl-5-hepten-2-ol. *Tetrahedron Lett.* **2003**, *44*, 715–719.

(87) Kita, Y.; Naka, T.; Imanishi, M.; Akai, S.; Takebe, Y.; Matsugi, M. Asymmetric Diels–Alder Reaction via Enzymatic Kinetic Resolution Using Ethoxyvinyl Methyl Fumarate. *J. Chem. Soc., Chem. Commun.* **1998**, 1183–1184.

(88) Akai, S.; Naka, T.; Omura, S.; Tanimoto, K.; Imanishi, M.; Takebe, Y.; Matsugi, M.; Kita, Y. Lipase-Catalyzed Domino Kinetic Resolution/Intramolecular Diels–Alder Reaction: One-Pot Synthesis of Optically Active 7-Oxabicyclo[2.2.1]heptenes from Furfuryl Alcohols and β-Substituted Acrylic Acids. *Chem. Eur. J.* **2002**, *8*, 4255–4264.

(89) Akai, S.; Tanimoto, K.; Kita, Y. Lipase-Catalyzed Domino Dynamic Kinetic Resolution of Racemic 3-Vinylcyclohex-2-en-1-ols/Intramolecular

Diels–Alder Reaction: One-Pot Synthesis of Optically Active Polysubstituted Decalins. *Angew. Chem., Int. Ed. Engl.* **2004**, *43*, 1407–1410.

(90) Takatori, K.; Hasegawa, K.; Narai, S.; Kajiwara, M. A Microwave-Accelerated Intramolecular Diels–Alder Reaction Approach to Compactin. *Heterocycles* **1996**, *42*, 525–528.

(91) Corey, E. J.; da Silva Jardine, P.; Mohri, T. Enantioselective Route to a Key Intermediate in the Total Synthesis of Forskolin. *Tetrahedron Lett.* **1988**, *29*, 6409–6412.

(92) Nagashima, S.; Kanematsu, K. A Synthesis of an Optically Active Forskolin Intermediate via Allenyl Ether Intramolecular Cycloaddition Strategy. *Tetrahedron: Asymmetry* **1990**, *1*, 743–749.

(93) Calvo, D.; Port, M.; Delpech, B.; Lett, R. Total Synthesis of Forskolin. Part III. Studies Related to an Asymmetric Synthesis. *Tetrahedron Lett.* **1996**, *37*, 1023–24.

15

Supported Chiral Catalysts

Many homogeneous catalytic asymmetric processes have now been developed that allow the assembly of complex molecules with high enantioselectivity and diastereoselectivity under mild conditions with low catalyst loadings. While catalysts used in these processes are readily employed in the laboratory, many challenges remain before they are suitable for applications on an industrial scale. A serious drawback of homogeneous catalysts is that they must ultimately be separated from the reaction products. Separation and recovery of homogeneous catalysts usually requires large amounts of solvent and is one of the major contributors to waste generation in chemical reactions.[1] Facilitating catalyst recovery would reduce waste generation and is a primary objective in green chemistry. In contrast to homogeneous catalysts, heterogeneous catalysts can be easily recovered from reaction mixtures. It has proven more difficult to develop highly enantioselective heterogeneous catalysts, however, because our understanding of how heterogeneous catalysts operate is rudimentary and they are considerably more challenging to study.

To take advantage of the best features of homogeneous and heterogeneous catalysts, chemists have investigated the attachment of homogeneous catalysts to soluble and insoluble supports. In addition to the features outlined above, other benefits of supported catalysts stem from the unique microenvironments defined by the support. The support matrix can impart greater catalyst stability, and the size of the microenvironment around the catalyst can influence both stereochemistry and catalyst activity. In some cases, the supported catalysts are more active than the homogeneous analogs, although this is not usually the case.

A number of reviews on supported catalysts, including asymmetric versions, have appeared.[2–20] In a very comprehensive survey,[21] supported chiral catalysts for a number of transformations (e.g., reduction, oxidation, and Diels–Alder) are presented. The truly practical utility of supported reagents and catalysts has been highlighted[22] in a 29-step total synthesis of the natural product epothiolone C, a potent anti-tumor agent, using solid-supported reagents and scavengers. At least one solid-supported reagent or scavenger was utilized in every step, with no less than 11 different types of supported reagents and scavengers being employed in the entire sequence. Use of the supported materials greatly facilitated purification, and the synthesis compares favorably to prior syntheses that employed traditional solution methods and purifications.

A unique feature of supported catalysts is the ability to use two different, and not necessarily mutually compatible, catalysts in one reaction vessel. These catalysts may be anchored to the same or to different supports. In either case, the support separates the two catalysts from each other while allowing soluble reaction components to undergo tandem or sequential processes (see Chapter 14). Another benefit of immobilized catalysts is their potential applications in continuous-flow processes. In principle, substrates could be subjected to a defined set of chemical transformations by simply passing them over a series of immobilized chiral catalysts. An advantage of supported catalysts is that the ligand, catalyst precursor, or catalyst can be prepared on the support medium, facilitating catalyst synthesis and purification while opening up combinatorial approaches.

This chapter begins with an overview of the most frequently employed immobilization methods for chiral catalysts, including covalent attachment to preformed supports, covalent incorporation as part of the support, and noncovalent incorporation into the support.[19]

15.1 Catalyst Supports—An Overview

Catalysts have been immobilized on a variety of supports, which can be roughly divided into polymeric organic supports and inorganic supports. Organic supports can be soluble or insoluble polymers or dendrimers. Soluble polymers are usually linear (no cross-linking). In theory, soluble supported catalysts are close approximations to homogeneous catalysts. The soluble support should have minimal impact on the reactivity and selectivity of the catalyst with respect to the unsupported analog. Removal of soluble supported catalysts from the reaction mixture is typically performed by precipitation and filtration. In contrast, cross-linked polymer supports are insoluble in organic solvents, facilitating catalyst removal by filtration. The reactivity and selectivity of such polymer-supported catalysts often depend on the nature of the reaction solvent. Solvents cause the polymer to swell, resulting in greater accessibility to inner surfaces of the support, increasing mass transport. Increasing the cross-linking of polymer supports makes the polymer more rigid, reducing swellability and mass transport, and limiting catalyst TOF.

Inorganic supports are typically porous materials with discrete surface areas, as exemplified by silica and alumina. Zeolites, which are crystalline materials consisting of well-defined channels and pores, have also been successfully employed as supports. Inorganic matrices are also mechanically rigid and not affected by solvent and temperature. In contrast, organic polymers are not rigid, and their shape and structures are strongly influenced by the solvent, temperature, and pressure.

15.1.1 Methods for Catalyst Heterogenization

The most common methods to generate polymer-supported catalysts are outlined in **Figure 15.1**. In **a–c**, catalysts are attached through covalent bonds to the support. This has been accomplished by grafting the catalyst to a preformed polymer bearing function groups (FG, **Figure 15.1a**), a variety of which are commercially available. Heterogenization of homogeneous catalyst by this technique often requires modification of the chiral ligand to incorporate a tether or linker. Alternatively, **Figure 15.1b** illustrates the generation of the supported catalyst by polymerization from monomers and an appropriately functionalized catalyst-bound polymer. An example of the latter would be the copolymerization of styrene and ligand or catalyst containing a styryl group. In **Figure 15.1c**, a catalyst or ligand is employed that contains two polymerizable groups, allowing incorporation of the active site into the backbone of the polymer or use as a cross-linking agent. By linking chiral ligands together such that they can bind two metal centers, coordination polymers can be generated (**Figure 15.1d**). Metal–ligand polymers of this type contain a high density of active sites.

Figure 15.2 illustrates other important methods that have been successfully used in the heterogenization of homogeneous catalysts. Addition of homogeneous catalysts containing functional groups to solid supports, such as silica, leads to the forma-

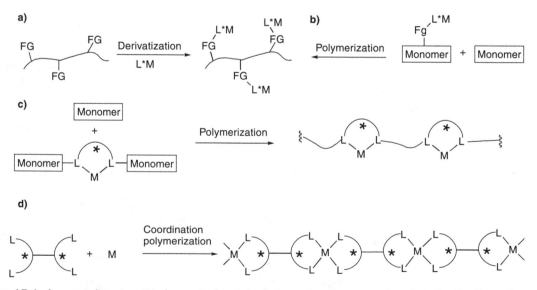

Figure 15.1. Approaches to attachment of catalysts to polymer supports: a) derivatization of preformed supports through pendent functional groups (FG), b) copolymerization of monomer and functionalized monomer carrying a chiral ligand or chiral catalyst, c) Incorporation of the active site into the backbone or cross-linker of the polymer, and d) Formation of coordination polymers.

tion of covalent linkages (**Figure 15.2e**). Reaction of charged catalysts with ion-exchange resins or ion-containing solids results in catalysts that are heterogenized by charge–charge attractions (**Figure 15.2f**). An ingenious approach to catalyst immobilization involves the synthesis of the catalyst inside support pores containing windows through which the catalyst, once assembled, cannot pass. This method, illustrated in **Figure 15.2g**, has been appropriately termed "ship in a bottle." An important attribute of the latter two methods is that they do not require modification of the homogeneous catalyst structure, unlike the methods in **Figure 15.1** and **Figure 15.2e**.

The attachment of homogeneous catalyst to supports often changes the properties of the catalyst. For example, functional groups on the support could interfere with catalytic activity. When a chiral catalyst is confined to a solid support, the shape and size of the channel or cavity can restrict the approach or orientation of the reaction partners, impacting the enantioselectivity of the system. The microenvironment of the catalyst can also affect the reaction rate by altering the local dielectric due to solvent exclusion or by interactions with the walls of the solid support. As mentioned earlier, mass transport issues arise when channel size is small or the polymer support is insufficiently swollen, restricting access of the reagents to the catalyst. As outlined in the sections below, there remain many challenges in the heterogenization of catalysts. Given the potential benefits of increased catalyst performance and recyclability, which can significantly alter the economics and environmental impact of the process, it is anticipated that considerable research effort will be invested in this area in the coming years.

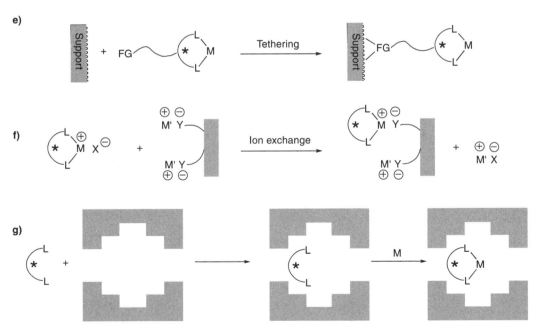

Figure 15.2. Heterogenization of homogeneous catalyst through e) tethering to solid supports, f) charge–charge attractions, and g) construction of the catalyst inside the pores of solid supports.

15.1.2. Chiral Catalysts on Polystyrene

The most popular organic polymer supports are polystyrene (PS) derivatives.[3] Polystyrenes are readily available, inexpensive, easily functionalized, and exhibit good chemical inertness and mechanical stability. These supports are generally prepared by the polymerization of styrene in the presence of a cross-linking reagent, such as divinylbenzene, and a functionalized styrene, to which the catalyst can be attached in a later step. The amount of cross-linking has a significant impact on the properties of the polymer. For example, polystyrene prepared with 2% divinylbenzene is a microporous gel. It must be used with an appropriate solvent to induce swelling and allow access to internal sites. Polar solvents, such as water and alcohols, can be problematic, because they do not adequately swell the polymer support, leading to poor site accessibility. In contrast, use of more than 10–15% cross-linker gives macroporous resins that exhibit little structural variation with organic solvents.

To improve the solvation properties of cross-linked polystyrene supports with polar solvents, poly(ethylene glycol) (PEG) moieties have been appended to the PS backbone. Two of the most common of this class are TentaGel and ArgoGel (**Figure 15.3**), both of which are commercially available. Catalysts are usually attached to polystyrene supports via covalent bonds, although ion-paring and entrapment have also been used.

Polymer-supported catalysts have proven essential in the hydrolytic kinetic resolution of certain epoxides by Co(salen) complexes. Epichlorohydrin, for example, undergoes partial racemization under the conditions employed during distillation away from the catalyst. A better method was required for the separation of the epichlorohydrin from the catalyst solution. As illustrated in **Figure 15.4**, hydroxymethylpolystyrene, derivatized as the 4-nitrophenyl carbonate, was combined with a hydroxy–salen ligand in the presence of base. Addition of a Co(II) salt to the PS-bound ligand formed the (salen)Co(II) complex, which underwent oxidation to the Co(III) derivative on exposure to oxygen. The resulting dark red beads were analyzed by elemental analysis and found to contain 160 μmol cobalt complex per gram of resin.

Results from the application of the resin-bound catalyst to the HKR of epichlorohydrin are illustrated in **Figure 15.4**. Using a catalyst loading of 0.25 mol%, the faster-reacting enantiomer was consumed in 3 h. The insoluble supported catalyst was

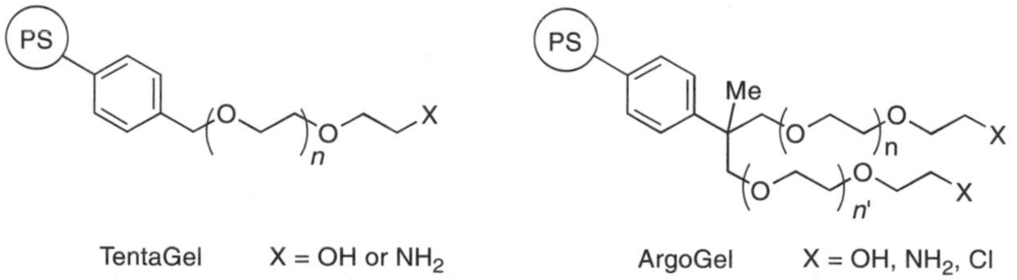

TentaGel   X = OH or NH$_2$ ArgoGel   X = OH, NH$_2$, Cl

Figure 15.3. Structures of TentaGel and ArgoGels (n, $n' \approx 70$).

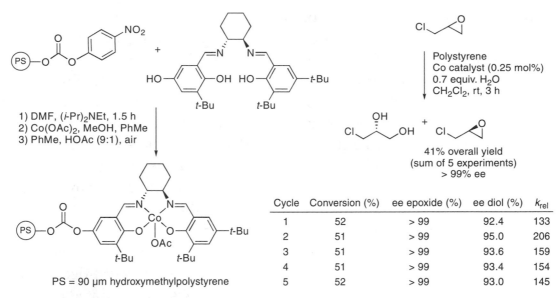

Cycle	Conversion (%)	ee epoxide (%)	ee diol (%)	k_{rel}
1	52	> 99	92.4	133
2	51	> 99	95.0	206
3	51	> 99	93.6	159
4	51	> 99	93.4	154
5	52	> 99	93.0	145

Figure 15.4. Polystyrene-supported chiral cobalt–salen catalysts in asymmetric hydrolytic kinetic resolution.

isolated by filtration and recycled four times without significant erosion of the selectivity factor.[23]

Recall from Chapter 7 that the mechanistic studies on the (salen)Co-catalyzed HKR indicated that the reaction is second-order in catalyst, with one molecule of catalyst activating the epoxide and the other the hydroxide nucleophile. The observation of an efficient kinetic resolution with the polymer-bound resin indicates that the support is sufficiently flexible such that two cobalt centers can interact in a cooperative fashion.[23]

15.1.3 Chiral Catalysts on Silica

Another common solid support for heterogenization of homogeneous catalysts is silica gel. Unlike swellable cross-linked polymer supports, silica gel is a rigid material with a porous structure that does not swell. Thus, it can be used in a variety of solvents over a wide range of temperatures and pressures. High catalyst loadings can be achieved with silica gel of high surface area (typically > 100 m^{2}g^{-1}) and pore sizes > 20 Å to allow facile diffusion of reagents to the active sites.[1] Silica gel can have high surface area (> 600 m^{2}g^{-1}) with pore sizes ranging from microporus to mesoporous (5–500 Å). The surface of silica gel contains Si–OH and Si–O–Si groups that can be derivatized. The pK_a of these Si–OH group is 7.1, which can interfere with reactions conducted requiring basic reagents.[24] The silica surface can be chemically capped, however, to render it inert and hydrophobic.[25]

For comparison of polystyrene and silica-gel supports, the discussion of the (salen)Co catalyst for the HKR reaction is continued. As shown in **Figure 15.5**, attachment of the chiral salen ligand to the silica gel is accomplished by condensation of the silyl ether with the silica support. Application of the supported catalyst in the HKR of the epoxy alcohol (**Figure 15.5**) indicated that a catalyst loading of 0.4 mol% resulted in 34% consumption of the fast-reacting enantiomer of the epoxy alcohol in 2.5 h.[23]

Silica-supported chiral catalysts are amenable to continuous-flow reaction systems, such as that illustrated in **Figure 15.6**. The epoxy alcohol and a THF/H$_2$O mixture were introduced into the reaction stream by means of a syringe pump. This solution was passed through a column packed with the silica-supported (salen)Co catalyst from **Figure 15.5** and collected as it eluted from the column. The results are illustrated in **Figure 15.6**. A second cycle was performed and gave similar results.[23]

Another advantage of rigid silica support is the ability to control the catalyst density, and therefore, the distance and degree of interaction between catalyst sites. Supported catalysts with well-defined distances between active sites allow site isolation, which can be useful in probing reaction mechanisms and catalyst decomposition routes. For example, by controlling the amount of catalyst incorporated onto a silica

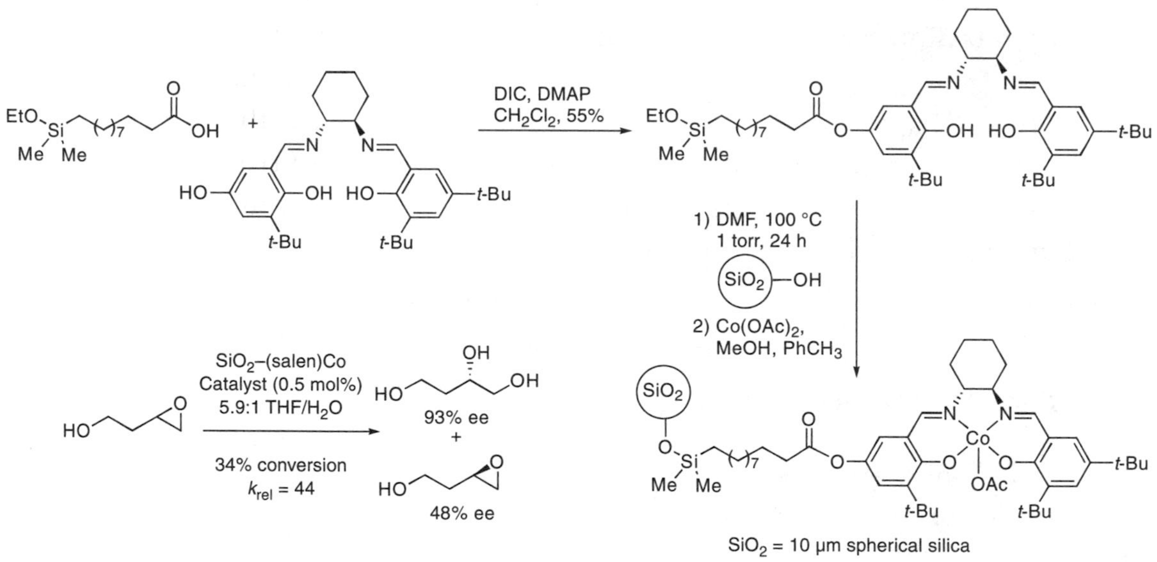

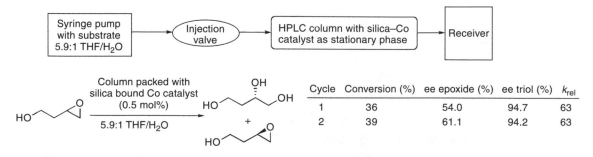

Figure 15.5. Synthesis of silica-supported chiral cobalt–salen catalysts (DIC= diisopropylcarbodiimide).

Cycle	Conversion (%)	ee epoxide (%)	ee triol (%)	k_{rel}
1	36	54.0	94.7	63
2	39	61.1	94.2	63

Figure 15.6. Continuous-flow reactor diagram. HKR of the epoxy alcohol in a continuous-flow reactor employing the silica-bound (salen)Co catalyst from **Figure 15.5** as stationary phase.

support, a clear relationship between local concentration and activity has been established in the cobalt–salen-catalyzed asymmetric HKR (**Figure 15.7**).[23] Based on the tether lengths, the surface area of the silica, and catalyst loadings on silica gel, which can be determined by elemental analysis, the probability of site interactions can be calculated.[26] A series of supported catalysts with decreasing distances between the active sites was prepared. As shown in **Figure 15.7**, when the distance between the active sites was large, no reaction occurred. Once a critical catalyst density is reached, cooperative epoxide-opening becomes more likely and the reaction proceeds readily. These data support the homogeneous kinetic results, which exhibit a second-order dependency on the concentration of catalyst. Notably, the same effect was not observed when polystyrene was employed as the solid support. The flexible nature of polystyrene allows the two cobalt complexes to interact, even when the complexes are widely separated along the polymer backbone.[23]

In the above case, the second-order dependency of the catalyst necessitates consideration of site proximity to insure that catalyst centers are sufficiently close to promote the asymmetric epoxide-opening reaction. In other systems, however, catalyst decomposition is a second-order process and site isolation can be used to prolong the lifetime of the catalyst.[27,28] In the asymmetric hydrogenation of methyl-acetamidecinnamate (MAC) with rhodium catalysts based on the bis(phosphine) (2S,4S)-4-diphenylphosphino-2-(diphenylphosphino methyl)pyrrolidine (PPM), the TOF exhibits a dependency on the site proximity (**Figure 15.8**). Initially, the rate of monomeric catalyst derived from BPMM and 0.5 equivalents of dimeric [Rh(cod)Cl]$_2$ was compared to the linked analogs. At constant catalyst loading, the relative catalyst rates were BPPM > PPM-C$_{12}$-PPM > PPM-C$_6$-PPM (**Figure 15.8**). Comparison of

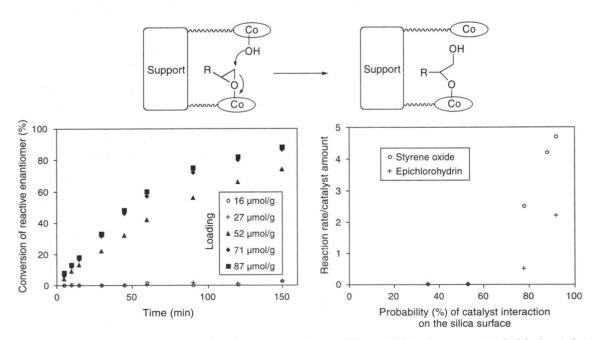

Figure 15.7. Relationship between local concentration and the activity of a supported chiral catalyst.

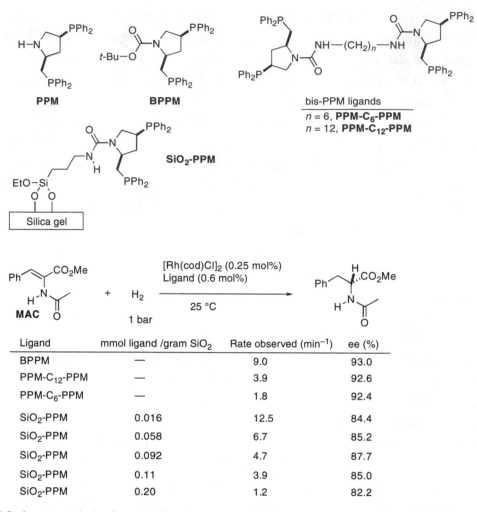

Ligand	mmol ligand /gram SiO₂	Rate observed (min⁻¹)	ee (%)
BPPM	—	9.0	93.0
PPM-C₁₂-PPM	—	3.9	92.6
PPM-C₆-PPM	—	1.8	92.4
SiO₂-PPM	0.016	12.5	84.4
SiO₂-PPM	0.058	6.7	85.2
SiO₂-PPM	0.092	4.7	87.7
SiO₂-PPM	0.11	3.9	85.0
SiO₂-PPM	0.20	1.2	82.2

Figure 15.8. Asymmetric hydrogenation catalyst deactivation is second-order in catalyst. Greater site isolation by decreasing the concentration of the catalyst on the support results in higher average catalyst TOF.

the results of the monomeric catalyst with the tethered versions indicates that as the local concentration increases, the catalyst activity decreases. This behavior is indicative of a bimolecular deactivation pathway, possibly involving dimer formation through bridging chlorides. Unlike the homogeneous catalysts, which can all form dimers, heterogenized analogs can be divided into those with catalysts that are sufficiently close to interact and those with catalysts that are isolated from neighboring catalysts. Catalysts in close proximity will likely form inactive dimers, whereas site-isolated catalyst will remain active. As the grams of catalyst per grams of silica increases, the percentage of those catalysts that can undergo the bimolecular deactivation increases and the overall activity of the catalyst decreases, as observed in **Figure 15.8**. Decreasing the concentration of the catalyst on the silica gel will lead to greater site isolation and greater activity.[28] This behavior can be modeled if the solid support is rigid and the individual catalytic centers are randomly distributed on the surface of the support.[29]

15.1.4 Chiral Catalysts on Mesoporous Silica

The IUPAC has divided porous materials into three size ranges: microporous (< 2 nm), mesoporous (2–50 nm), and macroporous (> 50 nm). Until recently, however, many porous materials consisted of disordered pore sizes with broad pore-size distributions. The synthesis of siliceous MCM-41,[30] the first regularly ordered mesoporous material with very narrow pore sizes was a major breakthrough in materials science. The honeycomb structure consists of an ordered hexagonal array of parallel cylindrical pores. Tethering catalysts inside MCM-41, with its regular pore dimensions, will enable studies of the relationship between pore size, substrate size, enantioselectivity, and the activity of catalysts anchored to the walls of the mesoporous support. In steps toward this goal, a Mn(salen) complex was tethered to MCM-41, as illustrated in **Figure 15.9**.[31] The Mn(salen)/MCM-41 was characterized by IR spectroscopy, which showed a characteristic absorption for the imine at 1635 cm^{-1} and was also characterized by EPR and UV–vis spectroscopies. The loading of the Mn(salen) was estimated to be 0.73 mmol/g support based on elemental analysis.

The Mn(salen)/MCM-41 was examined in the asymmetric epoxidation by stirring the immobilized catalyst (0.1 g) in dichloromethane (3 mL) with aqueous NaOCl (0.4 M, 2 mmol) and 1 mmol olefin substrate (**Equation 15.1**). In the heterogeneous dichloromethane/water system, the homogeneous catalyst epoxidized α-methylstyrene with an enantioselectivity of 56%. In comparison, Mn(salen)/MCM-41 exhibited lower activity in the 24 h reaction period, but higher enantioselectivity (72%). The lower activity in dichloromethane is due to slower diffusion of the substrate and oxidant

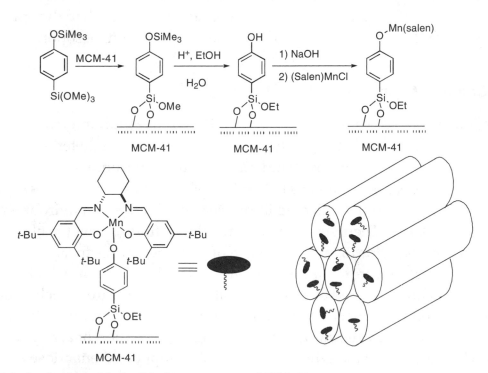

Figure 15.9. Anchoring of (salen)Mn to mesoporous MCM-41.

into the mesoporous support with the biphasic solvent system. The increase in the enantioselectivity of Mn(salen)/MCM-41 over the homogeneous catalyst is primarily attributed to the spatial constraints of the tether, the catalyst, and the size of the mesopores. Using ethanol as solvent, Mn(salen)/MCM-41 gave similar levels of enantioselectivity over three cycles, suggesting catalyst leaching was minimal. Interestingly, while the homogeneous catalyst readily epoxidized 1-phenylcyclohexene (93% conversion, 78% ee), no reaction with this substrate was observed with Mn(salen)/MCM-41. This result indicates that 1-phenylcyclohexene is too large to undergo reaction in the mesoporous material and that the Mn catalyst in Mn(salen)/MCM-41 remains bound to the support.[31]

Equation 15.1

$$Ph-\!\!\!<\!\!\! + NaOCl \xrightarrow{\text{Mn catalyst}} Ph-\!\!\!<\!\!\!\overset{O}{\triangleright}$$

Catalyst	Cycle	Solvent	Conversion (%)	ee (%)
ClMn(salen)	—	CH_2Cl_2	96	56
Mn(salen)/MCM-41	—	CH_2Cl_2	60	72
Mn(salen)/MCM-41	1	EtOH	99	70
Mn(salen)/MCM-41	2	EtOH	99	70
Mn(salen)/MCM-41	3	EtOH	98	73

The defined pore size of MCM-41 makes an ideal system to probe of the impact of the mesoporous support, the tether length, and the catalyst structure on the enantioselectivity and TOF of the catalyst system. Such studies will further our understanding of these interactions and allow a more rational approach to the optimization of the activity and enantioselectivity of catalysts supported on mesoporous supports.

15.1.5 Chiral Catalysts on Zeolites

Zeolites (molecular sieves) are a class of aluminosilicates with interconnected small cavities of 4–13 Å dimensions. Zeolites have been employed in heterogeneous catalysis due to their regular pore sizes and reactive microenvironments. The high reactivity arises from cations and anions in the channels, cavities, and walls of the zeolites, as well as the altered bulk dielectric in the nearly solvent-free environment of the channels and cavities. There has been a tremendous effort to develop chiral zeolites and related porous solids to take advantage of these properties in asymmetric catalysis. Typically, zeolites are synthesized by the assembly of an aluminosilicate matrix around a surfactant, followed by calcination at high temperature, which removes the surfactant. With chiral surfactants, the high temperature calcination destroys any chiral cavities generated by the surfactant, resulting in generation of achiral or racemic zeolites. Thus, alternate strategies to design zeolite-supported enantioselective catalysts have been employed, including assembling chiral catalysts inside the zeolite pores (see Section 15.3.4.a) and tethering catalysts to the walls of the zeolite channels and pores.

To compare the reactivity and enantioselectivity of homogeneous asymmetric hydrogenation catalysts with analogous catalysts tethered to silica and USY zeolite (pore diameter 12–30 Å), the catalysts in **Figure 15.10** were prepared.[32,33] Linking the catalyst to support was accomplished by controlled reaction of the silyl ether bonds of the tether with the surface Si-OH groups on either silica or the zeolite. The resulting supports were then washed to remove any nontethered catalyst. The spectroscopic and analytical properties of the supported complexes were similar to those of the homogeneous precursors.

These catalysts were then examined in the asymmetric hydrogenation of (Z)-α-N-acylcinnamic acid derivatives, as illustrated in **Figure 15.11**. In each case, the levels of enantioselectivity of the USY-zeolite-supported catalyst were higher than the silica-supported catalyst and the homogeneous catalyst. Additionally, in the hydrogenation of ethyl (Z)-α-benzolycinnamate, the time required to reach 50% conversion for the zeolite, silica, and homogeneous catalysts were roughly 2, 5, and 3 h, respectively. Based on these results, it was hypothesized that the nature of the zeolite support played a significant role in both the catalyst enantioselectivity and activity.[32] It is noteworthy that the zeolite-bound catalyst could be reused several times without loss of activity or enantioselectivity and no metal leaching was detected.

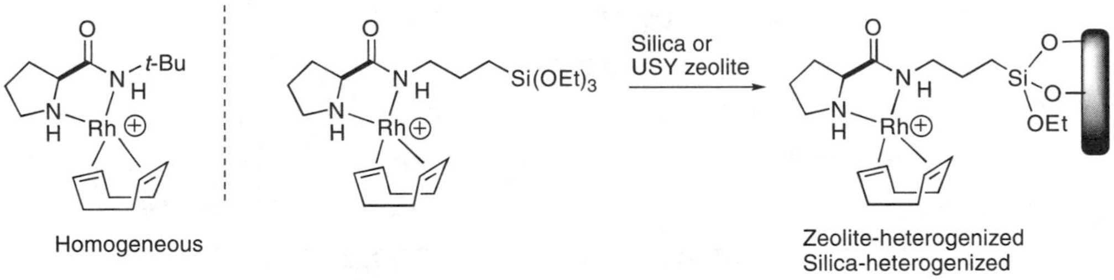

Figure 15.10. Proposed structure of the chiral catalyst anchored to a zeolite.

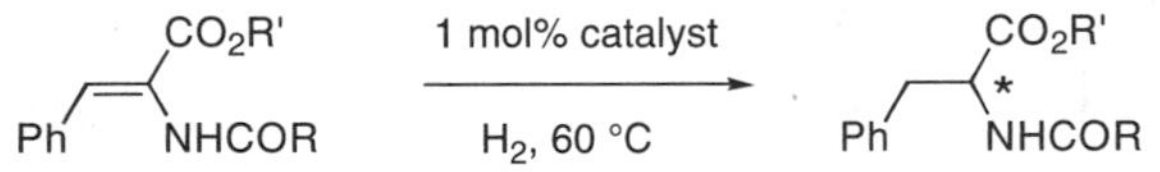

R'	R	Homogeneous	Silica	USY zeolite
H	Me	84.1	88.6	97.8
H	Ph	90.3	93.5	96.8
Et	Me	54.4	58.0	94.2
Et	Ph	85.6	92.2	99.0

Figure 15.11. Comparison of enantioselectivities employing homogeneous, silica-bound, and zeolite-bound catalysts in the enantioselective hydrogenation.

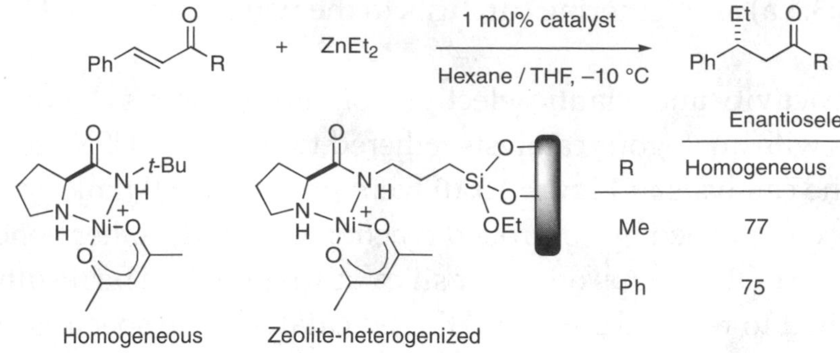

	Enantioselectivities (%)	
R	Homogeneous	USY zeolite
Me	77	91
Ph	75	95

Figure 15.12. Comparison between zeolite-supported and homogeneous catalysts in the conjugation addition of diethylzinc to enones.

Using the same ligand system with nickel in the conjugate addition of diethylzinc to enones (**Figure 15.12**), it was found that the zeolite-supported catalyst again gave higher enantioselectivities than the homogeneous system. In this case, however, the homogeneous system exhibited higher TOF.[34]

The examples above illustrate the positive impact that the zeolite support can impart on the enantioselectivity of catalysts. Further effort will be required in order to understand the cause of the increased enantioselectivity.

15.1.6 Chiral Catalysts Tethered to Nanoparticles

A central focus in this chapter has been the comparison of the attributes of homogeneous catalysts to those of their heterogenized counterparts. Laying between these two important classes are nanoparticles, which have also been referred to as nanoclusters, giant clusters, and colloids. Because of their small size, these clusters have very high surface areas. In some cases they are even soluble and have been called "semihomogeneous."[35] While many examples of catalytic nanoparticles have been reported,[36] very few examples of applications to asymmetric catalysis are known. Given the increasing interest in nanotechnology, it is anticipated that applications of nanoparticles to asymmetric catalysts will increase dramatically in the coming years.

While nanoparticles have high surface area and can be easily dispersed into organic solvents, recovery of the nanoparticle-supported catalysts is accomplished by settling or filtration, both of which can be complicated by the small size of the particles. A novel approach to the recovery of nanoparticle-supported catalyst is to immobilize the catalyst onto magnetically recoverable nanoparticles.[37]

Superparamagnetic materials are inherently nonmagnetic in the absence of a magnetic field. They are readily magnetized, however, in the presence of an external magnetic field. Immobilization of a phosphonic-acid-substituted ruthenium complex onto the superparamagnetic nanoparticles was performed as illustrated in **Figure 15.13**. A comparison of enantioselectivity between the unsupported catalyst and the nanoparticle-supported catalyst in the asymmetric hydrogenation of ketones indi-

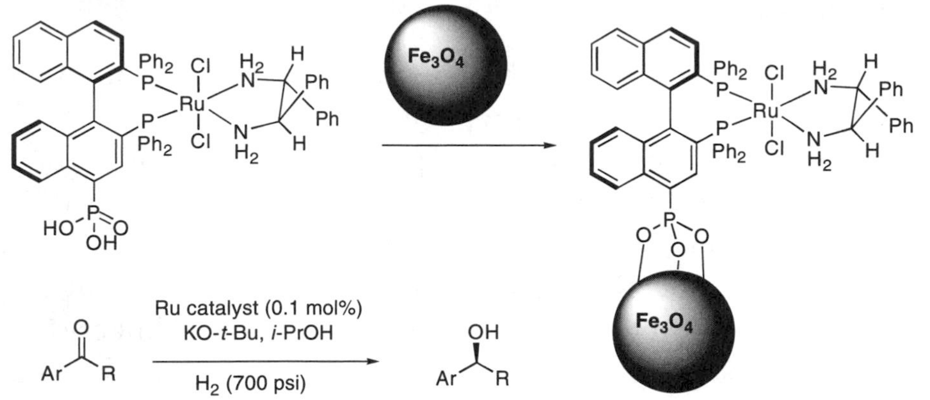

Ar	R	ee (%) Unsupported catalyst	ee (%) Supported catalyst
Ph	Me	87.0	87.6
1-Naphthyl	Me	98.0	98.0
4-C_6H_4-Cl	Me	75.9	76.6
Ph	Et	89.0	88.9

Figure 15.13. Catalyst immobilization onto superparamagnetic nanoparticles and application to the asymmetric hydrogenation of ketones.

cated that the catalysts exhibited virtually identical levels of enantioselectivity. The nanoparticle-supported catalyst was recovered by decanting the reaction mixture while attracting the nanoparticles with an external magnet. The supernatant was inactive in the hydrogenation reaction after separation from the nanoparticles, indicating little or no catalyst leaching occurred. The nanoparticle-supported catalyst exhibited no loss of enantioselectivity until the sixth cycle.[37]

15.2 Chiral Catalysts Incorporated as Parts of Solid Supports

The amount of a chiral catalyst that can be immobilized onto a preformed solid support is limited by the available surface area. In addition, functionalization of the interior pores of a polymer can be difficult. As a result, catalyst loadings on preformed solid supports are generally low. Higher catalyst densities can be obtained when the chiral catalyst is incorporated into the polymer backbone during formation of the polymer. Under these conditions, up to one catalyst unit per monomer is achievable. If greater spacing between catalyst units is desired then catalyst-substituted monomers can be mixed with inert monomers during the polymerization. The mechanical stability and pore size of the resultant polymers can also be modified by varying the amount of a cross-linking reagent.

15.2.1 Chiral Catalysts Incorporated into Polystyrene

Cross-linked polystyrene resins have been used successfully to support a variety of chiral catalysts. The methods for the preparation of PS-supported catalysts involve grafting the catalyst to a functionalized PS support, as outlined earlier, or copolymerization of styrene monomers carrying catalyst precursors or catalysts (**Figure 15.1b**). The ligand or catalyst must be stable to the conditions used in the free-radical polymerization process. In general, three monomers are used to form the support: the catalyst-bound styrene, styrene, and the divinylbenzene cross-linker (**Figure 15.14**). The properties of the polymeric matrix will depend on the degree and the nature of the of cross-linker, as outlined below.

Chiral oxazaborolidinone catalysts have been demonstrated to give high enantioselectivity in the asymmetric Diels–Alder reaction. The sulfonamide-based ligands from which these catalysts are generated are inert to the conditions of polymerization, making them ideal for copolymerization. The suspension polymerization outlined in **Figure 15.15** provided spherical beads with good swelling properties. The polymers were treated with $BH_3 \bullet SMe_2$ to generate the catalysts and used in the asymmetric Diels–Alder reaction between methacrolein and cyclopentadiene (**Figure 15.16**).[38]

Application of the PS-supported oxazaborolidinone in **Figure 15.15** resulted in formation of the Diels–Alder adducts with high *exo:endo* selectivity (> 90:10, **Figure 15.16**). Comparison of the enantioselectivities as a function of the cross-linker demonstrates the importance of this group. The PS cross-linked with divinylbenzene exhibited 65% ee, while catalyst derived from the hydrocarbon-spanned distyryl cross-linker gave 84% ee. This value is similar to that observed with the homogeneous analog (86% ee).[39,40] The oligo(oxyethylene) cross-linked polymers exhibited 92–95% enantioselec-

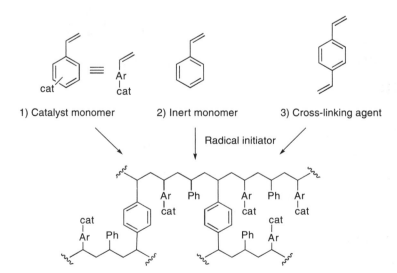

Figure 15.14. The three types of monomers used in making polystyrene-derived chiral catalyst polymers.

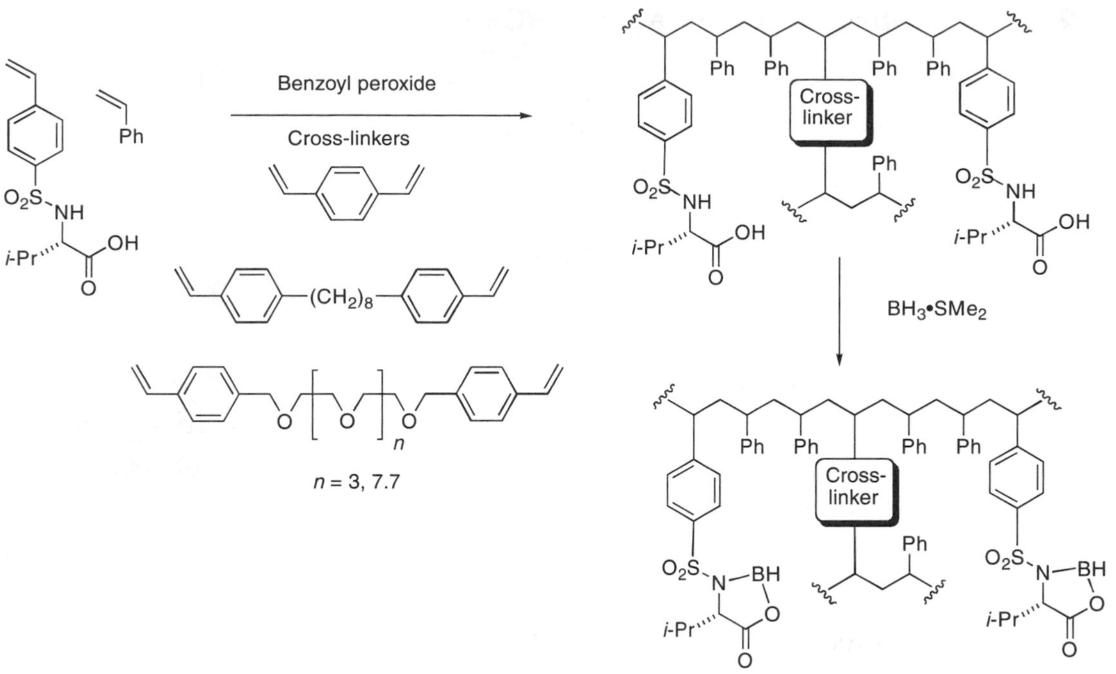

Figure 15.15. Generation of PS-supported oxazaborolidinone catalysts with various cross-linkers.

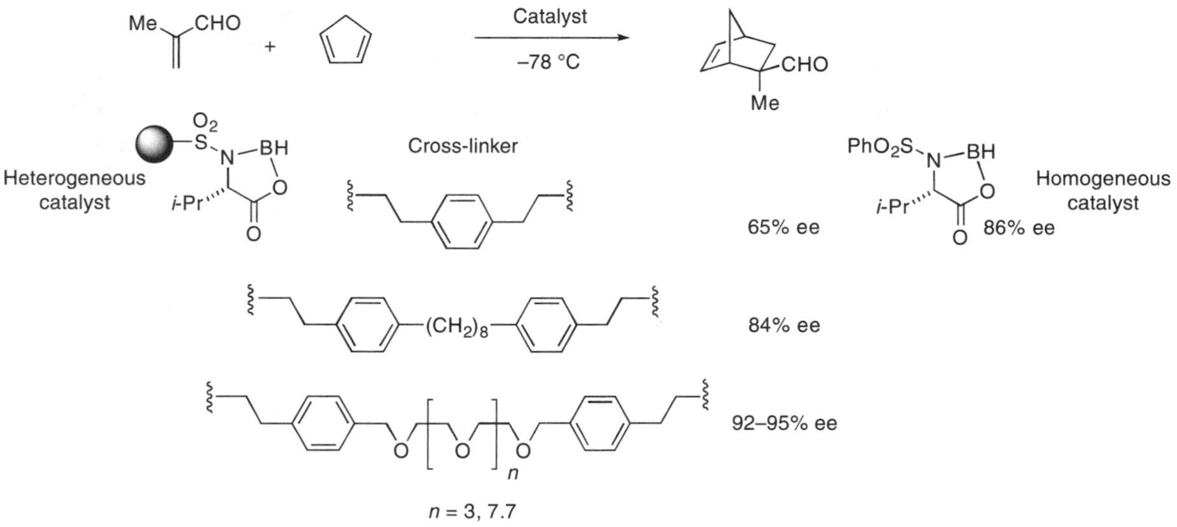

Figure 15.16. Use of PS-supported oxazaborolidinone catalysts with various cross-linkers in the asymmetric Diels–Alder reaction.

tivity at –78 °C. The oligo(oxyethylene) was used in a continuous-flow column in which the methacrolein and cyclopentadiene were passed through the catalyst mixture. The enantioselectivity of the continuous-flow system was found to be the same as the batch system.[38] Further investigation is necessary to understand the origin of the increase in enantioselectivity of the polymer-bound catalyst over the homogeneous analog.

15.2.2 Chiral Catalysts Incorporated by ROMP

Enantioselective olefin metathesis catalysts have been incorporated into polystyrene resins by using derivatized styrene monomers with good effect.[41] An alternative supported olefin metathesis catalyst has been reported wherein the polymer is formed via olefin metathesis.[4,42] Two norbornene units were attached to a chiral bisphenoxide, ring-opening metathesis polymerization (ROMP) with a ruthenium-based catalyst was employed to polymerize the chiral subunits, and then the molybdenum catalyst was generated on the support (**Figure 15.17**). Interestingly, the resultant molybdenum catalyst sites in the polymer do not cause metathesis reactions on the polymeric olefinic units. Rather, the chiral molybdenum polymer acts as an efficient and enantioselective ring-closing metathesis (RCM) catalyst for the desymmetrization and kinetic resolution of small-molecule triene and diene substrates, respectively. The conversion and enantioselection achieved with this polymer is comparable to that obtained with the corresponding monomeric homogeneous catalysts. In addition, removal of the catalyst by filtration is straightforward, which is particularly important when traces of molybdenum are unacceptable.

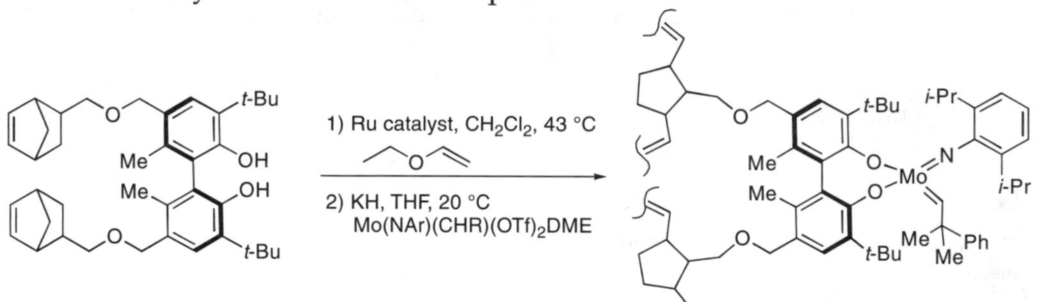

PolyMo

Substrate	Product[a]	Homogeneous catalyst Yield (%)	Homogeneous catalyst ee (%)	PolyMo Yield (%)	PolyMo ee (%)
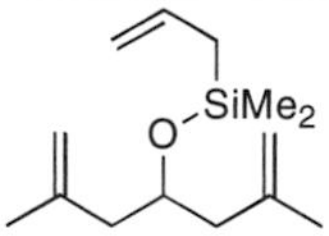	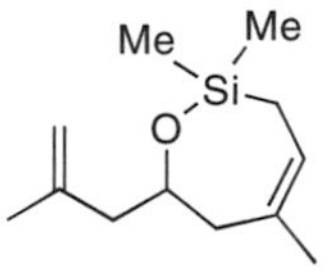	86	89	97	90
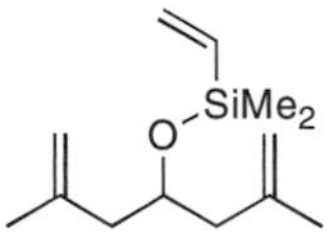	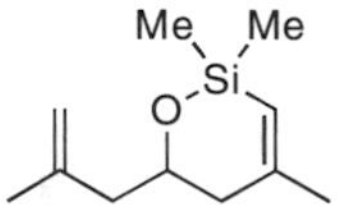			61	95
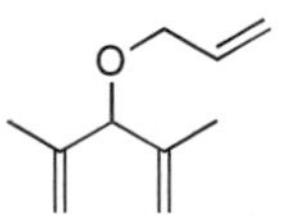	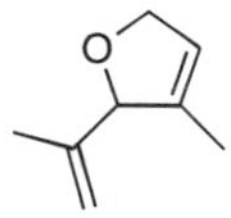	97	89	96	90
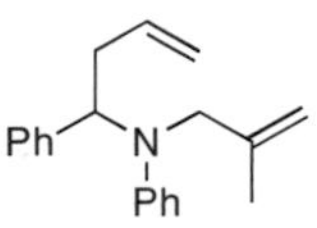	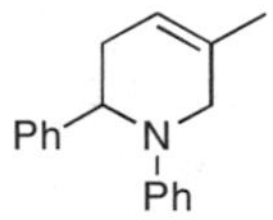	21	45	22	50

[a] Conditions: 2.8–5.0 mol% Mo, CH_2Cl_2, 30 min at 20 °C, 2 h at 45 °C

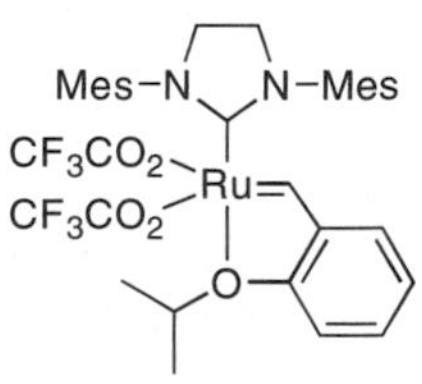

Ru ROMP catalyst

Figure 15.17. Chiral polymer RCM catalysts formed by ROMP.

15.2.3 Chiral Catalysts Incorporated into Other Organic Polymers

While many heterogenized asymmetric catalysts exhibit enantioselectivities approaching those observed in the parent homogeneous systems, the catalysts are often significantly less active because of the biphasic nature of many of these reactions. One strategy that has been successfully employed is to append the catalyst to a soluble polymeric support that can be easily precipitated upon completion of the reaction to facilitate catalyst isolation and recycling. In principle, such supported catalysts combine the high catalytic activity and enantioselectivity of homogeneous catalysts with the ease of catalyst isolation and recycling of heterogeneous systems. This class of supported catalysts has been called "one-phase catalysis–two-phase separation."[43]

Diastereomeric polymer-supported BINAP derivatives were generated by polycondensation of (R)- or (S)-5,5'-diamino-BINAP, terephthaloyl chloride, and (2S,4S)-pentanediol in the presence of pyridine (**Figure 15.18**). For the purpose of comparison, a monomeric model ligand was prepared by condensation of two equivalents of benzoyl chloride (PhCOCl) with (S)-5,5'-diamino-BINAP. The polymeric ligands were soluble in aprotic solvents such at toluene and dichloromethane, but insoluble in methanol. Conversion of the polymeric and monomeric ligands to the ruthenium-based precatalysts was performed by addition of the arene complex $[(\eta^6\text{-cymene})RuCl_2]_2$.

A comparison of the polymeric catalysts, the monomeric model system, and the (S)-BINAP analog was performed in the hydrogenation of the 2-arylacrylic acid (**Figure 15.18**). The hydrogenation product is naproxen, a common anti-inflammatory medication. The hydrogenation reactions were carried out in a mixture of toluene and methanol, in which the catalysts were completely soluble with a substrate:catalyst ratio of 200:1. The results are displayed in **Figure 15.18**. The diastereomeric catalysts derived from (R)- and (S)-5,5'-diamino-BINAP gave similar enantioselectivities to the monomeric model system and the BINAP-based catalyst (entries 1–4), indicating that the chiral diol in the polymer backbone had almost no impact on the enantioselectivity of the catalysts. Interestingly, $t_{1/2}$ of the polymeric catalyst was less than half that of the monomeric model system and a quarter of the BINAP-based catalyst. It is not clear why the polymer support exhibited a positive influence on the catalyst TOF.

Recovery and recycling of the polymer-supported catalysts was easily performed by addition of methanol to the reaction mixture after completion of the hydrogenation reaction. Filtration of the precipitated polymer and analysis of the filtrate by atomic absorption indicated that the ruthenium content was less than 16 ppb. In repeated application of the polymer-supported catalysts, no deterioration in activity or enantioselectivity was observed over 10 cycles.[43] These results attest to the potential benefits of supported catalysts on both catalyst recovery and TOF.

The same chiral diphosphine, (R)-5,5'-diamino-BINAP, was used in a similar fashion to prepare dendronized polymer-supported ligands and catalysts with chirality along the polymer backbone (**Figure 15.19**). Dendronized polymers are more easily prepared than dendrimers of higher generations.

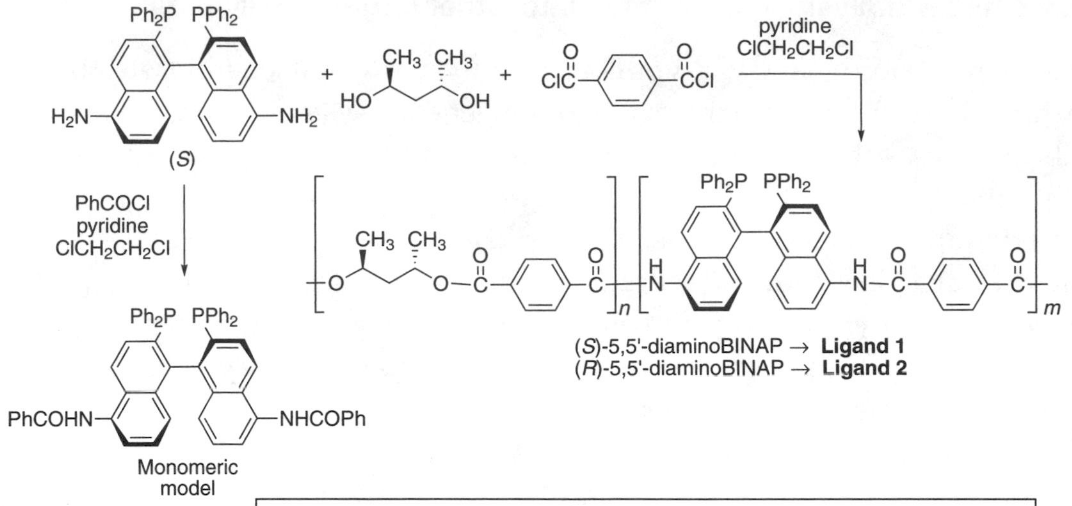

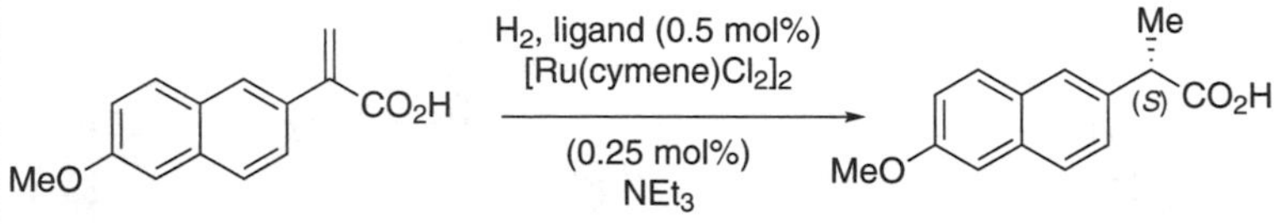

Figure 15.18. Synthesis of chiral polyester-supported BINAP polymer and application to the asymmetric hydrogenation of an acrylic acid derivative.

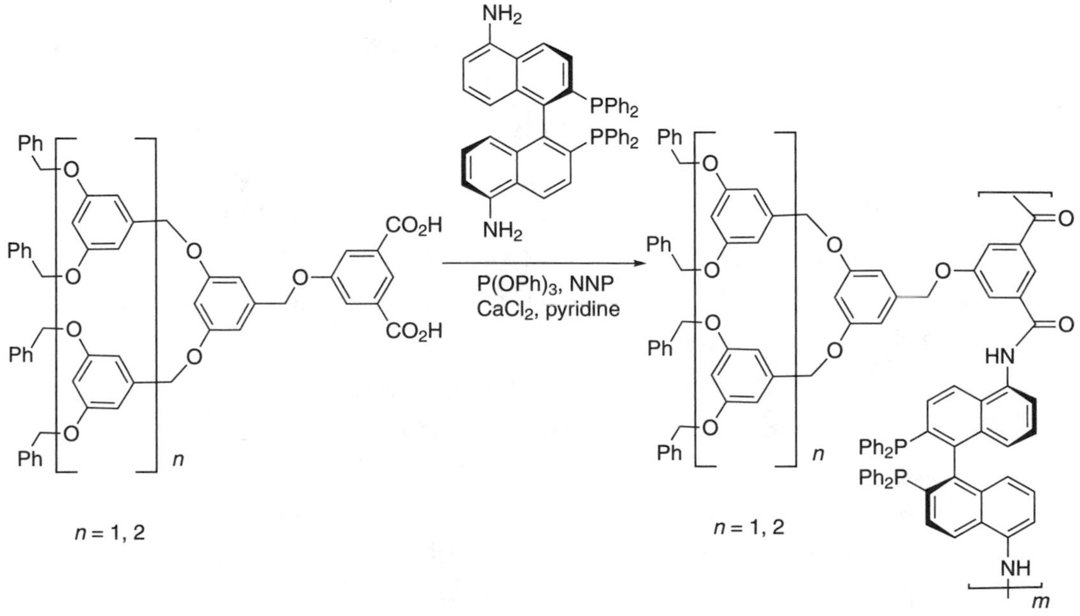

Entry	Ligand	Temp (°C)	Time (h)	Conv. (%)	ee (%)
1	Ligand 2	1	18	100	92.9 (*R*)
2	Ligand 1	1	12	97.4	93.6 (*S*)
3	Monomeric model	1	36	99.8	93.5 (*S*)
4	(*S*)-BINAP	1	48	94.7	93.5 (*S*)
5	Ligand 1	rt	4	95.4	87.7 (*S*)
6	Monomeric model	rt	4	64.2	89.2 (*S*)
7	(*S*)-BINAP	rt	4	56.5	88.7 (*S*)

Figure 15.19. Synthesis of dendronized poly(BINAP) ligands.

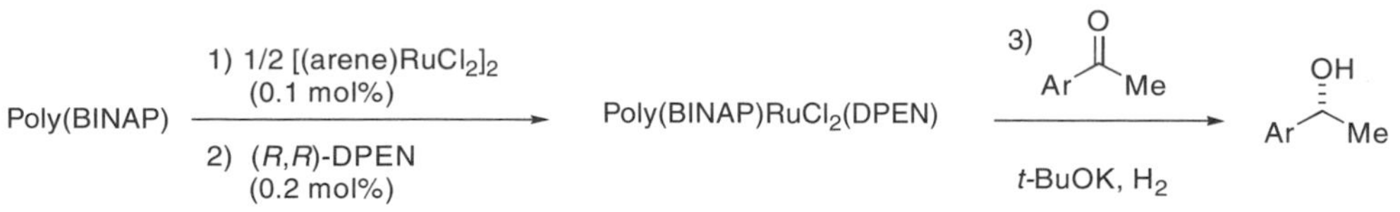

Poly(BINAP) = Dendronized polymer-supported BINAP

Ligand generation	Ar	ee (%)
$n = 1$	2-C_6H_4-Me	88
$n = 2$	2-C_6H_4-Me	92
$n = 1$	1-Naphthyl	88
$n = 2$	1-Naphthyl	92

Figure 15.20. Application of dendronized polymer-supported ligands in the asymmetric transfer hydrogenation reaction.

The dendronized polymers of **Figure 15.19** were then used in the asymmetric reduction of ketones, as illustrated in **Figure 15.20**. The reactions were conducted at 50 °C with a substrate:ruthenium ratio of 500:1. In this reaction, generation three, with $n = 2$, exhibited up to 4% higher enantioselectivity than generation two ($n = 1$).[44] The generation three catalyst was recycled three times without loss of enantioselectivity.

15.2.4 Catalytically Active Chiral Nanoparticles (Colloidal Catalysts)

While catalysts can be tethered to nanoparticle supports using techniques outlined throughout this chapter (see Section 15.1.6), the nanoparticles themselves can also be catalytically active. An example involving the synthesis of nanoparticles directly from molecular precursors is the decomposition of $Pd_2(dba)_3$ with hydrogen in THF in the presence of a sugar-based diphosphite ligand (L*) with atropisomeric biphenol units (Pd:L* = 5:1; **Figure 15.21**).[45] Under these conditions, small spherical particles with an average diameter of 4 nm formed. The reactivity and enantioselectivity of the chiral ligated palladium nanoparticles were then compared to a homogeneous complex, prepared from $[Pd(allyl)Cl]_2$ + 2 L*, in the allylic alkylation of racemic 3-acetoxy-1,3-diphenyl-1-propene.

Upon combination of 3-acetoxy-1,3-diphenyl-1-propene with dimethyl malonate under basic conditions, both the homogeneous and nanoparticle catalysts provided the allylic alkylation product with similar levels of enantioselectivity (**Figure 15.21**). The reactions displayed some important differences, however, most notably that the nanoparticle catalyst exhibited a dramatic change in rate at around 50% conversion. Such behavior in reactions of chiral racemic substrates is often indicative of a kinetic resolution (KR; see Chapter 7). Indeed, the ee of the remaining starting material in the nanoparticle-catalyzed reaction was found to be 89%, suggesting that an efficient KR had taken place. Decreasing the loading of the homogeneous catalyst resulted in TOFs similar to those observed in the nanoparticle system. The selectivity factor, s (= k_{rel}),

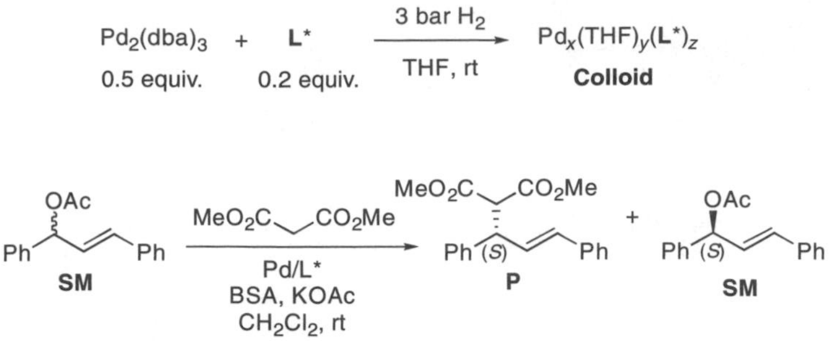

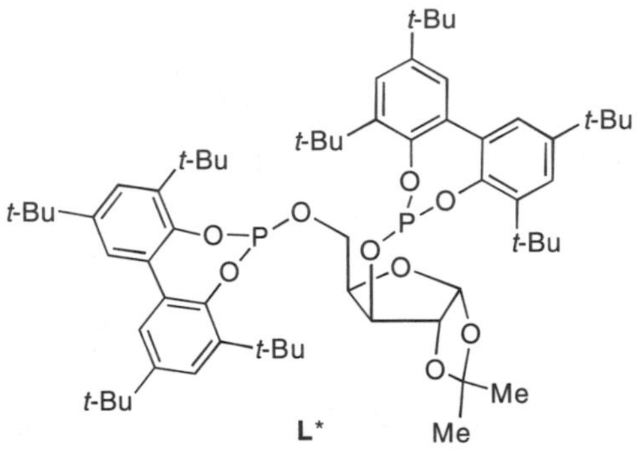

Figure 15.21. A colloidal chiral catalyst for asymmetric allylation.

Catalyst	SM:Pd:L*	Time (h)	Conv. (%)	ee P (%)	ee SM (%)
Colloid	100:1:0.2	24	56	97	89
	100:1:0.2	168	59	97	89
	100:1:1.05	168	61	97	89
Homogeneous	500:1:1	1.5	—	95	0
	2000:1:1	18	—	95	27
	10,000:1:1	48	—	95	40

for the homogeneous catalyst was 2, whereas for the nanoparicle catalyst it was significantly higher, ranging from 12–20.[45]

To determine if the nanoparticles were decomposing to a small amount of active homogeneous catalyst, several experiments were undertaken and the following observations were made: 1) no change in selectivity or KR was observed upon combining the nanoparticle catalyst with excess chiral ligand; 2) after the reaction of the nanoparticle catalysts had reached 50–60% conversion and the reaction had stopped, addition of a second dose of racemic substrate to the reaction mixture resulted in continued consumption of the (R)-starting material; and 3) addition of mercury or CS$_2$ to the nanoparticle catalyst inhibited catalysis, but had little or no impact on the homogeneous system. In many cases, mercury is known to poison colloidal catalysts. These experiments, and the significant differences in the selectivity factors between the colloidal and homogeneous systems, suggest that the nature of enantioselection is different, despite employing a common metal and chiral ligand. Further investigations are needed to determine if other chiral nanoparticle catalysts can be prepared and successfully employed in catalytic asymmetric reactions.

15.2.5 Chiral Catalysts Incorporated into Metal–Organic Coordination Networks

Several novel metal–organic coordination networks, with high stability, high porosity, and organic functionality have been reported.[46–49] These materials are rapidly constructed by combining simple bridging organic ligands with metal ions, as outlined in the supported catalyst overview (**Figure 15.1d**). The utility of achiral versions of these materials in the catalysis of organic reactions has been demonstrated.[50,51] By employing chiral bridging organic ligands to form chiral environments in the cavities of these materials, asymmetric catalysts can be constructed.[52,53] Due to the low solubility of metal–organic coordination networks in most media, these "self-supported" catalysts can be used to heterogenize homogeneous chiral catalysts to allow facile product purification and catalyst recycling.[54]

The idea behind metal–organic assemblies is to convert ligands used in homogeneous catalysis that traditionally contain a single binding site into ligands that possess two or more metal binding sites. Upon reaction with metal precursors, the metal binds the ligand to form metal–organic coordination networks (which have also been called metal–ligand polymers).[53] Metals in metal–organic coordination networks can serve as catalytically inactive bridges that connect the chiral metal complexes (M′, **Figure 15.22a**) or can serve as the bridges and active metal centers (**Figure 15.22b**). Examples are also known in which two different ligands must bind to each metal center to form the catalytically active metal–organic coordination network (**Figure 15.22c**).

An example of a coordination network from **Figure 15.22a** is described below. In such a system, a critical factor is the use of metal–ligand combinations that are self-selecting. The concept of self-assembly has been used to construct metal–organic coordination networks containing BINAP ligands (**Figure 15.23**).[55,56] For this support, phosphonate groups were added to the BINAP framework to act as linkers to anchor the catalyst to a zirconium–phosphonate support. Note that the character of the metals and ligands are orthogonal in that the ruthenium, phosphines, and amines are soft, and therefore matched, and the phosphonate and zirconium are hard (and also

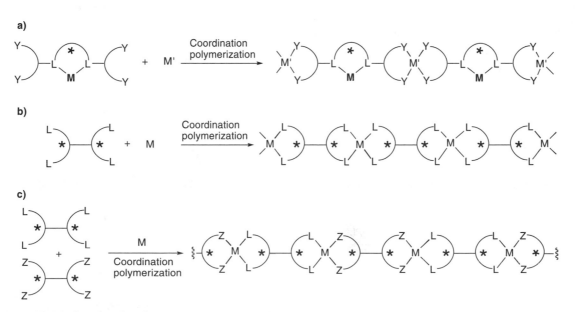

Figure 15.22. Synthesis of metal–organic coordination networks: (a) heterobimetallic system with M′ as a catalytically inactive linker, (b) M is the metal linker and the catalytically active site, and (c) heterotopic ligands linked by a metal complex.

matched). The ruthenium character, however, is mismatched with the phosphonate and the zirconium with the phosphines and amines.[57] Because the ruthenium adducts of chelating phosphine ligands are so stable, the preformed ruthenium complex can be subjected to the conditions used to generate the support. Thus, heating Zr(O-*t*-Bu)$_4$ in methanol with the ruthenium precursor generates the zirconium–phosphonate-supported catalyst. Characterization of the support revealed submicrometer particles comprised of a highly porous amorphous solid with a rather wide pore-size distribution.

The [(ZrO$_3$P)$_2$BINAP]RuCl$_2$(DPEN) catalyst exhibits exceptionally high activity and enantioselectivity in the hydrogenation of aromatic ketones (**Figure 15.24**).[55] For example, acetophenone was hydrogenated to 1-phenylethanol with 100% conversion and 96% ee when 0.1 mol% of the catalyst was employed. In comparison, the homogeneous monomeric (BINAP)RuCl$_2$(DPEN) precatalyst only provides ~80% ee under similar conditions. The implications of this confinement effect (see Section 15.4) are significant, in that it may be possible to improve the enantioselectivities of processes by the incorporation of catalyst into similar supports. Presumably, the extended chiral environment in the support, combined with a restricted reaction volume within the pores, contributes to this enantioselectivity improvement. Low catalyst loadings (0.005 mol%) could also be employed with TOF = 500/h (compared with 1250/h for the homogeneous system). Furthermore, the catalyst was very robust and could be successfully reused for six cycles with no deterioration in conversion or enantioselectiv-

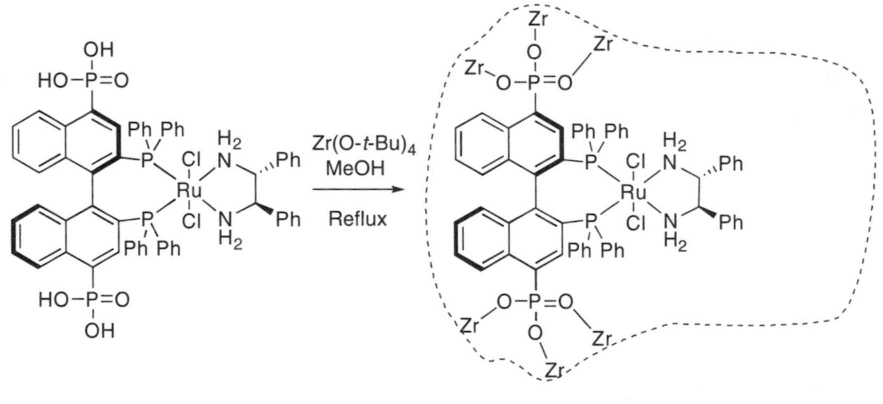

[(ZrO$_3$P)$_2$BINAP]RuCl$_2$(DPEN)

Figure 15.23. BINAP–DPEN ruthenium polymers with bridging zirconium phosphonate groups.

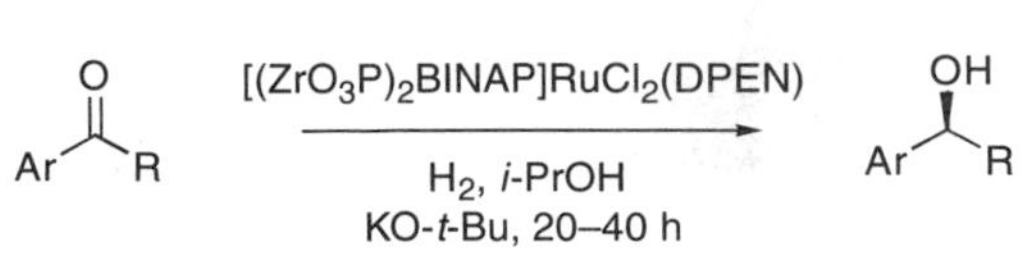

Ar	R	Ru (mol%)	KO-*t*-Bu (mol%)	ee (%)
Ph	Me	0.1	1	96
2-Naphthyl	Me	0.1	1	97
4-C$_6$H$_4$-OMe	Me	0.1	1	96
4-C$_6$H$_4$-Cl	Me	0.1	1	95
Ph	Et	0.1	1	93
1-Naphthyl	Me	0.005	0.02	99

Figure 15.24. Asymmetric hydrogenation with a chiral porous hybrid solid catalyst, [(ZrO$_3$P)$_2$BINAP]RuCl$_2$(DPEN), from **Figure 15.23**.

ity. It was determined that > 0.02% of the ruthenium metal leached into the organic phase during the hydrogenation reaction. Similar levels of enantioselectivity and activity have been reported with closely related catalysts in the asymmetric hydrogenation of β-ketoesters.[56]

A metal–organic coordination network of bis(BINOLate)-based catalysts was formed by the combination of dimeric BINOL units and either aluminum or titanium precursors (**Figure 15.25**).[58] Here, the metal functions as both a bridging unit and a catalytic center (**Figure 15.22b**). Characteristic IR peaks for the polymers are similar to those from the corresponding homogeneous complexes prepared from BINOL. The compositions of the metal–organic coordination networks were confirmed by elemental analysis.

The ALB coordination network displayed similar reactivity and selectivity compared to the homogeneous ALB catalyst in the conjugate addition of dibenzyl malonate (**Equation 15.2**). The supernatant from the heterogeneous catalyst ALB coordination network displayed no catalytic activity, confirming that the insoluble coordination network was catalytically viable. Recycling of the catalyst was possible by removal of the product-containing supernatant under an inert atmosphere and reapplication of the solvent and starting materials. While good activity was retained in subsequent cycles (59–86% yield), the enantioselection was eroded (second cycle = 87% ee).

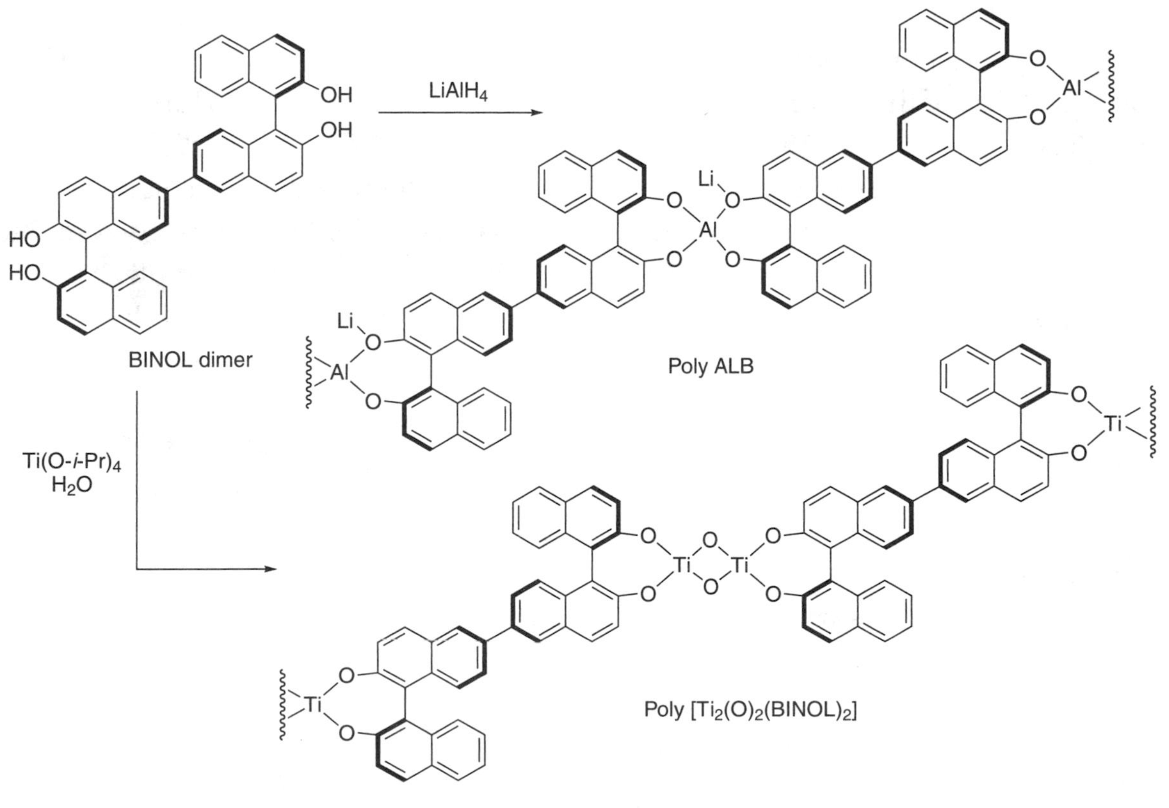

Figure 15.25. Synthesis of insoluble metal-bridged BINOL polymers.

Equation 15.2

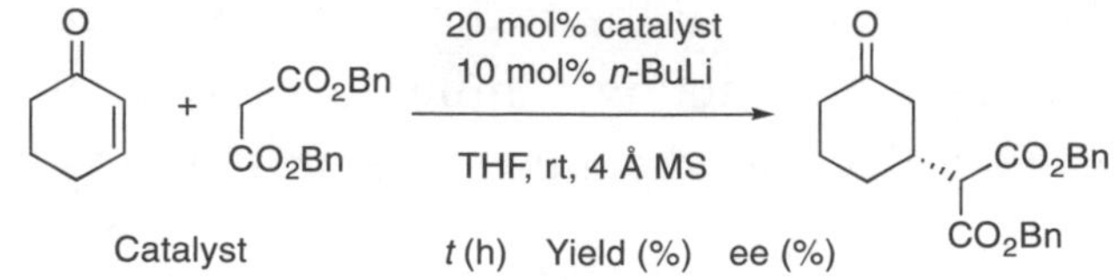

Catalyst	t (h)	Yield (%)	ee (%)
Heterogeneous ALB	48	86	96
Homogeneous ALB	12	100	97

In contrast to the homogeneous ALB catalyst, the heterogeneous $Ti_2O_2(BINOL)_2$ catalyst could be recycled in air. In addition, heterogeneous $Ti_2O_2(BINOL)_2$ displayed consistent activity (66–88% yield) and enantioselectivity (88–92% ee) over five cycles of use in the asymmetric carbonyl-ene reaction of α-methylstyrene with ethyl glyoxylate (**Equation 15.3**).[59] Overall, despite many advantages, these heterogeneous catalysts were typically less active (longer reactions times and/or lower conversions) than the corresponding homogeneous analogs.

Equation 15.3

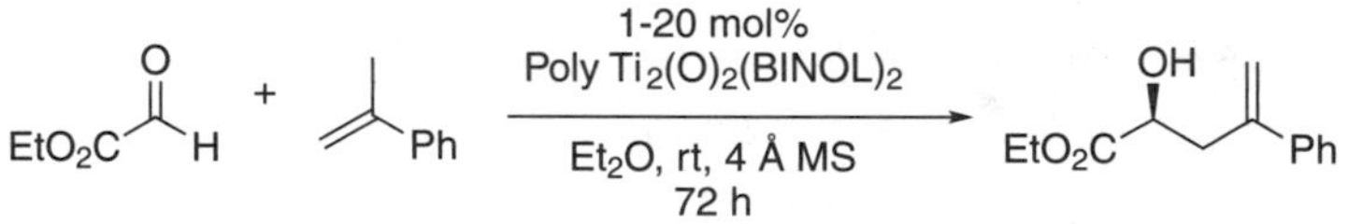

Catalyst	Yield (%)	ee (%)
Poly [$Ti_2(O)_2(BINOL)_2$]	81	90
$Ti_2(O)_2(BINOL)_2$	82	97

An interesting example of the programmed assembly of two different multitopic ligands to yield a metal–organic coordination network (**Figure 15.22c**) has also been reported.[60] One of the multitopic ligands is derived from BINAP units linked via a 1,4-phenylene spacer (**Figure 15.26**) and the second is composed of two DPEN derivatives separated by a diether spacer. The challenge in constructing the catalyst network is that there must be a strong preference for formation of the heteroligated combination over the catalytically inactive bis(diamine) or bis(diphosphine) complexes. Combining the ruthenium dimer, [($η^6$-arene)RuCl$_2$]$_2$, which contains a labile $η^6$-arene ligand, and the BINAP derivative at 100 °C results in formation of the bis(diphosphine)[RuCl$_2$(dmf)$_2$]$_2$. Addition of the linked diamine generates the heterogeneous-supported precatalyst, which has been referred to as "self-supported." Elemental analysis, [13]P CP-MAS, and FT-IR spectroscopy were consistent with the structure shown in **Figure 15.26**. The bulk material was composed of micrometer particles (1 μm) and shown to be amorphous.

Application of the supported P-phenyl derivative (Ar = Ph) to the hydrogenation of acetophenone provided the product with 78% enantioselectivity, which was similar to the homogeneous analog (BINAP)RuCl$_2$(DPEN) (80% ee).[61] The homogeneous system exhibited 96% enantioselectivity in the hydrogenation of acetophenone when Ar = 3,5-C$_6$H$_3$-Me$_2$. Based on this observation, the self-supported catalyst with

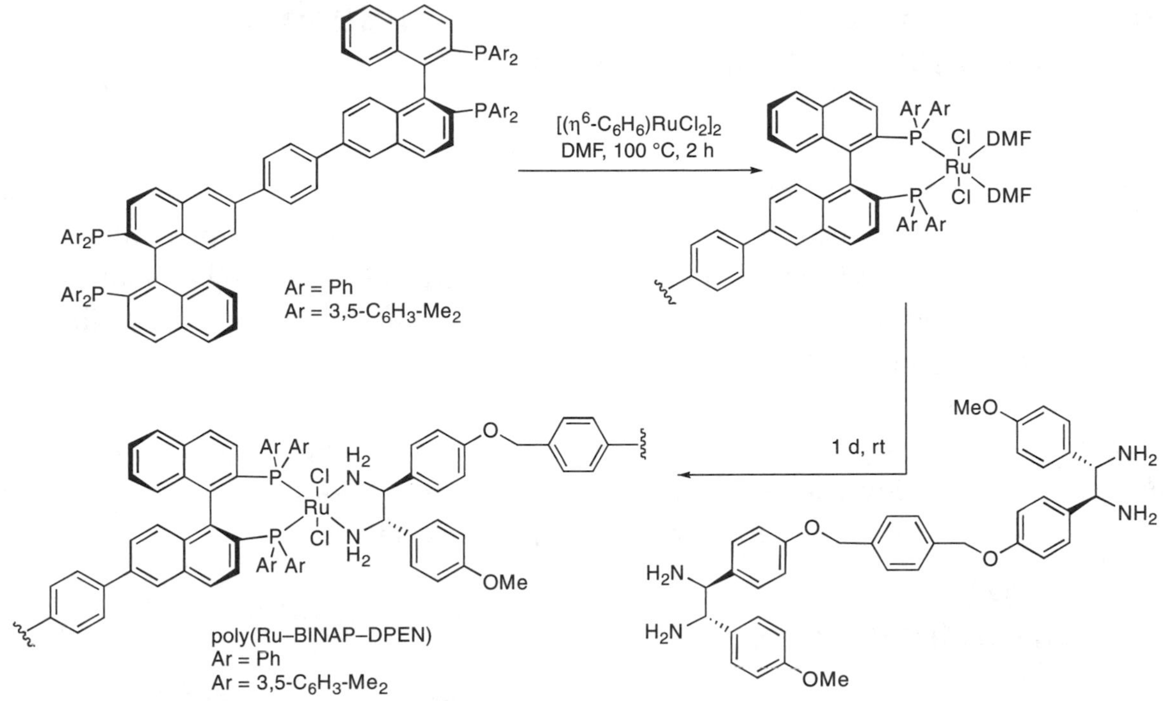

Figure 15.26. BINAP–DPEN ruthenium polymers via bridging BINAP and DPEN ligands.

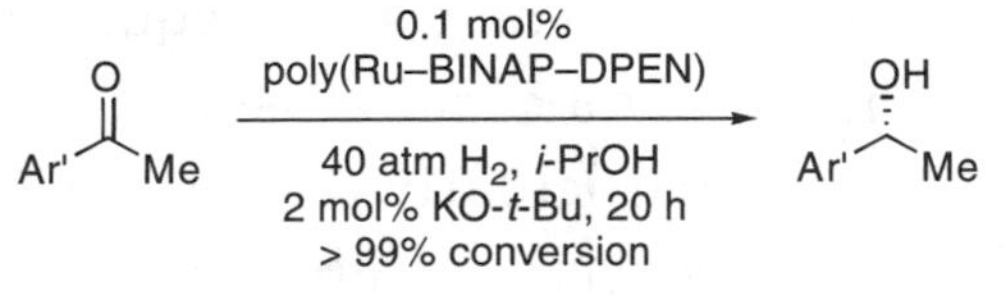

L*, Ar =	Ar'	ee (%)
Ph	Ph	78
3,5-C$_6$H$_3$-Me$_2$	Ph	97
3,5-C$_6$H$_3$-Me$_2$	2-Naphthyl	95
3,5-C$_6$H$_3$-Me$_2$	4-C$_6$H$_4$-OMe	96
3,5-C$_6$H$_3$-Me$_2$	4-C$_6$H$_4$-Cl	97
3,5-C$_6$H$_3$-Me$_2$	1-Naphthyl	98

Figure 15.27. Asymmetric hydrogenation with self-supported catalysts from **Figure 15.26**.

Ar = 3,5-C$_6$H$_3$-Me$_2$ was prepared. Use of this system in the analogous reaction resulted in slightly higher enantioselectivity than the homogeneous catalyst (97% ee; **Figure 15.27**). The insoluble self-supported catalyst was easily filtered from the reaction mixture and the filtrate was shown to be inactive in the hydrogenation of acetophenone. Furthermore, ruthenium leaching was found to be < 0.1 ppm. Both of these experimental observations are consistent with a heterogeneous catalyst. The self-supported catalyst was used in seven enantioselective hydrogenation cycles without loss of enantioselectivity or activity. Even at 0.01 mol% ruthenium, acetophenone was hydrogenated in 95% enantioselectivity with a TOF of approximately 500/h.[60]

With their simple assembly, high density of catalytic sites, stability, high catalytic efficiency, excellent enantioselectivity, facile recovery, and good recycling profiles, the use of chiral metal–organic coordination networks in asymmetric catalysis is bound to increase. One of the future challenges in this area will be to develop efficient syntheses of chiral ligands with multiple binding sites.

15.3 Chiral Catalysts Attached to Solid Supports by Noncovalent Interactions

The most common method to immobilize homogeneous catalysts onto supports is through covalent linkers, which are usually attached through the chiral ligand. While this method has met with success, it has several drawbacks. First, the chiral ligand must be modified to accommodate the tether. Not only does installing tethers require increased synthetic effort, the consequences of ligand modification on enantioselectivity are often unpredictable. To streamline the preparation of immobilized catalysts and to avoid potential problems caused by interference of the covalent tether, chemists have developed several noncovalent methods to heterogenize asymmetric catalysts involving ion-exchange, encapsulation, assembly of the catalyst inside the support, and construction of the support around the catalyst.

15.3.1 Absorbed Chiral Catalysts

Direct absorption of chiral ligands onto metal surfaces has not been found to be a general method to prepare highly enantioselective catalysts. First, the chiral ligand, also called a modifier, must adhere to the metal surface. Metal surfaces are not flat, but contain defects such as kinks, which may be chiral. There are an equal amount of these enantiomeric chiral kinks, however. The chiral modifier must either block one configuration of chiral metal sites (see Chapter 6, particularly the discussion of chiral poisoning) or accelerate the rate of the enantioselective pathway over the background reaction of the unmodified metal sites (i.e., ligand-accelerated catalysis).[62]

One notable exception is silica- or alumina-supported platinum modified by cinchona alkaloids for the asymmetric hydrogenation of α-functionalized ketones such as α-ketoesters, α-ketoamides, α-ketoacids, α-diketones, α-ketoacetals, and α-ketoethers, among others (**Figure 15.28**).[63–66] This catalyst system was introduced in 1979[67–69] and has been studied extensively in the ensuing years. The alkaloid-modified catalysts can be generated in situ by mixing the supported platinum with the alkaloid, followed by addition of the substrate and hydrogen. Reactions can be

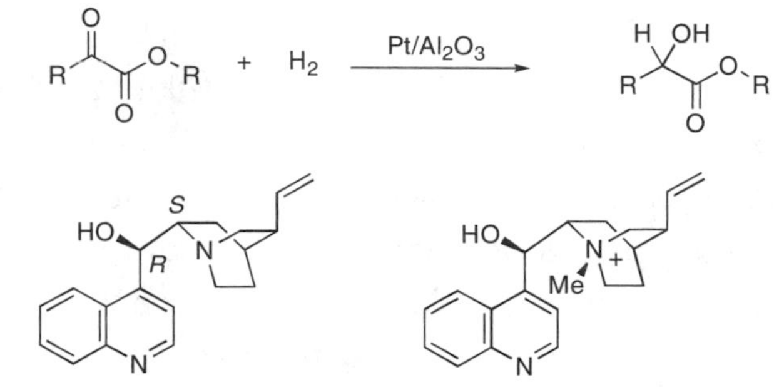

Figure 15.28. Alumina-supported platinum modified by cinchona alkaloids for the asymmetric hydrogenation of α-ketoesters. Alkylation of the basic nitrogen results in loss of enantioselectivity.

conducted at room temperature at hydrogen pressures of 1–100 bar. The amount of alkaloid necessary to achieve maximum enantioselectivity is quite low, estimated to be one adsorbed cinchona molecule for 10–20 surface platinum atoms.[70] Enantioselectivities as high as 90–95% have been achieved in this system.

The cinchona-modified platinum catalyst was applied to the synthesis of an intermediate for the ACE inhibitor benazepril (**Figure 15.29**, top). Hydrogenation of the diketoester was scaled-up to a production process of 10–200 kg (98% yield, 79–82% ee).[71] A more efficient route that is feasible on a large scale has been introduced.[65]

While considerable effort has been focused on understanding the mode of action of this unique catalyst system, several fundamental questions remain under debate. It is generally agreed that the alkaloid absorbs via the quinoline group, which is believed to be preferentially oriented flat on the platinum surface. In this orientation, the cinchonidine is tightly bound to the platinum surface via the π-system of the quinoline ring.[72] The crucial role of the quinuclidine nitrogen was revealed when experiments were conducted with N-alkylated derivatives (**Figure 15.28**), which exhibited complete loss of enantioselectivity.[73] Based on this and other mechanistic studies, a model for the rate acceleration and enantioselectivity has been proposed that involves protonation of the quinuclidine nitrogen by the acetic acid solvent. The protonated quinuclidine N–H is proposed to hydrogen-bond to the keto oxygen of the α-ketoester and preferentially orient the adsorbed substrate for reduction (**Figure 15.30**, left).[74] An alternative model involves hydrogen bonding to the ester carbonyl.[75]

A significantly different model is based on nucleophilic addition of the basic nitrogen to the keto group to generate a zwitterionic adduct between the cinchonidine and the pyruvate (**Figure 15.30**, right). The zwitterionic intermediate is stabilized under the acidic reaction conditions by hydrogen bonding (not shown). The C–N bond then undergoes reduction with inversion to provide the observed enantiomer of the product.[76] The divergent reaction mechanisms proposed for this important process reflect the difficulty associated with the study of absorbed catalysts.

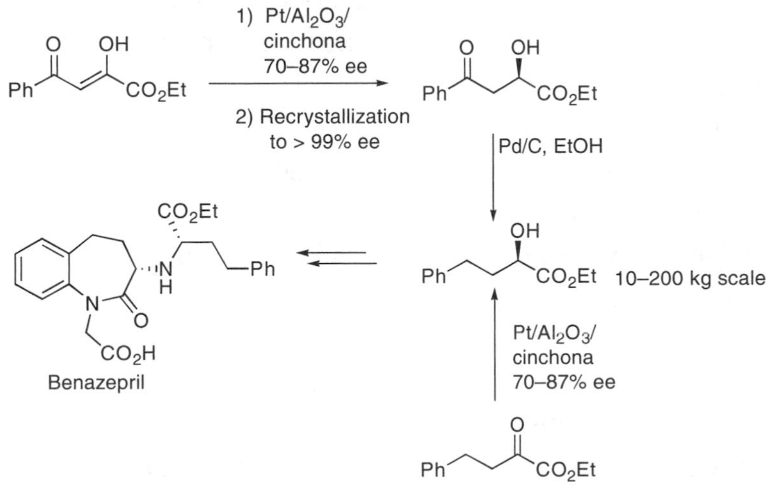

Figure 15.29. Application of the platinum–cinchona-catalyzed hydrogenation to the preparation of an intermediate in the ACE inhibitor benazepril.

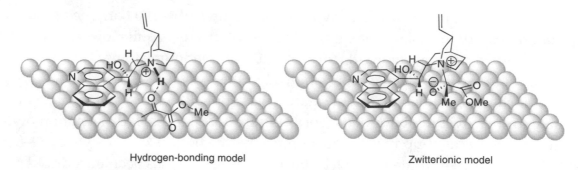

Hydrogen-bonding model   Zwitterionic model

Figure 15.30. Two models for the platinum–cinchonidine-catalyzed hydrogenation of pyruvate involving hydrogen-bond activation of the keto group (left) and a zwitterionic adduct that is subsequently reduced with inversion (right).

Several researchers have investigated the possibility of catalyst reuse and continuous reactions methods.[65] Catalyst reuse is found to be possible when fresh modifier is added. In continuous-flow reactors (e.g., fixed-bed reactors), the chiral modifier must be added in ppm concentrations to the feed to maintain high enantioselectivities.[77]

15.3.2 Heterogenization of Chiral Catalysts by Ionic Interactions

Many metal-based asymmetric catalysts are positively charged and contain noncoordinating or weakly coordinating anions. An effective method to anchor cationic catalysts to solid supports is through treatment with cation-exchange-type resins. Advantages of this method include 1) modification of the chiral ligands is unnecessary, 2) many ion-exchange resins are commercially available and inexpensive, 3) supported catalysts are easily removed from reaction mixtures and recycled, and 4) supported catalyst efficiency and enantioselectivity can be comparable to homogeneous analogs. Catalyst leaching, however, can be problematic if the reactions employ or generate salts.

Immobilization of cationic catalysts on cation-exchange resin can be straightforward. For example, treatment of the commercially available DOWEX 50WX2-100, a sulfonated polystyrene, with aqueous lithium hydroxide generates the cation-exchange beads. Stirring the lithiated resin with cationic rhodium complexes in methanol provides the immobilized catalysts (**Figure 15.31**).[78] The metal uptake of the resin was determined by analytical techniques to be 0.93% (w/w) with 68% of the initial rhodium complex absorbed onto the resin.

Comparison of the homogeneous and resin-bound catalysts in the asymmetric hydrogenation of methyl 2-acetamidoacrylate (MAA) was undertaken (**Equation 15.4**). In the homogeneous hydrogenation reactions with the PF_6^- salts, the DIOP-based rhodium complex exhibited marginal enantioselectivity (57%), while the TMBTP derivative catalyzed the reaction with excellent enantioselectivity (> 99%). Under identical reaction conditions, the immobilized catalysts exhibited similar levels of enantioselectivity and TOF to their homogeneous counterparts in the first cycle. A marked decrease in the activity of the resin-bound catalysts was observed in the sec-

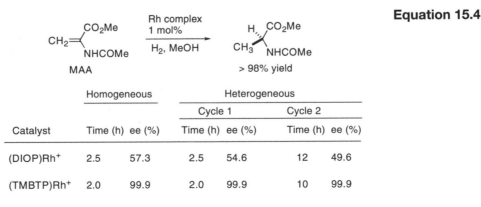

Figure 15.31. Anchoring of chiral catalyst precursors to a cation-exchange resin.

ond cycle, while a smaller change was found between the second and third cycle with the TMBTP-based catalyst. The enantioselectivity of the DIOP catalyst eroded slightly in the second cycle, but no change was observed with the TMBTP catalyst. Analysis of the reaction mixtures after removal of the catalyst beads indicated that > 2% rhodium leaching occurred in each cycle. To compare the solution structures of the homogeneous and heterogeneous catalysts, methanol solutions of $[(DIOP)Rh(NBD)]^+$ were pressurized with hydrogen. High-pressure NMR spectroscopy indicated that both the homogeneous and immobilized complexes formed $(DIOP)Rh(MeOH)_2^+$ after reduction of the NBD ligand. In situ low-temperature ^{31}P NMR monitoring of the asymmetric hydrogenation reactions with the homogeneous and heterogeneous catalysts indicated a common catalyst resting state, $(DIOP)Rh(MAA)_2^+$, with identical chemical shifts. Interaction of the immobilized catalyst with the anionic sulfonate groups had little impact on the spectroscopic features of the catalyst. Trace amounts of oxidized phosphine were also detected. Consistent with this mode of catalyst deactivation and rhodium leaching, removal of the supported catalysts from the reaction mixtures, followed by addition of substrate and pressurization with hydrogen, did not result in hydrogenation of the MAA.[78] This study not only demonstrates the ease of immobilization of charged transition-metal complexes onto ion-exchange resins and the benefits gained through this mode of heterogenization, it also illustrates the use of NMR spectroscopy to probe catalyst-support interactions.

Equation 15.4

Catalyst	Homogeneous		Heterogeneous			
			Cycle 1		Cycle 2	
	Time (h)	ee (%)	Time (h)	ee (%)	Time (h)	ee (%)
$(DIOP)Rh^+$	2.5	57.3	2.5	54.6	12	49.6
$(TMBTP)Rh^+$	2.0	99.9	2.0	99.9	10	99.9

Catalyst immobilization by ion-exchange methods is not limited to the use of polymer-based resins, but can also be performed with inorganic materials. An example of the latter is the immobilization of catalysts for the osmium-catalyzed Sharpless asymmetric dihydroxylation (AD) of olefins. The AD enables access to a variety of useful diols with high levels of enantioselectivity.[79] Despite all its attributes, the usefulness of the AD in large-scale pharmaceutical applications is limited by the high cost of the osmium precursors and the chiral ligand, the perceived toxicity of osmium, and potential difficulties of removing trace osmium in the products. To circumvent these issues, several immobilization strategies have been developed. The majority of these focus on immobilization of the chiral ligands on soluble[80–83] and insoluble polymers[84] or silica gel.[83,85,86] Recovery of the osmium using these approaches is challenging, because they depend on binding OsO_4 to the tethered ligands. The affinity of OsO_4 for the supported ligands is generally low, however, interfering with reuse of the valuable metal.

Alternative approaches to the heterogenization of the AD catalyst have focused on retaining the osmium metal on the support. A valuable precursor for osmium tetroxide is OsO_4^{2-}, which is commercially available as the potassium salt. The ion-exchange supports employed in this study were layered double hydroxides (LDH), a class of materials composed of alternating cationic and anionic layers (**Figure 15.32**). Cationic layers, $M(II)_{1-x}M(III)_x(OH_2)^{x+}$, are separated by anions and water. In this study, LDH crystals with $Mg_{1-x}Al_x(OH)(Cl)_x \bullet zH_2O$ with $x = 0.25$ were employed. The sizes of the crystals were 50–100 nm with an external surface area of around 100 m^2/g.[87] K_2OsO_4 was exchanged with the LDH to generate $LDH \bullet OsO_4$ with 0.975 mmol of Os per gram $LDH \bullet OsO_4$.[88,89] The $LDH \bullet OsO_4$ exhibited a broad band near 830–860 cm^{-1} in the FTIR spectrum for asymmetric O=Os=O absorption. Comparison with the sharp band at 819 cm^{-1} in the IR spectrum of K_2OsO_4 led to the conclusion that the heterogenized OsO_4^{2-} experiences only a weak interaction with the LDH. X-ray powder diffraction indicated that LDH and $LDH \bullet OsO_4$ exhibited similar spacing between layers, suggesting that the osmium salt did not intercalate and between the layers. Rather, the OsO_4^{2-} is located at the edges and on the surfaces.

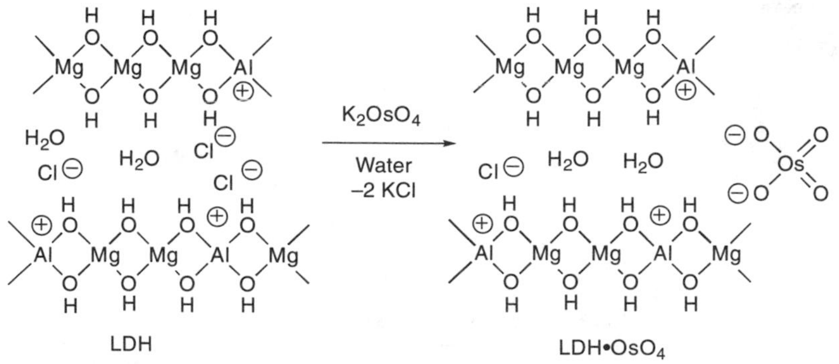

Figure 15.32. Ion-exchange of K_2OsO_4 into LDH to give $LDH \bullet OsO_4$.

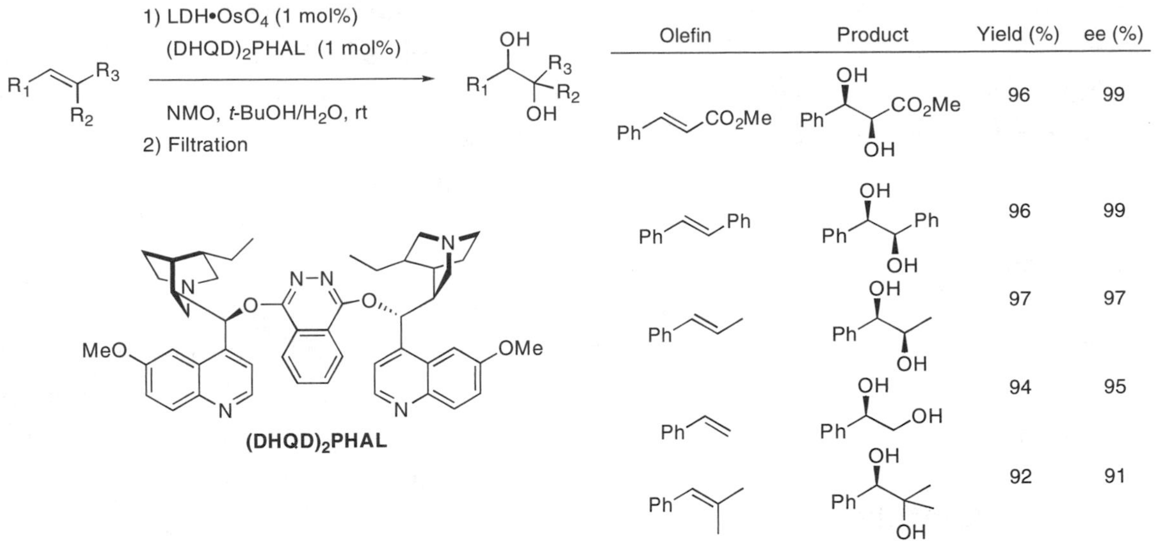

Figure 15.33. Asymmetric dihydroxylation of olefins with LDH•OsO$_4$ and (DHQD)$_2$PHAL.

Application of LDH•OsO$_4$ to the AD with the ligand (DHQD)$_2$PHAL is illustrated in **Figure 15.33**. The reactions were conducted under conditions similar to those used in the homogeneous AD and exhibited high levels of enantioselectivity. After completion of the reaction, the LDH•OsO$_4$ was recovered by filtration and the (DHQD)$_2$ PHAL by extraction with acid (95% ligand recovery). The filtrate was catalytically inactive and indicated that no leaching had occurred. The recovered LDH•OsO$_4$ was employed in four subsequent AD reactions without loss of activity. The high enantioselectivities in the LDH•OsO$_4$/(DHQD)$_2$PHAL system provides further support that the osmium(VI) did not intercalate into the LDH. The interlamellar space, which is the space between the cationic layers, is only about 3 Å. If the osmium had intercalated, it would be very difficult for the ligands to intercalate and bind to the metal due to its large volume. Experimental evidence supports the proposal that neutral OsO$_4$ is physisorbed onto the support during the Os(VI)/Os(VIII) redox cycle of the AD.[88]

The results outlined in this section indicate that simple ion-exchange techniques can be performed on supports with very different properties. Not only do these supports increase the practicality of the processes by facilitating catalyst handling and recovery, they also contribute to the development of environmentally benign chemical processes. We believe that these techniques will be of increasing importance for future catalytic asymmetric processes.

15.3.3 Immobilization by Anion Capture

Another noncovalent method to immobilize charged catalysts to silica currently functions specifically with sulfonate counterions. It has been demonstrated that hydrogen bonds form between complexes bearing sulfonate groups (L–SO$_3^-$) on ligands and surface silanols.[90] This technique required modification of the ligands to install the sulfonate group. A simpler version uses cationic metal complexes with trifluoromethanesulfonate (triflate, $^-$OTf) counterions combined with MCM-41. MCM-41 was chosen because it possesses large, tailorable, and well-defined pores, making it suitable for systematic studies. Furthermore, it is known to have a higher density of surface silanols than other silica-based supports. Combining [(R,R)-Me-(DuPHOS)Rh(COD)]OTf (**Figure 15.34**) with MCM-41 led to efficient incorporation of the rhodium catalyst, as judged by analytical techniques.[91]

To probe the association of the catalyst with the support, ^{13}P and ^{19}F NMR experiments were undertaken. Both the ^{13}P and ^{19}F NMR spectra of the immobilized catalyst exhibited significantly broadened resonances with respect to the homogeneous complex, indicative of restricted mobility of the precatalyst on the MCM-41. Addition of NaB(Ar$_F$)$_4$ {NaB(Ar$_F$)$_4$ = NaB[3,5-C$_6$H$_3$(CF$_3$)$_2$]$_4$} caused the release of [(R,R)-Me-(DuPHOS)Rh(COD)]B(Ar$_F$)$_4$ from the support, which again gave a sharp ^{31}P NMR spectrum. In contrast, the ^{19}F NMR spectrum of the support remained broadened and unchanged, consistent with no loss of triflate from the support. The absence of exchange of the silanol-bound triflate with the NaB(Ar$_F$)$_4$, and the loss of [(R,R)-Me-(DuPHOS)Rh(COD)]B(Ar$_F$)$_4$ from the support suggest that the triflate is tightly bound to the MCM-41. Cationic [(R,R)-Me-(DuPHOS)Rh(COD)]$^+$ associates with the MCM-41 support by charge–charge interaction with the triflate.[91]

The MCM-41-supported [(R,R)-Me-(DuPHOS)Rh(COD)]OTf was then compared to the homogeneous triflate salt in the hydrogenation of enamides (**Figure 15.34**). In each case, the supported catalyst gave higher enantioselectivities. There was also a significant difference in activity between the homogeneous and heterogeneous catalysts with the β,β-disubstituted substrates. For example, substrate **C** underwent hydrogenation in 16 h at 8 psi hydrogen with the supported catalyst, whereas the homogeneous analog exhibited only 26% conversion after 22 h at increased pressure (40 psi).

After completion of the hydrogenation of substrate **A**, the MCM-41-based catalyst reaction mixture was filtered and the filtrate found to be inactive toward hydro-

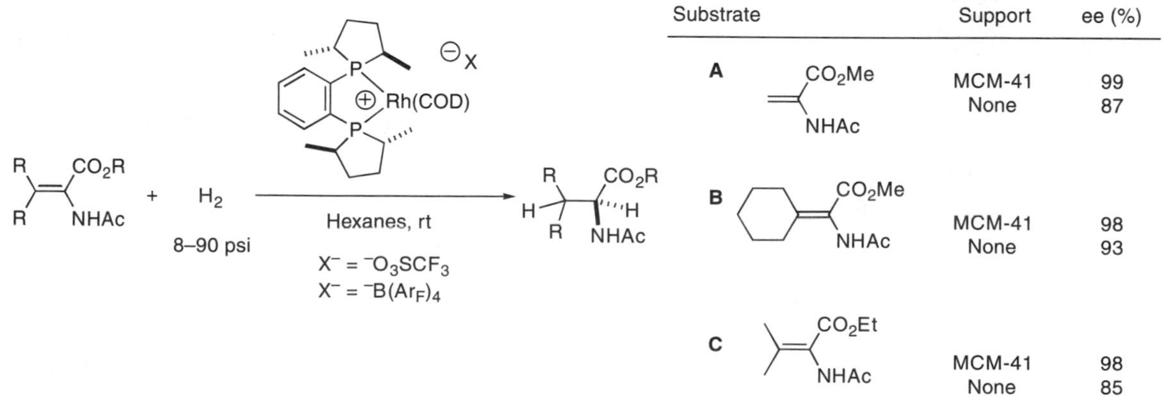

Substrate		Support	ee (%)
A	CO$_2$Me / NHAc	MCM-41	99
		None	87
B	CO$_2$Me / NHAc	MCM-41	98
		None	93
C	CO$_2$Et / NHAc	MCM-41	98
		None	85

Figure 15.34. Comparison of the asymmetric hydrogenation with a homogeneous catalyst and an MCM-41-based catalyst held to the support by hydrogen bonds between the support silanols and triflate counterions.

genation of additional substrate. Thus, little or no catalyst leaching had occurred. No loss of activity or enantioselectivity was observed on recycling the catalyst up to four times. This immobilization technique is currently limited to catalysts containing triflates and supports with a high density of silanol groups. Furthermore, the catalyst must be stable to the acidic silanol groups. Nonetheless, it appears to have great potential and these limitations will likely be reduced upon further investigations.

15.3.4 Microencapsulated Chiral Catalysts

The approaches to catalyst heterogenization outlined to this point have involved covalent tethering, physisorption, and ion-exchange techniques. Tethering requires structural modification of the catalysts, while ion-exchange and physisorption involve close association of the catalyst with the support, which can lead to changes in catalyst enantioselectivity relative to homogeneous counterparts. Encapsulation, on the other hand, does not involve modification of the homogeneous catalyst nor does it require direct interaction between the catalyst and support. For this reason, it represents the closest approximation to the homogeneous system.[19,20]

15.3.4.a Ship in a Bottle: Catalyst Assembly Inside a Support

In the first type of microencapsulation, the catalyst is constructed inside of a preformed cavity. In order to encapsulate a catalyst, the catalyst must be small enough to fit in the cavities of the support, but larger than channels to prevent its escape. The catalyst can be encapsulated by assembly of the catalyst from smaller precursors inside the preformed support, which is often referred to as the ship-in-a-bottle approach.[92] If the catalyst is to be assembled inside the support cavities, the reactions employed must be high yielding to avoid entrapment of byproducts in the support, which might also form viable catalysts that are less enantioselective. The most common supports for encapsulation of catalysts are zeolites (molecular sieves), which are microcrystalline and have regular structures. There are many naturally occurring and synthetic zeolites, offering a range of pore sizes.

An approach to encapsulation employed zeolite EMT, a partial structure of which is shown in **Figure 15.35**.[93] Zeolite EMT contains hypercages accessible through three pore windows of 0.69×0.74 nm and two circular apertures of 0.74 nm. The (salen) MnCl complexes used in this study have dimensions greater than 1.3 nm and, therefore, cannot escape the zeolite once encapsulated.

Beginning with dehydrated zeolite EMT, one equivalent of either (R,R)- or (S,S)-trans-1,2-diaminocyclohexane per EMT unit cell (repeat unit) was heated (**Figure 15.36**).[94] The salicylaldehyde (two equivalents) was then added, resulting in a color change of

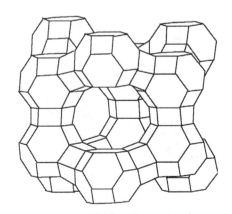

Figure 15.35. Partial structure of zeolite EMT. (Reproduced by permission of the PCCP Owner Societies with acknowledgment to Michael W. Anderson.)

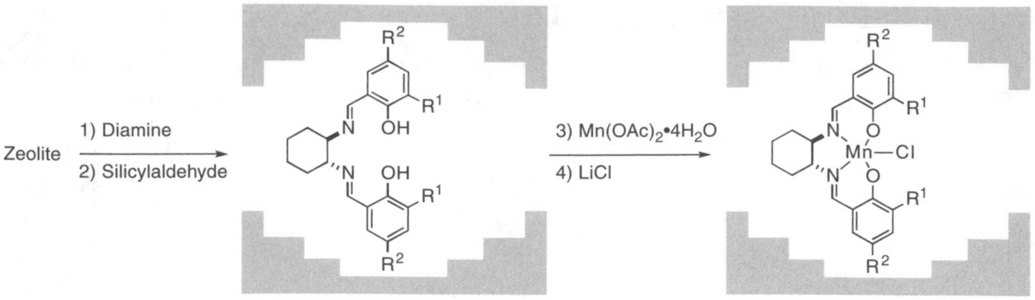

Zeolite $\xrightarrow{\text{1) Diamine}}$ $\xrightarrow{\text{3) Mn(OAc)}_2 \bullet 4\text{H}_2\text{O}}$
$\qquad\quad\text{2) Silicylaldehyde}\qquad\qquad\qquad\text{4) LiCl}$

$R^1 = R^2 = H$
$R^1 = t\text{-Bu}, R^2 = Me$

Figure 15.36. Ship-in-a-bottle approach to catalyst synthesis. The catalyst, (salen)MnCl, cannot escape through the pores.

the sieves to the characteristic yellow of the imine ligand. The metal precursor, $Mn(OAc)_2 \bullet 4H_2O$, was added, the resulting complex oxidized in air, and LiCl added to generate the (salen)MnCl complex. The zeolites were subjected to Soxhlet extraction to remove the excess reactants, ligand, and any nonencapsulated complex. The new materials were characterized by the imine C=N stretch in the FT-IR spectrum. Comparable stretching frequencies were measured for the interzeolite complexes and the homogeneous analog.[94] A model for the encapsulated (salen)MnCl complex inside the zeolite is illustrated in **Figure 15.37**.

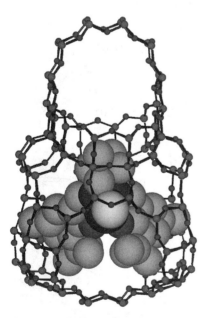

Figure 15.37. Calculated structure of (salen)MnCl encapsulated in a zeolite. (Professor Thomas Bein is acknowledged for permission to use this image.)

The encapsulated (salen)MnCl was examined in the asymmetric epoxidation reaction in **Figure 15.38**.[94] Addition of olefin and NaOCl to the supported catalyst resulted in epoxidation of the olefin. The enantioselectivities of the encapsulated and homogeneous catalysts were very similar, although the activity of the homogeneous catalyst was greater. The highest enantioselectivity with EMT-2 was achieved with cis-β-methylstyrene (88% ee with pyridine N-oxide additive). In this study, smaller olefins were observed to react faster with the encapsulated catalyst, consistent with the pore structure impacting diffusional access to the active sites. Very large olefins, such as cholesterol, did not undergo epoxidation, because they could not pass through the pores of the zeolite support. This result indicates that the epoxidation of the smaller substrates does occur inside the zeolite structure.[94] Furthermore, upon removal of the zeolites from the reaction mixture, epoxidation ceased, indicating that the epoxidation was occurring in the zeolite support. Optimization of this conceptually novel approach to catalyst immobilization represents a significant challenge, with many potential applications.

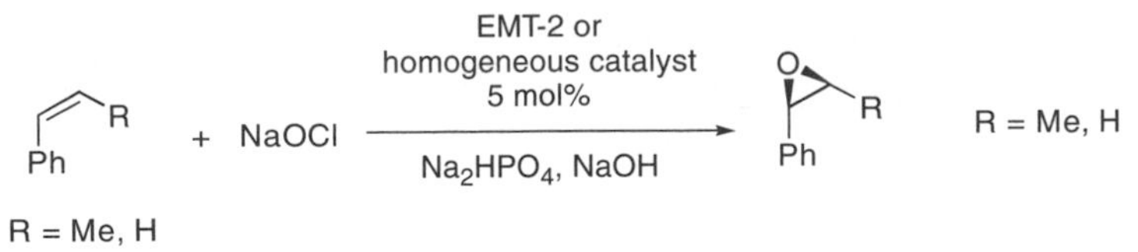

Figure 15.38. Comparison of EMT-2 and homogeneous (salen)MnCl complexes in the asymmetric epoxidation of olefins. The ee (%) with pyridine N-oxide was 88%.

	EMT-2		Homogeneous		
	Conversion (%)	ee (%)	Conversion (%)	ee (%)	Configuration
(styrene)	15	34	55	35	(S)
(β-methylstyrene)	15	80	85	80	(1S,2R)
Cholesterol	None	—	13	—	

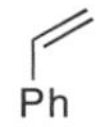

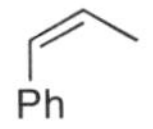

15.3.4.b Microencapsulation by Physical Envelopment in a Polymer

In this example of microencapsulation, a metal complex is physically enveloped by a polymer. In this case, when the polymer is dissolved or swollen, the metal complex can enter. Precipitation or shrinking of the polymer upon addition of an appropriate solvent traps the complex within the polymer.

The Sharpless AD will be highlighted in this section for the purpose of comparing microencapsulation of OsO_4 with the LDH ion-exchange discussed in Section 15.3.2. Polystyrene-microencapsulated OsO_4 (PS-MC•OsO_4) was prepared by combining a suspended 1% divinyl benzene cross-linked polystyrene resin in THF at 65 °C with OsO_4. After cooling, methanol was added and unencapsulated OsO_4 washed from the polymer beads. It was found that 1 g of polymer could uptake 170 mg of OsO_4, or 85% of the osmium employed in the immobilization process.[95] The resulting PS-MC•OsO_4 was determined to contain 10–20 nm capsules of the immobilized osmium. The PS-MC•OsO_4 was employed in the AD, using water as the solvent, with (DHQD)$_2$PHAL and Triton X-405, a nonionic surfactant, to aid in solubilizing the ligand and substrates (**Figure 15.39**). While enantioselectivities were moderate, it was found that PS-MC•OsO_4 could be recycled without loss of activity and only a small decrease in enantioselectivity (6%) over five cycles. No leaching was detected under the reaction conditions (**Figure 15.39**).

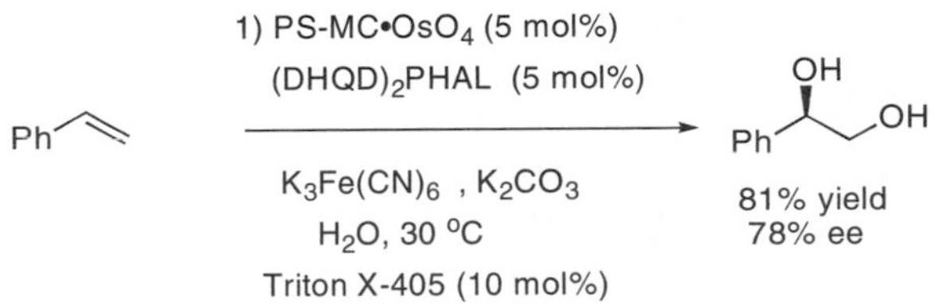

Figure 15.39. Asymmetric dihydroxylation of olefins with PS-Mc•OsO_4 and (DHQD)$_2$PHAL.

Interestingly, the catalyst activity and leaching depended on the amount of cross-linking. At low cross-linking, the activity was the highest, but leaching was elevated. PS-MC•OsO$_4$ with 2–5 mol% cross-linker exhibited lower activity but there was no detectable leaching of the osmium. It has been proposed that an interaction between the π-electrons of the PS and empty orbitals on the metal center contribute to the immobilization and retarding of the leaching.[96]

15.3.5 Assembling Supports Around Catalysts

A second strategy for catalyst encapsulation involves construction of the support around the catalyst. There are a number of important differences between this approach and the ship-in-a-bottle technique outlined above, despite the fact that both employ unmodified homogeneous catalysts as precursors. The sensitivity of most asymmetric catalysts precludes the high-temperature annealing of the support typically used in the synthesis of zeolites. The supports employed, therefore, are usually based on silica or polydimethylsiloxane (PDMS). Generation of a silica support around the catalyst can be accomplished by sol-gel processes, which consist of the controlled hydrolysis of tetraalkoxy silanes, Si(OR)$_4$, with water. The resulting silica contains pores and channels that vary significantly in size, giving rise to a range of catalyst sites.[19] Encapsulation of the catalyst in a smaller cavity, for example, will impact the enantioselectivity due to restricted catalyst conformations and substrate approaches and the reduced mass transport will decrease the TOF. Catalyst sites in larger cavities are anticipated to exhibit higher TOF, but might allow greater catalyst leaching.

Encapsulation in elastomeric PDMS has similar benefits and drawbacks to encapsulation with silica. The support is formed by cross-linking vinyldimethyl-terminated silicone polymer with a cross-linking reagent catalyzed by a platinum catalyst to promote the matrix formation. The asymmetric catalyst precursor, (MeDUPHOS)Rh(COD)$^+$ (COD = 1,4-cyclooctadiene) was introduced prior to the cross-linking (**Figure 15.40**). Slow evaporation of the solvent gives a membrane with the occluded catalyst.[97] The catalyst–polymer interaction is proposed to involve van der Waals forces and steric encapsulation.[16] Solvents that cause significant swelling expand the pores of the support, facilitating catalyst leaching. The best solvents are those in which the catalyst has limited solubility and that cause limited swelling. Furthermore, catalyst leaching can be reduced by increasing the size of the catalyst or the degree of support cross-linking.

The supported catalyst was examined in the asymmetric hydrogenation of methyl acetoacetate. Hydrogenation with homogeneous catalyst exhibited excellent enantioselectivity and TOF (**Figure 15.41**). The membrane-supported catalyst exhibited slightly reduced enantioselectivity and the TOF was about one order of magnitude lower. Importantly, no loss of activity or enantioselectivity was reported in the second cycle. Addition of more reagents and removal of the membrane resulted in no

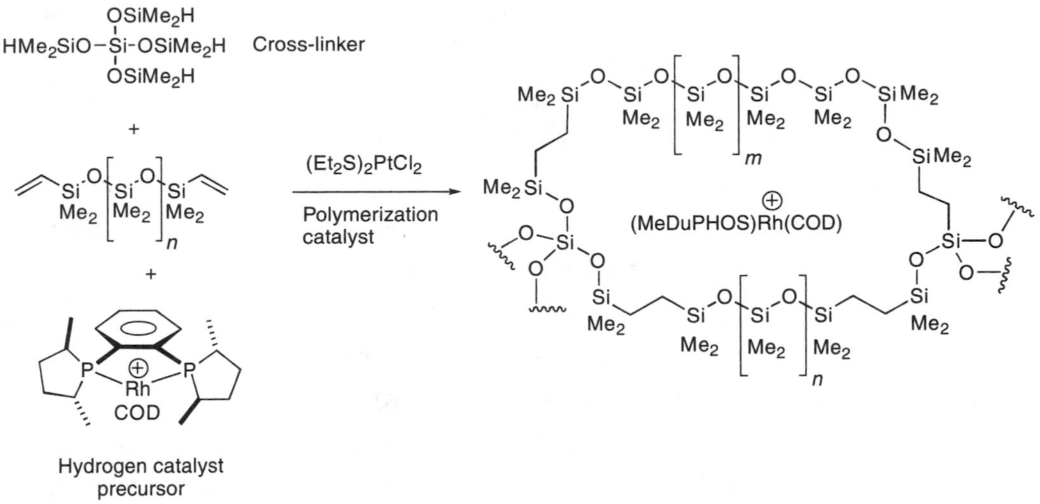

Figure 15.40. Encapsulation in PDMS to generate a solid-supported catalyst.

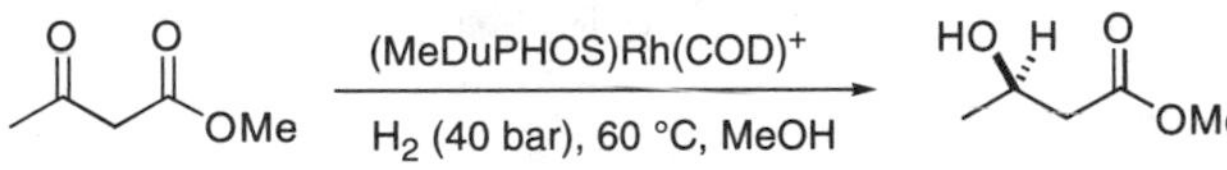

Medium	Run	TOF/h	ee (%)
Homogeneous	1	482	99
Membrane	1	28	90
Membrane	2	28	90

Figure 15.41. Comparison of a membrane-encapsulated chiral asymmetric hydrogenation catalyst (**Figure 15.40**) with the homogeneous analog.

further reaction, indicating that the catalyst remained imbedded in the support.[97] Similar results were obtained in the hydrogenation of a related substrate in water. The (MeDuPHOS)Rh cation is insoluble in water, minimizing leaching.[98]

15.4 Confinement Effects with Supported Catalysts

The impact of spatial restrictions of the solid support on the activity and selectivity (i.e., the enantioselectivity, diastereoselectivity, and chemoselectivity) of an immobilized catalyst is called the confinement effect. Confinement effects, both positive and negative, become more pronounced when the interaction of the catalyst and/or substrate with the support surface are stronger. Thus, constraining a catalyst in a small space will affect its conformation. Predicting how such interactions will impact enantioselectivity, however, is not at all straightforward at this time. As discussed in Chapter 1, the energy difference between diastereomeric transition states that lead to highly enantioenriched product is relatively small (< 4 kcal/mol at rt). These energy differences are similar to weak interactions between molecules or between molecules and surfaces, such as van der Waals forces, hydrogen bonding, and physical absorption. Not only can the support impart spatial restrictions, it can also induce electronic effects that increase or decrease polarity in the reaction cavity. If interactions between the catalyst or substrate and the support are on the same order of magnitude as those between the diastereomeric transition states, it is reasonable that such interactions will impact the magnitude of the enantioselectivity or even change the sense of the stereoinduction. Chemists are just beginning to design experiments to probe the nature of the interactions between supports and transition states. Given that it is currently difficult to predict enantioselectivity in homogeneous asymmetric catalysis, it is not surprising that such predictions with heterogenized systems are considerably more challenging.

While our understanding of the impact of the support on catalyst activity and stereoselectivity remains rudimentary, a great deal can be learned by designing straightforward experiments to measure the magnitude of such effects. Key to these experiments is the ability to systematically modify one variable at a time. A remarkable example that illustrates this point involved a study of support pore size on the

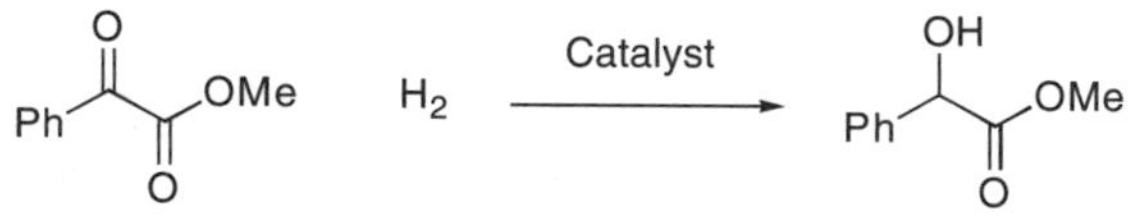

Catalyst	Silica pore size (Å)	Time (h)	Conv. (%)	ee (%)
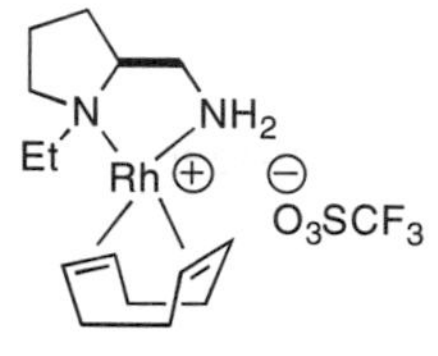 Homogeneous	—	2.0	62	0
Heterogeneous	38	0.5	83	82
Heterogeneous	60	0.5	67	65
Heterogeneous	250	0.5	45	0
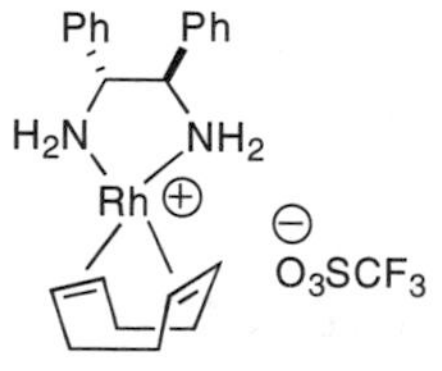 Homogeneous	—	2.0	70	0
Heterogeneous	38	0.5	78	50
Heterogeneous	60	0.5	60	68
Heterogeneous	250	0.5	39	0
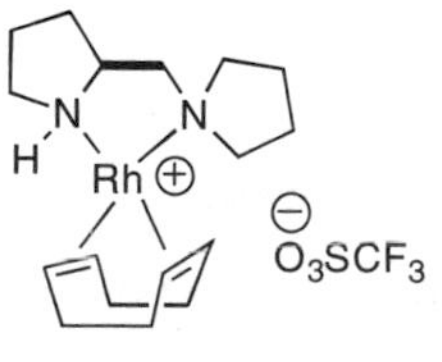 Homogeneous	—	0.5	46	53
Heterogeneous	38	0.5	93	85
Heterogeneous	60	0.5	63	72
Heterogeneous	250	0.5	61	65

Figure 15.42. Comparison of homogeneous catalysts with heterogenized analogs supported on silica with pore sizes of 38, 60, and 250 Å.

enantioselectivity of heterogeneous catalysts in the asymmetric hydrogenation of methyl benzoylformate to methyl mandelate (**Figure 15.42**).[99] Silicas with well-defined pore sizes of 38, 60, and 250 Å were used as supports for this study. The positively charged precatalysts with triflate counterions were immobilized by triflate capture through hydrogen bonding between the silanols and the triflate (see Section 15.3.3).[91] The homogeneous catalysts were initially examined and [Rh(COD)AEP][OTf] and [Rh(COD)DPEN][OTf] were found to exhibit no enantioselectivity under the conditions examined. [Rh(COD)PMP][OTf] exhibited moderate enantioselectivity (53%). Examination of the heterogenized derivatives in this reaction clearly indicated the positive impact of confinement on catalyst activity and enantioselectivity. Precatalysts bound to silica with smaller average pore size exhibited dramatic increases in enantioselectivity over the homogeneous catalysts and over heterogeneous catalyst on supports with larger pore sizes. These results illustrate the profound impact that can result from spatial constraint.

15.5 Impact of Tethers on Catalysis

Tethering catalysts to the support can often lead to unexpected changes in the activity and/or selectivity of the tethered catalyst with respect to the homogeneous analogs. Such effects are frequently attributed to interactions between the catalyst and support. Negative impacts of the support on catalysis can be minimized by screening different types of linkers and tether lengths between the catalyst and the support.

Intuition tells us that long, rigid linkers should direct the catalyst away from the support and better simulate homogeneous conditions. In contrast, shorter, more flexible tethers might be predicted to allow more catalyst-resin interactions, limiting access to the catalytic center and impacting enantioselectivity.[84,100] To explore these possibilities, Wang resin,[101] which consists of a cross-linked polystyrene core derivatized with 4-hydroxybenzyl alcohol, was monoesterified with a large excess of the diacid chlorides **A–C** (**Figure 15.43**). Quinine was introduced to react with the remaining acid chloride moiety to give supported quinine organocatalysts that were used in the catalytic asymmetric synthesis of lactams.[102] As outlined in **Figure 15.43**, the nature of the tether had a minor effect on product enantioselectivity. In contrast, the diastereomeric ratio dropped from ≥ 10:1 with the aryl-based tethers to 2:1 with the flexible C$_9$ tether. One possible explanation for the reduced diastereoselectivity is the greater residence time of the catalyst with the flexible tether near the sterically encumbered resin.

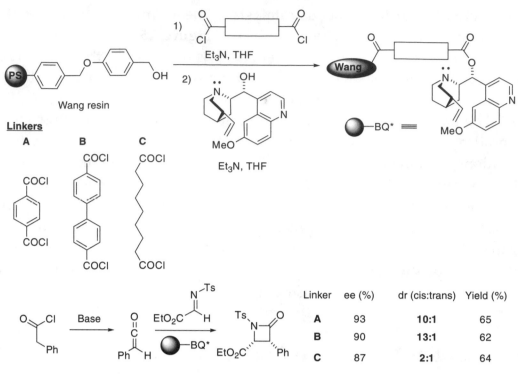

Figure 15.43. Dependence of enantioselectivity and diastereoselectivity on the tether in the supported quinine-catalyzed asymmetric synthesis of β-lactams.

Linker	ee (%)	dr (cis:trans)	Yield (%)
A	93	10:1	65
B	90	13:1	62
C	87	2:1	64

15.6 Enhancing Heterogeneous Catalysis Through Cooperative Catalyst-Surface Interactions

As discussed in this chapter, the role of the catalyst support has been largely that of an inert spectator intended to facilitate catalyst handling, recovery, and recycling. Unfortunately, the reality is that the heterogenization of homogeneous catalyst often has a negative impact on catalyst selectivity and activity. For this reason, chemists have chosen to approach this problem by minimizing the effect of the support on the catalyst performance. Perhaps a better approach, albeit significantly more challenging, would be to take advantage of the functionality of the support to enhance catalyst activity and selectivity in a similar manner that Nature uses protein secondary structure to facilitate reactions in metal-based enzymes.[103] The role of the support in such a cooperative catalyst–support interaction could take several different forms. For example, the support could behave as a ligand, as a hydrogen-bond donor, or as an internal Brønsted acid. Acting as a ligand, the support could stabilize unsaturated organometallic intermediates that are unstable in solution or facilitate the formation and stabilization of immobilized supramolecular assemblies. In other words, the surface actively participates and assists the catalytic reaction.

One important property of oxide supports like silica is that the surface allows supported complexes to form aggregates that are entropically prohibitive under homogeneous conditions. This behavior can contribute to the formation of new surface structures and reactivity or to the generation of less active clusters. In an effort to develop catalysts for the asymmetric coupling of 2-naphthols to form BINOLs enantioselectivily, the vanadium Schiff base complex was immobilized onto SiO_2. Characterization of the immobilized species by IR spectroscopy and EXAFS indicated that the coordinated water of the monomeric precursor was lost. Furthermore, it was found that surface silanols underwent exchange with the ligand phenoxides to anchor the vanadium to the surface. ESR spectroscopy indicated that two V(IV) centers were separated by 4 Å, indicative of association of the monomeric units (**Figure 15.44**). It is proposed that two vanadium complexes undergo surface association and form hydrogen bonds between the phenolic OH and carbonyl oxygen.[104]

The monomeric and silica-supported vanadium Schiff base complexes were examined in the oxidative coupling of 2-naphthol. The monomeric catalyst was inactive under the conditions in **Figure 15.45**. The silica-supported catalyst promoted the coupling reaction, albeit with low TOF. It was found that the enantioselectivity and activity depended on the surface density of vanadium species, consistent with an active dimeric catalyst.

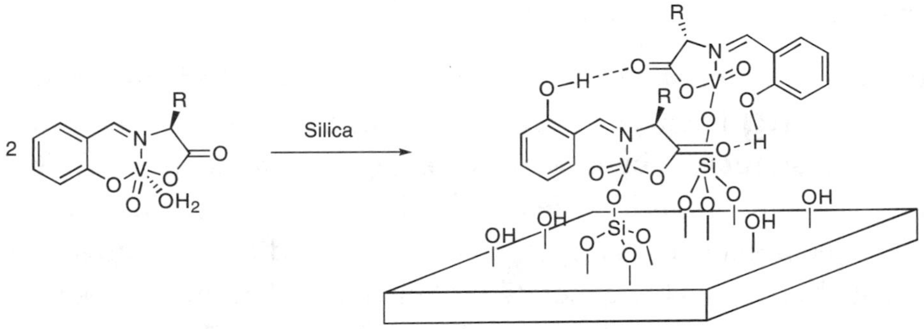

Figure 15.44. Reaction of a monomeric vanadium Schiff base adduct with silica and the proposed product dimer.

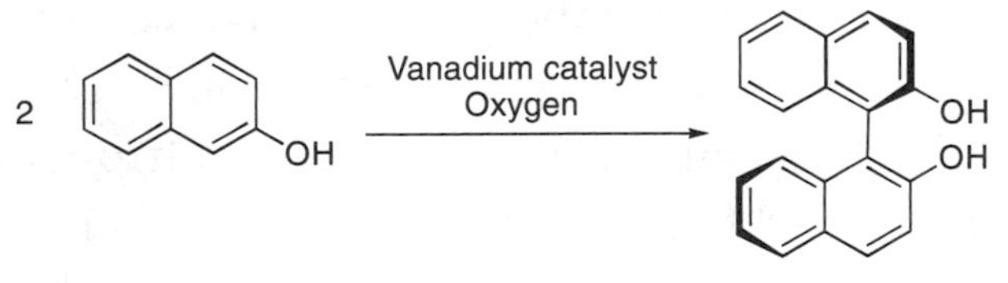

Vanadium loading (wt %)	Time (days)	Conversion (%)	ee (%)
Homogeneous	5	0	—
0.3	5	11	32
0.8	5	33	39
1.6	5	42	48
3.4	11	93	90

Figure 15.45. Coupling of 2-naphthol with homogeneous and heterogeneous catalysts from **Figure 15.44**.

In this example, the coordination sphere of the vanadium is modified upon immobilization, enhancing the activity in the asymmetric oxidative coupling reaction.

15.7 Multiple Catalysts on a Single Support

As highlighted in Chapter 14 on multistep catalytic asymmetric reactions, one of the frontiers in chemical synthesis is the development of asymmetric processes in which several steps can be performed in a single flask without isolation and purification of the intermediates. Such processes allow the generation of significant complexity in an efficient fashion. One promising approach to this challenge is to employ heterogeneous catalysts. If the catalysts are compatible, it may be possible to attach them to the same support. Alternatively, it may be necessary to insure that the catalysts remain isolated by tethering them to different supports and combining them in the reaction flask. Mutually incompatible heterogeneous catalysts that are on different supports cannot interact to an appreciable extent. The subsequent section, will outline methods to perform multicatalyst reactions with supported catalysts.

15.7.1 Dual Metal–Ligand Assemblies

One strategy to designing supported catalysts bearing two independent catalytically active sites is to take advantage of Pearson's rules, which state that hard electropositive metal centers prefer to bind to hard electronegative ligands, while soft metals tend to favor coordination to soft ligands.[57] The distinctively different coordination ability of hard metals, such as Al(III), Ti(IV), Zn(II), and Ln(III), and soft metals, like Ru(II) and Rh(III), permits the generation of orthogonal binding sites that can be incorporated into a single support, providing an opportunity for multifunctional catalyst design in asymmetric sequential reactions. In this example, the hard ligand BINOL forms very strong Ti–O bonds, whereas the soft ligand BINAP tightly binds to ruthenium(II) and not to the hard titanium center. Taking advantage of the different binding affinity of hard and soft metals, researchers have prepared an optically active BINOL/BINAP copolymer illustrated in **Figure 15.46**.[105] The copolymer is efficiently synthesized via a Suzuki coupling protocol using a triphenyl linker to space the BINOL and BINAP units apart from one another to prevent interference between the units. In doing so, the BINOL and BINAP units are randomly distributed along the polymer chain. Importantly the R-group-substitution of *n*-hexyl along the triphenyl linker was key to generating a sufficiently soluble polymer for homogenous asymmetric catalytic reactions.

Loading of 4 mol%, based on the repeating unit of the copolymer, was utilized in the sequential asymmetric reactions of *para*- and *meta*-acetylbenzaldehyde (**Figure**

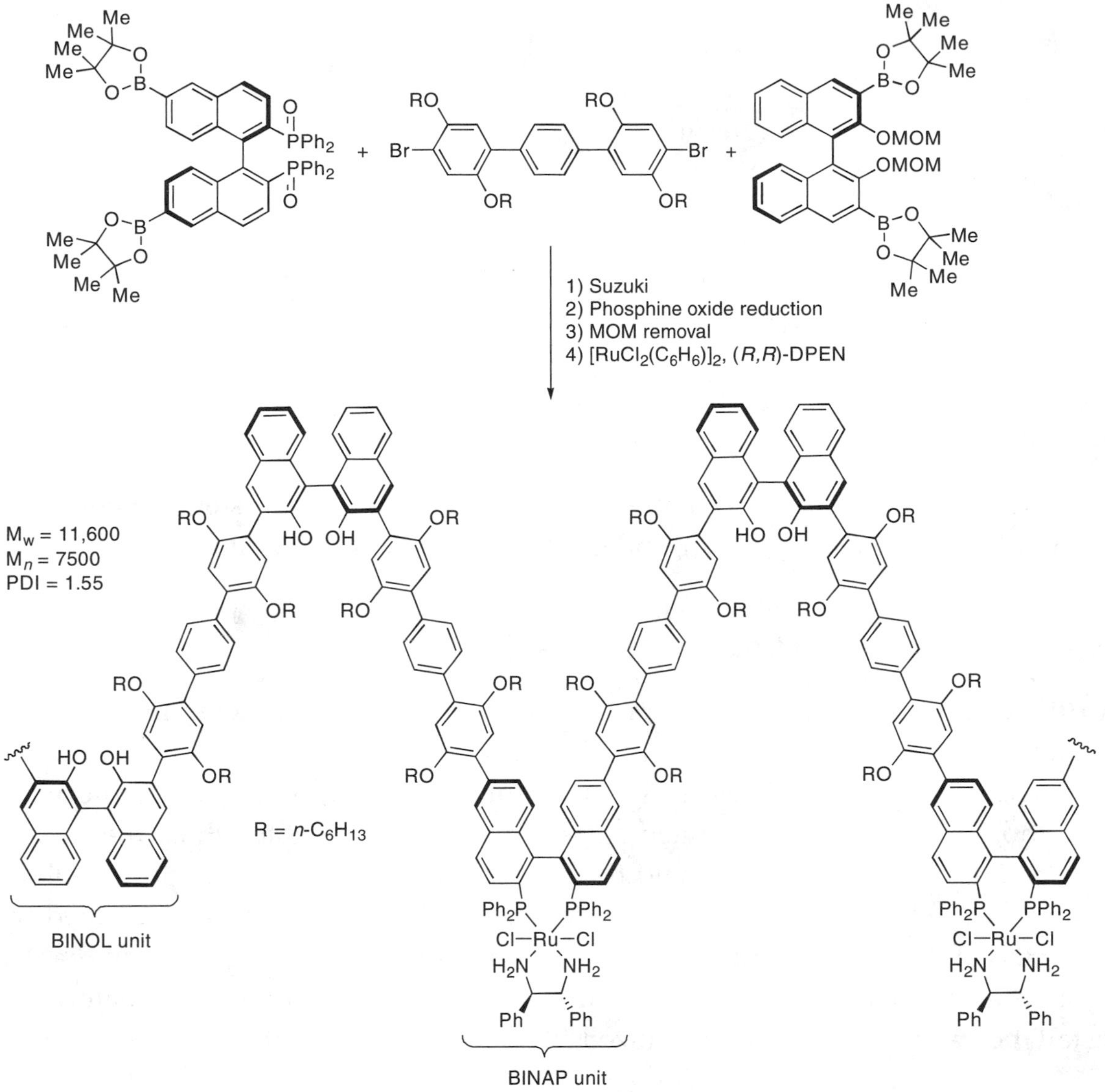

Figure 15.46. BINOL- and BINAP-containing polymeric catalysts.

15.47). Addition of Et$_2$Zn in toluene, followed by quenching with *i*-PrOH and transfer to a high-pressure apparatus utilizing H$_2$, yielded the diols as a result of asymmetric alkylation of the aldehyde (92–94% ee) and asymmetric hydrogenation of the ketone (75–86% de). The stereoselectivities of each transformation are similar to those encountered with the monomeric catalysts. As is the case for many soluble polymers, addition of the less-solubilizing solvent MeOH to a solution of the polymer in CH$_2$Cl$_2$ allows facile recovery via precipitation and filtration. The recovered catalyst was found to be almost as effective in the second cycle of the sequential reaction of *para*-acetylbenzaldehyde (**Figure 15.47**). Overall the procedure is operationally simple, complexity is built up rapidly in the creation of two independent asymmetric centers

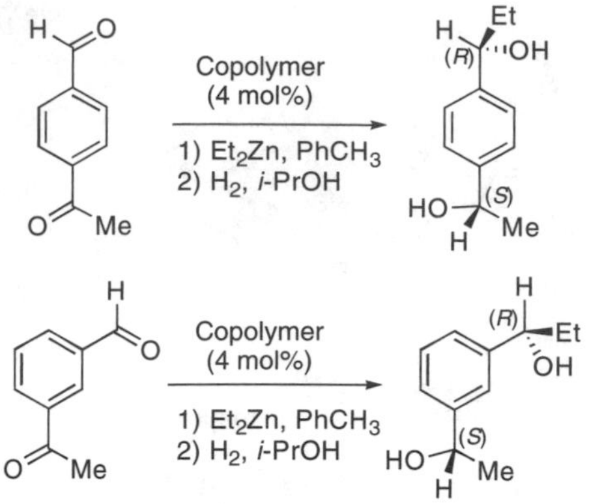

Cycle	Conv. (%)	ee (%)	de (%)
1	> 99	92	86
2	> 99	93	78
1	> 99	94	75

Figure 15.47. BINOL- and BINAP-containing polymeric catalysts in sequential aldehyde alkylation and ketone reduction.

(the selectivity of the second catalyst is independent of the first stereocenter, as the two reacting moieties are distal), and catalyst recovery is straightforward.

15.7.2 Supported Dual Catalyst Systems

A dual catalyst system has been prepared to catalyze the tandem Heck reaction/asymmetric dihydroxylation (AD).[106] The synthesis of this dual catalyst system is illustrated in **Figure 15.48**. Silica gel was functionalized with 3-mercaptopropyl trimethoxysilane, which generated surface mercapto groups to which the catalytically active metal complexes could be linked. Using a radical initiator, the DHQD analog was tethered to the derivatized silica gel. Elemental analysis indicated that 15.3 wt.% of the chiral ligand had been incorporated onto the silica gel and that free thiol groups remained, which was confirmed by IR spectroscopy of the material. To install the palladium, the functionalized silica gel was then treated with $PdCl_2$ in ace-

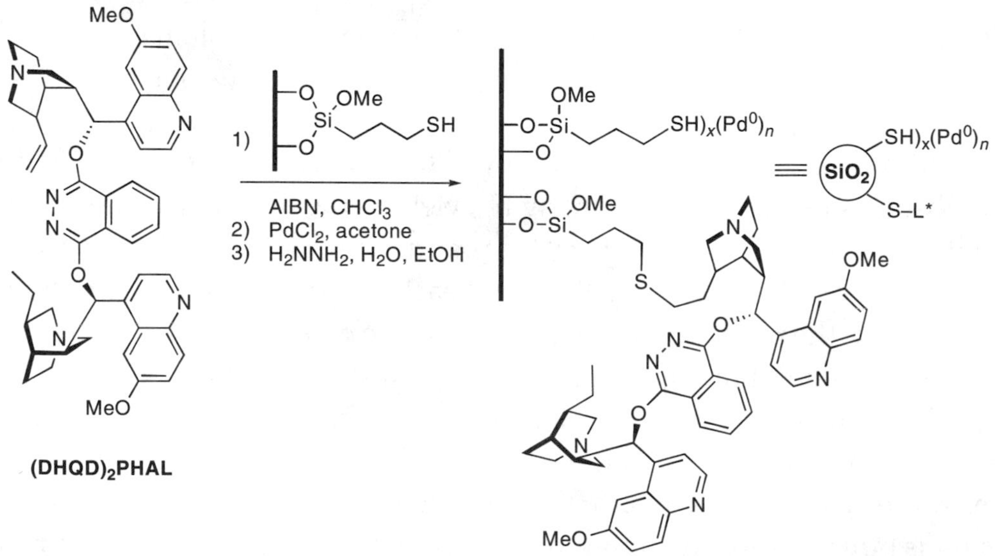

Figure 15.48. Synthesis of a silica-supported dual catalyst system.

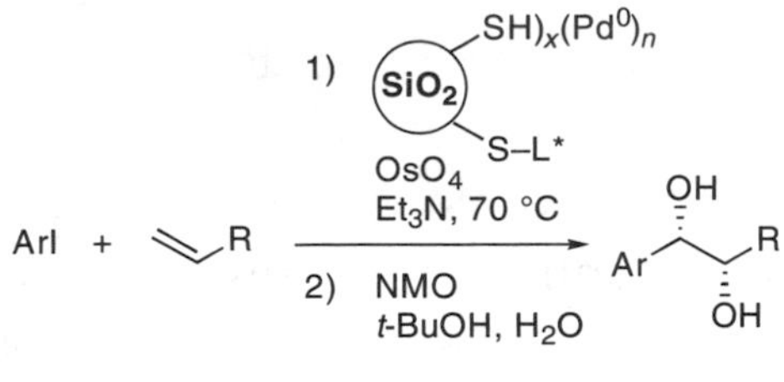

R	Product	Yield (%)	ee (%)
CO_2Me	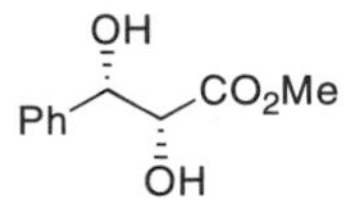	90	93
CO_2Et	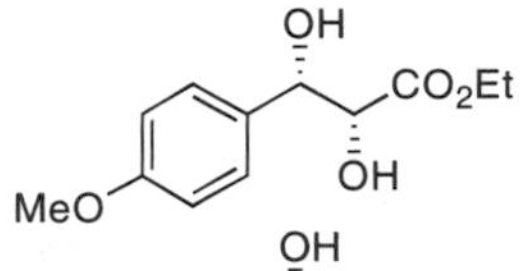	94	93
CO_2Et	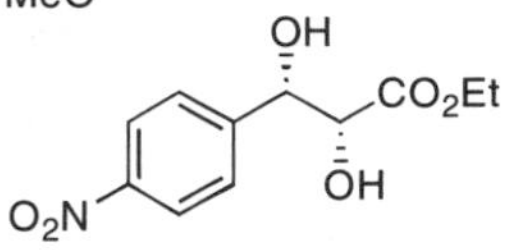	92	88
Ph	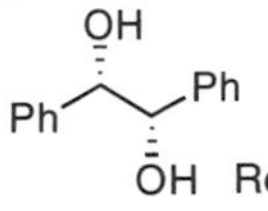	90	99
	Recovered catalyst	67	99
	Recovered catalyst + 0.3 mol% OsO_4	89	99

Figure 15.49. Silica-supported Pd and Os catalysts for sequential Heck and AD of alkenes.

tone under reflux. Reduction of the palladium with hydrazine hydrate in ethanol resulted in formation of a dispersion of palladium nanoparticles with size distributions of approximately 10–15 nm in diameter.

Application of this dual catalyst to the tandem Heck reaction/AD is illustrated in **Figure 15.49**. The silica-gel-supported palladium nanoparticles/linked PHAL(DHQD)$_2$ ligand was suspended in acetonitrile and treated with OsO_4, aryl iodide, olefin, and base and heated to 70 °C for eight hours to afford the trans olefin intermediate. After removal of the solvent, the N-methyl morpholine N-oxide oxidant for the AD was added in a mixture of t-BuOH/H$_2$O and the AD performed at rt. The yield and enantioselectivities of the diol products are impressive (**Figure 15.49**). While no change in activity of the palladium nanoparticles was observed on recycling, the activity in the AD dropped. This observation was attributed to leaching of the osmium from the support. Use of an additional 0.3 mol% OsO_4 with the recycled, supported catalyst restored the catalytic activity in the AD. This dual catalyst system has also been heterogenized onto microcrystalline MgO.[107] Heterogeneous supports with dual catalysts have the possibility to increase synthetic efficiency by reducing the number of isolation and purification steps. They are also easily recovered and can potentially be recycled, making such systems more economical in the long run.

15.8 Sequential Reactions Using Multiple Solid-Supported Catalysts

The ultimate goal of synthesis is to introduce methods to generate complex products of high purity, yield, and enantioselectivity and diastereoselectivity, from simple, readily available materials without isolation of intermediates. One can imagine that catalysts for such tandem syntheses would operate sequentially without interfering with each other. Of course, these catalysts should be easily recycled. One approach to this challenge would be to use a series of immobilized catalysts and reagents packed into connected columns. The starting material or materials could be continuously injected onto a column and undergo the required sequential transformations as the solution phase carries the intermediates past each catalyst. The enantioenriched product is then eluted at the bottom of the series of columns. While the ability to synthesize complex molecules using sequential columns seems distant, some progress toward this goal has been made, employing supported reagents and catalysts to make β-lactams.[102]

Enantioenriched β-lactams have long been recognized as important pharmacophoric components of penam antibiotics. The asymmetric Lewis base catalyzed Staudinger reaction, involving ketene-imine cyclization reactions, provides enantioenriched β-lactams and has been introduced in Chapter 12 on bifunctional catalysis. This reaction has been adapted to the sequential column technique to conduct four discrete reactions. The transformation is illustrated in **Figure 15.50**, along with

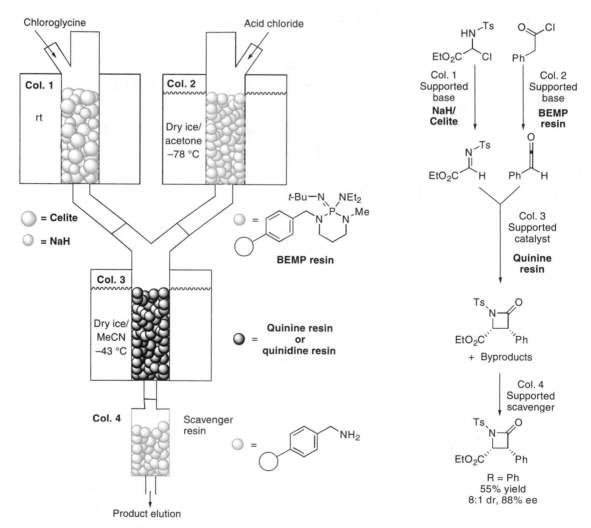

Figure 15.50. A series of supported catalysts and reagents for the synthesis and purification of chiral β-lactams.

a schematic of the reactor. Column 1 is loaded with NaH diluted with Celite, while column 2 is packed with BEMP resin, which contains a basic triaminophosphoramide imine. Column 3 contains the quinine catalyst tethered to the Wang resin with linker, as discussed in **Figure 15.43**. Finally, column 4 contains a scavenger resin to remove unreacted ketene and/or imine from the eluent.

The imine substrate is generated in column 1 by injecting a THF solution of the chloroglycine derivative that undergoes dehydrohalogenation with a stoichiometric heterogeneous strong base (NaH/Celite) at room temperature. Concurrently, a THF solution of the phenylacetylchloride was introduced onto the top of column 2, pre-cooled to $-78\ ^{\circ}C$. Percolation of the acid chloride through the BMEP resin results in dehydrohalogenation to generate the reactive ketene substrate. The ketene and imine flow onto column 3, which had been cooled to $-43\ ^{\circ}C$. In the presence of the supported catalyst, these reactants underwent conversion to the β-lactam over the course of 2 h. Finally, passage through the scavenger column and elution of the reaction mixture provided the enantioenriched β-lactam (90% ee, 10:1 cis:trans) in 62% yield.[102]

The potential advantages of using sequential columns are that reagents can be introduced onto a number of columns assembled in series or parallel to produce complex target molecules. The columns contain reactants or catalysts, many of which can be regenerated, stored, and reused for the same or related processes. Given the large number of supported asymmetric catalysts and processes outlined in this chapter, a vast number of transformations can be envisioned.

Outlook of Heterogeneous Asymmetric Catalysis

Despite a considerable body of study in synthesis, mechanism, and applications of supported enantioselective catalysts, this important field is still in its early stages of development. The simple adaptation of homogeneous asymmetric catalysts to heterogeneous catalysis by heterogenization generally results in a decrease in catalyst activity and/or enantioselectivity. Nonetheless, the potential benefits of heterogeneous catalysis in catalyst removal, isolation, and recycling will continue to drive research in this area. Promising subjects for future development include supporting catalysts through noncovalent interactions, because such methods do not require modification of the homogeneous catalysts. Investigation of catalytically active nanoparticles is also relatively unexplored and is expected to be an active area of research in the future.

One of the most important, and least understood, areas in heterogeneous catalysts is catalyst–support interactions. Biologically active catalysts are able to achieve extremely high activity and enantioselectivity by the precise organization of functional groups in and around the active site. At this time, it remains unclear if one can imitate nature's catalysts by heterogenization of asymmetric catalysts onto supports that not only facilitate the isolation and recycling of the catalysts, but also actively participate in the catalysis. Such synergistic interaction between the catalyst and support, or bifunctional behavior, could result in dramatic increases in activity and enantioselectivity (see Chapter 12 on bifunctional catalysis). Breakthroughs in this area could redefine enantioselective heterogeneous catalysis.

References

(1) Price, P. M.; Clark, J. H.; Macquarrie, D. J. Modified Silicas for Clean Technology. *J. Chem. Soc., Dalton. Trans.* **2000**, 101–110.

(2) Leadbeater, N. E.; Marco, M. Preparation of Polymer-Supported Ligands and Metal Complexes for Use in Catalysis. *Chem. Rev.* **2002**, *102*, 3217–3274.

(3) McNamara, C. A.; Dixon, M. J.; Bradley, M. Recoverable Catalysts and Reagents Using Recyclable Polystyrene-Based Supports. *Chem. Rev.* **2002**, *102*, 3275–3300.

(4) Barrett, A. G. M.; Hopkins, B. T.; Köbberling, J. ROMPgel Reagents in Parallel Synthesis. *Chem. Rev.* **2002**, *102*, 3301–3324.

(5) Dickerson, T. J.; Reed, N. N.; Janda, K. D. Soluble Polymers as Scaffolds for Recoverable Catalysts and Reagents. *Chem. Rev.* **2002**, *102*, 3325–2244.

(6) Bergbreiter, D. E. Using Soluble Polymers To Recover Catalysts and Ligands. *Chem. Rev.* **2002**, *102*, 3345–3384.

(7) Fan, Q.-H.; Li, Y.-M.; Chan, A. S. C. Recoverable Catalysts for Asymmetric Organic Synthesis. *Chem. Rev.* **2002**, *102*, 3385–3466.

(8) Rechavi, D.; Lemaire, M. Enantioselective Catalysis Using Heterogeneous Bis(oxazoline) Ligands: Which Factors Influence the Enantioselectivity? *Chem. Rev.* **2002**, *102*, 3467–3494.

(9) Song, C. E.; Lee, S.-g. Supported Chiral Catalysts on Inorganic Materials. *Chem. Rev.* **2002**, *102*, 3495–3524.

(10) Duchateau, R. Incompletely Condensed Silsesquioxanes: Versatile Tools in Developing Silica-Supported Olefin Polymerization Catalysts. *Chem. Rev.* **2002**, *102*, 3525–3542.

(11) Lu, Z.-l.; Lindner, E.; Mayer, H. A. Applications of Sol-Gel-Processed Interphase Catalysts. *Chem. Rev.* **2002**, *102*, 3543–3578.

(12) De Vos, D. E.; Dams, M.; Sels, B. F.; Jacobs, P. A. Ordered Mesoporous and Microporous Molecular Sieves Functionalized with Transition Metal Complexes as Catalysts for Selective Organic Transformations. *Chem. Rev.* **2002**, *102*, 3615–3640.

(13) Dupont, J.; de Souza, R. F.; Suarez, P. A. Z. Ionic Liquid (Molten Salt) Phase Organometallic Catalysis. *Chem. Rev.* **2002**, *102*, 3667–3692.

(14) Van Heerbeek, R.; Kamer, P. C. J.; van Leeuwen, P. W. N. M.; Reek, J. N. H. Dendrimers as Support for Recoverable Catalysts and Reagents. *Chem. Rev.* **2002**, *102*, 3717–3756.

(15) Roucoux, A.; Schulz, J.; Patin, H. Reduced Transition Metal Colloids: A Novel Family of Reusable Catalysts? *Chem. Rev.* **2002**, *102*, 3757–3778.

(16) Vankelecom, I. F. J. Polymeric Membranes in Catalytic Reactors. *Chem. Rev.* **2002**, *102*, 3779–3810.

(17) Saluzzo, C.; Lemaire, M. Homogeneous-Supported Catalysts for Enantioselective Hydrogenation and Hydrogen Transfer Reduction. *Adv. Synth. Catal.* **2002**, *244*, 915–928.

(18) Benaglia, M.; Puglisi, A.; Cozzi, F. Polymer-Supported Organic Catalysts. *Chem. Rev.* **2003**, *103*, 3401–3430.

(19) McMorn, P.; Hutchings, G. J. Heterogeneous Enantioselective Catalysts: Strategies for the Immobilisation of Homogeneous Catalysts. *Chem. Soc. Rev.* **2004**, *33*, 108–122.

(20) Li, C. Chiral Synthesis on Catalysts Immobilized in Microporous and Mesoporous Materials. *Catal. Rev.* **2004**, *46*, 419–492.

(21) Ley, S. V.; Baxendale, I. R.; Bream, R. N.; Jackson, P. S.; Leach, A. G.; Longbottom, D. A.; Nesi, M.; Scott, J. S.; Storer, R. I.; Taylor, S. J. Multi-Step Organic Synthesis Using Solid-Supported Reagents and Scavengers: A New Paradigm in Chemical Library Generation. *J. Chem. Soc., Perkin Trans. 1* **2000**, 3815–4195.

(22) Storer, R. I.; Takemoto, T.; Jackson, P. S.; Ley, S. V. A Total Synthesis of Epothilones Using Solid-Supported Reagents and Scavengers. *Angew. Chem., Int. Ed. Engl.* **2003**, *42*, 2521–2525.

(23) Annis, D. A.; Jacobsen, E. N. Polymer-Supported Chiral Co(Salen) Complexes: Synthetic Applications and Mechanistic Investigations in the Hydrolytic Kinetic Resolution of Terminal Epoxides. *J. Am. Chem. Soc.* **1999**, *121*, 4147–4154.

(24) Biernat, J. F.; Konieczka, P.; Tarbet, B. J.; Bradshaw, J. S.; Izatt, R. M. In *Separation and Purification Methods*; Wankat, P. C., van Oss, C. J., Henry, J. D., Eds.; Marcel Dekker: New York, 1994; Vol. 23, pp 77–348.

(25) Heckel, A.; Seebach, D. Preparation and Characterization of TADDOLs Immobilized on Hydrophobic Controlled-Pore-Glass Silica Gel and Their Use in Enantioselective Heterogeneous Catalysis. *Chem. Eur. J.* **2002**, *8*, 559–572.

(26) Collman, J. P.; Belmont, J. A.; Brauman, J. I. Silica Supported Rhodium Hydroformylation Catalysts. Evidence for Dinuclear Eliminations. *J. Am. Chem. Soc.* **1983**, *105*, 7288–7294.

(27) Vilim, J.; Hetflejs, J. Catalysis by Metal-Complexes. 50. Kinetics of Enantioselective Hydrogenation of α,α-Acetylaminocinnamic Acid-Catalyzed by a Rhodium Complex. *Collect. Czech. Chem. Commun.* **1978**, *43*, 121–133.

(28) Purgin, G. Immobilized Catalysts for Enantioselective Hydrogenation: The Effect of Site-Isolation. *J. Mol. Catal.* **1996**, *107*, 273–279.

(29) Grubbs, R. H.; Lau, C. P.; Cukier, R.; Brubaker, C. Polymer Attached Metallocenes. Evidence for Site Isolation. *J. Am. Chem. Soc.* **1977**, *99*, 4517–4518.

(30) Kresge, C. T.; Leonowicz, M. E.; Roth, W. J.; Vartuli, J. C.; Beck, J. S. Ordered Mesoporous Molecular Sieves Synthesized by a Liquid-Crystal Template Mechanism. *Nature* **1992**, *359*, 710–712.

(31) Xiang, S.; Zhang, Y. L.; Q., X.; Li, C. Enantioselective Epoxidation of Olefins Catalyzed by Mn(Salen)/MCM-41 Synthesized with a New Anchoring Method. *J. Chem. Soc., Chem. Commun.* **2002**, 696–2697.

(32) Corma, A.; Iglesias, M.; del Pino, C.; Sánchez, F. Optically Active Complexes of Transition Metals (Rh^I, Ru^{II}, Co^{II} and Ni^{II}) with 2-Aminocarbonylpyrrolidine Ligands. Selective Catalysts for Hydrogenation of Prochiral Olefins. *J. Organomet. Chem.* **1992**, *431*, 233–246.

(33) Corma, A.; Iglesias, M.; del Pino, C.; Sánchez, F. New Rhodium Complexes Anchored on Modified USY Zeolites. A Remarkable Effect of the Support on the Enantioselectivity of Catalytic Hydrogenation of Prochiral Alkenes. *J. Chem. Soc., Chem. Commun.* **1991**, 1253–1255.

(34) Corma, A.; Iglesias, M.; Martín, M. V.; Rubio, J.; Sanchez, F. Conjugate Addition of Diethylzinc to Enones Catalyzed by Homogeneous and Supported Chiral Ni-Complexes. Cooperative Effect of the Support on Enantioselectivity. *Tetrahedron: Asymmetry* **1992**, *3*, 845–848.

(35) Astruc, D.; Lu, F.; Aranzaes, J. R. Nanoparticles as Recyclable Catalysts: The Frontier Between Homogeneous and Heterogeneous Catalysis. *Angew. Chem. Int. Ed. Engl.* **2005**, *44*, 7852–7872.

(36) Grunes, J.; Zhu, J.; Somorjai, G. A. Catalysis and Nanoscience. *J. Chem. Soc., Chem. Commun.* **2003**, 2257–2260.

(37) Hu, A.; Yee, G. T.; Lin, W. Magnetically Recoverable Chiral Catalysts Immobilized on Magnetite Nanoparticles for Asymmetric Hydrogenation of Aromatic Ketones. *J. Am. Chem. Soc.* **2005**, *127*, 12486–12487.

(38) Kamahori, K.; Itsuno, S. Asymmetric Diels–Alder Reaction of Methacrolein with Cyclopentadiene Using Polymer-Supported Catalysts: Design of Highly Enantioselective Polymeric Catalysts. *J. Org. Chem.* **1996**, *61*, 8321–8324.

(39) Sartor, D.; Saffrich, J.; Helmchen, G. Enantioselective Diels–Alder Additions with New Chiral Lewis Acids Derived from Amino Acids. *Synlett* **1990**, 197–199.

(40) Sartor, D.; Saffrich, J.; Helmchen, G.; Richards, C. J.; Lambert, H. Enantioselective Diels–Alder Reactions of Enals: Fighting Species Multiplicity of the Catalyst with Donor Solvents. *Tetrahedron: Asymmetry* **1991**, *2*, 639–642.

(41) Hultzsch, K. C.; Jernelius, J. A.; Hoveyda, A. H.; Schrock, R. R. The First Polymer-Supported and Recyclable Chiral Catalyst for Enantioselective Olefin Metathesis. *Angew. Chem., Int. Ed. Engl.* **2002**, *41*, 589–593.

(42) Kröll, R. M.; Schuler, N.; Lubbad, S.; Buchmeiser, M. R. A ROMP-Derived, Polymer-Supported Chiral Schrock Catalyst for Enantioselective Ring-Closing Olefin Metathesis. *J. Chem. Soc., Chem. Commun.* **2003**, 2742–2743.

(43) Fan, Q.-h.; Ren, C.-y.; Yeung, C.-h.; Hu, W.-h.; Chan, A. S. C. Highly Effective Soluble Polymer-Supported Catalysts for Asymmetric Hydrogenation. *J. Am. Chem. Soc.* **1999**, *121*, 7407–7408.

(44) Deng, G.-J.; Yi, B.; Huang, Y.-Y.; Tang, W.-J.; He, Y.-M.; Fan, Q.-H. Dendronized Poly(Ru-BINAP) Complexes: Highly Effective and Easily Recyclable Catalysts for Asymmetric Hydrogenation. *Adv. Synth. Catal.* **2004**, *346*, 1440–1444.

(45) Jansat, S.; Gomez, M.; Philippot, K.; Muller, G.; Guiu, E.; Claver, C.; Castillon, S.; Chaudret, B. Case for Enantioselective Allylic Alkylation Catalyzed by Palladium Nanoparticles. *J. Am. Chem. Soc.* **2004**, *126*, 1592–1615.

(46) Yaghi, O. M.; Li, H.; Davis, C.; Richardson, D.; Groy, T. L. Synthetic Strategies, Structure Patterns, and Emerging Properties in the Chemistry of Modular Porous Solids. *Acc. Chem. Res.* **1998**, *31*, 474–484.

(47) Hagrman, P. J.; Hagrman, D.; Zubieta, J. Organic–Inorganic Hybrid Materials: From "Simple" Coordination Polymers to Organodiamine-Templated Molybdenum Oxides. *Angew. Chem., Int. Ed. Engl.* **1999**, *38*, 2639–2684.

(48) Blake, A. J.; Champness, N. R.; Hubberstey, P.; Li, W.-S.; Withersby, M. A.; Schroder, M. Inorganic Crystal Engineering Using Self-Assembly of Tailored Building-Blocks. *Coord. Chem. Rev.* **1999**, *183*, 117–138.

(49) Moulton, B.; Zaworotko, M. J. From Molecules to Crystal Engineering: Supramolecular Isomerism and Polymorphism in Network Solids. *Chem. Rev.* **2001**, *101*, 1629–1658.

(50) Sawaki, T.; Dewa, T.; Aoyama, Y. Immobilization of Soluble Metal Complexes with a Hydrogen-Bonded Organic Network as a Supporter. A Simple Route to Microporous Solid Lewis Acid Catalysts. *J. Am. Chem. Soc.* **1998**, *120*, 8539–8540.

(51) Sawaki, T.; Aoyama, Y. Immobilization of a Soluble Metal Complex in an Organic Network. Remarkable Catalytic Performance of a Porous Dialkoxyzirconium Polyphenoxide as a Functional Organic Zeolite Analogue. *J. Am. Chem. Soc.* **1999**, *121*, 4793–4798.

(52) Kesanli, B.; Lin, W. B. Chiral Porous Coordination Networks: Rational Design and Applications in Enantioselective Processes. *Coord. Chem. Rev.* **2003**, *241*, 305–326.

(53) Dai, L. X. Chiral Metal-Organic Assemblies—A New Approach to Immobilizing Homogeneous Asymmetric Catalysts. *Angew. Chem., Int. Ed. Engl.* **2004**, *43*, 5726–5729.

(54) Seo, J. S.; Whang, D.; Lee, H.; Jun, S. I.; Oh, J.; Jeon, Y. J.; Kim, K. A Homochiral Metal-Organic Porous Material for Enantioselective Separation and Catalysis. *Nature* **2000**, *404*, 982–986.

(55) Hu, A.; Ngo, H. L.; Lin, W. Chiral Porous Hybrid Solids for Practical Heterogeneous Asymmetric Hydrogenation of Aromatic Ketones. *J. Am. Chem. Soc.* **2003**, *125*, 11490–11491.

(56) Hu, A.; Ngo, H. L.; Lin, W. Chiral, Porous, Hybrid Solids for Highly Enantioselective Heterogeneous Asymmetric Hydrogenation of β-Keto Esters. *Angew. Chem., Int. Ed. Engl.* **2003**, *42*, 6000–6003.

(57) Pearson, R. G. Hard and Soft Acids and Bases. *J. Am. Chem. Soc.* **1963**, *85*, 3533–3539.

(58) Takizawa, S.; Somei, H.; Jayaprakash, D.; Sasai, H. Metal-Bridged Polymers as Insoluble Multicomponent Asymmetric Catalysts with High Enantiocontrol: An Approach for the Immobilization of Catalysts without Using any Support. *Angew. Chem., Int. Ed. Engl.* **2003**, *42*, 5711–5714.

(59) Guo, H.; Wang, W.; Ding, K. Assembled Enantioselective Catalysts for Carbonyl–Ene Reactions. *Tetrahedron Lett.* **2004**, *45*, 2009–2012.

(60) Liang, Y.; Jing, Q.; Li, X.; Shi, L.; Ding, K. Programmed Assembly of Two Different Ligands with Metallic Ions: Generation of Self-Supported Noyori-Type Catalysts for Heterogeneous Asymmetric Hydrogenation of Ketones. *J. Am. Chem. Soc.* **2005**, *127*, 7694–7695.

(61) Ohkuma, T.; Ooka, H.; Hashiguchi, S.; Ikariya, T.; Noyori, R. Practical Enantioselective Hydrogenation of Aromatic Ketones. *J. Am. Chem. Soc.* **1995**, *117*, 2675–2676.

(62) Berrisford, D. J.; Bolm, C.; Sharpless, K. B. Ligand-Accelerated Catalysis. *Angew. Chem., Int. Ed. Engl.* **1995**, *34*, 1059–1070.

(63) Baiker, A. Progress in Asymmetric Heterogeneous Catalysis: Design of Novel Chirally Modified Platinum Metal Catalysts. *J. Mol. Cat. A.* **1997**, *115*, 473–493.

(64) Mallat, T.; Orgimester, E.; Baiker, A. Asymmetric Catalysis at Chiral Metal Surfaces. *Chem. Rev.* **2007**, *107*, 4863–4890.

(65) Studer, M.; Blaser, H. U.; Exner, C. Enantioselective Hydrogenation Using Heterogeneous Modified Catalysts: An Update. *Adv. Synth. Catal.* **2003**, *345*, 45–65.

(66) Blaser, H.-U.; Pugin, B.; Spindler, F. Progress in Enantioselective Catalysis Assessed from an Industrial Point of View. *J. Mol. Catal.* **2005**, *231*, 1–20.

(67) Orito, Y.; Imai, S.; Niwa, S. Asymmetric Hydrogenation of α-Keto Esters Using a Platinum-Alumina Catalyst Modified with Cinchona Alkaloid. *J. Chem. Soc. Jpn.* **1980**, *37*, 670–672.

(68) Orito, Y.; Imai, S.; Niwa, S.; Nguyen, G. H. Asymmetric Hydrogenation of Methyl Benzoylformate Using Platinum-Carbon Catalysts Modified with Cinchonidine. *J. Synth. Org. Chem. Jpn.* **1979**, *37*, 173–174.

(69) Orito, Y.; Imai, S.; Niwa, S. Asymmetric Hydrogenation of Methyl Pyruvate Using a Platinum–Carbon Catalyst Modified with Cinchonidine. *J. Chem. Soc. Jpn.* **1979**, *37*, 1118–1120.

(70) Garland, M.; Blaser, H. U. A Heterogeneous Ligand-Accelerated Reaction: Enantioselective Hydrogenation of Ethyl Pyruvate Catalyzed by Cinchona-Modified Platinum/Aluminum Oxide Catalysts. *J. Am. Chem. Soc.* **1990**, *112*, 7048–7050.

(71) Studer, M.; Burkhardt, S.; Indolese, A. F.; Blaser, H.-U. Enantio- and Chemoselective Reduction of 2,4-Diketo Acid Derivatives with Cinchona Modified Pt-Catalyst—Synthesis of (R)-2-Hydroxy-4-Phenylbutyric Acid Ethyl Ester. *J. Chem. Soc., Chem. Comm.* **2000**, 1327–1328.

(72) Büergi, T.; Baiker, A. Heterogeneous Enantioselective Hydrogenation over Cinchona Alkaloid Modified Platinum: Mechanistic Insights into a Complex Reaction. *Acc. Chem. Res.* **2004**, *37*, 909–917.

(73) Blaser, H. U.; Jalett, H. P.; Monti, D. M.; Baiker, A.; Wehrli, J. T. Enantioselective Hydrogenation of Ethyl Pyruvate: Effect of Catalyst and Modifier Structure. *Surf. Sci. Catal.* **1991**, *67*, 147–55.

(74) Bonalumi, N.; Bürgi, T.; Baiker, A. Interaction between Ketopantolactone and Chirally Modified Pt Investigated by Attenuated Total Reflection IR Concentration Modulation Spectroscopy. *J. Am. Chem. Soc.* **2003**, *125*, 13342–13343.

(75) Lavoie, S.; Laliberté, M. A.; McBreen, P. H. Adsorption States and Modifier-Substrate Interactions on Pt(111) Relevant to the Enantioselective Hydrogenation of Alkyl Pyruvates in the Orito Reaction. *J. Am. Chem. Soc.* **2003**, *125*, 15756–15757.

(76) Vayner, G.; Houk, K. N.; Sun, Y. K. Origins of Enantioselectivity in Reductions of Ketones on Cinchona Alkaloid Modified Platinum. *J. Am. Chem. Soc.* **2004**, *126*, 199–203.

(77) Künzle, N.; Hess, R.; Mallat, T.; Baiker, A. Continuous Enantioselective Hydrogenation of Activated Ketones. *J. Catal.* **1999**, *186*, 239–241.

(78) Barbaro, P.; Bianchini, C.; Giambastiani, G.; Oberhauser, W.; Morassi Bonzi, L.; Rossi, F.; Dal Santo, V. Recycling Asymmetric Hydrogenation Catalysts by Their Immobilisation onto Ion-Exchange Resins. *J. Chem. Soc., Dalton Trans.* **2004**, 1783–1784.

(79) Kolb, H. C.; VanNieuwenhze, M. S.; Sharpless, K. B. Catalytic Asymmetric Dihydroxylation. *Chem. Rev.* **1994**, *94*, 2483–2547.

(80) Han, H.; Janda, K. D. Soluble Polymer-Bound Ligand-Accelerated Catalysis: Asymmetric Dihydroxylation. *J. Am. Chem. Soc.* **1996**, *118*, 7632–7633.

(81) Han, H.; Janda, K. D. Multipolymer-Supported Substrate and Ligand Approach to the Sharpless Asymmetric Dihydroxylation. *Angew. Chem. Int. Ed. Engl.* **1997**, *36*, 1731–1733.

(82) Bolm, C.; Gerlach, A. Asymmetric Dihydroxylation with MeO-Polyethyleneglycol-Bound Ligands. *Angew. Chem. Int. Ed. Engl.* **1997**, *36*, 741–743.

(83) Bolm, C.; Gerlach, A. Polymer-Supported Catalytic Asymmetric Sharpless Dihydroxylations of Olefins. *Eur. J. Org. Chem.* **1998**, 21–27.

(84) Kim, B. M.; Sharpless, K. B. Heterogeneous Catalytic Asymmetric Dihydroxylation: Use of a Polymer-Bound Alkaloid. *Tetrahedron Lett.* **1990**, *31*, 3003–3006.

(85) Bolm, C.; Maischak, A.; Gerlach, A. Asymmetric Dihydroxylation with Silica-Anchored Alkaloids. *J. Chem. Soc., Chem. Commun.* **1997**, 2353–2354.

(86) Motorina, I.; Crudden, C. M. Asymmetric Dihydroxylation of Olefins Using Cinchona Alkaloids on Highly Ordered Inorganic Supports. *Org. Lett.* **2001**, *3*, 2325–2328.

(87) Trifiro, F.; Vaccari, A. In *Comprehensive Supramolecular Chemistry*; Atwood, J. L., Macnicol, D. D., Davies, J. E. D., Vogtle, F., Eds.; New York: Oxford, 1996; Vol. 7, pp 251–291.

(88) Choudary, B. M.; Chowdari, N. S.; Kantam, M. L.; Raghavan, K. V. Catalytic Asymmetric Dihydroxylation of Olefins with New Catalysts: The First Example of Heterogenization of OsO_4^{2-} by Ion-Exchange Technique. *J. Am. Chem. Soc.* **2001**, *123*, 9220–9221.

(89) Choudary, B. M.; Chowdari, N. S.; Jyothi, K.; Kantam, M. L. Catalytic Asymmetric Dihydroxylation of Olefins with Reusable OsO_4^{2-} on Ion-Exchangers: The Scope and Reactivity Using Various Cooxidants. *J. Am. Chem. Soc.* **2002**, *124*, 5341–5349.

(90) Bianchini, D. G.; Burnaby, D. G.; Evans, J.; Frediani, P.; Meli, A.; Oberhauser, W.; Psaro, R.; Sordelli, L.; Vizza, F. Preparation, Characterization, and Performance of Tripodal Polyphosphine Rhodium Catalysts Immobilized on Silica via Hydrogen Bonding. *J. Am. Chem. Soc.* **1999**, *121*, 5961–5971.

(91) de Rege, F. M.; Morita, D. K.; Ott, K. C.; Tumas, W.; Broene, R. D. Non-covalent Immobilization of Homogeneous Cationic Chiral Rhodium–Phosphine Catalysts on Silica Surfaces. *J. Chem. Soc., Chem. Commun.* **2000**, 1797–1798.

(92) Herron, N. A Cobalt Oxygen Carrier in Zeolite Y. A Molecular "Ship in a Bottle." *Inorg. Chem.* **1986**, *25*, 4714–4717.

(93) Feijen, E. J. P.; De Vadder, K.; Bosschaerts, M. H.; Lievens, J. L.; Martens, J. A.; Grobet, P. J.; Jacobs, P. A. Role of 18-Crown-6 and 15-Crown-5 Ethers in the Crystallization of Polytype Faujasite Zeolites. *J. Am. Chem. Soc.* **1994**, *115*, 2950–2957.

(94) Ogunwumi, S. B.; Bein, T. Intrazeolite Assembly of a Chiral Manganese Salen Epoxidation Catalyst. *J. Chem. Soc., Chem. Commun.* **1997**, 901–902.

(95) Ishida, T.; Akiyama, R.; Kobayashi, S. A Novel Microencapsulated Osmium Catalyst Using Cross-Linked Polystyrene as an Efficient Catalyst for

Asymmetric Dihydroxylation of Olefins in Water. *Adv. Synth. Catal.* **2005**, *347*, 1189–1192.

(96) Kobayashi, S.; Akiyama, R. Renaissance of Immobilized Catalysts. New Types of Polymer-Supported Catalysts, "Microencapsulated Catalysts," Which Enable Environmentally Benign and Powerful High-Throughput Organic Synthesis. *J. Chem. Soc., Chem. Comm.* **2003**, 449–460.

(97) Vankelecom, I.; Wolfson, A.; Geresh, S.; Landau, M.; Gottlieb, M.; Hershkovitz, M. First Heterogenisation of Rh–MeDuPHOS by Occlusion in PDMS (Polydimethylsiloxane) Members. *J. Chem. Soc., Chem. Commun.* **1999**, 2407–2408.

(98) Wolfson, A.; Janssens, S.; Vankelecom, I.; Geresh, S.; Gottlieb, M.; Herskowitz, M. Aqueous Enantioselective Hydrogenation of Methyl 2-Acetamidoacrylate with Rh–MeDuPHOS Occluded in PDMS. *J. Chem. Soc., Chem. Commun.* **2002**, 388–389.

(99) Raja, R.; Thomas, J. M.; Jones, M. D.; Johnson, B. F. G.; Vaughan, D. E. W. Constraining Asymmetric Organometallic Catalysts Within Mesoporous Supports Boosts Their Enantioselectivity. *J. Am. Chem. Soc.* **2003**, *125*, 14982–14983.

(100) Pini, D.; Petri, A.; Nardi, A.; Rosini, C.; Salvadori, P. Heterogeneous Catalytic Asymmetric Dihydroxylation of Olefins with the OsO$_4$/Poly(9–Image-Acylquinine-co-Acrylonitrile) System. *Tetrahedron Lett.* **1991**, *32*, 5175–5178.

(101) Wang, S.-S. *p*-Alkoxybenzyl Alcohol Resin and *p*-Alkoxybenzyloxycarbonylhydrazide Resin for Solid Phase Synthesis of Protected Peptide Fragments. *J. Am. Chem. Soc.* **1973**, *95*, 1328–1333.

(102) Hafez, A. M.; Taggi, A. E.; Dudding, T.; Lectka, T. Asymmetric Catalysis on Sequentially-Linked Columns. *J. Am. Chem. Soc.* **2001**, *123*, 10853–10859.

(103) Notestein, J. M.; Katz, A. Enhancing Heterogeneous Catalysis Through Cooperative Hybrid Organic-Inorganic Interfaces. *Chem. Eur. J.* **2006**, *12*, 3954–3965.

(104) Tada, M.; Kojima, N.; Izumi, Y.; Taniike, T.; Iwasawa, Y. Chiral Self-Dimerization of Vanadium Complexes on a SiO$_2$ Surface for Asymmetric Catalytic Coupling of 2-Naphthol: Structure, Performance, and Mechanism. *J. Phys. Chem. B* **2005**, *109*, 9905–9916.

(105) Yu, H.-B.; Hu, Q.-S.; Pu, L. The First Optically Active BINOL–BINAP Copolymer Catalyst: Highly Stereoselective Tandem Asymmetric Reactions. *J. Am. Chem. Soc.* **2000**, *122*, 6500–6501.

(106) Choudary, B. M.; Chowdari, N. S.; Jyothi, K.; Kumar, N. S.; Kantam, M. L. A New Bifunctional Catalyst for Tandem Heck-Asymmetric Dihydroxylation of Olefins. *J. Chem. Soc., Chem. Commun.* **2002**, 586–587.

(107) Choudary, B. M.; Jyothi, K.; Roy, M.; Kantam, M. L.; Sreedhar, B. Bifunctional Catalysts Stabilized on Nanocrystalline Magnesium Oxide for One-Pot Synthesis of Chiral Diols. *Adv. Synth. Catal.* **2004**, *346*, 1471–1480.

16

Applications of Asymmetric Catalysis in Synthesis

16.1 Enantiomerically Pure Components

Enantiomerically pure small organic molecules are key starting material components in the synthesis of a host of more advanced structures. Applications for such materials occur in commodity chemical markets, including pesticides, fragrances, polymers, and materials. In addition, the fine-chemicals industry generates a diverse array of functionalized and unfunctionalized single enantiomer compounds on smaller scales for use in pharmaceutical synthesis and chemical research. Many of these compounds are obtained from natural sources. Such compounds are referred to as Nature's pool of chiral materials[1–3] which, by definition, is limited to those compounds that can be isolated in pure form from readily available natural sources. With the advent of many commercially available chiral catalysts that are effective over a range of substrates, the pool of chiral components has increased enormously.[4,5] Examples of small chiral molecules that are available in enantiomerically pure form via asymmetric catalysis include those outlined in the Introduction and **Figures 16.1–16.5**.

This ever-increasing array of small chiral components reduces the number of synthetic manipulations needed to utilize a component; components from natural sources are often subjected to many steps to adjust oxidation state, functionality, and protecting groups so that they may be employed as synthetic precursors. In addition, the greater diversity afforded by catalytic asymmetric methods means that fewer synthetic routes are discarded due to the lack of availability of appropriate precursors. Finally, the availability of these chiral precursors is driving changes in the construction of complex structures possessing multiple stereogenic units by facilitating the convergent assembly of multiple chiral components[6] versus the iterative diastereoselective introduction of stereogenic units (see Sections 16.1.2 and 16.1.3). The focus of this chapter will be the use of asymmetric catalysis in the synthesis of complex structures.

16.1.1 Generation of Enantiomerically Pure Components: Early–Stage Asymmetric Induction

The Introduction outlined some of the most successful examples of asymmetric catalysis in the generation of small/simple organic structures. In particular, catalytic asymmetric methods centered on reduction, oxidation, and cyclopropanation have been spectacularly successful. In general, catalytic asymmetric reduction (**Figure 16.1**) and oxidation (**Figure 16.2**) technologies[7–9] lead the field and a diverse array of enantiomerically pure small molecules are readily generated due to the high stereoselection, high reliability, wide substrate scope, and good turnover rates available with these methods. While many of these processes proceed with near perfect enantioselectivity (> 99% ee), lesser enantioselectivities (> 80% ee) are often acceptable, depending on the application or if further enrichment is straightforward. Notably, the catalyst productivity, given as turnover number (TON = mol product/mol catalyst) or as substrate/catalyst ratio (s/c), is the main determinant of cost in these processes. For hydrogenation reactions utilizing precious metal catalysts, a TON > 1000 for high-value products and > 50,000 for large-scale or less expensive products (catalyst reuse increases the productivity) is required.[7] The catalyst activity, given as average turnover frequency (TOF = mol product/mol catalyst-h^{-1}), affects the production capacity. For hydrogenation, a TOF > 500 h^{-1} for small -scale and 10,000 h^{-1} for large-scale products is desirable.[7] Due to lower catalyst costs and often higher added values, lower TON and TOF values are acceptable for enantioselective oxidation (**Figure 16.2**) and the few C–C bond-forming reactions that have been implemented on scale (**Figure 16.3**).

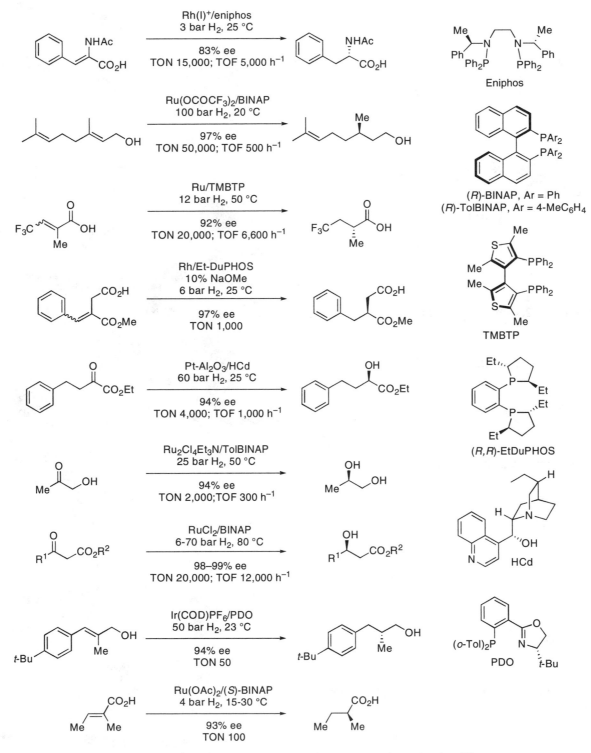

Figure 16.1. Simple components available on scale via asymmetric reduction.[7,8]

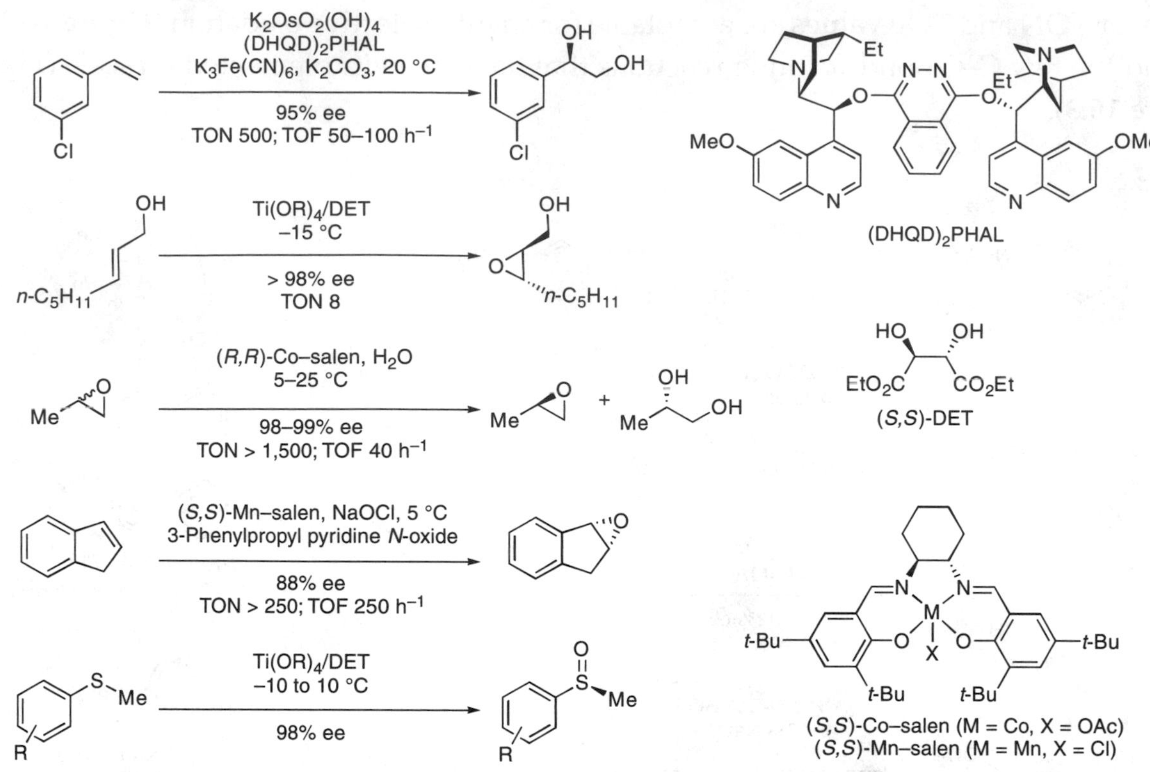

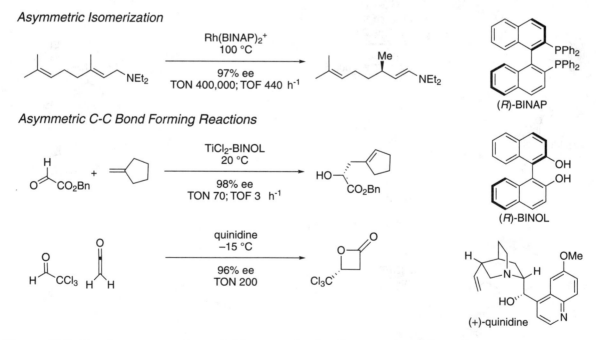

Figure 16.2. Simple components available on scale via asymmetric oxidation.[7]

Asymmetric Isomerization

Asymmetric C-C Bond Forming Reactions

Figure 16.3. Simple components available on scale via other processes.[7]

Examples of carbon–carbon bond-forming reactions that already possess many of these features and that are poised to make similar contributions include hydroformylation, olefin metathesis, π-allylation,[8–10] and conjugate additions[11–13] (**Figure 16.4**).

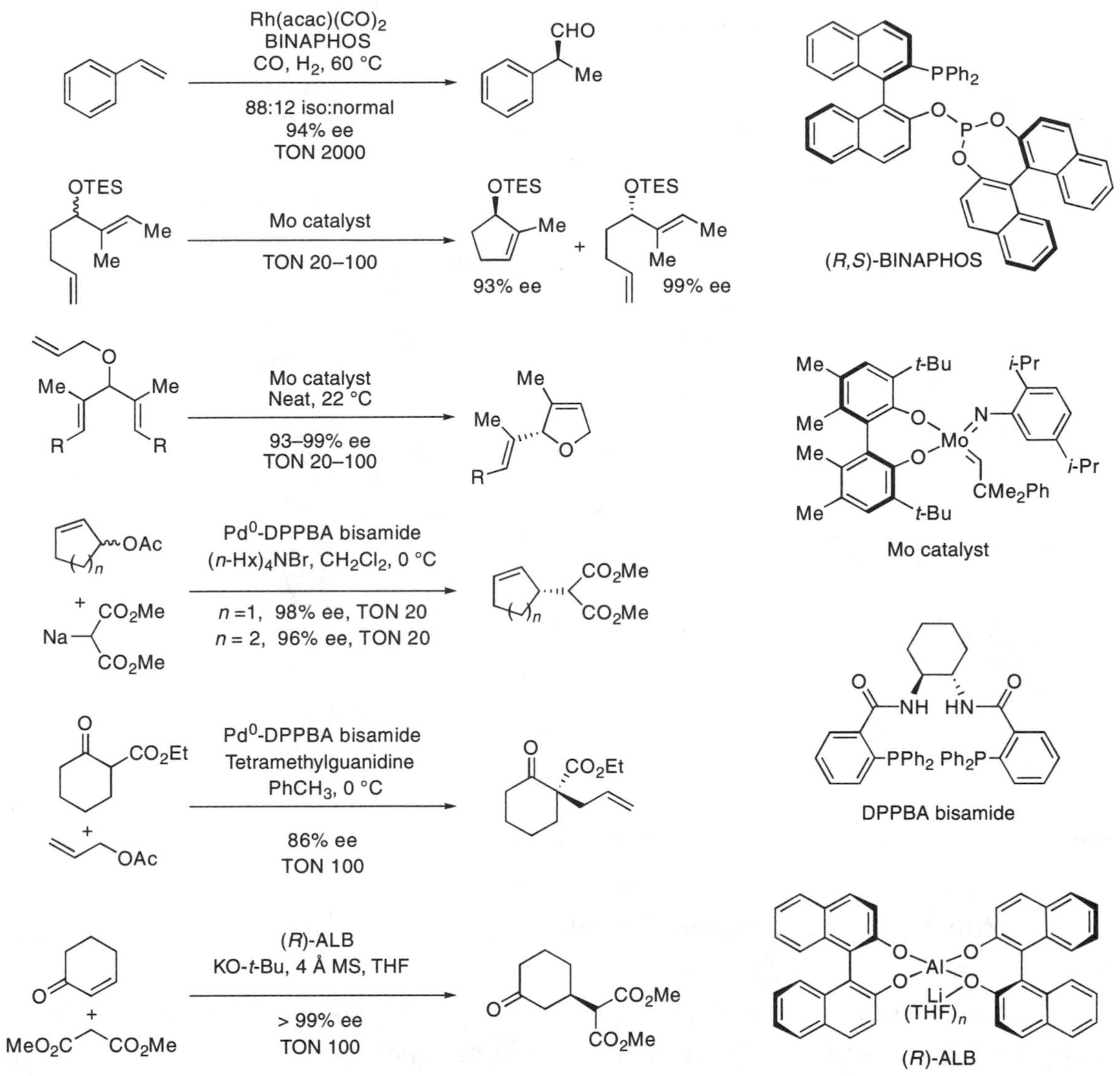

Figure 16.4. Simple components via asymmetric C–C bond-forming processes.[8–10,12]

Similarly, carbon–oxygen and carbon–nitrogen bond-forming reactions involving displacement rather than oxidation display favorable catalytic profiles and provide many useful functionalized small molecules (**Figure 16.5**).[10]

16.1.2 Small-Molecule Components to Generate Further Stereogenic Units

The field of asymmetric synthesis has benefited greatly from two major factors. The first is the diverse array of enantiomerically pure small molecules that are available from natural sources (i.e., amino acids, sugars, tartrate, etc.). The second is the concept of diastereoselective synthesis as first recognized by Emil Fischer,[14] wherein one stereogenic unit in a structure can be used to influence the creation of subsequent stereogenic units. With 2^n stereoisomers possible for molecules with n stereogenic

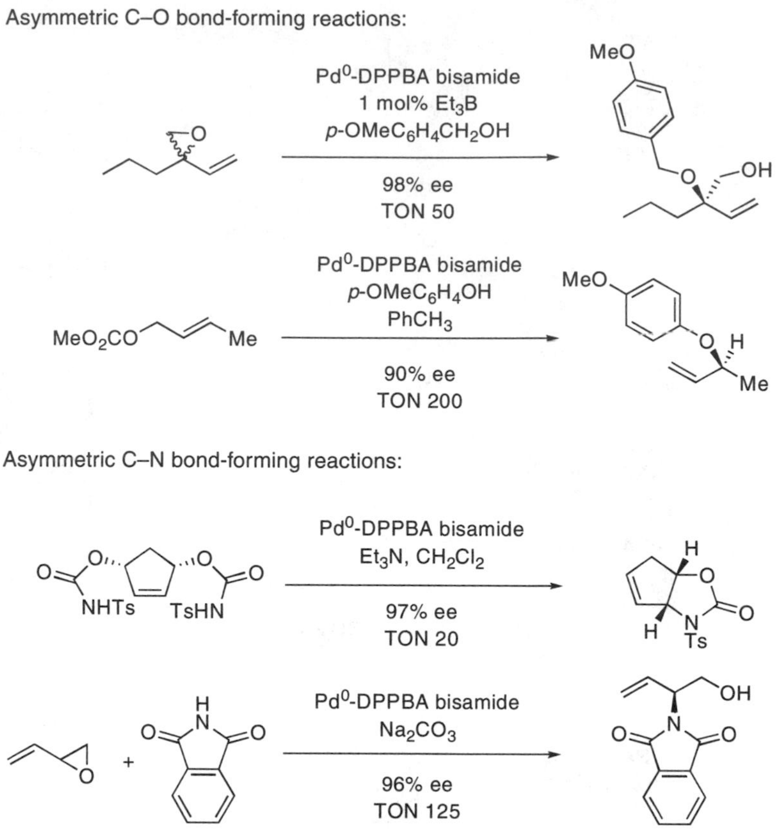

Asymmetric C–O bond-forming reactions:

Pd⁰-DPPBA bisamide
1 mol% Et₃B
p-OMeC₆H₄CH₂OH

98% ee
TON 50

Pd⁰-DPPBA bisamide
p-OMeC₆H₄OH
PhCH₃

90% ee
TON 200

Asymmetric C–N bond-forming reactions:

Pd⁰-DPPBA bisamide
Et₃N, CH₂Cl₂

97% ee
TON 20

Pd⁰-DPPBA bisamide
Na₂CO₃

96% ee
TON 125

Figure 16.5. Simple components via displacement processes.[10]

units, the construction of even moderately large structures would be hopeless without these contributions.

For example, one elegant synthesis of thienamycin commences with enantiomerically pure L-aspartic acid (**Figure 16.6**).[15] The stereogenic center from this compound is incorporated into the illustrated β-lactam. The enolate of this β-lactam is then employed in an acylation reaction. Due to the β-lactam stereogenic center, the enolate faces are diastereotopic and approach of the electrophile to the bottom face is favorable, which sets a second stereogenic center. In the next step, the extant stereocenters direct reduction to the bottom carbonyl diastereoface creating a third stereogenic center. Overall, the original stereogenic center from pure L-aspartic acid is used to establish the remaining two stereogenic centers of thienamycin. Early efforts in diastereoselective synthesis focused on the control of the relative stereochemistry within organized cyclic systems, wherein the stereochemical control elements are rigid and readily recognized. Later efforts have focused on the controlling features in conformationally flexible acyclic systems.[16–17] Today, both strategies are routinely employed in stereoselective synthesis. Regardless, diastereoselective strategies rely on a readily available supply of diverse enantiomerically pure chiral building blocks, since at least one stereogenic unit must be present to initiate such strategies.

The above example used an enantiomerically pure chiral building block from a natural source. The expanded pool of chiral materials[4,5] available via asymmetric

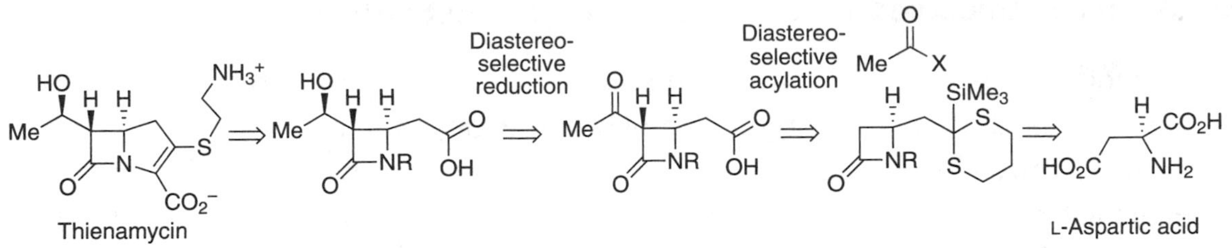

Figure 16.6. Diastereoselective retrosynthesis of thienamycin commencing with a single small chiral component.

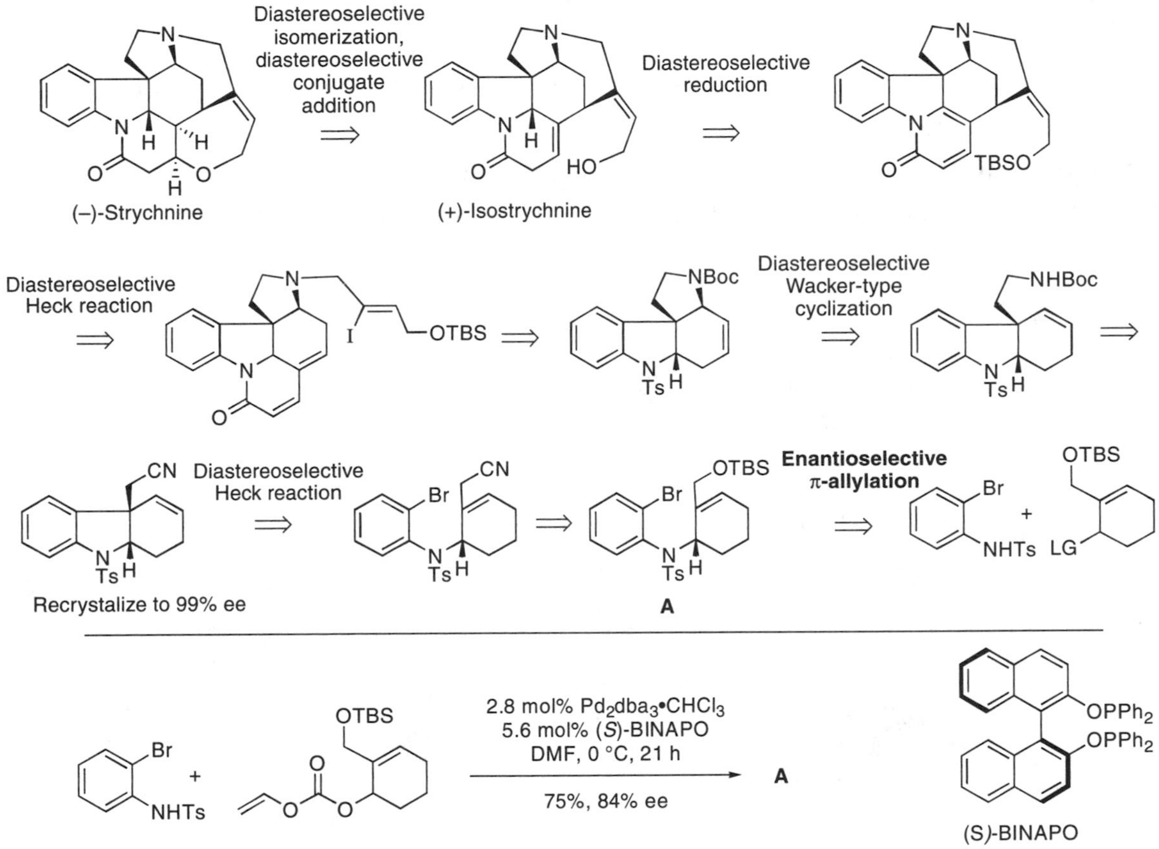

Figure 16.7. Diastereoselective synthesis of strychnine commencing with a single small chiral component made using asymmetric catalysis.

catalysis enables diastereoselective synthesis by permitting the design of distinct, more efficient, and more selective routes. For example, the total synthesis of the complex natural product strychnine,[18] illustrated in **Figure 16.7**, employs an enantiomerically pure component generated using asymmetric catalysis as the sole chiral starting material. The retrosynthetic analysis outlines a series of diastereoselective processes to afford the remaining stereogenic units from intermediate **A**. The stereogenic unit of **A** is, in turn, established using a catalytic asymmetric π-allylation reaction to unite two simple components.[19] In such an application, it is crucial that the catalytic asymmetric method be quite general to accommodate functionalized substrates that likely differ from those used in the initial development.

16.1.3 Small-Molecule Components in Convergent Synthesis

The expanded pool of chiral components[4,5] available via asymmetric catalysis also allows an increased use of convergent synthetic strategies in which more than one enantiomerically pure chiral component is employed as a starting material.[6] Analysis of target structures is then based upon identifying the chiral synthons ("chirons")[1,20–24] and relating them to available materials. For example, the synthesis of FR901464,[25,26] outlined in **Figure 16.8**, employs three separate chiral components, **A**, **B**, and **C**, all of which are generated using asymmetric catalysis. Propargylic alcohol **A** with one stereogenic unit is obtained via an asymmetric hydrogenation of an alkynyl ketone using a commercially available ruthenium catalyst.[27] Dihydropyrans **B** and **C**, with three and two stereogenic units, respectively, are assembled via asymmetric hetero-Diels–Alder reactions using a simple Schiff base chromium catalyst.[25,26] The final three stereogenic units found in FR901464 are generated via diastereoselective processes from **B** and **C**. The three elaborated components correspond to three distinct

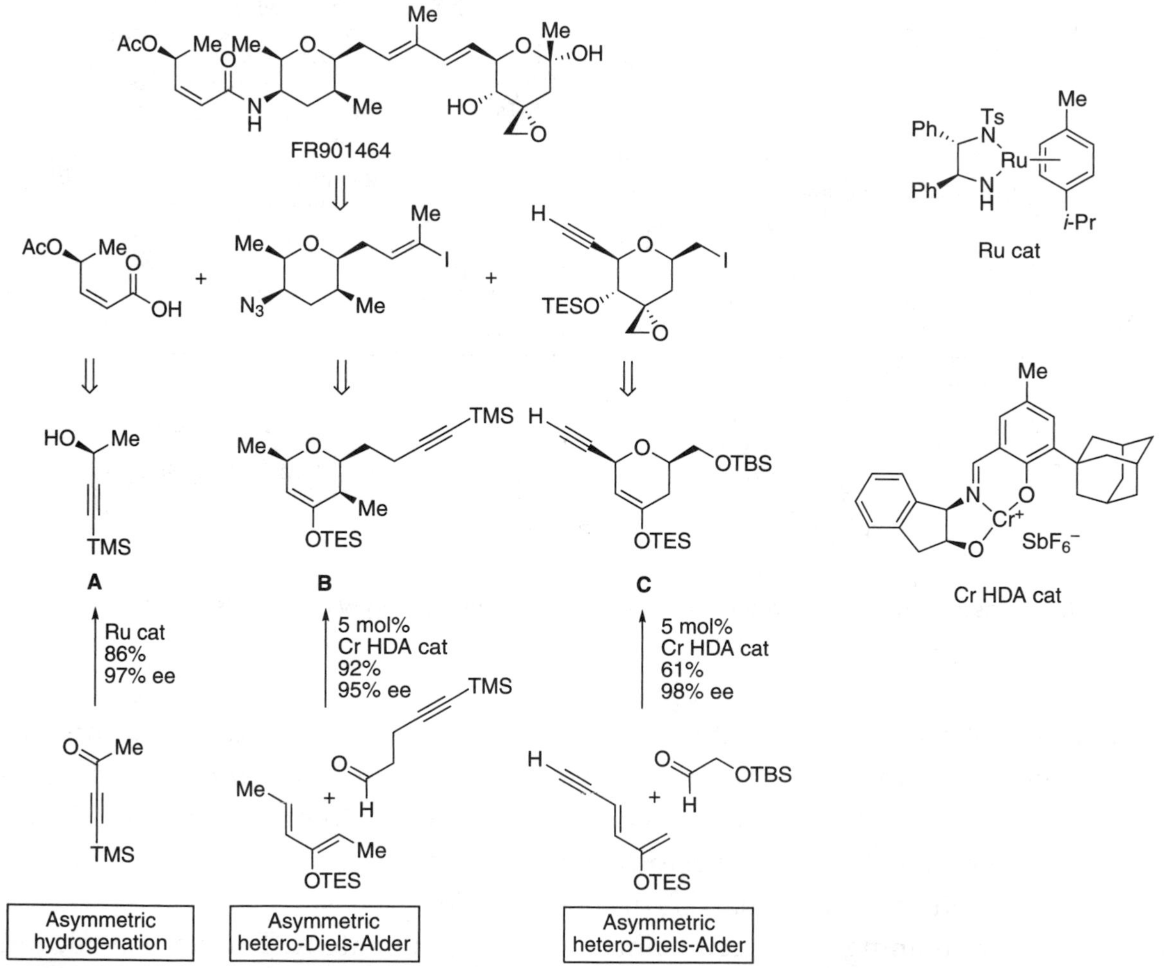

Figure 16.8. Convergent synthesis of FR901464 using chiral components generated via asymmetric catalysis.

stereochemical arrays in FR901464 and are subsequently united to provide the natural product.

Such a convergent approach is particularly powerful in the generation of stereochemical analogs. These analogs are useful in securing the stereochemistry of natural products and in identifying the key elements required for specific biological profiles. For example, two diastereomers of FR901464 were readily constructed using the same synthetic sequence (**Figure 16.9**). For **isomer 1**, the enantiomeric version of component **A** was employed, which was made by simply using the enantiomeric ruthenium catalyst. Similarly, the enantiomeric version of component **C** was employed in the synthesis of **isomer 2**. Subsequent characterization eliminated both of these isomers as candidates for the structure of FR901464.

The success of this convergent approach hinges on the presence of relatively isolated stereochemical arrays that can be synthesized independently and then assembled. If the stereochemical arrays are not isolated, then the assembly portion for different stereoisomers of the target structure may be complicated by diastereomeric effects (see the discussion of double diastereoselection in Chapter 13).

A second example of convergent synthesis using chiral components from asymmetric catalysis is illustrated in the synthesis of muconin in **Figure 16.10**.[28] Here, four different chiral components are employed, which contribute a total of four stereogenic units located at C4, C13, C22, and C36 in muconin. The remaining four stereogenic units are then set by diastereoselective processes. One of these chiral components is generated via an asymmetric catalytic hetero-Diels–Alder reaction of two simple achiral molecules. The remaining three chiral components are obtained via kinetic resolution of racemic starting materials. Specifically, hydrolytic epoxide-opening with cobalt–salen catalysts provides the requisite diol or epoxides. Even though only half of the chiral racemic epoxide is employed after the kinetic resolution, the low cost of these epoxides, which are obtained by epoxidation of the corresponding alkenes, the

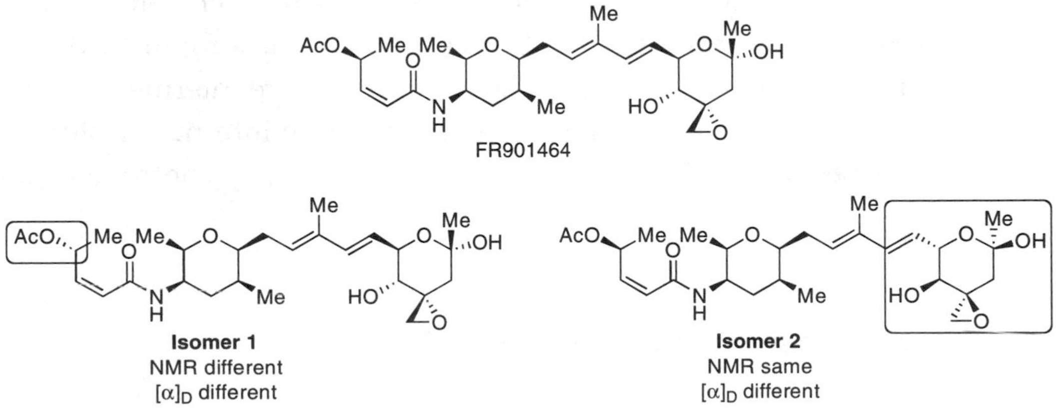

Figure 16.9. Flexibility of the convergent synthesis to FR901464 using chiral components generated via asymmetric catalysis.

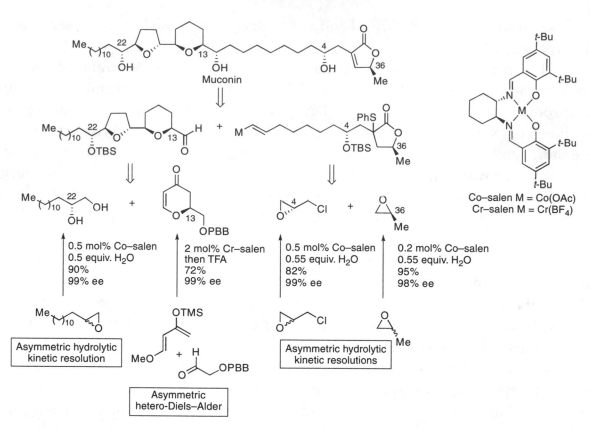

Figure 16.10. Convergent synthesis of muconin using chiral components generated via asymmetric catalysis.

low cost of the catalyst, and the efficiency of the resolutions, makes such processes very practical.[29]

A third example of convergent synthesis using chiral components from asymmetric catalysis is illustrated in the synthesis of epothilone in **Figure 16.11**.[30] Here, two different chiral components are employed that each contribute one stereogenic unit. These compounds are generated via asymmetric catalytic cyanosilylation and conjugate addition reactions. In the conjugate addition reaction, the stereochemistry is not generated in the C–C bond-forming process (i.e. at the β-carbon); instead, it arises due to an asymmetric protonation (i.e., at the α-carbon) of the intermediate enolate. The remaining stereogenic units are installed using simple internal diastereoselective processes (see Section 13.1) and a diastereoselective process controlled using an external chiral catalyst (see Section 13.3).

16.2 Late-Stage Asymmetric Induction

The utility of catalytic asymmetric methods in the late-stage introduction of stereogenic units in advanced, and valuable, synthetic intermediates[6] is appealing, in that lesser amounts of "high-cost" catalysts would be required compared to their appli-

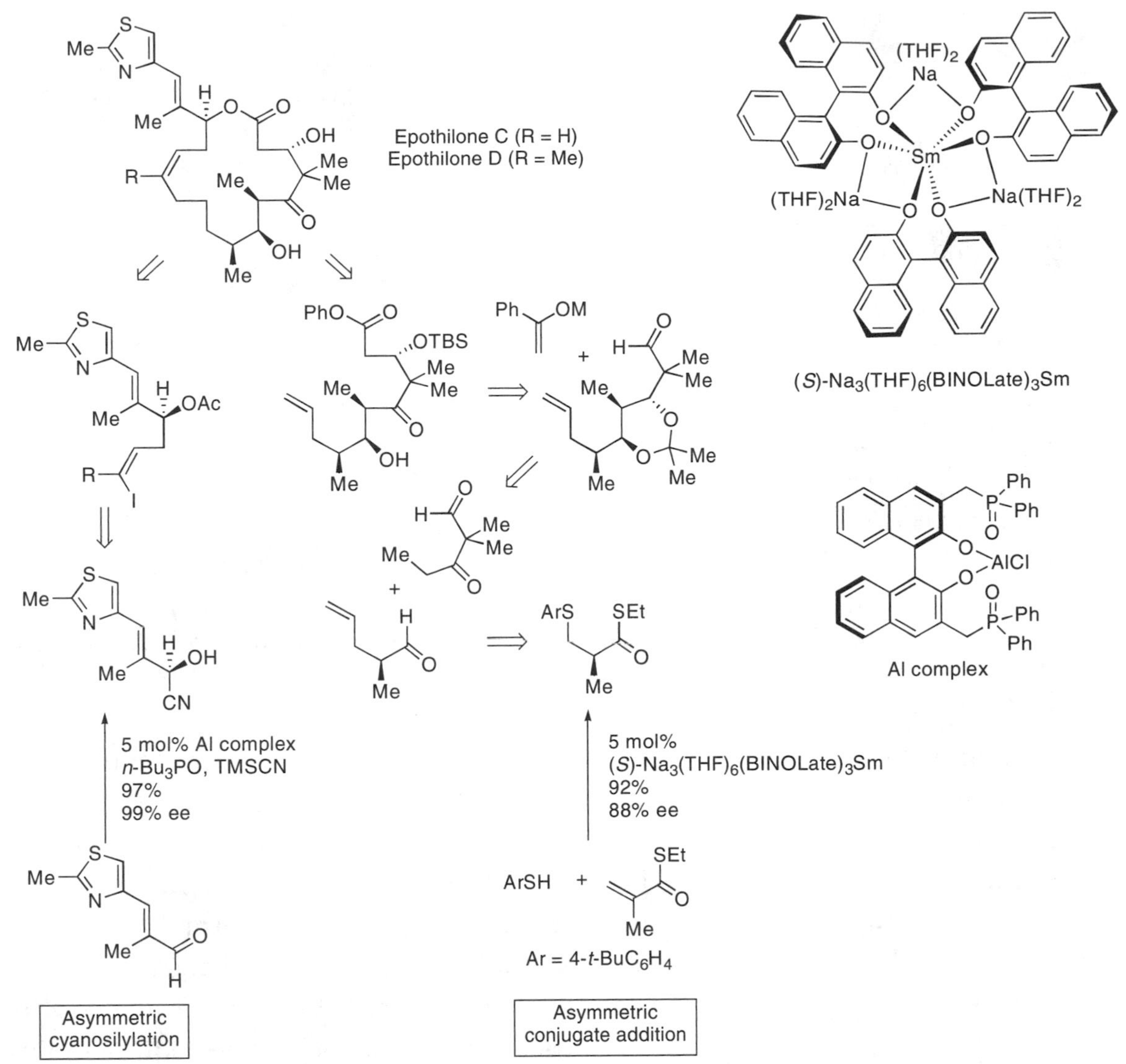

Figure 16.11. Convergent synthesis of epothilone using chiral components generated via asymmetric catalysis.

cation in the early stages of a synthetic endeavor. On the other hand, a high degree of reliability in the performance of the chiral catalyst is absolutely essential before committing to an involved synthesis of an advanced achiral precursor. Amongst the catalytic asymmetric methods, oxidation and reduction methods have emerged as particularly powerful in this genre.

16.2.1 Advanced Achiral Intermediates

In one synthesis of zaragozic acid,[31–33] an asymmetric dihydroxylation of the functionalized diene illustrated in **Figure 16.12** is employed to establish the C5 and C6 stereogenic units. The requisite diene is synthesized in four steps from the three components illustrated in **Figure 16.12**. The asymmetric dihydroxylation reaction has

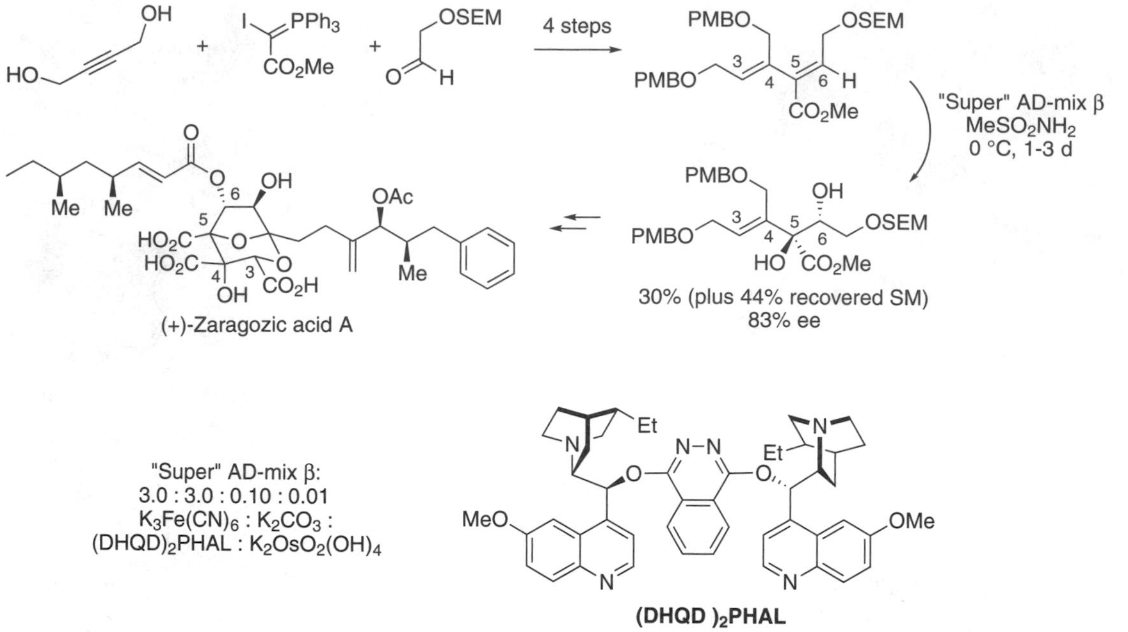

Figure 16.12. Asymmetric dihydroxylation of an advanced achiral intermediate to generate two new stereogenic units.

been thoroughly studied and exhibits a high degree of stereochemical regularity.[34] Even so, with such a functionalized diene neither the regiochemical nor the stereochemical course of the reaction is necessarily certain. In fact, a number of protecting-group combinations were surveyed before a selective process was identified. Aside from the starting material, eight products are possible in this transformation, including diols and tetrols. At first glance, dihydroxylation of the C3–C4 alkene is expected, as it is more electron-rich. In fact, dihydroxylation occurred at the C5–C6 alkene with complete regioselectivity and good enantioselectivity (83% ee). Molecular modeling results indicated that the ester group was perpendicular to the planar π-system of the diene. Moreover, the C–O bonds of the hydroxymethyl substituents were perpendicular to the plane of the diene, maximizing hyperconjugative electron withdrawal. As such, the C5–C6 bond with only one hydroxymethyl substituent is likely more electron-rich, contrary to superficial expectations.

An example of an asymmetric reduction on an advanced intermediate can be found in the enantioselective synthesis of (R)-denopamine, a β_1-receptor agonist used for the treatment of congestive heart failure (**Figure 16.13**).[35] Here, the extensive precedent of the chiral ruthenium catalysts proved reliable. Application of the *trans*-RuCl$_2$[(R)-Xyl-BINAP][(R)-DAIPEN] complex as the precatalyst provided the chiral alcohol in quantitative yield and 97% ee. Further removal of the amide, formation of the HCl salt and recrystallization, and hydrogenolysis of the benzyl group furnished (R)-denopamine hydrochloride in 94% yield. The use of low catalyst loading (substrate:catalyst = 2000:1) is key to the utility of this method due to the high cost of the catalyst.

Reactions aside from asymmetric oxidations and reductions that employ complex intermediates are less common. Such processes are less mature and further research

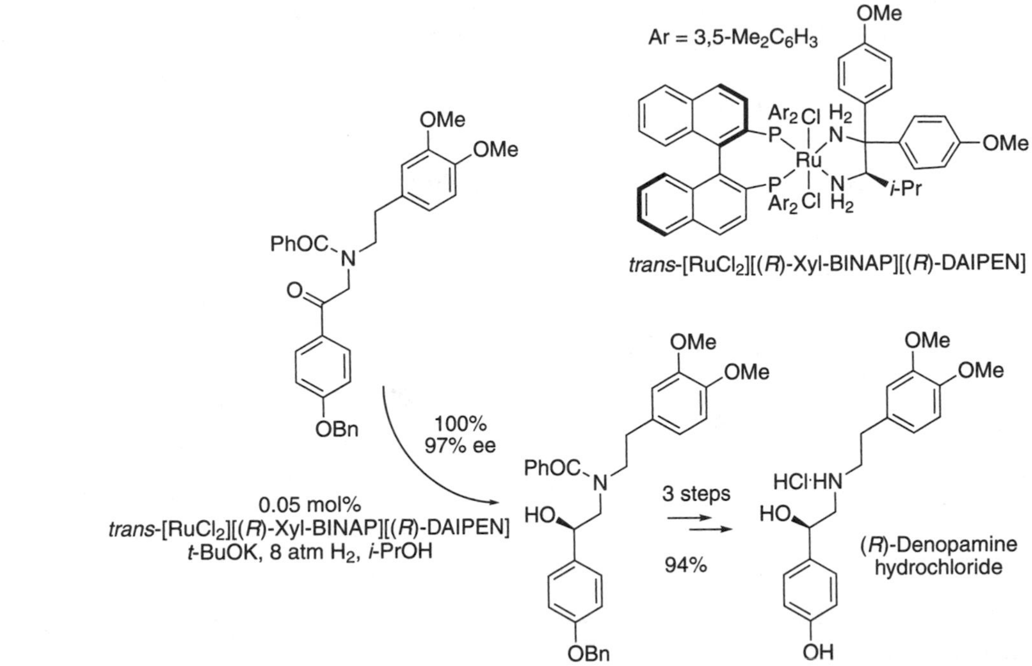

Figure 16.13. Asymmetric reduction of an advanced achiral intermediate.

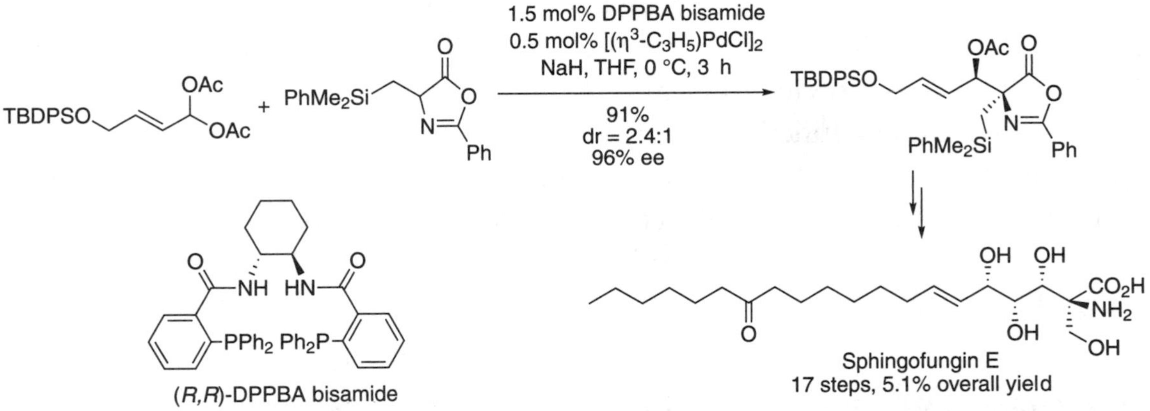

Figure 16.14. Enantioselective and diastereoselective catalytic C–C bond formation from moderately complex achiral intermediates.

will undoubtedly address concerns regarding substrate generality/predictability, functional-group tolerance, catalyst loadings, etc. An example can be found in the asymmetric π-allylation reaction of the allyl diacetate and the silyl azlactone illustrated in **Figure 16.14**.[36] The resultant adduct was formed in a 2.4:1.0 diastereomeric ratio with 96% ee. The selectivity-determining step involves both discrimination between the enantiotopic faces of the nucleophile and enantiotopic ionization of prochiral acetate leaving groups. The substantially lower disastereoselectivity observed with a silylmethylene (2.4:1.0 with 96% ee) versus a methyl (10.5:1.0 with 89% ee) analog highlights the difficulty in employing such transformations, as even minor modifications can lead to divergent results and can require rescreening reaction conditions and even catalysts. With this key carbon–carbon bond-forming transformation in hand, the synthesis of sphingofungin E was completed in 17 total steps and 5.1% overall yield.

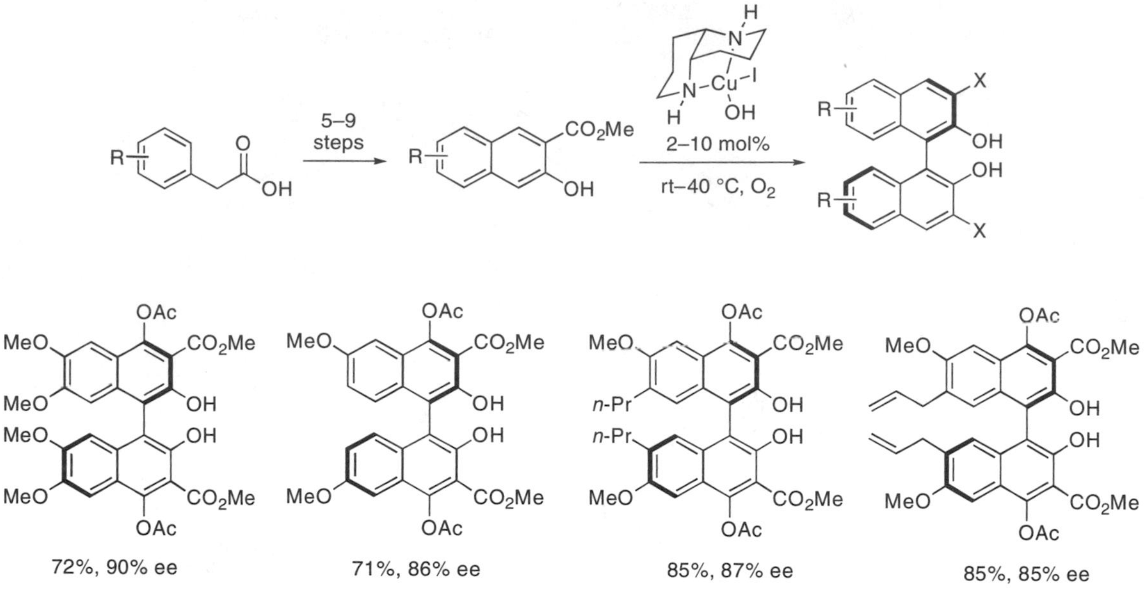

Figure 16.15. Asymmetric catalytic C–C bond formation with advanced achiral intermediates to generate axial stereogenic units.

Another example can be found in the oxidative biaryl coupling of achiral naphthols outlined in **Figure 16.15**. Here, advanced intermediates requiring 5–9 synthetic steps from commercial materials are found to reliably couple with good enantioselection (85–90% ee) to yield the chiral 1,1′-binaphthalenes.[37] The reliability of this coupling with highly functionalized naphthols to generate decasubstituted 1,1′-binaphthalenes has permitted the development of a general asymmetric approach to the perylenequinone natural products. The development of further reliable processes, especially those involving carbon–carbon bond-forming reactions, is key to advancing the utility of late-stage asymmetric transformations.

16.2.2 Multiple New Stereogenic Units from One Enantioselective Process

The processes in the above section all employ one catalytic asymmetric transformation with an advanced achiral intermediate that sets one or two stereogenic units. Due to the value of advanced intermediates, processes which can create larger stereochemical arrays are highly prized. Almost all such examples are found in desymmetrization reactions of meso compounds.[38,39] For example, the desymmetrization of the complex meso diene in **Figure 16.16** via Sharpless–Katsuki asymmetric epoxida-

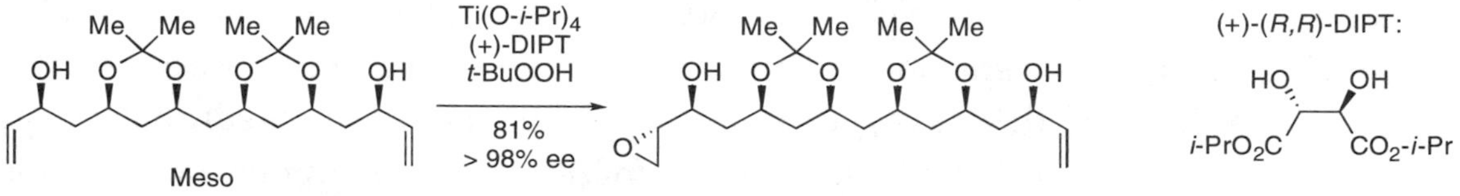

Figure 16.16. Desymmetrization of an advanced meso intermediate via enantioselectve epoxidation to generate seven stereogenic units.

tion generates an acyclic polyketide precursor with seven stereogenic units.[40] Good yield (81%) and very high selectivity (> 98% ee) were obtained. In this complex case, stoichiometric amounts of the chiral titanium complex were required. Catalyst systems that can function even more efficiently with such highly complex substrates is certainly an area of further study. Similar desymmetrization strategies with substoichiometric chiral catalysts have been applied in the synthesis of several natural products, including citreoviridin, FK506, and prostaglandin intermediates.[38] For a further discussion on enantioselective desymmetrization, see Chapter 10.

16.2.3 Multiple New Stereogenic Units from More Than One Reaction: Catalyzed Enantioselective and Diastereoselective Processes

When more than one selective process is catalyzed by a single chiral catalyst using an advanced achiral compound, larger stereochemical arrays can again be generated. For example, a double asymmetric epoxidation of the meso diene in **Figure 16.17** establishes a diastereomerically and enantiomerically pure compound with six stereogenic units;[41] the resultant bisepoxide was advanced to a C22–C34 fragment of halichondrin. This type of bidirectional strategy builds complexity efficiently by allowing the substrate assembly to proceed from two termini simultaneously.[40,42] For further examples of enantioselective desymmetrization, see Chapter 10.

Two factors contribute to the high enantiomeric excesses in such reactions. First, the inherent selectivities of the two asymmetric reactions on one substrate are multiplicative. Second, a stereochemical error in the first asymmetric reaction is negated in the second asymmetric reaction, thereby generating a separable diastereomer. These concepts have been examined in some detail and mathematical models put forth that are consistent with the observed results.[43,44] The basic premise is outlined in **Figure 16.18**. For the illustrated diene, there are four distinct alkene faces and their relationships are outlined below. If a directed epoxidation reaction is employed using Ti(O-i-Pr)$_4$ and tartrate (as was the case in **Figure 16.17**), and the X-group in **Figure 16.18**

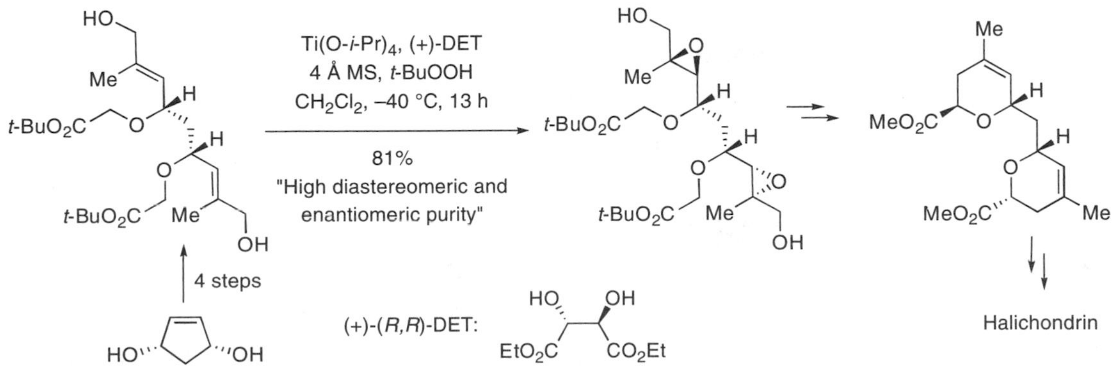

Figure 16.17. Desymmetrization of an advanced meso intermediate via asymmetric enantioselective and diastereoselective epoxidation to generate six stereogenic units.

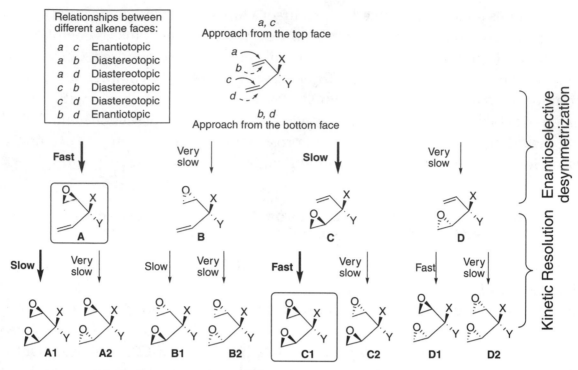

Figure 16.18. Combined enantiotopic group and diastereotopic face selectivity that can deliver products with very high enantiomeric excess. X = directing group, Y = non-directing group.

is the directing element, then only enantiomeric monoepoxides **A** and **C** should form in the first epoxidation (this assumes that the reaction is completely diastereoselective). If the reaction employs sufficient oxidant to also consume enantiomer **C**, then this enantiomer will be drained away selectively in the form of **C1** via a kinetic resolution (see Chapter 7). Upon termination of the reaction, *only* monoepoxide **A** and meso bisepoxide **C1** will be obtained. Since **A** and **C1** are different compounds, simple separation would then afford enantiomerically pure **A** as the major product. Thus, enantioselective desymmetrization of meso compounds can be a powerful technology, yielding products with near perfect enantiomeric excess, even when the asymmetric method employed is not inherently highly selective.

Achiral non-meso intermediates can also undergo multiple stereoselective reactions with single catalysts. For example, the analysis in **Figure 16.18** can be applied to the C–H insertion reaction of *N*-Boc pyrrolidine outlined in **Figure 16.19**.[45] Here, enantiotopic groups are differentiated rather than enantiotopic faces, but the same factors pertain. Of the four hydrogens available for the first insertion, there are only two enantiotopic sets (i.e., reaction at either H_C or H_B in the first insertion leads to the same product). As the first equation for the mono-insertion illustrates, the $Rh_2(S\text{-}DOSP)_4$ catalyst selects for pro-(*S*) hydrogens, yielding the product in 88% ee, which translates to a 94:6 enantiomeric ratio. If the same controlling factors are operative for a second sequential insertion (see second equation in **Figure 16.19**), then it should be possible to calculate the selectivity of the bis-insertion product using this 94:6 enan-

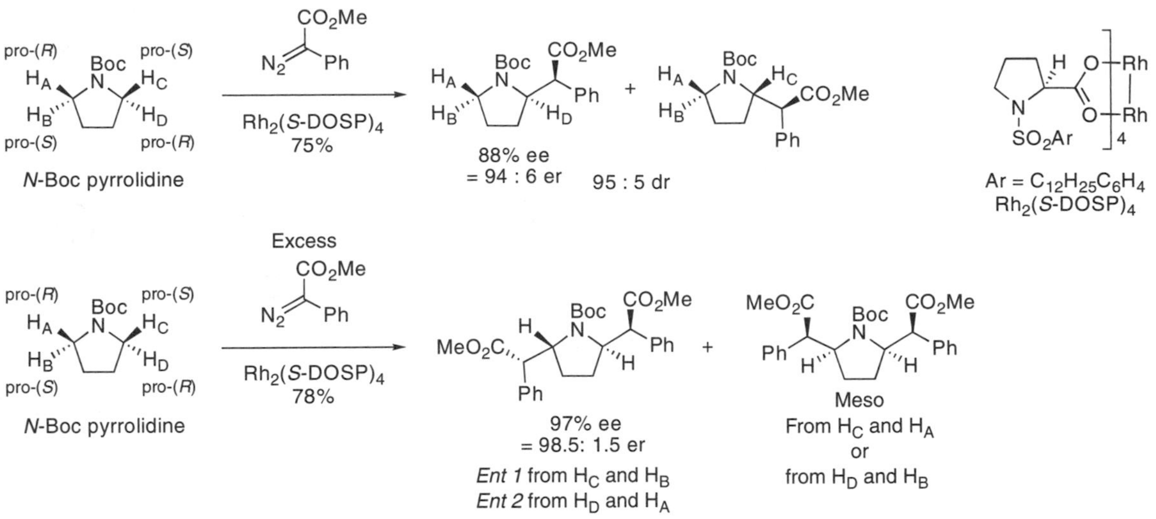

pro-(R) Boc pro-(S)
H_A N H_C
H_B H_D
pro-(S)   pro-(R)

N-Boc pyrrolidine

CO_2Me
N_2 Ph

$\xrightarrow{Rh_2(S\text{-DOSP})_4}$
75%

H_A Boc CO_2Me
N
H_B Ph
H_D

+

H_A Boc H_C
N CO_2Me
H_B Ph

88% ee
= 94 : 6 er   95 : 5 dr

Ar = $C_{12}H_{25}C_6H_4$
$Rh_2(S\text{-DOSP})_4$

Excess
CO_2Me
N_2 Ph

pro-(R) Boc pro-(S)
H_A N H_C
H_B H_D
pro-(S)   pro-(R)

N-Boc pyrrolidine

$\xrightarrow{Rh_2(S\text{-DOSP})_4}$
78%

MeO_2C H Boc CO_2Me
N   Ph
Ph   H

97% ee
= 98.5 : 1.5 er

Ent 1 from H_C and H_B
Ent 2 from H_D and H_A

+

MeO_2C Boc CO_2Me
N
Ph H   H Ph

Meso
From H_C and H_A
or
from H_D and H_B

Selectivity calculated based on ee of monoinsertion product:

1st reaction	2nd reaction		Calculated percentage		
	H_B	94%	94% x 94% = 88.4%		
H_C 94%	vs.				
	H_A	6%	94% x 6% = 5.6%	11.2%	*Ent 1:ent 2* = 99.5:0.5 er
vs.				Meso	99% ee
	H_B	94%	6% x 94% = 5.6%		
H_D 6%	vs.				
	H_A	6%	6% x 6% = 0.4%		

Figure 16.19. Selectivity enhancement from sequential enantioselective reactions with one catalyst.

tiomeric ratio. This exercise is outlined at the bottom of **Figure 16.19**. After the first insertion reaction, a 94:6 ratio of the pro-(S):pro-(R) insertion products is expected. Both of these mono-insertion products now possesses three possible protons for insertion. If H_C and H_D are disqualified due to the steric hindrance of the first insertion, then only H_A and H_B need to be considered. Applying a 94:6 ratio of pro-(S):pro-(R) selectivity to each of the mono-insertion adducts leads to the illustrated distribution, where the predicted selectivity for the non-meso bis-insertion product is 99% ee. The observed selectivity is slightly lower than calculated, which may arise from the stereogenic units in the mono-insertion product also influencing the stereochemistry of the second insertion (see the discussion of double diastereoselection in Chapter 13).

An extreme example of multiple stereoselective reactions with a single catalyst can be found when squalene is subjected to asymmetric dihydroxylation—no regioselectivity is observed and a mixture of diols forms due to reaction at more than one alkene position. However, the facial selectivity at each alkene is very high. This feature has been exploited in a synthesis of perhydroxysqualene (**Figure 16.20**).[46] After subjecting squalene to three rounds of asymmetric dihydroxylation and acetonide formation (to increase solubility), only one of the 36 possible stereoisomers was observed. This enantiomerically pure dodecanol was produced in an astounding 79% yield (> 96% per dihydroxylation). This 96% yield translates to ≥ 98% ee for the first dihydroxylation and

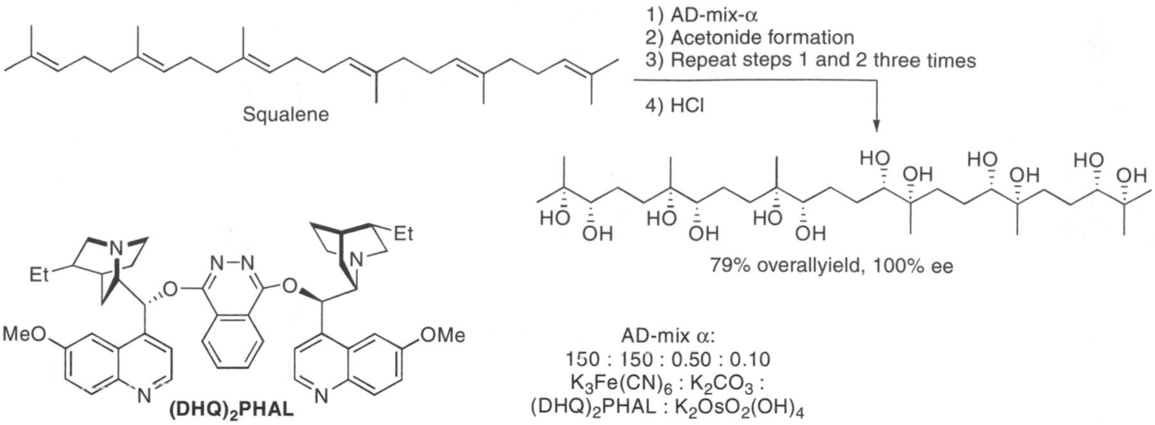

Figure 16.20. Stereoselective perhydroxylation of squalene to generate 12 stereogenic units.

≥ 98% de for the five subsequent diastereoselective dihydroxylations. Without the chiral ligand, all the possible stereoisomers were formed according to a statistical distribution. This example, in which 12 new stereogenic centers are established from a single catalyst, highlights the potential utility of chiral catalysts in generating enantiomerically pure compounds with multiple new stereogenic units.

The synthesis of the alkaloids quadrigemine C and psycholeine illustrates catalyst-controlled desymmetrization of meso compounds in total synthesis to dramatic effect (**Figure 16.21**).[47] The symmetrical meso diene substrate was assembled in 17 steps from commercial materials. This late-stage intermediate was then subjected to two iterative Heck cyclizations catalyzed by a chiral palladium–BINAP complex. The first enantioselective and the second diastereoselective Heck reactions proceeded with good selectivity to generate a product containing six of the eight stereogenic units present in the natural product. Notably, four of these units are all carbon-substituted quaternary stereocenters. Overall, the synthesis of quadrigemine C, which rigorously confirms its relative and absolute configuration, was executed in 20 linear steps with 2% overall yield from commercially available starting materials. The meso diene from **Figure 16.21** is one the most complex substrates ever subjected successfully to an enantioselective catalytic reaction. The use of a chiral catalyst to install the stereochemistry in the third-to-last synthetic step is an important departure from traditional methods for securing stereochemistry in total synthesis. The reliability and permissive scope of the Heck technology were key to this success; the future of such an approach will hinge on the development of catalysts with similar features.

In spite of the impressive accomplishments illustrated in the cases above, much work still remains in the field of asymmetric catalysis. Less expensive catalysts that are reliable at lower catalyst loadings and that can be employed with simple or complex substrates with a high degree of certainty are needed. For instance, the cyclization process illustrated in **Figure 16.22** is an example of the type of transformation

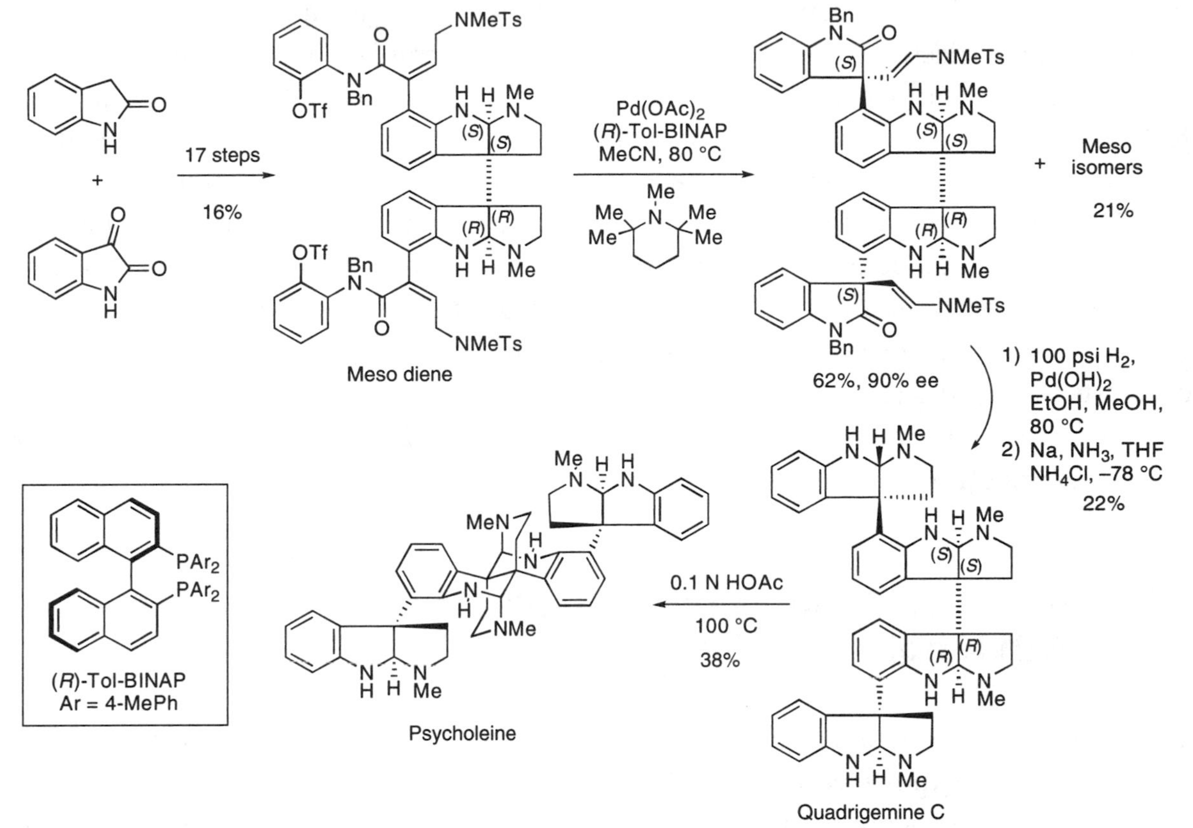

Figure 16.21. Asymmetric catalytic desymmetrization of an advanced achiral meso intermediate in the total synthesis of quadrigemine C and psycholeine.

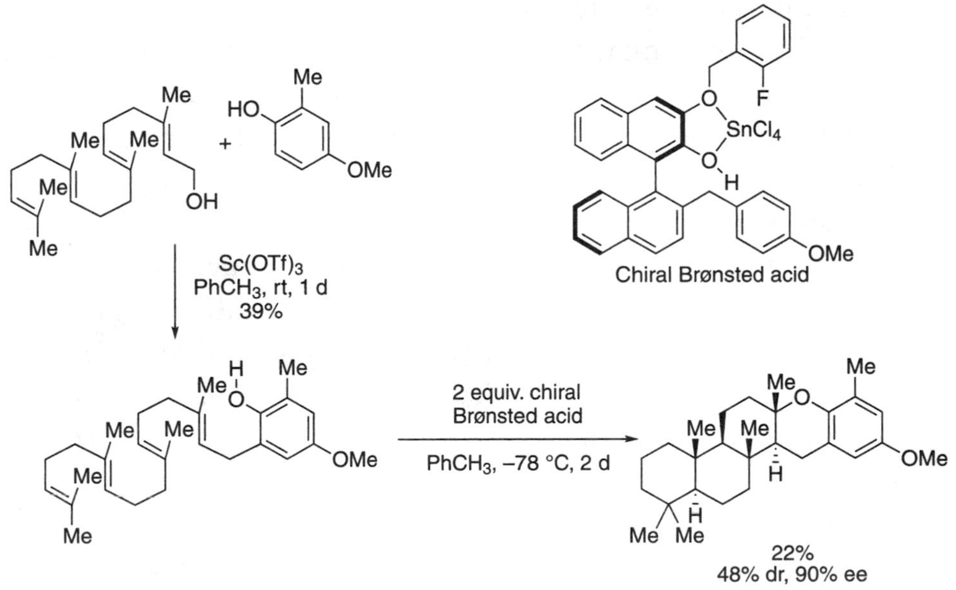

Figure 16.22. Complex stereoselective polycyclization using a stoichiometric chiral Brønsted acid to generate six stereogenic units.

that is most useful in asymmetric synthesis. Beginning with a multifunctional substrate, a highly diastereoselective and enantioselective transformation can be achieved using tin tetrachloride complexed with a chiral naphthol.[48] The tin tetrachloride binds to the oxygens of the napthol, increasing the acidity of the phenol. The resultant adduct selectively protonates one face of the dimethyl substituted alkene and triggers the stereoselective polycyclization to yield a synthetic analog of (–)-taondiol containing six stereogenic units with 90% ee and 48% de. This process is highly desirable from a synthetic point of view, as complexity is built rapidly and multiple bonds are constructed in a controlled manner in one event. Even so, the difficulty in devising catalysts that turn over in this process and that are compatible with oxygenated functionality highlights the fundamental principles that remain to be investigated and the development efforts that remain to be undertaken.

16.3 Synthesis Using Chiral Catalysts with Single Enantiomers

One strategy that is gaining popularity is the use of chiral catalysts with enantiomerically pure molecules (see Chapter 13). Since single-enantiomer substrates are employed, resolution is not the goal. Rather, the aim is to use chiral catalysts to establish stereocenters in spite of the stereochemistry already present. Given the fact that internal stereoselection (diastereoselection) works in many instances (see Section A.2.2 in Appendix A and Chapter 13) and does not require an additional chiral catalyst, one might wonder about the utility of such a strategy. There are several situations that would profit, including cases in which internal diastereoselection is weak or does not produce the desired stereoisomer, and cases in which it is desirable to synthesize the different diastereomers in pure form. This latter goal is especially important, as reliable external control with inexpensive and easily handled chiral catalysts will simplify synthetic endeavors undertaken by both experts and nonexperts.

In combining an enantiopure substrate with an enantiopure catalyst, double diasteroselection (see Chapter 13) may occur, where both the substrate and the catalyst stereochemistry contribute to the ultimate stereochemical outcome. For example, the catalytic efficiency of the BINAP–Ru catalyst is highly sensitive to the substitution patterns of the allylic alcohols. Even so, it has been applied to a chiral substrate in the synthesis of (3R,7R)-3,7,11-trimethyldodecanol, an intermediate for (R)-tocopherol (**Figure 16.23**).[49]

Another example of a catalytic asymmetric reduction with a chiral substrate can be found in the synthesis of taurospongin A (**Figure 16.24**).[50] Here, two additional stereochemical elements are present in the substrate, yet a highly selective hydrogenation occurs with the chiral catalyst.

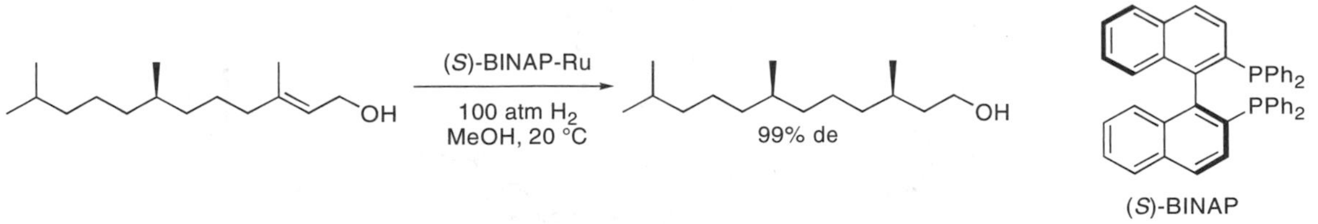

Figure 16.23. Catalytic asymmetric reduction of an advanced chiral intermediate to (3*R*,7*R*)-3,7,11-trimethyldodecanol.

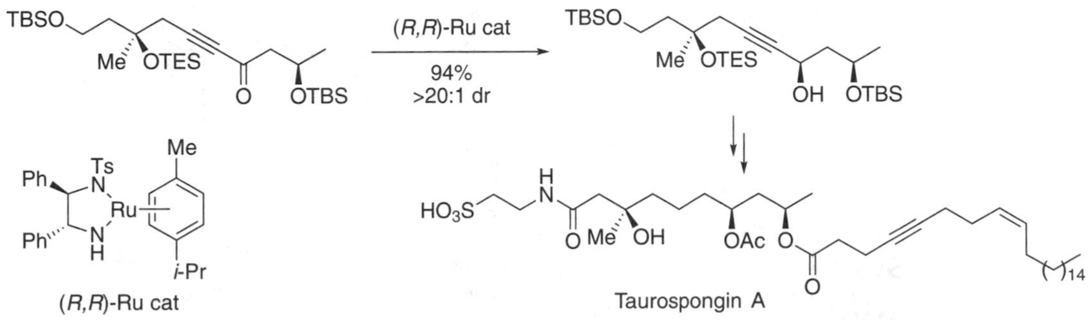

Figure 16.24. Catalytic asymmetric reduction of an advanced chiral intermediate.

In applying this strategy to advanced (and costly) intermediates, catalyst reliability is crucial. For example, there are several catalyst systems for catalytic asymmetric epoxidation, each of which has known strengths. The tartrate/titanium/*t*-BuOOH system works well with allylic alcohols,[51] the manganese–salens give good results with most cis and trisubstituted alkenes,[52] and the chiral dioxirane system provides high selectivity with trans and trisubstituted alkenes as well as allylic alcohols (**Figure 16.25**).[53] On this basis, it would appear that there are highly effective catalysts for most substrate types.

In the investigation of an efficient synthesis of cryptophycin 52, a cryptophycin variant for the treatment of solid tumors, an epoxidation of an advanced chiral intermediate was explored to install the sensitive epoxide pharmacaphore at a late stage (**Figure 16.26**).[54] Diastereoselective epoxidation with a variety of achiral oxidants was poor, leading researchers to anticipate that internal substrate control was weak. Chiral oxidants were explored in the hope of exerting external catalyst control and obtaining high diastereoselection. The inherent diastereoselection was 1.9:1.0, *syn:anti*, with achiral reagents, and the (*S*,*S*)-Mn–salen can override that to give 1.0:1.9, *syn:anti*. The best results were obtained with a chiral dioxirane catalyst derived from oxone and the ketone illustrated in **Figure 16.26** (3.5:1, *syn:anti*). However, under all of the asymmetric catalytic conditions explored, the conversion to the epoxide was poor and selectivity was nominal.

The problem of low diastereoselection encountered in the epoxidation of deoxocryptophycin 52 was solved by explorating alternative epoxidation substrates

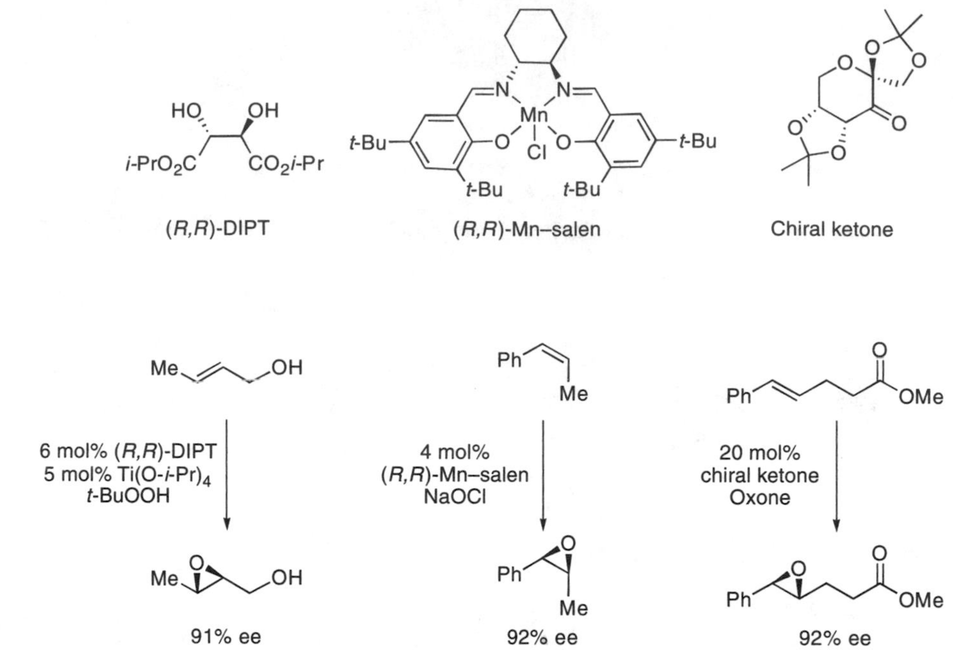

(R,R)-DIPT
(R,R)-Mn–salen
Chiral ketone
6 mol% (R,R)-DIPT
5 mol% Ti(O-i-Pr)4
t-BuOOH
4 mol%
(R,R)-Mn–salen
NaOCl
20 mol%
chiral ketone
Oxone
91% ee
92% ee
92% ee

Figure 16.25. Representative examples with different asymmetric epoxidation catalysts.

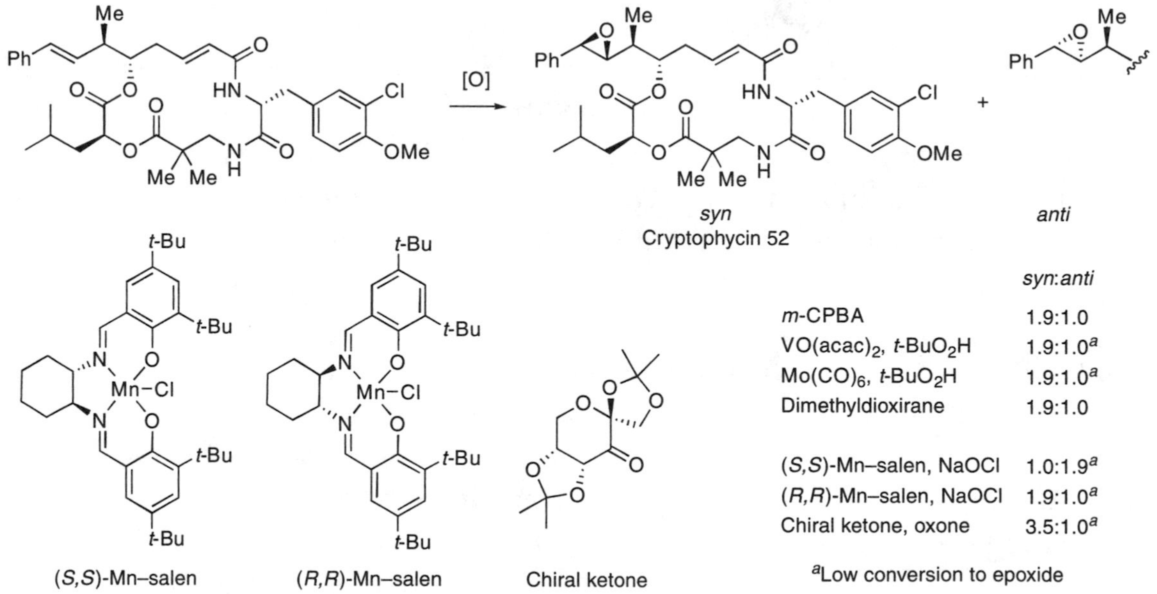

[O]
syn
Cryptophycin 52
anti
(S,S)-Mn–salen
(R,R)-Mn–salen
Chiral ketone
syn:anti
m-CPBA 1.9:1.0
VO(acac)2, t-BuO2H 1.9:1.0a
Mo(CO)6, t-BuO2H 1.9:1.0a
Dimethyldioxirane 1.9:1.0
(S,S)-Mn–salen, NaOCl 1.0:1.9a
(R,R)-Mn–salen, NaOCl 1.9:1.0a
Chiral ketone, oxone 3.5:1.0a
aLow conversion to epoxide

Figure 16.26. Poorly diastereselective epoxidations in crytophycin 52.

(**Figure 16.27**) in which the steric hindrance around the critical styrenyl group was reduced. With the earlier-stage intermediates **A–C**, higher conversions (> 95%) were obtained even with *m*-CPBA, but the stereoslection was still nominal (1:1 to 2:1). With the chiral dioxirane, conversions (70–95%) remained good and the stereoselectivities (5:1 to 9.5:1) improved. Ultimately, substrate **B** was selected for large scale synthesis of cryptophycin 52, as this was the latest intermediate that combined optimal conversion and selectivity. This situation highlights the difficulties in using chiral cata-

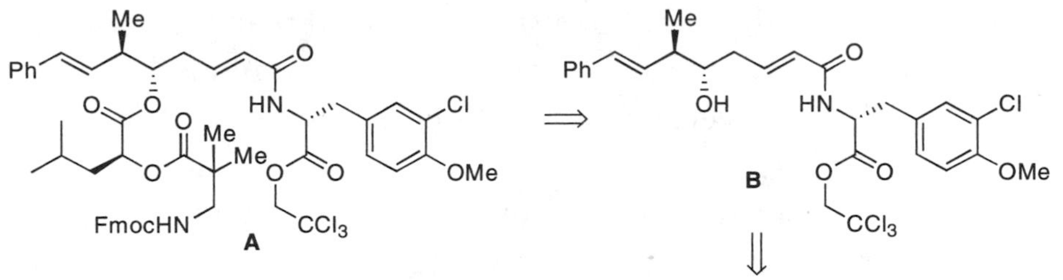

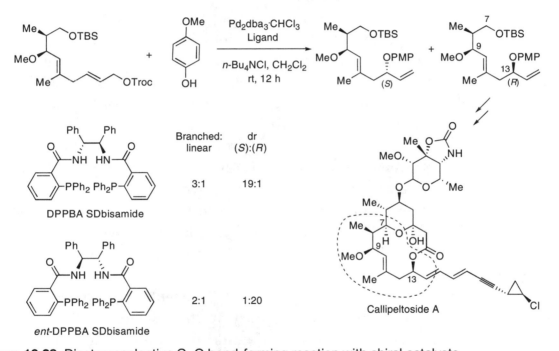

Figure 16.27. Solving diastereoselective epoxidation for crytophycin 52.

	m-CPBA		Chiral ketone/oxone	
	Conversion (%)	syn:anti	Conversion (%)	syn:anti
A	> 95	2:1	70–85	9.5:1
B	> 95	1.5:1	> 95	5:1
C	> 95	1:1	> 95	5:1

	Branched:linear	dr (S):(R)
DPPBA SDbisamide	3:1	19:1
ent-DPPBA SDbisamide	2:1	1:20

Callipeltoside A

Figure 16.28. Diastereoselective C–O bond-forming reaction with chiral catalysts.

lysts for diastereoselective transformations in complex compounds. Even highly stereoselective catalysts may not be as effective as anticipated from prior work or may require significant optimization in terms of reaction conditions and acceptable substrates. As such, a clear goal in asymmetric catalysis is the development of catalyst systems that do not suffer such limitations.

In addition to asymmetric reduction and oxidation, other catalytic asymmetric methods can be employed with advanced chiral substrates. For example, an asymmetric palladium-catalyzed allylic etherification was employed with a chiral substrate in the synthesis of callipeltoside A (**Figure 16.28**).[55] Upon screening a number of chiral bisphosphine ligands with the 4-methoxyphenol substrate, the best result was obtained with the DPPBA ligand that gave a 3:1 branched:linear ratio with excellent

(19:1) diastereoselectivity. In this case, the mechanism of reaction involves diastereo-facial exchange of the palladium–allyl complex through an η^1-allyl intermediate. Presumably, one of the interconverting diastereomeric complexes is more reactive and leads to product. Surprisingly, the stereochemistry was opposite that expected based on findings with related substrates. This result illustrates that caution must be exercised in extending prior stereochemical results to new substrates and that even distal stereogenic centers can have profound effects (see Chapter 13). However, a simple switch in the configuration of the ligand (i.e., *ent*-DPPBA) provided the required diastereoisomer with similar diastereoselectivity (20:1) and a somewhat reduced (2:1) branched:linear ratio from which the desired diastereomer was isolated in 51% yield. While ultimately solved, this situation highlights some of the problems encountered when using chiral catalysts with complex substrates.

The enantioselective synthesis of fostriecin[56] is a powerful example of the utility of asymmetric catalysis (**Figure 16.29**). One enantioselective reaction and then three diastereoselective reactions are employed, *all using chiral catalysts*. Notably, one chiral catalyst is used to establish each individual stereogenic unit. In principle, each stereogenic unit can be modified independently by the selection of different catalyst enantiomers, allowing the synthesis of all of the stereoisomers ($2^4 = 16$) of fostriecin via this route.

The first stereogenic center was installed via asymmetric ketone cyanation. The reaction was optimized fairly rapidly by examining two catalysts, temperature, and catalyst loading. Ultimately, the cyanation was performed using 5 mol% of the titanium catalyst illustrated in **Figure 16.29** on a 50-g scale without any difficulty (93% yield, 85% ee). The chiral ligand was recovered in 95% yield after silica-gel column chromatography.

The second stereochemical task was to enrich the enantiomeric excess and construct the C5 chiral carbon through a catalytic asymmetric allylation with the (*R*)–Tol-BINAP/AgF catalyst and an allyltrimethoxysilane donor (**Figure 16.29**). This provided the homoallylic alcohol in 28:1 diastereomeric ratio. The third stereogenic center was established via a direct aldol reaction between an ynone and an aldehyde with a lithium lanthanum–BINolate catalyst. From a stereochemistry standpoint, this portion was the most challenging, since the product was produced as a 3.6:1 diastereomeric mixture. In the final stereochemical task, a chiral ruthenium catalyst was used to effect asymmetric reduction of the ynone to provide a > 97:3 ratio of diastereomers.

Catalytic asymmetric resolutions have also been employed with advanced intermediates in complex molecules. For example, one strategy described for epothilone initially utilized the resolution outlined in **Figure 16.30**.[30] A relatively complex racemic aldehyde was employed in a direct catalytic asymmetric aldol reaction with acetophenone. After several attempts, a heteropolymetallic chiral catalyst gave the desired aldol product in 30% yield and in 89% ee, along with the diastereomer in 29% yield and in 88% ee. In this resolution, both of the enantiomers react, but they give rise to two *different* products (see the discussion of parallel kinetic resolution in Chapter 8), because the catalyst predominates, giving rise to the *Si*-facial-addition adduct regardless of the β-, γ-, δ-, and ε-stereocenters in the aldehyde. This approach was later supplanted by a nonresolution strategy (see **Figure 16.11**).[30]

In order to be useful with enantiopure substrates, chiral catalysts need to dominate the sterochemistry to be free from the effects of double diastereoselection (see

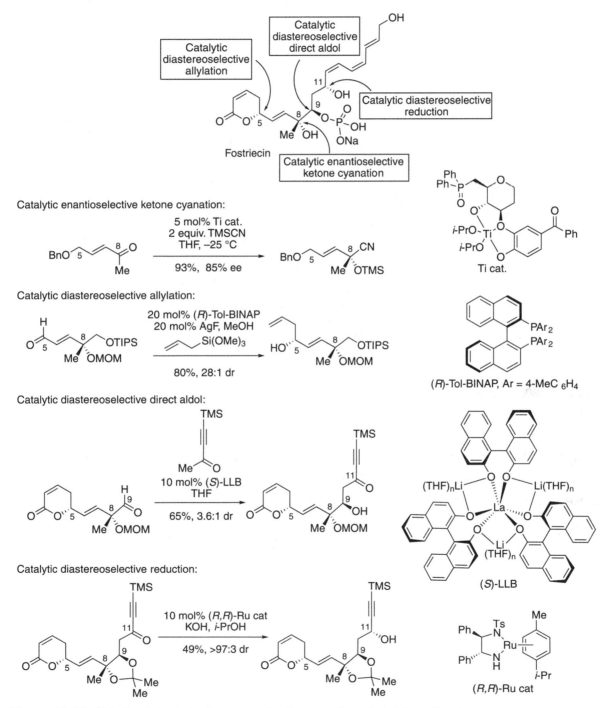

Figure 16.29. Chiral catalysts in diastereoselective reactions in total synthesis.

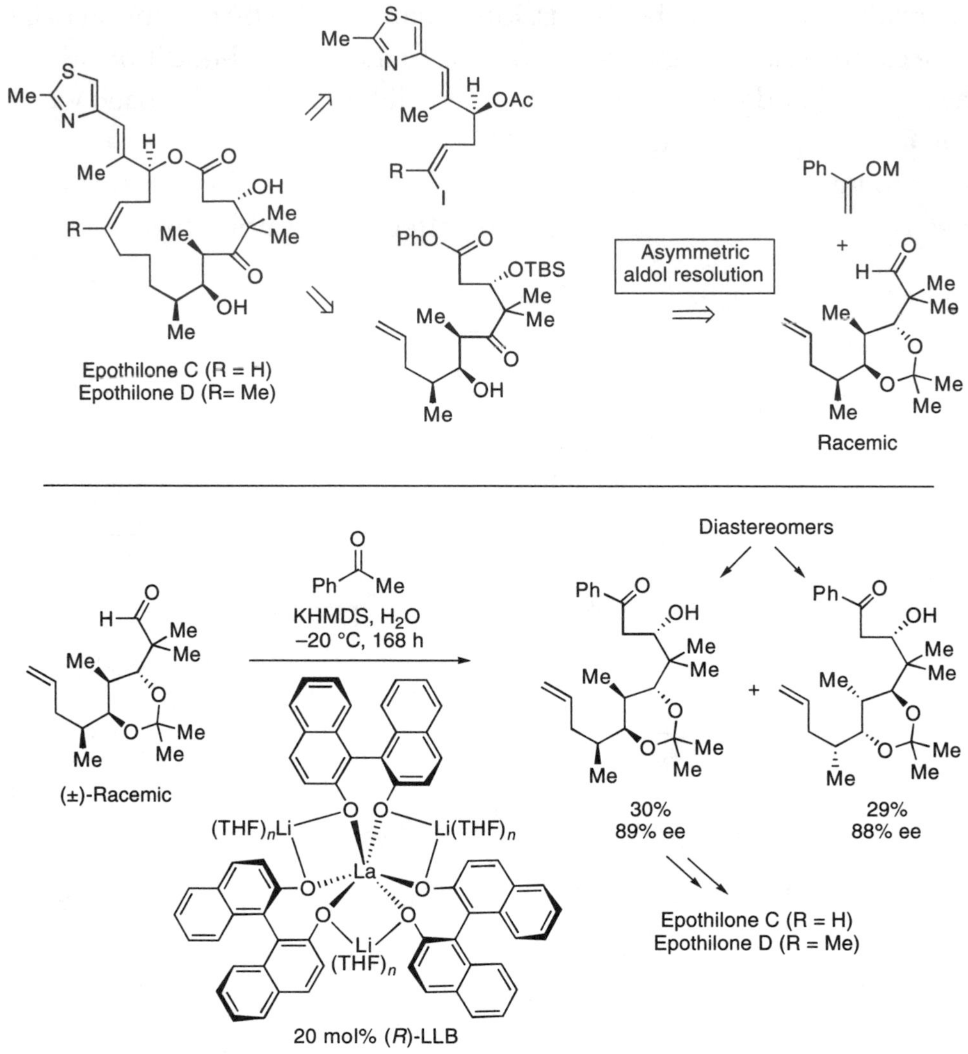

Figure 16.30. Parallel kinetic resolution in total synthesis.

Chapter 13). In addition, the chiral catalyst must be reliable across a range of substrates and tolerant of as many different functional groups as possible. Occasionally these goals are met, but further work is needed to discover ideal candidates for a wider array of chemistries.

References

(1) Hanessian, S. In *Organic Chemistry Series*; Baldwin, J. E., Ed.; Pergamon: New York, 1983; Vol. 3.

(2) Scott, J. W. In *Asymmetric Synthesis*; Morrison, J. D., Scott, J. W., Eds.; Academic Press: New York, 1984; Vol. 4, Chapter 1.

(3) Casiraghi, G.; Zanardi, F.; Rassu, G.; Spanu, P. Stereoselective Approaches to Bioactive Carbohydrates and Alkaloids With a Focus on Recent Syntheses Drawing from the Chiral Pool. *Chem. Rev.* **1995**, *95*, 1677–1716.

(4) Nugent, W. A.; RajanBabu, T. V.; Burk, M. J. Beyond Nature's Chiral Pool: Enantioselective Catalysis in Industry. *Science* **1993**, *259*, 479–483.

(5) Welch, C. J. Crawling Out of the Chiral Pool: The Evolution of Pirkle-type Chiral Stationary Phases. *Adv. Chrom.* **1995**, *35*, 171–197.

(6) Taylor, M. S.; Jacobsen, E. N. Asymmetric Catalysis in Complex Target Synthesis. *Proc. Natl. Acad. Sci. U.S.A.* **2004**, *101*, 5368–5373.

(7) Blaser, H. U.; Spindler, F.; Studer, M. Enantioselective Catalysis in Fine Chemicals Production. *Appl. Catal., A* **2001**, *221*, 119–143.

(8) Chapuis, C.; Jacoby, D. Catalysis in the Preparation of Fragrances and Flavours. *Appl. Catal., A* **2001**, *221*, 93–117.

(9) Farina, V.; Reeves, J. T.; Senanayake, C. H.; Song, J. J. Asymmetric Synthesis of Active Pharmaceutical Ingredients. *Chem. Rev.* **2006**, *106*, 2734–2793.

(10) Trost, B. M.; Crawley, M. L. Asymmetric Transition-Metal-Catalyzed Allylic Alkylations: Applications in Total Synthesis. *Chem. Rev.* **2003**, *103*, 2921–2943.

(11) Shibasaki, M.; Sasai, H.; Araixu, T. Asymmetric Catalysis with Heterobimetallic Compounds. *Angew. Chem., Int. Ed. Engl.* **1997**, *36*, 1236–1256.

(12) Xu, Y.; Ohori, K.; Ohshima, T.; Shibasaki, M. A Practical Large-Scale Synthesis of Enantiomerically Pure 3-[Bis(methoxycarbonyl)methyl]cyclohexanone via Catalytic Asymmetric Michael Reaction. *Tetrahedron* **2002**, *58*, 2585–2588.

(13) Ohshima, T.; Xu, Y.; Takita, R.; Shimizu, S.; Zhong, D.; Shibasaki, M. Enantioselective Total Synthesis of (–)-Strychnine Using the Catalytic Asymmetric Michael Reaction and Tandem Cyclization. *J. Am. Chem. Soc.* **2002**, *124*, 14546–14547.

(14) Fischer, E. Synthesen in der Zuckergruppe II. *Chem. Ber.* **1894**, *27*, 3189–3232.

(15) Salzmann, T. N.; Ratcliffe, R. W.; Christensen, B. G.; Bouffard, F. A. A Stereocontrolled Synthesis of (+)- Thienamycin *J. Am. Chem. Soc.* **1980**, *102*, 6161–6163.

(16) Hoveyda, A. H.; Evans, D. A.; Fu, G. C. Substrate-Directable Chemical Reactions. *Chem. Rev.* **1993**, *93*, 1307–1370.

(17) a) Bartlett, P. A. Stereocontrol in the Synthesis of Acyclic Systems–Applications to Natural Product Synthesis. *Tetrahedron* **1980**, *36*, 3–72.
b) Mikami, K.; Shimizu, M.; Zhang, H.-C.; Maryanoff, B. E. Acyclic Stereocontrol Between Remote Atom Centers via Intramolecular and Intermolecular Stereo-Communication. *Tetrahedron* **2001**, *57*, 2917–2951.

(18) Nakanishi, M.; Mori, M. Total Synthesis of (–)-Strychnine. *Angew. Chem., Int. Ed. Engl.* **2002**, *41*, 1934–1936.

(19) Mori, M.; Nakanishi, M.; Kajishima, D.; Sato, Y. A New and General Synthetic Pathway to *Strychnos* Indole Alkaloids: Total Syntheses of (–)-Dehydrotubifoline and (–)-Tubifoline by Palladium-Catalyzed Asymmetric Allylic Substitution. *Org. Lett.* **2001**, *3*, 1913–1916.

(20) Hanessian, S.; Ugolini, A.; Hodges, P. J.; Beaulieu, P.; Dube, D.; Andre, C. Progress in Natural Product Chemistry by the Chiron and Related Approaches—Synthesis of Avermectin B_{1a}. *Pure Appl. Chem.* **1987**, *59*, 299–316.

(21) Hanessian, S. Design and Implementation of Tactically Novel Strategies for Stereochemical Control Using the Chiron Approach. *Aldrichimica Acta* **1989**, *22*, 3–14.

(22) Hanessian, S.; Franco, J.; Larouche, B. The Psychobiological Basis of Heuristic Synthesis Planning—Man, Machine, and the Chiron Approach. *Pure Appl. Chem.* **1990**, *62*, 1887–910.

(23) Hanessian, S.; Franco, J.; Gagnon, G.; Laramee, D.; Larouche, B. Computer-Assisted Analysis and Perception of Stereochemical Features in Organic Molecules Using the CHIRON Program. *J. Chem. Inf. Comput. Sci.* **1990**, *30*, 413–425.

(24) Hanessian, S. Reflections on the Total Synthesis of Natural Products: Art, Craft, Logic, and the Chiron Approach. *Pure Appl. Chem.* **1993**, *65*, 1189–1204.

(25) Thompson, C. F.; Jamison, T. F.; Jacobsen, E. N. Total Synthesis of FR901464. Convergent Assembly of Chiral Components Prepared by Asymmetric Catalysis. *J. Am. Chem. Soc.* **2000**, *122*, 10482–10483.

(26) Thompson, C. F.; Jamison, T. F.; Jacobsen, E. N. FR901464: Total Synthesis, Proof of Structure, and Evaluation of Synthetic Analogues. *J. Am. Chem. Soc.* **2001**, *123*, 9974–9983.

(27) Matsumura, K.; Hashiguchi, S.; Ikariya, T.; Noyori, R. Asymmetric Transfer Hydrogenation of α,β-Acetylenic Ketone. *J. Am. Chem. Soc.* **1997**, *119*, 8738–8739.

(28) Schaus, S. E.; Brånalt, J.; Jacobsen, E. N. Total Synthesis of Muconin by Efficient Assembly of Chiral Building Blocks. *J. Org. Chem.* **1998**, *63*, 4876–4877.

(29) Keith, J. M.; Larrow, J. F.; Jacobsen, E. N. Practical Considerations in Kinetic Resolution Reactions. *Adv. Synth. Catal.* **2001**, *343*, 5–26.

(30) Sawada, D.; Kanai, M.; Shibasaki, M. Enantioselective Total Synthesis of Epothilones A and B Using Multifunctional Asymmetric Catalysis. *J. Am. Chem. Soc.* **2000**, *122*, 10521–10532.

(31) Nicolaou, K. C.; Yue, E. W.; Naniwa, Y.; De Riccardis, F.; Nadin, A.; Leresche, J. E.; La Greca, S.; Yang, Z. Zaragozic Acid A/Squalestatin S1: Synthetic and Retrosynthetic Studies. *Angew. Chem., Int. Ed. Engl.* **1994**, *33*, 2184–2187.

(32) Nicolaou, K. C.; Nadin, A.; Leresche, J. E.; La Greca, S.; Tsuri, T.; Yue, E. W.; Yang, Z. Synthesis of the First Fully Functionalized Core of the Zaragozic

Acids/Squalestatins. *Angew. Chem., Int. Ed. Engl.* **1994**, *33*, 2187–2190.

(33) Nicolaou, K. C.; Nadin, A.; Leresche, J. E.; Yue, E. W.; La Greca, S. Total Synthesis of Zaragozic Acid A/Squalestatin S1. *Angew. Chem., Int. Ed. Engl.* **1994**, *33*, 2190–2191.

(34) Kolb, H. C.; VanNieuwenhze, M. S.; Sharpless, K. B. Catalytic Asymmetric Dihydroxylation. *Chem. Rev.* **1994**, *94*, 2483–2547.

(35) Ohkuma, T.; Ishii, D.; Takeno, H.; Noyori, R. Asymmetric Hydrogenation of Amino Ketones Using Chiral RuCl$_2$(diphophine)(1,2-diamine) Complexes. *J. Am. Chem. Soc.* **2000**, *122*, 6510–6511.

(36) Trost, B. M.; Lee, C. *gem*-Diacetates as Carbonyl Surrogates for Asymmetric Synthesis. Total Syntheses of Sphingofungins E and F. *J. Am. Chem. Soc.* **2001**, *123*, 12191–12201.

(37) Mulrooney, C. A.; Li, X.; DiVirgilio, E. S.; Kozlowski, M. C. General Approach for the Synthesis of Chiral Perylenequinones via Catalytic Enantioselective Oxidative Biaryl Coupling. *J. Am. Chem. Soc.* **2003**, *125*, 6856–6857.

(38) Willis, M. C. Enantioselective Desymmetrisation. *J. Chem. Soc., Perkin Trans. 1* **1999**, 1765–1784.

(39) Rychnovsky, S. D. Oxo Polyene Macrolide Antibiotics. *Chem. Rev.* **1995**, *95*, 2021–2040.

(40) Schreiber, S. L.; Goulet, M. T.; Schulte, G. Two-Directional Chain Synthesis: The Enantioselective Preparation of Syn-Skipped Polyol Chains from *Meso* Precursors. *J. Am. Chem. Soc.* **1987**, *109*, 4718–4720.

(41) Burke, S. D.; Buchanan, J. L.; Rovin, J. D. Synthesis of a C(22)–C(34) Halichondrin Precursor via a Double Dioxanone-to-Dihydropyran Rearrangement. *Tetrahedron Lett.* **1991**, *32*, 3961–3964.

(42) Poss, C. S.; Schreiber, S. L. Two-Directional Chain Synthesis and Terminus Differentiation. *Acc. Chem. Res.* **1994**, *27*, 9.

(43) Smith, D. B.; Wang, Z. Y.; Schreiber, S. L. The Asymmetric Epoxidation of Divinyl Carbinols—Theory and Applications. *Tetrahedron* **1990**, *46*, 4793–4808.

(44) Schreiber, S. L.; Schreiber, T. S.; Smith, D. B. Reactions that Proceed with a Combination of Enantiotopic Group and Diastereotopic Face Selectivity Can Deliver Products with Very High Enantiomeric Excess: Experimental Support of a Mathematical Model. *J. Am. Chem. Soc.* **1987**, *109*, 1525–1529.

(45) Davies, H. M. L.; Venkataramani, C.; Hansen, T.; Hopper, D. W. New Strategic Reactions for Organic Synthesis: Catalytic Asymmetric C–H Activation α to Nitrogen as a Surrogate for the Mannich Reaction. *J. Am. Chem. Soc.* **2003**, *125*, 6462–6468.

(46) Crispino, G. A.; Ho, P. T.; Sharpless, K. B. Selective Perhydroxylation of Squalene: Taming the Arithmetic Demon. *Science* **1993**, *259*, 64–66.

(47) Lebsack, A. D.; Link, J. T.; Overman, L. E.; Stearns, B. A. Enantioselective Total Synthesis of Quadrigemine C and Psycholeine. *J. Am. Chem. Soc.* **2002**, *124*, 9008–9009.

(48) Ishibashi, H.; Ishihara, K.; Yamamoto, H. A New Artificial Cyclase for Polyprenoids: Enantioselective Total Synthesis of (–)-Chromazonarol, (+)-8-*epi*-Puupehedione, and (–)-11′-Deoxytaondiol Methyl Ether. *J. Am. Chem. Soc.* **2004**, *126*, 11122–11123.

(49) Takaya, H.; Ohta, T.; Sayo, N.; Kumobayashi, H.; Akutagawa, S.; Inoue, S.; Kasahara, I.; Noyori, R. Enantioselective Hydrogenation of Allylic and Homoallylic Alcohols. *J. Am. Chem. Soc.* **1987**, *109*, 1596–1597.

(50) Lebel, H.; Jacobsen, E. N. Enantioselective Total Synthesis of Taurospongin A. *J. Org. Chem.* **1998**, *63*, 9624–9625.

(51) Katsuki, T. In *Comprehensive Asymmetric Catalysis*; Jacobsen, E. N., Pfaltz, A., Yamamoto, H., Eds.; Springer-Verlag: Berlin, 1999; Vol. II, pp 621–648.

(52) Jacobsen, E. N.; Wu, M. H. In *Comprehensive Asymmetric Catalysis*; Jacobsen, E. N., Pfaltz, A., Yamamoto, H., Eds.; Springer-Verlag: Berlin, 1999; Vol. II, pp 649–677.

(53) Shi, Y. Organocatalytic Asymmetric Epoxidation of Olefins by Chiral Ketones. *Acc. Chem. Res.* **2004**, *37*, 488–496.

(54) Hoard, D. W.; Moher, E. D.; Martinelli, M. J.; Norman, B. H. Synthesis of Cryptophycin 52 Using the Shi Epoxidation. *Org. Lett.* **2002**, *4*, 1813–1815.

(55) Trost, B. M.; Gunzner, J. L.; Dirat, O.; Rhee, Y. H. Callipeltoside A: Total Synthesis, Assignment of the Absolute and Relative Configuration, and Evaluation of Synthetic Analogues. *J. Am. Chem. Soc.* **2002**, *124*, 10396–10415.

(56) Fujii, K.; Maki, K.; Kanai, M.; Shibasaki, M. Formal Catalytic Asymmetric Total Synthesis of Fostriecin. *Org. Lett.* **2003**, *5*, 733–736.

Appendix A

Terms and Enantioselective Processes in Asymmetric Catalysis

As a prerequisite to any useful discussion of asymmetric catalysis, a common means of describing various asymmetric groups and chirality types is required. In this appendix the definitions of chiral substances, their precursors, and their nomenclature are outlined. Different types of chiral compounds possessing a variety of stereogenic units, including central, axial, and planar chirality, are outlined. Achiral precursors that are prochiral, meaning that they can be transformed into enantiomerically pure chiral compounds, are delineated. The generation of pure enantiomers from racemic mixtures is also presented. Examples of the creation of each type of enantiomerically pure stereogenic unit are outlined. The appendix concludes with strategies for the creation of multiple stereogenic units in a controlled manner to generate single enantiomeric and diastereomeric compounds.

A.1 Definitions

Comprehensive and detailed analyses of stereochemistry have been published.[1–4] In the following sections, overviews of the main concepts as they pertain to catalytic asymmetric synthesis are provided.

A.1.1 Types of Chirality

A.1.1.a Stereogenic Units

Any object is considered chiral if it is nonsuperimposable with its mirror image. For example, (R)-alanine is chiral, as it is distinct from its mirror image, (S)-alanine (**Figure A.1**). In terms of symmetry, chiral structures are characterized by a lack of rotation–reflection axes (S-axes). For example, molecules belonging to the C_n or D_n space groups are chiral. Chiral molecules contain defined stereogenic units which confer this property. A stereogenic unit is the smallest portion of a structure that renders its mirror image nonsuperimposable. Notably, the presence of a stereogenic unit is in and of itself an insufficient condition for chirality. The most frequently encountered stereogenic unit is an sp^3 center with four different substituents. Compounds containing one such stereogenic unit are referred to as centrochiral and the center is referred to as a centrochiral unit or a stereogenic center. Other types of stereogenic units that cause molecules to exhibit axial chirality (axial chiral unit, stereogenic axis), as shown in BINAP,[5] helical chirality (helical chiral unit, stereogenic helix), as shown in a tetra-helicene, and planar chirality (planar chiral unit, stereogenic plane), as shown in a ferrocenyl phosphine,[6] also exist and play important roles in asymmetric catalysis.

If the nonsuperimposable mirror images of a compound are interconvertible at room temperature by means of a conformational change, then the compound will by definition exist as a racemic mixture (atropisomeric mixture, atropisomers-enantiomers that are conformational isomers) and exhibit no optical activity. Such compounds are often termed achiral, although they contain chiral conformations. For example, the mirror images of *cis*-decalin can interconvert by ring inversion (**Figure A.2**), which is viable at most temperatures used in asymmetric catalysis. Another example is seen in 1,1-binaphthyl, which can undergo facile rotation about the axial bond (**Figure A.2**).

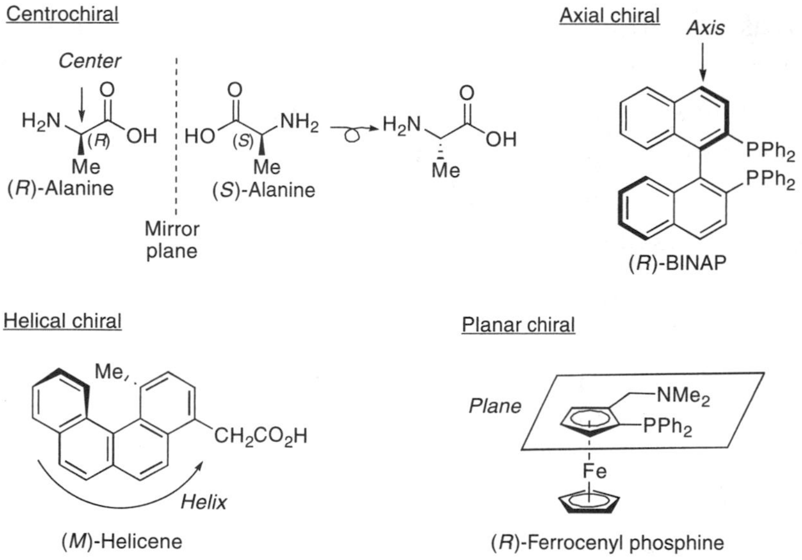

Figure A.1. Examples of different types of stereogenic units.

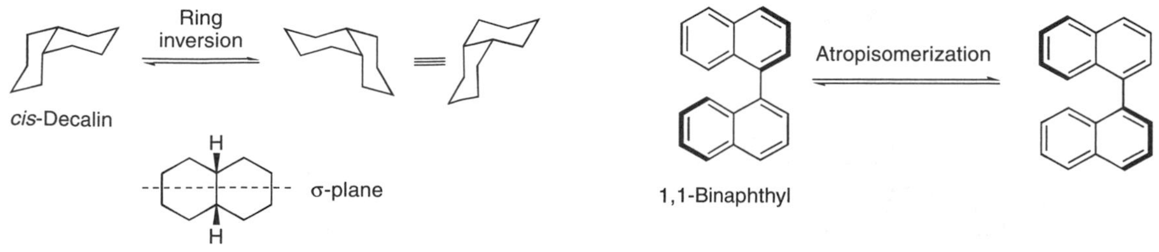

Figure A.2. Nonsuperimposable mirror-image compounds that are interconvertible at room temperature by means of a conformational change.

It is most convenient to determine if such compounds are chiral by examination of the most symmetric form of a structure. In the case of *cis*-decalin, internal σ-planes of symmetry are readily apparent in a planar representation, one of which is illustrated in **Figure A.2**. Thus, from this representation the molecule is classified as achiral.

A.1.1.b *Centrochirality*

The most common centrochiral unit is an *sp³* hybridized carbon center with four different substituents. The substituents can be arranged in two alternative ways to yield two forms of the molecule with the same composition that are nonsuperimposable mirror images. In **Figure A.3**, an example of a compound containing one centrochiral unit is shown with proline, one of the natural α-amino acids and a very useful catalyst in asymmetric synthesis.[7-10] Compounds with centrochiral centers are assigned configurations according to the Cahn–Ingold–Prelog convention.[11,12]

Spirocenters can also function as centrochiral units, as in olean, the olive fruit fly pheromone; the (R)-enantiomer is active against males and the (S)-enantiomer against females.[13,14] Compounds can contain more than one centrochiral unit. Two chiral ligands useful in asymmetric catalysis, (S,S)-t-Bu-Box[15,16] and (R,R)-Ph-bod*,[17] illustrate this principle by incorporating two centrochiral units. These compounds are attractive in asymmetric synthesis due to an element of C_2-symmetry. As a result, fewer distinct substrate coordination modes and transition states are possible, greatly simplifying any analysis (see Chapter 4 for further discussion). However, there is no requirement that useful compounds in asymmetric catalysis possess C_2-symmetry. For example, proline is quite useful.[7-10] In addition, quinine, which possesses five centrochiral units and is non-C_2-symmetric, has numerous applications in asymmetric catalysis.[18]

Centrochiral centers can originate from atoms other than carbon and examples of compounds with centrochiral silicon, nitrogen, phosphorous, and sulfur centers have been reported. For silicon, only a small number of silicon-centered chiral silanes[19-22]

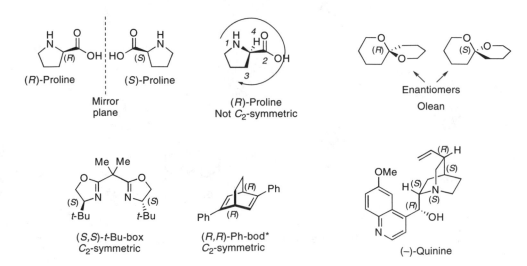

Figure A.3. Compounds containing one, two, and more centrochiral units.

have been generated (**Figure A.4**). Their utility is increasing; in particular, hydrosilane **B** provides high levels of selectivity (≥ 99:1 dr and 99% ee) in asymmetric hydrosilylation.[21,22]

In common practice, only units that are configurationally stable at room temperature are assigned as stereogenic units. Thus, few compounds containing a neutral sp^3 hybridized nitrogen with four different groups (here the nitrogen lone pair is considered as a group) as the only stereogenic unit are considered chiral, since rapid equilibration occurs between the two different forms of most trisubstituted amines (**Figure A.5**).[23] Amine derivatives with relatively stable nitrogen stereogenic units due to small rings include oxaziridines (**Figure A.5**). Alternatively bonding of the nitrogen lone pair to an electrophile can render the nitrogen stereogenic unit stable, as is the case for N-oxides and quaternary ammonium salts (**Figure A.5**). Enantiomerically pure versions of these compounds have found uses as chiral reagents and catalysts. Examples include the asymmetric oxidizing reagent Davis' oxaziridine,[24] prolyl N-oxide catalyst ligand,[25] and the asymmetric cinchonidium and spiro phase-transfer catalysts (PTCs),[26] where association of the stereogenic cationic N-center with an anionic substrate (i.e., an enolate) plays a key role. Metal complexation also can stabilize an N-centrochiral unit. In most cases, other stereogenic units present in an organic ligand cause a specific N-configuration to form upon metal complexation or equilibration via a relay effect (see Chapter 6). However, it is also feasible for an otherwise achiral amine ligand to form a stable chiral metal complex with an N-centrochiral unit, as is the case for the (R,R)-Pd complex[27] illustrated in **Figure A.5**.

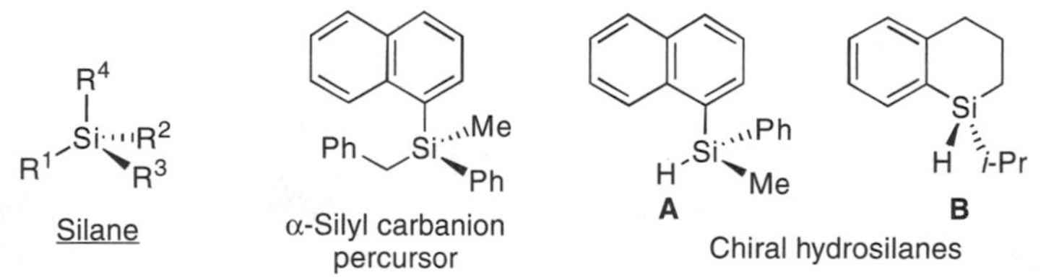

Figure A.4. Compounds containing centrochiral units at silicon.

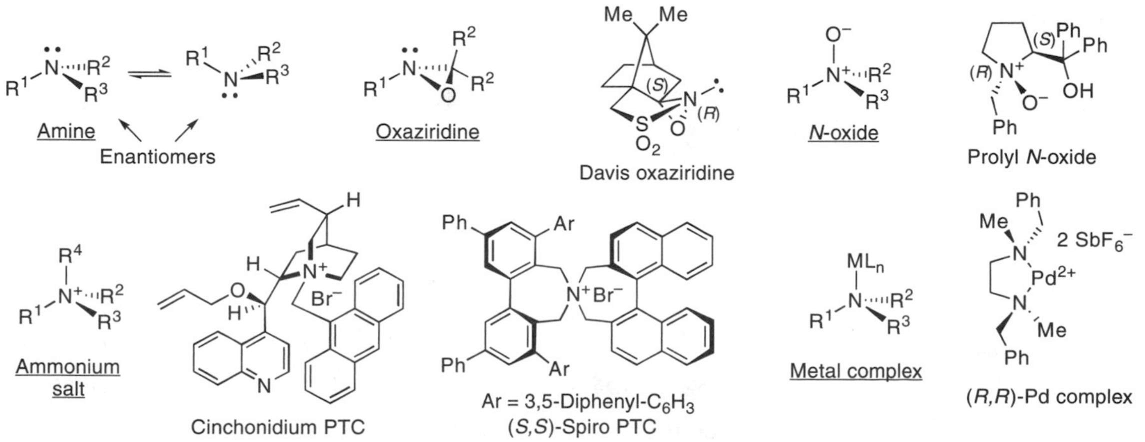

Figure A.5. Compounds containing centrochiral units at nitrogen.

The barrier to inversion at phosphorous and sulfur is usually considerably higher than in the case of nitrogen. Thus, such atoms can act as centrochiral units if four different groups are present (where the lone pair is again counted as a group). *P*-Chiral phosphine compounds[28–31] are particularly useful in transition-metal-catalyzed asymmetric reactions and include compounds such as CAMP and DIPAMP (**Figure A.6**). The corresponding phosphine oxides, phosphine sulfides, phosphonium salts, phosphine boranes, and phosphine metal complexes are also chiral (**Figure A.6**). Furthermore, these latter compounds can be generated from or converted to the corresponding chiral phosphines with excellent stereochemical fidelity.

Compounds with stereogenic centers at sulfur include the sulfoxides,[32,33] sulfoximines,[34,35] and sulfonium salts (**Figure A.7**).[36] Of these, the sulfoxides are of considerable importance in the pharmaceutical industry and in asymmetric synthesis. The top drug in sales in 2000 was Prilosec (US $6.2 billion) which contains a centrochiral unit in the form of a sulfoxide. This compound is a gastric proton-pump inhibitor used as an antiulcer agent and there was considerable interest in the single enantiomer formulation which is now sold under the name Nexium.[37,38] There are a wide

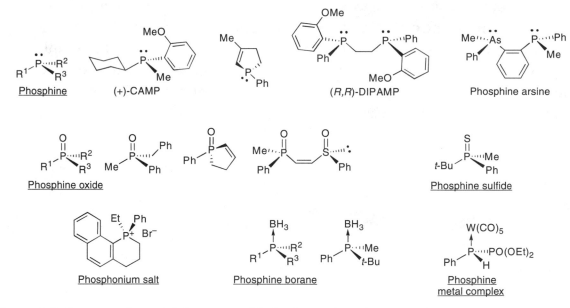

Figure A.6. Compounds containing centrochiral units at phosphorous.

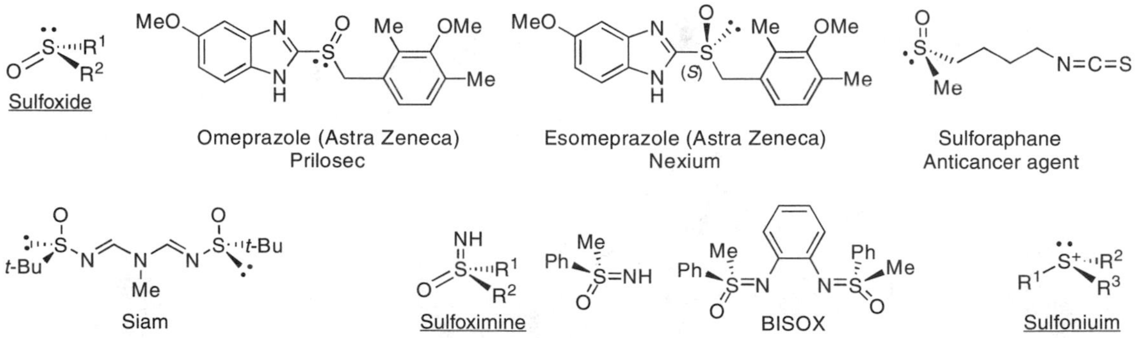

Figure A.7. Compounds containing centrochiral units at sulfur.

range of further industrially relevant chiral sulfoxides, including the anticancer agent sulforaphane illustrated in **Figure A.7**, as well as compounds that are flavors and aroma precursors, antibiotics, regulators of cholesterol catabolism, potassium-channel activators, calcium-channel antagonists, immunosuppressive agents, and platelet-adhesion inhibitors.[32] Unlike most other compounds containing heteroatom-centered chirality, highly selective methods for catalytic asymmetric sulfoxide synthesis have been developed[32] (see **Figure A.42** for an example).

Chiral sulfoxides and sulfoximines have also been employed as ligands in asymmetric catalysis (**Figure A.7**). Examples include the bis(sulfoxide) siam ligand and the bis(sulfoximine) BISOX ligands, which have been employed in copper-catalyzed Diels–Alder[39] and hetero-Diels–Alder reactions,[40] respectively.

Metal centers can also be centrochiral units,[41–45] even when more than four substituents are present (**Figure A.8**). While enantiomers are not observed in square-planar complexes with achiral ligands due to the symmetry, enantiomers are possible with the common regular geometries—namely, tetrahedral, trigonal bypyramidal, square pyramidal, and octahedral. Due to the greater number of substituents in the five- and six-coordinate geometries, a correspondingly larger number of stereo-isomers is possible, which varies depending on how many of the substituents are identical. For example, there are 20, 30, and 30 stereoisomers possible for the trigonal-bypyramidal, square-pyramidal, and octahedral geometries, respectively, if all the substituents are different. Nomenclature rules have been developed for these different isomers based upon clockwise (C) and anticlockwise (A) orientation of substituents but are seldom used in asymmetric catalysis. Some examples are given in **Figure A.8** and the rules are described in detail elsewhere.[43] Enantiomers are also possible with irregular geometries (i.e., distorted square planar, etc.).

As was the case for the tetrahedral centrochiral units described above, none of the ligands need to be chiral in order for metal centrochirality to occur. An example can be found in the acyl iron complex[46] illustrated in **Figure A.8**, which is regarded as

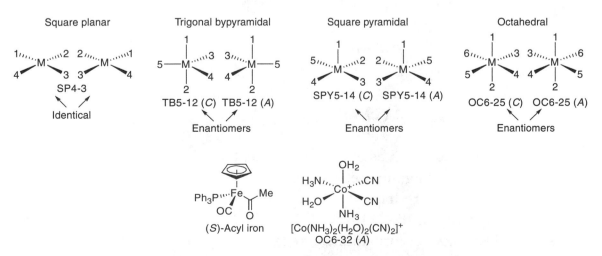

Figure A.8. Compounds containing centrochiral units at metal centers.

pseudotetrahedral and assigned the (*S*)-configuration (the cyclopentadienyl ligand is the highest priority group). Such complexes are configurationally stable and relatively inert to further complexation. Thus, their use in asymmetric synthesis has been predominantly as chiral auxiliaries—in this case, for asymmetric enolate alkylation.[46] Enantiomers within five-coordinate geometries are possible, but have fewer applications due to the dynamic flexibility of the coordination sphere (i.e., Berry pseudorotation), which allows facile interconversion between enantiomers. The octahedral complexes, however, are often configurationally stable and numerous instances of metal-centered chirality have been described. For example, the complex $[Co(NH_3)_2(H_2O)_2(CN)_2]^+$ has been isolated in enantiomerically pure form.[47] Note that metal centrochirality still occurs even when some of the ligands are identical. This configurational stability allows metal-centered chirality to impart selectivity in synthetic processes, as evidenced by the frequent appearance of octahedral metal complexes in asymmetric catalysis.

With chelating ligands, it is possible to have all the ligands identical (i.e., homoleptic system) while retaining metal centrochirality due to the helical arrangements of the chelating ligands. **Figure A.9** outlines the considerations in a generic system with nonsuperimposable mirror images that are assigned as the Δ (right-handed helical twist) and Λ (left-handed helical twist) configurations, respectively. The simplest variant of this phenomenon occurs with planar chelating ligands that are achiral and symmetric, as illustrated with $[Fe(bipyridine)_3]^{2+}$.[48]

Additional conformational isomerism is seen with ethylenediamine and other nonplanar chelating ligands (achiral and chiral), as illustrated in **Figure A.10**. Since each ethylenediamine can adopt either the λ (left-handed twist) or δ (right-handed twist) conformation when coordinated to a metal atom, stereoisomers arise with different ethylenediamine diamine conformations, as well as with different stereochemistries at the metal center. For example, there are four diastereomeric Δ-isomers along with the four corresponding Λ-enantiomers of the tris(ethylenediamine) $[Co(H_2NCH_2CH_2NH_2)_3]^{3+}$ complex:[49,50]

Δ (δ,δ,δ)	Λ (λ,λ,λ)
Δ (δ,δ,λ)	Λ (δ,λ,λ)
Δ (δ,λ,λ)	Λ (δ,δ,λ)
Δ (λ,λ,λ)	Λ (δ,δ,δ)

With achiral ligands such as ethylenediamine, the mixtures of these isomers are likely.[43,51] When a chiral ligand is employed, the λ- or δ-conformation is often predetermined, which greatly reduces the number of conformational isomers. For example, the axial chirality of the (*R*)-BINOL ligands in the tris(BINOLate) complex in **Figure A.10** dictate a λ-conformation at the lanthanide center.[52] As a result, only the Δ(λ,λ,λ) and the Λ(λ,λ,λ) stereochemistries are possible. These compounds are di-

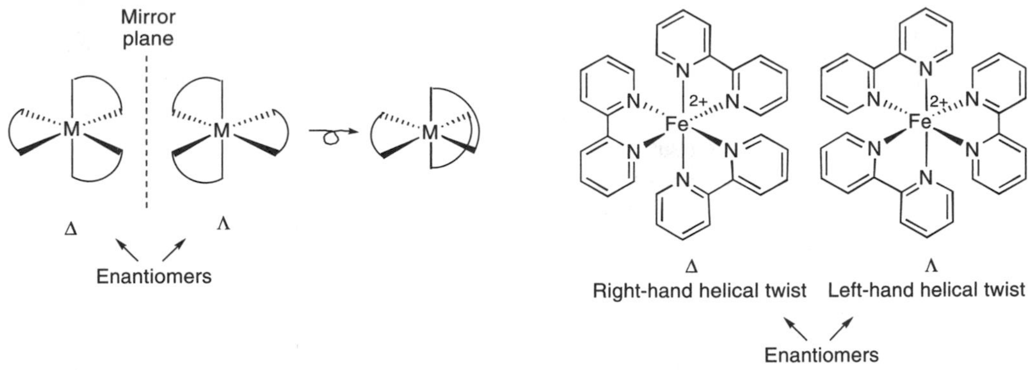

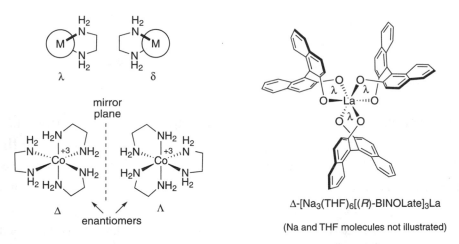

Figure A.9. Chiral metal complexes with achiral chelating ligands.

Δ-[Na₃(THF)₆[(R)-BINOLate]₃La

(Na and THF molecules not illustrated)

Figure A.10. Chiral metal complexes with conformationally "chiral" and chiral chelating ligands.

astereomeric and the more stable Δ(λ,λ,λ) predominates in this chiral catalyst. Control of these conformational isomers and how they pertain to selectivity in asymmetric catalysis is discussed further in Chapter 6. A further example of Δ- and Λ-isomers in asymmetric catalysis is presented in Chapter 12 (**Figure 12.50**).

In principle, metal centrochiral units would be most useful in asymmetric catalysis, as chemistry often occurs at the metal center itself. Two examples with only metal centrochirality are illustrated in **Figure A.11**. In the first, metal centrochirality occurs at the only metal center present (see Chapters 6 and 11 for reaction mechanism).[53] In the second, the centrochiral metal center comprises part of the ligand.[54] However, there are several major complicating factors with catalysts that only possess metal centrochiral units, including 1) efficient generation of metal centrochirality, 2) stability of metal centrochirality with respect to isomerization, and 3) the efficiency of stereochemical communication from a metal centrochiral unit to a reacting substrate. While the synthesis of metal catalysts with only metal centrochiral units has been documented (see **Figures A.9–A.11**), it is easier to employ a chiral ligand which in turn gives rise to ligand-centered chirality as well as metal-centered chirality. A second issue is that many metal complexes with metal centrochiral units are subject to

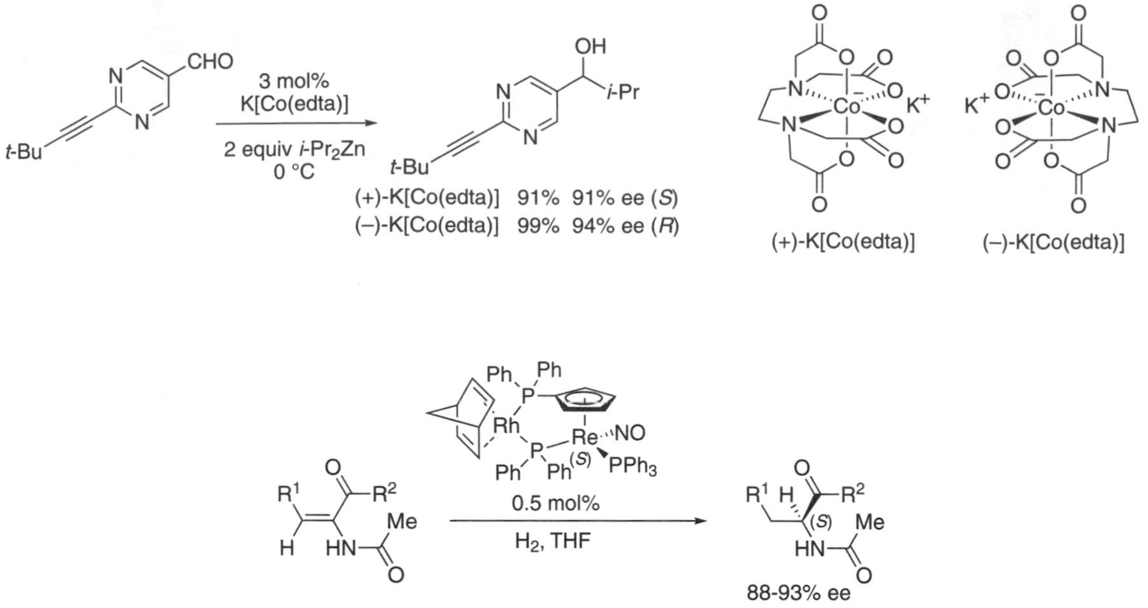

Figure A.11. Catalysts with metal centrochirality.

racemization either via low-barrier intramolecular intercoversions (i.e., square-pyramidal and trigonal-bipyramidal geometries) or via facile exchange through associative or dissociative pathways. Finally, metal centrochiral units are often distant from the prochiral reaction centers. For example, the oxygen of a carbonyl coordinated to the metal center would be most affected by local metal stereochemistry, whereas the more distal carbonyl carbon is more affected by the neighboring ligand portions. Thus, most chiral metal catalysts employ chiral ligands that create an extended asymmetric environment and which, upon substrate coordination, give rise to diastereomeric metal complexes of which one is more stable and/or more reactive. In such complexes, a metal centrochiral unit is often encountered, as illustrated in the Δ-Na$_3$(THF)$_6$[(R)-BINOL]$_3$La complex (**Figure A.10**).[52]

A.1.1.c Axial Chirality

Compound types exhibiting axial chirality include the biaryls, allenes, alkylidene cyclohexanes, and spiranes (**Figure A.12**). Although these compounds appear to be "extended" tetrahedrons, the loss of C_3-rotational symmetry means that it is no longer necessary that all four of the R-substituents be unique. It is sufficient that the pair at each end be different from each other. Assignment of configuration in axial chiral compounds follows a modified Cahn–Ingold–Prelog convention where near groups precede far groups. As illustrated in **Figure A.12**, the projections of the compounds along the chiral axis can be made from either the left side or right side without altering the final assignment. If the three highest priority groups are in a clockwise arrangement, then the (R)-configuration is obtained. If a counterclockwise arrangement occurs, then the (S)-configuration is obtained. The descriptors (aR) and (aS) can be used to distinguish axial chirality, but the prefix is optional.

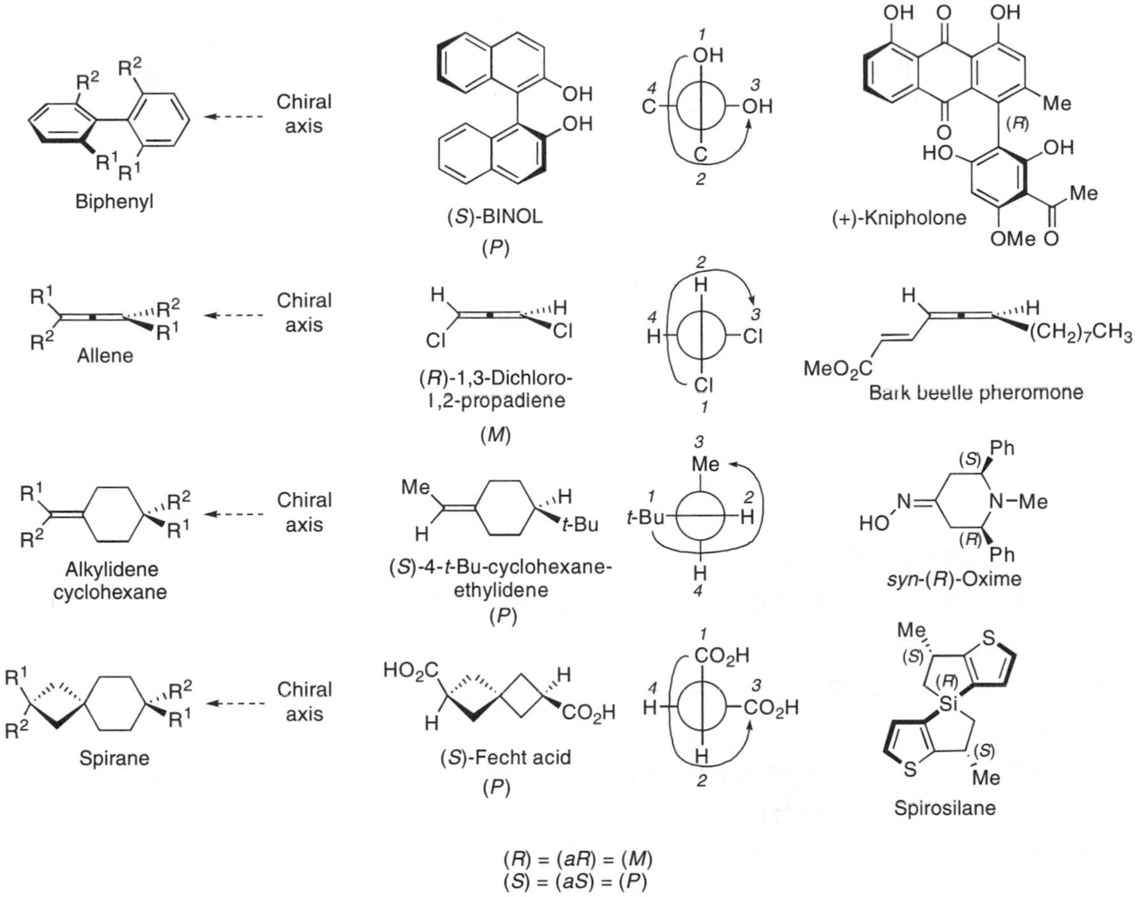

Figure A.12. Compounds exhibiting axial chirality.

Molecules with chiral axes may also be viewed as helices and their configurations denoted as (*P*) or (*M*). The symbol (*P*) stands for plus and denotes a right-handed twist (i.e., twist is clockwise moving along the axis of the helix away from the user). If the corresponding twist is counterclockwise (i.e., left-handed twist), then the symbol (*M*), which stands for minus, is used. For axial chiral compounds (**Figure A.12**), only the ligands of highest priority in the front (*1*) and back (*3*) are considered in the (*P*)- and (*M*)- assignment. If the turn from *1* to *3* is clockwise, the configuration (*P*) is assigned. If the turn is counterclockwise, (*M*) is assigned.

Next to centrochiral compounds, rotationally hindered biaryls, with their rigid chiral framework, are among the most successful reagents, ligands, and catalysts.[55–59] In addition, axial chirality is found in a number of symmetric and unsymmetric biaryl natural products,[60,61] such as the potent antimalarial (+)-knipholone (**Figure A.12**).[62–64] In this case, the biaryl bond axis is the stereogenic unit and is referred to as the axial chiral unit. Here, the two enantiomers are in principle directly convertible by a simple bond rotation and are conformational isomers. These compounds are referred to as atropisomers. If bond rotation is sufficiently hindered, the axial chiral unit can be stable at room temperature. The structural attributes needed to confer

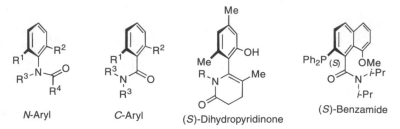

Figure A.13. Nonbiaryl atropisomeric compounds exhibiting axial chirality.

hindered rotation have been extensively examined and compiled.[3] Typically, sufficient hindrance is present if three of the four *ortho* positions of the biaryl are substituted.

Other atropisomeric compounds can contain stable axial chiral units, such as the *N*- and *C*-aryl amides[65–68] and the unusual dihydropyridinone[69] in **Figure A.13**. The application of such compounds in asymmetric catalysis is beginning to be explored. For example, the (*S*)-benzamide in **Figure A.13** has been employed in a chiral palladium catalyst in asymmetric α-allylation[70] and Heck[71] reactions. Atropisomerism can also occur about other types of bonds, such as those between sp^2–sp^3 and sp^3–sp^3 atoms, but compounds arising from such hindered rotation have not been the subject of much study in asymmetric catalysis.

Other compounds can exhibit axial chirality, including allenes, alkylidene cyclohexanes, and spiranes. Unlike the atropisomers discussed above, these compounds are not conformational isomers; conformational isomerism is not a required characteristic of axial chiral compounds. Allenes are a subclass of the cumulenes and any cumulene with an odd number of carbons in the alkene array is potentially chiral. Chiral allenes[72] have been found in several natural products,[73] including in the male bark beetle pheromone (**Figure A.12**). Alkylidene cyclohexanes (**Figure A.12**) are unusual in that the enantiomers can be generated by switching the substituents on the sp^3 carbon or by alkene isomerization. Apparent axial chirality is seen in cis-3,5- or cis-2,6-disubstituted alkylidene cyclohexanes, although configurational assignments are made for the two centrochiral units. An example is the *syn*-(*R*) oxime[74,75] illustrated in **Figure A.12**. Chiral spiranes that display axial chirality similar to allenes or alkylidene cyclohexanes, as exemplified by the Fecht acid,[76] are relatively rare compared to those such as spirosilane (**Figure A.12**). Compounds of this latter type appear to display axial chirality, but the stereochemical assignment is made using the centrochiral unit. Notably, the C_2-symmetric spirosilane was the first axial chiral spirane to be produced using a catalytic asymmetric process (see **Figure A.55**).[77]

A.1.1.d Helical Chirality

Compounds that exhibit helical chirality[78] do not contain a specific stereogenic unit such as a center, axis, or plane. Rather, the stereogenicity arises from the macroscopic

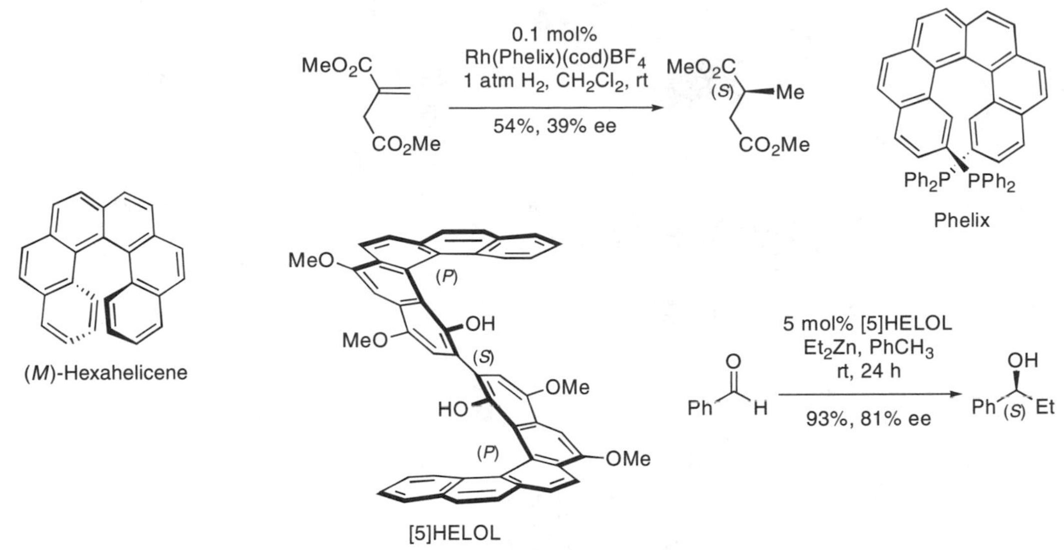

Figure A.14. Compounds exhibiting helical chirality.

structure itself. Helical chiral compounds are rare, with the helicenes such as the (M)-hexahelicine illustrated in **Figure A.14** comprising most known examples. Configurations of helical chiral compounds are assigned using the (P)- and (M)-descriptors, which denote plus and minus, respectively (see Section A.1.1.c). If the turn is clockwise, the configuration (P) is assigned; if the turn is counterclockwise, (M) is assigned. The utility of purely helical chiral compounds in asymmetric synthesis has not been widely explored, but examples using the diphosphine Phelix[79] and the diol [5]HELOL[80] are characterized by lengthy synthetic routes and low to moderate selectivity (**Figure A.14**). In the case of [5]HELOL, it is key that the two (P)-helicenes enforce the (S)-biaryl bond rather than vice versa. Macroscopic helical chirality, as induced by biaryl axial chirality in biaryl polymers, has been shown to amplify asymmetric induction in catalysis.[57]

A.1.1.e Planar Chirality

Figure A.15 illustrates several compounds exhibiting planar chirality.[81] While metal complexes[82] containing chiral planes such as the η^6-arene-Cr complex used to be considered curiosities, recent discoveries have changed that view.[83–90] The asymmetric environment afforded by an unsymmetrical plane has resulted in a number of excellent catalysts incorporating planar chirality. Examples include the ansa-bridged metallocenes, such as (ebthi)TiCl$_2$, which are useful polymerization and Lewis acid catalysts.[42] The planar chiral DMAP derivatives have found broad utility as chiral nucleophilic catalysts.[87–89] The asymmetric environment found in nonmetallocene derivatives, such as the paracyclophane[78] (R)-[2.2]PHANEPHOS, has also been shown to be useful in asymmetric catalysis.[91]

The definition of a chiral plane is less obvious than a chiral center or chiral axis. Compounds that contain neither a centrochiral nor axial chiral unit but that are

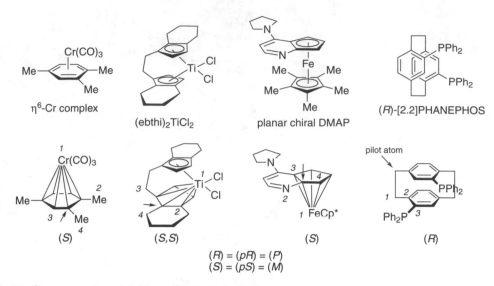

Figure A.15. Compounds exhibiting planar chirality.

chiral should be examined for a chiral plane comprised of as many of the atoms of the structure as possible. A plane can only be chiral if there is at least one substituent not contained in the plane. For example, in the η^6-arene-Cr complex, the chromium atom is not in the plane described by the arene ring.

The nomenclature of planar chiral compounds is complex.[3,92] For metallocene complexes, π-coordination is replaced by σ single bonds, as illustrated in **Figure A.15**. The configuration of the centrochiral unit is then determined from the atom of the highest precedence (marked by arrows) using the standard convention. An alternative protocol, resulting in the opposite configurational assignments, was described earlier,[93] but is not used here.

For nonmetallocenes such as [2.2]PHANEPHOS, the highest priority atom directly connected to, but outside of the plane containing the most atoms (i.e., the aromatic ring), is assigned as the pilot atom (marked by an arrow in **Figure A.15**). If the adjacent three atoms, chosen by precedence, are in a clockwise arrangement from the perspective of the pilot atom, then the (R)-configuration is assigned. If a counterclockwise arrangement is seen, then the (S)-configuration pertains. The descriptors (pR) and (pS) can be used to distinguish planar chirality, but the prefix is optional. Using the pilot atom and the three adjacent atoms, helical chirality can also be assigned using the (P)- and (M)-descriptors, which denote plus and minus, respectively (see Section A.1.1.c). If the turn from 1 to 3 is clockwise, the configuration (P) is assigned. If the turn is counterclockwise, (M) is assigned. Here, (pR) corresponds to (P) and (pS) to (M), which is opposite of the convention in axial chirality (see Section A.1.1.c).

A.1.1.f Enantiopure Compounds vs. Racemates

Compounds that are chiral, as described in the preceding sections, can exist in two enantiomeric forms. These enantiomers are constitutionally identical but differ only in the three-dimensional arrangement of their atoms. Specifically, enantiomers are mirror image forms that do not interconvert at room temperature.

If a compound exists as a single enantiomer, it is enantiomerically pure or homochiral. If a compound exists as a 1:1 mixture of the two possible enantiomers, it is racemic. The conversion of a single enantiomer into a racemic mixture is termed racemization, while the conversion of one enantiomer to the other is referred to as enantiomerization. Mixtures that are not 1:1 are referred to as scalemic or enantiomerically enriched. The ratio of two enantiomers can be quantified in terms of an enantiomeric excess (ee, **Equation A.1**) or an enantiomeric ratio (er, **Equation A.2**). Enantiomeric excess is popularly used and relates to optical purity as measured by optical rotation (**Equation A.1**). Enantiomeric ratios (**Equation A.2**), on the other hand, can be directly related to the difference in energy between the pathways to the separate enantiomers (see Chapter 1).[94] Since $\Delta G^{\ddagger}$ represents an activation energy, $\Delta\Delta G^{\ddagger}$ may not always correlate to enantiomeric ratios (see the discussion of the Curtin–Hammett paradigm in Chapter 1). This is especially true with computed transition states arising from different conformers of catalyst–substrate adducts. As such, ΔG ($\Delta G_{TS1} - \Delta G_{TS2}$) is used here to represent the energy difference between the highest transition state from each pathway (both pathways must commence from the same or isoenergetic species). The enantiomeric ratio can also be correlated to rates of formation of each enantiomer (again, rates must be measured from the same or isoenergetic starting materials). A ratio is typically used to represent the fraction in **Equation A.2** (e.g., 98:2) rather than a single number. Enantiomeric excesses or ratios can be determined by a number of analytical techniques, including chiral chromatography, NMR spectroscopy with chiral shift reagents, derivatization with chiral reagents, or optical rotation.

Equation A.1

$$\frac{\text{Enantomeric}}{\text{excess (\%)}} = \frac{(\text{Enantiomer 1}) - (\text{Enantiomer 2})}{(\text{Enantiomer 1}) + (\text{Enantiomer 2})} = \frac{\text{Optical}}{\text{purity (\%)}} = \frac{\text{Optical rotation of mixture}}{\text{Optical rotation of single enantiomer}}$$

Equation A.2

$$\frac{\text{Enantomeric}}{\text{ratio}} = \frac{(\text{Enantiomer 1})}{(\text{Enantiomer 2})} = \frac{k_{rel}(\text{Enantiomer 1})}{k_{rel}(\text{Enantiomer 2})} = e^{-\frac{\Delta G}{RT}}$$

$$\Delta G = G_{TS1} - G_{TS2}$$

Enantiomers have identical chemical and physical properties except in the presence of an external chiral influence such as plane-polarized light or another chiral molecule. For these reasons, preparative separation of enantiomers is not trivial. Thus, the generation of pure enantiomers via asymmetric catalysis is an important goal.

A.1.1.g More Than One Stereogenic Unit: Diastereomers

Stereoisomeric compounds that are not enantiomers are called diastereomers. Examples include alkenes such as *cis*-2-butene and *trans*-2-butene (**Figure A.16**). Diastereomeric compounds also arise when a compound contains at least two of the stereogenic units described in Sections A.1.1.b–A.1.1.e.

The number of stereoisomers for a given constitutional isomer is less than or equal to 2^n, where n is the number of stereogenic units. For example, tartaric acid, which contains two centrochiral units, gives rise to only three stereoisomers due to symmetry (**Figure A.16**). The isomers that are mirror images, (R,R)- and (S,S)-tartaric acid, are enantiomers. The (R,S)-compound is not the mirror image of either (R,R)- and (S,S)-tartaric acid and is hence a diastereomer of both. Not all diastereomers are chiral, even if they contain stereogenic units. This is highlighted by (R,S)-tartaric acid, which contains an internal plane of symmetry and is achiral. Such compounds are termed meso.

Another example of diastereomeric compounds can be found in the alkaloids quinine and quinidine (**Figure A.16**), which are employed frequently in asymmetric catalysis. These compounds are clearly diastereomers, but this arises solely due to the placement of the vinyl group. If this group were removed, the compounds would be enantiomers and, in fact, the compounds can be employed as pseudoenantiomers in asymmetric catalysis (for an example, see Chapter 14).

Not all diastereomeric compounds are due to the type of stereogenic units described in Sections A.1.1.b–A.1.1.e. For example, the 1,4-disubstituted cis and trans cyclohexane compounds illustrated in **Figure A.16** are clearly diastereomers, but do not contain such stereogenic units.

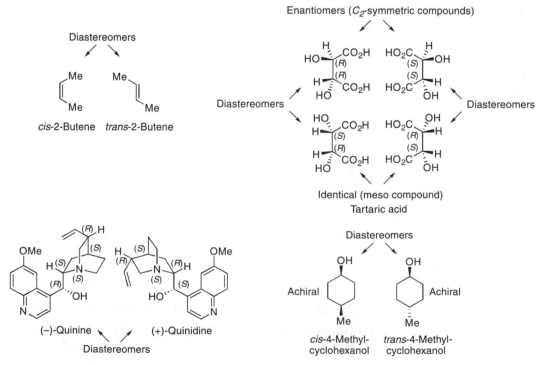

Figure A.16. Diastereomeric compounds.

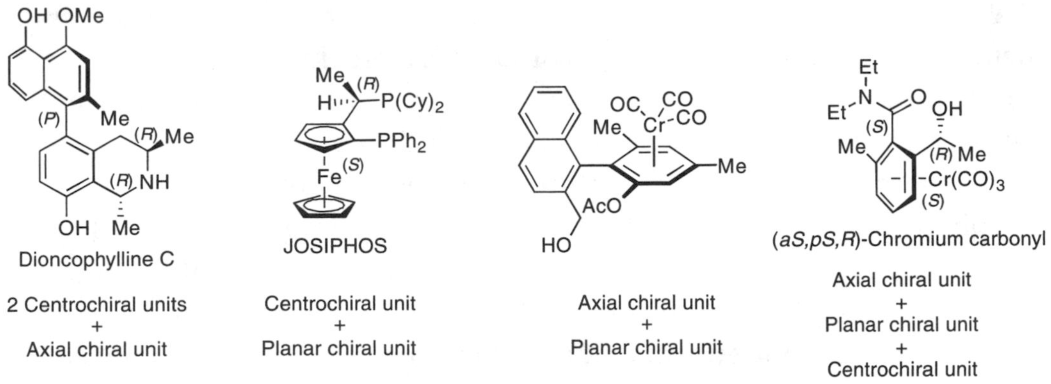

Figure A.17. Diastereomeric compounds contain different types of chiral units

The above cases illustrate diastereoisomerism arising from more than one centrochiral unit. However, any possible combination of stereogenic units can give rise to diastereomers (**Figure A.17**). Dioncophylline C, a naphthylisoquinoline alkaloid from the West African vine *Triphyophyllum peltatum* (Dioncophyllaceae),[95] is one natural-product example. This potent antimalarial agent[96] possesses two different kinds of stereogenic units, an axial chiral unit and two centrochiral units. Chiral ligands used in asymmetric catalysis, such as JOSIPHOS and related chiral ferrocene ligands,[83,84,97] also combine different stereogenic units. Three different types of chiral units can be found in the (*aS,pS,R*) chromium carbonyl.[98]

Similar to the case seen with enantiomers (see Section A.1.1.f), the amounts of diastereomers present in a mixture can be quantified using two different measures, the diastereomeric excess (de, **Equation A.3**) or the diastereomeric ratio (dr, **Equation A.4**). In contrast to enantiomeric excess, there is no physical measurement that correlates with diastereomeric excess.[94] Furthermore, diastereomeric excess is not usually useful if more than two stereogenic units are present in a molecule. As such, diastereomeric ratios are more commonly used in expressing the purity of diastereomeric mixtures.

Equation A.3

$$\text{Diastereomeric excess (\%)} = \frac{(\text{Diastereomer 1}) - (\text{Diastereomer 2})}{(\text{Diastereomer 1}) + (\text{Diastereomer 2})}$$

Equation A.4

$$\text{Diastereomeric ratio} = \frac{(\text{Diastereomer 1})}{(\text{Diastereomer 2})} = \frac{k_{rel}(\text{Diastereomer 1})}{k_{rel}(\text{Diastereomer 2})} = e^{-\frac{\Delta G}{RT}}$$

$$\Delta G = G_{TS1} - G_{TS2}$$

A.1.2 Prochiral Compounds

A compound is prochiral if replacement of one group or addition to one face gives rise to a chiral compound. In determining if a compound is prochiral or not, structures are typically examined for homotopic, enantiotopic, or diastereotopic groups or faces. Achiral structures with only homotopic groups or faces are not prochiral. Achiral structures with enantiotopic groups or faces are prochiral.

Two substituents, or groups, are homotopic (i.e., topologically equivalent) if they can be interchanged by rotation about any n-fold C_n-rotation axis to give a structure equivalent to the original. For example, the two methyl groups of *trans*-2,5-dimethylpiperidine are equivalent, as are the two hydrogens H_A and H_B of (*S,S*)-cyclohexanediamine (**Figure A.18**). Group homotopicity is easily determined by replacing each of the potential substituent in turn with an imaginary "Z" substituent. If replacement of first one substituent and then the other(s) leads to the same compound, then the substituents are equivalent (the replacement "Z" substituent must be achiral and different from any of the other substituents at the center in question).

Rapid conformational interconversion causes many groups to be homotopic even though, at first glance, they do not appear to be. For example, the H_A and H_B hydrogens in the Newman projection of *meta*-xylene do not appear equivalent (**Figure A.18**). However, all three methyl hydrogens are equivalent due to rapid rotation about the $C(sp^3)$–$C(sp^2)$ bond. This same phenomenon was observed with *trans*-2,5-dimethylpiperidine (**Figure A.18**).

Homotopic faces are characterized by a plane that contains a coplanar symmetry axis. Two faces of a molecule, usually double-bond faces, are equivalent or homotopic if they give rise to the same structure by addition of a reagent to either face. For example, reduction of cyclohexanone yields the same alcohol regardless of which face receives the hydrogen (**Figure A.19**). Similarly, hydroboration or dihydroxylation of cyclohexene from either alkene face yields the same product (**Figure A.19**).

If groups or faces are not equivalent or homotopic, then they are termed heterotopic. Heterotopic groups and faces can be divided into enantiotopic and diastereotopic classes (**Figure A.20**). Replacement of enantiotopic groups leads to enantiomeric structures, while replacement of diastereotopic groups leads to dia-

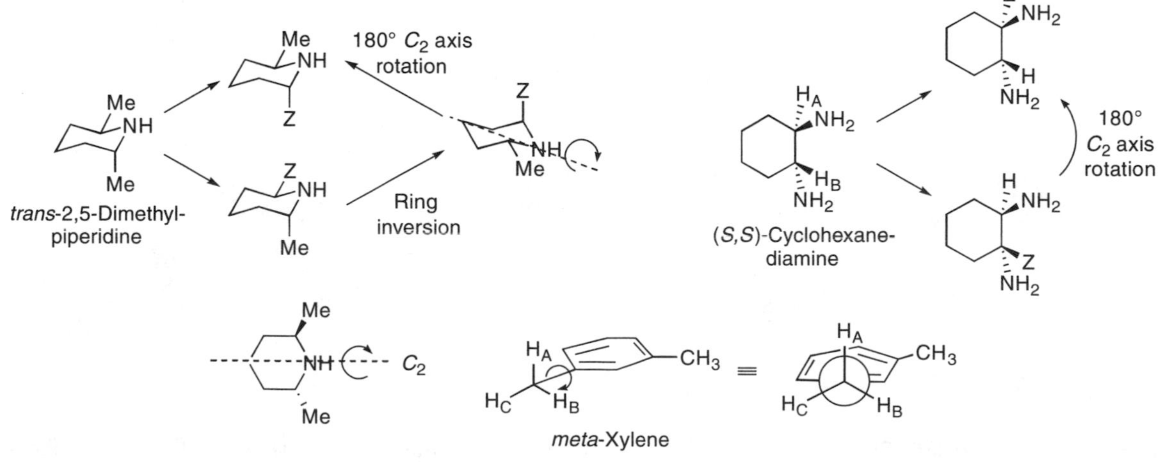

Figure A.18. Homotopic groups.

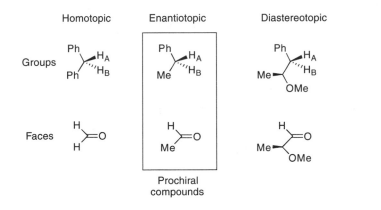

Figure A.19. Homotopic faces.

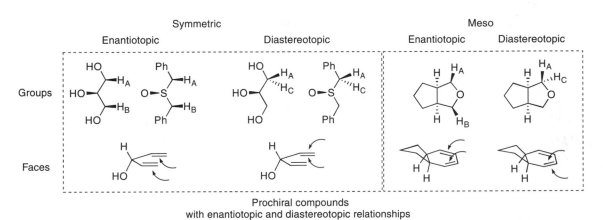

Figure A.20. Compounds with homotopic, enantiotopic, and diastereotopic groups and faces.

stereomeric structures. Similarly, additions to enantiotopic faces give rise to enantiomeric structures and additions to diastereotopic faces give rise to diastereomeric structures. Achiral compounds with enantiotopic groups or faces are thus termed prochiral, because they can be precursors in the formation of chiral compounds. Achiral symmetric compounds can possess both enantiotopic and diastereotopic groups and faces and are also prochiral compounds.

Enantiotopic groups and faces are not interchangeable by operation of any C_n (simple symmetry axis) symmetry element, but must be exchangeable by operation of a σ (plane of symmetry), i (center of symmetry), or S_n (alternating symmetry axis) element. Since chiral molecules cannot contain these latter symmetry elements, they cannot possess enantiotopic groups or faces.

A.1.2.a Compounds with Enantiotopic Groups

Enantiotopic groups are in mirror-image positions. They can be exchanged by a rotation–reflection axis. If replacement of first one group and then the other(s) leads to enantiomeric compounds, then the substituents are enantiotopic. For example, the illustrated hydrogens in *tert*-butyl ethyl ketone are enantiotopic (**Figure A.21**). Selective replacement of one of the H_A or H_B hydrogens would create a new centrochiral unit. Differentiation of enantiotopic groups can also give rise to other forms of chirality, as is the case for the biphenyl illustrated in **Figure A.21**, where the enantiomers are axial chiral.

Any type of group can be enantiotopic. In the case of the biphenyl compound in **Figure A.21**, the methyl groups Me_A and Me_B are enantiotopic. Lone pairs can also be enantiotiopic, as in the case of MeSEt (**Figure A.21**). Enantiotopic groups can be assigned as pro-(*R*) or pro-(*S*) by hypothetically assigning a higher priority to one of the groups. If assignment of a group to a higher priority results in the (*R*)-configuration, then that group is pro-(*R*), and if assignment results in the (*S*)-configuration, then it is pro-(*S*) (see **Figure A.21** for examples). Note that there is no direct correlation between pro-(*R*)/pro-(*S*) and the (*R*)/(*S*) product configuration from the reaction of prochiral substrates. This concept is illustrated in the substitution reactions of acetophenone dimethyl ketal,

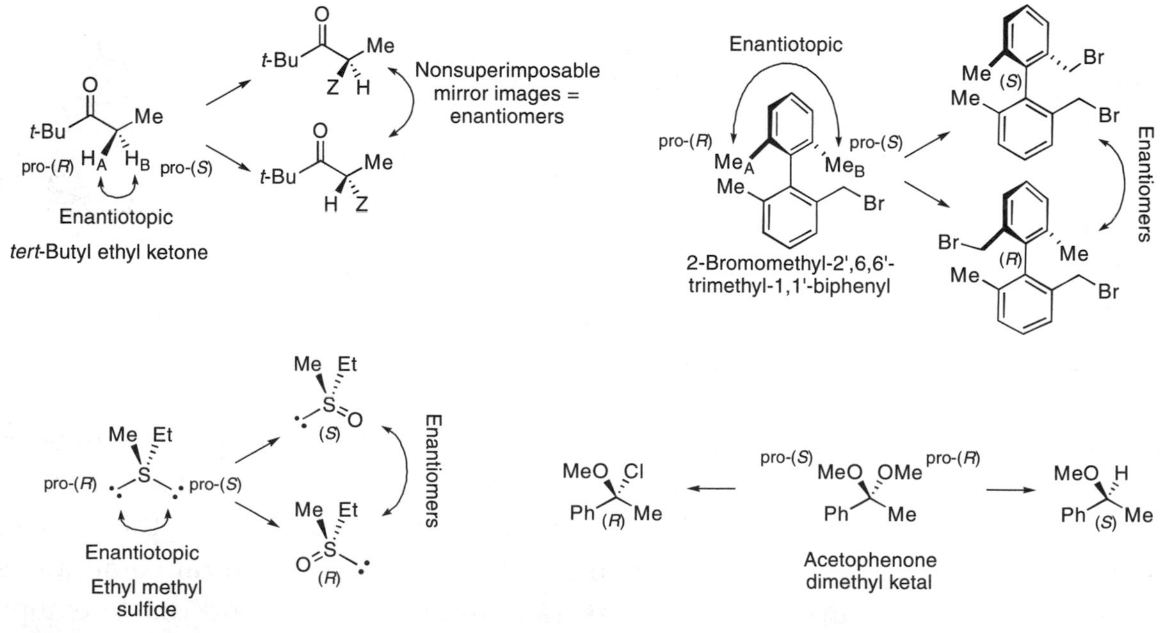

Figure A.21. Prochiral compounds with enantiotopic groups and no stereogenic units.

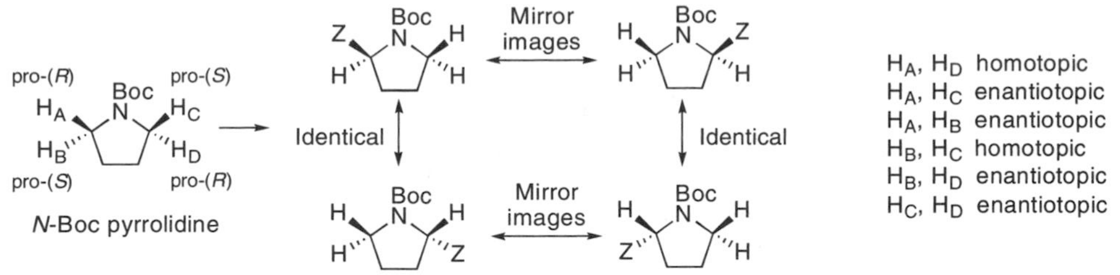

Figure A.22. A prochiral compound with more than two enantiotopic groups and no stereogenic units.

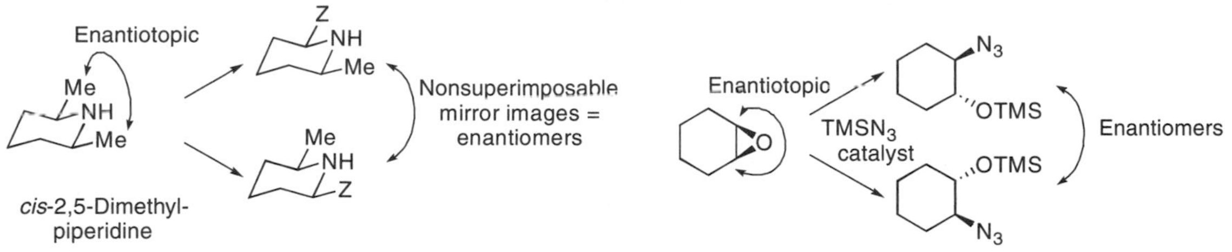

Figure A.23. Prochiral meso compounds with enantiotopic groups.

where the product configuration depends on the nucleophile employed and its relative ranking according to the Cahn–Ingold–Prelog convention (**Figure A.21**).

The arrangement of enantiotopic groups can be complex. For example, in *N*-Boc pyrrolidine (**Figure A.22**), there are four hydrogens that need to be considered. Amongst these, there are two homotopic relationships and four enantiotopic relationships. As a result, substitution of either H_A or H_D would lead to one enantiomer, whereas substitution of H_B or H_C would lead to the other enantiomer.

Compounds that are meso (i.e., that contain stereogenic units yet are achiral due to a symmetry plane or element) contain enantiotopic moieties. Most often, these are enantiotopic groups, as encountered in *cis*-2,5-dimethylpiperidine and cyclohexene oxide (**Figure A.23**). Selective reaction at one versus the other of these enantiotopic groups will lead to single enantiomers of a chiral compound. This process is termed desymmetrization[99–101] and is discussed further in Chapter 10.

A.1.2.b Compounds with Enantiotopic Faces

Similar criteria to those described above can be established for enantiotopic faces. If addition of the same achiral reagent to one face or another gives rise to enantiomeric products, then the faces are enantiotopic. For example, addition to acetophenone or *trans*-2-butene gives rise to chiral molecules containing one or two centrochiral units, respectively (**Figure A.24**). The prochiral face of the starting material at which addition occurs determines which product enantiomer is formed. Since the transition structures for either facial addition are identical with achiral reagents, both enantiomers are formed in equal amounts under these conditions. Chiral reagents or catalysts that can block addition to one face or direct addition to one face selectively can produce single-enantiomer products, because the transition structures leading to the different enantiomers are now diastereomeric (see Chapter 1).

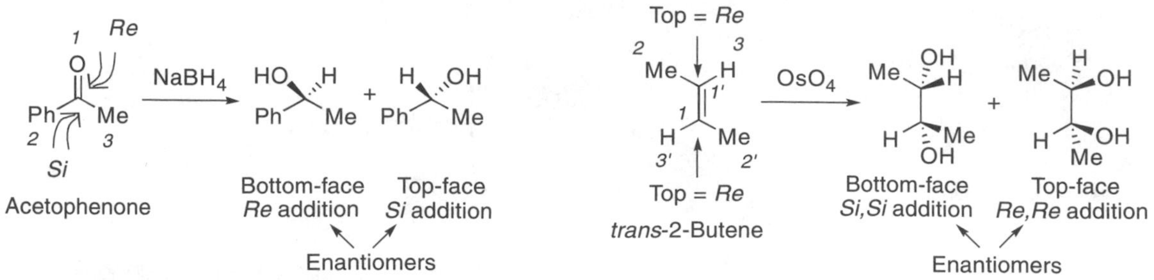

Figure A.24. Enantiotopic faces in the formation of centrochiral units.

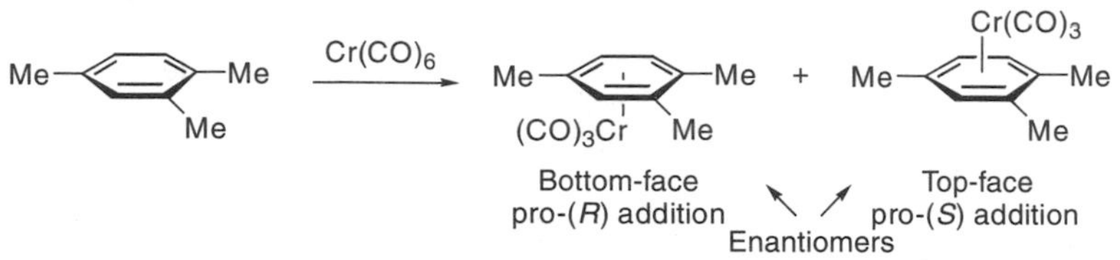

Figure A.25. Enantioface discrimination to generate a planar chiral unit.

The mode of facial attack can be described by assigning the priority of the substituents attached to the trigonal center undergoing reaction following the standard Cahn–Ingold–Prelog convention.[11,12] The face from which these three groups appear in a clockwise arrangement is termed the *Re* face, whereas the face that yields a counterclockwise arrangment is the *Si* face. Thus, in the case of acetophenone (**Figure A.24**), addition from the bottom occurs to the *Re* face and from the top to the *Si* face. Note that there is no direct correlation between (*R*)/(*S*) and *Re/Si*. If the reaction products are chiral, *Re/Si* should be used. If the reaction products are achiral (i.e. a meso product is generated), then *re/si* is appropriate.

If two new stereogenic centers are formed, then each of the two progenitor trigonal centers can be assigned as *Re* or *Si*. In the case of *trans*-2-butene both the upper and lower trigonal carbons of the alkene undergo *Re* additon when the reagent is added from the top face (**Figure A.24**). There is no requirement that the *Re/Si* descriptor be the same for both centers (see Section A.2.2.b for an example).

Prochiral enantioface discrimination can create compounds with axial and planar chiral units, as well as centrochiral units, as illustrated in **Figure A.24**. For example, the aromatic substrate in **Figure A.25** possesses enantiotopic faces. Chromiumhexacarbonyl addition gives rise to a planar chiral compound.

A.1.3 Compounds with Diastereotopic Groups and Faces

Diastereotopic groups are stereochemically distinct, yet are not mirror-image positions. They cannot be exchanged by any symmetry operation. If replacement of first one group and then the other(s) leads to the diastereomeric compounds, then substituents are diastereotopic. For example, the methylene protons of serine are diastereotopic (**Figure A.26**), because there is an extant stereogenic unit.

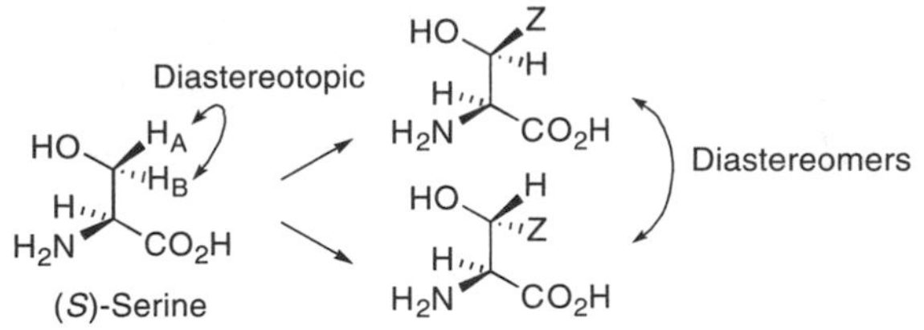

Figure A.26. Diasterotopic groups.

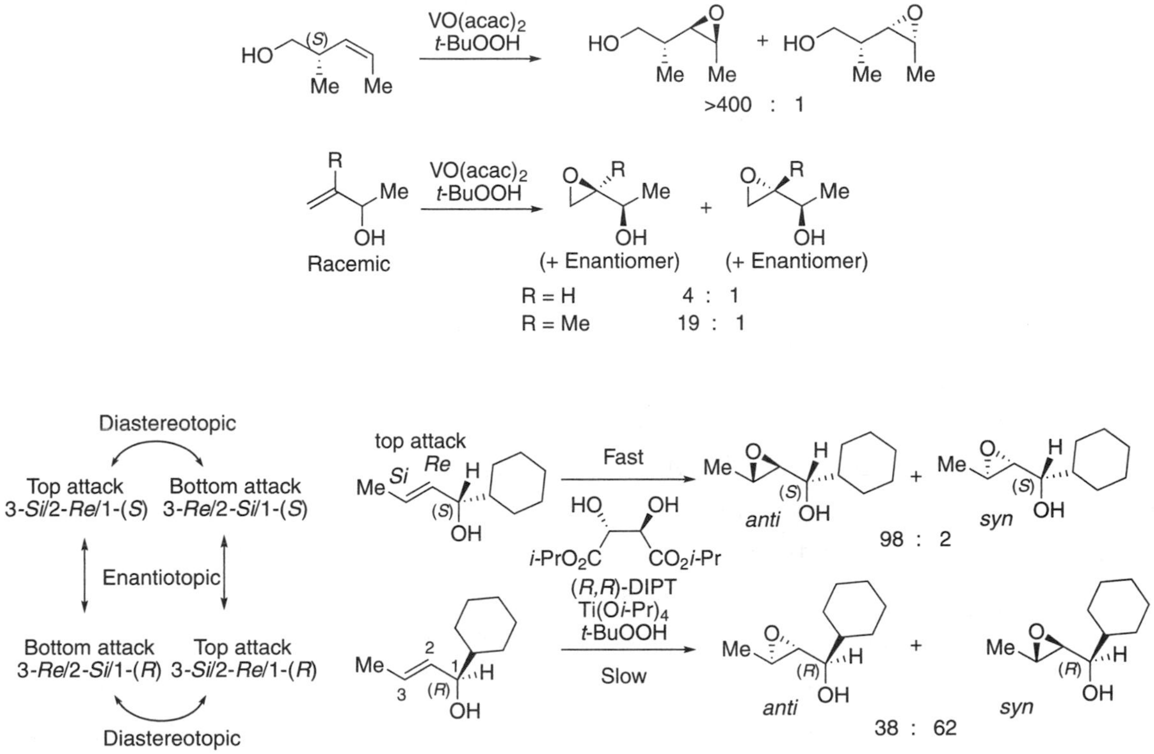

Figure A.27. Diastereotopic faces in single enantiomers and racemic mixtures.

Faces of a compound are diastereotopic if the relevant molecular plane is not planar symmetric and contains no symmetry axis. If addition of the same achiral reagent to one face or another gives rise to diastereomeric products, then the faces are diastereotopic. For example, epoxidation of enantiomerically pure (S)-2-methyl-3-cis-pentenol[102] gives rise to two new chiral molecules containing three centrochiral units (**Figure A.27**). Depending on which face of the starting material undergoes epoxidation determines which product diastereomer is formed. Hence, the two faces of the alkene are diastereotopic. Since the transition state structures for either facial addition are not identical with achiral reagents, the respective diastereomers can be (but are not required to be) formed in unequal amounts under these conditions. In this case, the diastereofacial selectivity is very high and a 400:1 ratio of product diastereomers is observed (**Figure A.27**, top).

In the epoxidation of the racemic 3-hydroxybutene (**Figure A.27**, middle),[103] similar considerations pertain, in that there are two diastereotopic faces of the alkene for each enantiomer of the starting material. Since the reagents are achiral, each enantiomer will react in a similar manner, favoring the same diastereotopic face (i.e., epoxidation *syn* to the alcohol). The relationships between enantiotopic and diasteretopic faces in racemic mixtures can be seen more clearly when chiral reagents or catalysts are employed with a racemic mixture. For example, epoxidation of a racemic mixture of (R)- and (S)-*trans*-1-cyclohexylbut-2-enol[104,105] gives rises to the *anti* (S)-product predominantly if the reaction is stopped at 50% conversion (see the discussion of kinetic resolution in Chapter 7). This result is a consequence of the diastereotopic and enantiotopic relationships between the alkene faces. Within each substrate, the two alkene faces are diastereotopic and respective addition leads to the diastereomeric *anti* and *syn* products (**Figure A.27**, bottom). Here the catalyst favors attack on the top alkene face (*Si/Re*) regardless of the centrochiral unit stereochemistry. Between the two substrates, the top and bottom alkene faces possess enantiotopic relationships. For example, top attack on the (S)-substrate gives the enantiomeric product relative to bottom attack on the (R)-substrate. Since the catalyst is chiral, these enantiotopic faces undergo epoxidation at different rates.

The symmetric achiral iron-dienyldialdehyde complex in **Figure A.28** illustrates another complicated scenario, where both enantiotopic and diastereotopic faces are present, but now in the same molecule.[106] Addition of the dialkyl zinc to the bottom face of either aldehyde results in *syn* delivery to the coordinated iron, whereas addition to the top face causes *anti* delivery. Thus, the top and bottom faces of each aldehyde are *diastereotopic*. Furthermore, addition from the bottom face to the left or right aldehyde moiety leads to different enantiomers. Thus, the two bottom aldehyde faces are *enantiotopic*. Overall, each face of each aldehyde is unique, leading to four possible products. Delivery of the dialkyl zinc is proposed to proceed via the more stable trans-aldehyde conformation from the same face as the iron carbonyl group due to dipole–dipole interactions. This control selects the diastereoface and restricts the possible products to the *syn* (R)- and *syn* (S)-enantiomers. The chiral catalyst adds another control element and restricts addition to the (S)-enantioface. The combination of these two control elements causes formation of the *syn* (S)-product [with an (R) planar chiral unit] with high enantioselectivity and diastereoselectivity. The minor enantiomer (*syn* (R)-product with an (S) planar chiral unit) is produced by bottom attack on the left carbonyl.

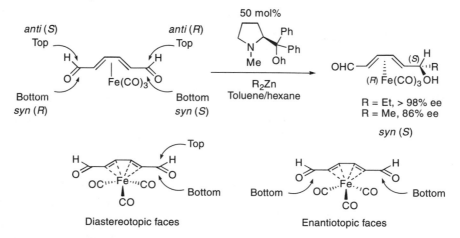

Figure A.28. Enantioface and diastereoface discrimination.

A.1.4 Describing Asymmetric Reactions

In discussing catalytic asymmetric reactions, there are different terms that allude to the net result (i.e., stereoselective) versus those that depend on the details of the process (i.e., stereospecific and chirality exchange). In this section these terms are defined and examples are provided to illustrate how the terms are used.

A.1.4.a Stereoselective vs. Stereospecific

A reaction is stereoselective if it causes preferential formation of one stereoisomer over another. Reactions can be diastereoselective (see the first[107] and second[108] reactions in **Figure A.29**), enantioselective (the third reaction[109] in **Figure A.29**), or both (the fourth reaction[109] in **Figure A.29**). A highly stereoselective reaction would lead to predominantly one compound with little of any other enantiomers or diastereomers.

A reaction is stereospecific if starting materials with different configurations lead to stereoisomerically distinct products. Accordingly, a stereospecific process

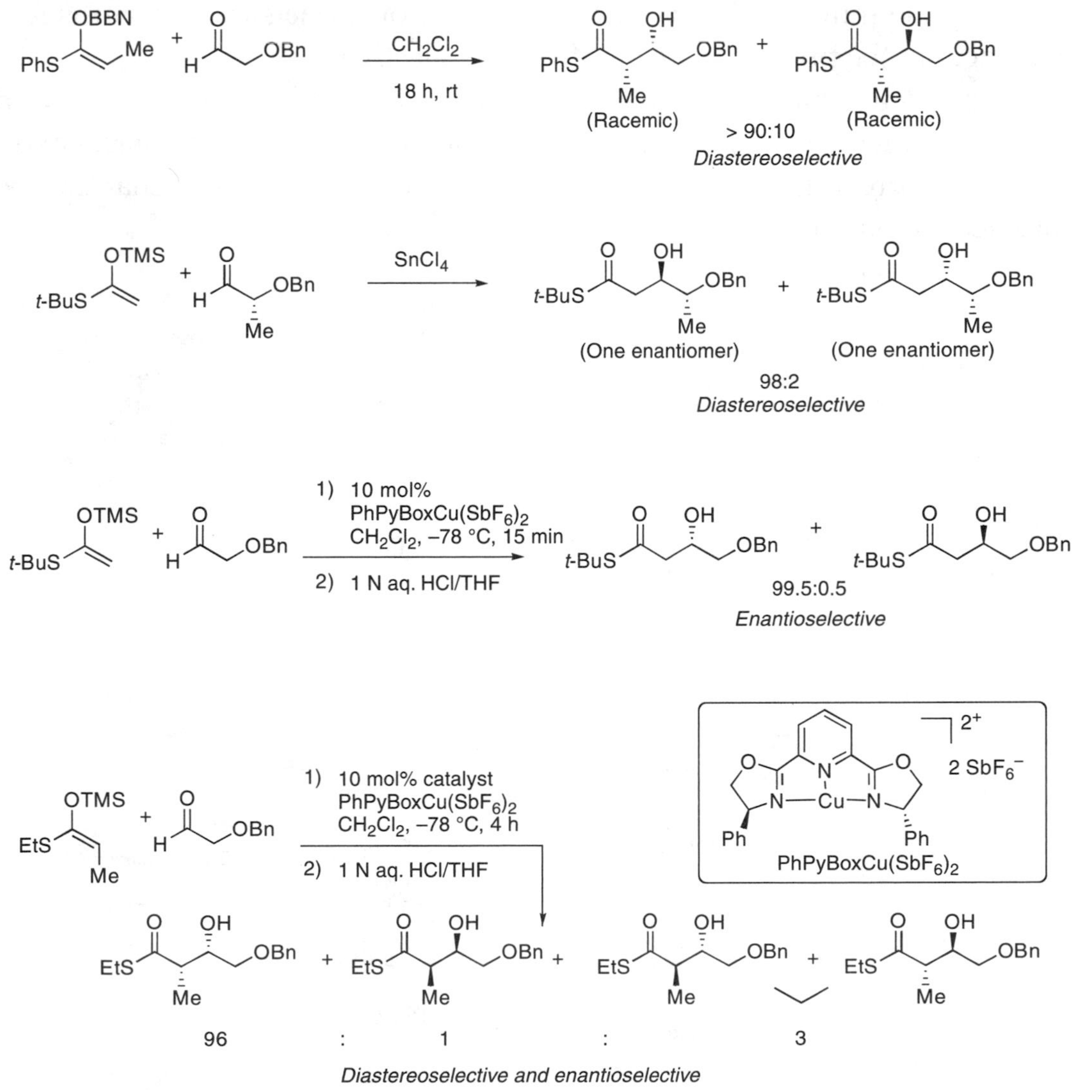

Figure A.29. Stereoselective processes.

commencing from a single stereoisomer (vs. a mixture) is necessarily stereoselective, but a stereoselective process in not necessarily stereospecific. An example would be the kinetic resolution illustrated in **Figure A.46**, which is stereospecific. In this example, the (S)-alcohol is acylated yielding the (S)-product. A counterexample would be the dynamic kinetic resolution illustrated in **Figure A.47**, which is not stereospecific. In contrast, this latter process is termed stereoconvergent, because more than one stereoisomeric starting material yields identical products. Another example of a stereoconvergent process is illustrated in the asymmetric hydrogenation of (E)- and (Z)-enamide mixtures with rhodium–DuPHOS catalysts (see the top of **Figure A.30**).[110] Both stereoisomeric starting materials converge to the same product. Thus, this process is not stereospecific, since different isomers do not yield different products, but it is stereoselective. This characteristic can be very useful if starting material isomers are difficult to separate. A related process that is both stereospecific and stereoselective is the hydrogenation of the tetrasubstituted alkenes (see the bottom of **Figure A.30**).[111] Here, the (E)- and (Z)-enamides yield different products (diastereomers) when the same catalyst is employed. Since only one stereoisomer is formed in each reaction, the processes are highly stereoselective.

Enantiospecific, enantioselective, diastereospecific, and diastereoselective have the corresponding meanings as applied to enantiomers and diastereomers. For example, each of the processes illustrated in the bottom of **Figure A.30** is enantiospecific, enantioselective, diastereospecific, and diastereoselective.

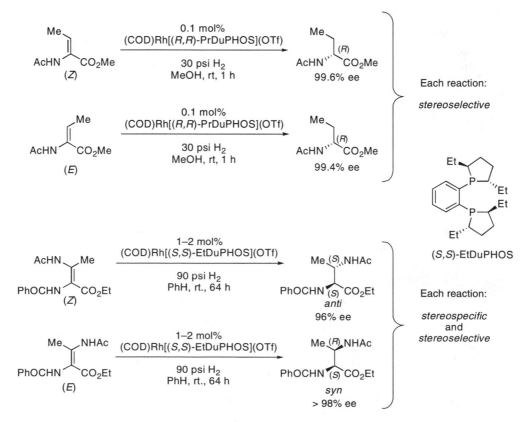

Figure A.30. Stereoselective and stereospecific processes.

In general, kinetic resolutions (see Chapter 7) are stereospecific and stereoselective (see **Figure A.46** for an example). On the other hand, dynamic kinetic resolutions (see Chapter 9) are stereoselective, but not stereospecfic, because both enantiomers convert into a single product (see **Figure A.47** for an example).

A.1.4.b Chirality Exchange

In some transformations, molecular chirality and enantiopurity can be conserved, even though events occur at the stereogenic unit. Chirality exchange or chirality transfer refers to the stereoselective formation of a new stereochemical element at the expense of another. Such reactions are also stereospecific. A classic example is the S_N2 reaction in which inversion occurs at the centrochiral unit of every molecule. If the starting material is enantio-enriched, then the product is chiral and nonracemic (**Figure A.31**).

In contrast, the S_N1 reaction does not conserve the stereochemistry of the original centrochiral unit; while the product in **Figure A.31** is still chiral, it is now racemic. This reaction is neither stereoselective or stereospecific. For the substrate in **Figure A.31**, E2 elimination results in a loss of centrochirality and the formation of an achiral compound; however, a new stereochemical element, the trans-alkene, is established. In this case, the reaction is stereoselective, but not stereospecific (i.e., the other enantiomer leads to the same trans-alkene).

Chirality exchange can also allow the transformation of one type of chirality unit for another. In the reaction of a propargylic mesylate with PhZnCl in the presence of catalytic [Pd(PPh$_3$)$_4$] (**Figure A.32**),[112] a centrochiral unit undergoes exchange to an axial chiral unit. This type of chirality exchange has also been referred to as self-immolative asymmetric synthesis.[113] This is an example of a completely stereospecific reaction (see Section A.1.4.a), since the same level of enantiomeric excess is observed in the starting material and the product. Here, the reaction proceeds via an allenyl-Pd

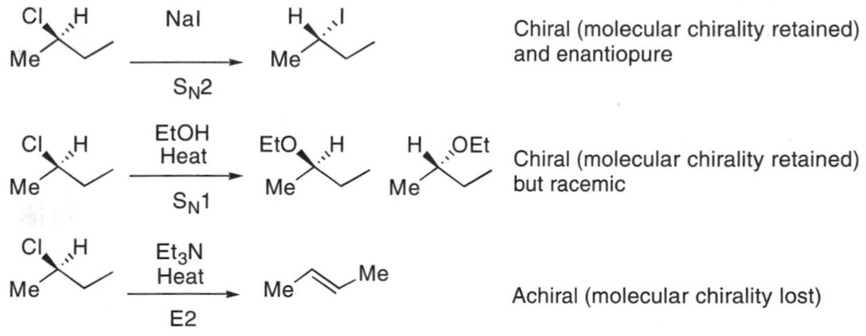

Figure A.31. Paradigms involving reaction at a stereogenic unit.

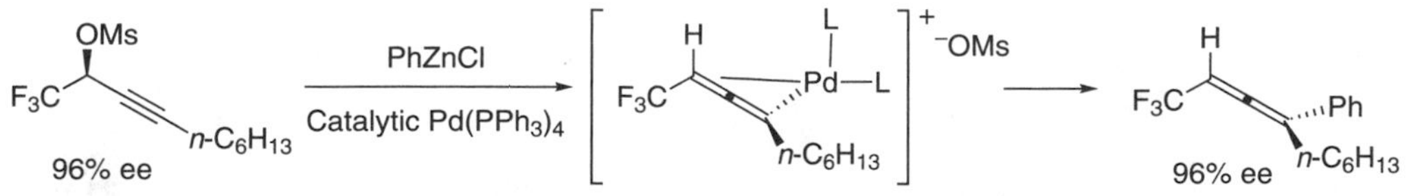

Figure A.32. Exchange of one centrochiral unit for one axial chiral unit.

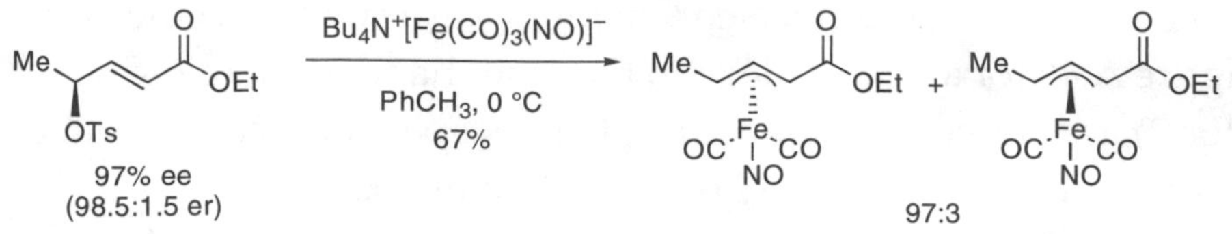

Figure A.33. Exchange of one centrochiral unit for one planar chiral unit.

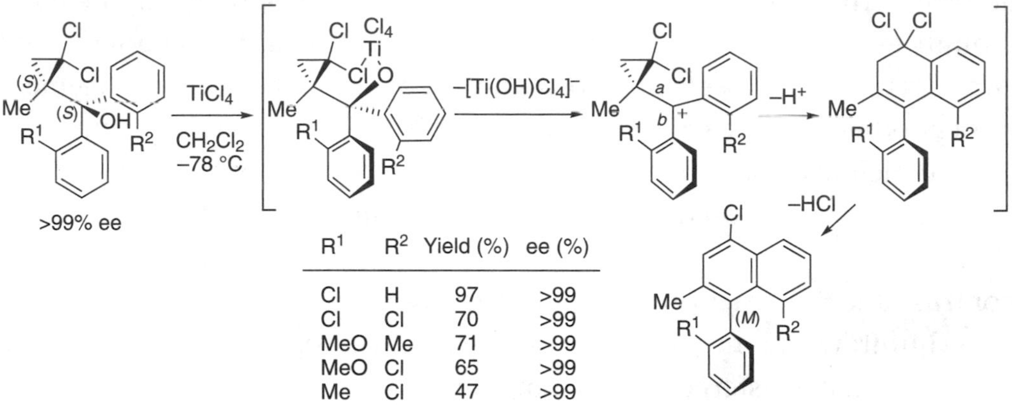

R^1	R^2	Yield (%)	ee (%)
Cl	H	97	>99
Cl	Cl	70	>99
MeO	Me	71	>99
MeO	Cl	65	>99
Me	Cl	47	>99

Figure A.34. Exchange of two centrochiral units for one axial chiral unit.

species that is configurationally stable and does not racemize via the corresponding propargylic-Pd intermediate due to the presence of the perfluorinated alkyl group.

In another example, the enantiopure allylic tosylate in **Figure A.33** undergoes stereoselective displacement to yield the η^3-iron complex arising from inversion in 94% ee.[114] In this case, a centrochiral unit undergoes exchange to a planar chiral unit. In principle, any combination of chiral units can be exchanged.

It is also possible to sacrifice more than one chiral unit to gain a new chiral unit. For example, a single-step chirality transformation from two centrochiral units to one axial chiral unit has been reported involving benzannulation of optically active aryl(aryl')-2,2-dichlorocyclopropylmethanols (AACMs) to chiral biaryls (**Figure A.34**)[115] Excellent levels of stereocontrol were observed, indicating a closely controlled chirality transfer. The proposed mechanism of the chirality transfer first invokes $TiCl_4$ chelation with the oxygen and chlorine of the substrate to give a rigid intermediate. Due to steric repulsion, the *ortho* substituent (R^1) orients itself away from the chelated titanium. Ionization of the OH group gives the cationic pre-(*M*) intermediate in which the rotation about bonds *a* and *b* is restricted. Subsequent highly regioselective Friedel–Crafts-type cyclization, followed by aromatization, yields the (*M*)-arylnaphthalene exclusively.

A.1.5 Other Selectivity in Asymmetric Catalysis: Regioselectivity and Chemoselectivity

In the course of an asymmetric catalytic reaction, other products may also form that are not stereoisomers. In this case, further descriptors are needed in addition to enan-

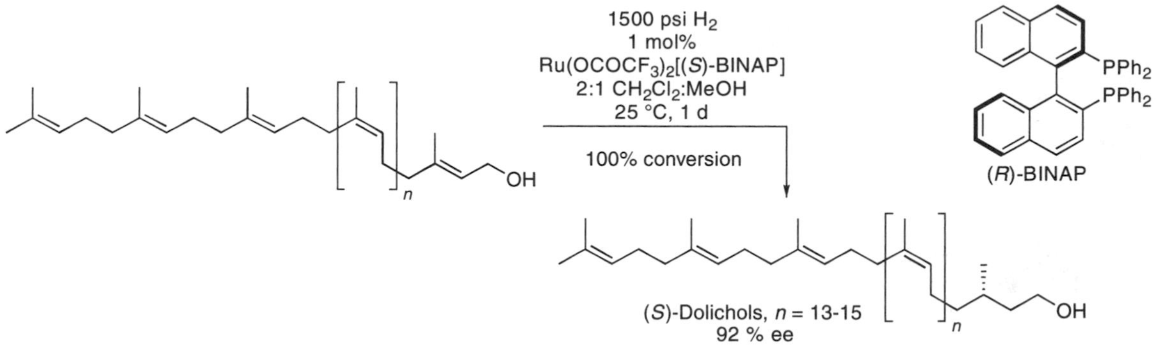

Figure A.35. Regioselective, enantioselective reduction of a polyene intermediate.

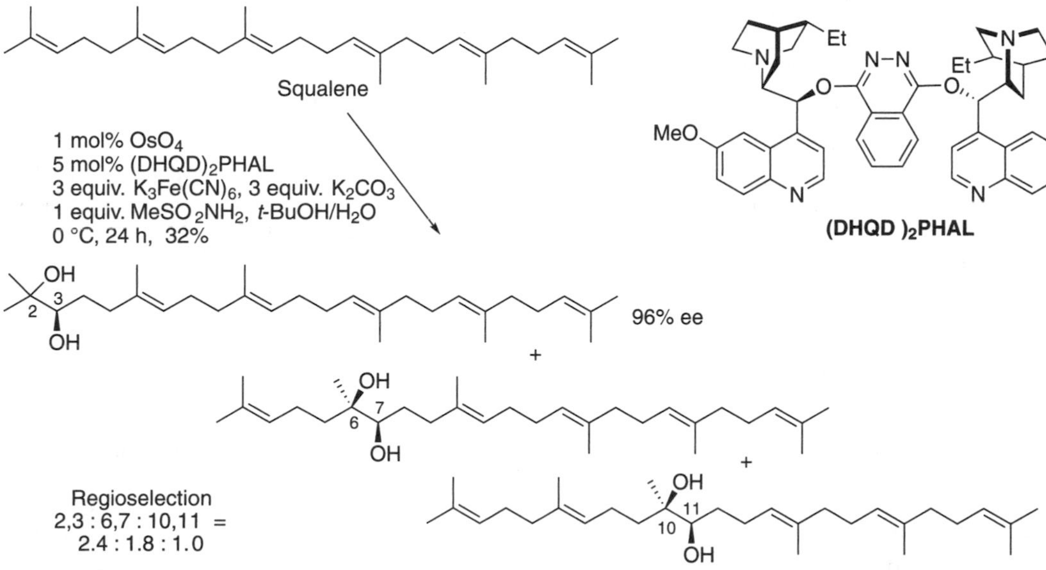

Figure A.36. Regioselective, enantioselective dihydroxylation of squalene.

tioselectivity and diastereoselectivity. Regioselectivity or site selectivity refers to the propensity for a reagent to react with different regions (sites) on one structure leading to constitutional isomers. This term is most often applied when the reacting sites are chemically similar but differ in location. For example, the alkenes in the polyene in **Figure A.35** are all highly similar, but only the alkene located in one region (near the hydroxyl) undergoes facile hydrogenation with a ruthenium–BINAP complex.[116] Thus, the reaction is highly regioselective. Reduction of one prochiral face of this particular alkene is more favorable, with the chiral catalyst causing the process to be enantioselective, as well.

With functional groups in highly similar environments, regioselection can be a significant challenge, as illustrated in the asymmetric dihydroxylation in **Figure A.36**.[117] Although high facial selectivity is observed, resulting in high enantioselection, the ability of the catalyst to distinguish very similar alkenes is moderate.

Examples illustrating diastereoselection, as well as regioselection and enantioselection, are shown in **Figure A.37**.[118] Here, a remarkably selective transformation is

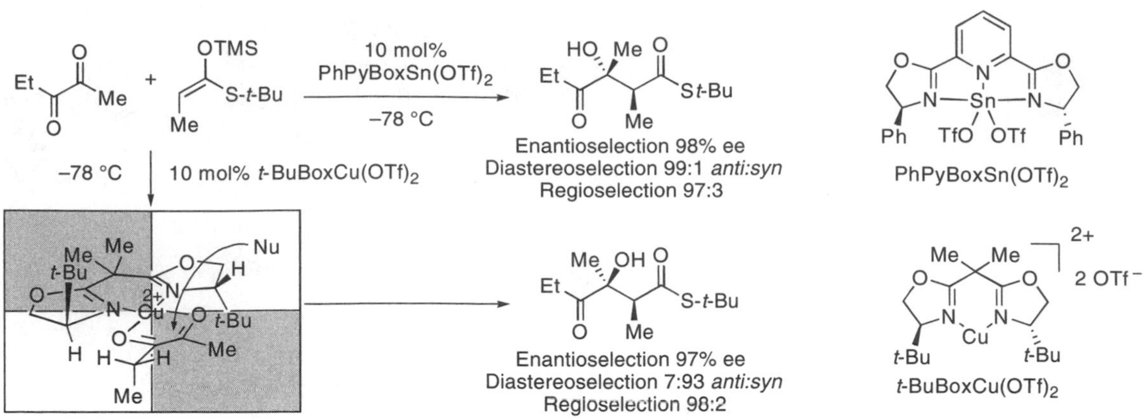

Figure A.37. Regio-, diastereo-, and enantioselective Mukaiyama aldol reaction.

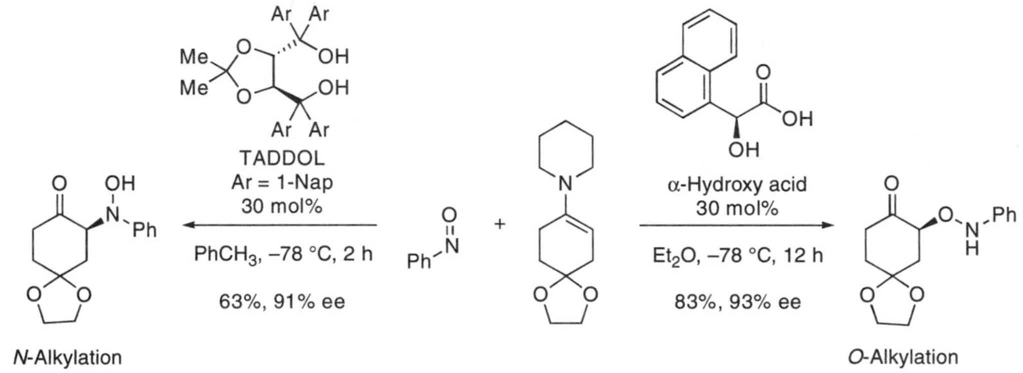

Figure A.38. Regioselectivity in the asymmetric nitroso aldol reaction using chiral Brønsted acids.

observed with reaction occurring mainly at the less-hindered carbonyl. The facial approach to both the 1,2-dicarbonyl and silyl ketene thioacetal is dictated by the catalysts, resulting in the observed diastereoselection and enantioselection. In both processes, a chelated dicarbonyl-catalyst adduct is presumed to form, as illustrated with the copper catalyst. Of the four quadrants (see Chapter 4) of approach to the carbonyl carbons, the upper left and lower right are blocked by the *tert*-butyl groups from the chiral ligand of the copper catalyst. Furthermore, the upper left *tert*-butyl group causes the terminal methyl group of the ethyl ketone to orient into the lower-left quadrant, which subsequently blocks approach to this quadrant. As a result, only the upper-right quadrant is sterically unencumbered, allowing approach of the silyl ketene thioacetal nucleophile, which is in accord with the observed product stereochemistry and regiochemistry.

Another example of an asymmetric regioselective process is seen in the enamine nitrosylation examples outlined in (**Figure A.38**).[119] In this work, high enantiomeric excesses are generally observed in the reactions of cyclohexene enamines bearing a piperidine-based amine moiety. The most gratifying aspect of this study was the exclusive formation of either regioisomer (*O*-alkylation vs. *N*-alkylation) in pure form depending on the Brønsted acid employed. With the TADDOL diol, Brønsted acid coordination of the nitrosobenzene oxygen presumably leads to N-alkylation. On the other hand, the α-hydroxy acid presumably coordinates with the nitrogen atom of nitrosobenzene. Upon nucleophilic attack at the oxygen atom, the O-adduct is realized.

Another type of selectivity is chemoselectivity, which occurs when reaction can occur with different chemicals (separate molecules) or with different chemical entities on one molecule. The latter type of process may also be termed regioselective or site selective (e.g., **Figure A.38)**, but chemoselective is more appropriate when the chemistry of two sites is very different.

Figure A.39 provides a schematic (coordination and activation of the zinc reagent by the chiral titanium catalyst is not shown) of a chemoselective, enantioselective process in the titanium–salen-catalyzed addition of diethyl zinc to α-ketoesters.[120,121] Here, two chemically distinct reaction pathways are possible, involving addition of an ethyl anion (*addition*) or a hydride anion (*reduction*) to the ketone carbonyl. In the absence of the catalyst, the *reduction* process is competitive and often accounts for the majority of the product. The chiral titanium–salen complex catalyzes the *addition* to a far greater extent than the *reduction*, generating a highly chemoselective asymmetric process.

Issues of chemoselectivity are particularly relevant in multicomponent reactions with their multiple reaction partners. For example, direct aldol and Mannich reactions will typically compete if imines and enol equivalents are not preformed, and their rates depend on the equilibrium ratio between the aldehyde and the imine (K_{eq}) and on their respective rate constants (k_{aldol} vs. $k_{Mannich}$) (**Figure A.40**).

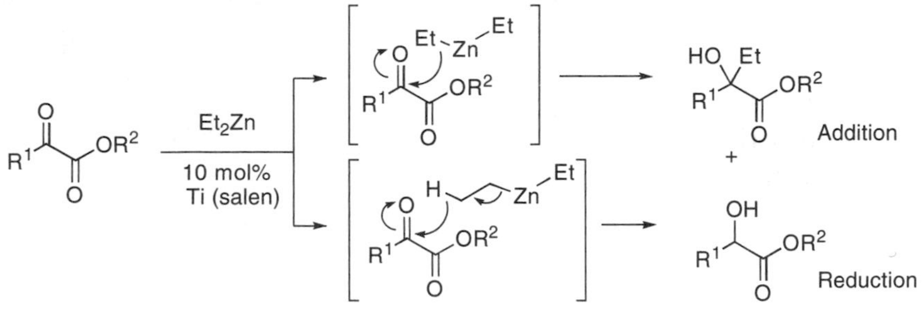

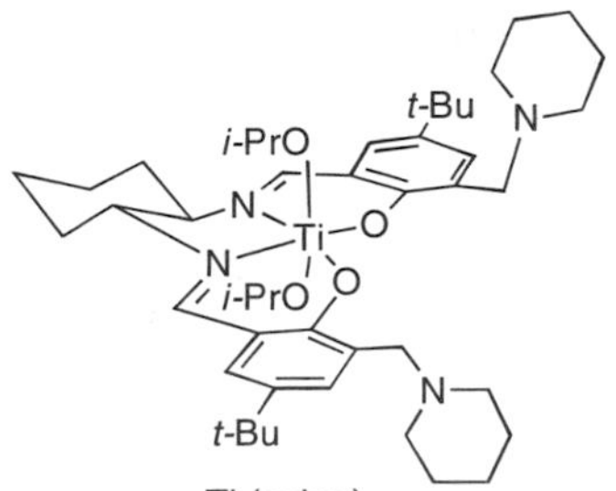

R[1]	R[2]	Addition: reduction	Addition ee (%)
Ph	Me	100:0	88
2-MeOC$_6$H$_4$	Me	100:0	85
PhCC	Me	100:0	80
Cy	Me	100:0	75
Me	i-Pr	100:0	72

Figure A.39. Chemoselectivity in asymmetric α-ketoester additions. The mechanism of this reaction is presented in **Figure 12.21.**

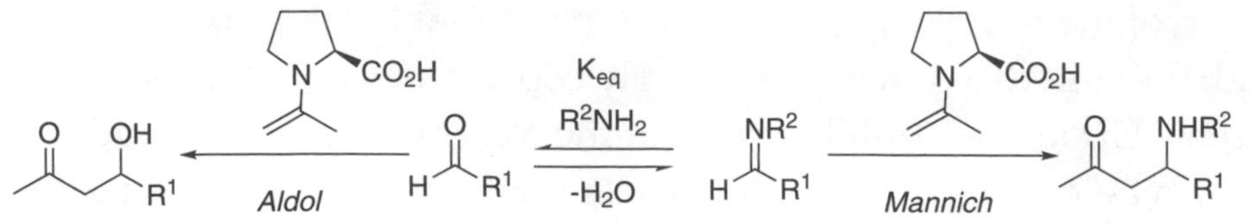

Figure A.40. Direct aldol vs. Mannich reactions.

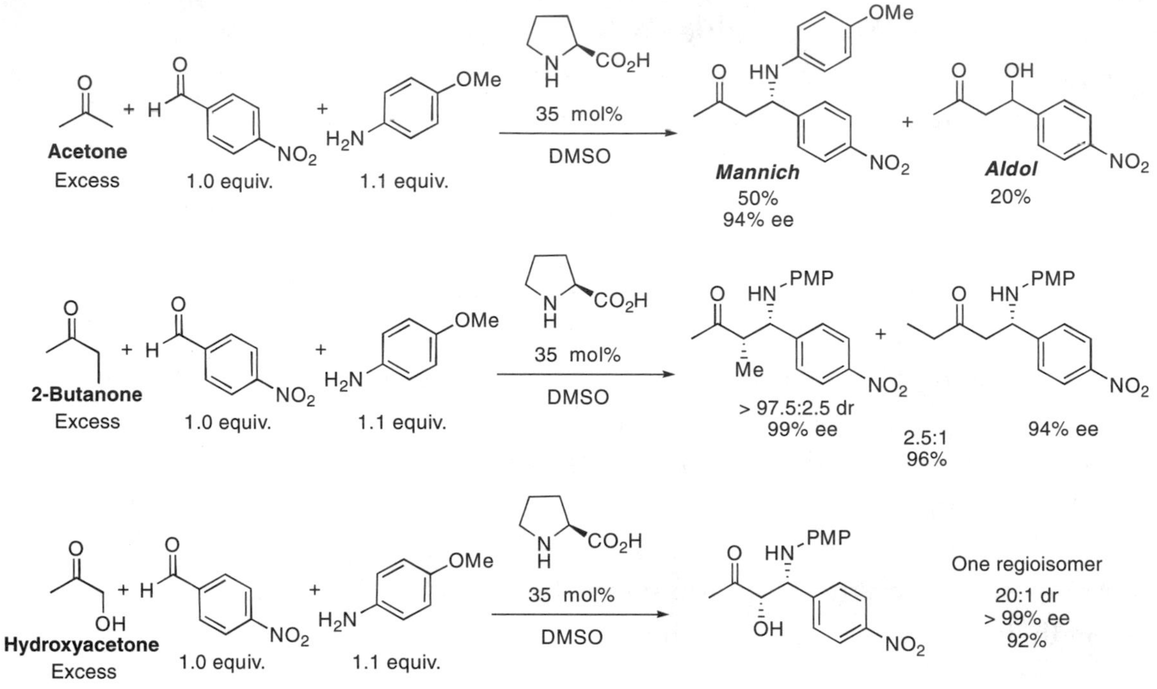

Figure A.41. Chemoselective, regioselective, diastereoselective, and enantioselective asymmetric Mannich reactions.

In spite of these challenges, catalytic asymmetric multicomponent Mannich reactions have been successfully developed.[122] For example, the proline-catalyzed reaction with acetone in **Figure A.41** leads to a 50:20 mixture of Mannich and aldol products. With 2-butanone and hydroxyacetone, this chemoselectivity is considerably improved, resulting in 92–96% yields of the Mannich product. With these ketones, regioselection, diastereoselection, and enantioselection all pertain. For butanone, the enantioselection (99% ee) and diastereoselection (97.5:2.5 dr) are very high, but the regioselection (2.5:1) is moderate. With hydroxyacetone, every aspect of the selectivity is excellent: the reaction is chemoselective (100:0, Mannich vs. aldol), regioselective (100:0 alkylation at the more-substituted vs. the less-substituted ketone α-position), diastereoselective (20:1, *syn* vs. *anti*), and enantioselective (> 99% ee).

Development of catalysts that can control all types of selectivity, including chemoselection, regioselection, diastereoselection, and enantioselection, is a continuing challenge. Such efforts will drive applications to highly functionalized structures.

A.2 Generation of Stereochemistry: Enantioselective Processes

While a wide variety of highly enantioselective catalytic processes have been developed, there are actually only a few fundamental strategies to generating enantiomerically pure chiral compounds. These include the differentation of prochiral groups, differentiation of prochiral faces, resolution of racemic mixtures of compounds with extant stereogenic units, and conversion of racemic mixtures of compounds into single enantiomers. Below these strategies are discussed in the context of creating either one or two new stereogenic units in a process, although it is entirely feasible to create more than two stereogenic units in one transformation. Selected examples are provided for the creation of single enantiomers incorporating several of the different types of stereogenic units, including centrochiral, axial chiral, and planar chiral compounds.

A.2.1. Generation of Single Enantiomers with One Stereogenic Unit

A.2.1.a. Transformation of Prochiral Compounds to Chiral Compounds

A.2.1.a.1. Differentiation of Enantiotopic Groups

Often, enantiotopic groups are two substituents at the same center of an otherwise achiral molecule. For example, in the prochiral precursor to esomeprazole, an antiulcer agent (see **Figure A.7**, Section A.1.1.b), the two lone pairs on the sulfur are the enantiotopic groups (**Figure A.42**). Selective oxidation of one lone pair versus the other by means of a chiral titanium–diethyl-tartrate complex provided esomeprazole with a centrochiral unit at the sulfur center in high selectivity.[123]

Whether a reaction involves the differentiation of enantiotopic groups or faces in a prochiral compound is not always clear. For example, in the overall process of the phase-transfer catalytic asymmetric alkylation in **Figure A.43**, the two enantiotopic hydrogens of the glycine imine appear to have been differentiated with a high degree of enantioselection.[124] However, it is most likely that an enolate initially forms

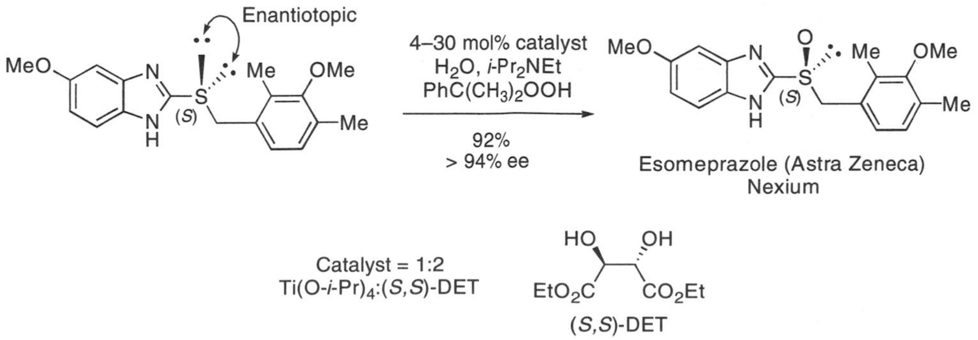

Figure A.42. Differentiation of enantiotopic groups.

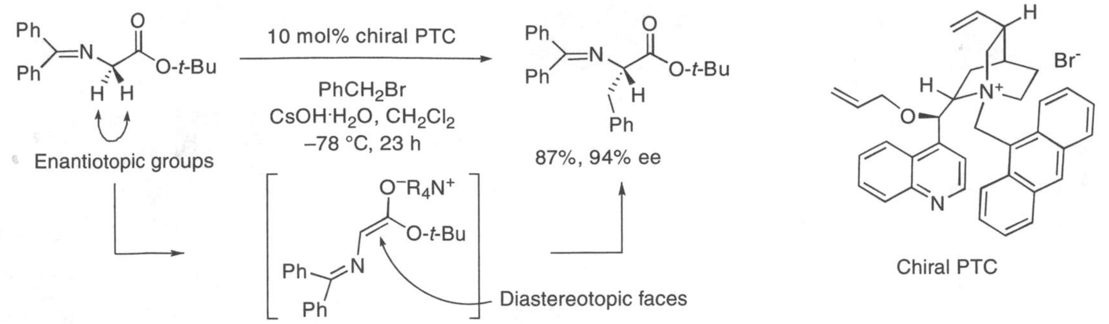

Figure A.43. Net differentiation of enantiotopic groups via diastereotopic faces.

outside of the sphere of the chiral phase-transfer catalyst (PTC). Subsequent exchange with the PTC then establishes the illustrated chiral enolate. From the point of view of stereochemical induction, the PTC establishes a bias by blocking one of the enantiotopic faces of the prochiral enolate to a greater extent. In other words, the enolate faces are diastereotopic in the chiral ammonium-enolate adduct. As a result, benzyl-bromide addition occurs predominantly on the unshielded face and this example becomes a problem in facial differentiation, not group differentiation.

Enantiotopic groups are not restricted to two substituents on the same center. For example, *N*-Boc pyrrolidine (see **Figure A.22** for an analysis) possesses enantiotopic groups on the same center, as well as between two different centers.

Desymmetrization reactions of achiral or meso compounds typically require the differentiation of enantiotopic groups (see Chapter 10). With achiral precursors, desymmetrization to products with only one stereogenic unit has been demonstrated on numerous occasions. For meso compounds, it is possible to create a compound with a single stereogenic unit, although most examples result in formation of two or more stereogenic units.[99–101] Such a transformation requires removal of at least one stereogenic unit since meso compounds, by definition, possesses more than one stereogenic unit (see Section A.1.1.h). An example can be seen in the desymmetrization of the enantiotopic alcohols in the diol in **Equation A.5**. Oxidation with the ruthenium catalyst leads to the product in 87% ee with a single stereogenic unit.[125]

Equation A.5

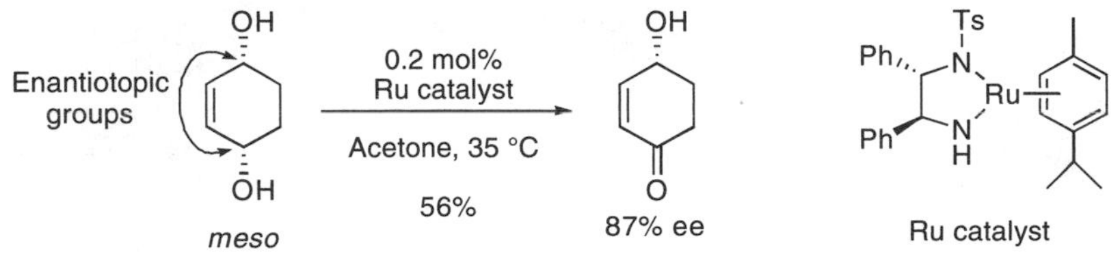

A.2.1.a.2. Addition to Enantiotopic Faces

In asymmetric catalysis, the most commonly used prochiral precursors possessing enantiotopic faces are carbonyl, imine, and alkene compounds. The first two allow formation of valuable functionalized compounds, including chiral alcohols and amines. Examples of selective additions to prochiral carbonyls, such as aldehydes, can be seen in **Figures A.28** and **A.29**. An example with imines is the asymmetric Strecker reaction catalyzed by the organocatalyst illustrated in **Figure A.44**. The urea catalyst activates HCN and delivers it selectively to one of the enantiotopic faces of the imine.[126] The resultant chiral α-aminonitriles are readily converted into chiral α-amino acids that are valuable components in pharmaceutical synthesis, total synthesis, and biological chemistry.

Since alkenes are readily available, they also serve as valuable prochiral precursors in asymmetric catalysis. Examples involving enantiofacial, selective alkene reduction (**Figure A.30**) or oxidation (**Figure A.27**) are common, but prochiral alkenes can also be employed in carbon–carbon bond-forming reactions. For example, in the transformation illustrated in **Figure A.45**,[127] it is proposed that an initial hydropalladation proceeds to give rise to an intermediate incorporating a chiral palladium species. This species can undergo intramolecular addition to either diastereoface of the alkene. Since the *Re* approach is lower energy, the (*S*)-enantiomeric product is formed predominantly. Here, the faces of the intermediate are diastereotopic because the intermediate contains a covalent bond to the chiral catalyst and is itself chiral. However, overall, the process results in the differentiation of the enantiotopic faces of the alkene in the achiral starting material.

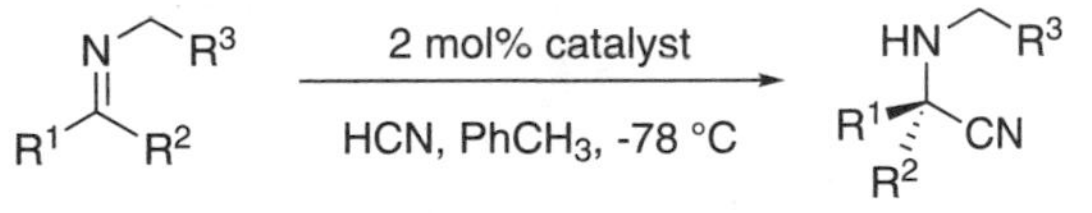

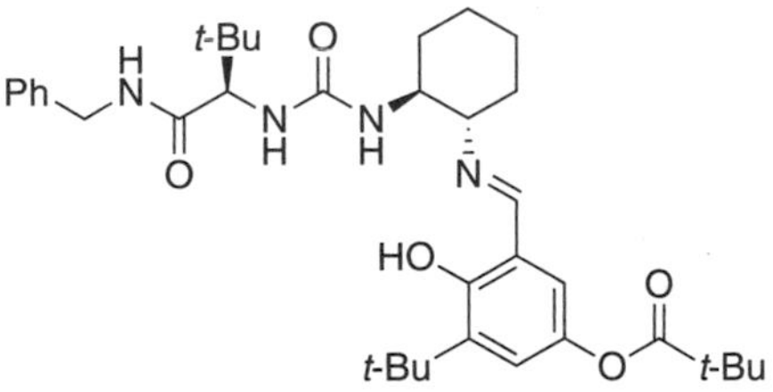

R¹	R²	R³	ee (%)
Ph	H	CHCH₂	95
t-Bu	H	Ph	96
Cy	H	Ph	87
n-Pent	H	Ph	78
Ph	Me	Ph	90

Figure A.44. Addition to a prochiral carbon–heteroatom double bond to generate one centrochiral unit.

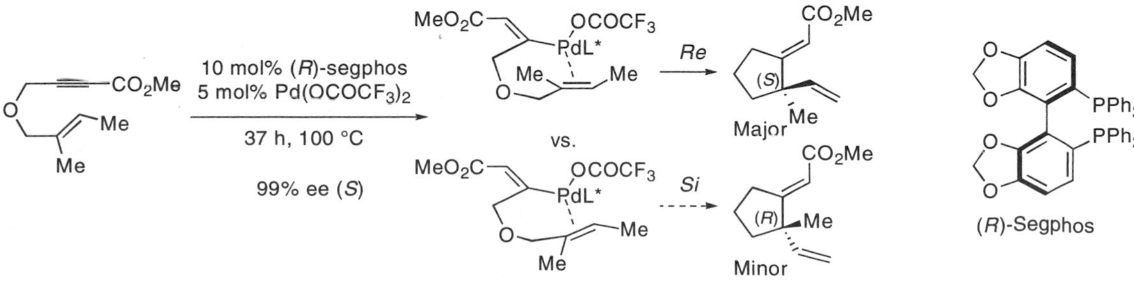

Figure A.45. Addition to a prochiral carbon–carbon double bond to generate one centrochiral unit.

A.2.1.b. Transformations of Racemic Compounds

Instead of beginning with achiral compounds and generating single enantiomers of compounds with new stereogenic units, an alternate approach commences with racemic mixtures of chiral materials with extant stereogenic units. Such racemic mixtures of compounds can be resolved by means of a chiral catalyst in a kinetic resolution process. In most such processes, a chiral catalyst is employed that catalyzes a transformation with one enantiomer of the racemic mixture more quickly than the other (see Chapter 2 for energy diagrams and Chapter 7 for further detail and related strategies).

Resolution of chiral alcohols via acylation (**Figure A.46**) is a classic version of such a kinetic resolution and was initially explored with enzymatic catalysts. Many highly successful small-molecule catalysts have also been developed for this process, including the biologically inspired peptide in **Figure A.46** that was identified via rapid combinatorial screening.[128] Notably, this peptide catalyst is general for a fairly broad array of chiral alcohols, in contrast to many enzymatic catalysts. The ability of this catalyst to resolve cyclohexyl ethanol and *sec*-butanol is especially unusual for a small-molecule catalyst. In the latter case, the catalyst undertakes the relatively subtle task of distinguishing between an ethyl versus a methyl group.

The disadvantage of kinetic resolution is that two materials (one enantiomer of the starting material and the product obtained from the other starting-material enantiomer) are left at the end, even if the reaction is perfect. As a consequence, the max-

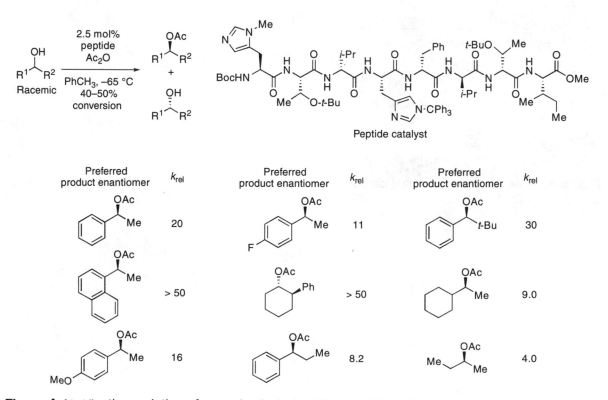

Figure A.46. Kinetic resolution of racemic alcohols with a peptide catalyst.

imum yield of any single enantiomer is 50%. In practice, the resolved starting material and the product must be separated, which is often difficult. While strategies have been developed to maximize the efficiency and utility of such processes,[129] a more efficient alternative lies in dynamic kinetic resolution (DKR; see Chapter 1 for energy diagrams and Chapter 9 for further detail and related strategies). In this strategy, a means to interconvert the starting material enantiomers is required. If such an interconversion is compatible with further reaction conditions, then all of a racemic starting material can be converted to a single product enantiomer.

For example, the chiral Grignard substrate illustrated in **Figure A.47** is not configurationally stable, resulting in rapid enantiomer interconversion. One of these enantiomers undergoes a coupling reaction more rapidly than the other when a chiral palladium catalyst is employed.[130] As the racemic Grignard substrate is "resolved" in the coupling reaction, Le Châtelier's principle becomes active: the substrate continually equilibrates to regenerate an equal amount of both enantiomers. As a result, both high yields and high levels of enantioselection can be achieved.

Dynamic kinetic resolution can occur via direct racemization, as is the case above, or via reaction of an achiral species that is accessible from both enantiomers. For example, the ketone substrate in **Figure A.48** is chiral and can be readily racemized via enolization/protonation. However, the asymmetric process[131] that is utilized to generate the subsequent quaternary center is a dynamic kinetic resolution only in the overall sense. Examination of the mechanism reveals that the stereochemical induction is established via differentiation of the two enantiotopic faces of an achiral enolate associated with the chiral phase-transfer catalyst (PTC), rather than by selective reaction with one enantiomer and equilibration of the starting ketone

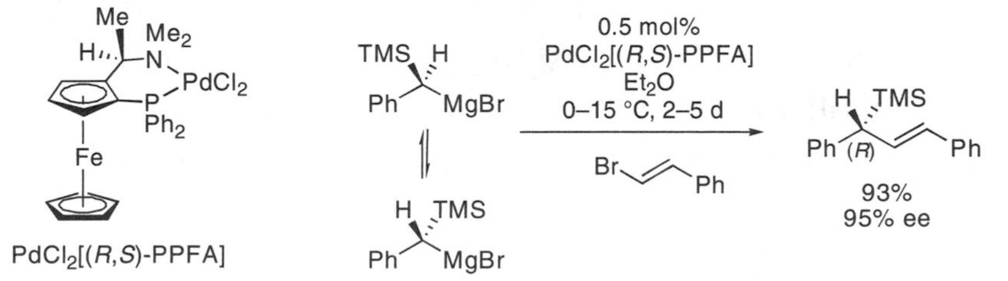

Figure A.47. Dynamic kinetic resolution (DKR) in cross-coupling.

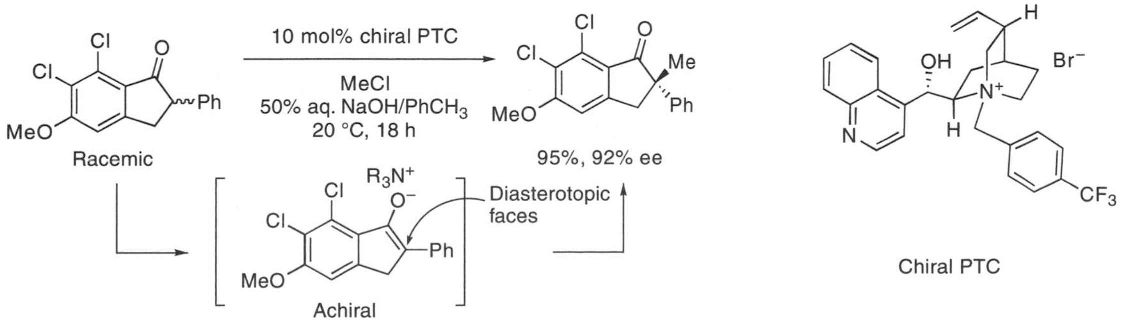

Figure A.48. Dynamic kinetic asymmetric transformation (DyKAT) via enolization.

enantiomers. Such a process is called a dynamic kinetic asymmetric transformation (DyKAT; see Chapter 1 for energy diagrams and Chapter 9 for further detail).

A.2.1.c. Creation of Central Chirality

Many different reaction types, including all of those encompassed in Sections A.2.1.a and A.2.1.b, have been employed to create single enantiomers of compounds with centrochiral units. The majority of these examples generate centrochiral units at *carbon* centers. In principle, the same methods can be used to establish centrochiral units at *silicon, nitrogen, phosphorous, sulfur* (see **Figure A.42**), and even *metal* centers. Examples of these methods can be found above and in Chapters 1–16.

A.2.1.d. Creation of Axial Chirality

There are four fundamentally different ways to establish axial chirality in atropoisomeric compounds: 1) creation of axial chirality at the same time the axial chiral bond is created, 2) enantiotopic differentiation of compounds with a prochiral axial unit, 3) resolution of nonequilibrating (kinetic resolution) or equilibrating (dynamic kinetic resolution) precursors containing axial chiral units, and 4) chirality exchange.

An example of the first of these can be found in asymmetric Suzuki couplings mediated by chiral palladium catalysts (**Figure A.49**).[132] Ironically, the chiral ligand that is used to create the new axial chiral unit is also an axial chiral compound. In this case the asymmetric induction presumably occurs during the reductive elimination step from the two illustrated intermediates. In this step, the biaryl bond is made while the stereochemical relationship between the two aromatic rings is established.

In oxidative binaphthol-coupling reactions, an axial chiral unit is also created during the course of biaryl bond formation (**Figure A.50**).[133,134] However, the biaryl bond terminated by two sp^2 centers is formed from two sp^3 hybridized centers via enolization of an intermediate. This intermediate is, in turn, formed by enantiofacial selective coupling to each of the aromatic substrates when bound to the chiral catalyst [see

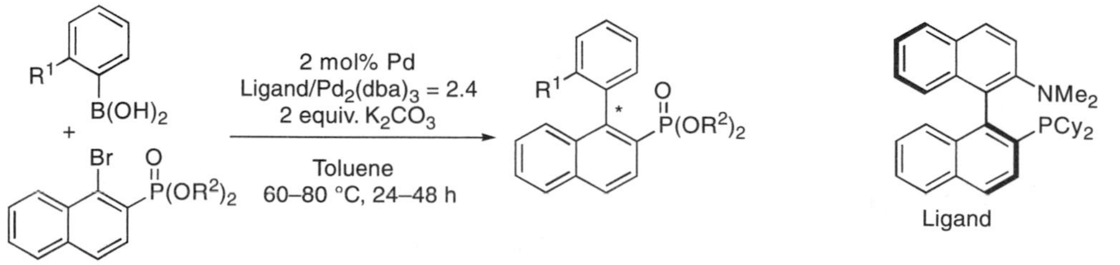

R^1	R^2	Yield (%)	ee (%)
Me	Et	100	87
Et	Et	96	92
i-Pr	Et	89	85
Ph	Et	74	74
Me	Me	91	84

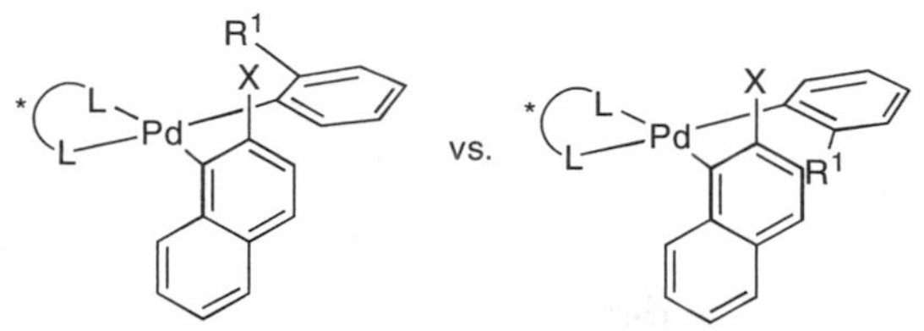

Figure A.49. Catalytic enantioselective Suzuki couplings generating axial chiral units.

the tetrahedral Cu(I) in **Figure A.50**]. Enolization is proposed to proceed with bond rotation in the direction of least steric resistance, resulting in the enantioinduction at the biaryl bond.

An example of prochiral enantiogroup differentiation (i.e., desymmetrization, see Chapter 10) to establish a chiral axis is illustrated below (**Figure A.51**).[135] In this instance, a secondary kinetic resolution of the monotriflate increases the observed enantioselectivity of this compound by draining away the minor (*R*)-component (see Chapters 7 and 8 for further discussion; see Chapter 16 for an additional example).[136,137]

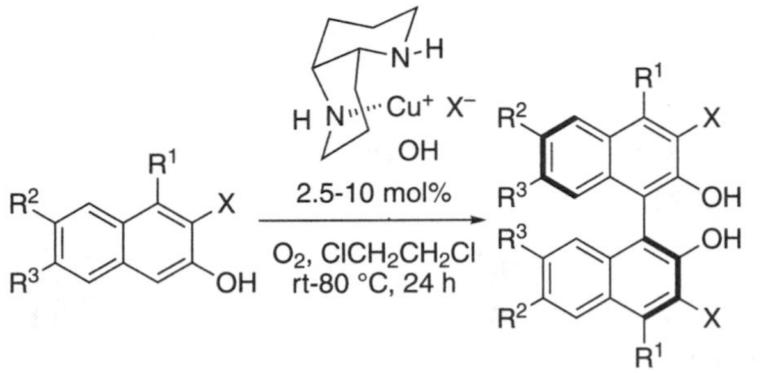

X	R^1	R^2	R^3	ee (%)
CO_2Me	H	H	H	90-93 (99)*
CO_2Me	H	H	Br	92
OBn	H	H	H	46
$CONR_2$	H	H	H	70-75
COAr	H	H	H	83-94
POR_2	H	H	H	92-96
SO_2(*p*-OMe-C_6H_4)	H	H	H	57 (98)*
SO_2(*p*-OMe-C_6H_4)	H	OMe	H	75
CO_2Me	OAc	OMe	OMe	90
CO_2Me	OAc	OMe	H	86
CO_2Me	OAc	OMe	*n*-Pr	87

*Triturated

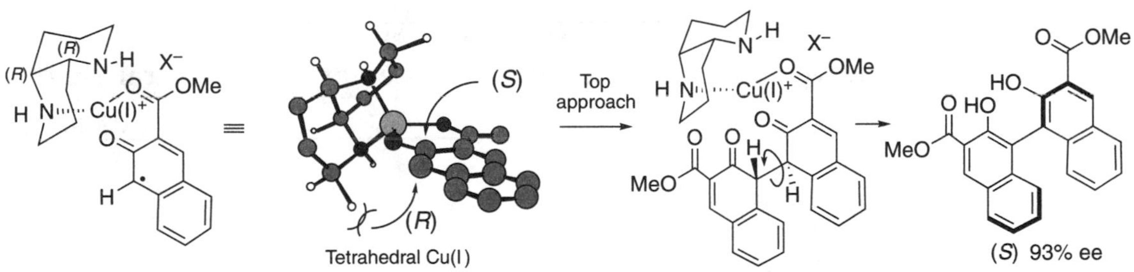

Figure A.50. Catalytic enantioselective oxidative naphthol coupling.

Conditions:

RMgBr
1 equiv. LiBr
$Et_2O/PhCH_3$
−20 to −30 °C

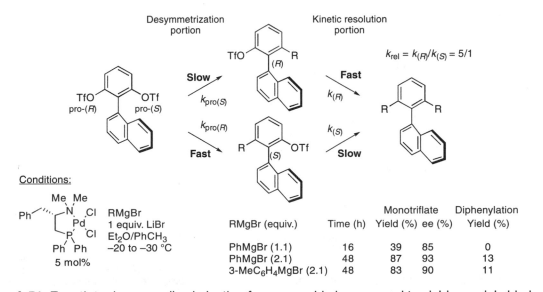

RMgBr (equiv.)	Time (h)	Monotriflate Yield (%)	Monotriflate ee (%)	Diphenylation Yield (%)
PhMgBr (1.1)	16	39	85	0
PhMgBr (2.1)	48	87	93	13
3-MeC_6H_4MgBr (2.1)	48	83	90	11

Figure A.51. Enantiotopic group discrimination from a prochiral compound to yield an axial chiral unit.

An example of catalytic asymmetric kinetic resolution in the formation of axial chiral compounds is illustrated in **Figure A.52**. Here atropisomerically enriched aryl amides are generated via the asymmetric dihydroxylation of one enantiomer of a racemic alkenyl aryl amide.[138]

The final method for creating axial chirality in an atropisomeric compound relies on chirality exchange (see Section A.1.4.b). An example can be found in **Figure A.34**.

Catalytic asymmetric methods have also been reported for the synthesis of non-atropisomeric compounds possessing axial chirality such as allenes.[139] For example, the chiral rhodium-catalyzed conjugate addition illustrated in **Figure A.53** provides the axial chiral allenyl enol ethers with good yield and selectivity.[140]

Catalytic asymmetric methods for the preparation of axial chiral alkylidene cyclo-hexanes and spiranes are rare. An example for the alkylidene cyclohexanes employs a catalytic asymmetric allylation (**Figure A.54**).[141] Under optimal conditions, the product was delivered in moderate yield (63%) and high selectivity both in terms of regioselection (96:4, internal:terminal alkene) and enantioselection (90% ee).

An example for the chiral spiranes can be found in the preparation of the chiral spirosilane (**Figure A.55**) from **Figure A.12** (see Section A.1.1.c) via an asymmetric intramolecular hydrosilation.[77] In this transformation, different types of asymmetric induction are excercised at two stages. In the first step, either of the two homotopic hydrogens at the silicon center can undergo oxidative addition with the chiral cata-lyst to yield the same intermediate. The silicon is not a stereogenic center at this stage but is prochiral. However, due to the extant stereocenter on the chiral rhodium

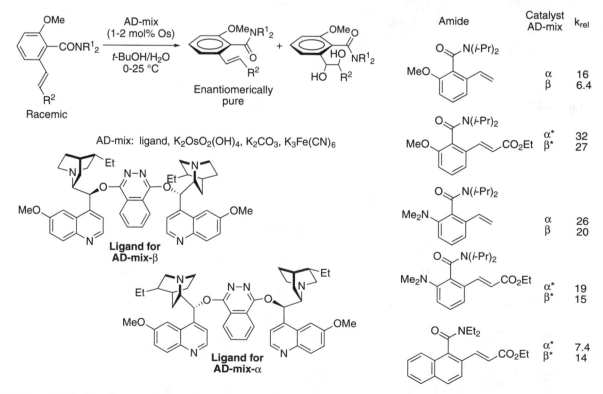

Figure A.52. Kinetic resolution of atropisomeric amides with the Sharpless asymmetric dihydroxylation (*MeSO$_2$NH$_2$ employed in the reaction).

ligand, the two thienyl groups on the silicon center are diastereotopic rather than enantiotopic. Reaction at the upper versus the lower thiophene results in subsequent diastereogroup selection. For reaction with the upper thiophene, a further stereo-chemical consideration arises with respect to the alkene face undergoing addition

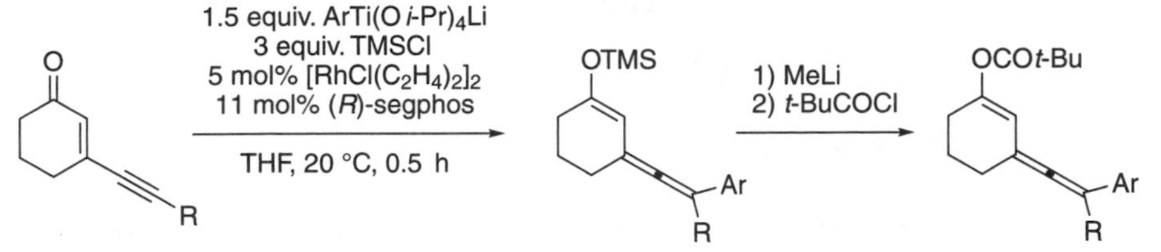

R	Ar	Yield (%)	ee (%)
n-Bu	Ph	85	92
n-Bu	4-FC$_6$H$_4$	85	91
n-Bu	4-MeOC$_6$H$_4$	83	93
Cyclohexyl	Ph	80	80
4-MeOC$_6$H$_4$	Ph	56	75

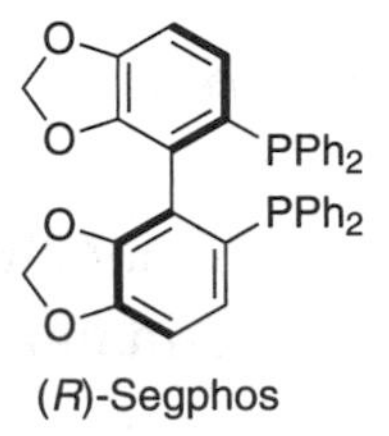

Figure A.53. Formation of axially chiral allenes via asymmetric catalysis.

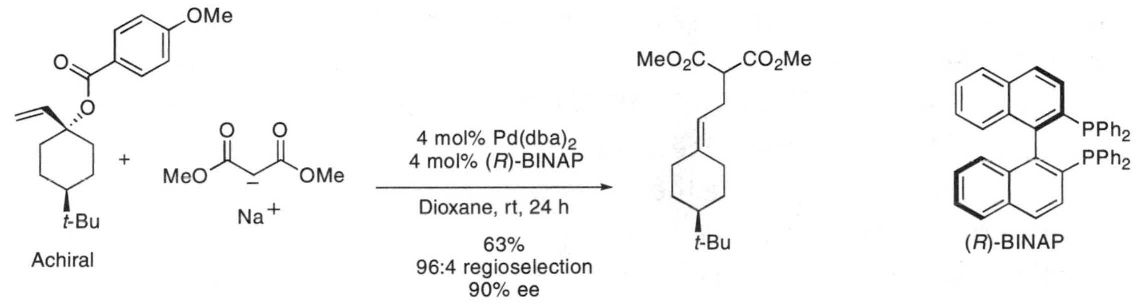

Figure A.54. Formation of an axial chiral alkylidene cyclohexane via asymmetric catalysis.

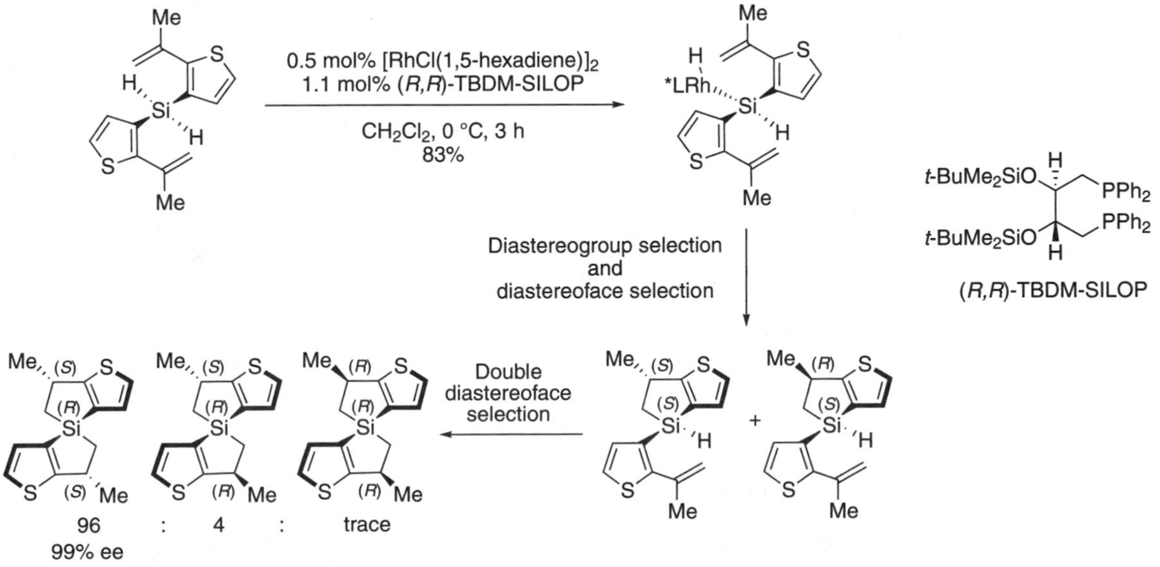

Figure A.55. Formation of an axial chiral spirosilane via asymmetric catalysis.

(i.e., diastereofacial selectivity, see Chapter 13). Thus, in this one step, the chiral ligand at the rhodium center exerts two types of stereocontrol in the migratory-insertion step. Subsequent stereospecific reductive elimination yields the illustrated diastereomeric silyl monohydrides with the (S,S)-diastereomer predominanting. A second oxidative addition/migratory-insertion/reductive-elimination sequence forms the second carbon–silicon bond diastereoselectively to yield the major product with high diastereoselection (96:4) and enantioselection (99% ee).

A.2.1.e. Creation of Planar Chirality

While methods for the synthesis of planar chiral compounds that also contain centrochiral units, such as JOSIPHOS (**Figure A.17**) and related chiral ferrocence ligands (e.g., **Figure A.47**),[83,84,97] are well-established, the catalytic asymmetric synthesis of chiral compounds containing only a planar chiral unit remains a significant challenge. Recently, planar chiral heterocycles have emerged as an important class of catalysts,[87–89] but these materials are typically obtained via chromatographic resolution. Direct catalytic asymmetric methods for the generation of planar chiral materials are rare[142,143] and fall into predominantly two classes—namely, asymmetric complexation and asymmetric substitution of the planar portion.

An example of the first of these methods, asymmetric complexation, is found in **Figure A.56**. Here, the planar chiral η^4-iron adduct is obtained with moderately good levels of enantioselection by complexation in the presence of a chiral catalytic azadiene and light.[144] In this reaction, the azadiene is proposed to act as an asymmetric tricarbonyliron transfer reagent via the η^4-adduct.

An example of the second method, asymmetric substitution on the planar portion, for the formation of planar chiral compounds, is illustrated in **Figure A.57**. Here, the enantiotopic chlorine groups of an achiral η^6-tricarbonylchromium complex are differentiated via a Suzuki coupling using a chiral palladium catalyst.[145] While levels of enantioselection are modest with alkenyl or aryl boronic acid coupling partners,

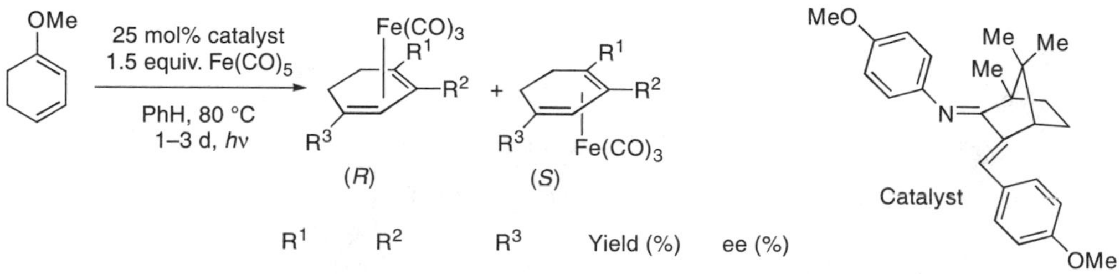

R^1	R^2	R^3	Yield (%)	ee (%)
OMe	H	H	97	86 (S)
O-i-Pr	H	H	78	79 (+)
OMe	H	Me	86	72 (S)
OMe	H	CH$_2$CO$_2$Me	93	50 (S)
H	CO$_2$Me	H	90	76 (−)

Figure A.56. Catalytic synthesis of planar chiral compounds via asymmetric metal complexation.

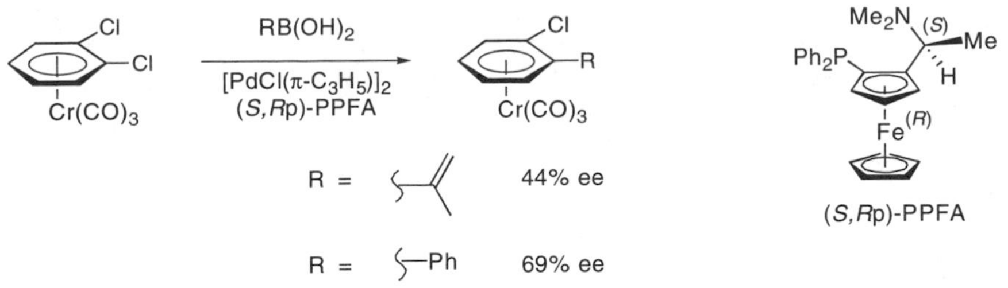

Figure A.57. Catalytic synthesis of planar chiral compounds via asymmetric substitution of the planar portion.

this approach holds considerable promise in the synthesis of this important class of compounds.

A.2.2 Generation of Single Enantiomers with More Than One Stereogenic Unit

In the above sections, examples of catalytic asymmetric methods for the creation of single enantiomers with one stereogenic unit were outlined. Fundamental strategies include enantiotopic group differentiation and desymmetrization, and enantiofacial discrimination, as well as the kinetic and dynamic kinetic resolution of racemic compounds with one stereogenic unit in which no additional stereogenic units are generated. All of these strategies can also be utilized to create single enantiomers with combinations of stereogenic units. The creation of two or more units in one catalytic asymmetric transformation is powerful and complements alternative diastereoselective methods in which additional stereogenic units are created iteratively (see Chapter 16).

A.2.2.a Differentiation of Enantiotopic Groups

In order to create more than one stereogenic center by differentiation of enantiotopic groups, at least two prochiral centers are needed. This is most often encountered in desymmetrization[99] (see Chapter 10) reactions involving enantiotopic groups on different centers (**Figures A.59–A.62**). In contrast, an example with enantiotopic groups on the same center is depicted in **Figure A.58**.[146] In this chiral rhodium-catalyzed asymmetric C–H insertion reaction, the two prochiral units reside on *separate* reaction partners. The hydrogens on the center adjacent to the OTBS of the alkene reaction partner are enantiotopic, giving rise to the first prochiral unit. The second prochiral unit is derived from the enantiotopic faces of the trigonal carbene that is obtained from the phenyldiazoacetate (in the rhodium carbenoid, these faces are diastereotopic due to the chiral ligand on rhodium). Each prochiral unit translates into one new stereogenic unit in the product.

An example involving a desymmetrization is shown in the chiral rhodium-catalyzed diazo insertions in **Figure A.59**.[147] Here, more than one stereogenic unit is generated from an achiral, non-meso compound via enantiotopic group differentiation.

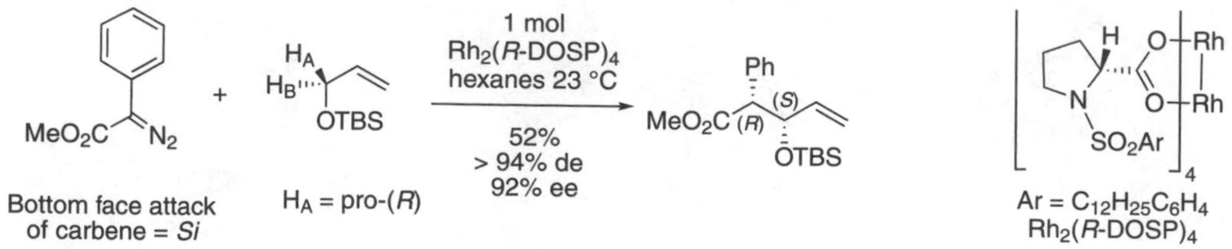

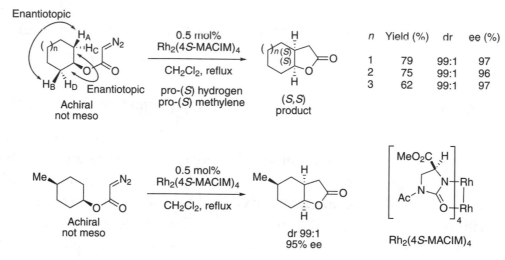

Figure A.58. Products with two stereogenic units from an achiral, non-meso compound via enantiotopic group differentiation and enantiofacial differentiation.

Figure A.59. Products with two stereogenic units from an achiral, non-meso compound via enantiotopic group differentiation.

In the first set of reactions, predominant insertion of the diazo into C–H bonds cis to the alkoxy tether is observed, giving rise to high cis:trans diastereoselection (99:1). Distinction of the two enantiotopic cis hydrogens by the chiral catalyst also gives rise to high enantioselection (96–97% ee), regardless of the ring size. In these cases, two stereogenic units are observed in the product, a consequence of two prochiral units in the starting material—namely, the illustrated enantiotopic hydrogens and the alkoxy carbon center. In the second equation in **Figure A.59**, three stereogenic units are obtained in the corresponding product due to the additional methyl-substituted center, which gives rise to a total of three prochiral units in the starting substrate.

Another desymmetrization reaction is depicted in **Figure A.60**. In this chiral rhodium-catalyzed desymmetrization reaction, the two prochiral units reside on *separate* reaction partners.[148] The hydrogens on the center adjacent to the nitrogen of the heterocyclic reaction partner are enantiotopic, giving rise to the first prochiral unit (reaction at the methylene on either side of the nitrogen involves the same enantiotopic relationships; see **Figure A.22** in Section A.1.2.a). The second prochiral unit is derived from the enantiotopic faces of the trigonal carbene that is obtained from the phenyldiazoacetate.

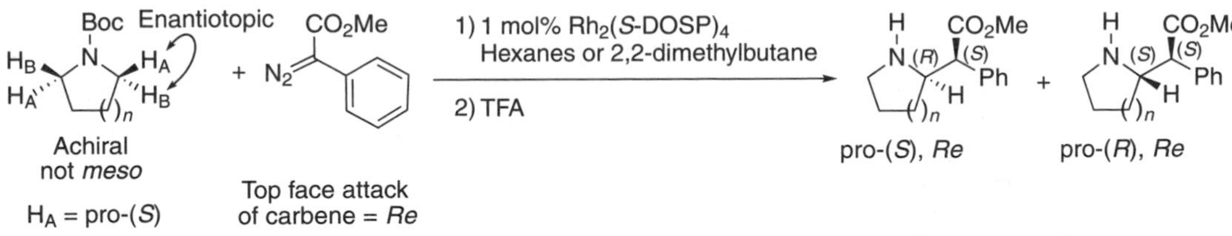

n	T (°C)	Yield (%)	dr	ee (%) of major
1	−50	72	96:4	94
2	−50	44	64:36	89
3	25	72	>95:5	92
4	25	74	>95:5	90

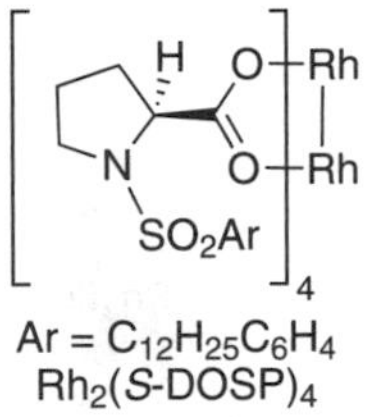

Figure A.60. Products with two stereogenic units from an achiral, non-meso compound via enantiotopic group differentiation and enantiofacial differentiation.

Desymmetrization by the differentiation of enantiotopic groups in meso compounds can be especially powerful.[99] Such transformations share many features with the examples in **Figures A.42** and **A.58–A.60**; the main difference is that stereogenic units already exist in the meso compounds, so breaking the symmetry of the structure readily manifests more than one new stereogenic unit. For example, the selective cleavage of meso anhydrides is a classic desymmetrization reaction. A recent catalytic asymmetric version that involves differentiation of the upper and lower carbonyls by a chiral catalyst is illustrated in **Figure A.61**.[149] More rapid insertion of the chiral palladium catalyst to the upper carbonyl causes the stereochemical differentiation. Subsequent transmetallation from the zinc reagent and reductive elimination yields the anhydride cleavage product concomitant with carbon–carbon bond formation. The resultant chiral compounds possess two or four centrochiral units.

The ability to reveal multiple stereocenters upon the desymmetrization of meso compounds renders this technique very powerful. The creation of structurally complex enantiomerically pure compounds from easy-to-construct achiral meso compounds is especially attractive.[99,150,151] For example, desymmetrization of the inositol intermediate in **Figure A.62** with different asymmetric phosphonylation catalysts gives rise to compounds with six new stereogenic centers with high yield and selectivity.[152,153] Prior syntheses of these important inositol-phosphate derivatives were considerably longer, requiring multiple protecting groups and resulting in overall low yields. For each transformation, the catalyst was identified relatively rapidly by combinatorial screening of peptide catalyst libraries. The resultant catalysts are proposed to function as transphosphonylation catalysts via initial formation of phosphonylated N-methylimidazole intermediates. The activated adducts then interact with the substrate via hydrogen bonding and the phosphonyl group is positioned for stereoselective transfer. Notably, the three hydroxyl groups available for substitution

Product	$R^2{}_2Zn$	Yield (%)	ee (%)	Product	$R^2{}_2Zn$	Yield (%)	ee (%)
	Ph_2Zn	69	91		Ph_2Zn	89	94
	Me_2Zn [25 mol% 4-$FC_6H_4(CHCH_2)$]	60	90		Ph_2Zn (40 °C)	74	89
	Ph_2Zn	83	97		Ph_2Zn (80 °C)	72	92

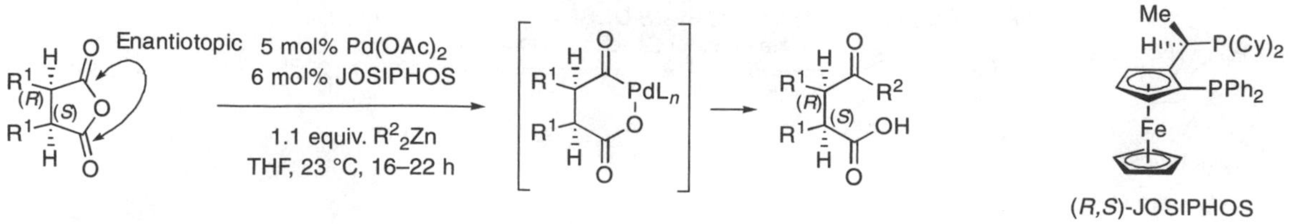

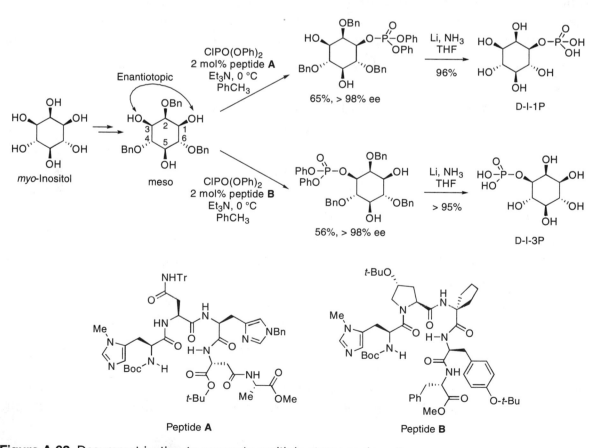

Figure A.61. Differentiation of enantiotopic groups via desymmetrization of meso compounds to provide products with two or more stereogenic units.

Peptide **A**

Peptide **B**

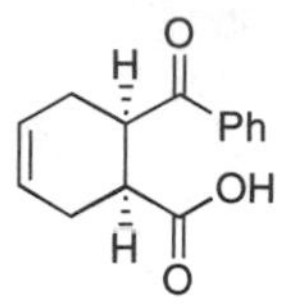

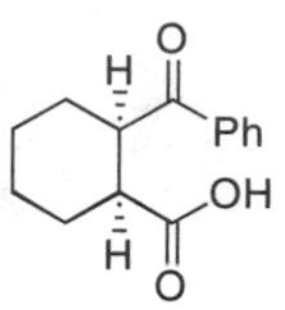

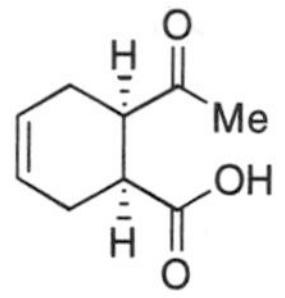

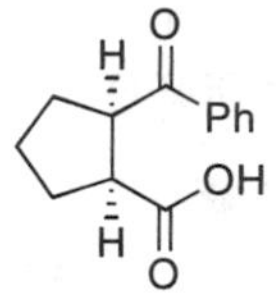

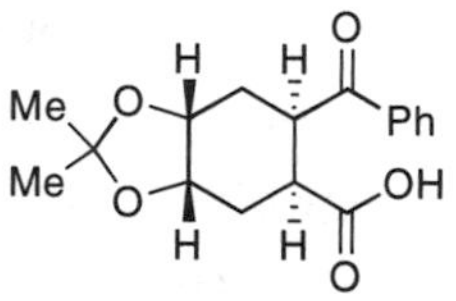

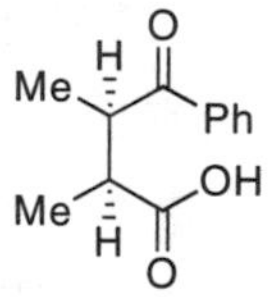

Figure A.62. Desymmetrization to generate multiple stereogenic units.

are stereochemically distinct. The C1 and C3 hydroxyls are enantiotopic and selective substitution of either leads to enantiomeric products. On the other hand, the C5 hydroxyl is diastereotopic with respect to the C1 and C3 hydroxyls and substitution results in a compound that is still meso. This type of catalyst, which results in the differentiation of three similar hydroxyl groups, addresses a key problem in asymmetric catalysis—namely, chiral catalysts that differentiate unique functional groups in polyfunctional molecules are highly desirable as they can be employed with a wider range of substrates, obviate the need for protection of other reactive functional groups,[154] and allow the late-stage functionalization of complex structures.

A.2.2.b Addition to Enantiotopic Faces

Enantiomerically pure compounds with more than one stereogenic unit are readily generated from achiral precursors via enantiofacial differentiation. Such transformations can involve one compound with prochiral faces (**Figure A.63**) or the combination of two or more compounds with prochiral faces (**Figures A.64** and **A.65**). With the large number of compounds possessing prochiral faces, including alkenes, dienes, carbonyls, imines, etc., this strategy is powerful and is employed frequently for the construction of chiral components with multiple stereogenic units.

The asymmetric hydrogenations of (E)- and (Z)-enamide substrates from **Figure A.30** are examined in more detail in **Figure A.63**.[110] In this case, the two diastereomeric starting materials expose different diastereofaces in the cis hydrogenation. The different facial approaches can be uniquely described using the Re and Si descriptors. For the first example, cis hydrogenation from the bottom face gives rise to an attack with Re approach to each trigonal carbon. For the second example, cis hydrogenation from the bottom face causes a Re approach to one trigonal carbon and a Si approach to the other. The products with the syn and anti descriptors are readily visualized (see **Figure A.63**). However, there is no uniform convention as to the staggered illustrations employed. Thus, syn and anti are relative terms that can change depending on the exact illustration used.

In the above two transformations, the stereofaces have changed due to the change in susbstrate geometry. The same set of stereochemical consequences with respect to

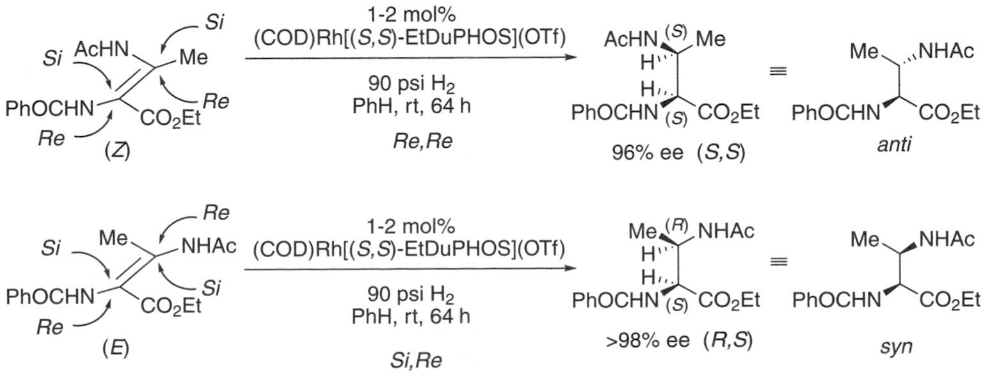

Figure A.63. Addition to a prochiral alkene to generate two centrochiral units.

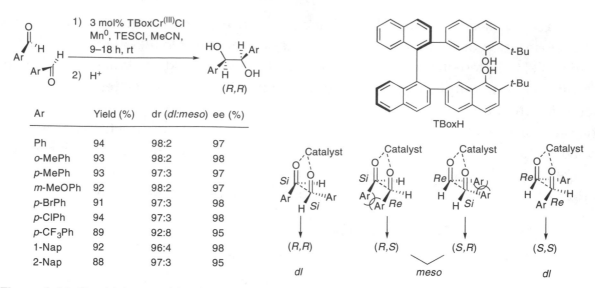

Ar	Yield (%)	dr (dl:meso)	ee (%)
Ph	94	98:2	97
o-MePh	93	98:2	98
p-MePh	93	97:3	97
m-MeOPh	92	98:2	97
p-BrPh	91	97:3	98
p-ClPh	94	97:3	98
p-CF₃Ph	89	92:8	95
1-Nap	92	96:4	98
2-Nap	88	97:3	95

Figure A.64. Combining two identical prochiral double bonds to generate two centrochiral units.

enantiofacial approach and product outcome could also have been realized by selective cis and trans hydrogenations starting from the same substrate.

Reactions that combine two different prochiral planar structures are a classic means for creating new compounds with multiple stereogenic units. The prochiral compounds undergoing reaction can either be identical (**Figure A.64**) or different (**Figure A.65**). In an example of the former, a chiral chromium catalyst is used to initiate the pinacol reaction between two identical aldehydes, leading to three possible stereoisomers (**Figure A.64**).[155] The resultant product, with two new centrochiral units, is formed with high diastereoselectivity and enantioselectivity. The diastereoselection is often a consequence that does not depend on the catalyst asymmetry. For example, most pinacol reactions give rise to high diastereoselection favoring the *dl*-isomer due to metal coordination with both carbonyls. As noted in **Figure A.64**, this metal organization causes disfavorable steric interaction between the aryl groups in the meso combinations. With diastereoselection secured, the main role of the chiral catalyst is to discriminate amongst the facial approaches of the *dl* combinations, eliminating one of two possibilites. In this example, approach to the *Si* face of each aldehyde is preferred (*Si,Si*) leading to the (*R,R*)-product predominantly.

When two different planar prochiral units undergo reaction, even more complexity can be created. For example, one of the Mukaiyama aldol reactions from **Figure A.29** is outlined in further detail in **Figure A.65**.[109] Since two planar prochiral units are present, four combinations are again possible. The diastereoselection is not a consequence of catalyst asymmetry, as racemic chelating Lewis acids give rise to predominantly *syn* adducts due to the illustrated steric interactions in the *anti* approaches. With the constraint for *syn* addition, the chiral catalyst then dicatates addition to the *Si* aldehyde face by sterically blocking the *Re* approach (see Chapter 2 for detailed stereochemical models).

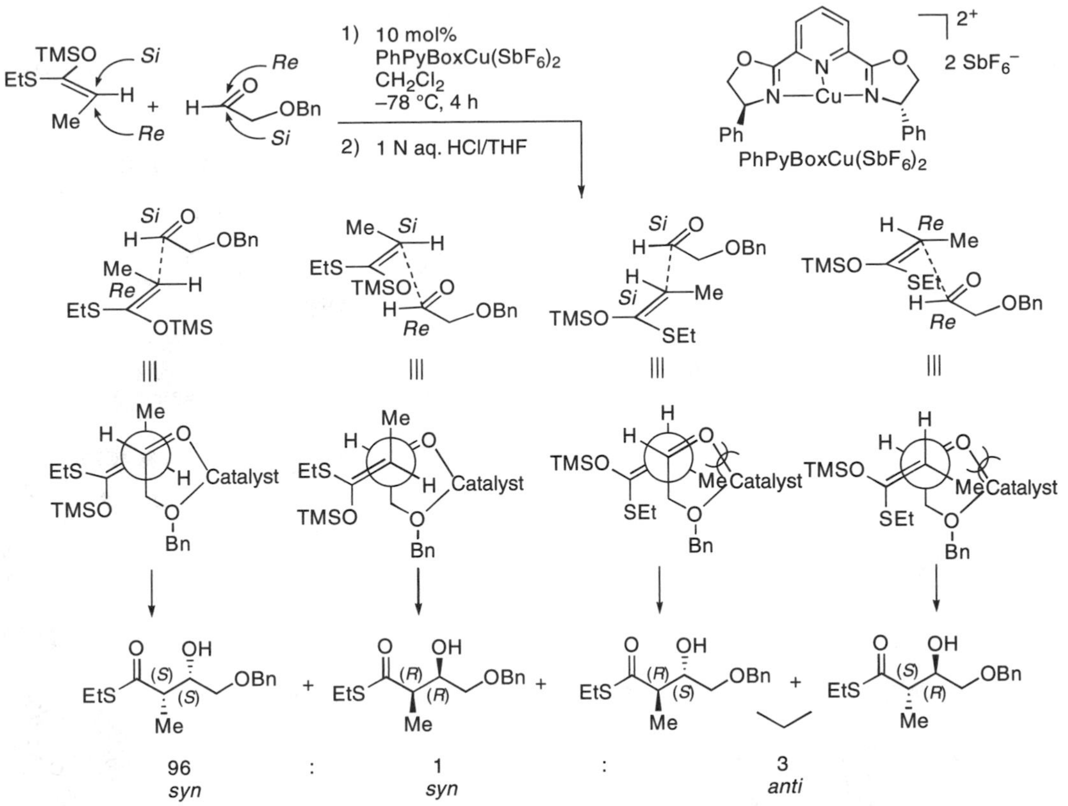

Figure A.65. Combining two different prochiral double bonds to generate two centrochiral units.

Combination of two planar prochiral units can also establish more than two stereogenic units in the products. For example, in the hetero-Diels–Alder reaction illustrated in **Figure A.66**, three centrochiral units can be established when a disubstituted dienophile such as dihydrofuran is employed.[156] The fixed cis stereochemistry of the dienophile only allows four possible stereochemical combinations even though there are three prochiral centers; in all cases the ring fusion is cis-substituted. While diastereoselection may be improved in the presence of a chiral catalyst (see Chapter 1), similar *endo* diastereoselection is observed here even with achiral catalysts. The primary role of the catalyst is then to dictate facial approach to the diene. In this case, *Si* approach is highly favorable (95% ee in the *endo* isomer).

A.2.2.c Transformations of Racemic Compounds

Enantiomerically pure compounds with more than one stereogenic unit are also generated from racemic precursors via resolution reactions (see Chapters 7–9) that establish additional stereogenic units. For example, the epoxidation of a racemic mixture of (R)- and (S)-trans-1-cyclohexylbut-2-enol from **Figure A.27**[104,105] gives rises to the *anti* (S)-product predominantly if the reaction is stopped at 50% conversion (**Figure A.67**). Since epoxidation is a cis stereospecific reaction, only two products can be produced from each starting material enantiomer. Under these conditions, the (R)-substrate is essentially unreactive. As such, only attack at the top face (*Si/Re*) versus

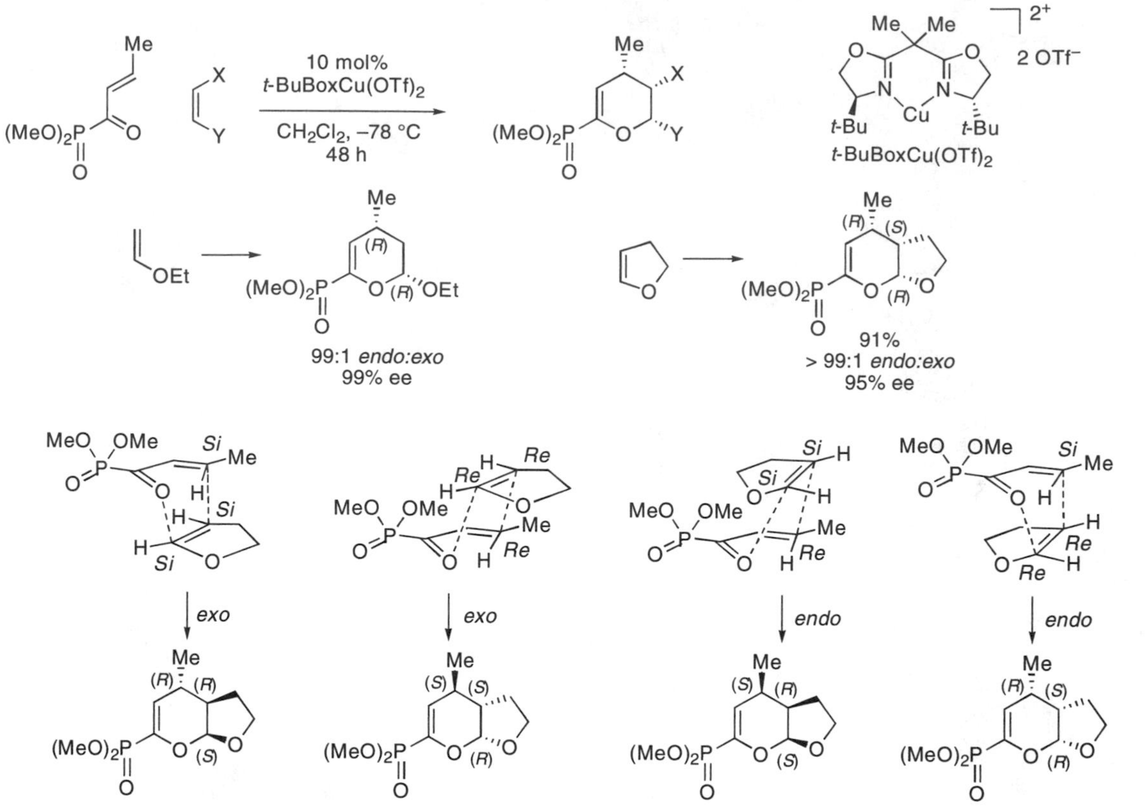

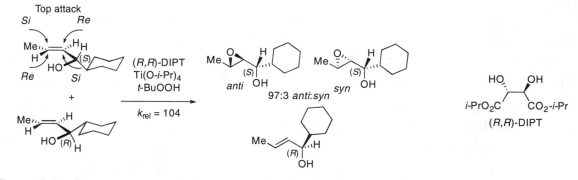

Figure A.66. Combining two different prochiral double bonds to generate multiple centrochiral units.

Figure A.67. Generation of multiple stereogenic units via kinetic resolution.

the bottom face (*Re/Si*) of the (*S*)-substrate needs to be considered. While a resolution of the starting material could have been achieved, even if no differentiation of these diastereofaces occurred, the product with three stereogenic units is also produced with useful levels of diastereocontrol (97:3, *anti:syn*).

Dynamic kinetic resolutions are more useful, as single enantiomers with multiple stereogenic units can be produced from racemic mixtures. For example, in the ruthenium–BINAP-catalyzed hydrogenation of the ketone in **Figure A.68**, the two ketone enantiomers can equilibrate via the enol. These hydrogenations proceed via the keto form in which the ester carbonyl acts as a donor tether, thus establishing a cis relationship between the delivered hydride and the ester. *Si* facial addition is favored by the catalyst and only the (*R*)-configured starting material can assist in the

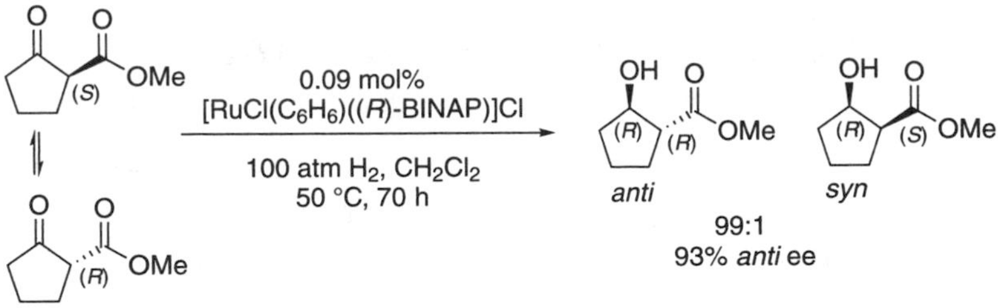

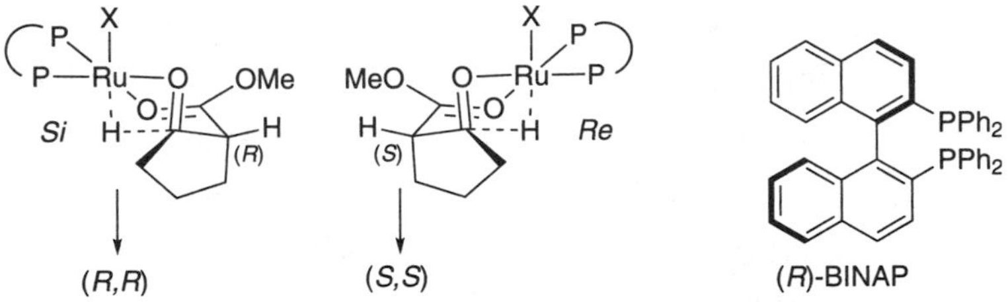

Figure A.68. Generation of multiple stereogenic units via dynamic kinetic resolution.

cis delivery of the ruthenium hydride to this prochiral face. As a result, the (R)-stereochemistry and Si face combination are favored, leading to the (R,R)-product.[157]

Kinetic resolutions that generate more than one stereogenic unit are not restricted to centrochiral units. For example, the dynamic kinetic resolution outlined in **Figure A.69** gives rise to products with a new centrochiral unit and a resolved axial chiral unit.[158] In the starting material, the chiral axis is not configurationally stable and the atropisomers equilibrate under the reaction conditions. With catalytic (S)-proline, the aldol reaction with acetone proceeds to the back (Si) face of the aldehyde. The Si facial approaches to the (R)- and (S)-configured atropisomers are diastereotopic and addition to the former is favored, resulting in a prepoderance of the anti (R,R)-product. At the product stage, the barrier to atropisomerization is higher and the chiral axis is configurationally stable.

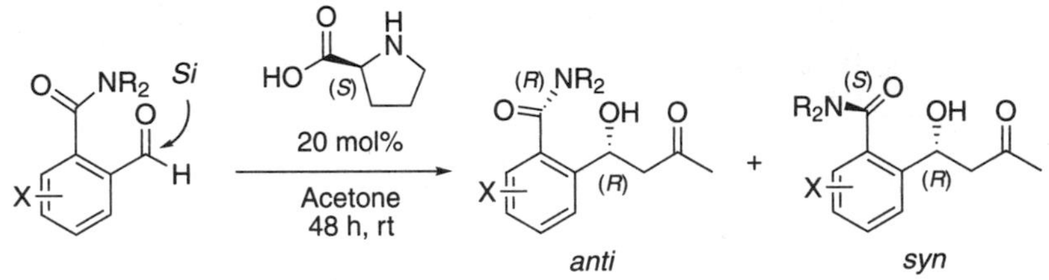

Substrate		Yield (%)	anti:syn	anti ee (%)
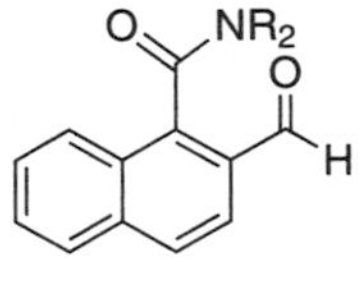	R = i-Pr	87	5.5:1	91
	R = Cy	89	4.8:1	92
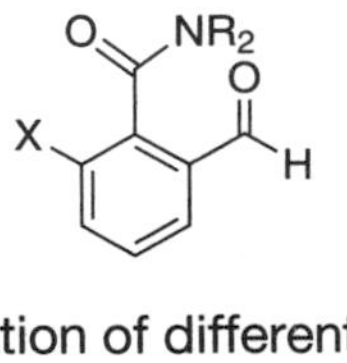	X = NMe₂	92	3.6:1	94
	X = CF₃	86	7.0:1	82
	X = Ph	100	3.0:1	90
	X = OMe	80	2.1:1	95

Figure A.69. Generation of different types of stereogenic units simultaneously via dynamic kinetic resolution.

A.3 Generation of Selected Stereoisomers

A key issue in asymmetric catalysis is the accessibility of all potential stereoisomers. Ideally, specific catalysts will produce each of the potential stereoisomers for a given process in pure form. For example, with a prochiral substrate, either enantiomeric product can be produced if both enantiomers of a chiral catalyst (e.g., an enantiopure amino acid) are equally accessible. On the other hand, obtaining the enantiomeric product can be problematic if the chiral catalyst is derived from a natural source where only one enantiomer is readily available. For example, in the asymmetric aerobic Wacker cyclization illustrated in **Figure A.70**, a (–)-sparteine–palladium complex is employed as the catalyst to good effect.[159] While a reliable supply of (–)-sparteine is available via isolation from natural resources, the analogous (+)-sparteine is not. Thus, the enantiomeric products from these reactions are not readily accessible. A great deal of effort has been expended to develop a scalable synthesis of (+)-sparteine or a readily available, reliable (+)-sparteine surrogate. The development of a surrogate appears promising;[160] however, the supply of the requisite precursor is limiting.

Many elegant solutions have been devised to address this problem. Most notably, the use of pseudoenantiomeric catalysts has proven highly successful. For example, in the Sharpless asymmetric hydroxylation, the discovery of the dihydroquinidine-(DHQD)- and dihydroquinine-(DHQ)-phthalazine-(PHAL)-derived osmium catalysts has proven very powerful, especially since both DHQD and DHQ are readily available.[161] Even though (DHQD)$_2$PHAL and (DHQ)$_2$PHAL chiral ligands are not enantiomeric, they provide almost equal levels of enantiomeric excess of the product enantiomers (**Figure A.71**). Their structures possess opposite configurations at the ethereal stereocenters and the same configurations in the azabicylooctanes. Apparently, the former acts as the dominant stereocontrol element.

Another solution has been to employ additives which alter the catalyst significantly or shift the mechanism of the reaction to provide the enantiomeric product. For example, one enantiomer of a chiral catalyst can provide either enantiomeric product in the Diels–Alder reaction from **Figure A.72** depending on which *achiral* additive is employed.[162] This approach is especially useful when the enantiomeric chi-

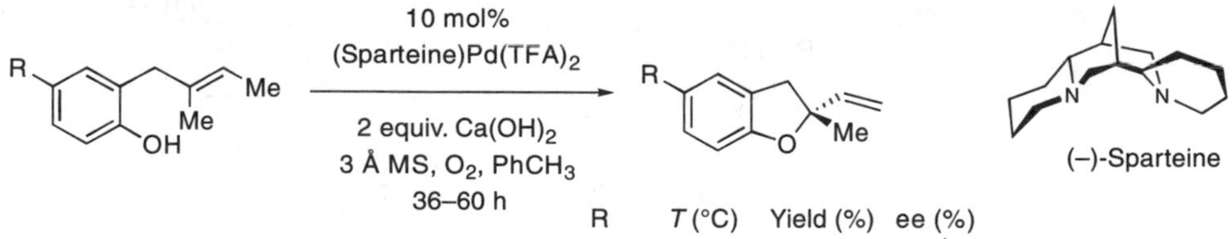

R	T (°C)	Yield (%)	ee (%)
H	80	87	81
OMe	55	57	90

Figure A.70. Asymmetric aerobic Wacker cyclization catalyzed by a palladium–sparteine complex.

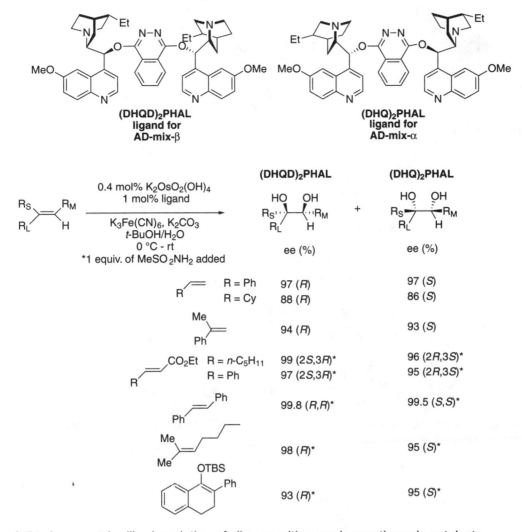

Figure A.71. Asymmetric dihydroxylation of alkenes with pseudoenantiomeric catalysts.

ral catalysts are unavailable or prohibitively expensive. For further discussion and examples, see Chapter 6.

When diastereoselection is also an issue, then at least three possible isomers can be produced. Ideally, the ability to obtain all compounds in pure form (> 90% product) by control of the reaction conditions and chiral catalyst is highly desirable. However, substrate control and catalyst control may both contribute (see Chapter 13). While the latter is more desirable in terms of synthetic flexibility and complete stereocontrol, it is not often achievable. For example, in the thermal reaction of a variety of dienes with the dienophile 3-propenoyl-2-oxazolidinone, the *endo* diastereomer tends to predominate (**Figure A.73**).[163] When a chiral bisoxazoline copper catalyst is employed this inherent selectivity is typically amplified (entries 1–4). For entries 1 and 2, the inherent differences in the two sets of diastereomeric pathways are relatively large with one pathway being much more favorable. In this case, the catalyst only modifies the diastereomeric ratio to a small extent and the other diastereomer is inherently inaccessible. In cases where the diene encounters additional steric interactions with the

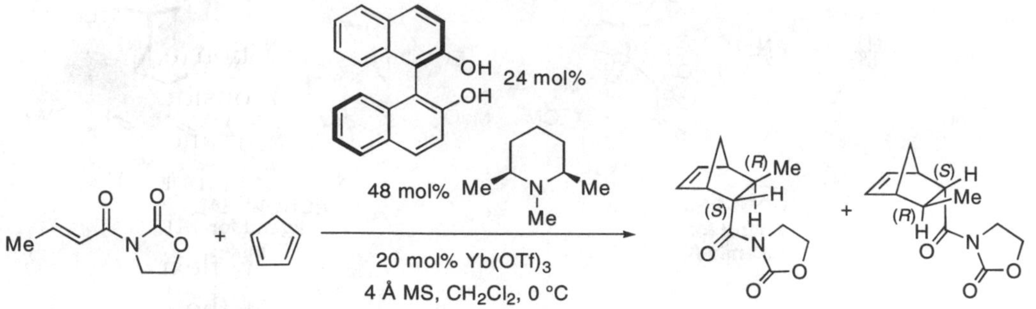

Entry	Additive	Yield (%)	endo:exo	(2S,3R):(2R,3S)
1	No catalyst aging	77	89:11	97.5:2.5
2	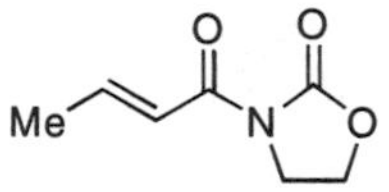	66	87:13	95:6
3	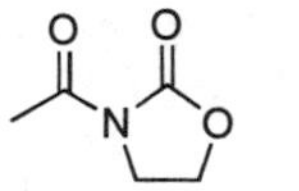	77	89:11	96.5:3.5
4	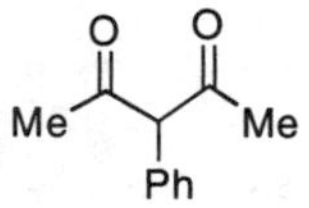	83	93:7	9.5:90.5

Figure A.72. Use of an additive with one enantiomer of a chiral catalyst to selectively generate enantiomeric Diels–Alder adducts.

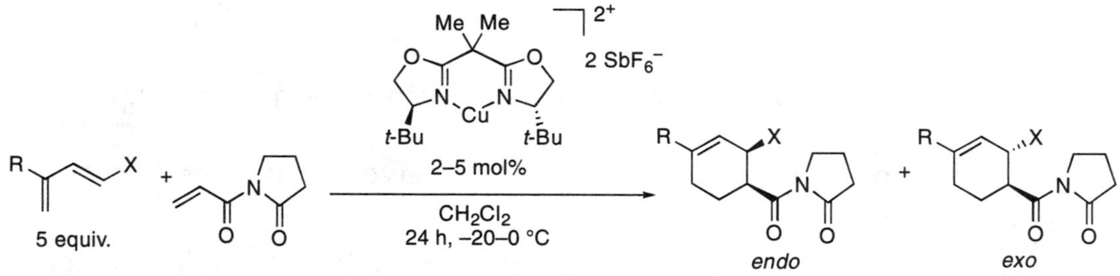

Entry	X	R	Yield (%)	endo:exo Thermal	Catalyzed	endo ee (%)	exo ee (%)
1	Ph	H	95	82:18	85:15	97	—
2	OAc	H	75	75:25	85:15	96	90
3	SPh	H	84	65:35	98:2	98	89
4	NHCbz	H	54	49:51	72:28	90	97
5	OAc	Me	57	60:40	27:73	—	98

Figure A.73. Diastereoselection vs. enantioselection in copper–bisoxazoline-catalyzed Diels–Alder reactions.

catalyst in the *endo* approach, the major diastereomer can switch from the *endo* to the *exo* compound in going from the thermal to the catalyzed reaction (entry 5). However, in all cases the changes are relatively small (especially when considered in the context of ΔG; see Chapter 1). Here substrate control is predominant and access to the *exo* products with high diastereoselectivity and enantioselectivity is typically difficult, whereas generation of the enantiomerically pure *endo* adducts is straightforward.

Other catalyst combinations have been devised that allow flexible diastereocontrol and enantiocontrol. Such a situation is ideal, as all of the stereoisomers are in principle accessible in pure form by selection of the appropriate catalyst. For example, in the additions of silyl ketene acetals to pyruvate esters, as catalyzed by various bisoxazoline-metal complexes, three of the four possible diastereomers can be generated in practically pure form (**Figure A.74**) by adjusting the ligand (bisoxazoline vs. pyridinebisoxazoline) and/or metal [Cu(II) or Sn(II)].[164,118] The fourth (*R,R*)-stereoisomer would be available from the enantiomeric Cu(II)–bisoxazoline complex. In these reactions, the catalyst is the predominant control element overriding any inherent substrate diastereoselection. As such, the catalyst provides stereocontrol on multiple levels (i.e., *Re* vs. *Si* facial selectivity for *both* the ketoester and the silyl ketene acetal substrates).

As the above paragraphs illustrate, there are several ways to modify catalysts to access the different stereoisomers of a reaction in pure form. Nonetheless, challenges remain when the enantiomeric catalyst is not readily available or when multiple diastereomers are possible.

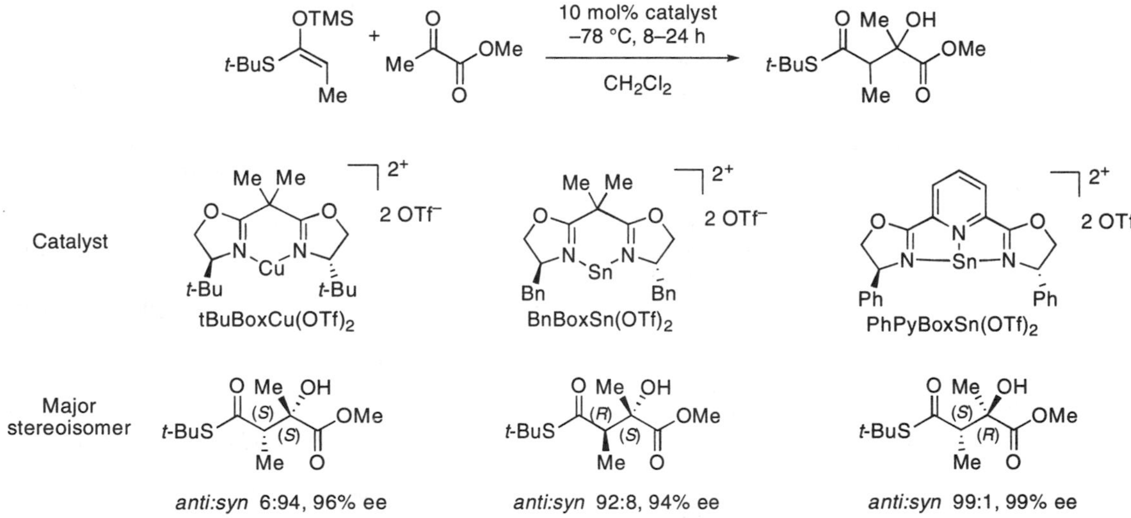

Figure A.74. Catalyst controlled diastereo- and enantioselective Mukaiyama aldol reactions between pyruvate esters and silyl ketene acetals.

References

(1) *Asymmetric Synthesis*; Aitken, R. A., Kilényi, S. N., Eds.; Chapman & Hall: New York, 1992.

(2) Koskinen, A. *Asymmetric Synthesis of Natural Products*; Wiley: New York, 1993.

(3) Eliel, E. L.; Wilen, S. H.; Mander, L. N. *Stereochemistry of Organic Compounds*; Wiley: New York, 1994.

(4) Nógrádi, M. *Stereoselective Synthesis*; VCH-Publishers: New York, 1995.

(5) Noyori, R.; Takaya, H. BINAP: An Efficient Chiral Element for Asymmetric Catalysis. *Acc. Chem. Res.* **1990**, *23*, 345–350.

(6) Hayashi, T.; Konishi, M.; Fukushima, M.; Mise, T.; Kagotani, M.; Tajika, M.; Kumada, M. Asymmetric Synthesis Catalyzed by Chiral Ferrocenylphosphine-Transition Metal Complexes. 2. Nickel- and Palladium-Catalyzed Asymmetric Grignard Cross-Coupling. *J. Am. Chem. Soc.* **1982**, *104*, 180–186.

(7) List, B. Proline-Catalyzed Asymmetric Reactions. *Tetrahedron* **2002**, *58*, 5573–5590.

(8) Movassaghi, M.; Jacobsen, E. N. The Simplest "Enzyme." *Science* **2002**, *298*, 1904–1905.

(9) Notz, W.; Tanaka, F.; Barbas, C. F., III. Enamine-Based Organocatalysis with Proline and Diamines: The Development of Direct Catalytic Asymmetric Aldol, Mannich, Michael, and Diels–Alder Reactions. *Acc. Chem. Res.* **2004**, *37*, 580–591.

(10) Sorensen, E. J.; Sammis, G. M. Chemistry: A Dash of Proline Makes Things Sweet. *Science* **2004**, *305*, 1725–1726.

(11) Cahn, R. S.; Ingold, C.; Prelog, V. Specification of Molecular Chirality. *Angew. Chem., Int. Ed. Engl.* **1966**, *5*, 385–415.

(12) Prelog, V.; Helmchen, G. Basic Principles of the CIP-System and Proposals for a Revision. *Angew. Chem., Int. Ed. Engl.* **1982**, *21*, 567–583.

(13) Mori, K.; Uematsu, T.; Yanagi, K.; Minobe, M. Synthesis of the Optically Active Forms of 4,10-Dihydroxy-1,7-dioxaspiro[5.5]undecane and Their Conversion to the Enantiomers of 1,7-Dioxaspirol[5.5]undecane, the Olive Fly Pheromone. *Tetrahedron* **1985**, *41*, 2751–2758.

(14) Haniotakis, G.; Francke, W.; Mori, K.; Redlich, H.; Schurig, V. Sexspecific Activity of (*R*)-(–)- and (*S*)-(+)-1,7-Dioxaspirol[5.5]undecane, the Major Pheromone of *Dacus oleae*. *J. Chem. Ecol.* **1986**, *12*, 1559–1568.

(15) Pfaltz, A. Design of Chiral Ligands for Asymmetric Catalysis. *Chimia* **2004**, *58*, 49–50.

(16) Johnson, J. S.; Evans, D. A. Chiral Bis(oxazoline) Copper(II) Complexes: Versatile Catalysts for Enantioselective Cycloaddition, Aldol, Michael, and Carbonyl Ene Reactions. *Acc. Chem. Res.* **2000**, *33*, 325–335.

(17) Tokunaga, N.; Otomaru, Y.; Okamoto, K.; Ueyama, K.; Shintani, R.; Hayashi, T. *C*₂-Symmetric Bicyclo[2.2.2]octadienes as Chiral Ligands: Their High Performance in Rhodium-Catalyzed Asymmetric Arylation of *N*-Tosylarylimines. *J. Am. Chem. Soc.* **2004**, *126*, 13584–13585.

(18) Kacprzak, K.; Gawronski, J. Cinchona Alkaloids and Their Derivatives: Versatile Catalysts and Ligands in Asymmetric Synthesis. *Synthesis* **2001**, 961–998.

(19) Maryanoff, C. A.; Maryanoff, B. E. In *Asymmetric Synthesis*; Morrison, J. D., Scott, J. W., Eds.; Academic Press: New York, 1984; Vol. 4, Chapter 5.

(20) Chan, T. H.; Wang, D. Chiral Organosilicon Compounds in Asymmetric Synthesis. *Chem. Rev.* **1992**, *92*, 995–1006.

(21) Oestreich, M.; Rendler, S. True Chirality Transfer from Silicon to Carbon: Asymmetric Amplification in a Reagent-Controlled Palladium-Catalyzed Hydrosilylation. *Angew. Chem., Int. Ed. Engl.* **2005**, *44*, 1661–1664.

(22) Rendler, S.; Oestreich, M.; Butts, C. P.; Lloyd-Jones, G. C. Intermolecular Chirality Transfer from Silicon to Carbon: Interrogation of the Two-Silicon Cycle for Pd-Catalyzed Hydrosilylation by Stereoisotopochemical Crossover. *J. Am. Chem. Soc.* **2007**, *129*, 502–503.

(23) Davis, F. D.; Jenkins, R. H. In *Asymmetric Synthesis*; Scott, J. W., Ed.; Academic Press: New York, 1984; Vol. 4, Chapter 4.

(24) Davis, F. A.; Reddy, R. T.; Han, W.; Reddy, R. E. Asymmetric Synthesis Using *N*-Sulfonyloxaziridines. *Pure Appl. Chem.* **1993**, *65*, 633–640.

(25) Shen, Y.; Feng, X.; Li, Y.; Zhang, G.; Jiang, Y. Asymmetric Cyanosilylation of Ketones Catalyzed by Bifunctional Chiral *N*-Oxide Titanium Complex Catalysts. *Eur. J. Org. Chem.* **2004**, 129–137.

(26) Maruoka, K.; Ooi, T. Enantioselective Amino Acid Synthesis by Chiral Phase-Transfer Catalysis. *Chem. Rev.* **2003**, *103*, 3013–3028.

(27) Pelz, K. A.; White, P. S.; Gagne, M. R. Persistent *N*-Chirality as the Only Source of Asymmetry in Nonracemic N₂PdCl₂ Complexes. *Organometallics* **2004**, *23*, 3210–3217.

(28) Valentine, D. R. In *Asymmetric Synthesis*; Morrison, J. D., Scott, J. W., Eds.; Academic Press: New York, 1984; Vol. 4, Chapter 3.

(29) Pietrusiewicz, K. M.; Zablocka, M. Preparation of Scalemic *P*-Chiral Phosphines and Their Derivatives. *Chem. Rev.* **1994**, *94*, 1375–1411.

(30) Crepy, K. V. L.; Imamoto, T. New *P*-Chirogenic Phosphine Ligands and Their Use in Catalytic Asymmetric Reactions. *Top. Curr. Chem.* **2003**, *229*, 1–40.

(31) Yamanoi, Y.; Imamoto, T. New Chiral Phosphine Ligands for Catalytic Asymmetric Reactions. *Rev. Heteroa. Chem.* **1999**, *20*, 227–248.

(32) Fernandez, I.; Khiar, N. Recent Developments in the Synthesis and Utilization of Chiral Sulfoxides. *Chem. Rev.* **2003**, *103*, 3651–3705.

(33) Allin, S. M.; Shuttleworth, S. J.; Page, P. C. B. Applications of Chiral Sulfoxides as Stereocontrol Elements in Organic Synthesis. *Organosulfur Chem.* **1998**, *2*, 97–155.

(34) Reggelin, M.; Zur, C. Sulfoximines. Structures, Properties, and Synthetic Applications. *Synthesis* **2000**, 1–64.

(35) Zhou, P.; Chen, B.-C.; Davis, F. A. Recent Advances in Asymmetric Reactions Using Sulfinimines (*N*-Sulfinyl Imines). *Tetrahedron* **2004**, *60*, 8003–8030.

(36) Barbachyn, M. R.; Johnson, C. R. In *Asymmetric Synthesis*; Morrison, J. D., Scott, J. W., Eds.; Academic Press: New York, 1984; Vol. 4, Chapter 2.

(37) Lindberg, P.; Brändstrom, A.; Wallmark, B.; Mattson, H.; Rikner, L.; Hoffman, K.-J. Omeprazole—The 1st Proton Pump Inhibitor. *Med. Res. Rev.* **1990**, *10*, 1–54.

(38) Carlsson, E.; Lindberg, P.; von Unge, S. Two of a Kind. *Chem. Br.* **2002**, *38*, 42–45.

(39) Owens, T. D.; Hollander, F. J.; Oliver, A. G.; Ellman, J. A. Synthesis, Utility, and Structure of Novel Bis(sulfinyl)imidoamidine Ligands for Asymmetric Lewis Acid Catalysis. *J. Am. Chem. Soc.* **2001**, *123*, 1539–1540.

(40) Bolm, C.; Simic, O. Highly Enantioselective Hetero-Diels–Alder Reactions Catalyzed by a C_2-Symmetric Bis(sulfoximine) Copper(II) Complex. *J. Am. Chem. Soc.* **2001**, *123*, 3830–3831.

(41) Brunner, H. Optical Induction in Organo-Transition-Metal Compounds and Asymmetric Catalysis. *Acc. Chem. Res.* **1979**, *12*, 250–257.

(42) Halterman, R. L. Synthesis and Applications of Chiral Cyclopentadienylmetal Complexes. *Chem. Rev.* **1992**, *92*, 965–994.

(43) Von Zelewsky, A. *Stereochemistry of Coordination Compounds*; Wiley: New York, 1996.

(44) Knof, U.; von Zelewsky, A. Predetermined Chirality at Metal Centers. *Angew. Chem., Int. Ed. Engl.* **1999**, *38*, 302–322.

(45) Brunner, H. Optically Active Organometallic Compounds of Transition Elements with Chiral Metal Atoms. *Angew. Chem., Int. Ed. Engl.* **1999**, *38*, 1194–1208.

(46) Liebeskind, L. S.; Welker, M. E.; Fengl, R. W. Transformations of Chiral Iron Complexes Used in Organic Synthesis. Reactions of η^5–CpFe(PPh$_3$)(CO)COCH$_3$ and Related Species Leading to a Mild, Stereospecific Synthesis of β-Lactams. *J. Am. Chem. Soc.* **1986**, *108*, 6328–6343.

(47) Shimba, S.; Fujinami, S.; Shibata, M. Preparation of a Novel Aminebromocyano[1,4,7-Triazacyclononane]Cobalt(III) Complex. *Chem. Lett.* **1979**, 783–784.

(48) Werner, A. Über Spiegelbild-Isomerie bei Eisenverbindungen. *Chem. Ber.* **1912**, *45*, 433–436.

(49) Bijvoet, J. M.; Peerdeman, A. F.; van Bommel, A. J. Determination of the Absolute Configuration of Optically Active Compounds by Means of X-Rays. *Nature (London)* **1951**, *168*, 271–272.

(50) Saito, Y.; Nakatsu, K.; Shiro, M.; Kuroya, H. Determination of the Absolute Configuration of Optically Active Complex Ion [Co(en)$_3$]$^{3+}$ by Means of X-Rays. *Acta Crystallog.* **1955**, *8*, 729–730.

(51) Cotton, F. A.; Wilkinson, G. *Advanced Inorganic Chemistry*; Wiley: New York, 1988.

(52) Sasai, H.; Arai, T.; Satow, Y.; Houk, K. N.; Shibasaki, M. The First Heterobimetallic Multifunctional Asymmetric Catalyst. *J. Am. Chem. Soc.* **1995**, *117*, 6194–6198.

(53) Sato, I.; Kadowaki, K.; Ohgo, Y.; Soai, K.; Ogino, H. Highly Enantioselective Asymmetric Autocatalysis Induced by Chiral Cobalt Complexes Due to the Topology of the Coordination of the Achiral Ligands. *J. Chem. Soc., Chem. Commun.* **2001**, 1022–1023.

(54) Kromm, K.; Osburn, P. L.; Gladysz, J. A. Chelating Diphosphines That Contain a Rhenium Stereocenter in the Backbone: Applications in Rhodium-Catalyzed Enantioselective Ketone Hydrosilylations and Alkene Hydrogenation. *Organometallics* **2002**, *21*, 4275–4280.

(55) Rosini, C.; Franzini, L.; Raffaelli, A.; Salvadori, P. Synthesis and Application of Binaphthylic C_2-Symmetry Derivatives as Chiral Auxiliaries in Enantioselective Reactions. *Synthesis* **1992**, 503–517.

(56) Noyori, R. In *Stereocontrolled Organic Synthesis*; Trost, B. M., Ed.; Blackwell Scientific Publications: Cambridge, MA, 1994; 1–15.

(57) Pu, L. 1,1'-Binaphthyl Dimers, Oligomers, and Polymers: Molecular Recognition, Asymmetric Catalysis, and New Materials. *Chem. Rev.* **1998**, *98*, 2405–2494.

(58) McCarthy, M.; Guiry, P. J. Axially Chiral Bidentate Ligands in Asymmetric Catalysis. *Tetrahedron* **2001**, *57*, 3809–3844.

(59) Brunel, J. M. BINOL: A Versatile Chiral Reagent. *Chem. Rev.* **2005**, *105*, 857–898.

(60) Bringmann, G.; Günther, C.; Ochse, M.; Schupp, O.; Tasler, S. Biaryls in Nature: A Multi-Facetted Class of Stereochemically, Biosynthetically, and Pharmacologically Intriguing Secondary Metaobolites. *Prog. Chem. Org. Nat. Prod.* **2001**, *82*, 1–249.

(61) Baudoin, O.; Gueritte, F. Natural Bridged Biaryls with Axial Chirality and Antimitotic Properties. *Stud. Nat. Prod. Chem.* **2003**, *29 (Part J)*, 355–417.

(62) Dagne, E.; Steglich, W. Knipholone: A Unique Anthraquinone Derivative from *Kniphofia foliosa*. *Phytochemistry* **1984**, *23*, 1729–1731.

(63) Bringmann, G.; Menche, D.; Bezabih, M.; Abegaz, B. M.; Kaminsky, R. Antiplasmodial Activity of Knipholone and Related Natural Phenylanthraquinones. *Planta Med.* **1999**, *65*, 757–758.

(64) Bringmann, G.; Menche, D. First, Atropo-Enantioselective Total Synthesis of the Axially Chiral Phenylanthraquinone Natural Products Knipholone and 6'-*O*-Methylknipholone. *Angew. Chem., Int. Ed. Engl.* **2001**, *40*, 1687–1690.

(65) Clayden, J. Non-biaryl Atropisomers: New Classes of Chiral Reagents, Auxiliaries, and Ligands? *Angew. Chem., Int. Ed. Engl.* **1997**, *36*, 949–951.

(66) Clayden, J. Stereocontrol with Rotationally Restricted Amides. *Synlett* **1998**, 810–816.

(67) Ahmed, A.; Bragg, R. A.; Clayden, J.; Lal, L. W.; McCarthy, C.; Pink, J. H.; Westlund, N.; Yasin, S. A. Barriers to Rotation about the Chiral Axis of Tertiary Aromatic Amides. *Tetrahedron* **1998**, *54*, 13277–13294.

(68) Clayden, J. Atropisomers and Near-Atropisomers: Achieving Stereoselectivity by Exploiting the Conformational Preferences of Aromatic Amides. *J. Chem. Soc., Chem. Commun.* **2004**, 127–135.

(69) Le Gac, S.; Monnier-Benoit, N.; Metoul, L. D.; Petit, S.; Jabin, I. Stereoselective Synthesis of New Classes of Atropisomeric Compounds Through a Tandem Michael Reaction-Azacyclization Process. Part 2. *Tetrahedron: Asymmetry* **2004**, *15*, 139–145.

(70) Dai, W.-M.; Yeung, K. K. Y.; Liu, J.-T.; Zhang, Y.; Williams, I. D. A Novel Class of Nonbiaryl Atropisomeric *P,O*-Ligands for Palladium-Catalyzed Asymmetric Allylic Alkylation. *Org. Lett.* **2002**, *4*, 1615–1618.

(71) Dai, W. M.; Yeung, K. K. Y.; Wang, Y. Q. The First Example of Atropisomeric Amide-Derived *P,O*-Ligands Used for an Asymmetric Heck Reaction. *Tetrahedron* **2004**, *60*, 4425–4430.

(72) Runge, W. In *The Chemistry of Allenes*; Landor, S. R., Ed.; Academic Press, New York, 1982; Vol. 2, Chapter 6.

(73) Hoffmann-Roder, A.; Krause, N. Synthesis and Properties of Allenic Natural Products and Pharmaceuticals. *Angew. Chem., Int. Ed. Engl.* **2004**, *43*, 1196–1216.

(74) Lyle, R. E.; Lyle, G. G. Resolution of 2,6-Diphenyl-1-methyl-4-piperidone Oxime, a Novel Example of Molecular Isomerism. *J. Org. Chem.* **1959**, *24*, 1679–1684.

(75) Lyle, G. G.; Pelosi, E. T. The Absolute Configuration of (+)-1-Methyl-2,6-diphenyl-4-piperidone Oxime. *J. Am. Chem. Soc.* **1966**, *88*, 5276–5279.

(76) Hulshof, L. A.; Wynberg, H.; van Dijk, B.; de Boer, J. L. Reassignment of the Chirality to a Series of 2,6-Disubstituted Spiro[3.3]heptanes by X-ray Methods and Implications Thereof on Empirical Rules and Theoretical Models. *J. Am. Chem. Soc.* **1976**, *98*, 2733–2740.

(77) Tamao, K.; Nakamura, K.; Ishii, H.; Yamaguchi, S.; Shiro, M. Axially Chiral Spirosilanes via Catalytic Asymmetric Intramolecular Hydrosilation. *J. Am. Chem. Soc.* **1996**, *118*, 12469–12470.

(78) Grimme, S.; Harren, J.; Sobanskib, A.; Vögtle, F. Structure/Chiroptics Relationships of Planar Chiral and Helical Molecules. *Eur. J. Org. Chem.* **1998**, 1491–1509.

(79) Reetz, M. T.; Beuttenmüller, E. W.; Goddard, R. First Enantioselective Catalysis using a Helical Diphosphane. *Tetrahedron Lett.* **1997**, *38*, 3211–3214.

(80) Dreher, S. D.; Katz, T. J.; Lam, K.-C.; Rheingold, A. L. Application of the Russig–Laatsch Reaction to Synthesize a Bis[5]helicene Chiral Pocket for Asymmetric Catalysis. *J. Org. Chem.* **2000**, *65*, 815–822.

(81) Schlögl, K. Planar Chiral Molecular Structures. *Top. Curr. Chem.* **1984**, *125*, 29–62.

(82) Paley, R. S. Enantiomerically Pure Planar Chiral Organometallic Complexes via Facially Selective π-Complexation. *Chem. Rev.* **2002**, *102*, 1493–1523.

(83) Togni, A.; Dorta, R.; Kollner, C.; Pioda, G. Some New Aspects of Asymmetric Catalysis with Chiral Ferrocenyl Ligands. *Pure Appl. Chem.* **1998**, *70*, 1477–1485.

(84) Togni, A.; Bieler, N.; Burckhardt, U.; Kollner, C.; Pioda, G.; Schneider, R.; Schnyder, A. Recent Studies in Asymmetric Catalysis Using Ferrocenyl Ligands. *Pure Appl. Chem.* **1999**, *71*, 1531–1537.

(85) Dai, L.-X.; Hou, X.-L.; Deng, W.-P.; You, S.-L.; Zhou, Y.-G. The Application of Ligands with Planar Chirality in Asymmetric Synthesis. *Pure Appl. Chem.* **1999**, *71*, 1401–1405.

(86) Dai, L.-X.; Tu, T.; You, S.-L.; Deng, W.-P.; Hou, X.-L. Asymmetric Catalysis with Chiral Ferrocene Ligands. *Acc. Chem. Res.* **2003**, *36*, 659–667.

(87) Fu, G. C. Enantioselective Nucleophilic Catalysis with "Planar-Chiral" Heterocycles. *Acc. Chem. Res.* **2000**, *33*, 412–420.

(88) Fu, G. C. Asymmetric Catalysis with "Planar-Chiral" Heterocycles. *Pure Appl. Chem.* **2001**, *73*, 347–349.

(89) Fu, G. C. Asymmetric Catalysis with "Planar-Chiral" Derivatives of 4-(Dimethylamino)pyridine. *Acc. Chem. Res.* **2004**, *37*, 542–547.

(90) Gibson, S. E.; Ibrahim, H. Asymmetric Catalysis Using Planar Chiral Arene Chromium Complexes. *J. Chem. Soc., Chem. Commun.* **2002**, 2465–2473.

(91) Pye, P. J.; Rossen, K.; Reamer, R. A.; Tsou, N. N.; Volante, R. P.; Reider, P. J. A New Planar Chiral Bisphosphine Ligand for Asymmetric Catalysis: Highly Enantioselective Hydrogenations Under Mild Conditions. *J. Am. Chem. Soc.* **1997**, *119*, 6207–6208.

(92) Solladié-Cavallo, A. In *Advances in Metal-Organic Chemistry*; Liebeskind, L. S., Ed.; JAI Press: London, 1989; Vol. 2, pp. 99–133.

(93) Schlögl, K. Stereochemistry of Metallocenes. *Top. Stereochem.* **1967**, *1*, 39–89.

(94) Gawley, R. E. Do the Terms "% ee" and "% de" Make Sense as Expressions of Stereoisomer Composition or Stereoselectivity? *J. Org. Chem.* **2006**, *71*, 2411–2416.

(95) Bringmann, G.; Rübenacker, M.; Weirich, R.; Aké Assi, L. Dioncophylline C from the Roots of *Triphyophyllum peltatum*, the First 5,1'-Coupled Dioncophyllaceae Alkaloid. *Phytochemistry* **1992**, *31*, 4019–4024.

(96) François, G.; Timperman, G.; Eling, W.; Aké Assi, L.; Holenz, J.; Bringmann, G. Naphthylisoquinoline Alkaloids Against Malaria: Evaluation of the Curative Potential of Dioncophylline C and Dioncopeltine A Against *Plasmodium berghei in vivo*. *Antimicrob. Agents Chemother.* **1997**, *41*, 2533–2539.

(97) Blaser, H.-U.; Brieden, W.; Pugin, B.; Spindler, F.; Studer, M.; Togni, A. Solvia Josiphos Ligands: From Discovery to Technical Applications. *Top. Catal.* **2002**, *19*, 3–16.

(98) Koide, H.; Uemura, M. Axially Chiral Benzamides: Diastereoselective Nucleophilic Additions to Planar Chiral (*N,N*-Diethyl-2-acyl-6-methylbenzamide)chromium Complexes. *Tetrahedron Lett.* **1999**, *40*, 3443–3446.

(99) Willis, M. C. Enantioselective Desymmetrisation. *J. Chem. Soc., Perkin Trans. 1* **1999**, 1765–1784.

(100) Spivey, A. C.; Andrews, B. I. Catalysis of the Asymmetric Desymmetrization of Cyclic Anhydrides by Nucleophilic Ring-Opening with Alcohols. *Angew. Chem., Int. Ed. Engl.* **2001**, *40*, 3131–3134.

(101) Chen, Y.; McDaid, P.; Deng, L. Asymmetric Alcoholysis of Cyclic Anhydrides. *Chem. Rev.* **2003**, *103*, 2965–2984.

(102) Mihelich, E. D.; Daniels, K.; Eickhoff, D. J. Vanadium-catalyzed Epoxidations. 2. Highly Stereoselective Epoxidations of Acyclic Homoallylic Alcohols Predicted by a Detailed Transition-State Model. *J. Am. Chem. Soc.* **1981**, *103*, 7690–7692.

(103) Tanaka, S.; Yamamoto, H.; Nozaki, H.; Sharpless, S. K. B.; Michaelson, R. C.; Cutting, J. D. Stereoselective Epoxidations of Acyclic Allylic Alcohols by Transition Metal-Hydroperoxide Reagents. Synthesis of *dl*-C_{18} *Cecropia* Juvenile Hormone from Farnesol. *J. Am. Chem. Soc.* **1974**, *96*, 5254–5255.

(104) Martín, V. S.; Woodard, S. S.; Katsuki, T.; Yamada, Y.; Ikeda, M.; Sharples, K. B. Kinetic Resolution of Racemic Allylic Alcohols by Enantioselective Epoxidation. A Route to Substances of Absolute Enantiomeric Purity? *J. Am. Chem. Soc.* **1981**, *103*, 6237–6240.

(105) Gao, Y.; Klunder, J. M.; Hanson, R. M.; Masamune, H.; Ko, S. Y.; Sharpless, K. B. Catalytic Asymmetric Epoxidation and Kinetic Resolution: Modified Procedures Including in situ Derivatization. *J. Am. Chem. Soc.* **1987**, *109*, 5765–5780.

(106) Takemoto, Y.; Baba, Y.; Honda, A.; Nakao, S.; Noguchi, I.; Iwata, C.; Tanaka, T.; Ibuka, T. Asymmetric Synthesis of (Diene)Fe(CO)₃ Complexes by a Catalytic Enantioselective Alkylation Using Dialkylzincs. *Tetrahedron* **1998**, *54*, 15567–15580.

(107) Christlieb, M.; Davies, J. E.; Eames, J.; Hooley, R.; Warren, S. The Stereoselective Synthesis of Oxetanes; Exploration of a New, Mitsunobu-Style Procedure for the Cyclisation of 1,3-Diols. *J. Chem. Soc., Perkin Trans. 1* **2001**, 2983–2996.

(108) Gennari, C.; Cozzi, P. G. Chelation Controlled Aldol Additions of the Enolsilane Derived from *tert*-Butyl Thioacetate : A Stereosetective Approach to 1β-Methylthienamycin. *Tetrahedron* **1988**, *44*, 5965–5974.

(109) Evans, D. A.; Kozlowski, M. C.; Murry, J. A.; Burgey, C. S.; Campos, K. R.; Connell, B. T.; Staples, R. J. C_2-Symmetric Copper(II) Complexes as Chiral Lewis Acids. Scope and Mechanism of Catalytic Enantioselective Aldol Additions of Enolsilanes to (Benzyloxy)acetaldehyde. *J. Am. Chem. Soc.* **1999**, *121*, 669–685.

(110) Burk, M. J.; Feaster, J. E.; Nugent, W. A.; Harlow, R. L. Preparation and Use of C_2-Symmetric Bis(phospholanes): Production of α-Amino Acid Derivatives via Highly Enantioselective Hydrogenation Reactions. *J. Am. Chem. Soc.* **1993**, *115*, 10125–10138.

(111) Robinson, A. J.; Stanislawski, P.; Mulholland, D.; He, L.; Li, H.-Y. Expedient Asymmetric Synthesis of All Four Isomers of N,N′-Protected 2,3-Diaminobutanoic Acid. *J. Org. Chem.* **2001**, *66*, 4148–4152.

(112) Konno, T.; Tanikawa, M.; Ishihara, T.; Yamanaka, H. Palladium-Catalyzed Coupling Reaction of Fluoroalkylated Propargyl Mesylates with Organozinc Reagents: Novel Synthesis of Optically Active Fluorine-Containing Trisubstituted Allenes. *Chem. Lett.* **2000**, 1360–1361.

(113) Mislow, K. *Introduction to Stereochemistry*; Benjamin: New York, 1965.

(114) Yamaguchi, H.; Nakanishi, S.; Takata, T. Synthesis of Planar Chiral η³-Allyldicarbonylnitrosyliron Complexes and Stereochemistry of the Complex Forming Reaction. *J. Organomet. Chem.* **1998**, *554*, 167–170.

(115) Nishii, Y.; Wakasugi, K.; Koga, K.; Tanabe, Y. Chirality Exchange from sp^3 Central Chirality to Axial Chirality: Benzannulation of Optically Active Diaryl-2,2-dichlorocyclopropylmethanols to Axially Chiral α-Arylnaphthalenes. *J. Am. Chem. Soc.* **2004**, *126*, 5358–5359.

(116) Imperiali, B.; Zimmerman, J. W. Synthesis of Dolichols via Asymmetric Hydrogenation of Plant Polyprenols. *Tetrahedron Lett.* **1988**, *29*, 5343–5344.

(117) Crispino, G. A.; Sharpless, K. B. Asymmetric Dihydroxylation of Squalene. *Tetrahedron Lett.* **1992**, *33*, 4273–4274.

(118) Evans, D. A.; MacMillan, D. W. C.; Campos, K. R. C_2-Symmetric Tin(II) Complexes as Chiral Lewis Acids. Catalytic Enantioselective Anti Aldol Additions of Enolsilanes to Glyoxylate and Pyruvate Esters. *J. Am. Chem. Soc.* **1997**, *119*, 10859–10860.

(119) Momiyama, N.; Yamamoto, H. Brønsted Acid Catalysis of Achiral Enamine for Regio- and Enantioselective Nitroso Aldol Synthesis. *J. Am. Chem. Soc.* **2005**, *127*, 1080–1081.

(120) DiMauro, E. F.; Kozlowski, M. C. Development of Bifunctional Salen Catalysts: Rapid, Chemoselective Alkylations of α-Ketoesters. *J. Am. Chem. Soc.* **2002**, *124*, 12668–12669.

(121) DiMauro, E. F.; Kozlowski, M. C. The First Catalytic Asymmetric Addition of Dialkylzincs to α-Ketoesters. *Org. Lett.* **2002**, *4*, 3781–3784.

(122) List, B.; Pojarliev, P.; Biller, W. T.; Martin, H. J. The Proline-Catalyzed Direct Asymmetric Three-Component Mannich Reaction: Scope, Optimization, and Application to the Highly Enantioselective Synthesis of 1,2-Amino Alcohols. *J. Am. Chem. Soc.* **2002**, *124*, 827–833.

(123) Cotton, H.; Elebring, T.; Larsson, M.; Li, L.; Sörensen, H.; von Unge, S. Asymmetric Synthesis of Esomeprazole. *Tetrahedron: Asymmetry* **2000**, *11*, 3819–3825.

(124) Corey, E. J.; Xu, F.; Noe, M. C. A Rational Approach to Catalytic Enantioselective Enolate Alkylation Using a Structurally Rigidified and Defined Chiral Quaternary Ammonium Salt Under Phase Transfer Conditions. *J. Am. Chem. Soc.* **1997**, *119*, 12414–12415.

(125) Hashiguchi, S.; Fujii, A.; Haack, K. J.; Matsumura, K.; Ikariya, T.; Noyori, R. Kinetic Resolution of Racemic Secondary Alcohols by Ru-II-Catalyzed Hydrogen Transfer. *Angew. Chem., Int. Ed. Engl.* **1997**, *36*, 288–290.

(126) a) Sigman, M. S.; Vachal, P.; Jacobsen, E. N., A General Catalyst for the Asymmetric Strecker Reaction. *Angew. Chem., Int. Ed.* **2000**, *39*, 1279–1281. b) Vachal, P.; Jacobsen, E. N., Enantioselective Catalytic Addition of HCN to Ketoimines. Catalytic Synthesis of Quaternary Amino Acids. *Org. Lett.* **2000**, *2*, 867–870.

(127) Hatano, M.; Terada, M.; Mikami, K. Highly Enantioselective Palladium-Catalyzed Ene-Type Cyclization of a 1,6-Enyne. *Angew. Chem., Int. Ed. Engl.* **2001**, *40*, 249–253.

(128) Copeland, G. T.; Miller, S. J. Selection of Enantioselective Acyl Transfer Catalysts from a Pooled Peptide Library Through a Fluorescence-Based Activity Assay: An Approach to Kinetic Resolution of Secondary Alcohols of Broad Structural Scope. *J. Am. Chem. Soc.* **2001**, *123*, 6496–6502.

(129) Keith, J. M.; Larrow, J. F.; Jacobsen, E. N. Practical Considerations in Kinetic Resolution Reactions. *Adv. Synth. Catal.* **2001**, *343*, 5–26.

(130) Hayashi, T.; Konishi, M.; Ito, H.; Kumada, M. Optically Active Allylsilanes. 1. Preparation by Palladium-Catalyzed Asymmetric Grignard Cross-Coupling and *anti* Stereochemistry in Electrophilic Substitution Reactions. *J. Am. Chem. Soc.* **1982**, *104*, 4962–4963.

(131) Dolling, U. H.; Davis, P.; Grabowski, E. J. J. Efficient Catalytic Asymmetric Alkylations. 1. Enantioselective Synthesis of (+)-Indacrinone via Chiral Phase-Transfer Catalysis. *J. Am. Chem. Soc.* **1984**, *106*, 446–447.

(132) Yin, J.; Buchwald, S. L. A Catalytic Asymmetric Suzuki Coupling for the Synthesis of Axially Chiral Biaryl Compounds. *J. Am. Chem. Soc.* **2000**, *122*, 12051–12052.

(133) Li, X.; Yang, J.; Kozlowski, M. C. Enantioselective Oxidative Biaryl Coupling Reactions Catalyzed by 1,5-Diazadecalin Metal Complexes. *Org. Lett.* **2000**, *3*, 1137–1140.

(134) Li, X.; Hewgley, J. B.; Mulrooney, C. A.; Yang, J.; Kozlowski, M. C. Enantioselective Oxidative Biaryl Coupling Reactions Catalyzed by 1,5-Diazadecalin Metal Complexes: Efficient Formation of Chiral Functionalized BINOL Derivatives. *J. Org. Chem.* **2003**, *68*, 5500–5511.

(135) Hayashi, T.; Niizuma, S.; Kamikawa, T.; Suzuki, N.; Uozumi, Y. Catalytic Asymmetric Synthesis of Axially Chiral Biaryls by Palladium-Catalyzed Enantioposi-

tion-Selective Cross-Coupling. *J. Am. Chem. Soc.* **1995**, *117*, 9101–9102.

(136) Kamikawa, T.; Uozumi, Y.; Hayashi, T. Enantioposition-Selective Alkynylation of Biaryl Ditriflates by Palladium-Catalyzed Asymmetric Cross-Coupling. *Tetrahedron Lett.* **1996**, *37*, 3161–3164.

(137) Kamikawa, T.; Hayashi, T. Enantioposition-Selective Arylation of Biaryl Ditriflates by Palladium-Catalyzed Asymmetric Grignard Cross-Coupling. *Tetrahedron* **1999**, *55*, 3455–3466.

(138) Rios, R.; Jimeno, C.; Carroll, P. J.; Walsh, P. J. Kinetic Resolution of Atropisomeric Amides. *J. Am. Chem. Soc.* **2002**, *124*, 10272–10273.

(139) Hoffmann-Röder, A.; Krause, N. Enantioselective Synthesis of and with Allenes. *Angew. Chem., Int. Ed. Engl.* **2002**, *41*, 2933–2935.

(140) Hayashi, T.; Tokunaga, N.; Inoue, K. Rhodium-Catalyzed Asymmetric 1,6-Addition of Aryltitanates to Enynones Giving Axially Chiral Allenes. *Org. Lett.* **2004**, *6*, 305–307.

(141) Fiaud, J. C.; Legros, J. Y. Substrate Leaving Group Control of the Enantioselectivity in the Palladium-Catalyzed Asymmetric Allylic Substitution of 4-Alkyl-1-vinylcyclohexyl Derivatives. *J. Org. Chem.* **1990**, *55*, 4840–4846.

(142) Bolm, C.; Muñiz, K. Planar Chiral Arene Chromium(0) Complexes: Potential Ligands for Asymmetric Catalysis. *Chem. Soc. Rev.* **1999**, *28*, 51–59.

(143) Knölker, H.-J. Efficient Synthesis of Tricarbonyliron-Diene Complexes—Development of an Asymmetric Catalytic Complexation. *Chem. Rev.* **2000**, *100*, 2941–2961.

(144) Knölker, H.-J.; Hermann, H.; Herzberg, D. Photolytic Induction of the Asymmetric Catalytic Complexation of Prochiral Cyclohexa-1,3-dienes by the Tricarbonyliron Fragment. *J. Chem. Soc., Chem. Commun.* **1999**, 831–832.

(145) Uemura, M.; Nishimura, H.; Hayashi, T. Catalytic Asymmetric Induction of Planar Chirality: Palladium-Catalyzed Asymmetric Cross-Coupling of *meso* Tricarbonyl(arene) Chromium Complexes with Alkenyl- and Arylboronic Acids. *J. Organomet. Chem.* **1994**, *473*, 129–137.

(146) Davies, H. M. L.; Antoulinakis, E. G.; Hansen, T. Catalytic Asymmetric Synthesis of *Syn*-Aldol Products from Intermolecular C–H Insertions Between Allyl Silyl Ethers and Methyl Aryldiazoacetates. *Org. Lett.* **1999**, *1*, 383–385.

(147) Doyle, M. P.; Dyatkin, A. B.; Roos, G. H. P.; Canas, F.; Pierson, D. A.; van Basten, A.; Mueller, P.; Polleux, P. Diastereocontrol for Highly Enantioselective Carbon–Hydrogen Insertion Reactions of Cycloalkyl Diazoacetates. *J. Am. Chem. Soc.* **1994**, *116*, 4507–4508.

(148) Davies, H. M. L.; Venkataramani, C.; Hansen, T.; Hopper, D. W. New Strategic Reactions for Organic Synthesis: Catalytic Asymmetric C–H Activation α to Nitrogen as a Surrogate for the Mannich Reaction. *J. Am. Chem. Soc.* **2003**, *125*, 6462–6468.

(149) Bercot, E. A.; Rovis, T. A Palladium-Catalyzed Enantioselective Alkylative Desymmetrization of *meso*-Succinic Anhydrides. *J. Am. Chem. Soc.* **2004**, *126*, 10248–10249.

(150) Rychnovsky, S. D. Oxo Polyene Macrolide Antibiotics. *Chem. Rev.* **1995**, *95*, 2021–2040.

(151) Ward, R. S. Nonenzymatic Asymmetric Transformations Involving Symmetrical Bifunctional Compounds. *Chem. Soc. Rev.* **1990**, *19*, 1–19.

(152) Sculimbrene, B. R.; Miller, S. J. Discovery of a Catalytic Asymmetric Phosphorylation Through Selection of a Minimal Kinase Mimic: A Concise Total Synthesis of D-*myo*-Inositol-1-Phosphate. *J. Am. Chem. Soc.* **2001**, *123*, 10125–10126.

(153) Sculimbrene, B. R.; Morgan, A. J.; Miller, S. J. Enantiodivergence in Small-Molecule Catalysis of Asymmetric Phosphorylation: Concise Total Syntheses of the Enantiomeric D-*myo*-Inositol-1-phosphate and D-*myo*-Inositol-3-phosphate. *J. Am. Chem. Soc.* **2002**, *124*, 11653–11656.

(154) Gani, D. An End to the Protection Racket. *Nature* **2001**, *414*, 703–705.

(155) Takenaka, N.; Xia, G.; Yamamoto, H. Catalytic, Highly Enantio- and Diastereoselective Pinacol Coupling Reaction with a New Tethered Bis(8-quinolinolato) Ligand. *J. Am. Chem. Soc.* **2004**, *126*, 13198–13199.

(156) Evans, D. A.; Johnson, J. S.; Olhava, E. J. Enantioselective Synthesis of Dihydropyrans. Catalysis of Hetero Diels–Alder Reactions by Bis(oxazoline) Copper(II) Complexes. *J. Am. Chem. Soc.* **2000**, *122*, 1635–1649.

(157) Noyori, R.; Ikeda, T.; Ohkuma, T.; Widhalm, M.; Kitamura, M.; Takaya, H.; Akutagawa, S.; Sayo, N.; Saito, T.; Taketomi, T.; Kumobayashi, H. Stereoselective Hydrogenation via Dynamic Kinetic Resolution. *J. Am. Chem. Soc.* **1989**, *111*, 9134–9135.

(158) Chan, V.; Kim, J. G.; Jimeno, C.; Carroll, P. J.; Walsh, P. J. Dynamic Kinetic Resolution of Atropisomeric Amides. *Org. Lett.* **2004**, *6*, 2051–2053.

(159) Trend, R. M.; Ramtohul, Y. K.; Ferreira, E. M.; Stoltz, B. M. Palladium-Catalyzed Oxidative Wacker Cyclizations in Nonpolar Organic Solvents with Molecular Oxygen: A Stepping Stone to Asymmetric Aerobic Cyclizations. *Angew. Chem., Int. Ed. Engl.* **2003**, *42*, 2892–2895.

(160) Dearden, M. J.; Firkin, C. R.; Hermet, J.-P. R.; O'Brien, P. A Readily-Accessible (+)-Sparteine Surrogate. *J. Am. Chem. Soc.* **2002**, *124*, 11870–11871.

(161) Kolb, H. C.; VanNieuwenhze, M. S.; Sharpless, K. B. Catalytic Asymmetric Dihydroxylation. *Chem. Rev.* **1994**, *94*, 2483–2547.

(162) Kobayashi, S.; Ishitani, H. Lanthanide(III)-Catalyzed Enantioselective Diels–Alder Reactions. Stereoselective Synthesis of Both Enantiomers by Using a Single Chiral Source and a Choice of Achiral Ligands. *J. Am. Chem. Soc.* **1994**, *116*, 4083–4084.

(163) Evans, D. A.; Barnes, D. M.; Johnson, J. S.; Lectka, T.; von Matt, P.; Miller, S. J.; Murry, J. A.; Norcross, R. D.; Shaughnessy, E. A.; Campos, K. R. Bis(oxazoline) and Bis(oxazolinyl)pyridine Copper Complexes as Enantioselective Diels–Alder Catalysts: Reaction Scope and Synthetic Applications. *J. Am. Chem. Soc.* **1999**, *121*, 7582–7594.

(164) Evans, D. A.; Burgey, C. S.; Kozlowski, M. C.; Tregay, S. W. C_2-Symmetric Copper(II) Complexes as Chiral Lewis Acids. Scope and Mechanism of the Catalytic Enantioselective Aldol Additions of Enolsilanes to Pyruvate Esters. *J. Am. Chem. Soc.* **1999**, *121*, 686–699.